The Age of
Dinosaurs in Russia
and Mongolia

The former Soviet Union and Mongolia cover a vast area of land and, in the past 100 years, many dozens of extraordinary dinosaurs and other fossil amphibians, reptiles, birds and mammals have been found in Mesozoic rocks in these territories. The Permo-Triassic of the Ural Mountains of Russia has produced hundreds of superb specimens, and many of the dinosaurs from Mongolia are unique.

This is the first compilation in any Western language of this large body of Russian, Mongolian, Kazakh, and Polish research and the first time so much of this research, previously unexplored by the West, has been introduced in English. *The Age of Dinosaurs in Russia and Mongolia* is written by a unique mix of Russian and Western palaeontologists, and provides an entrée to a range of fossil faunas, in particular reptiles, that have been little known outside Russia. It will undoubtedly become a major reference work for all vertebrate palaeontologists.

Michael J. Benton is Professor of Vertebrate Palaeontology in the Department of Earth Sciences at the University of Bristol and is the author of over 30 books on palaeontology.

Mikhail A. Shishkin is Chief of the Fossil Amphibian Laboratory at the Palaeontological Institute in Moscow.

David M. Unwin is Curator of Fossil Reptiles and Birds at the Museum für Naturkunde, Humboldt University in Berlin.

Evgenii N. Kurochkin is Chief of the Palaeoornithological Laboratory at the Palaeontological Institute in Moscow.

The Age of Dinosaurs in Russia and Mongolia

EDITED BY

Michael J. Benton
University of Bristol

Mikhail A. Shishkin
Palaeontological Institute, Moscow

David M. Unwin
Humboldt University, Berlin

Evgenii N. Kurochkin
Palaeontological Institute, Moscow

CAMBRIDGE
UNIVERSITY PRESS

PUBLISHED BY THE PRESS SYNDICATE OF THE UNIVERSITY OF CAMBRIDGE
The Pitt Building, Trumpington Street, Cambridge, United Kingdom

CAMBRIDGE UNIVERSITY PRESS
The Edinburgh Building, Cambridge CB2 2RU, UK
40 West 20th Street, New York NY 10011–4211, USA
477 Williamstown Road, Port Melbourne, VIC 3207, Australia
Ruiz de Alarcón 13, 28014 Madrid, Spain
Dock House, The Waterfront, Cape Town 8001, South Africa

http://www.cambridge.org

First published 2000
First paperback edition 2003

Typeface Monotype Janson 9.75/13pt.   *System* QuarkXPress®   [SE]

A catalogue record for this book is available from the British Library

Library of Congress Cataloguing in Publication data

The age of dinosaurs in Russia and Mongolia / edited by Michael J.
Benton . . . [et al.].
p. cm.
Includes bibliographical references.
ISBN 0 521 55476 4 hardback
1. Dinosaurs – Russia (Federation)   2. Dinosaurs – Mongolia.
I. Benton, M.J. (Michael J.)
QE862.D5A287   2000
567.9´0947–dc21   99-16743 CIP

ISBN 0 521 55476 4 hardback
ISBN 0 521 54582 X paperback

CONTENTS

Acknowledgements   *page* viii

List of contributors   ix

Preface   xiii

Introduction   xiv
Michael J. Benton, Mikhail A. Shishkin,
David M. Unwin and Evgenii N. Kurochkin

Conventions in Russian and Mongolian palaeontological
literature   xvi
Michael J. Benton

    Stratigraphic units in the Permo-Mesozoic of Russia and
    Middle Asia   xx

    Mongolian place names and stratigraphic terms   xxii

    Journals and series   xxix

    Transliterated names of Russian and Mongolian
    palaeontologists and geologists   xxxv

1 The history of excavation of Permo-Triassic vertebrates from
   Eastern Europe   1
   Vitalii G. Ochev and Mikhail V. Surkov

2 The amniote faunas of the Russian Permian: implications
   for Late Permian terrestrial vertebrate biogeography   17
   Sean P. Modesto and Natalia Rybczynski

3 Permian and Triassic temnospondyls from Russia   35
   Mikhail A. Shishkin, Igor. V. Novikov and Yurii M. Gubin

4 Permian and Triassic anthracosaurs from Eastern Europe   60
   Igor V. Novikov, Mikhail A. Shishkin and Valerii K. Golubev

Contents

5 The Russian pareiasaurs   71
    Michael S.Y. Lee

6 Mammal-like reptiles from Russia   86
    Bernard Battail and Mikhail V. Surkov

7 Tetrapod biostratigraphy of the Triassic of Eastern Europe   120
    Mikhail A. Shishkin, Vitalii G. Ochev, Vladlen R. Lozovskii
    and Igor V. Novikov

8 Early archosaurs from Russia   140
    David J. Gower and Andrei G. Sennikov

9 Procolophonoids from the Permo-Triassic of Russia   160
    Patrick S. Spencer and Michael J. Benton

10 Enigmatic small reptiles from the Middle–Late Triassic
    of Kirgizstan   177
    David M. Unwin, Vladimir R. Alifanov and Michael J. Benton

11 Mesozoic marine reptiles of Russia and other former
    Soviet republics   187
    Glenn W. Storrs, Maxim S. Arkhangel'skii and Vladimir M. Efimov

12 Asiatic dinosaur rush   211
    Edwin H. Colbert

13 The Russian–Mongolian expeditions and research in
    vertebrate palaeontology   235
    Evgenii N. Kurochkin and Rinchen Barsbold

14 The Cretaceous stratigraphy and palaeobiogeography of
    Mongolia   256
    The late Vladimir F. Shuvalov

15 Lithostratigraphy and sedimentary settings of the
    Cretaceous dinosaur beds of Mongolia   279
    Tom Jerzykiewicz

16 Mesozoic amphibians from Mongolia and the Central
    Asiatic republics   297
    Mikhail A. Shishkin

17 Mesozoic turtles of Middle and Central Asia   309
    Vladimir B. Sukhanov

Contents

18 The fossil record of Cretaceous lizards from Mongolia   368
    Vladimir R. Alifanov

19 Choristodera from the Lower Cretaceous of northern Asia   390
    Mikhail B. Efimov and Glenn W. Storrs

20 Mesozoic crocodyliforms of north-central Eurasia   402
    Glenn W. Storrs and Mikhail B. Efimov

21 Pterosaurs from Russia, Middle Asia and Mongolia   420
    David M. Unwin and Natasha N. Bakhurina

22 Theropods from the Cretaceous of Mongolia   434
    Philip J. Currie

23 Sauropods from Mongolia and the former Soviet Union   456
    Teresa Maryańska

24 Ornithopods from Kazakhstan, Mongolia and Siberia   462
    David B. Norman and Hans-Dieter Sues

25 The fossil record, systematics and evolution of
    pachycephalosaurs and ceratopsians from Asia   480
    Paul C. Sereno

26 Armoured dinosaurs from the Cretaceous of Mongolia   517
    Tat'yana A. Tumanova

27 Mesozoic birds of Mongolia and the former USSR   533
    Evgenii N. Kurochkin

28 Eggs and eggshells of dinosaurs and birds from the
    Cretaceous of Mongolia   560
    Konstantin E. Mikhailov

29 Mammals from the Mesozoic of Mongolia   573
    *Zofia Kielan-Jaworowska, Michael J. Novacek, Boris A. Trofimov
    and Demberlyin Dashzeveg*

30 Mammals from the Mesozoic of Kirgizstan, Uzbekistan,
    Kazakhstan and Tadzhikistan   627
    Alexander O. Averianov

    Index   653

ACKNOWLEDGEMENTS

We thank the Royal Society for funding the exchange visits and joint field expeditions of Russian and British palaeontologists from 1993 to 1996. Many people have helped the editors in their work. In particular, we thank the following for their reviews of individual chapters: J.D. Archibald, N.N. Bakhurina, M. Borsuk-Białynicka, E. Buffetaut, L. Chiappe, J.M. Clark, S.E. Evans, R.C. Fox, E. Frey, E. Gaffney, T. Holtz, J.A. Hopson, T. Jerzykiewicz, T.S. Kemp, Z. Kielan-Jaworowska, S. Macintosh, T. Maryańska, W.D. Maxwell, P. Meylan, A.R. Milner, R. Molnar, M. Norell, D.B. Norman, H. Osmólska, A.C. Panchen, J.M. Parrish, R.R. Reisz, B. Rubidge, K. Sabath, G.R. Warrington, D. Weishampel, P. Wellnhofer and W.A. Wimbledon.

We thank N.N. Bakhurina for translating some of the chapters, and for her extensive advice on questions concerning transliteration. We are also extremely grateful to Ariunchimeg Yarinpil for thorough checking of the Mongolian names. In Bristol, Pam Baldaro provided some of the diagrams, and Simon Powell printed many of the photographs. We are extremely grateful to Pat Rich for the supply of photographs of some of the Russian fossil reptiles scattered through the book. At Cambridge University Press, we are extremely grateful to Robin Smith and Tracey Sanderson for their encouragement, and for their understanding, with this book, and to Beverley Lawrence for her careful checking of the entire typescript.

LIST OF CONTRIBUTORS

Vladimir R. Alifanov
Paleontologicheskii Institut
Profsoyuznaya 123
117868 Moscow GSP-7, Russia

Maxim S. Arkhangel'skii
Geological Faculty
Saratov State University
Lenin Street 161
410750 Saratov, Russia

Alexander O. Averianov
Zoological Institute
Russian Academy of Sciences
Universitetskaya nab. 1
St Petersburg 199034, Russia

Natasha N. Bakhurina
Department of Earth Sciences
University of Bristol
Bristol BS8 1RJ, UK

Rinchen Barsbold
Palaeontological Centre
Mongolian Academy of Sciences
Enthaivan Avenue
Ulaanbaatar 51-63, Mongolia

Bernard Battail
Museum National d'Histoire Naturelle
Laboratoire de Paléontologie
8 rue Buffon
75005 Paris, France

Michael J. Benton
Department of Earth Sciences
University of Bristol
Bristol BS8 1RJ, UK

Edwin H. Colbert
Museum of Northern Arizona
3101 N. Fort Valley Road
Flagstaff
Arizona 86001, USA

Philip J. Currie
Royal Tyrrell Museum of Palaeontology
Drumheller
Alberta, Canada T0J 0Y0

Demberlyin Dashzeveg
Palaeontological Centre
Mongolian Academy of Sciences
Enthaivan Avenue
Ulanbataar 51-63, Mongolia

Mikhail B. Efimov
Paleontologicheskii Institut
Profsoyuznaya 123
117868 Moscow GSP-7, Russia

Vladimir M. Efimov
Paleontologicheskii Muzei
Undory
Ul'yanovskaya Oblast'
433312 Russia

List of contributors

Valerii K. Golubev
Paleontologicheskii Institut
Profsoyuznaya, 123
117868 Moscow GSP-7, Russia

David J. Gower
Department of Earth Sciences
University of Bristol
Bristol BS8 1RJ, UK
Present address:
Department of Zoology
Natural History Museum
Cromwell Road
London SW7 5BD, UK

Yurii M. Gubin
Paleontologicheskii Institut
Profsoyuznaya 123
117868 Moscow GSP-7, Russia

Tom Jerzykiewicz
210 Chinook Drive
Cochrane
Alberta
Canada, T0L 0W2

Zofia Kielan-Jaworowska
ul. Sadowa 9/1
PL-05-520 Konstancin-Jeziorna
Poland

Evgenii N. Kurochkin
Paleontologicheskii Institut
Profsoyuznaya 123
117868 Moscow GSP-7, Russia

Michael S. Y. Lee
Department of Zoology
University of Queensland
Brisbane
Queensland
4072 Australia

Vladlen R. Lozovskii
Moscow Geological Prospecting Institute
Miklucho-Maklay Str. 23
Room 5-62
117485 Moscow, Russia

Teresa Maryańska
Muzeum Ziemi
Polska Akademia Nauk
Al. Na Skarpie 20/26
00-488 Warszawa, Poland

Konstantin E. Mikhailov
Paleontologicheskii Institut
Profsoyuznaya 123
117868 Moscow GSP-7, Russia

Sean P. Modesto
Bernard Price Institute
University of Witwatersrand
1 Jan Smuts Avenue
Johannesberg 2050
South Africa

David B. Norman
Department of Earth Sciences
Sedgwick Museum
Downing Street
Cambridge CB2 3EQ, UK

Michael J. Novacek
Department of Vertebrate Paleontology
American Museum of Natural History
79th Street and West Central Park
New York
NY 10024, USA

Igor V. Novikov
Paleontologicheskii Institut
Profsoyuznaya 123
117868 Moscow GSP-7, Russia

List of contributors

Vitalii G. Ochev
Geological Faculty
Saratov State University
Lenin Street 161
410750 Saratov, Russia

Natalia Rybczynski
Biological Anthropology and Anatomy
Duke University
Box 90383
Durham
North Carolina 27708–0383, USA

Andrei G. Sennikov
Paleontologicheskii Institut
Profsoyuznaya 123
117868 Moscow GSP-7, Russia

Paul C. Sereno
Department of Anatomy
University of Chicago
1025 East 57 Street
Chicago, IL 60637, USA

Mikhail A. Shishkin
Paleontologicheskii Institut
Profsoyuznaya, 123
117868 Moscow GSP-7, Russia

Vladimir F. Shuvalov (the late)
formerly of
Institute of Lake Research
Russian Academy of Sciences
Sevast'yanova Street 9
196199 St. Petersburg, Russia

Patrick S. Spencer
Department of Earth Sciences
University of Bristol
Bristol BS8 1RJ, UK
Present address:
National Science Museum
Geology
3-23-1 Hyakunin-Cho
Shinjuku-Ku
Tokyo 169
Japan

Glenn W. Storrs
Cincinnati Museum of Natural History
1720 Gilbert Avenue
Cincinnati, OH 45202, USA

Hans-Dieter Sues
Department of Vertebrate Palaeontology
Royal Ontario Museum
100 Queen's Park
Toronto, Ontario M5S 2C6, Canada

Vladimir B. Sukhanov
Paleontologicheskii Institut
Profsoyuznaya 123
117868 Moscow GSP-7, Russia

Mikhail V. Surkov
Geological Faculty
Saratov State University
Lenin Street 161
410750 Saratov, Russia

Boris A. Trofimov
Paleontologicheskii Institut
Profsoyuznaya 123
117868 Moscow GSP-7, Russia

List of contributors

Tat'yana A. Tumanova
Paleontologicheskii Institut
Profsoyuznaya 123
117868 Moscow GSP-7, Russia

David M. Unwin
Department of Earth Sciences
University of Bristol
Bristol BS8 1RJ, UK
Present address:
Institut für Palaontologie
Museum für Naturkunde
Zentralinstitut der Humboldt-Universität zu Berlin
Invalidenstrasse 43
D-10115 Berlin, Germany

PREFACE

The fossil reptiles and amphibians of the Russian Permo-Triassic are world famous. The first specimens were recorded in the eighteenth century from the Permian of the Urals, during the reign of Peter the Great and, since then, some 150 species of small and large terrestrial tetrapods have now been recorded from a well-dated succession of faunas. The faunas include some Gondwanan forms, comparable with animals from South Africa, but others are unique to Russia. Marine reptiles have been found at many localities in the Russian marine Jurassic, but they are little known in the West. The dinosaurs and associated animals from the Cretaceous of Russia, Middle Asia, and Mongolia, are equally important. Some of the dinosaurs are like those from North America, but there are a number of important groups unique to Asia. Russian and Mongolian scientists have carried out a great deal of work on these ancient tetrapods, and yet they are little known in the West. Much of the work was published only in the Russian language, and the books and papers have rarely been translated.

Following *glasnost* and *perestroika*, true collaboration between Russian and Western scientists has again been possible, after a break of 75 years. One of the first major collaborations in palaeontology was the Joint Vertebrate Palaeontology Research Programme between the Palaeontological Institute, Moscow, and the University of Bristol, UK. This Programme was sponsored by the Royal Society and the Russian Academy of Sciences from 1993 to 1997, and some 25 palaeontologists took part. This book is one of the fruits of that collaboration. There are accounts of the history of collecting, the relevant stratigraphy and geological setting, and full accounts of the major animal groups, prepared by a mixture of Russian, Mongolian, Polish, and Western experts.

Introduction

MICHAEL J. BENTON, MIKHAIL A. SHISHKIN,
DAVID M. UNWIN AND EVGENII N. KUROCHKIN

There are many rich faunas of fossil amphibians, reptiles, birds, and mammals in the Permian and Mesozoic sediments of Russia, Mongolia, and the various republics of the Former Soviet Union (FSU). These include one of the best, or perhaps even *the* best, succession of Permo-Triassic continental tetrapods, consisting of some 150 described species known from 15 or more horizons. Many of the amphibians and reptiles may be compared with Gondwanan forms, from South Africa and elsewhere, but others are unique to Russia. These faunas have been described and summarized many times in Russian, but only sporadically in the Western literature (e.g. Efremov, 1940; Olson, 1957; Ochev and Shishkin, 1989; Nesov, 1992; Sennikov, 1996). The classic Permo-Triassic sequences of the Urals and of the Moscow Platform are treated in Chapters 1–9, and the unusual small forms from the Middle–Late Triassic of Kirgizstan in Chapter 10.

The record of marine tetrapods from Russia and surrounding countries of the FSU is extensive, but has never been reviewed, either in Russian or in any other language. Dozens of isolated finds of crocodilians, ichthyosaurs, and plesiosaurs have been reported from the banks of the Volga, and from marine Jurassic and Cretaceous sediments elsewhere, but these are sorely in need of revision (see Chapter 11).

The dinosaurs and other tetrapods from the Cretaceous of Mongolia, and contemporaneous units in Middle Asia and Russia, have been collected and studied by American, Russian, Polish, and Mongolian teams, since their discovery in the 1920s. Many of the available general accounts focus on the history of collecting and the arduous field conditions (e.g.

Andrews, 1932; Rozhdestvenskii, 1960; Colbert, 1968; Kielan-Jaworowska, 1969; Lavas, 1993; Novacek, 1996). Numerous papers and monographs have been devoted to the Mongolian faunas, and an enormous literature has developed. The chapters in this volume covering the Mongolian dinosaurs, and associated tetrapods, are the first comprehensive overview of these important faunas. Accounts of the various expeditions are given (Chapters 12 and 13), then a Russian and a Western view of the stratigraphy and geological setting of the dinosaur beds (Chapters 14 and 15), followed by accounts of the various tetrapod groups, from salamanders to mammals (Chapters 16–30).

References

Andrews, R.C. 1932. *The New Conquest of Central Asia. Natural History of Central Asia* 1. New York: American Museum of Natural History, 678 pp.

Colbert, E.H. 1968. *Men and Dinosaurs*. New York: Dutton; London: Evans, 283 pp.

Efremov, I.A. 1940. Kurze Übersicht über die Formen der Perm- und Trias-Tetrapoden-Fauna der UdSSR. *Centralblatt für Mineralogie, Geologie, und Paläontologie, Abteilung B* **1940**: 372–383.

Kielan-Jaworowska, Z. 1969. *Hunting for Dinosaurs*. Cambridge, Mass.: MIT Press, 177 pp.

Lavas, J.R. 1993. *Dragons from the Dunes*. Auckland: Academy Interprint Ltd., 138 pp.

Nesov, L.A. 1992. Mesozoic and Paleogene birds of the USSR and their paleoenvironments. *Natural History Museum of Los Angeles County, Science Series* **36**: 465–470.

Novacek, M.J. 1996. *Dinosaurs of the Flaming Cliffs*. New York: Doubleday, 369 pp.

Introduction

Ochev, V.G. and Shishkin, M.A. 1989. On the principles of global correlation of the continental Triassic on the tetrapods. *Acta Palaeontologia Polonica* **34**: 149–173.

Olson, E.C. 1957. Catalogue of localities of Permian and Triassic vertebrates of the territories of the U.S.S.R. *Journal of Geology* **65**: 196–226.

Rozhdestvenskii, A.K. 1960. *Chasse aux Dinosaures dans le Désert de Gobi.* Paris: Librairie Arthème Fayard.

Sennikov, A.G. 1996. Evolution of the Permian and Triassic tetrapod communities of Eastern Europe. *Palaeogeography, Palaeoclimatology, Palaeoecology,* **120**: 331–351.

Conventions in Russian and Mongolian palaeontological literature

MICHAEL J. BENTON

Introduction

We encountered a large number of problems in rendering Russian names of people, places, and geological units into English. We felt we should seek to standardize these throughout the book, and to do so according to a logical explicit system. (No doubt the astute reader will find some inconsistencies, despite all our best efforts.) At a basic level, standards of transliteration (rendering the Russian alphabet in Western letters) and translation had to be selected. However, the need for agreed standards goes beyond simple translation and transliteration, and includes standardization of spellings of place names, names of geological units, and names of journals.

We soon discovered that places, units, and people are not fixed entities when transliterated. As examples of the complexities that we faced, it is normal practice in Russian to treat the names of stratigraphic units as adjectives, and hence their endings are modified in agreement with the noun they qualify and their role in a sentence. The titles of journals are also debatable: the older ones went through many title changes, sometimes every two or three years, and from the nineteenth century until as late as the 1950s, many of them had both a Russian and a Western (French or German) title. Even the names of people are not fixed. Some palaeontologists have as many as three or more valid transliterated names. For example, Vitalii Ochev chose a Germanic transliteration of his surname, Otschev, for taxonomic usage, and his name is also transliterated in semi-Germanic fashion as Otchev or Otshev, while Ochev is the literal transliterated form that we choose. Traditionally, Russian scientists felt a

strong affinity with Germans and the German language, hence these forms of transliteration. We have, however, adopted English conventions which are now becoming the norm.

The rules of transliteration can be selected and fixed, and hence the transliterated names of Russian people and places are predictable. Journal titles can also be agreed and fixed. However, there is a further layer of complexity in rendering Russian stratigraphic terminology into English: the fundamental concepts of dividing up rock successions and spans of geological time into units are different from those used in the non-Russian-speaking world. This subject requires careful consideration.

Transliteration

We have selected a modern English-language system of transliteration in which the letters of the Cyrillic alphabet are given equivalents that, where possible, have the closest sound. There is little problem for the consonants, but the vowels and uniquely Russian characters are a little more difficult. The system here is based on that recommended by Zofia Kielan-Jaworowska (1993) and others, but with some modifications, and it has the advantage that the original Cyrillic spelling of a word can be reconstructed from the transliterated version. This will help in trying to trace specific publications, people, or places.

The scheme used here is not entirely unambiguous, recording the Russian letters 'E', 'Ë' and 'Э' as 'e', and 'И' and 'Й' as 'i', and both the soft and hard signs 'Ь' and 'Ъ' as '″', implying a modification to the preceding consonant. However, it is less ambiguous than the US

Library of Congress System, where, for example, 'Ю' is rendered as 'iu' and 'Я' is 'ia'. Our scheme involves some modifications to familiar forms of transliterated Russian names. For example, 'Yeltsin' beomes 'El'tsin' (it is spelled 'ЕЛЬЦИН').

А	а	a		К	к	k		Х	х	kh
Б	б	b		Л	л	l		Ц	ц	ts
В	в	v		М	м	m		Ч	ч	ch
Г	г	g		Н	н	n		Ш	ш	sh
Д	д	d		О	о	o		Щ	щ	shch
Е	е	e		П	п	p		Ъ	ъ	'
Ё	ё	e		Р	р	r		Ы	ы	y
Ж	ж	zh		С	с	s		Ь	ь	'
З	з	z		Т	т	t		Э	э	e
И	и	i		У	у	u		Ю	ю	yu
Й	й	i		Ф	ф	f		Я	я	ya

Russian stratigraphic approaches and terminology

There has been much confusion over comparisons of Russian and international systems of stratigraphy. Common Russian stratigraphic terms, such as svita, gorizont, and other subdivisions (e.g. podgorizont, nadgorizont), are sometimes anglicized, for example Gorizont as Horizon and Svita as Suite. Another solution has been to equate the Russian divisions with international units, for example, Gorizont with Horizon, and Svita with Formation. These approaches, however, mask some fundamental differences between the Russian and the international approaches to stratigraphy, and we prefer to retain transliterated versions of the Russian terms in order to avoid confusion.

The distinction between Russian and international approaches goes deeper than mere terminology, however, and we must seek to clarify the differences so that readers will understand the mix of systems used by different authors. In summary, the Russian stratigrapic approach generally used the concept of unified divisions of time, in which a specific time span, and the rocks of that age, are equated and treated as one. The Western, or now, international, system maintains a strict division between time units and rock units. Both Russian and international systems have been applied

in Mongolia (cf. Chapters 14 and 15), but only Russian systems in Russia and the FSU (cf. Chapters 2 and 7).

It should be noted, importantly, that the Russian system is evolving, and is approaching the international system (compare the 1959, 1977, and 1992 Russian stratigraphic codes). The current code (Zhamoida *et al.*, 1992) emphasizes the separation of geochronological and stratigraphic units, and distinguishes global, regional, and local systems. Many Russian stratigraphers work on international committees, and they interpret the Russian code in an international way. However, the pre-1990 literature, and some working Russian stratigraphers still apply the classic Russian system, and this must be outlined.

The classical Russian approach to stratigraphy was outlined in the 1959 code of practice (Interdepartmental Stratigraphic Committee, 1959, p. 26):

paleontologic criteria appear to be the most important and the most objective criteria for the distinction and especially for the correlation of the basic subdivisions of the stratigraphic and geochronologic scales, and the character and scale of the changes of fauna and flora serve as the principal basis for determining the taxonomic rank of the stratigraphic units, their 'hierarchical' intersubordination.

The text goes on to describe how the evolution of life means that fossil plants and animals occur in the rocks in predictably changing patterns with time, and hence stratigraphic boundaries should be established biostratigraphically. Lithological characters should be used only when fossils are absent. The account then (p. 27) describes the international system, with distinctive lithostratigraphic, biostratigraphic, and chronostratigraphic schemes, as untenable:

A unified stratigraphic scale ought to be accepted, based on the complex of historical-geological principles, on the distinction of definite steps in the history of the geological development of the Earth, and not on separate, arbitrarily selected characters of the rocks.

The stratigraphic units used at regional level are the gorizont and the lona, and in local description are, in descending order, complex, seriya, svita, and pachka.

According to the Russian Stratigraphic Code (Zhamoida, 1977, Art. IV.3), gorizonts are the main regional stratigraphic units, identified primarily from their palaeontological characteristics, and they do not pertain to lithostratigraphic units. The gorizont may unite several svitas, or parts of svitas, or deposits of different facies in various districts but clearly contemporaneous on the basis of included fossils. (In addition, and confusingly, the term 'marking gorizont' is sometimes used as a rough equivalent of the international 'horizon', in other words, to refer to a local rock unit that is characterized by a specific lithology or fossil; Zhamoida *et al.*, 1992, Art. VII.5, Recommendation 5A.)

Svitas, on the other hand, are largely lithostratigraphic units (Art. VII.5), given a locality name that is close to their characteristic exposure. The definition of a svita incorporates a mix of field lithological observations and biostratigraphic assumptions: 'In distinguishing a new svita, one ought without fail to establish at least an approximate, sufficiently proved correlation of it with the subdivisions of the unified [international] scale' (Interdepartmental Stratigraphic Committee, 1959, p. 34). The requirement for 'approximate correlation' has been deleted in the 1992 code, and the emphasis is on mappable lithological features (Zhamoida *et al.*, 1992, Art. V.10).

The stratigraphic system outside the Russian-speaking world is sometimes termed the dual classification, since it makes a clear distinction between 'rock units' and 'time units' (O'Rourke, 1976; Prothero, 1990). In other words, lithostratigraphy is quite distinct from chronostratigraphy and biostratigraphy. The normal procedure is for field geologists to name geological groups, formations, and members purely on the basis of mappable lithological features. Hence, the Otter Sandstone Formation in south Devon is a particular unit of red-coloured sandstone with channels and calcretes that may be mapped. It is defined by a type section where its lower contact with the Budleigh Salterton Pebble Bed Formation is observed, and by another type section where the overlying Mercia Mudstone Group caps it. The question of the age of the Otter Sandstone Formation is quite separate, and, in this case, it depends on the study of fossil fishes and tetrapod fossils which suggest a Mid Triassic Anisian age (Benton *et al.*, 1994). The lithostratigraphic definition of the Otter Sandstone Formation is fixed by field criteria, and this is unaffected by any independent chronostratigraphic determinations, where fossils, radiometric dates, and other evidence may be debated and discussed as hypotheses of age are considered. Changes in dating hypotheses do not affect the reality of the defined rock units.

The new understanding of the 'Svita', according to the 1992 code, allows for a closer equation with the 'Formation', since the emphasis is on local lithological and mapping criteria. In the future, as Russian practice more nearly approaches the international code, then equivalence may be assumed. For the present, though, until they are explicitly revised and redefined, classical svitas cannot be assumed to be purely lithostratigraphic units. The Gorizont, although treated in the 1992 code as a regional-scale stratigraphic unit, is largely geochronological, but applicable to particular rocks, a composite group/biochron. It remains problematic.

From a Western viewpoint, the classical Russian system incorporates a circularity. In other words, if stratigraphic units are defined, even in part, by fossils, how can the order of fossils be determined from stratigraphy? The dual approach cuts through the circularity, by defining stratigraphic units, and their relative sequence, in terms of lithological criteria and field relationships. The order of the fossils can then be extracted from the order of the rocks as determined independently of the fossils.

Perhaps the distinction between the two approaches, the Russian and the Western, can be traced to differences in philosophy. In an intriguing analysis, O'Rourke (1976) suggests that the unified Russian stratigraphic approach is a direct application of dialectical materialism in geology. This requires that time units, and the sedimentary rocks deposited in them, are treated as unified entities, material bodies. Dialectical materialism states that every material body originates in time, through a negation of negation, and therefore all rock bodies are chronostratigraphic units (Interdepartmental Stratigraphic Committee, 1959, p. 32). However, the dialectical materialist approach, as O'Rourke (1976, p. 47) notes, 'is ill-prepared to answer queries about how we obtain or verify a certain kind of knowledge'.

These comments could be seen as utter nonsense. Mikhail Shishkin, who wishes to emphasize the similarity of the current Russian and international stratigraphic approaches, notes that this analysis is quasi-scientific interpretation of a kind formerly touted by Marxist apologists in Russia. Although such arguments were made by philosophers, no practising Russian stratigrapher was ever affected by such considerations.

References

Benton, M.J., Warrington, G., Newell, A.J. and Spencer, P.S. 1994. A review of the British Middle Triassic tetrapod assemblages, pp. 131–160 in Fraser, N.C. and Sues, H.-D. (eds.), *In the Shadow of the Dinosaurs*. Cambridge: Cambridge University Press.

Interdepartmental Stratigraphic Committee 1959. Stratigraphic classification and terminology. *International Geology Review* 1 (2): 22–38. [Translation of 'Stratigraficheskaya klassifikatsiya i terminologiya', Moscow: State Scientific-Technical Publishing House for Literature on Geology and Mineral Resources, 1956.]

Kielan-Jaworowska, Z. 1993. Citing Russian papers. *News Bulletin of the Society of Vertebrate Paleontology* 159: 53–54.

O'Rourke, J.E. 1976. Pragmatism versus materialism in stratigraphy. *American Journal of Science* 276: 47–55.

Prothero, D.R. 1990. *Interpreting the Stratigraphic Record*. San Francisco: W.H. Freeman, 410 pp.

Zhamoida, A.I. (ed.) 1977. [*Code of Stratigraphy of the USSR.*] Leningrad: Vsesoyuznyi Geologicheskii Institut, 80 pp.

Zhamoida, A.I. *et al.* (eds.) 1992. [*Stratigraphic Code. Second Edition, Supplemented.*] Interdepartmental Stratigraphic Committee, St. Petersburg, 120 pp.

STRATIGRAPHIC UNITS IN THE
PERMO-MESOZOIC OF RUSSIA AND MIDDLE ASIA

In this list, current stratigraphic terms used to refer to major tetrapod-bearing subdivisions of the Permo-Mesozoic of Russia and the Middle Asian republics are included. We have chosen to retain an anglicized adjectival ending for Gorizont and Supergorizont terms. Svita names are given in adjectival form of the place name. Hence, we prefer the term Petropavlovskaya Svita, named after the village Petropavlovka, rather than 'Petropavlovsk', 'Petropavlovskian', 'Petropavlovska', or other variants. An exception to this rule is for names derived from non-Russian languages, such as Bashkir or Kazakh, which remain in the nominative. Examples are the Donguz Gorizont and the Karabastau and Yushatyr' svitas.

Admiralteistva Svita
Alamyshyk Svita
Astashikhian Member
Balabansai Svita
  Balobansay (= Balabansai)
Belebeiskaya Svita
Beleuta Svita
Bereznikovskaya Svita
Beshtyube Svita
  Beshtubinskaya (= Beshtyube)
Bissekty Svita
  Bogdinsk (= Bogdinskaya)
Bogdinskaya Svita
Bostobe Svita
  Bostobin (= Bostobe)
Bukobay Gorizont
Bukobay Svita
  Bysovsk (= Byzovskaya)
Byzovskaya Svita

Charkabozhskaya Svita
  Chodjakul (= Khodzhakul)
  Chodjakulsaiskaya (= Khodzhakul)
Dabrazinskaya Svita
Darbasa Svita
Donguz Gorizont
Dvuroginskian Gorizont
Eginsai Svita
Elton Gorizont
  Fedorovsk (= Fedorovskaya, Fedorovskian)
Fedorovskaya Svita
Fedorovskian Subgorizont
  Gam (= Gamskaya, Gamskian)
Gamskaya Svita
Gamskian Subgorizont
Ilekskaya Svita
Inder Gorizont
Intinskaya Svita
  Intinsk (= Intinskaya)
  Jalovatch (= Yalovach)
  Jushatyr' (= Yushatyr')
Karabastau Svita
Karakhskaya Svita
Keryamaiolskaya Svita
Khodzhakul Svita
Kopanskaya Svita
Krasnokamenskaya Svita
Kumanskaya Svita
Kutulukskaya Svita
Kzylsaiskaya Svita
Lestanshorskaya Svita
Lipovskaya Svita
Madygen Svita
Malokinel'skaya Svita

Stratigraphic units

Meshcherskii Gorizont
Mogoito Member
Moskvoretskaya Svita
Murtoi Svita
Nadkrasnokamenskaya Svita
Nizhneustinskaya Svita
Nyadeitinskaya Svita
   Petropavlovsk (= Petropavlovskaya)
Petropavlovskaya Svita
Pizhmomezenskaya Svita
Poldarsinskaya Svita
Ryabinskian Member
Rybinskaya Svita
Rybinskian Gorizont
Salarevskaya Svita
Severodvinskian Gorizont
Sheshminskian Gorizont
   Shilikhinsk (= Shilikhinskaya)
Shilikhinskaya Svita
Sludkinskaya Svita
Sludkian Gorizont

Staritskaya Svita
   Sukhonsk (= Sukhonskaya)
Sukhonskaya Svita
   Syninsk (= Syninskaya)
Syninskaya Svita
Syuksyukskaya Svita
Taikarshin Beds
Talkhabskaya Svita
Teryutekhskaya Svita
Ubukunskaya Svita
Urzhumian Gorizont
Ust'mylian Gorizont
Vetlugian Supergorizont
Vokhminskaya Svita
Vokhmian Gorizont
Vyatskian Gorizont
   Wetlugian (= Vetlugian)
Yalovach Svita
Yarenskian Gorizont
Yushatyr' Svita

MONGOLIAN PLACE NAMES AND STRATIGRAPHIC TERMS

Names of places and stratigraphic units are based on standard versions used by informed Mongolian and Western authors, expecially Gradziński *et al.* (1977), Jerzykiewicz and Russell (1991), and Dashzeveg *et al.* (1995). These works incorporate a number of changes in the transliterated forms of Mongolian place names, and these may look a little odd to people who have become accustomed to the older spellings. For example, 'Bayn Dzak' becomes 'Bayan Zag', 'Ulan Bator' becomes 'Ulaanbaatar', 'Dzun Bayan' becomes 'Züünbayan', 'Genghis Khan' becomes 'Chingis Khaan'. Some, fortunately, do not change: Choibalsan, Djadokhta, Gobi, Nemegt. In any case, we felt it was essential to attempt to standardize names, since different authors have evolved quite different systems. Who, but an expert, is to know that Hobur, Khoboor, Khobur, Khoobur, Khöövör, and Khovboor are one and the same place? Why the changes?

There are a number of reasons for the difficulties in transliterating Mongolian names. First is the fact that there are several Mongolian languages, second that there has been no single standardized alphabet for writing Mongolian, and third transliteration methods have sometimes proceeded directly from the Mongolian to English, but frequently have gone from Mongolian to Russian, and then to English.

Mongolian is the langauge spoken by most people in Mongolia and in Inner Mongolia (*Nei Mongol*), part of China. By origin it is one the languages of the Mongolian group of the Altaic family. In Mongolia today, one language is spoken, Mongolian, but many dialects, including Khalh, Buryad, Dörvöd, and Khalimag, while other languages of the Mongolian group are used in neighbouring regions of China, Russia, and Afghanistan. Much of the history of the Mongolian language has been oral, with constant evolution of local dialects, even though the population of Mongolia has never been large (it is just over 2 million today). The modern Mongolian language developed after the communist revolution in 1921 on the basis of the Khalh dialect. It consists of 46 phonemes (identifiable sounds), including 22 vowel phonemes.

Ten different scripts have been used to represent Mongolian on paper, and even today there is debate about which is most appropriate. The broad range of scripts, and their constant evolution reflect attempts by Mongolians to match the written language to the oral as closely as possible. Written texts on monuments from the seventh and eighth centuries are generally given in Chinese scripts, involving hundreds or thousands of individual ideograms. The Old Mongolian script, which evolved about 1000 years ago, and was perhaps borrowed by the Mongolians in the thirteenth century from an Aramaic source, the Uigur script, has letters that represent sounds, and which change depending on their position in a word. The script is written from the top downwards and from left to right. It has proved useful in representing words from all Mongolian dialects, and is still in use today. The Square script (hP'ags-pa) was invented between 1269 and 1368, on the order of Khubilai Khaan, for recording formal documents from a variety of languages, Tibetan, Sanskrit, Chinese, and Turkish. It consists of 44 letters, of which 30 are consonants. Further scripts include the Clear (Oirad) script invented in 1648, a modification of the Old Mongolian; the Soyombo and Horizontal-square scripts invented in 1686 to record holy texts in Mongolian, Tibetan, and Sanskrit; and the Vaghintara script invented in 1905, a further simplification of the Old Mongolian script, consisting of 36 letters, including eight vowels.

After the communist revolution in 1921, Mongolia came firmly under the influence of the USSR. Initially, the Mongolian script was used, but in the 1940s, a strong attempt at standardization was made with the introduction of a modified Cyrillic script, essentially the same as used in Russia, but with two additional letters for ö and ü, making a total of 35. The Cyrillic alphabet is still widely used, but after the democratic revolution in the early 1990s, official moves were made to reintroduce the classical Mongolian script.

In the face of continuing instability, it is no wonder that confusion reigns. Nevertheless, we have selected a single standard for transliteration, based on two works, the major official publication, *Information Mongolia*, prepared by the Mongolian Academy of Sciences, and published in 1990 [the book was published by the Pergamon Press in Oxford, and its owner, Robert Maxwell, is thanked for his 'far-sightedness'], and *A Modern Mongolian–English Dictionary* compiled by Altangerel Damdinsuren (1998). These books were recommended by Mongolian colleagues. In them, all Mongolian words and names are transliterated directly into English, using standard English consonants and vowels, but with the addition of the vowels 'ö', and 'ü', for the sounds 'ea' as in 'early' (but shorter, something between English 'o' and 'u') and 'o' as in 'who' respectively. Double vowels, 'aa', 'ee', 'ii', 'oo', 'öö', 'uu', and 'üü', indicate long vowels; these are rendered properly in the system adopted here. Additional consonants include 'kh/h', 'ts', 'ch' and 'sh', but the consonant often given as 'dz' (Russian spelling) is here rendered simply as 'z'.

Some standard geographic terms used in place names are listed, with the older transcription in parentheses: Aimag (aimak), major administrative division; Baruun (barun), right, Bulag (bulak), spring; Gol, river; Khudag (khudak), well; Nuruu (nuru), mountain range; Nuur (nur), lake; Ovoo (obo), heap, pile; Sum (somon), an administrative unit subordinate to an aimag; Teeg (teg), landform, any device to prevent things sliding; Tsagaan, white; Tsav (tsab), gorge; Ulaan (ulan), red; Uul (ula), mountain; Zoo (dzo), badland; Züün (dzun), left. Place names may take the form of several words. When these are converted into names of svitas or formations, the names are rendered as a single word.

Alag Teeg [locality]
Alag Tsav [locality]
Algui Ulaan Tsav [locality]
Alguiulaantsav Svita
Altai Sum
Altan Teeg [locality]
Altanteel Sum
  Altanulin (= Altanuul)
Altan Uul [locality]
Altanuul Svita
Amtgai [locality]
Andai Khudag [well]
Andaikhudag Formation/ Svita
Arts Bogd Ridge
Baga Mod Khudag [locality]
Baganuur [locality]
Baga Tariach [mountain]
Baga Zos Nuur [lake]
Bagazosnuur Svita
  Bain Chire (= Bayan Shiree)
  Baisheen (= Baishin)
Baishin Tsav [locality]
Baishin Tsav depression
Bakhar [locality]
Bambuu Khudag [locality]
  Barun (= Baruun)
Baruun Bayan cliffs
Baruunbayan Svita
Baruungoyot Formation/ Svita
Baruunurt [locality]
Bayandalai Sum
Bayankhongor
Bayan Mandahu [locality]
Bayan Mandahu basin
Bayanmönh Sum
  Bayan Munkh (= Bayanmönh)
Bayan Ovoo Uul [mountain]

Bayan Shiree cliffs [locality]

Bayanshiree Formation/ Svita

Bayan Tsav

Bayan Zag [locality]

Bayanzag Svita

 Bayn (= Bayan)

 Bayn Dzak (= Bayan Zag)

 Baynshin (= Baishin)

 Bayn Shire (= Bayan Shiree)

Beger Nuur depression

Berh Sum

 Berkhe Somon (= Berh Sum)

Böön Tsagaan [locality]

Bööntsagaan Gorizont

Bor Khovil [locality]

 Boro Khovil (= Bor Khovil)

Borzongiin Gobi [locality]

Bügiin Tsav [locality]

 Bugin Tsav (= Bügiin Tsav)

Builyastyn Khudag

Builyastyn Svita

 Bulagantu (= Bulgant)

 Buylyasutuin (= Builyastyn)

Bulgan Sum

Bulgant Svita

Bulgant Uul [mountain]

 Cav (= Tsav)

Choibalsan [locality]

Choibalsan Series

Choir [town]

Choir depression

Chono Kharaih [locality]

Chuluu Ungas Uul

Chuluut Uul

 Chzhirgalantuin (= Jargalantyn)

 Cis-Altai (= Pre-Altai)

 Dalandzadgad (= Dalanzadgad)

Dalanshandkhudag Formation/ Svita

Dalanzadgad (town)

 Darbi Somon (= Darvi Sum)

Darvi Sum

Davs (settlement)

Djadokhta Formation/ Svita

Dongsheng Formation/ Svita (Chinese)

Dornogov' Aimag

Dösh Uul (mountain)

Döshuul Formation/ Svita

Dundargalant Gorizont

Dundgov' Aimag

 Dushuul (= Döshuul)

 Dzabkhan (= Zavkhan)

 Dzagsokhairkhan (= Zogsookhairkhan)

 Dzamyn Khond (= Zamyn Khond)

 Dzergen (= Zereg)

 Dzhibkhalan (= Javkhlant)

 Dzhibkhalantu (= Javkhlant)

 Dzhirgalantum (= Jargalantyn)

 Dzost (= Zost)

 Dzun Bayan (= Züünbayan)

 Dzurumtai (= Zuramtai)

Ehingol depression

Elstiin [locality]

Erdenetsogtyn Gobi gorge

Erdene Uul [mountains]

Erenhot [locality]

 Ergelyeen Dzo (= Ergiliin Zoo)

Ergil Ovoo [locality]

Ergiliin Zoo [locality]

Ergiliinzoo Svita

 Flaming Cliffs (= Bayan Zag)

Galshar Sum

 Gashuni (= Gashuuny)

Gashuuny Khudag

Gilbent Ridge

 Gobi Altai (= Gov'altai)

Gobi Basin

Gobi Desert

Gov'altai

Guchin depression

Guchinus Sum

Gui Suin Gobi Depression (Chinese)

Guriliin Tsav [locality]

 Gurleen (= Guriliin)

Gurvan Ereen Nuruu [ridge]

Gurvanereen Svita

 Halhyn (= Khalkhyn)

 Halzan Hairhan (= Khalzan Khairkan)

 Hanbogd (= Khanbogd)

Mongolian place names

Hangai (= Khangai)
Har (= Khar)
Hatan Sudliin (= Khatan Suudal)
Hentii Mountains
Hermiin Tsav [locality]
Hetsüü Tsav [locality]
Hirgis Nuur [lake]
Hobur (= Höövör)
Höövör [locality]
Hötöl [locality]
Hovd (= Khovd)
Huachi Formation/ Svita (Chinese)
Huanhe Formation/ Svita (Chinese)
Hühteeg [mountain]
Hühteeg Gorizont
Hühteeg Svita
Hural (= Khural)
Hüren Dukh [locality]
Hürendukh Formation/ Svita
Hürmen Khudag [locality]
Hürmen Sum
Ih Bayan Uul [mountains]
Ih Ereen (mountain)
Ih Shunkht [locality]
Ih Zos Nuur [locality]
Ikh (= Ih)
Ikhe Dzosu Nuur (= Ih Zos Nuur)
Ihes Nuur [locality]
Ingeni Khovboor (= Ingenii Höövör)
Ingeni Khöövör (= Ingenii Höövör)
Ingenii Höövör [locality]
Ingenii Tsav [locality]
Iren Dabasu [lake; locality]
Iren Dabasu Formation (Chinese)
Jargalantyn Gol
Javkhlant Svita
Javkhlant Uul [locality]
Jibhalanta (= Javkhlant)
Jingchuang Formation/ Svita (Chinese)
Kalgan [town] (Chinese)
Khaichin Uul [mountain]
Khalkhyn Gol [river]
Khalzan Khairkhan [mountain]
Khamaran (= Khamaryn)

Khamareen (= Khamaryn)
Khamarin (= Khamaryn)
Khamaryn Us [locality]
Khamaryn Khural (= Khamaryn Us)
Khanbogd Sum
Khangai Mountains
Khar Hötöl Uul [mountain]
Khar Khutul (= Khar Hötöl)
Khar Us Nuur [locality]
Khara (= Har)
Khashaat [locality]
Khatan Suudal [locality]
Khaya Ulaan Nuur lake
Khentei (= Hentii)
Khentii (= Hentii)
Khermeen (= Hermiin)
Khermin (= Hermiin)
Khetzoo (= Hetsüü)
Khirgis (= Hirgis)
Khoboor (= Höövör)
Khobur (= Höövör)
Khoer (= Khoyor)
Kholboot Gol [river]
Kholboot Sair [locality]
Kholboot Svita
Kholbotu (= Kholboot)
Khongil Ovoo [mountain]
Khongil Tsav [locality]
Khoobur (= Höövör)
Khooldzin Plateau (Chinese)
Khoolsun (= Khulsan)
Khooren Dookh (= Hüren Dukh)
Khovboor (= Höövör)
Khovd [town]
Khoyor Zaan [locality]
Khukhtesk (= Hühteeg)
Khukhtyk (= Hühteeg)
Khulsan [locality]
Khulsangol Formation/ Svita
Khulsyn (= Khulsan)
Khural [locality]
Khuren [= Hüren]
Khurilt Ulaan Bulag [locality]
Khurmen (= Hürmen)

Khutoolyin (= Hötöl)
Khyra [locality] (Chinese)
Kobdo (= Khovd)
Kylodzhun [locality] (Chinese)
Lamawan Formation (Chinese)
Land Shan massif (Chinese)
Lehe Formation (Chinese)
Luohadong Formation (Chinese)
Mandalgov' [town]
Manlai Lake
Manlay (= Manlai)
Mergen [locality]
Mogoin Ulaagiin Hets [locality]
Mongol Altai Mountains
Mushgai Khudag well
Mushugai (= Mushgai)
Myangad (= Myangat)
Myangat Sum
Nalaih [locality]
Nalaikh (= Nalaih)
Naran Bulag [locality]
Naranbulag Svita
Naran Gol [locality]
Nemegt [locality]
Nemegt Basin
Nemegt Formation/ Svita
Nilgin (= Nyalga)
Nogon (= Nogoon)
Nogoon Tsav gorge [locality]
Nogoontsav Svita
Noyan (= Noyon)
Noyon Sum
Noyonsum Formation/ Svita
Nyalga [locality]
Ölgii Hiid [locality]
Olgoi (= Algui)
Ologoy (= Algui)
Ölzii Ovoo [mountain]
Ömnögov' Aimag
Ondai Sair (= Andai Khudag)
Ondaisair (= Andaikhudag) Formation/ Svita
Ondaisair Formation/ Svita
Öndörshil Sum

Öndör Ukhaa [mountain]
Öndörukhaa Svita
Ongon Ulaan [locality]
Ongon Ulaan Uul [locality]
Ongong (= Ongon)
Önjüül [locality]
Öösh Basin
Öösh Formation/ Svita
Ööshiin Nuruu [ridge]
Ööshiin Nuur [locality]
Ordos Basin
Orhon (= Orkhon)
Orkhon River
Osh (= Öösh)
Oshih (= Ööshiin)
Ovdog (= Övdög)
Övdög Khudag [locality]
Ovorhangai (= Övörkhangai)
Övörkhangai
Pre-Altai Gobi
Sainsar Bulag [locality]
Sainshand (town)
Sainshand Formation/ Svita
Sakhalak (= Sakhlag)
Sakhlag Uul [mountain]
Sangiin Dalai Nuur depression
Sayn Shand (= Sainshand)
Shabarakh Usu (= Bayan Zag)
Shaamar [locality]
Shabarakh Usu (= Bayan Zag)
Shamar (= Shaamar)
Shanh Sum
Shankh (= Shanh)
Sharga [locality]
Shariliin Formation/ Svita
Shar Teeg [locality]
Shar Tsav [locality]
Shiljust (= Shilüüt)
Shilt Uul [locality]
Shilüüt Uul [locality]
Shin Khuduk (= Shinekhudag)
Shine Khudag [locality]
Shinekhudag Gorizont
Shine Us Khudag

Mongolian place names

Shireegiin Gashuun [locality]
Shireegiin Gashuun Basin
   Shiregin Gashun (= Shireegiin Gashuun)
   Shirigin Gashoon (= Shireegiin Gashuun)
   Shirilin (= Shariliin)
   Shulutu Ula (= Chuluut Uul)
Sühbaatar [town]
Sümiin Nuur [lake]
   Tabun Tologoi (= Tavan Tolgoi)
Tamtsag Depression
Tariat Uul Ridge
   Taryatu (= Tariat)
Tatal Gol [locality]
Tatal Yavar [locality]
Tavan Tolgoi [locality]
   Tchono (= Chono)
   Tebshiin (= Tevshiin)
   Tebshün (= Tevshiin)
Tegaimiao Formation (Chinese)
Teel Ulaan Uul [mountain]
   Tel Ulan Ula (= Teel Ulaan Uul)
Tevsh Formation/ Svita
Tevshiin Gobi gorge
Tögrög [locality]
Tögrög Bulag [locality]
Tögrögiin Shiree [locality]
Tögrögiin Us [locality]
   Toogreek (= Tögrög)
   Toogreek Shire (= Tögrögiin Shiree)
Tormkhon Formation/ Svita
Trans-Altai Gobi
Tsagaan Gol [locality]
Tsagaangol Svita
Tsagaan Khushuu [locality]
Tsagaan Nuur depression
Tsagaan Teeg [locality]
Tsagaan Tsav well
Tsagaantsav Formation/ Svita
Tsagaantsav Gorizont
   Tsagan Tsab (= Tsagaan Tsav)
   Tsagan Ula (= Tsagaan Khushuu)
Tsakhiurt [locality]
Tsast Bogd Mountains
   Tsastoo Bogdo (= Tsast Bogd)

Tsetsen Uul [mountain]
   Tsogteen (= Tsogtyn)
Tsogtovoo Sum
Tsogtyn Gobi [gorge]
   Tugrig (= Tögrög)
   Tugrikin (= Tögrögiin)
   Tugrugiin Shireh (= Tögrögiin Shiree)
   Tüshilge (= Tüshleg)
Tüshleg [locality]
Tüshleg Ula [mountain]
   Ubur Khangai (= Övörkhangai)
   Udan Sayr (= Üüden Sair)
Uilgan river
Uilgan Svita
Ukhaa Tolgod [locality]
Ulaanbaatar [capital city]
   Ulan Bator (= Ulaanbaatar)
Ulaan Bulag [locality]
Ulaandel Svita
Ulaandel Uul [mountains]
Ulaan Nuur [lake]
Ulaan Nuur Basin
Ulaan Öösh [locality]
Ulaanöösh Svita
   Ulaan Sair (= Üüden Sair)
Ulaan Tolgoi [locality]
   Ulaan Tologoi (= Ulaan Tolgoi)
   Ulaanuush (= Ulaan Öösh)
   Ulan (= Ulaan)
   Ulan Osh (= Ulaan Öösh)
   Uldzhei (= Ölzii)
   Ülgii (= Ölgii)
   Ulugei (= Ölgii)
   Ülzii (= Ölzii)
   Umnogovi (= Ömnögov')
   Undershil (= Öndörshil)
   Under Ukhaa (= Öndör Ukhaa)
   Underukhiin (= Öndörukhaa)
   Undur (= Öndör)
   Undzhul (= Önjüül)
   Unjuul (= Önjüül)
Urilge Khudag [locality]
Üüden Sair [locality]
   Uvdeg (= Övdög)

Uverkhangai (= Övörkhangai)
Yagaan Khovil [locality]
Yagaan Shiree [locality]
Yavar [locality]
  Yavoor (= Yavar)
Yiginholo Formation (Chinese)
Yijun Formation (Chinese)
  Zaaltai (= Trans-Altai)
Zamyn Khond [locality]
  Zavhan (= Zavkhan)
Zavkhan River
Zereg Depression
Zereg Svita
Zhidan Group (Chinese)
  Zoost (= Zost)
Zost Uul
Zuramtai Depression
  Zurumtai (= Zuramtai)
Züün Bayan Cliffs
Züünbayan Svita

References

Academy of Sciences MPR. 1990. *Information Mongolia. The Comprehensive Reference Source of the People's Republic of Mongolia (MPR).* Oxford: Pergamon, 505 pp.

Damdinsuren, A. 1998. *A Modern Mongolian–English Dictionary.* Ulaanbaatar: Interpress, 654 pp.

Dashzeveg, D., Novacek, M.J., Norell, M.A., Clark, J.M., Chiappe, L.M., Davidson, A., McKenna, M.C., Dingus, L., Swisher, C. and Perle, A. 1995. Extraordinary preservation in a new vertebrate assemblage from the Late Cretaceous of Mongolia. *Nature* **374**: 446–449.

Gradziński, R., Kielan-Jaworowska, Z. and Maryańska, T. 1977. Upper Cretaceous Djadokhta, Barun-Goyot and Nemegt formations of Mongolia, including remarks on previous subdivisions. *Acta Geologica Polonica* **27**: 281–318.

Jerzykiewicz, T. and Russell, D.A. 1991. Late Mesozoic stratigraphy and vertebrates of the Gobi Basin. *Cretaceous Research,* **12**: 345–377.

JOURNALS AND SERIES

There follows an alphabetical listing of the Russian journals in which papers on Permian and Mesozoic tetrapods, and associated geology, have been described. Many of the journals have had a complex history, and it has been impossible to document the title changes in a fully researched bibliographic manner: titles and dates are merely indicative. So far as possible, details have been checked against serials in Russian libraries, and against the catalogues of the U.S. Library of Congress and the British Library. However, we felt that some standardization of nomenclature and spelling might be helpful.

The data have been derived first-hand from copies of journals and papers, from the various volumes of the *Bibliography of Vertebrate Paleontology* (BFV), and from the list of *Serial Titles held in the Library of the Geological Society* (London: The Geological Society, 1996). We have adopted the principle of using non-Russian (usually French) journal titles up to 1917. This approximates to the practice in Russia. Many of the key journals continued with double titles, often in Russian and in French, well after 1917, and the BFV tended to retain the use of French journal titles until well into the 1950s. Likewise, Russian sources often use the Russian journal titles for papers published before 1917. Our policy provides a working compromise.

[AN = Akademiya Nauk; RAN = Rossiiskaya Akademiya Nauk.]

Journals

Annuaire Géologique et Minéralogique de la Russie [St. Petersburg]

Vol. **1** (1895)–**17** (1917)

(= *Ezhegodnik po Geologii i Mineralogii Rossii*)

Annuaire de la Société Paléontologique de la Russie (see *Ezhegodnik Russkogo Paleontologicheskogo Obshchestva*)

Bulletin de l'Académie Impériale des Sciences de St. Pétersbourg

Series 3, Vol. **1** (1860) – Vol. **32** (1888)

Nov.Ser. **1**(33)–**4**(36) (1890–1894)

(= *Bulletin de la Classe Physico-Mathématique de l'Académie Impériale des Sciences de St. Pétersbourg*, 1836–1859)

(= *Bulletin Scientifique Public par l'Academie Impériale des Sciences de St. Pétersbourg*, Vol. **1–10**, 1837–1842)

(= *Izvestiya Imperatorskoi AN* [St. Petersburg], Series 5, Vol. **1** (1894)–Series 6, Vol. **4** (1917)

(= *Izvestiya Imperatorskoi AN. Fiziko-Matematicheskoe Otdelenie* [St. Petersburg], 1900–1908)

(= *Izvestiya Rossiiskoi AN. VI Seriya* [Petrograd], 1917–1925)

(= *Izvestiya AN SSSR. VI Seriya* [Leningrad], 1925–1927)

(= *Izvestiya AN SSSR. VII Seriya, Otdelenie Fiziko-Matematicheskikh Nauk* [Leningrad], 1928–1930)

(= *Izvestiya AN SSSR. VII Seriya, Otdelenie Matematicheskikh i Estestvennykh Nauk* [Leningrad], 1931–1935)

(= *Izvestiya AN SSSR. Seriya Geologicheskaya* [Moscow], 1936–1992)

(= *Izvestiya AN SSSR. Seriya Biologicheskaya* [Moscow], 1936–1992)

(= *Izvestiya AN Seriya Biologicheskaya* [Moscow], 1992–present day)

Bulletin de la Classe Physico-Mathématique de l'Académie Impériale des Sciences de St. Pétersbourg
Series 1, Vol. **1** (1836) – Vol. **7** (1842)
Series 2, Vol. **1** (1845) – Vol. **17** (1859)
(= *Bulletin de l'Académie Impériale des Sciences de St. Pétersbourg*, 1860–1894)

Bulletins du Comité Géologique, St. Pétersbourg
(see *Izvestiya Geologicheskogo Komiteta* [Leningrad])

Bulletin de la Société Impériale des Naturalistes de Moscou
Vol. **1** (1829) – Vol. **62** (1887) Nouvelle Série, Vol. **1** (1887)–**30** (1917)
(= *Byulleten' Moskovskogo Obshchestva Ispytatelei Prirody, Otdel Geologicheskii*, 1922–)

Byulleten' Moskovskogo Obshchestva Ispytatelei Prirody, Otdel Geologicheskii
Vol. **1** (1922)–present time
(= *Bulletin de la Société Impériale des Naturalistes de Moscou*, 1829–1917)

Doklady AN SSSR [Leningrad; then Moscow]
Novaya Seriya, Vol. **1** (1933) – Vol. **322** (1992)
(= *Doklady Rossiiskoi AN*, 1922–1925)
(= *Doklady RAN*, 1992–present time)

Doklady AN Tadzhikskoi SSR
Vol. **1** (1951) – **34** (1991)

Ezhegodnik po Geologii i Mineralogii Rossii Vol. **1** (1895) – **17** (1917)
(Parallel title = *Annuaire Géologique et Minéralogique de la Russie* [St. Petersburg]) Vol. **1** (1895) – **17** (1917)

Ezhegodnik Vserossiiskogo Paleontologicheskogo Obshchestva [Leningrad]
Vol. **1** (1916) – Vol. **9** (1931), vol. **34** (1991)–present time
(Parallel title = *Annuaire de la Société Paléontologique de la Russie*)
(= *Ezhegodnik Vsesoyuznogo Paleontologicheskogo Obshchestva*, 1932–1991)

Ezhegodnik TsNIGRI imeni Akademika F.N. Chernysheva [Leningrad]
Vol. **1** (1938)–present time

Ezhegodnik Vsesoyuznogo Paleontologicheskogo Obshchestva
Vol. **1** (1932) – **34** (1991)
(= *Ezhegodnik Vserossiiskogo Paleontologicheskogo Obshchestva*, 1916–1931)
(= *Ezhegodnik Vserossiiskogo Paleontologicheskogo Obshchestva*
Vol **34** (1991)–present time

Geologicheskii Vestnik [St. Petersburg]
Vol. **1** (1915)–

Geologicheskii Zhurnal [Kiev]
Vol. **1** (1934)–present time

Geologichnii Zhurnal
Vol. **1** (?1940)–**27** (1967)

Geologiya i Geofizika [Novosibirsk]
1961–present time (use year as volume number)

Gerpetologiya, Nauchnye Trudy Kubanskogo Gosudarstvennogo Universiteta

Gornyi Zhurnal [St. Petersburg to 1917; Moscow from 1922]
1825–1917, 1922–1929, 1934–1940, 1946–present time
(parallel title = *Journal des Mines, St. Pétersbourg*)
(= *Izvestiya Vysshikh Uchebnykh Zavedenii, Gornyi Zhurnal*)

Izvestiya Alekseevskogo Donskogo Politekhnicheskogo Instituta

Izvestiya AN Gruzinskoi SSR, Seriya Biologicheskaya [Tbilisi] Vol. **1** (1975)–**16**(5)(1990)
(= *Izvestiya AN Gruzii. Seriya Biologicheskaya* Vol. **16**(6) (1990)–present time)

Izvestiya AN Kazakhskoi SSR, Seriya Geologicheskaya
Vol. **1** (1940)–1991
(= *Izvestiya AN Respubliki Kazakhstan* 1992–present time)

Izvestiya AN SSSR. VII Seriya, Otdelenie Fiziko-Matematicheskikh Nauk [Leningrad]
1928–1930 [no volume numbers, use year and part numbers]
(= *Izvestiya AN SSSR. VII Seriya, Otdelenie Matematicheskikh i Estestvennykh Nauk* [Leningrad], 1931–1935)
(= *Izvestiya AN SSSR. Seriya Geologicheskaya* [Moscow], 1936–1962)

(= *Izvestiya AN SSSR. Seriya Biologicheskaya*
[Moscow], 1936–1962)
Izvestiya Biologicheskikh Nauk
Vol. **9** (1986)
Izvestiya Geologicheskogo Komiteta [Leningrad]
Vol. **1** (1882)–**48** (1929)
(Parallel title = *Bulletins du Comité Géologique, St.
Petersbourg*)
(= *Izvestiya Glavnogo Geologo-Razvedochnogo
Upravleniya*, Vol. **49** (1930)–Vol. **50** (1931))
(= *Izvestiya Vsesoyuznogo Geologo-Razvedochnogo
Ob'edineniya*, 1932–)
Izvestiya Imperatorskoi AN [St. Petersburg]
Vol. **1** (1894)–Vol. **17** (1902)
(= *Bulletin de l'Académie Impériale des Sciences de St.
Pétersbourg*, 1860–1894)
(= *Izvestiya Imperatorskoi AN. Fiziko-Matematicheskoe
Otdelenie* [St. Petersburg], Vol. **18** (1903)–Vol. **25**
(1906)
(= *Izvestiya Imperatorskoi AN. VI Seriya* [St.
Petersburg], Vol. **1** (1907)–Vol. **4** (1917)
(= *Izvestiya AN. VI Seriya* [Petrograd], 1917)
(= *Izvestiya Rossiiskoi AN. VI Seriya* [Petrograd],
1917–1924)
*Izvestiya Kazanskogo Filiala AN SSSR, Seriya
Geologicheskikh Nauk*
Vol. **1** (1950)–**10** (1963)
Izvestiya Rossiiskoi AN. VI Seriya [Petrograd]
1917–1924 [no volume numbers, use year and part
numbers]
(= *Izvestiya Imperatorskoi AN. VI Seriya* [St.
Petersburg], Vol. **1** (1907)–Vol. **4** (1917)
(= *Izvestiya AN SSSR. VI Seriya* [Leningrad],
1925–1927)
*Izvestiya Sibirskogo Otdeleniya Imperatorskogo Russkogo
Geograficheskogo Obshchestva* [Irkutsk]
Vol. **1** (1870)–**8** (1877)
(= *Izvestiya Vostochno-Sibirskogo Otdeleniya
Imperatorskogo Russkogo Geograficheskogo Obshchestva*
[Irkutsk]
Vol. **9** (1878)–**45** (1917))
(= *Izvestiya Vostochno-Sibirskogo Otdeleniya Russkogo
Geograficheskogo Obshchestva*

Vol. **46** (1921)–**57** (1937))
(= *Izvestiya Vostochno-Sibirskogo Otdeleniya
Geograficheskogo Obshchestva SSSR*
Vol. **58** (1954)–Vol. **69** (1976))
(= *Izvestiya Irkutskogo Gosudarstvennogo Nauchnogo
Muzeya*
Vol. **58** (1954)–Vol. **69** (1976))
*Izvestiya Vsesoyuznogo Geologo-Razvedochnogo
Ob'edineniya*
Vol. **51** (1932)–
(= *Izvestiya Geologicheskogo Komiteta* [Leningrad],
1883–1929)
(= *Izvestiya Glavnogo Geologo-Razvedochnogo
Upravleniya*, 1930–1931)
*Izvestiya Vysshikh Uchebnykh Zavedenii (Geologiya i
Razvedka)*
1958–1992 (use year as volume number)
Journal des Mines, St. Pétersbourg (see *Gornyi
Zhurnal*)
Materialy po Istorii Fauny i Flory Kazakhstana
Vols. **1** (1955), **2** (1958), **3** (1961)–**12** (1993)
Mémoires du Comité Géologique de St. Petersbourg
(see *Trudy Geologicheskogo Komiteta*)
*Mémoires de l'Académie Impériale des Sciences de St.
Pétersbourg*
Vol. **1** (1802)–**11** (1831)
(= *Mémoires de l'Académie Impériale des Sciences de St.
Pétersbourg. Séries 6. Sciences Mathématiques,
Physiques, et Naturelles*
Vol. **1** (1831)–**10** (1859))
Nauchnye Trudy Tashkentskogo, Universiteta
Vol. **237** (1964)
*Operativno-Informatsionnye Materialy k I Vsesoyuznomu
Soveshchaniyu po Paleoteriologii*
1989 (Moscow)
Otechestvennaya Geologiya (see *Sovetskaya Geologiya*)
Paleontologicheskii Zhurnal [Moscow]
1959–present time [no volume numbers, use year
and part numbers]
(available in English as *Paleontological Journal*)
Priroda [Leningrad-Moscow]
1912– present time [no volume numbers, use year
and part numbers]

Problemy Arktiki [Leningrad]
  Vol. **1** (1930)–Vol. **9** (1939)
Problemy Mongol'skoi Geologii
  Vol. **3** (1977), **5** (1982)
Problemy Paleontologii [Moscow]
  Vol. **1** (1936)–Vol. **5** (1939)
Problemy Sovetskoi Geologii (see *Sovetskaya
  Geologiya*)
Russkii Ornitologicheskii Zhurnal
  Vol. **1** (1992)–**4** (1996)
Sbornik Statei po Inzhenernoi Geologii
  Vol. **1** (1962)–
Sbornik Trudov Zoologicheskogo Muzeya MGU
  [Ulaanbaatar]
Selevinia [Almaty]
  Vol. **1** (1993)–present time
Soobshcheniya AN Gruzinskoi SSR [Tbilisi]
  Vol. **1** (1940)–**147** (1993)
Sovetskaya Geologiya [Moscow]
  1939–1992 [no volume numbers, use year and part
    numbers]
  (= *Problemy Sovetskoi Geologii*, Vol. **1** (1933)–Vol. **8**
    (1938)
  (= *Otechestvennaya Geologiya*, 1992–)
Teriologiya [Novosibirsk]
  Vol. **1** (1972)–present time
*Travaux de l'Institut Paléozoologique, Académie des Sciences
  de l'URSS*
  (see *Trudy Paleozoologicheskogo Instituta AN SSSR*)
Trudy Arkticheskogo Instituta AN SSSR [Leningrad]
  Vol. **1** (1930)–Vol. **140** (1939)
  (continued as *Trudy Arkticheskogo Nauchno-
    Issledovatel'skogo Instituta*, to Vol. **217** (1959))
Trudy Geologicheskogo Instituta AN SSSR [Leningrad]
  Vol. **1** (1932)–Vol. **9** (1939)
  (= *Trudy Geologicheskogo Muzeya AN SSSR*
    [Leningrad], 1926–1931)
  (= *Trudy, Ordena Trudovogo Krasnogo Znameni
    Geologicheskii Institut. AN SSSR* [Moscow],
    1939–1991)
Trudy Geologicheskogo Komiteta
  Vol. **1** (1883/4)–Vol. **20** (1902); Novaya Seriya, Vol.
    1 (1903)–Vol. **189** (1928)

(Parallel title = *Mémoires du Comité Géologique*)
*Trudy Geologicheskogo Muzeya imeni Petra Velikogo
  Imperatorskoi AN* [Petrograd]
  Vol. **1** (1907)–Vol. **8** (1915)
  (= *Trudy Geologicheskogo i Mineralogicheskogo Muzeya
    imeni Imperatora Petra Velikogo Imperatorskoi AN*
    [Petrograd], 1916)
  (= *Trudy Geologicheskogo i Mineralogicheskogo Muzeya
    imeni Imperatora Petra Velikogo Rossiiskoi AN*
    [Petrograd], 1918)
  (= *Trudy Geologicheskogo i Mineralogicheskogo Muzeya
    imeni Petra Velikogo Rossiiskoi AN* [Petrograd],
    1923–1926)
  (= *Trudy Geologicheskogo Muzeya AN SSSR*
    [Leningrad], 1926–1931)
  (= *Trudy Geologicheskogo Instituta AN SSSR*
    [Leningrad], 1932–1938)
  (= *Trudy, Ordena Trudovogo Krasnogo Znameni
    Geologicheskii Institut. AN SSSR* [Moscow],
    1939–1985)
Trudy Instituta Geologicheskikh Nauk AN SSSR
  Vol. **1** (1937)–**98** (1948)
  Vol. **101** (1948)–**165** (1955)
Trudy Instituta Zoologii Kazakhskoi SSR
  Vol. **1** (1953)–**45** (1990)
Trudy Mineralogicheskogo Obshchestva (see *Zapiski
  Vserossiiskogo Mineralogicheskogo Obshchestva*)
Trudy Mongol'skoi Komissii AN SSSR
  Vol. **1** (1932)–**33** (1937)
  Vol. **34** (1940)–**37** (1948)
  Vol. **38** (1949)–**44** (1953)
*Trudy Obshchestva Estestvoispytatelei Kazanskogo
  Universiteta*
  Vol. **1** (1871)–**67** (1964)
*Trudy, Ordena Trudovogo Krasnogo Znameni Geologicheskii
  Institut. AN SSSR* [Moscow]
  Vol. **9** (1939)–Vol. **398** (1985)
  (previously *Trudy Geologicheskogo Muzeya imeni Petra
    Velikogo Imperatorskoi AN* [Petrograd].
Trudy Paleontologicheskogo Instituta AN SSSR [Moscow]
  Vol. **8** (1937)–Vol. **249** (1991)
  (= *Trudy Paleozoologicheskogo Instituta AN SSSR*,
    1932–1937)

(= *Trudy Paleontologicheskogo Instituta RAN*, Vol. **250** (1992)–present time)

Trudy Paleozoologicheskogo Instituta AN SSSR [Leningrad]

Vol. **1** (1932)–Vol. **7** (1937)

(Parallel name = *Travaux de l'Institut Paléozoologique, Académie des Sciences de l'URSS*)

(= *Trudy Paleontologicheskogo Instituta AN SSSR*, 1937–1991)

Trudy Sankt-Peterburgskogo Obshchestva Estestvoispytatelei

Vol. **1** (1870)–**48** (1916), **91** (1994)

(= *Travaux de la Société des Naturalistes de Leningrad*, 1924–)

(= *Trudy Petrogradskogo Obshchestva Estestvoispytatelei*, 1922–1924)

(= *Trudy Leningradskogo Obshchestva Estestvoispytatelei*, 1924–1994)

Trudy Sovmestnoi Sovetsko-Mongol'skoi Nauchno-Issledovatel'skoi Geologicheskoi Expeditsii

Vol. **1** (1969)–Vol. **55** (1995)

Trudy Sovmestnoi Sovetsko-Mongol'skoi Paleontologicheskoi Ekspeditsii

Vol. **1** (1974)–Vol. **40** (1991)

(= *Trudy Sovmestnoi Rossiisko-Mongol'skoi Paleontologicheskoi Ekspeditsii*, Vol. **41** (1992)–**46** (1996))

Trudy VNIGRI (Vsesoyuznogo Nauchno-Issledovatel'skogo Geologo Razvedochnogo Instituta) [Leningrad]

Vol. **131** (1939)–Vol. **398** (1977); volumes unnumbered, 1977–1987.

(= *Trudy TsNIGRI (Tsentral'nogo Nauchno-Issledovatel'skogo Geologo-Razvedochnogo Instituta* [Leningrad and Moscow], Vol. **1** (1934)–Vol. **130** (1939)

Trudy Zoologicheskogo Instituta AN SSSR

Vol. **1** (1932)–**277** (1999)

(= *Annuaire du Musée Zoologique de l'Académie des Sciences de l'URSS*, – 1931)

Uchenye Zapiski Moskovskogo Universiteta, Otdel Estestvenno-Istoricheskii

Vol. **1** (1880)–**43** (1917)

Novaya Seriya, vol. **2** (1934)–**197** (1958)

Vestnik AN SSSR

Vol. **1** (1948)–present time

Vestnik Geologicheskogo Komiteta [Leningrad]

vol. **1** (1925)–present time

Vestnik Leningradskogo Universiteta. Seriya 3, Biologiya [*Sankt-Peterburgskogo*, after 1992]

Vol. **1** (1956)–present time

Vestnik Leningradskogo Universiteta. Seriya 7, Geologiya, Geografiya [*Sankt-Peterburgskogo*, after 1992]

Vol. **1** (1956)–present time

Vestnik Zoologii [Kiev]

Vol. **1** (1967)–present time

Voprosy Gerpetologii

1964, 1973, 1977, 1981, 1985, 1989 [no volume numbers, use year and part numbers]

Voprosy Geologii Azii

Vol. **1** (1954)–Vol. **2** (1955)

Voprosy Geologii Yuzhnogo Urala i Povolzh'ya

Vol. **1** (1964)–**4** (1967)/**22** (1981)

Voprosy Paleontologii

Numbered series in 1930s. After 1938, use year as volume number.

Yezhegodnik (see *Ezhegodnik*)

Zapiski Imperatorskogo Novorossiiskogo Universiteta [Odessa]

Vol. **1** (1867)–**104** (1906)

Zapiski Kievskogo Obshchestva Estestvoispytatelei

Vol. **1** (1870)–Vol. **27** (1929)

Zapiski Odesskogo Obshchestva Estestvoispytatelei

Vol. **1** (1872)–**42** (1918)

(= *Zapiski Novorossiiskogo Obshchestva Estestvoispytatelei* [Odessa])

Zapiski Sankt-Peterburgskogo Mineralogicheskogo Obshchestva

Vol. **1** (1830)–present time

Zapiski Vserossiiskogo Mineralogicheskogo Obshchestva [Moscow]

Vol. **62** (1933)–Vol. **67** (1938)

(= *Trudy Mineralogicheskogo Obshchestva* Vol. **1** (1830)–Vol. **2** (1842))

(= *Verhandlungen der Russisch-Kaiserlischen Mineralogischen Gesellschaft zu St. Petersburg*, 1842–1867)

(= *Verhandlungen der Kaiserlischen Gesellschaft für die Gesammte Mineralogie zu St. Petersburg*, 1862–1864)

(= *Zapiski Rossiiskogo Mineralogicheskogo Obshchestva*, Vol. 1 (1866)–Vol. 50 (1915))

(= *Zapiski Imperatorskogo Sankt-Peterburgskogo Mineralogicheskogo Obshchestva* [Petrograd], Vol. 51 (1918)–vol. 61 (1932))

(= *Zapiski Vsesoyuznogo Mineralogicheskogo Obshchestva* [Moscow and Leningrad], vol. 68 (1939)–vol. 123 (1994))

Zoologicheskii Zhurnal AN SSSR [*RAN* after 1990] Vol. 1 (1916)–present time

Publishers

Izdatel'stvo Akademii Nauk SSSR (Moscow)

Izdatel'stvo 'Metsniereba' Tbilisi

Izdatel'stvo Sankt-Peterburgskogo Universiteta

Izdatel'stvo Saratovskogo Universiteta

TRANSLITERATED NAMES OF RUSSIAN AND MONGOLIAN PALAEONTOLOGISTS AND GEOLOGISTS

In this list, we have followed a strict system of precise transliteration, using the system indicated earlier. This means that the spelling of names may differ from established forms, even forms preferred by the people themselves. However, it seemed preferable to adopt a single predictable system of transliteration, rather than to have many different schemes. The only exception is that certain forms used in formal taxonomic designations are retained, but only in such cases of attribution of authorship to a taxon name, even if they break the rules. These 'taxonomy-only' forms of names are indicated with an asterisk (*).

Afanas'ev, G.D.
   Afanasiev (= Afanas'ev)
Alifanov, V.R.
Amalitskii, V.P.
   Amalitskiy (= Amalitskii)
   Amalitsky (= Amalitskii)
   *Amalitzky (= Amalitskii)
Aref'ev, M.P.
   Arefiev (= Aref'ev)
Arkhangel'skii, M.S.
Auerbakh, I.B.
   Averianov (= Aver'yanov)
Aver'yanov, A.O.
Badamgarav, D.
Bakhurina, N.N.
Bannikov, A.F.
Barsbold, R.
Bayarunas, M.M.
Bazhanov, V.S.
   Beliajeva (= Belyaeva)

Belyaeva, E.I.
Blagonravov, V.A.
Blom, G.I.
Bogachev, V.V.
Bogdanova, T.N.
   Bogoljubow (= Bogolyubov)
   Bogolubov (= Bogolyubov)
   Bogolubow (= Bogolyubov)
Bogolyubov, N.N.
   Borissiak (= Borisyak)
Borisyak, A.A.
Borkin, L.J.
Brattseva, G.M.
Bugaenko, D.V.
Burakova, L.T.
Butsura, V.V.
Bystrov, A.P.
   Bystrow (= Bystrov)
Cherepanov, G.O.
Chkhikvadze, V.M.
Chudinov, P.K.
Danilov, A.I.
Darevskii, I.S.
   Darevsky (= Darevskii)
Dashzeveg, D.
   Davidov (= Davydov)
Davydov, V.A.
   Devjatkin (= Devyatkin)
Devyatkin, E.B.
Dmitriev, G.A.
Dmitriev, V. Yu.
Dmitrieva, E.L.
Dobruskina, I.A.

Doludenko, M.P.
Dombrovski, B.S.
  Dombrovsky (= Dombrovski)
Dovchin, N.
Dubeikovskii, S.G.
Durante, M.V.
Dzhalilov, M.P.
Efimov, M.B.
Efimov, V.M.
Efremov, I.A.
Efremova, T.I.
Eichwald, E.I. von
Eremin, A.V.
Fahrenkohl, A.
Favorskaya, T.A.
Fedorov, P.V.
Filin, V.R.
  Fischer de Waldheim (= Fischer von Waldheim)
Fischer von Waldheim, G.F.
Flerov, K.K.
Florentsov, N.A.
Frikh-Khar, D.I.
Gabunia, L.K.
  Garjainov (= Garyainov)
Garyainov, V.A.
Gavrilov, V.M.
Gekker, E.L.
Gekker, R.F.
Gerasimov, M.M.
Getmanov, S.N.
Glazunova, A.E.
Glazunova, K.P.
Glikman, A.L.
Glikman, L.C.
Golovneva, L.B.
Golubev, V.K.
Golubeva, L.P.
Goman'kov, A.V.
Gorbakh, L.G.
Gorbatkina, T.E.
Grishin, G.L.
Gubin, Y.M.
Gureev, A.A.
Hartmann-Weinberg, A.P.

Hofstein, I.D.
Holtman, E.D.
Ignat'ev, V.I.
  Ignatiev (= Ignat'ev)
Il'ichev, V.D.
Ivakhnenko, M.F.
Ivanov, A.Kh.
Ivanov, A.O.
Ivanov, V.G.
  Jakovlev (= Yakovlev)
  Jarkov (= Yarkov)
  Jaroshenko (= Yaroshenko)
  Julinen (= Yulinen)
Kabanov, V.A.
Kaie, A.
Kalandadze, N.N.
Kalantar', I.Z.
Kalugina, N.S.
  Karhu (= Karkhu)
Karkhu, A.A.
Kazanskii, P.
Kaznyshkin, M.N.
Kaznyshkina, L.F.
Khakimov, F.Kh.
Khand, E.
Khimenkov, V.G.
Khisarova, G.D.
Khosbayar, P.
Khozatskii, L.I.
Kiesielewski, F.Y.
  Kiprianoff (= Kipriyanov)
  Kiprijanoff (= Kipriyanov)
  Kiprijanov (= Kipriyanov)
Kipriyanov, W.A.
Klimov, P.N.
Kolesnikov, Ch.M.
  Konjukova (= Konzhukova)
Konzhukova, E.D.
Korabel'nikov, V.A.
Kordikova, E.
Kovaleva, N.P.
Kovalevskii, V.O.
Kramarenko, N.N.
Krasilov, V.A.

Krasovskaya, T.B.
  Krassilov (= Krasilov)
  Krassovskaya (= Krasovskaya)
Krupina, N.I.
Kuleva, G.V.
Kupletskii, B.M.
Kurochkin, E.N.
Kurzanov, S.M.
  Kusmin (= Kuz'min)
Kutorga, S.S.
Kuz'min, T.M.
Kuznetsov, V.V.
Kyansep-Romashkina, N.P.
Larishchev, A.A.
Lazurkin, D.V.
Lebedev, O.A.
Lebedev, V.D.
Lebedeva, Z.A.
Leonova, E.M.
Lopatin, A.V.
Lopato, A.Yu.
Lozovskii, V.R.
  Lozovskiy (= Losovskii)
  Lozovsky (= Lozovskii)
Luchitskaya, A.I.
Makarova, I.S.
Makulbekov, N.M.
Maleev, E.A.
Marinov, N.A.
  Martinene (= Mertinene)
Martinson, G.G.
Mashchenko, E.N.
Mazarovich, A.N.
Meien, S.V.
Menner, V.V.
Merkulova, N.N.
Mertinene, R.A.
  Meyen (= Meien)
Mikhailov, K.E.
Minikh, A.V.
Minikh, M.G.
Mitta, V.V.
Moiseenko, V.G.
Mokshantsev, K.B.

Molostovskii, E.M.
Mossakovskii, A.A.
Movshovich, E.V.
Murzaev, E.M.
Nagibina, N.S.
Naidin, D.P.
Nalbandyan, L.A.
Namsrai, T.N.
Narmandakh, P.
Nazarkin, M.V.
Nesov, L.A.
  Nessov (= Nesov)
Nikitin, S.N.
Nikolaeva, T.V.
Novikov, I.V.
Novodvorskaya, I.M.
Novokhatskii, I.P.
Novozhilov, N.I.
Nurumov, T.N.
Obruchev, V.A.
Ochev, V.G.
Orlov, Yu.A.
Orlovskaya, E.R.
  Otchev (= Ochev)
  *Otschev (= Ochev)
  Otshev (= Ochev)
Panteleev, A.V.
  Panteleyev (= Panteleev)
Perle, A.
Polubotko, I.V.
Ponomarenko, A.G.
Popov, Yu.A.
Potapov, A.Yu.
Potapov, D.O.
Potapova, O.P.
Pravoslavlev, P.A.
Prizemlin, B.V.
Rachevskii, I.P.
Rasnitsyn, A.P.
Rautian, A.S.
Reshetov, V.Yu.
  Riabinin (= Ryabinin)
  Rjabinin (= Ryabinin)
Rogovich, A.S.

Romanova, E.V.

Romanovskaya, G.M.

Rozanov, V.I.

Rozhdestvenskii, A.K.

   Rozhdestvenskiy (= Rozhdestvenskii)

   Rozhdestvensky (= Rozhdestvenskii)

Ryabinin, A.N.

Rychkov, P.I.

Sablin, M.B.

Saidakovskii, L.J.

Samoilov, V.S.

Selezneva, A.A.

Semeikhan, T.

Semenova, E.V.

Sennikov, A.G.

Sharov, A.G.

Shatkov, G.A.

Shelekhova, M.N.

Shilin, P.V.

Shimanskii, V.N.

Shishkin, M.A.

   Shiskin (= Shishkin)

   Shtukenberg (= Stuckenberg)

Shuvalov, V.F.

Sinitsa, S .M.

Sinitsyn, V.M.

   Sinitza (= Sinitsa)

Sintsov, I.F.

   Sintzov (= Sintsov)

   Sinzow (= Sintsov)

   Sinzoff (= Sintsov)

Skutschas, P.P.

Smagin, B.N.

Smirnov, V.N.

Sochava, A.V.

Sokolov, B.S.

Solonenko, V.P.

Solov'ev, V.K.

Solov'ev, A.N.

Solov'ev, N.S.

   Soloviev (= Solov'ev)

Stankevich, V.S.

Strok, N.I.

Stuckenberg, A.A.

   Stukenberg (= Stuckenberg)

Sukhanov, V.B.

Sukhov, I.

Sushkin, P.P.

Suslov, Y.V.

Sychevskaya, E.K.

   Sytchevskaya (= Sychevskaya)

Tatarinov, L.P.

   *Tchudinov (= Chudinov)

Trautschold, H.

Trofimov, B.A.

Trusova, E.K.

Tsaregradskii, V.

Tsybin, Yu.I.

Tumanova, T.A.

Turishchev, I.E.

Tverdokhlebov, V.P.

Tverdokhlebova, G.I.

Vakhrameev, V.A.

Vasil'ev, V.G.

Vavilov, M.N.

   Venjukov (= Venyukov)

Venyukov, P.N.

Vergai, I.F.

   Vergay (= Vergai)

Verzilin, N.N.

Vetrov, F.E.

Vislobokova, I.A.

   *Vjuschkov (= V'yushkov)

   Vjushkov (= V'yushkov)

Voinstvenskii, M.A.

Volkhonin, V.S.

Vorob'eva, E.I.

   Vorobjeva (= Vorob'eva)

   Vorobyeva (= Vorob'eva)

Voronin, Yu.I.

V'yushkov, B.P.

Yakobson, L.N.

Yakovlev, N.N.

Yanovskaya, N.M.

Yarkov, A.A.

Yaroshenko, O.P.

Yazikhov, P.M.
 Yefimov, M.B. (= Efimov)
 Yefimov, V.M. (= Efimov)
 Yefremov, I.A. (= Efremov)
 Yefremova, T.I. (= Efremova)
 Yeremin (= Eremin)
Yulinen, V.A.
Yur'ev, K.B.
 Yuriev (= Yur'ev)
Zaitsev, N.N.

Zaklinskaya, E.D.
 Zaytsev (= Zaitsev)
Zekkel, Ya D.
Zhamoida, A.I.
Zhegallo, V.I.
Zherikhin, V.V.
Zhuravlev, K.I.
Zonenshain, L.P.
Zykov, S.N.

The history of excavation of Permo-Triassic vertebrates from Eastern Europe

VITALII G. OCHEV AND MIKHAIL V. SURKOV

Introduction

Rich finds of tetrapods have been made in the mainly continental Upper Permian and Triassic deposits in the east of European Russia, west of the Ural Mountains. This territory, stretching from the Barents Sea to the Pre-Caspian, is covered with forests in the north and with steppe in the south. There are no vast badlands, yielding abundant fossil finds, as in South Africa or the Gobi Desert. The rocks are exposed only in river valleys and in ravines. Nevertheless, during almost two centuries of study, more than a thousand localities of Upper Permian and Triassic tetrapods have been discovered. They indicate a number of fossil faunas through time.

First discoveries in the Copper Sandstones (Late Permian)

The first of the local faunas to be discovered was one of the most ancient, the dinocephalian, from the early Kazanian to the early Tatarian (Late Permian). The localities of the dinocephalian fauna coincide mainly with the belt of Copper Sandstones, stretching for hundreds of kilometres along the western slope of the Ural Mountains. These finds were for a long time the oldest fossil reptiles from Europe. Their discovery was thanks to extensive mining works for copper, which were conducted in the eighteenth and nineteenth centuries. However, the finds which reached palaeontologists came only from a small number of mines in the present Orenburg Province, in Bashkortostan, and a very small number from Perm' Province. The distribution of the fossils probably reflects the presence of

people in charge of the mines who understood the great scientific significance of fossils.

The first indications of bones in the copper mines of the Cis-Urals are found in the works of participants of the 'Academical Expeditions' which were conducted by the Russian Academy of Sciences from 1765 to 1805 with the aim of studying the natural environment of Russia. However, the naturalists at that time did not fully understand the nature of these remains: the corresponding member of the Russian Academy of Science, P.I. Rychkov, mentioned his discovery of cupriferous fossil reptile bones (judging by their size, a dinocephalian) in the diary of his Orenburg travels (1770), but he took them for the remains of ancient mining workers.

The first scientific description of the remains of terrestrial vertebrates from the Copper Sandstones (of Perm' Province) was made by a Professor of the University of Saint Petersburg, S.S. Kutorga (1838). He established new taxa of predatory dinocephalians, *Brithopus* and *Syodon*, which were described on the basis of fragments of the humerus and tusk respectively. Kutorga took them for mammals, ascribing the first to the edentates, and the second to the pachyderms. Thus he was the first to notice the similarity between mammal-like reptiles and mammals.

Among a number of active collectors of fossils from the Copper Sandstones, there was a captain of the mining engineers corps, Sobolevskii, who gathered materials from the mines of Perm' Province. Especially notable was F. Wangenheim von Qualen, director of a number of mines in Ufa (Bashkortostan) and Orenburg Provinces, who assembled a large number of remains of fossil vertebrates. In the 1840s he began to publish a series of important essays in Russian scientific journals. Even before Murchison (1841) had distinguished the Permian System, he correlated the Copper Sandstones of the Cis-Urals with the Zechstein of Germany. We are indebted to him for the only data available now about the richest localities, as well as valuable information on the conditions of the burial of bones (Wangenheim von Qualen, 1845).

Materials from the collections of von Qualen were also studied by S.S. Kutorga and other Russian naturalists. Thus, a Professor of Moscow University, G.I. Fischer von Waldheim (1841) described a new genus of dinocephalian, *Rhopalodon*, and a corresponding member of the Academy of Science of Saint Petersburg, E.I. von Eichwald (1846), established the genus *Deuterosaurus*, and also described (Eichwald, 1848) the temnospondyl amphibian *Zygosaurus* on the basis of a complete skull.

A number of eminent foreign palaeontologists also studied the vertebrates from the Copper Sandstones. A significant part of von Qualen's collections was taken to Germany and later distributed to a number of museums, but most of these materials were destroyed during the Second World War. These specimens were studied by Hermann von Meyer. In 1866 he published a large monograph on terrestrial vertebrates from the Cis-Urals, in which he made many significant corrections to earlier researches. However, his understanding of the taxonomic content of the fauna was still not very clear: like the majority of naturalists at that time, he did not distinguish between amphibians and reptiles.

The material from the Copper Sandstones was first introduced to Richard Owen when it was brought from Russia to England after the 1841 expedition of Roderick I. Murchison (Figure 1.1). Owen took the Cis-Uralian dinocephalians for archosaurs. Later, he studied the materials in the British Museum, which had been collected in the Kargala mines in the Cis-Urals by the English company 'Russia Copper & Co.'. Owen (1876) gave the most complete analysis of this fauna: he compared the remains of the reptiles from the Copper Sandstones with the South African therio-

donts, a new order which had been established by him. Another English researcher, W.H. Twelvetrees, published in 1880–2 a number of articles on remains of the vertebrates from the Kargala mines, which he had visited. From here, he established the new temnospondyl genus *Platyops* on the basis of a skull. The monograph by H.G. Seeley (1894) was highly significant: he worked out in detail the collections of the Saint Petersburg Mining Institute and those of Kazan' University, and gave excellent drawings. However, Seeley's opinion that the Cis-Uralian reptiles were related to the Placodontia and Nothosauria delayed for a long time a correct understanding of the Copper Sandstones fauna.

The interest of foreign scientists in the fauna did not diminish at the beginning of the twentieth century. New reconstructions and descriptions were made on the basis of data from the literature, mainly from the work by Seeley. The eminent German palaeontologist Friedrich von Huene (1905) for the first time emphasized the similarity between the Cis-Uralian reptiles and the pelycosaurs. The famous English palaeontologist D.M.S. Watson (1914) concluded that all the genera of reptiles described from the Copper Sandstones belonged to the Dinocephalia. The last restudy of the original material in the Saint Petersburg Mining Institute was completed by F. von Nopcsa (1928). His work, although containing many mistakes, concluded with the correct assessment that the reptiles of the Cis-Urals were synapsids, and that they include forms intermediate between the pelycosaurs from the Early Permian of Texas, and the therapsids of South Africa.

With the gradual cessation of copper mining from the 1880s, and its termination at the beginning of the twentieth century, new discoveries of vertebrates from the Copper Sandstones of the Cis-Urals almost completely stopped. The main collections from this time were made in slag heaps of the old mines. Thus, the Russian geologist P.N. Venyukov found fragments of the jaws of dicynodonts in the slag heaps of Kargala mines, which were described posthumously by V.P. Amalitskii (1922) as the new genus *Venyukovia* and ascribed by him at first to the mammals. The first

Figure 1.1. The Gurmaya Hills of the South Ural Mountains, as seen from the steppes of Orenburg. This engraving shows the scene as witnessed by Sir Roderick Impey Murchison when he visited Russia in the early 1840s, and first recognized the Permian System. (From Murchison and de Verneuil, 1845.)

localities in natural exposures, not in mines, were found only in the 1890s by a Kazanian geologist, A.A. Stuckenberg (1898).

Amalitskii and the Late Permian of the North Dvina River

The main attention of Russian researchers at this time was drawn to new, younger (Late Tatarian) Permian faunas with pareiasaurs, gorgonopsians, and dicynodonts (Figure 1.2). We are indebted to a professor of Warsaw University, V.P. Amalitskii, for the discovery of this fauna on the River North Dvina (eastern Poland at that time was part of Russia). His work was an heroic episode in the history of Russian palaeontology. Amalitskii began to study the Permian deposits of the Middle Volga region and found bivalves, which

proved to be very similar to freshwater forms known from deposits of the same age from South Africa. It led Amalitskii to the suggestion that there should be other shared fossils, including the large reptiles (pareiasaurs and various mammal-like reptiles). This idea was supported by the fragmentary remains of dicynodonts, which were found by Amalitskii.

Amalitskii's idea was not greeted sympathetically at first, since it was in opposition to the generally accepted idea at the time that the animal and plant worlds were completely different in the northern and southern hemispheres in the Permian. But the enthusiasm of the scientist was strong and, with scanty resources, which he obtained from the Saint Petersburg Naturalists' Society, he began a field study along the banks of the Sukhona, North Dvina, and other smaller rivers.

Figure 1.2. A Late Permian fauna from the North Dvina River, containing reptiles excavated by V.P. Amalitskii. A gorgonopsian, *Inostrancevia*, attacks two large herbivorous pareiasaurs, *Scutosaurus*. (Restoration by A.P. Bystrov.)

These researches were conducted in difficult conditions. Amalitskii, with his wife, travelled along the northern rivers by boat under the open sky, and took shelter under the boat at night and in rainy weather. This continued each summer from 1895 to 1898. They became accustomed to midges, a very poor diet, and to the constant dampness and mists. After four years of determined searches, Amalitskii's efforts were rewarded. The fauna of Late Permian reptiles found by him, known before only from South Africa and India, became one of the greatest discoveries in palaeontology of the nineteenth and early twentieth century.

When Amalitskii delivered his first findings to a meeting of the Saint Petersburg Naturalists' Society, opinions changed, and he received a small grant for excavations, which were started in 1899. In the beginning the main work was conducted on the right bank of the River North Dvina, above Kotlas rail station, in the area called Sokolki. Here, striped marls with several large lenses of sand and sandstone are exposed, and in the lenses, he observed large spherical concretions, sometimes containing bones and remains of plants. Amalitskii chose one of these lenses for excavation. However, the absolutely vertical cliff did not allow access to the lens either from below or above. Only after removing sediment from above was it possible to get deeper into the sandstone to quarry the concretions from it. There were plenty of them, but often without fossils, and Amalitskii searched for a long time before he came across concretions with skeletons. To extract them, he dug a gallery 7 m long, 4 m wide, and the same in height. Altogether, he found 39 large bone-bearing concretions. All the collections were packed in 64 boxes, which filled two rail carriages and weighed 20 tonnes.

These were the results of the first organized excavations in the history of Russian palaeontology, and no-one doubted Amalitskii's success after that. A large sum of money, 50 000 roubles was given to him, which allowed him to continue excavation on a much more significant scale for many years. These resources also gave him the opportunity to organize the first palaeontological workshop in Russia, where the skeletons were prepared and mounted. This formed the nucleus of 'The North Dvina Gallery', created by Amalitskii. The present Palaeontological Museum of the Russian Academy of Science later developed from this gallery.

In the history of excavations by Amalitskii, there were many unexpected events and difficulties. The local people for a long time did not believe that he was searching for antediluvian animals, but thought that he was digging for gold. Only the finds of well-preserved jaws and skulls convinced them. The first season of excavation could have ended tragically, because of the appearance of the livestock plague, Siberian ulcer, in surrounding villages. A rumour spread that the professor was digging up an old cattle grave, and that the decayed corpses were spreading infection to the livestock. Fortunately, a veterinarian arrived and stopped the cattle plague, and the disturbances by the local people subsided.

The beginning of the First World War in 1914 marked a break in excavations on the North Dvina, and the unexpected death of Amalitskii stopped the research. The significant part of his results was published post-mortem, mainly in a special series by the Academy, *The North Dvina Excavations by Professor V.P. Amalitskii* (1921–7). These included preliminary descriptions and diagnoses of a number of new taxa of Permian tetrapods: the temnospondyl *Dvinosaurus*, the reptiliomorph amphibian *Kotlassia*, the gorgonopsian *Inostrancevia*, and others. The other materials of gorgonopsians were studied by a student of Amalitskii, Professor P.A. Pravoslavlev (1927). Academician P.P. Sushkin, known for his ornithological works, and since 1922 a curator of the North Dvina Gallery, used these collections for his classic works on the evolutionary morphology of vertebrates (Sushkin, 1926, 1927, 1936). These studies became widely known in the West.

The work of drawing geological maps, which was developing in Russia after the organization in 1922 of a Geological Committee, led to the discovery of numerous vertebrate localities in natural sections in the Permian and Triassic. Valuable finds were made in

European Russia by famous early twentieth century Russian geologists, N.G. Kassin, P.M. Zamyatin, V.A. Tsaregradskii, M.A. Zhirmunskii, and by many of their followers. The excavations and search expeditions started by the North Dvina Commission and Geological Museum, were particularly activated after the organization in 1930 by Academician A.A. Borisyak of the Palaeozoological (later, Palaeontological) Institute in the Academy of Sciences of the USSR. Before the Second World War, Permian and Triassic tetrapods were studied by a few researchers. A.P. Hartmann-Weinberg, who was the curator of the North Dvina Gallery after the death of P.P. Sushkin, worked on the rich materials discovered in the North Dvina, and she established (Hartmann-Weinberg, 1933) that the local pareiasaurs belonged to the new genus *Scutosaurus*. However, the main role at the beginning of the new stage of study belonged to a student of P.P. Sushkin, Professsor I.A. Efremov (Figure 1.3).

Ivan Antonovich Efremov (1907–72)

Efremov gained world fame as a vertebrate palaeontologist, the founder of a new science, taphonomy, and as a writer of science fiction, widely popular in Russia. His scientific activity began very early. In 1925, when he was 18 years old, he became a preparator in the Mining Museum in Leningrad and straightaway started independent expeditions. In 1926 he studied the conditions of burial of temnospondyl amphibians on the mountain Bolshoe Bogdo, near Lake Baskunchak in the Cis-Caspian, where at the end of the nineteenth century, I.B. Auerbakh, and subsequently M.M. Bayarunas, found amphibian bones in the marine Lower Triassic (see Efremov, 1928).

From 1927 to 1930, Efremov excavated localities yielding temnospondyls and small reptiles (archosaurs and others) in the continental Triassic on the Volga–Dvina watershed. As a result, very rich material was obtained and led him to the discovery of the hitherto unknown most ancient fauna of Early Triassic amphibians. Here a group of new genera was established by him, and partially by Professor A.I. Ryabinin,

Figure 1.3. Professor Ivan Antonovich Efremov (1907–72), most celebrated Russian vertebrate palaeontologist of the twentieth century.

named by Efremov the Neorachitomi. The description of the most richly represented form, *Benthosuchus sushkini*, completed later jointly with a skilled morphologist and excellent artist, Professor A.P. Bystrov (Efremov and Bystrov, 1940), was awarded a Diploma of the Linnean Society in London. The remains of small reptiles collected by Efremov from the Early Triassic of European Russia were described by Huene (1940).

In 1930, Efremov became head of the Ural–Dvina Expedition of the Palaeontological Institute, which embraced the study of the Permian and Triassic of the north of European Russia and the Cis-Urals, and from 1934 to 1939 he led the Volga–Kama Expedition. Now he gave his main attention to Permian vertebrates, first of all from the Cis-Uralian Copper Sandstones. His aim was to make new finds in the old mines and to study the conditions of deposition of the bones, and this study was continued up to 1939. He went down into the the old pits, often at considerable risk. His work at the Kargala mines and at the Akbatyrovo mines in the Kirov Province (former Vyatka Gouvernement) in 1934 did not lead to significant new

finds of bones, but they provided the first information on the conditions of burial of vertebrates in the Copper Sandstones.

Efremov summarized his research on tetrapods from the Copper Sandstones of the Cis-Urals in a fundamental monograph (Efremov, 1954). Having revised the systematic content of this fauna, he added to the list new genera of predatory dinocephalians and other therapsids.

Studies by geologists, and expeditions from the Palaeontological Institute of the Academy of Sciences, in the regions west of the zone of the Copper Sandstones led to the discovery of large new vertebrate localities of roughly the same geological age. A rich dinocephalian fauna was found near the village Isheevo in Tatarstan. From this locality, excavated from 1934 to 1939, complete skeletons of dinocephalians were obtained, similar to the latest forms from the Copper Sandstones. Efremov named this the Isheevo Dinocephalian Complex, to distinguish it from the Cis-Urals dinocephalian fauna. Yu.A. Orlov (1958) partially studied the findings from Isheevo, and described from this locality the carnivorous dinocephalians. Efremov (1946) established here a new genus of reptiliomorph amphibian, *Lanthanosuchus*, the analysis of which led him to distinguish the Subclass Batrachosauria.

Two further large localities of Permian vertebrates in the Cis-Urals were studied by Efremov, one low on the River Mezen' in northern Russia, and the other near the town Belebey in Bashkortostan. They were discovered by Ya.D. Zekkel in 1934–5, and by an assistant of Efremov, N.I. Novozhilov in 1937–8. In these localities, in contrast to the Dinocephalian complexes, the remains of small anapsid reptiles dominated. This gave Efremov the basis for the Mezen'–Belebey Cotylosaurian Complex, coeval with the Dinocephalian, but originally thought to be younger. In 1938–40, he established here the genera of anapsid reptiles *Nycteroleter*, *Nyctiphruretus*, *Rhiphaeosaurus*, an eotheriodont *Phthinosaurus*, and the pelycosaur *Mesenosaurus*.

Information on the younger North Dvina Pareiasaurian Complex was supplemented in the

1930s by the discovery of two new large localities: near the village of Il'inskoe on the Volga River near the town of Tetyushi in Tatarstan, and near the town of Kotel'nich on the Vyatka River in Kirov Province. However, the unique character of the tetrapod assemblages found here, which is slightly older than that from the North Dvina lenses, only became clear later.

Before the Second World War, Efremov summarized all the Permian and Triassic tetrapods from the former USSR (Efremov, 1940a, 1941). On the basis of his accumulated data, he created a zonal scheme for the stratigraphy of the continental Permian and Triassic on the basis of tetrapods (Efremov, 1937). It gained wide fame among geologists and was revised many times (Efremov, 1939, 1944, 1952) and is the basis of the current scheme.

Efremov had an early interest in the processes of burial of fossil vertebrates, and his wide experience of many cases led him to a number of important generalizations. In 1940 Efremov published an article on 'Taphonomy – the new branch of palaeontology' (Efremov, 1940b). In 1950, he published his monograph 'Taphonomy and the fossil record', in which he gave a broad outline of the study on burial of organic remains and formulated a number of generalizations. Although this book was published in Russian, it became widely known first in the USA, and in other Western countries.

Among researchers who worked with Efremov in the 1930s on Permian and Triassic tetrapods was Professor A.P. Bystrov. He combined the gifts of the morphologist and artist, and had created in 1935 reconstructions of the North Dvina fauna, and later of the dinocephalian fauna. He completed a number of studies on the histology of bones and teeth, the circulatory system of Palaeozoic amphibians (Bystrov, 1938, 1939, 1947) and, in the 1940s and early 1950s, he published detailed descriptions of some Permian amphibians and reptiles (e.g. Bystrov, 1944, 1957).

Researches in European Russia in the 1950s

Researches on Permian and Triassic tetrapods began in the late 1940s after a gap caused by the Second

Figure 1.4. Dr Valentin Petrovich Tverdokhlebov surveys Lower Triassic sediments at Petropavlovka, north of the Sakmara River, north-east of Orenburg. These river deposits have yielded isolated bones of temnospondyl amphibians, procolophonids, and other tetrapods. (Photograph taken on the 1995 Saratov–Bristol Expedition to the South Urals.)

World War. Many new localities were discovered by surveyors working on a new programme to create the States geological map of the USSR at a scale of 1:200 000. A number of geologists gained fame as fossil hunters. Those working mainly in the northern regions of European Russia were Dr G.I. Blom from Nizhnii Novgorod, and Professor V.I. Ignat'ev from Kazan' University. Significant finds were made in southern regions by Dr V.V. Butsura and Dr V.A. Garyainov (Figure 1.4).

Figure 1.5. Peter Konstantinovich Chudinov, one of the leading Russian experts on mammal-like reptiles. He began his studies in the 1950s. He is pictured here, ready to go fishing, accompanied by the camp cook's dog, Jimmy Carter. (Photograph taken on the 1995 Saratov–Bristol Expedition to the South Urals.)

At this time, new researchers were beginning to work on fossil tetrapods. In the Palaeontological Institute in Moscow these were E.D. Konzhukova, the wife of I.A. Efremov, and two of his young students, B.P. V'yushkov (who died tragically at the age of 32) and P.K. Chudinov (Figure 1.5). Konzhukova gave the first description (1953) of eryopoid temnospondyls from the coal-bearing deposits of the basin of the Pechora River near the city of Inta, found in a mine by the geologist G.A. Dmitriev and others. Now they are ascribed to the lowest part of the Upper Permian.

Later, Konzhukova (1955) established the presence in the southern Cis-Urals of the temnospondyl *Mastodonsaurus*, which suggested a Middle Triassic age. With this, she gave palaeontological evidence for the uppermost zone in the stratigraphic scheme of Efremov.

V'yushkov began his studies very early, at the age of 19–20, in 1945 with the discovery of the previously-mentioned Middle Triassic fauna with *Mastodonsaurus*, in southern Bashkortostan. He graduated from Saratov University, began to work at the Palaeontological Institute, and soon became the most energetic researcher in Efremov's laboratory. In 1947–9, V'yushkov and Konzhukova, as well as N.I. Novozhilov, conducted an expedition in the Orenburg region, where they excavated a locality of latest Permian (late Tatarian) batrachosaurs near the village of Pron'kino. In the same years, V'yushkov studied the still poorly known localities of the same age near the city of Gorky (now Nizhnii Novgorod), and also a number of other places in the European and Asian parts of the former USSR. V'yushkov's studies of these localities formed the basis for his establishment of the Gorky Batrachosaur Complex as part of Efremov's stratigraphic scheme, equivalent to the North Dvina Pareiasaur Complex. V'yushkov (1957b) described from here new genera of batrachosaurs, including the chroniosuchians. His study of theriodonts from Isheevo and from North Dvina (V'yushkov, 1955) supported Efremov's opinion that there was a significant gap in evolution between the Cis-Uralian Dinocephalian and Pareiasaur faunas.

We are indebted to V'yushkov and Chudinov (1956, 1957) for the first information on a number of groups of ancient tetrapods, previously unknown from Russia, whose remains had been collected by geologists and palaeontologists: unequivocal North American elements (captorhinids and caseids) from the Upper Permian of the northern regions, and procolophonids from the Lower Triassic of central and southern regions of the east of European Russia. In addition, V'yushkov (1957a), following finds in the basin of the Vetluga River made by the geologists G.I. Blom and V.I. Ignat'ev, recognized for the first time in

Figure 1.6. A scene from the early Triassic of the Urals Region, showing a large erythrosuchid, *Garjainia*, chasing some temnospondyls which are lingering at the water's edge. The theriodont *Silphedosuchus* is hunting some small prolacertiforms and procolophonids in the foreground. (Illustration by A.A. Prokhorov.)

Russia the amphibian *Tupilakosaurus*, which had previously been found in marine Induan deposits of Greenland. This was a critical stratigraphic marker, linking the continental basal Triassic of European Russia to the global standard stratigraphic scheme based on ammonoids.

However, V'yushkov's main interests were still in the Orenburg region of the Cis-Urals. In 1954, he conducted large excavations there using bulldozers, the first time this had been done in Russia, but commonplace later. Excavations near the village of Perovka, where the geologist P.N. Klimov had once found specimens of *Rhadiodromus* which were described by Efremov (1940c), yielded a diverse fauna of gigantic kannemeyeriid dicynodonts. A Lower Triassic locality near the village of Rassypnoe, found by V.A. Garyainov, produced the first whole skeletons of proterosuchian archosaurs (Figure 1.6). The untimely death of V'yushkov prevented him from completing these researches.

Efremov and V'yushkov (1955) published a catalogue of localities of Permian and Triassic terrestrial tetrapods from the territory of the USSR, which gave the results of the discoveries and researches of the first half of the twentieth century. This catalogue concluded the Efremov era in the study of the ancient tetrapods of Russia, and it was made available widely in English in an abbreviated version (Olson, 1957).

P.K. Chudinov, the last of Efremov's students, began research on the collections of small reptiles of the Mezen'–Belebey Cotylosaur Complex (Chudinov, 1955, 1957). However, his main achievement was the study of numerous early therapsids based on rich new material collected by him during excavations of the locality Ocher in Perm' Province. This was one of the greatest localities of Permian tetrapods, discovered in Upper Kazanian deposits, and it was the scene of perhaps the largest excavations in the history of Russian palaeontology. It was excavated by Chudinov during four field seasons (1952, 1957, 1958, 1960) with the use of bulldozers. The area of excavation reached 6000 square metres. This yielded an unknown number of individuals, but the known cranial remains indicate at least 50 animals. The results of the study of those remains were given by Chudinov (1983), who considered the morphology, phylogeny, and origin of the eotheriodonts, dinocephalians, and early dicynodonts, the Venyukoviidae. The vast Ocher fauna revealed one of the most important pages in the fossil record of the Late Permian of Russia.

1960–95: new specimens, revised biostratigraphy, and phylogeny

At the beginning of the 1960s, a new generation of researchers started to study ancient tetrapods, and the number of specialists grew with each decade. In addition, field geologists, primarily Prof. V.R. Lozovskii (Moscow Geological Prospecting Institute) and Dr V.P. Tverdokhlebov (Figure 1.4) contributed their efforts in the hunt for specimens and in establishing stratigraphies. From this time, Saratov University took an active part in the excavation and field study of localities, this work having been conducted earlier only by the Palaeontological Institute in Moscow. The largest excavations, using bulldozers, were conducted by V.G. Ochev (Figure 1.7) in the southern Cis-Urals. First of all, a group of taphonomically diverse large localities were dug out along tributaries of the Ural, the Donguz and Berdyanka Rivers in the Sol'-Iletsk district, Orenburg Province. This small territory with numerous finds of tetrapods was named the 'Sol'-Iletsk phenomenon'.

All these localities were considered earlier as Lower Triassic. As a result of the excavations it was possible to obtain, besides kannemeyeriid dicynodonts, theriodonts and basal archosaurs, complete skeletons of capitosauroid temnospondyls (in one of the localities on the River Berdyanka, a concentration of 20 skeletons was dug up; Figure 1.8), and for the first time in Russia mass remains of plagiosaurs were collected. On the basis of the temnospondyls, it was possible to date this assemblage as Middle Triassic, but older than the fauna with *Mastodonsaurus*, which had been found here earlier. This assemblage was named the *Eryosuchus* Fauna, after its dominant genus of capitosauroids.

As a result of overviews of the new material, mainly from the Moscow Syncline and the southern Cis-

Figure 1.7. Vitalii Georgievich Ochev, Professor at Saratov University, leading expert on the fossil reptiles and amphibians, and the biostratigraphy, of the Triassic of the south Urals. He began his studies in the late 1950s. (A) V.G. Ochev wields a pick at the Koltaevo III locality, site of numerous finds of dicynodonts in the 1960s. (B) V.G. Ochev and Misha Surkov (foreground) working at the Koltaevo III locality. (Photographs taken on the 1995 Saratov–Bristol Expedition to the South Urals.)

Urals, M.A. Shishkin (Palaeontological Institute) and V.G. Ochev (Saratov University) suggested in 1967 a more complete and detailed scheme of stratigraphy of the continental Triassic, based on tetrapods, in which a number of sequential faunas where distinguished: Lower Triassic Neorachitome (Zone V in the scheme of Efremov), *Parotosuchus* (Zone VI of Efremov), Middle Triassic *Eryosuchus* (not distinguished in Efremov's scheme), and *Mastodonsaurus* (Zone VII of Efremov). Some of these were subdivided in more detail. A similarly improved scheme for the Permian

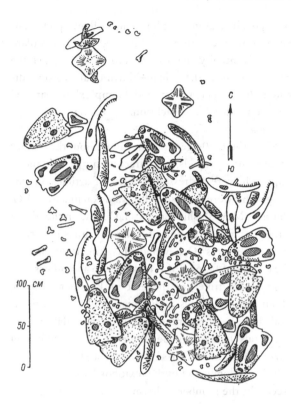

Figure 1.8. Map of a mass accumulation of temnospondyl (*Eryosuchus*) skulls and other elements (jaws, rhomboid interclavicle plates, ribs, vertebrae, and limb bones) from a site discovered on the Berdyanka River in the early 1960s by V.G. Ochev.

was suggested by Chudinov (1969). It was based mainly on three successive faunas: the Ocher and Isheevo, both dinocephalian (Zones I and II of Efremov) and the North Dvina or Sokolki, according to M.F. Ivakhnenko, the pareiasaurian (Zone IV of Efremov; in his opinion, a gap corresponded to Zone III).

Since the 1960s, further work has been carried out on the morphology of vertebrates, using the older specimens, as well as newly collected materials. L.P. Tatarinov (1974, 1976) described new therapsids. Among other new taxa of tetrapods established by him, was the first Upper Permian archosaur, *Archosaurus*, from the locality Vyazniki in the Oka River basin (Tatarinov, 1960).

M.A. Shishkin concentrated his attention on

amphibians. In publications on brachyopoids and plagiosaurs respectively, Shishkin (1973, 1987) considered the morphology and development of the head of temnospondyls, and the evolution of the middle ear and of the vertebral column in the lower tetrapods. He established for the first time the brachyopoid affinities of *Tupilakosaurus*, long regarded as a palaeontological puzzle. Shishkin described a number of new genera of temnospondyls. Of particular significance for the correlation of the continental and marine Triassic was his description of finds of temnospondyls from nearshore marine deposits: *Parotosuchus* from the Upper Olenekian from the Mangyshlak Peninsula, the benthosuchid *Benthosphenus* from the Lower Olenekian of the Far East, and the rhytidosteid *Boreopelta* from the Lower Olenekian of the River Olenek basin in Siberia. The interest of these findings were increased by Shishkin's (1994) identification of the Gondwanan genus *Rhytidosteus* in the *Parotosuchus* Fauna of the southern Cis-Urals.

V.G. Ochev studied the systematics and phylogeny of the largest group of Lower Triassic temnospondyls, the Capitosauroidea, and gave the results in two monographs (Ochev, 1966, 1972). Among a number of new taxa, the genus *Eryosuchus* was established, an index fossil of the new Middle Triassic fauna already noted. Later, Ochev focused his attention on the early archosaurs, described the first rauisuchid from the Triassic of Russia, *Vjushkovisaurus* (Ochev, 1982), and considered the phylogeny of proterosuchians in a special monograph (Ochev, 1991).

N.N. Kalandadze (1969, 1973) studied the extensive material of Middle Triassic dicynodonts of the southern Cis-Urals, which allowed him to recognize this fauna as one of richest in the world. A big event was the identification by Kalandadze (1975) of the first *Lystrosaurus* in Russia from materials which had been collected by G.I. Blom from the lowermost Triassic on the Vetluga River.

M.F. Ivakhnenko studied the fossil material of the most primitive reptiles and reptiliomorph amphibians. He published monographs on procolophonids (Ivakhnenko, 1979) and on the Permian reptiliomorph amphibians of the former USSR (Ivakhnenko, 1987).

In 1980, jointly with G.I. Tverdokhlebova from Saratov University, he published a monograph on Chroniosuchia. Subsequently, he identified some North America elements, such as the captorhinid *Rjabininus* in the Inta fauna on the River Pechora (Ivakhnenko, 1990). Ivakhnenko and Tverdokhlebova (1987) reported three skeletons of *Belebey*, described earlier from the Belebey locality, in the Upper Kazanian locality Krymskoe in Orenburg Province. Ivakhnenko (1990) suggested *Belebey* was related to *Bolosaurus* and he described a related genus, *Davletkulia*, from the southern Cis-Urals.

Since the second half of the 1970s, a number of new researchers at the Palaeontological Institute took an energetic part in field studies in the east of European Russia and in research on ancient tetrapods. S.N. Getmanov (1989) published his monograph on the Lower Triassic temnospondyl family Benthosuchidae on the basis of new materials, especially from the locality Tikhvinskoe, near Rybinsk, on the Upper Volga, which has yielded an astonishingly abundant collection of skulls. Yu.M. Gubin (1991) studied archegosauroid amphibians. M.A. Shishkin and I.V. Novikov, as a result of field work from 1984 to 1990, assembled new data on the little-known localities of Triassic tetrapods of the Timan–North Urals region. Novikov (1994) added to the systematic and stratigraphic distribution of Triassic amphibians and procolophonids. A.G. Sennikov (1995) studied the basal archosaurs of the eastern part of European Russia, and he revealed here for the first time Lower Triassic rauisuchids and Middle Triassic euparkeriids.

The vast amount of material accumulated by the 1990s indicated new conclusions on the evolution of the Cis-Uralian tetrapods and their stratigraphy. The largest excavations are being conducted now at an Upper Permian locality near Kotel'nich in Kirov Province, which is producing an abundance of pareiasaur skeletons. A rich assemblage of therapsids was also found there, the study of which has just started. This is a more ancient fauna of vertebrates than the North Dvina (Sokolki). The tetrapod complex with the oldest archosaur, *Archosaurus* (the locality of Vyazniki in Vladimir Province and others), was dated as uppermost Permian. This suggested the recognition of a specific stratigraphic horizon, with tetrapods transitional to the Triassic (Shishkin, 1990; Sennikov, 1991). Eventually, the work by Novikov (1994) on the assemblage from the Timan–North Urals region indicated the 'Tsyl'ma Tetrapod Complex' transitional between the 'Neorachitome' Fauna and the *Parotosuchus* Fauna.

Since the 1980s, general questions about the history of ancient tetrapods, a characteristic of Efremov's approach, were addressed. A series of articles on biogeography were published by N.N. Kalandadze, A.G. Rautian, M.A. Shishkin and V.G. Ochev, on the global stratigraphic correlation of Triassic faunas of terrestrial vertebrates (Ochev and Shishkin, 1989) and on the origins of faunas (M.F. Ivakhnenko and others).

In recent times, in the east of European Russia, about 1000 localities of Permian and Triassic tetrapods have been recorded. The growth of data on the Cis-Urals faunas can be observed in a number of review documents, in Efremov's and V'yushkov's (1955) catalogue and in the textbook by Orlov (1964). Recently, the number of Permian and Triassic genera described from Russia reached 150.

The lion's share of study of the Permian and Triassic tetrapods from the former Soviet Union still falls in the east of European Russia. The rest of this territory in general is poor in finds. The most significant were made in Central Asia. First were the large localities of discosauriscids discovered in the late 1950s and in the early 1960s in the Upper Carboniferous or Lower Permian of Kazakhstan and in the Lower Permian of Tadzhikistan. The largest excavation was conducted by an expedition of the Palaeontological Institute under the leadership of N.N. Kalandadze in 1975. Even more interesting is the Madygen locality in the Fergana Valley in Kirgizstan, first recognized for its rich plant remains. The palaeontologist, A.G. Sharov, when searching for fossil insects in the 1960s, found here, and described, the peculiar small reptiles *Longisquama* and *Podopteryx* (now *Sharovipteryx*; Figure 1.9). Continental deposits in Siberia have so far yielded only a few finds of temnospondyl larvae and a fragmentary dicynodont (probably *Lystrosaurus*). The Asian part of Russia still awaits more intensive researches.

Figure 1.9. The enigmatic small gliding reptile *Sharovipteryx* from the Middle-Late Triassic of Kirgizstan. (Reconstruction by A.A. Prokhorov.)

Acknowledgements

We thank Natasha Bakhurina for translating this article, and Mike Benton for editing it, and for supplying photographs.

References

Amalitskii, V.P. 1921–7. [I. Dvinosauridae. II. Seymouridae. V. The North Dvina therocephalian *Anna petri* gen. et sp. nov.] *Severo-Dvinskie Raskopki Professora V.P. Amalitskogo (AN SSSR)* **1**: 1–16; **2**: 1–14; **5**: 1–10.

Bystrov, A.P. [name often spelled Bystrow] 1935. Rekonstruktionsversuche einiger Vertreter der Nord-Dwina Fauna. *Travaux de l'Institut Paléozoologique, Académie des Sciences de l'URSS* **10**: 1–152.

—1938. Zahnstruktur der Labyrinthodonten. *Acta Zoologica* **19**: 387–425.

—1939. Blutgefässsystem der Labyrinthodonten (Gefässe des Kopfes). *Acta Zoologica* **20**: 125–155.

—1944. *Kotlassia prima* Amalitzky. *Bulletin of the Geological Society of America* **55**: 379–416.

—1947. Hydrophilous and xerophilous labyrinthodonts. *Acta Zoologica* **28**: 137–164.

—1957. [The pareiasaur skull.] *Trudy Paleontologicheskogo Instituta, AN SSSR* **68**: 3–18.

Chudinov, P.K. 1955. [Cotylosaurs from the Shikhovo–Chirki site.] *Doklady AN SSSR* **103**: 913–916.

—1957. [Cotylosaurs from the Upper Permian Red Beds of the Cis-Urals.] *Trudy Paleontologicheskogo Instituta AN SSSR* **68**: 19–87.

—1969. [On the stratigraphic distribution of Permian vertebrates in eastern European parts of the USSR.], pp. 96–105 in [*Questions on the Geology of the South Urals and Povolga Region*] Saratov: Izdatel'stvo Saratovskogo Universiteta.

—1983. [Early therapsids.] *Trudy Paleontologicheskogo Instituta AN SSSR* **202**: 1–230.

Efremov, I.A. 1928. [On the mode of deposition of labyrinthodonts in the Verfen deposits of the mountain Bolshoe Bogdo, Astrakhan Province.] *Trudy Geologicheskogo Muzeya AN SSSR* **3**: 9–14.

—1937. [On the stratigraphic divisions of the continental Permian and Triassic of the USSR, based on tetrapod faunas.] *Doklady AN SSSR* **16**: 125–132.

—1939. [On the evolution of Permian faunas of Tetrapoda of the USSR and on divisions of the continental Permian into stratigraphic zones.] *Izvestiya AN SSSR Seriya Biologiya* **2**: 272–289.

—1940a. Kurze Übersicht über die Formen der Perm- und Trias-Tetrapoden-Fauna der UdSSR. *Centralblatt für*

Mineralogie, Geologie, und Paläontologie, Abtheilung B **1940**: 372–383.

—1940b. [Taphonomy – a new branch of palaeontology.] *Izvestiya AN SSSR, Seriya Biologicheskaya* **3**: 405–413.

—1940c. [Preliminary description of new forms in the Permian and Triassic faunas of terrestrial vertebrates of the USSR. IV. Skeleton of a lystrosaur on the River Donguz, Chkalov region.] *Trudy Paleontologicheskogo Instituta AN SSSR* **10**: 73–81.

—1941. [Short survey of faunas of Permian and Triassic Tetrapoda of the USSR.] *Sovetskaya Geologiya* **5**: 96–103.

—1944. [Questions on the stratigraphy of the Upper Permian deposits of the USSR, based on terrestrial vertebrates.] *Izvestiya AN SSSR, Seriya Geologicheskaya* **6**: 52–60.

—1946. [On the Subclass Batrachosauria – a group of forms intermediate between amphibians and reptiles.] *Izvestiya AN SSSR, Seriya Geologicheskaya* **6**: 615–638.

—1950. [Taphonomy and the fossil record.] *Trudy Paleontologicheskogo Instituta AN SSSR* **24**: 1–178.

—1952. [On the stratigraphy of the Permian red beds of the USSR based on terrestrial vertebrates.] *Izvestiya AN SSSR, Seriya Geologicheskaya* **6**: 49–75.

—1954. [The fauna of terrestrial vertebrates in the Permian Copper Sandstone of the Western Cisuralian region.] *Trudy Paleontologicheskogo Instituta AN SSSR* **54**: 1–416.

— and Bystrov, A.P. 1940. [*Benthosuchus sushkini* Efr. – labyrinthodont from the Eotrias of the River Sharzhenga.] *Trudy Paleontologischekogo Instituta AN SSSR* **10**: 1–152.

— and V'yushkov, B.P. 1955. [Catalogue of the localities of Permian and Triassic terrestrial vertebrates in the territories of the USSR.] *Trudy Paleontologicheskogo Instituta AN SSSR* **46**: 1–147.

Eichwald, E.I. 1846. [*Geognosy of Russia.*] St Petersburg, 572 pp.

—1848. Ueber die Saurier der Kupferführenden Zechsteins Russlands. *Bulletin de la Société des Naturalistes de Moscou* **21**: 136–204.

Getmanov, S.N. 1989 [Triassic amphibians of the eastern European Platform.] *Trudy Paleontologicheskogo Instituta AN SSSR* **236**: 1–102.

Gubin, Yu.A. 1991. [Permian archegosauroid amphibians from the USSR.] *Trudy Paleontologicheskogo Instituta AN SSSR* **249**: 1–142.

Hartmann-Weinberg, A.P. 1933. Die Evolution der Pareiasauriden. *Travaux de l'Institut Paléozoologique, Académie des Sciences de l'URSS* **3**: 1–66.

Huene, F. von 1905. Pelycosaurier im deutschen Muschelkalk. *Neues Jahrbuch für Mineralogie, Geologie und Paläontologie (Beilage-Band)* **20**: 321–353.

—1940. Eine Reptilfauna aus der altesten Trias Nordrusslands. *Neues Jahrbuch für Mineralogie, Geologie und Paläontologie, Abteilung B* **84**, 1–23.

Ivakhnenko, M.F. 1979. [The Permian and Triassic procolophonians of the Russian platform.] *Trudy Paleontologicheskogo Instituta AN SSSR* **164**: 1–80.

—1987. [The Permian parareptiles of the USSR.] *Trudy Paleontologicheskogo Instituta AN SSSR* **223**: 1–160.

—1990. [The late Palaeozoic faunal assemblage of tetrapods from deposits of the Mezen' River.] *Paleontologicheskii Zhurnal* **1990**: 81–90.

—and Tverdokhlebova, I.G. 1980. [*Systematics, Morphology and Stratigraphic Significance of the Upper Permian Chroniosuchians from the East of the European Part of the USSR.*] Saratov: Izdatel'stvo Saratovskogo Universiteta, 69 pp.

—and — 1987. [A revision of the Permian bolosauromorphs of East Europe.] *Paleontologicheskii Zhurnal* **1987**: 98–106.

Kalandadze, N.N. 1969. [Triassic kannemeyeriids of the southern Urals.] *Byulleten' Moskovskogo Obshchestva Ispytatelei Prirody, Otdel Geologicheskii* **44**: 148.

—1973. [An outline of the phylogeny of the kannemeyeroids.] *Byulleten' Moskovskogo Obshchestva Ispytatelei Prirody, Otdel Geologicheskii* **48**: 151.

—1975. [The first find of a *Lystrosaurus* on the territory of the European part of the USSR.] *Paleontologicheskii Zhurnal* **1975**: 140–142.

Konzhukova, G.A. 1953. [Lower Permian terrestrial vertebrates of the northern Cis-Urals (Inta River basin).] *Doklady AN SSSR* **89**: 723–726.

—1955. [Permian and Triassic labyrinthodonts of the Volga and Ural regions.] *Trudy Paleontologicheskogo Instituta AN SSSR* **49**: 5–88.

Kutorga, S.S. 1838. *Beitrag zur Kenntniss der organischen Ueberreste des Kupfersandsteins am westlichen Abhange des Urals*. St Petersburg: N. Gretsch, 38 pp.

Meyer, H. von 1866. Reptilien aus dem Kupfer-Sandstein des West-Uralischen Gouvernements Orenburg. *Palaeontographica* **15**: 97–130.

Murchison, R.I. 1841. On the stratified deposits which occupy the northern and central regions of Russia.

Report and Proceedings of the British Association for the Advancement of Science **1841** (1840): 105–110.

—and de Verneuil, P.É.P. 1845. *The Geology of Russia in Europe and the Ural Mountains.* London and Paris, 700 pp.

Nopcsa, F. von 1928. Palaeontological notes on reptiles. II. On some fossil reptiles from the copper-bearing [Kazan] Permian strata of Russia. *Geologica Hungarica, Series Palaeontologica* **1**: 1–84.

Novikov, I.V. 1994. [Biostratigraphy of the continental Triassic of the Timan-North Uralian region on the tetrapod fauna.] *Trudy Paleontologicheskogo Instituta RAN* **261**: 1–139.

Ochev, V.G. 1966. [*Systematics and Phylogeny of Capitosauroid Labyrinthodonts.*] Saratov: Izdatel'stvo Saratovskogo Universiteta, 184 pp.

—1972. [*Capitosauroid Labyrinthodonts from the South-east European Part of the USSR.*] Saratov: Izdatel'stvo Saratovskogo Universiteta, 208 pp.

—1982. [Remains of pseudosuchians in the Lower Triassic of the southern Cis-Urals.] *Paleontologicheskii Zhurnal* **1982**: 96–102.

—1991. [On the History of the Evolution of Early Archosaurs.] *All-Union Institute of Scientific Information (VINITI), Deposited Manuscript* **2361**(**B91**): 1–78.

—and Shishkin, M.A. 1989. On the principles of global correlation of the continental Triassic on the tetrapods. *Acta Palaeontologica Polonica* **34**: 149–173.

Olson, E.C. 1955. Parallelism in the evolution of the Permian reptilian faunas of the Old and New Worlds. *Fieldiana, Zoology* **37**: 385–401.

—1957. Catalogue of localities of Permian and Triassic vertebrates of the territories of the U.S.S.R. *Journal of Geology* **65**: 196–226.

Orlov, Yu.A. 1958. [Carnivorous dinocephalians from the fauna of Isheevo (Titanosuchia).] *Trudy Paleontologicheskogo Instituta AN SSSR* **72**: 1–114.

—(ed.) 1964. [*Principles of Palaeontology, Amphibians, Reptiles, and Birds.*] Nauka, Moscow 722 pp.

Owen, R. 1876. Evidences of theriodonts in Permian deposits elsewhere than in South Africa. *Quarterly Journal of the Geological Society* **32**: 352–363.

Pravoslavlev, P.A. 1927. [III. Gorgonopsidae from the North Dvina Expedition of V. P. Amalitskii. IV. Gorgonopsid from the 1923 North Dvina Expedition (*Amalitzkia annae* gen. et sp. nov.).]. *Severo-Dvinskie Raskopki Professora V. P. Amalitskogo (AN SSSR)* **3**: 1–117; **4**: 1–20.

Seeley, H.G. 1894. Researches on the structure, organiza-tion, and classification of the fossil Reptilia. Part VIII. Further evidences of *Deuterosaurus* and *Rhopalodon* from the Permian rocks of Russia. *Philosophical Transactions of the Royal Society of London, Series B* **185**: 663–717.

Sennikov, A.G. 1991. [New stratigraphic horizon in eastern Europe.] *Priroda* **1991**(**4**): 121

—1995. [Early thecodonts of eastern Europe.] *Trudy Paleontologicheskogo Instituta RAN SSSR* **263**: 1–141.

Shishkin, M.A. 1973 [Morphology of early amphibians and problems of the evolution of lower tetrapods.] *Trudy Paleontologicheskogo Instituta AN SSSR* **137**: 1–260.

—1987. [The evolution of ancient amphibians (Plagiosauroidea).] *Trudy Paleontologicheskogo Instituta AN SSSR* **225**: 1–144.

—1990. [On the threefold subdivision of the Upper Tatarian Substage on the fauna of terrestrial verte-brates.] *Byulleten' Moskovskogo Obschestva Ispytatelei Prirody, Otdel Geologicheskii* **65**(**2**): 117.

—and Novikov, I.V. 1992. [Relict anthracosaurs in the Early Mesozoic of eastern Europe.] *Doklady RAN* **325**: 829–832.

—and Ochev, V.G. 1967. [Fauna of terrestrial vertebrates as guides to stratification of continental Triassic deposits of the USSR.], pp. 74–82 in [*Stratigraphy and Palaeontology of Mesozoic and Paleogene-Neogene Continental Deposits of Asiatic Parts of the USSR.*] Nauka, Leningrad.

Stuckenberg, A.A. 1898. Allgemeine geologische Karte von Russland. Blatt 127. *Trudy Geologicheskogo Komiteta* **16**: 1–362.

Sushkin, P.P. 1926. Notes on the pre-Jurassic Tetrapoda from Russia. I. *Dicynodon amalitzkii*, n. sp. II. Contributions to the morphology and ethology of the Anomodontia. III. On Seymouriamorphae from the Upper Permian of North Dvina. *Palaeontologia Hungarica* **1**: 323–344.

—1927. On the modifications of the mandibular and hyoid arches and their relations to the braincase in the early Tetrapoda. *Paläontologische Zeitschrift* **8**: 263–321.

—1936. Notes on the pre-Jurassic Tetrapoda from the USSR. III. *Dvinosaurus amalitski*, a perennibranchiate stegocephalian from the Upper Permian of North Dvina. *Travaux de l'Institut Paléozoologique, Académie des Sciences de l'URSS* **5**: 43–91.

Tatarinov, L.P. 1960. [Discovery of pseudosuchians in the Upper Permian of the USSR.] *Paleontologicheskii Zhurnal* **1960** (**4**): 74–80.

—1974. [The theriodonts of the USSR.] *Trudy Paleontologicheskogo Instituta AN SSSR* **193**: 1–252.

—1976. [*Morphology and Evolution of the Theriodonts and the General Problems of Phylogenetics.*] Nauka, Moscow, 258 pp.

Twelvetrees, W.H. 1880a. On a new theriodont reptile (*Cliorhizodon orenburgensis*, Twelvetr.) from the Upper Permian cupriferous sandstones of Kargalinsk, near Orenburg in south-eastern Russia. *Quarterly Journal of the Geological Society of London* **36**: 540–543.

—1880b. On a labyrinthodont skull (*Platyops rickardi*, Twelvetr.) from the Upper Permian cupriferous strata of Kargala, near Orenburg. *Bulletin de la Société des Naturalistes de Moscou* **55**: 117–122.

—1882. On the organic remains from the Upper Permian strata of Kargalinsk, in eastern Russia. *Quarterly Journal of the Geological Society of London* **38**: 490–501.

V'yushkov, B.P. 1955. [Theriodonts of the Soviet Union.] *Trudy Paleontologicheskogo Instituta AN SSSR* **49**: 128–175.

—1957a. [*Tupilakosaurus* – a new palaeontological riddle.] *Priroda* **1957** (9): 112–113.

—1957b. [New kotlassiomorphs from the Tatarian of the European part of the USSR.] *Trudy Paleontologicheskogo Instituta AN SSSR* **68**: 89–107.

—and Chudinov, P.K. 1956. [A contribution to the study of Triassic reptiles – *Microcnemus* and *Tichvinskia*.] *Doklady AN SSSR* **110**: 141–144.

—and — 1957. [The discovery of Captorhinidae in the Upper Permian rocks of the U.S.S.R.] *Doklady AN SSSR* **112**: 523–526.

Wangenheim von Qualen, F. 1845. Ueber einen im Kupfersandsteine der Westuralischen Formation entdeckten Saurierkopf. *Bulletin de la Société des Naturalistes de Moscou* **18**: 389–416.

Watson, D.M.S. 1914. The Deinocephalia, an order of mammal-like reptiles. *Proceedings of the Zoological Society of London* **1914**: 749–786.

2

The amniote faunas of the Russian Permian: implications for Late Permian terrestrial vertebrate biogeography

SEAN P. MODESTO AND NATALIA RYBCZYNSKI

Introduction

The Late Permian is a remarkable period in the history of vertebrate life on land. Spanning approximately 21 million years, this brief chapter of terrestrial vertebrate life is distinguished from the rest of the Palaeozoic by tremendous increases in species richness, distribution, and morphological diversity of parareptilian and non-mammalian synapsid amniotes; these vertebrates were postulated by Olson (1966) to form the oldest known herbivore-based terrestrial ecosystems. Intriguingly, diapsid reptiles continue to be uncommon during the Late Permian, a trend that belies their great evolutionary radiations in the succeeding Triassic Period, and their dominance of land, sea, and air environments throughout the Mesozoic Era.

Deposits of Permian age in southern Africa and European Russia document richly this crucial period in the evolution of large carnivorous and herbivorous terrestrial vertebrates. The Late Permian horizons of the Karoo Basin of South Africa have a long history of collection and, consequently, an impressive database on its vertebrates has been amassed (Kitching, 1977; Rubidge (Ed.), 1995). Accordingly, the Karoo has served as the basis of comparison with Upper Permian localities from elsewhere. However, the terrestrial record of the Karoo Permian is remarkable in that the vast majority of its amniotes have no close relatives from older North American strata; the single exception is the varanopseid eupelycosaur *Elliotsmithia longiceps* (Dilkes and Reisz, 1986). In strong contrast, many Late Permian Russian localities are characterized by the presence of tetrapods that are most closely related

to earlier forms from North America and Western Europe (Ivakhnenko, 1991a). Poorly fossiliferous Upper Permian facies in North America have been regarded to feature a similar mixture of typically Early Permian and Late Permian forms (Olson, 1962, 1986). However, the Late Permian materials from North America that were formerly attributed to therapsids (Olson and Beerbower, 1953; Olson, 1962, 1974) have been reidentified as a mixture of fragmentary caseid and sphenacodontid synapsid remains (Sidor and Hopson, 1995).

The Russian Permian faunas are, therefore, particularly important to the study of Permian terrestrial vertebrate biogeography and the evolution of terrestrial vertebrate environments during the Palaeozoic, since they appear to bridge a significant gap between the 'pelycosaur' faunas of the North American Early Permian and the therapsid-dominated faunas of the South African Late Permian. Traditional hypotheses regarded Laurussia as the centre of origin for most large amniote groups during the Permian (e.g. Boonstra, 1971; Olson, 1979), but, more recently, biogeographic inferences from phylogenetic work, suggest instead that amniotes may have been dispersing freely between Russia and Africa during the Permian (Rubidge and Hopson, 1990; Rubidge, 1993, 1995).

Despite pioneering efforts by Everett Olson to make knowledge of Russian Permian vertebrates more accessible to Western scientists (Olson, 1955, 1957, 1962), the significance of the Russian Late Permian faunas remains overshadowed by those of the South African Beaufort Series, an unfortunate situation exacerbated partly by the paucity of broad-scale lithostratigraphic studies of the Russian deposits. Continuing geological and palaeontological work at the Kotel'nich and Mezen' localities should elucidate further the stratigraphic relationships of the Russian faunas both to each other and to the synapsid-bearing horizons in North America and southern Africa. In this chapter, we present brief summaries of the amniote faunas from the Late Permian of Russia (Figure 2.1), compare them with related faunas in Africa and North America, and then examine the biogeographic implications that the phylogeny of one group, the therapsid clade Anomodontia (*sensu* Hopson, 1969), has for Permian terrestrial biogeography.

Stratigraphy of the Russian continental Late Permian

The Russian Late Permian was divided into several units in succession on the basis of the contained tetrapods by Efremov (1940), informally referred to as the dinocephalian (I–II), cotylosaurian (III), and pareiasaurian (IV) zones. A later version of this scheme (Efremov and V'yushkov, 1955) recognized only three successive units, two dinocephalian complexes and a pareaisaurian complex, with the assemblages of small captorhinids and procolophonids formerly grouped in zone III being recognized as coeval with the dinocephalian complexes. These units all equate with the Kazanian and Tatarian stages, and they were later found to be preceded by an unnamed temnospondyl fauna of Ufimian age.

This scheme, with minor modifications, was mostly followed by later authors. After the recovery by Chudinov in the late 1950s of the Ocher (Ezhovo) fauna, it became clear that the latter provides a standard for the early dinocephalian fauna of Efremov, which is succeeded by the late dinocephalian fauna from Isheevo (Lower Tatarian). Most authors (Ochev, 1976; Ochev *et al.*, 1979; Chudinov, 1969, 1983) discern three main events in the history of the Russian Permian biota, referred to as the Ocher (lower dinocephalian), Isheevo (upper dinocephalian), and pareiasaurian–gorgonopsian faunas. The Ocher fauna

Figure 2.1. Map of Cis-Uralian Russia showing distribution of Upper Permian localities (circles) discussed in the text. Locations of cities and other urban centres are indicated by squares. Map information courtesy of the National Geographic Society.

is immediately preceded by the poorly known Lower Kazanian assemblage ('Golyusherma complex') with the earliest therapsids. The assemblages of small reptiles from Mezen' and Belebey are often considered to be at least a partial correlate of the dinocephalian faunas (Ochev *et al.*, 1979), following Efremov and V'yushkov (1955).

Stratigraphically, the Cis-Uralian faunal succession mostly corresponds to the standard Late Permian stratigraphic stages, the Ufimian, Kazanian, and Tatarian, of which only the Kazanian includes marine facies in the stratotype. The detailed correlation of many local sections is obscure, and this makes uncertain the stratigraphic position of some localities. Particular problems still include, (1) the exact dating of the Mezen' fauna as compared to the dinocephalian units, and (2) an explanation of the dramatic faunal turnover in the Tatarian when the dinocephalian

18

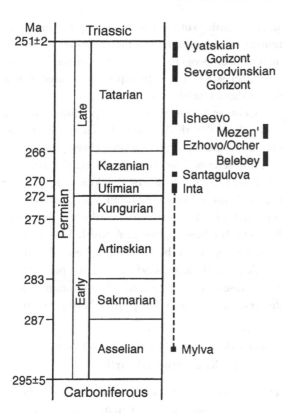

Figure 2.2. Approximate stratigraphic positions of Russian Permian assemblages, based on information from Chudinov (1983), Ivakhnenko (1991a,b), and Sennikov (1989). Dates for geological ages are from Ross *et al.* (1994).

more than two horizons (V.I. Golubev, pers. comm.). The oldest Russian amniotes are known only from fragmentary materials. Included here are the Ufimian-age Inta fauna, which consists of temnospondyls and small reptiles of North American Early Permian aspect, and a dinocephalian fauna of Early Kazanian age, from the lower horizons of the Copper Sandstones of Bashkortostan. Localities from younger horizons reveal a more complete view of the structure of terrestrial vertebrate communities during the Late Permian of Russia. The Late Kazanian Ocher or Ezhovo complex is characterized by numerous taxa of herbivorous and faunivorous therapsids. That fauna is apparently succeeded by a Tatarian-age therapsid complex at Isheevo, which features herbivorous therapsid taxa that suggest an increased Gondwanan influence. The Upper Tatarian Sokolki and Kotel'nich (Vyatka and North Dvina) vertebrate assemblages show the strongest ties with the Gondwanan fauna of South Africa. The far-north localities in the vast Mezen' basin are some of the most enigmatic Russian Permian localities; they preserve taxa with close relatives in both the Ezhovo and Isheevo faunas, together with a small faunivorous form and a large herbivore, taxa whose phylogenetic affinities lie with the Permo-Carboniferous synapsids from the Early Permian of North America.

Russian Late Permian amniote localities and their faunas

At least eight major amniote assemblages are recognized from the Late Permian of Russia (Figure 2.2). The age assignments follow those outlined by Chudinov (1965, 1983), with the exception of two Late Tatarian assemblages, which may be regarded provisionally as coeval (following an emendation of Chudinov's stratigraphic schemes by Sigogneau-Russell, 1989), although they may in reality represent

fauna was replaced by the pareiasaurian fauna. It is conceivable, as suggested by Efremov, that there is a large sedimentary gap in the Tatarian.

Inta fauna (Asselian (?) – Ufimian)

The Inta fauna, comprised of tetrapods from the localities of Inta, Pechora, and Us'va in northeastern Cis-Uralia, dated as Ufimian or even Asselian, is characterized largely by eryopoid temnospondyls of the genera *Intasuchus*, *Syndyodosuchus*, and *Clamorosaurus* (Gubin, 1983). The only amniote remains are fragmentary jaws of small, single-tooth-rowed captorhinid reptiles collected from Inta and Us'va. Ivakhnenko (1991a) erected the genus *Riabininus* for the Us'va captorhinid specimens, and suggested the Inta captorhinid specimens also belonged to this taxon. None of this material is distinguishable from basal captorhinid material from North America and Africa (Clark and Carroll, 1973; Gaffney and

McKenna, 1979; Modesto, 1996) and should be considered Captorhinidae *incertae sedis*.

Ivakhnenko (1991a) also associated a small bolosaurid maxilla, collected from the Asselian-age beds on the banks the Mylva River, with the Inta fauna. This specimen represents the oldest amniote material known from Eastern Europe. Although it was described as the second species of the genus *Bolosaurus* (Tatarinov, 1974a), *B. traati* does not share any apomorphies with the type species *Bolosaurus striatus* (Watson, 1954) that are not shared also with other Russian bolosaurs (Ivakhnenko and Tverdokhlebova, 1988; Ivakhnenko, 1991a). Accordingly, the Mylva species should be regarded provisionally as Bolosauridae *incertae sedis* until its holotype and the interrelationships of bolosaurs are reevaluated. Since the deposits at Mylva are separated from those at Inta and Us'va by at least 15 million years (Figure 2.2), the Mylva bolosaur should not be recognized as part of the Inta assemblage.

The earliest records of therapsid faunas (Early Kazanian)

Several localities from the Lower Kazanian beds of the Copper Sandstones of Bashkortostan preserve the oldest therapsid material in Russia, but most are known exclusively from isolated, fragmentary postcranial elements. Partial scapulae from the Bashkir Mines are assigned to the dinocephalian genus *Brithopus*, known best from the type species *Brithopus priscus* from an Upper Kazanian locality of the Copper Sandstones in Perm' Province (Olson, 1962; Chudinov, 1983). Four isolated femora represent holotypic material for an equal number of species that form the problematic synapsid group Phreatosuchidae (Efremov, 1954). These taxa include *Phreatosaurus bazhovi* and *Phreatosaurus menneri* from the Demsk Mines, and *Phreatosuchus qualeni* and *Phreatophasma aenigmaticum*, known only from the Santagulovo Mines. The last taxon was transferred to the synapsid clade Caseidae by Olson (1962), but in a recent discussion of caseid material from Russia, Ivakhnenko (1991a) concluded that *P. aenigmaticum* is not a caseid. The identity

and relationships of most of these Early Kazanian taxa remain indeterminable, and they provide little information on biogeographical aspects of therapsids prior to the 'classic' Russian therapsid-dominated faunas from Ezhovo and Isheevo.

Two other Kazanian sites have produced amniote materials of North American Permian aspect. The moradisaurine captorhinid reptile *Gecatogomphius kavejevi* was described on the basis of a partial mandible retrieved from the banks of the Vyatka River, near Gorka, Kirov Province; additional jaw material was collected just upriver from Gorka at Berezovye Polyanki (Ivakhnenko, 1991a). A North American distribution appears to be ancestral for moradisaurines (Dodick and Modesto, 1995), suggesting that *G. kavejevi* represents the second captorhinid lineage to have dispersed into Eastern Europe from North America.

Early dinocephalian (Ezhovo/Ocher) fauna (Late Kazanian – (?) earliest Tatarian)

The Ezhovo (Ocher) fauna, from Ezhovo in Perm' Province, includes several temnospondyls, including dissorophids, archegosaurids, and melosaurids, but it is best known for the earliest therapsids. The most conspicuous and abundant vertebrates at Ezhovo are dinocephalians of the family Estemmenosuchidae (Chudinov, 1965, 1983). Four estemmenosuchid taxa are known from adequate skeletal material: *Anoplosuchus tenuirostris*, *Estemmenosuchus mirabilis*, *Estemmenosuchus uralensis*, and *Zopherosuchus luceus* (Chudinov, 1983). These dinocephalians appear to constitute a clade of wholly endemic Russian herbivores. Other dinocephalian taxa known are the carnivores *Archaeosyodon praeventor* and *Chthamaloporus lenocinator* of the clade Anteosauria (*sensu* Hopson and Barghusen, 1986). Additional faunivorous therapsid taxa include *Biarmosuchus tener*, *Eotitanosuchus olsoni*, and *Ivantosaurus ensifer*. These three species are assigned currently to Biarmosuchia, which is in all likelihood a paraphyletic group (Hopson and Barghusen, 1986; Sigogneau-Russell, 1989).

The last taxon at Ezhovo to be considered here is the anomodont *Otsheria netzvetajevi* (Figure 2.3A). *O.*

netzvetajevi is traditionally grouped together with the genus *Venyukovia* (Hopson and Barghusen, 1986; King, 1988), but in a recent assessment of anomodont phylogeny (Rubidge and Hopson, 1990), *O. netzvetajevi* was recognized as the sister taxon of a clade containing *Venyukovia* and dicynodonts. The interrelationships of these and other basal anomodonts are examined in the final section of this chapter.

The Ezhovo fauna was recognized by Olson (1966) as the oldest known herbivore-based terrestrial community. However, this distinction may merely be one by default, in view of the rather scanty knowledge of terrestrial vertebrate faunas from older Russian and North American deposits. Regardless, the number of putative herbivorous taxa at Ezhovo is approached only by that at Kotel'nich among the succeeding Russian faunas. The presence of estemmenosuchids at Ezhovo and other Russian localities implies a strongly endemic component to dinocephalian diversity during the Late Kazanian. The presence of *O. netzvetajevi* in the same deposits, however, is suggestive that a minority of the Late Kazanian fauna of Russia were Gondwanan in origin, a postulate that is examined in greater detail below. The biogeographic implications of the faunivorous therapsid taxa at Ezhovo are, however, less clear, since the relationships of 'biarmosuchians' have yet to be established within Therapsida.

Belebey fauna (Late Kazanian)

Although its fauna is quite different from that known from Ezhovo, the small Kazanian-age locality of Belebey, Bashkortostan, contributes a significant reptilian component to the Ocher amniote assemblage (Tverdokhlebova and Ivakhnenko, 1985; Ivakhnenko and Tverdokhlebova, 1988). The two units are apparently coeval, and they are generally combined as one by Russian scientists. The Belebey assemblage includes platyoposaurid temnospondyls, as well as reptiles. The parareptile *Nycteroleter bashkyricus* and the bolosaurid *Belebey vegrandis*, represented by several specimens each, are the best-known members of the Belebey fauna. Our knowledge of *Nycteroleter* is based

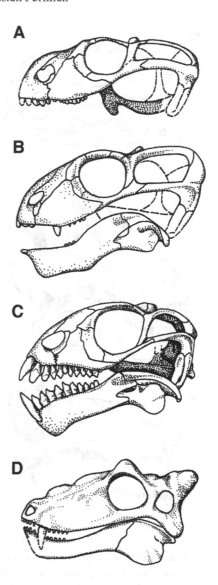

Figure 2.3. Skulls of Late Permian Russian therapsids in lateral aspect: A, *Otsheria netzvetajevi* from Ezhovo, × 0.66. B, *Ulemica invisa* from Isheevo, × 0.4. C, *Suminia getmanovi* from Kotel'nich, × 1. D, *Proburnetia viatkensis* from Kotel'nich, × 0.3. (A, after Chudinov, 1983; B, after Chudinov, 1983; C, after PIN 2212/32, 2212/33, and 2212/62; D, after Chudinov, 1983, and PIN 2416/1).

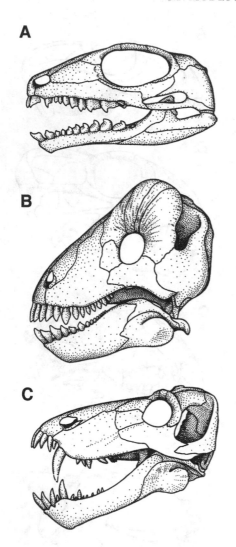

Figure 2.4. Skulls of representative Late Permian Russian amniotes in lateral aspect: A, *Belebey vegrandis* from Belebey, × 1.5. B, *Ulemosaurus svijagensis* from Isheevo, × 0.2. C, *Titanophoneus potens* from Isheevo, × 0.15. (A, after Ivakhnenko, 1991a; B, after Chudinov, 1983; C, after Orlov, 1958).

site ('*N.*' *kassini*) is probably not nycteroleterid (Ivakhnenko, 1991b).

Belebey vegrandis (Figure 2.4A) is the best-known bolosaurid reptile from Europe; referred material has been recovered also from the Krymskii locality in Orenburg (Ivakhnenko and Tverdokhlebova, 1988). Bolosaurid material collected recently from a locality near Dashankou, China, is closely comparable to this species (Li and Cheng, 1995). A second species, *B. maximi*, was described from a locality near Saray-Gir, Bashkortostan (Tverdokhlebova and Ivakhnenko, 1985). The holotype is a partial mandible that does not differ significantly in either size or discrete characters from *B. vegrandis*; *B. maximi* is probably a junior synonym of the type species. *Davletkulia gigantea*, the most recently described bolosaurid, is known only from a single large tooth from another locality in Bashkortostan (Ivakhnenko, 1991a). The phylogenetic affinities of bolosaurids are uncertain; the evolutionary implications of bolosaurid distributions must await a comprehensive anatomical and phylogenetic reevaluation of both North American and European bolosaurid material.

Materials of other amniotes at the locality are extremely limited in number. Fragmentary mandibles and a few postcranial elements are the only known material of the 'biarmosuchian' therapsid *Phthinosaurus borissiaki* (Sigogneau-Russell, 1989). Fragmentary, poorly preserved material of parareptilian nature was described originally as *Rhiphaeosaurus tricuspidens*. This taxon, interpreted originally as a relative of pareiasaurs (Chudinov, 1965; Ivakhnenko, 1987), is now considered to be a nycteroleterid (Lee, 1995), related closely to (if not referable to) the large nycteroleterid *Macroleter* (R. Reisz, pers. comm.).

Mezen' fauna (Early Tatarian)

A number of collecting sites along the banks of the Mezen' River and scattered through the vast Mezen' basin are traditionally referred to collectively as the Mezen' Group, Arkhangel'sk Province (Ivakhnenko, 1991b). The most common vertebrate at Mezen' is

almost exclusively on the species from Belebey, as the type species *Nycteroleter ineptus* is known from only three skulls from the Mezen' basin (Ivakhnenko, 1991b). Additional nycteroleterid material has been reported from both the Kotel'nich and Shikhovo–Chirki localities, although the material from the latter

the parareptile *Nyctiphruretus acudens*, which represents over two-thirds of the total number of specimens. *Macroleter poezicus* is the only other reptile that is collected in appreciable quantities from these deposits. A notable member of the fauna is the small faunivore *Mesenosaurus romeri*, which was described originally as a varanopseid synapsid (Efremov, 1937) but later interpreted as an archosaurian diapsid (Ivakhnenko and Kurzanov, 1979). Work in progress on this taxon and other varanopseids reaffirms the original interpretation (R. Reisz and D. Berman, pers. comm.). Other taxa described from the locality include the parareptiles *Nycteroleter ineptus* and *Lanthaniscus efremovi*, the basal therapsid *Biarmosuchus tagax*, and the enigmatic therapsid *Niaftasuchus zekkeli*. A skull attributable to the caseid synapsid *Ennatosaurus tecton*, known previously only from Karpogory (Ivakhnenko, 1991a), was recovered recently from the locality.

A related locality, Moroznitsa on the Pinega River, Arkhangel'sk Province, has yielded five skulls and associated postcrania of the caseid *Ennatosaurus tecton* (Ivakhnenko, 1991a). The closest relatives of this large herbivore are species of the genera *Cotylorhynchus* and *Angelosaurus* from the Permian of North America (Olson, 1968). This locality is dated as Early Tatarian (Chudinov, 1983), and indeed *Ennatosaurus* is known also from the Mezen' collecting area.

The age of the Mezen' deposits was uncertain; estimates ranged from Late Kazanian to the Early Tatarian (Ivakhnenko, 1991b), but most Russian geologists now accept an Early Tatarian age for the Nizhneustinskaya Svita, the main unit exposed. In the most recent review of the Mezen' basin localities, Ivakhnenko (1991b) argued that its fauna cannot be allied confidently with either Ezhovo or Isheevo, nor could it be interpreted confidently as a transitional assemblage. Close relatives of all taxa except *M. romeri* are found at other Russian Permian localities: the parareptile *Nycteroleter bashkyricus* from Belebey (Late Kazanian), *Lanthanosuchus efremovi* from Isheevo (Early Tatarian), and *Biarmosuchus tener* from Ezhovo (Late Kazanian).

Late Dinocephalian (Isheevo) fauna (Early Tatarian)

The late dinocephalian fauna is typified by, and mostly known from, the Isheevo assemblage from Tatarstan. The assemblage includes melosaurid temnospondyls and the seymouriamorph *Enosuchus*. As at Ezhovo, dinocephalians are the most conspicuous amniotes, but in strong contrast to this assemblage, estemmenosuchids are absent at Isheevo. The only herbivorous member of the Isheevo fauna is the tapinocephalid *Ulemosaurus svijagensis* (Figure 2.4B). Additional material of *U. svijagensis* is known from Malyi Uran, Orenburg Province (Olson, 1962). The genus *Ulemosaurus* has been regarded as a junior synonym of *Moschops* (Chudinov, 1983), although this synonymy is not universally accepted (Ivakhnenko, 1994). Interestingly, *U. svijagensis* is the only tapinocephalid known from Russia; all other members of this clade are known only from the *Eodicynodon–Tapinocaninus* and *Tapinocephalus* assemblage zones of South Africa. The remainder of the dinocephalian component at Isheevo consists of the anteosaurians *Syodon biarmicum*, *S. efremovi*, *Titanophoneus potens* (Figure 2.4C), and *Doliosauriscus janschinovi*, all of which are represented by excellent skeletal material (Orlov, 1958). These three carnivorous genera form a grade with respect to the South Africa genus *Anteosaurus* (Hopson and Barghusen, 1986). A second species of *Doliosauriscus* is known from Malyi Uran (Chudinov, 1983). Recently collected remains of faunivorous dinocephalians from Dashankou, China, are thought to be allied very closely with *Syodon* and *Titanophoneus* (Li and Cheng, 1995).

Non-dinocephalian materials from Isheevo include the holotype and referred material of the parareptile *Lanthanosuchus watsoni* and several anomodont skulls assigned formerly to the genus *Venyukovia* as *Venyukovia invisa*. The holotype of *L. watsoni* (PIN 127/1) was described by Olson (1962) as the most perfectly preserved skull of a Palaeozoic tetrapod; referred material is known from Malyi Uran (Efremov and V'yushkov, 1955; Ivakhnenko, 1980). The type and only known specimen of the anomodont *Venyukovia prima* is a fragmentary mandible from Kargala Mines, Orenburg (Chudinov, 1983). Accordingly, our understanding of the anatomy and phylogenetic relationships of *Venyukovia* have been based almost entirely on the material from Isheevo (Barghusen, 1976; Hopson and Barghusen, 1986; King, 1988, 1994: Rubidge and Hopson, 1990). However, the Isheevo specimens were suspected by Ivakhnenko (1994) to be 'sufficiently different' from *V. prima*, and he transferred them to the new genus *Ulemica* (Ivakhnenko, 1996) as *Ulemica invisa* (Figure 2.3B).

The Isheevo fauna can be recognized as what appears to be the ecological successor of the Ezhovo assemblage. As large herbivores, the estemmenosuchids appear to have been replaced by the tapinocephalid *Ulemosaurus svijagensis*, the faunivorous 'biarmosuchians' and the (presumably) basal anteosaurian dinocephalians replaced by more recently advanced anteosaurs, and *Otsheria netzvetajevi* replaced perhaps by a more recently derived anomodont. The presence of a lanthanosuchid parareptile is interesting, since the seymouriamorph *Enosuchus breviceps* is also known from Isheevo (Olson, 1962) and lanthanosuchid material is found with other temnospondyls at Malyi Uran, but not in the Mezen' region. The presence of *Ulemica invisa* and *Ulemosaurus svijagensis* suggests a slightly greater African influence at Isheevo, as compared with Ezhovo.

Pareiasaurian–gorgonopsian fauna (Late Tatarian)

In strong contrast to earlier assemblages, dinocephalians are completely absent in younger Tatarian horizons. Pareiasaurs, anomodonts, gorgonopsians, therocephalians, and cynodonts are the most conspicuous tetrapods in these horizons, whereas nonpareiasaurian reptiles are comparably rare, their presence indicated only by sparse, fragmentary remains.

The lower Kotel'nich fauna (Severodvinskian Gorizont)

The tetrapod fauna at Kotel'nich, on the Vyatka River in Kirov Province, is characterized by pareiasaurs,

anomodonts, and faunivorous therapsids. Recent work at Kotel'nich has produced a large number of tetrapod fossils, including a new species of basal anomodont and an unprecedented amount of dicynodont and therocephalian material (Ivakhnenko, 1994), some of which probably represents new taxa. Intriguingly, anamniote tetrapods are unknown as yet in the lower Kotel'nich assemblage, although the temnospondyl *Dvinosaurus* is recorded from the upper Kotel'nich assemblage.

The pareiasaur *Deltavjatia vyatkensis* is the largest amniote known from Kotel'nich. A second pareiasaur, *Scutosaurus rossicus* (Ivakhnenko, 1987), was apparently based upon juvenile specimens and is now recognized as a junior synonym of *D. vyatkensis* (Ivakhnenko, 1994). Ivakhnenko (1994) believed *D. vyatkensis* to be more closely related to the South African *Tapinocephalus* Zone genera *Bradysaurus* and *Embrithosaurus* than to the only other pareiasaur known from Russia, *Scutosaurus karpinskii*. However, recent phylogenetic work by Lee (1994; Chapter 5) suggests that the Kotel'nich taxon belongs to a clade which excludes *Bradysaurus* and *Embrithosaurus*, but is in turn excluded from a more recently derived pareiasaur clade containing *S. karpinskii*. Either phylogenetic scenario suggests that *S. karpinskii* and *D. vyatkensis* represent two separate invasions of Eastern Europe by pareiasaurs. Lee's (1994) pareiasaur phylogeny, in which the South African *Bradysaurus* and *Embrithosaurus* are the most basal pareiasaur genera, suggests further that pareiasaurs diversified originally in Africa, and later invaded Laurasian Pangaea. In this respect, pareiasaurian biogeography, as reconstructed from their phylogeny, resembles that suggested for tapinocephalid dinocephalians. Additional reptilian material is represented by a recently collected skull of a nycteroleterid (Ivakhnenko, 1994).

Anomodonts are well represented at Kotel'nich by two distinct dicynodonts and a basal anomodont ('dromasaur'). Material of a toothed dicynodont may be referable to the South Africa genus *Tropidostoma* (Ivakhnenko, 1994), whereas 15 recently collected skulls appear to belong to the genus *Dicynodon* (R. Reisz, pers. comm.). All basal dicynodonts are known exclusively from South Africa (King, 1988; Rubidge, 1990b), which implies that an African distribution is ancestral for dicynodonts. One of the most interesting therapsids described recently from Kotel'nich is the basal anomodont *Suminia getmanovi* (Ivakhnenko, 1994). The skull of *S. getmanovi* (Figure 2.3C) is only about 60 mm in length. It is distinguished from all other basal anomodonts by the presence of relatively large and closely spaced leaf-shaped teeth. Tooth occlusion has resulted in wear facets which bear distinct striations. In some specimens, all teeth are worn down almost to their necks, suggesting that tooth replacement was infrequent. Despite its elaborate dentition, *S. getmanovi*, like *Galeops* (King, 1994), shares many features of the masticatory apparatus with dicynodonts that were thought formerly to be characteristic of the latter. We examine the biogeographic implications of *S. getmanovi* and other basal anomodonts below.

The best known faunivorous therapsid is *Proburnetia viatkensis* (Figure 2.3D), known from the upper Kotel'nich assemblage. The holotype and only known specimen, a skull about 200 mm long, is preserved as impressions within a nodule (Tatarinov, 1968). *P. viatkensis* is thought to be the closest relative of the equally distinctive South African therapsid *Burnetia mirabilis* (Tatarinov, 1974b), but the only putative synapomorphy is the presence of the tuberosities; the anatomy of neither of the two is known well enough to make this sister-group relationship convincing. The remaining faunivores from Kotel'nich include the remains of small therocephalians and a large, euchambersiid-like therocephalian, all of which are undescribed (Ivakhnenko, 1994).

A number of faunivorous therapsid taxa are regarded as possibly coeval with the lower Kotel'nich assemblage. The gorgonopsian taxon, *Inostrancevia uralensis*, was erected for a partial braincase retrieved from the Blumental 3 locality, Orenburg. This may in fact be associated with the the Vyatskian assemblage (typified by Sokolki). Two cynodont taxa are of uncertain age assignment. A tiny, fragmentary cynodont jaw from Linovo in Kirov Region is the type and only known specimen of *Nanocynodon seductus*; this cynodont was assigned originally to the Galesauridae by Tatarinov (1968), but was later interpreted to be a juvenile procynosuchid (Hopson and Kitching, 1972). A second procynosuchid mandible, collected from the Blumental 3 site, is the holotype of *Uralocynodon tverdochlebovae* (Tatarinov, 1987); this specimen is also from a juvenile individual, and consequently the validity of *U. tverdochlebovae* is uncertain. The fragmentary, immature nature of the Russian procynosuchids complicates a biogeographic appraisal of Permian cynodonts. Although our knowledge of these basal cynodonts has been based almost exclusively on the southern African materials (Kemp, 1979), several well-preserved specimens of juvenile and adult procynosuchids have been collected from central Europe; a preliminary appraisal suggests that the European forms are indistinguishable from the African procynosuchids (H.-D. Sues, pers. comm.).

The Sokolki fauna (Lower Vyatskian Gorizont)

Sokolki, located near Kotlas on the bank of the Little Northern Dvina River, Arkhangel'sk Province, is the type and best known locality of the Vyatskian Gorizont. Its faunal complex is characterized largely by pareiasaurs, anomodonts, and gorgonopsians, but chroniosuchid and seymouriamorph anamniotes are also common. Among the latter include the genera *Karpinskiosaurus* and *Kotlassia* (Ivakhnenko, 1987), and the temnospondyl *Dvinosaurus* has been reported.

The pareiasaur *Scutosaurus karpinskii* is the largest herbivore collected from Sokolki. It appears to be more closely related to other pareiasaur taxa of Laurasian distribution than to *Deltavjatia vjatkensis* (Lee, 1994; Chapter 5). A second species (*S. turberculatus* Ivakhnenko, 1987) is now recognized as a junior synonym of *S. karpinskii* (Lee, 1994). In addition to the pareiasaur material, a partial reptilian jaw and isolated cranial bones with osteoderms have been collected from separate North Dvina sites. The former belongs to an indeterminate parareptile (Ivakhnenko, 1983), whereas the cranial elements appear to be nycteroleterid in nature (Ivakhnenko, 1991b).

The second herbivore at Sokolki is the dicynodont

Dicynodon trautscholdi (King, 1988). *Dicynodon* is primarily a southern African genus, with over 30 species having been described from South Africa, Tanzania, and Zambia. Some Chinese material has been assigned also to *Dicynodon*, but it is uncertain whether *D. trautscholdi* is related more closely to these Chinese forms or to the African members of the genus.

Faunivorous therapsids, comprised of gorgonopsians, therocephalians, and cynodonts, are more diverse throughout the Vyatskian Gorizont than are herbivores. The gorgonopsians *Inostrancevia latifrons* and its conspecific *I. alexandri* are the largest carnivores known from Sokolki. A slightly smaller species, *Pravoslavlevia parva*, has been collected from the same deposits (Tatarinov, 1974b). These gorgonopsians are grouped together in the family Inostranceviidae (Tatarinov, 1974b). A fourth gorgonopsian, *Sauroctonus progressus*, from Semin Ravin, Tetiyshi (Tatarinov, 1968), is believed to be more closely related to the South African genus *Scylacops* than to the inostranceviids (Sigogneau-Russell, 1989). This locality is sometimes assigned to the Vyatskian Gorizont, but may be older. Two theriodonts round out the complement of carnivores in the Vyatskian Gorizont. The therocephalian *Annatherapsidus petri* is thought to be a basal member of Euchambersiidae (*sensu* Hopson and Barghusen, 1986), a group with a predominantly African distribution. The cynodont member of the fauna is the most basal cynodont known, *Dvinia prima* (Tatarinov, 1968). Its closest relatives are procynosuchid cynodonts, which are known in Russia only from fragmentary juvenile specimens from Severodvinskian and Vyatskian Gorizont localities (Tatarinov, 1987); procynosuchids have also been collected from coeval deposits in southern Africa (Hopson and Kitching, 1972) and central Europe (Sues and Boy, 1988).

The Purly–Vyazniki fauna (latest Vyatskian Gorizont)

Purly, in Nizhni Novgorod Province, and Vyazniki, in Vladimir Province, have yielded the most westward deposits of the Vyatskian Gorizont. Unlike the other Late Tatarian localities in Russia, the Purly–Vyazniki faunal complex contains few gorgonopsians and, more significantly, neither pareiasaurs nor cynodonts are present. Instead, the fauna comprises dicynodonts, therocephalians, brachyopoid temnospondyls, chroniosuchids, and the only diapsid reptile found in Russia Permian deposits (Sennikov, 1989). The possible presence of procolophonid reptiles is suggested by fragmentary material. Sennikov (1989) regards the Purly–Vyazniki assemblage as the first appearance of the dicynodont-archosaur communities more characteristic of the succeeding Triassic.

Therocephalians from the Purly–Vyazniki assemblage include *Hexacynodon purlinensis*, represented only by fragmentary cranial elements and teeth from Purly (Tatarinov, 1974b) and *Moschowhaitsia vjuschkovi*, a taxon based on much more complete cranial material, from Vyazniki. The former taxon was assigned to Scylacosauridae by Tatarinov (1974b). The latter is regarded as a basal whaitsiid in as much as it retains a full marginal dentition (Tatarinov, 1974b), whereas African whaitsiids lack postcanine teeth.

Perhaps the most taxonomically significant tetrapod member of the Purly–Vyazniki fauna is the diapsid reptile *Archosaurus rossicus*. Described on the basis of associated premaxilla, skull roof, and presacral vertebrae (Tatarinov, 1960), this faunivorous reptile is distinguished among Late Permian reptiles as the oldest known archosaur (= archosauriform, *sensu* Sereno, 1991). Sennikov (1989) estimated that the skull was approximately 300 mm in length, which, by extrapolation, would make this diapsid the largest predator in the Purly–Vyazniki assemblage.

Basal anomodont phylogeny and biogeography

Our review of the major amniote assemblages of the Russian Permian, examined from a phylogenetic perspective using available amniote phylogenies (e.g. Hopson and Barghusen, 1986; Rubidge, 1991, 1994; Lee, 1994, 1996), suggests that some amniote groups were endemic to both Russia and southern Africa during the early stages of the Late Permian. The evidence is particularly suggestive for herbivorous Permian amniotes. Dinocephalians, pareiasaurs, and dicynodonts appear to have diversified initially in

Africa, with some lineages later dispersing into Eastern Europe. Estemmenosuchidae was an endemic Russian clade that appears to have been succeeded ecologically by tapinocephalid, pareiasaur, and dicynodont émigrés from Africa. Although anteosaurian phylogeny has been complicated by the uncertain position of the basal African species *Australosyodon nyaphuli* (Rubidge, 1994), what is known (Hopson and Barghusen, 1986) suggests that at least part of the evolutionary history of these dinocephalians occurred in Russia. On the other hand, taxa with close relatives from the Early Permian, such as *Mesenosaurus romeri*, *Ennatosaurus tecton*, *Gecatogomphius kavejevi*, and bolosaurids, appear to have dispersed into Russia from North America during Early to Mid-Permian times. Africa appears to have been less affected by these taxa with Early Permian affinities; only the varanopseid eupelycosaur *Elliotsmithia longiceps* and two captorhinid reptiles (Gaffney and McKenna, 1979; Ricqlès and Taquet, 1982) are known from Africa during the Late Permian.

The biogeography of the faunivorous therapsids and non-pareiasaurian parareptiles during the Late Permian is more obscure. Problems associated with the former group are discussed below. Although recent studies of the relationships of *Nyctiphruretus*, *Macroleter*, nycteroleterids, and lanthanosuchids have been undertaken (Laurin and Reisz, 1995; Lee, 1993, 1995; deBraga and Reisz, 1996), these parareptile taxa have not been examined together. Accordingly, we are reluctant to ascertain the role of Russia in the early evolution of these turtle relatives. Despite these problems, our review suggests strongly that endemic clades can be identified for both Russia and southern Africa during the Permian, and a Russian distribution was not ancestral for many Late Permian clades. These views contrast with traditional hypotheses that Late Permian amniotes dispersed into Africa from Eastern Europe, or the more recent hypothesis that faunal elements were dispersing freely between these regions during the Late Permian (Rubidge and Hopson, 1990; Ivakhnenko, 1994).

Early biogeographic hypotheses suggested that therapsids arose in the Laurussian portion of Pangaea during the Early Permian and later dispersed into Africa (Boonstra, 1971; Sigogneau and Chudinov, 1972). These ideas are probably based upon the perceived 'primitive' status of the Russian taxa, in combination perhaps with what was then regarded as an 'explosive' appearance for therapsids in the *Tapinocephalus* Zone of the South African Karoo (Boonstra, 1971). However, recent work in the Karoo has demonstrated the presence of an earlier terrestrial biozone (the *Eodicynodon–Tapinocaninus* Assemblage Zone: Rubidge, 1990a). The presence of basal taxa from five major therapsid clades (Barry, 1972; Rubidge *et al.*, 1983; Rubidge, 1991, 1994, 1995 (Ed.)) in the *Eodicynodon–Tapinocaninus* Zone does not support the earlier hypothesis that southern Africa was colonized by Russian émigrés during the Permian.

The distributions and evolution of anomodont therapsids did not figure largely in biogeographic hypotheses of therapsid evolution until recently. The description of *Patranomodon nyaphulii*, a small, toothed anomodont from the *Eodicynodon–Tapinocaninus* Zone, allowed Rubidge and Hopson (1990) to reexamine the interrelationships of the basal anomodonts. Their analysis indicated that the Russian taxa *Otsheria netzvetajevi* and *Ulemica invisa* (= '*Venyukovia invisa*') were not closely related, but formed a paraphyletic assemblage with the South African basal taxa *P. nyaphulii* and *Galeops whaitsi* (with respect to a monophyletic Dicynodontia). The distribution of the Russian and South African taxa on their cladogram (Figure 2.5) did not support the postulate that separate radiations occurred in northern and southern Pangaea, as suggested by previous phylogenetic studies (Hopson and Barghusen, 1986; King, 1988). Instead, Rubidge and Hopson (1990) concluded that anomodonts were dispersing freely between Europe and southern Africa in the Late Permian, an idea that has been expanded recently to encompass dinocephalians (Rubidge, 1993, 1995).

The cladistic analysis by Rubidge and Hopson (1990) is weak methodologically since it lacks a data matrix and outgroup taxa. The former oversight does not facilitate independent examination of their analysis, whereas the latter problem leads one to question

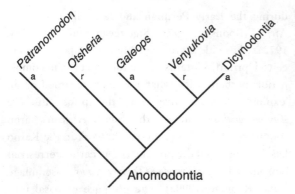

Figure 2.5. Interrelationships of basal anomodonts according to Rubidge and Hopson (1990). Distributions of taxa are indicated for each terminal taxon. *Venyukovia* in this tree is equivalent to *Ulemica* in the following figure. According to Rubidge and Hopson (1990), this phylogeny is suggestive of free faunal exchange between Eastern Europe and southern Africa. Note that *Suminia getmanovi* was not known when this cladogram was originally published. Abbreviations: r, Russian distribution; a, southern African distribution.

the polarity of their characters. Therefore, Rubidge and Hopson's (1990) study cannot be regarded as a rigorous treatment of the interrelationships of basal anomodonts. Accordingly, the hypothesis that there were two evolutionary radiations of endemic African and Russian anomodonts must be reexamined. Further, the recent discovery of *Suminia getmanovi* (Ivakhnenko, 1994) provides new anatomical information that may shed further light on anomodont phylogeny. With these goals in mind, we reanalysed the interrelationships of basal anomodonts, with the intention of mapping their geographic distributions onto the resultant phylogenetic tree(s).

The following taxa formed the ingroup in our analysis: *Patranomodon nyaphulii, Otsheria netzvetajevi, Ulemica invisa, Suminia getmanovi, Galeops whaitsi, Eodicynodon oosthuizeni, Pristerodon* sp., and *Dicynodon* sp. We did not include the basal anomodonts *Galepus jouberti* and *Galechirus scholtzi*, since the skulls of these basal forms are very poorly known (Brinkman, 1981) and we were unable to code them for most of our characters. Further, we were unable to examine the new

specimens attributed to *Galeops* by Rubidge and Hopson (1990), but we examined the holotype and latex casts made from it. Outgroup taxa included the dinocephalian *Titanophoneus potens*, the gorgonopsian *Gorgonops torvus*, and the therocephalian *Pristerognathus minor*. Since much of the palate, occiput, and mandible of basal therocephalians (including *P. minor*) is not well known, supplementary data were taken also from the more recently derived therocephalian *Moschorhinus kitchingi* (Mendrez, 1974). We consider this a preliminary study, since we were restricted to the use of cranial characters; among the 11 taxa examined here, we had access only to the postcrania of *G. whaitsi*. Sources for phylogenetic characters and anatomical data for all taxa are provided in Appendices 1 and 2, respectively. Thirty-eight characters were analysed by the branch-and-bound algorithm of PAUP 3.1.1. Characters were optimized using the delayed transformation algorithm, DELTRAN, and multistate characters were run unordered.

Three most parsimonious trees were discovered. Each tree has a length of 72 steps and a consistency index of 0.65, excluding uninformative characters. A strict consensus of these trees is shown in Figure 2.6. In all trees, *O. netzvetajevi, U. invisa*, and *S. getmanovi* form a monophyletic group defined by four unambiguous apomorphies. To this clade we attach provisionally the nomen Venyukovioidea, which was originally erected by Watson and Romer (1956) for the genera *Otsheria* and *Venyukovia*. The three most parsimonious trees differ only with respect to venyukovioid interrelationships. Venyukovioidea is the sister group of an unnamed clade consisting of *G. whaitsi* and the dicynodonts. The latter clade was recognized independently by Zanon (1987) and King (1994). Finally, Anomodontia is strengthened by the addition of several new characters; it is diagnosed here by 10 unambiguous synapomorphies. Optimization of the geographic distributions onto any of the three most parsimonious trees (Figure 2.6) suggests that the Russian basal anomodonts are the descendants of a single anomodont lineage that colonized Eastern Europe from Africa.

Unfortunately, only one extra step is required to

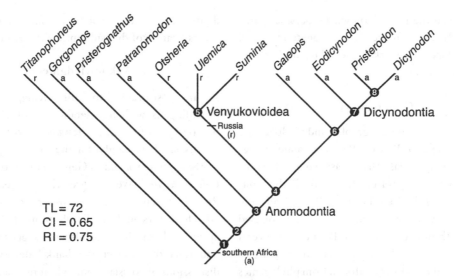

Figure 2.6. Interrelationships of basal anomodonts based on the present study. This cladogram is a strict consensus of the three most parsimonious trees. Distributions of taxa are indicated for each terminal taxon and are spelled in full at internodes for which the distribution is apomorphic. This phylogeny suggests that venyukovioids are the descendents of a single lineage that dispersed into Eastern Europe from Africa. Nodes are defined by the following unambiguous apomorphies (described in Appendix 1): Node 1: 4, 18, 27, 33; Node 2: 6, 25, 26, 34; Node 3 (Anomodontia): 3, 7, 11, 13, 14, 17, 23, 24, 28, 36; Node 4: 12, 21(2), 22; Node 5: 4, 16(2), 18, 19(2); Node 6: 2, 6(2), 10(2); Node 7 (Dicynodontia): 11(2), 21(3), 29, 35, 37; Node 8: 5, 16. Abbreviations as for Figure 2.5, plus: TL, tree length; CI, consistency index, RI, retention index.

alter dramatically the interrelationships of basal anomodonts, with 'venyukovioids' forming a paraphyletic group relative to the clade formed by *Galeops* and Dicynodontia. At two extra steps, *S. getmanovi* becomes the sister taxon of Dicynodontia. These slightly longer trees complicate a biogeographic appraisal of anomodont evolution. However, we would like to note that we were unable to examine original materials of *O. netzvetajevi* and *U. invisa*; we relied instead upon photographs, sketches, and information from the literature. Quite possibly, additional cranial synapomorphies may be discovered uniting venyukovioids using known material. Furthermore, the postcrania of both *O. netzvetajevi* and *U. invisa* are unknown, while that of *S. getmanovi* was described only briefly by Ivakhnenko (1994). Accordingly, venyukovioids may be united by additional, as yet unknown, apomorphies of the postcranial skeleton.

Lastly, a collateral hypothesis of our tree(s) is that an African distribution is ancestral for synapsids crownward of dinocephalians. That postulate, however, is based upon our use of a South African gorgonopsian and therocephalian, and is contingent upon our assumptions that the ancestral distribution for both Gorgonopsia and Therocephalia is Africa. No phylogeny is available for gorgonopsians, and so we cannot be sure whether the Russian gorgonopsians (for which detailed anatomical information was not available to us) are basal forms or more recently derived taxa; in the case of the former situation, an African distribution may be recognized as ancestral only for therocephalians and anomodonts among the therapsid taxa considered here. We are more confident that an African distribution will be demonstrated as ancestral for Therocephalia, since all basal therocephalians are known only from South Africa, except *Scylacosuchus orenbergensis* and the poorly known *Hexacynodon purlinensis*. Clearly, much more detailed anatomical and rigorous phylogenetic work is needed on all therapsid groups examined here.

Conclusions

The combined palaeobiogeographic and phylogenetic evidence available for Russian Permian amniotes suggests that, initially at least, there was recognizable provincialism among parareptiles and herbivorous therapsids during the Late Permian. Provincialism in Russia, however, appears to have been compromised increasingly throughout the Late Permian on several occasions by therapsid and pareiasaurian émigrés from Africa. Our conclusions, drawn from phylogenies of several groups (from the literature), and the anomodont phylogeny presented here, contrast with previous interpretations that therapsids were dispersing freely between Eastern Europe and Africa throughout the whole of the Late Permian. Furthermore, we believe that Africa was probably the centre of origin for pareiasaurs and most Permian therapsids crownwards of Dinocephalia.

The biogeographic implications of the distributions of faunivorous therapsid taxa in the Permian of Russia are indeterminate. We attribute this to the lack of resolution for basal therapsid interrelationships, specifically of those particularly problematic taxa, the 'biarmosuchians'. In a similar vein, no phylogenetic work has been done on gorgonopsians, and only limited, preliminary work has been published on therocephalians. Work on the former group will almost certainly have to await a rigorous review of the alpha-taxonomy of its approximately 80 recognized species. Most of the recent, rigorous phylogenetic studies on non-mammalian synapsids have focused almost exclusively on Mesozoic cynodonts, no doubt owing to the wide interest in mammalian origins and the acquisition of mammalian characters. This is unfortunate, since the evolutionary 'dead ends', the non-cynodont therapsids and stem-group parareptiles, were the dominant vertebrates in Permian terrestrial environments; it is in this evolutionary milieu that both the acquisition of mammalian characteristics by cyno-

donts and the taxonomic and morphological diversification of diapsid reptiles were initially staged.

Acknowledgments

We are most grateful to Gillian King, for generously providing us with her detailed notes, sketches, and photographs of *Otsheria netzvetajevi* and *Ulemica invisa*, to Mikhail Ivakhnenko for the loan of specimens of *Suminia getmanovi*, and to Gene Gaffney for the loan of *Galeops whaitsi*. We are greatly indebted to Robert Reisz for discussions of therapsid anatomy, phylogenetic characters, and the distributions of Permian synapsids and reptiles, and also for suggesting that we undertake this project. We thank Valerii Golubev for discussions of Russian anomodont taxonomy, Mikhail Shishkin for helpful comments on the stratigraphic sections of the paper, and Mike Benton for his consummate editing and organizational support. Bruce Rubidge kindly provided both thoughtful commentary on the manuscript and literature on taxa of which we were unaware. Comments by Jim Hopson also streamlined the final version of the paper. Diane Scott skillfully prepared the latex casts of *Galeops whaitsi*.

References

Barghusen, H.R. 1976. Notes on the adductor jaw musculature of *Venyukovia*, a primitive anomodont therapsid from the Permian of the USSR. *Annals of the South African Museum* **69**: 249–260.

Barry, T.H. 1972. Terrestrial vertebrate fossils from Ecca-defined beds in South Africa, pp. 653–659 in Haughton, S.H. (ed.) *Proceedings and Papers, 2nd IUGS Gondwana Symposium, 1970*, Pretoria, South Africa, CSIR.

Boonstra, L.D. 1971. The early therapsids. *Annals of the South African Museum* **59**: 17–46.

Brinkman, D.B. 1981. The structure and relationships of the dromasaurs (Reptilia: Therapsida). *Harvard Museum of Comparative Zoology, Breviora* **465**: 1–34.

Broom, R. 1932. *The Mammal-like Reptiles of South Africa and the Origin of Mammals*. London: H.F. & G. Witherby, 376 pp.

Chudinov, P.K. 1965. New facts about the fauna of the Upper Permian of the USSR. *Journal of Geology* **73**: 117–130.

—1969. [On the stratigraphic distribution of Permian vertebrates in eastern European parts of the USSR.], pp. 96–105 in [*Questions on the Geology of the South Urals and Povol'zhe.*] Saratov: Izdatel'stvo Saratovskogo Universiteta.

—1983. [Early therapsids.] *Trudy Paleontologicheskogo Instituta AN SSSR* **202**: 1–230.

Clark, J. and Carroll, R.L. 1973. Romeriid reptiles from the Lower Permian. *Bulletin of the Museum of Comparative Zoology* **144**: 353–407.

Cluver, M.A. and Hotton, N., III. 1981. The genera *Dicynodon* and *Diictodon* and their bearing on the classification of the Dicynodontia (Reptilia, Therapsida). *Annals of the South African Museum* **83**: 99–146.

deBraga, M. and Reisz, R.R. 1996. The Early Permian reptile *Acleistorhinus pteroticus* and its phylogenetic position. *Journal of Vertebrate Paleontology* **16**: 384–395.

Dodick, J.T. and Modesto, S.P. 1995. The cranial anatomy of the captorhinid reptile *Labidosaurikos meachami* from the Lower Permian of Oklahoma. *Palaeontology* **38**: 687–711.

Efremov, I.A. 1937. [New Permian reptiles of the USSR.] *Doklady AN SSSR* **19**: 771–776.

—1940. Kurze Übersicht über die Formen der Perm- und Trias-Tetrapoden-Fauna der UdSSR. *Centralblatt für Mineralogie, Geologie, und Paläontologie, Abteilung B* **1940**: 372–383.

—1954. [Fauna of terrestrial vertebrates from the Permian Copper Sandstones of the western Cis-Urals.] *Trudy Paleontologicheskogo Instituta AN SSSR* **54**: 1–416.

—and V'yushkov, B.P. 1955. [Catalogue of the localities of Permian and Triassic terrestrial vertebrates in the territories of the USSR.] *Trudy Paleontologicheskogo Instituta AN SSSR* **46**: 1–147.

Gaffney, E.S. and McKenna, M.C. 1979. A Late Permian captorhinid from Rhodesia. *American Museum Novitates* **2688**: 1–15.

Gauthier, J.A, Kluge, A.G. and Rowe, T. 1988. Amniote phylogeny and the importance of fossils. *Cladistics* **4**: 105–209.

Gubin, Y.M. 1983. The first eryopids from the Permian of the East European platform. *Paleontological Journal* **16**: 105–110.

Heever, J.A. van den and Hopson, J.A. 1982. The systematic position of 'Therocephalian B' (Reptilia: Therapsida). *South African Journal of Science* **78**: 424–425.

Hopson, J.A. 1969. The origin and adaptive radiation of mammal-like reptiles and non-therian mammals. *Annals of the New York Academy of Sciences* **167**: 199–216.

—and Barghusen, H.R. 1986. An analysis of therapsid relationships, pp. 83–105 in Hotton, N., III, MacLean, P.D., Roth, J.J. and Roth, E.C. (eds.), *The Ecology and Biology of Mammal-like Reptiles*. Washington, D.C.: Smithsonian Institution Press.

—and Kitching, J.W. 1972. A revised classification of cynodonts (Reptilia; Therapsida). *Palaeontologia Africana* **14**: 71–85.

Ivakhnenko, M.F. 1980. Lanthanosuchids from the Permian of the East European platform. *Paleontological Journal* **13**: 80–90.

—1983. New procolophonids from eastern Europe. *Paleontological Journal* **17**: 135–139.

—1987. [Permian parareptiles of the USSR.] *Trudy Paleontologicheskogo Instituta AN SSSR* **223**: 1–160.

—1991a. Elements of the Early Permian tetrapod faunal assemblages of Eastern Europe. *Paleontological Journal* **24**: 104–112.

—1991b. Late Palaeozoic tetrapod faunal assemblage from the Mezen' River Basin. *Paleontological Journal* **1991**: 76–84.

—1994. A new Late Permian dromasaurian (Anomodontia) from Eastern Europe. *Paleontological Journal* **28**: 96–103.

—1996. [Primitive anomodont-venjukoviid from the Upper Permian of eastern Europe.] *Paleontologicheskii Zhurnal* **1996** (3): 1–9.

—and Kurzanov, S.M. 1979. *Mesenosaurus*, a primitive archosaur. *Paleontological Journal* **12**: 139–141.

—and Tverdokhlebova, G.I. 1988. A revision of the Permian bolosauromorphs of Eastern Europe. *Paleontological Journal* **21**: 93–100.

Kemp, T.S. 1979. The primitive cynodont *Procynosuchus*: functional anatomy of the skull and relationships. *Philosophical Transactions of the Royal Society, Series B* **285**: 73–122.

King, G.M. 1981. The functional anatomy of a Permian dicynodont. *Philosophical Transactions of the Royal Society, Series B* **291**: 243–322.

—1988. Anomodontia, pp. 1–174 in P. Wellnhofer (ed.), *Handbuch der Paläoherpetologie, Teil 17C*. Stuttgart: Gustav Fischer Verlag.

—1994. The early anomodont *Venjukovia* and the evolution of the anomodont skull. *Journal of Zoology* **232**: 651–673.

Kitching, J.W. 1977. The distribution of the Karroo vertebrate fauna. *Bernard Price Institute for Palaeontological Research Memoir* 1: 1–131.

Laurin, M. and Reisz, R.R. 1995. A reevaluation of early amniote phylogeny. *Zoological Journal of the Linnean Society* 113: 165–223.

Lee, M.S.Y. 1993. The origin of the turtle body plan: bridging a famous morphological gap. *Science* 261: 1616–1620.

—1994. Evolutionary morphology of pareiasaurs. Ph.D. thesis, University of Cambridge.

—1995. Historical burden in systematics and the interrelationships of 'parareptiles'. *Biological Reviews* 70: 459–547.

—1996. Correlated progression and the origin of turtles. *Nature* 379: 812–815.

Li, J.-L. and Cheng, Z. 1995. A new Late Permian vertebrate fauna from Dashankou, Gansu with comments on Permian and Triassic vertebrate assemblage zones of China, pp. 33–37 in Sun, A. and Wang, Y. (eds.), *Proceedings of the Sixth Symposium on Mesozoic Terrestrial Ecosystems and Biota*. Beijing: China Ocean Press.

Mendrez, C.H. 1974. Étude du crane d'un jeune specimen de *Moschorhinus kitchingi* Broom, 1920 (?*Tigrisuchus simus* Owen, 1876), Therocephalia Pristerosauria Moschorhinidae d'Afrique australe (Remarques sur les Moschorhinidae et les Whaitsiidae). *Annals of the South African Museum* 64: 71–115.

Modesto, S.P. 1996. A basal captorhinid reptile from the Fort Sill fissures, Lower Permian of Oklahoma. *Oklahoma Geological Notes* 56: 4–14.

Ochev, V.G. 1976. [Stages in the history of Permian and Triassic tetrapods of the European parts of the USSR.], pp. 44–49 in [*Questions in Stratigraphy and Palaeontology*]. Saratov: Izdatel'stvo Saratovskogo Universiteta.

—, Tverdokhlebova, G.I., Minikh, M.G. and Minikh, A.V. 1979. [*Stratigraphic and Palaeontological Importance of Upper Permian and Triassic Vertebrates of the Eastern European Platform and the Cis-Urals.*] Saratov: Izdatel'stvo Saratovskogo Universiteta, 158 pp.

Olson, E.C. 1955. Parallelism in the evolution of Permian reptilian faunas of the Old and New Worlds. *Fieldiana: Zoology* 37: 385–401.

—1957. Catalogue of localities of Permian and Triassic terrestrial vertebrates of the USSR. *Journal of Geology* 65: 196–226.

—1962. Late Permian terrestrial vertebrates, USA and USSR. *Transactions of the American Philosophical Society, N.S.* 52, Part 2, 224 pp.

—1966. Community evolution and the origin of mammals. *Ecology* 47: 291–302.

—1968. The family Caseidae. *Fieldiana: Geology* 17: 223–349.

—1974. On the source of therapsids. *Annals of the South African Museum* 64: 27–46.

—1979. Biological and physical factors in the dispersal of Permo-Carboniferous terrestrial vertebrates, pp. 227–238 in Gray, J. and Boucot, A.J. (eds.), *Historical Biogeography, Plate Tectonics, and the Changing Environment*. Portland: Oregon State University Press.

—1986. Relationships and ecology of the early therapsids and their predecessors, pp. 47–60 in Hotton, N., III, MacLean, P.D., Roth, J.J. and Roth, E.C. (eds.), *The Ecology and Biology of Mammal-like Reptiles*. Washington, D.C.: Smithsonian Institution Press.

—and Beerbower, J.R. 1953. The San Angelo Formation, Permian of Texas, and its vertebrates. *Journal of Geology* 61: 381–423.

Orlov, Yu.A. 1958. [Carnivorous dinocephalians from the fauna of Isheev (Titanosuchia.)] *Trudy Paleontologicheskogo Instituta AN SSSR* 72: 1–114.

Ricqlès, A. de and Taquet, P. 1982. La faune de vertébrés de Permien Supérieur du Niger. I. Le captorhinomorphe *Moradisaurus grandis* (Reptilia, Cotylosauria) – Le crâne. *Annales de Paléontologie* 68: 33–63.

Ross, C.A., Baud, A. and Menning, M. 1994. Pangea time scale, p. 10 in Klein, G.D. (ed.), *Pangea: Palaeoclimate, Tectonics, and Sedimentation During Accretion, Zenith, and Breakup of a Supercontinent*. Boulder, Colorado: Geological Society of America Special Paper 288.

Rubidge, B.S. 1990a. A new vertebrate biozone at the base of the Beaufort Group, Karoo Sequence (South Africa). *Palaeontologia Africana* 27: 17–20.

—1990b. Redescription of the cranial morphology of *Eodicynodon oosthuizeni* (Therapsida: Dicynodontia). *Navorsinge van die Nasionale Museum Bloemfontein* 7: 1–25.

—1991. A new, primitive, dinocephalian mammal-like reptile from the Permian of southern Africa. *Palaeontology* 34: 547–559.

—1993. New South African fossil links with the earliest mammal-like reptile (therapsid) faunas from Russia. *South African Journal of Science* 89: 460–461.

—1994. *Australosyodon*, the first primitive anteosaurid dinocephalian from the Upper Permian of Gondwana. *Palaeontology* **37**: 579–607.

—1995. Did mammals originate in Africa? South African fossils and the Russian connection. *South African Museum, Sidney Haughton Memorial Lecture* **4**: 1–14.

—, (Ed.) (1995.) *Biostratigraphy of the Beaufort Group (Karoo Supergroup), South Africa.* Pretoria: Council of Geoscience, 72 pp.

—and Hopson, J.A. 1990. A new anomodont therapsid from South Africa and its bearing on the ancestry of Dicynodontia. *South African Journal of Science* **86**: 43–45.

Rubidge, B.S., Kitching, J.W., and van den Heever, J.A. 1983. First record of a therocephalian (Reptilia, Therapsida) from the Ecca of South Africa. *Navorsinge van die Nasionale Museum Bloemfontein* **4**: 229–235.

Sennikov, A.G. 1989. The role of the oldest thecodonts in the vertebrate assemblages of Eastern Europe. *Paleontological Journal* **22**: 74–82.

Sereno, P.C. 1991. Basal archosaurs: phylogenetic relationships and functional implications. *Journal of Vertebrate Paleontology Memoir* **2**: 1–53.

Sidor, C. and Hopson, J.A. 1995. The taxonomic status of the Upper Permian eotheriodont therapsids of the San Angelo Formation (Guadalupian), Texas. *Journal of Vertebrate Paleontology* **15** (3): 53A.

Sigogneau, D. and Chudinov, P.K. 1972. Reflections on some Russian eotheriodonts (Reptilia, Synapsida, Therapsida). *Palæovertebrata* **5**: 79–109.

Sigogneau-Russell, D. 1989. Theriodontia I, pp. 1–127 in Wellnhofer, P. (ed.), *Handbuch der Paläoherpetologie, Teil 17B/I.* Stuttgart: Gustav Fischer.

Sues, H.-D. and Boy, J.A. 1988. A procynosuchid cynodont from central Europe. *Nature* **331**: 523–524.

Tatarinov, L.P. 1960. [The discovery of pseudosuchians in the Upper Permian of the USSR.] *Paleontologicheskii Zhurnal* **1960**: 74–80.

—1968. Morphology and systematics of the Northern Dvina cynodonts (Reptilia, Therapsida; Upper Permian). *Postilla* **126**: 1–51.

—1974a. Discovery of a bolosaur in the Lower Permian of the USSR. *Paleontological Journal* **8**: 250–252.

—1974b. [Theriodonts of the USSR.] *Trudy Paleontologicheskogo Instituta AN SSSR* **143**: 1–252.

—1987. A new primitive cynodont from the Upper Permian of the southern Urals. *Paleontological Journal* **21**: 103–107.

Tverdokhlebova, G.I. and Ivakhnenko, M.F. 1985.

Nycteroleters from the Permian of Eastern Europe. *Paleontological Journal* **18**: 93–104.

Watson, D.M.S. 1954. On *Bolosaurus* and the origin and classification of reptiles. *Bulletin of the Museum of Comparative Zoology* **111**: 299–449.

—and Romer, A.S. 1956. A classification of therapsid reptiles. *Bulletin of the Museum of Comparative Zoology* **114**: 37–89.

Zanon, R.T. 1987. Phylogenetic relationships within Anomodontia. *Journal of Vertebrate Paleontology* **7** (3): 30A.

APPENDIX 1

Cranial characters used for cladistic analysis.

1. Caniniform present and long (0), maxillary teeth decrease gradually in size posteriorly (1), or caniniform present but short (2).

2. Incisors present (0) or absent and edentulous 'beak' formed by premaxilla and dentary present (1).

3. Fine serrations present (0), serrations absent (1), or coarse serrations present (2) on marginal teeth.

4. Premaxillary posterodorsal process elongate (0) or short (1).

5. Premaxillae sutured (0) or fused (1) together.

6. Antorbital region long (0), short (1), or greatly abbreviated (2).

7. Septomaxilla posterodorsal spur conspicuous and separates widely nasal from maxilla (0) or inconspicuous and nasal-maxillary suture well developed (1).

8. Frontal contribution large, equal to or greater than (0) or small, less than half of (1) that of the postfrontal.

9. Postorbital ventral portion sharply tipped (0) or expanded antero-posteriorly (1).

10. Postorbital-squamosal contact absent ventrally (0), present ventrally (1), or present ventrally with squamosal extending anteriorly beyond postorbital bar (2).

11. Squamosal anterior process parasagittally deep (0), narrow, rod-like (1), or dorso-ventrally compressed (2).

12. Squamosal without (0) or with (1) lateral fossa for origin of lateral slip of Mm. adductor mandibulae externus.

13. Squamosal posteroventral process absent (0) or present and extends ventrally to level of condyle (1).

14. Zygomatic arch roughly horizontal (0) or bowed dorsally (1).

15. Quadratojugal narrow dorsally (0) or dorsal part expanded transversely (1).

16. Parietals' contribution to skull table transversely broad as long (0), longer antero-posteriorly than broad (1), or shorter antero-posteriorly than broad (2).
17. Parietal posterolateral processes elongate (0) or short (1).
18. Pineal foramen raised on prominent boss (0) or opening flush with dorsal surface of skull roof (1).
19. Preparietal bone absent and interparietal suture well developed anterior to pineal foramen (0), preparietal bone present and interparietal suture either greatly reduced or absent anterior to pineal foramen (1), or preparietal bone absent and interparietal suture anterior to pineal foramen greatly reduced by frontals (2).
20. Tabular large and separates squamosal from supraoccipital (0) or small with supraoccipital-squamosal contact present (1).
21. Internal narial shelf absent (0), narrow narial shelf formed by maxilla and palatine only (1), narrow narial shelf formed by premaxilla, maxilla, and palatine (2), or narial shelf well developed, formed mostly by deep lingual shelves of premaxilla and maxilla (3).
22. Premaxilla-palatine contact absent (0) or present (1).
23. Ectopterygoid extends farther posteriorly than palatine (0) or vice versa (1) in palatal aspect.
24. Palatine posterior portion transversely broad (0) or narrow (1).
25. Palatine teeth present (0) or absent (1).
26. Pterygoid teeth present (0) or absent (1).
27. Epipterygoid separate from (0) or contacts (1) parietal.
28. Lateral pterygoid process large, conspicuous (0) or small and inconspicuous (1).
29. Pterygoids contact anteriorly (0) or vomer contributes to interpterygoid vacuity (1).
30. Parabasisphenoid excluded from (0) or reaches (1) interpterygoid vacuity.
31. Dentaries sutured (0) or fused (1) at symphysis.
32. Lateral dentary shelf absent (0) or present (1).
33. Dentary coronoid process absent (0) or present (1).
34. Mandibular fenestra absent (0) or present (1).
35. Surangular vertical lamina present and lateral to articular (0) or absent (1).
36. Coronoid bone present (0) or absent (1).
37. Prearticular with (0) or without (1) lateral exposure posteriorly.
38. Jaw articulation permitting strictly orthal closure (0) or parasagittal sliding action (1).

Sources: 1, 7, 12, 14, 18, 19, 21, 22, 25, 26, 31, 32, and 34 from Hopson and Barghusen (1986); 8 and 27 from Gauthier et al. (1988); 5, 21, 22, 28, 31, 32, and 38 from King (1988); 2, 6, 10, 15, 20, and 29 from Rubidge and Hopson (1990).

APPENDIX 2

Distribution of the cranial characters in eight anomodont genera and three outgroups

Titanophoneus	00000	00000	00000	00000	00000	00000	00000	000
Gorgonops	00010	00110	00000	00110	00000	01000	00100	000
Pristerognathus	00010	10100	0000?	10100	10001	11001	00110	010
Patranomodon	10110	11011	10111	01110	0011?	1?100	00010	100
Otsheria	10100	11000	1011?	21020	2?111	1?100	???1?	???
Ulemica	20100	11000	11111	2102?	21111	1?100	11110	110
Suminia	10200	11001	11111	21021	21111	1?101	01110	101
Galeops	?1??0	21012	1?01?	?????	21101	1?101	10110	101
Eodicynodon	01110	21012	21111	01111	30111	11110	11111	111
Pristerodon	01211	21002	21111	11111	31111	11110	11111	111
Dicynodon	01111	21012	21111	11111	31111	11110	11111	111

Sources: *Titanophoneus*, Orlov (1958); *Gorgonops*, Sigogneau-Russell (1989); *Pristerognathus*, Broom (1932) and van den Heever and Hopson (1982), with supplementary data from *Moschorhinus*, Mendrez (1974); *Patranomodon*, Rubidge and Hopson 1990; *Otsheria*, Chudinov (1983) and photographs of PIN 1758/5; *Ulemica*, Barghusen (1976), Chudinov (1983), King (1990) and photographs of PIN 157/1116 and 2793/1; *Galeops*, Brinkman (1981) and AMNH 5536; *Eodicynodon*, Rubidge (1990b); *Pristerodon*, BPI 3024; *Dicynodon*, Cluver and Hotton (1981), King (1981); *Suminia*, PIN 2212/32, 2212/33, and 2212/62.

3

Permian and Triassic temnospondyls from Russia

MIKHAIL A. SHISHKIN, IGOR V. NOVIKOV AND YURII M. GUBIN

Introduction

Among pre-Jurassic amphibians, temnospondyls are the predominant group, with a worldwide distribution by the earliest Triassic, and including the ancestors of modern frogs. They appeared first in the Early Carboniferous and declined during the Triassic; their latest record comes from the Early Cretaceous of Australia.

In appearance, temnospondyls varied from salamander-like to crocodile-like; different forms ranged in size from a few centimetres up to 3–4 m, thus including the largest amphibians that ever existed. Judging by their uniformly cone-shaped teeth, all temnospondyls were predators, primarily fish-eaters. As in modern amphibians, their life cycle was dependent on fresh water, but some Triassic groups, such as trematosaurs and plagiosaurs, invaded nearshore marine habitats, hence showing tolerance of high salinities when they were adult. During ontogeny, temnospondyls passed through a gill-breathing larval stage and underwent a gradual metamorphosis, which resembled the condition in modern urodeles. A number of forms, such as plagiosaurs, early brachyopoids and some dissorophoids, were neotenic and retained the gills in the adult.

Repository abbreviations

MB, Museum für Naturkunde, Humboldt Universität, Berlin; PIN, Paleontological Institute, Russian Academy of Sciences, Moscow; SGU, Saratov State University, Saratov; TsNIGRI, Tsentralny Nauchno-Issledovatelskii Geologo-Razvedochnyi Muzei, Saint Petersburg.

Temnospondyls and the amphibian radiation: a problem of interrelationships

Temnospondyls shared many primitive characters with other fossil amphibian groups (Ichthyostegidae, Acanthostegidae, Crassigyrinidae, Loxommatidae, Anthracosauromorpha, Seymouriamorpha) traditionally termed Labyrinthodontia. This latter taxon is now largely considered to be a paraphyletic grade; the former idea of the presence in its early members of the tympanic membrane as a common derived character is usually rejected. In line with this, the so-called otic notch in the skull roof, shared to various extents by all the labyrinthodont lineages[1], and once presumed to house the tympanum, is now regarded as either a rudiment of a spiracle, or a new evolutionary acquisition of some advanced forms (Carroll, 1980; Clack, 1983, 1987; Godfrey *et al.*, 1987; Milner, 1993b; Panchen, 1980, 1985; Smithson, 1985). This viewpoint is discussed below. The temnospondyls themselves are also regarded by some students as a grade of organization rather than a distinctive clade (Panchen, 1975, 1980; Milner, 1990, 1993b), but no consensus has been reached on that proposition.

Generalized labyrinthodont, or 'protetrapod', characters of temnospondyls include primarily the solid skull roof devoid of temporal fenestrations, the peculiar ornamentation of the dermal bones, the 'latitabular' pattern of the skull roof (with the tabular not contacting the parietal), the presence of otic notches and

---

[1] The only labyrinthodont group cited to be an exception, the Colosteidae, seems in fact to include the notch-bearing genus *Erpetosaurus*. Smithson's (1982) and Hook's (1983) assumption that this form is a trimerorhachoid is at variance with its morphology (Steen, 1931; Watson, 1956; cf. Foreman, 1990).

posttemporal fossae, the conical teeth with labyrinthine infolding of the dentine, usually three pairs of tusks on the palate, and the multipartite vertebral centra, primitively including the hypocentrum (intercentrum) as a main element, and the paired pleurocentra tending to meet its counterpart dorsally. The floor of the endocranium always shows an unossified zone between the basioccipital and basisphenoid, reminiscent of the separation into two parts of the rhipidistian braincase. In addition, the most primitive temnospondyls retain a separate intertemporal on the skull roof, the lacrimal extending from orbit to naris, a single concave occipital condyle, movable basipterygoid articulation, and anterior median contact between the pterygoids. All these characters were immediately inherited, or modified, from rhipidistian ancestors.

Of derived temnospondyl characters, very few can be effectively regarded as unique; but, combined together, these enable a rather clear characterization of the group. The most important of them (although also compound) is a dorsolaterally directed stapes inserted into the tympanum, firmly attached to the base of the oval fenestra, and showing reduced or lost cartilaginuous connection to the hyoid, a condition known elsewhere only in anurans (Tatarinov, 1962; Shishkin, 1973, 1975). Other apomorphies include the loss of pleurokinetism (with the cheek-table joint of the skull roof being replaced by a suture), the postparietals with occipital flanges sutured to the exoccipitals, the palate with interpterygoid vacuities, the parasphenoid cultriform process overlapping the vomers, the palatoquadrate otic process connected to the prootic (in contrast to anthracosaurs and amniotes, cf. Shishkin, 1973), the conspicuous parasphenoid wings underlying the basitrabecular processes, and the four-digit manus.

The main trends in temnospondyl evolution, largely known since Watson's (1919) classical account, and having proceeded in parallel within different lines, were caused by the return of the group in Late Permian and Triassic times to a purely aquatic, predominantly benthic, habitat. Particularly notable are the flattening of the skull, which resulted in expansion of the interpterygoid vacuities to retain the volume for housing the eyeballs and their muscles; progressive chondrification of the endocranium and replacement of the original

basipterygoid articulation by a firm pterygoid-parasphenoid suture; development of paired convex occipital condyles; reduction of the basioccipital and spread of the exoccipitals on to the otic capsules by means of giving off the paroccipital and subotic processes (unknown in other early tetrapods); development of the anterior palatal vacuity and palatal tooth rows; reduction of shagreen dentition of the palate; loss of the posterior projection of the jaw articulation beyond the level of the occiput; increase in length of the retroarticular process of the lower jaw; expansion and strengthening of the dermal shoulder girdle; and reduction of the vertebral pleurocentra to small plates of bone underlying the base of the neural arch. One further common trend, seen to start rather early in temnospondyl evolution, is re-integration of the vertebral hemicentra, from close association of the hypocentrum with the preceding pleurocentrum (inherited from rhipidistians) to a reverse condition, typical of temnospondyls (Shishkin, 1989a,b).

There are also a number of more specific trends, characterizing only particular temnospondyl lineages. These include, for instance, the obliteration of the otic notch, or its closure posteriorly (in many forms), loss of the lacrimal (rhytidosteids), or its fusion to the palatine (some brachyopoids), inclusion of the frontal in the orbital border (some dissorophoids and most capitosauroids), development of the pterygoid–exoccipital suture (many Triassic groups) or the pterygoid–basioccipital suture (tupilakosaurids), a trend towards reduction of the ectopterygoid (branchiosaurids and amphibamids), loss of the ectopterygoid tusks (intasuchids, archegosauroids, capitosauroids), or, more rarely, vomerine tusks (some plagiosaurs), and development of the dermal armour or dorsal scutes on the body (plagiosaurs, dissorophids). The vertebral structure is known to undergo re-modelling according to various patterns, such as stereospondylous, with the massive disc-shaped hypocentra (mastodonsaurs, metoposaurs, late cyclotosaurs and brachyopids), holospondylous, with hypo- and pleurocentra fused together (plagiosaurs), embolomerous, with both hemicentra discshaped (tupilakosaurids), or sub-gastrocentrous, approaching the reptile condition (doleserpetontids and some amphibamids).

An outline classification of temnospondyls, set out in terms of Linnaean taxa, has not been well-established. Most schemes published during the past 50 years (for a partial review, see Shishkin, 1984), follow, with various modifications, Romer's (1947, 1966) concept. One of the most recent revised versions of such a scheme is that of Carroll and Winer, (1977; see also Carroll, 1988). A sketchy attempt to outline the main lineages in temnospondyl evolution was made by Shishkin (1984, 1987). Cladistically, the interrelationships of temnospondyl families have been analysed by Milner (1990, 1993b), and his overall review of the group is yet to be published.

Detailed discussion on temnospondyl relationships, both with modern amphibians and among the amphibian-grade Palaeozoic tetrapods, is beyond the scope of this chapter, so only principal points will be outlined. In contrast to many recent workers who revive the taxon 'Lissamphibia' for all the living amphibians, and presume its origin from the Temnospondyli (see Milner, 1993b, for survey), we are confident that the latter gave rise only to a single living group, the Anura. Two key synapomorphies of temnospondyls and anurans preclude them from being related to the Caudata and Gymnophiona: (1) the presence of a tympanic sound-conducting system, with the 'dorsal' stapes separated from the hyoid, and (2) the pattern of the cranial arterial system (as far as it can be restored from impressions on the skull bones), in particular, the reduction of the stapedial artery, with transfer of its branches to the occipital artery, and mode of branching of the maxillary artery (Säve-Söderbergh, 1936; Shishkin, 1968, 1973, 1975; Tatarinov, 1962). We should stress that most characters commonly used to justify the concept of the 'Lissamphibia' either reflect common paedomorphic trends seen in recent amphibians (cf. Shishkin, 1973), or are soft (non-skeletal) structures which cannot be compared in fossils. The ancestors of caudates and gymnophionians most likely should be looked for among the Palaeozoic lepospondyl lineages, such as microsaurs (Carroll and Currie, 1975; Carroll and Holmes, 1980).

Another controversial problem is the position of temnospondyls with respect to the rest of the early amphibians. Re-assessment of the Labyrinthodontia as a grade based entirely on primitive characters (Smithson, 1982), and the Lepospondyli, uniting other Palaeozoic amphibians, as a polyphyletic assemblage of dwarf lineages (Thomson and Bossy, 1970), stimulated the search for a new concept of early tetrapod interrelationships, expressed primarily in cladistic terms. The most common re-arrangement, agreed by British authors (Smithson, 1985; Panchen and Smithson, 1988; Milner, 1993a,b; Ahlberg and Milner, 1994), and followed without further analysis in a number of accounts elsewhere (Gauthier et al., 1989; Lombard and Sumida, 1992; Bolt and Lombard, 1992), presumes that the whole tetrapod radiation falls into two principal clades, 'batrachomorphs' (Amphibia) and 'reptiliomorphs' (Amniota), or in these two and their common stem, comprising the most primitive labyrinthodont groups. The keystone of this concept is an assumption that extant amphibians and reptiles are both monophyletic and thus, providing the tetrapods as a whole are also monophyletic, all fossil groups should be members or sidelines of these clades.

Proceeding from this assumption, it was argued that temnospondyls and at least the main lepospondyl groups, i.e. microsaurs and possibly nectrideans[2], are stem-groups of the clade Amphibia, while anthracosaurs, seymouriamorphs, and perhaps more primitive labyrinthodonts, such as loxommatids and crassigyrinids, pertain to the clade Amniota. According to this hypothesis, temnospondyls, presumed to include the ancestors of living amphibians ('Lissamphibia'), share a sister-group relationships with microsaurs. As a consequence of all this revision, the systematic terminology for the early amphibian-grade tetrapods has been destabilized and no suitable alternative has so far been proposed.

In our opinion, this concept is too ill-supported to be accepted. We agree with Carroll and Chorn (1995) that, in the present state of knowledge, almost any pattern of relationship between the groups termed labyrinthodonts and lepospondyls can be imagined. As seen from the comments above, we reject the basic statement of the concept, the assumption that extant amphibians ('Lissamphibia') are more closely interrelated than any

---

[2] For reservations or alternative views regarding nectridean affinities within the basic tetrapod dichotomy, see Smithson (1985), Panchen and Smithson (1988)

one is to amniotes. Moreover, this statement is not in accord with other suggestions made by advocates of this view. For example, if Gymnophiona might have a microsaur ancestry (Milner, 1993b, p. 16), while microsaurs in turn may belong to the amniote clade (Panchen and Smithson, 1988, p. 12), then the hypothesis of a dichotomy of living Amphibia and Amniota would collapse. Likewise, with regard to extant amniotes, the assumption of their monophyly is seriously undermined by the unresolved problem of the sauropsid–theropsid split (Westoll, 1942; Tatarinov, 1958; Vaughn, 1962; Shishkin, 1973, 1975; cf. Panchen and Smithson, 1988, p. 25) and also by uncertainty about the level at which the living parareptiles (turtles) departed from the reptile stem. This divergence conflicts especially with the idea of two main tetrapod clades, if we accept Panchen's (1985) conclusion that the Seymouriamorpha (i.e. parareptile stem-group) is not related to anthracosaurs.

The number of derived characters currently available for analysis of early amphibian diversification is so limited that no one pattern of sister-group relationships can be substantiated as more convincing than the others. In the case of the presumed temnospondyl–microsaur linkage, their most frequently cited postulated synapomorphies (unique, or shared also with nectrideans and colosteids) are: (1) the exoccipital sutured to the skull roof, without the opisthotic intervening between them; (2) four (or fewer) digits in the manus; (3) waisted, not L-shaped, humerus (Panchen and Smithson, 1988; Milner, 1993b; Ahlberg and Milner, 1994). However, the lack of (1) and (2) in some nectrideans (Bossy, 1976; Panchen and Smithson, 1988) suggests that these characters could equally well have been acquired independently in the groups listed. The same may be suggested for (3), judging by the occurrence of the waisted humerus in caudates and earliest amniotes. On the other hand, while assessing microsaur relationships, it seems difficult to explain why the above features should be taken in preference to those derived characters which immediately link microsaurs with amniotes, without their being interlinked by anthracosaurs or any other labyrinthodonts (the separate supraoccipital, gastrocentrous vertebrae and, possibly, astragalus and calcaneum, cf. Panchen and Smithson, 1988; Smithson *et al.*, 1994).

In summary, available evidence is far from sufficient to indicate that any of the lepospondyl orders (regardless of their own interrelationships) are closely related to temnospondyls or any other labyrinthodont group. It is not ruled out that lepospondyls evolved prior to the rise of all or most labyrinthodont groups, from one or more stems for which the fossil record is currently unknown (cf. Carroll, 1995; Carroll and Chorn, 1995).

So it may be concluded that even being paraphyletic, the Labyrinthodontia still remains the best evidenced and most clearly demarcated taxonomic unit as compared with any other combination of primitive amphibian lines so far proposed. Hence, the idea of a temnospondyl–microsaur clade cannot be supported without stronger evidence.

Finally, it should be stressed that any solution of the problem of labyrinthodont interrelationships is heavily dependent on the interpretation of their middle ear structures. As noted above, most recent authors, and especially the supporters of a basic tetrapod dichotomy, deny the presence of a true otic notch and tympanum in primitive labyrinthodonts (or even in all labyrinthodonts except advanced temnospondyls and seymouriamorphs). The stapes in these forms is alleged to be a primitive hyomandibular-like structure, which was incapable of transmitting airborne vibrations and had some structural functions (Carroll, 1980; Clack, 1983, 1987, 1992; Smithson, 1982, 1985; Panchen, 1985; Panchen and Smithson, 1988). This viewpoint is widely used to strengthen the belief that labyrinthodont lineages had no shared derived characters. However, our interpretation of the primitive labyrinthont stapes is totally different (Shishkin, 1994a). In our opinion, it shows a well-marked indication of a distal hyotympanic expansion, comparable to that of paedomorphic temnospondyls, and thus corresponds precisely to what is expected in the transition from the rhipidistian hyomandibular to a dorsally directed temnospondyl stapes, connected to the tympanum. In other words, the primitive labyrinthodont stapes (known in *Acanthostega*, the colosteid *Greererepeton*, and the anthracosaur *Pholiderpeton*) invariably shows a specific derived condition which reflects the first steps towards the acquisition of a fully developed tympanic system. It is not immediately clear, of course, whether this condition is a true synapomorphy or arose in parallel within separate laby-rinthodont lineages; but at present the former possibil-ity can in no way be ruled out. In this connection, it worth remembering that none of the lepospondyl groups shows any evidence of the development, or former presence, of a tympanic system, a distinction that was traditionally used to separate the taxa Labyrinthodontia and Lepospondyli.

Permo-Triassic temnospondyl faunas of European Russia

In the Permo-Triassic tetrapod communities of East Europe, temnospondyls played an important role and achieved maximum diversity and abundance during the Early Triassic. Forms from Russia include over 40 genera in 20 families, half the familial diversity of the group worldwide (cf. Carroll and Winer, 1977; Milner, 1990). The great majority of finds come from European Russia (the Cis-Urals in a broad sense); geographically, the collecting areas extend from the Polar Urals to the Caspian basin. Only a few Triassic finds have been reported from northern and central Siberia and the Russian Far East. Stratigraphically, temnospondyl com-munities from Russia range from earliest Late Permian to Middle Triassic, and they show changes that corre-spond to turnovers in the whole coeval Cis-Uralian tetrapod fauna (cf. Chapters 2 and 7).

The earliest Late Permian (Ufimian) amphibian assemblage is represented by the terrestrial eryopoids, holdovers from the Early Permian Euramerican biota. These include eryopids and the endemic intasuchids (Konzhukova, 1956; Gubin, 1983, 1984). The succeed-ing Kazanian–Early Tatarian tetrapod faunas, primarily dominated by dinocephalian therapsids, show a broad expansion of the aquatic or semi-aquatic archegosaur-oids, unparalleled elsewhere (Gubin, 1991). These are represented by two main lineages, the alligator-like melosaurids and the long-snouted gavial-like archego-saurids, the latter probably being active swimmers. The bulk of this radiation derives from Early Permian archegosaurids of Europe; on the other hand, the origin of the Melosauridae from intasuchids is not ruled out. The terrestrial component of the discussed Cis-Uralian fauna is represented by large short-bodied dissorophids

(Gubin, 1980, 1987), which survived from the Early Permian Euramerican fauna and are close to the ances-try of frogs.

By the end of the Permian (Late Tatarian), which is marked in the Cis-Urals by the rise of the pareiasaurian tetrapod fauna, the variety of temnospondyls was reduced to a single primitive brachyopoid genus, the neotenic perennibranchiate *Dvinosaurus* (Bystrov, 1938; Shishkin, 1973). This form resembles the modern Japanese giant salamander, and is unquestionably a descendant of the North American Early Permian tri-merorhachoids, which were also a purely aquatic group. It is not ruled out that *Dvinosaurus*, or a closely related form, in turn gave rise to a single brachyopoid relict from the Middle Triassic of North America (*Hadrokkosaurus*). With regard to its reduced diversity, the Late Tatarian amphibian assemblage is comparable to that from the latest Permian of Gondwana, where most temnospondyls belong to just one or very few closely related genera (*Rhinesuchus* in a broad sense).

During the Triassic, the main changes of the Cis-Uralian amphibian assemblage follow the four succes-sive tetrapod faunas known for this area and termed by their predominant temnospondyl genera (for details, see Chapter 7). The earliest, *Benthosuchus– Wetlugasaurus* fauna (Induan–Early Olenekian), which immediately succeeded the Late Permian extinction, shows a marked increase in temnospondyl diversity to five families. All are represented by small- or medium-sized aquatic forms, indicating aridity of the land. The most peculiar temnospondyls of this time are the tupilakosaurids, small, probably limbless brachyopoids, which present the only temnospondyl group known to possess disc-shaped embolomerous vertebrae (Shishkin, 1973). Gondwanan influence is indicated by the lydekkerinid *Luzocephalus*, the single genus of this family recorded from Laurasia, and the rhytidosteoid *Boreopelta*, recov-ered from outside the Cis-Urals, in Siberia (Shishkin, 1980; Shishkin and Vavilov, 1985). Lydekkerinids are closely related to the earliest capitosauroids (Permian rhinesuchids), while rhytidosteoid ancestry is obscure. Yet the dominant earliest Lower Triassic Cis-Uralian temnospondyls were the crocodile-like Capitosauridae, derived from rhinesuchid-related Gondwanan ances-tors (Ochev, 1966, 1972), and their immediate endemic offshoot, the Benthosuchidae (Bystrov and Efremov, 1940; Getmanov, 1989; Novikov, 1994). In general, amphibians of this fauna were extremely abundant and their remains comprise over 90% of all tetrapod finds. This is paralleled in the Early Scythian Australian tetrapod fauna, but makes a contrast with other coeval Gondwanan communities, which were reptile-dominated.

The next (late Early Triassic) amphibian community belongs to the *Parotosuchus* faunal assemblage. It is represented mainly by descendents of former dominant forms. These include primarily the capitosaurid *Parotosuchus*, widespread in Europe, and trematosaurids, which are benthosuchid derivatives (Shishkin, 1960, 1980). The Cis-Uralian fossil record documents a nearly continuous morphological transition from advanced benthosuchids to trematosaurids (cf. Shishkin, 1980; Getmanov, 1989; Novikov, 1994). The latter bear a resemblance to the Permian archegosauroids, and obviously had similar adaptations; they are also peculiar for their trend to populate nearshore marine waters. Another benthosuchid offshoot are the poorly known yarengiids. The short-faced brachyopoids are represented by brachyopids (Shishkin, 1973), a family known elsewhere almost entirely from Gondwana. Remarkable is the first appearance of the Plagiosauridae, a group which became widespread in the European Middle to Late Triassic (Shishkin, 1987). Plagiosaurids are aberrant in many respects, particularly in the broadening of their skull, their enormous orbits, the development of dermal armour, and their unusual holospondylous vertebrae. At least some were neotenic. Their ancestry is obscure, and may be close to the Permian Zatrachydidae (Shishkin, 1987). Triassic members of the group are thought to have derived from the primitive Permian plagiosauroids of East Africa. The most spectacular evidence of Gondwanan influence on the Cis-Uralian *Parotosuchus* community is provided by the occurrence of the advanced South African rhytidosteid *Rhytidosteus* (Shishkin, 1994b).

Like the amphibians of the preceding fauna, the late Scythian temnospondyls were water-dwellers and dominated the coeval tetrapod assemblage. Their increase in body size relative to their forerunners is clear, and this trend is seen in succeeding communities.

The Middle Triassic temnospondyls of the Cis-Urals still remain abundant, but their diversity declines and is limited to two groups, the capitosauroids and plagiosauroids. In the earlier assemblage (Late Anisian (?)–Ladinian), the former are represented by the capitosaurid *Eryosuchus*, a large form with depressed skull and poorly ossified limb skeleton (Ochev, 1972); the plagiosaurids include two benthic genera, one of which is also known from the contemporaneous fauna of West Europe (Shishkin, 1987). Seasonal aridity of climate is evidenced by the discovery in the Ural basin of a mass burial of *Eryosuchus* individuals, which became trapped in a temporary deltaic pool, and died from overheating and oxygen deficiency during the drought (Figure 1.8; Ochev and Shishkin, 1967). In the late Middle Triassic (Late Ladinian) assemblage, capitosaurids are replaced by mastodonsaurids, primarily a giant form that is closely related to the West European *Mastodonsaurus* (Konzhukova, 1955). Like other late capitosauroids, it was a sort of 'living trap', which rested on the bottom of pools and rivers for long periods, and making snaps at passing prey. The accompanying plagiosaurids changed only at the species level, compared to their forerunners from the *Eryosuchus* fauna.

Later temnospondyls are so far unknown from Russia. The only recorded exception is a find of a single hypocentrum from Middle Jurassic nearshore marine deposits near Moscow. It is from a form similar to the brachyopid relict from the Jurassic of Mongolia (see Chapter 16).

Systematic survey

Class AMPHIBIA Linnaeus, 1758
Subclass TEMNOSPONDYLI Zittel, 1888
Superfamily ERYOPOIDEA Cope, 1882
Diagnosis. Skull with broad snout, preorbital part moderately elongate. Lacrimal excluded from orbital margin, but reaches naris or septomaxilla. Otic notch usually deep, partially bordered by supratemporal. Sensory grooves lacking. Basioccipital well developed, occipital condyle tripartite. No anterior palatal vacuities. Pterygoid contacts parasphenoid with narrow suture. Premaxillary–maxillary contact at level of anterior end of choana. No interchoanal teeth. Pterygoid

contacts vomer, both bones bearing shagreen dentition. Pterygoid flange not reduced.

Family ERYOPIDAE Cope, 1882

Diagnosis. Skull with convex side margins. Choanae broad. Palatine and ectopterygoid bearing no teeth other than tusk pairs. No, or few, parachoanal teeth on vomer. Vomerine tusks aligned subtransversally. Parasphenoid bears wide shagreen field and lacks muscular pockets.

Clamorosaurus Gubin, 1983

Diagnosis. Skull 200–230 mm long, with short preorbital part. Skull roof with shallow preorbital step in lateral view. Interfrontal may be present. Otic notch twice shorter than postorbital skull length. Pterygoid flange well expressed. Ectopterygoid tusk pair small; one parachoanal tooth on vomer.

Clamorosaurus nocturnus Gubin, 1983
See Figure 3.1.

Holotype and locality. PIN 1582/1, skull; Pechora River, Komi Republic.
Horizon[3]. Sheshminskian Gorizont, Ufimian, Upper Permian.

Clamorosaurus borealis Gubin, 1983

Holotype and locality. PIN 3950/1, skull; Inta city, Komi Republic.
Horizon. Intinskaya Svita, Ufimian, Upper Permian

Family INTASUCHIDAE Konzhukova, 1956

Diagnosis. Skull subtriangular with moderately narrow snout. Otic notch narrow, continuing anteriorly by groove on skull roof. Choanae narrow. Parachoanal tooth row well developed. Palatine tusks followed by tooth row extending to ectopterygoid. No tusks on ectopterygoid. Parasphenoid plate with muscular pockets; oval shagreen field at base of parasphenoid cultriform process. Vomerine tusks aligned anteroposteriorly.

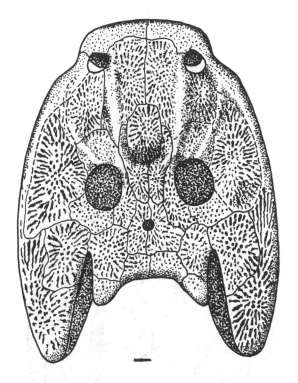

Figure 3.1. *Clamorosaurus nocturnus* Gubin; skull, dorsal view. Scale bar is 10 mm. (From Gubin, 1983.)

Intasuchus Konzhukova, 1956

Diagnosis. Skull elongate; preorbital ridges of skull roof well developed.

Intasuchus silvicola Konzhukova, 1956
See Figure 3.2.

Holotype and locality. PIN 570/1, skull; Inta city, Komi Republic.
Horizon. Intinskaya Svita, Ufimian, Upper Permian

Syndyodosuchus Konzhukova, 1956

Diagnosis. Skull with short preorbital part. Preorbital ridges of skull roof weakly expressed. Maxillary teeth arranged in pairs.

Syndyodosuchus tetricus Konzhukova, 1956

Holotype and locality. PIN 570/40, skull; Inta city, Komi Republic.
Horizon. Intinskaya Svita, Ufimian, Upper Permian.

---

[3] Horizons are shown for the whole range of each species, not just for the type specimen.

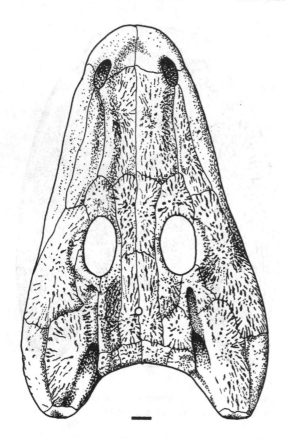

Figure 3.2. *Intasuchus silvicola* Konzhukova; skull, dorsal view. Scale bar is 10 mm. (From Gubin, 1984.)

Superfamily ARCHEGOSAUROIDEA Lydekker, 1885

[In this survey, the Archegosauroidea is considered to be a separate superfamily, following Efremov (1933), Konzhukova (1955) and Gubin (1991). Further analysis may support the alternative concept, that archegosauroids belong in Eryopoidea.]

Diagnosis. Skull triangular, with concave side margins; preorbital elongation moderate to strong. Lacrimal excluded from both orbital and narial margins. Septomaxilla usually absent. Sensory grooves poorly developed or absent. Choanae narrow. Anterior palatal vacuities usually present. Pterygoid contacts vomer. Pterygoid flange reduced. Parasphenoid plate with muscular pockets. Premaxilla contacts maxilla well in front of choana. Shagreen fields or rows on pterygoid and vomer; small oval shagreen field at base of cultriform process of parasphenoid. Palatine–ectopterygoid

tooth row present, no ectopterygoid tusks. Vomerine tusks aligned anteroposteriorly.

Family ARCHEGOSAURIDAE Lydekker, 1885

Diagnosis. Snout strongly elongated; ratio of skull length to skull width across centres of orbits is more than 2.1 in adults. Basioccipital well developed. Otic notches moderately deep. Occiput deep. Pterygo-parasphenoid suture weak or absent. Interchoanal vomerine teeth usually lacking.

Collidosuchus Gubin, 1986

Diagnosis. Skull more than 300 mm long; prenarial elongation moderate; snout with parallel side borders; choana drop-shaped; paired anterior palatal vacuities.

Collidosuchus tchudinovi Gubin, 1986

Holotype and locality. PIN 1758/334, incomplete skull; Ezhovo, Otcher district, Perm' Province.

Horizon. Upper Kazanian–lowermost Tatarian, Upper Permian

Platyoposaurus Lydekker, 1889

Diagnosis. Skull 240–400 mm, sometimes up to 700 mm long. Snout strongly elongated, narrow, with broadened terminal part. Nares far behind tip of snout. Interorbital distance equal to, or exceeds, diameter of orbit. Sensory grooves poorly developed. No parachoanal teeth. Parasphenoid plate subtriangular. Anterior palatal vacuity paired, broadly merged. Palatal shagreen extends to premaxilla.

Platyoposaurus rickardi (Twelvetrees, 1880)

Holotype and locality. Rickard's collection, Orenburg, poorly preserved skull (present location unknown); Rozhdestvenskii mine, Orenburg Province.

Horizon. Copper sandstone of Upper Kazanian, Upper Permian.

Platyoposaurus stuckenbergi (Trautschold, 1884)
See Figure 3.3.

Lectotype and locality. PIN 49/1, skull; Akbatyrovskii mine, Malmyzh district, Kirov Province.

Horizon. Upper Kazanian to Lower Tatarian; Upper Permian.

Platyoposaurus watsoni (Efremov, 1933)
Holotype and locality. PIN 2250/8, skull; Shikhovo-Chirki, Slobodskoy district, Kirov Province.
Horizon. Kazanian, Upper Permian.

Bashkirosaurus Gubin, 1981
Diagnosis. Close to *Platyoposaurus* in shape of skull. Orbits large; interorbital distance less than diameter of orbit. Shagreen on pterygoids much reduced. Palatal tooth row lacking.

Bashkirosaurus tcherdyncevi Gubin, 1981
Holotype and locality. PIN 164/70, incomplete skull; Belebey city, Bashkortostan Republic.
Horizon. Belebeiskaya Svita, Upper Kazanian, Upper Permian.

Family MELOSAURIDAE Fritsch, 1885
Diagnosis. Skull 250–300 mm, sometimes up to 500 mm long. Preorbital elongation moderate. Otic notch deep. Occiput moderately deep. Basioccipital considerably reduced. Short well-developed pterygoid– parasphenoid suture. Few interchoanal teeth.

Melosaurus Meyer, 1857
Diagnosis. Skull borders concave in postnarial area. Orbits rounded. Parasphenoid plate short. Anterior palatal vacuity paired. Parachoanal row includes two to six teeth.

Melosaurus uralensis Meyer, 1857
Holotype and locality. MB 334, skull; Sterlitamak district, Bashkortostan Republic.
Horizon. Kazanian, Upper Permian.

Melosaurus kamaensis Gubin, 1991
Holotype and locality. PIN 683/1, skull; Mamadysh district; Tatar Republic.
Horizon. Belebeiskaya Svita, Upper Kazanian, Upper Permian.

Melosaurus plathyrhinus Golubev, 1995
Holotype and locality. PIN 161/1, skull; Shikhovo-Chirki, Kirov Province.
Horizon. Upper Kazanian, Upper Permian.

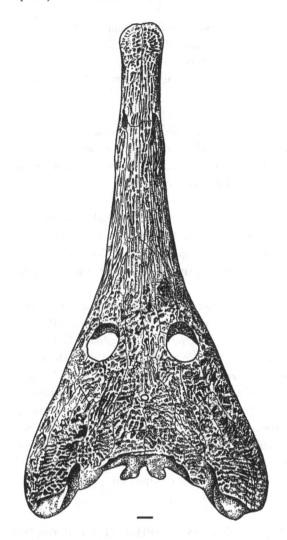

Figure 3.3. *Platyoposaurus stuckenbergi* (Trautschold); skull, dorsal view. Scale bar is 10 mm. (From Gubin, 1991.)

Konzhukovia Gubin, 1991
Diagnosis. Side borders of skull nearly straight. Orbits elongate. Parasphenoid plate moderately elongate. Anterior palatal vacuity paired. Two teeth in parachoanal row.

Konzhukovia vetusta (Konzhukova, 1955)
See Figure 3.4.
Holotype and locality. PIN 520/1 skull; Malyi Uran River, Orenburg Province.
Horizon. Lower Tatarian, Upper Permian.

Konzhukovia tarda Gubin, 1991

Holotype and locality. PIN 1758/253, skull; Ezhovo, Otcher district, Perm' Province.

Horizon. Upper Kazanian–lowermost Tatarian, Upper Permian.

Tryphosuchus Konzhukova, 1955

Diagnosis. Parasphenoid plate elongate. Anterior palatal vacuity unpaired. Ornament on palatal ramus of pterygoid. Two teeth in parachoanal row.

Tryphosuchus paucidens Konzhukova, 1955

Lectotype and locality. PIN 157/107, basi-parasphenoid complex; Isheevo, Apastov district, Tatar Republic.

Horizon. Sukhonskaya Svita, Lower Tatarian, Upper Permian.

Archegosauroid taxa based on fragmentary material or in need of re-evaluation (all from the Upper Permian). Platyoposauridae: *Platyoposaurus vyushkovi* Gubin, 1989, Orenburg Province, Upper Tatarian. Melosauridae: *Koinia* (*K. silantjevi*) Gubin, 1993, Vym' River, Vychegda basin, Komi Republic, Lower Kazanian; *Uralosuchus* (*U. tverdochlebovae*) Gubin, 1993, Ural basin, Orenburg Province, Upper Kazanian; *Melosaurus compilatus* Golubev, 1995, Golyusherma, Udmurt Republic, Lower Kazanian; *Tryphosuchus kinelensis* (Vjuschkov, 1955), Malaya Kinel River, Orenburg Province, Lower Tatarian.

Superfamily DISSOROPHOIDEA Williston, 1910

Diagnosis. Skull short or moderately elongate; orbits usually large. Lacrimal extending from orbit to naris. Jugal wedging out anteriorly, not reaching lacrimal. Otic notch exaggerated, extending from tabular to quadrate. Supratemporal included in otic notch margin. Presacral vertebral column reduced to 20–26 vertebrae.

Family DISSOROPHIDAE Williston, 1910

Diagnosis. Dermal ornament usually coarse-pitted. Palatine with small exposure on skull roof. Quadrate with dorsal process. Pterygoid and parasphenoid contacting by suture. Pterygoid with transverse flange. No teeth following tusk pairs on palatine and ectopterygoid.

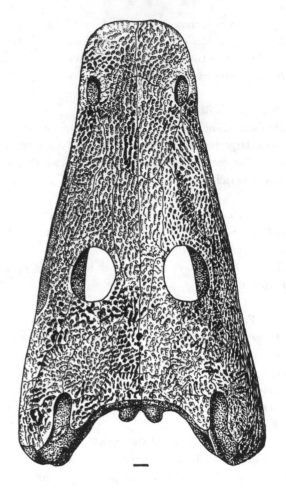

Figure 3.4. *Konzhukovia vetusta* (Konzhukova); skull, dorsal view. Scale bar is 10 mm. (From Gubin, 1991.)

Kamacops Gubin, 1980

Diagnosis. Skull 250–300 mm long. Otic notch closed by junction of dorsal quadrate process and tabular. Orbits slightly posterior to midlength of skull roof; orbital margins thickened and elevated. Septomaxilla present. Supraoccipital ossification paired. Choana elongate, bordered by parachoanal tooth row. Anterior palatal vacuity single. Shagreen dentition over whole palate. Axial dermal scutes present.

Kamacops acervalis Gubin, 1980

Holotype and locality. PIN 3817/1, incomplete skull; Erzovka, Kama River, Perm' Province.

Horizon. Belebeiskaya Svita, Upper Kazanian, Upper Permian.

Dissorophoid taxa based on fragmentary material or known from out-of-date descriptions. Dissorophidae: *Zygosaurus* (*Z. lucius*) Eichwald, 1848, Klyuchevskii mine, Bashkortostan; *Iratusaurus* (*I. vorax*) Gubin,1980, Belebey city, Bashkortostan (both Upper Kazanian, Upper Permian); *Alegeinosaurus* (?) sp. (cf. Gubin, 1993), Vym' River, Vychegda basin, Komi Republic, Lower Kazanian, Upper Permian. (?) Branchiosauridae: *Tungussogyrinus* (*T. bergi*) Efremov, 1939, Lower Tunguska River, Central Siberia, Dvuroginskian Gorizont, Lower Triassic. The attribution of this form to the Brachyopidae (presumed also to include plagiosaurs) on the presence of the broad cultriform process of the parasphenoid, and, allegedly, the solid vertebrae, as was suggested by the original description. However, recent re-investigation (Shishkin, 1998) revealed that, as in other branchiosaurs, the vertebral centra in *Tungussogyrinus* remained unossified. The expansion of the cultriform process in the Siberian genus also bears a resemblance to some typical European branchiosaurs, primarily the genus *Apatheon* Meyer (cf. Boy, 1986; Werneburg, 1989), on the alleged presence of the very broad cultriform process of the parasphenoid and the solid vertebrae; but this cannot be confirmed either from the holotype or from new material from the same provenance.

Superfamily CAPITOSAUROIDEA Watson, 1919
Diagnosis. Skull more or less elongate in facial region. Orbits rather close together. Supraorbital sensory groove out of lacrimal (except mastodonsaurids). Lacrimal does not reach orbit or naris. Supratemporal excluded from otic notch margin (except rhinesuchids). Basioccipital much reduced. Exoccipital extends into otic region. Opisthotic not exposed in occiput. Interpterygoid vacuities large. Pterygoid does not reach vomer. Pterygo-parasphenoid suture moderately elongate; quadrate ramus of pterygoid usually bears oblique crest. Anterior palatal vacuity, single or paired, always present (rudimentary in rhinesuchids). No ectopterygoid tusks; palatine tusks followed by long palatine–ectopterygoid tooth row. Inter- and parachoanal tooth rows well developed.

Family LYDEKKERINIDAE Watson, 1919
Diagnosis. Orbits close to midlength of skull roof. Lacrimal flexure of infraorbital groove gentle or step-shaped. Septomaxilla present. Frontal excluded from orbital margin. Preorbital projection of jugal very short. Lacrimal long. Anterior palatal vacuity single. Pterygoid oblique crest shallow or absent.

Luzocephalus Shishkin, 1980
Diagnosis. Skull about 170 mm long, with straight side borders. Orbitopineal distance long. No lateral projection of postorbital. Palatine contacts vomer laterally to choana. Pterygo-squamosal fissure open. No pterygoid oblique crest.
Comment. The genus is the only member of the Lydekkerinidae known from Laurasia.

Luzocephalus blomi Shishkin, 1980
See Figure 3.5.
Holotype and locality. PIN 3784/1, skull; Luza River, Kirov Province.
Horizon. Vokhmian Gorizont, Lower Triassic.

Family CAPITOSAURIDAE Watson, 1919
Diagnosis. Orbits in posterior half of skull roof. Nares elongate. Lacrimal flexure Z-shaped. No septomaxilla. Preorbital jugal projection long. Frontal usually enters orbital margin. Lateral projection of postorbital pronounced. Anterior palatal vacuity single. Palatal shagreen reduced or absent. Oblique crest well developed.

Wetlugasaurus Ryabinin, 1930
Diagnosis. Skull up to 200–220 mm long. Frontal excluded from orbital margin. Choanae moderately compressed. Parasphenoid ventral exposure does not reach level of choanae. Narrow shagreen field may present on pterygoid. Interchoanal tooth row arch-shaped. No medial parasymphysial teeth on lower jaw.

Wetlugasaurus angustifrons Ryabinin, 1930
See Figure 3.6.
Holotype and locality. TsNIGRI 3417/1, incomplete skeleton; Zubovskoe, Vetluga River, Nizhnii Novgorod Province.
Horizon. Sludkian Gorizont, Lower Triassic.

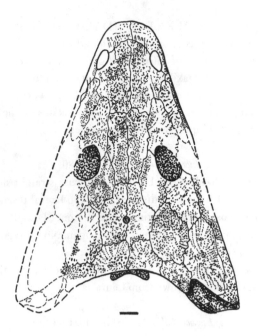

Figure 3.5. *Luzocephalus blomi* Shishkin; skull, dorsal view. Scale bar is 10 mm. (From Shishkin, 1980.)

Wetlugasaurus malachovi Novikov, 1990
Holotype and locality. PIN 4333/1, incomplete skull; Tsylma River, Pechora River basin, Komi Republic.
Horizon. Charkabozhskaya Svita, Ustmylian Gorizont, Lower Triassic.

Wetlugasaurus samarensis Sennikov, 1981
Holotype and locality. SGU 1277/1, skull; Shulaevka, Tarpanka River, Samara River basin, Orenburg Province.
Horizon. Kopanskaya Svita, Vokhmian Gorizont, Lower Triassic.

Parotosuchus Otschev et Shishkin, 1968
Diagnosis. Skull depressed, 250–450 mm long. Choanae slit-like. Parasphenoid ventral exposure extends to level of choanae. Interchoanal tooth row straight. Medial parasymphysial teeth present on lower jaw.

Parotosuchus orenburgensis (Konzhukova, 1965)
See Figure 3.7.
Holotype and locality. PIN 951/42, skull; Rossypnaya, Ural River, Orenburg Province.
Horizon. Petropavlovskaya Svita, Yarenskian Gorizont, Lower Triassic.

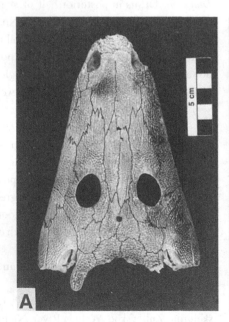
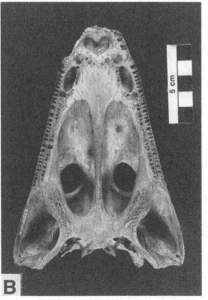

Figure 3.6. *Wetlugasaurus angustifrons* Ryabinin; skull in dorsal (A) and ventral (B) views. Scale bars are 50 mm.

Parotosuchus orientalis (Otschev, 1966)

Holotype and locality. PIN 4172/1 (originally SGU 104/222), incomplete skull; Kzyl-Say ravine, Kzyl-Oba River, Ural basin, Orenburg Province.

Horizon. Petropavlovskaya Svita, Yarenskian Gorizont, Lower Triassic.

Parotosuchus panteleevi (Otschev, 1966)

Holotype and locality. PIN 4173/54 (originally SGU 104/3518), skull fragment; Lipovskaya Balka, Don River, Volgograd Province.

Horizon. Lipovskaya Svita, Yarenskian Gorizont, Lower Triassic.

Parotosuchus sequester Shishkin, 1974

Holotype and locality. PIN 3300/1, incomplete skull; Dollapa, Mangyshlak Peninsula, Caspian basin[4].

Horizon. Upper Olenekian, Lower Triassic.

[4] Now the territory of Kazakhstan Republic.

Parotosuchus komiensis Novikov, 1986

Holotype and locality. PIN 3361/18, fragment of skull; Zheshart, Vychegda River, Komi Republic.

Horizon. Gamskaya Svita, Yarenskian Gorizont, Lower Triassic.

Eryosuchus Otschev, 1966

Diagnosis. Skull up to 500–600 mm long, close to that of *Parotosuchus* in structure and dentition. Tabular horns terminally broadened, showing trend to otic notch closure. Pterygo-parasphenoid suture elongate. Retroarticular process of lower jaw elongate, with dorsal depression. Vertebrae from rhachitomous to sub-stereospondylous.

Eryosuchus tverdochlebovi Otschev, 1966

Holotype and locality. PIN 4166/89 (originally SGU 104/3090), incomplete skeleton; Perovka, Donguz River, Ural basin, Orenburg Province.

Horizon. Donguz Gorizont, Middle Triassic.

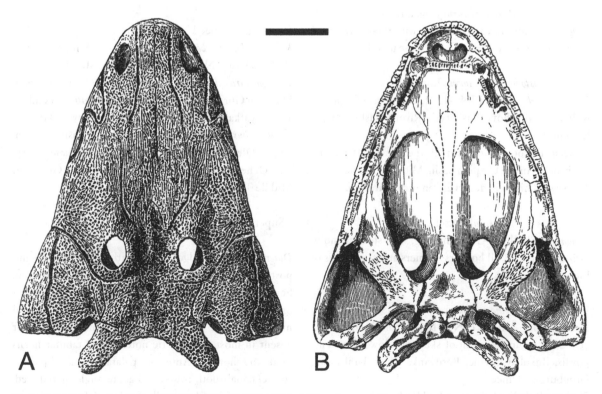

Figure 3.7. *Parotosuchus orenburgensis* Konzhukova; skull in dorsal (A) and ventral (B) views. Scale bar is 50 mm. (From Shishkin *et al.*, 1995.)

47

Eryosuchus garjainovi Otschev, 1966
See Figure 3.8.
Holotype and locality. SGU 104/3521, skull; Berdyanka River, Ural basin, Orenburg Province.
Horizon. Donguz Gorizont, Middle Triassic.

Eryosuchus antiquus Otschev, 1966
Holotype and locality. PIN 2973/65 (originally SGU 104/3515), lower jaw fragment; Karagachka River, Ural basin, Orenburg Province.
Horizon. Donguz Gorizont, Middle Triassic.

Family MASTODONSAURIDAE Lydekker, 1885
Diagnosis. Skull roof subtriangular, usually with prenarial perforations. Nares rounded. Supraorbital sensory groove passes across lacrimal. Frontal enters orbital margin. Anterior palatal vacuity paired. Pterygo-parasphenoid suture and parasphenoid plate elongate. Palatal dentition as in Capitosauridae. Vertebrae stereospondylous.

Mastodonsaurus Jaeger, 1828
Diagnosis. Skull up to 1.5 m long, with relatively broad snout. Orbits placed more or less close together.

Mastodonsaurus torvus Konzhukova, 1955
Holotype and locality. PIN 415/1, lower jaw fragment; Koltaevo, Bolshoy Yushatyr' River, Bashkortostan Republic.
Horizon. Bukobay Gorizont of the Cis-Urals and its equivalents in the Caspian basin, Middle Triassic.
Comment. This form deserves generic separation.

Bukobaja Otschev, 1966
Diagnosis. Skull 100–110 mm long, with slender snout. Orbits close to skull borders. Vomers and symphysis of lower jaw elongate.

Bukobaja enigmatica Otschev, 1966
Holotype and locality. PIN 4165/1 (originally SGU 104/245), snout fragment of skull with lower jaw symphysis; Bukobay ravine, Berdyanka River, Ural basin, Orenburg Province.
Horizon. Bukobay Gorizont, Middle Triassic.

Capitosauroid taxa based on incomplete material or

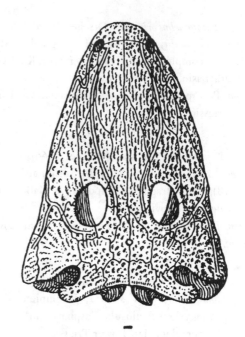

Figure 3.8. *Eryosuchus garjainovi* Otschev; skull, dorsal view. Scale bar is 10 mm. (From Shishkin *et al.*, 1995.)

those that need re-evaluation. Capitosauridae: *Komatosuchus* (*K. chalyshevi*) Novikov and Shishkin, 1992, Pechora basin, Nyadeytinskaya Svita, Middle Triassic; *Wetlugasaurus kzilsajensis* Otschev, 1966, Ural basin, Sludkian Gorizont, Lower Triassic; *W. vjatkensis* Gubin, 1987, Vyatka basin, Sludkian Gorizont, Lower Triassic; *Parotosuchus bogdoanus* Woodward, 1932, Caspian basin, Upper Olenekian, Lower Triassic; (?) *Cyclotosaurus* sp. (cf. Ochev, 1972), Ural basin, Bukobay Gorizont, Middle Triassic.

Superfamily TREMATOSAUROIDEA Watson, 1919
Diagnosis. Skull wedge-shaped with elongated pre- and postorbital areas. Orbits usually broadly separated. Sensory grooves well developed; supraorbital groove usually crossing lacrimal. Frontal excluded from orbital margin. Lateral projection of postorbital reduced or absent in most forms. Otic notches and tabular horns tend to shorten. Anterior palatal vacuity paired. Interchoanal tooth row forms acute angle or reduced. Ectopterygoid tusks usually present. Oblique crest of pterygoid tends to reduce. Shagreen field on pterygoid

well developed. Pterygoid–parasphenoid suture extends backwards to underlie anterior part of middle ear cavity.

Family BENTHOSUCHIDAE Efremov, 1937

Diagnosis. Snout elongated. Lacrimal flexure of infraorbital groove usually present. Pineal foramen in anterior half of orbito-occipital distance. Anterior (preotic) extension of ascending lamina weakly developed, not separating epipterygoid from basipterygoid articulation. Muscular crests on parasphenoid plate usually present. Elongation of pterygoid–parasphenoid suture weakly expressed.

Benthosuchus Efremov, 1937

Diagnosis. Primitive trematosauroids with skull up to 250–300 mm long. Orbits close together. Postorbital elongation of skull roof poorly developed. Jugal sensory groove along squamosal–quadratojugal suture. Lateral projection of postorbital well marked. Tabular horns rather long. Anterior palatal vacuities fused in central part or separated by thin partition. Interchoanal tooth row well developed, slightly wedged in between choanae. Ectopterygoid tusks absent or reduced. Pterygoid contacts palatine on ventral surface. Cultriform process of parasphenoid comparatively broad. Parasphenoid plate relatively short, with strong muscular crests. Subotic process of exoccipital (invading otic capsule) well developed. Lower jaw with marked angular bend and short retroarticular process. Posterior meckelian foramen small, oval in shape.

Benthosuchus sushkini (Efremov, 1929)
See Figure 3.9.

Holotype and locality. PIN 2243/1, skull; Vakhnevo, Sharzhenga River, Vologda Province. Holotype specimen is lost.
Horizon. Rybinskian Gorizont, Lower Triassic.

Benthosuchus uralensis (Otschev, 1958)

Holotype and locality. PIN 4167/1 (originally SGU 104/1), skull; Blumental ravine, Burtya basin, Orenburg Province.
Horizon. Kopanskaya Svita, Vokhmian Gorizont, Lower Triassic.

Benthosuchus bashkiricus Otschev, 1966

Holotype and locality. PIN 4168/1 (originally SGU 104/3810), skull fragment; Muraptalovo, Kazlair River, Bolshoi Yushatyr' basin, Bashkortostan Republic.
Horizon. Sludkian(?) Gorizont, Lower Triassic.

Benthosuchus korobkovi Ivachnenko, 1972

Holotype and locality. PIN 3200/1, skull; Tikhvinskoe, Volga River, Rybinsk district, Yaroslavl' Province.
Horizon. Rybinskaya Svita, Rybinskian Gorizont, Lower Triassic.

Benthosuchus bystrowi Getmanov, 1989

Holotype and locality. PIN 3783/1, skull; Makar'ev (?) district, Kostroma Province. Holotype is lost; exact locality unknown.
Horizon. Rybinskian(?) Gorizont, Lower Triassic.

Vyborosaurus Novikov, 1990

Diagnosis. Skull up to 400 mm long. Jugal sensory groove forming a bend on squamosal. Otic notches triangular and shallow, tabular horns short. Parasphenoid plate relatively short, without muscular crests. Sculpture on pterygoid predominates over shagreen. Subotic process of exoccipital poorly developed. Lower jaw with gentle angular bend and slightly elongated retroarticular process. Posterior meckelian foramen strongly elongated.

Vyborosaurus mirus Novikov, 1990

Holotype and locality. PIN 3360/9, lower jaw fragment; Vybor River, Mezen' basin, Arkhangel'sk Province.
Horizon. Ustmylian Gorizont, Lower Triassic.

Benthosphenus Shishkin, 1979

Diagnosis. Skull up to 200 mm long. Interchoanal tooth row deeply wedged in between choanae. Ectopterygoid tusks present. Pterygoid does not reach forward to palatine.
Comment. This genus is the only true benthosuchid recorded outside Europe.

Benthosphenus lozovskii Shishkin, 1979

Holotype and locality. PIN 3785/1, impression of anterior part of palatal surface; Russkii Island, Russian Far East.
Horizon. Lower Olenekian, Lower Triassic.

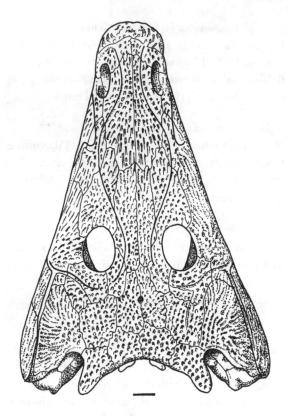

Figure 3.9. *Benthosuchus sushkini* Efremov; skull, dorsal view. Scale bar is 10 mm. (From Bystrov and Efremov, 1940.)

Thoosuchus Efremov, 1940

Diagnosis. Skull up to 150 mm long. Orbits broadly separated. Septomaxilla large and exposed on dorsal surface of skull. Postorbital elongation of skull roof moderate; pineal foramen in anterior third of orbito-occipital distance. Lateral projection of postorbital weak or absent. Tabular horns moderately elongated. Anterior palatal vacuities broadly separated. Interchoanal tooth row forms acute angle. No pterygoid–palatine contact. Tooth row between palatine and ectopterygoid tusks shortened. Cultriform process of parasphenoid very narrow. Parasphenoid plate elongate, with moderate muscular crests. Lower jaw with gentle angular bend and moderately pronounced retroarticular process. Medial process of supraangular poorly developed. Posterior meckelian foramen strongly elongate.

Thoosuchus yakovlevi (Ryabinin, 1927)
See Figure 3.10.
Holotype and locality. TsNIGRI 2169/1, preorbital skull fragment; Kormitsa River, Volga basin, Rybinsk district, Yaroslavl' Province.
Horizon. Rybinskian Gorizont, Lower Triassic.

Thoosuchus tardus Getmanov, 1989
Holotype and locality. PIN 4000/1, postorbital skull fragment; Chapayevka River, Samara basin, Samara Province.
Horizon. Rybinskian(?) Gorizont, Lower Triassic.

Thoosuchus tuberculatus Getmanov, 1989
Holotype and locality. PIN 4197/1, skull; Kamennyi Yar ravine, Sorochka River, Samara basin, Orenburg Province.
Horizon. Rybinskian Gorizont, Lower Triassic.
The benthosuchid genera listed below are very close to *Thoosuchus*, so that only their main differences from the latter are noted.

Prothoosuchus Getmanov, 1989
Diagnosis. Skull up to 70 mm long. Pineal foramen close to level of posterior margins of orbits. Cultriform process of parasphenoid comparatively broad.

Prothoosuchus blomi Getmanov, 1989
Holotype and locality. PIN 2423/1, incomplete skeleton; Mechet' ravine, Tavolzhanka River, Samara basin, Samara Province.
Horizon. Sludkian Gorizont, Lower Triassic.

Prothoosuchus samariensis Getmanov, 1989
Holotype and locality. PIN 3997/1, skull; Korneevskoe, Kalmanka River, Samara basin, Samara Province.
Horizon. Sludkian Gorizont, Lower Triassic.

Angusaurus Getmanov, 1989
Diagnosis. Skull up to 200 mm long. Pineal foramen in middle of orbito-occipital distance. Interchoanal tooth row strongly reduced. Few teeth on palatine following tusk pair. Retroarticular process of lower jaw elongate. Medial process of supraangular well developed.

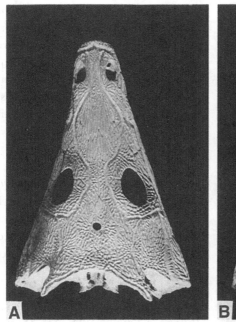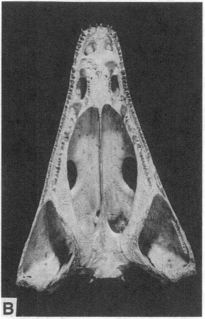

Figure 3.10. *Thoosuchus yakovlevi* (Ryabinin); skull in dorsal (A) and ventral (B) views. Skull is 150 mm long.

Angusaurus dentatus Getmanov, 1989
See Figure 3.11.
Holotype and locality. PIN 4196/1, skull; Logachevka, Bolshaya Pogromka River, Samara basin, Orenburg Province.
Horizon. Sludkian Gorizont, Lower Triassic.

Angusaurus succedaneus Getmanov, 1989
Holotype and locality. PIN 2428/1, incomplete skull; Borshchevka, Chapaevka River, Samara basin, Samara Province.
Horizon. Sludkian Gorizont, Lower Triassic.

Angusaurus tsylmensis Novikov, 1990
Holotype and locality. PIN 4333/6, skull; Tsylma River, Pechora basin, Komi Republic.
Horizon. Ustmylian Gorizont, Lower Triassic.

Trematotegmen Getmanov, 1982
Diagnosis. Skull up to 200 mm long. Pineal foramen in middle of orbito-occipital distance. Posterior border of lacrimal close to anterior margin of orbit. Postparietal wide. Tabular horns short.

Trematotegmen otschevi Getmanov, 1982
Holotype and locality. PIN 4200/1 (originally SGU 1599/1), incomplete skull; Panik creek, Buzuluk River, Orenburg Province.
Horizon. Rybinskian Gorizont, Lower Triassic.

Family YARENGIIDAE Shishkin, 1960
Diagnosis. Ornament on pterygoid predominates over shagreen. Parasphenoid plate ornamented. Pterygoid–parasphenoid suture extends far backward to underlie narrow floor of tympanic cavity (as in trematosaurids). Preotic extension of pterygoid ascending lamina shallow, not separating epipterygoid from basipterygoid articulation (as in benthosuchids). Exoccipital with reduced subotic process and elongated base. Anterior palatal vacuities fused in middle. Interchoanal tooth row forms acute angle.

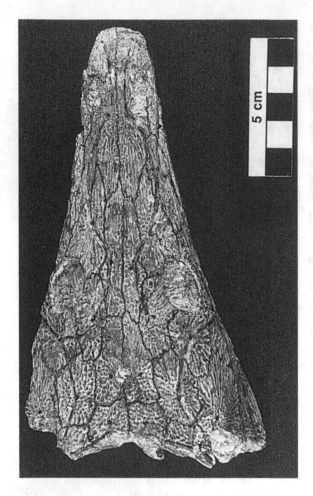

Figure 3.11. *Angusaurus dentatus* Getmanov; skull, dorsal view. Scale bar is 50 mm.

Yarengia Shishkin, 1960
Diagnosis. As for the family.

Yarengia perplexa Shishkin, 1960
Holotype and locality. PIN 1584/5, skull fragment; Yarenga River, Vychegda basin, Arkhangel'sk Province.
Horizon. Yarenskian Gorizont, Lower Triassic.

Family TREMATOSAURIDAE Watson, 1919
Diagnosis. Orbits broadly separated. No lacrimal flexure of infraorbital groove. Postorbital elongation strongly expressed; pineal foramen in posterior half of orbito-occipital distance. Anterior palatal vacuities broadly separated. Interchoanal tooth row reduced or absent.

Few enlarged teeth on palatine following tusk pair. Pterygoid–palatine contact usually absent. Cultriform process of parasphenoid very narrow. Pterygoid ascending lamina extends forward to separate epiptery-goid from basipterygoid articulation. Parasphenoid plate elongate, without muscular crests in palatal view. Occiput deep. Lower jaw with gentle angular bend and elongate retroarticular process. Posterior meckelian foramen strongly elongated.

Inflectosaurus Shishkin, 1960
Diagnosis. Skull up to 700 mm long. Pineal foramen close to middle of orbito-occipital distance. Orbits small. Quadrates behind the level of occipital condyles. Preotic extension of pterygoid ascending lamina strongly developed.

Inflectosaurus amplus Shishkin, 1960
Holotype and locality. PIN 2242/1, incomplete skull; Bolshoi Bogdo Mountain, Caspian basin, Astrakhan' Province.
Horizon. Yarenskian Gorizont, Lower Triassic.
Trematosauroid taxa based on incomplete material. Benthosuchidae: *Angusaurus weidenbaumi* (Kuzmin, 1935), Ples, Volga River, Sludkian Gorizont, Lower Triassic. Trematosauridae: (?) *Trematosaurus* sp., Lipovskaya Balka, Don River, Volgograd Province, Lipovskaya Svita, Yarenskian Gorizont, Lower Triassic.

Superfamily BRACHYOPOIDEA Lydekker, 1885
Diagnosis. Skull short, with laterally placed orbits; postorbital region longer than preorbital. Orbitopineal distance long. Sensory grooves well developed. Occipital surface slopes backward. Cheek region deep. No otic notch. Fissura pterygo-squamosa open. Quadrate wedge-shaped. Palate vaulted. Cultriform process of parasphenoid broad. Palatine and ectopter-ygoid tusks followed by few palatal teeth. Interchoanal and parachoanal teeth usually present. Lower jaw shallow, with well developed retroarticular and coronoid processes.

Family DVINOSAURIDAE Amalitskii, 1921
Diagnosis. Skull 200–220 mm long. Lacrimal reaches both to orbit and (usually) naris. Basioccipital strongly

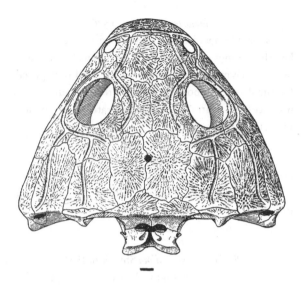

Figure 3.12. *Dvinosaurus primus* Amalitskii; skull, dorsal view. Scale bar is 10 mm. (From Shishkin, 1973.)

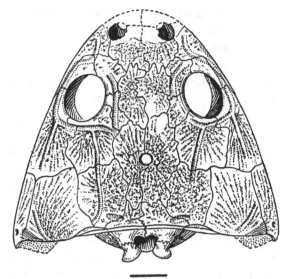

Figure 3.13. *Tupilakosaurus wetlugensis* Shishkin; skull, dorsal view. Scale bar is 10 mm. (From Shishkin, 1973.)

reduced. Paroccipital bar includes opisthotic. No anterior palatal vacuity. Pterygoid reaches vomer. No palatine–vomer contact laterally to choana. Movable articulation or short incipient suture between pterygoid and parasphenoid. Vertebrae rhachitomous.

Dvinosaurus Amalitskii, 1921
Diagnosis. As for family.

Dvinosaurus primus Amalitskii, 1921
See Figure 3.12.
Holotype and locality. PIN 2005/39, skull; Sokolki, North Dvina River, Arkhangel'sk Province.
Horizon. Upper Tatarian, Upper Permian.

Dvinosaurus egregius Shishkin, 1968
Holotype and locality. PIN 1100/23, skull; Vyazniki, Klyaz'ma River, Oka basin, Vladimir Province.
Horizon. Uppermost Tatarian, Upper Permian.

Dvinosaurus purlensis Shishkin, 1968
Holotype and locality. PIN 1538/18, lower jaw fragment; Purly, Pizhma River, Vyatka basin, Nizhnii Novgorod Province.
Horizon. Uppermost Tatarian, Upper Permian.

Family TUPILAKOSAURIDAE Kuhn, 1960
Diagnosis. Skull 80–100 mm long. Lacrimal small, bordering orbit and merged with palatine. Occipital articulation facet concave, single, dominated by strongly developed basioccipital. No opisthotic exposure in occiput. Pterygoid not reaching forward to palatine. Palatine contacts vomer laterally to choana. Anterior palatal vacuity paired. Pterygoid broadly sutured with parasphenoid and basioccipital. Vertebrae embolomerous.

Tupilakosaurus Nielsen, 1954
Diagnosis. As for the family.

Tupilakosaurus wetlugensis Shishkin, 1961
See Figure 3.13.
Holotype and locality. PIN 1025/1(1), skull; Spasskoe, Vetluga River, Nizhnii Novgorod Province.
Horizon. Vokhmian Gorizont, Lower Triassic.

Family BRACHYOPIDAE Lydekker, 1885
Diagnosis. Lacrimal small, bordering orbit, or absent (?). Basioccipital reduced. Occipital condyles paired, with convex articular surfaces. No opisthotic exposure

in occiput. Pterygoid not reaching forward to palatine. Vomer–palatine contact lateral to choana in many forms. Anterior palatal vacuities from paired to forming a single transverse slit. Pterygoid broadly sutured with parasphenoid and exoccipital. Vertebrae rhachitomous.

Batrachosuchoides Shishkin, 1966
Diagnosis. Skull about 150–180 mm long. Lacrimal merged with palatine. Anterior palatal vacuities partially confluent. Ectopterygoid tooth row long. Exoccipital–pterygoid contact narrow.

Batrachosuchoides lacer Shishkin, 1966
Holotype and locality. PIN 953/2, skull fragment; Okunevo, Fedorovka River, Vyatka basin, Kirov Province.
Horizon. Fedorovskaya Svita, Yarenskian Gorizont, Lower Triassic.

Batrachosuchoides impressus Novikov et Shishkin, 1994
Holotype and locality. PIN 4370/1; impression of skull roof fragment; Hei Yaga River, Pechora basin, Arkhangel'sk Province.
Horizon. Lower Lestanshorskaya Svita, Yarenskian Gorizont, Lower Triassic.

Superfamily PLAGIOSAUROIDEA Abel, 1919
Diagnosis. Skull short. Pineal foramen close to orbits. Dermal ornamentation from pitted to pustular. No pterygo-squamosal fissure. Interpterygoid vacuities large. Parasphenoid plate broad. Pterygoid sutured with parasphenoid. Trunk vertebral centra holospondylous, intrasegmental (formed by fusion of hypocentrum with preceding pleurocentrum).

Family PLAGIOSAURIDAE Abel, 1919
Diagnosis. Skull broadened transversely; facial region short, orbits very large. Supratemporal excluded from occipital border. Supratemporal–parietal contact much reduced. No otic notch and anterior palatal vacuity. Palate vaulted. Pterygoid does not reach palatine. Pterygo-parasphenoid suture long. Palatal tusks reduced in size or absent.

Plagiosternum Fraas, 1896
Diagnosis. Skull width up to 500–700 mm, exceeds skull length 2.5 times or more. Cheek margin with deep temporal embayment. Ornamentation of skull roof pitted. Lacrimal reaches both naris and orbit. Quadratojugal produces ventral supraquadrate projection. Interclavicle trapezoid.

Plagiosternum paraboliceps (Konzhukova, 1955)
Holotype and locality. PIN 415/5, skull fragment; Koltaevo I, Bolshoi Yushatyr' River, Bashkortostan Republic.
Horizon. Donguz Gorizont, Middle Triassic.

Plagiosternum danilovi Shishkin, 1986
See Figure 3.14A.
Holotype and locality. PIN 2867/17, skull fragment; Koltaevo IV, Bolshoi Yushatyr' River, Bashkortostan Republic.
Horizon. Bukobay Gorizont, Middle Triassic.

Plagioscutum Shishkin, 1986
Diagnosis. Skull moderately broad, with steep slope of cheek in occipital view. No embayment in lateral cheek margin. Quadratojugal lacks supraquadrate projection. Ornament pustulate. Interclavicle with long anterior process.

Plagioscutum ochevi Shishkin, 1986
See Figure 3.14B.
Holotype and locality. PIN 2430/80, clavicle; Donguz River, Ural basin, Orenburg Province.
Horizon. Donguz Gorizont, Middle Triassic.

Plagioscutum caspiense Shishkin, 1986
Holotype and locality. PIN 4121/11, clavicle fragment; Inder Lake, Caspian basin[5].
Horizon. Bukobay Gorizont of Cis-Urals and its equivalent in Caspian basin (upper part of Inder Gorizont), Middle Triassic.
Plagiosaurid taxa based on fragmentary material: *Melanopelta* (*M. antiqua*) Shishkin, 1967, Vyatka Basin, Yarenskian Gorizont, Lower Triassic (the earliest unquestionable plagiosaurid); *Aranetzia* (*A. improvisa*)

[5] Now the territory of Kazakhstan Republic.

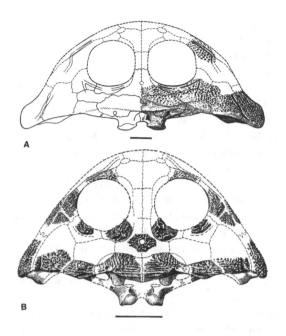

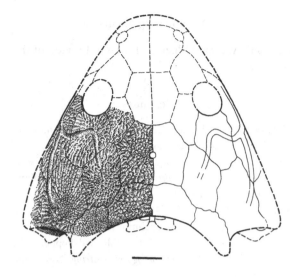

Figure 3.15. *Boreopelta vavilovi* Shishkin; skull, dorsal view. Scale bar is 10 mm. (From Shishkin and Vavilov, 1985.)

Figure 3.14. A, *Plagiosternum danilovi* Shishkin; skull, dorsal view. B, *Plagioscutum ochevi* Shishkin; skull, dorsal view. Scale bars are 50 mm. (From Shishkin, 1987.)

Novikov et Shishkin, 1992, Pechora basin, Krasnokamenskaya Svita, Middle Triassic.

Superfamily RHYTIDOSTEOIDEA Huene, 1920
Diagnosis. Skull subtriangular, usually short; orbits placed laterally. Ornamentation from pustular to pitted, showing radial pattern. Lacrimal absent. Otic notch reduced or absent. Palate shallow. Palatine–ectopterygoid tooth row usually present. Pterygoid-parasphenoid suture long.

Family RHYTIDOSTEIDAE Huene, 1920
Diagnosis. Palate moderately deep. Palatine–ectopterygoid tooth row irregular or incipient (not clearly differentiated from shagreen denticles). Shagreen on palate and coronoids, often extending to maxilla and dentary.

Rhytidosteus Owen, 1884
Diagnosis. Facial region of skull moderately elongate, snout slender. Choanae slit-shaped. Pterygoid contacts exoccipital. Precoronoid flattened, forming part of symphyseal plate.

Rhytidosteus uralensis Shishkin, 1994
Holotype and locality. PIN 2394/17, lower jaw fragment; Kzyl-Sai ravine, Kzyl Oba River, Ural basin, Orenburg Province.
Horizon. Petropavlovskaya Svita, Yarenskian Gorizont, Lower Triassic.
Comment. This form is notable for being the only Gondwanan tetrapod genus, other than the dicynodont *Lystrosaurus*, which has so far been recorded from the East European Lower Triassic.

Family PELTOSTEGIDAE Säve-Söderbergh, 1935
Diagnosis. Palate very shallow. Palatine–ectopterygoid tooth row well developed. Shagreen dentition strongly reduced or absent.

Boreopelta Shishkin, 1985
Diagnosis. Skull with short facial region. Orbitopineal distance long. Pterygo-squamosal fissure open. Pterygoid does not reach palatine and exoccipital.

Boreopelta vavilovi Shishkin, 1985
See Figure 3.15.
Holotype and locality. PIN 4115/1, skull fragment; Buur River, Olenek basin, northern Siberia.
Horizon. Teryutekh Svita, Lower Olenekian, Lower Triassic.

Acknowledgements

We thank Andrew Milner for his careful review of the manuscript.

References

Ahlberg, P.E. and Milner, A.R. 1994. The origin and early diversification of tetrapods. *Nature* **368**: 507–514.

Amalitzkii, V.P. 1921. [*Dvinosauridae. The North Dvina excavations by Prof. V.P. Amalitzkii. I.*]. Petrograd, Izdatel'stvo Akademii Nauk: 1–16.

Bolt, J.R. and Lombard, R.E. 1992. Nature and quality of the fossil evidence for otic evolution in early tetrapods, pp. 377–384 in Webster, D.B. Fay, R.R. and Popper, A.N. (eds.), *The Evolutionary Biology of Hearing*. New York: Springer.

Bossy, K.V. 1976. *Morphology, Paleoecology, and Evolutionary Relationships of the Pennsylvanian Urocordylid Nectrideans (Subclass Lepospondyli, Class Amphibia)*. Ph.D. thesis, Yale University, 370 pp.

Boy, J. 1986. Studien über die Branchiosauridae (Amphibia: Temnospondyli). 1. Neue und wenig ekannte Arten aus dem mittlereuropäischen Rotliegenden (?oberstes Karbon bis untere Perm). *Paläontologische Zeitschrift* **60**: 131–166.

Bystrov, A.P. 1938. *Dvinosaurus* als neotenische Form der Stegocephalen. *Acta Zoologica* **19**: 209–295.

—and Efremov, I.A. 1940. [*Benthosuchus sushkini* Efr. – a labyrinthodont from the Eotriassic of Sharzhenga River.] *Travaux de l'Institut Paléozoologique, Académie des Sciences de l'URSS* **4**: 289–299.

Carroll, R.L. 1980. The hyomandibular as a supporting element in the skull of primitive tetrapods, pp. 293–317 in Panchen, A.L. (ed.), *The Terrestrial Environment and the Origin of Land Vertebrates*. London: Academic.

—1988. *Vertebrate Paleontology and Evolution*. San Francisco: W.H. Freeman, 698pp.

—1995. Problems of the phylogenetic analysis of Paleozoic choanates. *Bulletin du Muséum National d'Histoire Naturelle, Paris, 4ème Série, Section C* **14**: 389–445.

—and Chorn, J. 1995. Vertebral development in the oldest microsaur and the problem of "lepospondyl" relationships. *Journal of Vertebrate Paleontology* **15**: 37–56.

—and Currie, P.J. 1975. Microsaurs as possible microsaur ancestors. *Zoological Journal of the Linnean Society* **57**: 229–247.

—and Holmes, R. 1980. The skull and jaw musculature as guides to the ancestry of salamanders. *Zoological Journal of the Linnean Society* **68**: 1–40.

—and Winer, L. 1977. Patterns of amphibian evolution, pp. 405–437 in Hallam, A. (ed.), in *Patterns of Evolution, as Illustrated by the Fossil Record*. Amsterdam: Elsevier. [Unpublished appendix: Classification of amphibians and list of genera, pp. 1–14.]

Clack, J.A. 1983. The stapes of the Coal Measure embolomere *Pholiderpeton scutigerum* Huxley (Amphibia: Anthracosauria) and otic evolution in early tetrapods. *Zoological Journal of the Linnean Society* **79**: 121–148.

—1987. *Pholiderpeton scutigerum* Huxley, an amphibian from the Yorkshire Coal Measures. *Philosophical Transactions of the Royal Society, Series B* **318**: 1–107.

—1992. The stapes of *Acanthostega gunnari* and the role of the stapes in early tetrapods, pp. 405–420 in Webster, D.B. (ed.), *The Evolutionary Biology of Hearing*. Berlin: Springer.

Efremov, J.A. 1929. *Benthosaurus sushkini*, ein neuer Labyrinthodont aus den permotriassischen Ablagerungen des Scharschenga-Flusses, Nord-Düna Gouvernement. *Bulletin de l'Académie des Sciences. URSS, Classe des Sciences Physico-Mathématiques*: 757–770.

—1933. Über die Labyrinthodonten der U.d.S.S.R. II. Permische Labyrinthodonten des früheren Gouvernements Wjatka. *Trudy Paleozoologicheskogo Instituta* **2**: 117–164.

—1937. [On the labyrinthodonts of the USSR. III. *Melosaurus uralensis* H.v.Meyer. IV. Notes on the lost forms *Zygosaurus* and *Chalcosaurus*]. *Trudy Paleontologicheskogo Instituta* **8**: 7–27.

—1939. First representative of Siberian early Tetrapoda. *Comptes-rendus des séances del'Académie des sciences del'URSS* **23**(1), 106–110.

—1940. [On the labyrinthodonts from Eotriassic deposits of the Upper Volga basin]. *Trudy Paleontologicheskogo Instituta AN SSSR* **10**(2): 6–23.

Eichwald, E.D. 1848. On *Zygosaurus lucius* from the Permian of Russia. *Bulletin de la Société Impériale des naturalistes de Moscou* **21**(3): 159.

Foreman, B.C. 1990. A revision of the cranial morphology of the Lower Permian temnospondyl amphibian *Acroplous vorax* Hotton. *Journal of Vertebrate Paleontology*, **10**: 390–397.

Gauthier, J., Cannatella, D., de Queiroz, K., Kluge, A.G. and Rowe, T. 1989. Tetrapod phylogeny, pp. 337–353 in Fernholm, B., Bremer, K. and Jörnvall, H. (eds.), *The Hierarchy of Life*. Amsterdam: Elsevier.

Getmanov, S.N. 1982. [A labyrinthodont from the Lower Triassic of Obshchy Syrt.] *Paleontologicheskii Zhurnal* **1982** (2): 98–104.

—1989. [Triassic amphibians of the East European platform.] *Trudy Paleontologicheskogo Instituta AN SSSR* **236**: 1–102.

Godfrey, S.J., Fiorillo, A.R. and Carroll, R.L. 1987. A newly discovered skull of the temnospondyl amphibian *Dendrerpeton acadianum* Owen. *Canadian Journal of Earth Sciences* **24**: 796–805.

Golubev, V.K. 1995. [New species of *Melosaurus* (Amphibia, Labyrinthodonia) from Kazanian deposits of the Kama River drainage area.] *Paleontologicheskii Zhurnal* **1995** (3): 86–97.

Gubin, Y.M. 1980. [New dissorophids from the Permian of Cis-Urals.] *Paleontologicheskii Zhurnal* **1980** (4): 82–90.

—1981. [A new platyoposaurid from Bashkiria]. *Paleontologicheskii Zhurnal* **1981** (2): 141–143.

—1983. [First eryopoids from the Permian of East European platform.] *Paleontologicheskii Zhurnal* **1983** (4): 110–115.

—1984. [On the systematic position of intasuchids.] *Paleontologicheskii Zhurnal* **1984** (2): 118–120.

—1986. [New data on the archegosauroids of the East European Platform]. *Paleontologicheskii Zhurnal* **1986** (2): 75–80.

—1987. [On the systematic position and dating of some labyrinthodonts from the Upper Permian of the western Cis-Urals.] *Paleontologicheskii Zhurnal* **1987** (1): 94–99.

—1989. [On systematic position of the labyrinthodont from the locality Malaya Kinel'.] *Paleontologicheskii Zhurnal* **1989** (4): 116–120.

—1991. [Permian archegosauroid amphibians of the USSR.] *Trudy Paleontologicheskogo Instituta AN SSSR* **249**: 1–141.

—1993. [New data on lower tetrapods from the Upper Permian of northern Cis-Urals and Obshchy Syrt.] *Paleontologischeskii Zhurnal* **1993** (4):97–105.

Hook, R.W. 1983. *Colosteus scutellatus* (Newberry), a primitive temnospondyl amphibian from the Middle Pennsylvanian of Linton, Ohio. *American Museum Novitates*, **2770**: 1–41.

Ivakhnenko, M.F. 1972. [A new benthosuchid from the Lower Triassic of the Upper Volga area]. *Paleontologicheskii Zhurnal* **1972** (4); 93–99.

Kalandadze, N.N., Ochev, V.G., Tatarinov, L.P., Chudinov, P.K., and Shishkin, M.A. 1968. [Catalogue of Permian and Triassic tetrapods of the USSR], pp. 73–92 in [*Upper Paleozoic and Mesozoic amphibians and reptiles of the USSR*]. Moscow: Nauka.

Konzhukova, E.D. 1955. [Permian and Triassic labyrinthodonts from the Volga basin and Cis-Urals.] *Trudy Paleontologicheskogo Instituta AN SSSR* **49**: 5–88.

—1956. [Fauna of Inta from the Lower Permian of the northern Cis-Urals.] *Trudy Paleontologicheskogo Instituta AN SSSR* **62**: 5–55.

—1965. [A new parotosaur from the Triassic of Cis-Urals]. *Paleontologicheskii Zhurnal* **1985** (1): 97–104.

Lombard, R.E. and Sumida, S.S. 1992. Recent progress in understanding early tetrapods. *American Zoologist* **32**: 609–692.

Lozovskii, V.R. and Shishkin, M.A. 1974. [The first labyrinthodont finds in the Lower Triassic of Mangyshlak]. *Doklady AN SSSR* **214** (1): 169–172.

Meyer, H. 1857. Über fossile Saurerierknochen des Orenburgischen Gouvernements. *Neues Jahrbuch für Mineralogie, Geologie und Palaeontologie* **1**: 539–543.

Milner, A.R. 1990. The radiations of temnospondyl amphibians., pp. 321–349 in Taylor, P.D. and Larwood, G.P. (eds.), *Major Evolutionary Radiations. Systematics Association Special Volume* 42. Oxford: Clarendon Press.

—1993a. Amphibian-grade Tetrapoda, pp. 665–679 in Benton, M.J. (ed.), *The Fossil Record 2*. London: Chapman & Hall.

—1993b. The Paleozoic relatives of lissamphibians. *Herpetological Monographs* **7**: 8–27.

Novikov, I.V. 1986. [A new species of *Parotosuchus* (Amphibia, Labyrinthodontia) from the Triassic of the Vychegda River Basin]. *Paleontologicheskii Zhurnal* **1986** (3): 129–131.

—1990. [New early Triassic labyrinthodonts of the central Tyman region]. Paleontologicheskii Zhurnal **1990** (1): 97–100.

—1994. [Biostratigraphy of the Triassic of the Timan-North Ural region on the tetrapod fauna.] *Trudy Paleontologicheskogo Instituta RAN* **261**: 1–139.

—and Shishkin, M.A. 1992. [New Mid Triassic labyrinthodonts from the Pechora Cis-Urals.] *Paleontologicheskii Zhurnal* **1992** (3): 71–80.

Ochev, V.G. 1958. [New data on the Triassic vertebrate fauna of the Orenburg Cis-Urals.] *Doklady AN SSSR* **122** (3): 485–488.

—1966. [*Systematics and Phylogeny of Capitosauroid Labyrinthodonts*.] Saratov: Izdatel'stvo Saratovskogo Universiteta, 184 pp.

—1972. [*Capitosauroid Labyrinthodonts from the Southeast of the European Part of the USSR.*] Saratov: Izdatel'stvo Saratovskogo Universiteta, 269 pp.

—and Shishkin, M.A. 1967. [Burial of ancient amphibians in Orenburzh'e.] *Priroda* **1967**(1): 79–85.

Panchen, A.L. 1975. A new genus and species of anthracosaur amphibian from the Lower Carboniferous of Scotland and the status of *Philodogaster pisciformis* Huxley. *Philosophical Transactions of the Royal Society of London, Series B* **269**: 581–637.

—1980. The origin and relationships of the anthracosaur amphibia from the late Palaeozoic, pp. 319–350 in Panchen, A.L. (ed.), *The Terrestrial Environment and the Origin of Land Vertebrates.* Academic: London.

—1985. On the amphibian *Crassigyrinus* Watson from the Carboniferous of Scotland. *Philosophical Transactions of the Royal Society, Series B* **309**: 505–568.

—and Smithson, T.R. 1988. The relationships of the earliest tetrapods, pp. 1–32 in Benton, M.J. (ed.), *The Phylogeny and Classification of the Tetrapods, Volume 1: Amphibians, Reptiles, Birds. Systematics Association Special Volume,* 35A. Oxford: Clarendon Press.

Romer, A.S. 1947. Review of the Labyrinthodontia. *Bulletin of the Museum of Comparative Zoology* **99**: 1–368.

—1966. *Vertebrate Paleontology.* 3rd edition. Chicago: The University of Chicago Press, 468 pp.

Ryabinin, A.N. 1927. [*Trematosuchus (?) yakovlevi* nov.sp. from Lower Triassic deposits of the Rybinsk vicinty]. *Izvestiya Geologicheskogo Komiteta SSSR* **45** (5): 519–527.

—1930. [*Wetlugasaurus angustifrons* nov.gen., nov.sp. from the Lower Triassic of Vetluga Land in northern Russia]. *Ezhegodnik Russkogo Paleontologicheskogo Obshchestva* **1930** (8): 49–76.

Säve-Söderbergh, G. 1936. On the morphology of Triassic stegocephalians from Spitzbergen and the interpretation of the endocranium in the Labyrinthodontia. *Kungliga Svenska Vetenskaps Akademiens Handlingar,* (3), **16** (1): 1–181.

Sennikov, A.G. 1981. [A new wetlugasaur from the Samara River basin]. *Paleontologicheskii Zhurnal* **1981** (2): 143–148.

Shishkin, M.A. 1960a. [A new Triassic trematosaurid *Inflectosaurus amplus.*] *Paleontologicheskii Zhurnal* **1960** (2): 130–148.

—1960b. [On the Yarengiidae, a new family of Triassic labyrinthodonts]. *Paleontologicheskii Zhurnal* **1960** (1): 97–106.

—1961. [New data on *Tupilakosaurus*]. *Doklady AN SSSR* **136** (4): 938–941.

—1966. [A brachyopid labyrinthodont from the Triassic of the Russian Platform]. *Paleontologicheskii Zhurnal* **1966** (2): 93–108.

—1967. [Plagiosaurs in the Triassic of the USSR]. *Paleontologicheskii Zhurnal* **1967** (1): 92–99.

—1968. On the cranial arterial system of labyrinthodonts. *Acta Zoologica,* **49**: 1–22.

—1973. [Morphology of early Amphibia and some problems of lower tetrapod evolution.] *Trudy Paleontologicheskogo Instituta AN SSSR* **137**: 1–260.

—1975. Labyrinthodont middle ear and some problems of amniote evolution. *Colloque international C.N.R.S. Problèmes actuels de paléontologie-évolution des vertébrés* **218**: 337–348.

—1980. [Luzocephalidae, a new family of Triassic labyrinthodonts.] *Paleontologicheskii Zhurnal* **1980**(1): 104–124.

—1984. [Amphibia. (Labyrinthodontia, Batrachomorpha).] pp. 124–133 in Tatarinov, L.P. and Shimanski, V.N. (eds.), [*Guide to the Systematics of Fossil Organisms.*] Moscow: Nauka.

—1986. [New data on plagiosaurs from the Triassic of the USSR]. *Byulleten' Moskovskogo Obshchestva Ispytatelei Prirody, Otdel Geologicheskii* **61** (3): 97–102.

—1987. [Evolution of early amphibians (Plagiosauroidea).] *Trudy Paleontologicheskogo Instituta AN SSSR* **225**: 1–143.

—1989a. The axial skeleton of early amphibians and the origin of resegmentation in tetrapod vertebrae. *Progress in Zoology* **35**: 180–185.

—1989b. On resegmentation of vertebrae in early tetrapods. *Acta Musei Reginae hradecensis S.A. Scientiae Naturales* **22**: 105–115.

—1994a. Multiple origin of the non-mammalian tympanic system: evidence or a priori belief? *Journal of Morphology* **220**(3): 393–394.

—1994b. [Gondwanan rhytidosteid (Amphibia, Temnospondyli) from the Lower Triassic of southern Cis-Urals.] *Paleontologicheskii Zhurnal* **1994** (4): 97–110.

—1998. *Tungussogyrinus,* a relict neotenic dissorophoid (Amphibia, Temnospondyli) from the Permo-Triassic of Siberia. *Paleontologicheskii Zhurnal* **1998**: 521–531.

—, and Lozovskii, V.R. 1979. [A labyrinthodont from the Triassic of the southern Maritime Land]. *Doklady AN SSSR* **246** (1): 201–205.

—, Ochev, V.G., Tverdokhlebov, V.P., Vergai, I.F., Goman'kov, A.V., Kalandadze, N.N., Leonova, E.M., Lopato, A.Yu., Makarova, I.S., Minikh, M.G., Molostovskii, E.M., Novikov, I.V. and Sennikov, A.G.

1995. [*Biostratigraphy of the Triassic of the Southern Cis-Urals.*] Moscow: Nauka, 206 pp.

—and Vavilov, M.N. 1985. [A find of rhytidosteid (Amphibia, Labyrinthodontia) from the Triassic of the USSR.] *Doklady AN SSSR* **282**: 971–975.

Smithson, T.R. 1982. The cranial morphology of *Greererpeton burkemorani* Romer (Amphibia: Temnospondyli). *Zoological Journal of the Linnean Society* **76**: 29–90.

—1985. The morphology and relationships of the Carboniferous amphibian *Eoherpeton watsoni* Panchen. *Zoological Journal of the Linnean Society* **85**: 317–410.

Smithson, T.R., Carroll, R.L., Panchen, A.L. and Andrews, S.M. 1994. *Westlothiana lizziae* from the Viséan of East Kirkton, West Lothian, Scotland, and the amniote stem. *Transactions of the Royal Society of Edinburgh: Earth Sciences* **84**: 383–412.

Steen, M. 1931. The British Museum collection of Amphibia from the Middle Coal Measures of Linton, Ohio. *Proceedings of the Zoological Society of London* **1930**: 849–891.

Tatarinov, L.P. 1958. [Evolution of sound-conducting apparatus of lower land vertebrates and the origin of reptiles.] *Zoologicheskii Zhurnal AN SSSR*, **37** (1): 57–73.

—1962. [Mode of functioning of sound-conducting apparatus of labyrinthodonts.] *Paleontologicheskii Zhurnal* **1962** (4): 21–30.

Thomson, K.S. and Bossy, K.V. 1970. Adaptive trends and relationships in early Amphibia. *Forma et Functio* **3**: 7–31.

Trautschold, H. 1884. Die Reste permischer Reptilien des paläontologischen Kabinets der Universität Kasan. *Bulletin de la Société Impériale des naturalistes de Moscou* **15**: 1–39.

Twelvetrees, W. 1880. On a labyrinthodont skull (*Platyops rickardi*, Twelvetr.) from the Upper Permian cupriferous strata of Kargalinsk near Orenburg. *Bulletin de la Société Impériale des naturalistes de Moscou* **55** (1): 117–122.

Vaughn, P.P. 1962. The Paleozoic microsaurs as close relatives of reptiles, again. *American Midland Naturalist* **67**: 79–84.

V'yushkov, B.P. 1955. [On the land tetrapod fauna from the River Malaya Kinel'.] *Trudy Paleontologicheskogo Instituta AN SSSR* **49**: 176–189.

Watson, D.M.S. 1919. The structure, evolution and origin of the Amphibia. *Philosophical Transactions of the Royal Society, Series B* **209**: 1–73.

—1956. The brachyopid labyrinthodonts. *Bulletin of the British Museum Natural History, Geology* **2**: 317–391.

Werneburg, R. 1989. Labyrinthodontier (Amphibia) aus dem Oberkarbon und Unterperm Mitteleuropas – Systematik, Phylogenie und Biostratigraphie. *Freuberger Forschungshefte* **436**: 7–57.

Westoll, T.S. 1942. Ancestry of captorhinomorph reptiles. *Nature* **149**: 667–668.

Woodard, A.S. 1932. *Text-book of Palaeontology by Karl Zittel*. II. London, MacMillan: 1–464.

4

Permian and Triassic anthracosaurs from Eastern Europe

IGOR V. NOVIKOV, MIKHAIL A. SHISHKIN, AND VALERII K. GOLUBEV

Introduction

Among the early amphibians informally termed laby-rinthodonts, the anthracosaurs played a far less important role than the temnospondyls (see Chapter 3) in terms of their variety, abundance and dispersal (both in time and space). However, this group is of particular interest for its proximity to amniote ancestry.

In recent schemes, the anthracosaurs in a strict sense are largely referred to as either the order Anthracosauromorpha (Ivakhnenko and Tverdokhlebova, 1980), or the suborder(order) Anthracosauria (Panchen, 1970, 1975, 1980, 1985; Holmes, 1984; Milner, 1993) or the suborder Anthracosauroideae (Smithson, 1985, 1986; Panchen and Smithson, 1988). The group is traditionally believed to be closely related to the Seymouriamorpha (which comprises another early amphibian lineage presumably linked to reptile ancestry); but this affinity has been questioned by Panchen (1985; Panchen, in Panchen and Smithson, 1988), who ascribed shared 'reptiliomorph' characters of anthracosaurs and seymouriamorphs to homoplasy or parallelism. Interrelationships of anthracosaurs with other amphibian-grade Palaeozoic lineages are still more uncertain, which reflects the lack of a clear general concept of early tetrapod radiation (Carroll and Chorn, 1995). As noted above (Chapter 3), attempts to arrange all these lineages within two principal tetrapod clades, Amphibia and Amniota (Smithson, 1985; Panchen and Smithson, 1988; Ahlberg and Milner, 1994) are not supported by the present authors.

Typical anthracosaurs are limited mainly to the Carboniferous and Early Permian of North America and Europe. The only late offshoot that survived into the Late Permian and Triassic is the poorly studied group Chroniosuchia, known almost entirely from European Russia. The only chroniosuchian record from elsewhere in the literature comes from the Late Permian of North China (Young, 1979).

Anthracosaurs so far recorded in the Permo-Triassic of European Russia are represented mainly by chroniosuchians. Early (typical) anthracosaurs are known from only two fragmentary finds, which represent the last survivors of Permo-Carboniferous lineages that persisted into the early Late Permian (Ufimian and Kazanian) of the Cis-Urals (Gubin, 1985,1988). It may also be mentioned that the primitive amphibian *Tulerpeton* from the Late Devonian of central Russia, regarded elsewhere as a stem-group tetrapod (Ahlberg and Milner, 1994), has been referred recently to the reptiliomorph clade (Lebedev and Coates, 1995), which should imply its close linkage to anthracosaurs; but we find no real evidence for this interpretation.

Chroniosuchians, which range in Russia from the latest Permian (Late Tatarian) to the late Middle Triassic, have been collected from many localities in the vast area between the Pechora and North Dvina basins in the north, and the Ural basin in the south (V'yushkov, 1957a,b; Ryabinin, 1962; Shishkin, 1962; Tverdokhlebova, 1967, 1968, 1972; Ivakhnenko and Tverdokhlebova, 1980; Shishkin and Novikov, 1992; Novikov and Shishkin, 1995, 2000). Finds from the Triassic are scarce and represented only by isolated bones, although geographically they are rather widespread.

Repository abbreviations

PIN, Paleontological Institute, Russian Academy of Sciences, Moscow; SGU, Saratov State University, Saratov; TsNIGRI, Tsentralny Nauchno-Issledovatelskii Geologo-Razvedochnyi Muzei, Saint Petersburg.

Anthracosaur characteristics

Anthracosaurs were crocodile-like piscivorous amphibians with a rather deep elongated skull from 50 to 500 mm long. During their history, they retained a resemblance to the most primitive temnospondyls in such characters as the extent of the lacrimal bone up to the naris (or septomaxilla), presence of an intertemporal in most forms, movable basipterygoid articulation (between braincase and upper jaw), narrow interpterygoid vacuities, pterygoids with median contact anteriorly, a single concave occipital condyle, occipital exposure of the opisthotic bone, and, usually, absence of a retroarticular process on the lower jaw. Peculiar for anthracosaurs also is the clear demarcation and loose joint between the skull roof and cheek, a condition inherited from rhipidistian fishes.

Derived characters of anthracosaurs include the 'angustitabular' skull roof (with tabular contacting parietal), usually long pointed tabular horns with an unornamented terminal part, loss of the posttemporal fossae, narrowing of the vomers, lack of well-defined tusk pairs on the vomers (except Gephyrostegidae) and on the lower jaw symphysis, and the presence of five digits on the manus, in contrast to four in temnospondyls. Many of these traits (except the shape of the tabular, the dentition and, to some extent, reduction of the posttemporal fossae) are also shared by seymouriamorphs. One more notable character, common to both anthracosaurs and seymouriamorphs, and stressed by many authors as a synapomorphy of the clade Amniota (Smithson, 1985; Panchen and Smithson, 1988), is the lack of contact between the exoccipital and skull roof; but there is no consensus about the significance of this distinction (cf. Carroll and Chorn, 1995).

As far as the vertebral pattern is concerned, the most peculiar derived character of anthracosaurs (shared again with seymouriamorphs) is the predominance of pleurocentra over hypocentra (intercentra), a trend which results in a nearly gastrocentrous (reptile-like) condition in some forms. However, the anthracosaur hypocentra, as compared to those in seymouriamorphs, do not show such strong reduction, and vary from the disc-shaped elements seen in typical embolomeres to the horseshoe- or crescent-shaped wedges in some groups such as gephyrostegids and proterogyrinids. In most typical anthracosaurs (infraorder Embolomeri) and their chroniosuchian descendants, the hypocentra have gently or markedly convex articulation surfaces, which acted like the ball of a 'ball-and-socket' joint between successive fully amphicoelous pleurocentra (cf. Panchen, 1970; V'yushkov, 1957a,b). Anthracosaurs resemble reptiles rather than temnospondyls in the firmer attachment of the neural arch to the pleurocentrum.

A number of evolutionary changes seen in some anthracosaur lineages are paralleled in temnospondyls. These include the withdrawal of the lacrimal from, and spread of the frontal to, the orbital margin, loss of the intertemporal and strong reduction of the endocranial ossifications (in Chroniosuchia), a trend to obliteration of the otic notch[1], and reduction of the sensory grooves in many forms. However, some modifications of the primitive labyrinthodont skull pattern seen in particular anthracosaur groups or genera are unknown in temnospondyls. These include fenestration of the skull roof, seen as preorbital fontanelles in chroniosuchids, perforation of the cheek in *Anthracosaurus* and perforation of the cheek/table joint in some bystrowianids (Panchen, 1977; Ivakhnenko and Tverdokhlebova, 1980).

---

[1] Most recent authors argue that the earliest, or even all, anthracosaurs lacked the true otic notch housing the tympanum and thus they were incapable of perceiving airborne vibrations (Clack, 1983; Panchen, 1977, 1985; Smithson, 1985; Panchen and Smithson, 1988). This belief is not shared in the present account (cf. Shishkin, 1994; see also Chapter 3).

Anthracosaur groups

Among early anthracosaurs, most diversity is seen in the infraorder Embolomeri (Panchen, 1975), which includes aquatic anguilliform animals with relatively small limbs and typically diplospondylous vertebrae, with a cylindrical pleurocentrum and disc-shaped hypocentrum. The elongated presacral division of the vertebral column included about 40 vertebrae. The embolomeres were obviously swamp-dwellers. Carboniferous forms are usually found associated with coal seams, and this is true for a single embolomere find from northern Russia.

Two other early anthracosaur groups, so far formally unnamed and typified by the families Eoherpetontidae and Gephyrostegidae (cf. Holmes, 1984; Panchen, 1985; Smithson, 1985), represent more terrestrial lineages, with no sign of sensory grooves on the skull, a relatively short body (usually about 25 presacral vertebrae), and markedly reduced ossification of the hypocentra. A presumed gephyrostegid find was reported from the Cis-Urals.

The latest anthracosaurs (Chroniosuchia) include two families, Chroniosuchidae and Bystrowianidae. They are readily distinguished from older groups by a row of dermal plates over the vertebral column, conspicuously ball-shaped hypocentra in most forms, fenestration of the skull roof (at least in the Chroniosuchidae) and some advanced characters in the skull roof pattern. The lack of well-developed sensory grooves in chroniosuchians suggests terrestrial or semiaquatic adaptations. The proportions of the body are not clear, but, judging from finds of associated series of dermal plates in the chroniosuchid *Chroniosaurus*, the number of presacral vertebrae was probably not as great as in embolomeres. Chroniosuchian remains come from deltaic, lacustrine, and alluvial facies.

In contrast to the early anthracosaurs, chroniosuchians have been little studied. The group was discovered by V'yushkov (1957a,b) in the Late Tatarian of the Cis-Urals and assigned by him to seymouriamorphs ('Kotlassiomorpha'). Shishkin (cf. Olson, 1965; Tverdokhlebova, 1967, 1968, p. 4) first identified chroniosuchians from the Triassic and suggested their

anthracosaur origin. The suborder Chroniosuchida was proposed by Tatarinov (1972) for chroniosuchids in a strict sense, while another family, the bystrowianids, was placed by him in the suborder Seymourida; both units were considered members of the Seymouriamorpha. Ivakhnenko and Tverdokhlebova (1980) extended the suborder Chroniosuchia (= Chroniosuchida) to include bystrowianids, and united it with the Anthracosauria (*sensu* Panchen, 1975) in the order Anthracosauromorpha.

In the present account, the Chroniosuchia is retained as a separate suborder only for convenience, to emphasize its specific distinctions from the rest of anthracosaurs (see below), which are referred to, following Smithson (1985), the suborder Anthracosauroideae. At the same time, it seems very likely that chroniosuchians are merely the aberrant descendants of the Embolomeri (cf. Holmes, 1984), as may be evidenced, in particular, by the similar structure of the tabulars and vertebrae in both groups. Based on this, it may be reasonable to include chroniosuchians in the Anthracosauroideae as an additional infraorder. Anthracosaurs as a whole, the suborders Anthracosauroidea and Chroniosuchia together, are termed Anthracosauromorpha in preference to Anthracosauria, since this name has been applied to different groups in the recent literature (cf. Panchen, 1975, 1980; Smithson, 1985, 1986; Gauthier *et al.*, 1988).

This chapter does not contain detailed information on other nonamniotes from Russia and adjacent areas of the former Soviet Union that stand close to amniote ancestry and represent the basal groups of the class Parareptilia Olson (cf. Ivakhnenko, 1987). Except for some discosauriscids (see below), all these forms come from the Late Permian of the East European Platform and Cis-Uralian Trough. In a recent survey by Ivakhnenko (1997), these are placed in two subclasses, the Seymouriamorpha Watson and Cheloniamorpha Ivachnenko.

According to Ivakhnenko's scheme, the local Seymouriamorpha belong to three families, the Discosauriscidae Romer, 1947, Leptorophidae Ivakhnenko, 1987, and Karpinskiosauridae Sushkin, 1925, the latter two being East European endemics. Among the Discosauriscidae, the genera recorded include *Discosauriscus* Kuhn, 1933 (Orenburg Province, Lower Tatarian, Upper Permian), *Utegenia* Kuznetzov and Ivachnenko, 1981 (Alma Ata Province, Kazakhstan,Upper Carboniferous or Lower Permian), and *Ariekanerpeton* Tatarinov, 1968 (Kuramin Ridge, Tadzhikistan, Lower Permian). The first of these genera is widespread elsewhere in the Lower Permian of central Europe, while the other two are endemic. The Leptorophidae includes *Leptoropha* Tchudinov, 1955 (Upper Kazanian[1]), *Biarmica* (Lower Kazanian), and *Raphanodon* (Upper Tatarian). The Karpinskiosauridae comprises *Kotlassia* Amalitzky, 1921, *Karpinskiosaurus* (Amalitzky, 1921) and *Buzulukia* Vjuschkov, 1957; all from the Vyatskian Horizon, Upper Tatarian.

Within the Cheloniamorpha, the nonamniote groups are united by Ivakhnenko (1997) in the suborder Nycteroleterina considered to be a member of the order Pareiasaurida. The families included in the Nycteroleterina are: Nycteroleteridae Romer, 1956, Tokosauridae Tverdochlebova et Ivachnenko, 1984, Rhipaeosauridae Tchudinov, 1955, and Lanthanosuchidae Efremov, 1940. The Nycteroleteridae includes the genera *Nycteroleter* Efremov, 1938 (Lower Tatarian), *Emeroleter* Ivachnenko, 1997 (Upper Tatarian), and *Bashkyroleter* Ivachnenko, 1997 (Upper Kazanian–Lower Tatarian). The Tokosauridae is typified by *Tokosaurus* (Upper Kazanian) and includes also *Macroleter* Tverdochlebova et Ivachnenko, 1984 (Lower Tatarian). The Rhipaeosauridae is limited to the type genus *Rhipaeosaurus* Efremov, 1940 (Upper Kazanian). The members of the Lanthanosuchidae are: *Lanthanosuchus* Efremov, 1946; *Lanthaniscus* Ivachnenko, 1980, and *Chalcosaurus* Meyer, 1866; all from the Lower Tatarian.

Systematic Survey

Class AMPHIBIA Linné, 1758
Order ANTHRACOSAUROMORPHA Ivachnenko et Tverdochlebova, 1980
Suborder ANTHRACOSAUROIDEAE Watson, 1929
Diagnosis. Amphibians with skull from 50 to over

[1] Here and below, the ranges shown for the listed parareptilian genera correspond to the age of the holotype(s) of their constituent species.

400 mm long. No osteoderm plates on dorsal side of body. No preorbital fenestrations in skull roof. Intertemporal retained. Frontal excluded from orbital margin. Vomers tuskless; pterygoid devoid of well-marked flanges (except for Gephyrostegidae). No anterior palatal vacuity. Braincase usually well ossified. Vertebral column with notochordal hypo- and pleurocentra. Hypocentrum ossification disc-shaped (with gently convex articulation surfaces) to horseshoe- or crescent-shaped.

Comment. Since the attribution of the 'old' anthracosaurs from Russia to particular infraorders of the Anthracosauroideae is debatable, diagnoses for groups below subordinal rank are given only for chroniosuchians.

(?)Infraorder EMBOLOMERI Cope, 1884
(?)Family EOGYRINIDAE Watson, 1929
Aversor Gubin, 1985

Diagnosis. Skull about 170 mm long. Dermal ornament formed by small radiating pits; no sensory grooves. Tusk pairs present on palatine and ectopterygoid but lacking on vomer. Pterygoid with dense shagreen dentition; pterygoid flange moderately expressed. Anterior and posterior meckelian foramina of lower jaw close together.

Aversor dmitrievi Gubin, 1985

Holotype and locality. PIN 570/50; skull fragment with associated lower jaw ramus; Inta River, Komi Republic.

Horizon. Intinskaya Svita, Ufimian, Upper Permian.

Comment. If correctly identified, this embolomere specimen is the youngest known member of the family Eogyrinidae, a group known otherwise largely from the Carboniferous (Milner, 1993).

Infraorder unspecified
(?)Family GEPHYROSTEGIDAE Jaekel, 1909

Comment. A short series of articulated dorsal vertebrae, showing cylindrical amphicoelous pleurocentra, small crescent-shaped intercentra, and deep, backwards-sloping neural arches was reported from the Copper Sandstone of the Kargala Mines (?Kazanian) on the Sakmara River, Orenburg Province (Gubin 1988). The proto-gastrocentrous pattern of these vertebrae suggests identification as Gephyrostegidae gen. indet., although, if this is the case, this record far post-dates any other gephyrostegid records (Milner, 1993).

Apart from this find, two Late Permian genera, formerly attributed to the non-anthracosauromorph groups, have recently been re-assessed by Golubev (1997) as members of the Gephyrostegidae. These are *Enosuchus* (*E. breviceps*) Konzukova, 1955, from the Lower Tatarian of Tatarstan Republic, and *Nyctiboetus* (*N. liteus*) Tchudinov, 1955, from the Upper Kazanian of Kirov Province. Both were long considered seymouriamorphs (cf. Tatarinov, 1972; Ivakhnenko, 1987). There is no consensus about the systematic position of these forms.

Suborder CHRONIOSUCHIA Tatarinov, 1972
Diagnosis. Tetrapods with skull from 50 to 300 mm long. Skull contour with antorbital step in side view. Sensory grooves absent or barely detectable. Skull roof lacking intertemporal, with median premaxillary fenestration and frontal entering orbital margin; palate with large anterior palatal vacuity (at least in Chroniosuchidae). Pterygoid flanges well developed. Braincase poorly ossified. Teeth conical, pointed, weakly infolded at base. A row of median ornamented osteoderm plates on dorsal side of body, each attached (via suture or ligament) to neural spine of underlying vertebra. Intercentra usually ball-shaped, with trend to obliteration of notochordal canal. Pleurocentra deeply amphicoelous, in most forms sutured to neural arches, sometimes indistinguishably fused to them.

Family CHRONIOSUCHIDAE Vjuschkov, 1957
Diagnosis. Skull deep, with markedly elongate preorbital part. Orbits round, facing dorsolaterally. Long preorbital fontanelles bordered by maxilla, lacrimal, prefrontal, and jugal. Otic notch positioned dorsolaterally or laterally (behind cheek). Quadrates well behind occipital margin of skull. Postorbital division of jugal short, not exceeding half of cheek length. Choanae long, slit-like. Infratemporal fossae narrow. Pterygoid flanges lying in palatal plane. Row of 3–4

tusk-like teeth on vomers, pairs or groups of small tusks on palatines and ectopterygoids. Osteoderm plates strongly expanded transversally, their median ventral processes not contacting neural spines. Each plate overlapped in squamate fashion by next posterior one. No paraneural holes or pits on neural arches. Intercentra from disc- or horseshoe-shaped, notochordal (in small forms) to ball-shaped, showing no notochordal perforation.

Chroniosuchus Vjuschkov, 1957
Diagnosis. Skull length up to 200–300 mm; in latest forms to 500 mm. Orbits relatively close together. Dermal ornament on skull roof and osteoderm plates pitted. Longitudinal crests of ornament on parietals lacking or retained only in juveniles. Hypocentra from disk-shaped (gently biconvex) or horseshoe-shaped in smaller forms to ball-shaped.

Chroniosuchus paradoxus Vjuschkov, 1957
See Figure 4.1.
Holotype and locality. PIN 521/6, articulated caudal vertebrae; Pron'kino, Orenburg Province.
Horizon. Kutuluk Svita, Vyatskian Gorizont, Upper Tatarian, Upper Permian.

Chroniosuchus licharevi (Ryabinin, 1962)
Holotype and locality. TsNIGRI 5813/1, lower jaw impression; Medvedkovo, Malaya North Dvina River, Arkhangel'sk Province.
Horizon. Salarevskaya Svita, Vyatskian Gorizont, Upper Tatarian, Upper Permian.
Comment. The form is a type of the genus *Jugosuchus* Ryabinin, here considered invalid.

Chroniosaurus Tverdokhlebova, 1972
Diagnosis. Skull from 70 to 200 mm long. Orbits relatively broadly separated. Ornament on skull and osteoderm plates largely tuberculate. Longitudinal crests on lacrimal, frontal, postfrontal, parietal, and postparietal well developed, continuing backwards on osteoderm plates. Intercentra change from horseshoe-shaped to nearly ball-shaped with growth. Neural arches not attached to pleurocentra in juveniles.

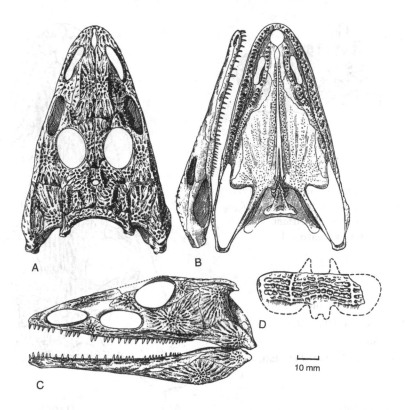

Figure 4.1. *Chroniosuchus paradoxus* Vjuschkov, 1957: A–C, reconstruction of skull and lower jaw, D, trunk osteoderm plate, No.521/3; A, D, from above, B, from below, C, side view. Pron'kino, Orenburg Province; Vyatskian Gorizont, Upper Tatarian, Upper Permian. (A–C, After Ivakhnenko and Tverdokhlebova, 1980.)

Chroniosaurus dongusensis Tverdochlebova, 1972
See Figure 4.2.
Holotype and locality. SGU 104B/198, skull; Donguz River, Ural basin, Orenburg Province.
Horizon. Malokinel'skaya Svita, Upper Tatarian, Upper Permian.

Chroniosaurus levis Golubev, 1998
Holotype and locality. SGU 104B/1102; osteoderm plate; Mutovino, Vologda Province.
Horizon. Severodvinskian Gorizont, Upper Tatarian, Upper Permian.

Jarilinus Golubev, 1998
Diagnosis. Skull more than 200 mm long. Orbits rather broadly separated. Skull roof bearing no ridges, with pitted dermal ornament. Osteoderm plates ornamented with meandering ridges of variable extent.

Jarilinus mirabilis (Vjuschkov, 1957)
See Figure 4.3.
Holotype and locality. PIN 528/1, skull table; Nizhnii Novgorod, Nizhnii Novgorod Province.
Horizon. Vyatskian Gorizont, Upper Permian, Upper Tatarian.

Uralerpeton Golubev, 1998
Diagnosis. Large chroniosuchid with narrowed osteoderm plates bearing pitted ornament. Ventral process of plate sutured with neural spine of vertebra.

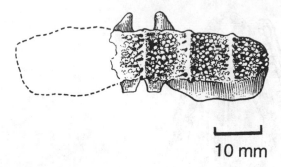

Figure 4.2. *Chroniosaurus dongusensis* Tverdochlebova, 1972: reconstruction of trunk osteoderm plate, from above. Donguz VI, Orenburg Province; Severodvinskian Gorizont, Upper Tatarian, Upper Permian. (After Ivakhnenko and Tverdokhlebova, 1980.)

Uralerpeton tverdochlebovae Golubev, 1998
Holotype and locality. PIN 1100/8, osteoderm plate; Vyazniki, Vladimir Province.
Horizon. Vyatskian Gorizont, Upper Tatarian, Upper Permian.

Suchonica Golubev, 1999
Diagnosis. Dermal plates narrow, with length exceeding width. Ventral process of plate not sutured to neural spine of vertebra. Dorsal surface of plate with ornament of pectinate type, devoid of longitudinal crests.

Suchonica vladimiri Golubev, 1999
Holotype and locality. PIN 4611/1, cervical osteoderm plate; Poldarsa, Vologda Province.
Horizon. Severodvinskian Gorizont, Upper Tatarian, Upper Permian.

Family BYSTROWIANIDAE Vjuschkov, 1957
Diagnosis. Skull of variable depth. No preorbital fontanelles. Otic notch shallow, broad, lying behind cheek. Quadrates close to level of occipital skull margin in top view. Postorbital division of jugal long. Intertemporal fossa broad, rounded. Pterygoid flanges steeply curved down. Osteoderm plates from weakly expanded to narrow (stretched along median axis); each plate usually bearing broadly separated anterior

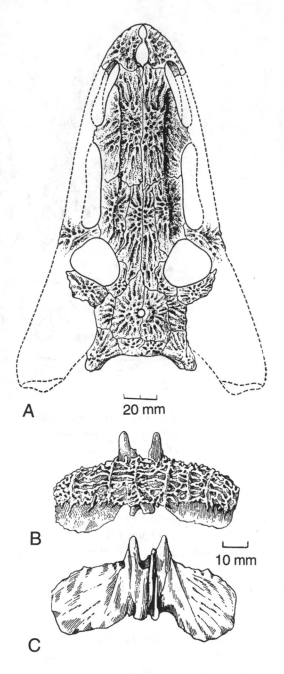

Figure 4.3. *Jarilinus mirabilis* (Vjuschkov, 1957): A, reconstruction of skull; B, C, osteoderm plate; A, B, from above, C, from below. Gorky I, Nizhnii Novgorod Province; Vyatskian Gorizont, Upper Tatarian, Upper Permian. (After Ivakhnenko and Tverdokhlebova, 1980.)

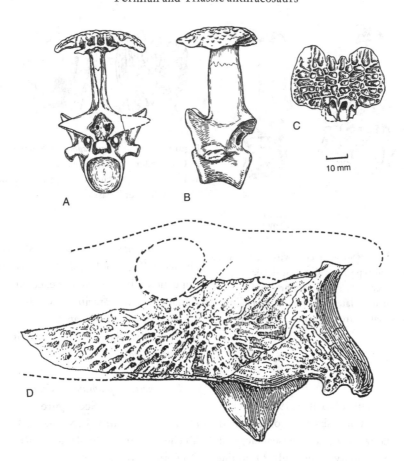

Figure 4.4. *Bystrowiana permira* Vjuschkov, 1957: A, B, trunk vertebra fused with osteoderm plate, holotype; C, osteoderm plate; D, reconstruction of posterior part of skull; A, from behind, B, D, from side, C, from above. Vyazniki, Vladimir Province; Vyatskian Gorizont, Upper Tatarian, Upper Permian. (A–C, After Ivakhnenko and Tverdokhlebova, 1980; D, modified from Ivakhnenko and Tverdoklebova, 1980.)

facets overlapped by preceding plate. Median ventral process of plate sutured with underlying neural spine. Paired deep paraneural canals and/or pits on anterior and posterior surfaces of neural arch. Intercentra ball-shaped, without continuous notochordal perforation.

Bystrowiana Vjuschkov, 1957

Diagnosis. Skull deep, up to 300 mm long. Suborbital division of jugal deep; postorbital division nearly separating squamosal from skull table. Skull roof with small fenestra between jugal and skull table. Zygapophysial facets of trunk vertebrae slope medially at angle of 20° to the horizontal plane in anterior view. Osteoderm plates well developed, broader than long, gently arched transversally. Plate locking mechanism elaborate: lateral portions of plate posterior margin overlap the next-following plate, while median portion of the same margin underlies paired anterior projection from following segment.

Bystrowiana permira Vjuschkov, 1957
See Figure 4.4.

Holotype and locality. PIN 1100/1, vertebra fused with osteoderm plate; Vyazniki, Vladimir Province.

Horizon. Vyatskian Gorizont, Upper Tatarian, Upper Permian.

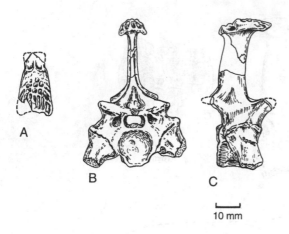

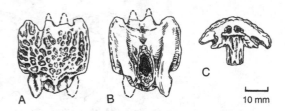

Figure 4.6. *Dromotectum spinosum* Novikov et Shishkin, 2000: A–C, osteoderm plate, holotype, from above (A), below (B) and behind (C). Mechet' ravine, Samara Province; Rybinskian Gorizont, Lower Triassic.

Figure 4.5. *Axitectum vjushkovi* Shishkin et Novikov, 1992: A, trunk osteoderm plate, holotype, from above; Spasskoe, Nizhnii Novgorod Province; B, C, sacral vertebra sutured with osteoderm plate and articulated with ribs, from behind (B) and from side (C); Konaki, Kirov Province; Vokhmian Gorizont, Lower Triassic.

Axitectum Shishkin et Novikov, 1992
Diagnosis. Tabular horn long, sharp, unornamented and directed posterolaterally. Zygapophyseal facets of trunk vertebrae slope medially at an angle 20° to the horizontal plane in anterior view. Osteoderm plates narrow and elongated (corresponding only to median portion of plate in *Bystrowiana*). Plate-locking mechanism simple, with posterior end of plate overlapped by next following segment.

Axitectum vjushkovi Shishkin et Novikov, 1992
See Figure 4.5.
Holotype and locality. PIN 1025/334, osteoderm plate; Spasskoe, Vetluga River, Nizhnii Novgorod Province.
Horizon. Vokhmian Gorizont, Lower Triassic.

Axitectum georgi Novikov et Shishkin, 2000
Holotype and locality. PIN 953/, osteoderm plate; Okunevo, Fedorovka River, Vyatka basin, Kirov Province.
Horizon. Yarenskian Gorizont, Lower Triassic.

Dromotectum Novikov et Shishkin, 1996
Diagnosis. Zygapophyseal facets of trunk vertebrae slope medially at angle not exceeding 30° to the horizontal plane in anterior view. Osteoderm plates well developed, broader than long, and strongly curved transversally. Plate-locking mechanism elaborate (as in *Bystrowiana*).

Dromotectum spinosum Novikov et Shishkin, 1996
See Figure 4.6.
Holotype and locality. PIN 2424/23, osteoderm plate; Mechet' ravine, Tavolzhanka River, Samara basin, Orenburg Province.
Horizon. Rybinskian Gorizont, Lower Triassic.

Synesuchus Novikov et Shishkin, 1996
Diagnosis. Skull depressed. Suborbital division of jugal narrow. Zygapophyseal facets of trunk vertebrae slope medially at angle of 40° to the horizontal plane. Osteoderm plates well developed, broader than long, gently curved transversally. Plate-locking mechanism as in *Bystrowiana* and *Dromotectum,* but devoid of anterior median projections from next-following segment.

Synesuchus muravjevi Novikov et Shishkin, 2000
See Figure 4.7.
Holotype and locality. PIN 4466/12, osteoderm plate; Bolshaya Synya River, Pechora basin, Komi Republic.
Horizon. Nadkrasnokamenskaya Svita, Bukobay Gorizont, Middle Triassic.

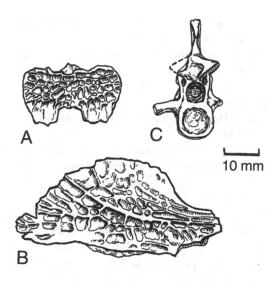

Figure 4.7. *Synesuchus muravjevi* Novikov et Shishkin, 2000. A, osteoderm plate, holotype, from above; B, jugal, PIN 4466/10, from side; C, trunk vertebra, PIN 4466/13, from behind. Bolshaya Synya River, Komi Republic; Nadkrasnokamenskaya Svita, Bukobay Gorizont, Middle Triassic.

Acknowledgements

We thank Alec Panchen and Andrew Milner for extremely helpful comments.

References

Ahlberg, P.E. and Milner, A.R. 1994. The origin and early diversification of tetrapods. *Nature* **368**: 507–514.

Carroll, R.L. and Chorn, J. 1995. Vertebral development in the oldest microsaur and the problem of "lepospondyl" relationships. *Journal of Vertebrate Paleontology* **15**: 37–56.

Clack, J.A. 1983. The stapes of the Coal Measure embolomere *Pholiderpeton scutigerum* Huxley (Amphibia: Anthracosauria) and otic evolution in early tetrapods. *Zoological Journal of the Linnean Society* **79**: 121–148.

Gauthier, J., Cannatella, D., de Querioz, K., Kluge, A.G., Rowe, T. 1989. Tetrapod phylogeny, pp. 337–353 in Fernholm, B., Bremer, K., Jörnvall, H. (eds.), *The Hierarchy of Life*. Amsterdam: Elsevier.

Golubev, V.K. 1997. [Subclass Anthracosauromorpha], pp. 20–22 in Ivakhnenko, M.F., Gubin, Y.M., Kalandaze, N.W., Novikov, I.V., Rautian, A.S. [*Permian and Triassic Tetrapods of Eastern Europe.*] Moscow: Geos.

—1998. [Revision of the Late Permian chroniosuchians (Amphibia, Anthracosauromorpha) of East Europe.] *Paleontologicheskii Zhurnal* **1996** (4): 68–77.

—1999. [A new narrow-armored chroniosuchian (Amphibia, Anthracosauromorpha) from the Late Permian of East Europe.] *Paleontologicheskii Zhurnal* **1999** (2): 43–50.

Gubin, Y.M. 1985. [First anthracosaur from the Permian of the East European Platform.] *Paleontologicheskii Zhurnal* **1985** (3): 118–122.

—1988. [On the land vertebrate fauna from the Copper Sandstones.] *Paleontologicheskii Zhurnal* **1988** (3): 116–119.

Holmes, R. 1984. The Carboniferous amphibian *Proterogyrinus scheelei* Romer, and the early evolution of tetrapods. *Philosophical Transactions of the Royal Society, Series B* **306**: 431–527.

Ivakhnenko, M.F. 1987. [Permian parareptiles of the USSR.] *Trudy Palaeontologicheskogo Instituta AN SSSR* **223**: 1–160.

—1997. [Class Parareptilia]., pp. 14–20 in Ivakhnenko, M.F., Golubev, V.K., Gubin, Y.M., Kalandadze, N.N., Novikov, I.V., Rautian, A.S. [*Permian and Triassic Tetrapods of Eastern Europe*]. Moscow: Geos.

Ivakhnenko, M.F. and Tverdokhlebova, G.I. 1980. [*Systematics, Morphology and Stratigraphic Significance of the Upper Permian Chroniosuchians from the East of the European Part of the USSR.*] Saratov: Izdatel'stvo Saratovskogo Universiteta, 69 pp.

Konzhukova, E.D. 1955. [Permian and Triassic labyrinthodonts from the Volga basin and Cis-Urals.] *Trudy Paleontologicheskogo Instituta AN SSSR* **49**: 5–88.

Lebedev, O.A. and Coates, M.I. 1995. The postcranial skeleton of the Devonian tetrapod *Tulerpeton curtum* Lebedev. *Zoological Journal of the Linnean Society* **114**: 307–348.

Milner, A.R. 1993. Amphibian-grade Tetrapoda, pp. 665–679 in Benton, M.J. (ed.), *The Fossil Record 2*. London: Chapman & Hall.

Novikov, I.V. and Shishkin, M.A. 1995. Palaeozoic relics in the Triassic tetrapod communities: the last anthracosaurian amphibians, pp. 29–32 in Ailing Sun and Yuanqing Wan (eds.), *Sixth Symposium on Mesozoic*

Terrestrial Ecosystems and Biota. Beijing: China Ocean Press.

—and—2000. [Triassic chroniosuchians (Amphibia, Anthracosauromorpha) and the evolution of the trunk dermal ossifications in the bystowianids.] *Paleontologicheskii Zhurnal*, in press.

Olson, E.C. 1965. Relationships of *Seymouria, Diadectes* and Chelonia. *American Zoologist* 5: 295–305.

Panchen, A.L. 1970. Anthracosauria, pp. 1–84 in Kuhn, O. (ed.), *Handbuch der Paläoherpetologie, Part 5A*. Stuttgart: Gustav Fischer.

—1975. A new genus and species of anthracosaur amphibian from the Lower Carboniferous of Scotland and the status of *Pholidogaster pisciformis* Huxley. *Philosophical Transactions of the Royal Society, Series B* 269: 581–640.

—1977. On *Anthracosaurus russeli* Huxley (Amphibia, Labyrinthodontia) and the family Anthracosauridae. *Philosophical Transactions of the Royal Society, Series B* 279: 447–512.

—1980. The origin and relationships of the anthracosaur amphibians from the late Palaeozoic, pp. 319–350 in Panchen, A.L. (ed.), *The Terrestrial Environment and the Origin of Land Vertebrates*. London: Academic.

—1985. On the amphibian *Crassigyrinus scoticus* Watson from the Carboniferous of Scotland. *Philosophical Transactions of the Royal Society, Series B* 309: 505–568.

—and Smithson, T.R. 1988. The relationships of the earliest tetrapods, pp. 1–32 in Benton, M.J. (ed.), *The Phylogeny and Classification of the Tetrapods, Volume 1: Amphibians, Reptiles, Birds, Systematics Association Special Volume*, 35A. Oxford: Clarendon Press.

Ryabinin, A.N. 1962. [A new stegocephalian from the Upper Permian of Malaya North Dvina River.] *Paleontologicheskii Zhurnal* 1962 (1): 140–143.

Shishkin, M.A. 1962. [On the systematic position and distribution of the genus *Jugosuchus*.] *Paleontologicheskii Zhurnal* 1962 (1): 143–145.

—1994. Multiple origin of the non-mammalian tympanic system: evidence or a priori belief? *Journal of Morphology* 220: 393–394.

—and Novikov, I.V. 1992. [Relict anthracosaurs in the Early Mesozoic of Eastern Europe.] *Doklady RAN* 325: 829–832.

Smithson, T.R. 1985. The morphology and relationships of the Carboniferous amphibian *Eoherpeton watsoni* Panchen. *Zoological Journal of the Linnean Society* 85: 317–410.

—1986. A new anthracosaur amphibian from the Carboniferous of Scotland. *Palaeontology* 29: 603–628.

Tatarinov, L.P. 1972. Seymouriamorphen aus der Fauna der USSR, pp. 70–80 in Kuhn, O. (ed.), *Handbuch der Paläoherpetologie, Part 5B*. Stuttgart: Gustav Fischer.

Tverdokhlebova, G.I. 1967. [On remains of genus *Chroniosuchus* from the Permian of the Orenburgian Cis-Urals.] *Izvestiya Vysshykh Uchebnykh Zavedenii. (Geologiya i Razvedka)* 1967 (9): 31–35.

—1968. [On the genera *Chroniosuchus* and *Jugosuchus* from the Upper Tatarian of the USSR.] pp. 11–15 in [*Upper Palaeozoic and Mesozoic Amphibians and Reptiles of the USSR*]. Moscow: Nauka.

—1972. [New batrachosaurian genus from the Upper Permian of the southern Cis-Urals.] *Paleontologischeskii Zhurnal* 1972 (1): 95–103.

V'yushkov, B.P. 1957a. [New kotlassiomorphs from the Tatarian of the European part of the USSR.] *Trudy Paleontologicheskogo Instituta AN SSSR* 68: 89–107.

—1957b. [New unusual animals from the Tatarian of the European part of the USSR.] *Doklady AN SSSR*, 113: 183–186.

Young, C.C. 1979. A new Late Permian fauna from Jiyuan, Honan. *Vertebrata Palasiatica* 17: 99–113.

5

The Russian pareiasaurs

MICHAEL S.Y. LEE

Introduction

Pareiasaurs are a highly distinctive and common component of the terrestrial faunas of the Late Permian of Russia. These anapsid-grade reptiles were large, sluggish herbivores with grotesquely ornamented skulls, stout bodies, massively ossified skeletons and heavy dermal armour. It is not surprising that even scientists have made unflattering comments about their appearance: for instance, Gadow (1909, p. 304) stated that '*Pareiasaurus* was clearly a very clumsy brute of most uncouth appearance'.

Recently, however, biologists have been treating the group rather more seriously. The position of pareiasaurs within the Reptilia has been discussed by Ivakhnenko (1987), Lee (1993, 1995), Laurin and Reisz (1995), and Spencer (1994). In all these works that pareiasaurs, procolophonoids, and turtles form a robust clade (Procolophonomorpha), to the exclusion of all other well-known amniotes (millerettids, mesosaurs, captorhinids, protorothyridids, diapsids, synapsids, and diadectomorphs). Laurin and Reisz suggest a procolophonoid-turtle pairing, while Ivakhnenko, Spencer and Lee all argue that pareiasaurs are the nearest relatives of turtles. The proposed morphological evidence linking turtles with lepidosaurs (de Braga and Rieppel, 1997) was shown to be weak (Wilkinson *et al.*, 1997; Lee, 1997a). However, there has recently been strong molecular evidence linking them with archosaurs (e.g. Zardoya and Meyer, 1998), an arrangement that appears to contradict all the morphological evidence.

A total of 15 species (in five genera) of pareiasaurs have been described from the Late Permian of Russia. Most of these descriptions, however, were published at a time when the biological species concept was not widely accepted, and palaeontologists felt compelled to represent every variation as a new species (Amalitskii, 1922; Hartmann-Weinberg, 1937; Efremov, 1940a), As with the South African pareiasaurs, therefore (see Lee 1997b), re-examination of the material and original descriptions (see below) results in a much lower estimate of species diversity: two (or three) species, in two genera. The definitely valid species of Russian pareiasaurs are *Deltavjatia vjatkensis*, a rather primitive and generalized form, and *Scutosaurus karpinskii*, a more derived form with a highly ornamented skull. Ivakhnenko (1987) has also suggested the existence of only two genera, although he divides *Scutosaurus* into several species. As will be discussed below, all the other named 'pareiasaurs' from Russia are either junior synonyms of these two species, or are not pareiasaurian. The relationships of the Russian taxa to other pareiasaurs will also be discussed.

Repository abbreviations

BMNH, British Museum (Natural History), London; PIN, Palaeontological Institute, Moscow; KM, Kotel'nich Museum, Kotel'nich, Russia; UMZC, University Museum of Zoology, Cambridge.

Systematic revision

Class REPTILIA Laurenti, 1768
Suborder PAREIASAURIA Seeley, 1888
Family PAREIASAURIDAE Cope, 1896
Diagnosis. Largest anapsid-grade reptiles; most 2–3 m

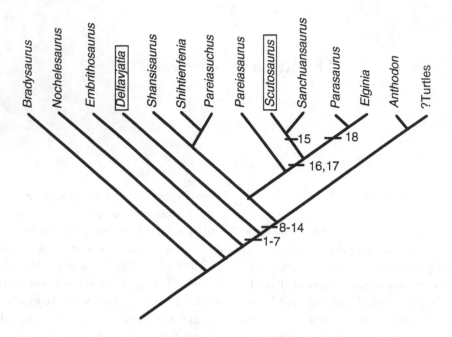

Figure 5.1. A phylogeny of pareiasaurs, based on Lee (1997c): Russian taxa are highlighted. For further details, including discussion of characters diagnosing the indicated clades, see text.

in body length; some late derived forms (*Anthodon, Nanoparia*) about 1 m long. Entire skeleton robust and heavily ossified. Skull heavily ornamented with bosses and ridges; large ventrolateral cheek flanges. Large medial flange of prefrontal forms antorbital buttress which contacts palatine ventrally; maxilla bears high anterodorsal process which almost excludes lacrimal from external naris. In palate, large foramen palatinum posterius; lateral margin of transverse flange of pterygoid with sharp ridge. Braincase with extremely thick floor; short blunt cultriform process; huge rounded basal tubera formed by basisphenoid and basioccipital; massive paroccipital processes. Stapes robust; lacks dorsal process and stapedial foramen. Lower jaw very stout; boss on ventral surface of angular; 'lateral shelf' formed by surangular and articular (Laurin and Reisz, 1995); coronoid process composed of coronoid bone only, fully exposed in lateral view. Teeth labiolingually compressed; 7–17 cusps along anterior and posterior edges. Vertebrae amphicoelous; small intercentra and wide ('cotylosaurian') neural arches.

Body short and stout, and tail relatively short; 19–20 presacrals, 4–6 sacrals, 15–50+ caudals. Scapula long and narrow; prominent acromion process; two coracoids; cleithrum retained only in early South African forms such as *Bradysaurus*. Forelimb short and very stout; manual phalangeal formula (when known), 23332. Pelvis with huge supraacetabular buttress and large anterior expansion of ilium. Hindlimb short and stout; pedal phalangeal formula (when known), 23343. Astragalus and calcaneum fused; metatarsals overlap each other proximally. Dermal ossifications present over dorsal region; gastralia absent. Conical bony studs present over limbs in many forms (e.g. *Scutosaurus, Elginia, Pareiasuchus, Anthodon*).

Deltavjatia Lebedev, in Ivakhnenko, 1987
Type species. Deltavjatia vjatkensis.
Synonym. Anthodon (partim) Owen, 1876.
Diagnosis. As for the type and only species.
Comments. The type species was originally described as a species of *Pareiasuchus*. Ivakhnenko (1987) sug-

Figure 5.2. Articulated skeleton of *Deltavjatia vjatkensis* (UMZC T1321). The pelvis and tail is missing, and much of the hindlimb is damaged.

gested that *Pareiasuchus vjatkensis* was actually unrelated to the other species of *Pareiasuchus*, and was more primitive in retaining closed interpterygoid vacuities. He erected the genus *Deltavjatia* to accommodate it. A recent phylogenetic analysis of the pareiasaurs indeed supports this interpretation (Figure 5.1). Thus, in order to maintain the monophyly of *Pareiasuchus*, Ivakhnenko's arrangement has been followed here.

Deltavjatia vjatkensis (Hartmann-Weinberg, 1937)
See Figures 5.2 and 5.3.
Pareiasuchus vjatkensis Hartmann-Weinberg, 1937
Anthodon rossicus Hartmann-Weinberg, 1937

Anthodon rossicus Hartmann-Weinberg, 1937; Efremov, 1940a

?*Anthodon chlynoviensis* Efremov, 1940a

Pareiasuchus vjatkensis Hartmann-Weinberg, 1937; Efremov, 1940b

Anthodon rossicus Hartmann-Weinberg, 1937; Efremov, 1940b

Pareiasuchus vjatkensis Hartmann-Weinberg, 1937; Olson, 1957

Anthodon rossicus Hartmann-Weinberg, 1937; Olson, 1957

Anthodon rossicus Hartmann-Weinberg, 1937; Kuhn, 1969

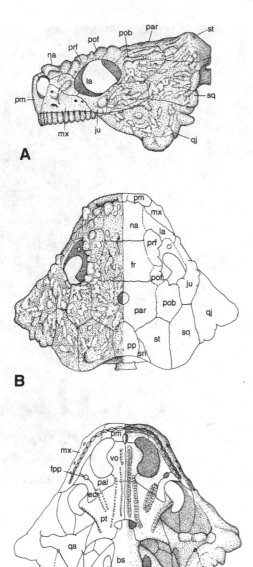

Figure 5.3. Skull of *Deltavjatia vjatkensis* in (A) lateral, (B) dorsal, and (C) ventral views. Reconstruction based on PIN 2212/1–3, and 2212/6. Skull breadth *c.* 0.3 m. Abbreviations: bo, basioccipital; bs, basisphenoid; ec, ectopterygoid; ex, exoccipital; fpp, foramen palatinum posterius; ju, jugal; la, lacrimal; mt, median tubercle; mx, maxilla; na, nasal; op, opisthotic; pal, palatine; par, parietal; pob, postorbital; pof, postfrontal; pp, postparietal; prf, prefrontal; pm, premaxilla; pt, pterygoid; qa, quadrate; qj, quadratojugal; sn, supernumerary; sq, squamosal; st, supratemporal; vo, vomer.

Pareiasuchus vjatkensis Hartmann-Weinberg, 1937; Kuhn, 1969

Pareiasaurus vjatkensis (Hartmann-Weinberg, 1937); Kuhn, 1969

Deltavjatia vjatkensis (Hartmann-Weinberg, 1937); Ivakhnenko, 1987

Scutosaurus rossicus (Hartmann-Weinberg, 1937); Ivakhnenko, 1987

Deltavjatia vjatkensis (Hartmann-Weinberg, 1937); Gao, 1989

Scutosaurus rossicus (Hartmann-Weinberg, 1937); Gao, 1989

Deltavjatia vjatkensis (Hartmann-Weinberg, 1937); Lee, 1993

Holotype and locality. PIN 2212/1, complete skull and mandible, Vyatka River, Kotel'nich, Kirov Province, Russia.

Horizon. Lower sections of the Upper Tatarian (Severodvinskian Gorizont), Upper Permian.

Referred specimens. PIN 2212/2, complete skull missing lower jaw; PIN 2212/6, complete skull and lower jaw; UMZC T1321, complete skull and lower jaw; vertebral column missing posterior sacrals and caudals; ribs; complete shoulder girdle and both forelimbs; partial left hindlimb; osteoderms; KMR (unnumbered), a skull, lower jaw, vertebral column (missing tail), and portions of both girdles, forelimbs and hindlimbs. Another complete skeleton has been excavated but at the time of writing, had yet to be prepared (N.N. Kalandadze, pers. comm., 1993). There are a further four skulls of this animal known. One is in the Palaeontological Institute, Moscow. The labels have been mixed up, but the catalogue number is probably PIN 2212/3. Another is in the private collection of Mr Terry Manning. Two more are currently undergoing preparation in the Department of Zoology, Erindale College, University of Toronto (Modesto, 1994). All referred material is from the type locality.

Diagnosis. Ventrolateral cheek flange terminates in a long, flattened, rectangular boss; occipital condyle projects far behind posterior border of postparietal; boss on postfrontal enlarged into a long conical horn.

Comments. The first two characters are clear autapomorphies of this taxon, and the third may also prove to be. Ivakhnenko (1987) suggested that all the bosses were pointed in this taxon: this does not apply to the most lateral cheek boss. The long conical postfrontal boss is known in the largest known specimen, currently in the private collection of Mr Terry Manning. This feature is present, but only incipiently, in other individuals. None of these three features is present in other pareiasaurs.

D. vjatkensis is a moderate-sized pareiasaur, with a body length of approximately 2 m. Much of this animal is known, but the pelvis, hind limb, and tail, are represented only by poorly preserved, or unprepared, material. *D. vjatkensis* possesses a rugose skull with large cheek flanges bearing distinct bosses – however, the bosses are not as sharp or as prominent as in *Scutosaurus karpinskii*. Other differences from *S. karpinskii* include: the absence of a maxillary horn immediately behind the external naris; the almost complete obliteration of the interpterygoid vacuities by the median union of the pterygoids; numbers of teeth in adults (approximately 14 alveoli per side in the upper jaw contains, approximately 12 in the lower); the presence of 9–11 cusps in all marginal teeth, arranged evenly around the crown; a smooth and featureless lingual surface of each marginal tooth; and, parallel medial rows of teeth on the pterygoids which are spaced far apart.

What is known of the postcranium conforms to the general family description given above. As in *S. karpinskii*, there are 19 presacral vertebrae, and the cleithrum is absent. Unlike *S. karpinskii*, the osteoderms are ornamented only with a central boss, and are restricted to a narrow sagittal band immediately above the vertebrae. There are no conical bony studs over the appendages. A new specimen of *D. vjatkensis* (N.N. Kalandadze, pers. comm., 1993) appears to have a tail as long as its trunk region – much longer than in *S. karpinskii*.

Nomenclature

The nomenclature of *D. vjatkensis* is particularly tortuous. The different names applied to this taxon, and evidence for synonymy, are outlined here.

Pareiasuchus vjatkensis. Hartmann-Weinberg (1937) made PIN 2212/1 the type of *Pareiasuchus vjatkensis*, supposedly related to *Pareiasuchus peringueyi* on the basis of the following features: 'low brain cavity', low supraoccipital, enlarged middle ear region, prominent pineal foramen, presence of a tubercle on the quadrate ramus of the pterygoid, absence of swelling on the distal end of the paroccipital process, and presence of a deep otic notch. However, the first five characters are found in all pareiasaurs, the sixth is primitive for pareiasaurs (the swelling is only found in *Bradysaurus baini*), and the last is not present in *Deltavjatia vjatkensis* or any other pareiasaur (e.g. Boonstra, 1934a). Thus, as also suggested by Ivakhnenko (1987), there was no evidence justifying the generic identification.

Pareiasuchus vjatkensis was also suggested to have some unique characters, but not the ones proposed in the above diagnosis. The lacrimal foramen was said to be current, and the transverse flange of the pterygoid was said to abut extensively against the lower jaw, forming a guide for propalinal movements. However, the lacrimal foramen is actually present, as in all other pareiasaurs. The extensive contact between the transverse flange of the pterygoid and the lower jaw is a taphonomic artefact: the lower jaw has simply been pushed dorsally into the palate. Many other pareiasaur skulls have been preserved in a similar manner (e.g. *Scutosaurus karpinskii* PIN 2005/1533). However, although the original characters proposed as diagnostic of *Deltavjatia vjatkensis* are doubtful, this taxon is a valid new species (see *Diagnosis* above).

Anthodon rossicus. Hartmann-Weinberg (1937) also made PIN 2212/2 the type of *Anthodon rossicus*. However, the characters uniting PIN 2212/2 with *Anthodon serrarius* are invalid, and the animal is in fact identical to *Deltavjatia vjatkensis* in all important respects. Hartmann-Weinberg suggested that PIN 2212/2 showed affinities with *A. serrarius* on the basis of the morphology of the paroccipital process, the position of the quadrate condyle, the orientation of the quadrate ramus of the pterygoid, the shape of the interpterygoid vacuity (misinterpreted as the choana), and the position of the palatal tooth rows. However, the nature of these similarities was not specified exactly. In fact, in the first three regions, *A. serrarius* and *D. vjatkensis* are identical: these characters therefore cannot be used to assign PIN 2212/2 to either taxon. *A. serrarius* does differ from *D. vjatkensis* in many other areas: for instance, there are 15 rather than 9 cusps on each maxillary tooth, the interpterygoid vacuities are more extensive, and the double median rows of palatal teeth are closer together. However, in all these traits, PIN 2212/2 exhibits the condition characteristic of *D. vjatkensis*. PIN 2212/2 was also united with *A. serrarius* because it only possessed about 11 teeth on each half of the upper jaw, asserted to be fewer than in *D. vjatkensis*, but the same number as in *A. serrarius*. However, adults of *A. serrarius* and *D. vjatkensis* both have about 14 teeth, while juveniles of both species have about 11 teeth. The size of PIN 2212/2 indicates that it is juvenile.

Hartmann-Weinberg (1937) suggested that PIN 2212/2 differed from *D. vjatkensis* in possessing basal tubera that are located more anteriorly, a slightly larger interpterygoid vacuity (mistakenly interpreted as a choana), and in lacking the pterygoid guide for propalinal movements of the mandible. However, the basal tubera are located in exactly the same position in both forms, the size of the interpterygoid vacuity in PIN 2212/2 is well within the range of variability exhibited by *D. vjatkensis*, and the 'pterygoid guide' supposedly present in *D. vjatkensis* is a preservational artefact (see above).

PIN 2212/2 was also described as differing from *Anthodon serrarius*, and, by implication, *Deltavjatia vjatkensis*, in having lost both the supratemporal bone and the bosses on the quadratojugal, in the sculpture of the skull roofing bones, and in the shape of the cheek flange. However, the supratemporal and the bosses on the quadratojugal are definitely present. PIN 2212/2 does differ from *A. serrarius* in having dermal sculpture consisting of coarse bosses and ridges, rather than small grooves, but in this feature it resembles *D. vjatkensis*. It does have rather small cheek flanges, but so do juveniles of all pareiasaurs (Lee, 1997b).

PIN 2212/2 is identical to *Deltavjatia vjatkensis* in all respects; in particular, it exhibits all three autapomorphies of the latter taxon, although, being a juvenile, the flat cheek boss and postorbital boss are weakly developed. Finally, PIN 2212/2 was found at the same site as the other specimens of *D. vjatkensis*, the only pareiasaur taxon known from this location. Thus, PIN 2212/2 is here referred to *Deltavjatia vjatkensis*, and *Anthodon rossicus* is made a junior synonym. The probability that all the pareiasaurs from Kotel'nich are the same species has been suggested before (Kalandadze *et al.*, 1968).

Kuhn's (1969) synonymies. Kuhn (1969) attempted to make *Deltavjatia vjatkensis* a junior synonym of two different taxa. He stated correctly that *Pareiasuchus vjatkensis* was identical with *Anthodon rossicus*, but did not give reasons, and made the former a junior synonym of the latter. This decision was technically illegal since *Pareiasuchus vjatkensis* has page priority over *Anthodon rossicus*. In the same work, Kuhn stated that '*Pareiasaurus*' *vjatkensis* was also identical to *Proelginia permicus*, and thus made the former a junior synonym of the latter as well. This synonymy is also invalid: '*Pareiasaurus*' *vjatkensis* is merely an illegitimate name change of *Pareiasuchus vjatkensis*, which in turn is a junior synonym of *Deltavjatia vjatkensis*, while *Proelginia permicus* is a junior synonym of *Scutosaurus karpinskii* (see below). These two Russian taxa are clearly distinct.

Anthodon chlynoviensis. *Anthodon chlynoviensis* is probably another junior synonym of *Deltavjatia vjatkensis*, as suggested by Ivakhnenko (1987). Efremov (1940a) founded this taxon on pareiasaur material from Kotel'nich, but did not nominate a type or give catalogue numbers of the relevant material. He then appended a paragraph mentioning that the Kotel'nich pareiasaurs had been named by Hartmann-Weinberg (1937) just before his paper went to press. As a result, Efremov made *Anthodon chlynoviensis* a junior synonym of one of Hartmann-Weinberg's taxa, *Anthodon rossicus* (=*Deltavjatia vjatkensis*). Although Efremov's material of *Anthodon chlynoviensis* can no longer be traced, it is probable that this taxon is also a junior synonym of *Deltavjatia vjatkensis*.

Scutosaurus Hartmann-Weinberg, 1930
Type species. Scutosaurus karpinskii.
Synonyms. Pariasaurus (partim) Watson, 1917,

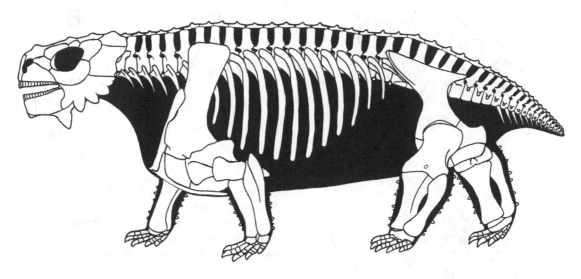

Figure 5.4. Skeletal restoration of *Scutosaurus karpinskii*, based on PIN 2005/1532. Total body length *c.* 3 m.

Pareiosaurus (partim) Amalitskii, 1922, *Proelginia* Hartmann-Weinberg, 1937.
Diagnosis. as for the type and only species.

Scutosaurus karpinskii (Amalitskii, 1922)
See Figures 5.4–5.6.

Pariasaurus karpinskyi Watson, 1917 (see *Comments*)
Pareiosaurus karpinskii Amalitskii, 1922
Pareiosaurus tuberculatus Amalitskii, 1922
Pareiosaurus elegans Amalitskii, 1922
Pareiosaurus horridus Amalitskii, 1922
Pareiasaurus karpinskii Amalitskii, 1922; Sushkin, 1927
Pareiasaurus karpinskii Amalitskii, 1922; Haughton, 1929
Pareiasaurus karpinsky Amalitskii, 1922; Hartmann-Weinberg, 1929, 1930
Pareiasaurus karpinskii Amalitskii, 1922; Haughton and Boonstra, 1930
Scutosaurus karpinskii (Amalitskii, 1922); Boonstra, 1932, 1934a, b
Scutosaurus karpinsky (Amalitskii, 1922); Hartmann-Weinberg, 1933
Pareiasaurus sp.; Hartmann-Weinberg, 1929, 1930
Scutosaurus karpinskii (Amalitskii, 1922); Hartmann-Weinberg, 1937
Proelginia permiana Hartmann-Weinberg, 1937

Scutosaurus karpinskii (Amalitskii, 1922); Efremov, 1940a, b, c
Scutosaurus permiana (Hartmann-Weinberg, 1937); Efremov, 1940a, b
Scutosaurus karpinskii (Amalitskii, 1922); Huene, 1944
Scutosaurus karpinskyi (Amalitskii, 1922); Gregory, 1946
Scutosaurus karpinskii (Amalitskii, 1922); Olson, 1957
Scutosaurus permianus (Hartmann-Weinberg, 1937); Olson, 1957
Scutosaurus karpinskii (Amalitskii, 1922); Bystrov, 1957
Scutosaurus karpinskii (Amalitskii, 1922); Kuhn, 1969
Scutosaurus karpinskyi (Amalitskii, 1922); Kuhn, 1969
Scutosaurus permicus (Hartmann-Weinberg, 1937); Kuhn, 1969
Scutosaurus karpinskii (Amalitskii, 1922); Ivakhnenko, 1987
Scutosaurus tuberculatus (Amalitskii, 1922); Ivakhnenko, 1987
Scutosaurus permianus (Hartmann-Weinberg, 1937); Ivakhnenko, 1987
?*Scutosaurus itilensis* Ivakhnenko, 1987
Scutosaurus karpinskii (Amalitskii, 1922); Gao, 1989
Scutosaurus karpinskii (Amalitskii, 1922); Lee, 1993, 1997b
Holotype and locality. PIN 2005/1532, skull, lower jaw

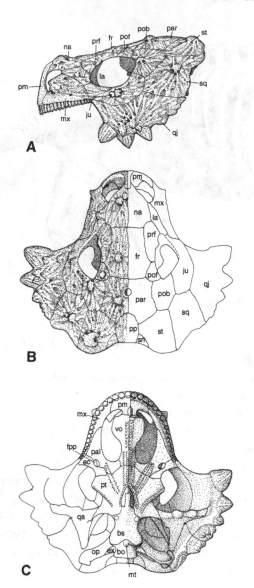

Figure 5.5. Skull of *Scutosaurus karpinskii* in (A) lateral, (B) dorsal, and (C) ventral views. Reconstruction based on PIN 2005/2471, 2005/1883, 156/1–3. Skull breadth *c.* 0.5 m. Abbreviations as for Figure 5.3.

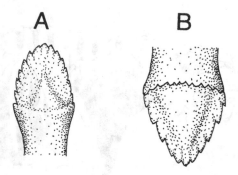

Figure 5.6. A, Lingual view of middle dentary tooth of *Scutosaurus karpinskii* showing the triangular lingual ridge. B, Lingual view of a maxillary tooth of *Scutosaurus* showing the cusped cingulum. Both after unnumbered isolated teeth from the Amalitskii collection (collection number: PIN 2005).

and complete postcranium missing only some manual and pedal elements; North Dvina, Arkhangel'sk Province, northern Russia.

Horizon. Upper section of the Upper Tatarian (lower Vyatskian Gorizont), Upper Permian.

Referred specimens. PIN 2005/1533, 2005/1535, 2005/1536, 2005/1537, 2005/1538: six skeletons including skull, lower jaw, complete vertebral column, ribs, most of shoulder and pelvic girdles and limbs. Osteoderms are only associated with the first specimen. The cast of PIN 2005/1535 in the BMNH has been extensively 'remodelled' and includes a string of caudals belonging to PIN 2005/1534. PIN 2005/1578: juvenile specimen, including skull and lower jaw, vertebral column (missing some caudals), parts of shoulder and pelvic girdles and limbs. PIN 2005/1534: skull and lower jaw, 17 presacrals, 5 sacrals and 14 caudals, pelvis and parts of both hindlimbs. PIN 2005/2471: partial skull including prepared braincase. PIN 11/215: skull and lower jaw. PIN 3919: braincase, and fragments of skull and lower jaw; osteoderms and numerous unidentifiable fragments. This number also includes a left dentary, left premaxilla, and palatal fragments from a second pareiasaur. PIN 2895: skull and lower jaw fragments. PIN 156/1, 156/2, 156/3: skulls of three juveniles, all missing lower jaws. PIN 156/305, 156/306: lower jaws of two juveniles. All from North Dvina and Il'inskoe, Upper Tatarian (lower Vyatskian Gorizont), Russia.

Scutosaurus karpinskii is the only pareiasaur known from North Dvina, and it is very common (see Figure 5.7). All the other pareiasaur material is identical to *S.*

Figure 5.7. Mounted skeleton of *Scutosaurus karpinskii*, from the collections of PIN. (Photograph courtesy of Pat Rich.)

karpinskii and can be referred tentatively to this species, despite often being too fragmentary to be diagnostic. Important specimens falling into this category are: PIN 2005/1885: weathered braincase. PIN 2005/2473: atlas–axis complex, sectioned sagitally. PIN 2005/1877: partial pes. PIN 2005/2489: phalangeal elements with associated sesamoids. PIN 2005/123: astragalo-calcaneum, sectioned. PIN 2005/2485: astragalocalcaneum. PIN 2005/2070: right fibula. PIN 2005/1543: interclavicle. PIN 2005/455: partial forelimb. PIN 11/28: intermedium and ulnare. PIN 2005/698: well-prepared osteoderm and rib.

Diagnosis. Large pareiasaur, 2.5 m in body length, with extremely rugose dermal sculpture, large cheek flanges with long pointed bosses, and osteoderms covering the entire body and limbs. Autapomorphies: small median tubercle on basioccipital between basal tubera; tips of teeth on upper jaw point slightly outwards; radiating ridges covering skull very coarse.

Comments. In all other pareiasaurs, the median basioccipital tubercle is absent, the teeth point either vertically or inwards, and the radiating ridges over the skull are less pronounced. *S. karpinskii* can also be readily identified because the bosses covering the skull, and on the cheek margins, are not drawn into long conical spines as in *Elginia*, but are more prominent than in all other pareiasaurs.

Scutosaurus karpinskii is represented by abundant,

though usually poorly prepared, material, and is the most completely known pareiasaur. It differs from *Deltavjatia vjatkensis* in several areas: the bosses on the skull, and in particular the edge of the cheeks, are much more prominent; there is a horn on the maxilla, immediately behind the external naris, and the interpterygoid vacuity is large; in adults, there are approximately 18 alveoli on each side of the upper jaw, and about 16 alveoli on each side of the lower jaw; the teeth are heterodont, each tooth on the upper jaw having 9–11 cusps, and each tooth on the mandible having 13–17 cusps; the lingual surfaces of the mandibular teeth possess a prominent triangular ridge; some marginal teeth, apparently randomly distributed throughout the upper and lower jaws, have a cusped cingulum on the lingual surface; and, the medial rows of palatal teeth are parallel and close together.

The postcranial anatomy also largely corresponds with the above family description. As in *Deltavjatia vjatkensis*, there are 19 presacral vertebrae. However, unlike *D. vjatkensis*, the osteoderms cover the entire dorsal region. Also, although most osteoderms are separate, a few are tightly sutured together. Presumably, the dermal armour was arranged as in the related *Elginia*, with isolated osteoderms over most of the dorsal region but suturally-united osteoderms over the shoulder and pelvic regions. Each osteoderm has a distinct central boss capped by a spine, surrounded by rugose, irregular radiating ridges. Small conical studs occur over the limbs.

Nomenclature

Watson (1917) and Amalitskii (1922). The North Dvina pareiasaurs were excavated around the turn of the century by Amalitskii (Buffetaut, 1987; Chapter 1 herein). However, he died before completing his work (see Woodward, 1918), and it was not until 1922 that his descriptions were published, posthumously. This delay caused nomenclatural problems, as Watson (1917) had in the meantime published a drawing of a cast of a scapulocoracoid of one of these pareiasaurs, labelling it '*Pariasaurus* (sic) *karpinskyi* Amalitskiy'. Watson's label should have priority over Amalitskii's description, but there are grounds for suppression because: (1) Watson clearly did not coin the name, as he credits Amalitskii as the source, (2) Watson's label consists only of a name, with no diagnosis or mention of types, and was clearly not meant to be a description of a new species, (3) if Watson's contribution is to be given formal recognition, the type of '*Pariasaurus karpinskyi*' would have to be Watson's material – a very inaccurate, extensively remodelled cast of part of a skeleton (PIN 2005/1535), (4) Watson misspelt the generic name, and his spelling of the specific name is also different from Amalitskii's eventual name, which is the spelling adopted by almost all subsequent works, and (5) all subsequent workers have attributed the taxon name to Amalitskii (1922).

Thus, I consider Amalitskii (1922) to be the first valid description of the Russian pareiasaurs. In this brief paper, he described four species, *Pareiosaurus* (sic) *karpinskii* (based on PIN 2005/1532, a different individual from the one represented by Watson's cast), *Pareiosaurus tuberculatus* (PIN 2005/1533), *Pareiosaurus horridus* (PIN 2005/1536), and *Pareiosaurus elegans* (PIN 2005/1536). Amalitskii was a geologist rather than a zoologist, and the four species he erected all appear to be conspecific. *Pareiosaurus elegans* was asserted to have a 'comparatively small head' (Amalitskii, 1922, p. 335), but the cranium of this animal is the same size as in all the other specimens of North Dvina pareiasaurs. *Pareiosaurus horridus* was said to be characterized by 'strong horn-like projections on the cheeks and on the lower jaw' (p. 335), but these are present in all the other specimens. *Pareiosaurus tuberculatus* was said to have boss-like osteoderms over the tail region: however, osteoderms are no longer associated with the tail of this or any other specimen, and this character is not useful. Hartmann-Weinberg (1929, 1930, 1933, 1937) later interpreted all four taxa as a single species. I agree with this assessment: I can detect no clear-cut differences in any of these features in the types or in any of the dozens of specimens of North Dvina pareiasaurs in the Palaeontological Institute, Moscow, that have since been prepared. The differences in skull shape and arrangement of bosses are very slight, and appear to have a continuous rather than discrete distribution. Similar variability occurs in other species of pareiasaurs (e.g. *Bradysaurus* speci-

mens in the South African Museum), and these differences can be ascribed to ontogenetic, tapho-nomic and individual variation.

Pareiosaurus and Pareiasaurus. Hartmann-Weinberg (1930) reduced all four species of Amalitskii to one, *Pareiosaurus* (sic) *karpinskii*, and corrected Amalitskii's spelling of the generic name, making *Pareiosaurus* a junior synonym of *Pareiasaurus*. However, she in turn consistently misspelt the specific name as *karpinsky*. Hartmann-Weinberg (1937) later realized this mistake, recognizing *karpinsky* as a junior synonym of *karpinskii*. Amalitskii's earlier error, though, is rather more troublesome. He clearly meant to refer the Russian forms to *Pareiasaurus*, but his misspelling meant that he inadvertently assigned his four species to a separate genus, *Pareiosaurus*. Nomenclatural difficulties arose when Hartmann-Weinberg suggested that *Pareiosaurus karpinskii* is sufficiently distinct from *Pareiasaurus serridens* to warrant generic separation. I would support this separation on cladistic grounds (Figure 5.1): the Russian forms appear only distantly related to *P. serridens*. Amalitskii's typographical error, therefore, had created a valid new generic name, but Hartmann-Weinberg (1930) instead synonymized *Pareiosaurus* with *Pareiasaurus* and then, in the same paper, erected another new genus for the Russian forms, *Scutosaurus*. While *Pareiosaurus* undoubtedly has priority, *Scutosaurus* should be retained. Amalitskii's name was obviously a misspelling (he also refers to Owen's species as '*Pareiosaurus*' *serridens*), and he did not intend to erect a new genus. Furthermore, 'pareiasaur' would have to be one of the most consistently misspelt names in biology: recognizing *Pareiasaurus* and *Pareiosaurus* as distinct but closely related genera would only create further chaos.

Scutosaurus itilensis. Ivakhnenko (1987) erected *Scutosaurus itilensis* based on PIN 3919. It was supposed to differ from *Scutosaurus karpinskii* in having more rounded cheek bosses, a larger middle ear cavity, and proportional differences in the skull (Ivakhnenko, 1987, pers. comm., 1993). However, the rounded cheek bosses in PIN 3919 are an artefact of weathering – one boss, on the left quadratojugal, is not damaged and is pointed and conical, as in *S. karpinskii*. The middle ear cavity is similar in size in PIN 3919 and *S. karpinskii*.

The proportional differences in the cranium cannot be confirmed because the material of PIN 3919 is so fragmentary that a confident reconstruction of the whole skull cannot be attempted: all the elements of PIN 3919 that are preserved, however, are identical to the corresponding elements in *S. karpinskii*. In particular, it possesses a cusped cingulum on some (but not all) marginal teeth, and a median tubercle between the basal tubera; these traits occur in no other pareiasaur. It also has the long conical cheek bosses (*contra* Ivakhnenko, 1987) and a horn on the maxilla. On this basis, *S. itilensis* is a junior synonym of *S. karpinskii*. Ivakhnenko (1987) also reinstated *Scutosaurus tuberculatus*, stating that the type (PIN 2005/1533) differed from all the other North Dvina pareiasaurs, which were referable to *S. karpinskii*. He claimed that, in *S. tuberculatus*, the nasal bosses were more prominent, dermal armour was present, and the neural spines were shorter. However, the size of the nasal bosses varies greatly within *S. karpinskii*, and PIN 2005/1533 does not appear unusual in this feature. Dermal armour was also present in most specimens of *S. karpinskii*, but most of it has been removed during preparation. For example, Amalitskii (1922) illustrates a block containing osteoderms associated with the type of *S. karpinskii* (PIN 2005/1532). This block can no longer be located.

Proelginia permiana. Hartmann-Weinberg (1937) erected *Proelginia permiana* on the basis of an isolated skull, PIN 156/2. Ivakhnenko (1987) lists PIN 156/1 as the type, but photographs in the original description clearly indicate that the type was PIN 156/2. It was supposed to differ from *Scutosaurus karpinskii* in the following features: (1) the pineal foramen is absent; (2) the 'otic notch' is weakly developed; (3) the interpterygoid vacuity (misinterpreted as a choana) is U-shaped rather than V-shaped; (4) the quadrate ramus of the pterygoid is directed laterally rather than posterolaterally, and the quadrate condyle is thus located more anteriorly; (5) the postorbital portion of the skull roof is elevated; (6) the dermal sculpturing consists of a honeycomb pattern rather than a system of bosses and radiating ridges; (7) the snout is shorter; (8) the supratemporal bosses (termed 'tubera tabularia') are larger but the other bosses on

the skull roof are weaker; and (9) the cheek flanges are smaller.

None of these alleged diagnostic characters of *Proelginia permiana* appears valid: (1) the parietal region is very badly mangled in PIN 156/2 and largely reconstructed in plaster – the absence or presence of a pineal foramen cannot be determined; (2) the otic notch is also weakly developed in all other pareiasaurs, including *Scutosaurus* (e.g. Boonstra, 1934a); (3) the V-shaped interpterygoid vacuity in some specimens of *Scutosaurus* is an artefact of over-preparation; (4) the quadrate ramus of the pterygoid has the same orientation (laterally and very slightly posteriorly) in both taxa; (5) the elevation of the postorbital portion of the skull roof is due to taphonomic distortion; (6) in both taxa, the dermal sculpturing consists of bosses and ridges, with occasional pits; (7) the snout is the same length in both taxa; (8) the supratemporal bosses are not especially large in PIN 156/2: in fact, all bosses in this specimen are slightly less developed than in the adult *Scutosaurus*; and (9) the cheek flanges are indeed smaller in *Proelginia permiana*. Thus, only the final two differences proposed by Hartmann-Weinberg actually exist: the poorly developed bosses and the cheek flanges. In addition, the horn on the maxilla is absent. For these reasons, Kalandadze *et al.* (1968) and Ivakhnenko (1987) regarded the animal as being related to *Scutosaurus karpinksii*, but as a distinct species, *S. permianus*. However, the linear dimensions of PIN 156/2 are about half those of typical skulls of adults of *Scutosaurus karpinskii*, and these differences might be ontogenetic. PIN 156/2 is otherwise identical to *S. karpinskii*; in particular, it possesses the diagnostic median tubercle between the basal tubera. *Proelginia permiana* might therefore be a junior synonym of *Scutosaurus karpinskii*. Kuhn (1969) suggested a rather different rearrangement: that *Pareiasaurus vjatkensis* (= *Deltavjatia vjatkensis*) is a junior synonym of *Proelginia permicus*. However, the two taxa are different, as the diagnoses of *Deltavjatia vjatkensis* and *Scutosaurus karpinskii* make clear. Furthermore, even if they were synonymous, *Pareiasaurus vjatkensis* has page priority over *Proelginia permiana*.

Shishkin (pers. comm., 1996) has, however, noted that the type of *Proelginia permiana* comes from a slightly older horizon than *Scutosaurus karpinskii,* and that all specimens from this horizon are small. This would suggest that the size differences are not ontogenetic, but taxonomic. However, as only three skulls are known from this horizon, all of slightly different sizes, the evidence for a real size difference is not overwhelming. For these reasons, a formal synonymy would at present be premature, and I will tentatively continue to recognize *P. permiana* as a distinct species.

Relationships of the Russian pareiasaurs to pareiasaurs from elsewhere

The phylogenetic relationships of the Russian pareiasaurs with other pareiasaurs are shown in Figure 5.1. This cladogram is based on a cladistic analysis of 129 informative osteological characters in the well-corroborated clade Procolophonomorpha (nycteroleterids, nyctiphruretids, procolophonids, *Owenetta*, *Barasaurus, Sclerosaurus,* lanthanosuchids, pareiasaurs, and turtles). Characters were polarized by outgroup comparison with millerettids and eureptiles (the two nearest outgroups: Laurin and Reisz, 1995). The following discussion will focus only on the immediate relationships of the two Russian pareiasaurs. Characters diagnosing other groupings are discussed in Lee (1997b).

Deltavjatia is an important animal because it appears to be transitional between the primitive early pareiasaurs and the derived younger forms. The earliest pareiasaurs are the three *Tapinocephalus* Zone forms from South Africa: *Bradysaurus, Nochelesaurus,* and *Embrithosaurus.* These are all large, heavily ossified, lightly armoured forms. All other pareiasaurs, including *Deltavjatia,* are from stratigraphically younger deposits in South Africa, Russia, China, Europe, and South America (Araújo, 1984).

The following characters are informative regarding the relationships of *Deltavjatia* with other pareiasaurs (Figure 5.1). The primitive state is listed as (0), the derived as (1).

1. Numbers of cusps on the teeth of the lower jaw: 11 or fewer (0), more than 12 (1).
2. Spacing of the medial rows of palatal teeth: far apart (0) (Figure 5.3B), closely spaced (1) (Figure 5.5B).
3. Interpterygoid vacuity: closed (0) (Figure 5.3B), extensive and slit-like (1) (Figure 5.5B).
4. Numbers of caudal vertebrae: 40 or more (0), fewer than 20 (1) (Figure 5.4).
5. Dermal armour: restricted to a narrow zone above the dorsal midline, or absent (0), covers the entire dorsum (1).
6. Osteoderms: when present at all, are isolated (0), are suturally united to each other above the shoulder and pelvic regions (1).
7. Conical bony studs over the appendages: absent (0), present (1).
8. Ventral view of the basisphenoid: greatly constricted (0), wide between the basicranial articulation and the occipital condyle (1) (Figures 5.3B and 5.5B).
9. Location of the basal tubera: closer to the occipital condyle than to the basicranial articulation (0), mid-way between the basicranial articulation and the occipital condyle (1) (Figures 5.3B and 5.5B).
10. Numbers of cusps on the marginal teeth: seven or fewer (0), nine or more (1) (Figure 6).
11. Spacing of the cusps around the crown: irregular, being close together near the apex and widely separated near the base (0), evenly spaced (1) (Figure 5.6).
12. Numbers of presacral vertebrae: 20 or more (0), 19 (1) (Figures 5.2 and 5.4).
13. Cleithrum: present (0), absent (1) (Figure 5.4). However, the cleithrum appears to have been lost convergently within procolophonids.
14. Large supernumerary element on the skull table: absent (0), present (1) (Figures 5.3A and 5.5A). This is not seen in turtles.

The three *Tapinocephalus* Zone pareiasaurs (*Bradysaurus, Nochelesaurus, Embrithosaurus*) and other procolophonomorphs (*sensu* Lee, 1995) show the primitive state for all these characters. *Deltavjatkia* also shows the primitive state for characters 1–7, but it shares the derived state of characters 8–14 with all later pareiasaurs. *Deltavjatia* bridges an otherwise rather large gap (Figure 5.1).

Scutosaurus, on the other hand, appears to be most closely related to *Sanchuansaurus* (Gao, 1989), known only from an isolated maxilla with teeth. These teeth, however, possess the cusped cingulum, otherwise found (among basal amniotes) only in *Scutosaurus* (Figure 5.1, character 15). *Sanchuansaurus* is different from *Scutosaurus* in that (1) all, rather than some, of the teeth possess the cusped cingulum, and (2) the two exits of the infraorbital canal are much further apart (see illustrations in Gao, 1989). The nearest relative of *Scutosaurus* and *Sanchuansaurus* appears to be *Elginia*. The derived characters supporting this arrangement (Figure 5.1, characters 16 and 17) are the small conical spine covering each osteoderm, and the horn-like cheek bosses. These characters are not found in any other pareiasaurs, although the latter feature has evolved convergently within (highly derived) procolophonids such as *Hypsognathus* (Colbert, 1946; Spencer, 1994). Although neither of these features is known in the German *Parasaurus*, this taxon can also be placed, tentatively, within this clade on the basis of one unique derived character which it shares with *Elginia*: the enlarged supernumerary element (Figure 5.1, character 18).

Material from the Russian Permian incorrectly assigned to Pareiasauria

Since most of the Russian taxa described as belonging to the Pareiasauria are not, in fact, pareiasaurs, a review of Russian pareiasaurs would be incomplete without a brief discussion of these taxa.

Parabradysaurus udmurticus

All that is known of this animal is a fragment of the lower jaw from the Mid Permian (Zone II) of Russia. Efremov (1954), who described the taxon, included it in the Pareiasauridae because it possessed multicuspid teeth. However, this resemblance is superficial, and later workers (e.g. Olson, 1962; Chudinov, 1983) placed it in the Dinocephalia, based on details of the tooth crowns and the shape of the dentary (King, 1988). *Parabradysaurus* was considered valid by Ivakhnenko (1995), and assigned by him to the eotheriodont family Rhopalodontidae.

Rhipaeosaurus tricuspidens

This taxon is based on an articulated specimen from the Mid Permian (Zone II) of Russia described briefly by Efremov (1940d) and interpreted as related to either *Nycteroleter* or *Nyctiphruretus*. Subsequently (Chudinov, 1955, 1957; Olson, 1957; Kuhn, 1969; Ivakhnenko, 1987) it was identified as a pareiasaurian.

R. tricuspidens was considered a pareiasaur relative by Chudinov (1955, 1957) largely on the basis of its weakly tricuspid teeth, which were interpreted as precursors of the crenulated multicuspid teeth of pareiasaurids. However, such teeth have evolved repeatedly in other lineages, including, significantly, nycteroleterids (Ivakhnenko, 1987), and there appears to be no other trait in *R. tricuspidens* which suggests pareiasaur affinities. Although Ivakhnenko (1987) suggests that nycteroleterids are closely related to pareiasaurs, a more recent analysis suggests that the two groups are only remotely related (Lee, 1995). The combination of traits found in *R. tricuspidens* – flat, triangular skull, tricuspid teeth, toothed transverse flange of the pterygoid, humerus with ectepicondylar foramen, tarsus with astragalus and calcaneum – is found elsewhere only in nycteroleterids. *R. tricuspidens* is slightly larger than most nycteroleterids, but there are no other differences. Furthermore, *R. tricuspidens* comes from the Zone II deposits of Russia, in which nycteroleterids are abundant. A comparison with all the other known groups of basal amniotes reveals that *R. tricuspidens* cannot be assigned to any of them, and it is almost certainly a large nycteroleterid.

Rhipaeosaurus talonophorus

This taxon was founded by Chudinov (1955) on a partial skull table and some postcranial remains from the Mid Permian (Zone II) of Russia, and interpreted as related to *Rhipaeosaurus tricuspidens* and pareiasaurs (Chudinov, 1955, 1957; Olson, 1957; Kuhn, 1969). However, subsequent discovery of more complete material has demonstrated that this taxon is a seymouriamorph (Ivakhnenko, 1987). Contrary to previous reconstructions (Chudinov, 1957; Kuhn, 1969;

Wild, 1985), the large postparietals on the skull table are not fused (Ivakhnenko, 1987), so the only character proposed to unite this species with pareiasaurs can be disregarded.

Leptoropha novojilovi

This taxon was founded by Chudinov (1955) on a partial skull roof from the Mid Permian (Zone II) of Russia. A mandible and some postcranial remains have been interpreted as coming from the same individual as the type, but this association is uncertain (Parrington, 1962). Subsequent discoveries have demonstrated that this taxon is a junior synonym of *R. talonophorus* (see above), and both have been transferred to the Seymouriamorpha (Ivakhnenko, 1987).

Acknowledgements

I thank Oleg Lebedev, Nick Kalandadze, Michael Ivakhnenko, Igor Novikov, and Andrei Sennikov of the Palaeontological Institute, Moscow, for doing their best under very trying conditions to make my three months in Russia both productive and enjoyable. They almost succeeded! I also thank the editors for inviting me to contribute to this volume, and Mikhail Shishkin and Desmond Maxwell for helpful reviews. I am grateful to Patrick Spencer, Robert Reisz, and my Ph.D. supervisor, Jenny Clack for general discussions about things pareiasaurian, Patrick Spencer and Des Maxwell for sharing unpublished material from their Ph.D. theses, and Ken Joysey for arranging the acquisition of the specimen shown in Figure 5.2. Financial support came from a Commonwealth Scholarship awarded by the Association of Commonwealth Universities and the British Council, the Cambridge Philosophical Society, and Queens' College (Cambridge), and in the final stages, an Australian Research Council Postdoctoral Research Fellowship. I am also grateful to everyone in Rick Shine's lab. (School of Biological Sciences, University of Sydney) for facilities and support.

References

Amalitskii, V.P. 1922. [Diagnoses of new forms of vertebrates and plants from the Upper Permian of North Dvina.] *Izvestiya AN SSSR, VI Seriya* 16: 329–340 [author name spelled Amalitzky].

Araújo, C.C. 1984. Sistematica e taxonomia dos pareiasaurios: historico perspectivas atuais. *Pesquisas, Porto Alegre* 16: 227–249.

Boonstra, L.D. 1932. The phylogenesis of the Pareiasauridae: a study in evolution. *South African Journal of Science* 29: 480–486.

—1934a. Pareiasaurian studies. Part IX. The cranial osteology. *Annals of the South African Museum* 31: 1–38.

—1934b. Pareiasaurian studies. Part X. The dermal armour. *Annals of the South African Museum* 31: 39–48.

Buffetaut, E. 1987. *A Short History of Vertebrate Palaeontology.* London: Croom Helm, 223 pp.

Bystrov, A.P. 1957. [The pareiasaur skull.] *Trudy Paleontologicheskogo Instituta AN SSSR* 68: 3–18.

Chudinov, P.K. 1955. [Procolophonia (cotylosaurs) of Russia.] *Doklady AN SSSR* 103: 913–916.

—1957. [Cotylosaurs from the Upper Permian redbed deposits of the PreUrals.] *Trudy Paleontologicheskogo Instituta AN SSSR* 68: 19–87.

—1983. [Early therapsids.] *Trudy Paleontologicheskogo Instituta AN SSSR* 202: 1–230.

Colbert, E.H. 1946. *Hypsognathus*, a Triassic reptile from New Jersey. *Bulletin of the American Museum of Natural History* 86: 225–274.

DeBraga, M. and Rieppel, O. 1997. Reptile phylogeny and the affinities of turtles. *Zoological Journal of the Linnean Society* 120: 281–354.

Efremov, I.A. 1940a. [Preliminary description of new Permian and Triassic Tetrapoda from USSR.] *Trudy Paleontologicheskogo Instituta AN SSSR* 10 (2): 1–140.

—1940b. Kurze Übersicht über die Formen der Perm- und Trias-Tetrapoden-Fauna der UdSSR. *Centralblatt für Mineralogie, Geologie, und Paläontologie, Abteilung B* 1940: 372–383.

—1940c. [Composition of the Northern Dvina Fauna of Permian Amphibia and Reptilia from the excavations of V.P. Amalitskii.] *Doklady AN SSSR* 27: 893–896.

—1940d. [New discoveries of Permian terrestrial vertebrates in Bashkiria and Chkalor Provinces.] *Doklady AN SSSR* 27: 412–415.

—1954. [The fauna of terrestrial vertebrates in the Permian Copper Sandstone of the Western Cisuralian region.] *Trudy Paleontologicheskogo Instituta AN SSSR* **54**: 1–416.

Gadow, H. 1909. *Amphibia and Reptiles*. London: MacMillan and Company.

Gao K.-Q. 1989. Pareiasaurs from the Upper Permian of north China. *Canadian Journal of Earth Sciences* **26**: 1234–1240.

Gregory, W.K. 1946. Pareiasaurs versus placodonts as near ancestors to the turtles. *Bulletin of the American Museum of Natural History* **86**: 276–323.

Hartmann-Weinberg, A.P. 1929. Über Carpus und Tarsus der Pareiasauriden. *Anatomischer Anzeiger* **67**: 401–428.

—1930. Zur Systematik der Nord-Duna-Pareiasauridae. *Paläontologische Zeitschrift* **12**: 47–59.

—1933. Die Evolution der Pareiasauriden. *Trudy Paleontologicheskogo Instituta AN SSSR* **3**: 3–66.

—1937. Pareiasauriden als Leitfossilien. *Problemy Paleontologii* **2/3**: 649–712.

Haughton, S.H. 1929. Pareiasaurian studies. Part II. Notes on some pareiasaurian brain-cases. *Annals of the South African Museum* **28**: 88–96.

—and Boonstra, L.D. 1930. Pareiasaurian studies. Part VI. The osteology and myology of the locomotor apparatus. A. Hind limb. *Annals of the South African Museum* **28**: 297–367.

Huene, F. von 1944. Pareiasaurierreste aus dem Ruhuhu-Gebiet. *Paläontologische Zeitschrift* **23C**: 386–410.

Ivakhnenko, M.F. 1987. [The Permian parareptiles of the USSR.] *Trudy Paleontologicheskogo Instituta AN SSSR* **223**: 1–160.

—1995. [New primitive therapsids from the Permian of East Europe.] *Paleontologicheskii Zhurnal* **1995** (4): 110–119.

Kalandadze, N.N., Ochev, V.G., Tatarinov, L.P., Chudinov, P.K., and Shishkin, M.A. 1968. [A Catalogue of the Permian and Triassic tetrapods of the USSR.] pp. 72–91 in [*Upper Palaeozoic and Mesozoic Amphibians and Reptiles of the USSR*.] Moscow: Nauka.

King, G.M. 1988. Anomodontia, pp. 1–174 in Wellnhofer, P. (ed.), *Handbuch der Paläoherpetologie*, 17C. Stuttgart: Gustav Fischer.

Kuhn, O. 1969. Cotylosauria, pp. 1–89 in Kuhn, O. (ed.), *Handbuch der Paläoherpetologie*, 6. Stuttgart: Gustav Fischer.

Laurin, M. and Reisz, R.R. 1995. A reevaluation of early amniote phylogeny. *Zoological Journal of the Linnean Society* **113**: 165–223.

Lee, M.S.Y. 1993. The origin of the turtle body plan: bridging a famous morphological gap. *Science* **261**: 1716–1720.

—1995. Historical burden in systematics and the interrelationships of parareptiles. *Biological Reviews* **70**: 459–547.

—1997a. Reptile relationships turn turtle. *Nature* **389**: 245–246.

—1997b. The species-level taxonomy of pareiasaurs: implications for Permian terrestrial palaeoecology. *Modern Geology* **21**: 231–298.

—1997c. Pareiasuar phylogeny and the affinities of turtles. *Zoological Journal of the Linnean Society* **120**: 197–280.

Modesto, S.P. 1994. [Untitled.] *News Bulletin, Society of Vertebrate Paleontology* **161**: 17.

Olson, E.C. 1957. Catalogue of localities of Permian and Triassic terrestrial vertebrates of the territories of the U.S.S.R. *Journal of Geology* **65**: 196–226.

—1962. Late Permian terrestrial vertebrates, USA and USSR. *Transactions of the American Philosophical Society, New Series* **52** (2): 1–224.

Parrington, F.R. 1962. Les relations des cotylosaurs diadectomorphes. *Colloques Internationaux du Centre National de la Recherche Scientifique, Paris* **104**: 175–185.

Spencer, P.S. 1994. *The Early Evolution of Amniota*. Ph.D. thesis, University of Bristol.

Sushkin, P.P. 1927. On the modifications of the mandibular and hyoid arches and their relations to the brain-case in the early Tetrapoda. *Paläontologische Zeitschrift* **8**: 263–321.

Watson, D.M.S. 1917. The evolution of the tetrapod shoulder girdle and fore-limb. *Journal of Anatomy* **7**: 1–63.

Wild, R. 1985. Ein Schadelrest von *Parasaurus geinitzi* H. v. Meyer (Reptilia; Cotylosauria) aus dem Kupferschiefer (Permian) von Richelsdorf (Hessen). *Geologische Blätter für Nordost-Bayern (Gedenkschrift B. v. Freyberg)* **34/35**: 897–920.

Wilkinson, M., Thorley, J. and Benton, M.J. 1997. Uncertain turtle relationships. *Nature* **387**: 466.

Woodward, A.S. 1918. Vladimir Prochorovich Amalitsky. Obituary. *Geological Magazine* **5**: 431–432.

Zardoya, R. and Meyer, A. 1998. Complete mitochondrial genome indicates diapsid affinities of turtles. *Proceedings of the National Academy of Sciences, USA* **95**: 14226–14231.

6

Mammal-like reptiles from Russia

BERNARD BATTAIL AND MIKHAIL V. SURKOV

Introduction

The mammal-like reptiles, a popular name for the Synapsida, excluding the Mammalia, are subdivided, in traditional systematics, into two major groups, the Pelycosauria and the Therapsida. The Pelycosauria, a heterogeneous set of early forms, display the basic characters of the synapsids, such as the presence of only one, lower, temporal fenestra; in Russia, the only well known pelycosaurs are the caseid *Ennatosaurus tecton* and the varanopsid *Mesenosaurus*.

The Therapsida are regarded by all recent authors as the sister-group of the sphenacodontid pelycosaurs, with which they share the following derived characters (according to Hopson and Barghusen, 1986, p. 86): (1) angular bone with reflected lamina, (2) zygomatic process of quadratojugal lost, replaced by process of squamosal, (3) canine teeth enlarged, transversely compressed, with mesial (anterior) and distal (posterior) cutting edges, (4) maxilla increased in height, eliminating lacrimal contact with external naris, (5) retroarticular process of articular turned downward, and (6) paroccipital process elongated and directed ventrolaterally.

The Therapsida are well defined: Hopson and Barghusen (1986) listed the following 11 cranial synapomorphies: (1) lateral temporal fenestra enlarged, (2) septomaxilla with posterodorsal process extending on to face between nasal and maxilla bones, (3) supratemporal bone lost, (4) squamosal with groove on posterior surface, (5) reflected lamina of angular deeply notched dorsally, (6) upper canine further increased in length, (7) dorsal process of premaxilla greatly elongated, (8) parietal foramen raised on a prominent boss, (9) maxilla increased further in height to eliminate contact of nasal and lacrimal bones, (10) interpterygoid vacuity reduced in size, (11) vomer transversely widened between internal nares.

Repository abbreviations

BMNH, British Museum (Natural History), London; LGM/Ch MP, Leningrad School of Mines (Tetrapods of the Copper-bearing Sandstones); LGU, Leningrad State University; PIN, Palaeontological Institute, Russian Academy of Sciences, Moscow; SGU, Saratov State University, Saratov.

The major groups of therapsids

Within the Therapsida, various groups have been defined, but their content and phylogenetic relationships are still a matter of debate.

Primitive carnivorous therapsids

The group Biarmosuchia, including the families Biarmosuchidae, Ictidorhinidae, Hipposauridae and Burnetiidae (Sigogneau-Russell, 1989), was once considered as possibly paraphyletic, 'characterized only by possession of primitive characters' (Hopson and Barghusen, 1986). The Biarmosuchia are now regarded as a probable monophyletic group, characterized by such apomorphies as: postcanine teeth with basal swelling and coarsely serrated margins (Hopson, 1991); distal carpals 4 and 5 fused (Sigogneau-Russell, 1989; Hopson, 1991); and, orbits greatly enlarged (Sigogneau-Russell, 1989; Battail, 1992). The

Biarmosuchia, sometimes considered as the sister-group of the more advanced carnivorous therapsids (Battail, 1992), are far more often considered as the sister-group of all remaining therapsids (Hopson and Barghusen, 1986; Kemp, 1988; Hopson, 1991).

The Phthinosuchia, indeed very poorly known, could be close to the Biarmosuchia (Kemp, 1982; Sigogneau-Russell, 1989); they have been tentatively put near to the Dinocephalia by Battail (1992).

It is equally difficult to judge the phylogenetic relationships of the Eotitanosuchia, known only by the skull without lower jaw of the type-specimen of *Eotitanosuchus olsoni* and by the maxilla of *Ivantosaurus ensifer*. *Eotitanosuchus* has been provisionally considered as the possible sister-group of the Gorgonopsidae by Kemp (1982, pp. 109–110). Similarly, both Sigogneau-Russell (1989) and Battail (1992) consider the Eotitanosuchia and Gorgonopsia as probable sister-groups. However, Kemp (1988) reconsidered his previous opinion on the phylogenetic position of *Eotitanosuchus*, and Hopson (1991) disputed the actual value of the few synapomorphies which might link the Eotitanosuchia and the Gorgonopsia.

The Gorgonopsia are undoubtedly the best documented group of primitive carnivorous therapsids. It is a very conservative group, characterized by many features, including: presence of a preparietal, which does not contact the pineal foramen; vomers fused, crested posteriorly and widened anteriorly; palatines meeting midventrally; fossa for lower canine confluent with internal naris; dentary displaying a narrow free-standing coronoid process; articular with a dorsal process; reflected lamina of the angular high and anteriorly situated; very large canines; precanine maxilla teeth lost; number of postcanine teeth reduced.

Dinocephalia and Anomodontia

The Dinocephalia are composed of 'large forms, both herbivores and carnivores, which tend towards pachyostosis of the skull bones. They are characterized by: incisor teeth with lingual heels, which intermesh when the jaws are closed; temporal fenestra expanded dorsoventrally; jaw adductor muscles invading the lateral and dorsal faces of the bones of the intertemporal regions of the skull' (Kemp, 1988, p. 8).

The content of the Anomodontia varies with different authors. In the original sense (Owen, 1859), the Anomodontia do not include the Dinocephalia. They are highly specialized herbivorous therapsids, the cranial synapomorphies of which are the following, according to Hopson and Barghusen (1986, p. 86): (1) zygomatic arch displaced dorsally, (2) dentaries fused at symphysis, (3) articular surface of lower jaw slopes steeply posteroventrally, (4) fenestra between dentary and angular bones, (5) facial region shortened, (6) lower canine reduced in size, (7) posterodorsal part of dentary with fossa for insertion of *M. adductor mandibulae externus*, (8) palatal portion of premaxilla greatly expanded posteriorly to closely approach palatine bone, (9) palatal teeth lost, and (10) coronoid bone lost.

With this definition, the Anomodontia include a few archaic forms, and a more advanced group known as the Dicynodontia, characterized by numerous apomorphies including in particular an edentulous beak, a secondary bony palate, and a lateral flange of the squamosal.

Watson and Romer (1956) introduced the Dinocephalia as a subgroup of the Anomodontia, broadening therefore considerably the extent of the latter. King (1988), in addition, extended the content of the Dicynodontia to all the Anomodontia *sensu* Hopson and Barghusen (1986). The Anomodontia are regarded by King as composed of two sister-groups, the Dinocephalia and the Dicynodontia *sensu lato*, which would be linked by four synapomorphies: (1) loss of coronoid bone(s), (2) non-terminal nostrils and long posterior spur of premaxilla (modified in higher dicynodonts), (3) grooved or troughed palatal exposure of vomers, and (4) reduction or loss of internal trochanter of femur. It appears, however, that the absence of the coronoid is not a constant feature, that characters (2) and (3) occur also in primitive therapsids, and that the internal trochanter of the femur is not reduced in certain primitive Dinocephalia.

Consequently, a special relationship between dinocephalians and dicynodonts seems to be only weakly supported (Hopson, 1991).

Higher carnivorous therapsids

It is now agreed that the two groups of advanced carnivorous therapsids, the Therocephalia and the Cynodontia, are sister-groups. Of the 11 synapomorphies listed by Hopson and Barghusen (1986), Hopson (1991) selected five which are unique to these two taxa among therapsids: (1) temporal roof completely eliminated so that the temporal fossa is completely open dorsally, (2) postorbital bone shortened so that it does not contact the squamosal medial to temporal fossa, (3) parietal expanded posteriorly on the midline behind the parietal foramen, increasing the length of the sagittal crest, (4) epipterygoid expanded anteroposteriorly, and (5) posteroventral portion of the dentary forming a thickened lower border that extends below the angular bone and supports the latter in a trough on its medial surface.

The diagnosis of the Therocephalia is based on only a few derived characters, but these are well defined and seem to be quite reliable, leaving no doubt about the monophyly of the group; the most obvious of the therocephalian features is perhaps the presence of a suborbital fossa, bounded by the palatine, the pterygoid and the ectopterygoid bones.

The Cynodontia (which, as a monophyletic group, must include the Mammalia) are defined by a very large number of apomorphies: cranial apomorphies, such as the contact between the postorbital and prefrontal bones on the orbital margin, excluding the frontal bone from the orbital rim, or the double occipital condyle; various mandibular apomorphies, expressing the huge development of the dentary, the reduction of the post-dentary bones, and changes in the jaw adductor muscle insertion; dental apomorphies, related to the high degree of specialization of the teeth, etc.

Most authors admit that the advanced carnivorous therapsids – Therocephalia and Cynodontia – share a few derived characters with the Gorgonopsia, for example a free-standing coronoid process of the dentary, or the reduction in size of the quadrate and quadratojugal, which are loosely set, without suture, in a depression on the anterior face of the squamosal.

The Therapsida in Russia

All major groups of therapsids, the Phthinosuchia, Biarmosuchia, Eotitanosuchia, Gorgonopsia, Dinocephalia, Anomodontia, Therocephalia, and Cynodontia are represented in Russia. The early forms of Late Permian age are better represented than the Triassic ones.

Order THERAPSIDA Broom, 1905
Suborder EOTHERIODONTIA Olson, 1962
Comment. The suborder Eotheriodontia is retained, with the same global content as in Kemp (1982, p. 348), only for the sake of convenience: it is a paraphyletic group which includes primitive carnivorous therapsids, the phylogenetic relationships of which still remain obscure and disputed.

Infraorder PHTHINOSUCHIA Romer, 1961
Comment. The Phthinosuchia were initially considered by Romer as the stem group of all other therapsids, playing therefore a major role in the evolution of the mammal-like reptiles. They were, however, not acknowledged as an independent infraorder by Olson (1962), who put them together with *Eotitanosuchus* and the families Biarmosuchidae and Brithopodidae in the infraorder Eotheriodontia, nov.; similarly, they were included by Chudinov (1983) in an order Eotheriodontia, to which belonged also the families Biarmosuchidae and Eotitanosuchidae. The Phthinosuchia are in fact very poorly known, and their phylogenetic position remains uncertain. Consequently, it is probably better to keep them, at least provisionally, as a separate infraorder, as did Sigogneau-Russell (1989). Two families are usually recognized, the Phthinosuchidae and the Phthinosauridae.

Family PHTHINOSUCHIDAE Efremov, 1954
Diagnosis. (Sigogneau-Russell, 1989, p. 4). Small temporal fenestra, only slightly larger than orbit, and

more developed dorso-ventrally than antero-posteriorly; incipient pachyostosis of dorsal border of orbit; occiput slightly inclined ventro-anteriorly, and consequently short basicranium and lower jaw articulation displaced anteriorly; lower jaw slender, without coronoid process.

Comment. Two genera, *Phthinosuchus* and *Phthinosaurus*, were originally attributed to the Phthinosuchidae by Efremov (1954). Tatarinov (1974) placed *Phthinosaurus* in a new family Phthinosauridae.

Phthinosuchus Efremov, 1954
See Figure 6.1B.

Type species. *P. discors* Efremov, 1954.

Holotype. PIN 1954/3, a badly preserved skull and lower jaw, lacking the anterior third. Copper Sandstones of Bashkortostan Republic; precise locality unknown; probably Early Tatarian, perhaps Late Kazanian.

Referred material. The only referred material is the fragmentary posterior part of a braincase.

Family PHTHINOSAURIDAE Tatarinov, 1974

Diagnosis. Much higher position of jaw articulation; development of low, but distinct coronoid process; mandibular fenestra below coronoid process (in comparison to Phthinosuchidae).

Phthinosaurus Efremov, 1940a

Type species. *P. borissiaki* Efremov, 1940a.

Holotype. PIN 164/7, a damaged left jaw, the sutures of which are often unclear; the teeth are all broken. Belebey, Bashkortostan Republic. Late Kazanian.

Referred material. Two dentary fragments.

Comment. *Phthinosaurus* has been described, figured and discussed many times (Efremov, 1940a, 1954; Watson, 1942; Olson, 1962; Tatarinov, 1974; Chudinov, 1983; Sigogneau-Russell, 1989), but its anatomy and precise phylogenetic position remain uncertain because of its incompleteness and poor state of preservation. Ivakhnenko (pers. comm., 1997) places *Phthinosaurus* in the Rhopalodontidae.

Infraorder BIARMOSUCHIA Sigogneau-Russell, 1989

Comments. According to Sigogneau-Russell (1989), the Biarmosuchia include four families of primitive therapsids, the Biarmosuchidae, Hipposauridae, Ictidorhinidae and Burnetiidae. The Biarmosuchidae are known only from Russia, the Hipposauridae and Ictidorhinidae only from South Africa, and the Burnetiidae are represented both in South Africa and in Russia. The family Niaftasuchidae, recently created by Ivakhnenko (1990), and attributed by him to the Tapinocephalia, should rather, in our opinion, also be integrated in the Biarmosuchia.

Family BIARMOSUCHIDAE Olson, 1962
Biarmosuchus Tchudinov, 1960
See Figure 6.1A.

Type species. *Biarmosuchus tener* Tchudinov, 1960.

Holotype. PIN 1785/2, a distorted skull with lower jaw, and anterior part of the postcranial skeleton. Ezhovo, Ocher district, Perm' Province. Late Kazanian–earliest Tatarian.

Referred material. *Biarmosuchus tener* is known from several skulls and postcranial skeletons, however often incomplete and not fully prepared.

Diagnosis. Skull characterized by slight ventral sinuosity; very narrow interorbital roof; long dorsal processes of premaxillae; postorbital does not reach ventral border of orbit; paroccipital process in contact with quadrate; parasphenoid keeled ventrally.

Synonyms. Olson (1962) suggested that *Biarmosaurus* Tchudinov, 1960 is a synonym of *Biarmosuchus*, and this was confirmed by Sigogneau and Chudinov (1972). *Biarmosaurus antecessor* Tchudinov, 1960 was based on a skull (PIN 1758/7) that is larger than the type specimen of *Biarmosuchus tener*, but it is otherwise very similar and comes from the same locality. It is a synonym of *Biarmosuchus tener* (Tchudinov, 1983). *Biarmosuchus tagax* Ivakhnenko, 1990, is based on a partial skull (PIN 3706/10) from the Upper Permian (Lower Tatarian?) of the Mezen' Basin, Arkhangel'sk Province. It is smaller than the type specimen of *Biarmosuchus tener*, but differs only very slightly from the latter. It is a probable junior synonym of *Biarmosuchus tener*.

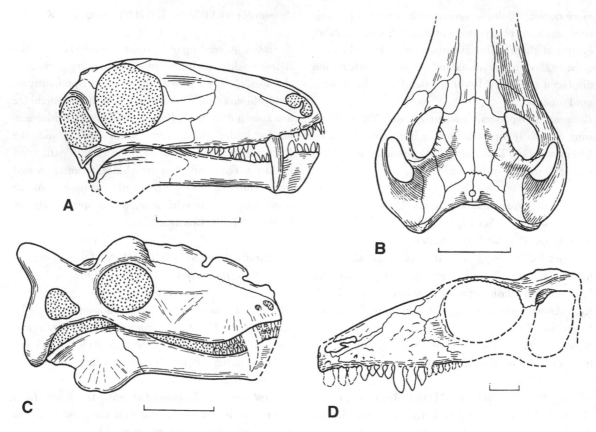

Figure 6.1. Basal therapsids from the Late Permian. A, *Biarmosuchus tener* Tchudinov, 1960, holotype, PIN 1758/2, skull, slightly restored, in lateral view. B, *Phthinosuchus discors* Efremov, 1954, holotype, PIN 1954/3, skull in dorsal view. C, *Proburnetia viatkensis* Tatarinov, 1968a, holotype, PIN 2416/1, skull in lateral view. D, *Niaftasuchus zekkeli* Ivakhnenko, 1990, holotype, PIN 3717/36, skull in lateral view. Scale bars, 50 mm, except in D, 10 mm. (A, after Sigogneau and Chudinov, 1972; B, C, after Tatarinov, 1974; D, after Ivakhnenko, 1990.)

Biarmosuchoides Tverdokhlebova et Ivakhnenko, 1994
Type species. Biarmosuchoides romanovi Tverdokhlebova et Ivakhnenko, 1994.
Holotype. SGU 104B/2051, an almost complete left dentary, the sole specimen. Locality Dubovka I, Orenburg Province, Novosergievka district. Early Tatarian.
Diagnosis. Single shallow, elongated lower jaw, with long symphysis; postcanine teeth slightly curved backwards.

Family BURNETIIDAE Broom, 1923
Diagnosis. Pachyostosis of skull bones; bony bosses over orbits and temporal fossae; orbits smaller than in

other Biarmosuchia, probably as a consequence of pachyostosis; paroccipital process in contact with quadrate; parasphenoid not keeled ventrally.
Comment. The Burnetiidae were usually included in the Gorgonopsia; however, Sigogneau-Russell (1989) indicated that they are close relatives of the Ictidorhinidae, and hence included them in the Biarmosuchia.

Proburnetia Tatarinov, 1968a
See Figure 6.1C.
Type species. Proburnetia viatkensis Tatarinov, 1968a.
Holotype. PIN 2416/1, a poorly preserved skull with lower jaw and a few cervical vertebrae. Kotel'nich,

90

Vyatka River basin, Kirov Province. Late Tatarian.
Comment. The type and only known specimen of *P. viatkensis* was described and figured by Tatarinov (1968a, 1974). *Proburnetia* is similar to the South African genus *Burnetia*, from which it differs essentially in skull proportions.

Family NIAFTASUCHIDAE Ivakhnenko, 1990
Comments. The family Niaftasuchidae was created by Ivakhnenko (1990) for a new genus and species, *Niaftasuchus zekkeli*, represented by only one specimen. It was attributed by him to the Tapinocephalia (the herbivorous dinocephalians, considered as a separate order), essentially on the basis of the dentition: in the Niaftasuchidae, the incisors are large and the canine is morphologically identical to the postcanine teeth. It must be noted, however, that there is no pachystosis in the Niaftasuchidae, and that the area of origin of the external adductor jaw muscles does not seem to extend much on to the dorsal surface of the postorbital; consequently, the Niaftasuchidae do not display the main synapomorphies of the Dinocephalia. They should rather, in our opinion, be regarded as Biarmosuchia, as they share with the latter two important features, big orbits and fused vomers.

Niaftasuchus Ivakhnenko, 1990
See Figure 6.1D.
Type species. Niaftasuchus zekkeli Ivakhnenko, 1990.
Holotype. PIN 3717/36, a small incomplete skull without lower jaw. Nyafta, basin of the Mezen' River, Arkhangel'sk Province. Late (?) Tatarian.
Diagnosis. Large orbits; small temporal fenestrae; skull roof broad; large pineal foramen; dentition includes, on each side, three incisors which were probably large (only alveoli preserved), four small precanine teeth, one canine, much bigger than precanine teeth, and series of smaller postcanine teeth; maxilla teeth transversely compressed, all morphologically identical.

Infraorder EOTITANOSUCHIA Boonstra, 1963
Diagnosis. Large forms of primitive carnivorous therapsids; temporal fossa larger than orbit; postorbital bar slightly twisted; dorsal processes of premaxillae elon-

gated; maxilla reaches maximum height in posterior part; short parietals; vomers incompletely fused; palatal dentigerous tuberosities well developed; behind their transverse flanges, pterygoids narrow and lateral borders of their quadrate rami almost parallel; small interpterygoid vacuity; paroccipital process contacts quadrate; incisors and postcanine teeth small; no precanine teeth; canines very large.
Comment. The Eotitanosuchia are known only from Russia.

Family EOTITANOSUCHIDAE Tchudinov, 1960
Eotitanosuchus Tchudinov, 1960
See Figure 6.2.
Type species. Eotitanosuchus olsoni Tchudinov, 1960.
Holotype. PIN 1758/1, a skull without lower jaw, crushed laterally, and posteriorly incomplete. Ezhovo, Ocher district, Perm' Province. Late Kazanian–earliest Tatarian.
Referred material. Two incomplete skulls, PIN 1758/85 and PIN 1758/319, from the same locality as the type specimen.
Comments. Eotitanosuchus olsoni was figured by Chudinov (1960, 1961, 1964a, 1983), Olson (1962), Sigogneau and Chudinov (1972), and Sigogneau-Russell (1989). The skull is large, with a long snout. The incisors are small. There are apparently no precanine teeth. In the type specimen, one large canine is present on either side. However, 'on the specimen PIN 1758/85 an obscure zone follows the functional canine, a zone which might correspond to a resorbed canine' (Sigogneau and Chudinov, 1972). There would be eight or nine postcanine teeth.

Ivantosaurus Tchudinov, 1983
Type species. Ivantosaurus ensifer Tchudinov, 1983.
Holotype. PIN 1758/292, two maxillae and one quadrate, belonging to the same individual. Ezhovo, Ocher district, Perm' Province. Late Kazanian–earliest Tatarian.
Comments. Ivantosaurus ensifer, known only from its fragmentary type specimen, is a very large form. The large canine is laterally compressed and curved; it is

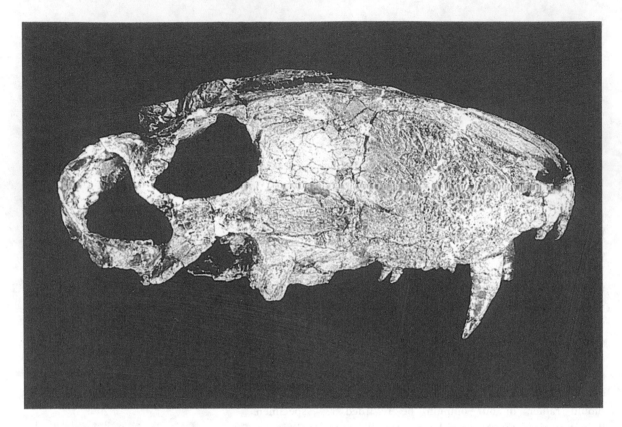

Figure 6.2. Skull of the Late Permian eotitanosuchian *Eotitanosuchus olsoni* Tchudinov, 1960, holotype, PIN 1758/1 in lateral view. Skull is about 150 mm long.

followed by a second, non functional, canine; 'we would have here, as in the Gorgonopsia, two alternately functional canines, each of them being individually replaced. However, the inclination of the canine towards the front and the structure of the quadrate do not seem to be of a gorgonopsian type' (Sigogneau-Russell, 1989).

Infraorder GORGONOPSIA Seeley, 1895
Comments. If the Ictidorhinidae and the Burnetiidae, which were often in the past regarded as Gorgonopsia, are considered in the Biarmosuchia, the Gorgonopsia are restricted to two families, the Watongiidae and the Gorgonopsidae. The Watongiidae include only one genus and species of primitive Gorgonopsia, *Watongia meieri*, known from only one very fragmentary skeleton from the Guadalupian (Late Permian) of

Oklahoma, USA (Sigogneau-Russell, 1989). On the other hand, the Gorgonopsidae are represented by very abundant material, attributed to many genera and species. Most Gorgonopsidae come from southern Africa (mainly South Africa, and also Zambia, Malawi and Tanzania), but a few are known from Russia. The Gorgonopsia are known only from the Upper Permian.

Family GORGONOPSIDAE Lydekker, 1890
Diagnosis. Medium to large carnivorous forms; temporal fenestrae much larger than orbits; intertemporal roof broad, without sagittal crest; incisors well developed; two upper canines, functioning alternately; postcanine teeth reduced both in size and in number.
Comment. The Gorgonopsidae represent a homogeneous, conservative family, which is difficult to classify.

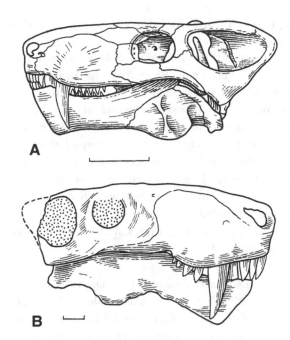

Figure 6.3. Gorgonopsians from the Late Permian. A, *Sauroctonus progressus* (Hartmann-Weinberg, 1938), holotype, PIN 156/5, skull in lateral view. B, *Inostrancevia alexandri* Amalitskii, 1922, holotype, PIN 2005/1578, skull in lateral view. Scale bars, 50 mm. (A, after Bystrov, 1955; B, after Pravoslavlev, 1927a,b.)

Sigogneau-Russell (1989) recognizes three subfamilies, the Gorgonopsinae, Rubidgeinae and Inostranceviinae.

Subfamily GORGONOPSINAE Sigogneau, 1970

Diagnosis. Zygomatic arches not thickened ventrally; posterior edge of cranial roof extends posteriorly beyond level of postorbital bar; preparietal usually present; posttemporal fossa orientated almost horizontally (Sigogneau-Russell, 1989, p. 66).

Sauroctonus Bystrov, 1955
See Figure 6.3A.

Diagnosis. Gorgonopsine of medium size; skull narrow posteriorly; orbits small; interorbital and intertemporal widths very narrow, narrower than in any other gorgonopsine; infraorbital and zygomatic arches rather slender; postorbital bar widens ventrally; trans-verse flanges of pterygoids with teeth; 4–6 postcanine teeth on maxilla.

Type species. Sauroctonus progressus (Hartmann-Weinberg, 1938).

Lectotype. PIN 156/5, a nearly complete skull. Semin Ravine, Il'inskoe, Tetyushi district, Tatarstan Republic. Severodvinskian Gorizont, Late Tatarian.

Referred material. PIN 156/6, a crushed skull and lower jaw with the anterior part of the skeleton, and many isolated bones; all the specimens come from the same locality as the type.

Comments. The species was described by Hartmann-Weinberg (1938) as *Arctognathus progressus*, reassigned by Efremov (1941) to *Inostrancevia progressa*, and by Bystrov (1955) to *Sauroctonus progressus*. It was redescribed or noted also by Efremov and V'yushkov (1955), Kalandadze *et al.* (1968), and Sigogneau-Russell (1989), and Tatarinov (1974) gave the most detailed description. *Sauroctonus* appears to be rather similar to the South African genus *Scylacops*.

Subfamily RUBIDGEINAE Sigogneau, 1970

Diagnosis. Skull usually broader than in Gorgonopsinae; postorbital bar very wide, and posterior edge of cranial roof does not extend beyond its level; infraorbital and zygomatic arches thick, and the latter with ventral expansion at posterior end; preparietal small or absent; posttemporal fossa slopes dorsally and laterally; dentary short and massive (Sigogneau-Russell, 1989, p. 101).

Niuksenitia Tatarinov, 1977a

Type species. Niuksenitia sukhonensis Tatarinov, 1977a.

Holotype. PIN 3159/1, an occiput with the otic region and the palate. Location 1.5 km from Navoloki, downstream, right bank of the Sukhona river, Nyuksenitsa district, Vologda Province. Sukhonskaya Svita, Late Tatarian.

Diagnosis. Medium-sized; transverse apophyses of pterygoids with teeth.

Comments. Little can be said of *Niuksenitia sukhonensis*, known only by its very incomplete type specimen. Tatarinov (1977a) compared it with the South African genus *Broomicephalus*. Sigogneau-Russell (1989) found also similarities with the South African genus *Rubidgea*. She did not exclude, however, that *Niuksenitia* could be in fact a close relative of the burnetiid *Proburnetia*.

Subfamily INOSTRANCEVIINAE Sigogneau-Russell, 1989

Diagnosis. Long snouts; wide interorbital and intertemporal spaces; cranial arches moderately developed, without thickening; posterior edge of cranial roof extends beyond level of postorbital bars; short parietal; preparietal present; no lower postcanines (Sigogneau-Russell, 1989, p. 111).

Inostrancevia Amalitskii, 1922
See Figure 6.3B.
Type species. *I. alexandri* Amalitskii, 1922.
Diagnosis. Large size; skull wide posteriorly; snout high and very long; small orbits; very wide intertemporal space; cranial arches slender; preparietal present; frontals do not reach orbital rim; palatal teeth reduced; transverse apophyses of pterygoids situated more anteriorly than orbits; very small interpterygoid fossa; very high mandibular symphysis; only four upper and three lower incisors on either side of skull.

Inostrancevia alexandri Amalitskii, 1922
Synonyms. *I. alexandri* (Efremov, 1940b, *pars*); *I. proclivis* Pravoslavlev, 1927a (V'yushkov, 1953; Efremov and V'yushkov, 1955).
Holotype. PIN 2005/1578, skeleton. Sokolki, basin of the Little Northern Dvina River, Arkhangel'sk Province, Kotlas district. Vyatskian Gorizont, Late Tatarian.
Referred material. One skeleton, one skull and one isolated frontal, from the same locality as the type specimen.
Diagnosis. Large size; relatively narrow occiput; fenestra ovale wide and rounded; transverse flanges of pterygoids with teeth.
Comment. Figured by Pravoslavlev (1927a), Tatarinov (1974), and Sigogneau-Russell (1989).

Inostrancevia latifrons Pravoslavlev, 1927b
Synonyms. *Amalitzkia vladimiri* Pravoslavlev, 1927a; *Inostrancevia vladimiri* (V'yushkov, 1953; Efremov and V'yushkov, 1955; Kalandadze *et al.*, 1968); *Amalitzkia annae* Pravoslavlev, 1927b.
Holotype. PIN 2005/1857, skull. Sokolki, basin of the Little Northern Dvina River, Arkhangel'sk Province, Kotlas district. Vyatskian Gorizont, Late Tatarian.
Referred material. One skull from the same locality as the type, and one incomplete skeleton from Zavrazhe.
Diagnosis. Extremely large; compared with *I. alexandri*, snout is lower and wider, parietal roof and occiput are wider, fewer teeth on transverse flanges of pterygoids, and palatal tuberosities less developed; fenestra ovale, wide and rounded.
Comment. Figured by Pravoslavlev (1927b) and Tatarinov (1974).

Inostrancevia uralensis Tatarinov, 1974
Holotype. PIN 2896/1, left half of the basioccipital portion of a braincase. Blyumental' 3, basin of the Ural River, Belyaevka district, Orenburg Province. Vyatskian Gorizont, Late Tatarian.
Referred material. An otico-occipital portion of a braincase, from the same locality as the type specimen.
Diagnosis. Smaller than *I. latifrons*; fenestra ovale in shape of a slot elongated transversally.

Pravoslavlevia Vjuschkov, 1953
Type species. *Pravoslavlevia parva* (Pravoslavlev, 1927a). See also V'yushkov (1953) (= *Inostrancevia parva* Pravoslavlev, 1927a; Efremov 1940b, 1941).
Holotype. PIN 2005/1859, a deformed skull. Sokolki, basin of the Little Northern Dvina River, Kotlas district, Arkhangel'sk Province. Vyatskian Gorizont, Late Tatarian.
Diagnosis. Small; snout long; skull does not widen posteriorly to orbits; small orbits; five upper and four lower incisors.

Suborder DINOCEPHALIA Seeley, 1895
Comments. No general agreement has yet been reached concerning the phylogeny of the Dinocephalia. All recent authors, however, recognize two main lineages: one includes the more plesiomorphic forms, which retained carnivorous adaptations, whereas the other consists of forms adapted to a herbivorous diet. But

some families remain difficult to locate in that scheme: the Estemmenosuchidae are included in the carnivorous Dinocephalia (Titanosuchia) by Chudinov (1983); they are put at the base of the herbivorous Dinocephalia (Tapinocephalia) by Hopson and Barghusen (1986); they are considered as the sister-group of all other Dinocephalia by King (1988). The Titanosuchidae are included among the carnivorous Dinocephalia by Chudinov (1983); are regarded as the sister-group of the more specialized herbivorous Tapinocephalidae by Hopson and Barghusen (1986); and King (1988) does not even keep a familial rank for the Tapinocephalidae, but considers them as a sub-family, the Tapinocephalinae, sister-group of the Titanosuchinae within the Titanosuchidae.

Recently, Ivakhnenko (1994, 1995) has expressed the view that the Estemmenosuchidae should be excluded from the Dinocephalia and brought together with the eotheriodonts; two main groups of Dinocephalia, the Titanosuchia and the Tapinocephalia, are retained, but the contents of these groups differ again sharply from the ones which had been proposed by previous authors.

In view of the discrepancies among the different classifications of the Dinocephalia, we have chosen here a simple, perhaps too simple, scheme, which retains only four families: the Estemmenosuchidae, the Anteosauridae, the Titanosuchidae, and the Tapinocephalidae. Two Russian genera, *Archaeosyodon* and *Microsyodon*, are attributed by Ivakhnenko (1995) to the Titanosuchidae; we consider however that they do not display typical apomorphies of the family, and should rather be attributed to the Anteosauridae; only the families Estemmenosuchidae, Anteosauridae and Tapinocephalidae are therefore present in Russia.

Family ESTEMMENOSUCHIDAE Tchudinov, 1960

Diagnosis. Skull heavily pachyostosed; often bears horn-like bosses; teeth on vomers.

Estemmenosuchus Tchudinov, 1960
See Figures 6.4 and 6.5A.
Type species. E. uralensis Tchudinov, 1960.

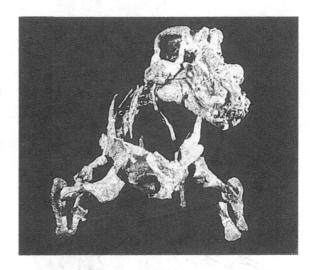

Figure 6.4. Anterior view of the skull and skeleton of the Late Permian dinocephalian *Estemmenosuchus uralensis* Tchudinov, 1968a,b. Skull is about 350 mm long.

Diagnosis. Massive skull; anterior parts of nasals forming unpaired boss on dorsal surface of snout; postorbitals and postfrontals forming pair of short outgrowths; jugal and squamosal forming very massive lateral outgrowth; large incisors; canines relatively short and thick; at least twenty small postcanine teeth.

Estemmenosuchus uralensis Tchudinov, 1960
Holotype. PIN 1758/4, a skull without lower jaw. Ezhovo, Ocher district, Perm' Province. Late Kazanian–earliest Tatarian.
Referred material. Very abundant, including more or less complete skulls and lower jaws, incomplete skeletons, and many isolated bones, all from the same locality as the type (complete list in Chudinov, 1983).
Diagnosis. Large species; length of skull exceeds width across jugal outgrowths; snout elongated and relatively narrow; postorbitals and postfrontals forming pair of short, undivided outgrowths, close to one another.

Estemmenosuchus mirabilis Tchudinov, 1968a
Synonym. Estemmenosuchus uralensis Tchudinov 1965, p. 122, *pars.*
Holotype. PIN 1758/6, incomplete skeleton with skull

B. BATTAIL & M.V. SURKOV

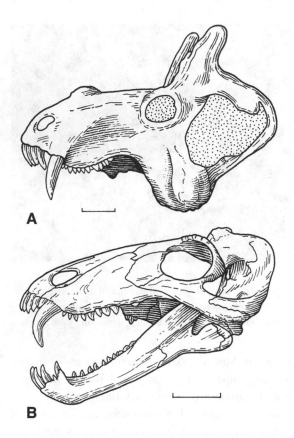

Figure 6.5. Dinocephalians from the Late Permian. A, *Estemmenosuchus mirabilis* Tchudinov, 1968a,b, holotype, PIN 1758/6, skull in lateral view. B, *Syodon efremovi* Orlov, 1940 (= *E. biarmicum* Kutorga, 1838), specimen PIN 157/2, skull, slightly restored, in lateral view. Scale bars, 50 mm. (A, after Chudinov, 1983; B, after Orlov, 1958.)

and lower jaw. Ezhovo, Ocher district, Perm' Province. Late Kazanian–earliest Tatarian.
Referred material. Left lower jaw with articulated quadrate; left and right lateral outgrowths. All the specimens come from the same locality as the type.
Diagnosis. Smaller than *E. uralensis*; skull length much less than width across jugal outgrowths; snout short and broad; postorbitals and postfrontals forming pair of bifurcated outgrowths, set far apart.

Molybdopygus Tchudinov, 1964b
Synonyms. Deuterosaurus Efremov, 1954, p. 103, *pars*; *Brithopus* Efremov, 1954, p. 223, *pars*.

Type species. Molybdopygus arcanus Tchudinov, 1964b (= *Deuterosaurus biarmicus* Efremov, 1954, p. 189, *pars*; = *Brithopus priscus* Efremov, 1954, p. 223, *pars*).
Holotype. PIN 2225/1, pelvic girdle and sacrum. Bolshoi Kityak, Malmyzh district, Kirov Province. Early Tatarian.
Referred material. Several pelvic bones, all from the Copper Sandstones of Bashkortostan Republic.
Diagnosis. Relatively small form; pelvis more massive than in *Estemmenosuchus*; skull unknown.

Anoplosuchus Tchudinov, 1968a
Type species. Anoplosuchus tenuirostris Tchudinov, 1968a.
Holotype. PIN 1758/79, an incomplete skeleton with a very deformed skull. Ezhovo, Ocher district, Perm' Province. Late Kazanian–earliest Tatarian.
Referred material. Many skull and jaw fragments from the Ocher district.
Diagnosis. Compared with *Estemmenosuchus*, skull more elongated; no cranial outgrowths; incisors about same size as canines; numerous small postcanine teeth; medium-sized form.

Zopherosuchus Tchudinov, 1983
Type species. Zopherosuchus luceus Tchudinov, 1983.
Holotype. PIN 1759/300, an incomplete skeleton with a very fragmentary skull, lacking the snout. Ezhovo, Ocher district, Perm' Province. Late Kazanian–earliest Tatarian.
Diagnosis. Small estemmenosuchid close to *Anoplosuchus*; no cranial outgrowths; temporal fenestrae short, but very developed dorso-ventrally; postcanine teeth less numerous than in *Estemmenosuchus* or *Anoplosuchus*.

Family ANTEOSAURIDAE Boonstra, 1954
Diagnosis. Carnivorous; skull long and narrow; tendency towards pachyostosis, affecting in particular orbital rims and parietals; raised alveolar margin in premaxillae; temporal fenestrae well open dorsally; heels developed on incisors; very large canines; postcanine teeth with bulbous crowns.
Comments. The Anteosauridae are understood here as including the Brithopodidae Efremov, 1954. The

96

family name Anteosauridae is preferred to the name Brithopodidae, because the genus *Brithopus* is very incompletely known.

Brithopus Kutorga, 1838

Synonyms. Orthopus Kutorga, 1838; *Rhopalodon* Fischer, 1841, *pars*; Fischer, 1845, *pars*; Eichwald, 1848, *pars*; Eichwald, 1860, *pars*; *Dinosaurus* Fischer, 1847; *Eurosaurus* Eichwald, 1860, *pars*; Meyer, 1866, *pars*).

Type species. Brithopus priscus Kutorga, 1838.

Diagnosis. Similar in size and general build to *Titanophoneus*, but differing by (probably) smaller number of anterior teeth; greater number of postcanine teeth (9–10 instead of 8); row of small teeth on each pterygoid flange; smaller palatal teeth on palatines; choanal depression deep; high supinator crest on humerus; powerful fourth trochanter and short adductor crest on femur (King, 1988, after Efremov, 1954).

Comments. Brithopus was described on the basis of very fragmentary, unsatisfactory material, from the Copper Sandstones of Cisuralia (precise localities often unknown). More detailed information can be found in Chudinov, 1983, pp. 79–80. The most complete study is by Efremov (1954), who recognized the following species:

Brithopus priscus Kutorga, 1838 (= *Orthopus primaevus* Kutorga, 1838; = *Rhopalodon murchisoni* Fischer, 1845; = *Dinosaurus murchisoni* Fischer, 1847; = *Eurosaurus verus* Meyer, 1866). Type specimen: PIN 296/14, distal part of humerus. Perm' Province. Late Kazanian.

Brithopus bashkyricus Efremov, 1954. Type specimen: PIN 294/14, incomplete scapula. Bashkortostan Republic. Late Kazanian.

Brithopus ponderus Efremov, 1954. Type specimen: LGM/ChMP 41, proximal part of humerus. Bashkortostan Republic. Late Kazanian.

Rhopalodon fischeri Eichwald, 1860; assigned to *Brithopus* by Efremov (1954). Type specimen: PIN collection, small fragment of lower jaw with teeth from a copper mine near Sterlitamak, Bashkortostan Republic. Late Kazanian.

Syodon Kutorga, 1838
See Figure 6.5B.

Synonym. Cliorhizodon Twelvetrees, 1880.

Type species. Syodon biarmicum Kutorga, 1838 (= *Cliorhizodon orenburgensis* Twelvetrees, 1880; = *Cliorhizodon efremovi* Orlov, 1940; = *Syodon efremovi* (Orlov, 1940); Orlov, 1958).

Holotype. LGM/ChMP 140/1, left upper canine. Copper Sandstones, Perm' Province. Early Tatarian.

Referred material. Partial skull from Kargala mines, Orenburg Province (BMNH R4055); skull with lower jaw from Isheevo, Tatarstan Republic; and various postcranial remains.

Diagnosis. Small form; skull moderately high; orbits relatively large; large pineal foramen opening in a boss; palatines and pterygoids with teeth; canines rounded and curved; postcanine teeth small, transversely compressed.

Comments. Detailed descriptions and figures of *Syodon* have been given by Efremov (1954), Orlov (1958), Olson (1962), Chudinov (1983), King (1988), and Ivakhnenko (1995). *Syodon* is considered by Ivakhnenko (1994, 1995) as the type genus of a new family, the Syodontidae, which would be close to the Titanosuchidae. A close relative of *Syodon*, *Australosyodon*, has recently been described from South Africa (Rubidge, 1994).

Titanophoneus Efremov, 1938
See Figures 6.6 and 6.7.

Type species. Titanophoneus potens Efremov, 1938.

Holotype. PIN 157/1, an almost complete skeleton, with skull and lower jaw. Kamennyi Ravine, 7 km from Isheevo, Tatarstan Republic. Early Tatarian.

Referred material. PIN collections, at least four other individuals from the same locality.

Diagnosis. Medium-sized; skull high and narrow; long snout; large pineal foramen opening in a boss; dorsal and posterior orbital borders thickened; palatines with teeth; transverse flange of pterygoids with two teeth; large canines, compressed transversely, with anterior and posterior keels; small postcanine teeth; body relatively lightly built.

Comment. Descriptions of *Titanophoneus potens* have

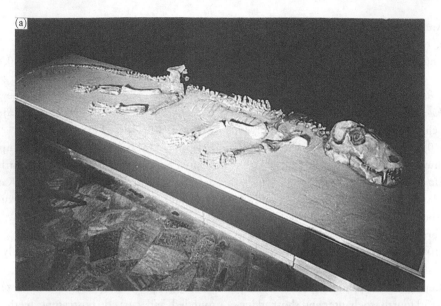

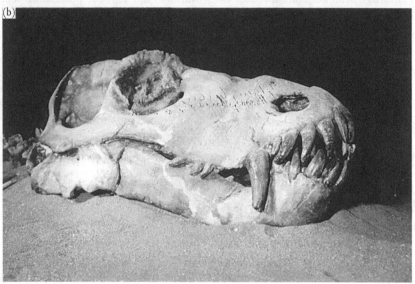

Figure 6.6. Skeleton (A) and skull (B) of the Late Permian dinocephalian *Titanophoneus potens* Efremov, 1938, holotype, PIN 157/1. Skeleton is about 3.5 m long.

been provided by Efremov (1938, 1940c, 1954), Orlov (1958), Olson (1962), Chudinov (1983), and King (1988).

Doliosauriscus Kuhn, 1961
See Figure 6.8A.
Synonym. Doliosaurus Orlov, 1958.

Type species. D. yanshinovi (Orlov, 1958) Kuhn, 1961.
Diagnosis. Large; skull high; small orbits; cranial roof thickened, with tuberosities; thick postorbital bar; palatines and pterygoids with small teeth; palatal depressions housing lower canines and first postcanine teeth; strong lower jaw, with massive symphysis; canines rounded.

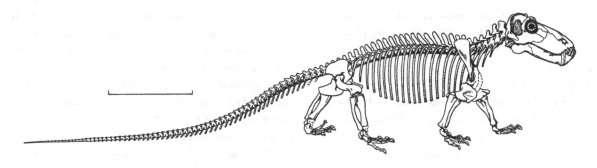

Figure 6.7. The Late Permian dinocephalian *Titanophoneus potens* Efremov, 1938, holotype, PIN 157/1. Skeleton, restored, in lateral view. Scale bar, 0.5 m. (After Orlov, 1958.)

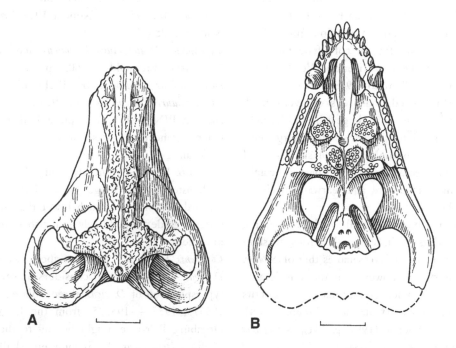

A   **B**

Figure 6.8. Anteosaurid dinocephalians from the Late Permian. A, *Doliosauriscus yanshinovi* (Orlov, 1958), holotype, PIN 157/3, skull in dorsal view. B, *Archaeosyodon praeventor* Tchudinov, 1960, holotype, PIN 1758/3, skull in ventral view. Scale bars, 50 mm. (A, after Orlov, 1958; B, after Chudinov, 1960.)

Doliosauriscus yanshinovi (Orlov, 1958)
Synonyms. Doliosaurus yanshinovi Orlov, 1958; = *Titanophoneus potens* Efremov 1940c, pp. 44–45, *pars*).
Holotype. PIN 157/3, skeleton with skull and lower jaw. Kamennyi Ravine, 7 km from Isheevo, Tatarstan Republic. Early Tatarian.

Referred material. Elements of forelimbs, figured by Olson (1962).
Diagnosis. Long-snouted.

Doliosauriscus adamanteus (Orlov, 1958)
Synonym. Doliosaurus adamanteus Orlov, 1958.

99

Holotype. PIN 519, distorted skull and partial skeleton. Right bank of Malyi Uran River, Orenburg Province.
Referred material. Many postcranial bones, described by Olson (1962).
Diagnosis. Short-snouted.
Comment. Doliosauriscus is considered by Ivakhnenko (1995, p. 99) as a junior synonym of *Titanophoneus.*

Notosyodon Tchudinov, 1968b
Type species. Notosyodon gusevi Tchudinov, 1968b.
Holotype. PIN 2505/1, an incomplete skull, without snout and lower jaw. Right bank of the Zhaksy-Karagala River, 7–8 km south of the village Turazhol, Aktyubinsk Province, Kazakhstan. Early Tatarian.
Referred material. Two teeth which belong probably to the type specimen, and a fragment of left dentary from the right bank of the Donguz River, Sol'-Iletsk district, Orenburg Province (Chudinov, 1968b, 1983).
Diagnosis. Medium-sized; massive skull; very thick interorbital and parietal areas; large orbits with thickened edges; high and massive parietal boss; large temporal fenestrae; interorbital portion of skull flat and straight, without median ridge; upper part of occiput strongly inclined downward and rearward; occipital condyle very large.
Comments. 'In its general overall configuration, the position of its orbits and its temporal cavities, the skull of *Notosyodon* very strongly resembles that of *Syodon*, from which it differs, however, in its considerably larger size and the much greater massiveness of its component elements' (Chudinov, 1968b, p. 4). According to Ivakhnenko (1995, p. 101), *Notosyodon* might be a junior synonym of *Syodon*.

Archaeosyodon Tchudinov, 1960
See Figure 6.8B.
Type species. Archaeosyodon praeventor Tchudinov, 1960.
Holotype. PIN 1758/3, deformed skull without lower jaw, and lacking the occipital region. Ezhovo, Ocher district, Perm' Province. Late Kazanian–earliest Tatarian.
Referred material. Many incomplete skulls and jaws from the same locality as the type (complete list in Chudinov, 1983, p. 89).

Diagnosis. Medium-sized; massive and relatively high skull; maxillae with sculptured surfaces; upper borders of orbits and skull roof thickened; short choanae occupying anterior position; many palatal teeth, situated on large prominences of palatines and pterygoid flanges; short, curved upper canines.
Comment. Archaeosyodon differs from *Syodon* in particular in its more anterior choanae, and in the greater development of the palatal tooth-bearing prominences. It is attributed to the Titanosuchidae by Ivakhnenko (1995).

Deuterosaurus Eichwald, 1860
Synonyms. Mnemeiosaurus Nopcsa, 1928; *Uraniscosaurus* Nopcsa, 1928.
Type species. Deuterosaurus biarmicus Eichwald, 1860 (= *Eurosaurus* Eichwald, 1860, p. 1613, *pars*; = *Deuterosaurus mnemonialis* Eichwald, 1860; = *Uraniscosaurus watsoni* Nopcsa, 1928).
Holotype. PIN 1954/1, incomplete skull and lower jaw. Copper sandstones, Bashkortostan Republic. Early Tatarian.
Diagnosis. Large; skull nearly as high as long; reduced pachyostosis; jaw hinge only slightly displaced forward, but situated much lower than upper tooth series; well developed heels on incisors; small postcanine teeth.
Comments. Efremov (1954) described a second species, *D. gigas*. Boonstra (1965) considers however that the type specimen of *D. gigas*, three isolated teeth (PIN 1955/3, 1955/4, 1955/5) from the Kargala mines, Orenburg Province, might belong to the tapinocephalid *Ulemosaurus*. An incomplete skull from the Copper Sandstones of Bashkortostan Republic (PIN 1954/2), attributed by Seeley (1895) to *D. biarmicus*, was renamed by Nopcsa (1928) *Mnemeiosaurus jubilaei*. However, the two species *D. biarmicus* and *M. jubilaei* are based on incomplete skulls, and the suites of characters given by Efremov (1954) to define the genera *Deuterosaurus* and *Mnemeiosaurus* are not mutually exclusive. There seem to be no good reason for keeping *Mnemeiosaurus* distinct from *Deuterosaurus* (King, 1988; Ivakhnenko, 1995). *Deuterosaurus* has been retained both by Chudinov (1983) and

Ivakhnenko (1995, p. 99) in the separate family Deuterosauridae.

Admetophoneus Efremov, 1954

Type species. *Admetophoneus kargalensis* Efremov, 1954.

Lectotype. PIN 1954/5, fragment of preorbital part of skull. Kargala mines, Orenburg Province. Early Tatarian.

Referred material. A fragment of palate and an incomplete humerus might belong to *A. kargalensis* (Chudinov, 1983).

Diagnosis. Large; upper canines strongly curved, compressed laterally, without keels; lower canines almost straight, their tips housed in palatal depressions; postcanine teeth very compressed transversely, bent backwards; palatal protuberances bearing ten teeth with short and blunt crowns.

Comments. This is a poorly known genus. The species *A. kargalensis* was initially erected by Efremov (1954, pp. 260–262) on the basis of a 'holotype' composed of two skull fragments belonging to two distinct individuals. Chudinov (1983, pp. 99–100) showed that they represent two different species, one of them belonging to the Estemmenosuchidae. The other was designated by him as lectotype of *A. kargalensis*.

Microsyodon Ivakhnenko, 1995

Type species. *Microsyodon orlovi* Ivakhnenko, 1995.

Holotype. PIN 4276/13, right maxilla. Golyusherma, Kama River, Udmurt Republic. Early Kazanian.

Referred material. An incomplete left maxilla from Kirov Province (Ivakhnenko, 1995).

Diagnosis. Very small; upper canine almost rounded in section, curved, with anterior and posterior keels; 10–11 postcanine teeth strongly compressed transversely, with cutting edges; small precanine tooth.

Comment. *Microsyodon orlovi* was attributed by Ivakhnenko (1995) to the Titanosuchidae.

Chthomaloporus Tchudinov, 1964b

Type species. *Chthomaloporus lenocinator* Tchudinov, 1964b.

Holotype. PIN 1758/17, pelvic girdle and sacral vertebrae. Ezhovo, Ocher district, Perm' Province. Late Kazanian–earliest Tatarian.

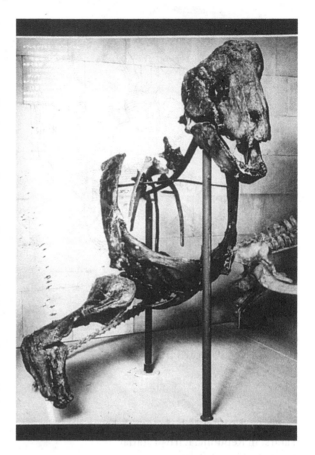

Figure 6.9. The Late Permian tapinocephalid dinocephalian *Ulemosaurus svijagensis* Ryabinin, 1938, lectotype, PIN 2207/2, skull and anterior skeleton. Skull is about 0.3 m long.

Diagnosis. Pelvic girdle lightly built; close proximity of acetabula; great height of ischiatic symphysis; well developed process on anterior edge of ilium; fused sacral vertebrae.

Family TAPINOCEPHALIDAE Gregory, 1926

Diagnosis. Extensive cranial pachyostosis, leading to reduction in size of temporal fenestrae; jaw hinge displaced forward; incisors with well developed heels; canines reduced; teeth interdigitating.

Ulemosaurus Ryabinin, 1938
See Figures 6.9 and 6.10.

Type species. *Ulemosaurus svijagensis* Ryabinin, 1938 (= *Moschops svijagensis* Tchudinov, 1983).

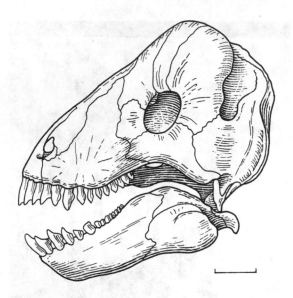

Figure 6.10. The Late Permian tapinocephalid dinocephalian *Ulemosaurus svijagensis* Ryabinin, 1938, lectotype, PIN 2207/2, skull in lateral view. Scale bar, 50 mm. (After Efremov, 1940a,b,c.)

Lectotype. PIN 2207/2, skull with lower jaw. Isheevo, Tatarstan Republic. Early Tatarian.

Referred material. Two other skulls from the same locality (syntypes).

Diagnosis. Very large; snout narrow and tapering; skull wide and very high in postorbital region; thick cranial roof; strongly developed pachyostosis of dorsal border of orbit; broad postorbital bar; temporal opening relatively large for tapinocephalid, and, hence, relatively narrow intertemporal region; large incisors; medium-sized canines; anterior postcanine teeth much larger than posterior ones.

Comments. *Ulemosaurus* was considered by Chudinov (1983) as a junior synonym of the South African genus *Moschops*; it appears however to be a far more archaic genus than the latter. Because of its primitive features, it is attributed by Ivakhnenko (1994) to a new family of primitive Tapinocephalia, the Ulemosauridae. In many respects, *Ulemosaurus* is similar to an even more primitive genus from South Africa, *Tapinocaninus* Rubidge, 1991.

DINOCEPHALIA *incertae sedis*
Phreatosuchus Efremov, 1954
Synonym. *Dinosaurus* Seeley, 1894, pp. 711–713, *pars*.
Type species. *Phreatosuchus qualeni* Efremov, 1954.
Holotype. LGM/ChMP 77, part of femur from Bashkortostan Republic. Kazanian.

Phreatosaurus Efremov, 1954
Type species. *Phreatosaurus bazhovi* Efremov, 1954 (= *Eurosaurus* Eichwald, 1860, *pars*; = *Rhopalodon* Seeley, 1894, *pars*).
Holotype. LGM/ChMP 75, part of femur from Bashkortostan Republic. Kazanian.

Phreatosaurus menneri Efremov, 1954
Holotype. LGM/ChMP 74, part of femur from Bashkortostan Republic. Kazanian.

Rhopalodon Fischer, 1841
Type species. *Rhopalodon wangenheimi* Fischer, 1881.
Holotype. Fragment of left ramus of lower jaw, with teeth; present location unknown. Bashkortostan Republic. Early Tatarian.
Comment. *R. wangenheimi* is considered to be a valid genus and species of Estemmenosuchidae by Chudinov (1983, pp. 108–110).

Parabradysaurus Efremov, 1954
Type species. *Parabradysaurus udmurticus* Efremov, 1954. (= *Rhopalodon murchisoni* in Stuckenberg, 1898, p. 71; = *Rhopalodon wangenheimi* in Stuckenberg, 1898, p. 301).
Holotype. TsNIGRI 2/1727, fragment of right ramus of lower jaw with teeth. Udmurt Republic. Late Kazanian. Originally described as a pareiasaur by Efremov (1954). *P. udmurticus* is considered to be a valid genus and species of Estemmenosuchidae by Chudinov (1983, pp. 110–111), and a rhopalodontid by Ivakhnenko (1996a).

Suborder ANOMODONTIA Owen, 1859
Comments. Within the Anomodontia (*sensu stricto*, not including the Dinocephalia), a specialized group, the Dicynodontia, can easily be recognized on the basis of its many synapomorphies, of which the most obvious

is probably the complete loss of both the upper and lower incisors. More primitive Anomodontia were traditionally split into two groups, the Dromasauria, known originally only from South Africa, and the Venyukoviamorpha, known only from Russia. Recent fossil finds and subsequent phylogenetic studies have shown, however, that both the Dromasauria and the Venyukoviamorpha are paraphyletic groups (Rubidge and Hopson, 1990).

The Dicynodontia, by far the most numerous of the Anomodontia, are known from the Late Permian to the Late Triassic. During the Late Permian, they became the dominant element of the herbivorous fauna. The Permo-Triassic boundary proved fatal to most taxa, and the numbers of genera were reduced more than ten times by the Early Triassic. There was another period of dicynodont diversification in the Middle Triassic, but this was much less pronounced than in the Late Permian.

The Dicynodontia are highly modified to a herbivorous mode of life. Their main features are: loss of teeth (except, usually, the upper tusks); development of characteristic horny beaks; temporal fenestrae elongated; occiput usually broad; jaw articulation lying at level of occipital condyle; preorbital region short and usually slightly inclined; zygomatic arch flaring laterally and emarginated ventrally.

In advanced dicynodonts, a secondary palate, formed by plates of the premaxillae and maxillae, is developed. There are no palatal teeth.

The first known anomodont remains from Russia, discovered at the beginning of the twentieth century in the Northern Dvina basin, were described by Amalitskii in 1922. Subsequently, anomodont remains were found in many other Permian and Triassic redbeds. The anomodont faunas of Orenburg Province and Bashkortostan Republic proved most abundant, yielding almost all the Triassic genera. At present, 13 anomodont genera in four families are known from Russia; most are endemic.

Anomodonts are a typical component of Late Permian and Middle Triassic tetrapod faunas in Russia, but they are either missing or rare in the Early Triassic. The Late Permian faunal complexes

(Ivakhnenko, 1992) are mainly characterized by anomodonts: *Otsheria* (Ocher complex), *Venyukovia* (Isheevo complex), *Suminia* and a dicynodont close to *Tropidostoma* (Kotel'nich complex). *Dicynodon* is a component of the Sokolki complex (uppermost part of the Upper Permian). Early Triassic dicynodonts of Russia are represented only by the genus *Lystrosaurus*, known by incomplete remains in the lowermost part of the *Tupilakosaurus* assemblage (Lozovskii, 1983; Ochev, 1992; Ochev and Shishkin, 1989). In the Middle Triassic, advanced dicynodonts became one of the main components of terrestrial vertebrate faunas. Dicynodonts of the *Eryosuchus* fauna are represented by six genera: *Calleonasus*, *Edaxosaurus*, *Rabidosaurus*, *Rhadiodromus*, *Rhinodicynodon*, and *Uralokannemeyeria*. The *Mastodonsaurus* fauna is characterized by only two genera: *Elephantosaurus* and *Elatosaurus*.

Primitive ANOMODONTIA
Family OTSHERIIDAE Tchudinov, 1960
Otsheria Tchudinov, 1960
See Figure 6.11A.

Type species. Otsheria netzvetajevi Tchudinov, 1960.
Holotype. PIN 1758/5, skull without lower jaw. Ezhovo, Ocher district, Perm' Province. Late Kazanian–earliest Tatarian.
Diagnosis. Small (skull length about 100 mm); snout high and narrow; parietal foramen very large; temporal fenestrae elongated; choanae large, wide; four teeth on premaxilla, nine on maxilla; in maxilla, fourth and fifth teeth slightly larger than others; no differentiated canine.

Family GALEOPIDAE Broom, 1912
Comment. The Galeopidae are understood here as including only the South African *Galeops* and the Russian *Suminia*: the other 'dromasaurs', *Patranomodon*, *Galechirus*, and *Galepus*, all from South Africa, appear to be at a comparatively more primitive grade (see Rubidge and Hopson, 1990).

Suminia Ivakhnenko, 1994
See Figure 6.11B.
Type species. Suminia getmanovi Ivakhnenko, 1994.

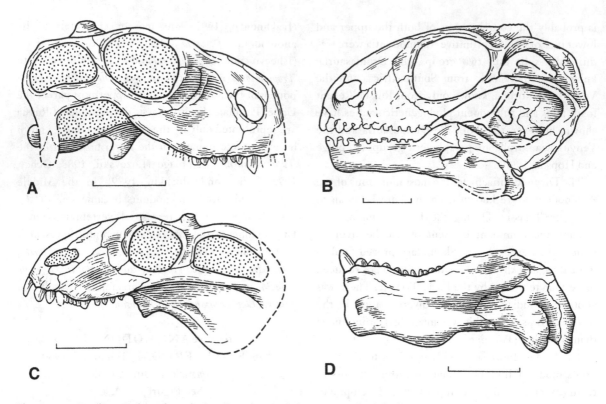

Figure 6.11. Basal anomodonts from the Late Permian. A, *Otsheria netzvetajevi* Tchudinov, 1960, holotype, PIN 1758/5, skull in lateral view. B, *Suminia getmanovi* Ivakhnenko, 1994, holotype, PIN 2212/10, skull in lateral view. C, *Venyukovia prima* Amalitskii, 1922, specimen PIN 2793/1, skull in lateral view. D, *Venyukovia prima* Amalitskii, 1922, specimen PIN 157/5, lower jaw in lateral view. Scale bars, 50 mm, except (B), 10 mm. (A, after Chudinov, 1960; B, after Ivakhnenko, 1994; C, D, after Chudinov, 1983.)

Holotype. PIN 2212/10, incomplete skeleton. Kotel'nich, Kirov Province. Earliest Late Tatarian.

Referred material. Several skulls and lower jaws, and isolated teeth, all from the same locality as the type.

Diagnosis. Very small (skull length up to 55 mm); skull high and short; zygomatic arch sharply bent upwards behind orbit; premaxilla in contact with palatine, and forming with the latter narrow palatal shelf; lateral depression on posterodorsal border of dentary for insertion of adductor jaw musculature; contact between premaxilla and palatine, and lateral depression on posterodorsal border of dentary, are derived characters shared with more advanced anomodonts (*Venyukovia* and the dicynodonts); incisors of decreasing size from front to rear; maxilla with small teeth; no

morphologically differentiated canine; limbs particularly long; digital formula, 2.3.3.3.3.

Family VENYUKOVIIDAE Efremov, 1940c
Venyukovia Amalitskii, 1922
See Figures 6.11C and D.

Synonym. Myctosuchus Efremov, 1937).

Type species. Venyukovia prima Amalitskii, 1922 (= *Venjukovia invisa* Efremov, 1940a, b, c).

Lectotype. PIN 48/1, a fragment of left ramus of lower jaw with teeth. Kargala mines (exact locality unknown), Orenburg Province. Early Tatarian.

Referred material. From Kargala mines: a lower jaw symphysis, PIN 48/2. From Isheevo, Tatarstan Republic: a preorbital part of skull with lower jaw, PIN 157/5 (holotype of *Venjukovia invisa* Efremov, 1940c);

lower jaw symphyses, PIN 157/6, 157/7, 157/8, 157/114; lower jaws, PIN 157/1111 and 157/1112; incomplete lower jaw, PIN 157/1113. From Novo-Nikolskoe: a skull without lower jaw, PIN 2793 (Efremov, 1940c; Chudinov, 1983).

Diagnosis. Rather small (skull length about 120 mm); preorbital part of skull elongated; lower jaw high and very massive; symphysial region subrectangular in dorsal view, and with chin processes directed downwards; opening between dentary and angular; front incisors chisel-shaped, followed by 5 or 6 small teeth and low, massive canine; postcanines small and blunt, arranged in two rows.

Comment. According to Rubidge and Hopson (1990), *Venyukovia* is the sister-taxon of the Dicynodontia.

Infraorder DICYNODONTIA Owen, 1859
Comment. Except for the primitive Anomodontia (removed from the Dicynodontia), the classification adopted here is taken from King (1988).

Superfamily PRISTERODONTOIDEA Cluver et King, 1983
Family DICYNODONTIDAE Owen, 1859
Subfamily DICYNODONTINAE Owen, 1859
Dicynodon Owen, 1845
Type species. Dicynodon lacerticeps Owen, 1845, from the Late Permian of South Africa.
Diagnosis. Medium-sized to large; dentition reduced to one pair of maxillary tusks; intertemporal region narrow; lower jaw with dentary ledge.
Comment. Dicynodon is cosmopolitan, known mainly from Gondwanaland (South Africa, in particular), but also from Laurasia.

Dicynodon trautscholdi Amalitskii, 1922
See Figure 6.12.
Synonyms. Gordonia annae Amalitskii, 1922; *Gordonia rossica* Amalitskii, 1922; *Oudenodon venyukovi* Amalitskii, 1922; *Dicynodon annae* (Amalitskii, 1922) Sushkin, 1926; *Dicynodon amalitzkii* Sushkin, 1926.

Subfamily KANNEMEYERIINAE Huene, 1948
Tribe LYSTROSAURINI Broom, 1903

Figure 6.12. The Late Permian dicynodont *Dicynodon trautscholdi* Amalitskii, 1922, skull in lateral view. The skull is about 0.2 m long.

Lystrosaurus Cope, 1870
See Figures 6.13 and 6.14A.
Type species. Lystrosaurus murrayi (Huxley, 1859) Broom, 1932, Early Triassic of South Africa.
Diagnosis. Medium-sized; snout short and very deep, sloping downward; orbits high, with nares immediately anterior to them.
Comment. Cosmopolitan: *Lystrosaurus* is known from South Africa, Antarctica, India, China (Xinjiang), and Russia.

Lystrosaurus georgi Kalandadze, 1975
Holotype. PIN 3447/1, anterior part of skull and substantial part of postcranial skeleton. Astashikha I, Vetluga River, Nizhnii Novgorod Province. Vokhmian Gorizont (Ryabinskian Member), Induan (Early Triassic).

Tribe SINOKANNEMEYERIINI King, 1988
Rhadiodromus Efremov, 1951
See Figure 6.14B.
Type species. Rhadiodromus klimovi (Efremov, 1940c) Efremov, 1951 (= *Lystrosaurus klimovi* Efremov, 1940c; = *Rhinocerocephalus cisuralensis* Vjuschkov, 1969).
Holotype. PIN 159/1, part of skull and incomplete skeleton. Donguz I, right bank of the Donguz river near Perovka, Sol'-Iletsk district, Orenburg Province. Donguz Gorizont, Middle Triassic.

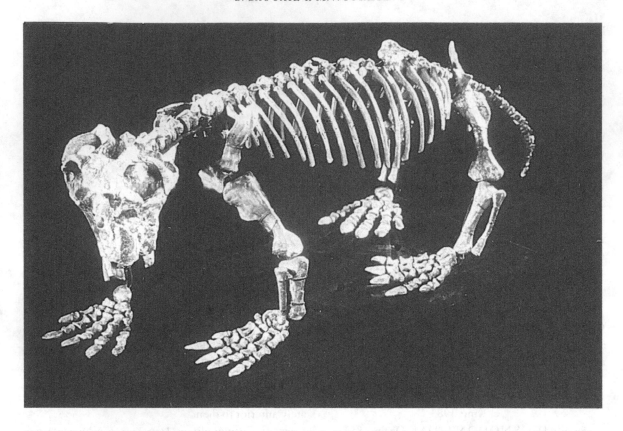

Figure 6.13. The Early Triassic dicynodont *Lystrosaurus georgi* Kalandadze, 1975, holotype, PIN 3447/1, skull and skeleton. The whole animal is about 1 m long.

Referred material. From the same locality as the type: an incomplete skull, PIN 952/111, and isolated bones, PIN 952/13, 18. From Koltaevo II (Bashkortostan Republic), skull fragments, PIN 2866/1. The skull of a large kannemeyriine from Berdyanka I (Orenburg Province), PIN 1579/14, belongs probably to the same species.

Diagnosis. Large; maximum length of skull, 450 mm; total length of skeleton, about 2 m (including skull and tail); skull very heavy; thickenings on premaxillae and nasals, and on maxillae around tusks, particularly conspicuous; sagittal crest narrow and relatively low; owerful tusks.

Uralokannemeyeria Danilov, 1971
See Figure 6.14C.

Type species. Uralokannemeyeria vjuschkovi Danilov, 1971.

Holotype. SGU D-104/1, a skull. Karagachka, Akbulak district, Orenburg Province. Donguz Gorizont, Middle Triassic.

Referred material. SGU D-104/2, an incomplete skull; an ilium and a tibia (Danilov, 1973).

Diagnosis. Large; snout wedge-shaped; skull roof flattened; rugose thickenings on nasal bones, maxillae, and superior orbital margins; large tusks directed obliquely, forward and downward; parietal crest massive, but low; occiput low, set at acute angle to dorsal surface of skull; secondary palate small.

Tribe KANNEMEYERIINI Lehman, 1961
Rabidosaurus Kalandadze, 1970
See Figure 6.15A.

Type species. Rabidosaurus cristatus Kalandadze, 1970.

Holotype. PIN 952/100, an almost complete skull.

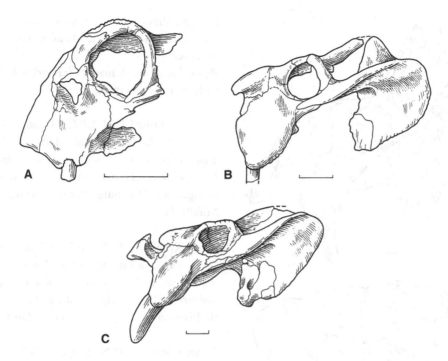

Figure 6.14. Skulls of dicynodonts from the Early and Middle Triassic, all in lateral view. A, *Lystrosaurus georgi* Kalandadze, 1975, holotype, PIN 3447/1. B, *Rhadiodromus klimovi* (Efremov, 1940a,b,c), PIN 952/111. C, *Uralokannemeyeria vjuschkovi* Danilov, 1971, holotype, SGU D-104/1. Scale bars, 50 mm. (A, after Kalandadze, 1975; B, after V'yushkov, 1969; C, after Danilov, 1971.)

Donguz I, Sol'-Iletsk district, Orenburg Province. Donguz Gorizont, Middle Triassic.

Referred material. PIN 2866/8, 9, fragments of skulls, from Koltaevo II (Bashkortostan Republic). Donguz Gorizont, Middle Triassic.

Diagnosis. Large; maximum length of skull, 570 mm; flat coarsely sculptured swellings on maxillae and premaxillae; narrow processes of postorbitals and thick parietals form huge sagittal crest, which slopes steeply posteriorly and dorsally at an angle of about 120° to dorsal surface of frontals; pineal opening huge; tusks large.

Tribe SHANSIODONTINI Cox, 1965
Rhinodicynodon Kalandadze, 1970
See Figure 6.15B.

Type species. Rhinodicynodon gracile Kalandadze, 1970.

Holotype. PIN 1579/50, a skull with lower jaw.

Berdyanka I, Sol'-Iletsk district, Orenburg Province. Donguz Gorizont, Middle Triassic.

Referred material. PIN 1579/51, 52, two skulls with postcranial remains; SGU 104/3885, a complete skeleton; all from the same locality as the type.

Diagnosis. Small (skull about 180 mm long); small, rounded thickening on each nasal bone; upper and posterior margins of orbit slightly thickened; interorbital distance much wider than intertemporal; sagittal crest low; parietal opening very small; maxillae swollen in region of tusk alveoli; tusks large and thick.

Tribe STAHLECKERIINI Lehman, 1961
Elephantosaurus Vjuschkov, 1969

Type species. Elephantosaurus jachimovitchi Vjuschkov, 1969.

Holotype. PIN 525/25, a fragment of the interorbital

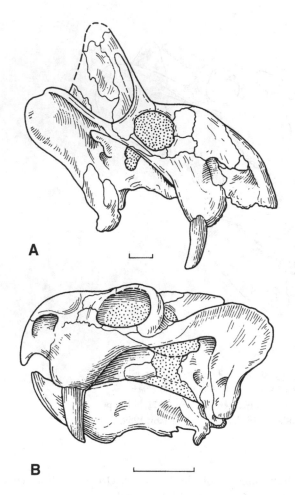

Figure 6.15. Skulls of dicynodonts from the Middle Triassic, both in lateral view. A, *Rabidosaurus cristatus* Kalandadze, 1970, holotype, PIN 952/100. B, *Rhinodicynodon gracile* Kalandadze, 1970, holotype, PIN 1579/50. Scale bars, 50 mm. (After Kalandadze, 1970.)

region of a skull roof. Koltaevo III, Bashkortostan Republic. Bukobay Gorizont, Middle Triassic.
Diagnosis. Giant form (skull probably up to 1 m long).
Comments. Elephantosaurus was regarded as close to *Stahleckeria* by V'yushkov (1969), but this was questioned by King (1988), who considers that the holotype and only known specimen is so fragmentary that its relationships cannot be determined.

Elatosaurus Kalandadze, 1985
Type species. Elatosaurus facetus Kalandadze, 1985.

Holotype. PIN 2867/1, a nasal bone. Koltaevo III, Bashkortostan Republic. Bukobay Gorizont, Middle Triassic.
Referred material. A maxilla fragment, found together with the type specimen.

Tribe PLACERIINI King, 1988
Edaxosaurus Kalandadze, 1985
Type species. Edaxosaurus edentatus Kalandadze, 1985.
Holotype. SGU D-104/4–1, an incomplete maxilla. Karagachka, Orenburg Province. Donguz Gorizont, Middle Triassic.

KANNEMEYERIINAE *incertae sedis*
Calleonasus Kalandadze, 1985
Type species. Calleonasus furvus Kalandadze, 1985.
Holotype. PIN 525/266, a left nasal bone. Koltaevo II, Bashkortostan Republic. Donguz Gorizont, Middle Triassic.
Referred material. PIN 525/267–270, several nasal bones from the same locality.

Suborder THERIODONTIA
Comments. The Theriodontia are understood here to include two sister-groups of higher carnivorous therapsids, the Therocephalia and the Cynodontia. It must be remembered, however, that many authors interpret the Theriodontia more broadly to include the Gorgonopsia, and sometimes also other groups of primitive carnivorous therapsids.

Infraorder THEROCEPHALIA Broom, 1903
Comments. The Therocephalia (including the Bauriamorpha of early authors) are undoubtedly a monophyletic group, but they are quite diverse in their structure and have been classified in many different ways (see Haughton and Brink, 1954; Watson and Romer, 1956; Mendrez, 1972, 1974, 1975; Tatarinov, 1974; Kemp, 1982; Hopson and Barghusen, 1986; Hopson, 1991). There is no agreement on the number of families (from eight according to Kemp, 1982, to 16 in Tatarinov, 1974) and on the content of each family. In the most recent work (Hopson, 1991), two major subgroups are recognized, the more primitive

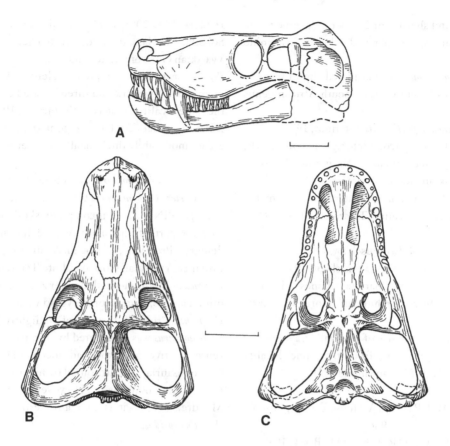

Figure 6.16. Therocephalians from the Late Permian. A, *Scylacosuchus orenburgensis* Tatarinov, 1968a, holotype, PIN 2628/1, skull in lateral view. B, C, *Annatherapsidus petri* (Amalitskii, 1922), holotype, PIN 2005/1993, skull in dorsal (B) and ventral (C) views. Scale bars, 50 mm. (A, after Tatarinov, 1968; B, C, after V'yushkov, 1955.)

Pristerosauria and the more derived Eutherocephalia; the Pristerosauria, however, might be paraphyletic, and the interrelationships among the four monophyletic subgroups of the Eutherocephalia (Hofmeyriidae, Euchambersiidae, Whaitsiidae, and the superfamily Baurioidea) remain unresolved (Hopson, 1991). Only a few Therocephalia are known from Russia.

PRISTEROSAURIA Hopson et Barghusen, 1986
Family SCYLACOSAURIDAE Broom, 1903
Comment. This family of primitive Therocephalia (Pristerosauria) is understood here as including the Pristerognathidae Watson et Romer, 1956. Russian genera are *Scylacosuchus*, as well as *Hexacynodon*

and *Porosteognathus*, both represented by fragmentary material.

Scylacosuchus Tatarinov, 1968a
See Figure 6.16A.
Type species. Scylacosuchus orenburgensis Tatarinov, 1968a.
Holotype. PIN 2628/1, an incomplete skeleton with most of the skull and lower jaw. Orenburg district, Orenburg Province. Severodvinskian Gorizont. Late Tatarian.
Diagnosis. Large form with heavy skull; snout long; intertemporal region with well developed sagittal crest; very narrow epipterygoid; 5 upper and 5 lower

incisors, long and sharp; 1 or 2 upper precanine teeth; very large canines; 6 upper and 4 lower postcanine teeth.

Comment. *Scylacosuchus* is considered by Tatarinov (1968a, 1974) to be close to the South African genus *Scylacosaurus*.

Hexacynodon purlinensis Tatarinov, 1974

Holotype. PIN 1538/6, incomplete right maxilla. Purly, Shakhun'ya district, Nizhnii Novgorod Province. Topmost Vyatskian Gorizont, late Tatarian.

Referred material. Fragment of dentary and five isolated teeth, from the same locality (Tatarinov, 1974, 1993).

Porosteognathus Vjuschkov, 1955
Porosteognathus efremovi Vjuschkov, 1955

Lectotype. PIN 157/19, parietal bones. Isheevo, Tatarstan Republic. Urzhumian Gorizont, Early Tatarian.

Referred material. Two pairs of parietals, fragments of jaws and left squamosal, from the same locality (V'yushkov, 1952, 1955; Tatarinov, 1974).

EUTHEROCEPHALIA Hopson et Barghusen, 1986
Family MOSCHORHINIDAE Brink, 1959

Synonym. Euchambersiidae *sensu* Hopson and Barghusen, 1986.

Diagnosis. Wide-snouted; no secondary palate; vomer very wide anteriorly.

Comment. According to Mendrez (1974), three subfamilies can be defined: the Annatherapsidinae, Moschorhininae, and Euchambersiinae.

Subfamily ANNATHERAPSIDINAE Kuhn, 1961

Diagnosis. Primitive: retain palatal teeth on pterygoids; complete dentition with well developed postcanine teeth; postfrontal present; no parietal foramen.

Annatherapsidus Kuhn, 1961
See Figure 6.16B, C.

Synonym. *Anna* Amalitskii, 1922.

Type species. *Annatherapsidus petri* (Amalitskii, 1922) Kuhn, 1961.

Holotype. PIN 2005/1993, a skull without lower jaw. Sokolki, Kotlas district, Arkhangel'sk Province. Vyatskian Gorizont, Late Tatarian.

Referred material. Incomplete skeleton, skull, incomplete skull, lower jaw and three isolated canines, from the same locality as the type (Tatarinov, 1974).

Diagnosis. Relatively large; zygomatic and postorbital arches moderately thick; small interpterygoid vacuity.

Chthonosaurus Vjuschkov, 1955

Type species. *Chthonosaurus velocidens* Vjuschkov, 1955.

Holotype. PIN 521/1, incomplete skull, lacking the anterior part of the snout, with a fragment of left dentary. Pron'kino, Sorochinsk district, Orenburg Province. Vyatskian Gorizont, Late Tatarian.

Diagnosis. Medium-sized; slender zygomatic and postorbital arches; large interpterygoid vacuity.

Comments. Probably in view of its lightly built skull, *Chthonosaurus* was considered by Tatarinov (1974) as a representative of the advanced Therocephalia (Scaloposauria), and attributed by him to a new family, the Chthonosauridae. However, according to Mendrez (1974), it is in many respects similar to *Annatherapsidus*.

Family WHAITSIIDAE Haughton, 1918

Diagnosis. Constriction of snout behind canines; tendency towards reduction in size of suborbital fenestra; small interpterygoid vacuity; reduced number of postcanine teeth; characteristic is presence of rudimentary secondary palate, formed by lateral processes of vomer contacting medial processes of maxillae.

Comment. Two subfamilies can be recognized, the more primitive Moschowhaitsiinae and the more derived Whaitsiinae (Tatarinov, 1963; Mendrez, 1974).

Subfamily MOSCHOWHAITSIINAE Tatarinov, 1963

Diagnosis. Retain surborbital fenestra; retain one or two precanine and some postcanine teeth.

Moschowhaitsia Tatarinov, 1963
See Figure 6.17A.

Type species. *Moschowhaitsia vjuschkovi* Tatarinov, 1963.

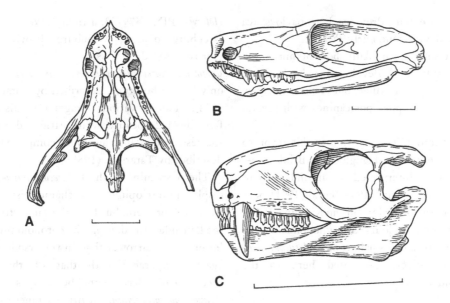

Figure 6.17. Therocephalians from the Late Permian (A), Early Triassic (B), and Middle Triassic (C). (A) *Moschowhaitsia vjuschkovi* Tatarinov, 1963, holotype, PIN 1100/20, skull in ventral view. (B) *Silphedosuchus orenburgensis* Tatarinov, 1977b, holotype, PIN 952/100, skull in lateral view. (C) *Nothogomphodon danilovi* Tatarinov, 1974, holotype, PIN 2865/1, skull in lateral view. Scale bars, 50 mm. (A, C, after Tatarinov, 1974; B, after Tatarinov, 1977b).

Holotype. PIN 1100/20, incomplete skull, lacking the otico-occipital region. Vyazniki-1, Vladimir Province. Topmost Vyatskian Gorizont, Late Tatarian.

Referred material. A few isolated skull and jaw fragments from the localities Vyazniki-1 and Vyazniki-2 (Tatarinov, 1974).

Diagnosis. Medium-sized; skull rather lightly built, broad in temporal region; no palatal teeth; upper dentition: 5 incisors, 1 very small precanine, large canine, 7 postcanines.

Comment. This therocephalian was described in great detail by Tatarinov (1963, 1974).

Viatkosuchus Tatarinov, 1995

Type species. Viatkosuchus sumini Tatarinov, 1995.

Holotype. PIN 2212/13, an almost complete but heavily deformed skull and lower jaw, with an important part of the postcranial skeleton. Kotel'nich-2, Vyatka River, Kirov Province. Late Tatarian.

Diagnosis. Medium-sized; skull rather lightly built, with temporal region slightly narrower than in

Moschowhaitsia; well developed, numerous palatal teeth, both on palatines and pterygoids; upper dentition: 5 incisors, 1 or 2 precanines, large canine, 7 postcanines.

Superfamily BAURIOIDEA Broom, 1911

Diagnosis. Advanced Therocephalia; maxillae play significant part in secondary palate.

Family ICTIDOSUCHOPSIDAE Hopson et Barghusen, 1986

Diagnosis. 'Palatal process of maxilla nearly, or just, contacts vomer, but sutural connection is lacking' (Hopson and Barghusen, 1986).

Silphedosuchus Tatarinov, 1977b
See Figure 6.17B.

Type species. Silphedosuchus orenburgensis Tatarinov, 1977b.

Holotype. PIN 952/100, skull with lower jaw, lacking the otico-occipital region. Rassypnaya, right bank of

the Ural River, Orenburg Province. Petropavlovskaya
Svita, Yarenskian Gorizont, Early Triassic.
Diagnosis. Very small form (skull about 30 mm long);
very lightly built; well developed palatal plates,
mainly constituted by maxillae, but leaving a narrow
slot in midline; last upper postcanines with several
cusps.
Comments. Tatarinov (1977b) attributes this taxon to
the family Silphedestidae Haughton et Brink, 1954.
However, whether interpreted as an independent
family (Haughton and Brink, 1954; Battail, 1991), or as
juvenile procynosuchids (Hopson and Kitching, 1972),
the silphedestids are undoubtedly cynodonts. On the
basis of the structure of its palate, *Silphedosuchus orenburgensis* is tentatively attributed here to the
Ictidosuchopsidae.

Family BAURIIDAE Broom, 1911
Diagnosis. Very advanced Therocephalia; complete
secondary palate, covering vomer; upper dentition:
only 4 incisors, no precanines, canine of moderate
size; postcanine teeth expanded tranversely; upper
and lower postcanine teeth intermeshing; herbivorous.

Subfamily NOTHOGOMPHODONTINAE
Tatarinov, 1974
Nothogomphodon Tatarinov, 1974
See Figure 6.17C.
Type species. Nothogomphodon danilovi Tatarinov, 1974.
Holotype. PIN 2865/1, anterior part of a skull with
lower jaw. Berdyanka-2, Orenburg district, Orenburg
Province. Donguz Gorizont, Middle Triassic.
Comments. Nothogomphodon was attributed by Tatarinov
(1974) to the new family Nothogomphodontidae,
because, compared with typical Bauriidae, it retains
archaic features (large canine, less modified postcanine teeth, complete postorbital bar). We consider
Nothogomphodon as representing a relatively primitive
subfamily within the Bauriidae, the Nothogomphodontinae.

Subfamily BAURIINAE Broom, 1911
Antecosuchus Tatarinov, 1973
Type species. Antecosuchus ochevi Tatarinov, 1973.

Holotype. PIN 1579/53, a right maxilla. Berdyanka 1,
Orenburg district, Orenburg Province. Donguz
Gorizont, Middle Triassic.
Comments. Antecosuchus ochevi was originally known
only by the holotype, described by Tatarinov (1973,
1974). Later, various fragments discovered at
Berdyanka 2 locality were attributed to the same
species; these include an almost complete left dentary,
described by Tatarinov (1988).

The postcanine teeth of *Antecosuchus ochevi* do not
display, in our opinion, the characteristic features of
traversodont teeth, but have the same structure as in
the Bauriidae. In addition, the coronoid process of the
lower jaw is narrower than in traversodontids, but is
again comparable with that of the Bauriidae.
Antecosuchus ochevi might be a close relative of
Traversodontoides wangwuensis, a Chinese form which
was also initially described as a traversodont cynodont, and subsequently attributed to the Bauriidae by
Sun Ailin (1981).

BAURIOIDEA *incertae sedis*
Scalopognathus Tatarinov, 1974
Scalopognathus multituberculatus Tatarinov, 1974
Holotype. PIN 3076/1, the posterior half of a right
ramus of a very small lower jaw. Srednyaya Makarikha,
90 km north-west of Inta, Inta district, Komi Republic.
The specimen comes from a bore hole, and was found
at a depth of 720.45 m. Vetlugian Supergorizont, Early
Triassic.
Diagnosis. Very light form; complex postcanine teeth,
with expanded crowns bearing many small, low cusps.
Comment. Tatarinov (1974) placed this genus in the
family Scalopognathidae. However, this is undoubtedly a baurioid but, as modern classifications of the
superfamily are based primarily on the structure of
the palate, its familial status remains uncertain.

Dongusaurus schepetovi Vjuschkov, 1964
Holotype. PIN 952/1, a right dentary, incomplete pos-
teriorly, and lacking teeth. Donguz-1, Orenburg dis-
trict, Orenburg Province. Donguz Gorizont, Middle
Triassic.
Comment. Dongusaurus schepetovi was attributed to the

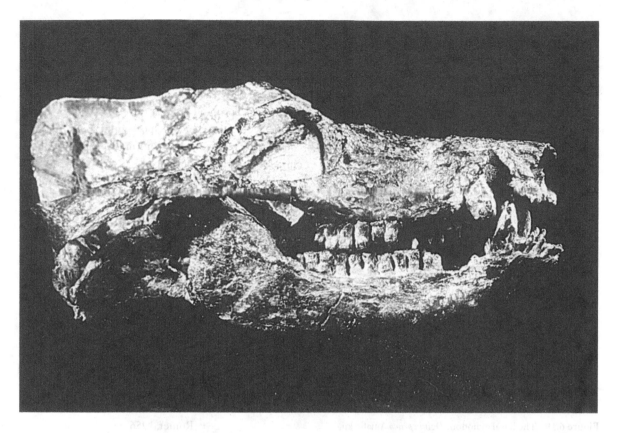

Figure 6.18. The basal cynodont *Dvinia prima* Amalitskii, 1922, from the Late Permian, skull in lateral view. Specimen PIN 2005/2469 (holotype of *Permocynodon sushkini* Woodward, 1932). The skull is about 120 mm long.

Bauriidae by V'yushkov (1964) and Tatarinov (1974). However, because of the very incomplete nature of the sole specimen, we prefer to consider it as Baurioidea *incertae sedis*.

Infraorder CYNODONTIA Owen, 1859

Comment. The Cynodontia are the most 'mammal-like' of the reptilian Synapsida. They are poorly represented in Russia, and mainly by archaic forms.

Family DVINIIDAE Tatarinov, 1968b.

Diagnosis. Primitive cynodonts with complete secondary palate. Primitive cynodont characters include: suture between prootic and epipterygoid, above trigeminal foramen, short; squamosal lightly built, leaving quadrate and quadratojugal well exposed; interpterygoid vacuity present; dentary relatively small; weak coronoid process; 6 incisors; in upper jaw, 1 very small vestigial precanine tooth. Autapomorphies of Dviniidae include: sagittal crest very high and long, without pineal foramen; postcanine teeth expanded transversely, with low main cusp and many small accessory cusps.

Comment. The Dviniidae are known only from Russia; they are represented by only one genus and species.

Dvinia Amalitskii, 1922

See Figures 6.18 and 6.19.

Synonym. *Permocynodon* Sushkin, 1927.

Type species. *Dvinia prima* Amalitskii, 1922 (= *Permocynodon sushkini* Woodward, 1932).

Holotype. PIN 2005/2465, anterior half of skull without lower jaw. Sokolki, Kotlas district,

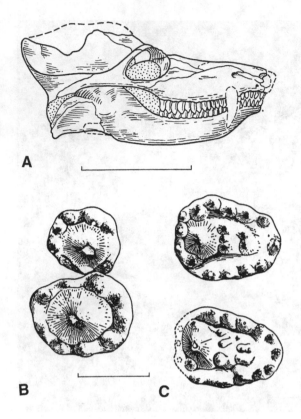

Figure 6.19. The basal cynodont *Dvinia prima* Amalitskii, 1922 from the Late Permian. A, Skull in lateral view, specimen PIN 2005/2469 (holotype of *Permocynodon sushkini* Woodward, 1932). B, C, Postcanine teeth in occlusal view (labial side right, anterior side down): upper postcanines, 12th and 13th right (B); lower postcanines, 11th and 13th right (C). Scale bars, 50 mm (A), 2 mm (B, C). (A, after Konzhukova, 1949; B, C, after Tatarinov, 1974.)

Arkhangel'sk Province. Vyatskian Gorizont, Late Tatarian.

Referred material. PIN 2005/2469, an almost complete skull with lower jaw (type specimen of *Permocynodon sushkini*), and PIN 2245/237, a right maxilla, both from the same locality as the holotype (Tatarinov, 1968b, 1974).

Diagnosis. Relatively small; snout narrow; suborbital and postorbital bars, and zygomatic arches slender; 6 incisors in lower jaw (the large number of lower incisors is considered by Hopson and Barghusen (1986) as a derived character).

Family **PROCYNOSUCHIDAE** Broom, 1938

Diagnosis. Primitive cynodonts; palatal plates of secondary palate do not meet in midline, and leave narrow slot; quadrate and quadratojugal well exposed; interpterygoid vacuity present; dentary relatively small; weak coronoid process; 6 incisors and 1 or 2 precanine teeth in upper jaw; postcanine teeth slightly compressed transversely, with 1 main cusp and lingual cingulum bearing small cusps.

Comment. The Procynosuchidae are known mainly from South Africa.

Uralocynodon Tatarinov, 1987

Type species. Uralocynodon tverdokhlebovae Tatarinov, 1987.

Holotype. SGU 10489/308, a left dentary. Blyumental' 3, Orenburg district, Orenburg Province. Kutuluk Svita, Late Tatarian.

Comment. Uralocynodon is obviously a close relative of *Procynosuchus*, a South African genus known also from Germany. It is smaller than the latter, and its coronoid process is narrower.

Family **THRINAXODONTIDAE** Watson and Romer, 1956

Diagnosis. Small, lightly built cynodonts; quadratojugal covered laterally by squamosal; secondary palate usually completely closed; small interpterygoid vacuity present at least in juveniles; dentary relatively small, but with well differentiated, high coronoid process; 4 upper and 3 lower incisors; no precanine teeth; postcanine teeth similar to Procynosuchidae, but more compressed transversely, and, hence, more sectorial.

Nanocynodon Tatarinov, 1968a
See Figure 6.20.

Type species. Nanocynodon seductus Tatarinov, 1968a.

Holotype. PIN 2415/1, an incomplete right dentary, Bol'shoe Linovo, Leninskoe district, Kirov Province. Vyatskian Gorizont, Late Tatarian. The specimen comes from a drill core, and was found at a depth of 85.1 m.

Diagnosis. Very small form; at least 10 lower postcanine teeth.

Comments. This form was attributed by Tatarinov

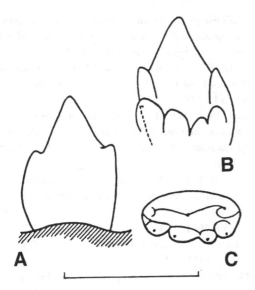

Figure 6.20. *Nanocynodon seductus* Tatarinov, 1968a, holotype, PIN 2415/1, 7th lower postcanine tooth, in labial (A), lingual (B), and occlusal (C) views. Scale bar, 2 mm. (After Tatarinov, 1974.)

(1968a, 1974) to the Galesauridae (*sensu lato*, including the Thrinaxodontidae), whereas Hopson and Kitching (1972) consider it as a probable juvenile procynosuchid. The postcanine teeth of *Nanocynodon* are very compressed transversely, distinctly sectorial, and they bear a narrow lingual cingulum with tiny cusps. Consequently, we think that *Nanocynodon* fits better in the Thrinaxodontidae than in the Procynosuchidae (Battail, 1991).

Family TRAVERSODONTIDAE Huene, 1936
Diagnosis. Very advanced gomphodont cynodonts; herbivorous; postcanine teeth greatly expanded transversely; crown-to-crown occlusion; upper and lower postcanine teeth with very different structures, but both with characteristic transverse ridges, longitudinal crests, and basins; lower postcanines much narrower transversely than upper ones.

Scalenodon Crompton, 1955
Type species. Scalenodon angustifrons (Parrington, 1946) Crompton, 1955, from the Middle Triassic Manda Formation of Tanzania.

Scalenodon boreus Tatarinov, 1973
Holotype. PIN 2973/1, upper postcanine tooth. Karagachka, Orenburg district, Orenburg Province. Donguz Gorizont, Middle Triassic.
Referred material. Another, very worn, postcanine tooth (Tatarinov, 1973, 1974).
Comment. The upper postcanine tooth of *Scalenodon boreus* is smaller than that of *S. angustifrons*, but has a similar structure; its longitudinal crest is however less distinct.

Conclusions

The Russian platform provides one of the best records of Permo-Triassic therapsids. The Russian therapsid faunas are comparable, in many respects, with the Southern African ones. Russia has yielded, however, a few more archaic forms, of Late Kazanian–Early Tatarian age, but is comparatively poorer in Early Triassic forms. In spite of the great distance between the two areas, they display many faunal similarities, evidencing the relative homogeneity of Pangaean faunas during Permo-Triassic times.

Acknowledgements

We thank Joséliane Maréchal and Danielle Quintus for practical help, Françoise Pilard who prepared the figures, and Mikhail Shishkin for his many valuable corrections.

References

Amalitskii, V.P. 1922. Diagnoses of new forms of vertebrates and plants from the Upper Permian of North Dvina. *Izvestiya AN SSSR, VI Seriya* **16**: 329–340 [author name spelled Amalitzky].

Battail, B. 1991. Les Cynodontes (Reptilia, Therapsida): une phylogénie. *Bulletin du Muséum National d'Histoire Naturelle, 4e Série, Section C* **13** (1–2): 17–105.

—1992. Sur la phylogénie des Thérapsides (Reptilia, Therapsida). *Annales de Paléontologie* **78** (2): 83–124.

Boonstra, L.D. 1954. The cranial structure of the titanosuchian: *Anteosaurus. Annals of the South African Museum* **42**: 108–148.

—1963. Early dichotomies in the therapsids. *South African Journal of Science* **59**: 176–195.

—1965. The Russian deinocephalian *Deuterosaurus*. *Annals of the South African Museum* **48**: 233–236.

Brink, A.S. 1959. Notes on some whaitsiids and moschorhinids. *Palaeontologia Africana* **6** (1958): 23–49.

Broom, R. 1903. On the classification of the theriodonts and their allies. *Reports of the South African Association for the Advancement of Science* **1**: 286–295.

—1911. On the structure of the skull in cynodont reptiles. *Proceedings of the Zoological Society of London* **1911**: 893–925.

—1912. On some fossil reptiles from the Permian and Triassic beds of South Africa. *Proceedings of the Zoological Society of London* **1912**: 859–876.

—1923. On the structure of the skull in the carnivorous dinocephalian reptiles. *Proceedings of the Zoological Society of London* **1923**: 661–684.

—1932. *The Mammal-like Reptiles of South Africa and the Origin of Mammals.* London: Witherby, 376 pp.

—1938. The origin of the cynodonts. *Annals of the Transvaal Museum* **19**: 279–288.

Bystrov, A.P. 1955. [A gorgonopsian from the Upper Permian beds of the Volga.] *Voprosy Paleontologii* **1955** (2): 7–18.

Chudinov, P.K. 1960. [Upper Permian therapsids from the Ezhovo locality.] *Paleontologicheskii Zhurnal* **1960** (4): 81–94.

—1961. Estemmenosuchidae. Eotitanosuchidae, pp. 79–83 in Piveteau, J. (ed.), *Traité de Paléontologie, Tome VI (I).* Paris: Masson.

—1964a. [Family Eotitanosuchidae.] pp. 247–249 in Orlov, Yu.A. (ed.), [*Elements of Palaeontology: Amphibians, Reptiles and Birds*]. Moscow: Nauka.

—1964b. [Contributions to the knowledge of dinocephalians of the USSR.] *Paleontologicheskii Zhurnal* **1964**: 85–98.

—1965. New facts about the fauna of the Upper Permian of the USSR *Journal of Geology* **75**: 117–128.

—1968a. [New dinocephalians from Ocher.] pp. 16–31 in [*Upper Palaeozoic and Mesozoic Amphibians and Reptiles of the U.S.S.R.*] Moscow: Nauka.

—1968b. A new dinocephalian from the Cisuralian region (Reptilia, Therapsida; Upper Permian). *Postilla,* **121**: 1–20.

—1983. [Early therapsids.] *Trudy Paleontologicheskogo Instituta AN SSSR* **202**: 1–230.

Cluver, M.A. and King, G.M. 1983. A reassessment of the relationships of Permian Dicynodontia (Reptilia, Therapsida) and a new classification of dicynodonts. *Annals of the South African Museum* **91**: 195, 273.

Cope, E.D. 1870. Remarks by Edward D. Cope at meeting May 6th 1870. *Proceedings of the American Philosophical Society* **11**: 419.

Cox, C.B. 1965. New Triassic dicynodonts from South America, their origins and relationships. *Philosophical Transactions of the Royal Society of London, Series B* **248**: 457–516.

Crompton, A.W. 1955. On some Triassic cynodonts from Tanganyika. *Proceedings of the Zoological Society of London* **125**: 617–669.

Danilov, A.I. 1971. [A new dicynodont from the Middle Triassic of Southern Cisuralia.] *Paleontologicheskii Zhurnal* **1971**: 132–135.

—1973. [Remains of the postcranial skeleton of *Uralokannemeyeria* (Dicynodontia).] *Paleontologicheskii Zhurnal* **1973**: 244–247.

Efremov, I.A. 1937. [On the stratigraphic subdivision of the continental Permian and Triassic of the USSR on the terrestrial vertebrate fauna.] *Doklady AN SSSR, Seriya Geologicheskaya* **16**: 121–126.

—1938. [Some new Permian reptiles of the U.S.S.R.] *Doklady AN SSSR* **19**: 771–776.

—1940a. [New discoveries of Permian terrestrial vertebrates in Bashkortostan and Chkalov Province.] *Doklady AN SSSR* **27**: 412–415.

—1940b. [Composition of the Northern Dvina Fauna of Permian Amphibia and Reptilia from the excavations of V.P. Amalitskii.] *Doklady AN SSSR* **27**: 893–896.

—1940c. [Preliminary description of new Permian and Triassic Tetrapoda from USSR.] *Trudy Paleontologicheskogo Instituta AN SSSR* **10** (2): 1–140.

—1941. [Short survey of faunas of Permian and Triassic Tetrapoda of the USSR.] *Sovetskaya Geologiya* **5**: 96–103.

—1951. [On the structure of the knee joint in higher dicynodonts.] *Doklady AN SSSR* **77**: 483–485.

—1954. [The fauna of terrestrial vertebrates in the Permian Copper Sandstone of the Western Cisuralian region.] *Trudy Paleontologicheskogo Instituta AN SSSR* **54**: 1–416.

—and V'yushkov, B.P. 1955. [Catalogue of the localities of Permian and Triassic terrestrial vertebrates in the territories of the USSR.] *Trudy Paleontologicheskogo Instituta AN SSSR* **46**: 1–147.

Eichwald, E. 1848. Über die Saurier des Kupferführenden

Zechsteins Russlands. *Bulletin de la Société Impériale des Naturalistes de Moscou* 21: 136–204.

—1860. *Lethaia Rossica ou Paléontologie de la Russie. Volume I. Ancienne Période.* Stuttgart: E. Schweizerbart.

Fischer von Waldheim, G. 1841. Notice sur le *Rhopalodon*, nouveau genre de sauriens fossiles du versant occidental de l'Oural. *Bulletin de la Société Impériale des Naturalistes de Moscou* 14: 460–464.

—1845. Beitrag zur naeheren Bestimmung des von Hrn. Wangenheim von Qualen abgebildeten und beschriebenen Saurier-Schaedels. *Bulletin de la Société Impériale des Naturalistes de Moscou* 18: 540–543.

—1847. Bemerkungen über das Schädel-Fragment, welches Herr Major Wangenheim von Qualen in dem West-Ural entdeckt und der Gesellschaft zur Beurteilung vorgelegt hat. *Bulletin de la Société Impériale des Naturalistes de Moscou* 20: 263–267.

Gregory, W.K. 1926. The skeleton of *Moschops capensis*, Broom, a dinocephalian reptile of South Africa. *Bulletin of the American Museum of Natural History* 56: 179–251.

Hartmann-Weinberg, A.P. 1938. [Gorgonopsians as time indicators.] *Voprosy Paleontologii* 1938 (4): 47–123.

Haughton, S.H. 1918. Investigations in South African reptiles and amphibians. 11. Some new carnivorous Therapsida, with notes upon the brain in certain species. *Annals of the South African Museum* 12: 175–216.

Haughton, S.H. and Brink, A.S. 1954. A bibliographic list of the Reptilia from the Karroo beds of Africa. *Palaeontologia Africana* 2: 1–187.

Hopson, J.A. 1991. Systematics of the nonmammalian Synapsida and implications for patterns of evolution in synapsids, pp. 635–693 in Schultze, H.-P. and Trueb, L. (eds.), *The Origin of Higher Groups of Tetrapods: Controversy and Consensus.* Ithaca: Cornell University Press.

—and Barghusen, H.R. 1986. An analysis of therapsid relationships, pp. 83–106 in Hotton, N., Maclean, P.D., Roth, J.J. and Roth, E.C. (eds.), *The Ecology and Biology of Mammal-like Reptiles.* Washington: Smithsonian Institution Press.

—and Kitching, J.W. 1972. A revised classification of cynodonts (Reptilia; Therapsida). *Palaeontologia Africana* 14: 71–85.

Huene, F. von 1936. *Die fossilen Reptilien des Südamerikanische Gondwanalandes. 2. Ordnung Cynodontia.* München: C.H. Beck.

—1948. Review of the lower tetrapods, pp. 65–106 in Du Toit, A.L. (ed.), *Robert Broom Commemorative Volume.* Cape Town: the Royal Society.

Huxley, T.H. 1859. On a new species of *Dicynodon* (*D. Murrayi*) from near Colesberg, South Africa; and on the structure of the skull in the dicynodonts. *Quarterly Journal of the Geological Society of London* 15: 649–658.

Ivakhnenko, M.F. 1990. [The late Palaeozoic faunal assemblage of tetrapods from deposits of the Mezen' River.] *Paleontologicheskii Zhurnal* 1990: 81–90.

—1992. [The Late Permian faunistic assemblages of tetrapods from Eastern Europe and their South Gondwan analogues.] In [*Palaeontology and Stratigraphy of Permian and Triassic Deposits of Northern Eurasia*]. Moscow: Paleontologicheskii Institut, Rossiiskoi Akademii Nauk.

—1994. A new Late Permian dromasaurian (Anomodontia) from Eastern Europe. *Paleontological Journal* 28: 96–103.

—1995. [Primitive titanosuchian dinocephalians the Late Permian of eastern Europe.] *Paleontologicheskii Zhurnal* 1995 (3): 98–105.

—1996a. New primitive therapsids from the Permian of Eastern Europe. *Paleontological Journal* 30: 337–343.

—1996b. Primitive anomodonts, venyukoviids, from the Late Permian of Eastern Europe. *Paleontological Journal* 30: 575–582.

Kalandadze, N.N. 1970. [New Triassic kannemeyeriids from Southern Cisuralia.] pp. 51–57 in Flerov, K.K. (ed.), [*Materials on the evolution of terrestrial vertebrates*]. Moscow: Nauka.

—1975. [The first lystrosaur find from the territory of the European part of the USSR.] *Paleontologicheskii Zhurnal* 1975 (4): 140–142.

—Ochev, V.G., Tatarinov, L.P., Chudinov, P.K. and Shishkin, M.A. 1968. [A Catalogue of the Permian and Triassic tetrapods of the USSR.] pp. 72–91 in [*Upper Palaeozoic and Mesozoic Amphibians and Reptiles of the USSR.*] Moscow: Nauka.

Kemp, T.S. 1982. *Mammal-like Reptiles and the Origin of Mammals.* London: Academic Press.

—1988. Interrelations of the Synapsida, pp. 1–22 in Benton, M.J. (ed.), *The Phylogeny and Classification of the Tetrapods. Volume 2. Mammals. Systematics Association Special Volume* 35B. Oxford: Clarendon Press.

King, G.M. 1988. Anomodontia, pp. 1–174 in Wellnhofer, P. (ed.), *Handbuch der Paläoherpetologie, Teil 17C.* Stuttgart: Gustav Fischer Verlag.

Konzhukova, E.D. 1949. [Concerning the morphology of *Permocynodon* and the evolution of the dental apparatus of the Cynodontia.] *Trudy Paleontologicheskogo Instituta, AN SSSR* **20**: 94–129.

Kuhn, O. 1961. *Die Familien der rezenten und fossilen Amphibien und Reptilien*. Bamberg: Verlagshaus Meisenback KG.

Kutorga, S.S. 1838. *Beitrag zur Kenntniss der organischen Ueberreste des Kupfersandsteins am westlichen Abhange des Urals*. St Petersburg: N. Gretsch, 38 pp.

Lehman, J.-P. 1961. Dicynodontia, pp. 287–351 in Piveteau, J. (ed.), *Traité de Paléontologie. VI (1). Mammifères. Origine Reptilienne. Evolution*. Paris: Masson.

Lozovskii, V.R. 1983. [On the age of the *Lystrosaurus*-bearing beds in the Moscow Syneclise.] *Doklady AN SSSR* **272**: 1433–1438.

Lydekker, R. 1890. *Catalogue of the Fossil Reptilia and Amphibia in the British Museum (Natural History). Part IV, Containing the Orders Anomodontia, Ecaudata, Caudata and Labyrinthodontia, and Supplement*. London: British Museum (Natural History).

Mendrez, C.H. 1972. On the skull of *Regisaurus jacobi*, a new genus and species of Bauriamorpha Watson and Romer 1956 (= Scaloposauria Boonstra 1953), from the *Lystrosaurus*-zone of South Africa, pp. 191–212 in Joysey, K.A. and Kemp, T.S. (eds.), *Studies in Vertebrate Evolution*. Edinburgh: Oliver & Boyd.

—1974. Étude du crâne d'un jeune spécimen de *Moschorhinus kitchingi* Broom 1920 (? *Tigrisuchus simus* Owen, 1876), Therocephalia Pristerosauria Moschorhinidae d'Afrique australe. *Annals of the South African Museum* **64**: 71–115.

—1975. Principales variations du palais chez les Thérocéphales sud-africains (Pristerosauria et Scaloposauria) au cours du Permien supérieur et du Trias inférieur, pp. 379–429 in *Problèmes actuels de Paléontologie – Evolution des Vertébrés*. Paris: Editions du C.N.R.S.

Meyer, H. von 1866. Reptilien aus dem Kupfer-Sandstein des West-Uralischen Gouvernements Orenburg. *Palaeontographica* **15**: 97–130.

Nopcsa, F. von 1928. Palaeontological notes on reptiles. II. On some fossil reptiles from the copper-bearing [Kazan] Permian strata of Russia. *Geologica Hungarica, Series Palaeontologica* **1**: 1–84.

Ochev, V.G. 1992. [On the second reliable anomodont find from the Lower Triassic of the Eastern European platform.] *Izvestiya Vysshikh Uchebnykh Zavednii (Geologiya i Razvedka)* **1992** (2): 132–133.

—and Shishkin, M.A. 1989. On the principles of global correlation of the continental Triassic on the tetrapods. *Acta Palaeontologica Polonica* **34**: 149–173.

Olson, E.C. 1962. Late Permian terrestrial vertebrates, USA and USSR. *Transactions of the American Philosophical Society* **52**: 3–224.

Orlov, Yu.A. 1940. [Titanosuchians of the Isheevo fauna.] pp. 266–267 in [*Abstracts of work of the Academy of Sciences of the USSR in 1940*]. Moscow: AN SSSR.

—1958. [Carnivorous dinocephalians from the fauna of Isheev (Titanosuchia).] *Trudy Paleontologicheskogo Instituta AN SSSR* **72**: 1–114.

Owen, R. 1845. Description of certain fossil crania discovered by A.G. Bain, Esq., in the sandstone rocks at the southeastern extremity of Africa, referable to different species of an extinct genus of Reptilia (Dicynodon), and indicative of a new tribe or suborder of Sauria. *Transactions of the Geological Society of London (2)* **7**: 59–84.

—1859. On the orders of fossil and recent Reptilia and their distribution in time. *Reports of the British Association for the Advancement of Science* **1859**: 153–166.

Parrington, F.R. 1946. On the cranial anatomy of cynodonts. *Proceedings of the Zoological Society of London* **116**:181–197.

Pravoslavlev, P.A. 1927a. [III. Gorgonopsidae from the North Dvina Expedition of V. P. Amalitskii.] *Severo-Dvinskie Raskopki Professora V.P. Amalitskogo (AN SSSR)* **3**: 1–117.

—1927b. [IV. Gorgonopsid from the 1923 North Dvina Expedition (*Amalitzkia annae* gen. et sp. nov.).] *Severo-Dvinskie Raskopki Professora V.P. Amalitskogo (AN SSSR)* **4**: 1–20.

Romer, A.S. 1961. Synapsid evolution and dentition, pp. 9–56 in Vandebroek, G. (ed.), *Colloque International sur l'Évolution des Mammifères Inférieurs et Non Spécialisés*. Brussels: Koninklijke Vlaamse Academie voor Wetenschappen, Letteren en Schone Kunsten van Belgie. Klasse der Wetenschappen.

Rubidge, B.S. 1991. A new primitive dinocephalian mammal-like reptile from the Permian of Southern Africa. *Palaeontology* **34**: 547–559.

—1994. *Australosyodon*, the first primitive anteosaurid dinocephalian from the Upper Permian of Gondwana. *Palaeontology* **37**: 579–594.

—and Hopson, J.A. 1990. A new anomodont therapsid from South Africa and its bearing on the ancestry of the Dicynodontia. *South African Journal of Science* **86**: 43–45.

Ryabinin, A.N. 1938. [Vertebrate fauna from the Upper Permian deposits of the Sviyaga basin: 1. A new dinocephalian, *Ulemosaurus svijagensis* n. gen. n. sp.] *Ezhegodnik TsNIGRI imeni Akademika F.N. Chernysheva* **1**: 4–40.

Seeley, H.G. 1895. Researches on the structure, organization and classification of the fossil Reptilia. IX, section 1. On the Therosuchia. *Philosophical Transactions of the Royal Society of London, Series B* **185**: 987–1018.

Sigogneau, D. 1970. *Révision Systématique des Gorgonopsiens Sud-africains*. Paris: Editions du CNRS (*Cahiers de Paléontologie*).

—and Chudinov, P.K. 1972. Reflections on some Russian eotheriodonts (Reptilia, Synapsida, Therapsida). *Palæovertebrata* **5**: 79–109.

Sigogneau-Russell, D. 1989. Theriodontia I, pp. 1–127 in Wellnhofer, P. (ed.), *Handbuch der Paläoherpetologie, Teil 17B/I*. Stuttgart: Gustav Fischer.

Stuckenberg, A.A, 1898. Allgemeine geologische Karte von Russland. Blatt 127. *Trudy Geologicheskogo Komiteta* **16**: 1–362.

Sun Ailin 1981. [Reidentification of *Traversodontoides wangwuensis* Young.] *Vertebrata PalAsiatica* **19**: 1–4.

Sushkin, P.P. 1926. Notes on the pre-Jurassic Tetrapoda from Russia. I. *Dicynodon amalitzkii*, n. sp. *Palaeontologia Hungarica* **1**: 323–327.

—1927. [*Permocynodon*, a cynodont from the Upper Permian beds of the Northern Dvina River.] *Trudy Paleozoologicheskogo Instituta*, **4**: 49–52.

Tatarinov, L.P. 1963. [A new Late Permian therocephalian.] *Paleontologicheskii Zhurnal* **1963** (4): 76–94.

—1968a. [New theriodonts from the Upper Permian of the USSR] pp. 32–45 in [*Upper Palaeozoic and Mesozoic Amphibians and Reptiles of the U.S.S.R.*] Moscow: Nauka.

—1968b. Morphology and systematics of the Northern Dvina cynodonts (Reptilia, Therapsida; Upper Permian). *Postilla* **126**: 1–51.

—1973. [Cynodonts of Gondwanan aspect in the Middle Triassic of the USSR.] *Paleontologicheskii Zhurnal* **1973** (2): 83–89.

—1974. [Theriodonts of the USSR.] *Trudy Paleontologicheskogo Instituta AN SSSR* **143**: 1–252.

—1977a. [A new gorgonopsian from the Upper Permian deposits of Vologda Oblast.] *Paleontologicheskii Zhurnal* **1977**: 97–104.

—1977b. [A new theriodont from the Lower Triassic of the Orenburg Province.] *Paleontologicheskii Zhurnal* **1977** (4): 86–91.

—1987. [A new primitive cynodont from the Upper Permian of Southern Cisuralia.] *Paleontologicheskii Zhurnal* **1987** (3): 110–114.

—1988. [On the morphology and systematic position of the gomphodont cynodont *Antecosuchus ochevi*.] *Paleontologicheskii Zhurnal* **1988** (2): 87–96.

—1993. [On crossopterygian characters in the skull structure of the Late Permian therocephalian *Hexacynodon purlinensis* (Reptilia, Theriodontia).] *Doklady RAN*, **332**: 124–126.

—1995. [*Viatkosuchus sumini* – a new therocephalian from the Upper Permian of the Kirov Province.] *Paleontologicheskii Zhurnal* **1995** (1): 84–96.

Tverdokhlebova, G.I. and Ivakhnenko, M.F. 1994. [New tetrapods from the Tatarian of Eastern Europe.] *Paleontologicheskii Zhurnal* **1994**: 122–126.

Twelvetrees, W.H. 1880. On a new theriodont reptile (*Cliorhizodon orenburgensis*, Twelvetr.) from the Upper Permian cupriferous sandstones of Kargalinsk, near Orenburg in south-eastern Russia. *Quarterly Journal of the Geological Society of London* **36**: 540–543.

V'yushkov, B.P. 1952. [On the relative age of the Isheevo and Northern Dvina faunas of terrestrial vertebrates from the Permian of the U.S.S.R.] *Doklady AN SSSR* **83**: 897–900.

—1953. [On the gorgonopsians of the Northern Dvina fauna.] *Doklady AN SSSR* **91**: 397–400.

—1955. [Theriodonts of the Soviet Union.] *Trudy Paleontologicheskogo Instituta AN SSSR* **49**: 128–175.

—1964. [Find of Triassic theriodonts in the USSR.] *Paleontologicheskii Zhurnal* **1964** (2): 158–160.

—1969. [New dicynodonts from the Triassic of Southern Cisuralia.] *Paleontologicheskii Zhurnal* **1969** (2): 99–106.

Watson, D.M.S. 1942. On Permian and Triassic tetrapods. *Geological Magazine* **79**: 81–116.

—and Romer, A.S. 1956. A classification of therapsids. *Bulletin of the Museum of Comparative Zoology at Harvard College* **114**: 35–98.

Woodward, A.S. 1932. *Zittel's Textbook of Palaeontology*, Volume 2. London: Macmillan, 464 pp.

7

Tetrapod biostratigraphy of the Triassic of Eastern Europe

MIKHAIL A. SHISHKIN, VITALII G. OCHEV, VLADLEN R. LOZOVSKII, AND IGOR V. NOVIKOV

Introduction

The Triassic tetrapods of Russia are known mostly from the European part of the country (Cis-Urals in a broad sense) where they form an almost continuous faunal succession ranging in age from the Induan to late Mid Triassic. Except for a very few records from nearshore marine facies, all the fossils come from continental deposits developed largely in the East European Platform and the Cis-Uralian Marginal Trough (Figure 7.1).

Triassic vertebrates were first reported from this area by Nikitin (1883), who recovered labyrinthodont teeth and dermal bones and dipnoan tooth plates from the Vetluga Basin. He was also the first to conclude that the rocks producing these fossils were Triassic in age. Further material from the Cis-Uralian Triassic was studied during the early twentieth century, principally by Yakovlev (1916), Sushkin (1927), Ryabinin (1930), Hartmann-Weinberg and Kuz'min (1936a,b), Kuz'min (1935, 1937), Efremov (1929, 1932, 1940), Bystrov and Efremov (1940), and Huene (1940).

A decisive step forward in these studies was made by Efremov who initiated a long-term programme of prospecting for Permo-Triassic vertebrate fossils in Russia. He first summarized the available data, and used them as a basis for a biostratigraphic zonation of the Cis-Uralian continental Triassic (Efremov, 1937, 1940; Efremov and V'yushkov, 1955). In Efremov's regional biozonal scheme spanning the Late Palaeozoic–Early Mesozoic, three units, the Zones V ('Neorhachitome'), VI (*Capitosaurus*), and VII (*Mastodonsaurus*) were recognized in Triassic sequences. Zones V and VI were placed in the Lower Triassic and Zone VII in the Middle–Upper Triassic.

Research during the last four decades has allowed a reappraisal and improvement of this scheme. According to modern views of the Triassic of the Cis-Urals, four principal successive faunas are discerned (Shishkin and Ochev, 1967, 1985; Ochev and Shishkin, 1989). They are named after their dominant amphibian genera and are as follows (in ascending order): (1) *Benthosuchus–Wetlugasaurus* fauna (Induan–Early Olenekian); (2) *Parotosuchus* (Late Olenekian); (3) *Eryosuchus* (Late Anisian(?)-Ladinian), and (4) *Mastodonsaurus* (Late Ladinian). Faunas (1) and (4) correspond to those of Efremov's Zones V and VII respectively, and (2) and (3) to the assemblages formerly attributed to Zone VI. The *Benthosuchus–Wetlugasaurus* fauna was shown to include three successive groupings (Shishkin and Ochev, 1967), the youngest of which may in turn be subdivided into two more biochrons or subgroupings (Novikov *et al.*, 1990).

The stratigraphic units that yield tetrapods also produce other fossils, including fish remains (among which dipnoan tooth plates are most common and dateable), molluscs, conchostracans, ostracods, insects, charophytes, sporomorphs, and plant macrofossils. In some cases, these fossils are important for dating and correlation, but in general the tetrapods are most useful.

The four principal faunas listed above (Figures 7.2 and 7.3) correspond to the following regional stratigraphic units: (1) Vetlugian Supergorizont, (2) Yarenskian Gorizont (Lower Triassic), (3) Donguz Gorizont, and (4) Bukobay Gorizont (both Middle Triassic). The Vetlugian is divided into the Vokhmian, Rybinskian, Sludkian, and Ustmylian Gorizonts. The two former are characterized by the *Tupilakosaurus–Luzocephalus* and *Benthosuchus* faunal

120

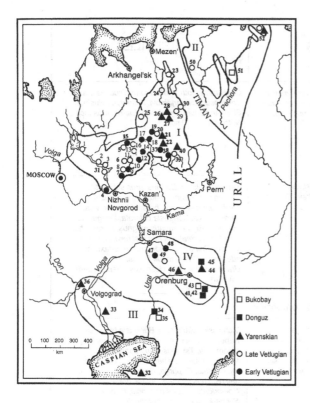

Figure 7.1. Distribution of the main tetrapod localities in the Triassic of European Russia. Localities are: 1, Tikhvinskoe (Parshino); 2, Krasne Pozhni, Ples, Semigore; 3, Reshma; 4, Gorbatovo; 5, Yuza I, II; 6, Berezniki; 7, Astashikha, Znamenskoe; 8, Shilikha; 9, Spasskoe; 10, Bolshaya Sludka; 11, Zubovskoe; 12, Vshivtsevo; 13, Kudanga; 14, Spasskoye–Semenovskoye; 15, Telyanino; 16, Vakhnevo; 17, Ananinskoye; 18, Podsaraitsa, Sholga; 19, Luza; 20, Chernyi Bor; 21, Mishakovskaya, Zanul'e, Keros; 22, Vaimos; 23, Mezen' (Vybor, Niz'ma); 24, Vashka; 25, North Dvina (Faustovo, Tarasovo, Permogor'e); 26, Lopatino; 27, Gam, Zheshart; 28, Yarenga; 29, Ors'yu; 30, Elva Vymskaya; 31, Kozlat'evo; 32, Mangyshlak; 33, Bolshoi Bogdo; 34, Azi Molla II; 35, Kara–Bolla–Kantemir; 36, Donskaya Luka; 37, Velikoretskoe; 38, Ryab'; 39, Okunevo (Fedorovka); 40, Teryukhan; 41, Karagachka; 42, Donguz X; 43, Bukobay I–V; 44, Petropavlovka; 45, Koltaevo; 46, Rossypnoe; 47, Yablonovi Vrag; 48, Zaplavnoe; 49, Mechet'; 50, Tzylma; 51, Synya; 52, Byzovaya; 53, Hey–Yaga. These localities divide into regions as follows: Moscow Syncline (1–17, 31), Volga–Ural Anticline (18–30, 37–40), Cis-Caspian Syncline (32–36), Cis-Uralian Trough (41–49), and Mezen' and Pechora Syncline (50–53).

groupings respectively, and the two latter by the *Wetlugasaurus* grouping with its two biochrons. Likewise, in the Yarenskian Gorizont, there are also two members: the Fedorovskian (lower) and Gamskian. Their faunal differences, currently under study, are outlined below.

Surveys of the principal tetrapod groups found in the Cis-Uralian Triassic include Ochev (1966, 1972), Shishkin (1973, 1987), Getmanov (1989), and Novikov (1994) for the amphibians, Ivakhnenko (1979) and Novikov (1994) for the procolophonids, Ochev (1991) and Sennikov (1995a, b) for the archosaurs, and Tatarinov (1974a) for the theriodonts. In addition, the compositions of individual faunas and localities are summarized by Efremov and V'yushkov (1955), Blom (1968), Kalandadze *et al.* (1968), Garyainov and Ochev (1962), Ochev *et al.* (1979), Ochev (1980), Novikov (1993, 1994), Shishkin *et al.* (1995), and Sennikov (1996). The abundance of fossils covered by these accounts is remarkably contrasted by the scarcity of finds of Triassic land vertebrates in the Asiatic part of the former Soviet Union; these finds were surveyed by Shishkin *et al.* (1986).

Lower Triassic

In the East European Platform, the most productive Lower Triassic tetrapod-bearing sites are scattered over the Moscow, Mezen', and Pechora synclines (basins of the Volga, Vyatka, North Dvina, Vychegda, Pechora, and Mezen' rivers) and the southern slope of the Volga–Ural Anticline (Ural–Samara watershed or the so-called Obshchy Syrt Upland). Many fossils have also been collected from the Don Basin and the Cis-Caspian Depression. In the Cis-Uralian Marginal Trough, most fossils come from the Ural Basin (Orenburg Province and Bashkortostan Republic); a number of finds were also made in the northern part of the trough area, which belongs to the Pechora Basin.

Benthosuchus–Wetlugasaurus *Fauna (Induan–Early Olenekian; Vetlugian Supergorizont)*

The tetrapods of the *Benthosuchus–Wetlugasaurus* fauna are known from hundreds of sites; they pertain to the

Series	Stage	Substage	Supergorizont	Gorizont	Fauna	Grouping	AMPHIBIANS	REPTILES
Middle Triassic	Ladinian			Bukobay	*Mastodonsaurus*		*"Mastodonsaurus" torvus, Bukobaja enigmatica, Plagiosternum danilovi, Plagioscutum caspiense, Synesuchus muravjevi, Cyclotosaurus (?)* sp.	*Chalishevia cothurnata, Malutinisuchus gratus, Energosuchus garjainovi, Jushatyria vjushkovi, Elephantosaurus vjushkovi, Elatosaurus facetus, Nothosaurus (?)* sp.
Middle Triassic	Anisian			Donguz	*Eryosuchus*		*Eryosuchus tverdochlebovi, E. garjainovi, E. antiquus, Komatosuchus chalyshevi, Bukobaja* sp., *Aranetsia improvisa, Plagiosternum paraboliceps, Plagioscutum ochevi*	*Kapes serotinus, Dongusaurus schepetovi, Nothogomphodon danilovi, Antecosuchus otschevi, Scalenodon boreus, Vjushkovisaurus berdjanensis, Dongusia colorata, Dongusuchus efremovi, Erythrosuchus magnus, Dorosuchus neoetus, Sarmatosuchus otschevi, Rhadiodromus klimovi, Rhinodicynodon gracile, Edaxosaurus edentatus, Calleonasus furvus, Uralokannemeyeria vjuschkovi, Rabidosaurus cristatus.*
				unconformity				
Lower Triassic	Olenekian	Upper		Yarenskian	*Parotosuchus*		*Parotosuchus orenburgensis, P. orientalis, P. sequester, P. panteleevi, P. komiensis, Yarengia perplexa, Inflectosaurus amplus, Trematosaurus (?)* sp., *Rhytidosteus uralensis, Melanopelta antiqua, Batrachosuchoides lacer, B. impressus, Axitectum georgi*	*Tichvinskia vjatkensis, Burtensia burtensis, Kapes amaenus, K.majmesculae, Orenburgia enigmatica, Tsylmosuchus donensis, Vytshegdosuchus zheshartensis,Garjainia triplicostata, G. prima, Gamosaurus lozovskii, Jaikosuchus magnus, Vitalia grata, Coelodontognathus donensis, C. ricovi, Silphedosuchus orenburgensis, Doniceps lipovensis.*
Lower Triassic	Olenekian	Lower	Vetlugian	Ustmylian	*Wetlugasaurus*	*Wetlugasaurus*	*Wetlugasaurus malachovi, Vyborosaurus mirus, Angusaurus tsylmensis.*	*Timanophon raridentatus, Orenburgia bruma, Lestanshoria massiva, Boreopricea funerea, Scharschengia* sp., *Tsylmosuchus jakovlevi, Microcnemus* sp.
Lower Triassic	Olenekian	Lower	Vetlugian	Sludkian	*Wetlugasaurus*	*Wetlugasaurus*	*Wetlugasaurus angustifrons, Angusaurus dentatus, A. weidenbaumi, A. succedaneus, Prothoosuchus blomi, P. samariensis, Benthosuchus bashkiricus.*	*Samaria concinna, Insulophon morachovskayae, Tichvinskia* sp., *Tsylmosuchus jakovlevi, Chasmatosuchus* sp., *Microcnemus* sp., *Scharschengia* sp., *Exilisuchus tubercularis.*
Lower Triassic	Olenekian	Lower	Vetlugian	Rybinskian	*Benthosuchus*	*Benthosuchus*	*Benthosuchus sushkini, B. korobkovi, B. bystrowi, Wetlugasaurus angustifrons, Thoosuchus yakovlevi, T.tardus, T.tuberculatus, Trematotegmen otschevi, Dromotectum spinosum.*	*Tichvinskia jugensis, Chasmatosuchus rossicus, Tsylmosuchus samariensis, Microcnermus efremovi, Scharschengia enigmatica.*
Lower Triassic	Induan		Vetlugian	Vokhmian	*Benthosuchus*	*Tupilakosaurus* *Luzocephalus*	*Tupilakosaurus wetlugensis, Luzocephalus blomi, Benthosuchus uralensis, Wetlugasaurus samarensis, Axitectum vjushkovi.*	*Phaanthosaurus ignatjevi, Contritosaurus simus, C.convector, Scalopognathus multituberculatus, Vonhuenia friedrichi, Blomia georgii, Blomosaurus ivachnenkoi, Lystrosaurus georgi, Microcnemus* sp., *Scharschengia* sp.

Figure 7.2. Succession of tetrapod assemblages in the Triassic of European Russia, showing stratigraphic nomenclature and names of amphibian and reptile taxa.

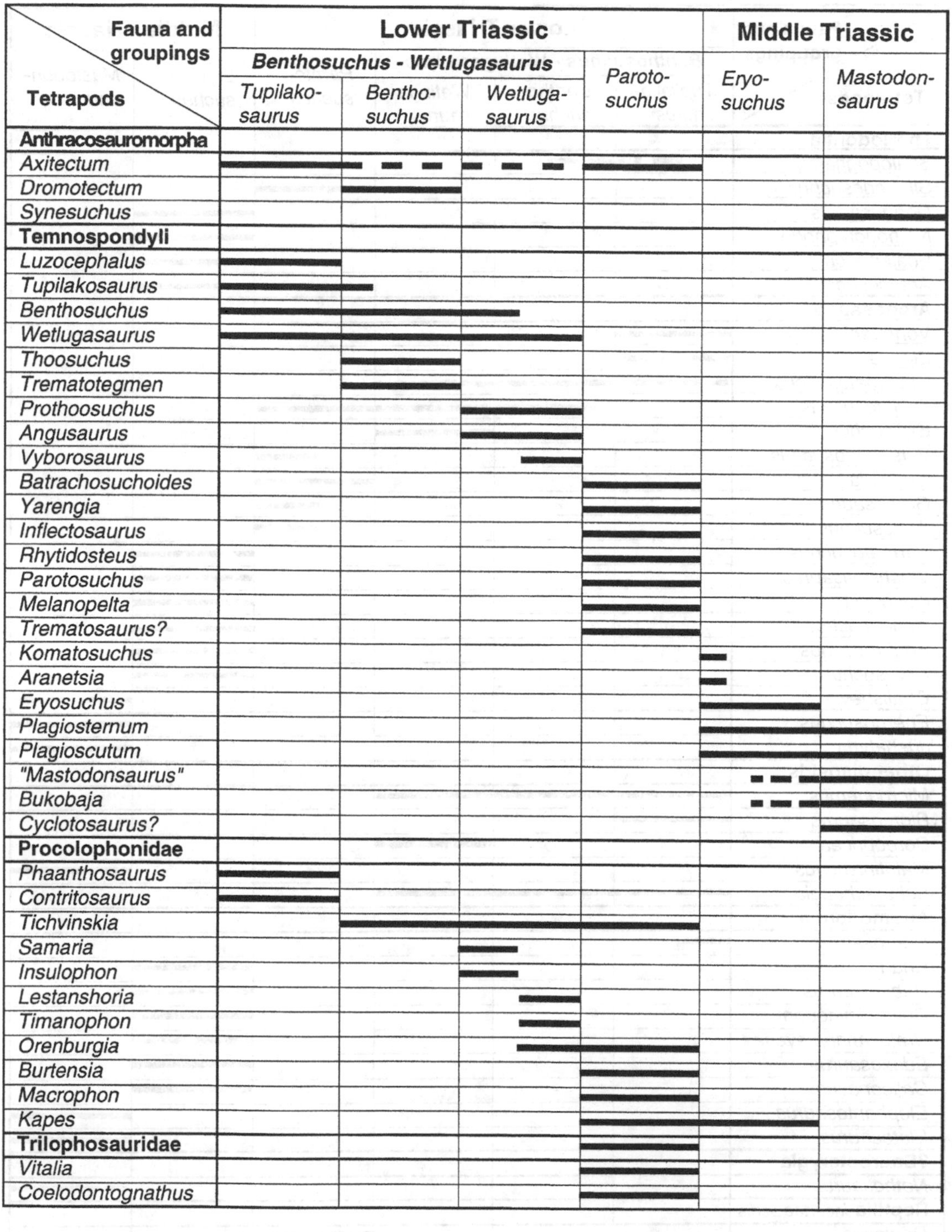

Fauna and groupings / Tetrapods	Lower Triassic				Middle Triassic	
	Benthosuchus - Wetlugasaurus			Paroto-suchus	Eryo-suchus	Mastodon-saurus
	Tupilako-saurus	Bentho-suchus	Wetluga-saurus			
Anthracosauromorpha						
Axitectum	████	▪ ▪ ▪	▪ ▪ ▪	████		
Dromotectum		████				
Synesuchus						████
Temnospondyli						
Luzocephalus	████					
Tupilakosaurus	████					
Benthosuchus	████	████				
Wetlugasaurus	████	████	████			
Thoosuchus		████				
Trematotegmen		████				
Prothoosuchus			████			
Angusaurus			████			
Vyborosaurus			████			
Batrachosuchoides				████		
Yarengia				████		
Inflectosaurus				████		
Rhytidosteus				████		
Parotosuchus				████		
Melanopelta				████		
Trematosaurus?				████		
Komatosuchus					██	
Aranetsia					██	
Eryosuchus					████	
Plagiosternum					████	████
Plagioscutum					████	████
"Mastodonsaurus"					▪▪ ████	████
Bukobaja					▪▪ ████	████
Cyclotosaurus?						████
Procolophonidae						
Phaanthosaurus	████					
Contritosaurus	████					
Tichvinskia		████				
Samaria			████			
Insulophon			████			
Lestanshoria			████			
Timanophon			████			
Orenburgia			████			
Burtensia				████		
Macrophon				████		
Kapes				████	████	
Trilophosauridae						
Vitalia				████		
Coelodontognathus				████		

Figure 7.3. Stratigraphic ranges of tetrapod genera recorded in the Triassic of European Russia.

Fauna and groupings / Tetrapods	Lower Triassic				Middle Triassic	
	Benthosuchus - Wetlugasaurus			Paroto-suchus	Eryo-suchus	Mastodon-saurus
	Tupilako-saurus	*Bentho-suchus*	*Wetluga-saurus*			
Theriodontia						
Scalopognathus	▬ ▬ ▬					
Silphedosuchus				▬▬▬		
Dongusaurus					▬▬▬	
Nothogomphodon					▬▬▬	
Scalenodon					▬▬▬	
Antecosuchus					▬▬▬	
Archosauria						
Vonhuenia	▬▬▬					
Blomia	▬▬▬					
Chasmatosuchus	▬ ▬ ▬ ▬	▬▬▬▬▬▬				
Tsylmosuchus		▬▬▬▬▬				
Exilisuchus			▬▬▬			
Vytshegdosuchus				▬▬▬		
Garjainia				▬▬▬		
Gamosaurus				▬▬▬		
Jaikosuchus				▬▬▬		
Sarmatosuchus					▬▬▬	
Vjushkovisaurus					▬▬▬	
Dongusia					▬▬▬	
Dongusuchus					▬▬▬	
Erythrosushus					▬▬▬	
Dorosuchus					▬▬▬	
Chalishevia						▬▬▬
Energosuchus						▬▬▬
Jushatyria						▬▬▬
Other diapsids						
Microcnemus	▬▬▬▬▬▬▬					
Blomosaurus	▬▬▬					
Boreopricea			▬▬▬			
Malutinisuchus						▬▬▬
Scharschengia	▬▬▬▬▬▬					
Anomodontia						
Lystrosaurus	▬▬					
Rhadiodromus					▬▬▬	
Rabidosaurus					▬▬▬	
Rhinodicynodon					▬▬▬	
Uralokannemeyeria					▬▬▬	
Edaxosaurus					▬▬▬	
Calleonasus					▬▬▬	
Elephantosaurus						▬▬▬
Elatosaurus						▬▬▬
?Sauropterygia						
Nothosaurus						▬▬▬
Reptilia *incertae sedis*						
Doniceps				▬▬▬		

Figure 7.3. (*cont.*)

Vetlugian, which lies unconformably on various levels of the Upper Tatarian (Upper Permian). The Vetlugian comprises fluvial–lacustrine sedimentary cycles which are dominated by floodplain, deltaic, and lake facies in the north of the platform, and by channel deposits (sandstone with conglomerate lenses) in the south and along the Urals.

The overwhelming majority of tetrapod remains were collected from the basal parts of cycles containing the coarsest fluvial sediments. The bones usually bear marks of water transport, and are isolated, fragmented, and often abraded to different degrees. Articulated skeletons are exceptionally rare, though complete skulls are not uncommon. Temnospondyl amphibian remains constitute about 90% of all finds, but reptiles outnumber amphibians in generic diversity (Ochev, 1992a). The former are primarily represented by procolophonids, archosaurs, and prolacertiforms. Therapsids are exceptionally rare. In the accompanying fish assemblage, the commonest and most readily identifiable genus is the dipnoan *Gnathorhiza*, ranging through the whole Vetlugian.

The three subdivisions (groupings) of the *Benthosuchus–Wetlugasaurus* fauna, along with the characteristics of the corresponding gorizonts of the Vetlugian (Figure 7.2), are now considered in turn, from oldest to youngest.

Tupilakosaurus–Luzocephalus *grouping (Vokhmian Gorizont)*

Stratigraphy and lithology. The Vokhmian Gorizont was established in the Moscow Syncline (Lozovskii, 1967; Blom *et al.*, 1982), and is represented there by the Vokhminskaya Svita. Its fossil-bearing equivalent in the Cis-Uralian Marginal Trough and Volga–Ural Anticline is the Kopanskaya Svita of the Ural Basin (Figures 7.4 and 7.5). In the Moscow Syncline, the Vokhmian usually consists of limnic cycles which comprise basal cross-bedded sands with conglomerate lenses, succeeded by reddish-brown or mottled clays with some mudcracks and paleosol horizons. Eastwards, toward the Urals, this facies gives way to coarser alluvial and proluvial deposits. In summary,

these sediments indicate deposition under arid conditions (Figure 7.5). The thickness of the Vokhmian is about 100 m on the East European Platform, and about 130 m in the Cis-Uralian Trough (Strok *et al.*, 1984; Tverdokhlebov, 1995).

Tetrapods. In the Moscow Syncline, Vokhmian assemblages are dominated by *Tupilakosaurus*, a small aberrant (probably eel-like) brachyopoid temnospondyl peculiar for its embolomerous vertebrae. Another, and much larger, index form is *Luzocephalus*, the only Laurasian lydekkerinid genus, recorded in European Russia from a single skull (Shishkin, 1980). Finds are rare, probably because of low-energy sedimentation in the area which caused taphonomic selection in favour of small bones.

By contrast, in the Volga–Ural Anticline, where the Vokhmian mainly comprises coarser channel deposits, the commonest amphibian is a primitive species of the medium-sized capitosaurid *Wetlugasaurus* (*W. samarensis*). A small needle-snouted trematosauroid comparable to the Indian *Gonioglyptus* (undescribed, Shishkin and Ochev, 1993b) is recorded from a single site, together with *W. samarensis* and *Tupilakosaurus*. In the adjacent area of the Cis-Uralian Trough, where similar lithologies occur in the Vokhmian, capitosauroid-related temnospondyls are represented only by a rare primitive benthosuchid, *Benthosuchus* (*Parabenthosuchus*) *uralensis*. In summary, the Vokhmian amphibian fauna seems to be regionally differentiated (Ochev, 1992a), which may be accentuated by taphonomic bias.

Along with the temnospondyls, relics of Tatarian anthracosaurs (Chroniosuchia) are represented by rare finds of isolated vertebrae and dorsal scutes. These pertain to two species of the bystrowianid *Axitectum* from the Vyatka–Vetluga Basin (Shishkin and Novikov, 1992; Novikov and Shishkin, 1995).

Among reptiles, the spondylolestine procolophonids with simple peg-like teeth (*Phaanthosaurus* and *Contritosaurus*) are known to be fairly abundant at some localities along the Vetluga River, but rather scarce on the Ural–Samara watershed. Rare remains of proterosuchian archosaurs, formerly assigned to *Chasmatosuchus*, the most common Vetlugian form,

Figure 7.4. Distal alluvial fan facies of the Kopanskaya Svita, Vokhmian Gorizont, at Astrakhanovka, Orenburg Province, South Cis-Urals. A, David Gower pauses while logging a section through coarse-grained channel sands of an advancing alluvial fan lobe. B, Glenn Storrs looks at some thin channel sandstones where isolated bones were found. (Photographs from the Saratov–Bristol Palaeontological Expedition of 1995.)

have been re-described as the new genera, *Vonhuenia* and *Blomia*. Both are rather small in size. Smaller diapsids, such as the prolacertiforms *Blomosaurus* and *?Microcnemus*, are not uncommon, but are represented mostly by isolated bone fragments that are poorly recognizable systematically (Sennikov, 1995a, b).

No therapsids other than the dicynodont *Lystrosaurus* have yet been reported. *Lystrosaurus* was, until recently, represented only by a single skeleton from the lowermost Vokhmian (Astashikhian Member) of the Vetluga basin (Kalandadze, 1975; Lozovskii, 1983). Later, *Lystrosaurus* bone fragments

Figure 7.5. Aeolian sandstone facies of the Kopanskaya Svita, Vokhmian Gorizont at Elshanka in the Obshchy Syrt region, South Cis-Urals. A, Darren Partridge looks at a section through several dune sets on the road above the village. The holes are nesting sites of house martins. B, Large-scale cross-bedding in the aeolian sandstones. (Photographs from the Saratov–Bristol Palaeontological Expedition of 1995.)

were found at the same locality at the base of the succeeding Ryabinskian Member of the Vokhmian, associated with *Tupilakosaurus* (Ochev, 1992b). Years of intensive collecting have yielded no further dicynodont remains.

In summary, the *Tupilakosaurus–Luzocephalus* assemblage is rather poor and clearly indicative of arid conditions (Ochev, 1992a; Shishkin and Ochev, 1993a, b). Apart from the lithological features of the Vokhmian, this is primarily evidenced by the small size of the constituent reptilian genera, except the semi-aquatic *Lystrosaurus* (which did not outlast the

Early Vokhmian) and the absence of other therapsids (Ochev, 1992a, 1995; cf. Robinson, 1971). On the other hand, strictly aquatic amphibians far exceed reptiles in both abundance and average body size. In all these respects, the fauna shows a marked contrast to that of the preceding Late Tatarian; also, these faunas have no genera and very few families (bystrowianid chroniosuchians, procolophonids, proterosuchids) in common. In spite of this, the Vokhmian assemblage retains the last signs of the worldwide tetrapod faunal uniformity which characterized the latest Permian; this phase of uniformity is designated the '*Lystrosaurus*-lydekkerinid episode' (Shishkin and Ochev, 1993a, b). However, in contrast to the Permian, in which the uniformity is represented by common reptile groups (pareiasaurs, gorgonopsians, theriodonts), the most cosmopolitan faunal components in the Induan are temnospondyls, such as lydekkerinids, tupilakosaurids, and *Gonioglyptus*-like trematosauroids. All these groups are recorded in the Cis-Urals and are known also from the coeval Gondwanan fauna of India (Shishkin, 1973; Tripathi, 1969). That Laurasian-Gondwanan faunal interconnections in the Induan were based mainly on aquatic or semiaquatic forms may be further exemplified by the occurrence of the dicynodont *Lystrosaurus* in the Vokhmian assemblage.

Dating and correlation. The correlation and dating of the Vokhmian tetrapod grouping is primarily based on its index temnospondyl genera, *Tupilakosaurus* and *Luzocephalus*. Both are known to occur in the nearshore marine Induan of East Greenland where they range from the *Glyptophiceras martini* and *Ophiceras commune* ammonite zones respectively to the *Proptychites rosenkrantzi* ammonite zone (Shishkin, 1961, 1980; Lozovskii, 1983). This is confirmed by a sporomorph assemblage from the Vokhmian, which resembles those from the marine Induan of East Greenland, Canada, and Pakistan (Golubeva *et al.*, 1985). In associated conchostracan assemblages, the spine-bearing *Vertexia tauricornis* is widespread, and provides a correlation with the Lower Buntsandstein of the Germanic Basin and the Upper Induan of East Greenland (Kozur *et al.*, 1983).

Benthosuchus *grouping (Rybinskian Gorizont)*

Stratigraphy and lithology. The stratotype of the Rybinskian Gorizont (and Rybinskaya Svita) was described in the western part of Moscow Syncline (Strok and Gorbatkina, 1974). Its middle and upper parts comprise thin-bedded grey and variegated silty clays with concretions containing temnospondyl remains. The presence of ornamented ostracod shells and abundant remains of the lycopsid plant *Pleuromeia* suggests sedimentation in the nearshore zone of a brackish-water basin, indicative of connection with the sea entering from the Baltic Syncline (Strok *et al.*, 1984). To the east, these sediments grade into channel and lacustrine facies (cross-bedded sands and clays) known as the Shilikhinskaya Svita, which also contains tetrapod fossils. The fossil-bearing equivalent of the Rybinskian in the Volga–Ural Anticline and the southern part of the Cis-Uralian Trough is the Staritskaya Svita, which consists largely of channel sandstones. The Rybinskian is 60 m thick in the Moscow Syncline, and reaches up to 200 m in the Cis-Uralian Trough.

Tetrapods. The beginning of the *Benthosuchus* grouping marks the most important event in the evolution of the Vetlugian fauna. *Luzocephalus* is not recorded from the Rybinskian, while *Tupilakosaurus* is known from just two finds of isolated vertebrae from the basal part of the unit. By contrast, benthosuchid temnospondyls increase in abundance, and include the new subfamily, Thoosuchinae. The latter is represented by the fairly common *Thoosuchus* and the more advanced *Trematotegmen*, known from a single specimen from the Ural–Samara area. Morphologically, Thoosuchinae show an obvious trend towards the trematosaurid condition (Shishkin, 1980; Getmanov, 1989). *Thoosuchus* is particularly abundant in the west of the Moscow Syncline, where Rybinskian sediments were deposited in brackish-water facies. Rare chroniosuchian, relict anthracosaur, remains are recorded in both the Vetluga and Samara basins (Shishkin and Novikov, 1992; Novikov and Shishkin, 1995).

Among reptiles, the spondylolestine procolophonids are replaced by true Procolophoninae with differentiated dentition (*Tichvinskia*). The proterosu-

Tetrapod biostratigraphy

chian archosaur *Chasmatosuchus* exceeds its Vokhmian forerunners in size and it has been found at a larger number of sites. Notable also is the earliest record of rauisuchian archosaurs (*Tsylmosuchus*). Other identifiable diapsids include *Microcnemus* and *Scharschengia*. The only known therapsid find is a theriodont, *Scalopognathus*, from the Pechora Syncline.

The wide distribution of the benthosuchids, trematosaur relatives which were active swimmers adapted to life in large water bodies, seems to indicate that stable and broadly connected basins had become established by Rybinskian time (Ochev, 1992a; Shishkin and Ochev, 1993b). The Rybinskian marks the maximum transgression of the Early Olenekian eustatic cycle traceable over the whole northern hemisphere (Lozovskii, 1989; Strok *et al.*, 1984). Along with some changes in lithology, primarily the spread of grey colour in some facies, this seems to indicate a somewhat milder setting as compared with Vokhmian time. The Rybinskian assemblage is peculiar in the scarcity of forms with close Gondwanan affinities. Both procolophonids and proterosuchian archosaurs are represented by lineages that evolved in parallel with those from southern Gondwana.

Dating and correlation. Assignment of the *Benthosuchus* grouping to the Early Olenekian (Lozovskii, 1967) is founded primarily on the following evidence. (1) In the brackish-water deposits in the west of the Moscow Syncline, this grouping is associated with *Pleuromeia*, a cosmopolitan lycopsid plant recorded nowhere before the Olenekian (Dobruskina, 1980, 1982). (2) The Rybinskian miospore assemblage is close to that from the Kumanskaya Svita of the eastern Cis-Caucasus (Aref'ev and Shelekhova, 1991), whose early Olenekian age is indicated by conodonts. (3) The genus *Benthosphenus* from the Russian Far East, the only true member of the Benthosuchidae recorded outside Europe, is accompanied by ammonites of the *Anasibirites nevolini* local zone, indicating a late early Olenekian age (Shishkin and Lozovskii, 1979).

These data show that the *Benthosuchus* grouping is roughly contemporaneous with the vertebrate assemblage from nearshore deposits of the Teryutekh Svita of northern Siberia (*Hedenstroemia hedenstroemi* local

zone, base of the Early Olenekian). This assemblage includes the rhytidosteid amphibian *Boreopelta* (Shishkin and Vavilov, 1985) which is not recorded in European Russia.

Wetlugasaurus *grouping (Sludkian and Ustmylian Gorizonts)*

Stratigraphy and lithology. The stratotype of the Sludkinskaya Svita, which gave its name to the gorizont, is exposed on the Vetluga River in the Moscow Syncline (Mazarovich, 1939). The upper part of the Sludkian Gorizont has been recognized recently on palaeontological evidence as a separate unit which is termed the Ustmylian Gorizont, and is typified by the upper member of the Charkabozhskaya Svita of the Pechora Syncline (Novikov *et al.*, 1990). In the Moscow Syncline, the Ustmylian is represented by the Bereznikovskaya Svita, and in the Mezen' Syncline by the Pizhmomezen'skaya Svita. In other tetrapod-bearing areas of the Cis-Urals, either only the Sludkian in a strict sense is detectable with confidence (Kzylsaiskaya Svita of Ural–Samara Basin), or the correlation with the two Upper Vetlugian gorizonts remains uncertain (lower part of Byzovskaya Svita in northern Cis-Urals).

In the Moscow and Pechora Synclines, the lithology of both gorizonts is rather uniform and suggests lacustrine–alluvial sedimentation in a lowland area. The deposits occur in cycles, in which variegated cross-bedded sands with clay gall lenses grade up into brownish and grey clay. In the Cis-Uralian Trough and the Volga–Ural Anticline, the channel facies, comprising sands and conglomerates, is predominant. The total thickness of the Sludkian and Ustmylian in the Moscow Syncline does not exceed 65 m; in the Cis-Uralian Trough, it does not exceed 235 m.

Tetrapods. The assemblages of the Sludkian and Ustmylian are dominated by late species of *Wetlugasaurus* (*W. angustifrons* and the more advanced *W. malachovi* respectively). In line with this, the *Wetlugasaurus* faunal grouping is subdivided into two biochrons, or subgroupings (Novikov *et al.*, 1990; Novikov, 1994). Both are characterized by the

129

advanced thoosuchine *Angusaurus*, closely approaching the trematosaurid condition, and by diapsid reptiles inherited from the preceding assemblage (*Microcnemus, Scharschengia, Chasmatosuchus, Tsylmosuchus*).

The lower (*Wetlugasaurus angustifrons*) biochron is known to include also such forms as the small thoosuchine *Prothoosuchus*, the procolophonids *Tichvinskia, Samaria*, and *Insulophon* (?), and the poorly known proterosuchian *Exilisuchus*, with the three latter being quite rare. However, the upper limit of occurrence of these forms in the Vetlugian remains uncertain. The upper (*Wetlugasaurus malachovi*) biochron is in most cases difficult to recognize other than by its index species. On the other hand, in the Pechora and Mezen' basins, where this biochron is known best, it has also yielded the advanced benthosuchid *Vyborosaurus*, the procolophonids *Orenburgia, Timanophon* (most common), and *Lestanshoria*, and the rauisuchid *Tsylmosuchus* which appears to predominate among the archosaurs. The prolacertiform *Boreopricea* is known from a single find from the same area. The distinction between the two faunal subdivisions may be further reinforced on the basis of the dipnoan assemblages associated with each (Minikh, 1977; Minikh and Minikh, 1985).

Dating and correlation. The Early Olenekian age of the *Wetlugasaurus* grouping immediately follows from its position between the well-dated Rybinskian Gorizont (Early Olenekian) and the Yarenskian (early and mid parts of Late Olenekian, see below). This is also consistent with the close similarity of the *Benthosuchus* and *Wetlugasaurus* groupings. The miospore assemblage from the Ustmylian Gorizont of the Pechora Basin belongs to the *Densoisporites nejburgii* complex (Golubeva *et al.*, 1985), indicative of the Olenekian.

Parotosuchus *fauna (Yarenskian Gorizont)*

Stratigraphy and lithology. The Yarenskian Gorizont (Blom *et al.*, 1982) is typified by sections in the Moscow and Mezen' Synclines. Here it includes two tetrapod-bearing units, the Fedorovskaya Svita and the overlying Gamskaya Svita, which have their stratotypes in the Vyatka and Vychegda basins respectively (Solov'ev, 1956; Blom, 1960; Lozovskii and Rozanov, 1969). In

this section, the predominant lithology is grey and blue clay, with siderite concretions and marl nodules, which includes subordinate layers of grey channel sands and sandstones. These sediments were laid down in a vast basin supplied by rivers which ran westward from the Urals and Timan uplifts. In the southern part of the Marginal Trough and the Volga–Ural Anticline, the Yarenskian (Petropavlovskaya Svita) is dominated by channel facies (Tverdokhlebov, 1995; Figure 7.6). The same holds north of the Marginal Trough (Pechora Basin), where the fossil-bearing Yarenskian is represented by the upper member of the Lestanshorskaya Svita. In the Cis-Caspian Depression, the Yarenskian consists of estuarine clay and limestone of the Bogdinskaya Svita (the first unit recognized as Triassic in Russia; cf. Chapter 1), which contains tetrapods associated with marine invertebrates.

The widespread grey colour of Yarenskian rocks, and the occurrence of siderite and kaolinite, point to increased humidity when compared to the Vetlugian. The thickness of the Yarenskian is up to 60 m in the Moscow Syncline, and 250 m in the Cis-Uralian Marginal Trough.

Bones from the Yarenskian are largely buried in coarse-grained channel facies; they are usually isolated, fragmented, and show other signs of transport. On the other hand, at a few sites in the Ural Basin, articulated skeletons have been recorded in mudrocks, indicative of burial in quiet water conditions. Accumulations of isolated bones in deltaic clay facies are also known from the Vychegda Basin. The fossil-bearing rocks in northern areas are peculiar for their essentially grey (sandstone) and dark (clay) colour, while the contained bones are usually black. This makes a contrast with the largely reddish colour which dominates both the sediments and the bones in the southern sites.

Tetrapods. The *Parotosuchus* fauna, specific to the Yarenskian Gorizont, is known from a more limited area than the Vetlugian assemblages. Most sites are in the east of the Moscow Syncline (Luza and Vyatka Basins), in the Mezen' Syncline (Vychegda Basin), and in the basins of the Pechora, Don, and Ural rivers.

The *Parotosuchus* fauna is markedly advanced, in

(A)

(B)

Figure 7.6. Channel facies of the Petropavlovskaya Svita, Yarenskian Gorizont, at the type locality, Petropavlovka, Orenburg Province, South Cis-Urals. A, David Gower looks at the channel lag at the base of a thick channel sand body, which is capped by finer-grained overbank deposits. Disarticulated bones occur in the channel lag. B, Glenn Storrs looks at cyclical fine-grained sandstone and mudstone beds close to the site where a well-preserved procolophonid skull was found. (Photographs from the Saratov–Bristol Palaeontological Expedition of 1995.)

comparison with the *Benthosuchus–Wetlugasaurus* fauna, and shares only a few genera with the latter. These include chroniosuchians, some procolophonids, and a rauisuchid. The main components of the *Parotosuchus* fauna are readily derivable from those of the preceding fauna. Among the temnospondyls, the index genus *Parotosuchus* is a descendant of *Wetlugasurus*, while benthosuchid offshoots are represented by two newly arising families, Yarengiidae (*Yarengia*) and Trematosauridae (*Inflectosaurus*). The revival of a Gondwanan influence is documented by records of the Brachyopidae (*Batrachosuchoides*), Rhytidosteidae (*Rhytidosteus*, previously known only from South Africa; Shishkin, 1994), and the early Plagiosauridae, related to the Upper Permian Peltobatrachidae of East Africa. The chroniosuchian anthracosaur *Axitectum* is inherited from the preceding (Vetlugian) fauna. Some of the procolophonids are also Vetlugian survivors (*Tichvinskia*, *Orenburgia*), while others are more advanced (*Burtensia*, *Kapes*). The first recorded trilophosaurids include *Coelodontognathus* and *Vitalia*. Archosaurs are represented by the erythrosuchids *Garjainia* (*Vjushkovia*) and *Gamosaurus*, and the rauisu-

chids *Tsylmosuchus*, *Vytshegdosuchus*, and *Jaikosuchus*. Except for *Garjainia*, all these archosaurs are based only on isolated bones. Other diapsids include large prolacertiforms known from a few fragments. Therapsids are extremely rare and known only from a single skull of the theriodont *Silphedosuchus* (Tatarinov, 1974b) and some galesaurid remains. In the accompanying fish assemblage, the dipnoan *Gnathorhiza* is replaced by *Ceratodus*.

The fauna of the Yarenskian Gorizont may be reasonably subdivided into two biochrons corresponding to the Fedorovskian and Gamskian members (Lozovskii and Rozanov, 1969; Minikh and Minikh, 1985; Minikh and Makarova, 1990; Lozovskii *et al.*, 1995). The earlier biochron is marked by the persistence of the lungfish *Gnathorhiza* (which co-exists with *Ceratodus*), as well as other Vetlugian survivors, the procolophonid *Tichvinskia*, and the rauisuchid *Tsylmosuchus*. The procolophonid *Burtensia* also characterises this unit. The later biochron, where *Gnathorhiza* disappears from the fish assemblage, includes the procolophonid genus *Kapes* and the rauisuchid *Vytshegdosuchus*. In the Don Basin, the latter unit

131

is also thought to include *Trematosaurus*, a temnospondyl genus otherwise unknown from Eastern Europe, but common in the middle Buntsandstein of the Germanic Basin.

The geographical differentiation of the Cis-Uralian fauna, which appeared less distinct after the Vokhmian, becomes clearer again. The Yarenskian assemblage of the Ural Basin differs from that of northern areas by its particular set of *Parotosuchus* species, the occurrence of the rhytidosteid amphibian, and the predominance of erythrosuchids among the archosaurs. On the other hand, northern areas are distinguished by the occurrence of the advanced procolophonid *Kapes* and the abundance of rauisuchid archosaurs.

A marked increase in the body size of Yarenskian amphibians (*Parotosuchus*, *Inflectosaurus*) and archosaurs (*Garjainia*, *Vytshegdosuchus*) supports the lithological evidence for a change towards a more humid climate compared to the Vetlugian. This conclusion is further reinforced by the replacement of the dipnoan *Gnathorhiza* by *Ceratodus*, based on biological differences of their living relatives (Ochev, 1992a). On the other hand, large- or medium-sized herbivores, indicative of abundant vegetation, are absent from the *Parotosuchus* community. The only evidence of their existence by that time in eastern Europe is the recent find (unpublished) of a fragment of a dicynodont lower jaw from the Yarenskian of the Don Basin.

Dating and correlation Key evidence for the late Olenekian age of the *Parotosuchus* fauna is provided by finds of its index genus in marginal marine deposits of the Caspian area. *Parotosuchus bogdoanus* has been found in the Cis-Caspian Depression (Bogdo Mountain), associated with ammonites of the *Tirolites cassianus* local zone, and *P. sequester* was recovered in the Mangyshlak Peninsula (western Kazakhstan) from beds containing the *Columbites karataucikus* ammonite fauna (Lozovskii and Shishkin, 1974). These two levels correspond to the *Tirolites harti* and *Columbites parisianus* Zones respectively of the Upper Olenekian of the Alpine standard scale. Miospore and charophyte assemblages are represented in the lower half of the range of the *Parotosuchus* fauna by the *Densoisporites nej-*

burgii and *Porochara triassica* complexes respectively. Both also occur in the Hardegsen beds of the Middle Buntsandstein of the Germanic Basin which contain the *Parotosuchus–Trematosaurus* amphibian assemblage (Blom *et al.*, 1982; Golubeva *et al.*, 1985; Saydakovsky and Kiesielewski, 1985). The upper part of the range of *Parotosuchus*, as represented by the Gamskaya Svita of the Mezen' Syncline, shows an *Aratrisporites*-dominated miospore assemblage, which is common in the Upper Olenekian worldwide (Yaroshenko *et al.*, 1991).

Middle Triassic

The Middle Triassic Cis-Uralian faunas range from the Late Anisian (?) to Late Ladinian. They have been collected mainly in the south of the Cis-Uralian Marginal Trough (Orenburg Province and Bashkortostan). A few finds are also known from isolated depressions in the northern part of the trough (Pechora Basin). On the East European platform, Middle Triassic tetrapods of the same faunal succession are known only in western Kazakhstan, where they are recorded from the Inder Lake area in the Cis-Caspian Depression, along the lower reaches of the Ural River. Like nearly all Middle Triassic tetrapod assemblages worldwide, these Cis-Uralian faunas show the rise and broad radiation of large herbivorous therapsids, kannemeyeroid dicynodonts, characterizing the global kannemeyeroid epoch (Ochev and Shishkin, 1989). Together with temnospondyls, these reptiles are the most abundant tetrapods; archosaurs also radiated. The Middle Triassic Cis-Uralian tetrapods constitute two successive communities, the *Eryosuchus* and *Mastodonsaurus* faunas, named after their marker amphibians, and corresponding to the Donguz and Bukobay Gorizonts respectively.

There is a sedimentary gap between the Yarenskian and Donguz deposits in the southern Cis-Urals, so that an earliest Anisian fauna comparable with the *Eocyclotosaurus* community of Euramerica seems to be lacking (Ochev and Shishkin, 1989). The only questionable analogue of this unit may be indicated by some sparse amphibian remains, the plagiosaurid *Aranetzia* and the capitosauroid *Komatosuchus*, from the

Figure 7.7. Fluviatile sandstones of the Donguz Svita, Donguz Gorizont, at the locality Koltaevo II, Bashkortostan Republic, South Cis-Urals. Vitalii G. Ochev (left) and Mikhail Surkov excavate in thick sandstones where several complete skeletons of dicynodonts had been found previously. (Photograph from the Saratov–Bristol Palaeontological Expedition of 1995.)

Nyadeytinskaya and Krasnokamenskaya Svitas of the northern Cis-Urals of the Pechora Basin (Shishkin and Novikov, 1992; Novikov, 1994; Chapter 3).

Eryosuchus *fauna (Donguz Gorizont)*

Stratigraphy and lithology. The Donguz Gorizont is established in the south of the Cis-Uralian Trough and is typified by the Donguz Svita, with its stratotype on the left Ural tributary bearing the same name (Garyainov *et al.*, 1967; Tverdokhlebov, 1995). The dominant lithology is variegated clay and silt, with occasional paleosol horizons; sandstones increase in importance to the north and east. The fossil-bearing facies are variable and indicate a vast deltaic plain with sediments accumulating in temporary channels, ponds, and muddy streams (Figure 7.7). In the Cis-Caspian Depression, the equivalent of the Donguz Gorizont is represented by the Elton Gorizont and the lower part of the Inder Gorizont, comprising brackish-water marine carbonates and clays (Blom *et al.*, 1982). In the northern part of the Cis-Uralian Trough,

possible correlatives of the Donguz Gorizont are the lower member of the Nyadeytinskaya Svita (Korotaikha Depression) and the Krasnokamenskaya Svita (Bolshaya Synia Depression); both contain sparse remains of post-Yarenskian tetrapods (Novikov, 1994). In the southern Cis-Urals, the Donguz Gorizont is 175–360 m thick. The tetrapods (*Eryosuchus* fauna) come from its middle and upper parts. On the evidence of the plagiosaurid amphibians, it seems possible to subdivide the tetrapod-bearing member into a lower (*Plagiosternum*) and upper (*Plagioscutum*) biochron (Shishkin, 1987).

Tetrapods. Compared to the Early Triassic, the generic diversity of temnospondyls in the *Eryosuchus* fauna is reduced to one third that of the reptiles (Ochev, 1992a). However, the dominant temnospondyl genera, the capitosauroid *Eryosuchus* and the plagiosaurid *Plagioscutum*, are fairly abundant; somewhat rare is *Plagiosternum*, recorded only from the early assemblage of the *Eryosuchus* fauna. All these forms evolved from Cis-Uralian late Olenekian forerunners (Shishkin, 1987; Ochev, 1992a), as did the only procolophonid, *Kapes*, which persisted from the *Parotosuchus* fauna (Novikov, 1994). Most archosaurs are thought also to have evolved *in situ*. These include the large proterosuchians *Erythrosuchus* and *Sarmatosuchus*, and the rauisichids *Vjushkovisaurus* and *Dongusuchus* (Ochev, 1991, 1992a; Sennikov, 1995a, b). On the other hand, the euparkeriid *Dorognathus* may reflect a Gondwanan influence.

The most striking feature of the *Eryosuchus* fauna is the radiation of therapsids, especially kannemeyeroid dicynodonts, which appear to have arrived from Gondwanaland. They include kannemeyeriids, most of the recorded dicynodont genera, (?) stahlekeriids (*Uralokannemeyeria*) and shansiodontids (*Rhinodicynodon*), all of them known only from the Cis-Urals (Kalandadze, 1995). Theriodonts are rather poorly known. These are primarily represented by bauriid therocephalians (*Dongusaurus*, *Nothogomphodon*) and a gomphodont cynodont assigned to the East African genus *Scalenodon*; the latter is identified from a single tooth (Tatarinov, 1973, 1974a). One more theriodont, *Anthecosuchus*, described by Tatarinov (1973) as a gomphodont, is believed by Battail (1984) to be a bauriid (see also Chapter 6 of this book).

The variegated colour of Donguz sediments may indicate seasonal increases of humidity (Ochev, 1992a). A trend towards a milder climate than that of the Late Olenekian is indicated by the large body size of the dominant temnospondyls, archosaurs, and dicynodonts; further, the diversity of herbivorous therapsids (kannemeyeroids, bauriids, and gomphodonts) suggests that there were areas of abundant vegetation. On the other hand, seasonal aridity is indicated by sedimentary features peculiar to arid swamps and soils (Tverdokhlebov, 1995), and the occurrence of 'dried pond' amphibian burials (Ochev and Shishkin, 1967; Chapter 3), the result of temporary droughts.

Dating and correlation In terms of main contstituent groups, the *Eryosuchus* fauna is similar to most other Middle Triassic tetrapod assemblages worldwide (Ochev *et al.*, 1964; Ochev and Shishkin, 1989). However, the stratigraphic position of such faunas in terms of the standard marine stages is hard to establish, so that comparison with the *Eryosuchus* fauna is not helpful for its dating. The only exception is the poor amphibian assemblage from the uppermost Muschelkalk of Crailsheim in the Germanic Basin which includes the plagiosaurids *Plagiosternum granulosum* and *Plagiosuchus pustuloglomeratus*, closely related to the Donguz forms *Plagiosternum paraboliceps* and *Plagiosuchus ochevi* respectively. The Upper Muschelkalk is equated by its invertebrate fossils with the early Late Ladinian of the Alpine section (Kozur, 1974), and a similar correlation and dating may be accepted for the *Eryosuchus* fauna (Shishkin and Ochev, 1992). However, since the deposits underlying the Upper Muschelkalk belong to marine facies, in which finds of continental vertebrates are unlikely, the range of the European plagiosaurid-*Eryosuchus* assemblage could extend down into the Anisian, though there is no evidence for this. In the Cis-Urals, a Late Anisian miospore assemblage has been identified from the lower part of the Donguz Gorizont, which is known to be barren of tetrapods. From the upper part of the unit, a Ladinian miospore assemblage has been reported, although the data do not come from the stratotype section (Makarova and Vergay, 1995).

Mastodonsaurus *fauna (Bukobay Gorizont)*

Stratigraphy and lithology. The term Bukobay Gorizont (Blom *et al.*1982; Tverdokhlebov, 1995) refers to the upper part of the Middle Triassic section of the southern Cis-Urals formed by the Bukobay Svita (with its stratotype in the basin of the Berdyanka River, a left tributary of the River Ural). The unit shows marked cyclicity. The basal cycle begins with greenish grey coarse- and medium-grained cross-bedded channel sandstones with gravels, clay galls and wood logs. It yields most of the tetrapod fossils known from the unit, and was formerly designated the Yushatyr Svita. It passes upwards into a variegated clay of deltaic and lacustrine origin. In successive cycles, the thickness of the basal sandstone layers progressively declines, while the clay becomes predominantly grey. The clay contains plants of the Ladinian–Carnian *Scytophyllum* flora. Other fossils include conchostracans, bivalves, and miospores. The thickness of the Bukobay Gorizont is up to 600m.

Outside the southern Cis-Urals, Bukobay equivalents containing the *Mastodonsaurus* fauna are known from two areas. A few remains have been recovered from sandstones and mudrocks of the Nadkrasnokamenskaya (Keryamayol) and Syninskaya svitas of the northern Cis-Urals (Novikov, 1994). In the Cis-Caspian Depression, more abundant fossils were recovered from a dark carbonate clay in the upper part of the Inder Gorizont in western Kazakhstan (Ochev and Smagin, 1974; Shishkin, 1987). These fossiliferous units show evidence of sedimentation in brackish-water basins connected to the sea.

Most bones collected from the Bukobay Gorizont and its correlatives in European Russia are disarticulated and often fragmented, evidence of active water transport. There have been rare finds of complete or fragmented skulls of amphibians. All the bones are black or dark brown, in contrast to the more variable, mostly brown to reddish, colour of finds from the Donguz Gorizont.

Tetrapods. The index genus of the Bukobay fauna is a giant mastodonsaurid temnospondyl which is currently referred to *Mastodonsaurus*, but deserves generic separation (Shishkin and Ochev, 1992). In the southern Cis-Urals this is the commonest form, and it may be associated with another mastodonsaurid, the small long-snouted *Bukobaja*. Mastodonsauridae are known only from Europe, and they first appear in the Early Anisian of the Germanic Basin. Other dominant components of the amphibian fauna are the plagiosaurs *Plagiosternum* and *Plagioscutum*, which are derivable from species of the same genera in the *Eryosuchus* assemblage. In the southern Cis-Urals, *Plagiosternum* is rather common, and *Plagioscutum* rare. By contrast, in the Cis-Caspian Depression (Inder Lake area), *Plagioscutum* outnumbers all other amphibians, while *Plagiosternum* is not recorded at all (Shishkin, 1987). A few bones of cyclotosaurid capitosauroids, a group otherwise typical of the Late Triassic of Eurasia, were recovered in both the southern and northern Cis-Urals. A noteworthy find is the last record of a chroniosuchian anthracosaur, *Synesuchus*, represented by dermal scutes from the northern Cis-Urals (Novikov and Shishkin, 1995).

The reptile component of the fauna has been poorly studied. It is dominated by the same groups as in the *Eryosuchus* fauna, large archosaurs and kannemeyeriid dicynodonts. Archosaurs include the giant erythrosuchid *Chalishevia* and the rauisuchids *Jushatyria* and *Energosuchus*; non-archosaurian diapsids are represented by the long-necked prolacertiform *Malutinisuchus* (Ochev, 1986, 1991; Sennikov, 1995a, b). Kannemeyeroids are represented by two genera identified from fragments, *Elephantosaurus* and *Elatosaurus* (Kalandadze, 1995). The former is the largest anomodont known from Eastern Europe, with skull roof bones up to 50 mm thick.

The predominance of the grey or variegated colour of the rocks, the abundance of plant remains, and the large body size of the main components of the tetrapod assemblage, including the terrestrial herbivores (dicynodonts), suggest an increase of humidity in Eastern Europe towards the end of the Mid Triassic.

Dating and correlation All the evidence suggests that the Bukobay Gorizont corresponds to the Lettenkeuper of the Germanic Basin, and that the mastodonsaurid-dominated faunas of both units are contemporaneous. As the Lettenkohle is correlated with the marine Late Ladinian of the Alpine Basin (Kozur, 1974), the same dating is accepted for the Eastern European *Mastodonsaurus* fauna. The correlation of the Bukobay Gorizont and its local equivalents with the Lettenkeuper is based on the following data: (1) the amphibian faunas of both units are similar and immediately succeed tetrapod assemblages, those of the Donguz Gorizont and topmost Muschelkalk, which are also correlatives (Shishkin and Ochev, 1992); (2) the plant assemblage from the Bukobay Gorizont belongs to the *Scytophyllum* Flora, which marks the Ladinian–Carnian in the northern hemisphere (Dobruskina, 1980, 1982); and, (3) the miospore assemblage of the Bukobay Gorizont is close to that of the Lettenkohle and to the uppermost Middle Triassic assemblage from the marine Masteksayan Gorizont of the Cis-Caspian section (Makarova and Vergay, 1995).

Late Triassic continental deposits are known in European Russia from the Cis-Uralian Marginal Trough, the Pechora Syncline, and some other areas. They contain plants and sporomorphs, but no tetrapod fossils have so far been recorded.

Acknowledgements

We thank Geoff Warrington and Andrew Milner for very helpful comments on the manuscript, and Pam Baldaro for redrafting the diagrams.

References

Aref'ev, M.P. and Shelekhova, M.N. 1991. [Palynological evidence for dating of Parshin Beds of the Lower Triassic of Moscow Syncline.] *Byulleten' Moskovskogo Obshchestva Ispytatelei Prirody, Otdel Geologicheskii* 66 (3): 73–77.

Battail, B. 1984. Les Cynodontes (Reptilia, Therapsida) et la paléobiogeographie du Trias. *Third Symposium on Mesozoic Terrestrial Ecosystems, Short Papers*, pp. 1–6, eds. W.-E. Reif & F. Westphal. Tubingen: ATTEMPTO Verlag.

Blom, G.I. 1960. [The Lower Triassic of the Volga-Vyatka Basin.] *Trudy Vsesoyuznogo Soveshchania po Utochneniu Stratigrafii Mezozoyskikh Otlozhenii Russkoi Platformy. I. Triasovaya Systema. Trudy VNIGNI* 29: 70–75.

—1968. [*Catalogue of Localities of Faunal Remains in the Lower Triassic Deposits of the Middle Volga and Kama Basins.*] Kazan: Izdatel'stvo Kazanskogo Universiteta, 375 pp.

—, Lozovskii, V.R., Minikh, M.G., Strok, N.I., Ochev, V.G., Tverdokhlebov, V.P. and Shishkin, M.A. 1982. [Moscow, Mezen' Syncline, Volga–Ural Anticline (Chart 1).] pp. 20–35 in *Reshenie Mezhvedomstvennogo Stratigraficheskogo Soveshchania po Triasu Vostochno-Evropeyskoi Platformy (Saratov, 1979).* Leningrad: Interdepartmental Stratigraphy Committee.

Bystrov, A.P. and Efremov, I.A. 1940. [*Benthosuchus sushkini* Efr. – a labyrinthodont from the Eotriassic of Sharzhenga River.] *Trudy Paleontologicheskogo Instituta AN SSSR* 10: 5–152.

Dobruskina, I.A. 1980. [Stratigraphic position of the plant-bearing Triassic sediments of Eurasia.] *Trudy Geologicheskogo Instituta AN SSSR* 346: 1–163.

—1982. [*The Triassic Floras of Eurasia.*] Moscow: Nauka, 196 pp.

Efremov, I.A. 1929. *Benthosaurus sushkini,* ein neuer Labyrinthodont aus den permotriassischen Ablagerungen des Scharschenga-Flusses, Nord-Düna Gouvernement. *Bulletin de l'Académie des Sciences, URSS, Classe de Sciences Physico-Mathématiques* 757–770.

—1932. [On the Permo-Triassic labyrinthodonts from the USSR. I. Labyrinthodonts of the Campylian beds of Bolshoi Bogdo mountain. II. On the morphology of the labyrinthodont *Dvinosaurus.*] *Trudy Paleozoologicheskogo Instituta AN SSSR* 1: 57–68.

—1937. [On the stratigraphic divisions of the continental Permian and Triassic of the USSR, based on tetrapod faunas.] *Doklady AN SSSR* 16: 125–132.

—1940. Kurze Übersicht über die Formen der Perm- und Trias-Tetrapoden-Fauna der UdSSR. *Centralblatt für Mineralogie, Geologie, und Paläontologie, Abteilung B* 1940: 372–383.

—and V'yushkov, B.P. 1955. [Catalogue of the localities of Permian and Triassic terrestrial vertebrates in the territories of the USSR.] *Trudy Paleontologicheskogo Instituta AN SSSR* 46: 1–147.

Garyainov, V.A., Kuleva, G.V., Ochev, V.G. and Tverdokhlebov, V.P. 1967. [*A Guide for an Excursion over the Upper Permian and Triassic Continental Formations of the Southeast of the Russian Platform and the Cis-Urals.*] Saratov: Izdatel'stvo Saratovskogo Universiteta, 148 pp.

—and Ochev, V.G. 1962. [*Catalogue of Vertebrate Localities in the Permian and Triassic Deposits of the Orenburg Cis-Urals.*] Saratov: Izdatel'stvo Saratovskogo Universiteta, 61 pp.

Getmanov, S.N. 1989. [Triassic amphibians of the East European Platform: family Benthosuchidae.] *Trudy Paleontologicheskogo Instituta AN SSSR* 236: 1–102.

Golubeva, L.P., Makarova, I.S., Romanovskaya, G.M. and Semenova, E.V. 1985. [Palynomorph complexes from the Triassic of East European Platform and their role for correlation of the coeval deposits.] pp. 77–85 in [*Triassic Deposits of the East European Platform.*] Saratov: Izdatel'stvo Saratovskogo Universiteta.

Hartmann-Weinberg, A. and Kusmin, T.M. 1936a. Untertriadische Stegocephalen der USSR. 1. *Lyrocephalus acutirostris* nov. spec. *Problemy Paleontologii* 1: 63–84.

—1936b. Untertriadische Stegocephalen der Oka-Zna Antiklinale. 2. *Capitosaurus volgensis* nov. sp. *Problemy Paleontologii* 1: 85–86.

Huene, F. von 1940. Eine Reptilfaune aus der ältesten Trias Nordrusslands. *Neues Jahrbuch für Geologie und Paläontologie, Abteilung B* 84: 1–23.

Ivakhnenko, M.F. 1979. [The Permian and Triassic procolophonians of the Russian platform.] *Trudy Paleontologicheskogo Instituta AN SSSR* 164: 1–80.

Kalandadze, N.N. 1975. [The first find of a *Lystrosaurus* on the territory of the European part of the USSR.] *Paleontologicheskii Zhurnal* **1975**: 140–142.

—1995. Synapsids, pp. 89–95 in Shishkin, M.A. (ed.), [*Biostratigraphy of the Continental Triassic of the Southern Cis-Urals.*] Moscow: Nauka.

—Ochev, V.G., Tatarinov, L.P., Chudinov, P.K. and Shishkin, M.A. 1968. Catalogue of the Permian and Triassic tetrapods of the USSR, pp. 72–91 in [*Upper Palaeozoic and Mesozoic Amphibians and Reptiles of the USSR.*] Moscow: Nauka.

Kozur, H. 1974. Biostratigraphie der germanischer Mitteltrias. Teil I. *Freiberger Forschungshefte* C **280**: 1–56.

—, Lozovskii, V.R., Lopato, A.Yu. and Movshovich, E.V. 1983. [Stratigraphic position of principal vertexiid localities in the Triassic of Europe.] *Byulleten' Moskovskogo Obshchestva Ispytatelei Prirody, Otdel Geologicheskii* **58** (5): 60–72.

Kuz'min, T.M. 1935. [Lower Triassic stegocephalians of the Oka-Tsna Anticline.] *Ezhegodnik Russkogo Paleontologicheskogo Obshchestva* **10**: 39–47.

—1937. Untertriadische Stegocephalen der Oka-Zna Antiklinale. 3. *Volgasaurus kalajevi* gen. et sp. nov. *Problemy Paleontologii* **2–3**: 621–648.

Lozovskii, V.R. 1967. [New data on stratigraphy of the Lower Triassic of the Moscow Syncline.] *Sbornik Statei po Inzhenernoi Geologii* **6**: 121–128.

—1983. [On the age of the *Lystrosaurus* beds in the Moscow Syncline.] *Doklady AN SSSR* **272**: 1433–1437.

—1989. Some peculiarities of development of East European and North American platforms during early Triassic. *28th International Geological Congress, Washington, D.C. Abstracts* **2**: 329–330 [author name spelled Lozovsky].

—, Novikov, I.V., Sennikov, A.G., Shishkin, M.A., and Minikh, M.G. 1995. [On the subdivision of the Early Triassic *Parotosuchus* Fauna of Eastern Europe]. pp. 20–21 in A.G. Sennikov (ed.): *All-Russian Conference. Palaeontology and Stratigraphy of Permian and Triassic Continental Deposits of Northern Eurasia.* Moscow, Paleontologicheskii Institut RAN: 1–48.

—and Rozanov, V.I. 1969. [Stratigraphy of the Triassic of the northern part of Moscow Syncline.] *Izvestiya Vysshykh Uchebnykh Zavedenii (Geologiya i Razvedka)* **1969** (10): 15–22.

—and Shishkin, M.A. 1974. [First labyrinthodont find from the Lower Triassic of Mangyshlak.] *Doklady AN SSSR* **214**: 169–172.

Makarova, I.S. and Vergay, I.F. 1995. [Miospores.] pp. 120–129 in Shishkin, M.A. (ed.), [*Biostratigraphy of the Continental Triassic of the Southern Cis-Urals.*] Moscow: Nauka.

Mazarovich, A.N. 1939. [On the Triassic of the Vetluga and Vyatka Basins.] *Uchenye Zapiski Moskovskogo Universiteta* **26** (1): 75–93.

Minikh, M.G. 1977. [*Triassic Dipnoan Fishes from the East of the European Part of the USSR.*] Saratov: Izdatel'stvo Saratovskogo Universiteta, 96 pp.

—and Minikh, A.V. 1985. [Subdivision of the Triassic deposits of the East European Platform on ichthyofauna.] pp. 44–51 in [*Triassic Deposits of the East European Platform.*] Saratov: Izdatel'stvo Saratovskogo Universiteta.

—and Makarova, I.S. 1990. [On the stratigraphic position of the Gam Svita in the Triassic of the Moscow Syncline.] *Trudy Vsesoyuznoy XI Geologicheskoi Konferentsii Komi ASSR, I, Syktyvkar.* 233–239.

Nikitin, S.N. 1883. [Geological account of the Vetluga district.] *Izdanie Sankt-Peterburgskogo Mineralogicheskogo Obshchestva* **1883**: 1–48.

Novikov, I.V. 1993. Triassic tetrapod assemblages of the Timan–North Urals region. *Bulletin of the New Mexico Museum of Natural History and Science* **3**: 371–373.

—1994. [Biostratigraphy of the continental Triassic of the Timan–North Uralian region on the tetrapod fauna.] *Trudy Paleontologicheskogo Instituta RAN* **261**: 1–139.

—, Lozovskii, V.R., Shishkin, M.A. and Minikh, M.G. 1990. [A new horizon in the Lower Triassic of the East European Platform.] *Doklady AN SSSR* **315**: 453–456.

—and Shishkin, M.A. 1995. Paleozoic relics in Triassic tetrapod communities: the last anthracosaur amphibians, pp. 29–32 in Sun, A. and Wang, Y. (eds.), *Sixth Symposium on Mesozoic Terrestrial Ecosystems and Biota. Short Papers.* Beijing: China Ocean Press.

Ochev, V.G. 1966. [*Systematics and Phylogeny of Capitosauroid Labyrinthodonts.*] Saratov: Izdatel'stvo Saratovskogo Universiteta, 184 pp.

—1972. [*Capitosauroid Labyrinthodonts from the Southeast of the European Part of the USSR.*] Saratov: Izdatel'stvo Saratovskogo Universiteta, 269 pp.

—1980. [On the localities of Middle Triassic tetrapods in the southern Cis-Urals.] pp. 38–50 in [*Questions on the Geology of the Southern Urals and Povolzh'e*]. Saratov: Izdatel'stvo Saratovskogo Universiteta, 148 pp.

—1986. [On Middle Triassic reptiles of the southern Cis-

Urals.] *Ezhegodnik Vsesoyuznogo Paleontologicheskogo Obshchestva* **29**: 171–180.

—1991. [On the History of the Evolution of Early Archosaurs.] *All-Union Institute of Scientific Information (VINITI). Deposited Manuscript* **2361** (**B91**): 1–78.

—1992a. [On the history of the Triassic vertebrates of the Cis-Urals.] *Byulleten' Moskovskogo Obshchestva Ispytatelei Prirody, Otdel Geologicheskii* **67** (4): 30–43.

—1992b. [On the second unquestionable find of an anomodont in the Lower Triassic of East Europe.] *Izvestiya Vysshykh Uchebnykh Zavedenii (Geologiya i Razvedka)* **1992** (2): 132–133.

—1995. On the relationship between the history of the Triassic tetrapods from Eastern Europe and climate evolution, pp. 43–46 in Sun, A. and Wang, Y. (eds.), *Sixth Symposium on Mesozoic Terrestrial Ecosystems and Biota. Short Papers.* Beijing: China Ocean Press.

—and Shishkin, M.A. 1967. [Burials of early amphibians in the Orenburg area.] *Priroda* **1967** (1): 79–85.

—and— 1989. On the principles of global correlation of the continental Triassic on the tetrapods. *Acta Palaeontologica Polonica* **34**: 149–173.

—, Shishkin, M.A., Garyainov, V.A. and Tverdokhlebov, V.P. 1964. [New data on stratification of the Triassic of Orenburg Cis-Urals on vertebrates.] *Doklady AN SSSR* **158**: 363–365.

—and Smagin, B.N. 1974. [On the locality of Triassic vertebrates near Inder Lake.] *Byulleten' Moskovskogo Obshchestva Ispytatelei Prirody, Otdel Geologicheskii* **49** (3): 74–81.

—, Tverdokhlebova, G.I., Minikh, M.G. and Minikh, A.V. 1979. [*Stratigraphic and Paleontological Significance of Upper Permian and Triassic Vertebrates of the Eastern European Platform and Cis-Urals.*] Saratov: Izdatel'stvo Saratovskogo Universiteta, 158 pp.

Robinson, P. 1971. A problem of faunal replacement on Permo-Triassic continents. *Palaeontology* **14**: 131–153.

Ryabinin, A.N. 1930. [A labyrinthodont stegocephalian *Wetlugasaurus angustifrons* nov. gen., nov.sp. from the Lower Triassic of Vetluga Land in northern Russia.] *Ezhegodnik Vserossiisskogo Paleontologicheskogo Obshchestva* **8**: 49–76.

Saydakovsky, L.J. and Kiesielewski, F.Y. 1985. [Significance of charophytes for stratigraphy of the Triassic of East European Platform.] pp. 67–77 in [*Triassic Deposits of the East European Platform.*] Saratov: Izdatel'stvo Saratovskogo Universiteta.

Sennikov, A.G. 1995a. [Diapsid reptiles from the Permian and Triassic of East Europe.] *Paleontologicheskii Zhurnal* **1995** (1): 75–83.

—1995b. [Early thecodonts of Eastern Europe.] *Trudy Paleontologicheskogo Instituta RAN* **263**: 1–141.

—1996. Evolution of the Permian and Triassic tetrapod communities of Eastern Europe. *Palaeogeography, Palaeoclimatology, Palaeoecology* **120**: 331–351.

Shishkin, M.A. 1961. [New data on *Tupilakosaurus.*] *Doklady AN SSSR* **136**: 938–941.

—1973. [Morphology of early amphibians and problems of the evolution of lower tetrapods.] *Trudy Paleontologicheskogo Instituta AN SSSR* **137**: 1–260.

—1980. [Luzocephalidae, a new family of the Triassic labyrinthodonts.] *Paleontologicheskii Zhurnal* **1980** (1): 104–124.

—1987. [The evolution of ancient amphibians (Plagiosauroidea).] *Trudy Paleontologicheskogo Instituta AN SSSR* **225**: 1–144.

—1994. [A Gondwanan rhytidosteid (Amphibia, Temnospondyli) in the Lower Triassic of the southern Cis-Urals.] *Paleontologicheskii Zhurnal* **1994** (4): 97–110.

—and Lozovskii V.R. 1979. [A labyrinthodont from the Triassic of the southern Maritime Land.] *Doklady AN SSSR* **246**: 201–205.

—, — and Ochev, V.G. 1986. [Review of Triassic land vertebrate localities in the Asiatic part of the USSR.] *Byulleten' Moskovskogo Obshchestva Ispytatelei Prirody, Otdel Geologicheskii* **61** (6): 51–63.

—and Novikov, I.V. 1992. [Anthracosaur relicts in the early Mesozoic of East Europe.] *Doklady RAN* **325**: 829–832.

—and Ochev, V.G. 1967. [Land vertebrate fauna as a basis for stratification of the continental Triassic of the USSR.] pp. 74–82 in Martinson, G.G. (ed.) [*Stratigraphy and Palaeontology of the Continental Mesozoic and Cenozoic of the Asiatic part of the USSR.*] Leningrad: Nauka.

—and— 1985. [Significance of land vertebrates for stratigraphy of the Triassic of East European Platform.] pp. 28–43 in [*Triassic Deposits of the East European Platform.*] Saratov: Izdatel'stvo Saratovskogo Universiteta.

—and— 1992. [On the age of the *Eryosuchus* and *Mastodonsaurus* faunas of East Europe.] *Izvestiya AN SSSR, Seriya Geologicheskaya* **7**: 28–35.

—and— 1993a. [On the spatial differentiation of the land

vertebrate fauna in the early Triassic.] pp. 101–109 in Sokolov, B.S. (ed.) [*Faunas and Ecosystems of the Geological Past.*] Moscow: Nauka.

—and— 1993b. The Permo-Triassic transition and the early Triassic history of the Euramerican tetrapod fauna. *Bulletin of the New Mexico Museum of Natural History and Science* 3: 435–437.

—, Ochev, V.G., Tverdokhlebov, V.P., Vergay, I.F., Goman'kov, A.V., Kalandadze, N.N., Leonova, E.M., Lopato, A.Yu., Makarova, I.S., Minikh, M.G., Molostovskii, E.M., Novikov, I.V. and Sennikov, A.G. 1995. [*Biostratigraphy of the Triassic of the Southern Cis-Urals.*] Moscow: Nauka, 206 pp.

—and Vavilov, M.N. 1985. [A find of a rhytidosteid (Amphibia, Labyrinthodontia) from the Triassic of the USSR.] *Doklady AN SSSR* 282: 971–975.

Solov'ev, V.K. 1956. [On the problem of the stratigraphy of the Lower Triassic of the Volga Basin.] *Doklady AN SSSR* 110: 430–433.

Strok, N.I. and Gorbatkina, T.E. 1974. [Stratigraphy of the Lower Triassic of the western and central parts of Moscow Syncline.] *Izvestiya Vysshikh Uchebnykh Zavedenii (Geologiya i Razvedka)* 7: 26–36.

—, Gorbatkina, T.E. and Lozovskii, V.R. 1984. [*The Upper Permian and Lower Triassic Deposits of Moscow Syncline.*] Moscow: Nedra, 139 pp.

Sushkin, P.P. 1927. On the modifications of the mandibular and hyoid arches and their relations to the braincase in the early Tetrapoda. *Paläontologische Zeitschrift* 8: 263–321.

Tatarinov, L.P. 1973. [Cynodonts of Gondwanan aspect in the Middle Triassic of the USSR.] *Paleontologicheskii Zhurnal* 1973 (2): 83–89.

—1974a. [Theriodonts of the USSR.] *Trudy Paleontologicheskogo Instituta AN SSSR* 143: 1–250.

—1974b. [A new theriodont from the Lower Triassic of the Orenburg district.] *Paleontologicheskii Zhurnal* 1974 (2): 86–91.

Tripathi, C. 1969. Fossil labyrinthodonts from the Panchet Series of the Indian Gondwanas. *Memoirs of the Geological Survey of India* 38: 1–53.

Tverdokhlebov, V.P. 1995. [Stratigraphy.] pp. 8–38 in Shishkin, M.A. (ed.) [*Biostratigraphy of the Continental Triassic of the Southern Cis-Urals*]. Moscow: Nauka.

Yakovlev, N.N. 1916. [The Triassic vertebrate fauna from variegated formation of the Vologda and Kostroma Gouvernements.] *Geologicheskii Vestnik* 3: 157–165.

Yaroshenko, O.P., Golubeva, L.P. and Kalantar, I.Z. 1991. [*Miospores and Stratigraphy of the Lower Triassic of the Pechora Syncline.*] Moscow: Nauka, 135 pp.

8

Early archosaurs from Russia

DAVID J. GOWER AND ANDREII G. SENNIKOV

Introduction

The Archosauria is a major clade of diapsid reptiles that first appeared in the uppermost Permian. Archosaurs radiated in the Triassic and eventually occupied every large vertebrate niche on land and in the air during the Mesozoic. The Archosauria is composed of groups as familiar and successful as the dinosaurs, birds, pterosaurs, and crocodiles, but also includes a number of other important groups that together form a paraphyletic assemblage informally termed the 'thecodontians'.

Recently (Gauthier, 1986; Sereno and Arcucci, 1990; Sereno, 1991; Parrish, 1992, 1993) the use of Archosauria has often been restricted to the crown group (the clade comprising all descendants of the most recent common ancestor of extant archosaurs, the birds and crocodilians). Here we use Archosauria in the traditional sense (see Juul, 1994; Benton, 1999), so that the non-crown-group taxa (currently understood to be proterosuchids, erythrosuchids, euparkeriids, and proterochampsids) formerly excluded by Gauthier's (1986) redefinition are included. The taxa lying within the crown group (= Gauthier's Archosauria) are referred to here as Avesuchia (after Benton, 1999).

In this chapter we review the early (Upper Permian to Middle Triassic) archosaurs of Russia. Based on current understanding, this includes proterosuchids and erythrosuchids (together forming the probably paraphyletic Proterosuchia), and possibly rauisuchids and a euparkeriid. To date, proterochampsids, parasuchians, ornithosuchids, aetosaurians, poposaurids, crocodylomorphs, pterosaurs, and dinosauromorphs have not been recorded from the Russian Permo-Triassic. A cervical vertebra from the Lower Triassic Vetlugian Supergorizont was originally described by Yakovlev (1916), in the first reference to Russian 'thecodontian' remains, as belonging to the prosauropod dinosaur *Thecodontosaurus*. This has recently been referred to the possible rauisuchid *Tsylmosuchus jakovlevi* by Sennikov (1990, 1995c). Efremov (1940) referred two additional elements from the same Supergorizont to *Thecodontosaurus*, but these were reassigned to the early archosaur *Chasmatosuchus rossicus* by Huene (1940).

The study of early archosaurs in Russia can be divided into three main periods. Material collected between 1920 and 1950 was described by F. von Huene (1940, 1960) and L.P. Tatarinov (1960, 1961). V.G. Ochev, based at Saratov State University, then carried out pioneering work in prospecting for specimens and describing new archosaur taxa. In the early 1980s, archosaur specimens were transferred to the collections of the Palaeontological Institute of the Russian Academy of Sciences, Moscow, and placed under new specimen numbers. Here, the search for and study of Russian 'thecodontians' has been continued by A. G. Sennikov. Previous works that have reviewed or discussed aspects of the Russian early archosaurs include Tatarinov (1961), Charig and Reig (1970), Charig and Sues (1976), Ochev (1991), and Sennikov (1993, 1995a, b, c, 1996).

Early archosaur remains have been discovered in all Permo-Triassic continental deposits from the Late Tatarian (Late Permian) to the Late Ladinian (Middle Triassic) of the eastern and central Russian platform, the Cis-Uralian trough, and the Caspian region (Figure 8.1). The remains have been recovered mostly

Period		Stage	Gorizont		Archosaurian Taxa
TRIASSIC	Middle	Ladinian	Bukobay Gorizont		*Chalishevia cothurnata* *Jushatyria vjushkovi* *Energosuchus garjainovi*
TRIASSIC	Middle	Anisian	Donguz Gorizont		*Sarmatosuchus otschevi* *Uralosaurus magnus* *Vjushkovisaurus berdjanensis* *Dongusuchus efremovi* *Dongusia colorata* *Dorosuchus neoetus*
TRIASSIC	Lower	Olenekian	Yarenskian Gorizont		*Gamosaurus lozovskii* *Garjainia prima* *Vjushkovia triplicostata* *Jaikosuchus magnus* *Tsylmosuchus donensis* *Vytshegdosuchus zheshartensis*
TRIASSIC	Lower	Olenekian	Vetlugian Supergorizont	Ustmylian Sludkian Rybinskian	*Chasmatosuchus rossicus* *Exilisuchus tubercularis* *Tsylmosuchus jakovlevi* *Tsylmosuchus samariensis*
TRIASSIC	Lower	Induan	Vetlugian Supergorizont	Vokhmian Gorizont	*Vonhuenia friedrichi* *Blomia georgii*
PERMIAN	Upper	Tatarian	Vyatskian Gorizont		*Archosaurus rossicus*

Figure 8.1. Stratigraphical distribution of early Russian archosaurs. Stratigraphic scheme is based on evidence given in Chapter 7.

from alluvial deposits, and they are typically isolated and often fragmentary. Despite the apparently high diversity of early Russian archosaurs, the very incomplete nature of most of the specimens means that many taxa are problematic and they are often overlooked by Western researchers.

Below, we consider each of the families in an increasingly crownward sequence. The order in which taxa are discussed within each family follows the chronological order in which they were erected. The systematic survey is complete in terms of mentioning all instances of referred material for all named taxa. The stratigraphy is based on data summarized in Shishkin *et al.* (1995) (see also Chapter 7).

Repository abbreviations

PIN, Palaeontological Institute Museum, Moscow; SGU, Saratov State University, Geological Collections.

Systematic survey

Family PROTEROSUCHIDAE Broom, 1906, or Huene, 1908?[1]

Proterosuchids include the earliest known archosaurs. Characterizing the family as a whole in terms of shared derived features has been problematic,

[1] See Charig and Sues (1976).

partly because a detailed investigation of the inter-relationships of all known proterosuchids has yet to be tackled (see Gower and Sennikov, 1997). Proterosuchids possess three derived features diagnostic of archosaurs – an antorbital fenestra, lateral mandibular fenestra, and laterosphenoid ossification. Despite being the earliest archosaurs and possessing many plesiomorphic features, proterosuchids appear to have been relatively specialized in having a conspicuously downturned, hooked snout.

Russian proterosuchids include *Archosaurus*, the oldest known archosaur, as well as many fragmentary and taxonomically problematic remains. The distinction between similar taxa based on fragmentary and sometimes mixed material has, in part, been based on the perception that two main morphs of proterosuchid can be recognized, namely short- and long-headed and necked (Sennikov 1994, 1995c).

Chasmatosuchus Huene, 1940

Type species. C. rossicus Huene, 1940
Comments. A large number of isolated and fragmentary remains of early archosaurs have been recovered from numerous localities in European Russia, from both the lower (Vokhmian Gorizont) and upper parts (Rybinskian, Sludkian, and Ustmylian Gorizonts) of the Vetlugian Supergorizont (Figure 8.1). Many of these remains have been placed in one of four species of *Chasmatosuchus*, but only the type species has withstood initial scrutiny. Huene suggested that *Chasmatosuchus* was comparable with *Chasmatosaurus*, Ochev (1978, 1979, 1991) considered the genus to be an early or ancestral erythrosuchid, while Charig and Sues (1976) and Sennikov (1995c) regarded it as a problematic proterosuchid.

Chasmatosuchus rossicus Huene, 1940

Holotype and locality. PIN 2252/381, two articulated posterior cervical vertebrae that bear well-defined facets for three-headed ribs; southern Cis-Urals.
Paratypes. PIN 32–2355/25 and PIN 160/9, two proximal halves of left tibiae.
Horizon. Rybinskian Gorizont, Early Olenekian.
Comments. '?*C. parvus*' was established at the same time

by Huene (1940), but with some uncertaintly. It was based on PIN 2252/382, an incomplete posterior cervical or anterior dorsal vertebra, also from the Rybinskian Gorizont of the southern Cis-Urals. It was synonymized with *C. rossicus* by Tatarinov (1961), but has since been referred to the prolacertiform *Microcnemus efremovi* (Huene, 1940) by Sennikov (1995c), as it is indistinguishable, other than in size, from other vertebrae of that taxon. Huene (1940) also described additional material as *Chasmatosuchus* sp. (see also Charig and Sues, 1976), and this was referred to *C. rossicus* by Sennikov (1995c).

'*C. ?vjushkovi*' was described (tentatively) by Ochev (1961), based on a small isolated left premaxilla (PIN 2394/4 – formerly SGU 104/45) from the lower part of the Yarenskian Gorizont. This was considered to be a juvenile of *Garjainia prima* by Tatarinov (1961) and Sennikov (1995c), but as generically indeterminate by Hughes (1963), Young (1964), and Charig and Reig (1970).

C. magnus was described by Ochev (1979) based on a holotype that is a cervical vertebra (PIN 951/65) from the upper Yarenskian Gorizont of the Orenburg region (Rassypnaya locality). The only other specimens are paratype vertebrae from Ustmylian and Yarenskian Gorizont deposits at different localities, and a fibula from the type locality that Ochev included, despite suggesting that it might belong to a lepidosaur. The holotype vertebra (Figure 8.8C) and a paratype neural arch has since been referred to *Jaikosuchus* by Sennikov (1990), who also referred the paratype material from northern Russia to *Tsylmosuchus jakovlevi* (Ustmylian Gorizont) and *Tsylmosuchus* sp. (lower Yarenskian Gorizont).

A basisphenoid was referred to *Chasmatosuchus* sp. by Ochev (1978), but has since been referred to *Vonhuenia* by Sennikov (1992).

Archosaurus Tatarinov, 1960
See Figure 8.2.

Type species. A. rossicus Tatarinov, 1960.
Holotype and locality. PIN 1100/55, an incomplete left premaxilla; Vyazniki locality, near Vyazniki, Vladimir Province.

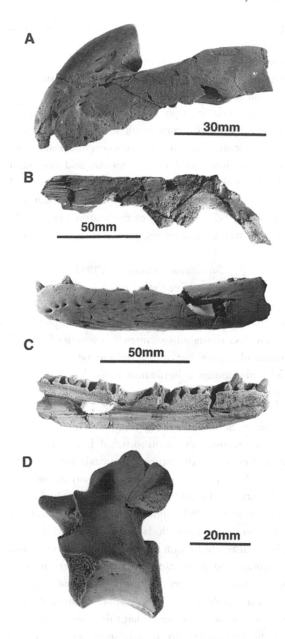

Figure 8.2. *Archosaurus rossicus* Tatarinov, 1960. A, holotype left premaxilla (PIN 1100/55) in lateral view; B, left side of skull roof (paratype PIN 1100/84) in dorsal view; C, left dentary (paratype PIN 1100/78) in lateral and medial views; D, cervical vertebra (paratype PIN 1100/66a) from left side.

Horizon. Upper part of Vyatskian Gorizont, Late Tatarian.

Paratypes. PIN 1100/84, a substantial part of the left side of the skull roof; PIN 1100/84a, an incomplete right squamosal; PIN 1100/85a, four isolated teeth; PIN 1100/66, three cervical vertebrae; PIN 1100, 67a, the proximal part of a right tibia; and PIN 1100/ 67, a sacral rib, all from the type locality. Also collected in 1955/6, with the type material described by Tatarinov (1960), is a left dentary (PIN 1100/78) that was only referred to *Archosaurus* by Sennikov (1988a) at a later date, and an undescribed clavicle (PIN 1100/427). Sennikov (1988a) also tentatively referred a trunk vertebra (PIN 1538/4) and left postorbital (PIN 1538/1), from similarly old collections from the Purly locality in the Nizhnii Novgorod region, to this taxon.

Comments. The exact number of taxa and individuals represented by the material cannot be judged with certainty. However, the premaxilla of *Archosaurus* is comparable only with that of the well known proterosuchids from South Africa and China (*Proterosuchus*/*Chasmatosaurus*) and the other elements are also very similar to those of these forms and therefore consistent with the understanding that the material represents a single taxon. Despite the lack of evidence for an antorbital or lateral mandibular fenestra, or a laterosphenoid, the known morphological features (in particular of the premaxilla and perhaps the presence of serrations on the teeth) probably make this the earliest known, and only Permian, archosaur globally. Parrington (1956) described some incomplete postcranial material from the Upper Permian Kawinga Formation of East Africa that, though probably indeterminate, could possibly be archosaurian (Charig and Sues, 1976, p. 17) . The identification of *Archosaurus* as a proterosuchid archosaur, however, has apparently never been seriously questioned.

Gamosaurus Ochev, 1979

Type species. *G. lozovskii* Ochev, 1979.

Holotype and locality. PIN 3361/13, an incomplete cervical vertebra; Zheshart locality, Vychegda River, Aikono district, Komi Republic.

Horizon. Upper part of Yarenskian Gorizont, Late Olenekian.

Paratypes. PIN 3361/14, 3361/94–96, cervical and dorsal vertebrae; same locality.

Comments. Gamosaurus has been distinguished from *Chasmatosuchus* on the basis of diapophyseal morphology (Ochev, 1979; Sennikov, 1995c). While considering it a proterosuchian, Ochev (1979) was uncertain about the familial affinities of this genus. Sennikov (1995c) referred additional vertebrae to this taxon, and has tentatively considered it to represent a proterosuchid (Sennikov 1990, 1995c).

Exilisuchus Ochev, 1979

Type species. E. tubercularis Ochev, 1979.

Holotype and locality. PIN 4171/25 (formerly SGU 104/2371), an incomplete left ilium; Kzyl-Sai III 2 locality, Akbulak district, Orenburg Province.

Horizon. Sludkian Gorizont, Early Olenekian.

Comments. Ochev considered this ilium to be of uncertain familial, sub-ordinal, and even ordinal affinity. Sennikov (1995c) admitted that assessing the systematic position of this material is highly problematic, but suggested it might represent an early, possibly proterosuchid, archosaur.

Vonhuenia Sennikov, 1992

Type species. V. friedrichi Sennikov, 1992.

Holotype and locality. PIN 1025/11, an anterior dorsal vertebra; Spasskoe I locality, Vetluga River, Nizhnii Novgorod Province.

Horizon. Vokhmian Gorizont, Induan.

Paratypes. Vertebrae (three apophyses clearly visible in one example) and fragmentary postcranial material from the same locality (see Sennikov, 1992), as well as a basisphenoid (PIN 1025/14) previously referred to *Chasmatosuchus* sp. (tentatively to *C. rossicus*) by Ochev (1978). Sennikov (1995c) has referred additional fragmentary material from the same locality to this taxon.

Blomia Sennikov, 1992

Type species. B. georgii Sennikov, 1992.

Holotype and locality. PIN 1025/348, a basisphenoid; Spasskoe I locality, Vetluga River, Nizhnii Novgorod Province.

Horizon. Vokhmian Gorizont, Induan.

Paratypes. Additional fragmentary postcranial material from the type locality (Sennikov, 1995c).

Comments. The basisphenoid differs from that of *Vonhuenia*, from the same locality. Both are of a form atypical among proterosuchids, where the basisphenoid is not horizontal and plate-like, and the basal tubera are above as well as behind the basipterygoid processes, as seen also in *Sarmatosuchus* (Gower and Sennikov, 1997). The unity of this material as a single taxon, as well as its affinities, are clearly problematic.

Sarmatosuchus Sennikov, 1994
See Figure 8.3.

Type species. S. otschevi Sennikov, 1994.

Holotype and locality. PIN 2865/68, an incomplete and largely disarticulated specimen, consisting of several cranial elements, cervical vertebrae, pectoral girdle, and limb elements; Berdyanka II locality, Sol'-Iletsk district, Orenburg Province.

Horizon. Donguz Gorizont, Anisian.

Comments. Sennikov (1994) originally identified the unique specimen as a proterosuchid, largely based on the presence of a downturned premaxilla in conjunction with the plesiomorphic form of many of the skull and postcranial elements. The reassessment by Gower and Sennikov (1997) highlighted further similarities with known proterosuchids, but also showed that the tall braincase, reduction in palatal teeth, short cervical vertebrae, and stratigraphic position all deviate from other known members of the Proterosuchidae as it is currently understood. The available morphological evidence strongly suggests that this taxon is indeed a Middle Triassic proterosuchid that shares a number of apparently homoplastic resemblances with various non-proterosuchid early archosaurs (Gower and Sennikov 1997). The material of *Sarmatosuchus* is important in indicating that the proterosuchids were morphologically a more diverse group than has previously been understood. That the earliest, most plesiomorphic known family of archosaurs still had living representatives at the time of the appearance of the

Figure 8.3. *Sarmatosuchus otschevi* Sennikov, 1994. Part of unique specimen (holotype PIN 2865/68). A, right premaxilla in lateral view; B, left scapula and coracoid in lateral view.

first dinosauromorphs, also requires a modification of existing views of the pattern of early archosaur evolution (see discussion below).

Family ERYTHROSUCHIDAE Watson, 1917

Comments. The erythrosuchids are a clade of large carnivorous archosaurs from the Lower/Middle Triassic (Parrish, 1992). Our understanding of them has been limited by similar problems to the proterosuchids – small amounts of often incomplete material that remains incompletely described. Five taxa have been described from Russia (but see also *Chasmatosuchus* above). These include a single Lower Triassic species, *Garjainia prima* (including *Vjushkovia triplicostata*) represented by some of the most extensive and well preserved early archosaur material, but this is in need of revision. The Middle Triassic has yielded two possible erythrosuchids, *Uralosaurus* and *Chalishevia*, which are known from very incomplete material. Their precise affinities must remain uncertain until more complete material is recovered.

Dongusia Huene, 1940

Type species. D. colorata Huene, 1940.

Holotype and locality. PIN 268/2, a dorsal vertebra; Donguz I locality, Sol'-Iletsk district, Orenburg Province.

Horizon. Donguz Gorizont, Anisian.

Comments. Huene (1940) assigned this single vertebra to a primitive thecodontian of the *Chasmatosaurus* group. Tatarinov (1961) considered it to belong to his expanded concept of *Erythrosuchus*, but most workers have regarded *Dongusia* and *D. colorata* as *nomina dubia* (e.g. Young, 1964; Charig and Reig, 1970; Charig and Sues, 1976). Charig and Reig (1970) and Sennikov (pers. obs.) consider PIN 268/2 to be similar in form to the dorsal vertebrae of rauisuchids from the East African Manda Formation in the Natural History Museum, London.

Garjainia Ochev, 1958

See Figures 8.4A, D.

Type species. G. prima Ochev, 1958.

Holotype and locality. PIN 2394/5 (formerly SGU 104/3–43), a single individual represented by an essentially complete articulated skull and mandible, and some presacral vertebrae (Ochev 1958), as well as both scapulae and coracoids and an interclavicle not mentioned by Ochev (1958); Kzyl-Sai II 2 locality, near the village Andreevka, Akbulak district, Orenburg Province.

Horizon. Yarenskian Gorizont, Late Olenekian.

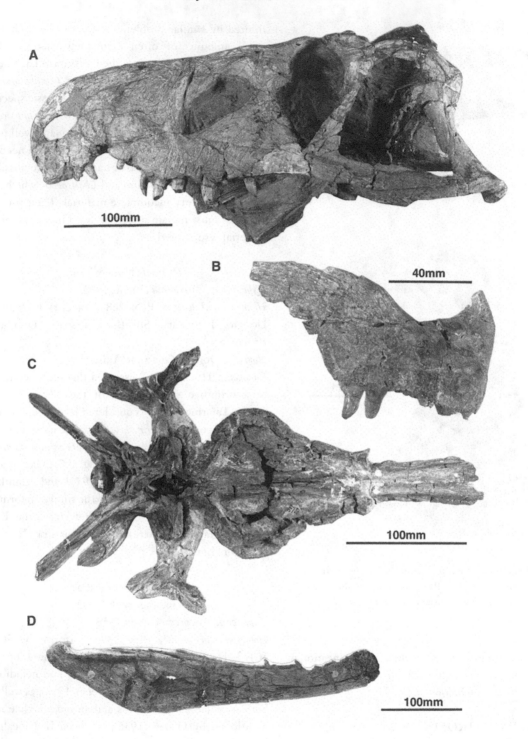

Figure 8.4. A, *Garjainia prima* Ochev, 1958, left side of skull of holotype (PIN 2394/5); B, *Vjushkovia triplicostata* Huene, 1960, lateral view of right premaxilla (PIN 951/63); C, *V. triplicostata*, ventral view of lectotype skull roof and braincase (PIN 951/59); D, *G. prima*, lateral view of right mandible of the holotype.

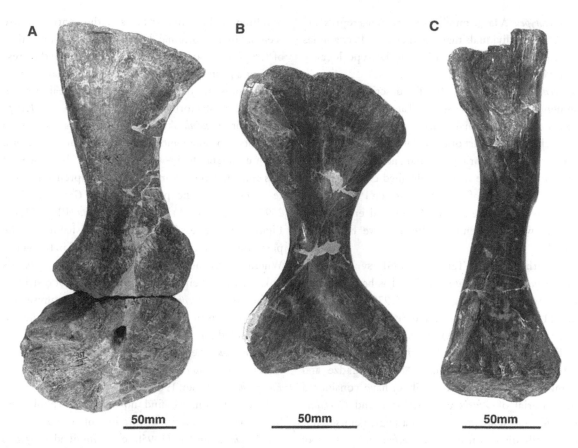

Figure 8.5. *Vjushkovia triplicostata* Huene, 1960. A, right scapula and coracoid (PIN 951/2) in lateral view; B, right humerus (PIN 951/36) in ventral view; C, right femur (PIN 951/27) in ventral view.

Comments. Ochev's original description was followed by a description of the palate (Ochev, 1975) and a more detailed consideration of further aspects of the morphology (Ochev, 1981). The holotype of *Chasmatosuchus vjushkovi* has sometimes been considered to belong to a juvenile of this species (see above). Sennikov (1995b, c) and Shishkin *et al.* (1995) also assign some isolated and fragmentary postcranial remains, including an incomplete femur, from various localities to *G. prima* or *Garjainia* sp.

Although originally described as belonging to the new family Garjainiidae (Ochev, 1958), Tatarinov (1961) highlighted similarities to *Erythrosuchus*, and even synonomized *Garjainia* with the South African genus, a synonomy rejected by subsequent authors (e.g. Young, 1964; Charig and Reig, 1970; Parrish, 1992; Sennikov, 1995b, c). *Garjainia* has long been considered to be one of the earliest (if not the earliest) and most plesiomorphic erythrosuchids – indeed, it was viewed as an intermediate between proterosuchids and later erythrosuchids (e.g. Young, 1964; Charig and Reig, 1970; Charig and Sues, 1976) before the paraphyly of proterosuchians was hypothesized.

Vjushkovia Huene, 1960
See Figures 8.4B, C and 8.5.
Type species. V. triplicostata Huene, 1960.
Lectotype and locality. PIN 951/59, a skull roof and occiput; Rassypnaya locality, Orenburg Province.
Horizon. Yarenskian Gorizont, Late Olenekian.

Paralectotypes. A large number of specimens representing several individuals that together include examples of nearly every element, all from the type locality (listed by Charig and Sues, 1976).

Comments. Huene (1960) briefly described selected elements from the type series, Tatarinov (1961) discussed some additional aspects of the morphology, and Ochev (1975) part of the palate. Clark *et al.* (1993) and Parrish (1992) briefly figured and commented on the braincase. More recently, detailed considerations of the morphology of the tarsus (Gower, 1996), braincase (Gower and Sennikov, 1996a), and endocranial cast (Gower and Sennikov, 1996b) have been presented.

As with *Garjainia*, Tatarinov (1961) synonomized *Vjushkovia* with *Erythrosuchus*. This has been rejected by Young (1964), Charig and Reig (1970), Charig and Sues (1976), Parrish (1992), and Sennikov (1995b, c). Since Tatarinov's (1961) work, some subsequent studies (Tatarinov reported *in litt.* in Charig and Sues, 1976; Ochev and Shishkin, 1988; Kalandadze and Sennikov, 1985; Sennikov 1995b, c) have considered the material of *Vjushkovia triplicostata* and *Garjainia prima* to represent two species of a single genus that is generically distinct from *Erythrosuchus*. In his study of erythrosuchid phylogeny, Parrish (1992) maintains the generic distinction. Furthermore, he challenges the more recent Russian conclusions by presenting a phylogenetic hypothesis in which a monophyletic *Vjushkovia*, including *V. triplicostata* and the Chinese taxon *V. sinensis* (Young, 1973a), shares a more recent common ancestor with both *Erythrosuchus* and *Shansisuchus* than with *Garjainia*.

A study of Parrish's (1992, table 2) data matrix shows that he considers *V. triplicostata* to differ from *Garjainia* in the absence of a pineal foramen (his character 13 – contradicting Ochev's original description of *Garjainia*), the absence of a posterodorsally expanded maxillary ramus (15), the presence of an antorbital fossa (16), the absence of intercentra (18), and a shorter posterior jugal process (19). Parrish's coding of his characters 16 and 19 is at variance with our observations of the material. The presence of intercentra (character 18) in *V. triplicostata* is suggested by strong

bevelling of the articular faces of the centra (clearly seen in Huene's original figures). The known maxillae of *V. triplicostata* are too incomplete to rule out the presence of a posteroventrally expanded maxillary ramus, and the presence or absence of a small pineal foramen cannot be satisfactorily determined in either the *V. triplicostata* or *Garjainia* skull roof remains. Parrish (1992) also lists two synapomorphies of the genus *Vjushkovia* as he understands it – the presence of three-headed ribs and a low posterior process of the premaxilla. On re-examination, the premaxillae of *Garjainia* (PIN 2394/5, Figure 8.4A) and *V. triplicostata* (PIN 951/63, Figure 8.4B) are essentially indistinguishable, and the posterior ascending process is clearly tall in both taxa. Whether three-headed ribs were present in *Garjainia* is unknown, but their presence is certainly not restricted to the genus *Vjushkovia*. Such ribs or apophyses for them are seen in a wide range of early archosaurian taxa including *Erythrosuchus* (D.J.G., pers. obs.), *Sarmatosuchus* (Sennikov, 1994; Gower and Sennikov, 1997), *Chasmatosuchus rossicus* (Huene, 1940), and *Energosuchus* (Ochev, 1986).

In conclusion, we find no sound morphological basis for the generic distinction of *Garjainia* and *Vjushkovia*. Sennikov (1995b, c) maintained *Garjainia prima* and *Garjainia* (= *Vjushkovia*) *triplicostata* as separate species on the basis of minor morphological details and their discovery from lower and upper parts respectively of the Petropavlovskaya Svita of the Yarenskian Gorizont. A re-examination by both of us concluded that the supposed morphological differences (except perhaps for a closer association between the teeth and surrounding alveolar bone in *triplicostata*) do not exist, and that all of the material probably represents but a single recognizable species (*G. prima*). A thorough revision of the osteology of this material would be worthwhile.

Uralosaurus Sennikov, 1995c
Type species. *U. magnus* Ochev, 1980.
Holotype and locality. PIN 2973/70 (formerly SGU 104/3516), a left pterygoid; Karagachka locality, Sol'-Iletsk district, Orenburg Province.
Horizon. Donguz Gorizont, Anisian.

Paratypes. PIN 2973/71 (formerly SGU 104/3516) and a right dentary; PIN 2973/72–79, isolated teeth, from the type locality. Further paratype material consists of four presacral vertebrae (PIN 952/95) and two dorsal vertebrae (PIN 2866/38, 39 – formerly SGU 104/3857, 58) from Donguz Gorizont sediments of the Donguz I and Koltaevo II localities respectively.

Comments. Ochev (1980) placed this material in a newly erected species of *Erythrosuchus* on the basis of the similarity between the known elements and those of *Garjainia prima*, *Vjushkovia triplicostata*, and *Erythrosuchus africanus* – and accepting Tatarinov's (1961) synonymy of the first two of these genera with the latter. Sennikov (1995c) maintained *magnus* as a recognizable species of erythrosuchid – but erected the genus *Uralosaurus* for it. Sennikov (1995c) also referred a vertebra and scapula from Koltaevo II, and teeth, a fragmentary jugal, and a caudal vertebra from Donguz I to this genus. The remains are very incomplete, and that they represent an erythrosuchid is not beyond doubt.

Chalishevia Ochev, 1980
See Figure 8.6.

Type species. *C. cothurnata* Ochev, 1980.

Holotype and locality. PIN 4356/1 (formerly SGU 104/385), a left maxilla and both nasals; Bukobay VII locality (incorrectly reported as Bukobay VI by Ochev, 1980), Sol'-Iletsk district, Orenburg Province.

Horizon. Bukobay Gorizont, Ladinian.

Paratypes. PIN 4366/3 (formerly SGU 104/3854), a tooth, and PIN 4366/2 (formerly SGU 104/3853), an incomplete right quadrate, from the Bukobay VII locality. An additional small fragment of nasal (PIN 2867/18 – formerly SGU 104/3862) from the Koltaevo III locality is also a paratype.

Comments. Sennikov (1995c) subsequently referred several vertebrae, an additional small nasal, and part of a surangular from the Bukobay I and V localities to this taxon. Apart from the distinctive nasals, referral of other isolated elements to the same taxon as the holotype cannot be certain. *Chalishevia* was originally referred by Ochev (1980) to the Erythrosuchidae, and while this has been followed by Sennikov (1995b, c),

Figure 8.6. *Chalishevia cothurnata* Ochev, 1980: lateral view of left maxilla and crushed left nasal of holotype (PIN 4366/1).

this genus has received little or no attention from Western students of early archosaurs.

The reconstruction by Ochev (1980, fig. 2a) depicts a dramatically tapered anterior part of the skull and a suggested very small premaxilla. It also shows two antorbital fenestrae. A re-examination of the material allows us to confirm the presence of two fenestrae, but also to modify Ochev's reconstruction. The area of the angle of the holotype left nasal has been crushed during preservation so that the snout should be taller and less tapered when reconstructed in lateral view. Evidence for this is found in the cracked and swollen nature of this part of the holotype left nasal, and in the much narrower angle between the maxillary and premaxillary rami than in the other known nasals.

The nasal and the region of the anterior antorbital fenestra are strikingly similar to that of *Shansisuchus* (Young, 1964) from the Middle Triassic of China (D.J.G., pers. obs.). The form of the maxilla–nasal articulation, the presence of the anteroventrally open 'sulcus' described by Ochev on the nasal above the anterior fenestra, and the facet for the posterior premaxillary process all correspond closely between the two taxa. In both, the anterodorsal edge of the maxilla is thin, but incompletely known, so that we are ignorant of the exact form of the anterior fenestra. A difference between the two taxa is possibly seen in the anteroventral corner of the antorbital fossa – much more acutely angled in *Chalishevia* – although the preservation of the surface of the Russian material is superior to that of the Chinese.

In other basal archosaurs with additional antorbital openings (e.g. some rauisuchians, crocodylomorphs), the additional openings are either only small 'subnarial foramina' or narrow slit-like gaps. A more substantial, subcircular opening formed between nasal (with a ventral 'sulcus'), maxilla, and premaxilla is known with certainty only in *Shansisuchus* and *Chalishevia*. The clearly derived morphology of the snout region of these two taxa may indicate a close relationship, and the conspicuously short cervical vertebrae in both is consistent with this hypothesis, but the fragmentary nature of the Russian taxon demands caution.

?Family EUPARKERIIDAE Huene, 1920

Comments. The euparkeriids are currently known reliably from only the type genus, *Euparkeria*, from the uppermost Lower or lower Middle Triassic of the *Cynognathus* Zone of South Africa. A single Russian genus has been claimed to represent an additional euparkeriid.

Dorosuchus Sennikov, 1989
See Figure 8.7.

Type species. D. neoetus Sennikov, 1989.

Holotype and locality. PIN 1579/161, a right ilium, femur, and tibia; Berdyanka I locality, Sol'-Iletsk district, Orenburg Province.

Horizon. Donguz Gorizont, Anisian.

Paratypes. PIN 1579/62, a braincase; PIN 1579/66, smaller left and right ilia; PIN 1579/63–64, sacral and caudal vertebrae, all recovered from the same block of siltstone as the holotype. An incomplete left ilium (PIN 952/200) from Donguz Gorizont deposits at the nearby Donguz I locality is also a paratype.

Comments. In his original description, Sennikov (1989) was mistaken in identifying the basioccipital–basisphenoid fossa (Gower and Sennikov, 1996a) as a medial eustachian foramen, and also in identifying the position of the foramina for the internal carotid arteries on the basisphenoid – they are in a posteroventral (Sennikov 1995c, fig. 19), not lateral position.

The pelvic girdle and limb of *Dorosuchus* closely resemble those of *Euparkeria* (Ewer, 1965), but there are problems with the hypothesis that the Russian taxon is a member of the same family. The family Euparkeriidae has not been satisfactorily characterized beyond *Euparkeria* itself. Supposed euparkeriids from China have probably been wrongly assigned. *Turfanosuchus dabanensis* (Young, 1973b) was originally identified as a euparkeriid, but the rotary, crocodilian-like astragalus–calcaneum joint (Parrish, 1993; D.J.G., pers. obs.) points against this. Another Chinese taxon referred to the Euparkeriidae, *Halazhaisuchus qiaoensis* (Wu, 1982), is represented by incomplete postcranial remains. It possesses both intercentra and well-developed dorsal osteoderms – a combination of plesiomorphic and derived features also found in euparkeriids – but the incompleteness of the material neither allows us to be sure of its affinities, nor furthers our understanding of euparkeriid morphology. A third Chinese taxon, *Turfanosuchus shageduensis* (Wu, 1982) is known from one poorly preserved specimen. Personal observation (D.J.G.) of the material found no support for the referral of this specimen to the genus *Turfanosuchus*. Indeed the material bears an equal, if not greater resemblance to *Halazhaisuchus* from the same formation (Lower Ehrmaying; Lower Triassic). Evidence for the presence of vertebral intercentra reported by Parrish (1993) for *T. shageduensis*, could not be found. Two further Chinese taxa placed in the Euparkeriidae by Sennikov (1989), *Wangisuchus* (Young, 1964) and *Xilousuchus* (Wu, 1981), are also incompletely known, but do not appear to share potential synapomorphies with *Euparkeria*. Further, material referred to *Wangisuchus* includes a calcaneum that was clearly part of a rotary, crocodilian-like ankle joint (Parrish, 1992, 1993; D.J.G., pers. obs.) – such as is unknown outside Avesuchia.

?Family RAUISUCHIDAE Huene, 1942

Comments. The 'rauisuchids' are inadequately understood in terms of their morphology, taxonomy, and phylogeny. There is uncertainty concerning the number of families that are represented by the taxa previously considered to be 'rauisuchids' (e.g. Parrish, 1993; Long and Murray, 1995; Benton and Gower, 1997) and their position within the Archosauria (Gower and Wilkinson, 1996). Russian 'rauisuchids'

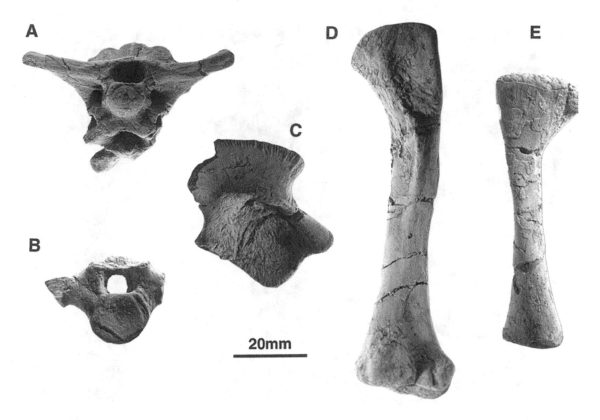

Figure 8.7. *Dorosuchus neoetus* Sennikov, 1989. A, occipital view of braincase (paratype PIN 1579/62); B, sacral vertebra (paratype PIN 1579/63) in anterior view; C, lateral view of incomplete right ilium of the holotype (PIN 1579/61); D, ventral view of right femur of the holotype; E, anterior dorsal view of right tibia of the holotype.

have mostly been identified with the understanding that those taxa that might be (e.g. Parrish, 1993) separated into prestosuchid and rauisuchid families are all covered by the latter. While accepting that 'rauisuchian' systematics is in need of revision (Gower, 2000), we use the term 'rauisuchids' rather loosely to refer to those avesuchians that might also be considered to be prestosuchids.

There are many incomplete, fragmentary, and isolated remains from the Lower and Middle Triassic of Russia that have been described as 'rauisuchids'. Until the variation in vertebral morphology within and between various clades of basal archosaurs is firmly established, the Russian 'rauisuchid' vertebrae will remain largely undiagnostic. While their status as 'rauisuchids' is not always clear, at least *Vytshegdosuchus*

bears an ilial swelling currently considered to be restricted to 'rauisuchians' (e.g. Parrish, 1993). In addition, the femora of *Vytshegdosuchus* and *Dongusuchus* are more derived than those of proterosuchians or euparkeriids. Long and Murray (1995, p. 117) accepted that at least some of the Russian material 'does appear to pertain to the Rauisuchia', and listed *Vytshegdosuchus*, *Dongusuchus*, *Jushatyria*, *Energosuchus*, *Tsylmosuchus*, and *Jaikosuchus* as Rauisuchia *incertae sedis*.

Vjushkovisaurus Ochev, 1982

See Figure 8.8A.

Type species. V. berdjanensis Ochev, 1982.

Holotype and locality. PIN 2865/62 (formerly SGU 104/3871), twelve vertebrae (five cervicals, and three dorsals are preserved as two short articulated series)

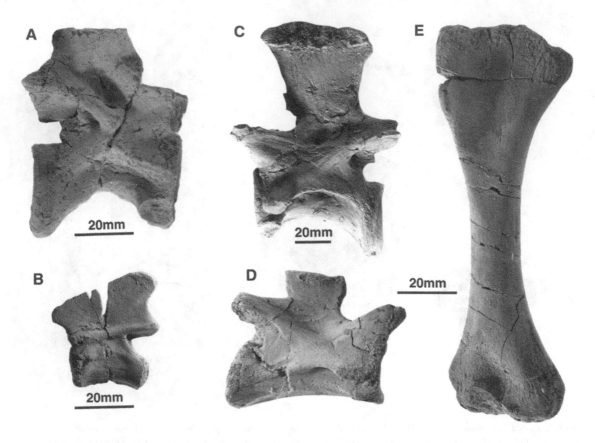

Figure 8.8. A, *Vjushkovisaurus berdjanensis* Ochev, 1982: right side of cervical vertebra (paratype PIN 2865/61); B, *Tsylmosuchus jakovlevi* Sennikov, 1990: left side of axis (paratype PIN 4339/1); C, *Jaikosuchus magnus* Ochev, 1979: left side of cervical vertebra (holotype PIN 951/65); D, *Energosuchus garjainovi* Ochev, 1986: right side of cervical vertebra (holotype PIN 4188/99); E, *E. garjainovi*: ventral view of right humerus (paratype PIN 4188/104).

and a complete left humerus; Berdyanka II locality, Sol'-Iletsk district, Orenburg Province.

Horizon. Donguz Gorizont, Anisian.

Paratypes. PIN 2865/61 (formerly SGU 104/3872), a vertebra, and some poorly preserved postcranial fragments, all from the type locality.

Comments. Two of the posterior cervical/anterior dorsal vertebrae bear well-defined facets for three-headed ribs (pers. obs.; Ochev, 1982) and the humerus is shorter and broader than in well known 'rauisuchids'. Ochev (1982) tentatively, and Sennikov (1990, 1995b, c) less tentatively, referred this taxon to the Rauisuchidae, but its systematic position remains uncertain. Juul (1994, p. 12) suggested that the

humerus of *Vjushkovisaurus* 'could belong to a rauisuchid or stagonolepid equally well'.

Jushatyria Sennikov in Kalandadze and Sennikov, 1985

Type species. J. vjushkovi Sennikov in Kalandadze and Sennikov, 1985.

Holotype and locality. PIN 2867/5, an incomplete left maxilla; Koltaevo III locality, Kumertau district, Bashkortostan Republic.

Horizon. Bukobay Gorizont, Ladinian.

Paratypes. PIN 2867/6, a cervical vertebra.

Comments. Additional fragmentary and isolated material from the Bukobay I locality was referred to this taxon by Sennikov (1995c). Sennikov (1995c) also transferred fragmentary postcranial material (from the Novo-Alexandrovka locality on the Berdyanka River) to *Jushatyria* that was originally described by Ochev (1986, introduction, fig. 1) as possibly representing a rauisuchid/*Prestosuchus*-like reptile. The referral of all of the material to a single taxon is problematic. The incomplete maxilla bears an antorbital fossa, but evidence for the additional, slit-like antorbital opening as reconstructed by Kalandadze and Sennikov (1985) is lacking. The referred femoral material is suggestive of a strongly sigmoidal element. While Kalandadze and Sennikov (1985) tentatively placed *Jushatyria* in the Rauisuchidae, its taxonomy, morphology, and affinities remain obscure.

Energosuchus Ochev, 1986
See Figure 8.8D, E.

Type species. *E. garjainovi* Ochev, 1986.
Holotype and locality. PIN 4188/99–100 (formerly SGU 104/383), a disarticulated elongate cervical vertebra with low neural spine; Bukobay V locality, near Mikhailovka, Sol'-Iletsk district, Orenburg Province.
Horizon. Bukobay Gorizont, Ladinian.
Paratypes. PIN 4188/100–122 (see Sennikov, 1995c, for former SGU numbers), vertebrae (one, now damaged, is shown in Ochev's figure 4 with three distinct apophyses on each side), a slender humerus, a radius, and an incomplete coracoid, all from the type locality.
Comments. Ochev (1986) designated two specimens as types, the holotype noted above, and PIN 4188/99 (formerly SGU 104/386), a further similar specimen. Sennikov (1995b, c) has subsequently referred only to PIN 4188/99 as the holotype – the only one that was figured by Ochev (1986, fig. 3, I). Sennikov (1995c) also referred an incomplete left femur (PIN 2867/19) from the Koltaevo III locality to this taxon.

Vytshegdosuchus Sennikov, 1988b
See Figure 8.9A, B.

Type species. *V. zheshartensis* Sennikov, 1988b.
Holotype and locality. PIN 3361/134, a right ilium; Zheshart locality, Vychegda River, Aikino district, Komi Republic.

Horizon. Upper Yarenskian Gorizont, Late Olenekian.
Paratypes. Paratypic femoral fragments are also from the Zheshart locality, while the other paratypes (fragments of pterygoid and additional postcranial material) were recovered from the Gam locality.
Comments. A cervical vertebra (PIN 3369/139), described in a separate paper (Sennikov, 1990), was referred to *Vytshegdosuchus* based on a match in size and location to the type material. Sennikov (1995c) has subsequently referred additional postcranial material from the Mezhog locality to this taxon. All three of the localities yielding material referred to *Vytshegdosuchus* are within a 40 km stretch of the Vychegda River.

The incomplete holotype ilium has a dorsal crest that is notably thickened just behind the anterior termination. The crest does not possess the anterior projection that is present in rauisuchians such as *Saurosuchus* (Sill, 1974), but the distinct swelling is currently considered to be characteristic of all known members of the Rauisuchidae (e.g. Parrish, 1993, p. 301). In ventral view, the proximal end of the femur is strongly asymmetrical, there is no intertrochanteric fossa, and the low fourth trochanter is positioned a short way down the shaft. This morphology is significantly derived over that seen in proterosuchids and erythrosuchids, and is consistent with rauisuchid affinity.

While the known material is clearly very incomplete, the morphology of *Vytshegdosuchus* is evidently more derived than previously described Lower Triassic archosaurs. Juul (1994, p. 12) considered it 'likely that *Vytshegdosuchus* is a rauisuchian'.

Dongusuchus Sennikov, 1988b
See Figure 8.9C–F.

Type species. *D. efremovi* Sennikov, 1988b.
Holotype and locality. PIN 952/15–1, a femur; Donguz I locality, Sol'-Iletsk district, Orenburg Province.
Horizon. Donguz Gorizont, Anisian.
Paratypes. PIN 952/15–2 to 6, femora, and PIN 952/84–1 to 7, epipodials, also from the Donguz I locality. An anterior cervical vertebra (PIN 2866/37 – formerly SGU 104/3880, see Ochev, 1986),

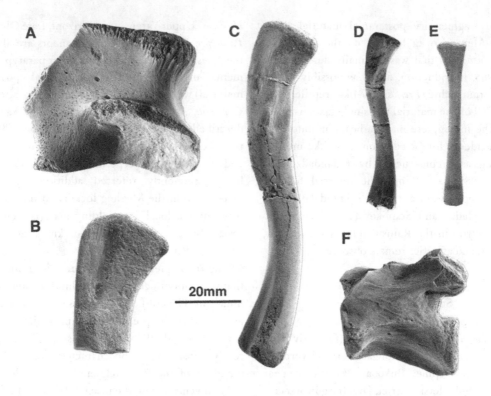

Figure 8.9. A, *Vytshegdosuchus zheshartensis* Sennikov, 1988b: lateral view of incomplete left ilium (holotype PIN 3361/134); B, *V. zheshartensis*: ventral view of proximal fragment of paratype left femur (PIN 3361/127); C, *Dongusuchus efremovi* Sennikov, 1988b: ventral view of holotype left femur (PIN 952/15–1); D, *D. efremovi*: ventral view of paratype left femur (PIN 952/15–2); E, *D. efremovi*: paratype left tibia (PIN 952/84–4); F, *D. efremovi*: left view of incomplete cervical vertebra (PIN 2866/37).

subsequently described by Sennikov (1990), was collected from the Koltaevo II locality in the Kumertau district of the Bashkortostan Republic. Sennikov (1995c) referred an additional cervical vertebra (PIN 4166/212) from the Donguz XII locality (some 100 m away from the type locality) to this taxon.

Comments. The vertebrae were referred to *Dongusuchus* on the basis that the localities are of the same age and geographically close, and their size and morphology were considered to be consistent with that of the type material (Sennikov, 1990).

While the referral of both the vertebrae and limb bones to a single taxon is problematic, some comments on the morphology of this very incomplete material can be made. The femora are markedly slender, and apparently more strongly and smoothly sigmoid in form than in any known non-crown-group

archosaur. The long axes of the proximal and distal ends are separated by an angle of about 70°. The proximal end bears an inturned head, lacks an intertrochanteric fossa, and has a low fourth trochanter positioned some way from the head. Where comparisons are possible (femora and vertebrae), the known material of *Dongusuchus* closely resembles that of *Vytshegdosuchus*.

Tsylmosuchus Sennikov, 1990

Comments. This genus includes three possible rauisuchid species based on incomplete vertebral remains from three separate horizons. The vertebrae are similar to those of 'rauisuchids' from the East African Manda Formation in the Natural History Museum, London (Sennikov, 1990, 1995c). Sennikov (1990) referred the lower Yarenskian Gorizont

Chasmatosuchus magnus paratypes (Ochev, 1979) to *Tsylmosuchus* sp.

Type species. *T. jakovlevi* Sennikov, 1990.

Tsylmosuchus jakovlevi Sennikov, 1990
See Figure 8.8B.

Holotype and locality. PIN 4332/1, a cervical vertebra; Cherepanka locality, Ust'-Tsylma district, Komi Republic.

Horizon. Ustmylian Gorizont, Early Olenekian.

Paratypes. Sennikov (1990) also referred to this taxon a large number of additional vertebrae, including the Ustmylian Gorizont *Chasmatosuchus magnus* paratypes (see above), and incomplete ilia. The material as a whole was collected from various localities in the Mezen' and Pechora River basins of northern European Russia (details in Sennikov, 1990).

Tsylmosuchus samariensis Sennikov, 1990

Holotype and locality. PIN 2424/6, a cervical vertebra; Mechet' II locality, Borskoi district, Obshchii Syrt region.

Horizon. Rybinskian Gorizont, Induan.

Paratypes. PIN 2424/7, 8, incomplete vertebrae from the type locality.

Tsylmosuchus donensis Sennikov, 1990

Holotype and locality. PIN 1043/42, a cervical vertebra; Donskaya Luka locality, Ilovlyanskii district, Volgograd Province.

Horizon. Yarenskian Gorizont, Late Olenekian.

Comments. Sennikov (1990) distinguished this taxon only tentatively.

Jaikosuchus Sennikov, 1990
See Figure 8.8C.

Type species. *J. magnus* Ochev, 1979.

Holotype and locality. PIN 951/65, a cervical vertebra, originally the holotype of *Chasmatosuchus magnus* (Ochev, 1979; see above); Rassypnaya locality, Orenburg Province.

Horizon. Yarenskian Gorizont, Late Olenekian.

Paratype. PIN 4187/25, a cervical neural arch; Donguz IX locality (Sennikov, 1990).

Comments. Sennikov (1990) referred two of the elements of Ochev's type material of *Chasmatosuchus magnus* to a new genus, based on their proposed rauisuchid, rather than proterosuchid, affinities.

Discussion

Interpreting the Russian early archosaur remains and placing them in a broader context is hampered by four related problems: their fragmentary nature, associated taxonomic problems, patchy knowledge of early archosaur morphology in general, and uncertainty surrounding the temporal correlation of Permo-Triassic faunas worldwide.

Our review shows that early Russian archosaurs consist largely of incomplete remains, and, on current understanding, represent genera that have not been discovered outside Russia. There have been several additional problems for interpretation of the material, and these mean that many of the taxon names discussed above are probably *nomina vana* and *nomina dubia*.

1. On occasion, dissociated material, sometimes from different localities, has been referred to single taxa, and at times as the type material.
2. An interpretive component is sometimes applied to the designation of taxa, for example, the perception that gracile and robust forms of certain taxa exist and can be distinguished. This has also led to the referral of material that is not always clearly associated or diagnostic.
3. At times, material has been referred to new taxa only tentatively.

Despite the problematic nature of much of the Russian 'thecodontian' material, we suggest that a significant part of it provides important information that prompts a modification of current understanding of the pattern of early archosaur evolution.

We perceive the current understanding (e.g. Charig and Reig, 1970; Charig and Sues, 1976; Benton and Clark, 1988; Sereno, 1991; Parrish, 1992) to be as follows. Archosaurs originated in the Late Permian, with the earliest recognizable taxa currently being the proterosuchids. These were specialized predators that persisted into the Early Triassic with little diversification or radiation. Erythrosuchids were larger predators that appeared in the Early Triassic and disappeared in the early part of the Middle Triassic – with little stratigraphic overlap with proterosuchids. The euparkeriids were a small group of archosaurs that were contemporaneous with erythrosuchids. These three families became extinct at the time of, or soon after the 'main' archosaur radiation. This radiation of the crown-group, the Avesuchia, occurred in the Middle Triassic, with the appearance of proterochampsids, suchians (*sensu* Benton and Clark, 1988), and dinosauromorphs.

With this framework in mind, the Russian archosaurs are important for the following reasons. *Archosaurus* remains the only robust, tangible evidence that archosaurs were present in the Permian. *Sarmatosuchus* is a Middle Triassic proterosuchid that extends the range of the family, and indicates that the group was more diverse than previously understood. *Garjainia* is perhaps the earliest known erythrosuchid, and one of the most completely known early archosaurs. *Chalishevia* is one of youngest known erythrosuchids (together with the possibly closely related Chinese *Shansisuchus*). The material of *Vytshegdosuchus* and *Dongusuchus* shows that possible rauisuchid, and more probably at least avesuchian archosaurs, were present in Russia during the Early Triassic. The presence of suchians in the Early Triassic of Russia is confirmed by our recent identification of an undescribed calcaneum that is interpreted as part of a rotary, crocodile-like (functionally) astragalus–calcaneum joint, among material collected from Yarenskian Gorizont deposits at the Donskaya Luka locality. The Russian archosaur fauna recovered from the Middle Triassic Donguz Gorizont includes a proterosuchid (*Sarmatosuchus*) and possible examples of euparkeriids (*Dorosuchus*), erythrosuchids (*Uralosaurus*), and rauisuchids (*Dongusuchus*). This builds on the recent work of Sereno (1991, fig. 26B) and Parrish (1992, 1993) in showing that the pattern of early archosaur evolution is more complex than previously thought, and suggests that the Anisian and Ladinian were host to a great diversity of taxa from which later archosaurian clades evolved. These modifications, summarized in Figure 8.10, should be taken into account in studies of the major faunal changes that occured in Permo-Triassic times.

Current knowledge of the details of basal archosaur phylogeny is perhaps improving, but many hypotheses of relationships are not robust (Gower and Wilkinson, 1996) and many taxa have yet to be included in phylogenetic analyses. This is directly related to our knowledge of morphology, which in turn affects the temporal correlation of Triassic faunas – which partly relies on the correlation of comparable vertebrates. As exemplified by the Russian forms, some of the less completely known archosaur material might yield important biogeographical and biostratigraphical information, even if not greatly advancing our understanding of morphology.

The apparent completeness of the succession of the continental Permo-Triassic in Russia, combined with the large temporal and spatial range represented, offers an important record of the early evolution of the Archosauria. We hope that new material can be collected, and suggest that our understanding of early archosaur evolution might be enhanced by the careful study of this material, rather than neglecting it because of its often problematic nature.

Acknowledgements

We thank J.M. Parrish for his careful review of an earlier draft of this chapter, and M.J. Benton for additional comments and advice. This chapter is the result of collaborative work between the two authors on early Russian archosaurs, partly conducted within the Bristol–Moscow Joint Palaeontological Research Programme. We thank the Royal Society and NERC for making it possible for A.G.S. to visit Britain and D.J.G. to visit Russia. D.J.G. was supported by a Royal Society Fellowship and NERC grant GR9/1569.

Note added in proof
Since this manuscript was written, Sennikov (1999) has described additional material, and erected the genus and species *Scythosuchus basileus*. (Sennikov, A.G. 1999. [The evolution of the postcranial skeleton in archosaurs in connection with new finds of the Rauisuchidae in the Early Triassic of Russia.] *Paleontologicheskii Zhurnal* 1999 (6): 44–56.)

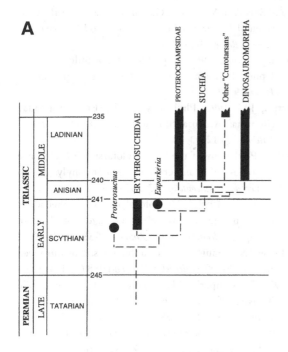

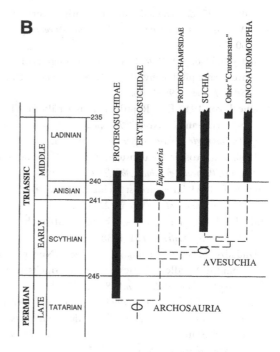

Figure 8.10. A, Pattern of early archosaur evolution based on Sereno (1991, fig. 26B). B, Revised pattern of biochronology incorporating information from Russian taxa discussed in text, and revised correlation of *Cynognathus* zone presented by e.g. Shishkin *et al.* (1995).

References

Benton, M.J. 1999. *Scleromochlus taylori* and the origin of dinosaurs and pterosaurs. *Philosophical Transactions of the Royal Society, Series B* **354**, 1423–1446.

—and Clark, J.M. 1988. Archosaur phylogeny and the relationships of the Crocodylia, pp. 295–338 in Benton, M.J. (ed.), *Phylogeny and Classification of Tetrapods. Vol. 1: Systematics Association Special Volume* 35A. Oxford: Clarendon Press.

—and Gower, D.J. 1997. Richard Owen's giant Triassic frogs: Middle Triassic archosaurs from England. *Journal of Vertebrate Paleontology* 17: 74–88.

Charig, A.J. and Reig, O.A. 1970. The classification of the Proterosuchia. *Biological Journal of the Linnean Society* 2: 125–171.

—and Sues, H.-D. 1976. Proterosuchia, pp. 11–39 in Kuhn, O. (ed.), *Handbuch der Paläoherpetologie*, Vol. 13. Stuttgart: Gustav Fischer, 137pp.

Clark, J.M., Welman, J., Gauthier, J.A. and Parrish, J.M. 1993. The laterosphenoid bone of early archosauriforms. *Journal of Vertebrate Paleontology* 13: 48–57.

Efremov, I.A. 1940. [Preliminary description of new forms in the Permian and Triassic faunas of terrestrial vertebrates of the USSR.] *Trudy Paleontologicheskogo Instituta AN SSSR* 10 (2): 1–156.

Ewer, R.F. 1965. The anatomy of the thecodont reptile *Euparkeria capensis* Broom. *Philosophical Transactions of the Royal Society of London, Series B* 248: 379–435.

Gauthier, J.A. 1986. Saurischian monophyly and the origin of birds. *Memoirs of the California Academy of Sciences* 8: 1–55.

Gower, D.J. 1996. The tarsus of erythrosuchid archosaurs (Reptilia), and implications for early diapsid phylogeny. *Zoological Journal of the Linnean Society* 116: 347–375.

—2000. Rauisuchian archosaurs (Reptilia, Diapsida): an overview. *Neues Jahrbuch für Geologie und Paläontologie, Abhandlungen*. In press.

—and Sennikov, A.G. 1996a. Morphology and phylogenetic informativeness of early archosaur braincases. *Palaeontology* 39: 883–906.

—and— 1996b. Endocranial casts of early archosaurian reptiles. *Paläontologische Zeitschrift* 70: 579–589.

—and— 1997. *Sarmatosuchus* and the early history of the Archosauria. *Journal of Vertebrate Paleontology* **17**: 60–73.

—and Wilkinson, M. 1996. Is there any consensus on basal archosaur phylogeny? *Proceedings of the Royal Society, Series B* **263**: 1399–1406.

Huene, F. von 1940. Eine Reptilfauna aus der ältesten Trias Nordrusslands. *Neues Jahrbuch für Geologie und Paläontologie, Abhandlungen* **1939**: 139–144.

—1960. Ein grosser Pseudosuchier aus der Orenburger Trias. *Palaeontographica, Abteilung A* **114**: 105–111.

Hughes, B. 1963. The earliest archosaurian reptiles. *South African Journal of Science* **59**: 221–241.

Juul, L. 1994. The phylogeny of basal archosaurs. *Palaeontologia Africana* **31**: 1–38.

Kalandadze, N.N. and Sennikov, A.G. 1985. [New reptiles from the Middle Triassic of the southern Cis-Urals.] *Paleontologicheskii Zhurnal* **1985** (2): 77–84.

Long, R.A. and Murray, P.A. 1995. Late Triassic (Carnian and Norian) Tetrapods from the Southwestern United States. *Bulletin of the New Mexico Museum of Natural History and Science* **4**: 1–254.

Ochev, V.G. 1958. [New data concerning the pseudosuchians of the USSR.] *Doklady AN SSSR* **123**: 749–751.

—1961. [New thecodont from the Triassic of the Orenburg region of the Cis-Urals.] *Paleontologicheskii Zhurnal* **1961** (1): 161–162.

—1975. [On the proterosuchian palate.] *Paleontologicheskii Zhurnal* **1975** (4): 98–105.

—1978. [On the morphology of *Chasmatosuchus*.] *Paleontologicheskii Zhurnal* **1978** (2): 98–106.

—1979. [New Early Triassic archosaurs from eastern European USSR.] *Paleontologicheskii Zhurnal* **1979** (1): 104–109.

—1980. [New archosaurs from the Middle Triassic of the southern Cis-Urals.] *Paleontologicheskii Zhurnal* **1980** (2): 101–107.

—1981. [On *Erythrosuchus* (*Garjainia*) *primus* Ochev.] *Voprosy Geologii Yuzhnogo Urala i Povolzh'ya* **22**: 3–22.

—1982. [Remains of pseudosuchians in the Lower Triassic of the southern Cis-Urals.] *Paleontologicheskii Zhurnal* **1982**: 96–102.

—1986. [On Middle Triassic reptiles of the southern Cis-Urals.] *Ezhegodnik Vsesoyuznogo Paleontologicheskogo Obshchestva* **29**: 171–180.

—1991. [*On the History of the Evolution of Early Archosaurs.*] Saratov: Izdatel'stvo Saratovskogo Universiteta, 77 pp.

—and Shishkin, M.A. 1988. Global correlation of the continental Triassic on the basis of tetrapods. *International Geology Review* **30**: 163–176.

Parrington, F.R. 1956. A problematic reptile from the Upper Permian. *Annals and Magazine of Natural History* **9**: 333–336.

Parrish, J.M. 1992. Phylogeny of the Erythrosuchidae (Reptilia: Archosauriformes). *Journal of Vertebrate Paleontology* **12**: 93–102.

—1993. Phylogeny of the Crocodylotarsi, with reference to archosaurian and crurotarsan monophyly. *Journal of Vertebrate Paleontology* **13**: 287–308.

Sennikov, A.G. 1988a. [The role of the oldest thecodontians in the vertebrate assemblages of Eastern Europe.] *Paleontologicheskii Zhurnal* **1988** (4): 78–87.

—1988b. [New rauisuchids from the Triassic of European Russia.] *Paleontologicheskii Zhurnal* **1988** (2): 124–128.

—1989. [New euparkeriid (Thecodontia) from the Middle Triassic of the southern Urals.] *Paleontologicheskii Zhurnal* **1989** (2): 71–78.

—1990. [New data on the rauisuchids of eastern Europe.] *Paleontologicheskii Zhurnal* **1990** (3): 3–16.

—1992. [Oldest proterosuchids from the Triassic of eastern Europe.] *Doklady RAN* **326**: 896–899.

—1993. Diapsids of the Permian and Triassic of Eastern Europe. *Bulletin of the New Mexico Museum of Natural History and Science* **3**: 429–430.

—1994. [The first Middle Triassic proterosuchid from eastern Europe.] *Doklady RAN* **336**: 659–661.

—1995a. [Diapsid reptiles from the Permian and Triassic of Eastern Europe.] *Paleontologicheskii Zhurnal* **1995** (1): 75–83.

—1995b. Diapsids, pp. 77–89 in Shishkin, M.A. (ed.) [*Biostratigraphy of the Continental Triassic of the Southern Cis-Urals.*] Moscow: Nauka.

—1995c. [Early thecodonts of Eastern Europe.] *Trudy Paleontologicheskogo Instituta RAN* **263**: 1–141.

—1996. Evolution of the Permian and Triassic tetrapod communities of Eastern Europe. *Palaeogeography, Palaeoclimatology, Palaeoecology* **120**: 331–351.

Sereno, P.C. 1991. Basal archosaurs: phylogenetic relationships and functional implications. *Journal of Vertebrate Paleontology Memoir* **2**: 1–53.

—and Arcucci, A. B. 1990. The monophyly of crurotarsal archosaurs and the origin of bird and crocodile ankle joints. *Neues Jahrbuch für Geologie und Paläontologie, Abhandlungen* **180**: 21–52.

Shishkin, M.A., Ochev, V.G., Tverdokhlebov, V.P., Vergay, I.F., Goman'kov, A.V., Kalandadze, N.N., Leonova, E.M., Lopato, A.Yu., Makarova, I.S., Minikh, M.G., Molostovskii, E.M., Novikov, I.V. and Sennikov, A.G. 1995. [*Biostratigraphy of the Triassic of the Southern Cis-Urals.*] Moscow: Nauka, 206 pp.

Sill, W.D. 1974. The anatomy of *Saurosuchus galilei* and the relationships of the rauisuchid thecodonts. *Bulletin of the Museum of Comparative Zoology, Harvard* **146**: 317–362.

Tatarinov, L.P. 1960. [Discovery of pseudosuchians in the Upper Permian of the USSR.] *Paleontologicheskii Zhurnal* **1960** (4): 74–80.

—1961. [The pseudosuchian material of the USSR.] *Paleontologicheskii Zhurnal* **1961** (1): 117–132.

Wu X.-C. 1981. [The discovery of a new thecodont from the north-east Shanxi.] *Vertebrata PalAsiatica* **19**: 122–132.

—1982. [Two pseudosuchian reptiles from Shan-Gan-Ning Basin.] *Vertebrata PalAsiatica* **20**: 291–301.

Yakovlev, N.N. 1916. [The Triassic vertebrate fauna from the variegated bed of the Vologda and Kostroma districts.] *Geologicheskii Vestnik* **2** (4): 157–165.

Young, C.C. 1964. [The pseudosuchians in China.] *Palaeontologia Sinica* **151**: 1–205.

—1973a. [On the occurrence of *Vjushkovia* in Sinkiang.] *Memoirs of the Institute of Vertebrate Palaeontology and Palaeoanthropology, Academia Sinica* **10**: 38–53.

—1973b. [On a new pseudosuchian from Turfan, Sinkiang.] *Memoirs of the Institute of Vertebrate Palaeontology and Palaeoanthropology, Academia Sinica* **10**: 15–37.

9

Procolophonoids from the Permo-Triassic of Russia

PATRICK S. SPENCER AND MICHAEL J. BENTON

Introduction

The Procolophonoidea is an important group of early amniotes currently placed in the subclass Parareptilia (Laurin and Reisz, 1995; Lee 1995). The procolophonoids arose in the Late Permian (Tatarian) and survived until the Late Triassic (Rhaetian or latest Norian; Benton, 1993), and during the Triassic formed a large component of many of the complex terrestrial assemblages of the period (Benton, 1983). The more derived Triassic forms had a world-wide distribution, with specimens known from European Russia, Western Europe, North America, South America, South Africa, Madagascar, Australia, China, and Antarctica.

The procolophonoids have been rather difficult to define, owing largely to dispute over the position of the primitive Late Permian genera *Barasaurus* and *Owenetta*. Ivakhnenko (1979), for example, placed both *Barasaurus* and *Owenetta* in the Russian taxon Nyctiphruretidae Efremov, 1938, while, Laurin and Reisz (1991) placed *Owenetta* in the Procolophonidae. Such discussions have largely ignored the status of the Owenettidae, a family erected by Broom (1939) on the basis of the monotypic South African species *Owenetta rubidgei*. Nevertheless, there is now evidence that *Barasaurus* and *Owenetta* may be closely related, comprising a morphologically more primitive taxon than the Procolophonidae (Meckert, 1993; Spencer, 1994). Lee (1995, 1997) has noted three postulated synapomorphies which support the inclusion of *Barasaurus* in the Owenettidae, and which distinguish the group from the Procolophonidae: postfrontal–supratemporal contact, median spur on back of skull table, and

absence of entepicondylar foramen. The validity of the Family Owenettidae, and its inclusion in Procolophonoidea as sister group to Procolophonidae, have been accepted in recent cladistic analyses of basal amniotes (Laurin and Reisz, 1995; DeBraga and Rieppel, 1997; Lee, 1997).

The monophyly of the Triassic family Procolophonidae was defined by Lee (1995) on the basis of two features: the exclusion of the parietal from the orbital margin, and the presence of enlarged palatal denticles, sparsely arranged in single rows. Unfortunately, however, the first character has a polymorphic distribution in many procolophonids, in which a small exposure of the parietal in the orbital margin may be present, and its polarity is presently undefined. The second character could apply to a plethora of basal amniotes (e.g. Captorhinidae, *Protorothyris*). Alternative potential synapomorphies supporting the Procolophonidae (Spencer, 1994) include: a splint-like postfrontal, confined to the orbital margin; postparietal much reduced or absent; fewer than four premaxillary teeth; fewer than nine maxillary teeth; a thickened layer of enamel restricted to the upper half of the crown of mid-row maxillary and dentary teeth; and lateral tooth base on the pterygoid and palatine reduced in length and aligned anteroposteriorly to anteromedially.

Lee (1995) defined the Procolophonoidea (Romer, 1956, *emend.* Lee, 1995), the most recent common ancestor of the Owenettidae and Procolophonidae, and all its descendants, on the basis of three unequivocal characters: posterior spur on prefrontal antorbital buttress; ventral embayment of the cheek; occipital flange of parietal. Other possible synapomorphies of procolophonoids (*sensu* Lee; Spencer, 1994) are: jugal–squamosal contact greatly reduced or absent, and elements of the scapulocoracoid unfused in mature specimens, a trait acquired convergently in mesosaurids and the 'protorothyridid' *Cephalerpeton*. In this chapter, we provisionally accept the Procolophonoidea as a taxon embracing the Owenettidae and Procolophonidae.

Within the Procolophonidae, three subfamilies of procolophonids, the Spondylolestinae, Procolophoninae, and Leptopleuroninae, are currently distinguished, based primarily on features of the skull and the marginal dentition (Ivakhnenko, 1979). Representatives of the first two subfamilies are found in the Permo-Triassic of Russia in four successive procolophonid assemblages which characterize the Triassic of the East European platform and Cis-Urals region (Chapter 7). The phylogeny of procolophonoids has been re-analysed in detail by Spencer and Sues (2000).

Russian procolophonids have been described in a number of papers (Chudinov and V'yushkov, 1956; Ochev, 1958, 1967, 1968; Ochev and Danilov, 1972; Ivakhnenko, 1973a,b, 1974, 1975, 1979, 1983; Novikov, 1991, 1994; Novikov and Orlov, 1992). Chudinov and V'yushkov (1956) established the genera *Phaanthosaurus* and *Tichvinskia*, based on partial skull and dentary remains from the Lower Triassic of various parts of European Russia. Ochev (1958) added the species *Tichvinskia burtensis*, later made the type of the genus *Burtensia* by Ivakhnenko (1975). Ivakhnenko also described a number of new genera, *Contritosaurus* Ivakhnenko, 1974, *Kapes* Ivakhnenko, 1975, *Macrophon* Ivakhnenko, 1975, *Orenburgia* Ivakhnenko, 1975, and *Microphon* Ivakhnenko, 1983, also based on incomplete skull remains. Novikov (1991) carried out further revisions of Russian procolophonids, and erected the genera *Timanophon* and *Lestanshoria* for new forms from the northern region of European Russia and *Samaria* for another specimen from the South Urals. Novikov and Orlov (1992) erected a further genus, *Insulophon*, for a specimen from Kolguev Island, north of the Arctic Circle. Two further genera were described as procolophonids, *Vitalia* Ivakhnenko,

1973a, and *Coelodontognathus* Ochev, 1967, but these are probably wrongly attributed to the group.

Russian procolophonid remains have been found in continental sediments, largely in fluvial settings. The elements are typically isolated and often abraded, and there is a preservational bias towards tooth-bearing bones in the museum collections. While this is true in general, some articulated skeletons have been recovered, most notably the type specimens of *Tichvinskia vjatkensis* (PIN 954/1; Figure 9.7) and *Timanophon burtensis* (PIN 3359/11). These specimens are preserved in a 'rolled-up' attitude which could indicate that the animals were located in burrows, or burrow systems. Supporting evidence for this suggestion comes from the recent discovery of flask-shaped burrow structures in the southern African *Lystrosaurus/Procolophon* assemblage zone that contained rolled-up, articulated skeletons of the procolophonine *Procolophon* (J. Welman, pers. comm. to P.S.S., 1991).

In the following overview, the Russian procolophonids are described in order of the two subfamilies, and in approximate stratigraphic order.

Repository abbreviations

PIN, Palaeontological Institute, Moscow; SGU, Saratov State University; TsNIGRI, Tsentralny Nauchno-Issledovatelskii Geologo-Razvedochnyi Muzei, Sankt Peterburg.

Systematic survey

Subclass PARAREPTILIA Olson, 1947
Suborder PROCOLOPHONOIDEA Romer, 1956
Family PROCOLOPHONIDAE Seeley, 1888
Subfamily SPONDYLOLESTINAE Ivakhnenko, 1979

Type genus. Spondylolestes Broom, 1937, *Lystrosaurus/ Procolophon* Zone, South Africa.
Diagnosis. The most primitive procolophonids; orbits elongated as a rule, not enlarged; teeth usually simple, conical, relatively weakly differentiated, more than ten on each jaw (Ivakhnenko, 1979).

Comments. Ivakhnenko (1979, p. 11) divided the Family Procolophonidae into two subfamilies, the Spondylolestinae and Procolophoninae, on the basis that the spondylolestines were more primitive than the procolophonines. His diagnosis of the Spondylolestinae, given above, differentiated this sub-family in broad terms from the procolophonine pro-colophonoids with their elongated orbits, their bicuspid marginal teeth, and the presence of fewer than ten teeth on each jaw. However, the features listed appear to be plesiomorphic for procolophonoids as a whole, and it is difficult to find autapomorphies supporting spondylolestine monophyly *sensu* Ivakhnenko (1979).

Ivakhnenko (1979, pp. 11–14) included a number of genera in the subfamily, some Russian (*Phaanthosaurus, Contritosaurus*), and others from South Africa (*Spondylolestes, Procolophonoides*), Brazil (*Candelaria*), and China (*Neoprocolophon*). These genera are all Early Triassic in age, except *Candelaria* from the Middle Triassic. *Procolophonoides* was erected by Ivakhnenko (1979, p. 13) for some South African materials previously assigned to *Procolophon*, but, according to him, not procolophonine.

The status of the Spondylolestinae is unclear. *Spondylolestes* appears to be a *nomen dubium*, and it may be difficult even to distinguish it as an amniote based on Broom's (1937) description and figures.

Phaanthosaurus Tchudinov and Vjuschkov, 1956
Phaanthosaurus ignatjevi Tchudinov and Vjuschkov, 1956
See Figures 9.1A–C and 9.2A, B.
Holotype and locality. PIN 1025/1, a dentary; Spasskoe village, Vetluga River, Nizhnii Novgorod Province.
Horizon. Vokhmian Gorizont, Lower Triassic.
Paratypes. PIN 1025/21, a dentary; PIN 1025/20, postdentary portion of a lower jaw; and further jaw fragments, all from around Spasskoe village.
Diagnosis. Eleven almost undifferentiated teeth on the lower jaw. Adductor notch narrow and long. Coronoid process of lower jaw massive and low (Ivakhnenko, 1979).
Comments. Phaanthosaurus is also characterized by the

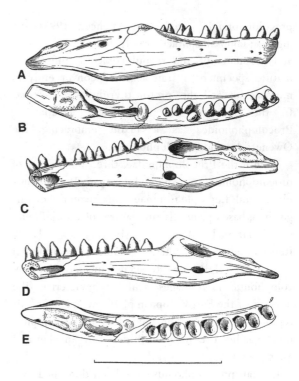

Figure 9.1. Lower jaws of the basal procolophonids *Phaanthosaurus ignatjevi* Tchudinov and Vjuschkov, 1956 (PIN 1025/1, 20; A–C) and *Contritosaurus simus* Ivakhnenko, 1974 (PIN 3355/1; D, E) in lateral (A), medial (B, D), and occlusal (C, E) views. Scale bars = 10 mm. (Modified from Ivakhnenko, 1979.)

tendency for the marginal teeth to form rare closely positioned pairs with replacement teeth, arranged diagonally (Figure 9.1B).

The manner of tooth wear in *Phaanthosaurus* and *Contritosaurus* is distinct. While most teeth show small terminal tooth-to-food wear facets, as in *Procolophon* for example, these are associated with extensive and steeply inclined facets. On the upper teeth these face inwards, and correspondingly on the lowers, these face outwards. This manner of wear in teeth that rarely occlude in a tooth-to-tooth fashion, as in all procolophonids (*contra* Gow, 1985) seems to be determined, as seen in unworn teeth, by the alternate displacement of the terminal cusp labially in uppers and lingually in lower teeth. It probably also corresponds to the differential thickness of the enamel of the tooth

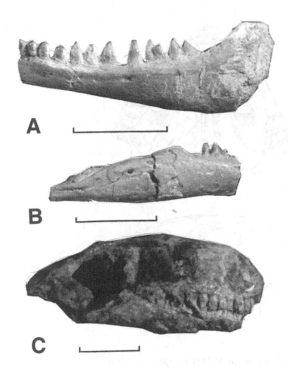

Figure 9.2. Partial lower jaws of *Phaanthosaurus ignatjevi* Tchudinov and Vjuschkov, 1956, PIN 1025/1 (A) and PIN 1025/20 (B), both in lateral view. C, Skull of *Contritosaurus simus* Ivakhnenko, 1974, PIN 3355/1, in lateral view. Scale bars = 5 mm.

crown, although this has yet to be determined by sectioning of the teeth. This character, not noted by Ivakhnenko or Novikov, seems to be a synapomorphy shared by *Phaanthosaurus* and *Contritosaurus*. Other similarities include: antorbital region of skull very high; maxillary lateral depression extends dorsally on to ventral part of nasal; anterodorsal region of prearticular, viewed medially, is bifurcated by the posteroventral ramus of the coronoid; and, base of each maxillary tooth has a small triangular distolingual flange, so that these teeth appear to be inclined anteroventrally in lingual view. Indeed, these taxa are so similar that they may well be conspecific, as is also suggested by their stratigraphic and geographic distribution.

Contritosaurus Ivakhnenko, 1974

Diagnosis. 'A very small procolophonid (length of skull reaching 2 cm). Skull high. Preorbital portion high and short. Orbits very large, rounded-trapezoidal, drawn out rearward. Fossa of lateral nasal gland very considerable, nostrils small. Lacrimal not extending to margin of nostril; postfrontal not fused with parietal. Upper margin of orbit formed by the joining of the prefontal and postfrontal bones. Teeth differentiated: front teeth (5–6 in upper jaw and 3 in lower jaw) conical, with slightly dilated and lingually recurved crowns, rear 7–8 teeth on both jaws with heavily dilated bases and pointed crowns, frequently with oblique wear surfaces. There are four rows of teeth on the palate, pterygoid, palatine, and two vomerine (perichoanal and medial); the perichoanal row of teeth is double.' (Ivakhnenko, 1974, p. 347).

Comment. Contritosaurus shares the same mode of tooth wear as *Phaanthosaurus*, but seems to lack the occasional diagonal pairing of marginal teeth. Most of the characters noted above are primitive. Ivakhnenko (1974, p. 347) notes that *Contritosaurus* is 'most similar to the genus *Phaanthosaurus*, but distinguished from it by the slightly shorter adductor fossa, relative to the length of the jaw, the longer and narrower retroarticular process, the thinner and higher coronoid process, and the lack of a crest on the outer face of the dentary.' In addition, the type specimen of *C. simus* (below) has a lacrimal with a narrow posteriorly directed process, resting on the dorsal surface of the extopterygoid, just medial to the foramen palatinum posterius, and a double row of vomerine teeth bordering the choana ('perichoanal tooth row' of Ivakhnenko, 1979). These features, however, are not preserved in any material assigned by Ivakhnenko to *Phaanthosaurus*. Ivakhnenko (1974) described two species of *Contritosaurus*, but these may be minor variants of a single species, *C. convector*.

Contritosaurus convector Ivakhnenko, 1974

Diagnosis. 'In contradistinction to *C. simus* the teeth in the upper jaw are slightly inclined, the adductor fossa is broader (the length of the adductor fossa is one-quarter greater than in *C. simus*, while its width is

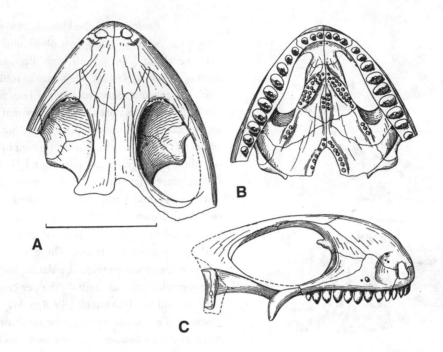

Figure 9.3. Holotype skull of *Contritosaurus simus* Ivakhnenko, 1974 (PIN 3355/1) in dorsal (A), ventral (B), and lateral (C) views. Scale bar = 10 mm. (Modified from Ivakhnenko, 1979.)

practically twice as great), the posterior margin of the dentary is sharply bent upwards, and of the two crests on the postdentary portion of the lower jaw, the outer crest is far more weakly expressed, while the other is completely absent.' (Ivakhnenko, 1974, p. 351).

Holotype and locality. PIN 3357/1, lower jaw fragment; Krasnie Baki village, Vetluga River, Nizhnii Novgorod Province.

Paratype. PIN 3357/2, a part of an upper jaw, from the same locality.

Horizon. Vokhmian Gorizont, Lower Triassic.

Contritosaurus simus Ivakhnenko, 1974
See Figures 9.1D, E, 9.2C and 9.3.

Diagnosis. [See *C. convector.*]

Holotype and locality. PIN 3355/1, an incomplete skull with lower jaw fragment; Lipovo village, Vetluga River, Nizhnii Novgorod Province.

Horizon. Vokhmian Gorizont, Lower Triassic.

Paratypes. PIN 2890/5, an incomplete skull from the Kasyanovtsy site in Kirov Province; PIN 3356/1, an incomplete skeleton and skull from the Sarafanikha site in Nizhnii Novgorod Province.

Microphon Ivakhnenko, 1983
Microphon exiguus Ivakhnenko, 1983
See Figure 9.4.

Diagnosis. 'Very small form, length of skull not more than 1 cm. Upper jaw short, maxilla with high ascending lamina. About 12 teeth in maxillary bone, teeth conical, with longitudinally compressed bases.' (Ivakhnenko, 1983, p. 136).

Holotype and locality. PIN 3583/31, lower jaw fragment; Donguz VI locality, Donguz River, Orenburg Province.

Horizon. Severodvinskian Gorizont, Upper Tatarian, Upper Permian.

Comments. It is unclear whether *Microphon* is a valid taxon or not. Among the characters listed by Ivakhnenko (1983, p. 136), only one seems to be acceptable, namely 'longitudinally compressed crowns'. The others are primitive for procolophonids

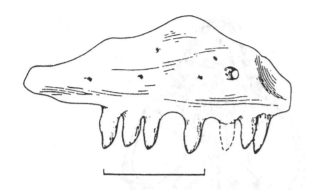

Figure 9.4. Partial right maxilla, holotype of *Microphon exiguus* Ivakhnenko, 1983 (PIN 3538/31) in lateral view. Scale bar = 25 mm. (Modified from Ivakhnenko, 1983.)

(presence of a maxillary depression) or for various basal amniotes (high maxillary process, prominent maxillary foramen: Lee, 1995). Ivakhnenko (1983) also argued that the small size of *Microphon*, the feature that gave rise to its name, is a further diagnostic feature. However, with only a single incomplete specimen, *Microphon* could be a juvenile of another taxon, although this is a unique Russian procolophonid record from the Late Permian.

The presence of single-cusped maxillary teeth is apparently primitive for procolophonoids, but would not exclude *Microphon* from the Spondylolestinae. Ivakhnenko (1983, p. 136) correctly points out similarities in the structure of the maxillary foramen, shared at least with *Phaanthosaurus*. Ivakhnenko (1983, p. 136) also lists 12 teeth as present in the maxilla of *Microphon*, according to him a spondylolestine feature, but one shared also with owenettids. However, his illustration (Figure 9.4 here) only shows seven teeth, although there may be space for 12.

Subfamily PROCOLOPHONINAE Seeley, 1888
emend. Ivakhnenko, 1979

Diagnosis. 'Skull usually high, orbits strongly elongated. Teeth differentiated into incisiforms and molariforms. Crowns of molariforms usually very complex, bicuspid. Ten or fewer teeth on the jaw' (Ivakhnenko, 1979, p. 14).

Comments. Ivakhnenko (1979, p. 14) erected this sub-

family to distinguish a group of procolophonids that he saw as distinctive from the spondylolestines. He included in the subfamily a number of Russian taxa (*Tichvinskia, Burtensia, Macrophon, Orenburgia, Kapes, Vitalia*), as well as some from South Africa (*Procolophon, Microtheledon, Thelegnathus, Myocephalus*), and Germany (*Anomoiodon, Koiloskiosaurus*). These taxa are all Early Triassic in age, except *Anomoiodon* and *Orenburgia*, which are Early to Middle Triassic.

Ivakhnenko's subfamilies Spondylolestinae and Procolophoninae were distinguished by him from a third, Leptopleuroninae, diagnosed as (Ivakhnenko, 1979, p. 21): 'Specialized procolophonids with strong elongation backwards of the orbits and with spines on the bones of the cheek complex. Teeth differentiated a little.' This subfamily, including *Leptopleuron* from the Late Triassic of Scotland and *Hypsognathus* from the Late Triassic of North America, is distinguished by clear synapomorphies (e.g., V-shaped incursion or embayment on the anterolateral surface of the jugal; strap-like process arising from medial side of descending process of prefrontal sutured to suborbital ridge on the frontal; pair of obtusely conical, laterally directed quadratojugal processes, each with an annular basal flange), unlike the Spondylolestinae and Procolophoninae, which are probably paraphyletic assemblages of outgroups to the Leptopleuroninae. Ivakhnenko (1979) included two additional Late Triassic taxa in Leptopleuroninae, *Sphodrosaurus* from North America and *Paoetodon* from China, but these are incorrectly assigned.

Tichvinskia Tchudinov and Vjuschkov, 1956
Diagnosis. Medium-sized procolophonines; interorbital depression slight; no lateral process on the quadratojugal; prominent supraorbital ridge formed by the frontal, postfrontal, and parietal; postfrontal fused to parietal; two conical teeth form the anterior part of the maxillary tooth row; a dorsomedially directed intermediate molariform (m7) sometimes forms the end of the dentary tooth row; posteromedial enamel ridge on the lower molariforms truncated dorsally by the distal occlusal basin (*emend.* from Ivakhnenko, 1979).

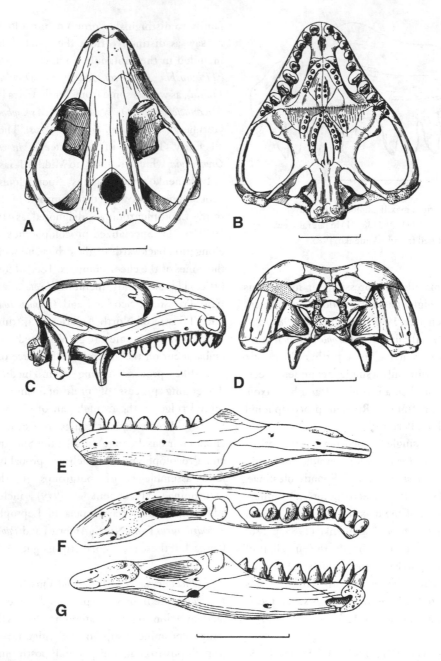

Figure 9.5. Skull (A–D) and lower jaw (E–G) of *Tichvinskia vjatkensis* Tchudinov and Vjuschkov, 1956 (PIN 954/1) in dorsal (A), ventral (B), lateral (C, E), occipital (D), occlusal (F), and medial (G) views. Scale bars = 10 mm. (Modified from Ivakhnenko, 1979.)

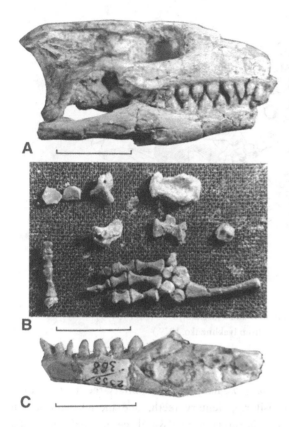

Figure 9.6. *Tichvinskia vjatkensis* Tchudinov and Vjuschkov, 1956, skull in lateral view, PIN 954/1 (A); skeletal remains, PIN 954/1, a partial foot and hindlimb and other elements (B); lower jaw of '*T. jugensis*' Vjuschkov and Tchudinov, 1956, PIN 2355/368 in lateral view (C). Scales bars = 10 mm.

Tichvinskia vjatkensis Tchudinov and Vjuschkov, 1956
See Figures 9.5, 9.6A, B and 9.7.

Diagnosis. 'Height of teeth exactly the same as the height of the dentary. Coronoid process of the lower jaw low and wide, adductor pit narrow. Post-dentary part of the mandible narrow and long.' (Ivakhnenko, 1979, p. 15).

Holotype and locality. PIN 953/1, skull; Okunevo, Fedorovka River, Vyatka River basin, Kirov Province.

Horizon. Yarenskian Gorizont, Lower Triassic.

Paratypes. Complete skeleton, series of skulls.

Comment. The holotype is lost. Ivakhnenko's (1973b) description was based on four skulls, including one

with an associated postcranial skeleton (PIN 954/1; Figures 9.6 and 9.7), and other tooth-bearing elements. The species almost certainly includes the genus *Burtensia* Ivakhnenko, 1975, founded on the species *T. burtensis* Otschev, 1958 (holotype, PIN 2394/12, formerly SGU 104/2, a dentary from Kzyl-Sai ravine at Andreevka settlement, Kzyl-Oba River, Ural River basin, Orenburg Province; Petropavlovskaya Svita, Yarenskian Gorizont, Lower Triassic; and paratypes, PIN 2394/11, right mandible from the type locality; PIN 4400/1, a right dentary, from the Meshcheryakovka locality; PIN 3359/11, a skull and lower jaw, from Pizhmo-Mezen' River, Arkhangel'sk Province). *Burtensia* was said to differ from *Tichvinskia* and other procolophonines by the 'small, relatively high dentary, differentiation of lower-jaw dentition, number and shape of quasi-molar teeth' (Novikov, 1991, p. 92). However, inspection of the specimens reveals no distinguishing characters in the dentition or skull shape (Spencer and Sues, 2000). Ivakhnenko (1975) referred another specimen (PIN 3359/11, a skull and lower jaw, from Pizhmo-Mezen' River, Arkhangel'sk Province) to *Burtensia*, but Novikov (1991) assigned it to a new genus, *Timanophon* (see below).

Tichvinskia jugensis Vjuschkov and Tchudinov, 1956
See Figure 9.6C.

Diagnosis. 'In lower jaw, height of double-peaked teeth is less than height of dentary. Coronoid process low, postdentary part of the jaw high and wide, adductor pit wide.' (Ivakhnenko, 1979, p. 16).

Holotype and locality. PIN 2252/308, skull fragment with lower jaw; Vakhnevo, Sharzhenga River, Vologda Province.

Horizon. Rybinskian Gorizont, Lower Triassic.

Paratypes. Parts of skulls.

Comment. On the basis of our re-examination of the currently limited material, this taxon is here distinguished only as Procolophonidae *incertae familiae.* It exhibits features that are absent in spondylolestine procolophonoids (*sensu* Ivakhnenko, 1979), including bicuspid molariform cheek teeth, and the lingual cusp of dentary molariforms positioned directly opposite

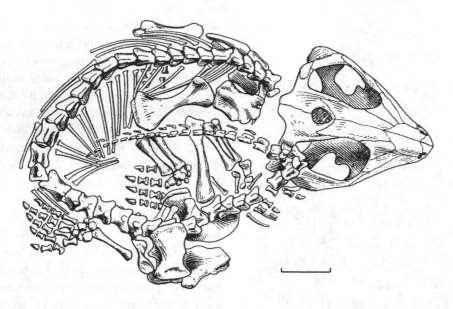

Figure 9.7. Skeleton of *Tichvinskia vjatkensis* Tchudinov and Vjuschkov, 1956 (PIN 954/1) in dorsal view, as found. Scale bar = 10 mm. (Modified from Ivakhnenko, 1979.)

the labial cusp, with both cusps of subequal size, but it lacks derived features of later procolophonoids, such as the Late Triassic leptopleuronines. Nevertheless, it is perhaps the sole published record from the Rybinskian Gorizont.

Kapes Ivakhnenko, 1975

Diagnosis. Maxillary teeth: four to five molariforms, one mesial intermediate molariform. Molariforms 1–4 of upper and lower jaws becoming progressively larger backwards. Lingual cusp in lower teeth subequal in height to labial cusp; labial and lingual cusps relatively close together.

Comments. Kapes was erected by Ivakhnenko (1975, pp. 87–88) for isolated remains of procolophonids that were larger than *Tichvinskia*, and differed by the number and shape of the teeth, and by the possession of a very large tooth on the lower jaw. This, and the characters listed in the diagnosis (Spencer and Storrs, 2000) justify the validity of this genus.

These medium to large procolophonines show a number of other characters (Novikov, 1991, p. 98) that are shared also with *Orenburgia, Samaria*, and *Lestanshoria*, namely: premaxillary teeth, at least one incisiform; dentary teeth, one mesial intermediate molariform (Figure 9.8), three molariforms, distal intermediate molariform absent or present, two incisors. Lower intermediate molariforms unicuspid, or bicuspid, with a labial and lingual cusp connected by a weak transverse crest. Molariforms 1–4 bicuspid and with shallow distal and mesial basins. Lower distal intermediate molariform absent, unicuspid, or bicuspid, lower than m4. Upper molariforms broader than long; bicuspid. Lower molariforms sub-conical, equidimensional to slightly longer than broad. Coronoid process not expanded transversely.

Macrophon is probably synonymous with *Kapes*, and it is possible that *Orenburgia, Samaria*, and *Lestanshoria* are also synonyms, although these last three lack the enlarged penultimate molariform of *Kapes*, and they may all be better assigned to *Orenburgia* as a distinct genus. When traits that are susceptible to intraspecific variation are excluded, these three genera are based primarily on primitive character states (Spencer and Storrs, 2000). Further study is necessary to confirm which, if any, of these genera are distinctive, and to

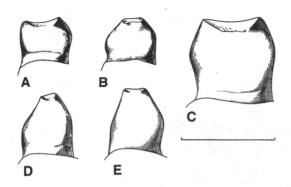

Figure 9.8. Molariform teeth, shown as the penultimate tooth on the left side, in posterior view, from the lower jaws of *Tichvinskia* (A), '*Burtensia*' (B), *Macrophon* (C), *Orenburgia* (D), and *Kapes* (E). Scale bar = 5 mm. (Modified from Ivakhnenko, 1979.)

determine which of the species that are synonymous with *Kapes* are distinct species of that genus.

Kapes amaenus Ivakhnenko, 1975
See Figure 9.9A, B.

Diagnosis. Adult mandible is 1.4–1.7 times the length of m3 beneath the distal end of m3. Upper part of molariform crowns compressed strongly distomesially; ratio of distance between labial and lingual cusp tips and maximum width of m3 is about 0.55 (Spencer and Storrs, 2000).

Holotype and locality. PIN 3361/2, dentary; Zheshart settlement, Vychegda River, Komi Republic.

Horizon. Gamskaya Svita, Yarenskian Gorizont, Lower Triassic.

Syntypes PIN 3361/4, right dentary; PIN 3361/15, 10, 11, fragments of left dentaries; PIN 3361/1, 6, 7, 12, fragments of right maxillae; PIN 3361/8, 9, 13, fragments of left maxillae; PIN 3361/14, a tooth, all from the Zheshart site (Novikov, 1991, p. 98).

Kapes majmesculae (Otschev, 1968)
See Figure 9.9C.

Diagnosis. Adult mandible is *c.* 2.0 times length of m3 beneath the distal end of m3. A weak cingulum occurs in middle height of all lower molariforms (Spencer and Storrs, 2000).

Holotype and locality. PIN 4365/5 (formerly SGU 104/3824), dentary; Petropavlovka, Sakmara River, Ural River basin, Orenburg Province.

Horizon. Yarenskian Gorizont, Lower Triassic.

Comment. This species was originally ascribed to *Tichvinskia* by Ochev (1968), and it was included in the new genus *Orenburgia* by Ivakhnenko (1975, p. 89). It was later re-assigned to *Kapes* by Ivakhnenko (1983), a view accepted by Novikov (1991) and Spencer and Storrs (2000).

Kapes serotinus Novikov, 1991

Diagnosis. Teeth widely spaced; ratio of maximum length of tooth row to maximum height of dentary is 2.7 (Novikov, 1991, p. 99).

Holotype and locality. PIN 1579/23, dentary; Berdyanka River, Ural River basin, Orenburg Province.

Horizon. Donguz Gorizont, Middle Triassic.

Comment. This specimen was originally described as *Tichvinskia* cf. *majmesculae* by Ochev and Danilov (1972), and was renamed as a distinct species by Novikov (1991, p. 99) on the basis of its less massive and widely spaced teeth. However, this is not a valid taxon, and is probably synonymous with *K. majmesculae* (Spencer and Storrs, 2000). The two diagnostic characters are both subject to individual variation in procolophonids. Indeed, spacing of the teeth varies within the holotypes of *K. serotinus* and *K. majmesculae* (Novikov, 1991, fig. 3c, e).

Macrophon Ivakhnenko, 1975
Macrophon komiensis Ivakhnenko, 1975
See Figures 9.8C and 9.9D.

Diagnosis. A very high posterior wall of the maxilla, and upper molariform teeth that have transverse axes directed distolabially and mesolingually. The anterior edge of the medial excavation of the ascending maxillary process is level with the anterior rim of the maxillary foramen, and faces posteroventrally (Spencer and Storrs, 2000).

Holotype and locality. PIN 3361/1, part of upper jaw; Zheshart settlement, Vychegda River, Komi Republic.

Horizon. Gamskaya Svita, Yarenskian Gorizont, Lower Triassic.

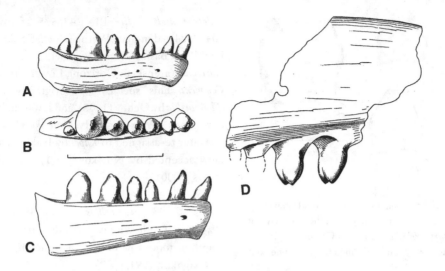

Figure 9.9. Anterior dentary fragments of *Kapes amaenus* Ivakhnenko, 1975 (PIN 3361/1; A, B), *Kapes majmesculae* (Otschev, 1968) (restored from SGU 104/3824 and PIN 4365/5; C), and partial left maxilla of *Kapes* (*Macrophon*) *komiensis* (Ivakhnenko, 1975) (PIN 3361/3; D) in lateral (A, C, D) and occlusal (B) views. Scale bars = 10 mm. (Modified from Ivakhnenko, 1979.)

Comments. This taxon is clearly like *Kapes* in its large size. However, it was said to differ from *Kapes* by the shape of its teeth (Ivakhnenko, 1975, p. 88), but that distinction is not clear. It is referred here to *Kapes* since it shares many of the characters noted as diagnostic of that genus by Novikov (1991), as indicated above. In addition, it shares some particular features with a new species of *Kapes* from the Otter Sandstone Formation of Devon, England (Spencer and Storrs, 2000): the upper molariform teeth are much broader than long, the M1–2 show a size increase distally, and the lingual cusp is higher than the labial cusp. The species *K. komiensis* (Ivakhnenko, 1975) appears to be valid, and distinct from other species of *Kapes*. It occurs at the same locality, and in the same horizon, as *Kapes amaenus*.

Orenburgia Ivakhnenko, 1975

Diagnosis. Medium-sized procolophonine. Skull up to 45 mm long, rounded-triangular, with concave lateral margins. Pineal foramen in front of posterior orbital margins, at the level of the middle of orbital length. Palate strongly curved longitudinally, with four short tooth rows. Interpterygoid notch narrow. Maxillary

teeth: two incisiforms, four molariforms; premaxillary teeth: three incisiforms; dentary teeth: two incisiforms, five molariforms gradually becoming smaller backwards. Crowns of molariforms bicuspid, widened transversely, with the maximal width at the middle of height (for anterior teeth) or lower third (for posterior). Incisiforms highest in dental tooth row. Coronoid process of lower jaw rounded. (Ivakhnenko, 1975, p. 89; Novikov, 1991, p. 94).

Comments. This genus was established for *Tichvinskia enigmatica* Tchudinov and V'yushkov, 1956 and *T. majmesculae* Otschev, 1968, the latter later (Ivakhnenko, 1983) re-assigned to *Kapes*. The genus differs from other procolophonines in the shape of the molariform teeth, and in the 'character of differentiation of dentition and in dominance of quasi-incisors'. It differs from *Tichvinskia*, and other taxa, in the 'location of pineal aperture and large size of buccal notch' (Novikov, 1991, p. 94).

Orenburgia shares affinities with *Lestanshoria*, for example in the height distribution of teeth along the dentary, which is very similar in both taxa. These two genera, and *Samaria*, all resemble *Kapes* in many

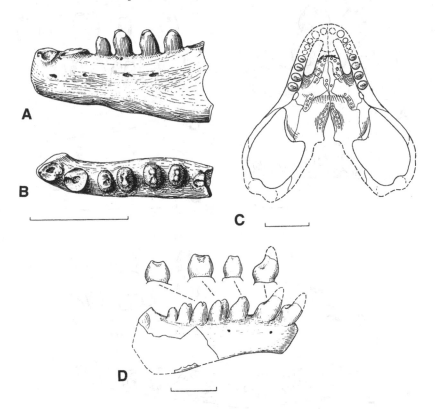

Figure 9.10. Dentary fragments (A, B, D) and partial skull (C) of *Orenburgia enigmatica* (Tchudinov and Vjuschkov, 1956) (A, B) and *Orenburgia bruma* Ivakhnenko, 1983 (C, D) in lateral (A, D) and occlusal/ ventral (B, C) views. A, B, PIN 1043/1; C, restored from PIN 3951/1 and 3952/2, D, PIN 4370/3. Scale bars = 10 mm. (Modified from Novikov, 1991.)

respects, and it could be that the small morphological differences are merely examples of individual variation within a single taxon (Spencer and Storrs, 2000).

O. enigmatica (Tchudinov and Vjuschkov, 1956)
See Figures 9.8D and 9.10A, B.
Diagnosis. 'Height of teeth similar to height of tooth-bearing element. Last tooth of lower jaw not much bigger than the second last.' (Ivakhnenko, 1979, p. 19).
Holotype and locality. PIN 1043/1, left dentary; Lipovaya Balka hollow, Don River basin, Volgograd Province.
Horizon. Lipovskaya Svita, Yarenskian Gorizont, Lower Triassic.
Comments. This species is probably a *nomen nudum*. The holotype dentary, the sole specimen, is damaged,

seemingly abraded during transport, and several of the teeth are broken (Figure 9.10A, B). It lacks diagnostic characters sufficient to distinguish it from other taxa. It seems very like *Tichvinskia vjatkensis* (cf. Figure 9.5E), except that the teeth are more widely spaced. It lacks the enlarged penultimate molariform of *Kapes* (cf. Figure 9.9A–C). The holotype has been lost.

O. bruma Ivakhnenko, 1983
See Figure 9.10C, D.
Diagnosis. Autapomorphies include a short jugal with dorsoventrally expanded anterior and posterior ends, the anterior portion of the tooth ridge on the palatine is edentulous and reaches the choana, and each vomer has a transverse expansion anteriorly bearing a diagonal row of three small teeth.

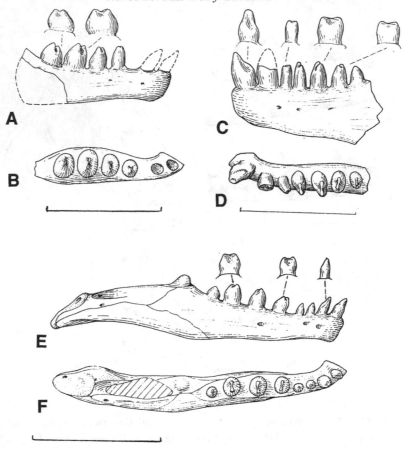

Figure 9.11. Lower jaws of *Samaria concinna* (Ivakhnenko, 1975), PIN 3362/1 (A, B), *Lestanshoria massiva* Novikov, 1991, PIN 4370/4 (C, D), and *Timanophon raridentatus* Novikov, 1991, PIN 3359/11 (E, F) in lateral (A, C, E) and occlusal (B, D, F) views. Scale bars = 10 mm. (Modified from Novikov, 1991.)

Holotype and locality. PIN 3952/1, incomplete skull; Cape Nikolaya, Admiralteistva Peninsula, Severnii Island, Novaya Zemlya Archipelago.

Horizon. Admiralteistva Svita, Ustmylian Gorizont, Lower Triassic.

Paratype. PIN 3952/2, an incomplete skull, from the same locality; PIN 4370/3, a partial dentary from Khei-Yaga River basin.

Comments. The distribution and shapes of dentary teeth are similar to *Kapes* (cf. Figure 9.9A–C), except that *O. bruma* lacks the enlarged penultimate molariform tooth. The ventral view of the skull (Figure 9.10C) is, in its preserved parts, similar to *Tichvinskia*

(Figure 9.5B). *O. bruma* is distinct from *Tichvinskia* in having widely spaced teeth, and from *Kapes* in lacking the enlarged penultimate molariform.

Samaria Novikov, 1991
Samaria concinna (Ivakhnenko, 1975)
See Figure 9.11A, B.

Diagnosis. Coronoid eminence very high; tooth ridge flanking interpterygoid vacuity with widely spaced denicles (Spencer and Sues, 2000).

Holotype and locality. PIN 3362/1, skull fragment with lower jaw; Markovka village, Soroka River, Samara River basin, Orenburg Province.

172

Horizon. Kzylsaiskaya Svita, Sludkian Gorizont, Lower Triassic.

Comments. This species was established as belonging to the genus *Orenburgia* by Ivakhnenko (1975, pp. 89–90), and was differentiated from the other species of that genus 'by the lesser height of the dentary relative to the height of the teeth and by the presence of a longitudinal depression on its anterior–inferior margin. Teeth slightly more obviously bicuspid.' It was made the type species of *Samaria* by Novikov (1991, pp. 101–102), and distinguished from other genera by the shape of the molariform teeth, by the reduction of palatal tooth rows, the relative height of the dentary, and possibly by the overall skull shape.

The taxon is most comparable to *Orenburgia* (Spencer and Storrs, 2000) from which it differs in only minor features that are subject to individual variation (cf. Figures 9.10D and 9.11A, B). Of the features listed in the diagnosis above, the coronoid is not preserved in the currently available material of *Orenburgia* for comparison. The dentary tooth morphology of *Samaria* is very similar to that of *Lestanshoria*.

Lestanshoria Novikov, 1991
Lestanshoria massiva Novikov, 1991
See Figure 9.11C, D.

Diagnosis. Medium-sized procolophonid. Skull short and massive (based on observation of lower jaw). Dentary teeth: two incisiforms, five molariforms (the third highest). Crowns of molariforms bicuspid, strongly widened, flattened longitudinally, with the maximal width at the middle of tooth height, narrowing slightly toward apex. Cusps of molariforms broadly placed. (Based on Novikov, 1991, pp. 102–103.)

Holotype and locality. PIN 4370/4, dentary; Lestanshor Creek, Khei-Yaga River, Nenetskii National District, Korotaikha River basin, Arkhangel'sk Province.

Horizon. Lestanshorskaya Svita, Ustmylian Gorizont, Lower Triassic.

Comments. Novikov (1991, p. 103) differentiated *Lestanshoria* from other procolophonids on the basis of the character of the differentiation of the teeth, the shape of the molariform teeth, the dominance of the incisiforms, and the purportedly more massive and

shorter skull. However, this genus appears to lack diagnostic criteria, and it strongly resembles *Orenburgia* (cf. Figure 9.10D), especially in the incremental height increase of the molariform teeth forwards to the mid row, the presence of five molariforms (inclusive of the mesial and distal intermediate molariforms), and the two massive anterior incisiform teeth (Novikov, 1991). *Lestanshoria* is probably a junior synonym of *Orenburgia*.

Timanophon Novikov, 1991
Timanophon raridentatus Novikov, 1991
See Figure 9.11E, F.

Diagnosis. Medium-sized procolophonine. Skull up to 30 mm long, egg-shaped, with almost straight lateral margins. Interorbital depression slight. Posterior margins of orbits and of pineal foramen at the same level. Frontal contributes to orbital margin. Postfrontal separated from parietal by suture. No lateral process on quadratojugal. Palate strongly curved longitudinally, with four short tooth rows. Interpterygoid notch wide. Maxillary teeth: two incisiforms, four molariforms; dentary teeth: three of four incisiforms, four molariforms (the second and third highest and equal in height). Crowns of molariforms bicuspid, widened transversely, with the maximal width on the middle of tooth height, narrowing evenly toward base and apex. Coronoid process of lower jaw rounded, broader than long. (Based on Novikov, 1991, p. 100.)

Holotype and locality. PIN 3359/11, incomplete skeleton with skull and lower jaw; Pizhmo-Mezen' River; Mezen' River basin, Arkhangel'sk Province.

Horizon. Pizhmomezen'skaya Svita, Ustmylian Gorizont, Lower Triassic.

Paratypes. Additional tooth-bearing elements from localities in the Mezen' River basin listed by Novikov (1991, p. 101).

Comments. Timanophon is superficially similar to *Orenburgia* (cf. Figure 9.10D), but it differs in the possession of three, rather than two, dentary incisiform teeth. However, *Timanophon* appears to represent a metataxon, being only distinguished by possession of a unique combination of primitive characters and autapomorphies (Spencer and Sues, 2000).

Insulophon Novikov, 1992

Insulophon morachovskayae Novikov, 1992

Diagnosis. Autapomorphies include: adductor fossa extremely narrow and straight; coronoid eminence with steeply angled posterior face formed by the posterodorsal process of the coronoid and surangular (Spencer and Sues, 2000).

Holotype and locality. TsNIGRI 842/10, incomplete skull with lower jaw fragment; borehole 23, Kolguev Island, Arkhangel'sk Province.

Horizon. Charkabozhskaya Svita, Sludkian(?) Gorizont, Lower Triassic.

Paratype. TsNIGRI 842/11, a partial disarticulated postcranial skeleton, from the type locality.

Comment. This specimen was obtained from an exploration borehole on Kolguev Island, in the Russian Arctic. Such boreholes had also yielded the prolacertiform *Boreopricea* from similar buried Triassic strata (Tatarinov, 1978; Benton and Allen, 1997). *Insulophon* is said to differ from other procolophonines in the form of the quasi-molars, in general skull shape, in the form of the coronoid process and of the adductor notch (Novikov and Orlov, 1992, pp. 182–183). Features of the palate and teeth seem, however, to be indistinguishable from *Orenburgia bruma* (cf. Novikov, 1994, figs. 10, 13).

Taxa provisionally removed from the Procolophonoidea

Vitalia Ivakhnenko, 1973a

Vitalia grata Ivakhnenko, 1973a

Diagnosis. 'Lower jaw low at the symphysis, teeth elongate-conical, with weakly expanded crowns, with blunt lateral cusps.' (Ivakhnenko, 1979, p. 21).

Holotype and locality. PIN 104/3105, lower jaw with teeth; Lipov hollow, Don River basin, Volgograd Province.

Horizon. Lipovskaya Svita, Yarenskian Gorizont, Lower Triassic.

Comments. Vitalia was described by Ivakhnenko (1973a) as a procolophonid, and specifically as a procolophonine (Ivakhnenko, 1979, p. 21). However, its dentition

is unusual, unlike any fully identified procolophonid. A new specimen (PIN unnumb.) displays a better preserved crown morphology than the holotype, and confirms its unusual nature.

The crown structure has no parallel in any definitely assignable procolophonid. There are two transversely expanded terminal cusps on each side of the teeth, set inside a flattened, transversely expanded terminal basin. The posteriormost two teeth are larger, and, viewed occlusally, are more equidimensional, with narrow mesial and distal hollows, which represent the parts of the terminal basin of more anterior teeth behind and in front of the central cusp.

Beyond exhibiting transversely expanded marginal teeth (seen in several other groups), there are no unequivocal characters supporting inclusion of *Vitalia* in Procolophonoidea. It is not possible to place *Vitalia* in another group: such obscure broad-toothed amniote remains are not uncommon in the Permo-Triassic. Some may belong to procolophonid relatives, others to trilophosaurids (archosauromorph diapsids), others to synapsids, or to other as yet incompletely known groups (e.g. Sues and Olsen, 1993).

Coelodontognathus Otschev, 1967

Coelodontognathus ricovi Otschev, 1967

Coelodontognathus donensis Otschev, 1967

Diagnosis. 'Skull 60 mm long. Lower jaw relatively low and elongate. Teeth transversely widened, and regularly serrated. First and last teeth of the lower jaw expanded and without serrations. In each half of the lower jaw, as well as, apparently, in the upper jaw, there were about 10 teeth.' (Ochev, 1967, p. 15)

Holotypes and locality. SGU 104/3101 (*C. ricovi*), right dentary; SGU 104/3103 (*C. donensis*), right dentary and nine teeth, and a further left and right dentary (SGU 104/3104, 3105); Lipovaya Balka, Don River basin, Volgograd Province.

Horizon. Lipovskaya Svita, Yarenskian Gorizont, Lower Triassic.

Comments. Coelodontognathus is almost certainly not a procolophonoid for basically the same reasons outlined for *Vitalia* above. The only character that seems to have been used to suggest procolophonid affinities is transversely expanded cheek teeth. However, expanded cheek teeth occur in other unrelated groups. The rest of the morphology of *Coelodontognathus* is quite unlike a procolophonoid.

Discussion

Over the years, 20 species and 14 genera of procoloph-onoids have been described from the Russian Permo-Triassic. Many of these are distinctive, and they attest to an important evolutionary radiation of the group in Eastern Europe. However, many of the taxa have been founded on rather incomplete materials, and comparisons among taxa have been difficult. Thorough revision is required, but a survey of the available specimens in PIN has suggested that the true diversity of described procolophonoids from Russia may be rather lower, at most six genera and 13 species (listed stratigraphically).

Upper Permian

Severodvinskian Gorizont

Microphon exiguus Ivakhnenko, 1983

Lower Triassic

Vokhmian Gorizont

Phaanthosaurus ignatjevi Tchudinov and Vjuschkov, 1956

Phaanthosaurus simus (Ivakhnenko, 1974) [incl. *Contritosaurus*; *C. convector*]

Rybinskian Gorizont

'*Tichvinskia jugensis*' Vjuschkov and Tchudinov, 1956

(Procolophonoidea *incertae familiae*)

Sludkian Gorizont

Orenburgia concinna Ivakhnenko, 1975 [incl. *Samaria* Novikov, 1991; *Lestanshoria massiva* Novikov, 1991; *Insulophon morachovskayae* Novikov, 1992]

Ustmylian Gorizont

Orenburgia bruma Ivakhnenko, 1983

Timanophon raridentatus Novikov, 1991

Yarenskian Gorizont

Tichvinskia vjatkensis Tchudinov and Vjuschkov, 1956

Kapes amaenus Ivakhnenko, 1975

Kapes majmesculae (Otschev, 1968)

Kapes komiensis (Ivakhnenko, 1975) [incl. *Macrophon* Ivakhnenko, 1975]

Orenburgia enigmaticus (Tchudinov and Vjuschkov, 1956)

Donguz Gorizont

Kapes majmesculae (Otschev, 1968) [incl. *Kapes serotinus* Novikov, 1991]

This tentative revision confirms that there was generally one taxon of procolophonid present in each gorizont in the Russian Upper Permian to Middle Triassic sequence (Chapter 7), except during the time of deposition of the Yarenskian Gorizont, when as many as four genera and six species may have occupied European parts of Russia.

Acknowledgements

We thank the Royal Society for funding of our visits to Moscow.

References

Benton, M.J. 1983. Dinosaur success in the Triassic: a non-competitive ecological model. *Quarterly Review of Biology* 58: 29–55.

—1993. Reptilia, pp. 681–715 in Benton, M.J. (ed.), *The Fossil Record 2*. London: Chapman & Hall.

—and Allen, R. 1997. *Boreopricea* from the Lower Triassic of Russia, and the relationships of the prolacertiform reptiles. *Palaeontology* 40: 931–953.

Broom, R. 1937. A further contribution to our knowledge of the fossil reptiles of the Karroo. *Proceedings of the Zoological Society of London, Series B* 107: 299–318.

—1939. A new type of cotylosaurian, *Owenetta rubidgei*. *Annals of the Transvaal Museum* 19: 319–321.

Chudinov, P.K. and V'yushkov, B.P. 1956. [New data on small cotylosaurs from the Permian and Triassic of the USSR.] *Doklady AN SSSR* 108: 547–550.

DeBraga, M. and Rieppel, O. 1997. Reptile phylogeny and the interrelationships of turtles. *Zoological Journal of the Linnean Society* 120: 281–354.

Ivakhnenko, M.F. 1973a. [New cotylosaurs of the Cis-Urals.] *Paleontologicheskii Zhurnal* 2: 131–134.

—1973b. Skull structure in the early Triassic procolopho-nian *Tichvinskia vjatkensis*. *Paleontological Journal* 7: 511–518.

—1974. New data on the early Triassic procolophonids of the USSR. *Paleontological Journal* 8: 346–351.

—1975. Early Triassic procolophonid genera of CisUral. *Paleontological Journal* 9: 86–91.

—1979. [Permian and Triassic procolophonids of the Russian platform.] *Trudy Paleontologicheskogo Instituta AN SSSR* 164: 1–80.

—1983. New procolophonids from eastern Europe. *Paleontological Journal* 17: 135–139.

Laurin, M. and Reisz, R.R. 1991. *Owenetta* and the origin of turtles. *Nature* 349: 324–326.

—and— 1995. A reevaluation of early amniote phylogeny. *Zoological Journal of the Linnean Society* 113: 165–223.

Lee, M.S.Y. 1995. Historical burden in systematics and the interrelationships of 'parareptiles'. *Biological Reviews* 70: 459–547.

—1997. Pareiasaur phylogeny and the origin of turtles. *Zoological Journal of the Linnean Society* 120: 197–280.

Meckert, D. 1993. A re-evaluation of the procolophonid *Barasaurus besairi* from the Upper Permian of Madagascar. *Journal of Vertebrate Paleontology* 13: 50A.

Novikov, I.V. 1991. New data on the procolophonids from the USSR. *Paleontological Journal* 25: 91–105.

—1994. [*Biostratigraphy of the Continental Triassic of the Timan-North Urals Region using the Tetrapod Fauna.*]. Moscow: Nauka, 139 pp.

—and Orlov, A.N. 1992. New Early Triassic vertebrates from Kolguyev Island. *Paleontological Journal* 26: 180–184.

Ochev, V.G. 1958. [New data on the Triassic vertebrate fauna of the Orenburg area of the CisUrals region.] *Doklady AN SSSR* 122: 485–488.

—1967. [A new species of procolophonid from the Triassic of the Don Basin.] *Izvestiya Vysshykh Uchebnykh Zavedenii (Geologiya i Razvedka)* 2: 15–20.

—1968. [A new representative of Triassic procolophonids of Bashkortostan]. *Ezhegodnik Vsesoyuznogo Paleontologicheskogo Obshchestva* 18: 298–301.

—and Danilov, A.I. 1972. [On the first discovery of a pro-colophonid in the Middle Triassic of the USSR.] pp. 81–84 in *Voprosy Geologii Yuzhnogo Urala i Povolzh'ya*. Saratov: Izdatel'stvo Saratovskogo Universitatea.

Spencer, P.S. 1994. *The early interrelationships and morphology of Amniota*. Unpublished PhD Thesis, University of Bristol.

—and Storrs, G.W. 2000. A reevaluation of small tetrapods from the Middle Triassic Otter Sandstone Formation of Devon, England. *Paleontology*. In press.

—and Sues, H.-D. 2000. Phylogenetic relationships within the Procolophonoidea (Amniota: Reptilia). Submitted.

Sues, H.-D. and Olsen, P.E. 1993. A new procolophonid and a new tetrapod of uncertain, possibly procolophonian affinities from the Upper Triassic of Virginia. *Journal of Vertebrate Paleontology* 13: 282–286.

Tatarinov, L.P. 1978. Triassic prolacertilians of the U.S.S.R. *Paleontological Journal* 12: 505–514.

10

Enigmatic small reptiles from the Middle–Late Triassic of Kirgizstan

DAVID M. UNWIN, VLADIMIR R. ALIFANOV AND MICHAEL J. BENTON

Introduction

Thick sequences of fluvial and lacustrine deposits at Madygen in Fergana, Kirgizstan provide an unparalleled record of Middle–Late Triassic continental floras and faunas of Middle Asia (Dobruskina, 1995). In addition to a wealth of plant fossils, these sediments have also produced vertebrate material, including the remains of two small diapsid reptiles, both discovered in 1965 by A.G. Sharov during expeditions organized by the Palaeontological Institute, Moscow, to collect fossil insects (Sharov, 1966). Rapid burial of the carcasses in fine-grained sediments led to the preservation of some integumentary structures.

One specimen, preserved with evidence of the integument and extensive flight membranes associated with the hind limbs, was named *Podopteryx* ('footwing') by Sharov (1971a). It was later (Cowen, 1981) renamed *Sharovipteryx*, since *Podopteryx* was found to be pre-occupied. Sharov interpreted *Sharovipteryx* as a small arboreal glider, an idea that was enthusiastically taken up by others (e.g. Halstead, 1975, 1982), as well as Gans *et al.* (1987), who redescribed *Sharovipteryx*. Initially, *Sharovipteryx* was thought to be a pseudosuchian (Sharov, 1971a) and possibly ancestral to pterosaurs (Halstead, 1975, 1979, 1982, 1989). Later, Gans *et al.* (1987) suggested that it may be a more primitive diapsid, an idea supported by Tatarinov (1989, 1994), who identified *Sharovipteryx* as a prolacertiform.

The second diapsid to be described from Madygen, *Longisquama insignis* (Sharov, 1970), is known from less complete material, but sufficient is preserved to show that this extraordinary animal bore a row of elongate frond-like scales upon its back. The anatomy and function of these structures remains unclear: Sharov (1970) supposed that they acted as parachutes, while Halstead (1975) and Haubold and Buffetaut (1987) suggested that they could be deployed laterally to form flight surfaces. *Longisquama* was identified as a pseudosuchian by Sharov (1970), but this idea has not yet been critically assessed.

Repository abbreviation

PIN, Palaeontological Institute, Russian Academy of Sciences, Moscow.

Geology

Both taxa were found at site 14 of Dobruskina (1995, fig. 9, T-14), at Dzhailyau-Cho, in the Madygen area, Lyailyakskii district, Osh Province, southern Fergana, in Kirgizstan (Figure 10.1). The remains were discovered in sediments of the Madygen Svita, a series of lacustrine and fluvial deposits about 500 m thick, consisting of intercalated sands, silts, and clays with discontinuous coal seams (Dobruskina, 1995). The Madygen Svita was previously thought to be Early Triassic (Sharov, 1970, 1971a), but more recent studies based on the extensive flora indicate a Middle or Late Triassic (Ladinian–Carnian) age (Dobruskina, 1970, 1976, 1980; Vakhrameev *et al.*, 1978). Dobruskina (1980) equates the flora with the Laurasian *Scytophyllum* Flora, derived from the Keuper strata of the Germanic basin.

Sharovipteryx and *Longisquama* were recovered from compact, light grey-yellow bedded shales in the 'upper member' of the Madygen Svita, about 50 m

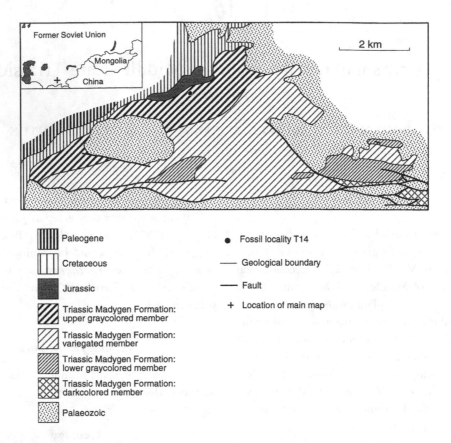

Paleogene

Cretaceous

Jurassic

Triassic Madygen Formation: upper graycolored member

Triassic Madygen Formation: variegated member

Triassic Madygen Formation: lower graycolored member

Triassic Madygen Formation: darkcolored member

Palaeozoic

● Fossil locality T14

—— Geological boundary

—— Fault

+ Location of main map

Figure 10.1. Geological map showing the site of discovery of *Sharovipteryx* and *Longisquama.* Inset, geographic location of main map. (Redrawn from Dobruskina, 1995.)

below the top of the unit (Dobruskina, 1995; Figure 10.1). Associated fossils include abundant plants, predominantly ferns, lycophytes, horsetails, cycads, ginkgos, and conifers (Dobruskina, 1995), bivalves, crustaceans, numerous insects, fishes, including the dipnoan *Asiatoceratodus* (Vorob'eva, 1967) and the saurichthyiform *Saurichthys* (Dobruskina, 1995), as well as a variety of other actinopterygians (Selezneva and Sychevskaya, 1989), a small poorly-preserved tetrapod, *Triassurus*, described as the earliest urodele (Ivakhnenko, 1978), and the skeleton of a cynodont, *Madygenia* (Tatarinov, 1980, 1994).

Systematic survey

Subclass DIAPSIDA Osborn, 1903
Order PROLACERTIFORMES Camp, 1945
Family PODOPTERYGIDAE Sharov, 1971a
(= SHAROVIPTERYGIDAE Tatarinov, 1989)
Sharovipteryx mirabilis (Sharov, 1971a)
See Figures 10.2 and 10.3.

Diagnosis. Pronounced caudal elongation of hyoids. Anterior development of preacetabular process of ilium. Process on distal end of femur; elongation of tibia, which is longer than the trunk. Sharov (1971a, p. 108) and Tatarinov (1989) cited long lists of supposedly diagnostic characters, but many of these are also

Figure 10.2. Holotype of *Sharovipteryx mirabilis* (PIN 2584/8), main slab. Scale bar divided into centimetres.

found in other diapsids. Tatarinov (1989) claimed that the relative shortness of the fore limbs, only one-third or one-quarter the length of the hind limbs, was also diagnostic of *Sharovipteryx*, but as the forelimbs have yet to be clearly identified this cannot be substantiated.)

Material and preservation. *Sharovipteryx* is represented by a single specimen, the holotype (PIN 2584/8), borne upon a slab and counterslab (Sharov, 1971a; Gans *et al.*, 1987). The skeleton is largely complete and almost fully articulated, though crushed in places. The skull is considerably compressed and visible in dorsal view, not ventral as Sharov (1971a) supposed (Gans *et al.*, 1987). The tip of the tail is lacking, but otherwise the spinal column is complete, though parts are not

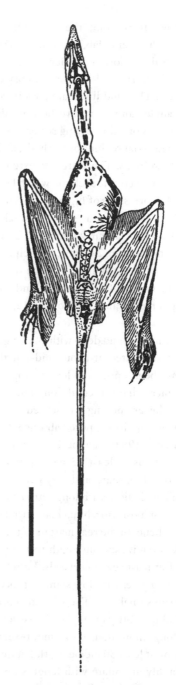

Figure 10.3. Preserved remains of the skeleton and impressions of the soft tissues of *Sharovipteryx*. Scale bar, 20 mm.

well preserved. If present, the forelimbs must be buried within the main slab, since structures previously identified as forelimb elements by Sharov (1971a) and Gans *et al.* (1987) are almost certainly remains of ribs. The hind limbs are largely complete, though the ankles and feet have become disarticulated. Impressions of skin bearing scales are common in the skull area, axial regions of the body, and around parts of the hind limbs. There are also extensive areas of sediment adjacent to the skeleton which bear superbly preserved impressions of flight membranes.

Anatomy. Sharovipteryx is a small animal with a snout–vent length of about 90 mm and a total body length of no more than 240 mm. Principal features include a narrow, deep skull with large orbit, a long neck and tail, remarkably elongate hind limbs and extensive flight membranes (Sharov, 1971a; Gans *et al.*, 1987).

The skull is narrow and deep with an elongate snout and large orbits located at about mid-length. There appear to have been upper and lower temporal openings, the typical diapsid condition, but the lower margin of the lower opening is obscured by the ramus of the lower jaw and its supposed absence (Gans *et al.*, 1987, Tatarinov, 1989) cannot be confirmed. The upper jaw contains at least 15 teeth per side. Each tooth is narrow, thin, sharp and very gently recurved. The mandible is shallow with long retroarticular processes and long posterior branchial cornua project caudally from beneath the rear margin of the skull.

The neck is equivalent in length to the trunk and composed of at least seven cervicals. Vertebrae three to seven are elongate, with low spinous processes and highly elongate spinal ribs. The sacrum consists of at least four, and possibly as many as six, vertebrae. The tail is very long (more than 1.5 × snout–vent length) and consists of at least 30 elements, the more distal of which are highly attenuate with lengths up to seven times their breadth. Gastralia, consisting of very thin fine rods of bone, are present in the trunk region.

Previous authors refer to the shoulder girdle (Sharov, 1971a) and very small forelimbs (Sharov, 1971a; Gans *et al.*, 1987; Tatarinov, 1994), but it is not clear to which elements they were alluding. A single

crescent-shaped bone lying to the left of the rostral end of the dorsal series might be remnants of a coracoid, but slender ossifications to the right of the dorsal series, identified by Sharov (1971a) as parts of the forelimb, appear to be the shafts of anterior dorsal ribs. Since the remains are not preserved on a single plane, but project into the sediment, we suppose that the forelimbs are buried within the main slab or, less likely, in the counterslab.

The ilium is remarkable for its elongate anterior process, but much of the rest of the pelvis is buried within the slab and few details are visible. Gans *et al.* (1987) refer to a possible epipubic element, but the identity of this structure is unclear and it might be a disarticulated rib.

The femur is remarkably long, reaching a length equivalent to that of the trunk, but straight, and with an unusual pulley-like process projecting from the distal end. The crus is also elongate, and slightly longer than the femur. The fibula is very slender, but reaches the ankle and is distinct from the tibia, except at the proximal end where the two appear to be fused.

Little remains of the ankle, except for a few disarticulated, isolated elements of uncertain identity. The metatarsus is short, only 25% the length of the femur. The pes contains five digits with a phalangeal formula of 2, 3, ?4, 2+, ?3. The digits are long and slender with intermediate elements reduced in length and elongate penultimate phalanges. There is an increase in length from the first to fifth digit, but digit one is not appreciably reduced in comparison to the other digits, as previous authors have stated, and digit five is of similar length to digit four. Digit five is unusual in that the first phalanx is equivalent in length to the fourth metatarsal.

Many of the appendicular elements are hollow, with bony trabeculae confined to the articular ends. It has been suggested (Sharov, 1971a) that the smooth inner surface of the bone walls indicates pneumatization of the bones, but pneumatophores are not evident on any element.

Impressions provide evidence of two types of external soft tissue: the integument, which usually bears small scales, or a distinctive diamond-shaped ornamentation; and wing membranes, impressions of which tend to be smoother and often exhibit folds and fine striae. The integument covering the skull, neck, trunk and extending up to the base of the tail bore small tubercular or keeled scales, while in places impressions of flat, imbricate scales are preserved along the margins of the hind limbs and toes.

The extent of the flight membranes is less certain. A large uropatagium was stretched between the hind limbs, attaching to the base of the tail as far as the seventh vertebra and along the posterior margin of the femur, tibia, and pes to the tip of the fifth toe. Sharov (1971a, fig. 5) also reconstructed a flight membrane attached to the body wall and extending forward from the femur to the fore limb. Though absent from their restoration, Gans et al. (1987) refer to 'half-moon'-shaped prefemoral folds, but conclude that these structures were not attached to the humerus. By contrast, Ivakhnenko and Korabelnikov (1987, fig. 260) accept Sharov's interpretation and even add a further membrane fringing the neck and extending to the base of the skull. New studies of the main slab and counterslab reveal impressions of a flight membrane anterior to the femur and extending laterally, at least as far as the knee. This observation provides some support for Sharov's reconstruction, but it is not possible to determine whether this membrane attached to the fore limb, or not.

Functional morphology. Sharov argued that *Sharovipteryx* was a small arboreal glider that flew from branch to branch using the tail as a counterweight and the head and body as a rudder. This idea has been widely accepted (e.g. Halstead, 1975; Gans *et al.*, 1987; Ivakhnenko and Korabel'nikov, 1987; Tatarinov, 1989; Wellnhofer, 1991), although Gans *et al.* (1987) opted for a somewhat different reconstruction of the flight apparatus, proposing that it consisted solely of a uropatagium, possibly assisted by a small canard wing supported by the fore limbs. Other possible functions of the membranes include camouflage and display (Gans *et al.*, 1987).

Phylogenetic relationships. Sharov assigned *Sharovipteryx* to the Pseudosuchia on the basis of a single character: the pronounced anterior development of the pre-

acetabular process of the ilium. The Pseudosuchia at the time was a broad grouping of 'advanced thecodontians', including rauisuchians, phytosaurs, aetosaurs, ornithosuchids, and others, whereas current views (Benton and Clark, 1988; Sereno, 1991) restrict the term to aetosaurs and rauisuchians. Charig *et al.* (1976) and Ivakhnenko and Korabel'nikov (1987) supported the idea of *Sharovipteryx* as an archosaur, as did Halstead, who went further and proposed that it was directly ancestral to pterosaurs (Halstead, 1975, 1979, 1982, 1989), basing this opinion on the presence, in both taxa, of wing membranes. Sharov (1971a) had already supposed a possible sister-group relationship with pterosaurs, but he also saw similarities with *Scleromochlus*, a small Late Triassic ornithodiran, and Benton (1993b, p. 698) listed Podopterygidae in Ornithodira.

Other workers (Gans *et al.*, 1987; Tatarinov, 1989, 1994; Wellnhofer, 1991) assigned *Sharovipteryx* to a much lower position within the diapsid tree. Gans *et al.* (1987) identify either Lepidosauria, or Protorosauria, as likely relatives, while Tatarinov (1989, 1994) argued that *Sharovipteryx* belonged within the Prolacertiformes on the basis of the following skull characters: lack of a preorbital opening, loss of the lower temporal bar, elongation of the nostrils, presence of a rudimentary coronoid process on the mandible, and the absence of a mandibular fenestra. However, only the first and last of these can be safely determined in *Sharovipteryx*, and neither character unites this taxon with prolacertiforms because they represent the primitive condition for diapsids (e.g. Benton, 1985; Evans, 1988; Carroll and Currie, 1991; Laurin, 1991).

Despite these difficulties, comparisons with recent cladistic analyses of diapsid relationships (Benton, 1985; Gauthier *et al.*, 1988; Evans, 1988; Carroll and Currie, 1991; Laurin, 1991; Benton and Allen, 1997) suggest that Tatarinov (1989, 1994) is essentially correct. *Sharovipteryx* exhibits two apomorphies of Prolacertiformes; (i) elongate cervical vertebrae (Benton, 1985; Chatterjee, 1986) with (ii) low neural spines (Benton, 1985; Chatterjee, 1986), and it may have a third: an incomplete lower temporal bar (Gow,

1975; Benton, 1985; Chatterjee, 1986), though this has yet to be confirmed. Other characters found in *Sharovipteryx* and some, though not all prolacertiforms, include a highly elongate femur (Chatterjee, 1986), tibia slightly longer than femur (Chatterjee, 1986), a feature which is also apomorphic for Ornithodira (e.g. Juul, 1994) and pes digit five with elongate proximal phalanx (Olsen, 1979; Chatterjee, 1986). Of the 48 characters of Prolacertiformes, or clades within Prolacertiformes, listed by Benton and Allen (1997), *Sharovipteryx* may be coded for eight. Derived prolacertiform characters are (numbers follow the list in Benton and Allen, 1997): cervicals longer than dorsals (19), cervical neural spines long and low (20), cervical ribs long and slender (22), femur straight (38), tibia longer than femur (39), and second phalanx of digit V of the foot long (47). *Sharovipteryx* codes as primitive for two prolacertiform characters: (?) seven or fewer cervical vertebrae (17), metatarsal IV less than three times the length of metatarsal V (45).

The hypothesis that *Sharovipteryx* is a prolacertiform, or a close outgroup of Prolacertiformes, is further supported by the presence in *Sharovipteryx* of apomorphies of Prolacertiformes + Archosauria (Benton, 1985) such as a long snout and narrow skull, recurved teeth, long, thin, tapering cervical ribs, and elongate transverse processes of the trunk vertebrae, but the absence of important archosaur apomorphies including an antorbital fenestra, mandibular fenestra and fourth trochanter on the femur.

? Subdivision ARCHOSAURIA Cope, 1869
Family LONGISQUAMIDAE Sharov, 1970
Longisquama insignis Sharov, 1970
See Figure 10.4.

Diagnosis. Elongate manus digit four, equivalent in length to the humerus. A series of elongate plume-like appendages inserting along dorsal mid-line of body. Occiput bears two tubercle-like structures. (The first two are clearly diagnostic, and the latter is a further possibility. Sharov (1970) cited numerous supposedly diagnostic characters, but they are present in other taxa.)

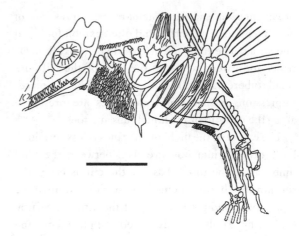

Figure 10.4. Holotype of *Longisquama insignis* (PIN 2584/4), main slab. Preserved remains of the skeleton and impressions of the soft tissues of *Longisquama*. Redrawn from Sharov (1970). Scale bar, 10 mm.

Material and preservation. The holotype (PIN 2584/4) consists of an incomplete skeleton, comprising the skull, neck and anterior half of the trunk, the pectoral girdle and forelimbs, and well preserved impressions of the integument. There are a further five specimens consisting of fragmentary remains of the plume-like dorsal appendages. All the remains are heavily crushed and details are poorly preserved.

Anatomy. *Longisquama* is a small reptile with a skull about 23 mm long and a total forelimb length of about 44 mm (Sharov, 1970).

The high skull has large orbits, small antorbital openings and, according to Sharov (1970, fig. 1), upper and lower temporal fenestrae. The teeth are small, conical, acrodont, and number 12–13 pairs in the upper jaw and 16–17 in the lower jaw. The neck is short and contains seven cervicals. Only the anterior-most dorsal vertebrae are visible and the rest of the spinal column is not preserved. Long, slender dorsal ribs are preserved in articulation with the trunk vertebrae, but cervical ribs seem to be lacking, though this may be because they are obscured or disarticulated rather than truly absent.

The shoulder girdle is composed of a long, narrow scapula, expanded at both ends, a short, rather rod-like coracoid, and a long crescent-shaped clavicle which articulates with a well developed interclavicle. The latter element has a rectangular anterior end and a broad stem tapering to a point distally. The humerus is slender, gently sigmoid and of similar length to the radius and ulna, both of which are long, thin and straight. The carpus consists of numerous small indeterminate elements which support five metacarpals. Metacarpals 1–4 are of similar size and about half the length of the forearm, while the fifth is much reduced. The fourth digit appears to contain five phalanges while the fifth has four, thus a phalangeal formula of 2, 3, 4, 5, 4 seems likely. In preserved digits the penultimate phalanx is elongate and supports a sharp-pointed and somewhat recurved claw.

The neck and ventral surface of the thorax appear to have been covered by long, overlapping, simple scales about 0.3 mm wide and up to 1.75 mm long. Similar scales fringe the anterior margin of the humerus and radius. Somewhat larger scales, up to 1.0 mm in width, fringe the posterior margin of the humerus and ulna. The most spectacular structures occur on the dorsal mid-line and consist of a series of paired appendages, apparently one pair per vertebra (Haubold and Buffetaut, 1987). Each paired appendage consists of two long plume-like structures, ranging from 100 to 150 mm in length and about 5–7 times longer than they are wide. The appendages decline slightly in size from front to back, with the longest occurring at the front (Haubold and Buffetaut, 1987). Each plume becomes gently expanded distally and slightly recurved, and is composed of finely folded anterior and posterior margins and a somewhat thickened medial region running from the base to the tip of the plume. Sharov (1970) argued that the paired plumes were joined along their anterior margins, and the posterior margin at the distal tip, but Halstead (1975) and Haubold and Buffetaut (1987) interpret the plumes as separate structures. They are generally thought to have been modified scales (Sharov 1970; Haubold and Buffetaut, 1987), but, unfortunately, they appear to have undergone some postmortem displacement and the nature of their insertion is unclear.

Functional morphology and ecology. It seems likely that *Longisquama* was a small arboreal insectivore (Ivakhnenko and Korabel'nikov, 1987). The numerous short, conical sharp-pointed teeth appear well suited for puncturing and dismembering the chitinous exoskeletons of insects, large numbers of which have already been reported from the Madygen Svita (overview in Dobruskina, 1995). The well developed shoulder girdle and proportions of the main forelimb elements suggest arboreal abilities (Sharov, 1970) and this is further indicated by the presence of elongate penultimate phalanges in the digits, which is a typical feature of climbers (Unwin, 1987, 1988).

The function of the dorsal appendages is uncertain. Sharov (1970) suggested that they might have functioned as parachutes, while Halstead (1975) and Haubold and Buffetaut (1987) went further and proposed that they could have been deployed in such a way as to form aerofoils, enabling the animal to glide from tree to tree. These authors argued that the overlapping plumes formed a flight surface which narrowed caudally, being broadest at the front, and could be raised and lowered by epaxial musculature.

Other functions are also possible. Sharov (1970) suggested that the appendages might have acted as insulation, by trapping air between the plumes, and Halstead (1975) supposed that they could have been erected as a crest and used to frighten predators. In a similar fashion, they would also have been very effective display structures, particularly if brightly coloured.

Phylogenetic relationships. Sharov (1970) assigned *Longisquama* to the Pseudosuchia on the grounds that it had an antorbital fenestra and a mandibular fenestra. These characters, if confirmed, would indicate that *Longisquama* belongs within the Archosauria, as others have accepted (e.g. Haubold and Buffetaut, 1987; Witmer, 1991), but they do not support any particular relationship with pseudosuchians (Charig *et al.*, 1976) or other archosaurian lineages. Benton (1993b, p. 698) listed *Longisquama* as an ornithodiran archosaur.

Assignment to Archosauria is problematic, however, since *Longisquama* is also said to have acrodont teeth and an ossified interclavicle, features that are more typical of lepidosaurs. Further, the key archosaur features are not entirely convincing: the antorbital fenestra is not certainly present (Figure 10.4), and the mandibular fenestra is, unusually, shown as located immediately below and behind the mandibular tooth row. It is neccesary to show that these two structures do not simply represent damage.

It has been argued by some that birds may be descended from 'pseudosuchians' and, as such, *Longisquama* has occasionally been incorporated into this hypothesis (see Witmer, 1991 for a review). Sharov (1970, 1971b) originated this idea by suggesting that the elongate scales represented an early stage in the evolution of feathers, an idea which was echoed by Halstead (1975) and Bakker (1975). Sharov (1970) also claimed that the clavicles resembled the avian furcula and may even be homologous with this structure. However, soon after *Longisquama* was first described, important new evidence supporting Huxley's contention (1868) that birds were descended from theropod dinosaurs began to emerge (see Witmer, 1991). In light of the widespread acceptance of this hypothesis, and the absence of any further evidence linking *Longisquama* and birds, the idea of *Longisquama* as an avian ancestor can now be safely abandoned.

Discussion

Diapsids have a good fossil record, but for taphonomic reasons much of it is dominated by medium to very large taxa. Small diapsids are relatively uncommon and examples with soft tissue preservation, such as those from Madygen, are extremely rare. The specimens are important geographically, since these are the only diapsids so far reported from the Triassic of Middle Asia. In addition, if confirmed as Middle–Late Triassic in age, *Sharovipteryx* and *Longisquama* also fall during an interval of major turnover among terrestrial faunas, when 'Palaeozoic' faunas of synapsids, rhynchosaurs, and basal archosaurs were replaced by 'modern' faunas of dinosaurs, pterosaurs, crocodylomorphs, basal lepidosaurs, turtles, and mammals (Benton, 1993a, 1994).

The Madygen diapsids have been widely ignored. Apart from two functional studies (Gans *et al.*, 1987;

Haubold and Buffetaut, 1987) and a few mentions in the semi-popular literature (e.g. Halstead, 1975, 1982, 1989; Cox, 1988) these taxa have been almost completely ignored and are not even listed in standard compendia such as Carroll (1988). New studies now under way (Unwin, in prep.) and greater ease of access to the original material may rectify this situation.

Acknowledgements

D.M.U. thanks N.N. Bakhurina, L.P. Tatarinov, V.U. Reshetov, M.F. Ivakhnenko, I.C. Barskov, A.S. Alekseev, and D. Yesin for all their help during his stay in Moscow. D.M.U. is grateful to the British Council for enabling him to carry out extended visits to the Former Soviet Union, and to M. Bird of the British Embassy, Moscow, for his assistance. Many thanks to R.R. Reisz, L.P. Tatarinov and M.F. Ivakhnenko for valuable discussion, to S. Powell (Department of Geology, Bristol University) for the photography, and to G.W. Storrs and M.F. Ivakhnenko for valuable comments on the manuscript. This research was supported by the Department of Geology, Bristol University, and a Royal Society University Research Fellowship awarded to D.M.U.

References

Bakker, R.T. 1975. Dinosaur renaissance. *Scientific American* **232**: 58–78.

Benton, M.J. 1985. Classification and phylogeny of the diapsid reptiles. *Zoological Journal of the Linnean Society* **84**: 97–164.

—1993a. Late Triassic extinctions and the origin of the dinosaurs. *Science* **260**: 769–770.

—1993b. Reptilia, pp. 681–715 in Benton, M.J. (ed.), *The Fossil Record 2*. London: Chapman & Hall.

—1994. Late Triassic to Middle Jurassic extinctions among continental tetrapods: testing the pattern, pp. 366–397 in Fraser, N.C. and Sues, H.-D. (eds.), *In the Shadow of the Dinosaurs*. Cambridge: Cambridge University Press.

—and Allen, J.A. 1997. *Boreopricea* from the Lower Triassic of Russia, and the relationships of the prolacertiform reptiles. *Palaeontology* **40**: 931–953.

—and Clark, J. 1988. Archosaur phylogeny and the relationships of the Crocodylia, pp. 295–338 in Benton, M.J. (ed.), *The Phylogeny and Classification of the Tetrapods. Systematics Association Special Volume*, 35A. Oxford: Clarendon Press.

Carroll, R. L. 1988. *Vertebrate Paleontology and Evolution.* New York: W.H. Freeman.

—and Currie, P.J. 1991. The early radiation of diapsid reptiles, pp. 354–424, in Schultze, H.-P. and Trueb, L. (eds.), *Origins of the Higher Groups of Tetrapods.* Ithaca, NY: Comstock.

Charig, A.J., Krebs, B., Sues, H.-D. and Westphal, F. 1976. *Handbuch der Paläoherpetologie. Teil 13, Thecodontia,* Stuttgart: Gustav Fischer, 137 pp.

Chatterjee, S. 1986. *Malerisaurus langstoni,* a new diapsid from the Triassic of Texas. *Journal of Vertebrate Paleontology* 6: 297–312.

Cowen, R. 1981. Homonyms of *Podopteryx. Journal of Paleontology* 55: 483.

Cox, C.B. 1988. Amphibians and reptiles, pp. 46–169 in Cox, C.B. (ed.), *Macmillan Illustrated Encyclopedia of Dinosaurs and Prehistoric Animals.* London: Guild Publishing.

Dobruskina, I.A. 1970. [The age of the Madygen Formation and the Permo-Triassic boundary in Middle Asia.] *Sovetskaya Geologiya* 1970 (12): 16–28.

—1976. [Correlation of the continental deposits of the Triassic.] *Sovetskaya Geologiya* 1976 (3): 34–45.

—1980. [Stratigraphic position of the Triassic plant-bearing beds of Eurasia.] *Trudy Paleontologicheskogo Instituta AN SSSR* 346: 1–160.

—1995. Keuper (Triassic) Flora from Middle Asia (Madygen, Southern Fergana). *Bulletin of the New Mexico Museum of Natural History and Science* 5: 1–49.

Evans, S.E. 1988. The early history and relationships of the Diapsida, pp. 221–260 in Benton, M.J. (ed.), *The Phylogeny and Classification of the Tetrapods. Systematics Association Special Volume*, 35A. Oxford: Clarendon Press.

Gans, C., Darevskii, I. and Tatarinov, L.P. 1987. *Sharovipteryx,* a reptilian glider? *Paleobiology* 13: 415–426.

Gauthier, J., Kluge, A.G. and Rowe, T. 1988. Amniote phylogeny and the importance of fossils. *Cladistics* 4: 105–209.

Gow, C.E. 1975. The morphology and relationships of

Youngina capensis Broom and *Prolacerta broomi* Parrington. *Palaeontologia Africana* 18: 89–131.

Halstead, L.B. 1975. *The Evolution and Ecology of the Dinosaurs.* London: Peter Lowe, 116 pp.

—1979. Pterosaurs, pp. 174–176 in Steel, R. and Harvey, A.P. (eds.), *The Encyclopaedia of Prehistoric Life.* London: Mitchell Beazely.

—1982. *Hunting the Past.* London: Hamish Hamilton, 208 pp.

—1989. *Dinosaurs and Prehistoric Life.* London: Collins, 240 pp.

Haubold, H. and Buffetaut, E. 1987. A new interpretation of *Longisquama insignis,* an enigmatic reptile from the Upper Triassic of Central Asia. *Comptes Rendus de l'Academie des Sciences Paris, Série II* 305: 65–70.

Huxley, T.H. 1868. On the animals which are most nearly intermediate between birds and reptiles. *Geological Magazine* 5: 357–365.

Ivakhnenko, M.F. 1978. Tailed amphibians from the Triassic and Jurassic of Middle Asia. *Paleontologicheskii Zhurnal* 1978 (3): 84–89.

—and Korabel'nikov, V.A. 1987. [*Life of the Past World.*] Moscow: Prosveshchenie Press, 253 pp.

Juul, L. 1994. The phylogeny of basal archosaurs. *Palaeontologia Africana* 31: 1–38.

Laurin, M. 1991. The osteology of a Lower Permian eosuchian from Texas and a review of diapsid phylogeny. *Zoological Journal of the Linnean Society* 101: 59–95.

Olsen, P.E. 1979. A new aquatic eosuchian from the Newark Supergroup (Late Triassic–Early Jurassic) of North Carolina and Virginia. *Postilla* 176: 1–14.

Selezneva, A.A. and Sychevskaya, E.K. 1989. [Triassic fishes from Madygen (Fergana).] *Byulleten' Moskovskogo Obshchestva Ispytatelei Prirody, Otdel Geologicheskii* 64: 131.

Sereno, P. C. 1991. Basal archosaurs: phylogenetic relationships and functional implications. *Journal of Vertebrate Paleontology* 11 (Supplement to no. 4): 1–53.

Sharov, A.G. 1966. [Unique discoveries of reptiles from Mesozoic beds of Central Asia.] *Byulleten' Moskovskogo Obshchestva Ispytatelei Prirody, Otdel Geologicheskii* 61: 145–146.

—1970. [Unusual reptile from the Lower Triassic of Fergana.] *Paleontologicheskii Zhurnal* 1970 (1): 127–131.

—1971a. [New flying reptiles from the Mesozoic of Kazakhstan and Kirgizstan.] *Trudy Paleontologicheskogo Instituta AN SSSR* 130: 104–113.

—1971b. Den Vorfähren der Vögel auf der Spur. *Presse Sowjetunion* 117: 5–6.

Tatarinov, L.P. 1980. [Towards a prehistory of mammals.] pp. 103–114 in Sokolov, B.S. (ed.), [*Palaeontology and Stratigraphy, 26th International Geological Congress*]. Moscow: Nauka.

—1989. [The systematic position and way of life of the problematic Upper Triassic reptile *Sharovipteryx mirabilis.*] *Paleontologicheskii Zhurnal* 1989 (2): 110–112.

—1994. Terrestrial vertebrates from the Triassic of the USSR with comments on the morphology of some reptiles, pp. 165–170 in Mazin, J.-M. and Pinna, G. (eds.), *Evolution, Ecology and Biogeography of the Triassic Reptiles. Paleontologia Lombarda, New Series,* 2.

Unwin, D. M. 1987. Pterosaur locomotion. Joggers or waddlers? *Nature* 327: 13–14.

—1988. New remains of the pterosaur *Dimorphodon* (Pterosauria: Rhamphorhynchoidea) and the terrestrial ability of early pterosaurs. *Modern Geology* 13, 57–68.

Vakhrameev, V.A., Dobruskina, I.A., Meien, S.V. and Zaklinskaya, E.D. 1978. *Paläozoische and Mesozoische Floren Eurasiens und die Phytogeographie dieser Zeit.* Jena: Gustav Fisher, 300 pp.

Vorob'eva, E.I. 1967. [Triassic *Ceratodus* from Southern Fergana with some comments on the systematics and phylogeny of ceratodontids.] *Paleontologicheskii Zhurnal* 1967 (4): 102–111.

Wellnhofer, P. 1991. *The Illustrated Encyclopedia of Pterosaurs.* London: Salamander, 192 pp.

Witmer, L.M. 1991. Perspectives on avian origins, pp. 427–466, in Schultze, H.-P. and Trueb, L. (eds.), *Origins of the Higher Groups of Tetrapods.* Ithaca, NY: Comstock.

11

Mesozoic marine reptiles of Russia and other former Soviet republics

GLENN W. STORRS, MAXIM S. ARKHANGEL'SKII, AND VLADIMIR M. EFIMOV

Introduction

Marine reptile remains have often been found in the extensive Mesozoic epicontinental marine units of the former Soviet Union, and in particular, the Russian Platform of the European part of Russia. These fossils include relatively common plesiosaurs, ichthyosaurs and mosasaurs, and rarer crocodilians and turtles. The Moscow Basin and the Ul'yanovsk, Samara, and Saratov regions of the Volga River Basin have been particularly productive. These areas not only contain large exposures of Upper Jurassic and Cretaceous sediments, but have proved relatively accessible to Russian workers over the years. Sadly, there has been little in-depth study and analysis of marine reptile fossils in the former Soviet Union in recent years, although a great many historical works have been devoted to them. In this article, we review the current state of knowledge of these interesting fossils and provide a basis for informed future study.

The quality of most Russian marine reptile holotypes is poor and most specimens have been recovered as float from the Volga and Moscow River Basins, or as chance occurrences in quarries and oil shale mines. Almost no deliberate excavations have been undertaken. Currently, the most productive Russian localities for plesiosaurs and ichthyosaurs are the phosphorite quarries near Voskresensk in the Moscow Region, quarries near the village of Sundokovo in Tataria, the vicinity of Kashpir in the Samara Region (Upper Jurassic to Aptian), the 30 km of Volga River shoreline from the village of Kriushi to Mordovo (Barremian to Albian), and the 24 km stretch of the Volga from Ul'yanovsk to Undory (Kimmeridgian to

Aptian). The phosphorites typically produce only fragmentary and disarticulated remains, but complete skeletons are common in the clays, shales, and marls of the Volga. There is good potential for discoveries in other quarries, but these are currently unmonitored. Historical localities of the Moscow Region generally no longer exist, as these quarries have been abandoned over time. The best potential for immediate results therefore lies with the excellent cliff-face exposures along the Volga.

Numerous Russian reports refer to localities in 'Povolzh'e' or 'Zavolzh'e'. These are Russian conventions for the Volga River Basin, Povolzh'e indicating the right bank of the river (Ul'yanovsk and Saratov shore; west bank), and Zavolzh'e the left (Samara shore; east bank). In the central Volga Basin near Ul'yanovsk, marine reptile remains have been found in all ammonite zones from the Callovian to the Albian. The greatest concentrations, however, occur in the Middle Volgian *Dorsoplanites panderi*, *Epivirgatites nikitini*, and Upper Volgian *Craspedites subdites* ammonite zones and in the Hauterivian *Speetoniceras versicolor* Zone. Uniform formational or 'svita' (suite) names have not been established generally for the Russian Platform marine rocks, and the use of local ammonite zones is preferred. The Russian Volgian essentially equals the Tithonian.

Lower Volga sediments producing ichthyosaurs, plesiosaurs, mosasaurs, and turtles in the Saratov and Penza Regions (Povolzh'e) are Senonian in age, typically Campanian and Maastrichtian. The precincts of the Serdoba River near the village of Malaya Serdoba, Penza, have been particularly rich in fragmentary remains of Campano-Maastrichtian age (pers. obs.,

M.S.A.). Jurassic and Cretaceous marine deposits containing fossil reptile remains are also found in some of the central Asian nations that were formerly republics of the Soviet Union, but these areas have been less well studied. Kazakhstan, Uzbekistan, and Azerbaijan are notable examples (Bazhanov, 1958; Rozhdestvenskii, 1973; Nesov and Krasovskaya, 1984; Glikman *et al.*, 1987).

Repository abbreviations

KGU, Geology and Mineralogy Museum, Kazan State University, Kazan'; MGRI, Moscow Geological Prospecting Institute, Vernadskii State Geological Museum, Moscow; PIN, Paleontological Institute, Russian Academy of Sciences, Moscow; PMK, Pugachev Regional Museum, Pugachev, Saratov region; POKM, Penza Regional Local History Museum, Penza, Penza Region; SGU, Paleontology Museum of the Department of Historical Geology and Paleontology, Saratov State University, Saratov; 'Simbirtsit', Paleontological collection of the 'Simbirtsit' Industrial Works (a free enterprise company), All-Russian Cultural Fund, Undory, Ul'yanovsk Region; TsGM, Central Geological Museum, St. Petersburg; UPM, Undory Palaeontological Musuem, Undory, Ul'yanovsk region; ZIN, Zoological Institute, Russian Academy of Sciences, St. Petersburg.

Systematic survey

DIAPSIDA Osborn, 1903
SAUROPTERYGIA Owen, 1860
PLESIOSAURIA de Blainville, 1835

Comments. Over two dozen plesiosaur species have been named, and numerous additional taxa recognized, from Russian sediments by Russian and German workers. Most of this effort was undertaken in the pre-revolutionary years of the nineteenth and early twentieth centuries (e.g. Fischer von Waldheim, 1845, 1846; Eichwald, 1865–1868; Kipriyanov, 1883; Ryabinin, 1909, 1915; Bogolyubov, 1911; Pravoslavlev, 1915, 1916). Zhuravlev (1941, 1943), Rozhdestvenskii

(1947), Menner (1948), Novozhilov (1948a, b, 1964), and Ochev (1976a, 1977), for example, have written on Russian plesiosaur remains more recently. Little study has been attempted in the past 20 years. Older workers (e.g. Bogolyubov, 1911; Menner, 1948) sometimes relied upon disassociated collections of bones for their type series. Where possible, holotypes in these cases have been selected on the basis of page priority.

Bogolyubov (1911) named numerous Russian plesiosaur species all of which, however, are indeterminate and must be considered *nomina dubia* (Table 11.1), even though some of these were retained by Pravoslavlev (1915) (see Welles, 1962). Indeed, virtually all of the 'species' unique to Russia are *nomina dubia*, and not a single complete skeleton has been described. Most are based upon isolated vertebrae and teeth and as such are non-diagnostic below the subordinal, or perhaps the familial level. Most, if not all, of Bogolyubov's (1911) holotypes, from the old museum of the Geological Cabinet of Moscow University, are now housed in MGRI. Some of these types have been identified (Table 11.1), but through neglect while under communist authority, parts of the collection are inaccessible. It is believed that a planned renovation project, now underway, will uncover the remaining specimens. Kipriyanov's (1882, 1883) material is presumably in the museum of the Academy of Sciences, St. Petersburg. The whereabouts of other collections, such as those of Eichwald and Fischer von Waldheim, are unknown at present.

Ostensibly, the genera *Cimoliasaurus, Colymbosaurus, Cryptoclidus, Elasmosaurus, Eretmosaurus, Georgiasaurus, Leutkesaurus, Liopleurodon, Muraenosaurus, Neopliosaurus, Peloneustes, Plesiosaurus, Pliosaurus, Polycotylus, Polyptychodon, 'Rhinosaurus', Scanisaurus, Simolestes, Spondylosaurus, Strongylokrotaphus,* and *Thaumatosaurus* are present in Russian rocks (Welles, 1962; Persson, 1963; Novozhilov, 1964). However, most generic identifications and assignments to previously known Western taxa have been based upon stratigraphical, rather than morphological information, with an historical readiness to name new species based upon geographic occurrence. Few of these identifications can be considered reliable. Indeed, most Western species

Table 11.1. *Compilation of plesiosaurian taxa based upon material from the former Soviet Union. Holotypes indicated by repository abbreviation or specimen number, where known*

Taxon	Holotype	Material	Locality	Horizon	Status
Colymbosaurus sklerodirus Bogolyubov, 1911	MGRI	fragmentary skeleton	Moscow Region	Volgian	Plesiosauria indet.
Cryptoclidus simbirskensis Bogolyubov, 1909	MGRI	vertebrae/limb frags.	Ul'yanovsk Region	Callovian–Oxfordian	Plesiosauria indet.
Elasmosaurus amalitskii Pravoslavlev, 1916	—	vertebral series	Don Region	Turonian	Elasmosauridae indet.
Elasmosaurus antiquus Dubeikovskii & Ochev, 1967	SGU 104a/17, 18, 19	cervical centra	Kama River Basin	Hauterivian	Elasmosauridae indet.
Elasmosaurus kurskensis Bogolyubov, 1911	MGRI [Kipr, 1882]	med. cervical centrum	Kursk Region	Cenomanian	Plesiosauria indet. [see Welles, 1962]
Elasmosaurus orskensis Bogolyubov, 1911	MGRI	cervical centra	Orenburg Region	Senonian	Elasmosauridae indet.
Elasmosaurus serdobensis Bogolyubov, 1911	MGRI	ant. cervical centrum	Penza Region	Campanian	Elasmosauridae indet.
?*Elasmosaurus sachalinensis* Ryabinin, 1915	—	phalanx	Sakhalin Island	Lower Senonian	Plesiosauria indet.
Eretmosaurus rzasnickii Menner, 1948	MGRI VI 61/1	cervical centrum	Vilyui River, Siberia	Middle Jurassic	Plesiosauria indet.
?*Eretmosaurus jakovlewi* Menner, 1948	MGRI VI 61/15	caudal centrum	Vilyui River, Siberia	Middle Jurassic	Plesiosauria indet.
Georgiasaurus (Georgia) penzensis (Ochev, 1976a)	POKM No. 11658	partial skull/skel.	Penza Region	Santonian	presumed valid
Leukesaurus [no sp.] Kipriyanov, 1883	ZIN?	teeth and vertebrae	Kursk Region	Cenomanian	?*Polyptychodon* sp.
Muraenosaurus elasmosauroides Bogolyubov, 1911	MGRI	cervical centrum	Moscow Region	Volgian	?Elasmosauridae indet.
Muraenosaurus kamensis Dubeikovskii & Ochev, 1967	SGU 104a/16 [lost]	cervical vertebrae	Kama River Basin	Volgian	?Elasmosauridae indet.
Muraenosaurus purbecki Bogolyubov, 1911	MGRI	centrum	Moscow Region	Volgian	Plesiosauria indet.
Neopliosaurus [no sp.] Sintzov, 1899	—	vertebrae/humeri	Penza Region	Senonian	?Polycotylidae indet.
Plesiosaurus belmerseni Kipriyanov, 1882 (emend. Bog. '11)	ZIN	cervical centrum	Penza Region	Campanian	Elasmosauridae indet. [see Persson, 1959]
Plesiosaurus nordmanni Eichwald, 1865	—	propodial fragment	Crimea	Neocomian	Ichthyosauria indet. [see Ryabinin, 1946b]
Pliosaurus giganteus Trautschold, 1860	—	tooth	Moscow Region	Oxfordian	*Liopleurodon* ?*ferox*
Pliosaurus rossicus Novozhilov, 1948a	PIN 304	partial skull/skel.	Chuvashia	Volgian	*Liopleurodon rossicus* [see Halstead, 1971]

Table 11.1. (*cont.*)

Taxon	Holotype	Material	Locality	Horizon	Status
Pliosaurus wosinskii Fischer von Waldheim, 1846	—	jaw fragment	Moscow Region	Kimmeridgian	?*Pliosaurus brachyspondylus*
Polycotylus brevispondylus Bogolyubov, 1911	MGRI	pectoral vertebra	unknown	Cenomanian	Polycotylidae indet.
Polycotylus donicus Pravoslavlev, 1915	—	cervical centra, etc.	Don Region	Senonian?	?Polycotylidae indet.
Polycotylus epigurgitis Bogolyubov, 1911	MGRI	posterodorsal centrum	Voronezh Region	Cenomanian	Plesiosauria indet.
Polycotylus ichthyospondylus var. *tanais* Bogolyubov, 1911	MGRI	vertebrae/propodial	Voronezh Region	Cenomanian	Polycotylidae indet.
Polycotylus orientalis Bogolyubov, 1911	MGRI (Sa109, 110 etc.)	centra/limb frags.	Orenburg Region	Senonian	?Polycotylidae indet.
Polycotylus ultimus Bogolyubov, 1911	MGRI	2 cerv. vertebrae	Penza Region	Campanian?	Polycotylidae indet.
Rhinosauriscus (*Rhinosaurus*) *jasykowii* (Fischer von Waldheim, 1847)	Original lost	skull	Ul'yanovsk Region?	Tatarian	Seymouriidae [see Rozhdestvenskii, 1973]
Scanisaurus (*Cimoliasaurus*) *nazarowi* (Bogolyubov, 1911)	MGRI	post. cervical centrum	Orenburg Region	Senonian	provisionally retained [see Persson, 1959]
Spondylosaurus fabrenkobli Fischer von Waldheim, 1846	—	vertebra	Moscow Region	Volgian	Plesiosauria indet.
Spondylosaurus frearsi Fischer von Waldheim, 1845	—	cervical centrum	Moscow Region	Kimmeridgian	?*Pliosaurus brachyspondylus*
Strongylokrotaphus (*Peloneustes*) *irgisensis* (Novozhilov, 1948a)	PIN 426	partial skull/skel.	Saratov Region	Volgian	*Pliosaurus irgisensis* [see Halstead, 1971]
Thaumatosaurus calloviensis Bogolyubov, 1911	MGRI	tooth	Moscow Region	Callovian	Pliosauridae indet. [but see Tarlo, 1960]
Thaumatosaurus mosquensis Kipriyanov, 1883	ZIN?	cervical vertebra	Moscow Region	Oxf.–Kimmeridgian	*Liopleurodon* ?*ferox*

Note:
Muraenosaurus kamensis and the ichthyosaur ?*Shastasaurus nordensis* inadvertently share the same catalogue number.

of 'Plesiosaurus' suffer from the same problem (Storrs, 1996) while Cimoliasaurus is apparently a nondiagnosable plesiosaur (?elasmosaur) and 'waste-basket taxon' (Williston, 1903). Fischer von Waldheim's (1847) Rhinosaurus, described as a plesiosaur, was based upon a Permian anthracosaur skull, the origin of which was perhaps misinterpreted through confusion within the containing collection, and is now known as Rhinosauriscus (Kabanov, 1959; Rozhdestvenskii, 1973). Rarer plesiosaur remains are known from some of the other former Soviet republics but most, if not all, of these are also generically indeterminate.

Few species of Russian plesiosaur are known from material of adequate quality to justify their retention (Table 11.1). Only three of these potentially distinct species are unique to Russia. The three include the best Russian specimens and are known from partial skulls, although they have been described only in a preliminary fashion, and a comprehensive review of each is required.

PLIOSAURIDAE Seeley, 1874a
Liopleurodon Sauvage, 1873

Diagnosis. Large pliosaur distinguished from other forms by relatively short and straight-sided mandibular symphysis bearing 5–7 pairs of teeth; dorsal aspect of symphysis tapers anteriorly to blunt V. This diagnosis conforms to the concept of Liopleurodon in Tarlo (1960), although the genus is founded upon a single tooth (Sauvage, 1873) that may prove problematic.

Liopleurodon rossicus (Novozhilov, 1948a)
See Figure 11.1.

Holotype and locality. PIN 304, most of a skull and a partial pectrum (scapulae and coracoid), with perhaps a few other elements associated; Buinsk Mine oil shales, Ibresi District, right bank of Volga (Povolzh'e), Autonomous Republic of Chuvashia, Russia.

Horizon. Middle Volgian (Dorsoplanites panderi Zone).

Comments. Originally described as 'Pliosaurus' rossicus Novozhilov, 1948a, this is a large, short-necked animal typical of the Pliosauridae. Most of what was once a complete skeleton was destroyed in 1938 during the process of mining the oil shales in which it was found,

the remainder being saved only by chance. Halstead (1971) described a number of additional bones as belonging to the holotype, but inadequate collection management practices have seemingly allowed confusion of at least some of the bones of this specimen with those of 'Strongylokrotaphus' irgisensis (discussed below). Ochev (pers. comm., 1995) insists that only the skull and pectrum of 'P.' rossicus were recovered. Novozhilov (1948a, b) described only the skull, and later (Novozhilov, 1964) figured the pectrum. The short dorsal blade and anteroventral ramus of the scapula suggest that the animal was immature. Halstead (1971) considered 'P.' rossicus to represent Liopleurodon largely on the basis of its short mandibular symphysis. It has, however, trihedral tooth cross-sections unlike the Oxfordian Liopleurodon ferox, but similar to Pliosaurus, a genus with a significantly longer symphysis. Tarlo's (1959) Kimmeridgian genus with trihedrally sectioned teeth and a short symphysis, Stretosaurus, was based upon incorrectly identified material and is hence invalid, much of the referred material probably belonging to Liopleurodon (Halstead, 1989). Pending new reviews of Pliosaurus and Liopleurodon, Halstead's interpretation of L. rossicus is accepted here. Points worth noting are that the bar between the naris and orbit is not as narrow as is suggested in the drawings in Novozhilov (1948a, 1964) and the naris itself is much smaller than shown. Furthermore, contrary to Novozhilov (1948a), there is no nasal bone or lacrimal.

Halstead (1971) also equated the enormous rostrum of 'Pliosaurus cf. P. grandis' (PIN 2440/1), described by Rozhdestvenskii (1947) from the left Volga bank (Zavolzh'e), Ozink Mine, Saratov Province, D. panderi Zone, with L. rossicus. This specimen too, was discovered complete, but only the rostrum, the proximal end of a humerus, a phalanx, and some rib fragments were saved from the mining operations (Rozhdestvenskii, 1947). The complete hind limb noted by Halstead (1971) undoubtedly belongs to 'Strongylokrotaphus' (PIN 426), as figured by Zhuravlev (1943) and Novozhilov (1964). The mandibular symphysis of PIN 2440/1 contains approximately six sharply trihedral teeth per ramus.

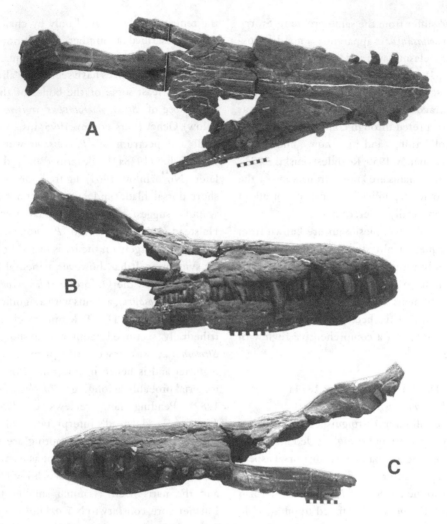

Figure 11.1. *Liopleurodon* (*Pliosaurus*) *rossicus* (Novozhilov, 1948a), PIN 304/1, from the middle Volgian oil shales of Buinsk Mine (*Dorsoplanites panderi* Zone), Ibresi District, Chuvashia. Skull and mandible in dorsal (A), right lateral (B), and left lateral (C) views. Scale bars, 100 mm. Compare with Novozhilov (1964).

Pliosaurus Owen, 1842

Diagnosis. Very large pliosaurid up to approximately 10 m, distinguished by long mandibular symphysis and trihedrally sectioned tooth crowns; approximately 11 pairs of caniniform teeth within symphysis; labial edges of mandibular symphysis approximately parallel.

Pliosaurus irgisensis (Novozhilov, 1948a)
See Figure 11.2.

Holotype and locality. PIN 426, a large partial skull, a partial vertebral column, and an articulated hind limb; oil shales of Savel'evsk Mine No. 1, near Gorny, about 35 km southwest of Pugachev, eastern Saratov Province, Russia.

Horizon. Savel'evsk oil shales, Volgian.

Comments. Most commonly known in Russian palae-ontological circles as '*Strongylokrotaphus*' *irgisensis*, this fossil was found during mining operations in 1933 (Zhuravlev, 1941, 1943; Novozhilov, 1948a, 1964). The

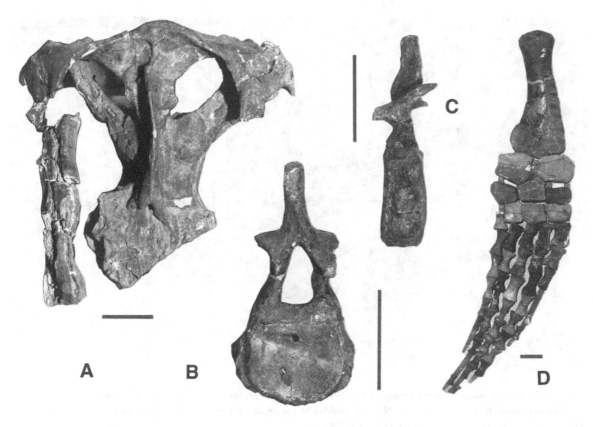

Figure 11.2. *Pliosaurus* (*Strongylokrotaphus*) *irgisensis* (Novozhilov, 1948a), PIN 426, from the Volgian oil shales of Savel'evsk Mine No. 1, near Gornyi, Pugachev District, Saratov Region. A, Skull roof, occiput, and right mandibular ramus in dorsal aspect (posterior at top). B, C, Anterior cervical vertebra in anterior and left lateral views, respectively. D, Right hind limb. Scale bars, 100 mm. Compare with Novozhilov (1964).

skeleton, as formerly exhibited in both the Pugachev Museum and PIN, was heavily restored and is less well preserved than appears in previously published photographs (Zhuravlev, 1941, 1943). All elements are now greatly affected by pyrite decay and the skull and mandible, in particular, are nearing destruction unless emergency conservation measures are soon employed, an unlikely scenario under present constraints.

Tarlo (1960) equated this taxon with *Pliosaurus*, noting that Novozhilov's (1948a) original assignment of this very large specimen to the relatively small *Peloneustes* was seemingly only on the basis of its elongated rostrum (of which only the extreme anterior tip, found in place, is preserved), and a misunderstanding of the characters of *Pliosaurus*. Indeed, *contra*

Novozhilov (1964), there appears little to differentiate the species from *Pliosaurus*, and his creation of a unique genus (*Stongylokrotaphus*) appears unjustified.

The cervical centra of *Pliosaurus irgisensis* are large and blocky, but very short, as in *Pliosaurus*, and their ribs are distinctly double-headed as is typical for pliosaurids. Novozhilov's (1964) placement of this species in the Polycotylidae (= 'Trinacromeriidae') is entirely inappropriate. The figures in Novozhilov (1948a, 1964) are misleading; the ovate 'fenestrae' seen lateral to the posterior ends of the parietal crest represent an oblique dorsal view through the distorted posttemporal fenestrae. The posterior interpterygoid vacuities are not visible dorsally. The apparent 'foramina' in the occiput of Novozhilov's (1964) figure represent the

Figure 11.3. *Georgiasaurus (Georgia) penzensis* (Ochev, 1976a), POKM 11658, from a building stone quarry in Santonian rocks at Zatolokino, Bekovo District, Penza Region. Natural mould of braincase, dorsal surface of palate, and rostrum. Total length of skull approximately 800 mm. Compare with Ochev (1976a).

splayed halves of the crushed foramen magnum with the exoccipitals forced laterally. There is no lacrimal preserved. Indeed, if once present, the orbits and anterior rami of the squamosals are now lost. Neither is Novozhilov's (1964) figure of the right hind limb entirely accurate, as comparison with Figure 11.2 indicates.

The specimen has been cited as one of the few known pliosaurs with gut contents, i.e. associated cephalopod hooklets (Zhuravlev, 1943; Gekker and Gekker, 1955). Zhuravlev (1943) also notes the association of an entire fish, shark teeth, and sand or gravel. Surely the shark teeth are the result of scavenging of the carcass by these animals, but the current state of conservation of this fossil reptile prevents any additional comment. Apparently, none of the associated remains has been preserved in the collections.

POLYCOTYLIDAE Williston, 1908
Georgiasaurus Ochev, 1977
See Figure 11.3.

Diagnosis. Relatively large polycotylid pliosaur distinguished by its size (4–5 m) and relatively elongate rostrum; details of palate and braincase noted by Ochev (1976a) may also be significant.

Georgiasaurus penzensis (Ochev, 1976a)

Holotype and locality. POKM 11658, the natural moulds of a skull and partial skeleton from a building stone quarry in the Penza Region, Russia.

Horizon. Upper Cretaceous (Santonian).

Comments. Georgiasaurus Ochev, 1977 replaces the preoccupied *Georgia* Ochev, 1976a. The snout of *Georgiasaurus penzensis* is narrow and elongate and the cervical ribs are single-headed as in other polycotylids. Ochev (1976a) estimates the total length of the animal in life at 4–5 m, relatively large for most polycotylids. Although the holotype material is poorly preserved, this taxon is provisionally retained on the basis of the adequate description by Ochev (1976a).

Other plesiosaur records

The primary contribution of Russian plesiosaur material is the supplementary information that it provides on the stratigraphic and geographic occurrence of the group, and particularly for suprageneric-level taxa. The earliest sauropterygian material from Russia was reported by Lazurkin and Ochev (1968) from the Ladinian of the left bank of the Taas-Krest River, Lena River Basin, near the Laptev Sea, northern Russia. This fossil, an isolated dorsal centrum (SGU 104a/15, now in PIN), was believed to represent a 'nothosaur', identified as *Nothosaurus*, but the presence of distinct foramina subcentralia, rare and irregular in stem-group nothosauriforms, suggests that the bone represents a primitive plesiosaur similar to *Pistosaurus* (Storrs, 1991). A similar centrum (PIN 4466/14) has been reported by Novikov (1993) from the late Ladinian of the Pechora River Basin, Bolshaya Synya tributary, as ?*Nothosaurus*.

Plesiosaurs are also extremely rare from the Lower Jurassic of Russia with but a single report of a vertebra from the Middle Liassic of the Jurung-Tumus Peninsula, Nordwick Bay, of the north Siberian Arctic (Ryabinin, 1939), but in the uppermost Middle Jurassic, and especially the Upper Jurassic, they have been widely encountered. The so-called cryptoclidids, a possible family of mesodiran, small-headed, many-toothed forms, have been reported from the Callovian–Oxfordian of Gorodishche near Ul'yanovsk (Bogolyubov, 1909) and the Vologda Region, and may exist in the middle Volgian of the Moscow Region (pers. obs., G.W.S., V.M.E.). Plesiomorphic elasmosaurs, reported as '*Muraenosaurus*' and '*Colymbosaurus*', are known in the Callovian of the Ryazan' Region, the Callovian or Oxfordian of Yaroslavl', and the Volgian of Moscow (especially Mnevniki and Schukino), the Kama River

Basin, and Ples in the Kostroma Region (Bogolyubov, 1911; Dubeikovskii and Ochev, 1967). Volgian to Valanginian records of '*Muraenosaurus*' come from Moscow and Orenburg, but these identifications have not been confirmed.

Pliosaurids, of the sort common in the English Upper Jurassic and represented in Russia by *Liopleurodon rossicus* and *Pliosaurus irgisensis*, are known from several localities in addition to those listed above. '*Pliosaurus*', perhaps representing both of the above taxa, has been reported from the Moscow Basin Oxfordian, Callovian, Kimmeridgian, and Volgian (Trautschold, 1860; Ryabinin, 1909; Bogolyubov, 1911), the Ryazan' Region Callovian, and the Volgian of the Samara, Ural'sk, western Kazakhstan, and Chuvashia regions (Rozhdestvenskii, 1947; Novozhilov, 1948a; Bazhanov, 1958). So-called *Peloneustes* is known from the Vologda–Vyatka railway (Callovian or Oxfordian), Kostroma Region, and the Upper Jurassic of Hooker Island, Franz-Josef Land (Ryabinin, 1909, 1936, 1939). Pliosaurids are also known from the Oxfordian of the Unzha River in the Kostroma region. '*Thaumatosaurus*' has been reported from the Callovian to Kimmeridgian of the Moscow Region (Bogolyubov, 1911), but *Thaumatosaurus* is based on a single tooth from the German Jurassic and is a *nomen dubium*. '*Polyptychodon*' has been reported from the Cenomanian of Kursk and of Podolia, Ukraine (Kipriyanov, 1883; Bogolyubov, 1911; Hofstein, 1961; Rozhdestvenskii, 1973).

Undoubted members of the Elasmosauridae occur in the Cenomanian of the Kursk and Saratov Regions and the Saratov Turonian and Campanian, but none is complete enough for generic identification (Kipriyanov, 1882; Pravoslavlev, 1918). Senonian elasmosaurs have come from the eastern Pre-Urals (Novokhatskii, 1954) and from the regions of Penza and Orenburg (Chkalov), around Malaya Serdoba (Campanian) and Konoplyanka, respectively (Bogolyubov, 1911, 1912; Pravoslavlev, 1918), and a possible elasmosaur was identified by Ryabinin (1915) from the lower Senonian of Sakhalin Island. Maastrichtian occurrences at the Adzhat River in north Kazakhstan and the Mugai River in the eastern

Pre-Urals were noted by Novokhatskii (1954). Upper Cretaceous (undifferentiated) elasmosaurs have been found along the Liska River, Don Region (Pravoslavlev, 1914, 1916, 1918), and in the Guberlin Mountains of the southern Pre-Urals (Rozhdestvenskii, 1973). Elasmosaur occurrences from the Russian Lower Cretaceous are known from the Hauterivian Vyatka-Kama phosphate deposit (Dubeikovskii and Ochev, 1967) and the Aptian of the Kara Sea (Ryabinin, 1939). Possible 'cimoliasaurs', or perhaps stem-group elasmosaurs, are reputed to occur in the Volgian of the Moscow Basin at Schukino and Tartarovo, the Albian? of the Moscow Basin (Berezniki), and the Senonian of Orenburg-Chkalov (Konoplyanka, Guberli), and the undifferentiated Upper Cretaceous glauconitic sandstone of the Charkov Region (Ryabinin, 1909; Bogolyubov, 1911).

Including *Georgiasaurus*, small, longirostrine 'pliosaurs', or polycotylids, have been identified in the Cenomanian of the Voronezh (Devitcha), Saratov, Volgograd, and Kursk regions, the Turonian of Saratov and the Tadzhik Depression of Uzbekistan (Talkhab Svita; Dzhalilov *et al.*, 1986), the Santonian of Penza (Ochev, 1976a), and the Senonian of Saratov (Campanian), Penza (Malaya Serdoba; Campanian) and Orenburg-Chkalov (Konoplyanka) (Bogolyubov, 1911, 1912; Sintsov, 1899). They are also known in the undifferentiated Upper Cretaceous of the Don Region's Liska River (Pravoslavlev, 1914), the Guberlin Mountains (Rozhdestvenskii, 1973) and the Lysogorsk District of the Saratov Region (Ochev, 1976a). Most of these fragmentary specimens have been identified as *Polycotylus*, but are generically indeterminate.

In addition, indeterminate plesiosaurs are known from the Upper Jurassic of the Vilyui River in Siberia (Menner, 1948), the Volgian of east Siberia, near Zhigansk, the polar Urals of the Lapinsk Region (Yakovlev, 1903; Ryabinin, 1939), and the Priural'sk District of west Kazakhstan (Bazhanov, 1958). They are also known from the undifferentiated Lower Cretaceous of the southwestern part of the Gissar Mountains near Baisun (Rozhdestvenskii, 1973), the Valanginian of the Lyapin area, north Urals (Ryabinin,

1939), and the Albian of Saratov (M.S.A., pers. obs.). In the Volga Basin near Ul'yanovsk, plesiosaurs range from Callovian to Aptian.

The Russian evidence of plesiosaur stratigraphic distribution agrees well with that suggested by occurrences in Western Europe, Australia, and the Americas. It appears that the Elasmosauridae, with possible representatives in the Late Jurassic, continued to exist until the latest part of the Cretaceous, as did the polycotylids, although the latter did not make their appearance until the latest Early Cretaceous. Cryptoclidids, if a legitimate clade, appear restricted to the Late Jurassic in Russia and elsewhere, although perhaps surviving locally into the Early Cretaceous. The giant Pliosauridae thrived in the Late Jurassic, but continued into the Early Cretaceous, seemingly becoming extinct soon thereafter.

ICHTHYOSAURIA de Blainville, 1835

Comments. The first Russian ichthyosaur remains were noted by Yazikhov (1832). Ichthyosaurs are commonly encountered in the Upper Jurassic, largely Oxfordian to Volgian, and Lower to lowermost Upper Cretaceous (Cenomanian) rocks of the Russian Platform. As elsewhere, their remains are more common in rocks of these ages than are those of plesiosaurs, and better material is available. Unfortunately, similar taxonomic chaos also exists. Over a dozen ichthyosaur species have been named from Russian specimens, and most have proved to be indeterminate. Likewise, historical works (e.g. Kipriyanov, 1881) have also made use of supposed type series with the implicit, but unproven, assumption of specific identity. The problematic taxa are included with potentially valid species in Table 11.2. Nesov *et al.* (1988) provide a stratigraphically organized list of ichthyosaur localities for the former Soviet Union, and Triassic records may be found in Callaway and Massare (1989).

Few remains of Russian ichthyosaurs are known from the Triassic and these are all of poor quality. Ryabinin (1946a) attributed a Norian vertebra from Kolymskii to *Shastasaurus* (*S. sieversi*), but Tatarinov (1964) believed that it could more probably be

Table 11.2. *Compilation of ichthyosaurian taxa based upon material from the former Soviet Union. Holotypes indicated by repository abbreviation or specimen number, where known*

Taxon	Holotype	Material	Locality	Horizon	Status
Brachypterygius zburavlevi Arkhangel'skii, 1998a	PIN 426/60–76	partial fore limb	Saratov Region	middle Volgian	?*Otschevia* sp.
Ichthyosaurus kurskensis Eichwald, 1853	—	teeth and vertebrae	Kursk Region	Neocomian	Ichthyosauria indet.
Ichthyosaurus nasimovii Fahrenkohl, 1856	—	2 vertebrae	Moscow Region	Volgian	Plesiosauria indet.
Ichthyosaurus steleodon Bogolyubov, 1909	MGRI	frag. jaws/vertebrae	Ul'yanovsk Region	Hauteriv.–Barremian	?*Platypterygius steleodon*
Ichthyosaurus volgensis Kazanskii, 1903	KGU	fragmentary skeleton	Ul'yanovsk Region	Volgian	?*Ophthalmosaurus* sp.
Khudiakovia callovensis Arkhangel'skii, 1999a	SGU 104a/27	partial fore limb	Saratov Region	Callovian	?*Ophthalmosaurus* sp.
Myopterygius kiprijanoffi Romer, 1968	ZIN? [Kipr., 1881]	partial skull	Kursk Region	up. Albian–Cenom.	*Platypterygius kiprijanoffi* [McGowan, 1972]
Ophthalmosaurus undorensis Efimov, 1991	'Simbirtsit' No. 140	fragmentary skeleton	Ul'yanovsk Region	upper Kimmerigian	presumed valid
Otschevia pseudoscythica Efimov, 1998	UPM No. 3/100	partial skeleton	Ul'yanovsk Region	lower Volgian	presumed valid
Paraophthalmosaurus saratoviensis Arkhangel'skii, 1998a	PMK No. 2836	frag. postcrania	Saratov Region	middle Volgian	*Paraophthalmosaurus* sp.
Paraophthalmosaurus saveljeviensis Arkhangel'skii, 1997	SGU 104a/23	partial skull/skeleton	Saratov Zavolzh'e	lower Volgian	provisionally retained
Platypterygius bannovkensis Arkhangel'skii, 1998b	SGU 104a/24	partial skull	Saratov Region	middle Cenomanian	provisionally retained
Plesiosaurus nordmanni Eichwald, 1865	—	propodial fragment	Crimea	Neocomian	Ichthyosauria indet. [Ryabinin, 1946b]
Plutoniosaurus bedengensis Efimov, 1997	UPM No. 2/740	skull/partial skeleton	Ul'yanovsk Region	upper Hauterivian	*Platypterygius bedengensis*
?*Shastasaurus nordensis* Polubotko & Ochev, 1972	SGU 104a/16	partial vert. series	Dyugadyak River Basin	Ladinian	Mixosauridae indet.
Shastasaurus sieversi Ryabinin, 1946a	—	vertebra	Iganii River Basin	Norian	Ichthyosauria indet.
Simbirskiasaurus birjukovi Ochev & Efimov, 1985	SGU 104a/22	frag. skull/skeleton	Ul'yanovsk Region	Hauterivian	presumed valid
Undorosaurus gorodischensis Efimov, 1996b	UPM EP-II-20 (527)	fragmentary skeleton	Ul'yanovsk Region	middle Volgian	presumed valid
Undorosaurus nesovi Efimov, 1996b	UPM EP-II-24 (785)	fragmentary skeleton	Ul'yanovsk Region	middle Volgian	*Undorosaurus gorodischensis*

Table 11.2. (cont.)

Taxon	Holotype	Material	Locality	Horizon	Status
Undorosaurus khorlovensis Efimov, 1999b	UPM EP–II–27 (870)	fragmentary skeleton	Moscow Region	middle Volgian	*Undorosaurus* sp.
Yasykovia yasykovi Efimov, 1999a	UPM EP–II–7 (1235)	fragmentary skeleton	Ul'yanovsk Region	upper Volgian	*Paraophthalmosaurus yasykovi*
Yasykovia mittai Efimov, 1999a	UPM EP–II–12 (4–M)	coracoids/ rt. scapula	Moscow Region	upper Volgian	Ichthyosauria indet.
Yasykovia sumini Efimov, 1999a	UPM EP–II–11 (3–M)	fragmentary skeleton	Moscow Region	upper Volgian	*Paraophthalmosaurus yasykovi*
Yasykovia kabanovi Efimov, 1999a	UPM EP–II–8 (1076)	fragmentary skeleton	Ul'yanovsk Region	middle Volgian	*Paraophthalmosaurus yasykovi*

Note:
? *Shastasaurus nordensis* and the plesiosaur *Muraenosaurus kamensis* inadvertently share the same catalogue number.

assigned to *Mixosaurus*. Callaway and Massare (1989) determined that this was not a mixosaur and that the species was a *nomen dubium*, reassigning the specimen to Shastasauridae *incertae sedis*. Even this assignment is too specific and the material is nondiagnostic beyond Ichthyosauria. Similarly, fragmentary specimens from northeastern Russia (Okhotsk Sea, Russkaya River, etc.) of the Carnian and Norian to Rhaetian discussed by Ochev and Polubotko (1964) are generically indeterminate. The Ladinian and Carnian Omolon Massif specimens of Polubotko and Ochev (1972) on the other hand, may represent *Mixosaurus* (their *?Shastasaurus nordensis*; Table 11.2) and *Cymbospondylus*, respectively (Callaway, pers. comm. 1996), although Mazin (1983) considered the former Mixosauridae *incertae sedis*. Nesov *et al.* (1988) also note problematic material ('*Shastasaurus* and *Mixosaurus*') from the Norian of the Magadan Region. Many very fragmentary individuals have been found in the undifferentiated Triassic of Chukotka District not far from Alaska (V.M.E.). The earliest potential ichthyosaurian occurrences in the former Soviet Union may be indeterminate Anisian and Induan specimens from the Omolon Massif, Volgograd Region (Ochev, 1976b; Polubotko and Ochev, 1972), although Shishkin and Lozovskii (1979) note indeterminate material from the 'Lower Triassic' of Russkii Island, as well as from the 'Middle Triassic' *Daonella* Shales of the Amur River estuary.

Although only a few ichthyosaur elements are known from the Lower and Middle Jurassic (e.g. Toarcian and Bajocian to Bathonian of Yakutia), Upper Jurassic finds abound. These latter are again primarily of Moscow and Volga Basin origin (Arkhangel'skii, 1999b; Arkhangel'skii *et al.* 1997), although indeterminate remains are known from Volgian sediments in western Kazakhstan (Bazhanov, 1958), Ul'yanovsk (Kabanov, 1959; Efimov, 1987), Moscow (Orlov, 1968), Saratov, and the Syzran' and Ukhta Regions (Nesov *et al.*, 1988). Various remains attributed to *Ichthyosaurus* have been reported from the Kimmeridgian of the Pechora River Basin (Ryabinin, 1912) and the Volgian of Moscow (Fahrenkohl, 1856; Trautschold, 1861) and Ul'yanovsk (Fischer von

Waldheim, 1847; Kazanskii, 1903), but it is unlikely that any of these specimens truly represent *Ichthyosaurus*, a Lower Jurassic genus long treated as a 'waste-basket taxon'. Indeed, the English holotype of *I. trigonus* Owen has been transferred (Huene, 1922) to *Macropterygius*, and *I. platyodon* Conybeare to *Temnodontosaurus* by McGowan (1974). Nevertheless, the Russian specimens assigned to these species (Ryabinin, 1912; Fischer von Waldheim, 1847, respectively) do not allow confident identification at even the generic level. *Ichthyosaurus nasimovii* Fahrenkohl, 1856 was founded upon two indeterminate plesiosaur vertebrae and is thus a *nomen dubium* (Eichwald, 1868; Kipriyanov, 1883; Bogolyubov, 1911).

ICHTHYOSAURIDAE Bonaparte, 1841
Ophthalmosaurus Seeley, 1874b

Diagnosis. Relatively large 'latipinnate' ichthyosaur, up to 4 m, characterized by extremely large orbits and very weakly developed or absent dentition; teeth certainly present only in anterior tips of jaws; humerus distinguished by presence of 3 distal facets of subequal size.

Ophthalmosaurus undorensis Efimov, 1991

Holotype and locality. 'Simbirtsit' 140, fragmentary skeleton; Volga River shore, found in 1982 near the village of Malye Undory, Ul'yanovsk Region, Russia.

Horizon. Aulacostephanus mutabilis Foraminiferan Zone, Upper Kimmeridgian.

Comments. Efimov (1991) described this species on the basis of a very fragmentary skeleton. Nevertheless, the character of the humerus clearly indicates that it belongs to *Ophthalmosaurus*. The material was considered to represent a new species partly on the basis of the very short vertebrae and on inferred differences in the scapula and forelimb between it and *O. icenicus* (Efimov, 1991). Provisionally retained here, this species is poorly characterized and requires additional study.

A number of Volgian-age ichthyosaur specimens perhaps referrable to *Ophthalmosaurus* are known. Moscow Region specimens (*?Ophthalmosaurus* sp.) include those noticed by Bogolyubov (1910a) and

Mitta (1984). Bogolyubov (1910a) also noted, but did not describe, a partial *Ophthalmosaurus* skeleton from the Volga River between Ul'yanovsk and Gorodische, which he provisionally assigned to the English species *O. icenicus.* Zhuravlev (1941, 1943) identified *Ophthalmosaurus* sp. in the Middle Volgian *Dorsoplanites panderi* and *Virgatites virgatus* zones of the Savel'evsk oil shale mines near Pugachev, but these may be misidentified (M.S.A.). Another unique Russian species, *Ichthyosaurus volgensis* Kazanskii, 1903 is based upon a fragmentary skeleton from the *V. virgatites* zone near Ul'yanovsk that, although probably specifically indeterminate, may represent *Ophthalmosaurus* (but see below). Indeed, Bogolyubov (1910a) assigned *I. volgensis* to the 'ophthalmosaurs'. Several species with possible 'ophthalmosaur' relationships have been described by Arkhangel'skii (1997, 1998a, 1999a) and Efimov (1999a). Lastly, a possible *Ophthalmosaurus* vertebra has come from undifferentiated rock on the left bank of the Volga near Kashpir and Syzran' (Bogolyubov, 1909).

Paraophthalmosaurus Arkhangel'skii, 1997
Diagnosis. Small ichthyosaur 2–2.5 m long with high skull, large orbits, and reduce dentition. Propodials with three distal facets as in *Ophthalmosaurus* but accessory facet reportedly bearing basal element of first digit rather than fifth.

Paraophthalmosaurus saveljeviensis Arkhangel'skii, 1997
Holotype and locality. SGU 104a/23, skull and limited anterior postcrania; brick factory spoil heap, Gornyi, eastern Saratov Province.
Horizon. lower Volgian.
Comments. Material variously assigned to *Ophthalmosaurus* and *Paraophthalmosaurus* (Arkhangel'skii, 1997, 1998a, 1999b; Bogolyubov, 1910a; Efimov, 1991; Zhuravlev, 1941, 1943) requires complete reexamination as the suspicion exists that the type of *Paraophthalmosaurus*, forelimb digit identifications notwithstanding, may merely represent a small or juvenile *Ophthalmosaurus.* However, until such time as proper comparisons are made, *P. saveljeviensis* is here retained. A second species, *P. saratoviensis*

Arkhangel'skii 1998a, is based upon insufficient material for diagnosis and is considered *Paraophthalmosaurus* sp.

Paraophthalmosaurus yasykovi (Efimov 1999a)
Holotype and locality. UPM No. EP-II-7 (1235), skull and partial skeleton; Detskii (Children's) Sanatorium area, Volga River, Ul'yanovsk Province.
Horizon. Craspedites subdites Zone, upper Volgian.
Comments. Described by Efimov (1999a) as a new genus, *Yasykovia*, this material is considered a junior synonym of *Paraophthalmosaurus* by Arkhangel'skii (1999b). The species is provisionally retained here. Three other specimens described by Efimov (1999a) as new species of *Yasykovia* are based upon suspect material, two of which are synonymized with *P. yasykovi* (see Table 11.2).

Undorosaurus Efimov, 1999b
Diagnosis. Moderate to large sized ichthyosaur (4–6 m) with massive vertebrae and stout ribs; broad, pentadactyl forelimbs with three humeral distal facets. Seemingly distinguished from *Ophthalmosaurus* by its large, strong teeth.

Undorosaurus gorodischensis Efimov, 1999b
Holotype and locality. UPM No. EP-II-20 (527), partial skeleton; Volga River at Gorodishche, Ul'yanovsk Province.
Horizon. Epivirgatites nikitini Zone, middle Volgian.
Comments. Contra Efimov (1999b), this taxon probably bears a close relationship with *Ophthalmosaurus* as evidenced by its relatively broad forelimb and three distal facets on the humerus. However, the stout dentition, in particular, argues that this may represent a valid taxon which is here retained. Other species of *Undorosaurus* described by Efimov (1999b) are likely variants of *Undorosaurus gorodischensis* or are *Undorosaurus* sp. (See Table 11.2).

LEPTOPTERYGIIDAE Kuhn, 1934
Platypterygius Huene, 1922
Diagnosis. 'Longipinnate' ichthyosaur of large size; skull usually over 1 m long in adults; full complement

of stout teeth; very long and slender rostrum and maxilla; relatively small orbit. See McGowan (1972) for full list of characteristics.

Platypterygius bannovkensis Arkhangel'skii, 1998b
Holotype and locality. SGU 104a/24, partial skull; Volga River right bank near Nizhnaya (Lower) Bannovka, Krasnoarmeisk District, Saratov Province.
Horizon. Middle Cenomanian.
Comments. Based upon only the rostrum and antorbital region of a deformed skull, the snout of this ichthyosaur appears quite low and long. It is not clear that the material is adequate for diagnosis, however, although the subgeneric name *Pervushoviasaurus* has been coined for this fossil (Arkhangel'skii, 1998b). Pending full revision of the Russian *Platypterygius* material, the species is tentatively retained.

Platypterygius bedengensis (Efimov, 1997)
Holotype and locality. UPM No. 2/740, skull and partial skeleton; Volga River right bank, 12 km north of Ul'yanovsk, Ul'yanovsk Province.
Horizon. *Speetoniceras versicolor* Zone, upper Hauterivian.
Comments. From the same stratigraphic horizon as *Platypterygius (Simbirskiasaurus) birjukovi* (Ochev and Efimov, 1985), Efimov (1997) distinguished this species largely by its different "narial" region. However, doubt exists regarding the true nature of the preorbital openings in *P. birjukovi* (see below). Well described by Efimov (1997) as the type of a new genus, *Plutoniosaurus*, this species was transferred to *Platypterygius* by Arkhangel'skii (1999b). It is provisionally retained here.

Platypterygius birjukovi (Ochev and Efimov, 1985)
Holotype and locality. SGU 104a/22, fragmentary skull and vertebral column; Volga River bank, between the Zakhar'evsk Mine and the Detskii (Children's) Sanatorium, 25 km north of Ul'yanovsk, Ul'yanovsk Province.
Horizon. *Simbirskites* Clay, upper Hauterivian.
Comments. The Cretaceous of Russia has its share of ichthyosaurs, ranging from the Berriasian to the Cenomanian, with numerous isolated occurrences in the upper Albian (Nesov *et al.*, 1988; Bardet, 1992). *Platypterygius (Simbirskiasaurus) birjukovi* (Simbirsk is the pre-revolutionary name of Ul'yanovsk) is known from only a single partial skeleton. The rostrum is elongate with the premaxillae extending slightly beyond the dentaries. If the fenestrae between the naris and orbit are not artefacts, these are notable characteristics. Ochev and Efimov (1985) suggest that they may have been associated with preorbital salt glands. The skull is high, but the orbits are stated to be subquadrate in shape. The maxilla is excluded from the external naris by the premaxilla and the jugal does not reach the premaxilla. Arkhangel'skii (1999b) places this species within *Platypterygius*.

Platypterygius kiprijanoffi (Romer, 1968)
See Figure 11.4.
Holotype and locality. A skull, the whereabouts of which are unknown; it is perhaps to be found in ZIN. Some postcranial elements may be associated; Kursk Region.
Horizon. Seversk Sandstone, 'Kursk Osteolite', upper Albian or Cenomanian.
Comments. *Platypterygius kiprijanoffi* (Romer, 1968) was originally described by Kipriyanov (1881) as *Ichthyosaurus campylodon* on the basis of a varied collection of material (Figure 11.4). Romer (1968) reassigned the material to *Myopterygius* (as had Huene, 1922), provided the new specific name, and designated the skull in the original Kipriyanov collection as the holotype without the benefit of a catalogue number or identified repository. McGowan (1972) and Kuhn (1946) have included the species within *Platypterygius*. This taxon has also been reported from the Lebedinsky and Stoylensky mines (upper Albian and Cenomanian) near Gubkin, Belgorod Region, and may be present in rocks of similar age in the Stanislavsk and Saratov Regions (Nesov, 1983; Nesov *et al.*, 1988; Rozhdestvenskii *et al.*, 1987).

Ichthyosaurus steleodon Bogolyubov, 1909 was based upon fragmentary jaws and vertebrae from the Ul'yanovsk Hauterivian or Barremian. This species is not attributable to *Ichthyosaurus*, but may also represent *Platypterygius* (Nesov *et al.*, 1988). Tatarinov (1964)

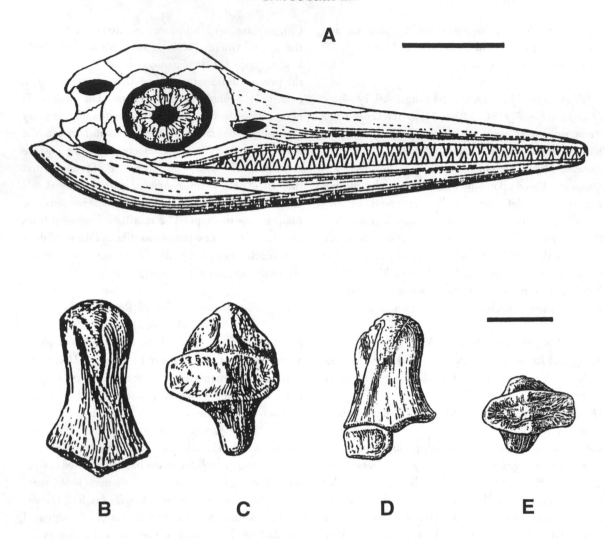

Figure 11.4. *Platypterygius* (*Myopterygius*) *kiprijanoffi* (Romer, 1968) from the upper Albian or Cenomanian Seversk Sandstone of Kursk (after Kipriyanov, 1881). A, Reconstructed skull and mandible. B, Right humerus in ventral (palmar) aspect. C, Right humerus in proximal aspect. D, left femur (with fibula) in ventral aspect. E, Left femur in proximal aspect. Scale bars, 150 mm.

suggested an affinity with *Myopterygius*, but there is scepticism about the validity of this genus (McGowan, 1972), most material now being referred to *Platypterygius*.

STENOPTERYGIIDAE Kuhn, 1934
Otschevia Efimov, 1998
Diagnosis. 'Longipinnate' ichthyosaur of moderate to large size with curved, stout teeth; epipodial and mesopodial bones robust; intermedium large and wedges between radius and ulna to narrowly contact humerus.

Otschevia pseudoscythica Efimov, 1998
Holotype and locality. UPM No. 3/100, a partial skull and skeleton with good forelimbs; Volga River right bank, 8 km east of Novaya Bedenga, Ul'yanovsk Province.
Horizon. Ilowaiskaya pseudoscythica Zone, lower Volgian.

Comments. The seemingly unique forelimb suggests that this is a distinct taxon with a relationship to the Stenopterygiidae. *Brachypterygius zhuravlevi* Arkhangel'skii, 1998a has been referred to *Otschevia* by Arkhangel'skii (1999b), although represented by extremely poor material. Arkhangel'skii (1999b) has also considered *Ichthyosaurus volgensis* Kazanskii, 1903 to represent *Otschevia*.

Other ichthyosaur records

Indeterminate ichthyosaur remains have been found in the Lower Cretaceous of Azerbaijan (Aptian) (Bogachev, 1962), and the Saratov, Samara (?Valanginian), Kirov (Valanginian), and Penza (Albian) Regions (Nesov *et al.*, 1988). *Plesiosaurus nordmani* Eichwald, 1865 is based upon a limb fragment from the Crimean (Biasala) Neocomian (Berriasian?; Bogolyubov, 1911) that Ryabinin (1946b) identified as ichthyosaurian. In the Upper Cretaceous, ichthyosaurs are known from Sakhalin Island, and from the Cenomanian of Saratov, the L'vov (Vinnitsa, Ukraine) and Tambov Regions, Cherkassia (Nesov *et al.*, 1988), and Moldavia (Moldova) (Sukhov, 1950). The Smolensk Region has produced material from undifferentiated Cretaceous sediments (Nesov *et al.*, 1988).

SQUAMATA Oppel, 1811
ANGUIMORPHA Fürbringer, 1900
MOSASAURIDAE Gervais, 1853

Comments. Mosasaurs of the former Soviet Union are known particularly from sediments of the Lower Volga Basin in Russia (Saratov and Penza Regions) and range in age from the Campanian to the Maastrichtian (Yarkov, 1993). The genus *Mosasaurus* has been reported from fragmentary remains, primarily vertebrae and teeth, from the Volga Region but also from the basins of the Don, Ural and Pechora rivers, Turgai, and Azerbaijan (Yakovlev, 1902, 1905; Khozatskii and Yur'ev, 1964). Not all of these reports are reliable beyond Mosasauridae indet. The one uniquely Russian species of *Mosasaurus, M. donicus* Pravoslavlev, 1914, being extremely fragmentary, cannot be diagnosed on

the basis of the available material. It is Campanian in age and comes from the Liska River Basin, Don Region. *Plioplatecarpus* is questionably present in Povolzh'e (Tsaregradskii, 1927; Khozatskii and Yur'ev, 1964; Yarkov, 1993). *Tylosaurus* (*Liodon*) *rhiphalus* (Bogolyubov, 1910b) was described from the Campanian of Orenburg Province, but is also probably not diagnostic of a valid taxon. *Tylosaurus* has also been reported from the Campanian of Saratov, Volgograd, and the Konoplanka River near Orsk, and from the Crimean Maastrichtian (Gorbakh, 1967). *Prognathodon* is thought to be present in the Campanian of Saratov and Penza, and in the Maastrichtian of the Crimea and Penza (Malaya Serdoba) (M.S.A.). *Platecarpus* and *Clidastes* reputedly have been found near Volgograd (Yarkov, 1993), but these reports are considered problematic at best. Similarly, *Clidastes* is questionably reported from the Campanian of Saratov and Penza (M.S.A.). Indeterminate mosasaur remains have also been found in the Lugansk and Serov regions, the eastern slope of the Urals, Karakalpakia, western Azerbaijan, and Kazakhstan (Novokhatskii, 1954; Gabunia, 1958; Rozhdestvenskii, 1973; Kovaleva *et al.*, 1982; Yarkov, 1993).

Dollosaurus Yakovlev, 1901
See Figure 11.5.

Diagnosis. A plioplatecarpine mosasaur characterized by no premaxillary rostrum anterior to the dentition; recurved maxillary teeth with single carina on back edge only; concave alveolar rim of dentary; strong symphysis; very large first two dentary teeth; transversely broad cervical and anterior dorsal vertebrae; well developed zygosphenes and zygantra; co-ossified chevrons (Yakovlev, 1901; Tsaregradskii, 1935).

Dollosaurus lutugini Yakovlev, 1901

Holotype and locality. A fragmentary skull and skeleton (whereabouts unknown); Donets River, Voroshilovgrad Region, Ukraine.
Horizon. Belemnitella mucronata Beds, Upper Campanian.
Comments. This species represents the only genus of mosasaur that has been described solely on the basis of

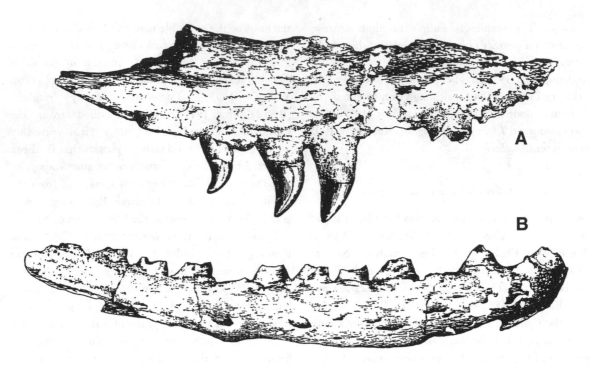

Figure 11.5. *Dollosaurus lutugini* Yakovlev, 1901 from the upper Campanian of the Ukraine (Voroshilovgrad Region) (after Tsaregradskii, 1935). A, Right maxilla. B, Right dentary in labial aspect. Estimated length of dentary 300 mm; maxilla to scale.

material from the former Soviet Union. Both Dollo (1924) and Russell (1967) accept *Dollosaurus* as a valid genus and have allied this animal with the Plioplatecarpinae. Russell (1967) considers it close to *Prognathodon*.

SERPENTES Linnaeus, 1758

Comment. Strictly marine fossil snakes are unknown from Russia, but a single occurrence of a primitive simoliophid is known from Albian– Cenomanian lagoonal deposits in Kzyl-Kum (Khodzhakul Village), Central Asia (Glikman *et al*, 1987).

ARCHOSAURIA Cope, 1869
CROCODYLIFORMES Benton and Clark, 1988

Comments. Few marine crocodyliform (crocodilian of traditional usage) remains are known from Russia and its former republics. Rare thalattosuchian fossils are known from the Russian Platform but are largely undescribed. For example, a vertebra and fifth meta-

tarsal of a metriorhynchid, perhaps *Dakosaurus*, were recovered from the uppermost Jurassic or lowermost Cretaceous of Khoroshevskii Island in the middle Volga Region (Ochev, 1981), and an indeterminate thalattosuchian has come from the Kimmeridgian of the Moscow Region (Efimov and Chkhikvadze, 1987). An unpublished partial thalattosuchian is also known from the Volgian of Gorodishche near Ul'yanovsk (V.M.E.). In Asia, teleosaurs have been reported, such as *Steneosaurus* from the Jurassic (Aalenian) Karakhskaya Svita of South Dagestan (Efimov, 1978), while *Teleosaurus* reputedly occurs in the uppermost Jurassic or lowermost Cretaceous of the Alaiskii Ridge, Fergana Basin (Efimov and Chkhikvadze, 1987). *Poekilopleuron schmidti*, from the Cenomanian of the Kursk Region, was described by Kipriyanov (1883) as a marine crocodile, but Efimov and Chkhikvadze (1987) believe that these remains represent a dinosaur.

Marine thoracosaurine remains are also rare in

Russia, but *Thoracosaurus* has been described from Maastrichtian marine rocks of the Inkerman Mines of the Crimea, near Sevastopol' (Borisyak, 1913; Ryabinin, 1946b), and Efimov and Chkhikvadze (1987) note a possible Maastrichtian thoracosaur from the Crimean village of Skalistoe. The remains described by Borisyak (1913) as *T. macrorhynchus* include a well preserved skull that Steel (1973) has assigned to *T. scanicus*, and Efimov (1988) to *T. borissiaki* (see Chapter 20, this volume). A possible, fragmentary thoracosaur has also come from the marine Cretaceous of Malaya Serdoba in the Penza Region (pers. comm., L.S. Glikman).

ANAPSIDA Williston, 1917
TESTUDINES Linnaeus, 1758

Comments. Marine turtles are rare and very poorly known in Russia and are only little more frequently seen in the rocks of the Soviet Union's former Asian republics. All known occurrences are Cretaceous. In Russia, the few remains of sea turtles are largely from the Upper Cretaceous sediments of the Lower Volga Basin (Campanian to Maastrichtian), such as Malaya Serdoba in the Penza Region. However, fragmentary chelosphargine protostegids are reported from the Aptian? to Cenomanian Lebedinsk and Soilensk quarries of the Belgorod Region (Nesov, 1985).

All the Asian localities represent abnormal marine conditions (deltaic to lagoonal), and have produced reports almost exclusively of toxochelyids. *Kirgizemys exaratus* Nesov and Khozatskii, 1978 is from the Albian Alamyshak Svita of Kirgizstan (ZIN T/F 491), and *Anatolemys maximus* (PIN 2398/501) and *A. oxensis* (ZIN RNT S 74–1) were described by Khozatskii and Nesov (1979) from the Turonian– Santonian of West Fergana and the upper part of the Khodzhakul Svita (Cenomanian) of Karakalpakia, respectively. *Oxemys gutta* Nesov, 1977 is from the Turonian (Beshtubinskaya Svita) of Lake Khodzhakul, Karakalpakia (TsGM 2/11478). The Kzyl-Kum Region of Karakalpakia and Uzbekistan has also produced remains of *Anatolemys* and *Kirgizemys* in Albian? to Turonian rocks (Nesov and Krasovskaya, 1984). A single desmatochelyid is known from the Santonian–Campanian of Uzbekistan (Glikman *et al.*, 1987).

Summary

Although the fossil record of Russian marine reptiles is currently poor, with very few well preserved specimens in existence, many fragmentary remains have been found. It is obvious that the vast territories of Russia and of her former satellite republics represent unrealized potential for the future discovery of important specimens. Fossil localities are concentrated in the central and southern parts of the Russian Platform (Jurassic and Cretaceous) and in the Asiatic former republics (Cretaceous), and reflect the palaeobiogeography of the times (Rozhdestvenskii, 1973). As now known, the marine reptiles of Russia comprise four to six families of plesiosaur (Pliosauridae, Elasmosauridae, Polycotylidae, Pistosauridae, ?Cryptoclididae, ?Cimoliasauridae), four families of ichthyosaur (Mixosauridae, Shastasauridae, Ichthyosauridae, Leptopterygiidae), perhaps two or three sub-families of Mosasauridae (Mosasaurinae, Plioplatecarpinae, ?Tylosaurinae), three families of turtle (Protostegidae, Toxochelyidae, Desmatocheyidae) and three families of crocodilian (Metriorhynchidae, Teleosauridae, Crocodylidae – Thoracosaurinae). A single snake (Simoliophidae) is known from lagoonal deposits.

Although few named Russian species may be regarded as valid, a picture of marine reptile stratigraphic distribution is emerging which correlates well with, and complements that suggested elsewhere. Plesiosaurs and ichthyosaurs are common in the Upper Jurassic and Lower Cretaceous. Plesiosaurs and mosasaurs dominate the Upper Cretaceous after the apparent extinction of ichthyosaurs in the Cenomanian. By far the greatest number of marine reptile remains have been recovered from Upper Jurassic rocks, followed by deposits of Senonian age.

Acknowledgements

The authors thank Mike Benton and David Unwin for their leadership in the establishment of the Joint

Moscow–Bristol Palaeontological Initiative which spawned this project and their invitation to contribute to this book. We are also indebted to Professor Vitalii Ochev for his enthusiastic help and guidance in the pursuit and completion of our studies. William Collier and Simon Powell provided photographic assistance, and we thank Mike Taylor and the late Jack Callaway for their review comments. G.W.S. thanks the Royal Society of London and the Natural Environment Research Council of the UK for financial support and the University of Bristol Department of Earth Sciences and the Museum of Natural History & Science at Cincinnati Museum Center for the use of their facilities.

References

Arkhangel'skii, M.S. 1997. [On a new genus of ichthyosaurs from the lower Volgian substage of the Saratov Zavolzh'e.] *Paleontologicheskii Zhurnal* 1997: 87–91.

—1998a. [On the remains of ichthyosaurs from the Volgian Stage of the Saratov Zavolzh'e.] *Paleontologicheskii Zhurnal* 1998: 87–91.

—1998b. [On the genus of ichthyosaurs *Platypterygius*.] *Paleontologicheskii Zhurnal* 1998: 65–69.

—1999a. [On an ichthyosaur from the Callovian Stage of the Saratov Povolzh'e.] *Paleontologicheskii Zhurnal* 1999: 88–91.

—1999b. [On the evolution of the skeleton of the fore fin of ichthyosaurs and the phylogeny of the group.] *Voprosy Paleontologii i Stratigrafii* 1999: 20–37.

—, Ivanov, A.V. and Popov, Y.V. 1997. [On the first reliable discovery of remains of the ichthyosaur *Platypterygius* in lower Aptian deposits of Povolzh'e.] *Uchenie Zapiski Geologicheskogo Faculteta Saratovskogo Gosudarstvennogo Universiteta* 1997: 57–59.

Bardet, N. 1992. Stratigraphic evidence for the extinction of the ichthyosaurs. *Terra Nova* 4: 649–656.

Bazhanov, V.S. 1958. [Concerning a pliosaur and ichthyosaur from the Upper Jurassic of western Kazakhstan.] *Materialy po Istorii Fauny i Flory Kazakhstana* 2: 72–76.

Benton, M.J. and Clark, J.M. 1988. Archosaur phylogeny and the relationships of the Crocodylia, pp. 295–338 in Benton, M.J. (ed.) *The Phylogeny and Classification of the Tetrapods. Volume 1. Amphibians, Reptiles, Birds.*

Systematics Association Special Volume 35A. Oxford: Clarendon.

Blainville, H.D. de 1835. Description de quelques espèces de reptiles de la Californie, précédée de l'analyse d'un système général d'Erpetologie et d'Amphibiologie. *Nouvelles Annales du Muséum (National) d'Histoire Naturelle, Paris* 3: 233–296.

Bogachev, V.V. 1962. [Discovery of Cretaceous ichthyosaurs in the Caucasus.] *Priroda* 1962: 119.

Bogolyubov, N.N. 1909. Sur quelques restes de deux reptiles (*Cryptoclidus simbirskensis* n. sp. et *Ichthyosaurus steleodon* n. sp.), trouvés par M. le Profes. A. P. Pavlow sur les bords de la Volga dans les couches mésozoiques de Simbirsk. *Annuaire Géologique et Minéralogique de la Russie* 11: 42–64.

—1910a. Sur les ichthyosaures portlandiens. *Bulletin de l'Académie Impériale des Sciences, St. Petersburg* 4: 469–476.

—1910b. Sur le restes des mosasauriens trouvés dans le gouvernement d'Orenburg. *Annuaire Géologique et Minéralogique de la Russie* 11: 42–64.

—1911. [*On the History of Plesiosaurs in Russia.*] Moscow: Imperial Moscow University Press.

—1912. Sur le présence de l'*Elasmosaurus* et du *Polycotylus* dans les dépôts de la Russie. *Annuaire Géologique et Minéralogique de la Russie* 14: 174–176.

Bonaparte, C.L. 1841. A new systematic arrangement of vertebrated animals. *Transactions of the Linnean Society of London* 18: 247–304.

Borisyak, A.A. 1913. Sur les restes d'un crocodile de l'étage supérieur du Crétacé de la Crimée. *Bulletin de l'Académie Impériale des Sciences, St. Pétersburg* 7: 555–558.

Callaway, J. and Massare, J. 1989. *Shastasaurus altispinus* (Ichthyosauria, Shastasauridae) from the Upper Triassic of the El Antimonio district, northwestern Sonora, Mexico. *Journal of Paleontology* 63: 930–939.

Cope, E.D. 1869. Synopsis of the extinct Batrachia, Reptilia and Aves of North America. *Transactions of the American Philosophical Society* 14: 1–252.

Dollo, L. 1924. *Globidens alabamaensis*, mosasaurien américain retrouvé dans le Craie d'Obourg (Sénonien supérieur) du Hainaut, et les mosasauriens de la Belgique en général. *Archives de Biologie* 34: 167–213.

Dubeikovskii, S.G. and Ochev, V.G. 1967. [On the remains of plesiosaurs from the Jurassic and Cretaceous deposits of the basin of the upper course of the River Kama.] *Voprosy Geologii Yuzhnogo Urala i Povolzh'ya* **4**: 97–103.

Dzhalilov, M.P., Holtman, E.V. and Khakimov, F.Kh. 1986. [First data on the find of bones of a Late Cretaceous pliosaurid (Sauropterygia, Reptilia) in the southwestern spur of the Gissar Range.] *Doklady AN Tadzhikskoi SSR* **29**: 553–556.

Efimov, M.B. 1978. [Survey of the fossil crocodiles of the USSR.] [Abstract] *Byulleten' Moskovskogo Obshchestva Ispytatelei Prirody, Otdel Geologicheskii* **53**: 157.

—1988. [Fossil crocodiles and champsosaurs of Mongolia and the USSR.] *Trudy Sovmestnoi Sovetsko–Mongol'skoi Paleontologicheskoi Ekspeditsii* **36**, 1–108.

—and Chkhikvadze, V.M. 1987. [Survey of the finds of fossil crocodiles in the USSR.] *Izvestiya AN Gruzinskoi SSR, Seriya Biologicheskaya* **13**: 200–207.

Efimov, V.M. 1987. [Marine reptiles in Mesozoic deposits of the Ul'yanovsk Region.] *Kraevedcheskie Zapiski Ul'yanovskogo Oblastnogo Kraevedcheskogo Myzeya* **7**: 60–66.

—1991. [On the first discovery of the ichthyosaur *Ophthalmosaurus* in Kimmeridgian deposits of the USSR.] *Paleontologicheskii Zhurnal* **1991**: 112–114.

—1997. [A new genus of ichthyosaurs from the Lower Cretaceous of the Ul'yanovsk Povolzh'e] *Paleontologicheskii Zhurnal* **1997**: 77–82.

—1998. [An ichthyosaur *Otschevia pseudoscythica* gen. et sp. nov. from Upper Jurassic deposits of the Ul'yanovsk Povolzh'e.] *Paleontologicheskii Zhurnal* **1998**: 82–86.

—1999a. [Ichthyosaurs of the new genus *Yasykovia* from the Upper Jurassic deposits of European Russia.] *Paleontologicheskii Zhurnal* **1999**: 92–100.

—1999b. [Ichthyosaurs of the Family Undorosauridae fam. nov. from Volgian Stage Upper Jurassic deposits of the European part of Russia] *Paleontologicheskii Zhurnal* **1999**: 54–61.

Eichwald, E.I. 1853. Einige palaeontologische Bemerkungen über den Eisensand von Kursk. *Bulletin de la Société Impériale des Naturalistes de Moscou* **2**: 209–231.

—1865–1868. *Lethaea Rossica ou Paléontologie de la Russie.* Stuttgart.

Fahrenkohl, A. 1856. Flüchtiger Blick auf die Bergkalk- und Jura-Bildung in der Umgebung Moskwas. *Zapiski, Vserossiiskoe Mineralogicheskoe Obshchestva* **1856**: 219–236.

Fischer von Waldheim, G.F. 1845. Notice sur le *Spondylosaurus*, genre de saurien fossiles de l'Oolithe de Moscou. *Bulletin de la Société Impériale des Naturalistes de Moscou* **18**: 343–351.

—1846. Notice sur quelques sauriens fossiles du Gouvernement de Moscou. *Bulletin de la Société Impériale des Naturalistes de Moscou* **19**: 90–107.

—1847. Notice sur quelques sauriens de l'Oolithe de Gouvernement de Simbirsk. *Bulletin de la Société Impériale des Naturalistes de Moscou* **20**: 362–370.

Fürbringer, M. 1900. *Beitrag zur Systematik und Genealogie der Reptilien.* Jena: Gustav Fischer.

Gabunia, L.K. 1958. [On the remains of mosasaurs from Upper Cretaceous deposits of the Caucasus.] *Soobshcheniya AN Gruzinskoi SSR* **20**: 561–564.

Gekker, E.L. and Gekker, R.F. 1955. [Remains of Teuthoidea from the White Jura and Lower Cretaceous of the Volga region.] *Voprosy Paleontologii* **1955** (2): 36–44.

Gervais, P. 1853. Observations relatives aux reptiles fossiles de France. *Comptes-rendus Hébdomadaires des Séances de l'Academie des Sciences, Paris* **36**: 374–377, 470–474.

Glikman, A.L., Mertinene, R.A., Nesov, L.A., Rozhdestvenskii, A.K., Khozatskii, L.I. and Yakovlev, V.N. 1987. [*Stratigraphy of the USSR. Cretaceous System. 2. Vertebrates.*] Moscow: Nedra.

Gorbakh, L.G. 1967. [First discovery of the remains of a mosasaur in the Crimea.] *Geologichnii Zhurnal* **27**: 93–96.

Halstead, L.B. 1971. *Liopleurodon rossicus* (Novozhilov) – a pliosaur from the lower Volgian of the Moscow Basin. *Palaeontology* **14**: 566–570.

—1989. Plesiosaur locomotion. *Journal of the Geological Society, London* **146**: 37–40.

Hofstein, I.D. 1961. [A tooth of a plesiosaur and fish from Cenomanian deposits of Podolia.] *Paleontologicheskii Sbornik L'vovskoi Geologicheskoi Obschestva pri L'vovskogo Universiteta* No. 1.

Huene, F. von 1922. *Die Ichthyosaurier des Lias und ihre Zusammenhänge.* Berlin: Gebrüder Borntraeger.

Kabanov, K.A. 1959. [Burial of Jurassic and Cretaceous reptiles in the region of Ulyanovsk.] *Izvestiya Kazanskogo Filiala AN SSSR, Seriya Geologicheskikh Nauk* **7**: 211–214.

Kazanskii, P. 1903. Ueber die *Ichthyosaurus*-Knochen aus dem Sysranischen Kreise des Gouvernement

Simbirsk. *Trudy Obshchestva Estestvoispytatelei Kazanskogo Universiteta* **37**: 1–33.

Khozatskii, L.I. and Nesov, L.A. 1979. [Large turtles from the Late Cretaceous of Central Asia.] *Trudy Zoologicheskogo Instituta AN SSSR* **89**: 98–108.

—and Yur'ev, K.B. 1964. [Family Mosasauridae.] pp. 475–481 in *Osnovy Paleontologii* 12. Rozhdestvenskii, A.K. and Tatarinov, L.P. (eds.), pp. 475–481. Moscow: Nauka.

Kipriyanov, W.A. 1881. Studien über die Fossilen Reptilien Russlands. I. Theil. Gattung *Ichthyosaurus* König aus dem Sewerischen Sandstein oder Osteolith der Kreidegruppe. *Mémoires de l'Académie Impériale des Sciences St. Pétersburg* (7) **28**: 1–103.

—1882. Studien über die Fossilen Reptilien Russlands. II. Theil. Gattung *Plesiosaurus* Conybeare aus dem Sewerischen Sandstein oder Osteolith der Kreidegruppe. *Mémoires de l'Académie Impériale des Sciences St. Pétersburg* (7) **30**: 1–57.

—1883. Studien über die Fossilen Reptilien Russlands. III. Theil. Gruppe Thaumatosauria n. aus der Kreide-Formation und dem Moskauer Jura. *Mémoires de l'Académie Impériale des Sciences St. Pétersburg* (7) **31**: 1–29.

Kovaleva, N.P., Nesov, L.A. and Favorskaya, T.A. 1982. [On the discovery of the remains of a gigantic marine lizard – a mosasaur in the Upper Cretaceous of Karakalpakia.] *Ezhegodnik Vsesoyuznogo Paleontologicheskogo Obshchestva* **25**: 262–265.

Kuhn, O. 1934. Ichthyosauria. *Fossilium Catalogus I: Animalia* **63**: 1–75.

—1946. Ein Skelett von *Ichthyosaurus* (*Platypterygius*) *hercynicus* n. sp. aus dem Aptium von Gitter. *Berichte der Naturforschenden Gesellschaft, Bamberg* **29**: 69–82.

Lazurkin, D.V. and Ochev, V.G. 1968. [First discovery of sauropterygian remains in the Triassic of the USSR.] *Paleontologicheskii Zhurnal* **1968**: 141–142.

Linnaeus, C. 1758. *Systema Naturae*. Stockholm: Laurentii Salvii.

Mazin, J.-M. 1983. Répartition stratigraphique et géographique des Mixosauria (Ichthyopterygia). Provincialité marine au Trias Moyen. In *Actes du Symposium Paléontologique Georges Cuvier*, Buffetaut, E., Mazin, J.-M. and Salmon, E. (eds.), pp. 375–387. Montbéliard.

McGowan, C. 1972. The systematics of Cretaceous ichthyosaurs with particular reference to the material from North America. *Contributions to Geology* **11**: 9–29.

—1974. A revision of the longipinnate ichthyosaurs of the Lower Jurassic of England, with descriptions of two new species. *Life Science Contributions, Royal Ontario Museum* **97**: 1–37.

Menner, V.V. 1948. [Remains of plesiosaurs from Middle Jurassic deposits of eastern Siberia.] *Trudy Instituta Geologicheskikh Nauk* **98**: 1–50.

Mitta, V.V. 1984. [On new discoveries of ichthyosaurs and plesiosaurs in the Moscow Region.] *Byulleten' Moskovskogo Obshchestva Ispytatelei Prirody, Otdel Geologicheskii* **59**: 131

Nesov, L.A. 1977. [On some special constructions of the skull of two Late Cretaceous turtles.] *Vestnik Leningradskogo Universiteta. Seriya 3, Biologiya* **4**: 45–48.

—1983. [Discovery of a jaw of a bony fish of the Family Aspidorhynchidae in Cretaceous deposits of the Belgorod Region.] *Ezhegodnik Vsesoyuznogo Paleontologicheskogo Obshchestva* **26**: 309–312.

—1985. [New data on the Late Cretaceous vertebrates of the USSR.] *Voprosy Gerpetologii* **1985**: 148–149.

—, Ivanov, A.O. and Khozatskii, L.I. 1988. [On the finds of the remains of ichthyosaurs in the USSR and the problem of faunal change in the middle Cretaceous.] *Vestnik Leningradskogo Universiteta. Seriya 7, Geologiya, Geografiya* **1988**: 15–25.

—and Khozatsky, L.I. 1978. [A turtle from the Early Cretaceous of Kirgizstan.] *Ezhegodnik Vsesoyuznogo Paleontologicheskogo Obshchestva* **21**: 267–279.

—and Krasovskaya, T.B. 1984. [Transformations in the complexes of turtles of the Cretaceous of Central Asia.] *Vestnik Leningradskogo Universiteta, Seriya 3, Biologiya* **1**: 15–25.

Novikov, I.V. 1993. Triassic tetrapod assemblages of the Timan–North Urals region. In *The Nonmarine Triassic*. Lucas, S.G. and Morales, M. (eds.), pp. 371–373. Albuquerque: New Mexico Museum of Natural History and Science.

Novokhatskii, I.P. 1954. [On the discovery of the remains of vertebrates in the Cretaceous deposits of the eastern Pre-Urals.] *Izvestiya AN Kazakhskoi SSR* **18**: 146–147.

Novozhilov, N.I. 1948a. [Two new pliosaurs from the lower Volgian tier of the Volga region (right bank).] *Doklady AN SSSR* **60**: 115–118.

—1948b. [On some peculiarities of the construction of the parietal bones in the Pliosauridae.] *Doklady AN SSSR* **60**: 285–288.

—1964. [Superfamily Pliosauroidea.] In *Osnovy Paleontologii* 12, Rozhdestvenskii, A.K. and Tatarinov, L.P. (eds.), pp. 327–332. Moscow: Nauka.

Ochev, V.G. 1976a. [A new pliosaur from the Upper Cretaceous of the Penza Region.] *Paleontologicheskii Zhurnal* 1976: 135–138.

—1976b. [An unusual tooth from the Lower Triassic of Donskaya Luka.] *Izvestiya Vysshikh Uchebnykh Zavedenii (Geologiya i Razvedka)* 19: 176–177.

—1977. [A substitution for the preoccupied name *Georgia penzensis*.] *Paleontologicheskii Zhurnal* 1977: 118.

—1981. [Marine crocodiles in the Mesozoic of Povolzh'e.] *Priroda* 1981: 103.

—and Efimov, V.M. 1985. [A new genus of ichthyosaur from the Ul'yanovsk Povolzh'e.] *Paleontologicheskii Zhurnal* 1985: 76–80.

—and Polubotko, I.V. 1964. [New finds of ichthyosaurs in the Triassic of the north-eastern USSR.] *Izvestiya Vysshikh Uchebnykh Zavedenii (Geologiya i Razvedka)* 1964: 50–55.

Oppel, M. 1811. *Die Ordnungen, Familien und Gattungen der Reptilien.* München.

Orlov, Yu. A. 1968. [*In the World of Ancient Animals.*] Moscow: Nauka.

Owen, R. 1842. Report on British fossil reptiles. Part II. *Report of the British Association for the Advancement of Science* 1841: 60–204.

Persson, P.O. 1959. Reptiles from the Senonian (U. Cret.) of Scania (S. Sweden). *Arkiv för Mineralogi och Geologi* 2: 431–478.

—1963. A revision of the classification of the Plesiosauria with a synopsis of the stratigraphical and geographical distribution of the group. *Lunds Universitets Årsskrift* 59: 1–60.

Polubotko, I.V. and Ochev, V.G. 1972. [New discoveries of ichthyosaurs in the Triassic of the northeastern USSR and some remarks on the conditions of their preservation.] *Izvestiya Vysshykh Uchebnykh Zavedenii (Geologiya i Razvedka)* 1972: 36–42.

Pravoslavlev, P.A. 1914. Restes d'un mosasaurien trouvé dans le Crétacé supérieur du bassin de la rivière Liski, Province du Don. *Izvestiya Alekseevskogo Donskogo Politekhnicheskogo Instituta* 3: 168–183. [In Russian with French résumé.]

—1915. Restes d'un jeune *Plesiosaurus* trouvés dans le Crétacé supérieur du bassin de la rivière Liski, Province du Don. *Annuaire Géologique et Minéralogique de la Russie* 17: 1–18.

—1916. [Elasmosaurs from the Upper Cretaceous deposits of the Don Region.] *Trudy Sankt-Peterburgskogo Obshchestva Estestvoispytatelei* 48: 153–322.

—1918. [Geological distribution of the elasmosaurs.] *Izvestiya Rossiiskoi AN. VI Seriya* 12: 2325–2343.

Romer, A.S. 1968. An ichthyosaur skull from the Cretaceous of Wyoming. *Contributions to Geology* 7: 27–41.

Rozhdestvenskii, A.K. 1947. [Discovery of a gigantic plesiosaur in the Volga region (left bank).] *Doklady AN SSSR* 56: 197–199.

—1973. [The study of Cretaceous reptiles in Russia.] *Paleontologicheskii Zhurnal* 1973: 90–99.

—, Nesov, L.A. and Khozatskii, L.I. 1987. [*Stratigraphy of the USSR. Cretaceous System. Reptiles.*] Moscow: Nedra.

Russell, D.A. 1967. Systematics and morphology of American mosasaurs. *Bulletin of the Peabody Museum of Natural History* 23: 1–241.

Ryabinin, A.N. 1909. Zwei Plesiosaurier aus den Jura- und Kreideablägerungen Russlands. *Mémoires du Comité Géologique, St. Petersburg* 43: 1–49.

—1912. Vertèbres d'un ichthyosaure provenant de Kimmeridge de Pecora. *Trudy Geologicheskogo i Mineralogicheskogo Muzeya imeni Petra Velikogo Imperatorskoi AN* 6: 43–47.

—1915. [Note on a plesiosaur from Sakhalin Island.] *Geologicheskii Vestnik* 1: 82–87.

—1936. [A vertebra of a plesiosaur from Franz-Josef Land.] *Trudy Arkticheskogo Instituta* 58: 143–146.

—1939. [New discoveries of Plesiosauria in the Soviet Arctic and a cervical vertebra of *Plesiosaurus latispinus* Owen from Lonely Island in the Kara Sea.] *Problemy Arktiki* 9: 49–54.

—1946a. [A vertebra of an ichthyosaur from the Upper Triassic of the Kolymskii District.] *Priroda* 1946: 57–58.

—1946b. [New finds of fossil reptiles in the Crimea.] *Priroda* 1946: 65–66.

Sauvage, H.E. 1873. Notes sur les reptiles fossiles. *Bulletin de la Société Géologique de France* 1: 365–380.

Seeley, H.G. 1874a. Note on some of the generic modifications of the plesiosaurian pectoral arch. *Quarterly Journal of the Geological Society of London* 30: 436–449.

—1874b. On the pectoral arch and fore limb of *Ophthalmosaurus*, a new ichthyosaurian genus from the Oxford Clay. *Quarterly Journal of the Geological Society of London* 30: 696–707.

Shishkin, M.A. and Lozovskii, V.R. 1979. [Labyrinthodonts from the Triassic of Lower Primor'ya.] *Doklady AN SSSR* **246**: 201–205.

Sintsov, J. 1899. Notizen über die Jura-, Kreide- und Neogen-Ablagerungen der Gouvernements Saratow, Simbirsk, Samara und Orenburg. *Zapiski Imperatorskogo Novorossiiskogo Universiteta (Odessa)* **77**: 1–106.

Steel, R. 1973. Crocodylia. *Handbuch der Paläoherpetologie* **16**: 1–116.

Storrs, G.W. 1991. Anatomy and relationships of *Corosaurus alcovensis* (Diapsida: Sauropterygia) and the Triassic Alcova Limestone of central Wyoming. *Bulletin of the Peabody Museum of Natural History* **44**: 1–151.

—1996. Morphological and taxonomic clarification of the genus *Plesiosaurus*, pp. 145–190 in *Ancient Marine Reptiles*, Callaway, J.M. and Nicholls, E.L. (eds.). New York: Academic Press.

Sukhov, I. 1950. [First discovery of an ichthyosaur in the Moldovan SSR.] *Priroda* **1950**: 66.

Tarlo, L.B. 1959. *Stretosaurus* gen. nov., a giant pliosaur from the Kimeridge [sic] Clay. *Palaeontology* **2**: 39–55.

—1960. A review of the Upper Jurassic pliosaurs. *Bulletin of the British Museum (Natural History), Geology* **4**: 145–189.

Tatarinov, L.P. 1964. [Subclass Ichthyopterygia. Ichthyopterygians, or ichthyosaurs.] In *Osnovy Paleontologii* 12, Rozhdestvenskii, A.K. and Tatarinov, L.P. (eds.), pp. 338–354. Moscow: Nauka.

Trautschold, H. 1860. Recherches géologiques aux environs de Moscou. Couche jurassique de Galiowo. *Bulletin de la Société Impériale des Naturalistes de Moscou* **33**: 338–361.

—1861. Recherches géologiques aux environs de Moscou. Couche jurassique de Mnioviki. *Bulletin de la Société Impériale des Naturalistes de Moscou* **34**: 64–94.

Tsaregradskii, V. 1927. [Remains of mosasaurs from the Saratov Region.] *Izvestiya Geologicheskogo Komiteta* **45**: 563–572.

—1935. [Detailed description of the mosasaur *Dollosaurus lutugini* Jak.] *Ezhegodnik Vsesoyuznogo Paleontologicheskogo Obshchestva* **10**: 49–54.

Wade, M. 1990. A review of the Australian Cretaceous ichthyosaur *Platypterygins* (Ichthyosauria, Ichthyopteryia). *Memoirs of the Queensland Museum* **28**: 115–137.

Welles, S.P. 1962. A new species of elasmosaur from the Aptian of Colombia and a review of the Cretaceous plesiosaurs. *University of California Publications in Geological Science* **44**: 1–96.

Williston, S.W. 1903. North American plesiosaurs. Part 1. *Field Columbian Museum Publication* **73**: 1–77.

—1908. North American plesiosaurs: *Trinacromerum*. *Journal of Geology* **16**: 715–736.

—1917. *Labidosaurus* Cope, a Lower Permian cotylosaur reptile from Texas. *Journal of Geology* **25**: 309–321.

Yakovlev, N.N. 1901. Restes d'un mosasaurien trouvé dans le Crétacé supérieur de sud de la Russie. *Bulletin du Comité Géologique, St. Petersburg* **20**: 507–520.

—1902. Ueber *Mosasaurus* aus den oberen Kreideablagerungen Süd-Russlands. *Zapiski Vserossiiskoe Mineralogicheskoe Obshchestva* **39**: 34.

—1903. Ueber *Plesiosaurus*-Reste aus der Wolga-Stufe an der Lena in Sibirien. *Zapiski Vserossiiskoe Mineralogicheskoe Obshchestva* **41**: 13–16.

—1905. Notes sur les mosasauriens. *Bulletin du Comité Géologique, St. Petersburg* **24**: 135–152.

Yarkov, A.A. 1993. [The history of study of mosasaurs in Russia and some remarks on their systematics.] In *Voprosy Stratigrafii Paleozoya, Mezozoyai, Kainozoya*, pp. 26–40. Saratov: Saratov University Press.

Yazikhov, P.M. 1832. [On the discovery of fossil remains of an ichthyosaur near the city of Simbirsk.] *Gornyi Zhurnal* **5**: 183.

Zhuravlev, K.I. 1941. [Ichthyosaurs and plesiosaurs from the fuel shales of Savel'evsk Shale Mine.] *Priroda* **1941**: 84–86.

—1943. [Finds of the remains of Upper Jurassic reptiles in the Savel'evsk Shale Mine.] *Izvestiya AN SSSR, Seriya Geologicheskaya* **5**: 293–306.

12

Asiatic dinosaur rush

EDWIN H. COLBERT[1]

Roy Andrews looks to Central Asia

The end of the great dinosaur rush along the banks of the Red Deer River in Alberta marked the close of a long and memorable chapter in the history of the search for dinosaurs. It was in effect a chapter covering four decades of exciting exploration in the western wilds of North America, beginning, as we have seen, with the intense rivalry between Marsh and Cope and their collectors in the Jurassic beds of Wyoming and Colorado, continuing with the turn-of-the-century diggings by the American Museum of Natural History at Como Bluff, continuing still with the massive quarry developed by Earl Douglass and his associates at what is now the Dinosaur National Monument, and culminating with the singularly romantic work by Barnum Brown and his rivals, the Sternbergs, who floated through the Cretaceous badlands of Alberta like hardy rivermen of a former age, in search of dinosaurian treasures. The discoveries made by the men who participated in the dinosaur hunts of western North America added new dimensions of unparalleled magnitude to our knowledge of the dinosaurs. They made North America in those days the centre of the world for dinosaur hunters – the region where dinosaurs were to be found in the greatest abundance and studied with the greatest facility.

Then in the early twenties of this century, the Asiatic dinosaur rush began, shifting the attention of dinosaur hunters (at least for the time being) from the Western to the Eastern Hemisphere. And it began by accident. For the initiation of this new phase of dinosaurian discovery and research was a side effect of other activities – something unexpected, something

that came as a very pleasant surprise indeed, and something that led to later expeditions and studies, the ends of which are still in the future.

This beginning had nothing to do with dinosaurs – the opening of the Asiatic dinosaur rush began as a search for the origin of man. Around the turn of the century Professor Henry Fairfield Osborn of the American Museum of Natural History had suggested that central Asia, at that time a vast *terra incognita* in the world of natural science, was the centre of origin for early man and for many of the various groups of mammals as well. This concept was elaborated a decade or so later by Dr. William Diller Matthew of the same institution in his classic work *Climate and Evolution*, in which he postulated that central Asia was the centre of origin for most of the mammals, basing his thesis upon his profound knowledge of the relationships and distributions of mammals, fossil and recent, throughout the world. In brief, he saw central Asia as a sort of palaeontological Garden of Eden for the ruling backboned, landliving animals of today. Therefore, said he, to get at the beginning of our modern world, let us look toward central Asia. This idea fired the imagination of Roy Chapman Andrews,

---

[1] This chapter is reproduced from Edwin Colbert's *Men and dinosaurs* (1968), chapter 8. We are extremely grateful to Dr Colbert for permission to include this chapter from his classic book, and for his help in obtaining the original photographs from the American Museum of Natural History. For the latter, we thank Dr Mike Novacek and the Archives of the American Museum of Natural History. The chapter has been modified only by the addition of references to the figures, and by an addendum to cover work since 1968. Russian and Mongolian names of places and people have been modified, where necessary, to the transliteration scheme employed here.

a member of the scientific staff of the museum in New York, and soon after the First World War he began to dream of an expedition on a grand scale to central Asia, particularly to Mongolia, to that Asiatic grassland and desert known as the Gobi.

The guideposts for such a venture were indeed vague. An American geologist, Raphael Pumpelly, had traversed central Asia and the Gobi in 1865, as had the German scholar Ferdinand Freiherr von Richthofen, in 1873. In later years there were some occasional incursions into the Gobi by other naturalists, but all these trips were of a reconnaissance nature, and with but one exception, there were no reports of fossils. The exception was the discovery in 1892 of a single fossil rhinoceros tooth along the old caravan route from Kalgan (now generally called Zhangjiakou), on the Chinese border, to Ulaanbaatar, the capital of Mongolia, by the Russian geologist and explorer Vladimir A. Obruchev. Andrews therefore necessarily envisaged an expedition into central Asia in search of early man with no more than the tooth of a long-extinct rhinoceros – a rhinoceros older by millions of years than the most ancient of men – to which he could point as evidence that there were *any* fossils in the Gobi.

It was a gamble, seemingly with the odds heavily against him, but not to be deterred, not even by the gloomy predictions of geologists who could visualize central Asia only as a vast, unfossiliferous waste of sand. He had the enthusiastic support of Professor Osborn and other knowledgeable authorities, and this was encouragement enough for him to make the gamble. From just such shots in the dark have come many of the great discoveries in science.

So Andrews went ahead with his plans and his organization. These are some of his own remarks about the problem, and about how he proposed to deal with it:

The main problem was to discover the geologic and paleontologic history of Central Asia; to find whether or not it had been the nursery of many of the dominant groups of animals, including the human race; and to reconstruct its past climate, vegetation and general physical conditions, particularly in relation to the evolution of man. It was necessary that a group of highly trained specialists be taken *together* into Central Asia in order that the knowledge of each man might supplement that of his colleagues. This was indeed the first expedition of such magnitude to employ these methods. The fossil history of Central Asia was completely unknown.

Mongolia is isolated in the heart of a continent; and there is not a single mile of railway in the country [this was written in 1929], which is nearly half as large as western Europe. The climate is extremely severe; the temperature drops to $-40°$ to $-50°$ and the plateau is swept by bitter winds from the Arctic. Effective paleontological work can be conducted only from the beginning of April to October. In the Gobi desert, which occupies a large part of Mongolia, food and water are scarce and the region is so inhospitable that there are but new inhabitants. The physical difficulties could only be overcome by some means of rapid transportation and that transportation the motor car successfully supplied. The automobiles could run into the desert, as soon as the heavy snows had disappeared, at the rate of one hundred miles a day, penetrate to the farthest reaches of Mongolia and return when cold made work impossible. Camels, which other explorers had used, average ten miles daily. Thus approximately ten years' work could be finished in one season (Andrews, 1929, p. 713).

Andrews had the audacity to think in big terms – at least in big terms for those unorganized days of the departed twenties, before government had rallied to the support of science with a capital S, as it has on a large and at times a lavish scale. As Andrews remarked, it was a venture to be undertaken by a group of specialists, working together in the Gobi. But how was this group of men, the scientists and technicians, their helpers and the helpers of the helpers, to be carried into a trackless and strange desert, even with automobiles available, in such a way that they would be able to concentrate their efforts and attention on the job at hand, the job of searching for and collecting fossils, of gathering and recording geological data, of collecting specimens of living animals and plants, and of making meteorological observations, without spending all their time and energy at the mere task of travelling and keeping alive? The answer was to use not only

Figure 12.1. The automobile caravan of the Central Asiatic Expedition of The American Museum of Natural History entering the Gobi Desert of Mongolia.

newfangled automobiles but also time-tested camels – in other words, to rely upon fast-moving cars for exploration and the transportation of personnel, combined with slow-moving camels for the hauling of supplies. It was planned to send a large camel caravan across the Gobi along certain predetermined routes. These camels would carry food and supplies of all sorts, including numerous five-gallon cans of gasoline and oil. The scientists and technicians and their assistants would travel in a fleet of sufficiently rugged cars (this was before the days of jeeps), traversing the desert in zigzag paths, sometimes together, sometimes in separate small parties, looking and searching and collecting. At stated intervals the cars would meet the caravan at designated rendezvous points, there to transfer gasoline and oil and food from the backs of the camels to the cars, and fossils and other objects from

the cars to the backs of the camels. The camels that had acquired loads of specimens in place of expendable supplies would return to a base, while the rest of the caravan would move on to the next rendezvous. And the cars would embark on new wandering courses across the rolling landscape. During one field season, for example, the expedition had a caravan of 125 camels that carried 1,000 gallons of gasoline, 100 gallons of oil, 3 tons of flour, 1½ tons of rice, and other food in proportion. This caravan left gasoline and food at two depots, and waited for the wandering cars at a well 800 miles out in the desert.

An interesting detail might be mentioned. The cars travelled across the undulating hills much as would ships across the long swells of a broad sea; consequently they were guided through the Gobi, back and forth, by the well-tested methods of navigation that

Figure 12.2. Frederick Morris, geologist of the Central Asiatic Expedition, at the plane table making a map. There is more to collecting dinosaurs than digging them out of the ground.

have been used by ships' captains ever since the invention of the compass and the sextant.

It worked – beautifully. And the first year was so successful that there followed four other expeditions. There would have been even more, but for internal political troubles that racked China during some of those years, and prevented the expedition from journeying into Mongolia.

So it was that on a spring morning in 1922 the first large-scale expedition of the Central Asiatic Expeditions of the American Museum of Natural History made its way out of Kalgan on the border, and entered Mongolia, to search for the birthplace of man in central Asia (Figure 12.1). The personnel of this expedition included, in addition to Andrews, the leader, Walter Granger, chief palaeontologist and second in command, and Professor C.P. Berkey and Frederick K. Morris (Figure 12.2), two outstanding geologists, who unravelled many of the hitherto unknown geological structures in this isolated land. Of particular importance to the success of the expedition was the representative for the Mongolian Government, T. Badmayapov, without whose help the party would have been unable to reach the inner recesses of Mongolia. There were in addition technicians and a group of Chinese and Mongol assistants – in all, some twenty-six persons. The route for the motorcars was from Peking to Kalgan, the gateway to Mongolia, then northwest toward the capital of Mongolia, Ulaanbaatar (sometimes Ulan Bator, or

Urga, the name then in frequent use), and from thence southwest into the desert (Figure 12.3).

Walter Granger

The key person in the assemblage of scientific talent – if the expedition were to accomplish its avowed purpose of finding the remains of ancient man – was Walter Granger, the palaeontologist (Figure 12.4). He was a good man to have along in every respect. He was a superb field palaeontologist, an expert with many years' experience in the search for fossil vertebrates, and a man with an unusually fine personality, a man who could work well with other people because he was a man other people universally liked.

Walter Granger was born in Vermont in 1872, and grew up in that wooded and rocky state. As a boy he was intensely interested in the wildlife around him. Through various circumstances he appeared at the American Museum of Natural History in 1890, a long, gangling youth not quite eighteen years old, to begin his lifelong association, and what was to prove a very distinguished career, with that institution. His beginnings at the museum were indeed lowly – he was a sort of handyman in the custodial department, who lent a hand in the Taxidermy Department. As he recalled, many years later:

In the taxidermy shop my task, aside from keeping the place clean, was to skin and preserve the birds, mammals and reptiles which died in the Central Park Zoo and elsewhere and I have never since been squeamish about odors!

All this time I had visions of future field work. My chance came in 1894 when I was sent west to collect mammals, and used the fossil collectors' camps as a base. Two years later, on the advice of Dr. Chapman [Frank Chapman, the noted ornithologist] I changed from the Department of Birds and Mammals to that of Vertebrate Paleontology, principally because of the opportunities it offered for field work in which I was getting more and more interested. During these years I have been absent from the field but very few seasons (Granger quoted by Simpson, 'Memorial', 1942, p. 160).

Young Granger quickly proved his worth as a palaeontologist not only in the field but also in the labora-

tory, so that by 1897 he was playing a leading role in the American Museum excavations for Jurassic dinosaurs along Como Bluff. During six seasons he participated in the dinosaur project in Wyoming, devoting his energies to the diggings at the Bone Cabin Quarry, which he helped to locate. In 1899 Dr. Wortman, who had been in charge of the dinosaur quarrying, left the American Museum to go to the Carnegie Museum in Pittsburgh, and from that time on, Granger took charge of the work.

Granger's association with dinosaurs came to an end in 1902. From then until he embarked upon the first of the Central Asiatic Expeditions he was primarily concerned with those ancient mammals that became the rulers of the land after the extinction of the dinosaurs. Granger's contributions in this field of study were of enormous importance; in fact, it was his work, carried on in collaboration with Dr. William Diller Matthew, that virtually defined the nature of mammalian life during the Paleocene epoch, the time of the first great evolutionary radiation of mammals. During about two decades, beginning in 1903, Granger made a solid reputation as a collector and student of the most primitive mammals – a reputation for which his name will be remembered as long as men study the annals of palaeontology.

Then in 1921 and 1922 Granger began a new scientific career – in Asia. On April 21, 1922, he, together with the other members of the Central Asiatic Expedition (Figure 12.5), left Kalgan on the first great journey into the unknown lands of Mongolia. It was a group of men who would share many experiences and even hardships before the end of the season. And long before the summer had run its course, every man of this historic group would be glad that Walter Granger was one of their party. A big, jovial person with a hearty laugh, Granger by his very presence, by the force of his remarkable personality, helped immeasurably to make the work of the expedition go smoothly. And Granger, by virtue of his expert knowledge of fossils and of how to look for fossils, contributed in a large degree to making the results of that first year in the Gobi most memorable indeed.

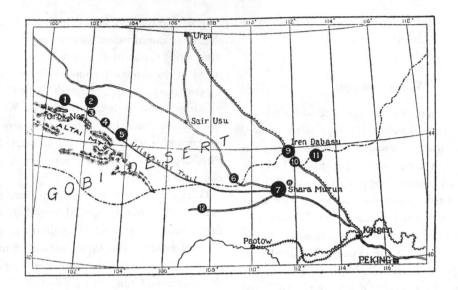

Figure 12.3. A map showing the areas of Mongolia explored by the Central Asiatic Expeditions of The American Museum of Natural History. Cretaceous dinosaurs were found at (4) the Osh Basin; (5) Shabarakh Us, or Bayan Zag; (9) Iren Dabasu; (12) Ongon. The capital city of Ulaanbaatar is designated on this map by its older name, Urga.

First discoveries

Before the first Central Asiatic Expedition had reached Ulaanbaatar, at a locality known as Iren Dabasu, less than half the distance to the Mongolian capital, fossil bones were discovered. But they were not the bones of early man – they were the bones of ancient mammals, and at one place, the bones of Cretaceous dinosaurs that had inhabited the world perhaps a hundred million years before the first man made his appearance on the earth. Such are the fortunes of Science; the search for one thing almost as often as not leads to the discovery of something else quite unexpected, quite unrelated to the goal that is being pursued, and frequently as important as that which was originally sought. Which does not imply that ancient dinosaurs are more important than ancient men, but they *are* none the less important.

Some hasty collecting was carried out, and then the cars went on their way to Ulaanbaatar. Not much time could be spent at Iren Dabasu, for this was a summer for reconnaissance, for tentative probing, a summer

during which much ground must be covered, a summer devoted to a general survey of the Mongolian scene (Figure 12.6).

From Ulaanbaatar, which was reached in due time, the motorcade turned toward the southwest and drove into the desert (Figure 12.7). Far out in Mongolia, at the end of many weeks of exploration to the west, the expedition finally turned eastward to begin the long journey back to the headquarters in Beijing. The party had driven to the east along an old trail for several days, when unexpectedly one afternoon the trail led the procession of cars to the edge of a large, eroded basin formed in red sandstones. Within this basin and around its edges were weathered natural monuments and sculptured cliffs, of which one large line of cliffs was particularly impressive, especially when the setting sun illuminated the red rocks so that they seemed to catch fire and glow against the darkening sky of evening (Figure 12.8). These became the 'Flaming Cliffs' of Shabarakh Us, as the locality was designated by the expedition (it was subsequently quoted as Bayn Dzak, but is, more correctly, Bayan

216

The Flaming Cliffs

Figure 12.4. Walter Granger, palaeontologist and second in command of the Central Asiatic Expedition, in a less than formal pose. He is applying shellac and rice paper to a fossil bone.

The next year arrived, as next years do, and again the expedition was in Mongolia. This time an adequate stop was made at Iren Dabasu, where bones and skeletons of Cretaceous dinosaurs were collected. The hurried work of the previous year had shown that the fossils of central Asia, if not the remains of ancient man, were none the less of great importance, for there were not only the bones of dinosaurs but also the bones of many other animals that were to add new and important facets to our knowledge of the history of life. Consequently the expedition this year included several trained palaeontologists and collectors, men who were abundantly able to cope with any fossils that might be found, under any conditions in which they might be found. There was Walter Granger again, to continue the direction of the palaeontological explorations, and with him were two of the American Museum's most able palaeontological field men, Peter Kaisen and George Olsen. In addition there was another fossil collector, Albert Johnson, who had worked with Barnum Brown along the Red Deer River. With such men at hand the weeks spent at Iren Dabasu were almost certain to be productive. And they were. Several quarries were opened and developed, from which a considerable collection of bones and skeletons was removed. As Andrews said in his account of the expedition 'bones of both flesh-eating and herbivorous dinosaurs, of several species and many individuals, were piled one upon the other in a heterogeneous mass.' It was a profitable dig palaeontologically speaking, but there were other things still to be done before the summer had run its course. The group was anxious to get back to Shabarakh Us (or Bayan Zag), to the Flaming Cliffs; it was an exciting spot, and it beckoned the men on to the west, into the heart of Mongolia.

To the west they journeyed, stopping for a rendezvous on the way with the camel caravan. On the afternoon of July 8,1923, the expedition was again at the Flaming Cliffs (Figure 12.10), setting up camp and looking forward with eager anticipation to the pleasant task of searching for fossils in this remote and starkly beautiful corner of central Asia.

Zag), a locality destined to become famous in the annals of dinosaur collecting. At the foot of these cliffs the fossil hunters found bones that were then unfamiliar to Granger and his associates, and fossilized eggshells that at the time were supposed to be portions of the eggs of large birds (Figure 12.9). Again, as at Iren Dabasu, there was little time; the expedition was forced to push on – to get out of the Gobi before the bad weather of autumn settled upon the land. But as the cars drove over to the east and the south, the men were already making plans for the next year.

Figure 12.5. Members of the Central Asiatic Expedition of The American Museum of Natural History, 1923. Middle row, from left to right: Walter Granger, Henry Fairfield Osborn, Roy Chapman Andrews (leader), Frederick K. Morris (geologist), Peter Kaisen, experienced dinosaur hunter who had spent many years with Barnum Brown in western North America. Back row, third from left, Albert Johnson, who worked with Brown along the Red Deer River in Alberta; third from right, George Olsen, who had also collected dinosaurs on previous occasions in North America.

It did not take long for the enthusiastic fossil hunters to locate the treasures for which they were searching. This year they could prospect and dig with some degree of understanding, because in the interval since their preliminary discoveries of the previous season some studies had been made of the few fossils that then had been collected, so there were now available clues as to the nature of these fossils. In short, the bones at Shabarakh Us were those of a very primitive horned dinosaur – one of the first of the horned dino-

saurs – and this ancestor of what was to be during late Cretaceous time a flourishing line of dinosaurian evolution had been christened *Protoceratops andrewsi* by Drs. W.K. Gregory and C.C. Mook of the American Museum (Figure 12.11). The name was an apt one: *protos* – the first, and *ceros* plus *ops* – horn plus face. (This compound suffix is commonly applied to the horned dinosaurs.) The trivial name *andrewsi* was of course in honour of Roy Andrews, the leader of the expedition.

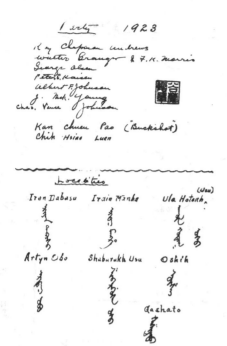

Figure 12.7. Central Asiatic Expedition trucks ploughing through the sand, with assistance from all except the canine.

Figure 12.6. The first page of Walter Granger's field book for 1923, showing the names of the scientific and technical members of the Central Asiatic Expedition to Mongolia, and the localities explored with their Mongolian designations.

The work at the Flaming Cliffs was the first phase of a collecting programme that extended through two seasons; the expedition returned to this locality again in 1925. And in the course of these two summers of work an unparalleled collection of *Protoceratops* was accumulated – skulls and bones and complete, articulated skeletons. All in all more than a hundred specimens of *Protoceratops* were collected, a series of dinosaurs that in numbers surpasses even the remarkable group of *Iguanodon* skeletons from Bernissart.

There was more than mere quantity to make the skulls and skeletons of *Protoceratops* significant, even though the dinosaurs excavated at the Flaming Cliffs probably exceeded in numbers of individuals belonging to a single species any collection of dinosaurs that had ever been made before. For instance, in this collection there were individuals in all stages of growth, from those newly hatched to fully adult animals. For

the first time in the history of the study of dinosaurs it was possible to get an idea as to how some of these ancient reptiles grew up. Strange as it may seem, there had been very few juvenile dinosaurs of any kind discovered before the collection was made at Shabarakh Us, and for that matter, very few have been discovered since. Most dinosaurs are known from the adults; it is therefore a rare privilege to be able to visualize growth in one species of dinosaur, as is so nicely illustrated by the display of a dozen *Protoceratops* skulls in the American Museum of Natural History.

It will be remembered that on the first brief visit to the Flaming Cliffs in 1922 some fragments of eggshell, supposedly of extinct birds, had been found. Soon after the beginning of the work at this place in 1923, George Olsen found a group of three eggs (Figure 12.12). Were these birds' eggs, or might it be possible that they were the eggs of dinosaurs, specifically of *Protoceratops*? This question, which immediately came to the mind of Walter Granger, raised hopes and expectations, and plunged the whole group of fossil hunters into a fever of egg-hunting excitement. The basin was searched with avidity and in detail, and as the days went by fossil eggs came to light in great numbers – as fragments, as complete eggs, as clusters of eggs, and even as nests of eggs.

These obviously were not the eggs of birds. They

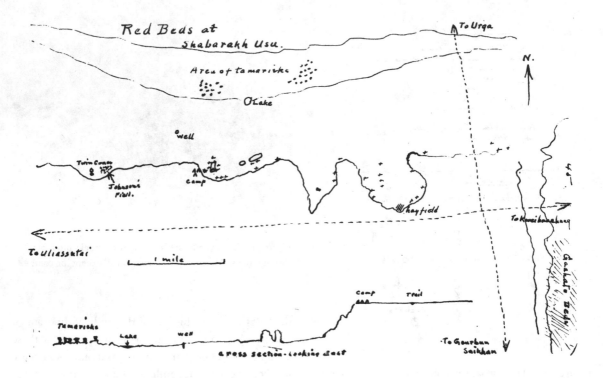

Figure 12.8. A page from the 1923 field book of Walter Granger, showing his map and diagram of the Shabarakh Us, or Bayan Zag, locality, where skeletons and eggs of the primitive horned dinosaur *Protoceratops* were excavated.

were quite elongated, shaped very much like the eggs of certain modern lizards, and their surfaces were roughly corrugated. Some of the clusters were especially revealing, for these were composed of circles of eggs. In one instance there is an inner circle of five eggs, a circle surrounding it with eleven eggs, and beyond that two peripheral eggs indicating an outer circle that originally may have contained as many as twenty eggs. Here the record is plainly preserved in the rock. Quite obviously a female dinosaur dug a large pit in the sand of an ancient Cretaceous plain, just as today her reptilian cousins, the gigantic marine turtles, dig pits in the sand on beaches of oceanic islands. And in this pit she deposited her eggs, perhaps turning around and around during the process, so that the eggs were laid in concentric circles. Then she carefully covered the eggs with sand, trusting to the heat of the sun to hatch them, in the same way as do the marine turtles or many modern crocodiles and other reptiles.

Fortunately for the fossil hunters at the Flaming Cliffs, and for the science of palaeontology, something happened so that the eggs never hatched. Perhaps there was a cool spell of weather, with the result that the sand never warmed up enough beneath the glowing rays of the sun to bring to completion the development of the embryos. Whatever the cause of the failure, the eggs remained in the ground; subsequently they cracked, and their cavities were filled with sand. Eventually they were fossilized, to preserve in a dramatic way the story of a dinosaur and her nest as it took place almost a hundred million years ago.

Granger and his associates assumed that the numerous eggs at the Flaming Cliffs were those of *Protoceratops*, and it is a logical assumption. Here in the ground were untold numbers of a small ancestral horned dinosaur, the adults of which were about six or eight feet in length, and here were untold numbers of fossil eggs of the right size to have been laid by the dinosaur *Protoceratops*, eggs about eight inches in

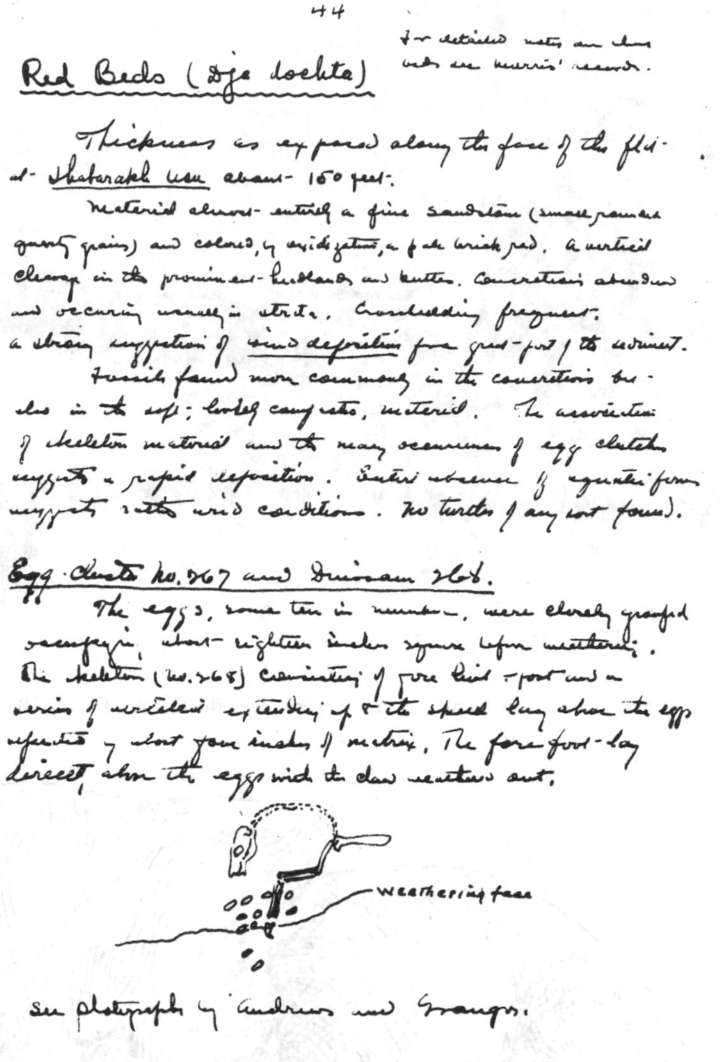
44

For detailed notes on these beds see Morris' record.

Red Beds (Djadochta)

Thickness as exposed along the face of the cliff at Shabarakh Usu about 150 feet.

Material almost entirely a fine sandstone (small rounded quartz grains) and colored, by oxidization, a pale brick red. A vertical cleavage in the prominent headlands and buttes. Concretions abundant and occurring usually in strata. Crossbedding frequent. A strong suggestion of wind deposition for greater part of the sediment.

Fossils found more commonly in the concretions but also in the soft; loosely compacted, material. The association of skeleton material and the many occurrences of egg clutches suggests a rapid deposition. Entire absence of aquatic forms suggests rather arid conditions. (No turtles of any sort found).

Egg cluster No. 267 and Dinosaur 268.

The eggs, some ten in number, were closely grouped occupying about eighteen inches square before weathering. The skeleton (No. 268) consisting of fore limb + foot and a series of vertebrae extending up to the skull lay above the eggs separated by about four inches of matrix. The fore foot lay directly above the eggs with the claws weathered out.

See photographs by Andrews and Granger.

Figure 12.9. Another page from the 1923 field book of Walter Granger, with notes concerning the occurrence of *Protoceratops*.

length. Why should not the fossil locality at Shabarakh Us, or Bayan Zag, be a burial ground for this particular dinosaur and for the eggs it laid, a palaeontological cemetery preserving not only all stages of growth, as has been mentioned, but also the egg that preceded the growth? The association of *Protoceratops* skeletons with the eggs seems supremely reasonable, and has been accepted by most palaeontologists, although some dissenters have denied that the eggs found at this locality belong to the dinosaur.

It should be added that in addition to the large quantities of eggs attributed to *Protoceratops* occurring at the Flaming Cliffs, several other types of fossil eggs also were found, though in quite limited numbers. Some of these eggs are round; others are elongated but rather small. How are they to be identified? If the most numerous eggs, the large elongated, rough-shelled eggs, are those of *Protoceratops*, then it follows that the other eggs may be associated with some of the other reptiles that were found at Shabarakh Us. But how are

Figure 12.10. The camel caravan of the Central Asiatic Expedition at the Flaming Cliffs of Bayan Zag, Mongolia.

Figure 12.11. The Cretaceous dinosaur *Protoceratops*. Restoration by Margaret Matthew Colbert.

Figure 12.12. George Olsen (left) and Roy Chapman Andrews (right) excavating a nest of dinosaur eggs at Shabarakh Us, or Bayan Zag, Mongolia.

any associations of eggs and bones to be made? It is a question as yet unanswered.

Perhaps a few words might be said about the animals that lived with *Protoceratops* so many million years ago. There were other dinosaurs, among them a large armoured dinosaur that has been named *Pinacosaurus*. Then there were some small, lightly built meat eating dinosaurs, *Saurornithoides* (birdlike dinosaur), *Velociraptor* (fast-running robber), and *Oviraptor* (egg stealer), as well as a large carnivore.

The discovery of *Oviraptor* provided a bit of unusual palaeontological drama for the fossil hunters. Its skeleton was exposed in the rock amid a cluster of *Protoceratops* eggs. Did this small and rather elegant little dinosaur perish in the act of plundering a *Protoceratops* nest? Perhaps – perhaps not. Perhaps the occurrence of this skeleton is entirely fortuitous. But

the vision of *Oviraptor* as a nest robber makes a pleasant bit of imaginative fancy and adds a certain touch of spice to the dinosaurian discoveries at the Flaming Cliffs; hence the name.

Some other reptiles were also found at the Flaming Cliffs: a crocodile and a pond turtle. This seems to indicate that the locality was a place of streams and perhaps small ponds, cutting through the sands across which *Protoceratops* wandered and congregated, to lay its eggs.

Our immediate interest is in the dinosaurs from Shabarakh Us, but it should be mentioned that famous though this place may be for the dinosaurs and the dinosaur eggs found there, it is equally if not even more important, palaeontologically speaking, by virtue of some tiny skulls of mammals, first collected at the Flaming Cliffs by the men of the Central Asiatic

Expeditions. Here again a great step forward in our knowledge of the history of life was made through the working of chance. Andrews had organized his elaborate venture in central Asia to search for the beginnings of ancient man, but instead his expedition had found dinosaurs. Andrews and Granger and their co-workers had returned to Shabarakh Us to collect dinosaurs – which they did on a grand scale. And to their surprise and delight they also found in these dinosaur beds the first placental mammals, small insect-eating animals related to modern shrews and hedgehogs. These ancient mammals are, in effect, representative of the ancestors of all of the dominant mammals that today rule the earth. There at the Flaming Cliffs, in rocks perhaps as much as a hundred million years in age, were the beginnings not only of shrews and hedgehogs but also in a sense of bats and lions, of rhinoceroses and cattle, of monkeys and apes and of ourselves. A few minute skulls and jaws were the 'haul' from the Flaming Cliffs, a collection to be carried in a cigar box. Yet these few fossils, seemingly so insignificant, represented one of the greatest accomplishments of the expedition. Here was visible proof of the genesis of a new era in the long story of life history. Here were the ancestors of the animals that eventually, and at no great interval in terms of geological time, were to replace the dinosaurs as the lords of the continents.

But let us get back to the dinosaurs. Certainly the discovery of dinosaur eggs at Shabarakh Us caught the fancy of the public, and important though the Cretaceous mammals might be, it was the collection of *Protoceratops* bones and eggs that gained for the expedition into Mongolia its widest acclaim. Dinosaur eggs became objects of lively interest in the news columns, and dinosaurs, already widely appreciated by people in many lands, became increasingly well established in the public mind.

The Central Asiatic Expedition collected dinosaurs at several other localities in Mongolia – such places as Öösh Nuur (Figure 12.13), Ondai Sair, and Ongon, each of these localities located, as are Iren Dabasu and Shabarakh Us, in an isolated basin, separate and disconnected from every other basin. But although

Figure 12.13. The Central Asiatic Expedition camp in the Öösh Basin, Mongolia, where remains of dinosaurs were found.

important finds were made at the various dinosaur localities, none could measure up to the discoveries made at the Flaming Cliffs.

The interim years – and their results

Such was the course of the Asiatic dinosaur rush – in its first phase. It is the story of three probing trips into Mongolia in 1922, 1923, and 1925. True enough, the Central Asiatic Expeditions ventured into the field again, in 1928 and in 1930, but after 1925 they never reached what was then called Outer Mongolia, where the dinosaurs are. Yet even though the men of the Central Asiatic Expedition were forced to limit their objectives and their activities in the later years of their work, they had already accomplished great things. In those three short summers, especially in 1923 and 1925, they had opened a new chapter in the buried record of the dinosaurs. They had established central Asia as a locale for the discovery of these extinct reptiles, comparable to western North America. They had made the first excavations in a palaeontologically unknown land, just as Marsh and Cope had done, a half-century earlier, in another palaeontologically unknown land.

The beginning of the Asiatic dinosaur rush was soon over; after 1925 there was a long lull in the search for dinosaurs in Mongolia. It was a lull occasioned by the militant march of world history, a procession of armies and battles across the face of Asia that crowded out the peaceful pursuits of fossil hunters and their ilk. This does not mean that the dinosaurs of central Asia were completely forgotten. In the practice of palaeontology there is always a necessary time lag between discovery and collecting on the one hand, and study and description on the other. This is the interval during which the fossils are 'prepared' in the laboratory, the long months during which they are chipped out of and freed from the encasing rock, and cleaned and hardened with various penetrating agents, to make of them significant objects suitable for study and interpretation. It is a tedious process – and it takes time.

Consequently the work on the dinosaurs from Mongolia continued for a number of years after the adventurous days of collecting had passed. After the fossils were prepared, there came the months and years of study and description, the end result being a series of technical publications by Granger and Matthew, Gregory, Mook, Gilmore, Brown, and Schlaikjer. This roster of names illustrates very nicely the change that had taken place since the days of Marsh and Cope; large projects were no longer the efforts of a single man and his subordinates. Many men were involved in bringing to light the dinosaurs of Mongolia – in the field, in the preparation laboratory, and at the study table. And the work at the study table, the hours of careful investigation from which flowed the stream of publications that are the ultimate goal of all serious research, revealed the nature and the significance of the dinosaurs that had been collected in Mongolia.

These scientific writings offered some clues as to the composition of dinosaurian life in central Asia during the Cretaceous period, when the gigantic reptilian rulers of the continents were at the climax of their evolutionary development. The clues to be read from the fossils indicated that in Cretaceous time this remote section of the world evidently had close connections with western North America, probably by way of a trans-Bering land bridge, so that dinosaurs and other land-living animals could wander back and forth from one region to the other. There would seem to have been connections with Europe, too, but the relationships with North America evidently were particularly strong.

Yet the evidence, satisfying though it might be, was still far from complete; indeed, it was, except for the abundantly recorded *Protoceratops*, based largely on scattered skeletons and skulls. Andrews, Granger, and their companions quite obviously had barely scratched the surface of the Mongolian dinosaur beds. This is not to be wondered at – they were the first palaeontological explorers in an interior subcontinent, a country half the extent of the continental United States.

Russian explorations

Naturally, the work of the Central Asiatic Expeditions attracted the attention of palaeontologists throughout the world. Consequently, when the turbulent days of the thirties and the cruel world conflict of the forties had run their successive courses eyes were again turned in the direction of Mongolia. But this time the view toward central Asia was not taken from America, but rather from a closer region.

Mongolia had come within the Russian sphere of influence as early as the mid-twenties; indeed, this was one reason why Andrews was unable to lead his party beyond the limits of Inner Mongolia after 1925. It was natural, therefore, that Russian palaeontologists should turn toward the fossiliferous green pastures of central Asia soon after the guns ceased to roar in 1945. Mongolia was the place for them to explore, and there they went. So began the second phase of the Asiatic dinosaur rush.

It began in 1946 and continued through 1947 and 1949 (Figure 12.14). It was, in effect, a modernized version of the Central Asiatic Expeditions of the twenties, with reliance placed upon heavy-duty trucks in place of camel caravans for the hauling of supplies, and upon jeep-like cars for scouting and exploration. It was a well-planned series of expeditions, based upon three years of preparations in Moscow, and crowned with success, the result of three summers of work in Mongolia.

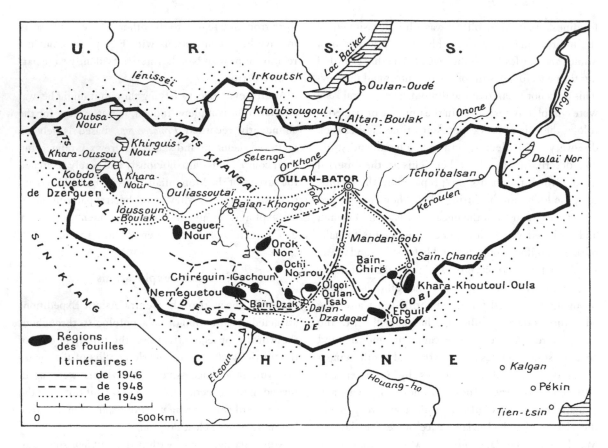

Figure 12.14. Map of Mongolia, showing the routes of the Russian expeditions of 1946, 1948, and 1949, and the fossil localities (in black) explored and excavated by those expeditions. Place names are given in French form.

This, being a Soviet effort, was strictly official. The expeditions were carried out under the auspices of the Academy of Sciences of the USSR, the central, controlling organization that is responsible for scientific research and technological development in Russia. The adviser for the expeditions was the late Academician Yu. A. Orlov, the dean of Russian palaeontologists, a gentle and charming man and an authority on fossil reptiles and mammals. (It should be said here that although the Academy of Sciences is a vast organization, employing thousands of people, it is ruled from the top by a small cadre of Academicians, in all, less than two hundred individuals. To be elected an Academician is perhaps the greatest honour that can come to a Russian scientist, an honour comparable to election in the United States to the National Academy of Sciences, in Britain to the Royal Society. Thus the expeditions to Mongolia had the personal attention from one of the elite in the scientific community of Russia.) The leader of the expeditions during the several field seasons was Dr. I. Efremov (Figure 12.15), renowned not only for the study of extinct animals, particularly reptiles, but also a very popular writer of fiction, a man with a large following of readers, most of whom are perhaps only slightly if at all aware of his reputation as a scientist. Efremov, a stocky individual with a keen mind and a restless temperament, was well fitted for leadership of such a large undertaking, and he led it with skill and vigour. Next in command to Efremov, also an outstanding authority on fossil reptiles, was A. Rozhdestvenskii (Figure 12.15), who has written a lucid and fascinating account

Figure 12.15. Palaeontologists, all specialists in the study of fossil reptiles, on the Russian Expedition to Mongolia in 1948: I.A. Efremov, leader of the expedition, in the foreground; A.K. Rozhdestvenskii, assistant leader to the left; E.A. Maleev to the right. The desert is a rugged place; it can be very cold at times.

of the Muscovite campaign in Mongolia. Of course, there were other participating palaeontologists, as well as a group of skilled technicians and assistants, together with the usual complement of truck drivers, mechanics, labourers, cooks, and other functionaries necessary to the success of the venture.

The second phase of the great Asiatic dinosaur rush began in 1946 with a preliminary reconnaissance journey through Mongolia during two summer months. The men on this first trip through the Gobi located various fossiliferous sites and made some preliminary excavations. Then, with the experience of the summer behind them, they were able to plan for a long and elaborate expedition during the next year, an expedition that began with the departure of some of its members from Moscow during the dead of winter, followed by two months or more of preparation at Ulaanbaatar before the caravan drove out of that city on March 18th (Figure 12.16). And the expedition remained in the field until the last possible moment, until the arctic winds from the north and the threat of impassable snows drove the group back to

Ulaanbaatar. This extended exploration into the Gobi was followed, two years later, by another long expedition that not only retraced the tracks of the previous explorations but also probed far to the west, to the western-most limits of Mongolia.

Of course, the Russians came into Mongolia from the north, entering Ulaanbaatar to find it metamorphosing into a modern city with shining new buildings, public squares and parks, a sizable university, and many other amenities, quite in contrast to the primitive Asiatic capital, hardly larger than a village, seen by Andrews and his companions in 1922. From Ulaanbaatar, Efremov and his caravan of five scout cars turned toward the south (Figure 12.17).

At a locality known as Bayan Shiree, in the direction of but several hundred kilometres to the northwest of, Iren Dabasu, where the Central Asiatic Expeditions had worked so successfully, the Soviet expedition discovered dinosaur bones, including an interesting skeleton of an armoured dinosaur. It was a significant discovery, but even greater things were to come.

The expedition then turned to the west – to Bayan

Figure 12.16. Twenty-five years later. The Russians follow trails first blazed by the Central Asiatic Expeditions of The American Museum of Natural History. Gone are the camel caravans; modern, powerful trucks now do the heavy hauling.

Figure 12.17. Even as late as 1946, the heavy Russian trucks were strange and terrifying objects to inhabitants of the far reaches of Mongolia – and to their camels and their dogs.

Zag (our familiar old locality of Shabarakh Us), and there continued the work that had been carried on amid such scenes of excitement and anticipation by the Central Asiatic Expeditions, more than a score of years earlier. Naturally, the Soviet Expedition uncovered more *Protoceratops* and more eggs. Bayan Zag proved to be as rich a hunting ground in the forties as it had been in the twenties, an illustration of the fact that

as the land erodes, more fossils come into view. Indeed, if the proper trained persons are not at hand to discover and excavate fossils, the ancient remains soon weather away into fragments and are lost. One may only speculate as to the vast numbers of fossils that have disappeared as a result of erosion at a locality such as Bayan Zag during the millennia before any men ever became palaeontologists; such thoughts may lead one along various stray philosophical paths. Our concern is dinosaurs, of which the Soviet palaeontologists discovered new as well as familiar forms at Bayan Zag. Here was found an armoured dinosaur, *Syrmosaurus*, that would appear to have been a connecting link between the plated dinosaurs so typical of Upper Jurassic deposits and the abundant armored dinosaurs that characterize the Upper Cretaceous beds of western North America. This was a most important new discovery, adding a link between two large groups of dinosaurs – one more building block in the evolutionary structure of the ancient reptiles.

The great Nemegt plain

The most important contributions of the Soviet expeditions perhaps were not so much in the work done at Bayan Shire and at Bayan Zag, as in the discovery and development of a series of fossil quarries in a great topographic basin some three hundred kilometres or more to the west of Bayan Zag, an area of Cretaceous rocks known as Nemegt. Much of the activity of these expeditions during three field seasons was centred in this fossiliferous region, and the efforts of the bone diggers were well repaid. It was an extraordinarily rich area, yielding many dinosaur skeletons, skulls, and bones. One of the main camps for the expeditions was at a spectacularly rich fossil locality, christened most appropriately 'The Dragon's Tomb' (Figure 12.18). For here were the bones of dragons – gigantic dragons that inhabited the earth in the distant past. There were other important dinosaurian graveyards scattered over a region more than a hundred kilometres in length within the Nemegt Basin.

As a result of the digging here, at least ten complete skeletons of dinosaurs were recovered, not to mention numerous other fossils representative of these ancient reptiles. There was a gigantic carnivorous dinosaur, closely related to if not identical with the well-known giant predator from the Cretaceous beds of Montana, *Tyrannosaurus*. There was also found the gigantic duck-billed dinosaur *Saurolophus*, hitherto known from North America. These dinosaurs, as well as various other dinosaurian genera, including some of the small

Figure 12.18. A partially exposed dinosaur skeleton at the 'Tomb of the Dragons', Nemegt Basin, Mongolia, with a member of the Russian Expedition happily seated on the hard sandstone ledge that protected the skeleton from erosion. (Courtesy of the American Museum of Natural History, from A.K. Rozhdestvenskii and I.A. Efremov.)

ostrich-like dinosaurs, indicate quite clearly the nature of the close connections that bound central Asia to western North America during Cretaceous times. As mentioned previously, the continental regions now separated by the Bering Straits were then one great tropical land, a broad lowland across which numerous dinosaurian giants and their lesser brethren wandered far and wide, from east to west, with great facility.

Thus the digs in southwestern Mongolia by Efremov, Rozhdestvenskii, and their fellow workers revealed a new and a fabulous hunting ground for dinosaurian treasures. The Nemegt Basin, in the heart of central Asia, now takes a position of first rank along with other classic dinosaur localities, especially those

Figure 12.19. Mongolian yurts (felt tents, supported by a lattice-like framework), the characteristic habitations of the Gobi nomads, with a truck of the Russian Expedition in the background.

with which it is intimately related, such as the Red Deer River in Alberta and the Hell Creek area in northern Montana. Barnum Brown and the Sternbergs had revealed a chapter in Cretaceous history by floating down a northern river to dig in its bordering cliffs; Efremov and his party, like Andrews and the Central Asiatic Expeditions two decades earlier, added still another chapter to this history by driving in their scout cars and trucks through the outer reaches of Mongolia, to dig in the broad basins of that interior land (Figure 12.19). In the two continents prodigious quantities of new and significant dinosaur bones were discovered – partly by luck, partly by intuitive judgment, and very much by a vast amount of hard, back-breaking work.

The Soviet expeditions shipped back to Moscow some 120 tons of fossil bones, a considerable proportion of which was excavated from the Nemegt Basin. Here is one measure of their success. And why should there have been such great numbers of dinosaur bones and skeletons deposited in the Nemegt Basin? The answer to this question was given by Efremov and one of his associates, Novozhilov, as a result of their study of the geology of Nemegt (Figure 12.20). Here, they maintain, are sediments laid down in the delta of an enormous ancient river, sediments that represent not only deposition by the river itself but

also accumulations of sands and muds on the bottom of ponds and lakes dotted about the broad delta. This delta, perhaps forty kilometres in width, and undoubtedly covered with lush vegetation, was the habitat of many dinosaurs – herbivorous dinosaurs that fed upon the abundant plant life at hand, and carnivorous dinosaurs that preyed upon the inoffensive plant-eaters. It was an ideal environment for the support of a large dinosaurian population; and where such a population existed there were bound to be deaths and burials. Some of these animals were preserved in the ever-accumulating flow of sands and muds, transported across the delta through numerous river channels or caught in the silt traps of small lakes and ponds. And so a record of the dinosaurian life of Mongolia was preserved, in the unlighted depths of sands and muds turned to rock; and so, as a result of long years of erosion, and of the efforts of some men in whom the bump of curiosity is strongly developed, this record was brought from its dark depths back to the light.

One interesting aspect of the Russian explorations in the Gobi, especially in the Nemegt Basin, is the manner in which the discovery of late Cretaceous dinosaurs, especially the giant duck-billed dinosaur *Saurolophus*, fulfilled in part a prophecy made by Professor Osborn in 1930.

There are still great unknown or unfossiliferous gaps to be filled in the prehistory of the ancient life of the Gobi Desert. Our explorations have as yet not revealed the closing periods of the Lower Cretaceous nor the closing period of the Upper Cretaceous in which large ceratopsians, like *Triceratops*, as well as large iguanodonts, like *Trachodon*, will doubtless be found (Osborn, 1930, p. 542).

Thus some of the gaps in the fossil record of Mongolia were filled, and the vision of central Asia as a great hunting ground in which would be found the fossilized remains of the last of the dinosaurs, as set forth by Osborn and Granger so many years ago, became a reality. It was the Nemegt Basin within which Upper Cretaceous dinosaurs came to light in remarkable numbers, and it was to the Nemegt Basin that the Russians returned, again and once again.

Figure 12.20. When the desert sun gets hot, the palaeontologists retreat to the comparatively cool shade beneath a truck, to study their maps and make plans: I.A. Efremov on the left, A.K. Rozhdestvenskii on the right, N.I. Novozhilov, chief of the prospecting group of the expedition, in the centre.

Mongols and Poles

And it was the Nemegt Basin that attracted other palaeontologists to its fossiliferous slopes, a decade and a half after the Soviet expeditions had completed their explorations. In 1964 what may be called the third phase of the Asiatic dinosaur rush began – this time with Mongols and Poles as the dinosaur hunters. When Roy Andrews led the pioneering American Museum expeditions into central Asia in the twenties, the Mongols were by and large nomadic shepherds, with their attention centred upon mutton and wool and camels and horses. Dinosaur skeletons were to them esoteric and perhaps incomprehensible objects. When Efremov led the elaborate Soviet expeditions into central Asia in the forties, Mongolia was in the throes of an industrial and cultural revolution. Ulaanbaatar was becoming a modern city, and many Mongols were turning away from the primitive life of their forebears, to participate in a twentieth-century world of sophisticated art, technology, and science.

The university in Ulaanbaatar was developing rapidly, and Mongolian scientists were making plans for the exploitation of their own scientific resources. Needless to say, one of the outstanding scientific resources of this nation is to be found in marvellous fossil beds of western Mongolia, with rich deposits of dinosaur skeletons. For the proper development of their plans, some outside help would be useful to the Mongolian palaeontologists.

So an agreement was made with Polish palaeontologists for joint work in Mongolia, and in 1964 the first Mongolian–Polish expedition entered the field. The Mongols and Poles concentrated a considerable amount of their efforts in the Nemegt Basin, where they, like their Russian predecessors, collected late Cretaceous dinosaurs, including a tyrannosaur skeleton. They also gave a great deal of attention to the famous Bayan Zag locality, the region of the Flaming Cliffs, which so long ago had been christened Shabarakh Us by the American Museum expedition, and there they collected *Protoceratops* skeletons and eggs as might be expected, a complete skeleton of the armoured dinosaur *Pinacosaurus*, and various other fossil reptiles. Of particular importance was their discovery at this locality of numerous Cretaceous mammals similar to the early mammals found here years before by Walter Granger and his associates, those little warm-blooded animals living with the dinosaurs, the ancestors of the many, varied mammals that became the rulers of the earth after the dinosaurs had become extinct. In addition to work at these two famous localities, the Mongolian–Polish expedition excavated excellent dinosaur skeletons and other remains at several other sites in Mongolia, adding to their collections still more tyrannosaur skeletons, as well as skeletons of armoured and duck-billed dinosaurs. The results of their work were spectacular, not so much because of completely new and unexpected discoveries, as for the fine fossils of known and expected dinosaurs and other extinct animals. It is always a good thing to get abundant and complete fossils for study and display; the worth of such collections should never be underestimated.

This latest phase of the extended Asiatic dinosaur hunt is but one part of a search for dinosaurs that began in Mongolia almost a half-century ago. It shows us that in our modern world, so reduced in size by the development of twentieth-century technology and especially by the speed of jet communications, there are even now faraway places where the romance of hunting for dinosaurs in an empty land still awaits the fossil hunter. It carries on the search for dinosaurs that was begun initially by accident and then very much by design by Roy Andrews, Walter Granger, and their associates in 1922. It continues the tradition of great dinosaur expeditions and excavations, begun almost a century ago by Marsh and Cope in the wild and then uninhabited western territories of North America.

Polish and American expeditions, 1960–2000[2]

Joint Russian–Mongolian expeditions continued from the 1940s to the present day (see Chapter 13). At the same time, other countries, notably Poland and the United States, have mounted major palaeontological expeditions to Mongolia. The Polish work has been reviewed by Kielan-Jaworowska (1969a, b, 1975) and Lavas (1993, chapter 5).

Polish involvement in Mongolia began in 1955, when Zofia Kielan (later Kielan-Jaworowska), a young palaeontologist from the Institute of Palaeozoology in Warsaw, Poland, visited Moscow. There she saw many of the remarkable fossil reptiles and mammals collected by the Russian expeditions from Mongolia, and she met the key palaeontologist from those expeditions, A.K. Rozhdestvenskii. At the time, Kielan was working on trilobites and Palaeozoic worms, but she made a complete switch in the 1960s to Mesozoic mammals, a field in which she has become a leading international expert (see Chapter 29).

In 1961, members of the Academies of Sciences of the COMECON countries met in Warsaw, and Polish representatives proposed to the chairman of the Mongolian Academy of Sciences, Professor Shyrendyb, that joint palaeontological expeditions should be organized. An agreement was signed in Ulan Baatar in 1962, and the first exploratory expedition set out in 1963, led by Julian Kulczycki on the Polish side, and Naydin Dovchin for the Mongolians. Larger expeditions ran in 1964 and 1965, and focussed mainly on Mesozoic localities, but also worked some Cenozoic sites.

Polish palaeontologists who worked in Mongolia included Zofia Kielan-Jaworowska (1964, 1965), the expedition leader, Kazimierz Kowalski (1964), Andrzej Sulimski (1964), Magdalena Borsuk-Białynicka (1964), Teresa Maryańska (1964, 1965), Halszka Osmólska (1965), Aleksander Nowiński (1965), Józef Kazmierczak (1965), Jerzy Malecki (1965), and Henryk Kubiak (1965), and the geologists Ryszard Gradziński and Jerzy Lefeld worked in Mongolia in 1964 and 1965. Mongolian palaeontologists included Demberlyin Dashzeveg (1964, 1965), Naydin Dovchin (1964, 1965), and Rinchen Barsbold (1965). These scientists were assisted by field technicians and drivers, which brought the total numbers to 16 in 1964 and 23 in 1965.

---

[2] This final section is added to Edwin Colbert's narrative to bring the story up to date. The treatment is brief, and reference is made to other recent accounts. M.J.B

Publications by the Polish scientists brought Mongolian dinosaurs to the notice of the world again. Major discoveries of the 1964 and 1965 expeditions included six incomplete skeletons of *Tarbosaurus*, two new sauropods, *Opisthocoelicaudia* based on a headless skeleton) and *Nemegtosaurus* (based on a skull), three skeletons of the new ornithomimosaur *Gallimimus*, the vast arms and hands of the new theropod *Deinocheirus*. Further highly important finds included some beautifully preserved therian mammals, *Zalambdalestes* and *Kennalestes* from the Djadokhta Formation of Bayan Zag.

Smaller Polish–Mongolian expeditions to Bayan Zag took place in 1967, 1968, and 1969, and then two further major expeditions were organized in 1970 and 1971, again focussing on the Bayan Zag and Nemegt Basin localities. Polish expedition members included Zofia Kielan-Jaworowska (1970, 1971), Teresa Maryańska (1970, 1971), Aleksander Nowiński (1970), Halszka Osmólska (1970), Hubert Szaniawski (1970), Adam Urbanek (1970), Andrzej Balinski (1971), Cyprian Kulicki (1971), and Andrzej Sulimski (1971). Mongolian scientists included Rinchen Barsbold (1970), Demberlyin Dashzeveg (1970), and Altangerel Perle (1970, 1971). The most important finds included four new pachycephalosaurs, *Tylocephale*, *Prenocephale*, *Homalocephale*, and *Goyocephale*, a new ankylosaur, *Saichania*, and the famous 'fighting dinosaurs', *Velociraptor* and *Protoceratops*, apparently locked in mortal combat, collected in 1971.

With these expeditions, Polish involvement in expeditions to Mongolia ended. The materials collected during the 1960s and 1970s continued to provide materials for extensive descriptions of dinosaurs and mammals by Polish and Mongolian scientists. At the same time, during the 1970s and 1980s, Russian–Mongolian joint expeditions took over as the main field campaign (see Chapter 13). Russian-sponsored work diminished after 1990.

Currently, a variety of agencies are sponsoring palaeontological expeditions in Mongolia, especially from the United States and from Japan. The most fruitful series of recent expeditions have been established by the American Museum of Natural History (AMNH), and these have run since 1990 (Novacek *et al.*, 1994; Novacek, 1996; Webster, 1996), which concentrated first on previously known sites in southern Mongolia. Sites visited include Höövör, Tatal Gol, Hermiin Tsav, Khulsan, Tögrögiin Shiree, Bayan Zag (Flaming Cliffs), and Khar Hötöl.

A key discovery on the 1993 AMNH expedition was the new locality Ukhaa Tolgod ('brown hills'), in the Nemegt Basin, where huge numbers of fossils were found in sediments that combine evidence for equivalence of age with the Djadokhta Formation (type locality: Bayan Zag) and the Barungoyot Formation. The fossils from Ukhaa Tolgod include over 400 specimens (Dashzeveg *et al.*, 1995) of dinosaurs (ankylosaur, *Bagaceratops*, carnosaur, troodontid, oviraptorids, *Velociraptor*), lizards (five species), a turtle (? *Basilemys*), birds (*Mononykus*), and mammals (14–15 species). The putative flightless bird *Mononykus* (Norell *et al.*, 1993; Perle *et al.*, 1993) and the discovery of the so-called 'egg thief' *Oviraptor* apparently incubating a nest of eggs, rather than stealing them (Norell *et al.*, 1994, 1995), have attracted a great deal of attention.

References

Andrews, R.C. 1929. Mongolia – explorations, pp. 713–715 in *Encyclopedia Britannica*, 14th edn., 15. New York: Encyclopedia Britannica.

—1932. *The New Conquest of Central Asia. Natural History of Central Asia* 1. New York: American Museum of Natural History, 678 pp.

Colbert, E.H. 1968. *Men and Dinosaurs*. New York: Dutton; London: Evans, 283 pp. [reprinted 1984, as *The Great Dinosaur Hunters and their Discoveries*, by Dover, New York].

Dashzeveg, D., Novacek, M.J., Norell, M.A., Clark, J.M., Chiappe, L.M., Davidson, A., McKenna, M.C., Dingus, L., Swisher, C. and Perle, A. 1995. Extraordinary preservation in a new vertebrate assemblage from the Late Cretaceous of Mongolia. *Nature* 374: 446–449.

Efremov, I.A. 1956. *Doroga Vetrov* [Road of the Wind]. Moscow: Trudrezervizdat, 360 pp.

Kielan-Jaworowska, Z. 1969a. *Hunting for Dinosaurs*. Cambridge, Mass.: MIT Press, 177 pp.

—1969b. Fossils from the Gobi Desert. *Science Journal* 5A: 32–38.

—1975. Late Cretaceous mammals and dinosaurs from the Gobi Desert. *American Scientist* 63: 150–159.

Lavas, J.R. 1993. *Dragons from the Dunes*. Auckland: Academy Interprint Ltd., 138 pp.

Norell, M.A., Chiappe, L.M. and Clark, J.M. 1993. A new limb on the avian family tree. *Natural History* 102: 38–43.

—, Clark, J.M., Dashzeveg, D., Barsbold, R., Chiappe, L.M., Davidson, A.R., McKenna, M.C., Perle, A. and Novacek, M.J. 1994. A theropod dinosaur embryo and the affinities of the Flaming Cliffs dinosaur eggs. *Science* 266: 779–782.

—, Clark, J.M., Chiappe, L.M. and Dashzeveg, D. 1995. A nesting dinosaur. *Nature* 378: 774–776.

Novacek, M.J. 1996. *Dinosaurs of the Flaming Cliffs*. New York: Doubleday, 369 pp.

—, Norell, M.A., McKenna, M.C. and Clark, J.M. 1994. Fossils of the Flaming Cliffs. *Scientific American* 271 (6): 60–69.

Osborn, H.F. 1930. Ancient vertebrate life of central Asia. Discoveries of the Central Asiatic Expeditions of the Museum of Natural History in the years 1921–1929, pp. 519–543 in *Livre Jubilaire, Centenaire de la Société Géologique de France*. Paris: Société Géologique de France.

Perle, A., Norell, M.A., Chiappe, L.M. and Clark, J.M. 1993. Flightless bird from the Cretaceous of Mongolia. *Nature* 362: 623–626.

Rozhdestvenskii, A.K. 1960. *Chasse aux Dinosaures dans le Désert de Gobi*. Paris: Librairie Arthème Fayard.

Simpson, G.G. 1942. Memorial to Walter Granger. *Proceedings of the Geological Society of America for 1941*, pp. 159–172.

Webster, D. 1996. Dinosaurs of the Gobi. *National Geographic* 1996 (7): 70–89.

13

The Russian–Mongolian expeditions and research in vertebrate palaeontology

EVGENII N. KUROCHKIN AND RINCHEN BARSBOLD

Introduction

Central Asia attracted the attention of palaeontologists after Tertiary mammals had been found in continental sediments. At first, Richthofen (1877) argued that Central Asia had been flooded by a huge sea which produced marine deposits, the so-called Khankha deposits. Borisyak (1915) predicted that Mongolia would be a storehouse of palaeontological treasures, based on his study of Tertiary mammals in adjacent Kazakhstan, and on the discovery of a brontothere tooth in Tertiary rocks on the Plateau Khooldzin, south of Iren Dabasu Lake, Inner Mongolia, China, by Vladimir Obruchev in 1892 (Obruchev, 1893). Borisyak expected rich finds of Tertiary mammals in Mongolia, since he considered that Central Asia was the centre of origin of Cenozoic mammals. These prognoses stimulated the Central Asiatic Expedition of the American Museum of Natural History in the 1920s, and the environs of Iren Dabasu Lake were the first area investigated; here the Expedition discovered a Cretaceous fauna of dinosaurs and Paleogene and Neogene mammals (Granger and Berkey, 1922).

Why is Mongolia so rich in remains of ancient vertebrates? During the past 200 Myr, since the early Mesozoic, the territory of Mongolia was never covered by the sea. The arid continental climate of the past 30 Myr has not encouraged the formation of a thick soil cover, nor the development of vegetation, and has generated strong erosion by water, wind, and temperature change. Rapid sedimentation, associated with significant water supplies and broken relief, occurred for significant periods of time in the Cretaceous, Paleogene, and Neogene over a large part of southern Mongolia, and this promoted the preservation of fossils. Moderate tectonic activity has generated small escarpments which reveal much of the thickness of the sedimentary succession. The absence of Pleistocene ice sheets meant that the ancient deposits were not erased. Thus, in Central Asia, there is a nearly continuous series of lake and river deposits containing continental biota, beginning in the Late Jurassic, and in some places in the Late Triassic.

The territory of Mongolia extends nearly 2400 km from east to west and nearly 1300 km from north to south. Most of Mongolia is a middle-level mountain plateau with average heights 1000–1200 m above sea level. In the southern half of the country, mountain ridges of the Mongolian and Gobi Altai with heights from 1500 to 4000 m above sea level, lie on either side of extended depressions, filled with Mesozoic and Cenozoic deposits, the products of erosion of the surrounding high countries (Murzaev, 1948). In these depressions the main palaeontological riches of Mongolia are buried.

Abbreviations

ARAS, Archives of the Russian Academy of Sciences; AS USSR, Academy of Sciences of the USSR; F, Fund of the ARAS; CAE, Central Asiatic Expedition of the American Museum of Natural History; L, List of the ARAS; MAS, Mongolian Academy of Sciences; MPE, Mongolian Palaeontological Expedition of the Academy of Sciences of the USSR; MPR, Mongolian Peoples' Republic; P, Page of the ARAS; PIN, Palaeontological Institute of the RAS; RAS, Russian

Academy of Sciences; RMPE, Joint Russian–Mongolian Palaeontological Expedition; SMPE, Joint Soviet–Mongolian Palaeontological Expedition; U, Deposition Unit of the ARAS.

The first discoveries of Mesozoic and Cenozoic vertebrate sites

From 1922 to 1930, the Central Asiatic Expedition of the American Museum of Natural History (CAE), under the leadership of Roy Chapman Andrews, worked in Mongolia. They found Late Cretaceous faunas of vertebrates (*Protoceratops*, small carnivorous dinosaurs, clutches of dinosaur eggs, mammals) at the Bayan Zag site in Southern Mongolia and Early Cretaceous dinosaurs (*Psittacosaurus*) and fishes in the Andai Khudag and Öösh sites in the Lakes Valley of Central Mongolia. The CAE also opened the Palaeocene Gashato site near Bayan Zag and the Oligocene Ardyn Ovoo (= Ergil Ovoo, = Ergiliin Zoo) site in south-eastern Mongolia and the Tatal Gol site in the Lakes Valley, which produced many new Paleogene mammals. Berkey and Morris (1927a, b) divided the Cretaceous deposits of Mongolia into 13 formations, and they considered that Late Cretaceous climates were arid and semi-desert, though in the Early Cretaceous epoch was more humid with many lakes, a wide river network, and rich vegetation.

Practically simultaneously with the CAE, Soviet geologists began searching for minerals in Mongolia as technical assistance to the young Mongolian Republic. The full story of the geological exploration of Mongolia and adjacent regions of China by Russian and Western scientists has been told by Marinov (1967). From this book, we will summarize the main projects by Soviet scientists.

From 1925 on 1932 geological expeditions of the Academy of Sciences of the USSR to Mongolia were led by I.P. Rachkovskii. These expeditions investigated rich Tertiary vertebrate sites in Western and Eastern Mongolia and Cretaceous dinosaur localities in Eastern Mongolia (Kupletskii, 1926; Lebedeva, 1926, 1934; Rachkovskii, 1928; Rachkovskii and Lebedeva, 1932; Belyaeva, 1937). In 1925, a geological expedition led by B.S. Dombrovski from the Far-Eastern University of Vladivostok conducted research in Central and Eastern Mongolia. It did some work on the new large dinosaur sites (Dombrovski, 1926). Further expeditions of Soviet scientists conducted researches in Mongolia from 1920 to 1930. The geographers and botanists A.D. Simukov, E.M. Murzaev, A.A. Yunatov, and B.M. Chudinov found Cretaceous dinosaurs in the Southern Gobi. In the Eastern Gobi, 20 sites of Paleogene mammals and Cretaceous dinosaurs were found by the geologists A.P. Chaikovskii, A.N. Alekseichik, N.I. Delnov, and Yu.S. Zhelubovskii (Marinov *et al.*, 1973), and later, some of these data were used by I. Efremov.

Mongolian Palaeontological Expedition (MPE) of the Academy of Sciences of the USSR, 1946–1949

At the end of 1940, the Scientific Committee of the MPR sent a letter to the director of the PIN, A. Borisyak, inviting him to organise a palaeontological expedition to Mongolia in 1941. The offer was agreed with the Praesidium of the AS USSR and by the Council of the Peoples' Commissars of the USSR, and money was allocated for the expedition. Yu.A. Orlov was nominated chief of the expedition, and I.A. Efremov vice-chief. Plans for the expedition included equipment for 10 people and an estimated duration of 3.5–4 months of field work, covering Southern Mongolia, the Trans-Altai and Middle Gobi, and Western Mongolia. Three GAS-AA motor trucks were received. However, because of a delay in the receipt of foreign passports for travel outside the USSR, the expedition could not leave Moscow at the end of May, as was planned, and it was postponed to 1942 (ARAS; F. 1712; L. 1; U. 18; pp. 1–38.). Then, in June 1941, Germany attacked the USSR, and clearly the work of the MPE had to be postponed.

As early as the end of 1945, Yu. Orlov submitted a request to run the Palaeontological expedition to Mongolia to the Praesidium of the AS USSR. This petition was accepted by the Council of Ministers of the USSR (Resolution N 2051 PC of February 16,

1946) and on March 28, 1946, the Praesidium of the AS USSR gave an order for the organization of the MPE for a period of seven months (ARAS; F. 1712; L. 1; U. 73.; pp. 1 and 8). The expedition left Moscow at the beginning of August, 1946. On August 10th it reached Ulaanbaatar, and on September 1st began to work at Bayan Zag. The leader of the MPE was now I.A. Efremov, and Yu.A. Orlov was its scientific adviser. The preparators J.M. Eglon and M.F. Lukiyanova also worked in Mongolia on the first expedition, as did the scientists K.K. Flerov, V.I. Gromov, and A.A. Kirpichnikov. The 1946 MPE worked for 2.5 months in the field. In Dalanzadgad, a forwarding base with stocks of petrol was created, from which forwarding routes extended over the whole of Southern and Eastern Mongolia. The motor vehicles travelled a total of 4700 km. The expedition returned to Ulaanbaatar on November 4th, and left for Moscow by rail on January 7th, 1947.

The MPE in 1946 carried out reconnaissance and prospecting on three main routes (Efremov, 1948, 1949; Orlov, 1952). North of Dalanzadgad, a northern group of routes included surveys at Bayan Zag, and the new Late Cretaceous sites Ulaan Öösh and Algui Ulaan Tsav were opened up. The western group of routes passed the foot of the Gilbent, Nemegt, and Altan Uul ridges and their surrounding depressions. In this direction, the rich Late Cretaceous dinosaur sites Nemegt and Altan Uul were opened, and the important Naran Bulag site with Paleogene mammals was found. The Late Cretaceous Shiregiin Gashuun site with crocodiles, north of Nemegt, and the Paleogene Gashato site were also surveyed. South of the Nemegt Ridge, in the region Noyon Sum, I.A. Efremov opened a section of continental Permian deposits, many kilometres long, with the remains of plants and trunks of cordaites (Efremov, 1952). On the eastern route from Dalanzadgad to Sainshand, new Late Cretaceous sites with dinosaurs were found at Bayan Shiree, Khamaryn Khural, Khar Hötöl, and Tüshleg, and excavations were also made at the Paleogene Ergil Ovoo locality, found by the CAE (Figure 13.1).

Despite the fact that the first MPE was essentially a reconnaissance trip, it was outstandingly successful,

both in terms of finds of Paleogene mammals and Late Cretaceous fossils. New large dinosaur sites were discovered in the Southern and Eastern Gobi. Particularly important were the finds in the Upper Cretaceous rocks of hadrosaurs and sauropods, as well as large terrestrial carnosaurs and ankylosaurs, generally unknown from the Old World, and abundant finds of fossil trees, crocodiles, and fishes. These fossils all suggested to the Soviet geologists that in the Late Cretaceous Central Asia was covered with extensive lakes, bogs, and large rivers, opposite to the usual view that these territories had experienced arid conditions since the Mesozoic.

The whole of 1947 was devoted to the preparation of the second expedition. A main task of this expedition was excavation at Nemegt and Bayan Shiree with large numbers of workers and large numbers of trucks. The trucks, equipment, and drivers were sent from Moscow to Ulaanbaatar in November 1947, and in December the expedition leaders arrived. The field base at Dalanzadgad was set up during the severe Mongolian winter with stocks of equipment and petrol, and one of the field teams began work in the Eastern Gobi in March, 1948. The second MPE ran for 11 months in 1947–1948, and participants included 16 employees of PIN and 10 hired hands (ARAS; F. 1712; L. 84; U. 28). I.A. Efremov continued as leader, and Yu.A. Orlov as scientific adviser. The scientists present were N. Novozhilov, J. Eglon, A. Rozhdestvenskii, and E. Maleev, the preparators were M. Lukiyanova and V. Presnyakov, and the drivers were V. Pronin, T. Bezborodov, N. Vylezhanin, I. Likhachev, and others.

The expedition left Ulaanbaatar for the Eastern Gobi on March 18th. Work at Bayan Shiree and Ergil Ovoo continued until April 20th, and at the same time remains of large sauropods, carnivorous dinosaurs, and turtles were found at the Lower Cretaceous Khar Hötöl locality, south of Sainshand. In the same region, fossil Cretaceous wood with huge vertical trunks of *Taxodium* were found. After April 20th, the main excavating team began work at Bayan Zag and, from the beginning of May, moved to the Nemegt locality. The skeletons of a huge hadrosaur (*Saurolophus angustiros-*

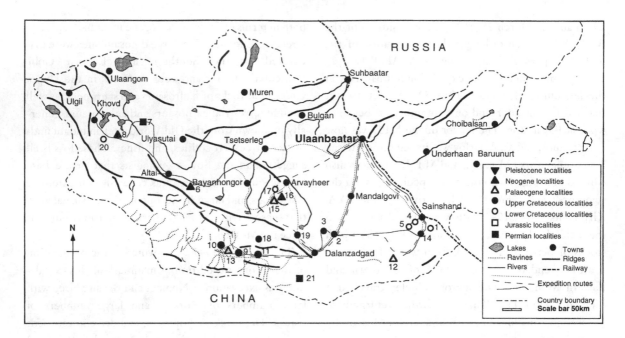

Figure 13.1. Mongolian Palaeontological Expedition of the Academy of Sciences of the USSR, 1946–1949. Map of the routes, localities, and newly discovered fossiliferous areas. Compiled by I. Efremov for Marinov (1967), with some additions. Localities: 1, Khamaryn Khural; 2, Algui Ulaan Tsav; 3, Ulaan Öösh; 4, Khar Hötöl; 5, Tushilge; 6, Beger Nuur; 7, Tsast Bogd; 8, Altanteel; 9, Altan Uul; 10, Tsagaan Uul; 11, Nemegt; 12, Ergil Ovoo; 13, Naran Bulag; 14, Bayan Shiree; 15, Tatal Gol; 16, Loo; 17, Andai Khudag; 18, Shireegiin Gashuun; 19, Bayan Zag; 20, Osh Nuur; 21, Noyon Sum.

tris), small carnivorous dinosaurs, isolated skulls of dinosaurs, huge turtles, crocodiles, and fishes were excavated. In Altan Uul, the Dragon's Tomb site was opened, where complete skeletons of *Tarbosaurus* and *Saurolophus* with the remains of fossilized skin were dug out. West of this site a new rich Late Cretaceous site, Tsagaan Uul (later named Tsagaan Khushuu), was opened. In Nemegt and Altan Uul, work continued nearly all summer and autumn. At the same time, a prospecting trip set out into western regions of the Southern Gobi, through the Trans-Altai Gobi, with a route through the Lakes Valley. Productive sites were not found south and west of Nemegt, but to the north, around the Mongol Altai Mountains, extensive outcrops of Lower Cretaceous sediments were exposed. North of Nemegt, at the bottom of the Ih Bayan Uul Mountains, vertebrates were found in the Eocene. The Lower Cretaceous Öösh Nuur and Andai Khudag localities, and the Oligocene Tatal Gol, and

Miocene Loo localities, which had been opened by the CAE, were surveyed in the Lakes Valley. At the end of the season, there was a further excavation at Ergil Ovoo in the Eastern Gobi. The 1948 MPE covered 14 000 km in total (Figure 13.1). Everyone returned to Moscow at the end of October (ARAS; F. 1712; L. 1; U. 97).

The second MPE obtained extraordinary materials of various dinosaurs, including complete skeletons of huge specimens 25–30 m in length, crocodiles, turtles, and mammals. In addition, detailed observations on the taphonomy of fossil-bearing river channels that ran into the large lake basins, and on the palaeogeography of dinosaur occurrences on the extensive lowlands of ancient Mongolia, covered with woods and bogs, and crossed by the rivers and covered by the lakes. It is necessary, however, to note that I.A. Efremov was wrong about the low palaeontological potential of the regions west of the Nemegt

Depression and south of the Gilbent Ridge, since some major discoveries were made there in the 1970s to 1990s.

In 1949, the MPE worked in the field from June 11th to September 23rd, and this time it consisted of 33 people, including the labourers. The scientific structure was the same as in 1948. The first trip was devoted to the search for vertebrates in the Beger Nuur Depression, the Zereg Depression, and the Gui Suin Gobi Depression. The rich Miocene Beger and Pliocene Altanteel mammal localities were opened up there, and Miocene vertebrates were also found at the western foot of the Jargalantyn Mountains, at the Öösh Ridge, 50 km from Khovd town. The Tertiary deposits here overlie Lower Cretaceous, where isolated remains of sauropods were found. In the region of the Tsast Bogd Mountains (Figure 13.1), continental Permian deposits were surveyed, but these yielded only plant remains. At the end of July, the expedition made short trips to the south-east in the region of Sainshand, to the Khar Hötöl locality, and to the south, to collect small dinosaurs from the Bayan Zag locality. After August 1st, large excavations were resumed in the Nemegt Depression at the Nemegt, Altan Uul, and Tsagaan Uul sites, and these yielded two skeletons of the large hadrosaur *Saurolophus*, two skulls, the skull of a young specimen, and blocks with fossilized skin of these dinosaurs. After the end of work on the Cretaceous sites, excavations continued at the Paleogene sites of Naran Bulag, Ulaandel Uul, and Tatal Gol, in the same region to the west, north, and east of the Nemegt Ridge. These yielded magnificent specimens of mammals, turtles, and fishes. On October 4th, all members of the expedition returned to Ulaanbaatar (ARAS; F. 1712; L. 1; U. 111).

In 1950, a last expedition to Mongolia was planned, but unexpectedly, in May, when the equipment had already been sent to Ulaanbaatar, the government of the USSR terminated the MPE and transferred all facilities, including trucks, to an agricultural expedition (ARAS; F. 1712; L. 1; U. 8 4). The reasons for this decision are still not clear. The sole surviving witness of those events, B. Trofimov, explains them as probably the result of diplomatic games connected with

the People's Republic of China, which had just been created.

In the end, then, the MPE worked for three seasons. Its main achievements were the excavation of diverse Late Cretaceous dinosaurs, especially in the Nemegt region, and of Palaeocene and Early Eocene vertebrates, the division of the Cretaceous faunas into three groups, the collection of new evidence about the palaeogeography and climate of Mongolia in the Cretaceous, and discovery of evidence against arid conditions (Efremov, 1948, 1949, 1950, 1952, 1953, 1954a, b, 1955; Efremov *et al.*, 1954). The history and results of the MPE have been described in detail by Efremov (1963), as well as by Rozhdestvenskii (1957, 1969), Chudinov (1987), and Lavas (1993).

Joint Soviet (Russian)–Mongolian Palaeontological Expedition

In 1964 the MAS invited B.A. Trofimov and P.K. Chudinov, employees of the PIN, to inspect the Bügiin Tsav locality, found by the Mongolian arats a little north of the Altan Uul Ridge. This Late Cretaceous locality has turned out to be one of the richest dinosaur sites in Mongolia, with many complete and partial skeletons of large and small dinosaurs exposed on the surface, although initial estimates of its richness (Chudinov, 1966; Trofimov and Chudinov, 1970) were rather exaggerated. Subsequently, large excavations at Bügiin Tsav by the SMPE and other expeditions produced magnificent turtles, interesting birds and mammals. Bügiin Tsav has now become an important palaeontological attraction for foreign tourists.

The Polish–Mongolian Palaeontological Expedition operated in Mongolia from 1963 to 1971, at first led by J. Kultchitskii, and then by Z. Kielan-Jaworowska. Rinchen Barsbold was Chief of the Expedition on the Mongolian side. These expeditions were notable for retrieving skeletons of large dinosaurs, for further developing Late Cretaceous sites in the Nemegt Depression, but especially for collecting large numbers of specimens of mammals and lizards at Bayan Zag and other localities of the Late

Figure 13.2. Excavations at the Tögrögiin Shiree locality, Ömnögov Aimag, Upper Cretaceous, Baruungoyot Svita, in September, 1969. From the left, M. Bragin, A. Tchangtoomoor, P. Chudinov, G. Namsray, N. Radkevich, A. Ponomarenko, A. Perle, I. Luk'yanov, P. Narmandakh and R. Barsbold. Photo by E. Kurochkin.

Cretaceous. The expedition discovered such important Late Cretaceous localities as Tögrögiin Shiree and Khulsan, and the finds of Pachycephalosauria, *Gobipteryx*, and Cretaceous bird embryos gave new directions to the palaeontology of Mongolia.

From 1967, the Joint Soviet–Mongolian Geological Expedition of the AS USSR and MAS began to work in Mongolia. The geologists G. Martinson, E. Devyatkin, A. Sochava and V. Shuvalov worked on palaeontological aspects of the Mesozoic and Cenozoic, discovered some new vertebrate localities, and collected a number of new fossil vertebrates. The Mongolian palaeontologists R. Barsbold, D. Dashzeveg, P. Khosbayar and T. Tomurtogoo began their scientific careers working with the Soviets on this expedition.

This expedition and the SMPE worked in close collaboration, and results were published in many joint papers and several monographs (Devyatkin, 1981; Martinson, 1982; Yanovskaya *et al.*, 1977).

At the end of 1960, Yu.A. Orlov addressed the Praesidium of the AS USSR with an offer to organize palaeontological researches in the system of the MAS. On a slip of paper, someone has written that there were no palaeontologists in Mongolia, although there is a palaeontological division in the State Museum. A proposal was made to organize joint palaeontological expeditions for five years in 1961–1965, and to use this to prepare two or three Mongolian palaeozoologists (ARAS; F. 1712; L. 1; U. 320; P. 1–2, P. 4–6). However, because of bureaucratic and political delays, this idea

Figure 13.3. R. Barsbold and G. Namsray prepare the skeleton of *Protoceratops* for plastering, at the Tögrögiin Shiree locality, September, 1969. Photo by E. Kurochkin.

was achieved only after the death of Yu.A. Orlov in 1966, when the Praesidium of the AS USSR made a decree on August 5th, 1968 about the organization of a Joint Soviet–Mongolian Palaeontological Expedition. A. Vologdin, Corresponding-member of the AS USSR, was nominated as chief of the SMPE and K. Flerov as scientific adviser (ARAS; F. 1712; L. 1; U. 320; P. 37–38). However, the main organizational work for the start of the SMPE were carried out by the director of the PIN, N. Kramarenko, and by the scientific researchers V. Zhegallo and Yu. Voronin. R. Barsbold was head of the SMPE on the Mongolian side.

In the following synopsis, attention will focus on the Mesozoic projects of the SMPE. The SMPE began work in 1969, when more than 40 people left Ulaanbaatar for the South Gobi in seven trucks and jeeps. The expedition was divided into an excavating team led by E.N. Kurochkin and an exploratory team led by V. Zhegallo, the Vice-Chief of the SMPE. Participants in the first season of field work included P. Chudinov, M. Shishkin, V. Sukhanov, A. Ponomarenko, M. Erbaeva, V. Reshetov, N. Kalandadze, R. Barsbold, A. Perle, E. Khand, P. Narmandakh, G. Namsray, and others, as well as many drivers, technicians, and students from Moscow and Ulaanbaatar. The main camp of the vertebrate teams was located at the Bayan Zag locality, but the main excavations took place at the Tögrögiin Shiree locality, about 40 km from Bayan Zag (Figures 13.2 and 13.3). In this season, the SMPE explored numbers of known Cretaceous, Paleogene, and Neogene localities in South, Central, and Western Mongolia. Field work finished by the middle of October, when strong night frosts start in the South Gobi (Figure 13.4). However, the discovery of the

Figure 13.4. Excavations at the Alag Teeg locality, Ömnögov' Aimag, Upper Cretaceous, Baruungoyot Svita, October 10th, 1969. From the left, G. Namsray, V. Reshetov, I. Luk'yanov, S. Kurzanov, and N. Radkevich. Photo by E. Kurochkin.

Lower Cretaceous Guchinus (then named Höövör) locality with a rich fauna of mammals (Figure 13.5) and lizards, and the Late Cretaceous Alag Teeg locality with numerous ankylosaurs, as well as the richest Eocene locality, Khaichin Uul II, made this beginning of the SMPE very successful, even though most of the Soviet participants had not had field experience in Mongolia before (Kurochkin *et al.*, 1970).

From 1970 to 1979, the SMPE prospected and excavated fossil vertebrates all over Mongolia for 2–4 months each year. There were 3–6 field crews, in total 30–40 people in 10–15 trucks (Figure 13.6). R. Barsbold remained the Chief of the SMPE from the Mongolian side all those years; on the Soviet side, after the first two years, N. Kramarenko headed the Expedition, and then Yu. Voronin, Yu. Popov, V. Sysoev, V. Reshetov and I. Manankov were consecutively heads of the SMPE and RMPE. Academician L.

Tatarinov was scientific adviser of the Expedition from 1975. Leaders of field teams were V. Reshetov, R. Barsbold, E. Kurochkin, V. Tverdokhlebov, S. Kurzanov, E. Dmitrieva, V. Ochev, N. Kalandadze, V. Zhegallo, D. Dashzeveg, E. Sychevskaya, I. Novodvorskaya, Yu. Tzybin and V. Yakovlev. In addition, many technicians and drivers from the PIN participated: V. Veselkin, V. Dorofeev, N. Radkevich, N. Frolkin, V. Chistoganov, V. Pronin, I. Likhachev, M. Bragin, L. Galukhina, I. Luk'yanov, and others. A number of Russians with Mongolian citizenship participated in the SMPE during the early years, and, during the first five years, many students from Saratov State University, Moscow University, and Perm' University took part in field work. Some of these students later became employees of the PIN, and others went to a variety of professional appointments throughout the USSR.

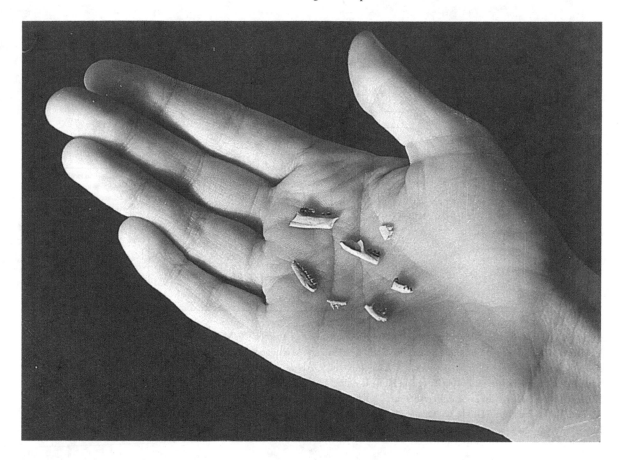

Figure 13.5. Jaw fragments of symmetrodonts and triconodonts found in 1969 at the Höövör locality, Övörkhangai Aimag, Central Mongolia, Lower Cretaceous, Andaikhudag Svita. Photo by E. Kurochkin.

Large excavations were carried out at the Lower Cretaceous Höövör and Hüren Dukh localities, and at the Upper Cretaceous Tögrögiin Shiree, Hermiin Tsav, Alag Teeg, Nogoon Tsav, Guriliin Tsav, Baishin Tsav, and Amtgai sites. Fishes, insects, and plants were collected widely in Lower Cretaceous deposits of Central and Western Mongolia. The East-Gobi team of the SMPE (E. Kurochkin) opened up in 1971 a completely new region of Lower Cretaceous deposits, the Züünbayan Svita, 150 km south-east of Sainshand. Here, in the localities Gashuuny Khudag, then renamed Khamaryn Us (Kalandadze and Kurzanov, 1974), and Tsakhiurt, good dinosaur specimens were found, including magnificent complete skeletons of *Psittacosaurus*. Champsosaurs were found for the first time in Asia at the Hüren Dukh site. Different groups of fishes were found at all horizons in the Cretaceous and Cenozoic. Early Cretaceous mammals and lizards were discovered, and thousands of bones were collected at Höövör. Clutches of huge sauropod eggs were collected at the Algui Ulaan Tsav locality (Figure 13.7). Rich faunas of vertebrates were discovered and excavated at the Late Palaeocene and Early Eocene Naran Bulag and Tsagaan Khushuu, at the Middle Eocene Khaichin Uul II–V, and at the Early Oligocene Ergileen Zoo (= Ergil Ovoo) localities (Figures 13.8 and 13.9). A number of fossil birds were found in the Lower Cretaceous and in all horizons of the Cenozoic. Rich localities with leaf floras of Late Palaeocene and Middle Eocene age from the Naran

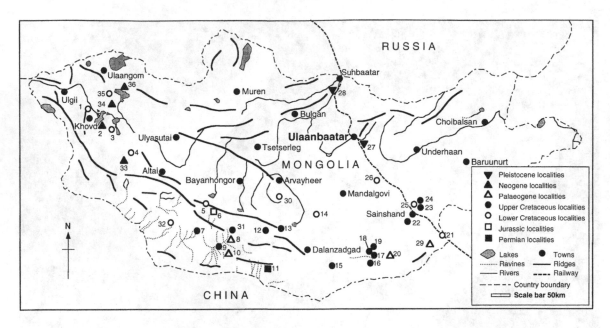

Figure 13.6. Joint Soviet–Mongolian Palaeontological Expedition of 1969–1979. Map of the localities, fossiliferous areas, new localities, and field teams. Compiled by E. Kurochkin. Localities: 1, Myangat; 2, Zagsokhairkhan; 3, Altanteel; 4, Gurvan Ereen; 5, Böön Tsagaan; 6, Bakhar; 7, Nogoon Tsav; 8, Khaichin Uul; 9, Hermiin Tsav; 10, Naran Bulag; 11, Sainsar Bulag; 12, Üüden Sair; 13, Tögrögiin Shiree and Alag Teeg; 14, Shine Khudag; 15, Shilt Uul; 16, Shar Tsav; 17, Baishin Tsav; 18, Amtgai; 19, Khongil Ovoo; 20, Ergiliin Zoo and Bayan Tsav; 21, Khamaryn Us; 22, Khongil Tsav; 23, Teel Ulaan Uul; 24, Baga Tariach; 25, Mogoin Bulag; 26, Hüren Dukh; 27, Nalaih; 28, Shaamar; 29, Khoyor Zaan; 30, Höövör; 31, Bügiin Tsav; 32, Dösh Uul; 33, Sharga; 34, Chono Kharaih; 35, Tatal Yavar; 36, Hirgis Nuur.

Bulag and Khaichin Uul sites have given important information on environments at that time (Figure 13.10).

The dinosaur finds were especially rich. In addition to *Protoceratops* (Figure 13.11), *Tarbosaurus*, and *Saurolophus*, genera that had been found before, the SMPE recovered new ankylosaurs, many different small and middle-sized theropods, Early Cretaceous iguanodontids, and complete skeletons of *Psittacosaurus*. New families of dinosaurs, the Oviraptoridae, Garudimimidae, Harpymimidae, Segnosauridae, Enigmosauridae, and Avimimidae were described on the basis of specimens collected by the SMPE (Perle, 1979; Barsbold, 1983; Kurzanov, 1987). Many sites also produced large collections of eggs and eggshells of dinosaurs and birds, some with embryos, and these provided the basis for an extensive systematic and structural study of eggshells

(Mikhailov, 1987, 1991, 1992, 1994). These results have created for Mongolia glory as the territory with the richest and most diverse fauna of dinosaurs in the world. The work of the SMPE focused on three questions: 1, the sequence of vertebrate faunas in the second half of the Mesozoic and in the Cenozoic; 2, faunal changes at the Mesozoic–Cenozoic boundary; and 3, discovery of new animals and new faunas (Barsbold *et al.*, 1971; Kramarenko, 1974).

The results of the expeditions were published in a series of transactions of the SMPE (Dmitrieva, 1971, 1977; Kurochkin, 1971; Sychevskaya and Lebedev, 1971; Zhegallo, 1971, 1978; Solov'ev and Shimanskii, 1978; Anonymous, 1979; Krasilov, 1982). After 10 years of work of the SMPE, 11 volumes of the transactions of the expedition were issued. Scientific sessions of the SMPE met annually, and more than 100 scientific papers were submitted by Mongolian and

Figure 13.7. N. Frolkin prepares some sauropod eggs from the Algui Ulaan Tsav locality, Dundgor Aimag, Upper Cretaceous, Nemegt Svita, 1969. Photo by E. Kurochkin.

Soviet experts. After 10 years of work, anniversary scientific conferences of the SMPE were held in Moscow and in Ulaanbaatar, at which 17 Soviet and six Mongolian scientists gave reports. At the Ulaanbaatar conference, a palaeontological exhibition was arranged, in which about 20 complete skeletons of Cretaceous dinosaurs and Mesozoic mammals were displayed; these are now on show at the State Museum of Mongolia or in the collection of the Geological Institute in Ulaanbaatar.

During the next 10 years (1980–1989), the SMPE continued more detailed development of certain sites (Figures 13.12–13.14). Rinchen Barsbold took a large Mongolian team in several trucks to Mesozoic sites, and D. Dashzeveg each season headed a separate

group on the Paleogene sites in the East, Central, and South Gobi. S. Kurzanov or Yu. Gubin headed the Mesozoic teams, investigating Cretaceous sites in the East, Central, and South Gobi, and V. Reshetov continued to explore Paleogene and Mesozoic localities in the South and Central Gobi. N. Bakhurina and E. Sychevskaya, with separate teams, began collecting pterosaurs and fishes, and with great success. Some seasons, the palaeobotany team of N. Makulbekov worked at Mesozoic and Paleogene sites. Two of the most interesting discoveries were the Upper Cretaceous Üüden Sair locality, south of the eastern end of the Arts Bogd Ridge, and the Upper Jurassic locality Shar Teeg, south-east of Altai Sum (Figure 13.12). The first produced avimimid remains, and the

Figure 13.8. The combined field team of the Joint Soviet–Mongolian Palaeontological and Geological Expedition at the Sevkhul Khudag camp site in July, 1970, Ergiliin Zoo locality, Dornogov' Aimag. From the left, E. Devyatkin, I. Liskun, D. Dashzeveg, I. Kuzikov, M. Sytin, V. Kocherzhenko, E. Kurochkin, N. Radkevich, Z. Shalneva, A. Saitsev, M. Borisoglebskaya, V. Zhegallo, A. Tchangtoomoor and V. Kutyrkin. Photo by E. Kurochkin.

second turns out to be the richest Late Jurassic site for plants, insects, fishes, reptiles, and amphibians, including one of the latest temnospondyls. The SMPE had occasionally surveyed the Upper Permian deposits south of Noyon Sum, but for the first time in 1989 the team of S. Kurzanov found terrestrial tetrapods at the Sainsar Bulag site.

Publications from this decade of the SMPE include the monographs of Yanovskaya (1980) on brontotheres, Badamgarav and Reshetov (1985) on the Paleogene of Southern Mongolia, Sychevskaya (1986, 1989) on fossil fishes, Kurzanov (1987) on avimimids, Tumanova (1987) on ankylosaurs, and Efimov (1988) on crocodilians. Collected papers on the fossil vertebrates of Mongolia were also published (Trofimov, 1971; Tatarinov, 1979, 1981, 1983; Kurochkin, 1988).

Thirty-nine volumes of transactions of the SMPE were published up to 1989, of which 19 were devoted wholly or mainly to vertebrates. In 1989, a conference was held in Ulaanbaatar to mark the twentieth jubilee of the SMPE (Anonymous, 1989), and a further large palaeontological exhibition was presented by the Soviet side of the SMPE, with many new mounted skeletons of dinosaurs and fossil mammals.

Since 1990, the work of the SMPE decreased sharply for financial and political reasons. Only one or two small teams work each year for short spells collecting vertebrates. From the 1992 expedition onwards, the programme was named the Russian–Mongolian Palaeontological Expedition. Forty-five volumes of the transactions of the RMPE have been published up to 1995.

Figure 13.9. Field camp of the Joint Soviet–Mongolian Palaeontological Expedition at the Sevkhul Khudag site in 1970, Ergiliin Zoo locality, Dornogov' Aimag. The Lower Oligocene Ergiliinzoo Svita outcrops in the background with the Hetsüü Tsav beds (upper cover sandstone) at the top of the outcrop, the alluvial upper member of the Ergiliin Zoo beds, and the lacustrine lower member of the Ergiliinzoo beds at the base of the outcrop. Photo by E. Kurochkin.

Main results of the RMPE

From the beginning, the SMPE/RMPE has been a complex expedition, involving clashes of individual personalities, and a complex set of geological and palaeontological objectives. In addition to the teams seeking fossil vertebrates, there were also large groups working on marine faunas from the Precambrian to the Upper Palaeozoic. Perhaps, over its long span, the RMPE was the largest set of expeditions in the history of palaeontology, in terms of the numbers of employees, the technical support (numerous motor vehicles, bulldozers, explosives and compressors), and the materials obtained.

Several major achievements of the RMPE can be noted.

1. A great diversity of faunas and floras existed during the Mesozoic and Cenozoic in Mongolia.
2. Dinosaur evolution can now be viewed in a new light, with Central Asia as a major region for their evolution. Three groups of theropods were endemic to Mongolia, the Oviraptoridae, Deinocheiridae and Therizinosauridae (Segnosauria) (Barsbold, 1983; Barsbold *et al.*, 1989). New data were obtained for the study of hadrosaurs, ankylosaurs, psittacosaurs, and protoceratopsians. RMPE specimens formed the basis of 25 new species of dinosaurs, and several higher taxa, as well as 19 forms of dinosaur eggshells, which had some stratigraphic value (Mikhailov, 1991, 1992).
3. Data were obtained on other groups of reptiles: turtles (Sukhanov, this volume), crocodiles

Figure 13.10. Excavations in deposits of the lacustrine lower member of the Ergiliin Zoo of the Lower Oligocene Ergiliinzoo Svita in the Novozhilov Hills site, Ergiliin Zoo locality in 1971. Photo by E. Kurochkin.

(Efimov and Storrs, this volume), pterosaurs (Unwin and Bakhurina, this volume), and lizards (Alifanov, this volume).

4. Important bird specimens were found from the Lower Cretaceous to the Upper Neogene. Lower Cretaceous birds and a number of bird feathers from many localities demonstrated the early beginning of modern birds and the existence of Enantiornithes in the Lower and Upper Cretaceous of Mongolia (Kurochkin, 1995, 1996). Hesperornithids were also discovered in the Cretaceous. Among Cenozoic birds, especially rich collections were obtained from the Palaeocene, Lower Oligocene (Kurochkin, 1981), and Upper Miocene (Kurochkin, 1985).

5. Extensive data were obtained on mammals of the Cretaceous and Paleogene. The RMPE discovered a magnificent fauna of Lower Cretaceous mammals in the Höövör locality where placental insectivores prevailed, but also multituberculates, triconodonts, symmetrodonts, and pantotheres (Trofimov, 1978, 1980, 1981; Kielan-Jaworowska *et al.*, 1987, and this volume; Dashzeveg *et al.*, 1989; Kielan-Jaworowska and Dashzeveg, 1989). The outstanding discovery was a complete skeleton of the new Late Cretaceous marsupial *Asiatherium reshetovi* from the Üüden Sair locality (Trofimov and Szalay, 1994). The RMPE also found the rich Late Palaeocene fauna at Tsagaan Khushuu and the Middle Eocene fauna of Khaichin Uul in the South Gobi (Reshetov, 1979; Badamgarav and Reshetov, 1985). This work also provided information for correlation of the Paleogene in Central Asia and North America. The RMPE provided evidence that, in the Aptian–Albian, Central Asia was one of major centres of the adaptive radiation of placental mammals.

6. The RMPE investigated huge areas of fossiliferous mudstone and bituminous shale deposits of the great Lower Cretaceous lakes of Central and Western Mongolia. Detailed inventories of these localities was conducted mainly by the palaeoen-

Figure 13.11. Monolith with a skeleton of *Protoceratops andrewsi* from the Tögrögiin Shiree locality, Ömnögov' Aimag, Upper Cretaceous, Baruungoyot Svita, 1969. Photo by E. Kurochkin.

tomologists of the SMPE (Yu. Popov and A. Ponomarenko). These sites yielded rich collections of insects, fishes, plants, and the very important finds of birds, and tens of their feathers. Monographs were published on the fauna, ecosystems, geology, and palaeogeography of the Manlai, Gurvan Ereen, Myangat, Böön Tsagaan, Kholboot, and other localities (Kalugina, 1980; Rasnitsyn, 1986; Sinitsa, 1993).

7. The collection of vertebrates and plants at the same Upper Cretaceous and Paleogene sites, and the discovery of the most ancient angiosperms in the Neocomian were also major achievements of the RMPE (Krasilov, 1982; Makulbekov, 1988).

8. The RMPE found a number of new vertebrate

localities dating from the Late Jurassic to the Pleistocene, many in regions which had not been investigated before (Figures 13.2 and 13.3). A number of sites were opened for palaeontology by the RMPE, although information on finds of bones came first from Mongolian arats or geologists. Most important were the Early Cretaceous Höövör and Hüren Dukh localities, where the expedition worked in 1969 and 1970. In 1971 the East-Gobi team found a new area with the Züünbayan Formation, south-east of Sainshand, where the rich Khamaryn Us (= Gashuuny Khudag) locality with complete skeletons of *Psittacosaurus* and new ankylosaurs and sauropods (Kurzanov and Kalandadze, 1974). In 1971 also the

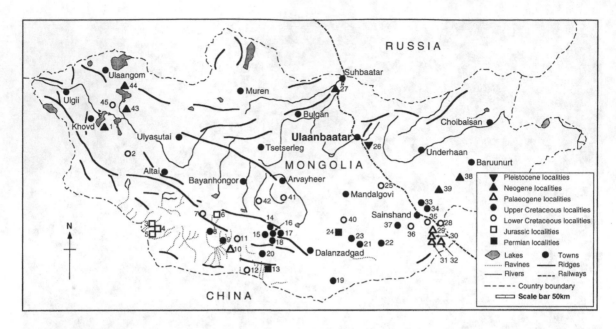

Figure 13.12. Joint Soviet (Russian)–Mongolian Palaeontological Expedition in 1980–1995. Map of the localities, fossiliferous areas, new localities, and field teams. Compiled by E. Kurochkin. Localities: 1, Yavar; 2, Gurvan Ereen; 3, Shar Teeg; 4, Khatan Sudal; 5, Elstiin; 6, Bakhar; 7, Kholboot; 8, Nogon Tsav; 9, Hermiin; 10, Tsagaan Khushuu; 11, Altan Uul; 12, Ih Shunkht; 13, Sainsar Bulag; 14, Yagaan Shiree; 15, Bor Khovil; 16, Zamyn Khond; 17, Khongil; 18, Üüden Sair; 19, Shilt Uul; 20, Khulsan; 21, Baishin Tsav; 22, Ölgii Khiid; 23, Amtgai; 24, Tavan Tolgoi; 25, Hüren Dukh; 26, Nalaih; 27, Shaamar; 28, Tsakhiurt; 29, Mergen; 30, Tsagaan Tsav; 31, Alag Tsav; 32, Khoyor Zaan; 33, Baga Tariach; 34, Teel Ulaan Uul; 35, Khamaryn Khural; 36, Khongil Tsav; 37, Khar Hötöl; 38, Baruunurt; 39, Indyn Uul; 40, Manlai; 41, Höövör; 42, Builyastyn Khudag; 43, Zavkhan; 44, Hirgis Nuur; 45, Tatal Gol.

RMPE opened the rich Late Cretaceous Baishin Tsav locality and two smaller ones, Amtgai and Shar Tsav, in the same area in the eastern region of the South Gobi, with vertebrates of the Bayanshiree Svita (Turonian–Coniacian). Important also was the find in 1970 of the huge Hermiin Tsav locality of Barungoyotian age (Campanian) on the eastern edge of the Trans-Altai Gobi, where various dinosaurs, lizards, mammals, and birds were found later. The Late Jurassic locality Shar Teeg, in the south-western corner of Mongolia, found by the RMPE in 1984, is a very large locality, with lacustrine and alluvial facies, where various animals and plants were found.

9. The excavations of the RMPE produced not only major scientific materials, but also many fine museum skeletons for exhibition, as seen in the halls of the Museum of the Palaeontological Institute in Moscow and the State Museum in Ulaanbaatar.

10. A further important achievement of the RMPE has been in training; through the expeditions, whole new generations of Russian and Mongolian palaeontologists who have received experience of field work, scientific work, and joint scientific co-operation. Many dissertations were based on materials extracted by the RMPE, including 12 Candidates of Sciences (Ph.D.) and three Doctors of Sciences (D.Sc.) at the PIN RAS and eight dissertations of Candidates of Sciences and two Doctors of Sciences at the Geological Institute of MAS. The training of Mongolian palaeontologists since the MPE in the late 1940s has been very important. For example, B. Luvsandazan began as a student on one of I. Efremov's expeditions, and he

Figure 13.13. Field camp at the Baishin Tsav locality, east of Ömnögov' Aimag, in August 1985, Upper Cretaceous, Baruungoyot Svita. Photo by S. Kurzanov.

later became an Academician and Director of the Geological Institute of MAS, and scientific adviser to the RMPE. In the years of the RMPE, a large group of excellent palaeontologists has developed in the MAS, and they now conduct independent research on various animal groups and engage in independent international co-operation.

Acknowledgements

We thank many employees of the PIN and GI for advice and corrections in the process of preparing of this paper, and especially of B. Trofimov, S. Kurzanov, and Yu. Gubin. The help of the staff of the ARAS was also very important. We thank Mike Benton for revisions and help with the English.

References

Anonymous 1979. [*Principal Results of the Investigations of the Joint Soviet–Mongolian Palaeontological Expedition in 1969–1979. Abstracts of Papers.*] Moscow: Paleontologicheskii Instituta, 20 pp.

—1989. [*Principal Results of the Investigations of the Joint Soviet–Mongolian Palaeontological Expedition in 1969–1988. Abstracts of Papers.*] Moscow: Paleontologicheskii Instituta, 47 pp.

Badamgarav, D. and Reshetov, V.Yu. 1985. [Palaeontology and stratigraphy of the Paleogene of the Trans-Altai Gobi.] *Trudy Sovmestnoi Sovetsko-Mongol'skoi Paleontologicheskoi Ekspeditsii* **25**: 1–104.

Barsbold, R. 1983. [Carnivorous dinosaurs from the Cretaceous of Mongolia.] *Trudy Sovmestnoi Sovetsko-Mongol'skoi Paleontologicheskoi Ekspeditsii* **19**: 1–120.

Figure 13.14. Excavations at the Khongil locality, Ömnögov' Aimag, in 1985 for the skeleton of an ankylosaur, Upper Cretaceous, Baruungoyot Svita, 1985. From the left, S. Kurzanov, K. Mikhailov and G. Vinogradov. Photo by S. Kurzanov.

—, Kurzanov, S.M., Perle, A., and Tumanova, T.A. 1989. [Some results of the study of dinosaurs from Mongolia, pp. 10–12 in [*Principal Results of the Investigations of the Joint Soviet-Mongolian-Palaeontological Expedition in 1969–1988. Abstracts of Papers*]. Moscow: Paleontologicheskii Institut.

—, Voronin, Yu.I., and Zhegallo, V.I. 1971. [The work of the Soviet-Mongolian Palaeontological Expedition in 1969–1970.] *Paleontologicheskii Zhurnal* **2**: 139–143.

Belyaeva, E.I. 1937. [Materials characteristic of the Upper Tertiary faunas of mammals of North-Western Mongolia.] *Trudy Mongol'skoi Komissii AN SSSR* **33** (9): 1–52.

Berkey, C.P. and Morris, F.K. 1927a. Climatic pulsations in Mongolia. *Bulletin of the Geological Society of America* **38**: 211–212.

—and— 1927b. *Geology of Mongolia. Natural History of Central Asia. Vol. II.* New York: American Museum of Natural History, 475 pp.

Borisyak, A.A. 1915. Sur les restes d'*Epiaceratherium turgiacum. Bulletin de l'Académie Impériale des Sciences de St. Pétersburg* (6)**9**: 781–787.

Chudinov, P.K. 1966. [A unique site for late Cretaceous reptiles in Bayan Khongor Aimag.] pp. 74–78 in Marinov, N.A. (ed.), *Materialy po Geologii Mongol'skoi Narodnoi Respubliki.* Moscow: Nedra.

—1987. [*Ivan Antonovich Efremov. 1907–1972.*] Moscow: Nauka, 224 pp.

Dashzeveg, D., Reshetov, V.Yu. and Trofimov, B.A. 1989. [The early stages of evolution of mammals of Mongolia.] pp. 5–6 in [*Principal Results of the Investigations of the Joint Soviet-Mongolian-Palaeontological*

Expedition in 1969–1988. Abstracts of Papers]. Moscow: Paleontologicheskii Institut.

Devyatkin, E.V. 1981. Cenozoic of Inner Asia. *Trudy Sovmestnoi Sovetsko-Mongol'skoi Nauchno-Issledovatel'skoi Geologicheskoi Ekspeditsii* 2 7: 1–196.

Dmitrieva, E.L. 1971. [Neogene gazelles of Western Mongolia,] pp. 124–131 in Trofimov, B.A. (ed.), *Fauna Mezozoya i Kainozoya Mongolii. (Trudy Sovmestnoi Sovetsko-Mongol'skoi Nauchno-Issledovatel'skoi Geologicheskoi Ekspeditsii* 3: Moscow: Nauka

—1977. [Neogene antelopes from Mongolia and adjacent territories.] *Trudy Sovmestnoi Sovetsko-Mongol'skoi Paleontologicheskoi Ekspeditsii* 6: 1–119.

Dombrovski, B.S. 1926. [*Preliminary Report on Geological Investigations in the Mongolian People Republic in 1925. Part I. Investigations in the Highlands of Mongolia.*] Vladivostok.

Efimov, M.B. 1988. [The fossil crocodiles and champsosaurids of Mongolia and the USSR.] *Trudy Sovmestnoi Sovetsko-Mongol'skoi Paleontologicheskoi Ekspeditsii* 36: 1–108.

Efremov, I.A. 1948. [First Mongolian Palaeontological Expedition of the Academy of Sciences of the USSR.] *Vestnik AN SSSR* 1: 47–58.

—1949. [Preliminary results of activity of the First Mongolian Palaeontological Expedition of the Academy of Sciences of the USSR, 1946.] *Trudy Mongol'skoi Komissii AN SSSR* 38: 1–49.

—1950. [Taphonomy and the fossil record.] *Trudy Paleontologicheskogo Instituta AN SSSR* 24: 1–178.

—1952. [Questions on the development of the continental Upper Palaeozoic of Central Asia.] *Doklady AN SSSR* 85: 627–630.

—1953. [Questions about studies of dinosaurs (on materials from the Mongolian expeditions of the Academy of Sciences of the USSR).] *Priroda*, **1953** (6): 26–37.

—1954a. [Some observations on questions of the historical development of the dinosaurs.] *Trudy Paleontologicheskogo Instituta AN SSSR* 48: 125–141.

—1954b. [Palaeontological research in the Mongolian People Republic: results of the expeditions of 1946, 1948, and 1949.] *Trudy Mongol'skoi Komissii AN SSSR* 59: 3–32.

—1955. [Burial of dinosaurs in Nemegt (South Gobi, MPR).] pp. 789–809 in *Voprosy Geologii Azii*. Moscow: Izdatel'stvo Akademii Nauk SSSR.

—1963. [Perspectives on the development of palaeontological researches in Mongolia.] pp. 82–92 in *Materialy po Geologii Mongol'skoi Narodnoi Respubliki*. Moscow: Gostoptekhizdat.

—, Novozhilov, N.I., and Rozhdestvenskii, A.K. (eds.) 1954. [Collected papers on the palaeontology of the Mongolian Peoples' Republic.] *Trudy Mongol'skoi Komissii AN SSSR* 59: 1–55.

Granger, W., and Berkey, C.P. 1922. Discovery of Cretaceous and older Tertiary strata in Mongolia. *American Museum Novitates* 42: 1–7.

Kalandadze, N.N. and Kurzanov, S.M. 1974. [Lower Cretaceous localities of terrestrial vertebrates of Mongolia.] *Trudy Sovmestnoi Sovetsko-Mongol'skoi Paleontologicheskoi Ekspeditsii* 1: 9–18.

Kalugina, N.S. (ed.) 1980. [The Early Cretaceous Manlai Lake.] *Trudy Sovmestnoi Sovetsko-Mongol'skoi Paleontologicheskoi Ekspeditsii* 43: 1–115.

Kielan-Jaworowska, Z. and Dashzeveg, D. 1989. Eutherian mammals from the Early Cretaceous of Mongolia. *Zoologica Scripta* 18: 347–355.

—, Dashzeveg, D., and Trofimov, B.A. 1987. Early Cretaceous multituberculates from Mongolia and a comparison with Late Jurassic forms. *Acta Palaeontologica Polonica* 32: 3–47.

Kramarenko, N.N. 1974. [On work of the Joint Soviet-Mongolian Palaeontological Expedition.] *Trudy Sovmestnoi Sovetsko-Mongol'skoi Paleontologicheskoi Ekspeditsii* 1: 9–18.

Krasilov, V.A. 1982. Early Cretaceous flora of Mongolia. *Palaeontographica, Abteilung B* 181: 1–43.

Kupletskii, B.M. 1926. [On the geology of Eastern Mongolia. North Mongolia. I.] pp. 31–50 in *Predvaritelnie Otcheti Geologicheskoi, Geokhemicheskoi i Pochvenno-Geografischeskoi Ekspeditsii o Rabotakh, Proizvedennykh v 1925 godu*. Leningrad: Izdatel'stvo AN SSSR.

Kurochkin, E.N. 1971. [On the avifauna of the Mongolian Pliocene.] pp. 58–67 in Trofimov, B.A. (ed.), *Fauna Mezozoya i Kainozoya Mongolii. (Trudy Sovmestnoi Sovetsko-Mongol'skoi Nauchno-Issledovatel'skoi Geologicheskoi Ekspeditsii* 3: Moscow: Nauka.

—1981. [New forms and the evolution of two families of archaic gruiforms in Eurasia.] *Trudy Sovmestnoi Sovetsko-Mongol'skoi Paleontologicheskoi Ekspeditsii* 15: 59–85.

—1985. [Birds of Central Asia in the Pliocene.] *Trudy*

Sovmestnoi Sovetsko-Mongol'skoi Paleontologicheskoi Ekspeditsii 26: 1–120.

—(ed.) 1988. [Fossil reptiles and birds of Mongolia.] *Trudy Sovmestnoi Sovetsko-Mongol'skoi Paleontologicheskoi Ekspeditsii* 34: 1–112.

—1995. Synopsis of Mesozoic birds and early evolution of Class Aves. *Archaeopteryx* 13: 47–66.

—1996. A new enantiornithid of the Mongolian Late Cretaceous, and a general appraisal of the Infraclass Enantiornithes (Aves). *Palaeontological Institute of the Russian Academy of Sciences, Special Issue.* Moscow: Palaeontological Institute, 60 pp.

—, Kalandadze, N.N. and Reshetov, V.Yu. 1970. [First results of the Soviet-Mongolian Palaeontological Expedition.] *Priroda* 1970 (4): 115.

Kurzanov, S.M. 1987. [Avimimidae and the problem of the origin of birds.] *Trudy Sovmestnoi Sovetsko-Mongol'skoi Paleontologicheskoi Ekspeditsii* 31: 1–95.

—and Kalandadze, N.N. 1974. [Lower Cretaceous sites of terrestrial vertebrates of Mongolia.] *Trudy Sovmestnoi Sovetsko-Mongol'skoi Paleontologicheskoi Ekspeditsii* 1: 288–295.

Lavas, J.R. 1993. *Dragons from the Dunes. The Search for Dinosaurs in the Gobi Desert.* Auckland: Academy Interprint, 138 pp.

Lebedeva, Z.A. 1926. The detailed geological works in North-Western Mongolia, pp. 76–79, 319–325 in *Otchet o Deyatel'nosti AN SSSR 1925 godu.* Leningrad: Izdatel'stvo AN SSSR.

—1934. [On the geology of the Gurvan Saykahan mountain group in Gobi Altai.] *Trudy Mongol'skoi Komissii AN SSSR* 18: 1–74.

Makulbekov, N.M. 1988. [Paleogene flora of South Mongolia.] *Trudy Sovmestnoi Sovetsko-Mongol'skoi Paleontologicheskoi Ekspeditsii* 35: 1–96.

Marinov, N.A. 1967. [*Geological Explorations of the Mongolian Peoples' Republic.*] Moscow: Nedra, 843 pp.

—, Zonenshain, L.P. and Blagonravov, V.A. (eds.) 1973. [*Geology of the Mongolian Peoples' Republic. Vol. I. Stratigraphy.*] Moscow: Nedra, 583 pp.

Martinson, G.G. 1982. [Upper Cretaceous molluscs of Mongolia.] *Trudy Sovmestnoi Sovetsko-Mongol'skoi Paleontologicheskoi Ekspeditsii* 17: 1–82.

Mikhailov, K.E. 1987. Principal structure of the avian eggshell: data of SEM studies. *Acta Zoologica Cracovienza* 30 (5): 53–70.

—1991. Classification of fossil eggshells of amniotic vertebrates. *Acta Palaeontologica Polonica* 36 (2): 21–39.

—1992. The microstructure of avian and dinosaurian eggshell: phylogenetic implications. *Papers in Avian Paleontology. Natural History Museum of Los Angeles County, Science Series* 36: 361–373.

—1994. [Eggs of theropodan and protoceratopsian dinosaurs from the Cretaceous deposits of Mongolia and Kazakhstan.] *Paleontologicheskii Zhurnal* 2: 81–95.

Murzaev, E.M. 1948. [*Mongolian Peoples' Republic. Physico-Geographical Description.*] Moscow: State Publishing for Geography Literature, 314 pp.

Obruchev, V.A. 1893. [Some words on the geological texture of Eastern Mongolia along the caravan road from Kjakhta to Kalgan.] *Izvestiya Vostochno-Sibirskogo Otdeleniya Imperatorskogo Russkogo Geograficheskogo Obshchestva* 24 (3–4): 104–108.

Orlov, Yu.A. 1952. [Works of Soviet palaeontologists in Central Asia.] *Priroda* 6: 78–87.

Perle, A. 1979. [Segnosauridae, new family of carnivorous dinosaurs from the Upper Cretaceous of Mongolia.] *Trudy Sovmestnoi Sovetsko-Mongol'skoi Paleontologicheskoi Ekspeditsii* 8: 45–55.

Rachkovskii, I.P. 1928. [Exploration of the Mongolian and Tannu-Tuva Peoples' Republics.] pp. 260–264 in *Otchet o Deyatel'nosti AN SSSR in 1925 godu. Part II. Otchet o Nauchnykh Komandirovkakh i Ekspeditsiikh.* Leningrad: Izdatel'stvo AN SSSR.

—and Lebedeva, Z.A. 1932. [Short report on results of the work of the geological team of the expedition of the Academy of Sciences of the USSR and Scientific-Exploration Committee of the Mongolian Peoples' Republic in 1931.] *Trudy Mongol'skoi Komissii AN SSSR* 6: 1–28.

Rasnitsyn, A.P. (ed.) 1986. [Insects in early Cretaceous biocoenoses of Mongolia.] *Trudy Sovmestnoi Sovetsko-Mongol'skoi Paleontologicheskoi Ekspeditsii* 28: 1–214.

Reshetov, V.Yu. 1979. [Early Tertiary Tapiroidea of Mongolia and the USSR.] *Trudy Sovmestnoi Sovetsko-Mongol'skoi Paleontologicheskoi Ekspeditsii* 11: 1–144.

Richthofen, F.V. 1877. *China. Ergebnisse eigener Reisen und darauf gegründeter Studien. Band I.* Berlin.

Rozhdestvenskii, A.K. 1957. [Brief results of the study of fossil vertebrates of Mongolia on the data of the Mongolian Palaeontological Expedition of the Academy of Sciences of the USSR in 1946–1949.] *Vertebrata PalAsiatica* 1: 169–183.

—1969. [*On the Trail of the Dinosaurs of the Gobi.*] Moscow: Nauka, 293 pp.

Sinitsa, S.M. 1993. [Jurassic and Lower Cretaceous of

Central Mongolia.] *Trudy Sovmestnoi Rossiiskoi Paleontologicheskoi Ekspeditsii* **42**: 1–238.

Solov'ev, A.N. and Shimanskii, V.N. (eds.) 1978. [*The Development and Replacement of Organic Fossils at the Boundary of the Mesozoic and Cenozoic.*] Moscow: Nauka, 136 pp.

Sychevskaya, E.K. 1986. [Paleogene freshwater fish fauna of the USSR and Mongolia.] *Trudy Sovmestnoi Sovetsko-Mongol'skoi Paleontologicheskoi Ekspeditsii* **29**: 1–157.

—1989. [Neogene freshwater fish faunas of Mongolia.] *Trudy Sovmestnoi Sovetsko-Mongol'skoi Paleontologicheskoi Ekspeditsii* **39**: 1–144.

—and Lebedev, V.D. 1971. [Freshwater Neogene fish faunas of the Basin of the Great Lakes.] pp. 49–57 in Trofimov, B.A. (ed.), *Fauna Mezozoya i Kainozoya Mongolii. Trudy Sovmestnoi Sovetsko-Mongol'skoi Nauchno-Issledovatel'skoi Geologicheskoi Ekspeditsii* **3**: 49–57.

Tatarinov, L.P. (ed.) 1979. [Fauna of the Mesozoic and Cenozoic of Mongolia.] *Trudy Sovmestnoi Sovetsko-Mongol'skoi Paleontologicheskoi Ekspeditsii* **8**: 1–150.

—(ed.) 1981. [Fossil vertebrates of Mongolia.] *Trudy Sovmestnoi Sovetsko-Mongol'skoi Paleontologicheskoi Ekspeditsii* **15**: 1–131.

—(ed.) 1983. [Fossil reptiles of Mongolia.] *Trudy Sovmestnoi Sovetsko-Mongol'skoi Paleontologicheskoi Ekspeditsii* **24**: 1–136.

Trofimov, B.A. (ed.). 1971. [Mesozoic and Cenozoic faunas of Western Mongolia.] *Trudy Sovmestnoi Sovetsko-Mongol'skoi Nauchno-Issledovatel'skoi Geologicheskoi Ekspeditsii* **3**: 1–134.

—1978. [First triconodonts from Mongolia.] *Doklady AN SSSR* **243**: 213–216.

—1980. [Multituberculata and Symmetrodonta from the Lower Cretaceous deposits of Mongolia.] *Doklady AN SSSR* **251**: 209–212.

—1981. [Eutheria from the Lower Cretaceous of Mongolia.] *Doklady AN SSSR* **261**: 111–114.

—and Chudinov, P.K. 1970. [New data about the vertebrate sites of Mongolia.] pp. 152–156 in Flerov, K.K. (ed.), *Materialy po Evolutsii Nazemnykh Pozvonochnykh Mongolii*. Moscow: Nauka.

—and Szalay, F.S. 1994. New Cretaceous marsupial from Mongolia and the early radiation of Metatheria. *Proceedings of the National Academy of Sciences, USA* **91**: 12 569–12 573.

Tumanova, T.A. 1987. [The armoured dinosaurs of Mongolia.] *Trudy Sovmestnoi Sovetsko-Mongol'skoi Paleontologicheskoi Ekspeditsii* **32**: 1–80.

Yanovskaya, N.M. 1980. [The brontotheres of Mongolia.] *Trudy Sovmestnoi Sovetsko-Mongol'skoi Paleontologicheskoi Ekspeditsii* **12**: 1–219.

—, Kurochkin, E.N. and Devyatkin, E.V. 1977. [Locality Ergiliin Zoo as stratotype of the Lower Oligocene in South-Eastern Mongolia.] pp. 14–33 in Trofimov, B.A. (ed.), *Fuana, Flora i Biostratigrafiya Mezozoya y Kainozoya Mongolii. Trudy Sovmestnoi Sovetsko-Mongol'skoi Nauchno-Issledovatel'skoi Geologicheskoi Ekspeditsii* **4**: 14–33.

Zhegallo, V. I. 1971. [Hipparions from Neogene deposits of Western Mongolia and Tuva.] pp. 98–119 in Trofimov, B.A. (ed.), *Fauna Mezozoya i Kainozoya Mongolii. Trudy Sovmestnoi Sovetsko-Mongol'skoi Nauchno-Issledovatel'skoi Geologicheskoi Ekspeditsii* **3**: Moscow: Nauka.

—1978. [The hipparions of Central Asia.] *Trudy Sovmestnoi Sovetsko-Mongol'skoi Paleontologicheskoi Ekspeditsii* **7**: 1–155.

14

The Cretaceous stratigraphy and palaeobiogeography of Mongolia

VLADIMIR F. SHUVALOV[1]

Introduction

Mongolia is rich in Cretaceous localities yielding dinosaurs and other vertebrates. The American Central Asiatic Expedition from 1923 to 1925 discovered fossils in the Öösh Basin, Bayan Zag (Flaming Cliffs), Iren Dabasu, and Ongon Ulaan Uul (see Chapter 12). Further work from 1946 to 1949, by the Russian Academy of Science, led to the discovery of the new localities Nemegt (Trans-Altai Gobi) and Ulaan Öösh (North Gobi), and dinosaurs were excavated at Bayan Zag and other localities (see Chapter 13). The Polish–Mongolian Palaeontological Expedition of the 1960s (Kielan-Jaworowska and Barsbold, 1972), and further Russian and American–Mongolian expeditions also met with success.

The first stratigraphic scheme was produced by Berkey and Morris (1927). Efremov (1954) divided the Mongolian bone-bearing deposits into the Lower Cretaceous sequences of Ondai Sair or Andaikhudag, and Öösh, and the Upper Cretaceous sequences of Baruun Bayan, Ulaan Öösh, Algui Ulaan Tsav, Bayan Zag (Djadokhta), and others. The stratigraphy of the Cretaceous of Mongolia, and the East Gobi in particular, was revised by Mongolian and Russian oil geologists in the 1950s (Turishchev, 1954; Marinov, 1957; Vasil'ev *et al.*, 1959). The majority of researchers distinguished three svitas in the Lower Cretaceous, the Shariliin, Tsagaantsav and Züünbayan, and two in the Upper Cretaceous, the Sainshand and Bayanshiree. A similar scheme was suggested by Florentsov and Solonenko (1963) for the Gobi Altai, the svitas Tormkhon, Tevsh and Kholboot, virtually repeating the svitas Shariliin, Tsagaantsav, and Züünbayan, with

division of the last into upper and lower parts. For the Upper Cretaceous the svitas Builyastyn and Bulgant were proposed. The latter includes mainly acid volcanic rocks, the Tevsh, basic. At the same time, further local schemes were also proposed, but these were in principle little different from the scheme of Vasil'ev *et al.* (1959).

A new scheme for division of the Upper Cretaceous was suggested by Martinson *et al.* (1969) and by Gradziński *et al.* (1977). For southern regions of Mongolia, these authors distinguished the Sainshand, Bayanshiree, Baruungoyot, and Nemegt formations. For East Mongolia, instead of the Baruungoyot and Nemegt, the Javkhlant Svita of concurrent age (Late Senonian) was suggested, deposits of which had hitherto been mistakenly included in the Paleogene (Vasil'ev *et al.*, 1959; Sochava, 1975). The Sainshand Svita was later rejected (Verzilin, 1979b; Shuvalov, 1982), since the type section on the mountain Khar Hötöl Uul in South-east Mongolia has not yielded fossils, and basalts from the same section have not been dated radiometrically.

New stratigraphic schemes have been proposed for the Lower Cretaceous of a number of regions of Mongolia (Khosbayar, 1972; Martinson and Shuvalov, 1973, 1976). The Shariliin Svita of the East Gobi and the corresponding Tormkhon Svita of central Mongolia were assigned to the Upper Jurassic (Kimmeridgian–Tithonian) by Shuvalov (1969). Further revisions led to new stratigraphic schemes for the Cretaceous of Central, South (Shuvalov, 1970, 1975a; Barsbold, 1972) and other regions of Mongolia, followed by a scheme for the whole of Mongolia (Nagibina *et al.*, 1977). This scheme, in its most recent state (Shuvalov, 1994) is the basis of Table 14.1.

---

[1] Vladimir Shuvalov died in 1999.

Table 14.1 *Scheme of stratigraphy of the Cretaceous deposits of Mongolia*

System	Division	Stage	Gorizont	Svita						
				East Gobi	Trans-Altai Gobi	Central Mongolia	South Mongolia	West Mongolia	North-east Mongolia	North Mongolia
Cretaceous	Upper	Maastrichtian			Nemegt					
		Campanian–Santonian			Baruungoyot					
		Santonian–Cenomanian			Bayanshiree					
	Lower	Albian–Aptian	Hühteeg	Hühteeg	Baruunbayan Döshuul	Khulsangol	Ulaandel	Zereg	Bagazosnuur	
		Barremian–Hauterivian	Shinekhudag	Shinekhudag	Altanuul	Andaikhudag	Tsagaangol	Gurvanereen	Shinekhudag	
		Valanginian–Berriasian	Tsagaantsav	Tsagaantsav		Öndörukhaa			Tsagaantsav	Uilgan

Source: From Shuvalov, 1994

Character and dating of the Cretaceous formations of Mongolia

Lower Cretaceous

Tsagaantsav (Berriasian–Valanginian)

The oldest horizon of the Cretaceous of Mongolia, the Tsagaantsav, occurs in nearly all regions of Mongolia, but most widely in the centre and south of the country. It is represented by conglomerates (below), sandstones, clays, limestones, marls, in some places with layers of volcanics and tuffs up to 700–800 m and more thick (Figure 14.1). Basic volcanics, and rarely acid volcanics, and tuffs are characteristic of the Tsagaantsav Gorizont in South-east, Central, and North-east Mongolia, but not in the west or north (Shuvalov, 1975a,b, 1982). Sections of the Tsagaantsav Gorizont of the Sangiin Dalai Nuur depression have been given by Devyatkin *et al.* (1975) and Bakhurina (1983).

Dinosaurs are rare in the Tsagaantsav Gorizont. The westernmost dinosaur locality is the Sangiin Dalai Nuur depression, south of Hirgis Nuur lake, where, according to Bakhurina (1983), remains of the dinosaur *Psittacosaurus,* numerous pterosaurs *Dsungaripterus weii* and *D. parvus,* the tooth of a carnivorous dinosaur, the scapula and rib of a large dinosaur, and the jaw of a paramacellodid lizard have been found (Bakhurina, 1983; Bakhurina and Unwin, 1995). The psittacosaurs here are among the oldest known, being generally characteristic of the Aptian–Albian (Shuvalov, 1982; Sereno, 1990). To the north, from the mountain Bayan Ovoo Uul, in many-coloured sandstones, siltstones and clays dated as lower Gurvanereen Svita (Khosbayar, 1972), remains of molluscs and ostracods, characteristic of the Tsagaantsav Gorizont, have also been collected (Devyatkin *et al.,* 1975).

In the region of the Gurvan Ereen range, deposits of the lower part of the Gurvanereen Svita (included in the Tsagaantsav Gorizont) are observed with a thickness of more than 150–200 m (Devyatkin *et al.,* 1975). They are represented mostly by clays, siltstones and sandstones, mainly grey in colour, and lying on red-

Figure 14.1. Intercalation of clays and marls near the mountain Öndör Ukhaa (north-west part of Gobi Altai). Öndörukhaa Svita, Tsagaantsav Gorizont. (Photo by V.F. Shuvalov.)

coloured coarse sediments of Late Jurassic age (Kimmeridgian–Tithonian). Here, and south of Darvi Sum, numerous ostracods and conchostracans characteristic of the Tsagaantsav Gorizont have been found, as well as remains of the primitive chondrostean fish *Stichopterus* sp., characteristic of the lower part of the Early Cretaceous (Yakovlev, 1986). Devyatkin *et al.* (1975) report corixid heteropteran insects similar to *Baissocovixa jacijewski* Popov from the Lower Cretaceous of Transbaikalia, Siberia, and higher in the section, numerous coprolites, probably of dinosaurs.

In Central Mongolia, in the Öndörukhaa Svita

(Shuvalov, 1975a), remains of dinosaurs are also very rare. They have been found near Tsogtovoo Sum. Sections of the svita were given by Florentsov and Solonenko (1963) and Shuvalov (1975a, 1982, 1987). Among molluscs, most characteristic for this svita are representatives of *Arguniella*, as well as *Limnocyrena kweichowensis* (Grab.), *L. wangshihensis* (Grab.) and others (Martinson, 1975; Shuvalov, 1975a). Numerous ostracods (*Cypridea trita* Lüb., *C.priva* Lüb., *C. remota* Lüb.) and conchostracans (*Brachigrapta kansuensis (Chi), Bairdestheria halobiformis* Kob. et Kus., and others) are characteristic of the Tsagaantsav Gorizont of Mongolia and of the Early Cretaceous of China and Transbaikalia (Shuvalov, 1975a; Shuvalov and Stepanov, 1970).

The same formation occurs in the south-east of the Gobi Altai, south of Khangai, in particular in the lower part of the Ondaisair Svita of the Tsagaantsav Gorizont, in North Gobi and other places (Shuvalov, 1975a). In South Trans-Altai and East Gobi, the composition of the Tsagaantsav Gorizont is in general similar to Central Mongolia. But, in some places there are thick horizons of tuffs of acid volcanics, which contain zeolites (mainly clinoptilolite, containing more than 20%), especially in the East and South Gobi: on Tüshleg Mountain, near Tsagaan Tsav well, north of the Khanbogd Sum, and in other places. In Central Mongolia, north of the ridge Tariat Uul (Gobi Altai), silts of the Öndörukhaa Svita (or Tevsh Svita) contain 14.81–15.81% of phosphide (Florentsov and Solonenko, 1963). Tsagaantsav deposits in North-east and East Mongolia have a similar content. Here deposits of the Tsagaantsav Svita were included in the Choibalsan series by Nagibina and Badamgarav (1975). The volcanics of the Tsagaantsav Svita have been described by Frikh-Khar and Luchitskaya (1978). In North Mongolia, deposits of the Tsagaantsav Svita are known in the basin of the Uilgan river, where the remains of molluscs have been found, but volcanics are absent.

The Tsagaantsav Svita is dated as Berriasian–Valanginian on the basis of numerous finds of fossils, and K–Ar and Rb–Sr dating of the volcanics (119–141 Myr) (Devyatkin *et al.* 1990; Solov'ev *et al.*, 1977;

Verzilin, 1979a; Shuvalov, 1987, 1994). See also Table 14.2.

Shinekhudag Gorizont (Hauterivian–Barremian)

This widely distributed horizon includes the Shinekhudag, Andaikhudag, the upper part of Altanuul, and other svitas (Table 14.1). The Shinekhudag Svita occurs in Central and South-east Mongolia, but it is not known yet in North Mongolia. The unit is composed everywhere of sandstones, argillites, marls and clays together with bituminous shales ('fish shales'), which are the most characteristic facies of these deposits (Figure 14.2). The first deposits were described by Berkey and Morris (1927) from south of Khangai (near the Andai Khudag well) area as a member of the Ondaisair Formation. Detailed descriptions of the deposits were given by Marinov (1957), Vasil'ev *et al.*(1959), and Shuvalov (1975a).

Remains of dinosaurs in Shinekhudag deposits are extremely rare. South of Khangai the Americans found remains of *Protiguanodon mongoliensis* and *Psittacosaurus mongoliensis.* However, everywhere in the Shinekhudag sequences there are ostracods, conchostracans, fishes and molluscs (Devyatkin *et al.*, 1990; Shuvalov, 1975a; Martinson and Shuvalov, 1973; Shuvalov, 1980). Fishes are represented mostly by *Lycoptera fragilis* Huss., but not by *L. middendorfii* Müll., as at Andai Khudag (Berkey and Morris, 1927). The latter form, judging by its most frequent occurrence, is characteristic of the Tsagaantsav Gorizont (Martinson, 1975; Shuvalov, 1975a). Other fossils from Khangai include the insects *Indasia reisi* Cockerell, *Ephemeropsis trisetalis* Eich., and others, and numerous insects were also collected from the Andaikhudag Svita of the Andai Khudag well region. Here, remains of dinosaurs also occur in higher (Aptian–Albian) horizons of the Lower Cretaceous (Khulsangol Svita) (Shuvalov, 1975a).

In West Mongolia, sections of the upper part of the Gurvanereen Svita (Khosbayar, 1972) can be observed to the south-west of Darvi Sum and near Myangat Sum. In Central Mongolia, deposits of the Andaikhudag Svita, belonging to the same horizon,

Figure 14.2. Rhythmical intercalation of dolomites and bituminous shales on the river Kholboot Gol (north-west part of the Gobi Altai). Andaikhudag Svita, Shinekhudag Gorizont. (Photo by V.F. Shuvalov.)

occur at Kholboot Sair, where their thickness reaches 700 m; also near the Erdene Uul mountains (west and east), near Shanh Sum (basin of the Orkhon river), in Tevshiin Gobi gorge, to the west of Mandalgov' town and in other places of Gobi Altai, Khangai and North Gobi (Shuvalov, 1975a).

In the Trans-Altai Gobi, Shinekhudag deposits are known only near Altan Uul mountain (Shuvalov, 1993). In South Gobi they can be observed in the upper part of the Tsagaangol Svita (Shuvalov, 1982). In East Gobi the Shinekhudag Svita (lower part of the Züünbayan Svita according to Vasil'ev *et al.*, 1959) is most completely represented near the Shine Us Khudag well, near Öndörshil Sum, near Tüshleg mountain, in the Baishin Tsav depression (extreme south-west of the region) and in a number of other locations. In the east and north-east of the country, similar deposits occur at Nyalga, Choibalsan, Ih Zos Nuur, and in other depressions. Nearly everywhere they contain characteristic ostracods, conchostracans, fishes, and other organisms (Martinson and Shuvalov, 1973; Martinson, 1975; Nagibina and Badamgarav, 1975; Shuvalov, 1980).

The conchostracans *Bairdestheria sinensis* (Chi), *B. mattoxi* Kras., *Pseudograpta asanoi* (Kob. et Kus.), which occur low in the Züünbayan Svita (Vasil'ev *et al.*, 1959) and in the Shinekhudag Svita of the East Gobi (Martinson and Shuvalov, 1973) give a Hauterivian–Barremian age for the Shinekhudag Svita.

Hühteeg Gorizont (Aptian–Albian)

This unit is richer in dinosaurs. It embraces the Hühteeg, Khulsangol, Zereg, and other svitas in different regions of Mongolia (Table 14.1). The Hühteeg Gorizont is dominated by grey-coloured sandstones, and in the east and centre of Mongolia a characteristic feature is coal (Övdög Khudag, Nalaih, Baganuur, and other localities). Horizons of basic volcanics of different thickness (up to 100 m) occur, as well as oval sandstone concretions (Figure 14.3), partial tree trunks, and other water-borne organic materials (Martinson, 1975; Martinson and Shuvalov, 1973; Shuvalov, 1980). The thickness of the Hühteeg Svita sometimes reaches 400–500 m (west slope of the mountain Tsetsen Uul in Gobi

Table 14.2. *Stratigraphical position and radioactive age of the Cretaceous deposits of Mongolia (constants according to 1976, Australia)*

Division	Stage	Regional svitas, biostratigraphical horizons, location, rocks	Radioactive age, Myr, laboratory
Upper	Campanian	Baruungoyot Svita	
		1. South Gobi, Chuluut Uul mountain, upper part of svita, basalt	75 ± 7 (IGEM)
		2. North Mongolia, Tsagaan Nuur depression, basaltic sheet on the Lower Cretaceous	80 (IG)
	Cenomanian	Bayaanshiree Svita	
		3. South Gobi, Dösh Uul mountain (80 km south of Bayandalai Sum), upper part of svita, basalt	90 ± 10 (IGEM)
		4. Central Mongolia, Chuluu Ungas Uul mountain (south of Arts Bogd ridge), upper part of svita, basalt	92 ± 9 (IGEM)
		5. Altitude 1640 m (south of Arts Bogd ridge), basalt	94 ± 3 (IGEM)
		6. Top of Zoost Uul mountain, upper part of svita, basalt	94 ± 6 (IGEM)
		7. Altitude to the south of Khalzan Khairkhan mountain (upper layer), basalt	93 ± 4.5 (IEC)
		8. Altitude to the south of Khalzan Khairkhan mountain (lower layer), basalt	95 ± 4.5 (IEC)
		9. Separate hill west of Mushgai Khudag well, lower part of svita, basalt	99 (IGEM)
		10. Mountain Khalzan Khairkhan (south of ridge Arts Bogd), lower part of svita, basalt	101 ± 7 (IGEM)
		11. Altitude 1122.5 m (south of lake Ulaan Nuur), lower part of svita, basalt	101 ± 10 (IGEM)
Lower	Albian–Aptian	Baruunbayan Svita	
		12. Trans-Altai Gobi, 10 km east of mountain Dösh Uul, upper part of svita, basalt	104 ± 13 (IG)
		13. Central Mongolia, well Hürmen Khudag (south of ridge Arts Bogd), basalt	108 ± 7 (IGEM)
		14. Trans-Altai Gobi, mountain Dösh Uul, lower part of svita, basalt	113 (IG)
		15–16. Altitude 1072 m, lower part of svita, basalt	110 ± 6 112 (IGEM)
		Hüteeg Gorizont Döshuul Svita	
		17. South Gobi, Tsagaan Gol dry river-bed, lower part of svita, basalt	116 ± 8 (IGEM)
	Valanginian–Berriasian	**Tsagaantsav Gorizont** Tsagaangol Svita	
		18–19. South Gobi, Nogoon Tsav, lower part of svita, basalt	122 ± 16 129 ± 12 (IG)
		Tsagaantsav Svita	
		20. East Mongolia, Galshar Sum, low part of svita, trachydacite	133 ± 6 (Pb–Sr) (IGEM)
		21. Khalkhyn Gol river, upper part of svita, liparite	120 ± 5 (Rb–Sr) (IGEM)
		22. Bayanmunkh Sum, middle part of svita, liparite	127 ± 6 (Rb–Sr) (IGEM)
		23. Berkhe Sum, middle part of svita, liparite	128 ± 6 (Rb–Sr) (IGEM)
		24. Bayanmönh Sum, lower part of svita, basalt	134 ± 7 (IGEM)
		25. Sümiin Nuur lake, lower part of svita, basalt	138 ± 7 (IGEM)

Table 14.2. (*cont.*)

26. Ulaan Nuur lake (south of Khar Airak Sum), liparite, perlite	126 ± 5 (Rb–Sr) (ARGI)
27. Galshar Sum, lower part of svita, trachydacite	133 ± 6 (Rb–Sr) (ARGI)
28. Khaya Ulaan Nuur lake, lower part of svita, basalt	125 ± 7 (IGEM)
29. Ölzii Ovoo mountain, lower part of svita, basalt	129 ± 8 (IGEM)
30. Ölgii Khiid, middle part of svita, dacite	123 ± 4 (IGEM)
31. Khar Hötöl Uul mountain, lower part of svita, basalt	128 ± 9 (IGEM)
32. Ölgi Khiid (north-west of ruins of monastery), lower part of svita, basalt	131 (IGEM)
33. same place, upper part of svita	123 ± 4 (IGEM)
Öndörukhaa Svita	
Central Mongolia	
34. Erdene Uul mountain, lower part of svita, basalt	132 (PGS UG)
35. Bulgant Uul, lower part of svita, basalt	127 ± 10 (IGEM)
36–37. Öösh Nuur ridge, lower part of svita (lower and upper layers of basalt correspondingly)	141 ± 8 (IGEM) 126 ± 9 (IGEM)
38. Zost Uul mountain, upper part of svita, basalt	119 ± 6 (IGEM)

Notes:
Laboratory abbreviations are: ARGI, All-Russian Geological Institute, Saint Petersburg; IEC, Institute of the Earth's Crust, Siberian Branch of the Russian Academy of Sciences, Irkutsk; IG, Institute of Geology, Georgian Academy of Sciences, Tbilisi; IGEM, Institute of Ore Deposits, Mineralogy, and Geochemistry, Russian Academy of Sciences, Moscow; PGS UG, Production Geological Society, Ural Geology, Ekaterinburg

Altai, Central Mongolia); more often it is less (150–200 m).

Höövör, situated on the north of Gobi, 15 km from Guchinus Sum, in Övörkhangai Aimag, shows a detailed section of the sediments (Table 14.3). The thickness of this section is more than 100 m; altogether the thickness of Aptian–Albian deposits in this region is more then 500 m (Shuvalov, 1974). Lacustrine and lacustrine-fluvial (deltaic) facies predominate here. Besides the vertebrate fauna, remains of molluscs and conchostracans have also been collected from this section (Shuvalov, 1974).

The richest dinosaur localities are in the regions of the gorges Hüren Dukh, Khamaryn-Khural, Hühteeg (East Gobi), Höövör, Andai Khudag (Central Mongolia), Ih Ereen (South Gobi), and others. The dinosaurs of Höövör are most commonly *Psittacosaurus mongoliensis*, as well as sauropods and small theropods (Kalandadze and Kurzanov, 1974). Other vertebrate fossils from Höövör include turtles of the families Sinemydidae, Dermatemydidae, and *Mongolemys* sp. (Kalandadze and Kurzanov, 1974). Mammals are represented by multituberculates (the plagiaulacid *Aguinbaatar dmitrievae* Trofimov, 1980), specialized triconodonts (*Gobiconodon borissiaki* Trofimov, 1978; *Guchinodon hoburensis* Trofimov, 1978), a new symmetrodont (*Gobiodon infinitus* Trofimov, 1980) and two new genera of archaic Insectivora (Eutheria), similar to the Upper Cretaceous *Kennalestes* and *Zalambdalestes* from Bayaan Zag.

The Hüren Dukh locality coincides with the north-west part of the Choir depression, located about 60 km to the south of the railway station Choir (Figure 14.4). The rocks occur in a monocline dipping 7–12° southeast. The sediments, mainly lacustrine sandstones and clays, are about 130 m thick in all. Low in the sequence, close to the underlying Palaeozoic, deltaic facies of sandstones and gravels can be observed. Dinosaurs from this sequence include *Psittacosaurus* sp. and *Iguanodon orientalis* (?) (Rozhdestvenskii, 1955; Norman, 1996).

At the Andai Khudag (Ondai Sair) locality,

Figure 14.3. Concretions of compact sandstone included in friable sandstones of Khulsangol Svita near the mountain Sakhlag Uul, next to Ööshiin Nuruu cliffs (Pre-Altai part of the North Gobi). (Photo by V.F. Shuvalov.)

described earlier, the Khulsangol Svita has yielded remains of turtles and dinosaurs, among them *Psittacosaurus mongoliensis*, sauropods and theropods (Kalandadze and Kurzanov, 1974; Shuvalov, 1975a).

The Khamaryn Khural locality, in South-east Gobi, 40 km south-east of the town Sainshand has yielded fragmentary remains of crocodiles and dinosaurs (*Iguanodon orientalis*, sauropods, theropods) from sandstones and gravels of the Hühteeg Svita (Kalandadze and Kurzanov, 1974). From the same svita the remains of psittacosaurs have been found in the south from the mountain Khar Hötöl Uul (Verzilin, 1979b; Shuvalov, 1982). Near the mountain Hühteeg (East Gobi), in the stratotype of the svita, the remains of dinosaurs and turtles have also been found (Martinson and Shuvalov, 1973; Shuvalov, 1980).

In South Gobi the remains of dinosaurs have been recovered in the region of the mountain Ih Ereen (south-west of the town Dalanzadgad). Here, in the sandstones and clays of the Ulaandel Svita, we have found the remains of *Iguanodon orientalis* (teeth and jaws) and turtle and dinosaur eggshells (Shuvalov, 1982). North-east of Noyon Sum, low in the svita, bones of *Psittacosaurus* sp. were found (Shuvalov, 1982). The Ulaandel Svita of South Gobi (region Hürmen Sum) has also yielded abundant molluscs and conchostracans characteristic of the Hühteeg Gorizont (Shuvalov, 1982), as well as insects. Oval concretions of sandstone and wood are found in many regions in the Khulsangol Svita, but especially in North and East Gobi (mountain Erdene Uul, Erdenetsogtyn Gobi gorge, mountain Hühteeg and in many other places). In the stratotype of the Hühteeg Svita near the mountain Hühteeg, there are abundant remains of stromatolites (Sochava, 1977). Coprolites from the Zereg Svita of West Mongolia may belong to champsosaurs (Efimov, 1983).

Dinosaurs occur also in red-coloured deposits of the Baruunbayan Svita (Aptian–Albian) (Shuvalov, 1982), which were earlier ascribed to the Sainshand Svita (Albian–Cenomanian) or generally to the Upper

Table 14.3. *The Hühteeg Svita in the Höövör section, recorded from the bottom*

1. Sandstones loose and poorly cemented, of grey, greyish-brown and greenish-grey colour with single interlayers of dark-grey and greenish-grey sandy clays, containing the remains of dinosaurs and turtles. Numerous disc-shaped and spherical ferrous sandstone concretions.	15 m
2. Clays of dark-grey and greenish-grey colour with interlayers of grey and pinkish-grey sandstone.	6 m
3. Sandstones of different cementation, of grey, greenish-grey and yellowish-grey colour, intercalated with clays and single horizons of intraformational gravels, which include the remains of dinosaurs and turtles.	35–40 m
4. Sandstones of yellowish-grey and greenish-grey colour with single interlayers of solid argillite-like clays of brownish, greyish-brown, and dark grey colour.	5 m
5. Sandstones of different grain sizes, of grey and yellowish-grey colour, with interlayers of brownish, greyish-brown, greenish-grey, and dark grey-coloured clays and intraformational gravels and conglomerates (up 0.4 m), which include the dinosaur remains.	20 m
6. Sandstones loose, of yellowish-grey colour, cross-bedded, with remains of turtles, lizards, and mammals.	up to 3m
7. Sandstones of different grain size and cementation, of greenish-grey, yellowish-grey, and reddish greyish-brown colour, with interlayers of sandstone concretions and sandy clays (up 0.6 m). In the sandstones were found remains of turtles and dinosaurs.	20m

Source: From Shuvalov, 1974.

Cretaceous (Vasil'ev *et al.*, 1959; Martinson *et al.*, 1969; Barsbold, 1972, 1983; Shuvalov, 1975a; Sochava, 1975). The Baruunbayan Svita is as much as 250–300 m thick, and it is composed predominantly of poorly sorted sediments from alluvial fans, and it is capped by basalts, mainly in the upper part of the svita, in Trans-Altai Gobi and Gobi Altai, and on the Arts Bogd ridge, in the south. In one of the most complete sections in the cliffs Baruunbayan and Züün Bayan in North Gobi (Table 14.4), the Baruunbayan Svita rests on sediments of the Khulsangol Svita, in places directly on the Palaeozoic. The section is overlain unconformably by Quaternary grey boulders and pebbles. The thickness of the svita here is about 200 m. Sections like this have also been observed in the west (Ulaanöösh, Algui Ulaan Tsav, and others).

Among dinosaurs found by us and other researchers in the Baruunbayan Svita, iguanodonts and small psittacosaurs are commonest (Rozhdestvenskii, 1971, 1974; Barsbold, 1983). In addition, bones of sauropods were found, as well as turtle eggs, tortoise shells (South Gobi), and others in the gorges Ulaanöösh, Baruun Bayan, Züün Bayan (see also Maleev, 1952; Efremov, 1954; Rozhdestvenskii, 1955; Shuvalov, 1975a; Barsbold, 1983; Figure 14.5). Numerous oval eggs, up to 200 mm in diameter, and with multicanal

shells, *Faveoloolithus ningxiaensis* (Figure 14.5) occur rarely isolated, as well as in clutches of up to ten and more (Sochava, 1969; 1975; Shuvalov, 1975a). This kind of egg was found in the Algui Ulaan Tsav gorge (South Gobi) and north of Nogoon Tsav gorge (south of the mountain Dösh Uul, Trans-Altai Gobi). The Baruunbayan Svita has also yielded the charophyte alga *Mesochara tursoni*, the mollusc *Oxynaia sainshandica*, ostracods, and conchostracans (Barsbold, 1972; Martinson, 1975).

The Hühteeg Svita is dated as Aptian–Albian on the basis of fossils from several localities, and radiometric dating. The Höövör turtles and mammals suggest an Aptian–Albian age (Shuvalov, 1974). The turtles are different from Upper Cretaceous forms and similar to those known from the Aptian–Albian of Trans-Altai and East Gobi (Dösh Uul, Khar Hötöl, Hühteeg). The Hüren Dukh succession is dated by molluscs and dinosaurs. *Iguanodon orientalis* is similar to a specimen previously described from the Khamaryn Khural region (East Gobi) from the Aptian–Albian (Vasil'ev *et al.*, 1959). The molluscs are characteristic of the Aptian–Albian (Martinson, 1975). The spore-pollen assemblage, according to G.M. Brattseva (pers. comm.), is also Aptian–Albian (Shuvalov, 1974). In the Trans-Altai Gobi, Hühteeg deposits ascribed to the

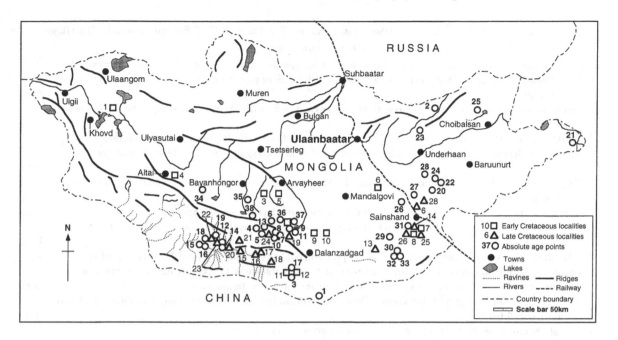

Figure 14.4. The main localities of dinosaurs and locations of samples of Cretaceous volcanics used for radioactive dating. *1–12, Early Cretaceous dinosaur localities.* 1, Tatal Yavar (West Mongolia, Sangiin Dalai Nuur depression, Tsagaantsav Gorizont); 2, Ööshiin Nuruu (Pre-Altai Gobi, Tsagaantsav Gorizont); 3, Andai Khudag (South Khangai, Shinekhudag and Hühteeg gorizonts); 4, Gurvan Ereen (West Mongolia, region of Darvi Sum, Hühteeg Gorizont); 5, Höövör (South Khangai, depression Guchin, Hühteeg Gorizont); 6, Hüren Dukh (Middle Gobi, Hühteeg Gorizont); 7, Khamaryn Khural (East Gobi, Hühteeg Gorizont); 8, Khar Hötöl Uul (East Gobi, Hühteeg Gorizont); 9, Algui Ulaan Tsav (Middle Gobi, Baruunbayan Svita); 10, Ulaanöösh, Baruun Bayan and Züün Bayan (Middle Gobi, Baruunbayan Svita); 11, Tsagaan Gol (South Gobi, Baruunbayan Svita); 12, Ih Ereen (South Gobi, Hühteeg Gorizont). *13–28, Late Cretaceous dinosaur localities.* 13, Baishin Tsav (East Gobi, Bayanshiree and Baruungoyot svitas); 14, Khar Hötöl Uul (East Gobi, Bayaanshiree Svita); 15, Altan Uul (South Gobi, Bayanshiree Svita); 16, Nemegt (South Gobi, Baruungoyot and Nemegt svitas); 17, Khulsan (South Gobi, Baruungoyot Svita); 18, Ukhaa Tolgod (South Gobi, Baruungoyot Svita); 19, Bayan Zag and Tögrögiin Shiree (South Gobi, Baruungoyot Svita); 20, Hermiin Tsav (Trans-Altai Gobi, Baruungoyot Svita); 21, Bügiin Tsav (South Gobi, Nemegt Svita); 22, Nogoon Tsav (Trans-Altai Gobi, Nemegt Svita); 23, Ingenii Tsav (Trans-Altai Gobi, Nemegt Svita); 24, Üüden Sair (South Gobi, Baruungoyot and Nemegt svitas); 25 Bayan Shiree (East Gobi, Bayanshiree Svita); 26, Khongil Tsav (East Gobi, Bayanshiree and Baruungoyot svitas); 27, Teel Ulaan Uul (East Gobi, Baruungoyot Svita); 28, Baga Tariach (East Gobi, Baruungoyot Svita). Locality numbers of samples used in radiometric dating are given in Table 14.2, which also contains localities numbered above 28.

Döshuul Svita (Martinson and Shuvalov, 1976), contain a horizon of basic volcanics (basalts), whose radiometric age is 110–113 Myr (Table 14.2; Shuvalov and Nikolaeva, 1985).

The Baruun Bayan basalts yield a radiometric age of Aptian–Albian (Shuvalov, 1982; Devyatkin *et al.*, 1990). The charophyte alga *Mesochara tursoni* from the Baruunbayan Svita was recorded earlier from the Aptian of Hungary (Kyansep-Romashkina, 1982), and

the mollusc *Oxynaia sainshandica* is known only from the Aptian–Albian (Martinson, 1975).

Upper Cretaceous

The Upper Cretaceous deposits of the Gobi region of Mongolia are represented by three successive svitas, formed on a platform (Nagibina *et al.*, 1977; Shuvalov, 1975b) and in an arid climate (Sochava,

Table 14.4. *The sequence of the Baruunbayan Svita (Aptian–Albian) in the cliffs of Baruun Bayan and Züün Bayan, recorded from below*

1. Conglomerate-breccias of red greyish-brown colour, with lenses of gravel, sandstone and, detrital clay of the same colour.	30 m
2. Intercalation of conglomerate-breccias with sandstones and clays, with common lens-shaped beds 0.2–5 m in thickness.	40 m
3. Boulder-pebble conglomerates with lens-shaped interlayers of red-greyish-brown and ochre-yellow sandstones, detrital clays and gravels. Thickness of interlayers of clays and sandstones is 3–4 m, of gravels 0.3 m.	30 m
4. Clay of red greyish-brown colour, sandy, solid, with lens-shaped interlayers of small-pebble conglomerates, 0.1–1 m in thickness.	15 m
5. Conglomerates, grey and greyish-brownish-red, with lenses of clays and sandstones.	5 m
6. Detrital clay, reddish greyish-brown with lens-shaped interlayers of conglomerates and sandstones up to 2 m in thickness.	37 m
7. Clay, dark grey and grey, merging into yellowish-pink and reddish greyish-brown, with interlayers of small-pebble conglomerates and ochre-yellow sandstone, up to 1m in thickness. In clays are marl concretions with remains of molluscs and ostracods, coalified plants remains, and ones of dinosaurs.	8 m
8. Intercalation of detrital greyish-brownish yellow clays with sandstones, gravels and conglomerates. Thickness of interlayers 0.1–1.5 m.	6 m
9. Conglomerates, red greyish-brown with lenses of sandstones and clays.	5 m
10. Sandstone, red greyish-brown with inclusion of rare pebbles.	3 m
11. Clay, red greyish-brown, with lens-shaped interlayers of pinkish-grey conglomerate.	22 m

1975; Shuvalov, 1985): Bayanshiree (Cenomanian–Santonian), Baruungoyot (Santonian–Campanian), and Nemegt (Maastrichtian). The first two svitas are most widespread, occurring practically in all depressions in the Gobi. The Nemegt Svita occurs mainly in Trans-Altai Gobi, as well as in the south-west of East Gobi (Baishin Tsav depression, south of the Arts Bogd ridge, Üüden Sair, and others), Central Mongolia, and a number of other places (Shuvalov and Stankevich, 1977). Dinosaurs are recorded throughout (Barsbold, 1983; Tatarinov, 1983), but are particularly abundant in the Nemegt Svita (localities Nemegt, Bügiin Tsav, Altan Uul, and others). Remains of conchostracans, molluscs, turtles, and mammals occur everywhere (Martinson, 1975; Shuvalov and Chkhikvadze, 1975).

Berkey and Morris (1927) distinguished a number of formations in the Upper Cretaceous. They named the Djadokhta Formation with dinosaurs, dinosaur eggs, and mammals, in the Bayan Zag cliffs (or Shabarak Us gorge), north-east of Bulgan Sum. Originally, this formation was ascribed to the lower part of the Upper Cretaceous. Later, Morris (1936) suggested a younger age, in the upper part of the Upper Cretaceous, on the basis of analysis of the fauna from the sandstones of Bayan Zag. The age of the Bayan Zag deposits is still debated (Martinson *et al.*, 1969; Sochava, 1969, 1975; Barsbold, 1972, 1983; Kielan-Jaworowska, 1975; Martinson, 1975; Shuvalov, 1975a), but neither the dinosaurs nor the mammals will give an unequivocal answer. Taking into account the data of various researchers (Verzilin, 1979a, 1980; Kolesnikov, 1982; Jerzykiewicz *et al.*, 1993), we present our own scheme of the stratigraphy of the Upper Cretaceous of the Gobi regions of Mongolia (Shuvalov, 1982, 1994).

Bayanshiree Svita (Cenomanian–Santonian)

In sections of the Bayanshiree Svita in East and South-east Mongolia its two-part structure is clear: in lower parts are usually grey sandstones and gravels with isolated interlayers of clays and conglomerates, in upper

Figure 14.5. Large spherical dinosaur eggs with multicanal shell structure from the Baruunbayan Svita (locality Algui Ulaan Tsav, Middle Gobi). (Photo by V.F. Shuvalov.)

Figure 14.6. Clays, sandstones and gravels in the upper part of Bayanshiree Svita (region of Altan Uul mountain, South Gobi). (Photo by V.F. Shuvalov.)

parts, multicoloured clays and sandstones (Figure 14.6). In the south and south-west, lower parts of the svita are usually formed from multicoloured, and upper parts by red-coloured deposits, but here also the coarsest facies are found in lower parts (Shuvalov, 1993). Its total thickness in West and Trans-Altai Gobi is 150–200 m. In South-east Mongolia, Sochava (1969) argued that the Baruungoyot and Sainshand svitas were equivalent to the Bayanshiree, but this is not possible since the Baruungoyot Svita is an independent unit in both regions (Martinson, 1975; Shuvalov, 1975b). In the west, north, and north-east of the country, neither the Bayanshiree Svita, nor the younger units of the Upper Cretaceous have yet been found (Shuvalov, 1982; Devyatkin et al., 1975; Martinson, 1975). Sections of the Bayanshiree Svita have been described many times (Barsbold, 1972, 1983; Martinson, 1975; Shuvalov, 1975a, 1982).

Dinosaurs have been recorded from various horizons and various regions: (1) a skeleton of the ankylosaur *Talarurus plicatospineus* from the upper part of the svita at Bayan Shiree cliffs (South-east Gobi), USSR Palaeontological Expedition of 1946 (Efremov, 1949; Maleev, 1952; Sochava, 1975); (2) various dinosaurs, turtles (*Amida orlovi*, and others), and crocodiles (*Shamosuchus major*), found later in Khongil Tsav and to the north from the mountain Khar Hötöl Uul (Martinson et al., 1969; Sochava, 1975; Efimov, 1983); (3) large numbers of dinosaur remains from sandstones at Baishin Tsav and other locations in the south-west of East Gobi, including ornithomimosaurs, hadrosaurs, and segnosaurs (Barsbold, 1983); (4) the hadrosaur *Arstanosaurus* sp. from the lower part of the svita in the north from Altan Uul mountain in Trans-Altai Gobi, a form known hitherto only from the Bayanshiree Svita of East Gobi (Shuvalov et al., 1991).

The Bayanshiree Svita embraces deposits dated from Cenomanian to lower Santonian (Shuvalov, 1982, 1994). In Central Mongolia, covering basalts yield absolute ages, measured by K–Ar, of 101–92 Myr (Table 14.2), in other words, Cenomanian (Baskina et al., 1978; Naidin, 1981; Harland et al., 1990; Shuvalov and Nikolaeva, 1985; Devyatkin et al., 1990). Upper horizons have been dated as Coniacian–Santonian (Martinson, 1975) on the basis of comparison of the molluscs *Sainshandia robusta, S. sculpturata, Pseudohyria turischewi, P. tuberculata* and others, with specimens from the Yalovach Svita of Fergana and the Bostobe Svita of the Aral Sea region.

Baruungoyot Svita (Santonian–Campanian)

The mainly red sandstones and mudstones of the Baruungoyot Svita occur in all regions of the Gobi. The Baruungoyot Svita unites the red sediments of the Djadokhta Formation (Berkey and Morris, 1927; Kielan-Jaworowska, 1968, 1970, 1975; Kaie and Devyatkin, 1969), the Djadokhta Svita (Barsbold, 1972, 1983), or the Bayanzag Svita (Khand, 1974), the multicoloured sediments of the Tögrög Bulag and multicoloured sediments to the west from Bayan Zag (Barsbold, 1972, 1983; Khand, 1974; Tverdokhlebov and Tsybin, 1974), and also the multicoloured sediments of the regions of Altan Uul, Nemegt, Khulsan, Bügiin Tsav, Hermiin Tsav, Bambuu Khudag, Ingenii Höövör, Shireegiin Gashuun and many others in Trans-Altai and South Gobi (Figure 14.7). The recently described (Dashzeveg et al., 1995) red-coloured sandstones from Ukhaa Tolgod (north of the settlement Davs in South Gobi Aimag) also belong here, and those on the watershed between the ridges Gilbent and Sevrei (pers. obs.). In South and East Gobi, the Baruungoyot Svita unites red sandstones and mudstones in the regions of Bayandalai Sum, south-east of the town Dalanzadgad, at Baishin Tsav, Javkhlant Uul, near the mountains Teel Ulaan Uul and Baga Tariach, in the depression Tamtsag, in the gorge Borzongiin Gobi, near Khanbogd Sum, in the upper part of the section of red sediments in the gorge Khongil Tsav, south-west of the town Züün Bayan, and in many other places in the Gobi (Figure 14.4; Martinson et al., 1969; Martinson, 1975; Shuvalov, 1975a; Sochava, 1975).

The Baruungoyot Svita cannot readily be subdivided, since it is a uniform unit which often extends laterally for tens and hundreds of kilometres without change (red sandstones and mudstones with interlay-

Figure 14.7. Sandstones of the Baruungoyot Svita on the southern slope of the Nemegt ridge (South Gobi). (Photo by V.F. Shuvalov.)

ers of conglomerates). The svita varies from 30–50 m to 100 m in thickness (Martinson, 1975; Shuvalov, 1975a; Sochava, 1975).

Numerous dinosaur remains have been found, mainly at Bayan Zag, Hermiin Tsav, Khulsan, and other places in the South and Trans-Altai Gobi: *Protoceratops andrewsi, Velociraptor mongoliensis, Oviraptor philoceratops, Syrmosaurus viminicadus* (Berkey and Morris, 1927; Morris, 1936; Barsbold, 1983). Large numbers of mammal remains have also been found in the Baruungoyot Svita, initially *Djadochtatherium mat-thewi, Zalambdalestes lechei, Deltatheroides cretacius*, and others (Berkey and Morris, 1927; Kielan-Jaworowska, 1970, 1971, 1975; Barsbold, 1972, 1983; Shuvalov, 1975a). They occur not only at Bayan Zag, but in other regions in the south and Trans-Altai Gobi (Hermiin Tsav, Khulsan, and others). Other common fossils include turtles (*Lindholmemis*), lizards, crocodiles, and

dinosaur eggs (Berkey and Morris, 1927; Kielan-Jaworowska, 1968; Martinson, 1975; Shuvalov and Chkhikvadze, 1975). Invertebrates (ostracods, molluscs, and others) are rare. Ostracods (*Cypridea, Rhinocypris,* and others) were found in the Trans-Altai Gobi, in the badlands of Altan Uul (eastern) and Bügiin Tsav (Stankevich and Sochava, 1974), and *Gobiocypris* was found at Tögrögiin Shiree, west of Bayan Zag (Khand, 1974).

Dating of the Barungoyot Svita is difficult: most researchers give a Senonian, most likely Campanian, age. Kielan-Jaworowska (1968, 1970) lowered the age of the Djadokhta to Coniacian–Santonian on the basis of mammal remains from Bayan Zag and other places. However, molluscs, ostracods, and turtles do not allow an age greater than Santonian. In addition, K–Ar dates from basalts in the middle and upper parts of the svita (in the gorge Borzongiin Gobi and others) are 78–80

Figure 14.8. Clays, sandstones and marls of the Nemegt Svita (Ingenii Tsav gorge, Nogoon Tsav locality, Trans-Altai Gobi). (Photo by V.F. Shuvalov.)

Myr (Campanian). Hence, we date the Baruungoyot Svita as Santonian–Campanian (Shuvalov and Nikolaeva, 1985).

Nemegt Svita (Maastrichtian)

The youngest svita of the Cretaceous of Mongolia, though not widely distributed, has yielded the largest number of dinosaur remains and other fauna and flora. The Nemegt Svita consists mainly of grey and yellow greyish-brown sandstones and mudstones with inter-layers of red mudstones (Figure 14.8) and widespread intraformational gravels and conglomerates (Sochava, 1975; Verzilin, 1980, 1982), mainly grey and yellow-ish-grey. The Nemegt Svita occurs mainly in Trans-Altai Gobi, in South-west Mongolia, in the south-west of East Gobi (Baishin Tsav), south of the ridge Arts Bogd, and elsewhere in the Gobi (Shuvalov, 1976, 1982, 1985; Shuvalov and Stankevich, 1977). In Trans-Altai Gobi the richest localities for dinosaurs and other associated fauna and flora are in the badlands Nemegt, Altan Uul, Tsagaan Khushuu, Bügiin Tsav, Ingenii Tsav, and others (Figure 14.4; Efremov, 1949, 1955; Barsbold, 1972, 1983; Sochava, 1975).

Dinosaurs in Mongolia are most abundant in the Nemegt Svita. Taxa discovered by various expeditions in Trans-Altai Gobi include the tyrannosaurid *Tarbosaurus*, the ornithomimid *Gallimimus*, the dromaeosaurid *Adasaurus*, the deinocheirids *Therizinosaurus* and *Deinocheirus,* and others (Barsbold, 1983). Other faunal elements include charophyte algae, molluscs, ostracods, conchostracans, turtles, crocodiles, lizards, mammals, and eggs of dinosaurs and turtles. The thickness of deposits of the Nemegt Svita is more than 40–50 m (Martinson, 1975; Shuvalov, 1982).

The Nemegt Svita is dated as Maastrichtian on the basis of a variety of fossil evidence (Martinson *et al.*, 1969; Martinson, 1975; Shuvalov, 1976, 1982, 1994). The left-coiled gastropods *Mesolanistes* are also known from Maastrichtian deposits of North America and China, as are some conchostracans, ostracods, and charophytes (Martinson, 1975). The absolute age of the middle and upper parts of the underlying Baruungoyot Svita (75–80 Myr) confirms the Maastrichtian age. Formerly, Barsbold (1983) dated the dinosaur-rich lower horizons of the Nemegt Svita as Upper Campanian–Lower Maastrichtian, and the less rich upper layers as Maastrichtian, sometimes distinguished as a separate Nogoontsav Svita.

Conditions of deposition of the Cretaceous of Mongolia

Geologists and palaeontologists have presented different views on the conditions of deposition of the Mongolian continental Cretaceous (summarized in Shuvalov, 1982, 1994; see also Chapter 15).

Lower Cretaceous (Neocomian)

Deposits of the Tsagaantsav Svita in depressions began as early as the end of the Upper Jurassic. The Early Neocomian (earliest Cretaceous) represents the time of the greatest distribution in Mongolia of lake basins and active volcanism (Shuvalov, 1982, 1994). In the centre, east, and south of Mongolia significant sequences of basic, and in some places, acid volcanics and tuffs accumulated, partly underwater (Vasil'ev *et al.*, 1959). For this reason, rare zeolites of lake-tuffogenic type were formed.

At the end of the Neocomian, in Shinekhudag time, denudation and flattening dominated everywhere. Western and northern regions of Mongolia at this time, and before, had undergone general uplift, which led to cessation of sediment accumulation in a number of Jurassic depressions. There was a relatively calm tectonic regime at the end of Shinekhudag time and little volcanic activity. Shinekhudag sediments indicate lagoonal conditions in all depressions, and hydro-gen sulphide contamination in bottom layers (Sochava, 1975; Shuvalov, 1982). Temperate-humid climates prevailed (Shuvalov, 1982, 1985).

The wide distribution of Neocomian lake basins is reflected on palaeogeographic maps (Shuvalov, 1982, 1985, 1994). In the west of Mongolia, where volcanic activity did not occur at the beginning of the Cretaceous, Neocomian basins were different from those in other regions, because of their isolation from the rest of territory, as well as with a more arid climate, as shown by lithologies and faunas (Khosbayar, 1972; Devyatkin *et al.*, 1975; Ponomarenko and Popov, 1980).

Lower Cretaceous (Aptian–Albian)

The beginning of Aptian–Albian time (Hühteeg Svita) in Mongolia was marked by a more humid climate, activation of tectonic movements, complication of relief, and reduction in the rate of sediment accumulation in most depressions in West and North Mongolia. In the west, sediment accumulated only in the Ihes Nuur (region of Gurvan Ereen ridge) and Zereg (near Altanteel Sum) depressions, in their lowest south-eastern parts. In South Mongolia, on the other hand, the number of depression zones, and their sizes, increased. These include the Trans-Altai zone of depression (Düshuul, Ingenii Höövör, Shireegiin Gashuun depressions) extending 300 km from east to west, and the Zuramtai depression. To the south of Khangai and Hentii, Höövör, Hüren Dukh and many other depressions appeared. Among these, the Nyalga, Choibalsan, Tamtsag, and some other older depressions, expanded and became more complicated in their outlines. On the other hand, in South-east Mongolia, the area of such depressions as Züün Bayan, Baishin Tsav, and Khar Hötöl reduced (Shuvalov, 1982).

Warm humid climates dominated at the beginning of the Aptian–Albian, and this led to intensive denudation and the erosion of large river channels which fed the depressions with clastic material. In depressions of the humid zone, which included the northern, eastern, south-eastern and central regions of

Mongolia, sediments accumulated in lakes and rivers. The lakes were significantly smaller than those of the Neocomian, and they were often boggy around their margins, which led to the formation of coal (Övdög Khudag, Jargalantyn, Baganuur). At the end of this time, tectonic movements intensified, and this led to the accumulation of significant thicknesses (up to 300–500 m) of sandstones, gravels, and conglomerates with well-rounded pebbles, quite often with the remains of trunks and fragments of petrified trees (Erdenetsogtyn Gobi gorge, north of Khar Hötöl Uul mountain).

The Aptian–Albian lakes of the arid zone of Mongolia (west, south-west, and south regions) apparently had no outflows; temporary connections were formed only during periods of flooding. The lakes were characterized by variable depths, sizes, and hydrodynamical regimes; many were shallow (Martinson and Shuvalov, 1976). Their distinguishing feature was high salinity (Kolesnikov, 1982).

Volcanic activity in the Aptian–Albian of Mongolia occurred only in the north-east (near the lake Baga Zos Nuur), in Gobi Altai (in the region of Jargalantyn Gol, south of Arts Bogd ridge), and in Trans-Altai Gobi (near the mountain Dösh Uul). Fragments of basalts of this age still remain exposed in some places (Shuvalov, 1982).

In the humid zone, forest landscapes dominated, vegetated with heat-loving conifers, mainly pines (*Picea*, *Pinus*, *Cedrus*), as well as araucarians, marsh cypresses, and others. Angiosperms and some species of ferns occurred in all areas, but ginkgos were much less abundant (Larishchev, 1955; Brattseva and Novodvorskaya, 1975). In the semiarid warm zone in the south and south-west, the character of the vegetation was generally similar to previous times. Sinitsyn (1962) suggested that landscapes then looked like modern subtropical savannas with isolated woods of oasis type.

At the end of the Aptian–Albian (Baruunbayan time), the most important tectonic event in this story occurred: the area of active downwarp of South and South-east Mongolia (the Gobi Downwarp) became separated from the area of uplift of West, North, and North-east Mongolia (the North Mongolian Uplift). Since this time, sediments accumulated only in South and South-east Mongolia, that is, in the Gobi regions. This marked the establishment of the basic elements of the platform structure of Mongolia, which finally took shape in the Upper Cretaceous (Shuvalov, 1975b).

At the same time, the warm humid climate was replaced by hot arid and semiarid conditions, which became characteristic of the following Cretaceous stages. This is shown by the deposits of the Baruunbayan Svita – red, mainly alluvial, conglomerates, breccia, sandstones and detrital clays of marginal parts of depressions. In central parts in some places there were apparently lakes, shown by remains of molluscs, ostracods, and charophytes (Shuvalov, 1982) in some sections (Baruun Bayan, Züün Bayan). Downwarping of the Gobi was still in progress, associated with fractured basalt eruptions. The remains of the basalt cover have been observed by us only in the north-west in the Ingenii Höövör depression (mountain Dösh Uul), in the south on the Arts Bogd ridge and in some other places (Figure 14.4; Table 14.2). Apparently, dinosaurs and their eggs were associated with the coastal zone of the Gobi lakes, which were characterized by irregular levels and slightly increased salinity. Around the lakes on the plains of the Gobi, apparently, dry savannas and semi-deserts dominated, and forests were mainly present on watersheds in northern and western parts of Mongolia (Shuvalov, 1982).

Upper Cretaceous

In the Late Cretaceous, sediments accumulated only in the Gobi regions of Mongolia. This marks the beginning of a long period of levelling of the relief of Mongolia, mainly by denudation in the north and west, and accumulation in the south, south-east, and extreme east of the country (Shuvalov, 1975b; Nagibina *et al.*, 1977). The slow sagging of the Gobi part of Mongolia, and periodic humidification, especially at the beginning of Bayanshiree time, contributed to the origin of large inland water basins, which attracted dinosaurs and other reptiles (Rozhdestvenskii, 1971, 1974; Shuvalov, 1982, 1985).

The lakes apparently reached their largest size in the Santonian, at the time of the greatest depression of the Gobi, when they covered up to 50% of its area and were connected with each other, and possibly also with the marine basins of China and Central Asia, which is shown by the discovery of remains of sharks, some turtles, and ostracods (Shuvalov, 1982). It is not out of the question that at this time the Gobi basin was connected with the world ocean (Nagibina *et al.*, 1977). In the Campanian slow uplift of all of Mongolia began, including the Gobi (although relatively it was lowered), and lake basins were reduced (Shuvalov, 1982). The all-Gobi lake apparently did not exist any more, but there were separate large basins which often reduced in area, and became more saline. Wide beach zones appeared, where dinosaurs and turtles laid eggs.

There has been debate over the sedimentary setting of the sediments of the Baruungoyot Svita, whether they were largely lacustrine or fluvial/aeolian. The lacustrine interpretation was suggested by Verzilin (1980, 1982), and Rozhdestvenskii (1971, 1974) argued that the dinosaurs lived near water, and Shuvalov and Chkhikvadze (1975) noted that the turtles were mainly lacustrine, as were the molluscs and ostracods of this unit, and the units before and after (Khand, 1974; Stankevich and Sochava, 1974; Martinson, 1975, 1982; Kolesnikov, 1982). The alternative proposal, a mainly fluvial or even aeolian origin of the sandstones of the Baruungoyot Svita, and the included Djadokhta Formation, as well as of correlated formations (Ukhaa Tolgod in Trans-Altai Gobi, Bayan Mandahu in North China), was presented by Gradziński (1970), Gradziński and Jerzykiewicz (1972), Sochava (1975), and Barsbold (1983) (see Chapter 15). As is widely known, similar dinosaurs, their eggs, and mammals occur in the Djadokhta Formation and its analogues (Berkey and Morris, 1927; Kielan-Jaworowska, 1975; Jerzykiewicz and Russell, 1991). However, it is practi-

cally impossible to imagine how herbivorous dinosaurs could have been inhabitants of a desert and have been a source of food for predatory dinosaurs too. They required adequate food, and this suggests the vegetation of lake shores and lakes. The work of the palaeobotanists N.M. Makulbekov, I.A. Shilkina, P.I. Dorofeev, V.A. Krasilov and others in Trans-Altai Gobi and other regions, summarized by Martinson (1975, 1982) and Shuvalov (1982, 1985), testify to the wide distribution of vegetation near the water basins of the Gobi in the Campanian and Maastrichtian.

At the end of the Late Cretaceous (Maastrichtian), in connection with continuing uplift and with possible increasing aridity, the area of the lake basins reduced even more, which is seen in the distribution of sediments, mostly in Trans-Altai Gobi, and rarely in other regions. These sediments, containing abundant ostracods, conchostracans, and charophytes, also appear to be lacustrine (Martinson, 1975, 1982; Kolesnikov, 1982; Shuvalov, 1982; Verzilin, 1982), and there is also taphonomic evidence to support this view (Tverdokhlebov and Tsybin, 1974; Efremov, 1955). Not all researchers agree, however. Verzilin (1982) suggested that turbidity flows deposited some of the lake sediments, and these may have been triggered by increased seismicity of the South Gobi. In our opinion, however, the lacustrine turtles and other fauna could quite readily have been buried in mud flows (Shuvalov and Chkhikvadze, 1975). At the same time, it is impossible to agree with the view of Gradziński (1970), Gradziński and Jerzykiewicz (1972), and Sochava (1975), that these deposits formed in arid-zone rivers. We accept the idea of a hot and arid climate at this time, but the climate was not identical everywhere.

Observations on the modern lakes of Mongolia support the arid-land river point of view. Lakes today are concentrated in the arid zone of West Mongolia, the Great Lakes Depression. This apparent anomaly is explained by the fact that the Great Lakes Depression is surrounded by the high mountain systems of Khangai and Altai, from which the largest rivers of this region, the Zavkhan, Khovd, and others, flow. Thus, rapid evaporation from the surface of the lakes is balanced by the influx of river water (Shuvalov, 1985), and aridity and lake development are fully compatible. The uplift of the Gobi regions of Mongolia no doubt led to a reduction in the area of lake basins at the end of the Late Cretaceous, and to the extinction of dinosaurs as a result of habitat loss.

Conclusion

The Lower Cretaceous deposits of Mongolia are much thicker than those of the Upper Cretaceous (2000–3000 m against 200–300 m). This may be explained by the fact that the former were deposited during tectonic activity, and the latter during a time of platform development. The thicknesses of volcanics (in the Lower Cretaceous they are in some places over 200–300 m) confirms this. In addition, Lower Cretaceous deposits are distributed everywhere in Mongolia, but Upper Cretaceous sediments are found only in the Gobi.

There is no agreement on the genesis of the deposits and the climate during the Cretaceous, although there is no doubt that Late Cretaceous climates were generally more arid than those of the Early Cretaceous. Those who interpret the majority of the Upper Cretaceous sediments as lacustrine see evidence for extensive sedimentary basins and substantial vegetation in and around them to provide food for the herbivorous dinosaurs.

There are more than 30 radiometric dates (K–Ar and Rb–Sr) on volcanics of the Cretaceous of Mongolia (Table 14.2). The majority come from the Tsagaantsav Svita, where volcanics are most common, and they, and abundant fossils, confirm a Lower Neocomian Age (Shuvalov, 1987). There are also Aptian–Albian (Hühteeg Gorizont and Baruunbayan Svita), Cenomanian (Bayanshiree Svita), and Campanian (Baruungoyot Svita) dates on basalts. These latter radiometric dates are important since they come from volcanics associated with red beds where organic remains are rare. The radiometric dates have been checked by measurements on acid and basic volcanics, and on minerals and gross samples from the same svitas, and they appear to be reliable (Afanas'ev and Zykov, 1975; Solov'ev *et al.*, 1977; Zhamoida, 1977; Frikh-Khar and Luchitskaya, 1978; Naidin, 1981; Shuvalov, 1982, 1987; Shuvalov and Nikolaeva, 1985; Devyatkin *et al.*, 1990).

Acknowledgements

I thank Z. Kielan-Jaworowska, E.V. Devyatkin, E.N. Kurochkin, and D.A. Subetto for constant help in preparing material for publication. I thank the leadership of the Joint Soviet–Mongolian Geological and Palaeontological Expeditions for the opportunity to take part in field work for many years. The palaeontologists G.G. Martinson, E.K. Trusova, I.V. Stepanov, I.Yu. Neustrueva, N.P. Kyansep-Romashkina, S.M. Kurzanov, and V.M. Chkhikvadze, and the field associates N.N. Verzilin, A.V. Sochava, and R. Barsbold have also been most helpful. Absolute ages of volcanics from the Cretaceous of Mongolia were obtained in the laboratories of IGEM (M.M. Arakelyants) and IG (M.M. Rubinshtein), to whom the author is very grateful. I express my special gratitude to T.V. Nikolaeva (St. Petersburg State University) for constant help with the manuscript and in field studies, and to N.N. Bakhurina for translation of the manuscript.

References

Afanas'ev, G.D. and Zykov, S.N. 1975. [*Geochronological Scale of the Phanerozoic in the Light of New Data on Decay Constants.*] Moscow: Nedra, 156 pp.

Bakhurina, N.N. 1983. [The Early Cretaceous locality of pterosaurs from West Mongolia.] *Trudy Sovmestnoi Sovetsko-Mongol'skoi Paleontologicheskoi Ekspeditsii* **24**: 126–129.

—and Unwin, D. 1995. A survey of pterosaurs from the Jurassic and Cretaceous of the Former Soviet Union and Mongolia. *Historical Biology* **10**: 197–245.

Barsbold, R. 1972. [*Biostratigraphy and Freshwater Molluscs of the Upper Cretaceous of the Gobi part of the MPR.*] Moscow: Nauka, 88 pp.

—1983. [Carnivorous dinosaurs of the Cretaceous of

Mongolia.] *Trudy Sovmestnoi Sovetsko-Mongol'skoi Paleontologicheskoi Ekspeditsii* **19**: 1–120.

Baskina, V.A., Volchanskaya, I.K., Kovalenko, V.I. and others. 1978. [Potassium alkaline volcano-plutonic complex Mushgai Khudag in the south of MPR and associated mineralization.] *Sovetskaya Geologiya* **1978** (4): 86–89.

Berkey, C. and Morris, F. 1927. *Geology of Mongolia. Natural History of Central Asia. Vol. II.* New York: American Museum of Natural History, 475 pp.

Brattseva, G.M. and Novodvorskaya, I.M. 1975. [Spores and pollen from the Lower Cretaceous deposits of Hüren Dukh locality.] *Trudy Sovmestnoi Sovetsko-Mongol'skoi Paleontologicheskoi Ekspeditsii* **2**: 205–209.

Dashzeveg, D., Novacek, M.J., Norell, M.A., Clark, J.M., Chiappe, L.M., Davidson, A., McKenna, M.C., Dingus, L., Swisher, C. and Perle, A. 1995. Extraordinary preservation in a new vertebrate assemblage from the Late Cretaceous of Mongolia. *Nature* **374**: 446–449.

Devyatkin, E.V., Martinson, G.G., Shuvalov, V.F., and Khosbayar, P. 1975. [Stratigraphy of the Mesozoic of West Mongolia.] *Trudy Sovmestnoi Sovetsko-Mongol'skoi Geologicheskoi Ekspeditsii* **13**: 25–41.

—, Nikolaeva, T.V. and Shuvalov, V.F. 1990. [Structural-geomorphological position and main stages of activity of the basalt magmatism in Mongolia in Mesozoic and Cenozoic.] pp. 126–134 in *Geodynamics of Intercontinental Mountain Regions.* Novosibirsk: Nauka.

Efimov, M.B. 1983. [The Champsosauridae of Central Asia.] *Trudy Sovmestnoi Sovetsko-Mongol'skoi Paleontologicheskoi Ekspeditsii* **24**: 67–75.

Efremov, I.A. 1949. [Preliminary results of activity of the First Mongolian Palaeontological Expedition of the Academy of Sciences of the USSR, 1946.] *Trudy Mongol'skoi Komissii AN SSSR* **38**: 1–49.

—1954. [Palaeontological research in the Mongolian Peoples' Republic: results of the expeditions of 1946, 1948, and 1949.] *Trudy Mongol'skoi Komissii AN SSSR* **59**: 3–32.

—1955. Burial of dinosaurs in Nemegt (South Gobi, MPR), pp. 789–809 in *Voprosy Geologii Azii.* Moscow: Izdatel'stvo Akademii Nauk SSSR.

Florentsov, N.A. and Solonenko, V.P. (eds.) 1963. [*The Gobi-Altai Earthquake.*] Moscow: Izdatel'stvo Akademii Nauk SSSR, 391 pp.

Frikh-Khar, D.I. and Luchitskaya, A.I. 1978. [*The Late*

Mesozoic Volcanics and Associated Hypabyssal Intrusions of Mongolia.] Moscow: Nauka, 167 pp.

Gradziński, R. 1970. Sedimentation of dinosaur-bearing Upper Cretaceous deposits of the Nemegt Basin, Gobi Desert. *Palaeontologia Polonica* **21**: 147–229.

—, and Jerzykiewicz, T. 1972. Additional geographical and geological data from the Polish–Mongolian Palaeontological Expeditions. *Palaeontologia Polonica* **22**: 17–32.

—, Kielan-Jaworowska, Z. and Maryańska, T. 1977. Upper Cretaceous Djadokhta, Barun-Goyot and Nemegt Formations of Mongolia, including remarks on previous subdivisions. *Acta Geologica Polonica* **27**: 281–318.

Harland, W.B., Armstrong, R.L., Cox, A.V., Craig, L.E., Smith, A.G. and Smith, D.G. 1990. *A Geologic Time Scale 1989.* Cambridge: Cambridge University Press, 263 pp.

Jerzykiewicz, T., Currie, P. L., Eberth, D.A., Johnston, P.A., Koster, E.H., and Zheng, J.-J. 1993. Djadokhta Formation correlative strata in Chinese Inner Mongolia: an overview of the stratigraphy, sedimentary geology, and palaeontology and comparisons with the type locality in the Pre-Altai Gobi. *Canadian Journal of Earth Sciences* **30**: 2180–2195.

—, and Russell, D.A. 1991. Late Mesozoic stratigraphy and vertebrates of the Gobi Basin. *Cretaceous Research* **12**: 345–377.

Kaie, A. and Devyatkin, E.V. 1969. [Morphostructural studies of quartz grains from the sands of the Mesozoic-Cenozoic deposits of Mongolia.] pp. 67–85 in *Litologiya i Poleznye Iskopaemye* **5**.

Kalandadze, N.N. and Kurzanov, S.M. 1974. [Lower Cretaceous localities of terrestrial vertebrates of Mongolia.] *Trudy Sovmestnoi Sovetsko-Mongol'skoi Paleontologicheskoi Ekspeditsii* **1**: 9–18.

Khand, E. 1974. [The Late Cretaceous genus *Gobiocypris* gen. nov. in Mongolia.] *Trudy Sovmestnoi Sovetsko-Mongol'skoi Paleontologicheskoi Ekspeditsii* **1**: 265–267.

Khosbayar, P. 1972. [*Stratigraphy of the Mesozoic of West Mongolia and the History of its Geological Development at that Time.*] [Summary of doctoral thesis.] Moscow: GIN RAN, 35 pp.

Kielan-Jaworowska, Z. 1968. Preliminary data on the Upper Cretaceous eutherian mammals from Bayn Dzak, Gobi Desert. *Palaeontologia Polonica* **19**: 171–191.

—1970. New Upper Cretaceous multituberculate genera

from Bayn Dzak, Gobi Desert. *Palaeontologica Polonica* 21: 35–49.

—1971. Skull structure and affinities of the multituberculates. *Palaeontologia Polonica* 25: 5–41.

—1975. Preliminary descriptions of two new eutherian genera from the Late Cretaceous of Mongolia. *Palaeontologia Polonica* 33: 5–16.

—and Barsbold, R. 1972. Narrative of Polish–Mongolian Palaeontological Expeditions 1967–1971. *Palaeontologia Polonica* 27: 5–13.

Kolesnikov, Ch.M. 1982. [Biogeochemical study of the hydrochemistry and thermodynamics of the Cretaceous lagoon basins of Mongolia.] pp. 101–125 in Martinson, G.G. (ed.), *Mesozoic Lake Basins of Mongolia*. Leningrad: Nauka.

Kyansep-Romashkina, N.P. 1982. [Distribution of charophyte algae in the Mesozoic lake basins of Mongolia and conditions of their growth.] pp. 158–193 in Martinson, G.G. (ed.), *Mesozoic Lake Basins of Mongolia*. Leningrad: Nauka.

Larishchev, A.A. 1955. [On the composition of the flora in the Mesozoic forests of the Gobi in Mongolia.] *Byulleten' Moskovskogo Obshchestva Ispytatelei Prirody, Otdel Geologicheskii* 6: 97–98.

Maleev, E.A. 1952. [Some comments on the geological age and stratigraphical distribution of the armoured dinosaurs of Mongolia.] *Doklady AN SSSR* 85: 893–896.

Marinov, N.A. 1957. [*Stratigraphy of the Mongolian People's Republic.*] Moscow, 268 pp.

Martinson, G.G. 1975. [On the question of the principles of stratigraphy and correlation of the continental formations of Mongolia.] *Trudy Sovmestnoi Sovetsko-Mongol'skoi Paleontologicheskoi Ekspeditsii* 13: 7–24.

—1982. [General problems of palaeolimnological studies in Mongolia.] pp. 5–17 in Martinson, G.G. (ed.), *Mesozoic Lake Basins of Mongolia*. Leningrad: Nauka.

—and Shuvalov, V.F. 1973. [Stratigraphical division of the Upper Jurassic and Lower Cretaceous of South-east Mongolia.] *Izvestiya AN SSSR, Seriya Geologicheskaya* 10: 139–143.

—and— 1976. [Stratigraphy and fossil molluscs of the Lower Cretaceous lacustrine deposits of the Trans-Altai Gobi in Mongolia.] pp. 20–50 in *Fossil Freshwater Molluscs and their Significance for Palaeolimnology*. Leningrad: Nauka.

—, Sochava, A.V. and Barsbold, R. 1969. [On the strati-

graphical division of the Upper Cretaceous deposits of Mongolia.] *Doklady AN SSSR* 189: 1081–1084.

Morris, F.K. 1936. Central Asia in Cretaceous time. *Bulletin of the Geological Society of America* 47: 1477–1534.

Nagibina, M.S. and Badamgarav, Zh. 1975. [Stratigraphy of the Late Mesozoic formations of North-east Mongolia.] *Trudy Sovmestnoi Sovetsko-Mongol'skoi Paleontologicheskoi Ekspeditsii* 13: 198–225.

—, Shuvalov V.F. and Martinson G.G. 1977. [The main features of stratigraphy and of history of development of the Mesozoic structures of Mongolia.] *Trudy Sovmestnoi Sovetsko-Mongol'skoi Paleontologicheskoi Ekspeditsii* 22: 76–91.

Naidin, D.P. 1981. [Geochronology of the Mesozoic.] *Itogi Nauki i Tekhniki. VINITI. Stratigrafiya. Palaeontologiya* 11: 34–73.

Norman, D.B. 1996. On Mongolian ornithopods (Dinosauria: Ornithischia). 1. *Iguanodon orientalis* Rozhdestvensky 1952. *Zoological Journal of the Linnean Society* 116: 303–315.

Ponomarenko, A.G. and Popov, Yu.A. 1980. [On palaeobiocoenoses of the Early Cretaceous lakes of Mongolia.] *Paleontologicheskii Zhurnal* 1980 (3): 3–13.

Rozhdestvenskii, A.K. 1955. [New data on psittacosaurs-Cretaceous ornithopods.] *Voprosy Geologii Azii* 2: 783–788.

—1971. [The study of dinosaurs of Mongolia and their role in the division of the continental Mesozoic.] *Trudy Sovmestnoi Sovetsko-Mongol'skoi Geologicheskoi Ekspeditsii* 3: 21–23.

—1974. [The history of the dinosaur faunas of Asia and of other continents, and questions of palaeogeography.] *Trudy Sovmestnoi Sovetsko-Mongol'skoi Paleontologicheskoi Ekspeditsii* 1: 107–131.

Sereno, P.C. 1990. Psittacosauridae, pp. 579–592 in Weishampel, D.B., Osmólska, H. and Dodson, P. (eds.), *The Dinosauria*. Berkeley: University of California Press.

Shuvalov, V.F. 1969. [On the Upper Jurassic red-coloured continental deposits of Mongolia.] *Doklady AN SSSR* 189: 1088–1091.

—1970. [The age of the Tsagaantsav Gorizont of Mongolia in the light of new radiometric data.] *Izvestiya AN SSSR, Seriya Geologicheskaya* 10: 68–77.

—1974. [On the geological structure and age of the Höövör and Hüren Dukh localities.] *Trudy Sovmestnoi Sovetsko-Mongol'skoi Paleontologicheskoi Ekspeditsii* 1:

296–304.

—1975a. [The stratigraphy of the Mesozoic of Central Mongolia.] *Trudy Sovmestnoi Sovetsko-Mongol'skoi Paleontologicheskoi Ekspeditsii* **13**: 50–112.

—1975b. [The structures of the platform stage of development of Mongolia (Late Cretaceous-Paleogene).] *Trudy Sovmestnoi Sovetsko-Mongol'skoi Geologicheskoi Ekspeditsii* **11**: 243–259.

—1976. [The Upper Senonian of the south-east of Mongolia.] *Izvestiya AN SSSR, Seriya Geologicheskaya* **2**: 58–62.

—1980. [The Jurassic and Lower Cretaceous lacustrine deposits of the East Gobi and distribution of associated fossil fauna and flora.] pp. 91–118 in *The Limnobiosis of the Ancient Lake Basins of Eurasia.* Leningrad: Nauka.

—1982. [Palaeogeography and history of the development of the lake systems of Mongolia in Jurassic and Cretaceous time.] pp. 18–80 in Martinson, G.G. (ed.), *Mesozoic Lake Basins of Mongolia.* Leningrad: Nauka.

—1985. [The lake basins of the arid and humid regions of Mongolia in the Late Mesozoic.] pp. 39–61 in Martinson, G.G. (ed.), *Palaeolimnology of Lakes in Arid and Humid Zones.* Leningrad: Nauka.

—1987. [The age of the Tsagaantsav Gorizont of Mongolia as shown by radioisotope data.] *Izvestiya AN SSSR, Seriya Geologicheskaya* **10**: 68–77.

—1993. [Upper Jurassic and Neocomian deposits in Trans-Altai Gobi (Mongolia).] In *Stratigraphy. Geological Correlation* **1** (3): 76–81.

—1994. [Palaeogeography of the lakes of Mongolia in the Mesozoic.] pp. 148–181 in Sevast'anov, D.V., Shuvalov, V.F. and Neustrueva, I.Yu. (eds.), *Limnology and Palaeolimnology of Mongolia.* Leningrad: Nauka.

—and Chkhikvadze, V.M. 1975. [New data on the Late Cretaceous turtles of South Mongolia.] *Trudy Sovmestnoi Sovetsko-Mongol'skoi Paleontologicheskoi Ekspeditsii* **2**: 214–229.

—, Devyatkin, E.V. and Semeikhan, T. 1991. [On the age of the gold-bearing conglomerates of the Trans-Altai Gobi (Mongolia).] *Doklady AN SSSR* **320**: 1207–1211.

—and Nikolaeva, T.V. 1985. [On the age and territorial distribution of Cenozoic basalts in the south of Mongolia.] *Vestnik Leningradskogo Universiteta, Seriya 7, Geologiya, Geografiya* **14**: 52–59.

—and Stankevich, E.S. 1977. [Late Cretaceous ostracods and stratigraphy of the Baishin Tsav region of South-east Mongolia.] *Trudy Sovmestnoi Sovetsko-Mongol'skoi Paleontologicheskoi Ekspeditsii* **4**: 127–136.

—and Stepanov, I.V. 1970. [New data on stratigraphy of the Khara-Airak region of the East Gobi.] *Trudy Sovmestnoi Sovetsko-Mongol'skoi Geologicheskoi Ekspeditsii* **2**: 20–27.

Sinitsyn, V.M. 1962. [*Palaeogeography of Asia.*] Moscow-Leningrad, 267 pp.

Sochava, A.V. 1969. [Dinosaur eggs from the Upper Cretaceous of the Gobi.] *Paleontologicheskii Zhurnal* **1969** (4): 65–68.

—1975. [Stratigraphy and lithology of the Upper Cretaceous deposits of South Mongolia.] *Trudy Sovmestnoi Sovetsko-Mongol'skoi Paleontologicheskoi Ekspeditsii* **13**: 113–182.

—1977. [The Early Cretaceous stromatolites of Mongolia.] *Trudy Sovmestnoi Sovetsko-Mongol'skoi Paleontologicheskoi Ekspeditsii* **4**: 145–160.

Solov'ev, N.S., Shatkov, G.A., Yakobson, L.N. and others 1977. [The Pri-Argun Mongolian volcanic belt.] *Geologiya i Geofizika* **3**: 20–31.

Stankevich, E.S. and Sochava, A.V. 1974. [Ostracods of the Senonian of Mongolia.] *Trudy Sovmestnoi Sovetsko-Mongol'skoi Paleontologicheskoi Ekspeditsii* **1**: 268–289.

Tatarinov, L.P. (ed.) 1983. [Fossil reptiles of Mongolia.] *Trudy Sovmestnoi Sovetsko-Mongol'skoi Paleontologicheskoi Ekspeditsii* **24**: 1–136.

Turishchev, I.E. 1954. [The Lower Cretaceous deposits of the South-east part of Mongolia.] *Doklady AN SSSR* **99**: 445–448.

Tverdokhlebov, V.P. and Tsybin, Yu.I. 1974. [Genesis of the Upper Cretaceous dinosaur localities Tögrögiin Us and Alag Teeg.] *Trudy Sovmestnoi Sovetsko-Mongol'skoi Paleontologicheskoi Ekspeditsii* **1**: 314–319.

Vasil'ev, V.G., Volkhonin, V.S., Grishin, G.L., Ivanov, A.Kh., Marinov, I.A. and Mokshantsev, K.B. 1959. [*Geological structure of the Peoples' Republic of Mongolia (Stratigraphy and Tectonics).*] Leningrad: Gostoptekhizdat, 492 pp.

Verzilin, N.N. 1979a. [Genesis of the Upper Cretaceous deposits of South Mongolia on the basis of taphonomic observations.] *Vestnik Leningradskogo Universiteta, Seriya 7, Geologiya, Geografiya* **12** (2): 7–13.

—1979b. [Cretaceous deposits of Khar Hötöl and the problem of Sainshand Svita.] *Byulleten' Moskovskogo*

Obshchestva Ispytatelei Prirody, Otdel Geologicheskii **54**: 123.

—1980. [Special features of deposition of sediments in the territory of South Mongolia in Late Cretaceous time.] *Vestnik Leningradskogo Universiteta, Seriya 7, Geologiya, Geografiya* **14** (6): 18–27.

—1982. [Palaeolimnological significance of a peculiarity in the texture of the Upper Cretaceous deposits of Mongolia.] pp. 81–100 in Martinson, G.G. (ed.), *Mesozoic Lake Basins of Mongolia*. Leningrad: Nauka.

Yakovlev, N.N. 1986. [Fishes.] *Trudy Sovmestnoi Sovetsko-Mongol'skoi Paleontologicheskoi Ekspeditsii* **28**: 178–181.

Zhamoida, A.I. (ed.) 1977. [*Code of Stratigraphy of the USSR.*] Leningrad: Vsesoyuznyi Geologicheskii Institut, 80 pp.

15

Lithostratigraphy and sedimentary settings of the Cretaceous dinosaur beds of Mongolia

TOM JERZYKIEWICZ

Introduction

There are two conflicting approaches to, and interpretations of, the stratigraphic record of the Cretaceous period in Mongolia. The earlier approach stems from the pioneer work of American palaeontologists and stratigraphers in Mongolia (Granger and Gregory, 1923), and relies on the 'formation' as the basic stratigraphic unit (Berkey and Morris, 1927; Gradziński and Jerzykiewicz, 1974a; Gradziński *et al.*, 1977). A later method, employed largely by authors publishing in Russian, utilizes the 'svita' as a basic stratigraphic category (cf. Chapter 14). The 'formation' and 'svita' cannot be compared. The former is entirely a lithostratigraphic term, but the latter is an eclectic concept combining the lithologic and temporal aspects of a stratigraphic classification assuming isochronous boundaries throughout the entire extent of the 'svita'.

Application of these different concepts has led to considerable confusion in stratigraphic terminology (see Gradziński *et al.*, 1977, and Jerzykiewicz and Russell, 1991, for discussion), and contributed to a controversy about the sedimentary environment of the dinosaur-bearing strata of Mongolia. The debate centres around the question of whether a 'lacustrine' or 'aeolian' depositional model better describes the Cretaceous stratigraphic record of Mongolia. It seems clear that both lacustrine and aeolian processes are recognizable, but there is no agreement about which of these processes predominated in any area and at any time during the Cretaceous. To resolve this dispute the following essential questions should be addressed.

1. How extensive were lacustrine and aeolian processes in the Cretaceous of Mongolia, and what was the size and duration of lakes and deserts?
2. What were the geological mechanisms controlling the size and longevity of the lakes and deserts? What was the prime geological control upon the stratigraphic record – base-level change in the lakes (related or unrelated to world sea-level changes), tectonic subsidence/uplift, or sediment supply?
3. What was the impact of the changing environment upon the biota, especially the vertebrate fauna? How did these postulated 'humid' and 'arid' environments coexist in Central Asia throughout the Cretaceous, providing conditions suitable for the rich and diversified aquatic and terrestrial faunas?

The purpose of this chapter is to address these problems, some of which must remain unsolved either because of lacunas in the stratigraphic record of Mongolia or gaps in our knowledge. This compilation is constrained by the stratigraphic, sedimentological and palaeontological data gathered throughout Mongolia from the Khangai Mountains in the north to the Yellow River valley in the south. The emphasis is on the Upper Cretaceous Djadokhta, Baruungoyot and Nemegt formations which have produced most of the terrestrial vertebrate faunas (e.g. Granger and Gregory, 1923; Kielan-Jaworowska, 1970, 1974; 1981; Osmólska and Roniewicz, 1970; Maryańska and Osmólska, 1975; Rozhdestvenskii, 1977; Maryańska, 1977; Clemens and Kielan-Jaworowska, 1979; Osmólska, 1980, 1987; Norell *et al.*, 1994; Dashzeveg *et al.*, 1995; Gambaryan and Kielan-Jaworowska, 1995; see also Chapters 16–30 in the present publication).

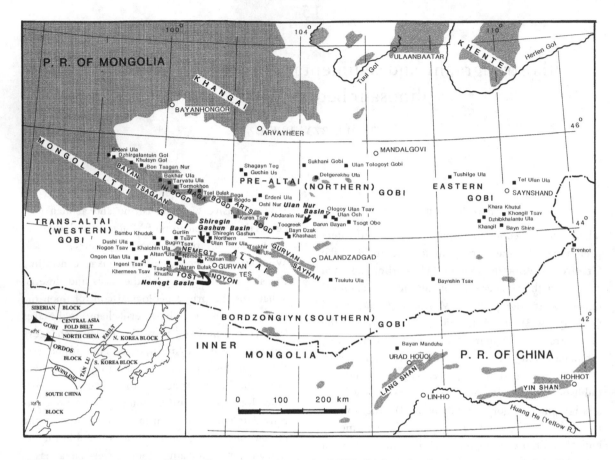

Figure 15.1. Location of fossiliferous Cretaceous sections in the Gobi Basin. Some locality names are given their traditional spellings. (Modified from Jerzykiewicz and Russell, 1991.)

In conclusion, I outline a new depositional model, which resolves aspects of the long-standing debate between proponents of 'lacustrine' and 'desert' origins of the dinosaur-bearing Cretaceous strata of Mongolia (Jerzykiewicz, 1996b). The model synthesizes sedimentary processes recorded in the stratigraphic sections, and suggests that the Okavango Oasis of the Kalahari Desert is a close contemporary analogue of the Cretaceous vertebrate beds of Mongolia.

Geological and palaeogeographic setting

Cretaceous vertebrate-bearing strata of Mongolia infill the Gobi Basin, and a part of the Ordos Basin, which extend between the Khangai and Mongol Altai mountains in the northwest and the Yellow River

valley in the southeast. Most of this area lies within the People's Republic of Mongolia, except its southeastern periphery, which belongs to Chinese Inner Mongolia (Figure 15.1).

The Gobi Basin came into existence only in the Jurassic, long after the Permian collision between Siberia and north China. The basin was created by regional extension behind an active plate-margin, where the Pacific sea floor was subducted under the Asian continent (Hsu, 1989). Jurassic volcanics and intrusives were accompanied by coarse clastics and followed, over the regional unconformity (Morris, 1936), by a succession of predominantly fine-grained lacustrine, fluvio-lacustrine, alluvial plain and aeolian sediments of Cretaceous age. These vertebrate-bearing strata were deposited mainly during the last

Figure 15.2. Palaeogeographical setting of the Gobi Basin during the Late Cretaceous (Senonian Stage, 90–80 Myr). Outlines of Cretaceous continental areas (dotted) are superimosed on the present outlines of continents. Note the difference between the position of the Gobi and Ordos Basins located far away from the seashore and the position of the Alberta Basin located in a coastal area. (Modified from Jerzykiewicz, 1996a; outlines simplified from H.G. Owen in Howarth, 1981.)

stage of tectonic evolution of the Gobi and Ordos Basins, which commenced in the Cretaceous and spanned the Paleogene (Jerzykiewicz, 1995). During that time the Gobi and Ordos Basins were separated by the Lang Shan massif and partitioned into component half grabens. The Gobi Basin was fragmented into the Ulaan Nuur, Shireegiin Gashuun, Nemegt, and Bayan Mandahu Basins (Figure 15.1).

Palaeogeographically, the vertebrate habitats of the Gobi and Ordos basins developed many hundreds of kilometres away from the coastline in the middle of the Asiatic mainland (Figure 15.2). The influence of the sea upon sedimentation within the Ordos and Gobi Basins was minimal as they represent typical inland basins (Sun *et al.*, 1989).

Stratigraphy and sedimentation

Background and methodology

The sedimentary style of the Cretaceous in Mongolia is typical of an extensional inland basin. Sedimentation and erosion were controlled by changes in the base level, related to subsidence/uplift, and climate, rather than eustasy. The maximum lateral extent of lithostratigraphic units is restricted by the magnitude of depositional/erosional processes which diminished throughout the Cretaceous continental rift basins of Central Asia. The large, open lacustrine basins of the Early Cretaceous broke up into a system of restricted fault-controlled halfgrabens by the Late Cretaceous. These were ephemerally and spasmodically infilled with dinosaur-bearing sediments (Jerzykiewicz, 1995, 1996b).

The stratigraphic record of the Cretaceous in Mongolia is discontinuous because of these changes. Localized, tectonically controlled and episodic sedimentation in the Late Cretaceous was punctuated by lacunas marked by semiarid paleosols and reworked dunes. The overall pattern of sedimentation strongly suggests that the Late Cretaceous stratigraphic record in Mongolia is incomplete and laterally more restricted than the Early Cretaceous. This should be recognized when attempts are made to correlate the dinosaur-bearing Cretaceous either within the Gobi Basin or outside it, using either international marine stages or eustatic sequences.

The Cretaceous dinosaur-bearing deposits of Mongolia have been classified using two fundamentally different approaches to stratigraphy. The one used here is known to the international community of stratigraphers as the dual classification (e.g. Krumbein and Sloss, 1963, p. 22), and it makes a clear distinction between 'rock units' and 'time units'. The other method, not applied here, combines lithological and temporal aspects of strata into 'svitas' (cf. Chapter 14). A comprehensive discussion of the differences between these two stratigraphic methods was given by O'Rourke (1976), and the rationale behind the dual classification of the Cretaceous in Mongolia was presented by Gradziński et al. (1977) and Jerzykiewicz and Russell (1991).

Lithostratigraphy

The Cretaceous of Mongolia is represented by an unconformity-bounded succession of continental clastic deposits and rare volcanic rocks (Morris, 1936; Sochava, 1975; Gradziński et al., 1977; Martinson, 1982; Samoilov et al., 1988; Jerzykiewicz and Russell, 1991). A composite stratigraphic/sedimentological column of the Cretaceous formations of Mongolia is shown in Figure 15.3. The boundary between the Lower and the Upper Cretaceous appears to be at the base of thick accumulations of conglomerates of the Sainshand Formation in the

Gobi Basin, and the Dongsheng Formation in the Ordos Basin.

The Lower Cretaceous is represented by shales, claystones, and mudstones interbedded with thin siltstones, fine-grained sandstones, limestones, and marls. These open-lacustrine sediments form the bulk of the Andaikhudag, Ondaisair, Khulsangol and Döshuul formations in the Gobi Basin. Marginal-lacustrine and fluvio-lacustrine channel deposits are subordinate (Figure 15.3). The open-lacustrine to fluvio-lacustrine origin of these formations is suggested by the presence of invertebrate faunas of molluscs and ostracodes (Vasil'ev et al., 1959; Shuvalov, 1975a,b; Martinson, 1971).

Coeval strata of the Ordos Basin (Zhidan Group) consist of perennial lacustrine shales infilling the largely axial part of the basin (Huachi and Huanhe formations), and marginal-lacustrine facies which consist of fluvio-lacustrine, deltaic, and aeolian facies (Lehe, Luohadong, Jingchuang, Lamawan and Yiginholo formations; Figure 15.3).

The Upper Cretaceous dinosaur-bearing red beds form the bedrock in most of the Gobi Basin, and extend south of the Yellow River where they have been reported in the northern part of the Ordos Basin (the Dongsheng and Tegaimiao formations in Figure 15.3).

The Sainshand Formation is a succession of alluvial fan breccias and conglomerates passing upwards into sandstones and lacustrine mudstones and/or paleosols. Basalts up to 30 m thick are present in some localities (Samoilov et al., 1988). Lacustrine mudstones yielded a molluscan assemblage dominated by gastropods which occur widely in Albian to Cenomanian strata in eastern Asia (Barsbold, 1972; Makulbekov and Kurzanov, 1986). The type locality at Sainshand is located in East Gobi, but correlative strata are also well preserved along faults bordering the Ulaan Nuur Basin of the North Gobi (Figure 15.1).

The Bayanshiree Formation consists of alternating varicoloured claystones and fluvial sandstones and conglomerates. The stratigraphic position and correlation of the Bayanshiree Formation is somewhat controversial, and depends upon where its upper boundary is placed. At the Khar Hötöl locality this

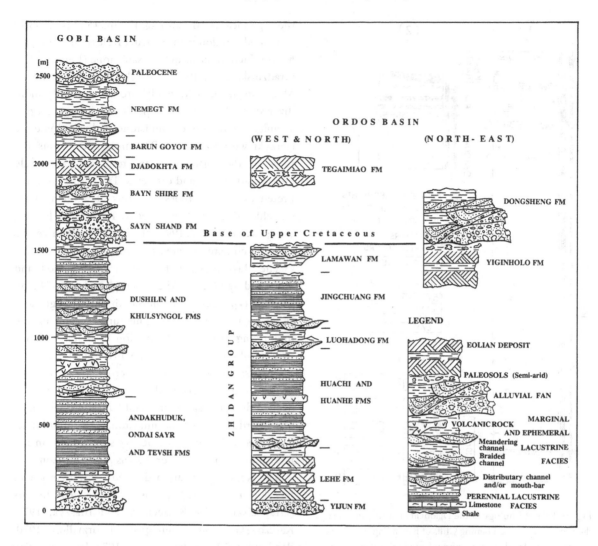

Figure 15.3. Composite stratigraphic column of the Cretaceous in Mongolia. Superposition relations of the formations, and facies, are inferred from Gradziński (1970), Gradziński and Jerzykiewicz (1974a, b), Shuvalov (1975a), Sochava (1975), Gradziński *et al.*, (1977), Jerzykiewicz and Russell (1991), and Jerzykiewicz (1995). Mongolian formation names are given traditional spellings (Barun Goyot = Baruungoyot; Bayn Shire = Bayanshiree; Sayn Shand = Sainshand; Dushilin = Döshuul; Khulsyngol = Khulsangol; Andakhuduk = Andaikhudag; Ondai Sayr = Ondaisair).

boundary is at the base of a persistent layer of calcrete (Javkhlant Svita; cf. Sochava, 1975; Martinson 1982), which has been interpreted by Jerzykiewicz and Russell (1991) as the boundary between the Bayanshiree and Djadokhta Formations (Figure 15.3). The Bayanshiree Formation has been correlated in South Mongolia from the Shireegiin Gashuun Basin (Martinson, 1982) with the Iren Dabasu Formation

near Erenhot in East Gobi (Figure 15.1; Jerzykiewicz and Russell, 1991; Currie and Eberth, 1993).

The Djadokhta Formation yielded the first Cretaceous mammals, nests of dinosaur eggs, and numerous well-preserved dinosaur skeletons, notably *Protoceratops*, *Pinacosaurus*, *Oviraptor*, and *Velociraptor* (Berkey and Morris, 1927; Lefeld, 1971; Gradziński *et al.*, 1977; Chapters 16–30). The dominant lithology at

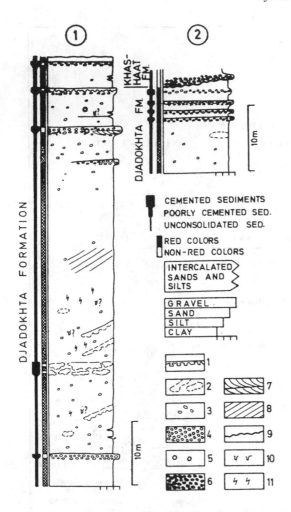

Figure 15.4. Lithology of the Djadokhta Formation: (1) at the type locality, the Flaming Cliffs of Bayan Zag, and (2) at the stratotype locality of the upper boundary of the formation at Khashaat (from Gradziński *et al.*, 1977). Refer to Figure 15.1 for location of the sections. Symbols and patterns: 1, calcrete horizons in structureless sandstone; 2, zones of calcareous cementation; 3, calcrete concretions; 4, conglomerate consisting of redeposited calcrete concretions; 5, exotic pebbles; 6, coarse gravels; 7, large-scale cross-stratification; 8, very large-scale cross-stratification; 9, erosional surface; 10, burrows; 11, other trace fossils. The substantial blank spaces between symbols indicate structureless sandy background.

the type locality (Bayan Zag; Figure 15.4) is a poorly cemented, predominantly structureless, fine- to very fine-grained, reddish-orange sandstone (Lefeld; 1971; Gradziński *et al.*, 1977). Correlative strata at Bayan Mandahu in Inner Mongolia are lithologically more diversified than at the type locality, and contain significant amounts of mudstone and conglomerate (Jerzykiewicz *et al.*, 1993). The most conspicuous features of the Djadokhta sandstone are large-scale cross-stratification and calcretes. The former is interpreted by most authors in terms of aeolian dunes (Lefeld, 1971; Jerzykiewicz *et al.*, 1989, 1993; Eberth, 1993; Fastovsky *et al.*, 1997), the latter is indicative of a semiarid climate (see next section).

The Baruungoyot Formation was defined and described by Gradziński and Jerzykiewicz (1974a, b), with the type section at Khulsan in the Nemegt Basin (Figures 15.1 and 15.5). This unit resembles the underlying Djadokhta Formation in that the dominant lithology is fine-grained, poorly cemented structureless sandstone. Large-scale cross stratification of aeolian origin is very well developed. Cross- to flat-bedded fluviatile sandstone, and laterally discontinuous interdune lacustrine mudstone are subordinate. Reworked calcrete debris occurs sporadically in the interdune channels, rather than as *in situ* horizons.

The Nemegt Formation, defined by Gradziński and Jerzykiewicz (1974a), is most fossiliferous (Gradziński, 1970; Szczechura and Blaszyk, 1970; Karczewska and Ziembinska-Tworzydlo, 1970; Barsbold, 1972; Gradziński *et al.*, 1977; Verzilin, 1979; Osmólska, 1980; Jerzykiewicz and Russell, 1991; Chapters 16–30). It has been thoroughly described in the Nemegt Basin at Nemegt (type locality) and Altan Uul, and at Hermiin Tsav in the Trans-Altai Gobi (Gradziński, 1970; Gradziński and Jerzykiewicz, 1972; Gradziński *et al.*, 1977; Verzilin, 1979). At Nemegt, the formation consists of upward-fining successions of channel sandstone and mudstone (Figure 15.6), typical of point-bar/overbank accumulations, and interpreted as products of a meandering fluvial environment (Gradziński, 1970). Correlative strata at Altan Uul and other localities of the Nemegt Basin (Figure 15.1) were interpreted as lacustrine, on the

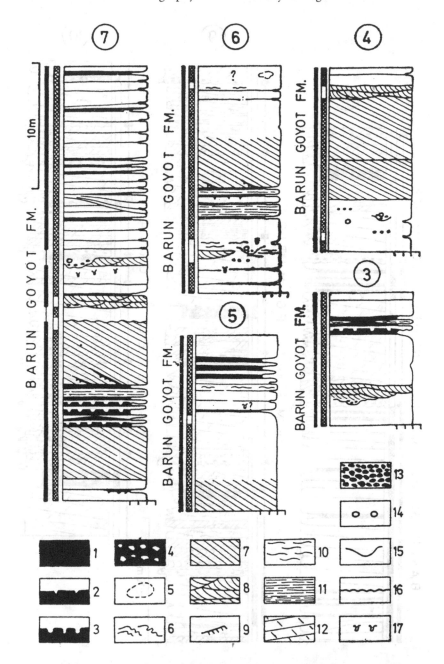

Figure 15.5. Lithology of the Baruungoyot Formation at the type and reference sections in the Nemegt Basin. The Khulsan locality: (3) and (5) Central Cliffs; (4) Western Sair; (6) Northern Cliffs. The Nemegt locality: (7) Southern Monadnocks. Symbols and patterns: 1, mudstone; 2, mud cracks; 3, load casts; 4, calcareous concretions in mudstone; 5, large calcareous concretions; 6, deformational structures; 7, large-scale, planar cross-stratification; 8, large-scale wedge-shaped and trough cross-stratification; 9, small-scale cross-lamination; 10, wavy lamination; 11, horizontal lamination; 12, inclined bedding with internal structures; 13, intraformational conglomerate; 14, exotic pebbles; 15, channel-like erosional surface; 16, flat erosional surface; 17, burrows; other symbols as in Figure 15.4. (From Gradziński *et al.*, 1977.)

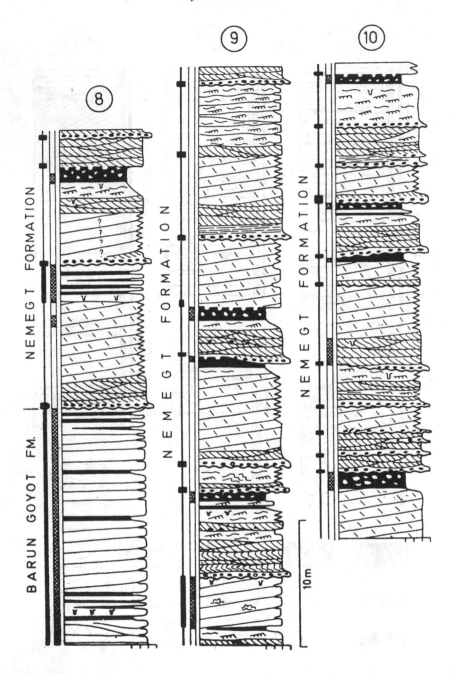

Figure 15.6. Lithology of the Nemegt Formation at the type and reference sections of the Nemegt locality (8) Red Walls (the Baruungoyot/Nemegt formations stratotype); (9) Central Sair; (10) Northern Sair (from Gradziński *et al.*, 1977). For symbols and patterns, see Figures 15.4 and 15.5.

basis of rich bivalve faunas recovered mainly from channel lag deposits (Verzilin, 1979).

Age and correlation

Lack of marine fossils in the Mongolian Cretaceous prevents direct biostratigraphic correlation with the international marine stages of the Cretaceous Period. International correlations are also severely hampered by the largely endemic nature of Mongolian Cretaceous vertebrate assemblages (Jerzykiewicz and Russell, 1991; Chapters 16–30).

Nonmarine fossils have been used to correlate the major lithostratigraphic units of the Gobi Basin with the international standard scale (e.g. Martinson, 1971, 1982; Barsbold, 1972; Kalugina, 1980; Krasilov, 1980, 1982; Ponomarenko and Kalugina, 1980; Makulbekov and Kurzanov, 1986). All these papers, in a more or less direct manner, refer to sections containing stratigraphic interdigitations of terrestrial and marine fossils, either in Asia or North America. Ages of the Lower Cretaceous formations in Mongolia (Figure 15.3) – Valanginian, Hauterivian, Barremian, Aptian, Albian – were estimated largely on the basis of plant megafossils and molluscs (Vasil'ev, *et al.*, 1959; Shuvalov, 1975a; Krasilov, 1982; Martinson, 1982). For instance, the Cenomanian age of the Sainshand Formation was inferred from the resemblance of its molluscan assemblages to those of Cenomanian age in eastern Asia (Makulbekov and Kurzanov, 1986).

Ages of the Upper Cretaceous formations of Mongolia, inferred from comparisons of the vertebrates with those in North American nonmarine units (which are constrained by ammonite zonation and palynology and calibrated by chronostratigraphic methods), are as follows: Bayanshiree Formation (Late Cenomanian–Coniacian to ?Early Santonian), Djadokhta Formation (Mid-Campanian), Baruungoyot Formation (Mid–Late Campanian), and Nemegt Formation (Late Campanian–Early Maastrichtian) (Gradziński *et al.*, 1977; Fox, 1978; Lillegraven and McKenna, 1986; Jerzykiewicz and Russell, 1991). Biostratigraphic methods based on Cretaceous mammals (Kielan-Jaworowska, 1974,

1981; Clemens and Kielan-Jaworowska, 1979; Lillegraven and McKenna, 1986; Lillegraven and Ostresh, 1990; Nesov *et al.*, 1998; Chapters 29 and 30), may lead to refined ages and allow more precise correlation between the Cretaceous vertebrate-bearing strata of Central Asia and North America.

Strata of Late Maastrichtian age have not been documented in Mongolia. The Cretaceous/Tertiary boundary is at an erosional unconformity, marked by calcrete paleosol in the Naran Gol section of the Nemegt Basin (Gradziński *et al.*, 1968, 1977; Jerzykiewicz, 1995). Palaeontological evidence is not sufficient to assess the magnitude of the hiatus between the Upper Cretaceous and Palaeocene in the Gobi Desert.

Sedimentary facies and environment

The Cretaceous strata of Mongolia consist of a wide spectrum of continental facies. Two extremes of this spectrum are: (1) open lacustrine and associated marginal lacustrine and fluvial facies, and (2) aeolian dunes and associated interdune facies including calcretes and other features of subaerial deposition and erosion. The former facies association predominates in the Lower, and the latter ones in the Upper Cretaceous. Lacustrine facies and fossils were described from the Lower Cretaceous formations of the Gobi in the context of the 'Manlai Lake' (Kalugina, 1980) and from correlative strata of the Zhidan Group in the Ordos Basin (Jerzykiewicz, 1995).

Aeolian and associated facies of subaerial deposition interfingering with water-lain interdune/ephemeral facies were documented from the Upper Cretaceous Djadokhta and Baruungoyot formations, and correlative strata (Berkey and Morris, 1927; Lefeld, 1971; Gradziński and Jerzykiewicz, 1974b; Jerzykiewicz *et al.*, 1989, 1993; Eberth, 1993; Fastovsky *et al.*, 1997). The most conspicuous wind-formed feature is large-scale tabular- or wedge-planar cross stratification which is diagnostic of accumulation on slopes of aeolian dunes (Figure 15.7A). Detailed sedimentological studies of the Upper Cretaceous in

Figure 15.7. Aeolian dune deposits and *Protoceratops* buried in 'standing' pose. A, Large-scale cross-stratified sandstone interpreted as transverse dunes, Baruungoyot Formation at Khulsan. B, C, Skeleton of 'standing' *Protoceratops* in aeolian sandstone of the Djadokhta Formation at Tögrög. Oblique and side views of the same specimen excavated during the Polish–Mongolian Expedition in 1971.

Mongolia suggest the presence of various types of aeolian dunes including linear, transverse, and parabolic forms (Gradziński and Jerzykiewicz, 1974b; Jerzykiewicz *et al.*, 1993; Eberth, 1993; Fastovsky *et al.*, 1997). Some of these forms at Tögrög (Fastovsky *et al.*, 1997) have been interpreted as a 'dry delta' in a 'shallow-water wave dominated environment' by Tverdokhlebov and Tsybin (1974).

Associated with the large-scale cross-stratified aeolian sandstone are structureless sandstones which form the bulk of the Djadokhta and Baruungoyot formations (Figures 15.4 and 15.5). The structureless sandstones have been interpreted as vertically aggraded, windblown sand trapped by growing plants, and thus similar in origin to loess (Berkey and Morris, 1927). However, these sandstones are coarser than loess, up to 25% dust, and contain randomly distributed granules and pebbles, some of which have frosted and pitted surfaces. Sometimes they show traces of internal stratification, which can be inferred from the orientation of large clasts of redeposited calcrete, or bone debris. The polymodal grain-size distribution, disrupted traces of stratification, and the association of the structureless sandstone with aeolian dunes and calcretes, suggest a complex origin. This deposit is interpreted as having originated from slumping of the slipfaces of aeolian dunes and/or infilling of interdune areas by high-energy sand storms (Jerzykiewicz *et al.*, 1989, 1993). This interpretation is consistent with the fact that many dinosaur skeletons were found in the structureless sandstone, some of them in poses suggesting that the animals were encased within the sediment at the time of their death (Figure 15.7B, C). They may have been trapped and died while attempting to free themselves from the sand during or shortly after a sandstorm (Jerzykiewicz *et al.*, 1989, 1993; Fastovsky *et al.*, 1997).

Interdune facies consist of channelized sandstone and conglomerate, sheet flood sandstone of fluvial/alluvial fan origin, and interstratified sandstones and mudstones of ephemeral lake origin (Gradziński and Jerzykiewicz, 1974b; Jerzykiewicz *et al.*, 1993; Eberth, 1993). Fluvial deposits forming fining-upward cycles, interpreted as point bar deposits (Figure 15.6; Gradziński, 1970), and fluvio-lacustrine facies (Verzilin, 1979) have been described from the Nemegt Formation.

Djadokhta strata (Figure 15.8) are dominated by laterally persistent aeolian and structureless sandstones, which are cyclically interbedded with interdune deposits and calcrete horizons. Calcretes occur either as *in situ* nodules and hardpans, or as redeposited debris in interdune channels, which suggests that a great deal of penecontemporaneous erosion.

The North Canyon section (Figure 15.8) contains six calcrete horizons analogous to those described by Blumel (1982) and Summerfield (1982) from the Kalahari beds of Namibia and Botswana. In that area, calcrete forms in a multistage process involving an alternation of climatic conditions, predominantly semiarid to subhumid. Diagenesis takes place during the subaerial stage in semiarid to arid conditions (Blumel, 1982, fig. 1).

This kind of facies pattern (Figure 15.8) strongly suggests climatic oscillations. Wet periods within an overall semiarid climate are marked by calcretes, and intervening dry periods are recorded by aeolian dunes. The calcretes, extending laterally for several hundred metres, show a distinct relief, indicating syndepositional topography (Eberth, 1993). An abundance of rhizoliths and endogenic invertebrate traces in some layers underlying the calcretes (Figure 15.8) suggests that they were formed in organic-rich and moist subenvironments (Jerzykiewicz *et al.*, 1993). The sedimentological characteristics of the Djadokhta strata, especially grain size, paleosols and related topography, and colour, makes them very similar to the Kalahari beds (Thomas and Shaw, 1991; McCarthy and Ellery, 1995).

The extensive calcretes and thick layers of redeposited calcrete debris in the Djadokhta and Bayanshiree formations indicate time gaps of unknown magnitude. The overall pattern of sedimentation of the Upper Cretaceous strata of the Gobi in general, and the above formations in particular, was discontinuous and episodic (see below).

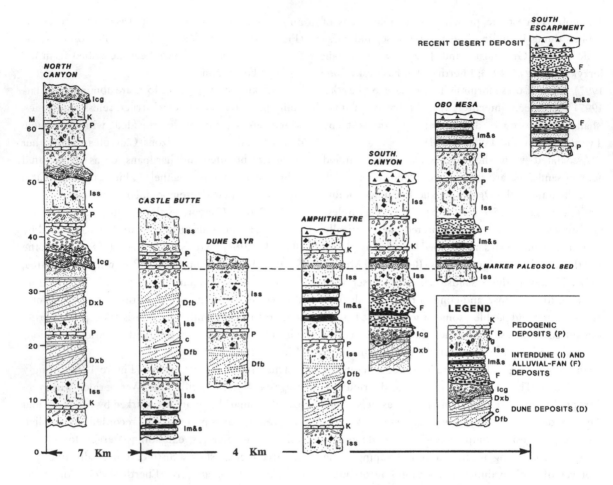

Figure 15.8. Sedimentary facies in the Djadokhta Formation correlative strata at Bayan Mandahu, Inner Mongolia. Symbols: Dxb, large-scale, tabular and/or wedge planar sets of cross-stratified sandstone of aeolian origin; Dfb, structureless and/or faintly bedded sandstone, containing inclined layers of indurated, cemented sandstone (c); Im&s, alternating sandstone and mudstone of lacustrine origin; Iss, structureless sandstone; P, calcrete deposits (g, concretion, r, rhizocretion, K, hardpan); Icg, intraformational conglomerate consisting of redeposited calcrete debris. F, extraformational conglomerate of alluvial fan origin. (From Jerzykiewicz *et al.*, 1993.)

An environmental model for the Cretaceous dinosaur beds of Mongolia

A number of palaeoenvironments have been proposed for specific dinosaur-bearing sites in Mongolia: open lake (Kalugina, 1980; Verzilin, 1982; Shuvalov, 1982; Chapter 14), lakeshore (Lefeld, 1971), lacustrine delta (Jerzykiewicz, 1995), meandering-channel/flood plain (Gradziński, 1970), aeolian/alluvial plain with ephemeral lakes (Gradziński and Jerzykiewicz, 1974b; Jerzykiewicz *et al.*, 1993), transitional alluvial fan-desert (Eberth, 1993), and braid-plain to aeolian (Fastovsky *et al.*, 1997). All of these settings may be identified in the Cretaceous of Mongolia, but none of them fully explains the ecological diversity and faunal abundance of the dinosaur environments. This is particularly true of the Late Cretaceous, which was characterized by rapid lateral and vertical facies changes,

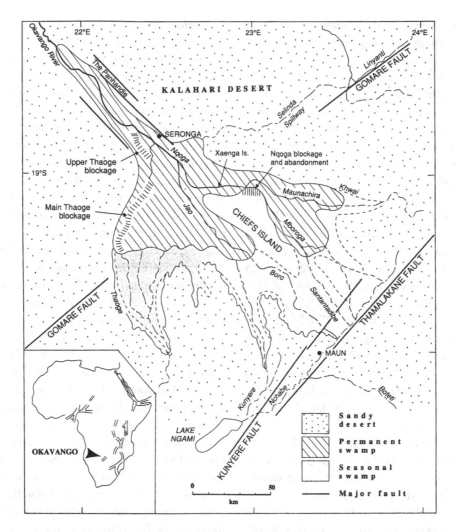

Figure 15.9. Map of the Okavango Oasis in the Kalahari Desert, a contemporary analogue of the Cretaceous vertebrate-bearing environments of Mongolia. The inset shows the location of the Okavango in relation to the other African rifts. The main features of the Okavango Delta are modified after Hutchins *et al.* (1976), and McCarthy *et al.* (1988, *in* Thomas and Shaw, 1991).

and by the occurrence of both terrestrial and aquatic fauna (Kielan-Jaworowska, 1970, 1974, 1981; Osmólska and Roniewicz, 1970; Barsbold, 1972, 1983; Maryańska and Osmólska, 1975; Maryańska, 1977; Rozhdestvenskii, 1977; Elzanowski, 1977; Gradziński *et al.*, 1977; Osmólska, 1980, 1982, 1987; Ponomarenko and Kalugina, 1980; Martinson, 1982; Jerzykiewicz *et al.*, 1993; Currie and Eberth, 1993; Norell *et al.*, 1994; Dashzeveg *et al.*, 1995; Chapters 16–30).

The Okavango Delta (Figure 15.9) and related ephemeral rivers and lakes/pans of the Kalahari Desert (Hutchins *et al.*, 1976; Thomas and Shaw, 1991; McCarthy and Ellery, 1995) are probably the closest contemporary analogues of the late Cretaceous environments of Mongolia (Jerzykiewicz, 1996b). The Okavango combines aspects of a life-supporting oasis (for most of the time), with a life-threatening desert during episodic droughts. This area, comparable in size to the Late Cretaceous of Mongolia (cf. Figures 15.1 and 15.9), supports a large number of animals, including grazing mammals, crocodiles, and other reptiles and birds (Thomas and Shaw, 1991). The Okavango is within the African rift system, and it experiences a semiarid climate, where evapotranspiration exceeds rainfall. The area is subject to seasonal flooding from subtropical Angola to the northwest. Water is distributed by meandering and anastomosing channels through the permanent and seasonal swamps, and disappears in a system of faults which abruptly terminate the delta. However, during seasonal flood events, water from the Okavango River may reach farther southeast into distal areas of the ephemeral drainage which contains lakes, pans, and playas, such as the Ngami (Figure 15.9), the Etosha, and the Makgadikgadi (Thomas and Shaw, 1991). The Okavango is surrounded by relic aeolian-dune topography (linear, transverse, and barchanoid forms) which is stabilized by vegetation and duricrust (Blumel, 1982; Summerfield, 1982). However, during prolonged droughts, the area is subject to dust and sand storms. The last cataclysmic drought in 1987 killed hundreds of thousands of animals.

The African rift system appears to provide modern analogues of the Early and Late Cretaceous environments of Mongolia. Some of the rifts in Africa contain large lakes in their axial parts (e.g. Lake Tanganyika), while others, exemplified by the Okavango Delta, have no large reservoirs of water and are drained by distributaries into subterranean seepage along major faults (Figure 15.9). The former environment is comparable with the Early Cretaceous perennial lakes, and the latter conforms more closely with the Late Cretaceous drainage system (Jerzykiewicz, 1995, 1996b). Timing and magnitude of tectonic events exert an overriding geological control upon the drainage systems of the African rifts (Frostick and Reid, 1987). Tectonically induced changes of climate cause great changes in the area and volume of lakes, some being temporarily reduced to swamps, or even obliterated. The climatic changes are of various frequencies from long-term changes (millions of years) to short-term cycles (thousands of years). The chronology of climatic

fluctuations in the Okavango–Makgadikgadi area since 50 000 years BP includes at least four phases of relative humidity, and intervening phases of relative aridity. The former were expressed by high levels of the water table and development of lakes, and the latter by increasing intensity of aeolian processes (Thomas and Shaw, 1991). At present, the Okavango is in a period of a relative aridity, and it is subject to seasonal droughts.

The stratigraphic record of the Cretaceous in Mongolia contains two different types of sediments, 'cyclic' and 'episodic'. They differ in mode of deposition, rate of accumulation, frequency of occurrence, and preservation potential (cf. Dott, 1988). High-frequency cyclic sedimentation is exemplified by repetitions of fining-upward successions of meandering-river origin or aeolian palaeodunes at individual localities (Figures 15.6 and 15.8). Long-term cyclicity, on the other hand, was responsible for changing the pattern of sedimentation in the Nemegt and Shireegiin Gashuun basins from the aeolian-dominated Djadokhta, through the aeolian/fluvial Baruungoyot, to the fluvially-dominated Nemegt Formation during the late Campanian to early Maastrichtian timespan. It is worth noting, however, that the Okavango accommodates aeolian-dominated, fluvially-dominated, and intermediate depositional facies within one environment. Episodic sedimentation is exemplified by abrupt changes in the Baruungoyot Formation at Nemegt (e.g. Gradziński and Jerzykiewicz, 1974b, fig. 10). There, the aeolian dunes are capped by flash-flood deposits, indicating sudden inundation of the dune field by a flood event. Instantaneous events interpreted in terms of sandstorms are suggested by numerous skeletons of *Protoceratops* trapped in the aeolian sandstone (Figure 15.7B, C, and by other cases of sudden death including the 'fighting dinosaurs', and the mass grave of *Pinacosaurus* babies (see Jerzykiewicz *et al.*, 1993, figs. 9, 11).

Both the gradual–cyclic and the instantaneous–episodic modes of accumulation of the wind-blown and water-laid sediments are consistent with the Okavango model. This composite model consists of two contrasting depositional settings or metastable states, (1) subaerial/aeolian, and (2) subaqueous/fluvio-lacustrine. Changes from one to the other might have been gradual or instantaneous. The 'switching mechanisms' may have been allogenic–extrabasinal (predominantly tectonic) or autogenic–intrabasinal (combined tectonic and climatic). It is impossible to be sure which of these controls was most important in specific dinosaur beds of Mongolia.

Acknowledgement

Thanks to Bill Wimbledon for helpful comments, and to Mike Benton for editing.

References

Barsbold, R. 1972. [*Biostratigraphy and Freshwater Molluscs of the Upper Cretaceous of the Gobi part of the MPR.*] Moscow: Nauka, 88 pp.

—1983. [Carnivorous dinosaurs of the Cretaceous of Mongolia.] *Trudy Sovmestnoi Sovetsko-Mongol'skoi Paleontologicheskoi Ekspeditsii* 19: 1–120.

Berkey, C. and Morris, F. 1927. *Geology of Mongolia. Natural History of Central Asia. Vol. II.* New York: American Museum of Natural History, 475 pp.

Blumel, W.D. 1982. Calcretes in Namibia and SE-Spain, relations to substratum, soil formation and geomorphic factors, pp. 67–82 in Yaalon, D.H. (ed.), *Aridic Soils and Geomorphic Processes. Catena Supplement* 1.

Clemens, W.A. and Kielan-Jaworowska Z. 1979. Multituberculata, pp. 99–149 in Lillegraven, J.A., Kielan-Jaworowska, Z. and Clemens, W.A. (eds.), *Mesozoic Mammals: the First Two-thirds of Mammalian History.* Berkeley: University of California Press.

Currie, P.J. and Eberth, D.A. 1993. Paleontology, sedimentology and palaeoecology of the Iren Dabasu Formation (Upper Cretaceous), Inner Mongolia, People's Republic of China. *Cretaceous Research* 14: 127–144.

Dashzeveg, D., Novacek, M.J., Norell, M.A., Clark, J.M., Chiappe, L.M., Davidson, A., McKenna, M.C., Dingus, L., Swisher, C. and Perle, A. 1995. Extraordinary preservation in a new vertebrate assemblage from the Late Cretaceous of Mongolia. *Nature* 374: 446–449.

Dott, R.H., Jr. 1988. An episodic view of shallow marine clastic sedimentation, pp. 3–12 in de Boer, P.L. (ed.), *Tide-influenced Sedimentary Environments and Facies.* Amsterdam: D. Reidel.

Eberth, D.A. 1993. Depositional environments and facies transitions of dinosaur-bearing Upper Cretaceous redbeds at Bayan Mandahu (Inner Mongolia, People's Republic of China). *Canadian Journal of Earth Sciences* 30: 2196–2213.

Elzanowski, A. 1977. Skulls of *Gobipteryx* (Aves) from the Upper Cretaceous of Mongolia. *Palaeontologica Polonica* 37: 153–165.

Fastovsky, D.E., Badamgarav, D., Ishimoto, H., Watabe, M. and Weishampel, D.B. 1997. The paleoenvironments of Tugrikin-Shireh (Gobi Desert, Mongolia) and aspects of the taphonomy and paleoecology of *Protoceratops* (Dinosauria: Ornithischia). *Palaios* 12: 59–70.

Fox, R.C. 1978. Upper Cretaceous terrestrial vertebrate stratigraphy of the Gobi Desert (Mongolian People's Republic) and Western North America, pp. 577–594 in Stelck, C.R. and Chatterton, B.D.E. (eds.), *Western and Arctic Canadian Biostratigraphy. Geological Association of Canada Special Paper* 18.

Frostick, L.E. and Reid, I. 1987. Tectonic control of desert sediments in rift basins ancient and modern, pp. 53–68 in Frostick, L. and Reid, I. (eds.), *Desert Sediments: Ancient and Modern. Geological Society of London Special Publication* 35.

Gambaryan, P.P. and Kielan-Jaworowska, Z. 1995. Masticatory musculature of Asian taeniolabidoid multituberculate mammals. *Acta Palaeontologica Polonica* 40: 45–108.

Gradziński, R. 1970. Sedimentation of dinosaur-bearing Upper Cretaceous deposits of the Nemegt Basin, Gobi Desert. *Palaeontologia Polonica* 21: 147–229.

—and Jerzykiewicz, T. 1972. Additional geographical and geological data from the Polish-Mongolian Palaeontological Expeditions. *Palaeontologia Polonica* 22: 17–32.

—and— 1974a. Sedimentation of the Barun Goyot Formation. *Palaeontologia Polonica* 30: 111–146.

—and— 1974b. Dinosaur- and mammal-bearing aeolian and associated deposits of the Upper Cretaceous in the Gobi Desert (Mongolia). *Sedimentary Geology* 12: 249–278.

—, Kazimierczak, J. and Lefeld, J. 1968. Geographical and

geological data from the Polish-Mongolian Palaeontological Expeditions. *Palaeontologica Polonica* 1: 33–82.

—, Kielan-Jaworowska, Z. and Maryańska, T. 1977. Upper Cretaceous Djadokhta, Barun-Goyot and Nemegt Formations of Mongolia, including remarks on previous subdivisions. *Acta Geologica Polonica* 27: 281–318.

Granger, W. and Gregory, W.K. 1923. *Protoceratops andrewsi*, a pre-ceratopsian dinosaur from Mongolia. *American Museum Novitates* 42: 1–9.

Howarth, M.K. 1981. Palaeogeography of the Mesozoic, pp. 197–220 in Cocks, L.R.M. (ed.), *The Evolving Earth.* London: British Museum (Natural History).

Hsu, K.J. 1989. Origin of sedimentary basins of China, pp. 207–227 in Zhu, X. (ed.), *Chinese Sedimentary Basins.* Amsterdam: Elsevier.

Hutchins, D.G., Hutton, L.G., Hutton, S.M., Jones, C.R. and Loenhert, E.P. 1976. A summary of the geology, seismicity, geomorphology and hydrogeology of the Okavango Delta. *Botswana Geological Survey Department Bulletin* 7: 1–27.

Jerzykiewicz, T. 1995. Cretaceous vertebrate-bearing strata of the Gobi and Ordos basins – a demise of the Central Asian lacustrine dinosaur habitat, pp. 233–256 in Chang, K.-H. (ed.), *Environmental and Tectonic History of East and South Asia, Proceedings of 15th International Symposium of Kyungpook National University.*

—1996a. Late Cretaceous dinosaurian habitats of western Canada and central Asia – a comparison from a geological standpoint, pp. 63–83 in Sahni, A. (ed.), *Cretaceous Stratigraphy and Palaeoenvironments. Geological Society of India, Memoir* 37.

—1996b. Cretaceous of the Gobi and Ordos basins. The stratigraphic record of tectonically controlled demise of the lacustrine dinosaur habitat of central Asia. *Geological Society of America, Abstracts with Programs* 28 (7): 68.

—, Currie, P.L., Eberth, D.A., Johnston, P.A., Koster, E.H. and Zheng, J.-J. 1993. Djadokhta Formation correlative strata in Chinese Inner Mongolia: an overview of the stratigraphy, sedimentary geology, and palaeontology and comparisons with the type locality in the pre-Altai Gobi. *Canadian Journal of Earth Sciences* 30: 2180–2195.

—, Currie, P.J., Johnston, P.A., Koster, E.H. and Gradziński, R. 1989. Upper Cretaceous dinosaur-bearing eolia-

nites in the Mongolian Basin. *28th International Geological Congress, Washington, D.C. Abstracts* 2: 122–123.

—and Russell, D.A. 1991. Late Mesozoic stratigraphy and vertebrates of the Gobi Basin. *Cretaceous Research* 12: 345–377.

Kalugina, N.S. (ed.) 1980. [Early Cretaceous Lake Manlai]. *Trudy Sovmestnoi Sovetsko-Mongol'skoi Paleontologicheskoi Ekspeditsii* 13: 1–93.

Karczewska, J. and Ziembinska-Tworzydlo, M. 1970. Upper Cretaceous Charophyta from the Nemegt Basin, Gobi Desert. *Palaeontologia Polonica* 21: 121–144.

Kielan-Jaworowska, Z. 1970. New Upper Cretaceous multituberculate genera from Bayan Dzak, Gobi Desert. *Palaeontologia Polonica* 21: 35–49.

—1974. Multituberculate succession in the Late Cretaceous of the Gobi Desert (Mongolia). *Palaeontologia Polonica* 30: 23–44.

—1981. Evolution of the therian mammals in the Late Cretaceous of Asia. Part IV. Skull structure in *Kennalestes* and *Asioryctes. Palaeontologia Polonia* 42: 25–78.

Krasilov, V.A. 1980. [The fossil plants of Manlai]. *Trudy Sovmestnoi Sovetsko-Mongol'skoi Paleontologicheskoi Ekspeditsii* 13: 40–42.

—1982. Early Cretaceous flora of Mongolia. *Palaeontographica, Abteilung B* 181: 1–43.

Krumbein, W.C. and Sloss, L.L. 1963. *Stratigraphy and Sedimentation.* San Francisco: W.H. Freeman, 660 pp.

Lefeld, J., 1971. Geology of the Djadokhta Formation at Bayan Dzak (Mongolia). *Palaeontologica Polonica* 25: 101–127.

Lillegraven, J.A. and McKenna, M.C. 1986. Fossil Mammals from the 'Mesaverde' Formation (Late Cretaceous, Judithian) of the Bighorn and Wind River Basins, Wyoming, with Definitions of Late Cretaceous North American Land-Mammal 'Ages'. *American Museum Novitates* 2840: 1–68.

—and Ostresh, L.M., Jr. 1990. Late Cretaceous (earliest Campanian/Maastrichtian) evolution of western shorelines of the North American Western Interior Seaway in relation to known mammalian faunas, pp. 1–29 in Bown, T.M. and Rose, K.D. (eds.), *Dawn of the Age of Mammals in the Northern Part of the Rocky Mountain Interior, North America. Geological Society of America, Special Paper* 243.

Makulbekov, N.M. and Kurzanov, S. M. 1986. [The biogeographical relationships of the Late Cretaceous biota of Mongolia.] *Trudy Sovmestnoi Sovetsko-Mongol'skoi Paleontologicheskoi Ekspeditsii* 29: 106–112.

Martinson, G.G. 1971. [Fresh-water molluscs from Albian deposits of the Bayan Khongor county.] *Trudy Sovmestnoi Sovetsko-Mongol'skoi Paleontologicheskoi Ekspeditsii* 3: 7–13.

—1982. [The Upper Cretaceous molluscs of Mongolia.] *Trudy Sovmestnoi Sovetsko-Mongol'skoi Paleontologicheskoi Ekspeditsii* 17: 5–76.

Maryańska, T. 1977. Ankylosauridae (Dinosauria) of Mongolia. *Palaeontologia Polonica* 37: 65–151.

—and Osmólska, H. 1975. Protoceratopsidae (Dinosauria) of Asia. *Palaeontologia Polonica* 33: 133–181.

McCarthy T.S. and Ellery, W.N. 1995. Sedimentation on the distal reaches of the Okavango Fan, Botswana, and its bearing on calcrete and silcrete (ganister) formation. *Journal of Sedimentary Research* A65: 77–90.

Morris, F.K. 1936. Central Asia in Cretaceous time. *Bulletin of the Geological Society of America* 47: 1477–1534.

Nesov, L.A., Archibald, D.J. and Kielan-Jaworowska, Z. (1998). Ungulate-like mammals from the Late Cretaceous of Uzbekistan and a phylogenetic analysis of Ungulatomorpha, in Beard, C. and Dawson, M. (eds.), *The Dawn of Asian Mammals. Bulletin of the Carnegie Museum of Natural History* 34: 40–88.

Norell, M.A., Clark, J.M., Dashzeveg, D., Barsbold, R., Chiappe, L.M., Davidson, A.R., McKenna, M.C., Perle, A. and Novacek, J. 1994. A theropod dinosaur embryo and the affinities of the Flaming cliffs dinosaur eggs. *Science* 266: 779–782.

O'Rourke, J.E. 1976. Pragmatism versus materialism in stratigraphy. *American Journal of Science* 276: 47–55.

Osmólska, H. 1980. The Late Cretaceous vertebrate assemblages of the Gobi Desert, Mongolia. *Mémoires de la Société Géologique de France* 59: 145–150.

—1982. *Hulsanpes perlei* n.g. n.sp. (Deinonychosauria, Saurischia, Dinosauria) from the Upper Cretaceous Barun Goyot Formation of Mongolia. *Neues Jahrbuch für Geologie und Paläontologie, Monatshefte* **1982**: 440–448.

—1987. *Borogovia gracilicrus* gen. et sp. n., a new troodontid dinosaur from the Late Cretaceous of Mongolia. *Acta Palaeontologica Polonica* 32: 133–150.

—and Roniewicz, E. 1970. Deinocheridae, a new family of theropod dinosaurs. *Paleontologia Polonica* 21: 5–19.

Ponomarenko, A.G. and Kalugina, N.S. 1980. [The general characteristics of insects at the Manlai locality.] *Trudy Sovmestnoi Sovetsko-Mongol'skoi Paleontologicheskoi Ekspeditsii* 13: 68–81.

Rozhdestvenskii, A.K. 1977. The study of dinosaurs in Asia, pp. 102–119 in Singh, S.N. (ed.), *Jurij Alexandrovich Orlov Memorial Volume. Journal of the Palaeontological Society of India* 20.

Samoilov, V.S., Ivanov, V.G. and Smirnov, V.N. 1988. Late Mesozoic ritogenic magmatism in the northeastern part of the Gobi Desert (Mongolia). *Soviet Geology and Geophysics* 29: 13–21.

Shuvalov, V.F. 1975a. [The stratigraphy of the Mesozoic of Central Mongolia.] *Trudy Sovmestnoi Sovetsko-Mongol'skoi Paleontologicheskoi Ekspeditsii* 13: 50–112.

—1975b. [The structures of the platform stage of development of Mongolia (Late Cretaceous-Paleogene).] *Trudy Sovmestnoi Sovetsko-Mongol'skoi Geologicheskoi Ekspeditsii* 11: 243–259.

—1982. [Palaeogeography and history of the development of the lake systems of Mongolia in Jurassic and Cretaceous time.] pp. 18–80 in Martinson, G.G. (ed.), *Mesozoic Lake Basins of Mongolia.* Leningrad: Nauka.

Sochava, A.V. 1975. [Stratigraphy and lithology of the Upper Cretaceous deposits of South Mongolia.] *Trudy Sovmestnoi Sovetsko-Mongol'skoi Paleontologicheskoi Ekspeditsii* 13: 113–182.

Summerfield, M.A. 1982. Distribution, nature and probable genesis of silcrete in arid and semi-arid southern Africa, pp. 37–65 in Yaalon, D.H. (ed.), *Aridic Soil and Geomorphic Processes. Catena Supplement* 1.

Sun, Z., Xie, Q. and Yang, J. 1989. Ordos Basin – a typical example of an unstable cratonic interior superimposed basin, pp. 63–75 in Zhu, X. (ed.), *Chinese Sedimentary Basins.* Amsterdam: Elsevier.

Sczczechura J. and Blaszyk J. 1970. Fresh-water Ostracoda from the Upper Cretaceous of the Nemegt Basin, Gobi Desert. *Palaeontologia Polonica* 21: 107–118.

Thomas, D.S.G. and Shaw, P.A. 1991. *The Kalahari Environment.* Cambridge: Cambridge University Press, 284 pp.

Tverdokhlebov, V.P. and Tsybin, Yu.I. 1974. [Genesis of the Upper Cretaceous dinosaur localities Tögrögïn Us and Alag Teeg.] *Trudy Sovmestnoi Sovetsko-Mongol'skoi Paleontologicheskoi Ekspeditsii* 1: 314–319.

Vasil'ev, V.G., Volkhonin, V.S., Grishin, G.L., Ivanov, A.Kh., Marinov, N.A. and Mokshantsev, K.B. 1959. [*Geological structure of the People's Republic of Mongolia (Stratigraphy and Tectonics).*] Leningrad: Gostoptekhizdat, 492 pp.

Verzilin, N.N. 1979. [Genesis of the Upper Cretaceous deposits of South Mongolia on the basis of taphonomic observations.] *Vestnik Leningradskogo*

Universiteta, Seriya 7, Geologiya, Geografiya **12** (2): 7–13.

—1982. [Palaeolimnological significance of a peculiarity in the texture of the Upper Cretaceous deposits of Mongolia.] pp. 81–100 in Martinson, G.G. (ed.), *Mesozoic Lake Basins of Mongolia.* Leningrad: Nauka.

16

Mesozoic amphibians from Mongolia and the Central Asiatic republics

MIKHAIL A. SHISHKIN

Introduction

Mesozoic amphibians of Mongolia are very poorly known. They come from several sites in the south and southwest of the country, and include a few members of two batrachomorph groups (*sensu* Säve-Söderbergh, 1934), temnospondyls and anurans. Compared with the Mongolian finds, those from the Asian part of the former Soviet Union are more numerous and diverse, although represented almost entirely by isolated fragments. These belong to temnospondyls (very rare), anurans, and urodeles. The collecting area producing them is broad and extends between the longitudes of the Aral Sea and Balkhash Lake. The main groups of localities are in (1) the central and southwestern Kyzylkum Desert, south of the Aral Sea, Uzbekistan; (2) the northern Cis-Aralian area, Kazakhstan; (3) Karatau Lake, southern Kazakhstan; and (4) the Fergana Depression (Kirgizstan and Tadzhikistan).

Although temnospondyls were the dominant amphibian group in the Triassic, they have not yet been found in the earliest tetrapod-bearing Triassic deposits of Mongolia. These are developed in the southern Gobi Desert, and form the upper part of the Noyonsum Svita (Gubin and Sinitsa, 1993). In the overlying Dalanshandkhudag Svita, containing a sparse Triassic flora, an indeterminate capitosauroid has been found (Zaitsev *et al.*, 1973). The only other temnospondyl find recorded in Mongolia comes from the Jurassic and belongs to the aberrant brachyopid *Gobiops* (Shishkin, 1991). In regions of former Soviet Central Asia, similar brachyopid remains have been found in the Jurassic of the Fergana Depression (Nesov, 1988, 1990; Nesov and Fedorov, 1989).

Several Mesozoic anurans have been reported from Mongolia, all from the Cretaceous of the Gobi Desert and adjacent areas. In the Central Asiatic republics, anurans have been reported from Uzbekistan, and they are also limited to the Cretaceous. The urodele fossils range in age from Middle Jurassic to Late Cretaceous, those from the Jurassic having been recovered from the Fergana Depression and southern Kazakhstan.

The bulk of data on Mesozoic amphibians from the former Soviet Union has been summarized in several accounts (Nesov, 1981, 1988; Roček and Nesov, 1993), but this has never been done for the coeval finds from Mongolia. For this reason, the latter are emphasized in the present survey. It should also be stressed that most of the amphibian taxa described from the Asiatic republics are based on fairly incomplete material, and so should be re-evaluated. These taxa are only conventionally considered here to be valid, on the assumption that future revision will establish their true status.

Repository abbreviations

LU, St Petersburg University; PIN, Paleontological Institute, Moscow; TsNIGRI, Tsentralny Nauchno-Issledovatelskii Geologo-Razvedochni Muzei (Chernyshev's Central Museum of Geological Exploration), St Petersburg; ZIN, Zoological Institute, St Petersburg; ZPAL, Paleontological Institute, Warsaw, Poland.

Systematic survey

Subclass TEMNOSPONDYLI Zittel, 1888
Superfamily CAPITOSAUROIDEA Watson,
1919
Capitosauroid indet.

Specimen and locality. PIN 3350/1, impression of fragment of interclavicle; southern Gobi Desert, near Noyon Sum, Mongolia.

Horizon. Dalanshandkhudag Formation (Upper Olenekian to Upper Triassic).

Discussion. A fragment has been collected by the Soviet–Mongolian Geological Expedition (Zaitsev *et al.*, 1973). The specimen cannot be identified beyond Capitosauroidea gen. indet.; the pattern of its ornamentation suggests that its age could be anything from Upper Olenekian to Late Triassic. Some of the accompanying plant fossils are known to be most common in the upper half of the Triassic.

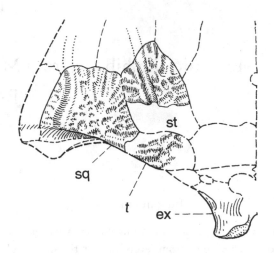

Figure 16.1. *Gobiops desertus* Shishkin, 1991. Restoration of postorbital part of the skull roof (left half). Abbreviations: ex, exoccipital; sq, squamosal; st, supratemporal; t, tabular. (From Shishkin 1991.)

Family BRACHYOPIDAE Lydekker, 1885
Gobiops Shishkin, 1991

Diagnosis. Medium-sized form with total body length up to 1–1.5 m. Judging by structure of squamosal and exoccipitals, skull was strongly depressed, while its width exceeded that in other brachyopids. Occiput with very gentle slope backwards. Exoccipitals with large jugular notch instead of foramen, and vast ventral exposure caused by strong reduction of parasphenoid body. Atlas of typically brachyopid pattern, elongate rostrocaudally. Trunk vertebrae stereospondylous, with disc-shaped hypocentra (intercentra) perforated by notochordal canal.

Gobiops desertus Shishkin, 1991
See Figures 16.1 and 16.2.

Holotype and locality. PIN 4174/102, a left squamosal; Shar Teeg locality; Trans-Altai Gobi, southwestern Mongolia.

Horizon. Details unknown, Upper Jurassic.

Discussion. Gobiops desertus is one of a few post-Triassic temnospondyl holdovers so far recorded, and the only taxonomically recognizable temnospondyl from Mongolia (Shishkin, 1991). The Shar Teeg locality

was discovered by the Soviet–Mongolian Paleontological Expedition. The material is largely isolated dermal bones and fragments, as well as disarticulated hypocentra. Apart from this amphibian, the local fossil biota includes insects (belonging to 11 orders), dipnoans (Krupina, 1994), palaeoniscoid fishes, turtles, crocodiles (Efimov, 1988), dinosaurs, and symmetrodont mammals (Tatarinov, 1994).

Compared to its Triassic brachyopid forerunners, *Gobiops* is remarkable for the extreme flattening of its head, which approached the condition seen in late plagiosaurs, and provides evidence of its benthic habitat. Another unusual character of *Gobiops* is the stereospondylous pattern of its hypocentra, which contrasts with the rhachitomous condition (with crescent-shaped hypocentra) known in earlier brachyopids.

Ferganobatrachus Nesov, 1990

Diagnosis. Dermal bones ornamented with pits and ridges. Hypocentra rhachitomous, or substereospondylous, with very deep, semi-closed notochordal notch. Dorsal process of clavicle short.

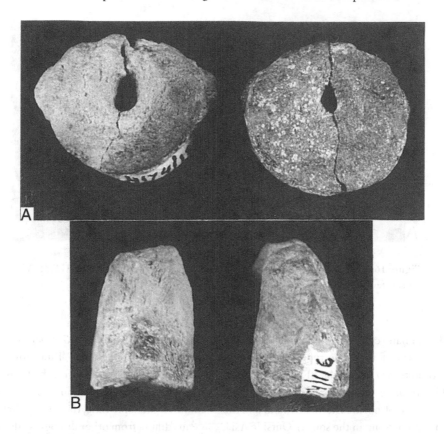

Figure 16.2. *Gobiops desertus* Shishkin, 1991. Two trunk hypocentra: A, anterior view; B, lateral view. Scale × 2.

Ferganobatrachus riabinini Nesov, 1990

Holotype and locality. TsNIGRI 6/12217, left clavicle; Sarykamyshsai locality, near Tashkumyr, Fergana Depression, Kirgizstan.

Horizon. Balabansai Svita, Callovian, Upper Jurassic.

Discussion. Ferganobatrachus is a poorly known and ill-determined form which may be close to, or congeneric with, *Gobiops.* Nesov (1990) attributed *Ferganobatrachus* to the Capitosauridae mostly on the grounds of the shape and strong ossification of the hypocentra, which are said to imply 'possible affinity with Mastodonsauridae and Cyclotosauridae'. However, the type material, a series of isolated bones and fragments, shows no capitosauroid characters. On the other hand, the strong dorsoventral compression of the axis hypocentrum figured by Nesov (1990, Fig. 1z, hypocentrum) strongly suggests an assignment to Brachyopidae.

Relict temnospondyls from Central Asia and their record in adjacent regions

Brachyopid relicts related to *Gobiops* seem to have been rather common in the Jurassic of Asia. Among these, the best known is *Sinobrachyops placenticephalus* from the Middle Jurassic of Sichuan Province of China (Dong, 1985). In the Central Asiatic republics, similar forms are represented by two records from the brackish-water estuarine deposits of the Fergana Depression, *Ferganobatrachus riabinini* (see above) from the Callovian, and an isolated trunk hypocentrum, indistinguishable from those of *Gobiops* (pers. obs.), from

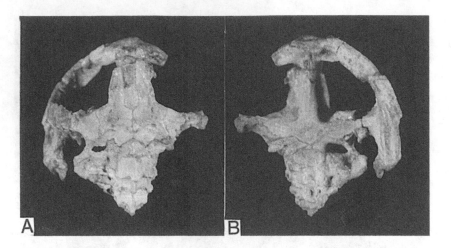

Figure 16.3. *Gobiates kermeentsavi* Špinar and Tatarinov, 1986. Type skull, PIN 3142/1: A, dorsal view; B, ventral view. Scale × 2.

the Upper Bathonian (cf. Nesov and Fedorov, 1989). Further hypocentra of the same type were recently reported by Buffetaut *et al.* (1994a, b) from the continental Middle Jurassic sediments of Thailand (Phu Kradung Formation of northeastern Thailand and the rocks west of Thung Song in the south). Outside Asia, the only record of a *Gobiops*-like hypocentrum comes from the coastal marine Middle (?) Jurassic of the Moscow vicinity (unpublished).

Order ANURA Rafinesque, 1815
Family GOBIATIDAE Roček and Nesov, 1993
Gobiates Špinar and Tatarinov, 1986
See Figure 16.3.

Diagnosis. Medium-sized frogs with broad skull and almost circular orbits; frontoparietals with median suture posteriorly and long central fontanelle anteriorly. Most of dermal roofing bones sculptured with widely spaced pits. Orbital margin of maxilla moderately concave. Maxillary arch complete; zygomaticomaxillary process of maxilla not developed. Squamosal not meeting frontoparietal; quadratojugal delineated by suture from quadrate; exoccipital and prootic not fused. Pectoral girdle arciferal, scapula long, lacking anterior lamina. Vertebrae perichordal, possibly amphicoelous.

Discussion. The genus *Gobiates* was provisionally placed by Špinar and Tatarinov (1986) in the Discoglossidae, primarily on the basis of possessing a long outstanding prootic and a simply built sphenethmoid. At the same time, they noted that the Mongolian genus differs from other discoglossids, both recent and fossil, by its pitted (not pustular) dermal sculpture and frontoparietal fontanelle. The resemblances to the Pelobatidae, in the shape of the frontoparietals, the presence of foramina for occipital arteries in these bones, and the structure of maxilla and squamosal are mostly considered by Špinar and Tatarinov (1986) to be correlated with parallel broadening of the skull in both groups. It was suggested that *Gobiates* may represent the early evolutionary line of the *Discoglossus*-group discoglossids, possibly related to *Scotiophryne* (Cretaceous of North America), *Opisthocoelellus* (Eocene–Miocene of Europe), and *Latonia* (Late Cretaceous–Miocene of Eurasia).

On the other hand, a separate family, Gobiatidae, was proposed for *Gobiates* by Roček and Nesov (1993), who described a number of new species of *Gobiates* and a further genus, *Gobiatoides*, from the Late Cretaceous (Coniacian) of the central Kyzylkum Desert, Uzbekistan. It is worth noting, however, that these alleged *Gobiates* finds are represented by isolated bones, mostly fragments of maxillae and squamosals and some postcranial elements, and so it seems open to question whether the attribution of any of them to *Gobiates* is really justified. This doubt is especially strengthened by the presence of the zygomatico-maxillary process of the maxilla on the holotypes of several forms described by Roček and Nesov (1993, figs. 3, 4, pp. 13, 15–17) as *Gobiates*, which contrasts with the lack of this structure in both the Mongolian species (cf. Špinar and Tatarinov, 1986, p. 121). Besides, as all the ten species of *Gobiates* and *Gobiatoides* erected by Roček and Nesov are based on isolated fragments of the same provenance (Bissekty Svita of the central Kyzylkum Desert), their taxonomic validity seems rather questionable, regardless of their affinities.

The postcranial characters included by Roček and Nesov in the diagnosis of the family Gobiatidae cannot be inferred from the available *Gobiates* specimens from Mongolia (bicondylar articulation between sacrum and urostyle; sacral vertebrae with scarcely expanded transverse processes; absence of lateral epicondyle of humerus; prominent crista femoris; possible presence of at least one pair of transverse processes on urostyle).

The Gobiatidae is thought by Roček and Nesov (1993) to be closely related to the Leiopelmatidae and Discoglossidae, from which they differ by the narrow sacral diapophyses and the absence of the lateral epicondyle of the humerus. All three families share a number of plesiomorphies, such as free ribs, paired frontoparietal, tooth-bearing maxilla and premaxilla, and, probably, a urostyle with at least one pair of transverse processes. The ancestry of both Gobiatidae and Discoglossidae may be looked for among those Jurassic leiopelmatids related to *Vieraella* and *Notobatrachus*.

A possible record of Gobiatidae outside Central Asia, according to Roček and Nesov (1993), is material from the (?) Early Cretaceous (Comanchean) of North America (Winkler *et al.*, 1990).

Gobiates kermeentsavi Špinar and Tatarinov, 1986
Holotype and locality. PIN 3142/1, skull; Hermiin Tsav, Gobi Desert, Mongolia.

Horizon. Baruungoyot Formation (Middle Campanian); Upper Cretaceous.
Discussion. The original material was collected by Mongolian geologists, and it includes a well preserved type skull with the first three presacral vertebrae, and two further incomplete skulls. Roček (in Roček and Nesov, 1993; p.11) believed *G. kermeentsavi* to be a junior synonym of *Gobiates sosedkoi* (Nesov, 1981), a species based on a single fragment of frontoparietal (cf. below).

Gobiates leptocolaptus (Borsuk-Białynicka, 1978)
Eopelobates leptocolaptus Borsuk-Białynicka, 1978, p. 57, figs 1, 2, pl. 15.
Holotype and locality. ZPAL MgAb-III/1, incomplete skull with mandible and a part of pectoral girdle; Hermiin Tsav, Gobi Desert, Mongolia.
Horizon. Baruungoyot Formation (Middle Campanian); Upper Cretaceous.
Discussion. This species is the first frog ever described from the Cretaceous of Central Asia. The holotype, collected by the Polish–Mongolian Paleontological Expedition, was originally assigned (Borsuk-Białynicka, 1978) to the pelobatid genus *Eopelobates*. It was later referred by Špinar and Tatarinov (1986) to *Gobiates* because of its similarity to *G. kermeentsavi* in the shape of the skull, the presence of a quadratojugal-quadrate suture, and the lack of a zygomatico-maxillary process of maxilla.

Gobiates sosedkoi (Nesov, 1981)
Eopelobates sosedkoi Nesov, 1981, p. 71.
Holotype and locality. ZIN PHA K77–5, fragment of right frontoparietal; Dzharakhuduk, central Kyzylkum Desert, Uzbekistan.
Horizon. Middle part of Bissekty Svita, Coniacian, Upper Cretaceous (for holotype). The full range is reported (Roček and Nesov, 1993) as ?Comanchean–Coniacian.
Discussion. Roček (in Roček and Nesov, 1993, p. 11) argued that *Gobiates kermeentsavi* from Mongolia (see above) is a junior synonym of *G. sosedkoi*, but the material available for the latter species seems too limited to substantiate such an identification. As noted

above, all the forms described by Roček and Nesov (1993) from the central Kyzylkum Desert as different species are rather questionable both in terms of their validity and the attribution to the discussed genus. Placing these forms, starting with *G. sosedkoi*, under their original names is only provisional and requires re-evaluation.

Gobiates bogatchovi Roček and Nesov, 1993
Holotype and locality. TsNIGRI 3/12936 (=LU-N5/107), part of maxilla and prearticular associated with quadratojugal–quadrate complex and ptery-goid; Dzharakhuduk, central Kyzylkum Desert, Uzbekistan.
Horizon. Middle part of Bissekty Svita, Coniacian, Upper Cretaceous (for holotype). The full range is reported as ?Late Turonian–Coniacian.

Gobiates dzhyrakudukensis Roček and Nesov, 1993
Holotype and locality. TsNIGRI 5/12936 (=LU-N6/357); part of maxilla; Dzharakhuduk, central Kyzylkum Desert, Uzbekistan.
Horizon. Middle part of Bissekty Svita, Coniacian, Upper Cretaceous.

Gobiates fritschi Roček and Nesov, 1993
Holotype and locality. TsNIGRI 8/12936 (=LU-N5/143); part of maxilla; Dzharakhuduk, central Kyzylkum Desert, Uzbekistan.
Horizon. Middle part of Bissekty Svita, Coniacian, Upper Cretaceous.

Gobiates tatarinovi Roček and Nesov, 1993
Holotype and locality. TsNIGRI 9/12936 (=LU-N6/405); part of maxilla; Dzharakhuduk, central Kyzylkum Desert, Uzbekistan.
Horizon. Middle or upper part of Bissekty Svita, Coniacian, Upper Cretaceous.

Gobiates spinari Roček and Nesov, 1993
Holotype and locality. TsNIGRI 11/12936 (=LU-N5/137), left squamosal; Dzharakhuduk, central Kyzylkum Desert, Uzbekistan.

Horizon. Middle part of Bissekty Svita, Coniacian, Upper Cretaceous.

Gobiates asiaticus Roček and Nesov, 1993
Holotype and locality. TsNIGRI 14/12936 (=LU-N6/370), fragment of right squamosal; Dzharakhuduk, central Kyzylkum Desert, Uzbekistan.
Horizon. Middle part of Bissekty Svita, Coniacian, Upper Cretaceous.

Gobiates kizylkumensis Roček and Nesov, 1993
Holotype and locality. TsNIGRI 16/12936 (=LU-N6/363), fragment of right squamosal; Dzharakhuduk, central Kyzylkum Desert, Uzbekistan.
Horizon. Middle part of Bissekty Svita, Coniacian, Upper Cretaceous.

Gobiates furcatus Roček and Nesov, 1993
Holotype and locality. TsNIGRI 17/12936 (=LU-N5/165), fragment of left maxilla; Dzharakhuduk, central Kyzylkum Desert, Uzbekistan.
Horizon. Middle part of Bissekty Svita, Coniacian, Upper Cretaceous.

Gobiatoides Roček and Nesov, 1993
Diagnosis. Maxilla with smooth dorsal suface, very shallow at lowest point of orbital margin. Orbital margin of maxilla deeply concave and paralelled on inner side of bone by a ridge extending from processus frontalis.

Gobiatoides parvus Roček and Nesov, 1993
Holotype and locality. TsNIGRI 30/12936 (=LU-N6/344), fragment of right maxilla; Dzharakhuduk, central Kyzylkum Desert, Uzbekistan.
Horizon. Middle part of Bissekty Svita, Coniacian, Upper Cretaceous.

Cretasalia Gubin, 1999
Diagnosis. Skull pointed anteriorly. No dermal sculpture on skull bones except anterior margin of frontal process of maxilla. Maxilla very shallow, with smooth inner surface above horizontal lamina. Orbital embay-

ment of maxilla weakly expressed; frontal process deep.

Cretasalia tsybini Gubin, 1999

Holotype and locality. PIN 3142/399, skull with part of skeleton; Hermiin Tsav, Southern Gobi Aimag, Gobi Desert, Mongolia.
Horizon. Campanian–Maastrichtian, Upper Cretaceous.

Family DISCOGLOSSIDAE Günther, 1858
Kizylkuma Nesov, 1981

Diagnosis. Maxilla tooth-bearing, not ornamented; its postorbital part only moderately exceeding orbital margin in depth. Horizontal lamina of maxilla with abrupt posterior end. Tooth row terminating at level of posterior end of horizontal lamina. Fossa cubitalis ventralis humeri absent or shallow.
Discussion. Duellman and Trueb (1986) placed this genus in the Pelobatidae.

Kizylkuma antiqua Nesov, 1981

Holotype and locality. ZIN, PHA K77–10, left maxilla; Dzharakhuduk, central Kyzylkum Desert, Uzbekistan.
Horizon. Upper Turonian–Coniacian (for holotype). Full range reported as Upper Turonian–Santonian, Upper Cretaceous.

Aralobatrachus Nesov, 1981

Diagnosis. Large-sized anuran (with estimated length of maxilla about 18 mm); postorbital division of maxilla with longitudinal grooves and/or groove-like pits on outer surface; tooth row extending behind level of posterior end of horizontal lamina.
Discussion. Duellman and Trueb (1986) placed this genus in the Pelobatidae.

Aralobatrachus robustus Nesov, 1981

Holotype and locality. ZIN, PHA 477–7, fragment of right maxilla; Dzharakhuduk, central Kyzylkum Desert, Uzbekistan.
Horizon. Lower part of Bissekty Svita, Upper Turonian, Upper Cretaceous.

Itemirella Nesov, 1981

Diagnosis. Maxilla extending behind posterior end of tooth row, not ornamented, with shallow embayment of dorsal edge. Horizontal lamina of maxilla strongly projected and gradually wedged out posteriorly.
Discussion. This genus is not reviewed in Roček and Nesov's (1993) recent account on the Cretaceous anurans of Central Asia, but they accept its validity since they compare *Saevesoederberghia egredia* (see below) with *Itemirella* (Roček and Nesov, 1993, p. 28).

Itemirella cretacea Nesov, 1981

Holotype and locality. ZIN PHA K77–6, fragment of right maxilla; Dzharakhuduk, central Kyzylkum Desert, Uzbekistan.
Horizon. Lower part of Bissekty Svita, Upper Turonian, Upper Cretaceous.

Saevesoederberghia Roček and Nesov, 1993

Diagnosis. Posterior end of maxillary orbital margin with medially directed process. Backward extent of tooth row as in *Kizylkuma*. Posterior end of horizontal maxillary lamina terminating abruptly.

Saevesoederberghia egredia Roček and Nesov, 1993

Holotype and locality. TsNIGRI 136/12936 (= LU-N6/375), fragment of right maxilla; Dzharakhuduk, central Kyzylkum Desert, Uzbekistan.
Horizon. Middle part of Bissekty Svita, Coniacian, Upper Cretaceous.

Procerobatrachus Roček and Nesov, 1993

Diagnosis. Dorsal edge of maxilla (including orbital margin) straight, with no sign of zygomatico-maxillary process. Horizontal lamina of maxilla only slightly projecting medially, with gradual posterior termination.

Procerobatrachus paulus Roček and Nesov, 1993

Holotype and locality. TsNIGRI 138/12936 (= LU-N6/412), fragment of right maxilla; Dzharakhuduk, central Kyzylkum Desert, Uzbekistan.

Horizon. Middle part of Bissekty Svita, Coniacian, Upper Cretaceous.

Estesina Roček and Nesov, 1993

Diagnosis. Maxilla with well developed frontal process and horizontal lamina widely rounded anteriorly; tooth row extending backward posterior to termination of horizontal lamina.

Estesina elegans Roček and Nesov, 1993

Holotype and locality. TsNIGRI 139/12936 (= LU-N5/172), fragment of right maxilla; Dzharakhuduk, central Kyzylkum Desert, Uzbekistan.

Horizon. Middle part of Bissekty Svita, Coniacian, Upper Cretaceous.

Altanulia Gubin, 1993

Diagnosis. Large frogs. Maxilla up to 20 mm long, bearing 45–47 teeth. Posterior part of maxilla deep, with longitudinal wedge-shaped labial depression; pterygoid tubercle of maxilla well expressed; frontal process in anterior part of bone.

Discussion. The genus differs from *Bombina* in the number of teeth and the shape of the maxilla, from *Discoglossus* and *Eodiscoglossus* in the depth of the maxilla and the well-developed pterygoid tubercle, from *Gobiates* in its size, the relative length of the maxilla, and the depression in its posterior part, and from *Scotiophryne* in the reduced number of teeth, the increase of maxillary depth posteriorly, and the presence of a labial depression on this bone.

Altanulia alifanovi Gubin, 1993

Holotype and locality. PIN 553/300, an isolated maxilla; Altan Uul II, southern Gobi Desert, Mongolia.

Horizon. Nemegt Formation (Upper Campanian–Lower Maastrichtian), Upper Cretaceous.

Discussion. The holotype of *A. alifanovi*, collected by the Soviet–Mongolian Palaeontological Expedition, is the only find hitherto known for the genus and species. The bone is ascribed to the Discoglossidae on the basis of its shape and the structure of the lingual surface. Among known Mesozoic forms, *Altanulia* may be most closely related to *Kizylkuma* Nesov (see above).

Eodiscoglossus sp.

Locality. Höövör, southeast from Guchinus Somon; Övörkhangai Aimag, Mongolia.

Horizon. Aptian–Albian, Lower Cretaceous.

Discussion. The earliest anuran ever recorded from Mongolia, based on two fragments of left maxillae collected by the Soviet–Mongolian Palaeontological Expedition (Gubin, 1993). The attribution of these finds to the Jurassic genus *Eodiscoglossus,* formerly reported from Spain and Britain (Vergnaud-Grazzini and Wenz, 1975; Evans *et al.,* 1990) requires harder evidence, since the Mongolian specimens are so incomplete.

Anura indet.

Apart from the anuran taxa from the Cretaceous of central Kyzylkum Desert, thought by Roček and Nesov (1993) to be distinctive and determinable, and listed in the above review, the latter authors cited a number of other finds from the same area, noted as cf. *Gobiates,* cf. *Gobiatoides,* cf. *Kizylkuma,* cf. *Aralobatrachus,* and indeterminate Gobiatidae and Discoglossidae. Nesov (1988) also mentions finds of Discoglossidae? and Pelobatidae? from neighbouring areas of the southwestern Kyzylkum Desert, Karakalpakia (lower or middle part of the Khodzhakul Svita, Upper Albian, Lower Cretaceous; upper part of the same unit, Lower Cenomanian, Upper Cretaceous). *Eopelobates* sp. is reported by Nesov (1988) from Kansai, western Fergana, Tadzhikistan (upper part of Yalovach Svita, Lower Santonian, Upper Cretaceous).

Roček and Nesov (1993, p. 10, fig. 18A,F,G) also noted and figured some axial elements of an anuran from the Upper Cretaceous (Djadokhta Formation) of the Ulaan Sair locality in the Gobi Desert, Mongolia.

A genus *Bissektia* Nesov, 1981, with one species, *B. nana,* from the Upper Cretaceous of the central Kyzylkum Desert, based on an isolated maxilla, was originally described by Nesov as a urodele, but was later re-evaluated as Anura inc. sed. (cf. Nesov, 1981, 1988, p. 478). In Roček and Nesov's (1993) survey of the Kyzylkum anurans, this form is not mentioned.

Order CAUDATA Oppel, 1911
Suborder KARAUROIDEA Estes, 1981
Family KARAURIDAE Ivakhnenko, 1978
Karaurus Ivakhnenko, 1978
Diagnosis. Total length about 200 mm. Skull large, depressed, ornamented with pits. Lacrimal and quadratojugal present. Premaxilla broad, without posterior processes. Vomeropalatine tooth row set on conspicuous ridge. Angular merged to prearticular. Body short (15 presacral vertebrae).

Karaurus sharovi Ivakhnenko, 1978
See Figure 16.4.
Holotype and locality. PIN 2585/2, skeleton; Mikhailovka village, Karatau Ridge, southern Kazakhstan.
Horizon. Karabastau Svita, Kimmeridgian, Upper Jurassic.

Kokartus Nesov, 1988
Diagnosis. Skull roof with ornamented bulges. Teeth sharp, curved inside. Trunk vertebrae without hypapophyses. Differs from *Karaurus* in the straighter ridge of the lateral process of the squamosal, a longer and narrower posterolateral process of the frontal, and a narrower parasphenoid and femur.

Kokartus honorarius Nesov, 1988
Holotype and locality. TsNIGRI 1/11998, frontal; Kyzylsu River (near Niczke Spring), Kugart Basin, 100 km east-southeastward from Tashkumyr, Kirgizstan.
Horizon. Bathonian, Middle Jurassic.

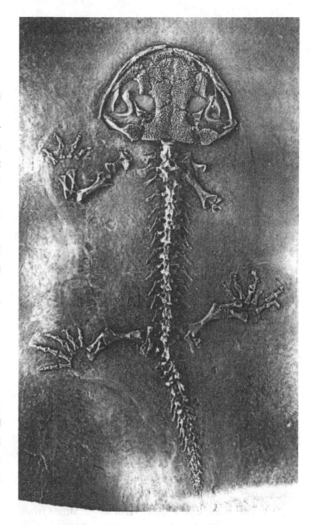

Figure 16.4. *Karaurus sharovi* Ivakhnenko, 1978. Type skeleton in dorsal view.

Suborder AMBYSTOMATOIDEA Noble, 1931
Family SCAPHERPETONTIDAE Auffenberg and Goin, 1959
Subfamily EOSCAPHERPETONTINAE Nesov, 1981
Eoscapherpeton Nesov, 1981
Diagnosis. Small forms. Maxilla long, abruptly reduced in depth anteriorly. Ascending process of maxilla shallow, extended anteroposteriorly. Premaxilla and maxilla with lateral outgrowth bearing ventral ridge parallel with tooth row. Frontal broad,

contacting nasal and prefrontal. Parietal long, with stout arched ridge on ventral side. Meckelian groove of dentary long, not reaching symphysis. Teeth thin, rounded in cross section, arranged in a single irregular row. Atlas with strongly developed hypapophysis projecting ventrally. Ilium deep.

Eoscapherpeton asiaticum Nesov, 1981
Holotype and locality. ZIN, PHA K77–1, left maxilla; Dzharakhuduk, central Kyzylkum Desert, Uzbekistan.
Horizon. Species reported from the whole range of

Bissekty Svita, Upper Turonian–Coniacian, Upper Cretaceous. (Holotype comes from the lower part of section, Upper Turonian.)

Eoscapherton superum Nesov, 1997
Holotype and locality. TsNIGRI 232/12177, atlas; Kansai, western Fergana, Tadzhikistan.
Horizon. Yalovach Svita; Santonian, Upper Cretaceous.
Discussion. Validity of the species is uncertain. The name was originally coined as a *nomen nudum* by Nesov and Udovichenko (1986). Nesov (1988) and Aver'yanov (1999) designate the type specimen as *E.* sp.

Horezmia Nesov, 1981
Diagnosis. Similar to *Eoscapherpeton* in structure of maxilla, premaxilla, and dentary. Atlas with large depressed triangular area ventrally to intercotylar tubercle and narrow lateral outgrowths; hypapophysis directed posteriorly rather than ventrally. Neural spine of atlas deep, sloped backwards.

Horezmia gracile Nesov, 1981
Holotype and locality. ZIN, PHA K77–2, atlas; Khodzhakul Lake, southwestern Kyzylkum Desert, Karakalpakia, Uzbekistan.
Horizon. Species reported from Khodzhakul Svita and lower part of Beshtyube Svita, Upper Albian to ?Lower Turonian, Cretaceous. (Holotype comes from lower part of Beshtyube Svita, ?Lower Turonian, Upper Cretaceous.)

Suborder PROTEIDA Cope, 1866
Family BATRACHOSAUROIDIDAE
Auffenberg, 1958
Mynbulakia Nesov, 1981
Diagnosis. Maxilla short, with rather deep ascending process and short posterior (pterygoid) process bent medially near its end. Dentary attaining maximum depth immediately posterior to tooth row. Meckelian groove long, strongly narrowed anteriorly, not reaching symphysis. Teeth rather large, transversely expanded, with nearly blunt apex. Anterior facets of atlas rounded; intercotylar tubercle large, only

slightly projecting forward. Trunk vertebrae elongate, with ventral median ridge reduced or absent. Neural spines shallow, zygapophyses conspicuously extended anteroposteriorly. Femur short, stout, with strong plate-like trochanter ridge.

Mynbulakia surgayi Nesov, 1981
Holotype and locality. ZIN, PHA K77–3, left maxilla; Dzharakhuduk, central Kyzylkum Desert, Uzbekistan.
Horizon. Middle part of Bissekty Svita, Coniacian, Upper Cretaceous.
Discussion. Another presumed species of *Mynbulakia*, *M. nongratis* Nesov, originally described from the Cenomanian–?Lower Turonian of southwestern Kyzylkum Desert (Nesov, 1981) was later implicitly rejected by Nesov (1988, p. 481).

Suborder inc. sed.
Family ALBANERPETONTIDAE Fox and Naylor, 1982
Nukusaurus Nesov, 1981
Diagnosis. Very small forms with shallow dentary. Meckelian groove passing to canal slightly anterior to hind end of tooth row. Dentary teeth increasing in size from behind to a point near midlength of tooth row. Pleurodont condition of teeth is weakly expressed. Joint of dentaries strengthened by knobs and pits on symphysial surfaces.
Discussion. The genus was originally assigned by Nesov (1981) to the Protosirenidae Estes, and later transferred (Nesov, 1988) to the Albanerpetontidae, following Fox and Naylor's (1982) concept, which excludes this group from the Caudata. Milner (1988, 1993) designates the Albanerpetontidae as 'Lissamphibia *incertae sedis*' and leaves open the problem of possible positions of this group inside or outside the caudates. Gardner and Aver'yanov (1998) consider *Nukusaurus* as a non-diagnosable albanerpetontid, and designate this generic name as *nomen dubium*

Nukusurus insuetus Nesov, 1981
Holotype and locality. ZIN, PHA K77–4, left dentary; Chelpyk, southwestern Kyzylkum Desert, Karakalpakia, Uzbekistan.

Horizon. Upper part of Khodzhakul Svita or basal part of Beshtyube Svita, Cenomanian–?Lower Turonian, Upper Cretaceous.

Nukusurus sodalis Nesov, 1997

Holotype and locality. TsNIGRI 241/12177; dentary; Dzharakhuduk, central Kyzylkum Desert, Uzbekistan. *Horizon.* Middle part of Bissekty Svita, Coniacian, Upper Cretaceous.

Caudata indet.

Triassurus sixtelae Ivakhnenko, 1978, type species of the genus *Triassurus* and of the family Triassuridae Ivakhnenko, 1978, is a very poorly preserved impression of a small tetrapod skeleton (PIN 2584/10) from the Madygen Svita (Upper Triassic) of southern Kirgizstan. Restoration of the specimen (Ivakhnenko, 1978, Fig. 2) and its very attribution to the Caudata seem open to debate.

Bishara backa Nesov, 1997, type species of *Bishara* Nesov, 1997 from the Bostobe Svita (Santonian-Lower Campanian, Upper Cretaceous) of Baybishe, Kazakhstan, was originally mentioned as a *nomen nudum* (Nesov and Udovichenko, 1986), and further attributed by Nesov (1997) to the (?)Albanerpetontidae, Gardner and Aver'yanov (1998) re-evaluated it as a salamander. The holotype and only specimen (TsNIGRI 240/12177, atlas) is lost.

In addition, further fragmentary finds of Mesozoic urodeles from the Asian part of the former Soviet Union are reported by Nesov (1988) as not identified at the species or higher level. These include the following.

Albanerpetontidae: Tashkumyr, northern Fergana, Kirgizstan; Balabansai Svita, Callovian, Upper Jurassic; Dzharakhuduk, central Kyzylkum Desert, Uzbekistan; middle part of Bissekty Svita, Coniacian, Upper Cretaceous.

Eoscapherpetontinae: Shakh-Shakh, northeast of Cis-Aral region, Kazakhstan; lower part of Bostobe Svita, Santonian, Upper Cretaceous; *Eoscapherpeton* sp. and (?) prosirenid: Baybishe, northeast of Cis-Aral region, Kazakhstan; middle or upper part of Bostobe Svita, Santonian–?Campanian, Upper Cretaceous.

Acknowledgements

I thank Andrew Milner and Jean-Claude Rage for helpful comments on an earlier version of the manuscript.

References

Aver'yanov, A.O. 1999. [Annotated list of taxa described by L.A. Nesov]. *Trudy Zoologicheskogo Instituta RAN* **227**: 6–37.

Borsuk-Białynicka, M. 1978. *Eopelobates leptocolaptus* sp.n – the first Upper Cretaceous pelobatid frog from Asia. *Palaeontologia Polonica* **38**: 57–63.

Buffetaut, E., Raksaskulwong, L., Suteethorn, V. and Tong, H. 1994a. First post-Triassic temnospondyl amphibians from the Shan-Thai block: intercentra from the Jurassic of peninsular Thailand. *Geological Magazine* **131**: 837–839.

—, Tong, H. and Suteethorn, V. 1994b. First post-Triassic labyrinthodont amphibian in South East Asia: a temnospondyl intercentrum from the Jurassic of Thailand. *Neues Jahrbuch für Geologie und Paläontologie, Monatshefte* **1994**: 385–390.

Dong, Zh. 1985. A Middle Triassic labyrinthodont (*Sinobrachyops placentiephalus* gen. et sp. nov.) from Dashanpu, Zigong, Sichuan Province. *Vertebrata PalAsiatica*, **23**: 301–305.

Duellman, W.E. and Trueb, L. 1986. *Biology of Amphibians.* New York: McGraw-Hill.

Efimov, B. 1988. [On the fossil crocodiles of Mongolia and the Soviet Union.] *Trudy Sovmestnoi Sovetsko-Mongol'skoi Paleontologicheskoi Ekspeditsii* **34**: 81–90.

Evans, S.E., Milner, A.R. and Mussett, F. 1990. A discoglossid frog from the Middle Jurassic of England. *Palaeontology* **33**: 299–311.

Fox, R.C. and Naylor, B.G. 1982. A reconsideration of the relationships of the fossil amphibian *Albanerpeton*. *Canadian Journal of Earth Sciences* **19**: 118–128.

Gardner, J.D. and Aver'yanov, A.O. 1998. Albanerpetontid amphibians from the Upper Cretaceous of Middle Asia. *Acta Palaeontologica Polonica* **43**: 453–467.

Gubin Y.M. 1993. [Cretaceous anurans of Mongolia.] *Paleontologicheskii Zhurnal* **1993** (1): 51–56.

—1999. [The gobiatids (Arura) from the Upper Cretaceous locality Hermiin Tsav (Gobi, Mongolia)]. *Paleontologicheskii Zhurnal* **1999** (1):

76–87.

—and Sinitsa, S.M. 1993. Triassic terrestrial tetrapods of Mongolia and the geological structure of the Sain-Sar-Bulak locality. *New Mexico Museum Natural History and Science Bulletin* 3: 169–170.

Ivakhnenko, M.F. 1978. [Urodele amphibians from the Triassic and Jurassic of Middle Asia.] *Paleontologicheskii Zhurnal* 1978 (3): 84–89.

Krupina, N.I. 1994. [First finds of dipnoans from Mongolia.] *Paleontologicheskii Zhurnal* 1994 (2): 75–80.

Milner, A. R. 1988. The relationships and origin of living amphibians, pp. 59–102 in Benton, M.J. (ed.), *The Phylogeny and Classification of the Tetrapods. Volume 1. Amphibians, Reptiles, and Birds. Systematics Association Special Volume* 35A. Oxford: Clarendon Press.

—1993. Amphibian-grade Tetrapoda, pp. 666–679 in Benton, M.J. (ed.), *The Fossil Record 2*. London: Chapman & Hall.

Nesov, L.A. 1981. [Urodele and anuran amphibians from the Cretaceous of Kyzylkum.] *Trudy Zoologicheskogo Instituta AN SSSR* 101: 57–58.

—1988. Late Mesozoic amphibians and lizards of Soviet Middle Asia. *Acta Zoologica Cracoviensis* 31: 475–486.

—1990. [Late Jurassic labyrinthodont (Amphibia, Labyrinthodontia) among other relict groups of vertebrates of northern Fergana.] *Paleontologicheskii Zhurnal* 1990 (3): 82–90.

—1997. [*None-marine vertebrates from the Cretaceous of Middle Asia.*] St. Petersburg: Izdatel'stvo Sankt-Peterburgskogo Universiteta, 218 pp.

—and Fedorov, P.V. 1989. [Vertebrates from the Jurassic, Cretaceous and Palaeogene of northeastern Fergana and their significance for refining the age of deposits and the setting of the past. 1. Jurassic and early Cretaceous.] *Vestnik Leningradskogo Universiteta, Seriya 7, Geologiya, Geografiya* 14 (2): 20–30.

—and Udovichenko, N.I. 1986. [New finds of vertebrate remains from the Cretaceous and Palaeogene of Middle Asia.] *Voprosy Paleontologii* 9: 129–136.

Roček, Z. and Nesov, L.A. 1993. Cretaceous anurans from Central Asia. *Palaeontographica, Abteilung A* 226: 1–54.

Säve-Söderbergh, G. 1934. Some points of view concerning the evolution of the vertebrates and the classification of this group. *Arkiv Zoologii* 26A (17): 1–20.

Shishkin, M.A. 1991. [Labyrinthodont from the Late Jurassic of Mongolia.] *Paleontologicheskii Zhurnal* 1991 (1): 81–95.

Špinar, Z.V. and Tatarinov, L.P. 1986. A new genus and species of discoglossid frog from the Upper Cretaceous of the Gobi Desert. *Journal of Vertebrate Paleontology* 6: 113–122.

Tatarinov, L.P. 1994. [On an unusual mammal tooth from the Jurassic of Mongolia.] *Paleontologicheskii Zhurnal* 1994 (2): 97–105.

Vergnaud-Grazzini, C. and Wenz, S. 1975. Les discoglossidés du jurassique supérieur du Montsech (Province de Lerida, Espagne). *Annales de Paléontologie (Vertébrés)* 61: 19–36.

Winkler, D.A., Murry, P.A., and Jacobs, L.L. 1990. Early Cretaceous (Comanchean) vertebrates of central Texas. *Journal of Vertebrate Paleontology* 10: 95–116.

Zaitsev, N.S., Mossakovskii, A.A., Durante, M.V. and Shishkin, M.A. 1973. [Reference section of the Upper Palaeozoic and Triassic continental deposits of southern Mongolia with the first finds of labyrinthodonts.] *Izvestiya AN SSSR, Seriya Geologiya* 7: 133–134.

17

Mesozoic turtles of Middle and Central Asia

VLADIMIR B. SUKHANOV

Introduction

Mongolia is located in the heart of Central Asia, a vast continent which has existed uninterrupted for a long period of time and was, without doubt, the centre of origin and evolution of many vertebrate groups including the turtles. Good exposures of Mesozoic deposits, numerous discoveries and the often exceptional preservation of the remains of Mesozoic turtles from the territory of Mongolia, the Middle Asian republics and China have added much new information to our knowledge of the systematics and evolution of this group. This chapter presents the results of a preliminary study of the fossil turtles of Mongolia based on remarkably complete material collected in 1946–1949 by the Mongolian Palaeontological Expedition of Academy of Sciences of the USSR, under the leadership of I.A. Efremov, and from 1969 to the present day by the Joint Soviet–Mongolian (now Russian–Mongolian) Palaeontological Expedition (JSMPE). This study was carried out by the author, much of it in collaboration with Dr P. Narmandakh from the Geological Institute of the Mongolian Academy of Sciences.

Turtles are a common component of the so-called 'dinosaur faunas' of Central Asia. Frequently, in respect of total numbers, they are the dominant group of fossil reptiles. In some horizons there are accumulations of tens or even hundreds of shells or complete skeletons, often representing only a single genus. It is sufficient to mention the Mongolian turtle cemeteries of Bambuu Khudag, Nogoon Tsav, Tsagaan Khushuu and Höövör, while at Shar Teeg some beds are formed almost entirely of single fragments of turtles. At other localities turtles are rarer, occurring as single complete skeletons or just shells. In the latter case the remains usually represent large turtles, highly specialized for life in the water. Differences in the preservation of turtles result not only from local taphonomic conditions, but also from the character of the faunas themselves, which reflect differences in the physico-geographical conditions and in the age of the deposits.

Repository abbreviations

CCMGE, Chernyshev Central Museum of Geological Exploration, St. Petersburg; GIN, Geological Institute, Mongolian Academy of Sciences, Ulaanbaatar; IVPP, Institute of Vertebrate Paleontology and Paleoanthropology, Chinese Academy of Sciences, Beijing; PIN Palaeontological Institute, Russian Academy of Sciences, Moscow; ZIN, Zoological Institute, Russian Academy of Sciences, St. Petersburg.

The history of study

The study of fossil turtles from Central Asia began in the 1920s with the Central-Asian Expeditions of the American Museum of Natural History and the Swedish–Chinese Expeditions. The American expeditions proceeded along Lake Valley through the South Gobi, visiting Bayan Zag and Ondai Sair, and into the Eastern Gobi (Ergil Ovoo) and adjoining regions of China (Inner Mongolia). Material collected by this expedition was studied by Matthew and Granger (1923) and Gilmore (1931, 1934). The Swedish–Chinese expedition worked at Gansu and in

Inner Mongolia, China, collecting material which was later described by Wiman (1930) and Bohlin (1953). The American expedition found only fragmentary material in the Mesozoic deposits, their identification was dubious and comparisons were only made with American forms. However, the identification amongst these fragments of turtles belonging to the Adocidae (included at that time in the Dermatemydidae) in deposits ascribed to the Lower Cretaceous was interesting, since, previously, these turtles had only been known from the Upper Cretaceous (Campanian) and the Palaeogene of North America. The collections of the Swedish–Chinese expeditions were of considerable significance despite the fragmentary nature and poor preservation of the remains and, more importantly, their uncertain age. These materials showed, for the first time, the high degree of endemism of Mesozoic Central Asian turtles, and their distant relationship to North American forms. Practically all the turtles including *Sinemys*, *Sinochelys*, *Tsaotanemys*, *Peishanemys* and *Yumenemis* proved to represent new genera. Only *Osteopygis* was common to Mongolia and America, though later (Nesov and Khozatskii, 1973; Sukhanov and Narmandakh, 1974) it was shown that the remains ascribed to this turtle represented a completely different genus and family.

The study of the fossil turtles of Central Asia was largely halted during the Second World War, and the only work to be published was a description of *Manchurochelys* (Endo and Shikama, 1942) from the Late Jurassic of China. Subsequently, and until quite recently, turtles from Middle Asia, Mongolia and China were studied separately.

Fossil turtles from Middle Asia are housed in PIN (Moscow), the Zoological Institute (St. Petersburg), RAN, the Central Museum of Geological Exploration (St. Petersburg) and in the Institute of Zoology of the Academy of Sciences, Alma-Ata, Kazakhstan. The material was found in four main regions: Southern Kazakhstan, Karakalpakia and Central Kyzylkum (Uzbekistan), the Fergana Valley (Kirgizstan, Uzbekistan and Tadzhikistan) and Transbaikal (Russia) (Nesov, 1984).

Mesozoic turtles from the Middle Asian republics

were first described by Ryabinin (1935, 1948) and Khozatskii (1957) and the latter's students, Kuznetsov (Kuznetsov, 1976; Kuznetsov and Shilin, 1983; Kuznetsov and Chkhikvadze, 1984) and, most importantly, Nesov (Nesov, 1977a, b, c, d, 1978, 1981a, b, 1984, 1986a, b, 1995; Nesov and Khozatskii, 1973, 1977a, b, 1978, 1980, 1981a, b; Khozatskii and Nesov 1977, 1979; Nesov and Krasovskaya, 1984; Nesov and Kaznyshkin, 1985). Nesov carried out many studies of Middle Asian Mesozoic turtles (Nesov, 1984a, b, 1988; Nesov and Krasovskaya, 1984; Nesov and Golovneva, 1983; Nesov and Mertinene, 1982), their phylogenetic relationships and the distribution of some groups of turtles (Nesov, 1976, 1977a, b, 1981a, b, 1986a, b; Nesov and Yulinen, 1977; Nesov and Khozatskii, 1981a, b; Kaznyshkin *et al.*, 1990). Recently, Kaznyshkin began investigating Jurassic turtles (Nesov and Kaznyshkin, 1985; Kaznyshkin, 1988; Kaznyshkin *et al.*, 1990) and Kordikova (1992a, b; 1994a, b) published work on fossil trionychids.

Following the Second World War, the study of Chinese Mesozoic turtles began with work on the Jurassic turtles from Sechuan (Young and Chow, 1953) and Lower Cretaceous turtles from Shandun (Chow, 1954). The greatest contributions to the study of Chinese Mesozoic turtles were made by Yeh Hsiang-K'uei (Yeh, 1963, 1965, 1966, 1973a, b, 1974a, b, 1982, 1983a, b, 1986a, b, 1988, 1990). A new stage in the study of Chinese Mesozoic turtles began with the establishment of the Sino-Canadian 'Dinosaur Project'. This joint project led to the discovery of new forms and additional material of previously described turtles (Brinkman and Peng, 1993b; Brinkman and Nesov, 1993), and new detailed revisions of some important Central Asian groups such as the so-called Chinese plesiochelyids (Peng and Brinkman, 1993) and turtles from the family Sinemydidae (Brinkman and Peng, 1993a). The description of the skull of the Lower Cretaceous macrobaenid *Dracochelys* (Gaffney and Yeh, 1992) should also be included here.

Despite excellent collections made by Efremov's expeditions in the 1940s, the study of fossil turtles from Mongolia was much delayed. Fossil material was given to L. Khozatskii to study, but the first publications, including the description of *Mongolemys elegans* (Khozatskii and Młynarski, 1971), a new Late Cretaceous turtle based mainly on material collected by the Joint Polish–Mongolian Expedition, and a new species of *Trionyx* from the early Upper Cretaceous (Khozatskii, 1976) appeared only in the 1970s. The most interesting component of the Efremov collection, *Mongolochelys efremovi*, will be published shortly (Khozatskii, in press) following preparation of the paper by L.A. Nesov from rough notes, drawings and photographs left by Khozatskii.

Since the mid-1960s the study of Efremov's material was undertaken by V.B. Sukhanov, but for ethical reasons little progress was made. Following the start of the Joint Soviet–Mongolian Paleontological Expedition in 1969 new material was collected and descriptions of new turtles from the territory of Mongolia began to appear (Sukhanov and Narmandakh, 1974, 1975, 1976, 1977, 1983; Narmandakh, 1985). Most of the work on Mongolian turtles was conducted by the author and P. Narmandakh, but some fragmentary remains from this region, now housed in St. Petersburg and Tbilisi, were described by the late L.A. Nesov (Nesov and Verzilin, 1981) and V. Chkhikvadze (Shuvalov and Chkhikvadze, 1975, 1979, 1986; Chkhikvadze, 1976, 1981, 1987; Chkhikvadze and Shuvalov, 1980, 1988). This material is of particular interest because it was collected from regions and deposits not investigated by the SSMPE. However, a large percentage of the taxa described are based on fragmentary or non-diagnostic material. A comparatively small collection of fossil turtles, now located in Poland, was recovered by the Polish–Mongolian Palaeontological Expedition (1963–1965) under the leadership of Kielan-Jaworowska. This material was described by Młynarski and others (Młynarski, 1969, 1972; Młynarski and Narmandakh, 1972; Khozatskii and Młynarski, 1971).

Preliminary studies of fossil material from the territory of Mongolia have led to the identification of 52 species, 37 genera and 12 families of fossil turtles, from 14 biostratigraphical horizons ranging in age from Upper Jurassic to Pliocene. Nine of these horizons are found in the Mesozoic (two Jurassic, two Lower Cretaceous and five Upper Cretaceous) and, apart from single horizons in the Miocene and Pliocene (which have each yielded just one species of terrestrial turtle), each contains a special complex of turtles. The greatest diversity (10–12 species), is found in faunal complexes at the end of the Late Cretaceous (Nemegt Svita) and in the Late Palaeocene. Mesozoic horizons have yielded 34 species representing 24 genera distributed among nine families. Five species and three genera of one family are Jurassic, five–six species and four genera are from the Early Cretaceous and 22 species and 15 genera from the Late Cretaceous. By comparison, Mesozoic deposits of the Middle Asian republics have produced 22 species in 16–17 genera from eight families (two species and two genera from the Jurassic, three genera and four species from two Early Cretaceous families and 13–14 genera and 16 species in six families from the Late Cretaceous). All the Late Cretaceous Middle Asian forms are from the Cenomanian to early Santonian, while the Mongolian turtles (9–10 genera and 13 species) are from the Campanian–Maastrichtian interval.

From the Mesozoic of China 21 genera and 27 species have been described (five to eight genera and 12 species in five families from the Jurassic, seven to nine genera and nine species in four families from the Early Cretaceous and four to six genera and six species in four or five families from the Late Cretaceous), though of these, seven genera and seven species are based on very poor material. In spite of the provisional nature of the taxonomic numbers given above, which is related to the fragmentary nature of the material and the different approaches of specialists, they show that, in terms of diversity, Mongolian Mesozoic turtles clearly outnumber those from adjoining regions. More importantly, they fill many gaps, especially in the stratigraphic record, in our knowledge of the faunas of turtles from Central Asia.

It should be emphasized that, in contrast to taxa collected from adjoining regions, the anatomy of Mongolian turtles is known in considerable detail. This is mainly a result of the good preservation and completeness of Mongolian taxa, many of which are represented by multiple specimens. This permits detailed morphological study and, naturally, the recognition of new diagnostic features, enabling the distinction of new taxa or elevation in rank of previously described taxa. It is also worth emphasizing one important feature of work on fossil turtles. Specific features of species (identified by morphological comparisons) are often exhibited by single fragments and can easily be established, especially where multiple specimens have been collected. They also enable direct comparison with analogous elements of turtles from the same or other deposits. However, the identification of taxa at the generic level is much more difficult, owing to the frequent repetition of the same modifications of bony and horny elements of shells in different taxa. Hence, the degree of substantiation (or rather the degree of validity) of turtle taxa directly depends on the quality (i.e. completeness and preservation) of material. Even if remains are fragmentary (e.g. single plates of the shell) where series of specimens representing one taxon exist it is possible to produce accurate reconstructions of the shell, or its parts. By contrast, single plates are inadequate as holotypes since they do not exhibit specific features of species and genera.

Systematic review

Jurassic turtles

The oldest known turtles from Central Asia are Jurassic in age and mainly from China where they have been recovered from lacustrine and river deposits dating back to the Early Jurassic (Yeh, 1988). Five families have been identified, although two of these are represented only by single discoveries. In Mongolia there are only two localities of Jurassic turtles. Two bone-bearing horizons, probably of different age, as they represent two cycles of basin development with an interval between them, in Upper Jurassic lacustrine deposits at Shar Teeg in the Transaltai Gobi have both produced turtles, as has the locality of Önjüül in Central Mongolia. These turtles, of which five species in three genera have been identified, all seem to belong within one family (Sukhanov and Narmandakh, in press).

In Middle Asia, numerous remains of turtles have been found in Jurassic brackish water deposits in the Fergana depression (Kirgizstan). With the exception of some small, but as yet undescribed Middle Jurassic turtles (Kaznyshkin et al., 1990, p. 186), all these remains were initially ascribed to *Xinjiangchelys latimarginalis*, also known from China (Nesov and Kaznyshkin, 1985; Kaznyshkin, 1988; Kaznyshkin et al., 1990).

Jurassic turtle faunas are composed of three groups of turtles: the Middle Jurassic Chengyuchelyidae; the upper Middle Jurassic to Lower Cretaceous Xinjiangchelyidae and the Upper Jurassic–Lower Cretaceous Sinemydidae. Dominant among these are the Xinjiangchelyidae, but each group is restricted to a comparatively short time interval and all of them are exclusively Central Asian.

Family CHENGYUCHELYIDAE Yeh, 1990
Type genus. Chengyuchelys Young and Chow, 1953. Middle Jurassic of Sechuan, China.

This family is represented by quite small turtles up to 300 mm in length. Three species have been recognized (Yeh, 1963, 1982, 1990; Fang, 1987), but, unfortunately, all specimens originate from one region and sediments of the same age, and are incomplete and poorly preserved. Only the presence of mesoplastra and at least three pairs of inframarginals provide systematic information about this family and show, in particular, its primitive nature compared to other turtles from Central Asia (an exception to this is the Upper Cretaceous Mongolochelyidae; see below). Other characteristics of this family include a sutural connection between the carapace and plastron, presence of medial narrowing of the narrow mesoplastra and overlap of the hypoplastra near the midline by the anals. The diagnosis given by Yeh (1990), is unsatisfactory, because, on this basis, the family includes *Xinjiangchelys*, which, following more precise descriptions (Peng and Brinkman, 1993), does not possess mesoplastra and thus should be assigned to another family (Kaznyshkin et al., 1990).

Family XINJIANGCHELYIDAE Nesov, 1990

Type genus. Xinjiangchelys Yeh, 1986. Middle–Upper Jurassic, Central Asia.

Turtles of this family are diverse and widely distributed across Middle Asia, Mongolia and China. Fossil material includes complete and fragmentary shells, and skulls in a few forms. Thus, until recently, the Xinjiangchelyidae was the best studied group of Asian Jurassic turtles.

Xinjiangchelys Yeh, 1986 (Figure 17.1), was established on the basis of a fairly complete, but poorly preserved shell of *X. junggarensis* from the upper part of the Middle Jurassic (Bathonian–Callovian) of the Dzhungarian depression. However, Yeh (1986a, b) mistook cracks in the plastron as evidence of the presence of mesoplastra and may also have been mistaken regarding the existence of only ten pairs of peripherals and eleven pairs of marginals. Later, material from Fergana was assigned to this species, and it was synonymized with *Plesiochelys latimarginalis* Young and Chow, 1958, from Sechuan, known to be widely distributed and to have a very variable shell morphology (Kaznyshkin, 1988). *Xinjiangchelys latimarginalis* was proposed as the type species for *Xinjiangchelys*, which in turn was established by Nesov (Kaznyshkin *et al.*, 1990) as the type genus for the Xinjiangchelyidae, though this created considerable difficulties for later researchers. Following the study of new material from Sinzyan and Sechuan (Peng and Brinkman, 1993) it was suggested that not all the Middle Asian material can be identified as *X. latimarginalis*. Only two incomplete shells (one of these representing a juvenile), can be ascribed to this species, while the rest of material, including an incomplete skull, was identified as *Xinjiangchelys* sp. Recently, Nesov (1995) proposed a new name for this species, *X. tianshanensis* (Figure 17.1C). At the same time, the type specimen of *X. junggarensis* is distinctly different from the rest of the material ascribed to *X. latimarginalis* and, at best, must be situated at the very edge of the potential limits of this species (Peng and Brinkman, 1993). Peng and Brinkman (1993) recently published more detailed descriptions of *X. latimarginalis* (Figure 17.1A, B), but the skull of this form is as yet unknown, and the

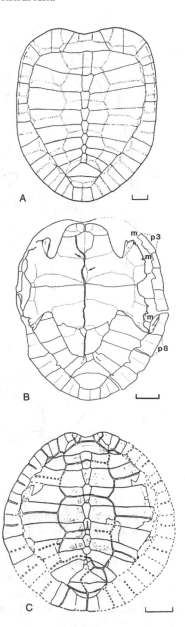

Figure 17.1. *Xinjiangchelys* Yeh, 1986. (A, B) *X. latimarginalis* (Young and Chow, 1953), IVPP V9537–1, Qigu Formation, Upper Jurassic, Pingfengshan locality, Xinjiang, China. (A) reconstruction of the carapace, (B) ventral view of the shell. (C) *X. tianshanensis* Nesov, 1995, holotype, CCMGE 1/12526, Balabansai Formation, Middle Jurassic, (Callovian), Sarykamyshsai locality, Fergana, Kirgizstan. (A and B, after Peng and Brinkman, 1993; C, after Nesov, 1995). Abbreviations: m, musk ducts; p, peripheral. Scale bar = 50 mm.

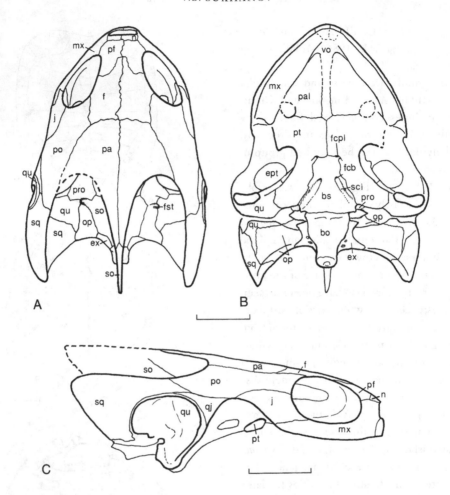

Figure 17.2. *Annemys levensis* Sukhanov and Narmandakh, in press, PIN 4636/4–1, from the Upper Jurassic of Shar Teeg, Transaltai Gobi, Mongolia. Skull in (A) dorsal view, (B) ventral view, and (C) right lateral view. Abbreviations: bo, basioccipital; bs, basisphenoid; ex, exoccipital; f, frontal; fcb, foramen caroticum basisphenoidale; fpcl, foramen posterius canalis carotici laterale; fst, foramen stapedio-temporale; j, jugal; mx, maxilla; n, nasal; op, opisthotic; pa, parietal; pal, palatine; pf, prefrontal; po, postorbital; pro, prootic; pt, pterygoid; qj, quadratojugal; qu, quadrate; sci, sulcus caroticus internus; so, supraoccipital; sq, squamosal; vo, vomer. Scale bar = 10 mm.

diagnosis of the genus given by these authors includes only characteristics of the family.

Xinjiangchelyids are middle-sized turtles up to 300–350 mm long. The skull, which is known for the genus *Annemys* (Figure 17.2), has large temporal and cheek emarginations, small nasals, and large prefrontals that contact along the midline. There is no interpterygoid vacuity, but the pterygoids contact along the

midline anteriorly, though posteriorly they do not reach the basioccipital or exoccipitals. There are no internal carotid canals anterior to the foramen caroticum basisphenoidale. The internal carotid arteries are situated in an open ventral groove formed by the basisphenoid. The neck vertebrae have free ribs and exhibit a transitionary condition from amphycoely to a more complex form. The shell is relatively low and wide, round posteriorly and with a wide nuchal incision anteriorly. The plastron was connected to the carapace by ligaments. Anterior and posterior processes of the axillary and inguinal buttresses are correspondingly engaged in grooves on the peripherals. The neurals are relatively narrow with a tendency to partial reduction of the hind-most elements (seventh–eighth), leading to the appearance of individual variation in the contact along the midline of the seventh pair of costals. The first and second neurals (or only the latter) are rectangular, the rest are hexagonal with short antero-lateral sides, reduction of which can lead to a pentagonal shape. The second to seventh peripherals are thickened, and their free edges bend upwards somewhat. The seventh to eleventh peripherals are enlarged in height (medio-laterally). There is one cervical scute: its width is noticeably larger than its length. The marginals overlap the distal ends of costals two to four. The epiplastra bear well developed dorsal processes and there are no mesoplastra. Pairs of gular and intergular scutes and exit foramina for canals of the musk gland are present. The middle sulcus behind the entoplastron is meandering. The first dorsal rib is reduced and reaches only midway toward the axillary buttress, while the free part of the first costal rib engages in a groove on the third peripheral.

So, on the one hand, Xinjiangchelyidae is characterized by a very primitive skull morphology, similar to that of the European Pleurosternidae, but distinguished from them by a relatively short basisphenoid which does not extend far forward, thus permitting the pterygoids to contact along the midline. On the other hand, the shell is evolutionarily advanced, corresponding to the condition in Plesiochelyidae, but distinguished by features such as the connection of the carapace and plastron, the unpaired cervical scute and the anteriorly thickened high posterior peripheral.

In addition to the type genus, Xinjiangchelyidae also includes *Shartegemys* (Sukhanov and Narmandakh, in press), from the upper part of the (?)Middle Jurassic (Figure 17.3), *Annemys* (Sukhanov and Narmandakh, in press) from the Upper Jurassic of the Transaltai Gobi (Figure 17.4) and possibly

Undjulemys (Sukhanov and Narmandakh, in press; Figure 17.5) from the Upper Jurassic of Central Mongolia.

Shartegemys (Figure 17.3) exhibits a number of distinctive characters: narrow neurals, extremely wide but short vertebrals (especially the first one) and correspondingly short pleurals, a very wide but short cervical scute, relatively wide bridges and anals, which do not reach the hypoplastra, and a femoro-anal sulcus that runs antero-medially. Particularly distinctive for this genus are the enlarged epiplastra, connected to the entoplastron and the hypoplastra via a suture, and directed antero-medially rather than transversely as in other representatives of this family. In this respect *Shartegemys* is similar to the macrobaenids.

Annemys (Figure 17.4), is characterized by wider neurals, longer pleurals and narrower middle vertebrals (second and third) which are almost square in shape. In addition, the twelfth pair of marginals is lower and not overlapped by the second suprapygal, the bridges are narrower, the anals overlap the hypoplastra and, as in *Xinjiangchelys*, the femoro-anal suclus has a characteristic knee-shaped curve.

Undjulemys (Figure 17.5) is known only on the basis of two incomplete impressions of the carapace. This is a small turtle wherein the border fontanelles are retained in the carapace. It is distinguished by wide neurals and low peripherals and marginals. The general shape of the carapace is reminiscent of *Manjurochelys* (especially *M. liaoxiensis*, Ji, 1995) from the Late Jurassic of North-Eastern China.

As already mentioned, the greater part of the material from the Middle Jurassic of Central Asia and initially identified as *Xinjiangchelys latimarginalis* (Kaznyshkin, 1988; Kaznyshkin *et al.*, 1990), has more recently been considered as *X.* sp. (Peng and Brinkman, 1993) or *X. tianshanensis* (Nesov, 1995). However, it is possible that restudy of this material may lead to its identification as a new genus, as might also be the case for the Chinese turtle *X. chungkingensis* (Young and Chow, 1953). A number of other genera can also be provisionally assigned to the Xinjiangchelyidae. These include the poorly known forms *Jastmelchyi* (Chkhikvadze, 1987)

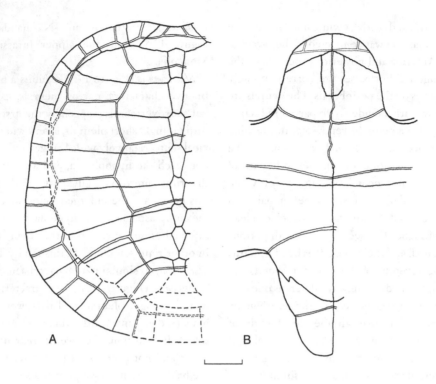

Figure 17.3. *Shartegemys laticentralis* Sukhanov and Narmandakh, in press, PIN 4636, from the Upper Jurassic of Shar Teeg, Transaltai Gobi, Mongolia. Reconstruction of the shell based on single plates and an almost complete plastron in (A) dorsal view and (B) ventral view. Scale bar = 20 mm.

('*Tienfucheloides*' *jastmelchyi* (Chkhikvadze, 1981) according to Nesov (1987b)) from the Early Cretaceous of Mongolia and *Tienfuchelys* (Young and Chow, 1953) from the Late Jurassic of Sechuan (China), known only from isolated fragmentary remains.

Family SINEMYDIDAE Yeh, 1963

Type genus. Sinemys Wiman, 1930. Late Jurassic–Early Cretaceous of Central Asia (Shandong, Gansu, Inner Mongolia, Sechuan)

The Sinemydidae are small, aberrant turtles from 100 to 200 mm long and with a strongly flattened carapace. The family was established by Yeh (1963) for two forms: *Sinemys lens* Wiman, 1930 and *Manchurochelys manchoukuoensis* Endo and Shikama (1942) from Upper Jurassic deposits of Eastern and North-Eastern China

(Shandong and Laioning respectively). These turtles are recognized by the presence of a reduced crest-shaped plastron with fontanelles and a narrow, elongate posterior lobe. Recently, new material of *S. lens* has been found and two new sinemydids, *Sinemys gamera* from the Early Cretaceous of Inner Mongolia (Brinkman and Peng, 1993b) and *Manchurochelys liaoxiensis*, from the Late Jurassic of Liaoning (Ji, 1995), have been described. This new material, which includes skulls, has broadened our knowledge of *Sinemys*, and has revealed some of its exceptional features. It has also become clear that *Sinemys wuerhoensis* from the Early Cretaceous of Dzhungaria (Yeh, 1973a, b) does not belong to this genus, or to this family (Brinkman and Peng, 1993b). The new material casts doubt on earlier suggestions (Chkhikvadze, 1976, 1987) that the Sinemydidae should include some

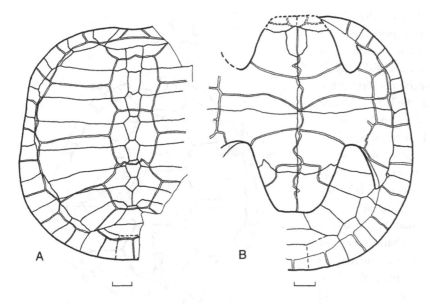

Figure 17.4. *Annemys latiens* Sukhanov and Narmandakh, in press, holotype, GIN, Mongolian Academy of Sciences. Shell in (A) dorsal view and (B) ventral view. Scale bars = 20 mm.

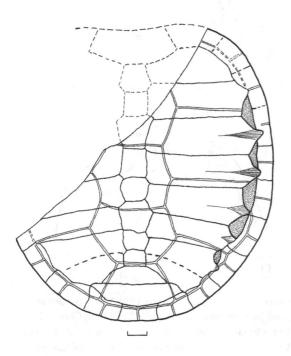

Figure 17.5. *Undjulemys platensis* Sukhanov, 1997, holotype, PIN 4637/1, Upper Jurassic, Önjüül locality, Central Aimag, Mongolia. Reconstruction of the carapace. Scale bar = 10 mm.

macrobaenid turtles, an idea that, despite the existence of objections (Khozatskii and Nesov, 1979) has been uncritically accepted by some authors (McKenna *et al.*, 1987; Gaffney and Meylan, 1988). Brinkman and Peng (1993b) united *Sinemys* and *Manchurochelys* in the Sinemydidae on the basis of two synapomorphies: the presence of a first and second suprapygal, the former being small, and the elongate posterior lobe of the plastron, which has a narrow base and almost parallel sides. However, there are striking differences between these taxa and *Manchurochelys* is probably an early representative of another family, the Macrobaenidae, which achieved peak diversity in the Early Cretaceous (see below).

The Sinemydidae is defined here as follows: the posterior portion of the carapace is reduced as a result of the disappearance of the pygal and the decrease in size of the twelfth pair of peripherals, leading to the exposure of the second suprapygal on the posterior edge of the carapace; the first suprapygal is very small, while the second is large and wide; the shape of the neurals is variable: usually the second and third are rectangular while the rest are hexagonal, often with short postero-lateral sides; in place of the eighth neural there are two small plates that prevent contact between the eighth costals; and in contrast to the usual condition in turtles where the free ends of the ribs of the last costal are engaged in special pockets within each peripheral, the distal ends of the last costals are inclined between the corresponding and adjoining peripherals. The strengthening of the seventh peripheral and the development on this plate of a special process ('spine' according to Brinkman and Peng, 1933b) that projects outwards and backwards is highly characteristic for this family. In addition, the antero-lateral peripherals have sharp external edges, that is to say, the thickened edges that are characteristic of Xinjiangchelyidae and Macrobaenidae are missing, as is the groove on the dorsal surface of the peripherals; there is no cervical scute and the first vertebral reaches the anterior edge of carapace; the twelfth pair of marginals appear to be small triangular plates and do not contact along the midline, as a result of which the fifth vertebral also reaches the posterior edge of the carapace; the plastron has wide bridges and becomes narrow and more wedge-shaped anteriorly, but long with almost parallel lateral edges posteriorly; the epiplastra are tiny triangular elements, connected posteriorly, perhaps loosely, only with the entoplastron. The latter element is unusual, consisting of a bony plate, that in external (ventral) view appears very short and widened transversely, and which completely separates the epiplastra from the hyoplastra. The central and marginal fontanelles are always present in the plastron, and the skull (see *Sinemys lens*, Figure 17.6) has a very large temporal emargination, extending far forward beyond the auditory capsules and connected to the cheek emargination because of the absence of contact between the postorbital and the quadrate, and, possibly, because of the absence of the quadrato-jugal. The nasals are large and the prefrontals have only a small entry on to the skull roof and do not contact along the midline. The basisphenoid extends far forward between the pterygoids, though it is not clear if it separates them completely, as in Pleurosternidae, or if the pterygoids have a small contact along the midline. The incisura columellae

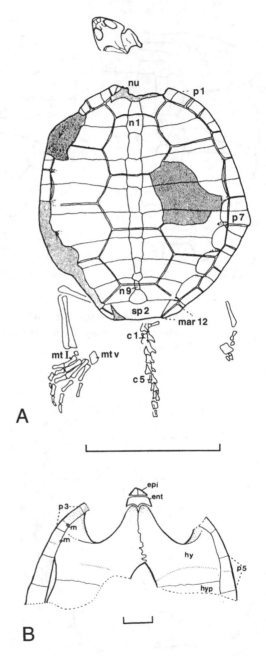

Figure 17.6. *Sinemys lens* Wiman, 1930, Mengyin Formation, Upper Jurassic, Ningjiagou locality, Shandong, China. (A) IVPP V9533–3, almost complete skeleton in dorsal view. (B) IVPP V9533–1, anterior part of a plastron. (A and B, after Brinkman and Peng, 1993b). Abbreviations: epi, epiplastron; ent, entoplastron; hy, hyoplastron; hyp, hypoplastron; m, musk duct; p3, third peripheral, p5, fifth peripheral. Scale bar: A = 50 mm, B = 10 mm.

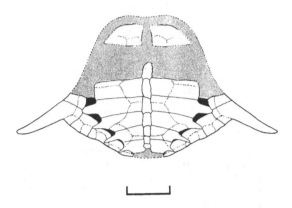

Figure 17.7. *Sinemys gamera* Brinkman and Peng, 1993, holotype, IVPP V9532–1, Laolonghuoze locality, Ordos, Inner Mongolia, China. Reconstruction of the carapace. (After Brinkman and Peng, 1993b.). Scale bar = 50 mm.

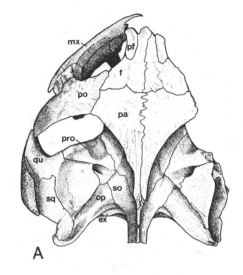

A

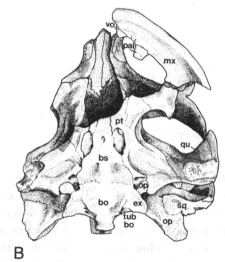

B

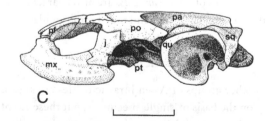

C

Figure 17.8. *Sinemys gamera* Brinkman and Peng, 1993, IVPP V9532–11, Laolonghuoze locality, Ordos, Inner Mongolia, China. The skull in (A) dorsal, (B) ventral, and (C) left lateral view. (After Brinkman and Peng, 1993b). Abbreviations as for Figure 17.2 except: tub. bo, tuberculum basiooccipitale. Scale bar = 10 mm.

auris is closed. The postero-medial part of the pterygoid does not extend very far backwards, has no contact with the basioccipital, and does not cover the processus interfenestralis of the opisthotic ventrally.

Following this definition of the group it includes, for now, only one genus, *Sinemys*, containing three species, all from China: *S. lens* (Figure 17.6) from the Late Jurassic–Early Cretaceous; *S. gamera* (Figures 17.7, 17.8) from the Aptian and ?*S. efremovi* (Khozatskii, 1996) recently described on the basis of collections made by Russian geologists in 1940–1942 from deposits of ?Barremian age in Dzhungaria. *Tienfucheloides undatus* Nesov, 1978, from the early Late Cretaceous of Karakalpakia (Uzbekistan), described on the basis of single shell plates, is reminiscent of *Sinemys* in some details (Nesov, 1978, 1988), but there are few grounds for ascribing this taxon to the Sinemydidae (Brinkman and Peng, 1993a, b).

Unfortunately, the skull of *Manchurochelys* is not yet known, but its shell (Figure 17.9) is strikingly different from that of *Sinemys*: the neural plates in the middle part of the series are hexagonal with short sides antero-lateraly, but not postero-lateraly; the eighth neural is normal and not divided into two; the first suprapygal and the pygale are relatively large, and the eleventh pair of peripherals are large, as is usual in turtles, so that unlike *Sinemys* the posterior part of the

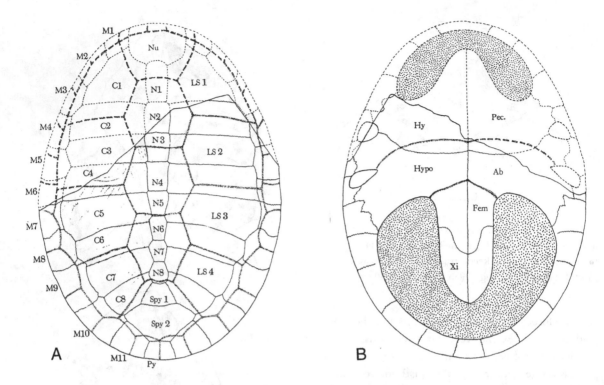

Figure 17.9. *Manchurochelys manchoukuoensis* Endo and Shikama, 1942, holotype, National Museum of Manchoukuo n.3898, Yixian Formation, Upper Jurassic, Zaocishan locality, Laioning, China. (A) carapace and (B) plastron. (After Endo and Shikama, 1942).

carapace does not appear to be shortened; the distal ends of the posterior costals enter grooves on the peripherals; and the cervical scute and twelfth pair of marginals are well developed and in contact along the midline, so that neither the first nor the fifth vertebrals exit onto the edge of the carapace. The majority of these features of *Manchurochelys* are also characteristic of the Xinjiangchelyidae and the Macrobaenidae.

Other Jurassic turtles from Asia

Two other groups of Asian Jurassic turtles are known only on the basis of single specimens, but these are of particular interest because they may be the earliest representatives of families which later became widely distributed. 'Plesiochelys' tatsuensis Yeh, 1963 (referred to as *Ferganemys* by Nesov and Yulinen, 1977; Nesov and Khozatskii, 1981a) from Sechuan, China, possibly belongs to the Adocidae, as may undescribed remains

(cf. Shachemydinae) from the Callovian of Fergana, Kirgizstan (Nesov, 1984). *Sinaspideretes wimani* (Young and Chow, 1953), also from Sechuan, China, and long thought to be a trionychid (Meylan and Gaffney, 1992), may be an early representative of the Carettochelyidae.

Yaxartemys longicauda (Ryabinin, 1948) from Upper Jurassic (Kimmeridgian) lacustrine deposits of South Kazakhstan, was initially ascribed to the Thalassemydidae (Ryabinin, 1948; Sukhanov, 1964), but the two known specimens are both juveniles and it may be preferable to consider them as incertae sedis (Nesov, 1984).

Early Cretaceous turtles

Early Cretaceous deposits are widely distributed in Central Asia, but attempts at dating them are very contradictory. In Mongolia, turtles are found at three different levels. The upper-most level, represented by deposits in the region of lake Böön-Tsagaan, contains only indeterminate fragmentary remains of macrobaenid turtles. The two remaining levels are each characterized by their own complex of turtles: the lower-most is known on the basis of a turtle fauna from the locality of Höövör, while the intermediate is represented by a turtle fauna from Hüren Dukh. Traditionally, these two levels were correlated with the Hühteeg Gorizont, dated as Aptian–Albian (Shuvalov and Chkhikvadze, 1979; Shuvalov, 1982), but it should be noted that the dominant form from the lowest level, *Hangaiemys hoburensis*, was also reported by these authors from the Tsagaantsav (Valanginian–Hauterivian) and Shinekhudag (Barremian–Hauterivian) Svitas, raising doubts as to the identification of these turtles and the assignment of deposits from various localities to a particular horizon. Moreover, there is an alternate interpretation of the age of these deposits, wherein the lowest and middle levels are assigned to the Neocomian and the upper one to the Aptian. Chinese researchers (e.g. Dong, 1995) continue to refer to two faunal assemblages in the Early Cretaceous of Northern China: an earlier Psittacosaur–Pterosaur assemblage (Neocomian–Early Aptian), which has been correlated with the Neocomian–Aptian Tsagaantsav Gorizont of Mongolia (Jerzykiewicz and Russell, 1991) and a later Probactrosaurian assemblage (Late Aptian–Albian). In earlier interpretations the entire Early Cretaceous was characterized as the Psittacosauridae-Pterosauria fauna (Chen, 1983).

The only Neocomian turtles from the territory of the Former Soviet Union are from Buryatia and from four assemblages found in Middle Asia (Nesov, 1984), the oldest of which is dated as Upper Aptian and the remainder as Upper Albian, though one of the latter, an assemblage which characterises the Alamyshyk Svita at the locality of Kylodzhun, South-Eastern Fergana, was recently redated as Lower–Middle Albian (Nesov, 1995).

At the start of the Cretaceous there were significant changes in turtle assemblages forming Central Asian vertebrate faunas, though representatives of the Sinemydidae are still found in the Neocomian of China. Early Cretaceous turtles were of only small to middle size (Nesov, 1984). The dominant group, both in numbers and in diversity, was the Macrobaenidae, which replaced the more primitive Xinjiangchelyidae and may well be descended from them, probably first appearing in the Late Jurassic. Morphologically, it is easy to derive macrobaenids from xinjiangchelyids and, as already mentioned, *Manchurochelys* from the Late Jurassic of China may represent an intermediate form. Macrobaenids seem to have reached peak diversity in the Early Cretaceous, where they are represented by five or six genera. From the Upper Cretaceous only one genus with two species, from the Cenomanian–Early Santonian of Middle Asia, has been reported (Nesov, 1984), though the group does survive into the Late Palaeocene.

A second group of Early Cretaceous turtles, the Sinochelyidae, are known only from Mongolia and China. However, they are extremely rare by comparison to the macrobaenids which can be present in mass burials.

Adocid turtles, though rare, distinguish Early Cretaceous Middle Asian turtle faunas from those of Mongolia and China. One exception is the locality of Kylodzhun, South-Eastern Fergana, where, in deposits ascribed to the Early–Middle Albian, remains of the adocid *Ferganemys verzilini* (Figure 17.17) are far more common than remains of the macrobaenid *Kirgizemys exaratus*. Two genera of adocids are known from the Early Cretaceous of Middle Asia, though one of these is represented only by fragmentary remains. Adocids are important in the Late Cretaceous of Central Asia and North America (see below), but only one indeterminate fragmentary remain, a bony plate from the carapace, has so far been found in the Early Cretaceous of Mongolia. This specimen was identified as *Adocus* sp. on the basis of the characteristic sculpture on the surface of the bony plate (Gilmore, 1931). Remains of Early Cretaceous adocids have not yet been found in China, though if the assignment of '*Plesiochelys*' *tatsuensis* from the Late Jurassic of China to *Ferganemys* (Nesov and Yulinen, 1977) is accepted, it seems possible that adocids were present in China in the Early Cretaceous.

The next major change in Middle Asian turtle faunas took place in the late Early Cretaceous (Albian), with a number of new families appearing alongside the macrobaenids and adocids. The Carettochelyidae are represented by remains of *Kyzilkumemys* sp., and the Lindholmemydidae, by fragments of *Mongolemys* cf. *occidentalis*, and the shells of two turtles initially identified as *Mongolemys* sp. (Sukhanov and Narmandakh, 1974) from deposits dated as Aptian–Albian in age and not far from the famous locality of Höövör. The Trionychidae are represented by '*Trionyx*' *kyrgyzensis* (Nesov, 1995; Figure 17.19) and two species of *Aspideretes* (Yeh, 1965), though these are now thought to be representatives of two different genera: *Axestemys* Hay, 1899, also known from the Palaeocene of North America, and *Paraplastomenus* Kordikova, 1991 (Kordikova, 1994a), found also in the Late Cretaceous–Miocene of Kazakhstan. The late Early Cretaceous also saw the first appearance of the Nanhsiungchelyidae in the form of fragmentary remains of '*Basilemys*' sp. (Chkhikvadze, 1976) from Lower Cretaceous deposits of Shandong, first reported by Wiman (1930).

Family MACROBAENIDAE Sukhanov, 1964

Type genus. Macrobaena Tatarinov, 1959. Late Palaeocene, Central Asia.

This family was established on the basis of *Macrobaena mongolica* (Tatarinov, 1959) when it became clear that the latter could not be retained in the Baenidae (Sukhanov, 1964). Later, two new Early Cretaceous genera, *Kirgizemys* (Nesov and Khozatskii, 1973, 1978, 1981b) and *Hangaiemys* (Sukhanov and Narmandakh, 1974, in press), were also ascribed to this family (according to Nesov a subfamily in the Toxochelyidae). Numerous characteristics of the Macrobaenidae became known following the redescription of *Macrobaena* (Sukhanov and Narmandakh, 1976), though the skull of this genus needs further description, and detailed accounts of *Hangaiemys* have yet to be published.

Practically all researchers agree on the morphological uniformity of macrobaenids, but assume that this group is united only by plesiomorphic features (a comparatively low shell, a crest-shaped plastron and a loose connection between the carapace and plastron), which are also characteristic for a number of other eucryptodirans. Thus, Brinkman and Peng (1993a) consider macrobaenids to be paraphyletic and, according to Gaffney and Meylan (1988) they are primitive representatives of the Suborder Polycryptodira. However, Macrobaenidae and Xinjiangchelyidae share many apomorphic features, which separate them from the Plesiochelyidae. These apomorphies include a loose connection between the carapace and the plastron, thickening of the lateral edges of the second to seventh peripherals and enlargement of the seventh to eleventh peripherals (Peng and Brinkman, 1993), but no overlap of the marginals on the first five costals in macrobaenids.

Macrobaenidae is characterized by the following apomorphies, which separates it from the Xinjiangchelyidae: the formation between the basisphenoid and prootic dorsally, and the pterygoid ventrally, of an almost completely enclosed canal (for the internal carotid artery) that has a posterior opening located toward the posterior margin of the pterygoid, and a special opening (the foramen basisphenoidale of Brinkman and Nicholls, 1993) that connected the canal to the palatal surface of the skull. The same condition is also present in the Sinemydidae and some other groups of turtles including the Adocidae, the Nanhsiungchelyidae and the Lindholmemydidae, thus at the present time it is difficult to decide whether this condition is apomorphic or plesiomorphic for macrobaenids. Other macrobaenid apomorphies include: the appearance of completely formed articular surfaces on the cervical vertebrae, and one double cervical; the reshaping, reorientation and reduction in size of the epiplastra and reduction of their dorsal processes; and reduction in the number of gulars to a single pair.

The relationships of Macrobaenidae to later groups of turtles is not yet clear. Nesov and Khozatskii (1978) considered macrobaenids to be a subfamily of the Toxochelyidae, assuming that only a complex of generalised features distinguished them from the latter . But recent studies of the skull of macrobaenids has led to the discovery of important differences between macrobaenids and toxochelyids that are related to the formation of the canal for the internal carotid artery. It has also been stated that chelydrids are the direct descendants of macrobaenids (Chkhikvadze, 1973; de Broin, 1977), but features of the shell, for example, the absence of costiform processes of the nuchal, argue against this opinion. Chkhikvadze (1973, 1977) included some macrobaenids in his group 'Sinemydidae', but this is contradicted by the recent redescription of *Sinemys* (Brinkman and Peng, 1993b).

Generally, macrobaenids are small to medium sized (200–500 mm long) with relatively small, low skulls about 20–25% of carapace length. The skull (Figure 17.10), based mainly on *Hangaiemys* and *Macrobaena*, is wide and subtriangular. It has a temporal emargination of moderate size, with only a small contact between the squamosal and parietal, and a large cheek emargination with a very short contact between the jugal and quadratojugal. The prefrontals are large and in contact along their entire midline, while the frontals are reduced in size. The incisura collumella auris is not closed, distinguishing macrobaenids from sinemydids. The palate is of primitive type with a large posterior palatine foramen. There is a large epipterygoid. The pterygoids contact along the midline anteriorly, and the basioccipital posteriorly. The posterior foramen for the internal carotid canal is located posteriorly, and enclosed between the pterygoid ventrally and the prootic dorsally. Anteriorly, where the internal carotid artery divides into cranial and palatal branches, the canal carrying the artery opens ventrally on to the palatal surface of the skull via a special opening in that part of the pterygoid which, in most turtles, covers the basisphenoid and forms the base of the canal. Thus, there is a slot or an opening (foramen basisphenoidale), providing a view into the basisphenoid which contains a passage (foramen caroticum basisphenoidale) for the internal carotid artery leading into the brain cavity. This opening also reveals, anteriorly and somewhat laterally, between the basisphenoid and pterygoid, the lateral carotid foramen for the palatal artery. The trabeculae of the basisphenoid are short and widely spaced and the

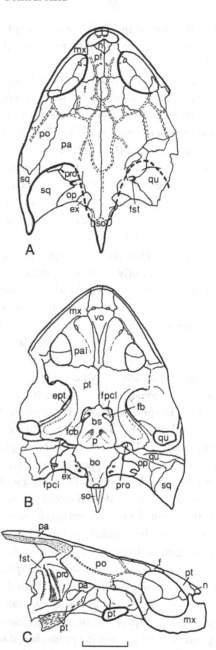

Figure 17.10. *Hangaiemys hoburensis* Sukhanov and Narmandakh, 1974, Döshuul Svita, Lower Cretaceous (Aptian–Albian), Höövör locality, Northern Gobi, Mongolia. Skull in (A) dorsal, (B) ventral and (C) left lateral view. (A) and (B) based on PIN 3334–4 and 3334–36, (C) based on PIN 3334–6. Abbreviations as in Figures 17.2 and 17.8 except for: fb, foramen basisphenoidale; fpci, foramen posterius canalis carotici interni; p, pit. Scale bar = 10 mm.

dorsum sellae is high and does not overhang the sella turcica. The stapedio-temporal foramen is well developed.

Neck mobility in macrobaenids is relatively small. The cervical vertebrae are short and high with large ventral keels and well developed articular surfaces on the centrum. The fourth cervical has convex condyles anteriorly and posteriorly: cervicals anterior to this vertebra are opisthocoelous, whilst the remaining cervicals are procoelous. All cervical vertebrae, except for the first, have double-headed free ribs. The rib of the first dorsal, as a rule, reaches only about the midlength of the first costal.

The shell is of mesochelidian type, relatively low and rounded posteriorly. Anteriorly, it has a shallow nuchal emargination, often outlined from above by a shallow medial groove. As a rule, the peripheral fontanelles are absent in adults, but there is a normal complement of bony elements. The nuchal has no costiform processes and the neurals are relatively narrow. The first or third may be quadrangular, while the rest are hexagonal, with short antero- or posterolateral edges. Despite occasional traces of the reduction of the eighth neural, the last costals do not contact along the midline. There are two large suprapygals and the postero-lateral peripherals are enlarged. The cervical is large with a trapezoidal shape and the first and fifth vertebrals are wide and hexagonal, while the second, third and fourth are relatively narrow, the width more or less equal to the length. The pleural–marginal sulcus passes along the edge of the shell or partially unites with the costal–peripheral suture, and only the twelfth marginal overlaps the second suprapygal. The plastron is relatively short, only 60–70% of carapace length, and has wide axillary and inguinal notches, giving it a crest-shaped outline. The plastron is connected to the carapace by ligaments and has special sockets on the peripherals for anterior and posterior buttress horns and smaller projections on the lateral edge of the bridge parts of the hyo- and hypoplastra. The bridges are relatively narrow, only 30–35% of plastron length. The epiplastra are narrow and elongated and have short contacts with each other. They are oriented latero-caudally

from the midline and embrace the sides of the hyoplastra. The entoplastrons were elongate and wedge-shaped, and probably had a mobile connection to the epiplastra. Special processes of the hyoplastra go under the edge of the entoplastron, entering grooves on their ventral surface. Occasionally, the central fontanelle may be retained in the plastron in adults. The xiphiplastra are large and embrace the hypoplastra from behind and partially from the sides. There is probably only one pair of gular scutes. The midline sulcus is straight and the humero-pectoral sulcus enters behind the entoplastron. The femoro-anal sulcus reached the hyoplastra medially and there are four pairs of inframarginals.

Hangaiemys from Mongolia (Sukhanov and Narmandakh, 1974) is the most numerous and best researched Early Cretaceous macrobaenid. Skulls of this form often exhibit excellent preservation (Figure 17.10) and a series of complete and fragmentary shells (Figure 17.11B) and other elements of the postcranial skeleton are now known. It is primarily on this form that we judge the whole group, not the later and more specialized type genus, *Macrobaena*. Two species have been described: *H. hoburensis* (Sukhanov and Narmandakh, 1974; Figure 17.11A) and *H. leptis* (Sukhanov and Narmandakh, in press; Figure 17.11B) from two horizons subsequent in age. Neither of these species is large, with shells only 250–350 mm long. The shell, as in the majority of macrobaenids, is oval, the width no more than 70% the length, and, with the exception of radial wrinkles which sometimes appear near the anterior and lateral edges of some scutes, lacks sculpture.

Kirgizemys is a typical turtle from the Early Cretaceous deposits of Middle Asia (Nesov and Khozatskii, 1977a, b). Two forms are known: the type species *K. exaratus* (Figure 17.12) from the Albian of Kirgizstan and *K. dmitrievi* from the Neocomian of Buryatia (Nesov and Khozatskii, 1981a, b). Despite the numerous remains, complete shells of this genus have not yet been found, but published reconstructions seem to be correct. *Kirgizemys* is medium-sized (250 mm long) with significant relief on the shell: in addition to very fine, meandering sulci on the carapace,

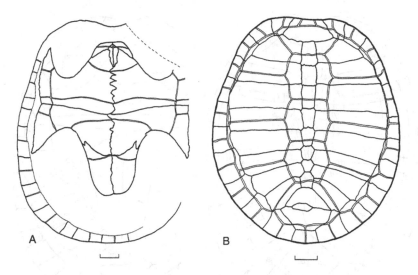

Figure 17.11. *Hangaiemys* Sukhanov and Narmandakh, 1974. (A) *H. hoburensis* Sukhanov and Narmandakh, 1974, reconstruction of the plastron based mainly on the holotype, PIN 3334–1, with epiplastra and entoplastron based on PIN 3334–4, Döshuul Svita, Lower Cretaceous (Aptian–Albian), Höövör locality, Northern Gobi, Mongolia. (B) *H. leptis* Sukhanov and Narmandakh, in press, holotype carapace, GIN 25/85, Upper part of Khulsangol Svita, Early Cretaceous (Albian), Hüren Dukh locality, Middle Gobi, Mongolia. Scale bar = 20 mm.

large folds connecting with the anterior and lateral edges of the central and pleural scutes are clearly visible. This genus is also distinguished from *Hangaiemys* by the presence of a wider cervical, small peripheral fontanelles in the carapace and rather large lateral fontanelles in the plastron. To this genus Nesov and Khozatskii (1978) assigned another species, *K. kansuensis* (Bohlin, 1953), known from fragmentary remains from Chia-yü-kuan locality, Gansu, and originally described under the name '*Osteopygis*'. This decision was based on similarities in the cross-sections of the anterior and bridge marginal scutes: they have upwardly-reflected thickened free edges, lacking the sharply expressed outline seen in *Hangaiemys hoburensis*. However, the marginal scutes of *H. leptis*, from Mongolia, have the same type of edge and, obviously, this character is not representative at the generic level. Corresponding remains of turtles described from the territory of Mongolia as *H. kansuensis* (Shuvalov and

Chkhikvadze, 1979) could belong to the latter species.

Asiachelys, from the Early Cretaceous of Mongolia, is known only on the basis of one plastron and some fragments of the carapace (Sukhanov and Narmandakh, in press; Figure 17.13). These remains seem to indicate that the shell was up to 300 mm in length and more rounded in shape when compared to most other macrobaenids. The most distinctive features are the width of the shell and the presence of middle vertebrals which are twice as wide as they are long. In addition, adults have large, round central and smaller semicircular lateral fontanelles, and small rhomb-shaped fontanelles at the boundary between the hypo- and xiphiplastron. Other characteristics include wide bridges, about 75% of the half-width of the plastron, and an anterior lobe of the plastron that is approximately 1.6 times shorter than the posterior lobe.

Ordosemys from the Early Cretaceous of Inner

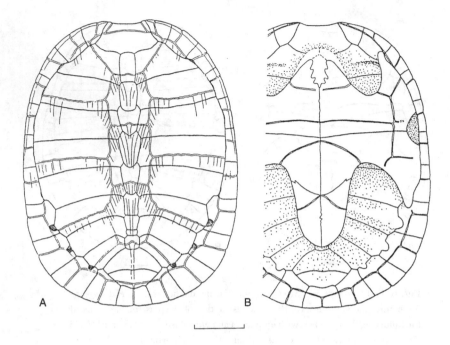

Figure 17.12. *Kirgizemys exaratus* Nesov and Khozatskii, 1978, Alamyshyk Formation, Lower Cretaceous, (Early–Middle Albian), Kylodzhun locality, South-Eastern Fergana, Kirgizstan. Reconstruction of the shell, based on single plates, in (A) dorsal and (B) ventral view. (After Nesov, 1988a.) Scale bar = 40 mm.

Mongolia (Brinkman and Peng, 1993a; Figure 17.14) is similar to *Asiachelys* in terms of the size and the shape of the shell (the width is similar to the length), the wide vertebrals and presence, in adults, of fontanelles in the plastron and wide bridges. However, *Ordosemys* is distinguished by the significantly smaller size of the central and lateral fontanelles, the first of which is only a little larger than the hypo-xiphiplastron fontanelle, while the last is long and narrow. Specific features of *Ordosemys* are the relatively wide and short nuchal, the presence of a preneural, which is individually variable in other macrobaenids (e.g. *Anatolemys*), and the presence of wide alveolar surfaces in the upper jaw, a feature that is reminiscent of *Macrobaena* from the Late Palaeocene of Mongolia. Contrary to the opinion of Peng and Brinkman (1993b), the elongate first dorsal rib of *Ordosemys*, which reaches to the distal extension of the axillary buttress, is not a general characteristic of macrobaenids because in other Asian representatives of this family it reaches only to the mid-point of the first costal, where it is connected with the latter by a broad base and a suture.

Parathalassemys, with its wide vertebrals and central fontanelle in the plastron, was described on the basis of fragments of one specimen from the Early Cenomanian of Kyzylkum, Uzbekistan (Nesov and Krasovskaya, 1984, table 4, figs. 9–12) and apparently belongs to the same branch of macrobaenids as the two genera mentioned above. This genus is distinguished by its relatively large size, with a carapace up to 500 mm long. Initially, *Parathalassemys* was ascribed, though without any evidence, to the Thalassemydidae, a family which is of doubtful validity (see Gaffney and Meylan, 1988).

Dracochelys from the Early Cretaceous of Sinkiang, China, is known only on the basis of the skull (Gaffney and Yeh, 1992; Figure 17.15). This was a large turtle, twice the size of *Hangaiemys*, with a condylar–basal

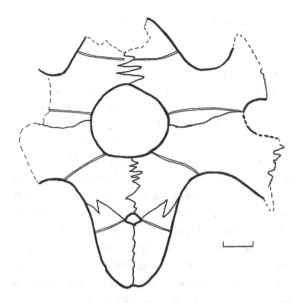

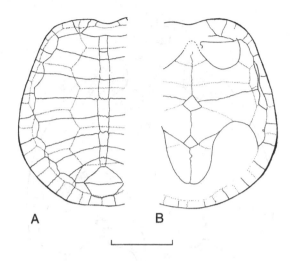

A   B

Figure 17.14. *Ordosemys leios* Brinkman and Peng, 1993. Reconstruction of the shell in (A) dorsal and (B) ventral view. Based on the holotype IVPP V9534–1 and IVPP V9534–3 and IVPP V9534–11 from the Lower Cretaceous of Laolonghuoze locality, Ordos, Inner Mongolia, China. (After Brinkman and Peng, 1993a.) Scale bar = 50 mm.

Figure 17.13. *Asiachelys perforata* Sukhanov and Narmandakh, in press, holotype plastron, GIN 25/87, Upper part of Khulsangol Svita, Early Cretaceous (Albian), Hüren Dukh locality, Middle Gobi, Mongolia. Scale bar = 20 mm.

skull length of 90 mm, compared to 45 mm in the latter. The skull of *Dracochelys* is similar to that of *Hangaiemys*, but has a more extended basisphenoid, which is arrow-shaped in ventral view, not sub-rectangular, and correspondingly, there is a very short medial contact between the pterygoids. In addition, the basisphenoid is noticeably broadened posteriorly, ending in blunt postero-lateral processes. A specific characteristic of these turtles is the presence on the upper jaw of tooth-shaped bony cusps between the premaxilla and maxilla.

Macrobaenids persisted into the Late Cretaceous where they are represented by middle-sized and large shells (up to 600–700 mm in length) of the turtle *Anatolemys* (Khozatskii and Nesov, 1977). *Anatolemys maximus*, the type species (Figure 17.16), comes from Late Turonian–Santonian deposits of the Fergana Valley, Tadzhikistan, while *Anatolemys oxensis* is from the Cenomanian–Turonian of Karakalpakia, Uzbekistan (Khozatskii and Nesov, 1977, 1979; Nesov, 1977d). *Anatolemys* has some features that are not char-

acteristic of macrobaenids, including the particular shape of the nuchal which has almost parallel lateral edges, a rectangular-shaped cervical and large, wide xiphiplastra, connected to the hypoplastra by a direct transverse suture. In addition, *Anatolemys* is distinguished from other Early Cretaceous macrobaenids by the shorter, straighter free end of the rib on the first costal, in which respect it is closer to *Macrobaena*.

Macrobaenids are not known from the Late Cretaceous of Mongolia and China, but appear again in the Late Palaeocene, represented by *Macrobaena* from lacustrine deposits of the Naranbulag Svita in South Gobi.

Family SINOCHELYIDAE Chkhikvadze, 1970
Type genus. *Sinochelys* Wiman, 1930. Lower Cretaceous (Neocomian) of China.

The family name was first published in 1970, but justifications and definitions of this taxon only came later (Chkhikvadze, 1976, 1983, 1985). Chkhikvadze suggested a new reconstruction of the plastron for the type genus (1983, fig. 55) on the basis of photographs

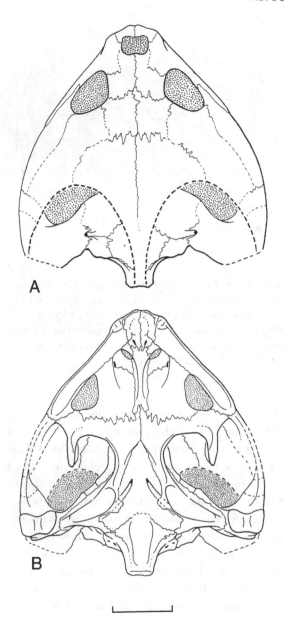

Figure 17.15. *Dracochelys bicuspis* Gaffney and Yeh, 1992, holotype, IVPP V4075, Upper part of Tugulo Series, Lower Cretaceous, Wuerho district, Xinjiang, China. Partially restored skull in (A) dorsal, and (B) ventral view. (After Gaffney and Yeh, 1992.) Scale bar = 20 mm.

published by Wiman (1930). He also believed that *Scutemys tecta*, described from the same locality by Wiman (1930), was a junior synonym of *Sinochelys*. In addition to the type genus, Chkhikvadze proposed that the family should include *Peishanemys* Bohlin, 1953, '*Nesovemys*' Chkhikvadze, 1985, and, probably, *Heishanemys* Bohlin, 1953, though the latter was described on insufficient material. '*Nesovemys*' was established for *Peishanemys testudiformis* Nesov, 1981, from the Early Cretaceous of Mongolia. However, Nesov (1986a, b) believed that this new genus could not be substantiated because the features used by Chkhikvadze are highly variable among turtles. Nesov suggested that *Peishanemys* should be placed in a separate family, the Peishanemydidae, believing that *Sinochelys* could not be used as a type genus because of its poor preservation and errors in the original definition of Sinochelyidae (Nesov and Verzilin, 1981). There may be some truth in this opinion, but it cannot be accepted because Sinochelyidae has priority.

Sinochelyids range up to 300 mm in length and have a relatively small skull (up to about 25% the length of the carapace) that is short, wide and low. The temporal emargination is weakly developed, there is no cheek emargination and the stapedio-temporal foramen is well developed. The rami of the lower jaw are short, while the symphysis is relatively long. According to Zangerl (1969), the shell is of mesochelidian type. The carapace is wide and relatively low, with a nuchal emargination. There are eight neurals, the first three pairs having short postero-lateral sides, and two large suprapygals. The anterior peripherals are thickened, the cervical is very small and sub-quadrilateral, and the first to fourth vertebrals are relatively narrow, the width more or less equal to the length. The pleuro-marginal sulcus occurs on the peripherals, but the eleventh and twelfth marginals far overlap the second suprapygal. The plastron is large and connected with the carapace by a suture. The buttresses are weakly developed, while the bridges are wide, corresponding to the short, wide lobes of the plastron. The anterior lobe has a straight anterior edge and slightly thickened epiplastral lips, while the posterior lobe appears to

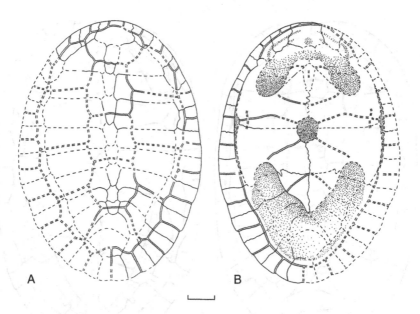

Figure 17.16. *Anatolemys maximus* Khozatskii and Nesov, 1979, holotype, PIN 2398/501. Reconstruction of the shell, on the basis of the anterior part of the carapace (holotype specimen) and isolated plates, in (A) dorsal, and (B) ventral view. (After Nesov, 1986a.) Scale bar = 60 mm.

bear a shallow, wide, anal emargination. In external view the entoplastron is very large and rhomboid in shape, the length more or less equal to the width, and reaches posteriorly to the level of the axillary notch. The epiplastra are also large and, unusually for turtles, the epihyoplastral sutures run antero-laterally from the lateral corners of the entoplastron. The xiphiplastra are large and wide and the hypo-xiphiplastral suture runs transversely. There are two pairs of scutes on the gular region of the anterior lobe of the plastron. The gulars are small, sub-quadrilateral, and located at the antero-lateral corners of the epiplastra. The intergulars are large, sometimes unpaired and overlap the entoplastron. The humero-pectoral sulcus crosses behind the entoplastron and the pectorals widen towards the midline. The scute sulci on the dorsal surface of the plastron are inset from its edge. There may have been small caudal scutes, at least in an extremely rudimentary state, as in *Peishanemys* (see Chkhikvadze, 1985). Finally, there are four well developed pairs of inframarginals, the largest being the third pair.

Sinochelyidae includes, as already mentioned, two genera: *Sinochelys* from the Neocomian Mengyin Formation of Shandong, China, and *Peishanemys* (Figure 17.18) represented by two species, one from Gansu, China (Bohlin, 1953), the second from the upper part of the Döshuul Svita of the Transaltai Gobi (Nesov and Verzilin, 1981), dated as Aptian/Albian (Shuvalov, 1982) and from Aptian–Albian deposits of the Chingshan Formation (Chen, 1983) in Shandong (Chow, 1954). All three species are known from single discoveries. The Chinese material has not been described in detail and many aspects remain uncertain (Chkhikvadze, 1983, 1985). Existing material clearly demonstrates morphological affinities between *Peishanemys* and *Sinochelys*. At the same time, distinctions between these taxa, at the generic level, if they exist at all, have yet to be demonstrated. Descriptions of *Peishanemys* (Bohlin, 1953; Chow, 1954) were not accompanied by comparison with *Sinochelys*, though Nesov (in Nesov and Verzilin, 1981) considered that comparison was not possible because of the poor preservation of *Sinochelys*.

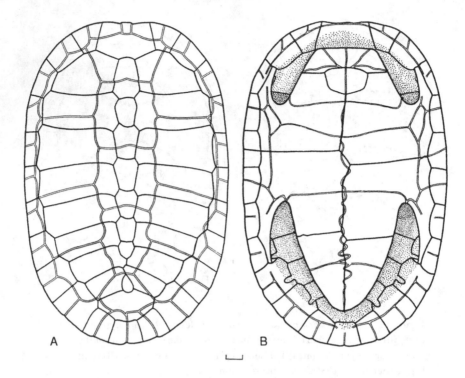

A   B

Figure 17.17. *Ferganemys verzilini* Nesov and Khozatskii, 1977, holotype, ZIN PNT n. F
67–7, Alamyshyk Formation, Lower Cretaceous, (Early–Middle Albian), Kylodzhun
locality, South-Eastern Fergana, Kirgizstan. Reconstruction of the carapace (A) and plastron
(B) based on holotype (an incomplete plastron) and isolated shell plates. (After Nesov and
Khozatskii, 1977.) Scale bar = 15 mm.

Chkhikvadze (1985) also failed to identify significant
differences as the majority of characters he cited can
be attributed to individual variation.

The systematic position of the Sinochelyidae is
unclear. Chkhikvadze (1985) considered sinochelyids
as most likely to be the ancestors of Testudinidae,
based on the probable presence, in both taxa, of rudi-
ments of caudal scutes. However, other serious argu-
mentation in favour of this idea has not been brought
forward. Nesov (in Nesov and Verzilin, 1981) considered that the Sinochelyidae belonged in the
Testudinoidea, and that although they may be close to
some groups of testudinoids, they cannot be their
ancestors because of the presence of some specialized
features. General features of 'terrestrial' organisation
shared by Sinochelyidae and Testudinidae, according
to Nesov and Verzilin (1981), include the general

shape of the shell, a shortened plastron, the probable
presence of osteoderms in the skin, a relatively short
humerus and femur, and the morphology of their
proximal epiphyses, although the degree of reduction
of the intertrochanteric groove on the femur, resulting
in the confluence of both trochanters, seems to have
been exaggerated by these authors. Similarities shared
with Platysternidae include significant development
of the skull roof in the temporal region, small development of the frontal, the morphology of the lower
jaw, with its slanting anterior region and strengthened
symphysis, and the development of a nuchal emargination in the carapace. The Sinochelyidae are distinguished from the Testudinidae by the morphology of
the temporal region of the skull and lower jaw, the
presence of two pairs of gular scutes, a complete set of
inframarginals and the weak thickening of the lips of

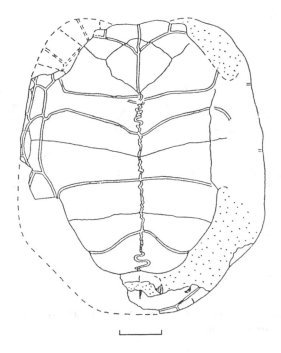

Figure 17.18. *Peishanemys testudiformis* Nesov, 1981, holotype plastron, PIN 46335/1 (ZIN, PNT M77–3), Döshuul Svita, Lower Cretaceous (Aptian–Albian), Dösh Uul II locality, Transaltai Gobi, Mongolia. (After Nesov and Verzilin, 1981.) Scale bar = 40 mm.

the epiplastron. They are distinguished from the Platysternidae by the sutural connection between the carapace and plastron, the morphology of the posterior part of the carapace, the greater doming of the shell, the different shape of the epiplastra and the presence of two pairs of gular scutes. It is also clear that neither genus assigned to the Sinochelyidae can be ascribed to the Dermatemydidae, as now defined.

The identity of the ancestors of the Sinochelyidae also needs to be solved. It is unlikely that they will be found in the Xinjiangchelyidae, Macrobaenidae or Sinemydidae which have a different shell construction with a loose connection between the carapace and plastron, narrow bridges, elongate epiplastra that are obliquely directed, a relatively small, elongate entoplastron and a low skull with strongly reduced temporal roof. At the same time, the skull is only known in one sinochelyid, *Peishanemys testudiformis*, and has not

been fully described, as the palatal region remains unprepared. All the groups mentioned are united only by plesiomorphic features such as the presence of two pairs of gulars and a complete set of well developed inframarginals.

Late Cretaceous turtles

Late Cretaceous deposits are widely distributed in southern Mongolia. The oldest (Cenomanian–Turonian) are found in the Eastern Gobi localities of Baishin Tsav, Amtgai and Khar Hötöl. The youngest occur in the South and Transaltai Gobi and include the Campanian localities of Bayan Zag, Yagaan Khovil, Zamyn Khond, Khulsan and Ukhaa Tolgod and the Maastrichtian localities of Nemegt, Altan Uul, Tsagaan Khushuu, Bambuu Khudag, Nogoon Tsav, Bügiin Tsav, Guriliin Tsav, Khaichin Uul and many others. Generally, these deposits have been assigned to three successive svitas: the Bayanshiree (Cenomanian–Santonian), the Baruungoyot (Santonian–Campanian) and the Nemegt (Maastrichtian) (Shuvalov, 1982). In the first two svitas, vertebrate assemblages divide them into lower and upper parts, though the division of the Baruungoyot into the lower or Djadokhta, and upper, strictly speaking, Baruungoyot (Shuvalov and Chkhikvadze, 1975; Jerzykiewicz and Russell, 1991) has recently been the subject of new discussions (Novacek *et al.*, 1994).

Turtles from the Upper Cretaceous deposits of Mongolia can be assigned to five separate complexes, containing different genera and species. However, the first four have similar faunal characters and are dominated by the Adocidae and Nanhsiungchelyidae, both represented, as a rule, by large turtles with moderately convex shells, and most likely adapted to life in streams. Later, in the Campanian, where, as a rule, turtles are found only as single specimens, these taxa are joined by the Lindholmemydidae, also water turtles of small to middle size with thickened plates and strengthened buttresses of the plastron. The fifth complex, represented by turtles often found in mass burials, is characteristic of the youngest deposits (ascribed to the Nemegt Svita), and differs sharply

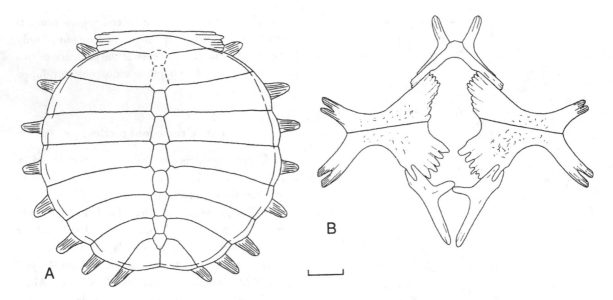

Figure 17.19. '*Trionyx*' *kyrgyzensis* Nesov, 1995, Alamyshyk Formation, Lower Cretaceous (Early–Middle Albian), Kylodzhun locality, South-Eastern Fergana, Kirgizstan. (A) Carapace and (B) plastron. Scale bar = 20 mm.

from the older complexes in that the dominant forms belong to the Mongolochelyidae, large flattened turtles, most probably inhabitants of still water basins, and the Lindholmemydidae. By contrast, the Adocidae and Nanhsiungchelyidae are rare, and turtles of the family Haichemydidae appear for the first time. Analogues of the fifth complex are not found in Middle Asia or China. Trionychids are found in all the complexes, but are most common in the Bayanshiree and Nemegt assemblages.

Nesov (1984) considered the Late Cretaceous turtles of Middle Asia to belong to five successive assemblages. The first is dated, according to Nesov, as ?Albian–Cenomanian, the second as Upper Cenomanian and, probably, Lower Turonian, the third as Upper Turonian–Coniacian and the fourth as Santonian–Campanian. A separate assemblage from Late Cretaceous deposits of South Kazakhstan is dated as Upper Santonian–Campanian, but, apart from mentions of unidentified middle-sized turtles and remains of very large chelonioids, it is poorly known. The first four complexes are undoubtedly similar in that they all contain adocids, usually the

dominant form, and nanhsiungchelyids, known only as rare fragments from the Cenomanian. In contrast to contemporaneous assemblages from Mongolia, carettochelyids are rare in the Cenomanian and macrobaenids and adocids form dominant grous, represented by the rather small *Ferganemys* in Cenomanian–Lower Turonian deposits and the larger *Anatolemys*, found in all Late Cretaceous assemblages. The aberrant Late Cretaceous turtle *Shachemys* is unique to Middle Asia and placed by Nesov (1977c) in a special subfamily, the Shachemydinae, in the Adocidae. The Lindholmemydidae and Trionychidae are ubiquitous, though, unlike Mongolia, the former, unlike the latter, never become dominant. Nesov noted a significant change in turtle assemblages at the Early to Late Turonian boundary. Many genera disappear, diversity is reduced and, instead of small, thin-shelled forms such as *Ferganemys*, *Anatolemys oxensis*, *Adocus kizylkumensis* and *Kizylkumemys*, there appear large turtles (*Anatolemys maximus*, *Shachemys*, *Adocus aksary*, *A. foveatus* and large forms of *Tryonix*) and forms such as *Lindholmemys* with thick, strong shells. The most famous localities for Late Cretaceous Middle Asian turtles and other vertebrates are Kansai in Western Fergana, Tadzhikistan, and Shakh-Shakh in South Kazakhstan, which produce Early Santonian–Campanian assemblages.

Upper Cretaceous deposits of China have yielded small numbers of turtles, mostly from the early Late Cretaceous and often described on the basis of poor material, which makes it difficult to compare them with Mongolian remains. Researchers from China divide the Chinese deposits into three biostratigraphic horizons, named after the dominant dinosaurs found therein. These are: the *Bactrosaurus* assemblage, found in the Irendabasu Formation of Inner Mongolia, which has yielded the trionychid *Khunnuchelys* from the locality of Erenhot (Brinkman and Nesov, 1993), the *Protoceratops* assemblage from the localities of Chia-yü-kuan and Hui-hui-p'u in Gansu (Bohlin, 1953), and the *Nanhsiungosaurus* assemblage. The first two assemblages are correlated with the Bayanshiree and Baruungoyot Gorizonts of Mongolia, though it is possible that at localities in Gansu, deposits which have produced *Microceratops sulcidens* and turtles (Maryańska and Osmólska, 1975) may be older than the Djadokhta of Mongolia and, according to Nesov (1984), may be Cenomanian–Lower Turonian in age and thus equivalent to the lower part of the Bayanshiree Svita. The third assemblage is found in deposits of the Nanxiong Formation of the Nanhsing group and the Guandun and Subashi Formations of the Turfan depression, Xinjiang. This assemblage, which commonly contains turtles such as *Nanhsiungchelys*, is dated as Campanian–Maastrichtian, but has nothing in common with the Nemegt assemblage of Mongolia.

Family ADOCIDAE Cope, 1870

Type genus. Adocus Cope, 1870. Late Cretaceous–(Campanian–Maastrichtian)–Palaeocene of North America. It has been suggested that this genus also occurs in Central Asia from the Upper Albian to Eocene (Gilmore, 1931; Nesov, 1977c, 1995; Khozatskii and Nesov, 1977; Sukhanov, 1978), but this is disputable.

Adocidae was established by Cope, but was subsequently included in Dermatemydidae Gray, 1870

(Hay, 1908), after similarity was found with *Dermatemys*. This point of view became wide spread and appeared in a number of fundamental works (Williams, 1950) and surveys (Romer, 1956; Sukhanov, 1964; Młynarski, 1969). However, descriptions of the skulls of the adocids *Ferganemys* (Nesov, 1977c) and *Adocus* (Meylan and Gaffney, 1989) confirmed the independent status of Adocidae. Various diagnoses of this family have been given (Nesov, 1977c; Chkhikvadze, 1987), depending on the size of the group. According to Nesov (1977c) the Adocidae must unite *Adocus* (including *Alamosemys* and *Zygoramma*, according to Meylan and Gaffney, 1989) *Ferganemys*, and, separately, *Shachemys*. However, Meylan and Gaffney (1989) identified the latter as Eucryptodira, incertae sedis. At the same time, Chkhikvadze (1987), and Gaffney and Meylan (1988) included in the Adocidae turtles which belong to the Nanhsiungchelyidae.

Here, Adocidae is considered to include the North American *Adocus* (including *Alamosemys* and *Zygoramma*) and the Central Asian forms *Ferganemys*, *Adocoides*, *Shineusemys*, *Mlynarskiella*, '*Plesiochelys*' *tatsuensis* and, doubtfully, *Shachemys*. A number of other species from Central Asia, currently assigned to *Adocus*, are problematic. '*Adocus*' *orientalis* Gilmore, 1931, from the Upper Eocene of Irdyn Manga, Inner Mongolia, China, and the Zaisan depression of Eastern Kazakhstan (Chkhikvadze, 1973, 1976) and from the Lower Oligocene of Ergiliin Zoo, Eastern Gobi, Mongolia, is only represented by plastra, which lack characteristic features of *Adocus* (Khozatskii and Nesov, 1977). Other species, including '*Adocus*' *kizylkumensis* from the Upper Albian of Karakalpakia (Nesov, 1981a, 1984), '*Adocus*' *askari* Nesov (Nesov and Krasovskaya, 1984) from the Upper Turonian Coniacian of Central Kyzylkum, Uzbekistan, and '*Adocus*' *foreatus* (Nesov and Khozatskii, 1977a) from the Santonian of Western Fergana, Tadzhikistan, are based on fragmentary remains. These remains are insufficient as a basis for particular species, but exhibit characters of adocids, such as sculpturing and significant overlap of the lateral marginals on the costals, though, as is now clear, the latter feature is not restricted to *Adocus*.

Adocids achieved their greatest diversity in Central Asia, first appearing in the Late Jurassic (Nesov and Khozatskii, 1981a) and persisting until the Early Oligocene, whereas in North America they appear only in the Campanian and disappear towards the end of the Palaeocene. Asian adocids reached peak diversity at the start of the Late Cretaceous, but became relatively rare by the Maastrichtian. In Mongolia, adocids are as yet unknown from the Maastrichtian, but fragments of indeterminate adocids have been found in the Palaeocene of Naran Bulag and more complete remains of '*Adocus*' *orientalis* are known from the Upper Eocene of China and the Lower Oligocene of Mongolia. In Middle Asia, adocids are usually found in brackish water deposits of estuaries, which also produce sharks and skates. In North America they are found in marginal marine deposits, and in Mongolia and adjoining regions in sediments of inner continental freshwater basins. In the latter case, adocids are predominantly found in regions which experienced a sub-arid climate and were dominated by dune landscapes and fluvial systems. There was a sharp decrease in diversity following the appearance of large lake type basins in the Maastrichtian.

Early adocids were small, about 150 mm long, with a relatively small skull that reached a maximum of about 20% of carapace length. Later adocids reached large sizes, up to 500–700 mm in length. The head was covered with horny scutes, but the bones of the skull roof are smooth, without sculpture. The temporal emargination is deep, and the contact between the squamosal and postorbital is lost. The parietal participates in the processus trochlearis oticum, which is wide and projects strongly. The frontal forms a large part of the orbital margin. The basisphenoid is separated from the vomer and palatine by midline contacts of the pterygoids. The posterior foramen for the internal carotid canal is displaced far posteriorly and formed only by the pterygoid. The foramen caroticum basisphenoidale is relatively large, as is the stapediotemporal foramen, and the diameter of the canal for the internal carotid, which pierces the basisphenoid, is approximately equal to the diameter of the lateral carotid canal. The incisura collumella auris is narrow

and characteristically bowed, but not enclosed, and basal tubercles are present. All the cervical vertebrae are opisthocoelous and the centra of the dorsal vertebrae, together with the heads of the corresponding ribs, are moderately reduced. The humerus is about 17% the length of the carapace and has a relatively small head and thin shaft. The forearm is about 70% of humerus length. The manus is typical for freshwater turtles with a phalangeal formula of 2–3–3–3–3, and all the terminal phalanges are clawed. The shell is elongate and moderately convex, but without keels. The carapace has a sutural connection with the plastron and the plastral buttresses are thin, directed obliquely antero-posteriorly, rather than upward, and contact only the peripherals. The plates forming the shell are covered with a fine sculpture in the shape of small tubercles (which form rows whose orientation can change depending on their location), separated by shallow pits (from 6–10 per cm) forming a punctate sculpture. Moreover, the plates are penetrated by very small pores for blood vessels (averaging up to 30–35 pores per 0.25 cm^2), which may or may not coincide with the sculpturing.

The nuchal lacks costiform processes and the neural formula is: 6-4-6-6-6-6-0-0 or 6-4-6-6-6-5-0, that is, the posterior costals may contact each other. There are two suprapygals: the first is small, the second very large. The anterior peripherals have a height which slightly exceeds the width of the free margin. Further posteriorly, the height increases, reaching a maximum in the sixth and seventh peripherals, which are 1.5 times the height of the anterior plates: subsequently, the height decreases again. The posterior (eleventh and twelfth) marginals and often the lateral marginals, from the fourth or fifth, overlap the corresponding costals and second suprapygal to some extent. The plastron is wide and relatively short (65–75% of carapace length), with an almost complete set of plastral plates, lacking only the mesoplastra. The anterior lobe of the plastron has, as a rule, a straight anterior margin, as does the posterior lobe, which can also be rounded, but lacks an anal notch. The anterior margins of the epiplastra are not thickened. The entoplastron is large, its width exceeding its

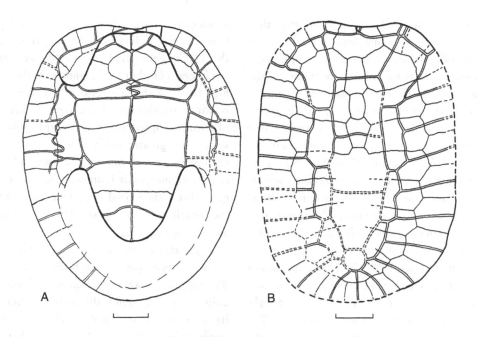

Figure 17.20. *Adocoides amtgai* (Narmandakh, 1985), upper part of Bayanshiree Svita, Upper Cretaceous, (Turonian–Lower Santonian), Amtgai locality, Eastern Gobi, Mongolia. (A) Holotype plastron, PIN 3648–2. (Based on Narmandakh, 1985.) (B) Incomplete carapace, PIN 3648–3. Scale bar: A = 50 mm, B = 60 mm.

length; posteriorly it reaches the level of the axillary notch. The skin-scute sulci approach the free margin of the plastron. There are a couple of gulars, though these are always separated by the intergulars which may, sometimes, be unpaired. The pectorals are usually narrow, and shorter in length near the midline, broadening somewhat medially. The abdominals are the largest plastral scutes and, occasionally, the femorals approach them along the midline. The anals do not overlie the hypoplastron and the medial sulcus is gently meandering. There is a complete row of four (sometimes three) pairs of inframarginals.

Meylan and Gaffney (1989) consider the adocids to be relatively primitive representatives of the superfamily Trionychoidea, including them in the epifamily Trionychoidae, the sister group to the Kinosternoidae. At the same time, the Adocidae were identified as the sister group to Nanhsiungchelyidae and Trionychia (Carettochelyidae and Trionychidae). Furthermore, the Adocidae share some features in common with Testudinoidea as Nesov (1977c) has emphasised, including adocids in the latter taxon.

Adocoides Sukhanov and Narmandakh, in press (Figure 17.20) was established on the basis of *Adocus amtgai* Narmandakh, 1985, from the upper part of the Bayanshiree Svita of Eastern Gobi, Mongolia. This is a middle-sized turtle, up to 450 mm long, with a maximum skull length of 75 mm. The carapace is oval and elongate. The height (length) of the peripherals increases from front to back, reaching a maximum at the seventh where the height is 1.5 times the length of the free margin and 1.6 times the average height of the first peripheral. Subsequently, peripheral height decreases, with the eleventh the lowest. The cervical is absent, or remains an insignificant rudiment displaced to the anteriormost margin of the carapace. The second to fourth vertebrals are relatively wide (width equals length), and the fifth is small. The first and second marginals are low, 2.5 times wider than they are tall. By contrast, the lateral and posterior marginals are very tall, 2–2.2 times higher than they are wide and they overlap one third of the length of the corresponding costals and second suprapygal. There is a sudden increase in height beginning at the fourth marginal, which has an unusual triangular shape (as does the twelfth marginal), unlike *Adocus*, where it is square. Correspondingly, the first pleural has an unusually elongate shape with an obliquely oriented long axis. The remaining pleurals are extremely short, their width exceeding three times their length. The skin-scute sulci overlap the ventral surface of the carapace, remote from the free margin and approaching the free margin only anteriorly along the midline. The plastron is large, reaching 75% the length of the carapace, compared to 65–70% in *Adocus*, and almost reaches the anterior margin of the carapace. The anterior lobe is shorter than the posterior and has a straight anterior margin, while the latter is rounded. The intergulars are paired and completely separate the gulars, exceeding them in size and reaching to the entoplastron. Altogether, the gulars and the intergulars occupy a little more than half the surface of the epiplastra. The pectorals are very narrow, reaching or slightly overlapping the entoplastron. The abdominals are correspondingly large and 1.5 times longer than the femorals near the midline. There are three pairs of large inframarginals, four if an additional small, asymmetrically located scute appears. The sulcus separating the marginals and inframarginals is strongly meandering. The inframarginals reach a line connecting the centres of the axillary and inguinal notches medially and the margins of the plastron laterally.

Adocoides differs from *Adocus* in the complete, or almost complete, reduction of the cervical and the appearance of a contact between the first marginals, the relatively wide second to fourth vertebrals, the significantly higher lateral and posterior (fourth to twelfth) marginals, the very short second to fifth pleurals, the very narrow fifth vertebral, the larger size of the plastron, the smaller size of the intergulars and gulars in relation to the surface of the epiplastron, the greater narrowing of the pectorals and corresponding greater length of the abdominals along the midline, and the greater width of the inframarginals as a result

of which the marginals only slightly overlap the plastron. *Adocoides* differs from *Ferganemys* in the width of the skull, its generally larger size, the greater height of the lateral peripherals (beginning at the fourth), the high lateral marginals, reduction of the cervical, the relatively smaller size of the outer surface of the epiplastron, occupied jointly by the intergular and gular scutes, the greater narrowing of the pectorals and expansion of the abdominals, and the width and shape of the inframarginals. It differs from *Mlynarskiella* in its general size, the paired intergular scutes, which reach the entoplastron, and the smaller size of the gulars relative to the intergulars.

Mlynarskiella (Shuvalov and Chkhikvadze, 1986) from the upper part of the Bayanshiree Svita of the Transaltai Gobi was described on the basis of a single epiplastron. This genus differs from other adocids in its small size (the shell is about 150–180 mm long), the unpaired intergulars, which do not reach the entoplastron, and the relatively large gulars.

Shineusemys (Sukhanov and Narmandakh, in press; Figure 17.21) from the lower part of the Bayanshiree Svita of the Eastern Gobi, Mongolia, is a medium-sized turtle with a shell up to 300 mm in length. The neurals are relatively narrow and the first suprapygal has a high and narrow trapezoid shape. The second is noticeably wider: its width is 2.4 times its length and it is twice as wide as the first suprapygal. There is a cervical and the plastron is long, reaching 80% the length of the carapace. The anterior lobe is wide and short, and truncated anteriorly, while the posterior lobe is elongate with a narrower base and rounded posteriorly. Anteriorly, the plastron almost reaches the edge of the carapace: posteriorly it terminates at 80% the length of the carapace. Correspondingly, the axillary notch is short and narrow, the inguinal notch long and wide. The bridges are relatively narrow (up to 45% the width of the plastron and 35% of its length, compared to 50–57.5% and 40% in *Adocoides*). The epiplastra are large, unlike those of *Adocoides*, have a long contact with the entoplastron and are half the length of the interepiplastral suture. The entoplastron is large, with a broad rhombic shape and the length is only 70% of the width. There are gulars and inter-

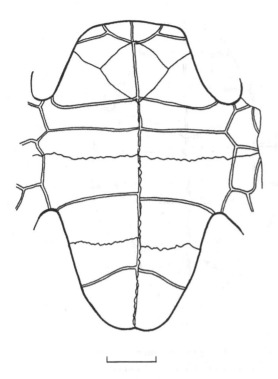

Figure 17.21. *Shineusemys plana* Sukhanov and Narmandakh, in press, holotype plastron, PIN 4636–1, lower part of Bayanshiree Svita, Upper Cretaceous (Cenomanian–Lower Turonian), Shine Us Khudag locality, Eastern Gobi, Mongolia. Scale bar = 40 mm.

gulars which, together, occupy no more than half the surface of the epiplastra. The intergulars separate the gulars and reach posteriorly to the entoplastron. The pectorals are narrow, but widen medially, and slightly overlap the entoplastron. The abdominals are relatively narrower than in other adocids, and their medial length is less than that of the femorals. There are four pairs of inframarginals: the first is small, while the second to fourth are almost equal in size and relatively large and wide. The marginals slightly overlap the plastron.

Ferganemys (Nesov and Khozatskii, 1977a; Figure 17.17) is known from the Albian of Fergana and the Cenomanian of the Central Kyzylkum. According to Nesov (Nesov and Yulinen, 1977; Nesov and Khozatskii, 1981a), '*Plesiochelys*' *tatsuensis* from the Upper Jurassic of Sechuan, China, should also belong

to this genus. Judging by the illustrations published by Yeh (1963, Pl. I: figs. 3–4; Figure 17.7) the assignment of the latter taxon to Adocidae seems correct, but there is no reason for its synonymy with *Ferganemys*. The evidence in favour of uniting this genus and *Shachemys* in the subfamily Shachemydinae is very weak (see Meylan and Gaffney, 1989).

Ferganemys is represented by a small Early Cretaceous species, *F. verzilini* (Figure 17.17), up to 200 mm in length, and a middle-sized species, *F. itemirensis* from the Cenomanian, which has a relatively elongate, narrow skull (maximum width only 65% of skull length) and a body length of up to 400 mm (Nesov, 1988, fig. 5). The suture between the prefrontal and the frontal is not transverse. The ventral edge of the upper jaw is even, without a toothed margin and the alveolar surface is narrow and groove-shaped, without additional crests. The posterior palatine foramen is of moderate size and situated between the palatine and pterygoid: the latter does not contact the vomer. The bony plates of the shell are thin. The height of the anterior peripherals is a little less than their width along the free edge. There is an increase in height posteriorly, reaching a maximum in the region of the ninth peripheral, where the height is a little greater than the width. The cervical is well developed, long and narrow. The width of the second to fourth vertebrals more or less equals the length, but the fifth vertebral is wider. The pleuro-marginal sulcus almost always lies on the edges of the peripherals. The posterior marginals (eleven and twelve) are relatively high and overlap the second suprapygal and, to a small extent, the eighth costal. The skin-scute sulcus is remote from the posterior edge of the inner surface of the carapace and does not approach this margin along the midline, as in *Shineusemys*. The plastron terminates well short of the anterior edge of the carapace. Its anterior lobe is short and truncated, the posterior lobe narrows. The bridges are wide, about 50% the width of the plastron. There are two pairs of gulars: together they occupy almost the entire surface of the epiplastra. In addition, the gulars are larger than the intergulars, which reach or slightly overlap the entoplastron. The pectorals are, relatively speaking, wider than in

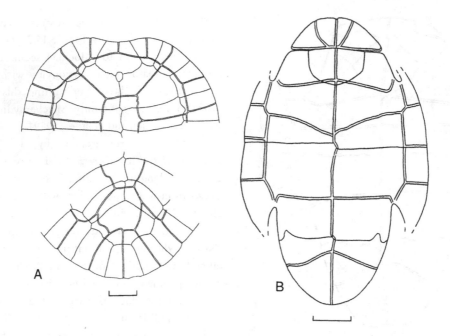

Figure 17.22. *Shachemys baibolatica* Nesov, 1984. Reconstruction of the anterior and posterior parts of the carapace (A) and plastron (B), based on isolated shell plates from the upper part of the Taikarshin Beds, Upper Cretaceous (Upper Turonian–Coniacian), Dzharakhuduk locality, Central Kyzylkum, Uzbekistan. (After Nesov, 1986a.) Scale bar = 40 mm.

Adocus and *Adocoides*, anteriorly they occasionally reach the entoplastron. The abdominals and femorals are more or less equal in length along the midline and there are four pairs of inframarginals. The latter are narrow, and relatively closer to the lateral edge of the plastron than they are to a line connecting the centres of the axillary and inguinal notches. The marginals virtually do not overlap the plastron.

Shachemys (Kuznetsov, 1976; Figure 17.22) is a middle-sized turtle, 300–400 mm in length, with a flattened carapace and wide shallow nuchal emargination. The surface is smooth, without keels and the bony plates of the shell are thin. Sculpture is practically absent, except for a weakly expressed tuberculation, and stroke-like furrows, but a system for blood vessels is present in the shell elements and the canal openings are significantly wider than in other adocids. The nuchal has a rounded posterior edge without costiform processes. All, or almost all, the neurals are

lost and all the costals contact along the midline. There is one large suprapygal. The anterior peripherals are thickened, unlike those to the posterior which are very thin. The buttresses are thick, but short, and articulate with a massive third peripheral and possibly the seventh. The cervical is absent and the first marginals have a long midline contact, occupying more than one third of the length of the nuchal. The first vertebral is wide, trapezoid-shaped and has a long basis anteriorly. The remaining vertebrals are noticeably narrower, and longer than wide. The pleuro-marginal sulcus runs along the peripheral plates. Only the eleventh and twelfth marginals show a sharp increase in height and reach over almost one half the length of the eighth costal and the posterior third of the suprapygal. The skin-scute sulcus approaches the peripherals along their ventral surface and near the very edge of the plastron. The bridges reach about 42% the width of the plastron, which is large. The anterior and posterior lobes are very wide, the former 1.5 times longer than the latter, and rounded, and the axillary and inguinal notches are correspondingly narrow. The epiplastra are large and sub-triangular and their contacts with the hyoplastra and entoplastra form a straight transverse line, which corresponds to the straight anterior edge of the wide entoplastron. The area occupied by the gulars and inter-gulars coincides completely with the outer surface of the epiplastra. All this suggests a hinge-like connection. The humero-pectoral sulcus enters the entoplastron or reaches almost to its centre. The pectorals are wide, and of similar width to the femoral scutes; posteriorly they approach the midline coming close to the hyo-hypoplastron suture. The medial sulcus on the plastron is practically straight and, contrary to Meylan and Gaffney (1989), there are four large pairs of inframarginals. The complete reduction of the neurals, absence of the cervical and the development of a hinge in the anterior lobe of the plastron reflect the special nature of *Shachemys*.

Family NANHSIUNGCHELYIDAE Yeh, 1966

Type genus. Nanhsiungchelys Yeh, 1966. Nanxiong Formation (Nanhsing Group), Upper Cretaceous (Maastrichtian), Guandun, China.

Yeh (1966) established this family for *Nanhsiungchelys*, but did not make any comparisons with *Basilemys*, from North America, even though fragments identified as *Basilemys* sp. had previously been reported from Asia by Tokunaga and Shimizu in 1926. Subsequently, similar fragmentary remains from the territory of Middle Asia, Mongolia and China were ascribed to *Basilemys* (Sukhanov and Narmandakh, 1975; Nesov, 1981b, 1984; Chkhikvadze, 1987), although, as has become clear recently, there are no grounds for this. In 1972 Młynarski described a new turtle, *Zangerlia*, from Mongolia, considering it a side branch of the Dermatemydidae with adaptations to a terrestrial way of life analogous to those of the Testudinidae. In 1977, following their detailed description of *Basilemys orientalis*, Sukhanov and Narmandakh compared *Nanhsiungchelys*, *Zangerlia* and *Basilemys* and included them all in a single genus,

Basilemys, in the family Dermatemydidae. Meylan and Gaffney (1989) rejected this, considering the Nanhsiungchelyidae as a distinct family in the epifamily Trionychoidae and a sister group to *Peltochelys*, Carettochelyidae and Trionychidae. Nanhsiungchelyidae includes the following genera: *Nanhsiungchelys*, *Zangerlia*, *Hanbogdemys* and *Bulganemys* from Asia and *Basilemys* from North America. The two species *B. sinuosa* and *B. praeeclara* should probably be placed in separate genera. All Asian remains previously identified as *Basilemys* sp. should be treated as Nanhsiungchelyidae gen. et sp. indet.

The history of this family is extremely short: the earliest representatives appear in the Late Albian of Middle Asia (Nesov, 1984) and the lower part of the Bayanshiree Svita (Cenomanian–Turonian) of Mongolia (Shuvalov and Chkhikvadze, 1979), but their remains are unknown in Middle Asia from the Late Turonian onwards. In Mongolia they are known from the upper part of the Baruungoyot Svita (Campanian), but Maastrichtian age remains are extremely rare (Młynarski and Narmandakh, 1972). In China, the Nanhsiungchelyidae are known only from the Maastrichtian, whereas in North America they occur in the Campanian and remained quite diverse in the Maastrichtian (Langston, 1956).

The characters of the Nanhsiungchelyidae are not yet completely clear because the skull is so far known only in *Nanhsiungchelys* (Yeh, 1963). However, Jerzykiewicz *et al.* (1993) recently reported on the discovery of several skulls of Asian *Basilemys*-like turtles and these may provide much more detailed information for this family.

Nanhsiungchelyids were large, up to 1 m long, with large heads up to one third the length of the carapace. Bones which covered the skull, lower jaw, shell and, in some cases, the ischial symphysis of the pelvis bore a sculpture composed of large cells (up to 3–4 per centimetre), separated by crests with pyramidal elevations at the points where they connected (termed large-celled, or 'pock-mark' sculpture).

The skull was covered with horny scutes. In *Nanhsiungchelys* the temporal region is well developed, the cheek emargination is completely absent, with the temporal emargination only weakly represented. There is an extensive contact between the parietal and squamosal. The rostral region of the skull is narrow and elongate as a result of the development of the prefrontals and the maxillae which form a distinctive short, wide tube. Consequently, the oral opening is displaced onto the ventral side of the head, and the orbits are located almost at the mid-length of the skull. In addition, they face laterally and are not visible from above. The vomer seems to have been reduced (Meylan and Gaffney, 1989) and does not contact the basisphenoid. In addition, the pterygoids seem not to contact along the midline and the palatines contact the basisphenoid. The external pterygoid processus is absent.

The carapace is oval, the width reaching 70–90% of the length, and has a large nuchal notch. The nuchal has costiform processes and the neurals form a complete row (neural formula: 6-4-6-6-6-6-6-6), such that the posterior costals do not meet along the midline. There are two suprapygals: usually the first is relatively small, while the second is large and wide. The plastron is large (85–95% of carapace length) with broad bridges and connected to the carapace by a suture. The buttresses are short and thick, reaching only the peripheral plates. The epiplastra are strongly thickened along their anterior edges which form epiplastral lips. The entoplastron is large and usually hexagonal, the width exceeding the length; posteriorly it may extend beyond the level of the axillary notch. The xiphiplastra are thickened along their lateral edge. There are two pairs of gular scutes. Between them the intergulars may merge into a single scute, while the gulars themselves may be more or less reduced. The humerals are wide laterally, but narrow near the midline, while the pectorals are wide medially, where they extend far over the entoplastron, but narrow laterally. The marginals also extend far over the plastron, but the inframarginals are more or less reduced. Except on the midline near the xiphiplastra, the skin-scute sulcus of the anterior and posterior lobes of the plastron is located far from the free edge.

The forelimbs are strong: the humerus reaches 23–27% the length of the carapace, the forearm is shortened (the radius is only 40% the length of the humerus) and the metacarpals and digital phalanges are extremely shortened, but terminate in large phalanges, 65% the length of the forearm and five times longer than the penultimate phalanges. The phalangeal formula is 2-2-2-2-1.

Młynarski (1972) considered, without doubt, that *Zangerlia* was a terrestrial turtle. He cited the presence of a convex shell (though without estimating the degree of convexity), the shortened phalanges in the manus digits, the coincidence of the pleuromarginal sulcus and costo-peripheral suture in the holotype of *Zangerlia testudinimorpha* (though this may be explained by the fact that this turtle is a juvenile: the length of the shell is approximately 270 mm), and the incomplete formation of the inner blade of the peripheral plates (their height increases with age). It was later argued by Sukhanov and Narmandakh (1977) that the presence of powerful forelimbs, the construction of the humerus and range of its possible movements, and the construction of the pelvis and its position in relation to the carapace in *Hanbogdemys orientalis* is not consistent with a terrestrial mode of life. These turtles were probably specialized swimmers, using the forelimbs to move on the bottom and to cling to the substrate under the conditions of strong currents (Nesov, 1981b). The aqueous mode of life also corresponds to a significant flattening of the shell in all American species of *Basilemys* (the height is not more than 25% of shell width), though Asian forms do have higher shells.

Hanbogdemys (Sukhanov and Narmandakh, in press; Figure 17.23), established on the basis of *Basilemys orientalis* (Sukhanov and Narmandakh, 1975), from the upper part of the Bayanshiree Svita (Turonian–Santonian) of the Eastern Gobi, Mongolia, represents the earliest nanhsiungchelyid identified to the generic level. Older findings are fragmentary and not determinable beyond the family level. *Hanbogdemys* was a large turtle, up to 700 mm long, with a high shell (40% the maximum width of carapace) that was also oval and somewhat elongate (width = 80% of shell length). The sickle-shaped nuchal notch is relatively narrow, up to 30% the width of the carapace and short (5% the

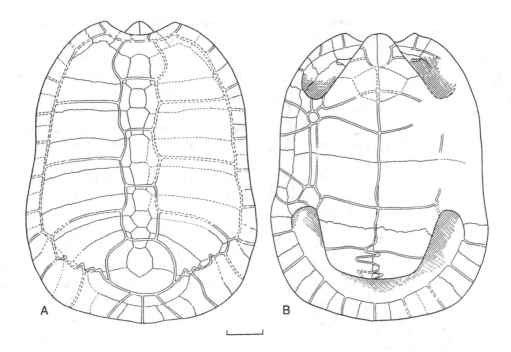

Figure 17.23. *Hanbogdemys orientalis* (Sukhanov and Narmandakh, 1975). Reconstruction of the carapace (A) and plastron (B) based on PIN 3458–3 from the upper part of the Bayanshiree Svita, Upper Cretaceous (Late Turonian–Early Santonian) of Baishin Tsav, Eastern Gobi, Mongolia. (After Sukhanov and Narmandakh, 1977.) Scale bar = 80 mm.

length of the carapace). Asian nanhsiungchelyids are united by this character and the shape of the first peripherals, wherein the angular free edge, marking the lateral edge of the notch, projects forward, unlike the condition in American forms where the edge of the carapace is rounded. The carapace bears a longitudinal groove, but a medial keel is absent. In the area of the bridges the free edge of the shell forms an acute angle in cross-section. The nuchal is relatively narrow, 29% the width of the carapace, and the costiform processes reach the second peripheral. The first suprapygal is small and heptagonal, the second very large and half-moon shaped. The ninth and tenth peripherals are relatively high, the eleventh and twelfth lower. The cervical is relatively large and wide. The first vertebral narrows anteriorly and does not reach the limits of the nuchal, the second to fourth are narrow and the fifth is large and rounded. The first marginal has an unusual shape, extending along the edge of the nuchal

notch and clasps the corner of the free edge of the first peripheral, mentioned above. The pleuro-marginal sulcus probably coincides with the costo-peripheral suture. The tenth to twelfth marginals are noticeably enlarged in height and the last two extend far onto the suprapygal. The anterior lobe of the plastron is wedge-shaped and the epiplastra are large with a long interepiplastral suture, more or less equal to the epi-entoplastral suture, but significantly longer than the epi-hyoplastral suture. The entoplastron is large, wide and hexagonal and the hypoplastra are longer than the remaining plastral scutes. The xiphiplastron is relatively short and the hypo-xiphiplastral suture is located behind the level of the inguinal notch. The bridges are not particularly wide (45% the length of the plastron). The axillary buttress reaches the third peripheral, the inguinal buttress to the seventh peripheral. The large unpaired intergular reaches the entoplastron posteriorly and pushes the small gulars

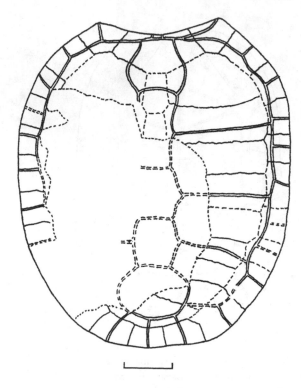

Figure 17.24. *Bulganemys jaganchobili* Sukhanov and Narmandakh, in press. Reconstruction of the carapace based on the holotype shell (GIN, Mongolia) from the Baruungoyot Svita, Upper Cretaceous (Late Santonian) of Yagaan Khovil, Southern Gobi, Mongolia. Scale bar = 50 mm.

towards the lateral edges of the epiplastra. The pectorals overlap the rear third of the entoplastron. The medial sulcus on the plastron is almost straight, but strongly meanders in the region of the xiphiplastra. The sixth and seventh marginals overlap the plastron significantly, almost reaching the level of the axillary and inguinal notches, and the plastral end of the sixth marginal is strongly broadened. There is a complete row of inframarginals: the first and second are small, the third is narrow and long and the fourth is rather large.

Bulganemys (Sukhanov and Narmandakh, in press; Figure 17.24) is based on an incomplete carapace and bridge parts of a plastron from the lower part of the Baruungoyot Svita (Santonian) of South Gobi, Mongolia. These small turtles, up to 450 mm long, do not have a longitudinal groove or a medial keel on their high (up to 50% of the width) and wide carapace (width = 90% of the length). The nuchal notch is wide (40% the width of the carapace) but not particularly deep (7% the length of the carapace). In the area of the bridges the free edge of the shell has an obtuse angle in cross-section. *Bulganemys* differs from *Hanbogdemys* in its smaller size and shell proportions, being relatively higher and broader. In addition, the nuchal notch is relatively wider, the cervical is smaller in size, the second vertebral is narrower and longer, and the first marginal is low and has a very long free edge. The pleuro-marginal sulcus remains within the limits of the peripherals, but the eleventh and twelfth marginals are higher than in *Hanbogdemys* and far overlap the second suprapygal. The axillary buttress reaches the second peripheral, the inguinal buttress reaches the eighth peripheral, and the plastral end of the sixth marginal is slightly broadened. The fourth marginal overlaps the plastron to a greater extent that the others and there seems to have been a complete row of inframarginals.

Zangerlia (Młynarski, 1972; Figure 17.25) is of medium to large size, and up to 700 mm long. The shell is convex and wide (width about 90% the length) and has a longitudinal groove and a weakly expressed medial keel. *Zangerlia* is sharply distinguished from other members of the Nanhsiunchelyidae by the narrow (30% of maximal width of the carapace) and deep nuchal notch (10–11% of the length of carapace), which forms an almost right-angled corner, and by the wide straight anterior edge of the trapezoidal anterior lobe of the plastron. In addition, the shortened hypoplastra are practically excluded from the posterior lobe of the plastron. The xiphiplastra are correspondingly very large and their length along the midline is greater than in all other plastral scutes. Also characteristic of this genus are the wide, but short and, probably, unpaired intergulars, which displace the small triangular gulars into the antero-lateral corners of the epiplastra. Moreover, the shared boundary of the intergulars and the gulars is almost perpendicular to the midline and situated in front of the entoplastron. Two species have been described, both from the

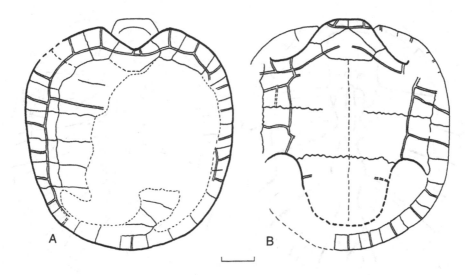

Figure 17.25. *Zangerlia dzamynchondi* Sukhanov and Narmandakh, in press. Reconstruction of the incomplete carapace (A) and plastron (B) based on the holotype, PIN 4698–1, from the lower part of the Baruungoyot Svita, Upper Cretaceous (Late Santonian) of Zamyn Khond, Southern Gobi, Mongolia. Scale bar = 100 mm.

upper part of the Baruungoyot Svita (Campanian). *Zangerlia testudinimorpha* (Młynarski, 1972: fig. 1, pl. 28) was founded on the basis of incomplete juvenile specimens from the localities of Khulsan and Nemegt in the Transaltai Gobi and *Z. dzamynchondi* (Sukhanov and Narmandakh, in press) is based on incomplete shells of large turtles from Zamyn Khond, South Gobi.

Nanhsiungchelys is a large turtle, up to 1 m in length, from the Upper Cretaceous Nanhsiung Group (Maastrichtian) of Guandun, China (Yeh, 1966, figs. 1–3, pls. 1–4). The shell is elongate (width 70% of the length) and moderately convex, and this turtle is distinguished by its gigantic nuchal notch (almost 50% the width of the carapace and 10% its length) that includes the first (as in other members of this family), the second and possibly even the third peripheral. The neurals were very wide. Yeh (1966) reported an octagonal neural, but this may have been the barrel-shaped second neural. The anterior lobe of the plastron was wedge-shaped and rounded anteriorly and the same length as the posterior lobe. The epiplastra are elongated along the free edge and form a very short epi-

hyoplastral suture, a short inter-epiplastral suture and very long epi-entoplastral sutures. The entoplastron is very large, hexagonal, and slightly wider than long, as in other representatives of the family. The intergular is very large, unpaired and extends posteriorly over the entoplastron to no less than one third of its length. The gulars seem to be displaced toward the lateral edge of the epiplastra. The pectorals broaden medially and almost reach the mid-length of the entoplastron, but the pectoral-abdominal sulcus occupies a transverse position and does not approach the hyo-hypoplastral suture. The inframarginals appear to be absent.

Family CARETTOCHELYIDAE Boulenger, 1887

The Carettochelyidae is composed of two sub-families. The Carettochelyinae Boulenger, 1887 is represented by genera from the Eocene of Western Europe, the ?Eocene of China, and extant forms from New Guinea and Australia while the fossil record of the Anosteirinae extends back to the Late Cretaceous.

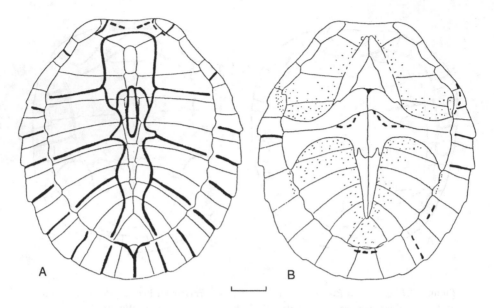

Figure 17.26. *Kizylkumemys schultzi* Nesov, 1977. Reconstruction of the carapace (A) and plastron (B), based on isolated shell plates from the upper part of the Khodzhakul Svita and the lower part of the Beshtyube Svita, Late Cretaceous (Cenomanian–Early Turonian), Khodzhakul'sai, Sultan-Uvais ridge, Karakalpakia, Uzbekistan. Scale bar = 40 mm.

Subfamily ANOSTEIRINAE Lydekker, 1889

Type genus. Anosteira Leidy, 1871. Middle Eocene–?Oligocene of North America and the ?Middle Eocene–Early Oligocene of Asia. Five species are known, four from China.

Characters of this group, according to Nesov (1976) are: the presence of horny sulci on the carapace; the distance between the anterior and posterior points of connection of the carapace with the plastron is noticeably less than the length of the suture between the hyoplastron and the hypoplastron; the xiphiplastra are narrow, their midline length more than double their width; the facets for the eighth cervical vertebrae lie far from the posterior edge of the nuchal; the symphysis of the lower jaw is short; and the widest part of the alveolar surface is situated posteriorly and takes the form of a shelf composed of the dentary and the coronoid.

In addition to *Anosteira*, the subfamily includes two monotypic genera: *Pseudanosteira* (Clark, 1932) from the Late Eocene of North America and *Kizylkumemys*

(Nesov, 1977a) based on fragmentary remains from deposits in Karakalpakia, Uzbekistan, dated as Cenomanian–Early Turonian and, consequently, the oldest reliably known anosteirines. However, it should be noted that *Sinaspideretes wimani* (Young and Chow, 1953) from the ?Late Jurassic of China might represent an even earlier carettochelyid (Meylan and Gaffney, 1992).

Kizylkumemys (Nesov, 1977a; Figure 17.26) is a small turtle, 250–350 mm long and represented by a single species, *K. schultzi* (including *Anosteira shuvalovi* Chkhikvadze, 1979), from the Cenomanian–Early Turonian of Middle Asia and the lower part of the Bayanshiree Svita, Mongolia. *Kizylkumemys* is characterized by the location of the second vertebral in the boundaries of the neural plates, a strongly developed spine on the carapace, the presence of small spines on the free edge of the fifth and sixth peripherals, and the narrowness of the posterior lobe of the plastron (Nesov, 1977a). To this genus Nesov (1981a, 1984) ascribed fragments found in Middle Asian deposits,

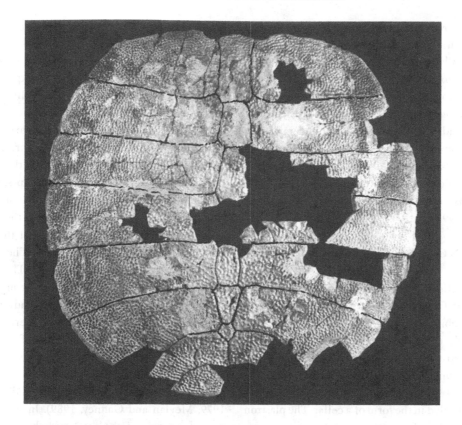

Figure 17.27. *Amyda orlovi* Khozatskii, 1976. Incomplete carapace of the holotype, PIN 557–1/1, from the lower part of the Bayanshiree Svita, Late Cretaceous (Cenomanian) of Bayan Shiree, Eastern Gobi, Mongolia. × 0.5.

dated as Late Albian–Late Turonian, and in the collections of the JSMPE there are naturally articulated, but incomplete shells of *Kizylkumemys* from the upper part of the Bayanshiree Svita (Upper Turonian–Santonian).

Family TRIONYCHIDAE Bell, 1828

Remains of trionychids are ubiquitous in Cretaceous deposits of Central Asia, from the Albian onwards. However, for the most part, these findings are fragmentary, or consist only of the carapace without the plastron: more complete remains, in which the shell and skull are associated, are extremely rare. There are such specimens in the collections of the JSMPE, but they have not yet been prepared. Practically all the remains which have been described so far have been

conditionally ascribed to extant genera. Brinkman and Nesov (1993) described a new genus based on an isolated skull, but without the shell this find is not very informative. Thus, despite the existence of rich collections, which contain the remains of trionychids from Middle Asia, Mongolia and China, few have yet been described. Khozatskii (in press) has reviewed trionychids from the Cretaceous of Mongolia, on the basis of material collected over many years, but has not formally diagnosed what are undoubtedly morphologically different species.

Two species of trionychid have been described from Mongolia. '*Amyda*' *orlovi* (Khozatskii, 1976; Figure 17.27) is based on an almost complete carapace from the Bayanshiree Svita of the Eastern Gobi and '*Amyda*' *menenri* (Chkhikvadze and Shuvalov, 1988) is based on fragmentary remains from the Nemegt Svita (Maastrichtian) of the Transaltai Gobi. The complete shell of a small trionychid, probably belonging to the genus *Platypeltis*, has also been reported (Merkulova, 1978), but not yet described. Two trionychids, *Palaeotrionyx riabinini* and *Trionyx riabinini* (Kuznetsov and Chkhikvadze, 1987) have been found in Upper Cretaceous deposits (Upper Turonian–Santonian) of Southern Kazakhstan. Recently it has been suggested that they should be assigned to *Axestemys* (Hay, 1889) and *Paraplastomenus* Kordikova, 1991, respectively (Kordikova, 1994a, b).

Family MONGOLOCHELYIDAE Sukhanov and Pozdnjakov in press

Type species. Mongolochelys Khozatskii, 1997. Nemegt Svita (Upper Cretaceous: Maastrichtian) Transaltai and South Gobi, Mongolia.

This family contains a single genus, *Mongolochelys* (Figure 17.28) with two species. *Mongolochelys* was a large turtle, up to 800 mm long, with a relatively flattened shell and a skull in which the roof was strongly expanded in the form of a collar. The plastron was strongly reduced, but retains mesoplastra, and was connected with the carapace via ligaments. The first remains of this turtle, including several complete carapaces, fragments of plastra and complete skulls, some in natural articulation with the shell, were discovered by Efremov's expeditions in 1946–1949. Delay in the publication of descriptions of *Mongolochelys* led to various speculations regarding this turtle: early reports identified it as a North American genus, *Baena*, or even the marine form *Dermochelys*. It was also identified with *Yümenemys*, based on fragmentary remains from China, and there was even an attempt to unite it with *Meiolania* (from Australia), *Neurankylus* (from America) and *Kallokibotion* (from Europe), into the family Meiolanidae (Chkhikvadze, 1987). Between 1969 and 1980 large quantities of additional material representing *Mongolochelys* were collected by the JSMPE and detailed descriptions of the shell and skull, based on these remains, will be published shortly.

Characteristics of *Mongolochelys* are as follows. The skull (Figure 17.28) reaches about one quarter the length of the carapace and lacks cheek and temporal emarginations. The region of the latter is occupied by a considerable posterior and medial expansion of the squamosals, which meet along the midline behind the caudal end of the crista supraoccipitalis, forming a peculiar collar covering the anterior part of the neck. The skull is covered by symmetrically arranged, regularly shaped horny scutes. Underneath these scutes, in the area of the 'collar', there are lump-shaped bony thickenings resembling the more developed 'horns' of Meiolaniidae. The nasals are large and the prefrontals open widely onto the dorsal surface of the skull, but do not contact along the midline. Their ventrally directed processes contact the vomer. The frontals do not enter the edge of the orbit and the jugal contacts the quadrate ventral to the small quadratojugal. An interpterygoid vacuity is absent and there are no teeth on the pterygoid. The anterior part of the pterygoid forms a horizontal plate and the postero-lateral corner lacks the vertically expanded thickness, considered to be a characteristic feature of Cryptodira (Gaffney, 1979; Meylan and Gaffney, 1989). In this region the medial margins of the pterygoids have an extensive midline contact in front of the basisphenoid. Posteriorly, the pterygoids form only part of the base of the middle ear cavity and do not cover the prootic when viewed from beneath. A large section of the internal carotid artery lay exposed in a small groove on the basisphenoid before entering a canal, via the foramen caroticum basisphenoidale, which leads through the basisphenoid into the area of the sella turcica. Immediately in front of the paired foramen caroticum basisphenoidale, between the pterygoids and basisphenoid, lie another pair of openings, the foramina caroticum laterale for the palatine arteries. An epipterygoid is present. The basisphenoid rostrum is long, the trabeculae melding together in front of the sella turcica. A processus trochlearis oticum is also present, though only weakly expressed.

The neck is relatively short and the vertebral centra bear well developed articular surfaces. The third and eighth cervical are convex at both ends, while the sixth is concave at both ends. The articular surface of the

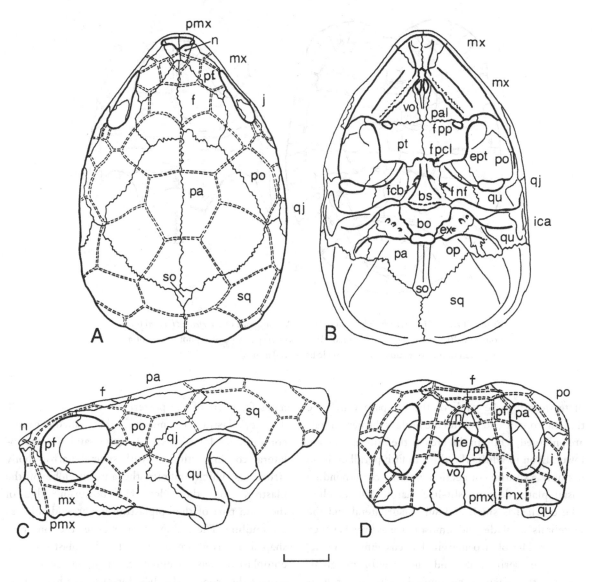

Figure 17.28. *Mongolochelys efremovi* Khozatskii, 1997, holotype, PIN 551–459, Nemegt Svita (Late Cretaceous–Maastrichtian), Nemegt, Transaltai Gobi, Mongolia. Skull in (A) dorsal, (B) ventral, (C) lateral and (D) anterior view. Abbreviations as in Figures 17.2 and 17.8 except for: ept, epipterygoideum; fe, fissura ethmoidalis; fnf, foramen nervi facialis; fpp, foramen palatinum posterius; ica, incisura columellae auris; pmx, premaxillare. Scale bar = 20 mm.

first thoracic centrum faces forward. The cervical ribs are well developed and double-headed.

The shell (Figure 17.29) is low (maximum height equivalent to about 15% of the length), and wide (width reaches 80% of the length). The carapace is sub-oval with a large nuchal notch in front. The nuchal lacks costiform processes and there are nine neurals and nine pairs of costals. The first suprapygal is small, the second very large, and the pygal is short and wide. The first costal is narrow, the ninth wide and no smaller than the previous costal. The peripherals are thickened and massive, in contrast to the bony

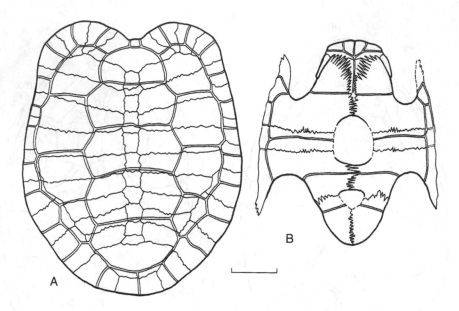

Figure 17.29. *Mongolochelys* Khozatskii, 1997. (A) Carapace of *M. efremovi* based on the holotype, PIN 551–459. (B) Plastron of *Mongolochelys* sp., based on a photograph of a specimen in the collections of GIN. Scale bar = 100 mm.

plates forming the vault of the carapace which are thin. In front of and behind the bridges the peripherals are enlarged, while in the area of the bridges they are lower. Even in adults the third to fifth peripherals do not have a sutural connection with the corresponding costal plates, leaving slit-like marginal fontanelles. The cervical is small and sub-quadrilateral and the vertebrals are wide, and almost twice as broad as they are long. The pleuro-marginal sulcus runs, as a rule, along the peripherals and the twelfth marginals overlap the second suprapygal. The plastron is strongly reduced, leaving a large central and small, narrow xiphiplastral fontanelles. The plastron is connected to the carapace by ligaments and the axillary and inguinal buttresses, whose anterior and posterior processes reach the second and eighth peripherals respectively. The bridges are relatively wide, reaching 40% the width of the plastron. The anterior and posterior lobes are relatively short and narrow toward their external margins. The former has a more or less straight edge, the latter is rounded. The anterior lobe reaches the nuchal notch, but the posterior lobe termi-

nates far short of the posterior margin of the shell. The epiplastra are massive, their medio-caudal corners bear long dorsal processes, and the caudo-lateral corners articulate with special narrow bony structures which embrace the anterior lobe of the plastron from the sides. The posterior edge of the main part of the epiplastron is perpendicular to the midline. The endoplastron overlies the long ray-shaped anterior processes of the hyoplastron. The entoplastron has an elongate, triangular shape, the base at the front, and it lies dorsal to the hyoplastra, preserving the original position characteristic for the reptilian episternum. The hyo- and hypoplastra meet along the midline, both anterior and posterior to the central fontanelle, via ray-like processes which interfinger between each other like thorns. The xiphiplastra contact each other behind the second fontanelle via similar, but shorter processes. A pair of gulars and intergulars are present. The intergulars reach the entoplastron, completely separating the gulars, which are elongated transversely and displaced laterally. There are four pairs of narrow inframarginals, pressed to the boundary between the plastron and the carapace.

In Mongolia the Mongolochelyidae are found only in the Nemegt Svita and seem to be absent from contemporaneous or earlier sequences of adjoining regions in Middle Asia and China. The origins of this family are unclear. The presence of mesoplastra in the Mongolochelyidae and in the Chengiuchelyidae from the Middle Jurassic of China, is not sufficient to indicate a direct connection between these two, and besides, no other Central Asian turtles have mesoplastra. *Mongolochelys* presents a unique combination of primitive and derived (apomorphic) characters. Primitive features include: the absence of a vertical expansion on the external process of the pterygoid and absence of a bony canal for the internal carotid artery; the presence of the dorsal processes on the epiplastra; the presence of mesoplastra, nine pairs of costals, nine pairs of neurals, seven pairs of plastral scutes, a complete row of inframarginals and the structure and relative position of the entoplastron in relation to other elements of the plastron. Autapomorphies include: the expansion of the squamosals to form a 'collar' behind the supraoccipital; the development of a contact between the jugal and quadrate, leading to the disappearance of the cheek emargination; melding of the trabeculae anterior to the sella turcica; the formation of special bony elements in the latero-caudal corners of the epiplastra; the presence of large fontanelles in the plastron and small fontanelles in the carapace; the connection of single elements of the plastron along the midline via interfingering of special processes; and the connection of the plastron and carapace by ligaments. The absence of both an interpterygoid vacuity and teeth on the vomer and pterygoids shows that *Mongolochelys* is not related to turtles of the Triassic type, although in a number of respects (the weak development of the caudo-medial processes of the pterygoids, the number of neurals and costals and the dorsal processes of the epiplastra) it resembles the Early Jurassic turtle *Kayentachelys* from North America. At the same time, the structure of the canal for the internal carotid artery, presence of dorsal processes on the epiplastra, and connection of the cara-

pace and plastron by ligaments is also found in the Late Jurassic Central Asian Xinjiangchelyidae, but they do not have mesoplastra, and the skull roof is extremely reduced. Similarities can also be found between *Mongolochelys* and *Kallokibotion* from the Early Cretaceous of Europe, but pleurosternids from the Late Jurassic and Early Cretaceous of Western Europe show the closest level of development, exhibiting the primitive condition for the morphology of the palatal surface of the skull and retaining mesoplastra.

Family LINDHOLMEMYDIDAE Chkhikvadze, 1970

Type genus. Lindholmemys Ryabinin, 1935. Upper Cretaceous (Cenomanian–Santonian) of Kyzylkum, Middle Asia.

The Lindholmemydidae includes: *Lindholmemys*, *Mongolemys*, *Gravemys*, *Hongilemys* and *Tsaotanemys*. The Lindholmemydidae is a large and diverse group of Asian turtles, apparently amphibious, and first known from the late Early Cretaceous. They reach peak diversity in the late Late Cretaceous, accounting for the majority of turtle remains in the Nemegt Svita (Maastrichtian) of Mongolia. These turtles are also the most abundant elements in the Late Palaeocene Naranbulag Svita of Mongolia, and it has been observed that this group appears to have 'not noticed' the boundary between the Cretaceous and Palaeogene (Sukhanov, 1978). During the Cretaceous the Lindholmemydidae occupied a niche which, in the Cenozoic, appears to be firmly engaged by the Emydidae (in the widest taxonomic sense).

The Lindholmemydidae are the only group of Asian Cretaceous turtles where the shell consists of a well developed carapace and plastron, connected by a strong suture and strengthened by a buttress, which extends far dorsally along the inner surface of the corresponding costals.

The name for the family was suggested, without characterization, by Chkhikvadze (1970) for two genera of turtles: *Lindholmemys* Ryabinin, 1935 and *Mongolemys* Khozatskii and Młynarski, 1971. These were previously ascribed, on the basis of the presence of a complete row of inframarginals (Khozatskii and

Młynarski, 1971) to the Dermatemydidae (Williams, 1950; Romer, 1956; Sukhanov, 1964; Młynarski, 1969), a group already considered as a 'waste-basket' taxon. A short characterization of the family was published by Chkhikvadze in 1975 (Shuvalov and Chkhikvadze, 1975) and repeated, practically without any changes, by Chkhikvadze in 1987. On the basis of superficial similarities in the mosaic of bony and epidermal elements of the shell, Chkhikvadze (1981) identified the Lindholmemydidae as a group of late, rather highly evolved plesiochelids, and ancestors of the family Platysternidae. In connection with this, Chkhikvadze's diagnosis included features which distinguished Lindholmemydidae from Plesiochelyidae (specifically European turtles, though connections were implied with a little known Chinese form '*Plesiocheylis*' *chungkingensis*, recently ascribed to a completely different family, the Xinjiangchelyidae) and from Platysternidae, the position of which in turtle systematics is still disputed: are they related to the Chelydridae (Gaffney, 1975a, b; Gaffney and Meylan, 1988), or to a completely different lineage of turtles, the Testudinoidea? Apart from these two families the Lindholmemydidae were not compared with other turtles. Other authors (e.g. Nesov and Khozatskii, 1980; Nesov, 1981b) continued to defend the idea that *Lindholmemys* and *Mongolemys* belonged to the Dermatemydidae. This was based on the seemingly narrow stapedio-temporal foramen in the skull, implying a trionychoid type of blood supply to the head, rather than the 'arrangement seen in Chelydroidea and Testudinoidea, and these taxa were assigned to the subfamily Lindholmemydinae in the Dermatemydidae (Nesov, 1986a, b).

Detailed study of the skull and construction of the shell of *Mongolemys*, a typical lindholmemydid represented by a good series of complete and fragmentary skulls including separate skull bones, nearly one hundred more or less complete shells and numerous fragments of separate shell elements and postcranial bones, show that this group of turtles has nothing in common with the Dermatemydidae, in the modern sense (Nesov, 1977c; Meylan and Gaffney, 1989), the Trionychoidea or the Platysternidae.

Lindholmemydidae are small- to medium-sized turtles, 150–400 mm long, with a relatively small skull no more than one fifth the length of the carapace. The skull of *Mongolemys* (Figure 17.30) is wide and relatively short, with a maximum width nearly 80% of the condylar–basal length. The orbits are large and face outward and upward. The nasals are lacking and, in contrast to *Dermatemys* and *Platysternon*, the prefrontals do not contact along the midline, but are separated by narrow, forward directed processes of the frontals. The ethmoidal fissure is 'key-hole shaped': very narrow ventrally, but widens dorsally. The frontal reaches the edge of the orbit, preventing contact between the prefrontal and postorbital. The temporal emargination is deep and extends beyond the processus trochlearis oticum to level with the posterolateral corner of the external process of the pterygoid. However, a small contact remains between the squamosal and postorbital. The cheek emargination is also deep and almost reaches the level of the middle of the orbits. The quadrato-jugal forms a small, 'T'-shaped element, and does not prevent contact between the postorbital and the quadrate. The jugal is large and almost excluded from the margin of the orbit. A powerful, medially-directed process of the jugal contacts ventrally with both the external and the maxillary process of the pterygoid. A ventrally-directed process of the parietal forms the posterior margin of the foramen interorbitalis, displacing the small epipterygoid toward the anterior margin of the trigeminal foramen and contacting ventrally with the pterygoid, but not the palatine, which (unlike *Dermatemys* and *Platysternon*) does not take part in the formation of the brain case or have a corresponding dorsal crest. The stapedio-temporal canal is well developed and the columella auris incisure is not enclosed.

The maxilla has an even, slightly rounded ventral margin formed by a tall and sharp labial crest lacking any tooth-like swellings. The alveolar surface is relatively narrow and formed mainly from the maxillary and premaxillary. A lingual crest is absent and there are no commissural crests. The premaxillae are connected along the midline by a suture and are moderately developed in an antero-posterior direction.

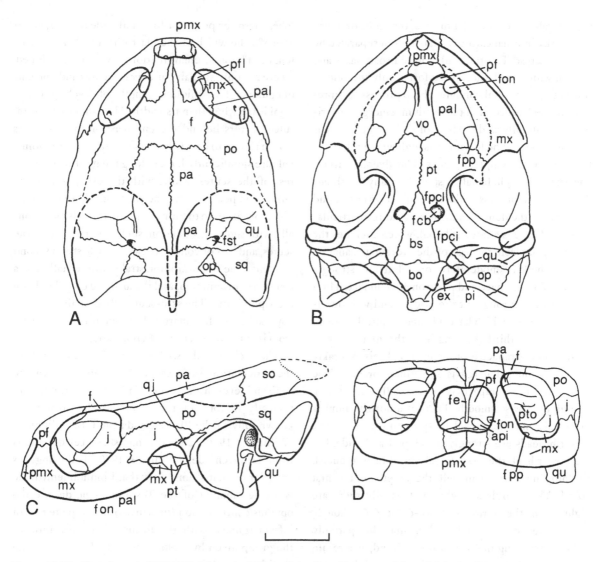

Figure 17.30. *Mongolemys* sp., PIN 4693–1 from the Nemegt Svita (Late Cretaceous–Maastrichtian), of Nogoon Tsav, Transaltai Gobi, Mongolia. Skull in (A) dorsal, (B) ventral, (C) lateral and (D) anterior view. Abbreviations as in Figures 17.2, 17.8 and 17.28 except for: api, apertura narium interna; fon, foramen orbito-nasale; pi, processus interfenestralis; pto, processus trochlearis oticum. Scale bar = 10 mm.

Coupled prepalatinal foramina are located between the premaxillae and the vomer, and the entire ventral surface of each premaxilla is included in the alveolar surface of the upper jaw. The posterior palatal foramen is situated between the palatine, the maxillary and pterygoid, but not separated from the latter by contact between the first two. The pterygoid waist is wide. The medial margin of the pterygoid does not contact the basisphenoid throughout its length, but leaves a special opening, the foramen basisphenoidale, which opens in the canalis carotici interni directly ventral to the divergence of the cerebral carotid artery and the palatal artery. Posterior to this the internal carotid canal passes between the pterygoid and basisphenoid. The foramen posterius canalis carotici interni is far to the rear. The dorsum sellae projects considerably over the sella turcica, the anterior openings of the internal carotid canals are not separated by a sagittal crest. The trabeculae have broad bases and remain wide. The basioccipital forms only the ventral third of the occipital condyle. Sharp processes, representing the basioccipital tuberculum, emerge posteriorly and laterally, and the base of the basal tubercle is well developed. The lower jaw is high with a relatively long, but small symphysial beak. The alveolar surface consists of a single labial crest without any 'toothing'. The coronoid process of the coronoid does not extend higher than the dentary and there is no retroarticular process. There is a large splenial, which excludes the small angular from the base of the Meckelian canal.

The shell is moderately or relatively strongly convex. The carapace and plastron are connected by a strong suture throughout the bridges and well developed buttresses, which have a contact equivalent to no less than one third the length of the first and fifth costals. The carapace is oval and lacks keels. A weakly developed nuchal notch is sometimes present. The surface of the carapace is occasionally sculptured, though otherwise smooth. There is a normal number of costals and the nuchal lacks costiform processes. The rib heads are strongly developed and wide, but the free rib of the first dorsal vertebra is reduced, reaching no more than half the length of the first costal. The neurals are wide and usually there are eight with the formula 4-6-6-6-6-6-6-6, though, rarely, the seventh and eighth may by partially reduced resulting in 4-6-6-6-6-6-6-5 and, as a result, the eighth costals may contact along the midline. As a rule, there are two suprapygals, rarely one. Their shape and size is varied, but usually the first is narrow and trapezoidal while the second is wide. The peripherals in the anterior part of the carapace are relatively massive and in the region of the bridges the free margin forms a smooth angle. There is one well developed cervical and the anterior marginals (first to third) are low. The pleuromarginal sulcus runs along the edge of the peripherals, or, if starting from the bridge region and more caudally, it follows the costal–peripheral suture. The plastron is large, reaching 85–90% the length of the carapace, with a wide, short anterior lobe, narrower posterior lobe and wide bridges, about 55–65% the width of the plastron and 40–45% of its length. The anal emargination is weakly developed. There are only six pairs of plastral scutes and one pair of gulars. The midline length of each gular is significantly less than its width. The humero-pectoral sulcus enters behind the entoplastron, which varies from a rhomb to a wide hexagonal shape. The abdominals are significantly larger along the midline than the rest of the scutes, while, with the exception of the gulars, the pectorals are the shortest. There is a complete row of three to four pairs of marginals, occasionally with signs of reduction, such as narrowing of the scutes, and loss of contact between the first and second element. The hypo-xiphiplastral suture usually has a transverse orientation, and the anals lie completely on the xiphiplastra. The skin-scute sulcus on the plastron approaches the free margin, but sometimes departs far from it in the region of the hypoplastra.

The neck formula is: (2(, (3(, (4),)5),)6),)7(, (8), and the articulation surfaces of the joints between the sixth and seventh and seventh and eighth vertebrae are typically double. In recent turtles this formula is known only in the Platysternidae and Testudinoidea (Williams, 1950) and distinguishes Testudinoidea from Trionychoidea (Williams, 1950). Following this consideration, it seems likely that Lindholmemydidae is a true member of the Testudinoidea, despite the obvious contradiction (presence of a complete row of inframarginals) with the formal characterization of the group given by Gaffney and Meylan (1988). The skull has a testudinoid appearance and does not have any apomorphic features which would place Linholmemydidae closer to the Dermatemydidae, whereas there are numerous differences. In fact, there is not a single characteristic that is specific for Dermatemydidae, as well as for Trionychoidea in general (see Meylan and Gaffney, 1989), that can be observed in *Mongolemys*. Moreover, *Mongolemys* does not have a single apomorphic feature of *Platysternon* or Platysternini in general, including *Chelydropsis* from the Oligocene–Pliocene, or the recent form *Macroclemys* (Gaffney and Meylan, 1988). At the same time, an analysis of skull morphology in *Mongolemys*, based on 48 characters used by Hirayama (1984) in a study of the skull of Batagurinae, shows that only three of them can be said to have achieved an advanced state. Moreover, current study of Lindholmemydidae shows that many specific characters (e.g. presence of a foramen basisphenoidale, a large splenial and others), must be considered plesiomorphic. This may also be true for the presence of a deep temporal emargination, since this is characteristic for Xinjiangchelyidae and for Macrobaenidae. The following may be considered as apomorphies: the development of a deep cheek emargination with the partial reduction of the quadratojugal; the general strengthening of the shell by thickening of the costal elements; the development of an interdigitating sutural connection between the carapace and plastron; and strengthening of the buttresses.

Mongolemys (Khozatskii and Młynarski, 1971; Figures 17.30 and 17.31) is best known from remains of the type species, *M. elegans*, found in the Nemegt Svita (Maastrichtian) of the Transaltai Gobi. This turtle is up to 250–350 mm long, with a moderately convex shell (height about 40% of the width), composed of thickened costal elements and without a nuchal notch. Marginal and plastral fontanelles are absent, even in juvenile specimens only 70–100 mm long. The surface of the shell is almost smooth, bearing only a net of thin branching sulci, probably connected with microvessels, or has a well developed sculpture, consisting of small tubercles and crests, though specimens bearing this type of sculpture should probably be ascribed to a separate species. There are two suprapygals and, while the shape of the area occupied by these two plates is approximately equal, the transverse suture between them is varied in location. This results in mutual changes in their shape and size. Thus, when the first is narrow and trapezoidal the second is wide and hexagonal and, vice versa, when the first is large and hexagonal, the second is narrower, with a convex lens shape. The cervical scute is wide, but short and sub-quadrangular, and the first vertebral is wide, its antero-lateral corners overlapping the boundaries of the nuchal and contacting the second marginals. The second and third vertebrals are sub-quadrangular and the margi-

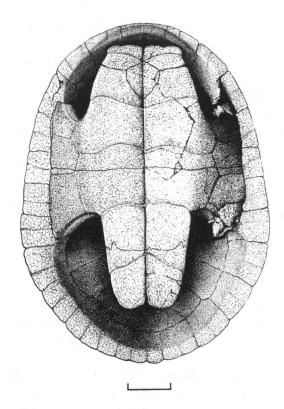

Figure 17.31. *Mongolemys elegans* Khozatskii and Młynarski, 1971. Holotype plastron, PIN 551–422/1 (=ZIN. RN. T/M – 46.1), from the Nemegt Svita, Late Cretaceous (Maastrichtian) of Nemegt, Transaltai Gobi, Mongolia. Illustration by S.M. Shteinberg, after a reconstruction by L.I. Khozatskii. Scale bar = 20 mm.

nals are relatively low, the pleuro-marginal sulcus approaching the costal–peripheral suture only in the region of the bridges. The twelfth marginals are subtriangular and the contact between them is short, so that the fifth vertebral closely approaches the posterior margin of the carapace. The plastron has no anal emargination and the bridges are wide, reaching 110–120% of the half-width of the plastron. The buttresses reach to the distal third of the costals. The gulars are small, transversely elongate and posteriorly they contact or slightly overlap the entoplastron. The inframarginals are large, there are three, sometimes four pairs and their lateral boundary is coincident with the suture between the carapace and plastron.

Other species of *Mongolemys* include: *M. turfanensis* from the Upper Cretaceous–Palaeocene deposits of Sechuan (Yeh, 1974a, b), and *M. tatarinovi* and *M. reshetovi*, both from the Late Palaeocene of Mongolia (Sukhanov and Narmandakh, 1976). New, as yet undescribed, material of *M. tatarinovi* suggests that this species may have to be placed in a separate genus. A fifth species, presently referred to as *Mongolemys* sp., is also known from the late Early Cretaceous (Sukhanov and Narmandakh, 1974).

A number of other species assigned to *Mongolemys* do not belong in this genus. '*Mongolemys*' *planicostatus* from the Late Cretaceous of the Pre-Amur, China, was described on the basis of a single first costal as *Aspideretes* sp. (Ryabinin, 1930) and later assigned by Nesov (1981a) to *Mongolemys*. However, taking into account the diversity of Lindholmemydidae and the almost total absence of distinctive features of this particular element of the shell this reassignment seems unjustified.

'*Mongolemys*' *australis* (Yeh, 1974a) from the Palaeocene of Guandun, China, was correctly assigned by Chkhikvadze (1976) to a distinct genus, *Elkemys*, on the basis of the presence of an anal notch in the plastron and the significant overlap of the plastral scutes on the inner surface of the plastron. This taxon should be included in the family Emydidae.

Tsaotanemys was established on the basis of fragmentary material from the badlands of Chia-yü-kuan in Gansu, China, from deposits of uncertain age (Bohlin, 1953). Judging by the turtles and dinosaurs recovered from these deposits, they are probably Aptian–Albian (Nesov and Mertinene, 1982) to Santonian (Maryańska and Osmólska, 1975) in age, but not Maastrichtian, in which the majority of the mass discoveries of *Mongolemys elegans* have been made. Even accepting the probable inaccuracy of the reconstruction of the plastron of *Tsaotanemys*, given by Bohlin (1953, fig. 62), there are no grounds for synonymizing this genus with *Mongolemys*, as proposed by Chkhikvadze (1976, 1987). *Tsaotanemys* is smaller, only up to 150 mm long, the bony elements of the shell are much thicker, the sculpturing is coarser, there is a shallow nuchal notch and the anterior lobe of the plastron is shaped differently. Moreover, it seems likely that *Tsaotanemys* was established on insufficient type material.

The remaining representatives of the Lindholmemydidae are based on incomplete and rather poor material. *Lindholmemys* (Ryabinin, 1935; Figure 17.32) is known from the early Late Cretaceous (Cenomanian–Early Santonian) of Middle Asia, where it is represented by at least two species (Ryabinin, 1935; Nesov and Khozatskii, 1980). *L. elegans* from Kyzylkum, Uzbekistan, is represented by one almost complete shell (the holotype), two isolated skulls, as yet undescribed (Nesov, 1986a), and by several insignificant fragments. *L. gravis* occurs in the Yalovach Svita (Turonian–Santonian) of South Kazakhstan and was described on the basis of separate plates, none of which possess any specific features.

Lindholmemys is rather small, up to 250 mm long, and has a notably convex shell, the height reaching 56% of the width. The buttresses are very powerfully developed. The axillary buttress extends behind the middle of the first costal and usually reaches the end of the first dorsal rib; and the inguinal buttress reaches higher than the mid-length of the fifth and sixth costals. The bony plates of the shell are unusually thick and the carapace has only a very small nuchal emargination. The pygal narrows posteriorly and there is only one suprapygal. The cervical scute is large and trapezoidal in shape, while the first vertebral is relatively narrow, failing to extend laterally beyond the boundary of the nuchal, or to contact the second marginals. The pleuro-marginal sulcus runs close to the costal–peripheral suture, except for the first to third peripherals, where it runs nearer to the mid-height of these plates. The plastron has an anal emargination and there are three inframarginals. The inframarginals are narrow and displaced laterally far behind the line that connects the centres of the axillary and inguinal notches, though only the third inframarginal extends behind the peripherals. The second and third inframarginals probably lost contact with each other.

Hongilemys is known from the Upper Cretaceous deposits of Mongolia (Sukhanov and Narmandakh, in

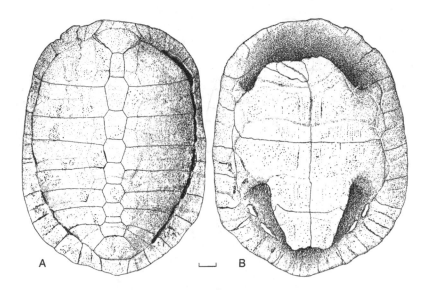

Figure 17.32. *Lindholmemys elegans* Ryabinin, 1935. Holotype carapace (A) and plastron (B) from the Taikarshin Beds (Late Cretaceous: Late Turonian–Coniacian) of Dzharakhuduk, Central Kyzylkum, Uzbekistan. Illustration by S.M. Shteinberg, after a reconstruction by L.I. Khozatskii. Scale bar = 20 mm.

press; Figure 17.33). *H. martinsoni* (Shuvalov and Chkhikvadze, 1975) is known from levels ascribed to the upper part of the Bayanshiree Svita (Upper Turonian–Santonian), while *H. kurzanovi* (the type species) comes from the lower part of the Baruungoyot Svita (Santonian). This genus, like *Lindholmemys*, has strengthened buttresses and the bony elements of the shell are very thick. It also has an anal emargination of the plastron and narrow inframarginals, but, unlike *Lindholmemys*, the shell is slightly less convex, there is a small nuchal emargination, two suprapygals, and the pleuro-marginal sulcus always runs along the peripherals.

Gravemys is known from the Nemegt Svita (Maastrichtian) of the Transaltai Gobi (Sukhanov and Narmandakh, 1976, 1983; Figure 17.34). The shell is moderately convex (height = 45% of the width) and bears a characteristic sculpture consisting of tubercles, aligned in rows, or small crests with a clearly expressed orientation. There is a small nuchal emargination, two suprapygals and the bridges are large, reaching 45% the length of the plastron. The cervical

scute is large and sub-rectangular. Unusually, the first vertebral is rounded anteriorly, while the second and third are sub-rectangular and narrow, the width noticeably less than the length. The first marginals are pentagonal and, with the exception of the first three marginal scutes, the pleuro-marginal sulcus runs close to the costal–peripheral suture. The plastron has narrow, wedge-shaped lobes. The anterior lobe is almost half the length of the posterior, but has a wider base. There is an anal emargination and the epiplastra are large and pentagonal while the entoplastron is hexagonal. The buttresses are less well developed than in *Lindholmemys*, but more so than in *Mongolemys*. The gular plates overlap about one third of the entoplastron and the pectorals are narrower than the humerals. The pectoral-abdominal sulcus is flexed backward behind the level of the axillary notch. *Gravemys* is distinguished from other genera by the shape and size of the epiplastra and the presence of four, not three, large inframarginals, which are displaced laterally and reach far over the peripheral plates.

355

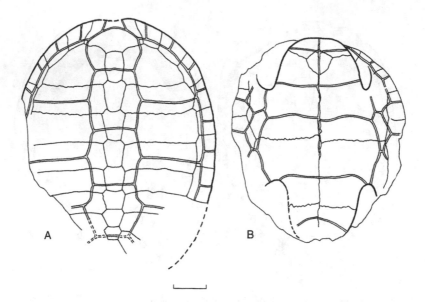

Figure 17.33. *Hongilemys kurzanovi* Sukhanov and Narmandakh, in press. Carapace (A) and plastron (B) based on the holotype, PIN 4695–1 and PIN 4695–2, from the upper part of the Baruungoyot Svita, Late Cretaceous (Campanian) of Khongil Tsav, Southern Gobi, Mongolia. Scale bar = 30 mm.

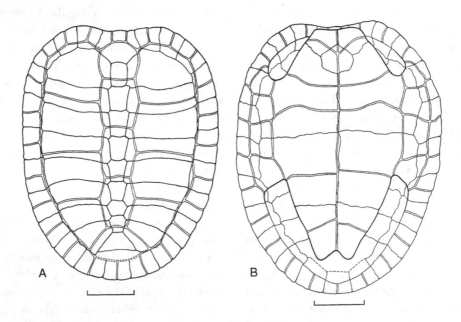

Figure 17.34. *Gravemys barsboldi* Sukhanov and Narmandakh, 1974. Reconstruction of the carapace (A) and plastron (B) based on the holotype (housed in GIN) from the Nemegt Svita, Late Cretaceous (Maastrichtian) of Hermiin Tsav, Transaltai Gobi, Mongolia and on PIN 4064–1 from the Nemegt Svita at Bambuu Khuduag. Scale bar = 50 mm.

356

Family HAICHEMYDIDAE Sukhanov and
Narmandakh, in press.

Type genus. *Haichemys* Sukhanov and Narmandakh, in
press. Nemegt Svita (Upper Cretaceous: Maastrichtian), Transaltai Gobi, Mongolia.

The Haichemydidae are found in the Late
Cretaceous to Late Palaeocene of Central Asia where
they are represented by two genera, one of which,
from the Palaeocene, has yet to be described. These
turtles are extremely rare: among the mass collections
of turtles from the Nemegt Svita only one practically
complete specimen with part of the skull, the holotype of *Haichemys ulensis*, has been found. Palaeocene
representatives of the family have been found in the
Naranbulag Svita of Transaltai Gobi and are represented by three almost complete specimens, though,
unfortunately, without skulls.

Haichemydids were small turtles, up to 250 mm
long, with a flattened shell, the height no more than
30% of the width, and a relatively small skull only
20–25% the length of the shell. The bony elements of
the shell are thin and the carapace is relatively wide
and round, the width reaching 85% of the length. It
has sharp marginal edges, even in the region of the
bridges, and small fontanelles.

A nuchal emargination is absent and there is the
usual number of bony plates. The neural formula
varies from the standard 4-6-6-6-6-6-6-(4)-6 (though
in some cases the antero-lateral sides are very short)
up to 4-8-4-6-6-6-6-(4)-6. Between the seventh and
eighth neurals there is an additional small, short,
transversely elongated element. There are two large
suprapygals: the first is trapezoid and relatively
narrow, the second is wider. The cervical is large and
the first, fourth and fifth vertebrals are wide, their
width greater than their length, while the second and
third are narrower (width = length). The sulcus
between the fourth and fifth vertebrals crosses the
eighth neural and the pleuro-marginal sulcus is
confined to the peripheral plates, though in the
regions of the bridges it is coincident with the
costal–peripheral suture. The plastron is large with a
wide, relatively short anterior lobe and a narrower, but
longer posterior lobe. An anal emargination is present

and the connection between the carapace and plastron
seems to have been weak and formed by ligaments.
The axillary buttresses reach the caudal margin of the
second peripheral and, in the largest specimens from
the Palaeocene, there is a small contact with the inner
surface of the first costal. The inguinal buttresses
contact only with the seventh peripheral and the
bridges are wide, reaching about 40% the length of
the plastron. The epiplastra are relatively small and
sub-rectangular. Their connection with the hypoplastron is via tooth-shaped processes on the latter which
overlap, externally, special little grooves along the
posterior margins of the epiplastra. The entoplastron
is relatively small, only 12–13% the length of the plastron. The connection between the right and left halves
of the plastron is relatively weak, and consists of interlocking processes. A narrow central fontanelle seems
to have been present even in adult stages in the
Palaeocene form. The hypo-xiphiplastral suture is
transverse and there is one pair of transversely elongated gulars. The humero-pectoral sulcus enters
behind the entoplastron, but the anals do not overlap
the hypoplastra. The pectorals, abdominals and femorals are roughly equal in medial length. There is a
complete row of three inframarginals. They are large
and their lateral boundary is coincident with the
contact between the carapace and plastron. The skin-
scute sulci on the plastron approach the free margin.

Externally, the numbers and shapes of the bony and
horny elements of the shell of haichemydid turtles
resemble the condition in lindholmemydids, but there
are striking differences in the general construction of
the shell and the contacts between individual bony
elements. Unfortunately, only a small fragment of the
skull is preserved in the type species. From this it
seems likely that the foramen basisphenoidale was
absent in these turtles and thus the entire internal
carotid canal was enclosed in bone. In addition, peculiarities of shell construction and the presence of a
complete row of inframarginals suggests possible relationships with the extant turtle *Platysternon*. However,
there are some important differences concerning the
relative size of the skull, the size and shape of the epiplastra, the width of the bridges (which in *Platysternon*

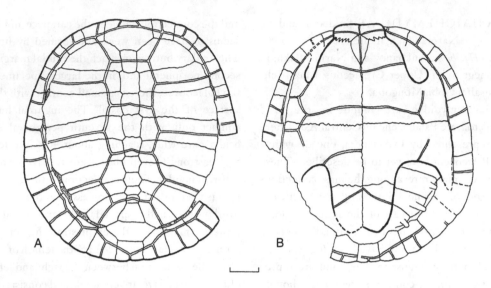

Figure 17.35. *Haichemys ulensis* Sukhanov and Narmandakh, in press, holotype, PIN 4691–1 from the Nemegt Svita, Late Cretaceous (Maastrichtian) of Khaichin Uul, Transaltai Gobi, Mongolia. Reconstruction of the carapace (A) and plastron (B). Scale bar = 20 mm.

attain no more than 20% the length of the plastron), the length of the buttresses, and the size and shape of the anterior and posterior lobes of the plastron. This shows that haichemydids and platysternids cannot be included in the same family. Various remains of Tertiary turtles from Kazakhstan, including material from the Late Palaeocene (Chkhikvadze, 1989), *Planiplastron* from the Late Oligocene (Chkhikvadze, 1971), and *Kazachemys* from the Miocene (Chkhikvadze, 1989), have been ascribed to the Platysternidae, but probably belong rather to the Haichemydidae.

Haichemys (Sukhanov and Narmandakh, in press; Figure 17.35) is known only from the Nemegt Svita of the Transaltai Gobi. *Haichemys* was a small turtle, up to 200 mm long. The shell lacked a nuchal emargination and there was weak scalloping of the posterior margin of the carapace. Small marginal fontanelles occur in the carapace, mainly in its posterior third, but in contrast to the larger Palaeocene form there are no fontanelles in the plastron. The sub-parallel antero-lateral sides of the nuchal, relatively wide neurals and relatively short first neural with correspondingly greater narrowness of the first costal are characteristic of *Haichemys*. The cervical scute is short, its length only one third of its width, and the first vertebral is wide, but noticeably shorter than the succeeding vertebrals. The marginals are relatively low, especially the first, which is sub-triangular. The gular edges of the epiplastral have thickened lips and the axillary buttress hardly contacts the first costal.

Discussion

In this review of Middle and Central Asian turtles special attention was given to the broad morphological characteristics, not diagnoses, of the main groups of turtles, based on reasonably complete material. In many respects this review is built on new, unpublished work based mainly on the study of Mongolian fossil turtles. The nature of this review is not accidental. A key turning point in the macro-systematics of turtles can be attributed to Gaffney (1975a, b, 1984; Gaffney and Meylan, 1988; Meylan and Gaffney, 1989; Gaffney *et al.*, 1991), and stems from the study of the skulls of mainly modern turtles and earlier success in elucidating the nature of the blood supply to the turtle's head by McDowell and Albrecht. The systematic scheme devised from these studies initially included all modern groups of turtles, but very few fossil turtles. The application of cladistic methods led to further profound developments in turtle systematics and clearly established the need for a new task – to extend this system to fossil turtles. Aside from the rarity of skulls in fossil turtles, especially in association with the skeleton, the main obstacle to this task is our inadequate knowledge of the morphology of fossil and modern forms, and the difficulty of reducing the morphology to restricted sets of taxonomic characters, many of which seem to be affected by homoplasy. Thus, much more detailed descriptions and analyses of turtle morphology are now a primary need and this will also require restudy and revision of many previously described forms and groups, including extant taxa.

The necessity for highly detailed studies of the anatomy of fossil turtles with a special accent on skull morphology is self-evident, and has already resulted in the appearance of many articles and monographs by Gaffney and others, dedicated to morphological descriptions of forms important for understanding the evolution of turtles. Pre-eminent among these are *Proganochelys* and *Australochelys* from the Triassic, *Kayentachelys* from the Early Jurassic, *Glyptops*, *Pleurosternon* (= *Mesochelys*), *Plesiochelys*, *Portlandemys* and *Solnhofia* from the Late Jurassic, *Dorsetochelys* from the Early Cretaceous, *Kallokibotion* from the Late Cretaceous, Baenidae, *Neurankylus* and *Boremys*, from the Cretaceous and Palaeogene and *Meiolania* from the Pleistocene. With the exception of *Meiolania*, from Australia, these are all West European and North American forms which belong either to turtles of the Triassic type, or to primitive cryptodirans such as the Pleurosternidae and Plesiochelyidae, or are clearly rare aberrant forms such as *Kallokibotion*. Asian forms already described include *Sinemys* (Brinkman and Peng, 1993b) from the Late Jurassic and Early Cretaceous, *Dracochelys* (Gaffney and Yeh, 1992) representing the Cretaceous macrobaenids, the trionychidid *Khunnuchelys* (Brinkman and Nesov, 1993), and

the American form *Adocus* (Meylan and Gaffney, 1989) from a group of undoubtedly Asian origin.

The long isolation and stable conditions of the Asian continent, which includes the territory of modern Central Asia, inevitably led to the development, within its boundaries, of an autochthonous vertebrate fauna including turtles. During the Mesozoic, as a rule, Asian turtles were found in the lakes and rivers of inner-continental basins, sometimes appearing in marginal marine environments such as estuaries in the western limits of Middle Asia. As a result, there are a large number of endemic Asian groups in the Mesozoic, but an almost complete absence of purely marine forms and representatives of the Pleurodira. The Asian fossil record also contains the earliest representatives of a number of families (Adocidae, Nanhsiungchelyidae, Carettochelyidae, Trionychidae) all of which are also known from other continents. It also includes primary stages in the major lineages of the Cryptodira (Chelydroidea, Testudinoidea and Trionychoidea) represented by the endemic Asian families Xinjiangchelyidae, Macrobaenidae, Sinemydidae, Lindholmemydidae and Haichemydidae. It should also be added that the earliest representatives of some widely distributed modern families, such as the Emydidae (in its broadest sense) and Testudinidae are also found in this region, in the Palaeocene of Mongolia and China. Finally, we cannot name a single group of Mesozoic Asian turtles whose origins can be traced to other continents. The only exception may be the Mongolochelyidae, which, despite its late occurrence, occupies a special place among Mongolian turtles because it is extremely primitive.

So far, however, apart from fragments described as *Proganochelys nuchae* from Thailand (Broin, 1984), the earliest stages of turtle evolution on the Asian Continent, which presumably took place in the Late Triassic and Early to Middle Jurassic, are not yet known. The Middle Jurassic Chenguichelyidae reveal nothing regarding the origin of Asian turtles and, in any case, require redescription and, preferably, new material.

Late Jurassic Asian turtles were diverse, and

probably numerous, and all belong to endemic families (Xinjiangchelyidae, Sinemydidae and Macrobaenidae), or families which clearly have an Asian origin, such as the Adocidae (see above for further comments regarding putative Jurassic records for this family). In Western Europe and North America, Late Jurassic turtles are represented by the Pleurosternidae, the Plesiochelyidae (which probably includes turtles previously assigned to the 'Thalassemydidae', see Gaffney and Meylan, 1988) and primitive pleurodires such as *Platychelys*. These turtles seem to be associated with coastal-marine deposits and the Plesiochelyidae (excluding Thalassemydidae for the moment), are often referred to as marine turtles (Gaffney *et al.*, 1991). In addition, with the exception of clearly specialized 'marine' forms, the shell is well consolidated in all cases (for example, the carapace is connected to the plastron by a suture), there is only moderate development or absence of the temporal emargination in the skull and mesoplastra are present in the Pleurosternidae and *Platychelys*.

By contrast, the majority of Late Jurassic and Early Cretaceous Asian turtles (Xinjianchelyidae, Macrobaenidae and Sinemydidae) have a completely different shell construction with no mesoplastra and the plastron is ligamentously attached to the carapace. They also have deep temporal and cheek emarginations, features traditionally considered to be advanced characters. Where they occur in some European turtles (represented by lack of sutures between the carapace and plastron and the appearance of fontanelles in the shell) they are attributed to profound marine specialization.

Middle Jurassic turtles have recently been found in continental deposits in the vicinity of Moscow, associated with a diverse flora and fauna containing temnospondyls, crocodiles and fishes. Interestingly, they exhibit a number of characters typical of the 'Triassic type' (an open interpterygoid vacuity, comparable in size to that in *Proganochelys*, but without any teeth on the jaws or palate), and characters found in Late Jurassic Asian forms: a loose connection between the carapace and plastron and a deep temporal emargination in the skull. In addition, the plastron contains

mesoplastra and a large central fontanelle, while in the carapace there are small marginal fontanelles. These middle-sized turtles, up to 400 mm long, are now under study and probably represent a new family of Middle Jurassic turtles. However, it is too early to discuss possible relationships between these turtles and those Central Asian forms with a similar type of shell construction, or questions concerning the probable independent disappearance of the interpterygoid vacuity in different lineages of turtles.

In any case (with the exception of the Mongolochelyidae), in the most morphologically primitive Asian turtles, the Xinjiangchelyidae, the internal carotid artery is not enclosed in bony canals before entering the basisphenoid, resembling the state observed in Pleurosternidae, though without the separation of the pterygoids by the basisphenoid. In the Macrobaenidae and Sinemydidae the bony canals are present, and the posterior foramen for one internal carotid canal is displaced backwards to the posterior margin of the pterygoid, while anteriorly, in the region of separation between the cranial branch of the internal carotid artery and the palatal artery, there remains a more or less large opening which connected the bony canal of the carotid artery with the palatal surface of the skull. Interestingly, this opening, which can be considered as the plesiomorphic condition, is retained not only in late representatives of Macrobaenidae, such as *Macrobaena* from the Late Palaeocene, but is present also in other Asian groups of turtles, such as the Lindholmemydidae (*Mongolemys*), where this region of the skull has been researched, and is even found in some American turtles of Asian origin such as *Adocus* (Meylan and Gaffney, 1989). The wide distribution of this character and its temporal persistence is a reminder of distant relationships among Asian turtles and, at the same time, confirms the independent formation of a bony canal for the internal carotid artery in these turtles, in contrast to the American Baenidae (Brinkman and Nicholls, 1993).

All three groups of early Central Asian turtles, Xinjiangchelyidae, Macrobaenidae and Sinemydidae, are undoubtedly related to each other and the origins of the latter two families should probably be sought among the former. Their relationship to the Chelydridae, the first reliable representative of which appears only in the Palaeocene of North America, is a more complex question. They share the same general construction of the shell and the skulls are similar, although in the Chelydridae, even in the earliest forms, there is a completely enclosed bony canal for the internal carotid artery and no hint of a foramen basisphenoidale (possibly as in the Haichemydidae from the Late Cretaceous of Asia). For the present, the inclusion of all these turtles, together with Platysternidae, in one group, the Chelydroidea, does not seem unreasonable. At the same time, a relationship between the Asian groups mentioned above and the coastal-marine Toxochelyidae, first known from the Late Cretaceous of North America, does not look so likely. Even in the earliest forms, the skull has enclosed bony canals for the carotid artery, which suggests relationships with the marine West European Plesiochelyidae, and not with the continental Asian Macrobaenidae, with which they have sometimes been associated. However, all these questions are only likely to be resolved following the publication of detailed descriptions of the skulls of *Annemys*, *Hangaiemys* and *Macrobaena*. Macrobaenidae reached their greatest diversity in the Early Cretaceous and persisted into the Palaeocene, though their role in vertebrate faunas seems to have decreased. Late Cretaceous and Palaeocene macrobaenids were larger and clearly more specialized than Early Cretaceous forms.

New groups of turtles appear in Asia at the Early–Late Cretaceous boundary. These include the extant and widely distributed trionychids, known mainly from Asia, and followed later by the Carettochelyidae, represented by the Anosteirinae, known also from Europe. Both these families are associated with aqueous habitats and are active swimmers. Other water turtles which began to play a significant role in Late Cretaceous vertebrate faunas of Asia include the Adocidae and Nanhsiungchelyidae, though they are specialized in a different direction: walking on the bottom of lakes and streams. The shell construction was different in these turtles, the cara-

pace and plastron were connected by a suture, the bones of the shell were relatively thin, and the buttresses in the plastron were weakly developed and did not reach the costals. In the early Late Cretaceous these turtles achieved relatively large size. The Adocidae survived in Mongolia until the Early Oligocene, but the Nanhsiungchelyidae became extremely rare in the late Late Cretaceous. Adocids and nanhsiungchelyids appeared in North America in the Campanian, but did not achieve great diversity: only two genera have been reported: *Adocus* and *Basilemys*, and the latter is not known to have survived beyond the end of the Cretaceous. However, adocids were important in the Late Cretaceous and, in the form of *Adocus*, also in the Palaeocene faunas of North America where they equalled in numbers the turtles of local origin, the Baenidae. Notably, American faunas of this age also contain diverse marine turtles and representatives of the Pleurodira.

Another important lineage of turtles, the Testudinoidea, represented by the Sinochelyidae and the Lindholmemydidae, also appeared in Central Asia in the earliest Late Cretaceous, or possibly even slightly earlier. Remains of sinochelyids are extremely rare and more material is required to establish the systematic position of these turtles firmly. The Lindholmemydidae occupy a special place among Asian turtles because of their diversity, their occurrence in mass burials in the Late Cretaceous, and because they cross the Cretaceous–Palaeogene boundary. These are also the first Central Asian turtles that show a tendency to develop a stronger shell with thickened bony scutes, strong sutures which fasten the carapace and plastron together, and strongly developed buttresses. Despite the presence of a number of primitive characters (e.g. a complete row of inframarginals and a large splenial), the skull is similar to those of modern batagurins, though differing from them in the depth of the cheek emarginations and the presence of a foramen basisphenoidale. Recently, the Lindholmemydidae have been considered to lie near the base of Testudinoidea, and, at the same time, it is clear that they have no relation to the Dermatemydidae. Relationships with the

Platysternidae also seem extremely unlikely, since in all its basic characters the development of the shell in lindholmemydids seems to be evolving in a different direction.

In the Late Palaeocene of Mongolia lindholmemydids and emydids (in their broadest sense) occur in about equal numbers, though the former are more taxonomically diverse. It should be noted here that *Mongolemys tatarinovi* (Lindholmemydidae) and *Pseudochrysemys gobiensis* (Emydidae), both originally described on the basis of fragments from the Palaeocene (Sukhanov and Narmandakh, 1976), are now known from much more complete, but as yet unstudied remains, which may eventually elucidate the relationships of these two families.

The Haichemydidae, which have yet to be properly studied, are extremely interesting because of the construction of their shell, which has a weakened connection between the carapace and plastron and is thus similar to the majority of older Central Asian turtles (Xinjiangchelyidae, Sinemydidae and Macrobaenidae). From a morphological point of view this also renders them more likely to be ancestors to the Platisternidae, rather than the Lindholmemydidae.

In Mongolia, the latest Cretaceous is represented by deposits of the Nemegt Svita, which have no analogues in other regions of Central Asia. Primarily this is connected with the sudden appearance of abundant and large water turtles belonging to the Mongolochelyidae. These turtles occur in the same localities as *Mongolemys* and have a relatively flat shell and reduced plastron that, again, was connected to the carapace by ligaments. The many autapomorphies of *Mongolochelys* (discussed above) stand in contrast to the extremely primitive state of the palatal surface of the skull, equivalent to the stage of development seen in Pleurosternidae and Xinjiangchelyidae, and a number of primitive features of the shell (discussed above). The origins of this apparently relict group of Mongolian turtles is not clear: they stand apart from all other Central Asian turtles and do not seem to be related to any of them. The secondary overgrowth in the temporal region of the skull roof is observed in

some other turtles (e.g. *Kallokibotion* and the Meiolanidae), but is constructed according to different principles in *Mongolochelys* and is thus unique to this taxon.

Acknowledgements

The present study was supported by the Russian Foundation for Fundamental Research (RFFR) within the framework of project 96–04–50822 'Early stages in the evolution of reptiles and birds'. The original text was translated by N. Bakhurina and extensively edited by D. Unwin.

References

Bohlin, B. 1953. Fossil reptiles from Mongolia and Kansu. Report from the scientific expedition to the north western provinces of China under the leadership of Dr. Sven Hedin. *The Sino-Swedish expedition, VI. Vertebrate Palaeontology* 6: 1–113.

Brinkman, D.B. and Nesov, L.A. 1993. *Khunnuchelys* gen. nov., a new trionychid (Testudines: Trionichydae) from the Late Cretaceous of Inner Mongolia and Uzbekistan. *Canadian Journal of Earth Sciences* 30: 2214–2223.

—and Nicholls E.L. 1993. The skull of *Neurankylus eximus* (Testudines: Baenidae) and a reinterpretation of the relationships of this taxon. *Journal of Vertebrate Paleontology* 13: 273–281.

—and Peng Jiang-Hua 1993a. *Ordosemys leios*, n. gen., n. sp., a new turtle from the Early Cretaceous of the Ordos Basin, Inner Mongolia. *Canadian Journal of Earth Sciences* 30: 2128–2138.

—and— 1993b. New material of *Sinemys* (Testudines, Sinemydidae) from the Early Cretaceous of China. *Canadian Journal of Earth Sciences* 30: 2139–2152.

Broin F. de 1977. Contribution à l'étude des Chéloniens. Chéloniens continentaux du Crétacé et du Tertiare de France. *Mémoires du Museum National d'Histoire Naturelle, Série C* 38: 1–336.

—1984. *Proganochelys ruchae* n. sp., chélonien du Trias supérieur de Thailande. *Studia Geologie Salamanticensia, Volume Especial 1: Studia Palaecheloniologica* 2: 88–97.

Chen, Pei-Ji 1983. A survey on the non-marine Cretaceous in China. *Cretaceous Research* 4: 123–143.

Chkhikvadze, V.M. 1970. [Classification of the subclass Testudinata.] *16th Scientific Session of the Institute of Palaeobiology, Academy of Sciences, Georgia,* 7–8.

—1971. New turtles from the Oligocene of Kazakhstan and the taxonomic status of some species from Mongolia. *Soobshcheniya AN Gruzinskoi SSR* **62** (2): 489–492.

—1973. [*Tertiary turtles from the Zaisan depression.*] Izdatel'stvo 'Metsniereba', Tbilisi, 101 pp.

—1976. [New dates on fossil turtles from Mongolia, China and Western Kazakhstan.] *Soobshcheniya AN Gruzinskoi SSR* **82**: 746–748.

—1977. [Fossil turtles of the family Sinemydidae.] *Izvestiya AN Gruzinskoi SSR, Seriya Biologicheskaya* **3**: 265–270.

—1981. [On the question of the origin of bigheaded turtles.] In [*General questions of palaeobiology*]. Izdatel'stvo 'Metsniereba', Tbilisi: 131–146.

—1983. [*Fossil turtles from the Caucasus and northern regions of the Black Sea.*] Izdatel'stvo "Metsniereba", Tbilisi, 149 pp.

—1985. [Fossil turtles of the family Sinochelyidae.] *Vestnik Zoologii* **1**: 40–44.

—1987. Sur la classification et charactères de certaines tortues fossiles d'Asie, rares et peu etudiées. *Studia Geologie Salamanticensia, Volume Especial, 2. Studia Palaecheloniologica* **2**: 55–85.

—1989. [*Neogene turtles of the USSR.*] Izdatel'stvo 'Metsniereba', Tbilisi, 1–104.

—and Shuvalov, V. F. 1980. [On the question of the origin of three-clawed turtles.] *Soobshcheniya AN Gruzinskoi SSR* **100**: 502–503.

—and— 1988. The first Cretaceous chelonians in the Ekhingol Basin (Mongolia). *Acta Zoologica Cracoviensia* **31**: 509–512.

Chow, M.C. 1954. Cretaceous turtles from Laiyang, Shantung. *Acta Paleontologica Sinica* **2**: 395–408.

Clark, J. 1932. A new anosteirid from the Uinta Eocene. *Annals of the Carnegie Museum* **21**: 161–170.

Cope, E.D. 1870. On *Adocus*, a genus of Cretaceous Emydidae. *Proceedings of the American Philosophical Society* **11**: 295–298.

Dong, Zhiming 1995. The dinosaur complexes of China and their biochronology, pp. 91–96 in Ailing Sun and Yuanqing Wang (eds.), *Sixth Symposium on Mesozoic Terrestrial Ecosystems and Biota, Short Papers.* Beijing.

Endo, R. and Shikama, T. 1942. Mesozoic reptilian fauna in the Jehol Mountainland, Manchoukuo. *Bulletin Central National Museum, Manchoukuo* **3**: 1–23.

Fang, Q.R. 1987. A new species of Middle Jurassic turtles from Sichuan. *Acta Herpetologica Sinica* **6** (1): 65–69.

Gaffney, E.S. 1975a. A phylogeny and classification of the higher categories of turtles. *Bulletin of the American Museum of Natural History* **155**: 387–436.

—1975b. Phylogeny of the chelydrid turtles: a study of shared derived characters in the skull. *Fieldiana Geology* **33**: 157–178.

—1979. The Jurassic turtles of North America. *Bulletin of the American Museum of Natural History* **162**: 91–136.

—1984. Historical analysis of theories of chelonian relationships. *Systematic Zoology* **33**: 283–301.

—and Meylan, P.A. 1988. A phylogeny of turtles, pp. 157–219 in Benton, M.J. (ed.), *The phylogeny and classification of the tetrapods. Vol. 1: Amphibians, reptiles, birds. Systematics Association Special Volume 35A.* Oxford: Clarendon Press.

—, Meylan, P.A. and Wyss, A.R. 1991. A computer assisted analysis of the relationships of the higher categories of turtles. *Cladistics* **7**: 313–335.

—and Yeh, Hsiang-k'uei 1992. *Dracochelys*, a new cryptodiran turtle from the Early Cretaceous of China. *American Museum Novitates* **3048**: 1–13.

Gilmore, C.W. 1931. Fossil turtles of Mongolia. *Bulletin of the American Museum of Natural History* **59**: 213–257.

—1934. Fossil turtles of Mongolia. Second contribution. *American Museum Novitates* **689**: 1–14.

Hay, O.P. 1908. The fossil turtles of North America. *Publications of the Carnegie Institution, Washington* **75**: 1–568.

Hirayama, R. 1984. Cladistic analysis of batagurine turtles (Batagurinae: Emydidae: Testudinoidea); A preliminary Result. *Studia Geologie Salamanticensia, Volume Especial 1: Studia Palaecheloniologica* **1**: 141–157.

Jerzykiewicz, T., Koster, E.H. and Jia, Jian-Zheng 1993. Djadokhta Formation correlative strata in Chinese Inner Mongolia; an overview of the stratigraphy, sedimentary geology, and paleontology and comparisons with the type locality in the Pre-Altai Gobi. *Canadian Journal of Earth Sciences* **30**: 2180–2195.

—and Russell, D.A. 1991. Late Mesozoic stratigraphy and vertebrates of the Gobi Basin. *Cretaceous Research* **12**: 345–377.

Ji, S. 1995. Fossil Reptiles. Order Testudines, pp. 140–146 and 202–203 in *Fauna and stratigraphy of the Jurassic-*

Cretaceous in Beijing and the adjacent areas. Beijing: Seismic Publishing House.

Kaznyshkin, M.N. 1988. [Late Jurassic turtles from Northern Fergana (Kirgizskaya SSR).] *Vestnik Zoologii* 1: 26–32.

—Nalbandyan, L.A. and Nesov, L.A. 1990. [Turtles from the Middle and Late Jurassic of Fergana (Kirgizskaya SSR).] *Ezhegodnik Vsesoyuznogo Paleontologicheskogo Obshchestva* 33: 185–204.

Khozatskii, L.I. 1957. [Freshwater turtles from the Upper Cretaceous of Fergana.] *Doklady AN Tadzhikskoi SSR* 22: 19–21.

—1976. [A new representative of the trionychids from the Late Cretaceous of Mongolia.] *Gerpetologiya. Nauchnye Trudy Kubanskogo Gos. Universiteta,* 218: 3–19.

—1997. [Large turtles from the Late Cretaceous of Mongolia.] *Russian Journal of Herpetology* 4: 148–154.

—2000 [Turtles-trionychids of the Cretaceous of Mongolia.] (in press).

—and Młynarski, M. 1971. Chelonians from the Upper Cretaceous of the Gobi desert, Mongolia. *Acta Palaeontologica Polonica* 25: 191–144.

—and Nesov, L.A. 1977. [Turtles of the genus *Adocus* from the Late Cretaceous of the SSSR.] *Trudy Zoologicheskogo Instituta AN SSSR* 74: 116–118.

—and— 1979. [Large turtles of the Late Cretaceous of Middle Asia.] In [Ecology and the systematics of amphibians and reptiles.] *Trudy Zoologicheskogo Instituta AN SSSR* 89: 98–108.

Kordikova, E. 1992a. [On the time and ways of prochoresis of the fossil trionychids from Kazakhstan.] *Soobshcheniya AN Gruzinskoi SSR* 145: 647–650.

—1992b. [Review of fossil trionychids from the Soviet Union.] *Izvestiya AN Gruzinskoi SSR, Seriya Biologicheskaya* 18: 131–141.

—1994a. On the systematics of fossil trionychids in Kazakhstan. *Selevinia* 2: 3–10.

—1994b. Review of fossil trionychid localities in the Soviet Union. *Courier Forschungs-Institut, Senckenberg* 173: 341–358.

Kuznetsov, V.V. 1976. [Freshwater turtles from the Senonian deposits of north-east Pre-Aralia.] *Paleontologicheskii Zhurnal* 1976 (4): 125–127

—and Chkhikvadze, V.M. 1987. [Late Cretaceous trionychids from the locality of Shak-Shak, in Kazakhstan. In [*Materials on the history of the fauna and flora of Kazakhstan*] 9: 33–39. Alma-Ata.

—and Shilin, P.V. 1983. [Late Cretaceous turtles from Baibishe (north-east Pre-Aralia).] *Izvestiya AN Kazakhskoi SSR, Seriya Biologicheskaya* 1983 (6): 41–44.

Langston, W.Jr. 1956. The shell of *Basilemys varialosa* (Cope). *Bulletin of the National Museum of Canada* 142: 155–165.

McKenna, M.C., Hutchison, J.H. and Hartman, J.H. 1987. Paleocene vertebrates and nonmarine Mollusca from the Goler Formation, California, pp. 31–41 in Cox, B.F. (ed.) *Basin Analysis and Paleontology of the Paleocene and Eocene Goler Formation, El Paso Mountains, California.* Los Angeles: Society of Economic Paleontologists and Mineralogists, Pacific Section.

Maryańska, T. and Osmólska, H. 1975. Protoceratopsidae (Dinosauria) of Asia. *Acta Palaeontologica Polonica* 33: 133–182.

Matthew, W.D. and Granger, W. 1923. The fauna of the Ardyn Obo formation. *American Museum Novitates* 98: 1–5.

Merkulova, N.N. 1978. [New *Trionyx* from Nemeget (MPR).] *Byulleten' Moskovskogo Obshchestva Ispytatelei Prirody, Otdel Geologicheskii* 53 (3): 156.

Meylan, P.A. and Gaffney, E.S. 1989. The skeletal morphology of the Cretaceous cryptodiran turtle *Adocus*, and the relationships of the Trionychoidea. *American Museum Novitates* 2941: 1–60.

—and— 1992. *Sinaspideretes* is not the oldest trionychid turtle. *Journal of Vertebrate Paleontology* 12: 257–259.

Młynarskii, M. 1969. *Fossile Schildkröten.* Die Neue Brehm-Bucherei, 396 pp.

—1972. *Zangerlia testudimorpha* n. gen., n. sp. a primitive land tortoise from the Upper Cretaceous of Mongolia. *Acta Palaeontologica Polonica* 27: 85–97.

—and Narmandakh, P. 1972. New turtle from the Upper Cretaceous of the Gobi desert, Mongolia. *Acta Palaeontologica Polonica* 27: 95–102.

Narmandakh, P. 1985. [A new species of *Adocus* from the Late Cretaceous of Mongolia.] *Paleontologicheskii Zhurnal* 1985 (2): 85–93.

Nesov, L.A. 1976. [On the systematics and phylogeny of two clawed turtles.] *Vestnik Leningradskogo Universiteta* 1976 (9): 7–17.

—1977a. [A new genus of two clawed turtle from the Upper Cretaceous of Karakalpakia.] *Paleontologicheskii Zhurnal* 1977 (1): 103–114.

—1977b. [On some particulars of the skull of two Late Cretaceous turtles.] *Vestnik Leningradskogo Universiteta* 1977 (21): 45–48.

—1977c. [Construction of the skull in Early Cretaceous

turtles of the family Adocidae.] *Trudy Zoologicheskogo Instituta AN SSSR* **74**: 75–79.

—1977d. [Turtles and some other reptiles from the Cretaceous of Karakalpakia.] pp. 155–156 in [*Questions of herpetology. Sixth All Union Herpetological Conference*]. Leningrad: Izdatel'stvo Nauka.

—1978. [Archaic Late Cretaceous turtles from Western Uzbekistan.] *Paleontologicheskii Zhurnal* **1978** (4): 101–105.

—1981a. [On turtles of the family Dermatemydidae from the Cretaceous of the Amur river basin, and some other rare discoveries of the remains of ancient turtles of Asia.] pp. 69–73 in [*Herpetological research in Siberia and the Far East.*] Leningrad.

—1981b. [On the phylogenetic relationships of some families of terrestrial turtles.] in [*Life on ancient continents, its establishment and development.*] *Trudy Vsesoyuznogo Paleontologicheskogo Obshchestva* **1981**: 133–141.

—1984. Data on Late Mesozoic turtles from the USSR. *Studia Geologie Salamanticensia, Volume Especial 1: Studia Palaecheloniologica* **1**: 215–223.

—1986a. Some Late Mesozoic and Paleocene turtles of Soviet Middle Asia. *Studia Geologie Salamanticensia, Studia Palaecheloniologica* **2**: 7–22.

—1986b. [On the level of morphology and phylogenetic relationships in the evolution of marine turtles.] pp. 179–186 in [*Morphology and evolution of animals*]. Moscow: Izdatel'stvo Nauka.

—1987a. [Results of research on Cretaceous and early Palaeogene mammals on the territory of the USSR.] *Ezhegodnik Vsesoyuznogo Palaeontologicheskogo Obshchestva* **30** (1987): 199–218.

—1987b. On some Mesozoic turtles of the Soviet Union, Mongolia and China, with comments on systematics. *Studia Palaeocheloniologica* **2** (4): 87–102.

—1988. [Vertebrate assemblages from the Late Mesozoic and Paleocene of Middle Asia.] in [*The establishment and evolution of continental biotas.*] *Trudy Vsesoyuznogo Paleontologicheskogo Obshchestva* **1988**: 93–101.

—1995. On some Mesozoic turtles of the Fergana depression (Kirgizstan) and Dzhungar Alatau ridge (Kazakhstan). *Russian Journal of Herpetology* **2**: 134–141.

—and Golovneva, L.B. 1983. [Changes in vertebrate assemblages in the Cenomanian–Santonian (Late Cretaceous) of Kyzylkum.] pp. 126–134 in [*Palaeontology and the evolution of the biosphere.*] Leningrad.

—and Yulinen, V.A. 1977. [On the phylogenetic relation-

ships and history of distribution of some families of continental turtles.] in [*Life on ancient continents, its establishment and development.*] *Trudy Vsesoyuznogo Paleontologicheskogo Obshchestva* **1977**: 54–56.

—and Kaznyshkin, M.N. 1985. [Lungfish and turtles from the Late Jurassic of North Fergana (Kirgiz SSR).] *Vestnik Zoologii* **1**: 33–39.

—and Khozatskii, L.I. 1973. [Early Cretaceous turtles from South-Eastern Fergana.] pp. 132–133 in [*Questions of Herpetology.*] *Doklady III Vsesoyuznoi Gerpetologicheskoi Konferentsii.* Leningrad: Izdatel'stvo 'Nauka'.

—and— 1977a. [Freshwater turtle from the Early Cretaceous of Fergana.] *Ezhegodnik Vsesoyuznogo Paleontologicheskogo Obshchestva* **20**: 248–262.

—and— 1977b. [Mesozoic turtles of the USSR.] pp. 157–159 in [*Questions of herpetology.*] *Doklady IV Vsesoyuznoi Gerpetologicheskoi Konferentsii.* Leningrad: Izdatel'stvo 'Nauka'.

—and— 1978. [Early Cretaceous turtles of Kirgizstan.] *Ezhegodnik Vsesoyuznogo Paleontologicheskogo Obshchestva* **21**: 267–279.

—and— 1980. [Turtles of the genus *Lindholmemys* from the Late Cretaceous of the USSR.] *Ezhegodnik Vsesoyuznogo Paleontologicheskogo Obshchestva* **23**: 250–264.

—and— 1981a. [The history of some groups of turtles in connection with the destiny of continents.] pp. 153–160 in [*Palaeontology, palaeobiogeography and mobilism.*] Magadan.

—and— 1981b. [Early Cretaceous turtles of Transbaikal.] pp. 74–78 in [*Herpetological research in Siberia and the Far East.*] Leningrad.

—and Krasovskaya, T.B. 1984. [Transformations in the composition of turtle assemblages in the Cretaceous of Middle Asia.] *Vestnik Leningradskogo Universiteta* **3**: 15–25.

—and Mertinene, R.A. 1982. [Teeth of cartilaginous fish as a source of information concerning the age of deposits of Cretaceous estuaries in Middle Asia.] in [*Palaeontology and detailed stratigraphic correlation.*] *Tezisy dokladov XVIII Vsesoyuznogo Paleontologicheskogo Obshchestva*, Tashkent **1982**: 56–57.

—and Verzilin, N.N. 1981. [Remains of turtles from Aptian-Albian deposits of the Transaltai Gobi in Mongolia and conditions of their burial.] in [*Fossil vertebrates of Mongolia.*] *Trudy Sovmestnoi Sovetsko-Mongol'skoi Paleontologicheskoi Ekspeditsii* **15**: 12–26.

Novacek, M. J., Norell, M., McKenna, M. C. and Clark, J.

1994. Fossils of the Flaming Cliffs. *Scientific American* **271** (6): 36–43.

Peng, Jiang-Hua and Brinkman, D.B. 1993. New material of *Xinjiangchelys* (Reptilia: Testudines) from the Late Jurassic Qigu Formation (Shishugou Group) of the Pingfengshan locality, Junggar Basin, Xinjiang. *Canadian Journal of Earth Sciences* **30**: 2013–2026.

Ryabinin, A.N. 1930. On the age and fauna of the dinosaur beds on the Amur River. *Zapiski Imperatorskogo Sankt-Peterburgskogo Mineralogicheskogo Obshchestva* **59** (2): 41–51.

—1935. [Remains of turtles from the Upper Cretaceous deposits of the Kyzylkum desert.] *Trudy Paleontologicheskogo Instituta AN SSSR* **4**: 69–78.

—1948. [Turtles from Jurassic of Kara-Tau.] *Trudy Paleontologicheskogo Instituta AN SSSR* **15**: 94–98.

Romer, A.S. 1956. *The osteology of the reptiles.* Chicago.

Shuvalov, V.F. 1982. [Palaeogeography and history of development of the lake systems in Mongolia in Jurassic and Cretaceous time.] pp. 18–80 in Martinson, G.G. (ed.) [*Mesozoic lake basins of Mongolia. Palaeogeography, lithology, palaeobiogeochemistry and palaeontology.*] Leningrad: Nauka.

—and Chkhikvadze, V.M. 1975. [New data on Late Cretaceous turtles of South Mongolia.] in [*Fossil faunas and floras of Mongolia.*] *Trudy Sovmestnoi Sovetsko-Mongol'skoi Paleontologicheskoi Ekspeditsii* **2**: 214–229.

—and— 1979. [On the stratigraphic and systematic position of some freshwater turtles from new Cretaceous localities in Mongolia.] in [*Mesozoic and Cenozoic faunas of Mongolia.*] *Trudy Sovmestnoi Sovetsko-Mongol'skoi Paleontologicheskoi Ekspeditsii* **8**: 58–76.

—and— 1986. [On the composition and age of deposits of the Ehingol depression in Mongolia and the first findings of fossil turtles there.] *Soobshcheniya AN Gruzinskoi SSR* **124**: 433–436.

Sukhanov, V.B. 1964. [Subclass Testudinata.] pp. 354–438 in Orlov, Yu.A. (ed.) [*Osnovy paleontologii. Zemnovodnye, presmykayushchiesya i ptitsy.*] Moscow: Nauka.

—1978. [Subclass Testudinata.] pp. 84–102 in *Razvitie i smena organicheskogo mira na rubezhe mezozoya i kainozoya. Pozvonochnye.* Moscow: Nauka.

—and Narmandakh, P. 1974. [A new Early Cretaceous turtle from continental deposits of the Northern Gobi.] In [*Fauna and biostratigraphy of the Mesozoic and Cenozoic of Mongolia.*] *Trudy Sovmestnoi Sovetsko-Mongol'skoi Paleontologicheskoi Ekspeditsii* **1**: 192–220.

—and— 1975. [Turtles of the group *Basilemys* (Chelonia, Dermatemydidae) in Asia]. in [*Fossil faunas and floras of Mongolia.*] *Trudy Sovmestnoi Sovetsko-Mongol'skoi Paleontologicheskoi Ekspeditsii* **2**: 94–101. Moscow, Izdatel'stvo 'Nauka'.

—and— 1976. [Paleocene turtles from Mongolia.] in [Palaeontology and biostratigraphy of Mongolia.] *Trudy Sovmestnoi Sovetsko-Mongol'skoi Paleontologicheskoi Ekspeditsii* **3**: 107–133.

—and— 1977. [The shell and limbs of *Basilemys orientalis* (Chelonia, Dermatemydidae).] in [*Fauna, flora and biostratigraphy of the Mesozoic and Cenozoic of Mongolia.*] *Trudy Sovmestnoi Sovetsko-Mongol'skoi Paleontologicheskoi Ekspeditsii* **4**: 57–80.

—and— 1983. [A new genus of Late Cretaceous turtle from Mongolia.] in [Fossil reptiles of Mongolia.] *Trudy Sovmestnoi Sovetsko-Mongol'skoi Paleontologicheskoi Ekspeditsii* **24**: 44–66.

—and— [New forms of turtles from Mesozoic deposits of Mongolia.] *Paleontologicheskii Zhurnal* (in press).

Tatarinov, L.P. 1959. [New turtle of the family Baenidae from the Lower Eocene of Mongolia.] *Paleontologicheskii Zhurnal* **1**: 100–113.

Tokunaga, S. and Shimizu, S. 1926. The Cretaceous Formation of Futaba in Iwaki and its fossils. *Journal of the Faculty of Science, University of Tokyo, Section 2* **1**: 181–212.

Williams, E.E. 1950. Variation and selection in the cervical central articulation of living turtles. *Bulletin of the American Museum of Natural History* **94**: 505–562.

Wiman, C. 1930. Fossile Schildkröten aus China. *Palaeontologia Sinica, series C* **76**: 1–56.

Yeh, Hsiang-K'uei 1963. Fossil turtles of China. *Palaeontologia Sinica, series C* **18**: 1–112.

—1965. New materials of fossil turtles of Inner Mongolia. *Vertebrata PalAsiatica* **9**(1): 47–69.

—1966. A new Cretaceous turtle from Nansiung, Northern Kwangtung. *Vertebrata PalAsiatica* **10** (2): 191–200.

—1973a. [Chelonian fossils from Wuerho.] *Memoirs of the Institute of Vertebrate Paleontology and Paleoanthropology* **11**: 8–11.

—1973b. Discovery of *Plesiochelys* from the Upper Lufeng series, Oshan, Yunnan and its stratigraphical significance. *Vertebrata PalAsiatica* **11** (2): 160–163.

—1974a. Cenozoic chelonian fossils from Nanhsiung, Kwangtung. *Vertebrata PalAsiatica* **12** (1): 26–42.

—1974b. A new fossil dermatemydid from Sinkiang. *Vertebrata PalAsiatica* 12 (4): 257–261.

—1982. Middle Jurassic turtles from Sechuan, S.W. China. *Vertebrata PalAsiatica* 20 (4): 282–290.

—1983a. A turtle carapace from the Late Jurassic of Weiyuan, Sichuan. *Vertebrata PalAsiatica* 21 (3): 188–192.

—1983b. A Jurassic turtle from Chenxi, Hunan. *Vertebrata PalAsiatica* 21 (4): 286–291.

—1986a. A Jurassic turtle is discovered for the first time in the Xinjiang Uygur Autonomus Region. *Vertebrata PalAsiatica* 24 (3): 171–181.

—1986b. New material of *Plesiochelys radiplicatus* with preliminary discussion of related problems. *Vertebrata PalAsiatica* 24 (4): 269–273.

—1988. Early records of fossil turtles in China. *Acta Zoologica Cracoviensis* 31 (2): 451–455.

—1990. Fossil turtles from Dashanpu, Zigong, Sechuan. *Vertebrata PalAsiatica* 28 (4): 304–311.

Young, Chung-chien and Chow, Min-chen 1953. New fossil reptiles from Szechuan, China. *Acta Scientia Sinica* 2 (3): 216–243.

Zangerl, R. 1969. The turtle shell, pp. 311–339 in Gans, C. (ed.) *Biology of the Reptilia, Volume 1, Morphology.* New York: John Wiley and Sons.

18

The fossil record of Cretaceous lizards from Mongolia

VLADIMIR R. ALIFANOV

Introduction

Lizards, represented by several thousand extant species, are one of the most successful groups of modern reptiles. The first lizards of modern type are known from the Middle Jurassic (Evans, 1993) and the first representatives of modern families appear in the Late Cretaceous (Alifanov, 1989a). By the Late Cretaceous the diversity, at the family level, of Asian lizards had reached its apogee, with both Mesozoic and modern families found together following the beginning of the Asian–American faunal interchange.

In this chapter, extraordinarily rich and diverse lizard complexes from the Lower and Upper Cretaceous deposits of the Gobi Desert in the People's Republic of Mongolia, known mainly from palaeontological discoveries in the South Gobi, are briefly outlined. Lizard remains from the Upper Cretaceous of this region include thousands of specimens, many of them well preserved, often with articulated skulls and sometimes with postcranial skeletons. This concentration of fossil material is unusual for lizards because of their small size and kinetic skulls, which renders them particularly susceptible to rapid disarticulation after death.

The most important aspect of the Cretaceous lizards of Mongolia is their surprisingly high taxonomic diversity. Almost twenty families, representing all infraorders of lizards, have so far been described, and some of these families contain large numbers of genera and species. The origin of this abundance is enigmatic and requires explanation. A brief comparative analysis of Early and Late Cretaceous lizard faunas helps to clarify some of the main stages in the evolution of lizards on the territory of Central Asia.

The first fossil Mongolian lizards were described by Gilmore (1943) on the basis of several rather poorly preserved specimens collected from Upper Cretaceous and Palaeogene localities of the Gobi Desert by expeditions of the American Museum of Natural History in the years 1923 to 1930. Sulimski (1972, 1975, 1978, 1984) and later Borsuk-Białynicka (1984, 1985, 1987, 1988, 1990), Borsuk-Białynicka and Moody (1984) and Borsuk-Białynicka and Alifanov (1991) described numerous new Upper Cretaceous taxa from the Gobi Desert of Mongolia, based on specimens collected by the Polish–Mongolian Expeditions (1965–1972). Recently, Alifanov (1988, 1989a, b, 1991, 1993a, 1996) described further new Cretaceous taxa found by the Soviet (Russian)–Mongolian expeditions which began in 1969. These papers also included information on fossil remains from the Lower Cretaceous and the Palaeogene (Alifanov, 1993b, c). More recently, a brief review of fossil lizards from the Upper Cretaceous Djadokhta and Baruungoyot Svitas of Mongolia was conducted by Borsuk-Białynicka (1991) and further remains were collected by the Joint expedition of Mongolian Academy of Sciences and American Museum of Natural History (Dashzeveg *et al.*, 1995).

The fossil material referred to in the systematic section of this paper was collected from several highly productive localities in the Mongolian Cretaceous, also famous for their fossil mammals and dinosaurs (see Chapters 12–13, 22–26, 29). The Upper Cretaceous lizard material comes from three consecutive svitas (Gradziński *et al.*, 1977; Jerzykiewicz and

Russell, 1991): the Djadokhta (localities of Bayan Zag, Tögrögiin Shiree, Zamyn Khond), the Baruungoyot (localities of Nemegt, Hermiin Tsav and Khulsan) and the Nemegt (localities of Bügiin Tsav, Guriliin Tsav and Tsagaan Khushuu), the terminal Cretaceous sequence and in which lizard remains are rare. The oldest fossil material discussed in this work comes from the locality of Höövör. Deposits at this site are thought to belong to the Hühteeg Svita, originally dated as Aptian–Albian (Shuvalov, 1974). Fossil material from this locality is disarticulated and was collected by screen washing.

Repository abbreviations

AMNH, American Museum of National History, New York; PIN, Palaeontological Institute, Russian Academy of Sciences, Moscow; ZPAL, Palaeobiological Institute, Polish Academy of Sciences, Warsaw.

Systematic palaeontology

IGUANIA Cope, 1864

The families Iguanidae, Agamidae and Chamaeleonidae are traditionally included within this infraorder. Moody (1980), Estes *et al.* (1988), Etheridge and Queiroz (1988) and Frost and Etheridge (1989) have noted the possibility that the first two families may be paraphyletic and propose that they be divided into several separate family groups for the Agamidae (*sensu lato*) (Moody, 1980) and Iguanidae (*sensu lato*) (Frost and Etheridge, 1989).

'IGUANIA' indet.

Remarks. Numerous unpaired and rugose iguanian-like frontals with a clear incision on their parietal border for the parietal opening have been found in late Lower Cretaceous deposits at Höövör. The exact systematic position of these remains is still unclear, but they appear to represent the earliest iguanian lizards yet known.

PHRYNOSOMATIDAE Fitzinger, 1843
Record. This family is represented by two genera from Khulsan: *Polrussia mongoliensis* Borsuk-Białynicka and

Alifanov, 1991 (holotype: ZPAL MgR-I/119), is represented by a partly broken skull with mandible (Figure 18.1A–D) and *Igua minuta* Borsuk-Białynicka and Alifanov, 1991 (holotype: ZPAL MgR-I/60) is based on a skull with a damaged preorbital region and mandible.

Polrussia is distinguished by its wide and flat skull with short snout, large orbits, largely unreduced splenial and conical teeth. *Igua* has wide and non-rugose parietals, fusion of the dentary tube anterior to the splenial, which reaches the mid-level of the dentary, clearly tricuspid teeth and very long quadrates.

Remarks. The two Mongolian genera were described by Borsuk-Białynicka and Alifanov (1991) as members of the Iguanidae *sensu lato*, without clear definition of their relationships to extant lizards. In my opinion, both genera belong most likely to the family Phrynosomatidae because of similarities between the construction of the posterior processes of the dentary with the same element in the extant forms *Phrynosoma* and *Sceloporus*. In these taxa the dentary has a distinct surangular process which does not cross the level of the anterior angular opening, and an elongate angular process. Occasionally, there is an extensive incision between these processes.

Among Asiatic lizards the genera *Anchaurosaurus* and *Xihaina*, described by Gao and Hou (1995) from the Upper Cretaceous of Inner Mongolia (China), can also be preliminarily assigned to this group. A sceleporine lizard has also been reported from the Upper Cretaceous of North America (Denton and O'Neill, 1993).

PRISCAGAMIDAE Borsuk-Białynicka and Moody, 1984 (=AGAMIDAE, PRISCAGAMINAE Borsuk-Białynicka and Moody, 1984)
Record. Six genera belonging to the subfamily Priscagamidae have been described from the Upper Cretaceous of Mongolia. *Priscagama gobiensis* Borsuk-Białynicka and Moody, 1984, is based on a holotype (ZPAL MgR/III-32) consisting of a damaged skull with mandible (Figure 18.2A–C) from the locality of Hermiin Tsav, and other specimens have been found

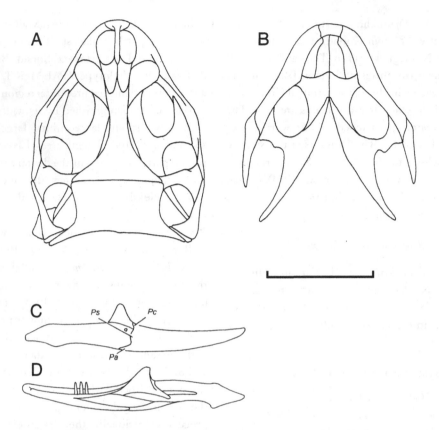

Figure 18.1. *Polrussia mongoliensis* Borsuk-Białynicka and Alifanov, 1991: reconstruction of the skull in (A) dorsal and (B) ventral view, and the right mandible in (C) lateral and (D) medial view. Abbreviations: Pa, angular process; Pc, coronoid process; Ps, surangular process. Scale bar = 5 mm.

at Bayan Zag. *Chamaeleognathus iordanskyi* Alifanov, 1996, is known from a very well preserved skull with mandible (holotype: PIN 3142/345) and one other specimen collected from Hermiin Tsav. The only known specimen of *Cretagama bialynickae* Alifanov, 1996 also consists of a well preserved skull and mandible (holotype: ZPAL MgR/III-32). *Flaviagama dzerzhinskii* Alifanov, 1989b (holotype: PIN 3143/101) is based on a well preserved skull with mandible from Tögrögiin Shiree and *Morunasius modestus* Alifanov, 1996 (holotype: PIN N 3142/317) on a well preserved skull and mandible from Hermiin Tsav. *Phrynosomimus asper* Alifanov, 1996 (holotype: PIN 3142/318), is represented by a skull, lacking palatal bones, and a fragmentary mandible, also from Hermiin Tsav.

In addition to these records, Nesov (1988) reported the discovery of representatives of Priscagamidae in the Upper Cretaceous of Middle Asia and *Priscagama* has also been found in Inner Mongolia, China (Gao and Hou, 1995).

Remarks. The Priscagamidae is a group of Upper Cretaceous Asiatic acrodontan iguanians, that are distinguished by an unusual combination of derived and plesiomorphic characters: the bones of the skull roof are sculptured; the maxillae contact behind the premaxilla; the postfrontal is completely reduced; the labial process of the coronoid covers the posterior part of the dentary; the splenial is large; and the dentition is agamid-like: canine-like teeth are present in hatchlings and the post canine-like teeth are sub-triangular

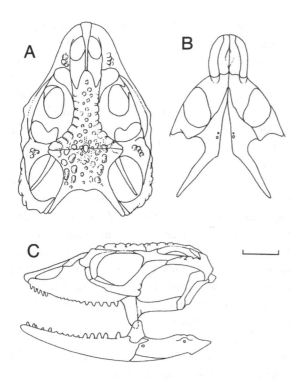

Figure 18.2. *Priscagama gobiensis* Borsuk-Białynicka and Moody, 1984: reconstruction of the skull in (A) dorsal, (B) ventral and, with the left mandible in (C) left lateral view. Scale bar = 5 mm.

in form (conical with an expanded base) and do not undergo replacement.

Initially, this group was established as a separate subfamily, the Priscagaminae, in the family 'Agamidae' by Borsuk-Białynicka and Moody (1984). Alifanov (1989b) noted the possible non-agamid nature of priscagamids and proposed that they be recognized as a separate family, and subsequently argued (Alifanov, 1996) that their closest relationships lay with the extant Hoplocercidae (*sensu* Frost and Etheridge, 1989). The Priscagamidae–Hoplocercidae group has some characters also found in the Chamaeleonidae including: sculptured bones of the skull roof, posterior expansion of the surangular process of the dentary above the anterior surangular foramen, and absence of the coronoid process of the dentary. Thus, the agamid-like condition of the priscagamid dentition may be homoplastic.

The family Priscagamidae was divided by Alifanov (1996) into two subfamilies: Priscagaminae Borsuk-Białynicka and Moody, 1984, including *Priscagama*, *Chamaeleognathus*, and *Cretagama*, and Flaviagaminae Alifanov, 1996, comprising *Flaviagama*, *Morunasius* and *Phrynosomimus*. The Priscagaminae are distinguished by caudally lengthened and dorsally oriented nares, lack of contact between pterygoids and vomers, the wide and short labial process of the coronoid, isolation of the nasals following contact between the ascending process of the premaxilla and the frontal, and the presence of four or more canine-like teeth. Flaviagamines are distinguished by their larger parietal foramen, the narrow and elongate labial process of the coronoid and the presence of, at most, three canine-like teeth.

HOPLOCERCIDAE Frost and Etheridge, 1989
Record. Four species of this family are known from Mongolia. *Pleurodontagama aenigmatodes* Borsuk-Białynicka and Moody, 1984, is based on a single specimen from Hermiin Tsav (holotype: ZPAL MgR-III/35), consisting of a fairly complete skull with mandible. *Gladidenagama semiplena* Alifanov, 1996, from the same locality, is based on a single specimen (PIN 3142/319), consisting of a fairly complete skull with mandible. The holotype of *Mimeosaurus crassus* Gilmore, 1943, consists of a left maxilla and associated incomplete jugal and ectopterygoid (AMNH 6655) from Bayan Zag, which has yielded further remains of this lizard including isolated maxillae and dentaries (Borsuk-Białynicka and Moody, 1984). *Mimeosaurus tugrikinensis* Alifanov, 1989 was established upon a single specimen, consisting of a disarticulated skull and fragmentary mandible (PIN 3143/102, Figure 18.3A, B) from Tögrögiin Shiree.

Pleurodontagama aenigmatodes and *Mimeosaurus* cf. *tugrikinensis* have also been reported from Djadokhta sediments of Inner Mongolia, China (Gao and Hou, 1995), and further *Mimeosaurus* material from China was referred by Gao and Hou (1995) to *M. crassus*. However, judging from their illustrations (Gao and Hou 1995, fig. 8c), this material is more similar to *M. tugrikinensis* in that the first two teeth are enlarged and 'canine-like' and there are clear diastemas between

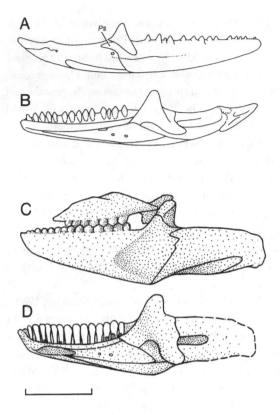

Figure 18.3. *Mimeosaurus tugrikinensis* Alifanov, 1989b: reconstruction of the right mandible in (A) lateral and (B) medial view. *Isodontosaurus gracilis* Gilmore, 1943 (ZPAL MgR-II/39): reconstruction of the left mandible in (C) lateral view and the right mandible in (D) medial view. Abbreviations: Ps, surangular process. Scale bar = 5 mm.

to a separate subfamily, the Pleurodontagaminae by Alifanov (1996). This taxon is characterized by contact between the maxillae behind the premaxilla, contact between the frontal and the dorsal process of the maxilla, large and wide vomers, a long anterior process of the pterygoids that sometimes reaches the vomers, labio-lingual compression of the upper parts of the tooth crowns, and the enlargement of one or two maxillary teeth which become caniniform.

The first hoplocercid to be recovered from the Late Cretaceous of Asia, *Mimeosaurus crassus*, was assigned by Gilmore (1943) to the Chamaeleonidae. Later, this lizard was redescribed by Borsuk-Białynicka and Moody (1984) and placed in their Priscagaminae together with the poorly known *Pleurodontagama*. The latter has some general similarity to *Priscagama*, though not in the dentition which is pleurodont and undergoes replacement.

Mimeosaurus tugrikinensis Alifanov, 1989 was declared by Gao and Hou (1995) with reference to the International Code of Zoological Nomenclature (ICZN; Ride *et al.*, 1985) as an invalid taxon for which 'Alifanov (1989) provided no diagnosis' (Gao and Hou, 1995, p. 73). However, according to Article 13a of the ICZN, establishment of a taxon requires a 'description or diagnosis'. Alifanov's erection of *Mimeosaurus tugrikinensis* conforms to this requirement, since a description was given under the special heading 'description', both in the original Russian version of the paper and in the English translation.

ISODONTOSAURIDAE Alifanov, 1993a stat. nov. (=AGAMIDAE, ISODONTOSAURINAE Alifanov, 1993 b)

Record. In the Upper Cretaceous of Mongolia this family is represented by a single form, *Isodontosaurus gracilis* Gilmore, 1943, based on a fragmentary right ramus of a mandible from Bayan Zag (AMNH N 6647). A partially damaged and compressed skull with mandibles (ZPAL MgR-II/39, Figure 18.3C, D) also appears to belong to this species. Remains of *Isodontosaurus* have also been found in Djadokhta sediments of the Gobi Desert in China (Gao and Hou, 1995) and further remains of isodontosaurids, with

adjacent teeth. Resolution of this problem may be possible after further preparation of the Chinese specimens and re-evaluation of all *Mimeosaurus* material.

A series of as yet undescribed remains belonging to this subfamily has also been collected from the Upper Cretaceous of Middle Asia (Nesov, 1988).

Remarks. The Hoplocercidae, erected by Frost and Etheridge (1989), is a relatively small family of modern lizards now restricted to South America. They resemble the Priscagamidae, but are distinguished from them by their pleurodont non-agamid-like dentition, with tooth crowns that have narrow bases and are non-conical.

The four Mongolian taxa listed above were assigned

various forms of teeth, are also known from the Upper Palaeocene and Middle Eocene of Mongolia. The latter are similar to Chinese finds, including *Qianshanosaurus* from the Palaeogene (Hou, 1974) and *Creberidentat* from the Middle Eocene (Li, 1991), both of which probably belong to the Isodontosauridae.

Remarks. Previously, the systematic status of *Isodontosaurus* and *Qianshanosaurus* was uncertain, and Estes (1983) listed them as 'Incertae sedis'. Alifanov (1993a) showed that the Isodontosauridae are Late Cretaceous Asiatic iguanians with an agamid-like skull and mandibles. The bones of the skull roof are smooth, the coronoid has no labial process, the surangular process of the dentary extends above the anterior surangular foramen, the coronoid process of the dentary expands on to the labial surface of the coronoid and the anterior end of the angular is situated below the lower border of the splenial. The dentition is pleurodont, or hyperpleurodont, and shovel-like (teeth have narrow bases and labio-lingually flattened apices), but there are no fang-like teeth and there is no tooth replacement. The absence of maxillary fangs and presence of narrow-based teeth shows some similarities to the dentition of the extant lizard *Uromastyx*.

Isodontosauridae was first described by Alifanov (1993b) as a subfamily within the Agamidae *sensu lato*. However, the hyperpleurodont dentition with unusual shovel-like teeth (similar to those of pleurodontagamine hoplocercids) does not permit unification of *Isodontosaurus* with known agamoid taxa and thus it is placed within its own family.

SCINCOMORPHA Camp, 1923

Scincomorpha is a rather heterogeneous and controversial lizard group originally proposed by Camp (1923) for uniting Amphisbaenia and Cope's Leptoglossa. At present there is no satisfactory phylogenetic concept for the numerous fossil and living families assigned to this most problematic taxon and cranial characteristics valid for all members of this group seem impossible to find. Rieppel (1988), Estes (1983), Estes *et al.*, (1988) and Presch (1988) give further, more extended accounts of this problem.

TEIIDAE Gray, 1827

Records. A maxilla and dentary, identified as cf. *Leptochamos*, was collected at the locality of Tsagaan Khushuu.

Remarks. The Teiidae is distinguished by a complex of cranial characters including: a dorsal process on the squamosal, fused frontals, lateral contact between the ectopterygoids and palatines medial to the supradental ridge of the maxilla, expansion of the medial process of the postfrontal behind the postorbital to reach the parietal, a deep incision between the surangular and angular processes of the dentary, which is short, and a series of bicuspid teeth.

The extant subfamily Teiinae is known mainly from South America and was present there from at least the Late Cretaceous (Valencia *et al.*, 1990). North American Upper Cretaceous representatives of this family were placed by Denton and O'Neill (1995) in a new subfamily, Chamopinae, separating them from the extant Teiinae and Tupinambinae.

MACROCEPHALOSAURIDAE Sulimski, 1975

Record. The type genus *Macrocephalosaurus* is represented by three Mongolian species: *M. ferrugenosus* Gilmore, 1943 (holotype: AMNH 6520, based on a single incomplete skull with part of the right ramus of the mandible from Bayan Zag; *M. gilmorei* Sulimski, 1975 (holotype: ZPAL MgR-III/18, a complete skull with lower jaw from Hermiin Tsav); and *M. chulsanensis* Sulimski, 1975 (holotype: ZPAL MgR-I/14, a complete skull and postcranial skeleton from Khulsan). The latter species is also represented by many other remains.

The following macrocephalosaurids have also been described from Hermiin Tsav, Mongolia: *Darchansaurus estesi* Sulimski, 1975 (holotype: ZPAL MgR-III/6, based on a complete skull with lower jaw) (Figure 18.4A–D); *Erdenetosaurus robinsonae* Sulimski, 1975 (holotype: ZPAL MgR-III/19, based on a single skull with lower jaw); and *Cherminsaurus kozlowskii* Sulimski, 1975 (holotype: ZPAL MgR-III/24 based on a skull with lower jaw).

Remarks. Macrocephalosaurids are unusual Upper Cretaceous Asiatic lizards with a massive skull,

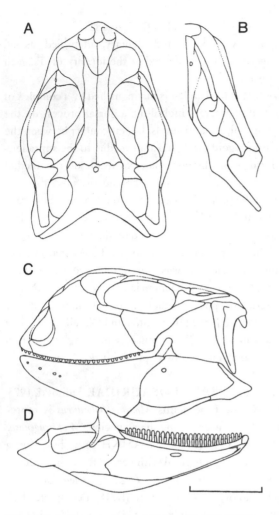

postfrontal; transversely situated articular processes of the postorbital; paired frontals; a massive medial (dorsal) process on the squamosal which partly or completely fuses with the supratemporal; contact between the ectopterygoids and palatines medial to the supradental ridge of the maxillae; contact between the pterygoids and vomers; reduction or complete loss of the infraorbital fenestrae; a well-developed surangular process on the dentary which extends posteriorly above the anterior surangular foramen; a large splenial; and subpleurodont dentition.

Sulimski (1975) identified the Macrocephalosauridae as a family within the Scincomorpha, but Estes (1983) united it with the modern Teiidae. Alifanov (1993a) re-established this family and erected a new subfamily, the Mongolochamopinae, though this has now been recognized as a separate family (see below). Wu *et al.* (1996) proposed a relationship between Macrocephalosauridae and Amphisbaenia within the Scincomorpha, but this has been questioned by Gao and Hou (1996).

MONGOLOCHAMOPIDAE Alifanov, 1993a stat. nov. (MACROCEPHALOSAURIDAE, MONGOLOCHAMOPINAE Alifanov, 1993a)

Record. More than ten genera belonging to this family are known from the Cretaceous of Mongolia. Seven of these genera have been recovered from Hermiin Tsav: *Mongolochamops reshetovi* Alifanov, 1988 (holotype: PIN 3142/304, based on a fragmentary skull and mandible); *Altanteius facilis* (Alifanov, 1988) (holotype PIN 3142/306, based on a skull, lacking the parietal, and the associated left ramus of the mandible); *Gobinatus arenosus* Alifanov, 1993a (holotype: PIN 3142/308, known from a complete skull with mandible); *Parameiva oculea* Alifanov, 1993a (holotype: PIN 3142/310, a skull without brain case, but with a mandible; *Prodenteia ministra* Alifanov, 1993a (holotype: PIN 3142/324, based on articulated bones of the skull roof and a mandible); *Piramicephalosaurus cherminicus* Alifanov, 1988 (holotype: PIN 3142/310, represented by articulated bones of the anterior part of skull, a

Figure 18.4. *Darchansaurus estesi* Sulimski, 1975 (ZPAL MgR-III/6): reconstruction of the skull in (A) dorsal, (B) ventral and, with the left mandible, in (C) left lateral view, and the left mandible in (D) medial view. Scale bar = 20 mm.

Iguana-like teeth with 4–6 small cusps, and trunk vertebrae with clear zygosphenes and zygantra. They exhibit a complex of primitive and derived cranial characters including: a wide upper temporal fenestra; a parietal fenestra on the fronto-parietal suture; a V- or U-shaped posterior edge of the nasals; contact between the prefrontal and nasal; lateral expansion of the postorbital which develops bifurcated articular surfaces forming a mutual cruciform contact with the

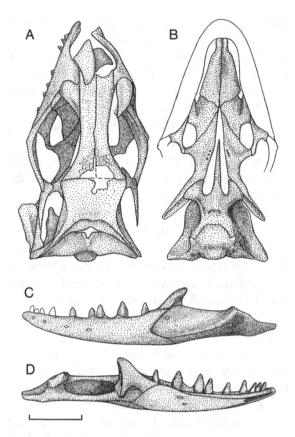

Figure 18.5. *Barungoia vasta* Alifanov, 1993a (PIN 4487/2): skull in (A) dorsal and (B) ventral view and left mandible in (C) lateral and (D) medial view. Scale bar = 5 mm.

mandible and isolated vertebrae); and *Tchingisaurus multivagus* Alifanov, 1993a (holotype: PIN 3142/309, based on a mandible).

Khulsan has produced just two taxa: *Barungoia vasta* Alifanov, 1993a (holotype: PIN 4487/2, a complete skull with mandible) (Figure 18.5A–D) and *Gurvansaurus canaliculatus* Alifanov, 1993a (holotype: PIN 4487/3, based on a right dentary with dentition). Tögrögiin Shiree has also produced two genera: *Dzhadochtosaurus giganteus* Alifanov, 1993a (holotype: PIN N3142/103, based on a complete skull with mandible) and *Gurvansaurus potissimus* Alifanov, 1993a (holotype: PIN 3143/104, a skull without roof, but with mandible). Finally, one taxon is known from

Bayan Zag: *Conicodontosaurus djadochtaensis* Gilmore, 1943 (holotype: AMNH 6519 represented by a fragmentary skull and mandible). Mongolochamopids from Mongolia are distinguished from one another by their size, the proportions of separate skull bones and parts of the skull, and the numbers and morphology of the teeth.

Mongolochamopid lizards have also been reported from the Nemegt Formation (Alifanov, 1993a) and Höövör (Alifanov, 1993b, c).

Remarks. The Mongolochamopidae are Cretaceous Asian–American lizards that, like the Macrocephalosauridae, have double processes on the postfrontal and postorbital which form a mutual cruciform: a feature that is unique among lizards. The Mongolochamopidae are distinguished from the Macrocephalosauridae by their small size, a contact between the frontals and maxillae, a large infraobital fenestra, loss of the pterygoid–vomer contact, a well-developed angular process on the dentary, and teeth with two symmetrical lateral shoulders or additional denticles. This family also shows secondary similarities to macroteid genera, in particular the tricuspid subpleurodont teeth. However, mongolochamopids are distinguished by the presence of additional symmetrical cusps and the absence of bicuspid teeth.

Initially, these lizards were described by Alifanov (1993a) as a subfamily within the Macrocephalosauridae. They are raised to family rank here on the basis of their distinctive morphological characteristics and their more ancient and diverse distribution.

The Upper Cretaceous Chinese lizard *Chilingosaurus*, described by Dong (1965) and placed by Estes (1983) within 'Incertae sedis', is clearly a mongolochamopid lizard, as is *Buckantaus*, described by Nesov (1985) from the Late Cretaceous of Middle Asia (Alifanov, 1993a). The North American genera *Gerontoseps*, *Socognathus* and *Sphenosiagon*, described by Gao and Fox (1991) and originally assigned to the Teiidae, also belong in the Mongolochamopidae (Alifanov, 1993a). Another North American form *Prototeius*, described by Denton and O'Neill (1993), probably also belongs in this family.

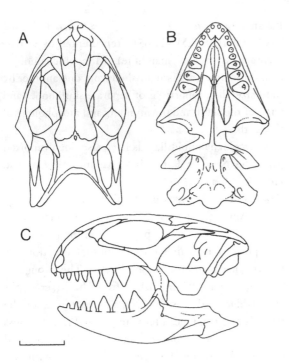

Figure 18.6. *Adamisaurus magnidentatus* Sulimski, 1972: reconstruction of the skull in (A) dorsal, (B) ventral and, with the left mandible, in (C) left lateral view. Scale bar = 5 mm.

ADAMISAURIDAE Sulimski, 1978

Record. At present this family is represented in the Mesozoic of Mongolia by a single genus and species, *Adamisaurus magnidentatus* Sulimski, 1972. The holotype (ZPAL MgR-II/80), recovered from Bayan Zag, consists of a skull with both mandibles (Figure 18.6A–C) and numerous other remains including skulls, sometimes associated with skeletons, are known from Djadokhta and Baruun Goyot localities. Moreover, remains of *Adamisaurus magnidentatus* have also been found from Djadokhta sediments in the Gobi Desert of China (Gao and Hou, 1995).

Remarks. The Adamisauridae is a distinctive Late Cretaceous family, described by Sulimski (1978) on the basis of a single genus that possesses paired frontals, a subtriangular postfrontal, a very large coronoid process on the dentary that is expanded on the labial surface of the coronoid, absence of the labial process of the coronoid, a surangular process on the dentary

that is developed above the anterior surangular foramen, a large angular process on the dentary, subacrodont teeth that are slightly expanded transversely and have swollen bases, and teeth that are sharp and increase in size caudally. In addition, the dentition includes not more than eight teeth.

Like the Macrocephalosauridae this family was placed in synonymy with the Polyglyphanodontinae (Teiidae) by Estes (1983). Actually, there is some similarity between the Polyglyphanodontidae and the Adamisauridae, described from the Upper Cretaceous of North America by Gilmore (1940, 1942). They both have agamoid-like subacrodont teeth with transversely expanded tooth bases and an agamid-like dentary with posterior processes. However, there are many differences between the Adamisauridae and the Polyglyphanodontidae and they cannot be united within a single family.

The North American lizard *Peneteius*, described by Estes (1969a), has a small series of subacrodont teeth that increase in size posteriorly and have bulbous bases as in *Adamisaurus*. Thus, the inclusion of *Peneteius* in Adamisauridae seems reasonable. However, *Peneteius* is distinguished by its transversely expanded teeth that have a notch in their dorsal border, resulting in double apices. This unusual structure of the teeth is convergent with the dentition of the extant forms *Teius* and *Dicrodon*. Like *Callopistes* and some species of *Cnemidophorus* the anterior teeth of *Teius* and *Dicrodon* have additional cusps, though these are displaced lingually on the posterior teeth in the latter genus.

?EICHSTAETTISAURIDAE Kuhn, 1958

Record. The only Mongolian record is *Globaura venusta* Borsuk-Białynicka, 1988 (holotype: ZPAL MgR-III/40), based on a complete skull with mandible (Figure 18.7A–C) from Hermiin Tsav. Numerous remains of *Globaura* have also been found at the type locality, Bayan Zag and Khulsan. This lizard is characterized by its large brain case, a maxilla–frontal contact, paired premaxillae, reduced lacrimals, a wide upper temporal fenestra, fused and narrow frontals and a scincid-like dentary.

Remarks. The Family Eichstaettisauridae was des-

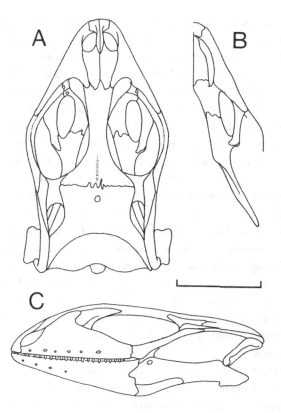

Figure 18.7. *Globaura venusta* Borsuk-Białynicka, 1988: reconstruction of the skull in (A) dorsal, (B) ventral and, wth left mandible, in (C) left lateral view. Scale bar = 5 mm.

The reconstruction of the skull roof of *Eichstaettisaurus* by Estes (1983) exhibits many similarities to *Globaura*: a short snout, large orbits, small, separate postorbitals and postfrontals, small nasals, and an ectopterygoid–palatine contact above the supradental ridge of the maxillae. *Globaura* itself is a true scincomorphan, as Borsuk-Białynicka (1988) has shown, and this is confirmed by further scincid characters including: a labial process of the coronoid overlapped anteriorly by the coronoid process of the dentary so that the lateral exposure of this process is limited to a narrow wedge between the dentary and surangular; a postfrontal with two, short, lateral processes; the frequently poor development of the medial process of the postorbital, situated between the lateral processes of the postfrontal; and the massive angular process of the dentary. The stability of skull features in Jurassic and Cretaceous forms and the unique complex of primitive and derived characters of Eichstaettisauridae Kuhn, 1958, permit inclusion, tentatively of *Globaura* in this family.

CARUSIIDAE Borsuk-Białynicka, 1987
(=CAROLINIDAE Borsuk-Białynicka, 1985)

Record. Two forms belonging to this subfamily are known from Mongolia. *Carusia intermedia* Borsuk-Białynicka, 1985 (holotype: ZPAL MgR-III/34) is based on a skull without the upper temporal bars, but with both mandibles (Figure 18.8A, B) from Hermiin Tsav, and *Shinisauroides latipalatum* Borsuk-Białynicka, 1985 (holotype: ZPAL MgR-I/58) is based on a skull without the parietal region, but with a mandible, from the Nemegt.

Shinisauroides differs from *Carusia* in that it has a wider snout and palatines, more massive ectopterygoids, fused vomers, an unusually strongly expressed osteodermal sculpture on the skull roofing bones and more numerous teeth. These taxa are also represented by a small series of further remains from Mongolia.

Remarks. Carusiidae is an Upper Cretaceous scincomorphan family distinguished by osteodermal sculpture on the bones of the skull roof, a short and vaulted snout, paired premaxillae, reduced lacrimals, a large upper temporal fenestra, partly or completely fused

cribed by Kuhn (1958) on the basis of a single specimen of *Eichstaettisaurus* from the Upper Jurassic of Germany. It was later synonymized with Ardeosauridae and recognized as gekkotan by Hoffstetter (1964, 1966). The history of these problematic families was well described by Estes (1983). All species of Ardeosauridae *sensu* Hoffstetter are represented by materials of different preservation and seen in dorsal aspect only, which obscures some skull characters. In my opinion, there are no formal arguments supporting the hypothesis of gekkotan relationships for both *Ardeosaurus* and *Eichstaettisaurus*. On the contrary, both genera demonstrate several non-gekkotan features, including the position of the ectopterygoid and the lateral process of the palatine dorsal to the supradental ridge of the maxilla, the complete temporal bar, a large jugal and the presence of a parietal foramen.

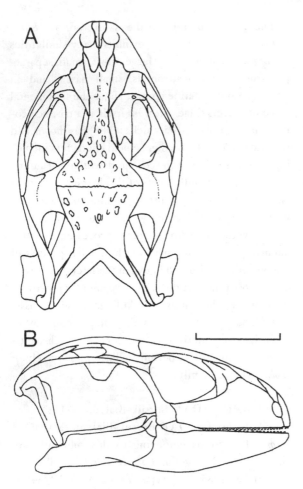

Figure 18.8. *Carusia intermedia* Borsuk-Białynicka, 1985: reconstruction of the skull in (A) dorsal and, with the right mandible, in (B) right lateral view. Scale bar = 10 mm.

between this family and the Late Jurassic European lizard *Ardeosaurus*, known from the dorsal aspect only. The latter was described on the basis of some incomplete fossil finds and recognized as a gekkotan by Camp (1923), Hoffstetter (1964, 1966), Mateer (1982) and Estes (1983). The best preserved specimen of *Ardeosaurus*, described by Mateer (1982), exhibits rugose skull roofing bones, an elongate parietal, paired premaxillae and frontals, expansion of the medial process of the postorbital anterior to the postfrontal, massive upper temporal bars, location of the parietal foramen in the centre of the parietal, and large adductor fossae. Most of these features are also characteristic for carusiid genera, with only the paired frontals distinguishing *Ardeosaurus* from other carusiids.

Poorly preserved remains of *Carusia* and possibly also of *Shinisauroides* have been found in Inner Mongolia, China (Gao and Hou, 1996). Gao and Hou (1996) proposed on the basis of their material that morphological differences between *Carusia* and *Shinisauroides* reflect sexual dimorphism. However, their formal conclusion does not take into consideration the absence of distinctive sexual dimorphism in the cranial features of the majority of modern lizard species and they do not state which genus should be regarded as male and which as female.

The scincomorphan *Contogenys* was described by Estes (1969b) from the Upper Cretaceous of North America on the basis of a fragmentary dentary which has teeth similar to those of *Carusia* and may thus belong to the Carusiidae.

PARAMACELLODIDAE Estes, 1983a
Record. Undescribed remains, represented by disarticulated bones, including numerous maxillae and dentaries bearing teeth, from the Early Cretaceous of Höövör are tentatively referred to this family. So far, however, paramacellodid-like lizards have not been reported from the Upper Cretaceous deposits of Mongolia.

Other Asian paramacellodids include *Sharovisaurus* from the Upper Jurassic of Kazakhstan (Hecht and Hecht, 1984) and *Mimobecklesiosaurus* from the Upper Jurassic of China (Li, 1985). *Changetisaurus*, a distinc-

postorbitals and postfrontals, the position of the medial contact of the processes of the ectopterygoid and palatine dorsal to the supradental ridge of the maxilla, unpaired and narrow frontals with closed olfactory canal, a large parietal with descending finger-like processes, a wide labial process on the coronoid that is overlapped by the coronoid process of the dentary, the exceptionally large angular and surangular processes of the dentary, and the exceptionally large, and skull-like hyperpleurodont teeth.

The relationships of Carusiidae to other scincomorphans is unclear, but there are some similarities

tive paramacellodid-like genus represented by an articulated skull and compound rectangular body osteoscutes from the Callovian of Fergana, Kirgizstan, was described by Nesov (Fedorov and Nesov, 1992) as a dorsetisaurid.

Remarks. Paramacellodids are Jurassic-Cretaceous cordyloid-like scincomorphans distinguished by compound rectangular body osteoscutes. Unfortunately, the skull is only poorly known. Paramacellodidae was proposed by Estes (1983) to distinguish some problematic forms that 'on present material cannot be referred to the Holocene family' (Estes, 1983, p. 115).

SLAVOIIDAE Alifanov, 1993c

Record. Two genera have been described from the Upper Cretaceous of Mongolia: *Slavoia darevskyi* Sulimski, 1984 (holotype: ZPAL MgR-I/8, based on a skull and complete postcranial skeleton from Khulsan) (Figure 18.9A–C) and *Eoxanta lacertifrons* Borsuk-Białynicka, 1988 (holotype: ZPAL MgR-III/37, represented by a complete skull with both mandibles from Hermiin Tsav). *Slavoia* is the commonest lizard from Bayan Zag, Khulsan and Hermiin Tsav and remains of *Slavoia*-like and *Eoxanta*-like genera are also known from the Early Cretaceous of Mongolia (Alifanov, 1993b, c).

Remarks. The *Acontias*-like features of the skull in *Slavoia* (wide snout, large upper temporal fenestrae, nasals and frontals, contact between the prefrontal and postfrontal above the orbits, small postfrontals, the reduced size of the orbit and teeth which lack additional anterior denticles) clearly distinguish it from *Eoxanta*. A distinctive feature of *Eoxanta* is its large and long postfrontal the enlargement of which obliterates the upper temporal fenestra.

The Slavoiidae, unique to the Cretaceous of Asia, was erected by Alifanov (1993c) and distinguished by an unusual complex of primitive and derived characters including: paired premaxillae and frontals, wide nasals and frontals, the position of the short medial process of the postorbital between the lateral processes of the postfrontal, the development of subolfactory processes of the frontals, reduced lacrimals, small orbits, contact between the lateral process of the

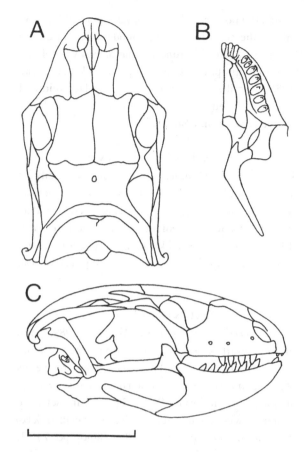

Figure 18.9. *Slavoia darevskyi* Sulimski, 1984: reconstruction of the skull in (A) dorsal, (B) ventral and, with the right mandible in (C) right lateral view. Scale bar = 5 mm.

palatine and the massive ectopterygoid medial to the supradental ridge of the maxilla, the massive angular process of the dentary, overlapping of the labial process of the coronoid by the coronoid process of the dentary, opening of the Meckelian canal in front of the partially reduced splenial and the presence of no more than fifteen conical teeth.

The structure of the posterior border of the dentary shows that the Slavoiidae are undoubtedly scincomorphans. The absence of osteodermal sculpture on the skull roofing bones and the location of the contact between the lateral process of the palatine and the ectopterygoid, medial to the supradental ridge of the maxilla, is similar to that in extant Xantusiidae, though the postorbital–postfrontal contact is comparable to the situation in some modern genera of the Scincidae. Sulimski proposed that *Slavoia* was related to the Gymnophthalmidae, while Borsuk-Białynicka (1988) described *Eoxanta* as a non-teiid lacertoid and correctly noted, for the first time, the similarity between *Slavoia* and *Eoxanta*.

GEKKOTA Cuvier, 1817

This infraorder includes lizards that are usually of small size and with a homeomorphan skull structure. Gekkota *sensu* Kluge (1987) includes Gekkonidae, Diplodactilidae and Eublepharidae, while Estes *et al.* (1988) divided the Gekkota into Pygopodidae and Gekkonidae, noting the possible paraphyly of the latter.

'GEKKONIDAE' Gray, 1825

Record. Hoburogekko sukhanovi Alifanov, 1989a (the specific epithet is emended from the original '*suchanovi*' a lapsus calami) from the Lower Cretaceous of Mongolia, has been described on the basis of a fragmentary skull (holotype: PIN 3334/ 500) and separate dentaries with an unfused Meckelian canal. *Gobekko cretacicus* Borsuk-Białynicka, 1990 (holotype: ZPAL MgR-II/4), from the Upper Cretaceous of Mongolia, is based on one of three fragmentary skulls, with mandibles, from Bayan Zag. *Gobekko* is characterized by paired bones of the skull roof (except for the premaxillae), the paired, M-shaped posterior edge of the wide parietal and unpaired vomers.

ANGUIMORPHA Fürbringer, 1900

This taxon was proposed by Fürbringer (1900) in synonymy with Diploglossa of Cope (1900). Later, Camp (1923) united Cope's Diploglossa with Fürbringer's Platynota (Varano–Dolichosauria) and Mosasauria in his Anguimorpha. The modern conception of this taxon includes Varanoidea and Anguoidea (McDowell and Bogert, 1954; Gauthier, 1982; Pregill *et al.*, 1986; Estes *et al.*, 1988).

VARANIDAE Hardwike and Gray, 1824
(= MEGALANIDAE Fejervary, 1935)

Record. The following taxa have been described from the Upper Cretaceous of Mongolia: *Telmasaurus grangeri* Gilmore, 1943 (holotype: AMNH 6645), based on the parietal region of a skull from Bayan Zag; *Saniwides mongoliensis* Borsuk-Białynicka, 1984 (holotype: ZPAL MgRI/72), based on a complete skull with mandible from Khulsan; and *Estesia mongoliensis* Norell *et al.*, 1992 (holotype: M 3/14), based on a skull and left ramus of the lower jaw, also from Khulsan. *Cherminotus longifrons* Borsuk-Białynicka, 1984 (holotype: ZPAL MgR-III/59) was established on a complete skull, lacking the upper temporal bars, but associated with the mandibles (Figure 18.10C–E) from Hermiin Tsav, which has also yielded vertebrae of *Saniwa* sp. *Varanus*-like vertebrae have been found in the Nemegt Formation (Alifanov, 1993b) and varanids have also been reported from the Palaeogene of Mongolia (Alifanov, 1993b, c).

Remarks. In *Saniwides*, and possibly in *Telmasaurus*, the frontals are excluded from the posterior border of the bony nares by nasal–maxilla contacts. In both cases, the palatines and pterygoids bear many teeth and the fused postorbitals and postfrontal, including the frontals, take part in forming the upper border of the orbits. The largest form, *Estesia*, described recently by Norell *et al.* (1992), and the smallest, *Cherminotus*, exhibit the alternative condition for the features mentioned above, and have distinct similarities in cranial morphology. *Cherminotus* was originally assigned to Lanthanotidae by Borsuk-Białynicka (1984), but new remains do not confirm this conclusion.

Derived osteological character states of Varanidae include: the absence of regular body osteoscutes or rugosity on the skull roof bones, posterior retraction of the bony nares, the posterior position of the nasal processes of the maxilla, well-developed subolfactory processes, intramandibular mobility (streptognathy), anterior expansion of the labial process of the coronoid and fusion of the upper borders of the labial and antero-medial processes of the coronoid, and in the vertebrae precondylar constriction leads to the formation of flanges on the condyle.

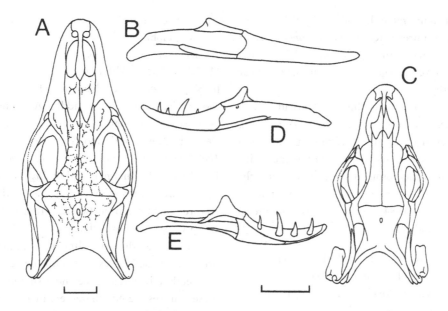

Figure 18.10. *Proplatynotia longirostrata* Borsuk-Białynicka, 1984: reconstruction of the skull in (A) dorsal view and the right mandible in (B) lateral view. *Cherminothus longifrons* Borsuk-Białynicka, 1984: reconstruction of the skull in (C) dorsal view and the left mandible in (D) lateral and (E) medial view. Scale bar = 5 mm.

The Varanidae currently occur in the Old World and Australia, but they had a broader distribution in the Palaeogene and were also found in North America, Europe and Central Asia. The oldest reliable evidence of this family comes from the Late Cretaceous of Central Asia, but at this time varanids seem to have been absent from North America. *Paleosaniwa*, known on the basis of fragmentary remains from North America, has sometimes been identified as a varanid (Estes, 1983), but it lacks precondylar constriction of the trunk vertebrae, a characteristic of true varanids and is more likely to be a 'necrosaurid'.

'NECROSAURIDAE' Hoffstetter, 1943
(=PARASANIWIDAE Estes, 1964)

Record. The following forms have been described from the Upper Cretaceous of Khulsan, Mongolia: *Proplatynotia longirostrata* Borsuk-Białynicka, 1984 (holotype: ZPAL MgR-I/68), based on a complete skull with mandible (Figure 18.10A, B); *Gobiderma pulchra* Borsuk-Białynicka, 1984 (holotype: ZPAL

MgR-III/64), represented by a complete skull with mandible; and *Parviderma inexacta* Borsuk-Białynicka, 1984 (holotype: ZPAL MgR-I/43), based on a damaged skull and mandible. Necrosaurid lizards, as yet undescribed, have also been found in the Nemegt Formation and the Upper Palaeocene of Mongolia (Alifanov, 1993b, c).

The necrosaurids from Khulsan are distinguished from one another by skull dimensions and the proportions of individual bones. *Parviderma* is distinguished by its fused and narrow frontals, while *Gobiderma* has non-imbricate rounded or polygonal osteoscutes and parietals with flattened lateral borders. The structure of the dentary and teeth of *Proplatynotia* shows clear similarity to that of *Colpodontosaurus* from the Upper Cretaceous of North America.

Remarks. The Necrosauridae are predatory anguimorphans from the Cretaceous of Asia–America. They exhibit a combination of primitive and derived characters including the unusual form of the body osteoscutes, osteodermal sculpture on the skull roof bones, distinct upper temporal arches, absence of intramandibular mobility, unretracted nares and *Anguis*-like or *Varanus*-like teeth. In addition, all Asiatic 'necrosaurids' have pterygoids and palatines that bear numerous teeth.

Estes (1983) synonymized his Parasaniwidae with Necrosauridae, erected earlier by Hoffstetter (1943), who also believed in the unity of these groups, and proposed (Estes, 1983) that helodermatids may have been derived from necrosaurid stock. More recently, Necrosauridae *sensu lato* has been interpreted by Borsuk-Białynicka (1984) as a paraphyletic group within the Varanoidea.

ANGUIDAE Gray, 1825 (=BAINGUIDAE
Borsuk-Białynicka, 1984)

Record. Bainguis parvus Borsuk-Białynicka, 1984 (holotype: ZPAL MgR-II/46) from Bayan Zag, and represented by a damaged skull with both mandibles is the only record of this family in Mongolia.

Remarks. Bainguis was initially established in a separate family, Bainguidae, by Borsuk-Białynicka (1984), but later, on the basis of similarities in the configuration of the osteodermal sheets on the bones of the skull roof, it was assigned, apparently correctly, to the Anguidae (Borsuk-Białynicka, 1991).

DORSETISAURIDAE Hoffstetter, 1967

Record. Dorsetisaurus, represented by numerous isolated maxillae and dentaries is known only from Höövör.

Remarks. The Dorsetidauridae are poorly known anguimorphans from the Late Jurassic of North America and Europe and the Early Cretaceous of Europe and Asia. They are distinguished by frontal scutellation, a low and elongate braincase, wide upper temporal fenestrae, and lancet-like teeth (see Estes, 1983 for further details). So far, no *Dorsetisaurus*-like lizards have been found in the Upper Cretaceous deposits of South Gobi, Mongolia.

HODZHAKULIIDAE Alifanov, 1993c

Record. Numerous isolated maxillae, premaxillae and dentaries of a single taxon, *Hodzhakulia* sp. have been collected from Höövör. So far, hodzhakuliid lizards

have not been found in the Upper Cretaceous deposits of South Gobi, Mongolia.

Remarks. Hodzhakulia, the type genus of the family Hodzhakuliidae (Alifanov, 1993c), was first described by Nesov (1985) on the basis of isolated dentaries from the Aptian–Albian of Middle Asia, but was not assigned to any particular taxon within Lacertilia. Recently, Nesov and Gao (1993) have suggested that *Hodzhakulia* may be related to amphisbaenians. However, additional material of *Hodzhakulia* from the Lower Cretaceous of Mongolia shows the following features: unpaired premaxillae; a thin and long posterior process of the maxillae; varanid-like structure of dentary; tall teeth with cylindrical or antero-posteriorly compressed bases and multiple resorbtion pits; and teeth apices that have short shoulders. Many of these features are not characteristic of Amphisbaenia. Alternatively, the tall teeth of *Hodzhakulia* show some similarities to those of *Litakis*, a problematic form assigned by Estes to Anguimorpha *incertae sedis* (1964) or Eolacertilia (1983).

?XENOSAURIDAE Cope, 1886
(=SHINISAURIDAE Ahl, 1929)

Record. Deposits at Höövör have yielded several anguoid-like dentaries with conical teeth, that have been assigned, on a preliminary basis, to Xenosauridae. By contrast, xenosaurids have not been reported from the Upper Cretaceous of South Gobi, Mongolia.

Remarks. Fragmentary remains of *Oxia karakalpakiensis* Nesov, 1985 from the Lower Cretaceous of Uzbekistan were referred to the Xenosauridae by Nesov and Gao (1993). Apart from this record and the Höövör material, the earliest xenosaurids, a small group of extant Asian–American anguimorphans, are known also from the Upper Cretaceous and Palaeogene of North America.

PARAVARANIDAE Borsuk-Białynicka, 1984

Record. A single genus and species, *Paravaranus angustifrons* Borsuk-Białynicka, 1984 (holotype: MgR-I/67, a partly damaged skull and mandible), known only from Khulsan. Additional remains of *Paravaranus* are held in the collections of PIN.

Remarks. The Paravaranidae is a somewhat doubtful monotypic family characterized by the following features: absence of a subolfactory process; squamosal with an Iguania-like dorsal process; parietals bearing concavities limited by ridges; Y-shaped vomers; strongly toothed pterygoids; slender and low mandibles; and pointed and recurved pleurodont teeth. Borsuk-Białynicka (1984) proposed that *Paravaranus* belonged within Anguimorpha, but, in my opinion, there is some evidence to suggest a possible relationship with mosasaurs. This is supported by a complex of common cranial characters: restricted bony nares; unpaired premaxillae, frontals, parietals and nasals; a deep division between the vomers; toothed palatines; and the presence of dorsal processes of the squamosal.

Discussion

Patterns of diversity

Representatives of almost 20 families of lizards, the total number of families known from Cretaceous deposits, are listed in the survey above. Some details concerning the origin of this diverse assemblage can be obtained by comparing lizard assemblages collected from localities in the southern part of the Gobi Desert in the Mongolian People's Republic and representing the following intervals: the Early and Late Cretaceous, the pre-Maastrichtian and the Maastrichtian, and the Cretaceous and the Palaeogene (see Tables 18.1–18.3).

Table 18.1 compares the diversity of lizards, at the family level, known from the Djadokhta and Baruungoyot Formations (pre-Maastrichtian) with that known from the younger Nemegt Formation (Maastrichtian). Only four families are found in the Nemegt, compared with 15 in older deposits, indicating a decline in lizard diversity in the Maastrichtian of Mongolia.

Although the history of many of the families discussed above is disputable, there is clear evidence of significant levels of extinction of lizards in Asia before the beginning of the Cenozoic. The vast majority (11 out of 16) of families of Late Cretaceous lizards

Table 18.1. *Comparison of the fossil record of lizards from Upper Cretaceous deposits of the Gobi Desert, Mongolia.*

Pre-Maastrichtian (Djadokhta and Baruungoyot formations)	Maastrichtian (Nemegt Formation)
†Adamisauridae	—
Anguidae	—
†Carusiidae	—
†?Eichstaettisauridae	—
'Gekkonidae'	—
Hoplocercidae	—
†Isodontosauridae	—
†Macrocephalosauridae	—
†Mongolochamopidae	†Mongolochamopidae
†'Necrosauridae'	†Necrosauridae
†Paravaranidae	—
Phrynosomatidae	—
†Priscagamidae	—
†Slavoiidae	—
Varanidae	Varanidae
—	Teiidae

Notes:
† = extinct taxon.

became extinct at this time and only three families (Varanidae, Isodontosauridae, 'Necrosauridae') crossed the Cretaceous–Tertiary boundary in Asia, though these families (Varanidae), or their descendants (Gekkonidae, Anguidae) are still found in Asia today. The results of preliminary studies of lizards from the Mongolian Palaeogene (Tables 18.2 and 18.3) confirm this pattern. The Palaeogene yields eight families, half of them (Arretosauridae, Uromasticidae, Agamidae and Lacertidae) appearing for the first time in Asia. In the case of the Agamidae and the Uromasticidae, and possibly also the Arretosauridae, this may reflect a local radiation in Asia during the latest Cretaceous–earliest Palaeocene. The Anguidae, well known from the Late Cretaceous and Palaeogene of North America, were widely distributed during this interval, while the Lacertidae may have arrived from an ancient centre of diversity apparently located in Europe.

Comparisons between lizard assemblages from the

Table 18.2. *Comparison of the fossil record of lizards from Cretaceous and Palaeogene deposits of the Gobi Desert, Mongolia*

Lower Cretaceous	Upper Cretaceous	Palaeogene
—	†Adamisauridae	—
—	—	Agamidae
—	Anguidae	Anguidae
—	—	Arretosauridae
—	†Carusiidae	—
†Dorsetisauridae	—	—
†?Eichstaettisauridae	†?Eichstaettisauridae	—
'Gekkonidae'	'Gekkonidae'	—
—	Hoplocercidae	—
†Hodzhakuliidae	—	—
'Iguanidae' indet.	—	—
—	†Isodontosauridae	†Isodontosauridae
—	—	Lacertidae
—	†Macrocephalosauridae	—
?†Mongolochamopidae	†Mongolochamopidae	—
—	†'Necrosauridae'	†'Necrosauridae'
—	†Paravaranidae	—
?†Paramacellodidae	—	—
—	Phrynosomatidae	—
—	†Priscagamidae	—
†Slavoiidae	†Slavoiidae	—
—	Teiidae	—
—	—	Uromasticidae
—	Varanidae	Varanidae
?Xenosauridae	—	—

Notes:

† = extinct taxon.

Table 18.3. *Comparison of the fossil record of lizards in the Palaeogene of the Gobi Desert, Mongolia*

Late Palaeocene (Tsagaan Khushuu)	Early Eocene (Tsagaan Khushuu)	Middle Eocene (Khaichin Uul II)	Early Oligocene (Ergiliin Zoo, Khoyor Zaan)
Agamidae	Agamidae	Agamidae	—
—	—	?†Arretosauridae	?†Arretosauridae
—	Anguidae	—	Anguidae
†Isodontosauridae	—	†Isodontosauridae	—
—	—	—	Lacertidae
†'Necrosauridae'	—	—	—
—	Varanidae	Varanidae	Varanidae
—	Uromasticidae	Uromasticidae	Uromasticidae

Notes:

† = extinct taxon.

Early Cretaceous (Höövör) and the Late Cretaceous (Table 18.2) suggest a modest level of extinction at the family level. Only the Paramacellodidae, the Hodzhakuliidae and the problematic Dorsetisauridae disappeared completely, while the single extant family, the Xenosauridae, is not known from post-Early Cretaceous deposits in Asia.

Palaeobiogeography

Three families found in the Early Cretaceous of Mongolia (?Eichstaettisauridae, Paramacellodidae and Dorsetisauridae) are also known from the Late Jurassic or Early Cretaceous of Europe. This suggests a connection between and the possible development of Cretaceous lizards of Asia from Jurassic lizards of Panlaurasia. These families together with the extant and cosmopolitan 'Gekkonidae', also known from the Early Cretaceous of Asia, can be named Laurasian relics.

Most Early Cretaceous lizard families from Asia appear to be endemic to this region at this time. This is consistent with the supposed isolation of Asia during this interval, which, according to Russell (1993), began in the Middle Jurassic. If this is correct, then the Mongolochamopidae, Slavoiidae, Xenosauridae and Hodzhakuliidae presumably had an Asiatic origin during the Late Jurassic–Early Cretaceous, and may be named Late Jurassic–Early Cretaceous derivatives of Asia. At least some 'iguanoid' groups may also have originated in Asia during this interval.

Table 18.2 shows that 12 lizard families (which can be linked in three groups: the Hoplocercidae–Teiidae–Anguidae, the Isodontosauridae–Priscagamidae–Macrocephalosauridae–Varanidae–Paravaranidae and the Phrynosomatidae–Adamisauridae–Carusiidae–'Necrosauridae'), first appear in the Late Cretaceous. Members of the first group are usually referred to as American lizards and are currently found in North and South America. The Teiidae have been reported from the Late Cretaceous of North and South America, but the Anguidae and Hoplocercidae are known only from the Late Cretaceous of North and South America respectively. By contrast, several families from the second group (Isodontosauridae, Priscagamidae, Macrocephalosauridae, Paravaranidae and Varanidae) are found only in the Late Cretaceous of Asia. The third group includes families (Phrynosomatidae, Adamisauridae, Carusiidae and 'Necrosauridae') that occur in both Asia and America.

The appearance of American and Asian–American families in Asia indicates the establishment of an Asian–American connection toward the end of the Late Cretaceous (Russell, 1993). However, the evidence from lizards, and other taxa, of endemism in Asia in the Late Cretaceous shows that this connection was subsequently lost, leading to temporary isolation of Asia (Alifanov, 1993c), prior to the Palaeogene consolidation of all northern continents.

Faunal interchange between North America and Asia in the Cretaceous led to the introduction of the Asian lizard families Mongolochamopidae and Xenosauridae into North America. These families do not appear to have reached South America, suggesting that this continent was isolated from North America by the beginning of the Late Cretaceous Asian–American interchange. However, the Americas must have been connected prior to this, since only this configuration explains the distribution of Teiidae, Hoplocercidae and Anguidae in Asia and the Americas, and the Asiatic occurrences of Mongolochamopidae and Xenosauridae, which are otherwise only found in North America. This suggests that the Early Cretaceous lizards from Höövör existed before the beginning of the Cretaceous Asian–American connection, while Late Cretaceous lizards from the Djadokhta, Baruungoyot and Nemegt formations existed after this event. This palaeozoogeographic hypothesis is also supported by studies of other tetrapod groups (Kalandadze and Rautian, 1992) and can be used to establish the palaeozoogeographic origins of Late Cretaceous lizard groups. For example, the Asian Cretaceous endemics Isodontosauridae, Priscagamidae, Macrocephalosauridae and Varanidae are derived from the Late Cretaceous of Asia. By contrast, Teiidae and Hoplocercidae, which are so far unknown in the Early Cretaceous of Asia are possibly derived from the Neocomian of America or, perhaps more precisely, North America, since the nearest relations of all the American lizard families that appear in the Late Cretaceous of Asia originally had a Panlaurasian distribution.

The origin of the Asian–American Phrynosomatidae, Carusiidae, Adamisauridae and 'Necrosauridae', which are unknown in South America or from the rich assemblage found at Höövör, possibly happened in North America after the Neocomian, following the loss of the first inter-American connection and just before the beginning of interchanges between Asia and America.

Conclusion

The material described above fills an important gap in our knowledge of lacertilian evolution during the Mesozoic. The establishment of rich Cretaceous lizard assemblages on the territory of Central Asia is connected with the protracted and relatively stable existence of a local fauna beginning in the Middle Jurassic, when lizard families first appeared. By the Maastrichtian, the diversity of Asiatic lizards had declined and was never restored to this level, hence the Late Cretaceous lizards of the Gobi Desert would appear to represent the apogee of lacertilian diversity in this region.

The origin of some endemic Asiatic groups took place after the Jurassic disjunction of Panlaurasia which led to the isolation of Central Asia. An Asian–American contact in the late Early Cretaceous temporarily suspended this isolation and led to interchange between the two contrasting assemblages of Asia and Northern America, as a result of which some American forms appeared in Asia. By contrast to the Asiatic groups, which only appear in North America during the Late Cretaceous, the American migrants reached South America too. This indicates the possibility of direct inter-American faunal interchange before the establishment of a connection between Central Asia and North America in the Cretaceous. During the Late Cretaceous renewed isolation of Asia led to the appearance of new Asiatic families both before and after the Maastrichtian crisis.

Thus, Cretaceous lizards from the Gobi Desert of Mongolia can be characterized palaeozoogeographically as follows: Laurasian relics (?Eichstaettisauridae, Paramacellodidae, Dorsetisauridae and 'Gekkonidae'); Late Jurassic–Early Cretaceous Asian forms (Slavoiidae, Mongolochamopidae, Xenosauridae and Hodzhakuliidae); Late Cretaceous Asian forms (Varanidae, Isodontosauridae, Priscagamidae, Macrocephalosauridae and Paravaranidae); American migrants (Teiidae, Hoplocercidae and Anguidae); and Asian-American Forms (Phrynosomatidae, Carusiidae, Adamisauridae and 'Necrosauridae').

Acknowledgements

I am grateful to Dr D. Unwin, Bristol University, for his advice and comments on several earlier versions of this paper, as well as Dr M. Borsuk-Białynicka from the Institute of Paleobiology, Polish Academy of Sciences and Dr S. Evans University College, London for their critical remarks on this paper. I thank Dr M. Borsuk-Białynicka for the loan of fossil material from collections of the Polish–Mongolian expedition and Dr C. Bell, Berkeley, California, Dr V. Orlova, Zoological Museum, Moscow State University, and Dr V. Sukhanov, Paleontological Institute, Russian Academy of Sciences, for the loan of comparative osteological material and their kind assistance.

References

Alifanov, V.R. 1988. [New lizards (Lacertilia, Teiidae) from the Upper Cretaceous of Mongolia]. *Trudy Sovmestnoi Sovetsko-Mongol'skoi Paleontologicheskoi Ekspeditsii* **34**: 90–100.

—1989a. [More ancient gekkos (Lacertilia, Gekkonidae) from the Lower Cretaceous of Mongolia]. *Paleontologicheskii Zhurnal* **1**: 124–126.

—1989b. [New priscagamids (Lacertilia) from the Upper Cretaceous of Mongolia]. *Paleontologicheskii Zhurnal* **4**: 73–87.

—1991. [A revision of *Tinosaurus asiaticus* Gilmore (Agamidae)]. *Paleontologicheskii Zhurnal* **3**: 115–119.

—1993a. [New lizards of the family Macrocephalosauridae (Sauria) from the Upper Cretaceous of

Mongolia and some critical comments on the classification of Teiidae (*sensu* Estes, 1983a)]. *Paleontologicheskii Zhurnal* 1: 57–74.

—1993b. Some peculiarities of the Cretaceous and Palaeogene lizard faunas of the Mongolian People's Republic. *Kaupia. Darmstadter Beiträge zur Naturgeschichte* 3: 9–13.

—1993c. [Upper Cretaceous lizards of Mongolia and the first inter-American contact]. *Paleontologicheskii Zhurnal* 3: 79–85.

—1996. [The lizard families Priscagamidae and Hoplocercidae (Sauria, Iguania): phylogenetic position and new representatives from the Late Cretaceous of Mongolia]. *Paleontologicheskii Zhurnal* 3: 100–118.

Borsuk-Białynicka, M. 1984. Anguimorphans and related lizards. *Palaeontologia Polonica* 46: 5105.

—1985. Carolinidae, a new family of xenosaurid-like lizards from the Upper Cretaceous of Mongolia. *Acta Palaeontologica Polonica* 30: 151–176.

—1987. *Carusia*, a new name for the Late Cretaceous lizards from the Upper Cretaceous of Mongolia. *Acta Palaeontologica Polonica* 32: 153.

—1988. *Globaura venusta* gen. et sp. nov. and *Eoxanta lacertifrons* gen. et sp. n. – non-teiid lacertoids from the Late Cretaceous of Mongolia. *Acta Palaeontologica Polonica* 33: 211–248.

—1990. *Gobekko cretacicus* gen. et sp. n., a new gekkonid lizard from the Cretaceous of the Gobi Desert. *Acta Palaeontologica Polonica* 35: 67–76.

—1991. Questions and controversies about saurian phylogeny, a Mongolian perspective, pp. 9–10 in Kielan-Jaworowska, Z., Heintz, N. and Nacrem, H.A. (eds.), *5th Symposium on Mesozoic Terrestrial Ecosystems and Biota (Extended Abstracts): Contributions of the Palaeontological Museum, University of Oslo*, 364.

—and Alifanov, V.R. 1991. First Asiatic 'iguanid' lizards in the Late Cretaceous of Mongolia. *Acta Palaeontologica Polonica* 36: 325–342.

—and Moody, S.M. 1984. Priscagaminae, a new subfamily of the Agamidae (Sauria) from the Late Cretaceous of the Gobi Desert. *Acta Palaeontologica Polonica* 29: 51–81.

Camp, C.L. 1923. Classification of the lizards. *Bulletin of the American Museum of Natural History* 48: 289–481.

Cope, E.D. 1900. The crocodilians, lizards and snakes of North America. *Annual Reports of US National Museum, for 1898*: 153–1294.

Dashzeveg, D., Novacek, M.J., Norell, M., Clark, J.M., Chiappe, L., Davidson, A., McKenna, M., Dingus, L., Swisher, C. and Perle. A. 1995. Extraordinary preservation in a new vertebrate assemblage from the Late Cretaceous of Mongolia. *Nature* 374: 446–448.

Denton, R. and O'Neill, R.C. 1993. "Precocious" squamates from the Late Cretaceous of New Jersey, including the earliest record of a North American iguanian. *Journal of Vertebrate Paleontology* 13: 32A–33A (supplement).

—and— 1995. *Prototeius stageri*, gen. et sp. nov., a new teiid lizard from the Upper Cretaceous Marshalltown Formation of New Jersey, with a preliminary phylogenetic revision of the Teiidae. *Journal of Vertebrate Paleontology* 15: 235–253.

Dong, Z. 1965. A new species of *Tinosaurus* from Lushich, Honan. *Vertebrata PalAsiatica* 9: 79–82.

Estes, R. 1964. Fossil vertebrates from the Late Cretaceous Lance Formation Eastern Wyoming. *University of California Publications in Geological Sciences* 49: 1–180.

—1969a. Relationships of two Cretaceous lizards (Sauria, Teiidae). *Museum of Comparative Zoology, Breviora* 317: 1–8.

—1969b. A scincoid lizard from the Cretaceous and Paleocene of Montana. *Museum of Comparative Zoology, Breviora* 331: 1–9.

—1983. Sauria terrestria, Amphisbaenia. In: *Handbuch der Paläoherpetologie*, Part 10A. Stuttgart: Gustav Fisher Verlag.

—, de Queiroz, K. and Gauthier, J. 1988. Phylogenetic relationships within Squamata, pp. 119–281 in Estes, R. and Pregill, G.K. (eds.), *Phylogenetic relationships of the lizard families. Essays commemorating Charles L. Camp.* Stanford University Press.

Etheridge, R. and de Queiroz, K. 1988. A phylogeny of Iguanidae, pp. 283–368 in Estes, R. and Pregill, G.K. (eds.), *Phylogenetic relationships of the lizard families. Essays commemorating Charles L. Camp.* Stanford University Press.

Evans, S.E. 1993. Jurassic lizard assemblages. *Revue de Paléobiologie.* Vol. spéc. (7): 55–65.

Fedorov, P.V. and Nesov, L.A. 1992. [Lizard from the Middle and Late Jurassic border of Northeast Fergana]. *Vestnik Sankt-Peterburgskogo Universiteta* 21: 9–15.

Frost, D.R. and Etheridge, R. 1989. A phylogenetic analysis and taxonomy of Iguanian lizards (Reptilia:

Squamata). *The University of Kansas, Museum of Natural History, Miscellaneous Publication* 81: 1–65.

Fürbringer, M. 1900. Zur vergleichenden Anatomie des Brustschulterapparates und der Schultermuskeln. *Jenaische Zeitschrift* 34: 217–718.

Gao, K. and Fox, R. 1991. New teiid lizards from the Upper Cretaceous Oldman Formation (Judithian) of southeastern Alberta, Canada, with a review of the Upper Cretaceous record of teiids. *Annals of the Carnegie Museum* 60: 145–162.

—and Hou, L. 1995. Iguanians from the Upper Cretaceous Djadokhta Formation, Gobi Desert, China. *Journal of Vertebrate Paleontology* 15: 57–78.

—and— 1996. Systematics and taxonomic diversity of squamates from the Upper Cretaceous Djadokhta Formation, Bayan Mandahu, Gobi Desert, People's Republic of China. *Canadian Journal of Earth Sciences* 33: 578–598.

Gauthier, J. 1982. Fossil xenosaurid and anguid lizards from the Early Eocene Wasatch Formation, southeast Wyoming, and a revision of the Anguioidea. *Contribution to Geology, University of Wyoming* 21: 7–54.

Gilmore, C.W. 1940. New fossil lizards from the Upper Cretaceous of Utah. *Smithsonian Miscellaneous Collections* 99: 1–3.

—1942. Osteology of *Polyglyphanodon*, an Upper Cretaceous lizard from Utah. *Proceedings of the United States National Museum* 92: 229–265.

—1943. Fossil lizards of Mongolia. *Bulletin American Museum of Natural History* 81: 361–384.

Gradziński, R., Kielan-Jaworowska, Z. and Maryańska, T. 1977. Upper Cretaceous Djadokhta, Barun Goyot and Nemegt formations of Mongolia, including remarks on previous subdivisions. *Acta Geologica Polonica* 27: 281–318.

Hecht, M.K. and Hecht, B.M. 1984. [A new lizard from Jurassic deposits of Middle Asia]. *Paleontologicheskii Zhurnal* 3: 135–138.

Hoffstetter, R. 1943. Varanidae et Necrosauridae fossiles. *Bulletin du Museum d'Histoire Naturelle, Paris* 15: 134–141.

—1964. Les Sauria du Jurassique supérieur et specialement les Gekkota de Bavière et de Mandchourie. *Senckenbergiana Biologica* 45: 281–324.

—1966. A-propos des genres *Ardeosaurus* et *Eichstaettisaurus* (Reptilia, Sauria, Gekkonoidea) du Jurassique Supérieur de Franconie. *Bulletin de la Société Geologique de France* 8: 592–595.

Hou, L. 1974. Paleocene lizards from Anhui, China. *Vertebrata PalAsiatica* 3: 193–202.

Jerzykiewicz, T. and Russell, D.A. 1991. Late Mesozoic stratigraphy and vertebrates of the Gobi Basin. *Cretaceous Research* 12: 345–377.

Kalandadze, N.N. and Rautian, A.S. 1992. [The systematics of mammals and historical zoogeography] pp. 44–152 in Rossolimo, O. (ed.), *Phylogenetics of mammals. Sbornik Trudov Zoologicheskogo Museya MGU* 29.

Kluge, A.G. 1987. Cladistic relationships among the Gekkonoidea. *Museum of Zoology, University of Michigan, Miscellaneous Publication* 173: 1–54.

Kuhn, O. 1958. Ein neuer Lacertilier aus dem frankischen Lithographieschiefer. *Neues Jahrbuch für Geologie und Paläontologie, Monatshefte.* 1958: 437–440.

Li, J. 1985. A new lizard from Late Jurassic of Subei, Gansu. *Vertebrata PalAsiatica* 1: 13–18.

—1991. Fossil reptiles from Hetaoyuan Formation, Xichuan, Henan. *Vertebrata PalAsiatica* 3: 199–203.

Mateer, N. 1982. Osteology of the Jurassic lizard *Ardeosaurus brevipes* (Meyer). *Palaeontology* 3: 461–469.

McDowell, S.B.J. and Bogert, C.M. 1954. The systematic position of *Lanthanothus* and the affinities of the anguinomorphan lizards. *Bulletin of the American Museum of Natural History* 105: 1–105.

Moody, S. 1980. *Phylogenetic and historical biogeographical relationships of the genera in the Agamidae (Reptilia: Lacertilia).* Unpublished PhD thesis, University of Michigan.

Nesov, L.A. 1985. [Rare osteichthyans, terrestrial lizards and mammals of the coastal lake zone and the coastal plain from the Cretaceous of Kyzylkum]. *Ezhegodnik Vsesoyuznogo Paleontologicheskogo Obshchestva* 28: 199–219.

—1988. Late Mesozoic amphibians and lizards of Soviet Middle Asia. *Acta Zoologica Cracovensia* 31: 475–486.

—and Gao, K. 1993. Cretaceous lizard from the Kyzylkum Desert, Uzbekistan. *Journal of Vertebrate Paleontology* 13: 51A (supplement).

Norell, M.A., McKenna, M.C. and Novacek, M.J. 1992. *Estesia mongoliensis*, a new fossil varanoid from the Late Cretaceous Barun Goyot Formation of Mongolia. *American Museum Novitates* 3045: 1–24.

Pregill, G.K., Gauthier, J.A. and Greene, H.G. 1986. The evolution of helodermatid squamates, with description of a new taxon and an overview of Varanoidea. *Transactions of the San Diego Society of Natural History* **21**: 167–202.

Presch, W. 1988. Phylogenetic relationships of the Scincomorpha, pp. 472–492 in Estes, R. and Pregill, G.K. (eds.), *Phylogenetic relationships of the lizard families. Essays commemorating Charles L. Camp.* Stanford University Press.

Ride, W.D.L., Sabrosky, C.W., Bernardi, G. and Melville, R.D. 1985. *International Code of Zoological Nomenclature.* Third Edition. International Trust for Zoological Nomenclature in Association with British Museum (Natural History), London, 338 pp.

Rieppel, O. 1988. The classification of the Squamata, pp. 262–293 in Benton, M.J. (ed.). *The phylogeny and classification of the tetrapods. Vol. 1: Amphibians, reptiles, birds.* Oxford: Clarendon Press.

Russell, D.A. 1993. The role of Central Asia in dinosaurian biogeography. *Canadian Journal of Earth Sciences* **30**: 2002–2012.

Shuvalov, V.F. 1974. [On the geology and age of the Höövör and Hüren Dukh localities in Mongolia] *Trudy Sovmestnoi Sovetsko-Mongol'skoi Paleontologicheskoi Ekspeditsii* **34**: 296–304.

Sulimski, A. 1972. *Adamisaurus magnidentatus* n. gen. et n. sp. (Sauria) from the Upper Cretaceous of Mongolia. *Palaeontologia Polonica* **27**: 33–40.

—1975. Macrocephalosauridae and Polyglyphanodontidae (Sauria) from the Late Cretaceous of Mongolia. *Palaeontologia Polonica* **33**: 25–102.

—1978. New data on the genus *Adamisaurus* Sulimski, 1972 (Sauria) from the Upper Cretaceous of Mongolia. *Palaeontologia Polonica* **38**: 43–56.

—1984. New Cretaceous scincomorph lizard from Mongolia. *Palaeontologia Polonica* **46**: 143–155.

Valencia, J., Covacevich, V., Marshall, L., Rivano, S., Charrier, R. and Salinas, P. 1990. Registro fossil mas antiguo de la familia Teiidae de la formacion de Colimapu (cretaceo temprano) en las termas del Flaco Chile central. *II Congresso Latinoamericano Herpetologica. (Libro de Resumenes).* Merida: 75.

Wu, X.-Ch., Brinkmann, D.B. and Russell, A.R. 1996. *Sineoamphisbaenia hexatabularis,* amphisbaenian (Diapsida: Squamata) from the Upper Cretaceous redbeds at Bayan Mandahu (Inner Mongolia, People's Republic of China), and comments on the phylogenetic relationships of the Amphisbaenia. *Canadian Journal of Earth Sciences* **33**: 541–577.

19

Choristodera from the Lower Cretaceous of northern Asia

MIKHAIL B. EFIMOV AND GLENN W. STORRS

Introduction

The Choristodera are a relatively enigmatic group of piscivorous diapsid reptiles that are superficially similar to the Crocodyliformes in habitus. The group has typically been known from fossils belonging to the Champsosauridae from the Upper Cretaceous and Palaeogene of North America and Europe (Brown, 1905; Cope, 1876; Dollo, 1884; Erickson, 1972, 1985, 1987; Fox, 1968; Gao and Fox, 1998; Parks, 1927; Russell, 1956; Russell-Sigogneau and Russell, 1978; Sigogneau-Russell, 1981a), but more plesiomorphic taxa have recently been identified in the Upper Triassic of Europe (Huene, 1935; Storrs, 1993, 1994, 1999; Storrs and Gower, 1993; Storrs et al., 1996), the Middle and Upper Jurassic of Europe and North America (Evans, 1989, 1990, 1991), and the Oligocene (Stampian) of France (Hecht, 1992). Currently, their remains and our knowledge of their evolutionary history are restricted to the Laurasian continents.

Carroll and Currie (1991), Evans (1988), Evans and Hecht (1993), Gao and Fox (1998), and Gauthier et al. (1988) have discussed the possible position of Choristodera within Diapsida, with some disagreement regarding a potential relationship to Archosauromorpha. The most recent analyses (Evans and Hecht, 1993; Gao and Fox, 1998) conclude that Choristodera, once variously identified as a group within Rhynchocephalia (Huene, 1935; Hoffstetter, 1955) or 'Eosuchia' (Romer, 1956), a known paraphyletic assemblage, is most likely an early diapsid stem-group and sister taxon to the Neodiapsida of Benton (1985).

Recent explorations by Russian, Chinese, and Canadian scientists have recovered fossils of choristoderes, generally regarded as champsosaurs or crown-group choristoderes, in the Lower Cretaceous terrestrial deposits of Mongolia and Central Asia (Brinkman and Dong, 1993; Efimov, 1975, 1979, 1983, 1988, 1996; Evans and Manabe, 1999; Sigogneau-Russell, 1981b; Sigogneau-Russell and Efimov, 1984; Figure 19.1). These occur with some frequency at several localities, but are as yet limited in their taxonomic diversity. Four putative genera of Asian choristoderes are known: *Ikechosaurus*, *Irenosaurus*, *Khurendukhosaurus*, and *Tchoiria*. Five species have been described from Mongolia and the former Soviet Union (Table 19:1). Most are generally similar in known morphology, although they differ in size and other particulars. They seem to have also been similar in ecological habit. Outside China, material is known from Hüren Dukh in east-central Mongolia, Tüshleg and Khamaryn Khural in south-eastern Mongolia, Lake Gusinoe in Buryatia, Russia, and possibly Khodzhakul in Uzbekistan (Nesov, personal communication, 1995).

In Mongolia and Buryatia, the remains of choristoderes are usually found in the occasional mass death assemblages of palaeolakes and in attritional deltaic facies. They occur there together with fossils of dinosaurs, pterosaurs, turtles, fish, and other organisms. Although choristodere remains may be locally abundant, many are fragmentary and/or have yet to be prepared and studied. This paper briefly summarizes the history of discovery, occurrence, and material of the described Mongolian choristoderes and of the single specimen from the Trans-Baikal region of Russia.

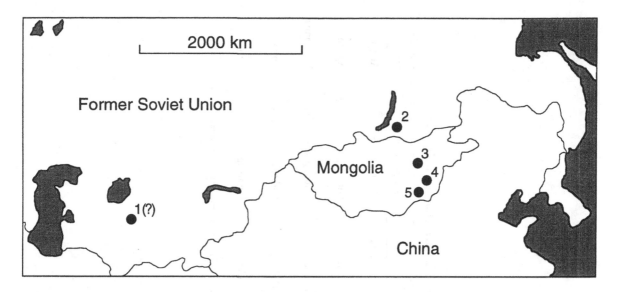

Figure 19.1. The primary localities for choristoderans as noted in the text and Table 19.1. 1, Khodzhakul; 2, Lake Gusinoe; 3, Hüren Dukh; 4, Tüshleg; 5, Khamaryn Khural.

Repository abbreviation

PIN, Paleontological Institute, Russian Academy of Sciences, Russia.

Systematic survey

CHORISTODERA Cope, 1876
Champsosauridae Cope, 1884
Tchoiria Efimov, 1975
See Figures 19.2–19.5.

Diagnosis. Champsosaur of moderate size with total body length of 1–1.5 m. Distinguished from *Champsosaurus* primarily on the basis of the greater numbers of marginal teeth, a broader rostrum, shorter mandibular symphysis, and relatively solid occiput. Distinct from both *Simoedosaurus* and *Ikechosaurus* in having subcircular tooth bases.

Tchoiria namsarai Efimov, 1975
Holotype and locality. PIN 3386/1, partial skull, mandible, and anterior portion of postcranial skeleton; Hüren Dukh, Central-Gobi Aimag, Mongolia.
Horizon. Hühteeg Gorizont, Züünbayan Svita, Aptian, Lower Cretaceous.

Discussion. This, the first Asian choristodere known, was described as *Tchoiria namsarai* by Efimov (1975) on the basis of a good partial skull (Figures 19.2 and 19.3) and skeleton (Figures 19.4 and 19.5) from the Aptian of the Gobi Desert. It is typical of the sort of active aquatic ambush predator envisioned by Evans and Hecht (1993) and Erickson (1985) and exemplified by *Champsosaurus*. As is also typical for champsosaurs, an elongate rostrum with numerous conical teeth served as a trap for fish, its primary food source, and the animal was propelled by lateral undulations of the strong tail. In general, champsosaurs can be considered to have been largely aquatic (Erickson, 1972, 1985). In addition to the holotype of *Tchoiria*, fragmentary postcranial remains are known from the type locality of Hüren Dukh and other localities in central Mongolia.

Tchoiria differs from *Champsosaurus* in the possession of some potentially plesiomorphic features relative to later champsosaurs, such as its slightly broader and shorter rostrum, the posterior displacement of the mandibular articulation, the more numerous marginal teeth, and the rather shortened lower jaw symphysis (Figure 19.2). Evans and Hecht (1993) and Gao and

391

Table 19.1. *Compilation of choristoderan taxa based upon material from the Lower Cretaceous of Mongolia and Buryatia*

Taxon	Holotype	Material	Locality	Gorizont	Stage
Ikechosaurus magnus (Efimov, 1979)	PIN 559/501	partial jaws/skeleton	Khamaryn Khural	Hühteeg	Albian
Irenosaurus egloni (Efimov, 1983)	PIN 3386/2	fragmentary skeleton	Hüren Dukh	Hühteeg	Aptian
Khurendukhosaurus bajkalensis Efimov, 1996	PIN 2234/201	scapulocoracoid/rib	Lake Gusinoe	Ubukunskaya Svita	Neocomian
Khurendukhosaurus orlovi Sigogneau-Russell & Efimov, 1984	PIN 3386/3	fragmentary skeleton	Hüren Dukh	Hühteeg	Aptian
Tchoiria namsarai Efimov, 1975	PIN 3386/1	partial skull/skeleton	Hüren Dukh	Hühteeg	Aptian

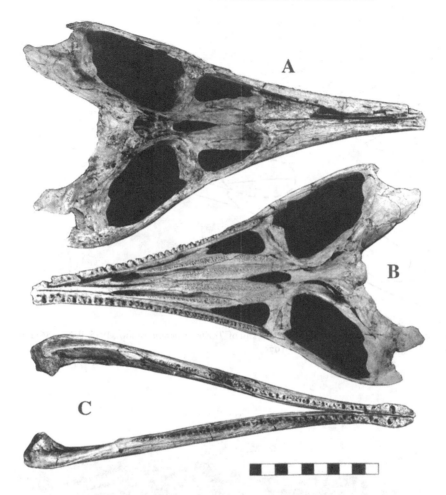

Figure 19.2. *Tchoiria namsarai* Efimov, 1975, PIN 3386/1, holotype skull and mandible from the Aptian of Hüren Dukh, Central Gobi Aimag, Mongolia. Partial skull in (A) dorsal and (B) palatal aspect. Mandible in (C) occusal aspect. Scale bar = 100 mm.

Fox (1998) provide two possible scenarios for character polarity and in-group relationships in Choristodera. *Tchoiria* also exhibits a wide entrance of the basisphenoid on to the posterolateral edge of the large interpterygoid vacuity, rudimentary or absent posttemporal fenestrae, and a longer, more open entepicondylar foramen (Figure 19.5).

Most of the holotype skull of *Tchoiria namsarai* is preserved matrix-free and uncrushed, although the skull roof, rostral extremity, and dorsal rami of the squamosals are lacking (Figure 19.2). It resembles the skull of *Ikechosaurus* Sigogneau-Russell, 1981, in its general snout shape and seemingly short interorbital distance, conditions presumably derived relative to, for example, *Simoedosaurus* (Brinkman and Dong, 1993). The rostrum of *Tchoiria*, and that of *Ikechosaurus*, is rather more elongated than that of *Simoedosaurus*. *Tchoiria*, however, has ovate tooth bases and a relatively large interpterygoid vacuity, as well as an apparently narrower interorbital area than that seen in *Ikechosaurus*.

Three species of *Tchoiria* have been described

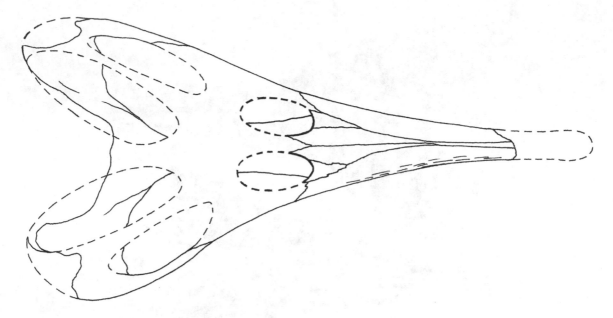

Figure 19.3. Schematic reconstruction of the holotype skull of *Tchoiria namsarai* Efimov, 1975, PIN 3386/1, in dorsal aspect, indicating extent of missing sections (after Efimov, 1975).

Figure 19.4. Block containing articulated gastralia from the holotype of *Tchoiria namsarai* Efimov, 1975, PIN 3386/1. Scale bar = 20 mm.

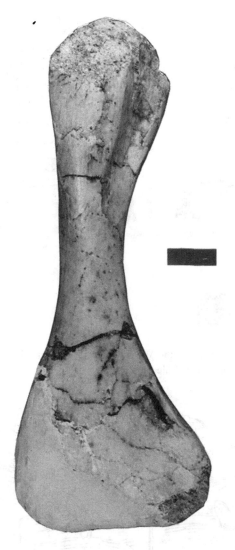

Figure 19.5. Left humerus of the holotype of *Tchoiria namsarai* Efimov, 1975, PIN 3386/1. Note elongate entepicondylar foramen. Scale bar = 10 mm.

(Efimov, 1979, 1983), but all save the type species of the genus, *T. namsarai*, have subsequently been referred to other genera (Efimov, 1983, 1988).

Ikechosaurus Sigogneau-Russell, 1981

Discussion. Ikechosaurus was first described (Sigogneau-Russell, 1981a, b) from the Lower Cretaceous of China (Inner Mongolia) following the reidentification of a rostral fragment once considered crocodylian (Young, 1964). *Ikechosaurus sunailinae*, the resulting species, is now known from a moderate amount of good quality referred material from China (Brinkman and Dong, 1993).

Ikechosaurus magnus (Efimov, 1979)
See Figure 19.6.

Holotype and locality. PIN 559/501, partial mandible and portions of postcranial skeleton; Khamaryn Khural, Eastern-Gobi Aimag, south-eastern Mongolia.

Horizon. Upper Hühteeg Gorizont, Züünbayan (?Sainshand) Svita, Albian, Lower Cretaceous.

Discussion. The second known species of *Ikechosaurus* was created with the transfer of *I. magnus* (Efimov, 1979) from the genus *Tchoiria* by Efimov in 1983. This reassignment was made on the basis of the rectangular, transversely elongate tooth bases of both *I. sunailinae* Sigogneau-Russell, 1981, and *I. magnus* (Efimov, 1979) (Figure 19.6). *Ikechosaurus magnus* is a very large animal, up to 3 m in length, and is also the geologically youngest (Albian) of the known Mongolian choristoderes. Only the holotype of *I. magnus*, from Khamaryn Khural, can be certainly ascribed to this species.

The nasal of *Tchoiria* only just reaches the prefrontal, in contrast to that of *Ikechosaurus* as seen in the Chinese specimens, although this may be supposed to be a variable feature. Additionally, the tooth bases of *Tchoiria* are ovate, not sub-rectangular as in *Ikechosaurus*. Otherwise, the two genera are difficult to distinguish and require renewed study, as noted by Brinkman and Dong (1993). We likewise tentatively retain *Ikechosaurus* until this can be accomplished. *Ikechosaurus* and *Simoedosaurus* share transversely elongate tooth sockets and apparently fused postorbital/postfrontals. Evans and Hecht (1993) link these two genera by virtue of these features and a presumed abbreviated rostrum, although this latter character is not apparent in the new material reported by Brinkman and Dong (1993), nor in the Mongolian specimen. Gao and Fox (1998) link *Ikechosaurus* and *Simoedosaurus* as a sister group to *Tchoiria* within Simoedosauridae (excluding *Champsosaurus*).

Figure 19.6. Holotype jaw of *Ikechosaurus magnus* (Efimov, 1979), PIN 559/501, from the Albian of Khamaryn Khural, Eastern-Gobi Aimag, Mongolia in (A) lateral aspect, anterior to left, and (B) alveolar aspect. Scale bar = 50 mm.

Figure 19.7. Holotype cervical vertebrae of *Khurendukhosaurus orlovi* Sigogneau-Russell and Efimov, 1984, PIN 3386/3, from the Aptian of Hüren Dukh, Mongolia, in left lateral view. 2× natural size.

CHORISTODERA *incertae sedis*
Khurendukhosaurus Sigogneau-Russell and Efimov, 1984
See Figure 19.7.

Discussion. Khurendukhosaurus is known from a variety of fossils occurring at Hüren Dukh in central Mongolia and, as the only Russian choristodere, material from the Lake Gusinoe locality (Neocomian) of the Trans-Baikal region in Buryatia (Figure 19.1). It is a generally small animal, up to 1 m in length, and supposedly with a relatively high and narrow trunk and an elongate cervical region. Typically, the coracoid and scapula are fused, as are the neural arches with the body of the vertebrae, and the vertebral centra are elongate. *Khurendukhosaurus* appears to represent a relatively plesiomorphic choristodere, possibly a stem-group member.

Khurendukhosaurus orlovi Sigogneau-Russell and Efimov, 1984

Holotype and locality. PIN 3386/3, fragmentary postcranial skeleton; Hüren Dukh, Central-Gobi Aimag, Mongolia.

Horizon. Hühteeg Gorizont, Züünbayan Svita, Aptian, Lower Cretaceous.

Discussion. Aside from the holotype, *Khurendukhosaurus orlovi* is known from a number of postcranial remains. The vertebrae are unusual for champsosaurian-grade choristoderes (Figure 19.7), but there are clear similarities to the putative plesiomorphic genus *Pachystropheus* (Storrs, 1993, 1994, 1999; Storrs and Gower, 1993; Storrs *et al.*, 1996). For example, the centra are generally elongate and amphi- to platycoelous. Both the ventral longitudinal keel and the transverse processes are pronounced. The neural spines are also similar, in being anteroposteriorly elongate and having distal tips that are transversely expanded and rugose, reinforcing the suggestion that *Khurendukhosaurus* is a relatively plesiomorphic choristodere. The cervical centra, with lengths of approximately 2.5 times their height, are similar also to the condition found in *Cteniogenys* (Evans, 1991) and *Lazarussuchus* (Hecht, 1992), while a marked ventral keel is also seen in *Cteniogenys* (Evans, 1991), *Champsosaurus* (Erickson, 1972), and *Ikechosaurus* (Brinkman and Dong, 1993). The proportions of the dorsal centrum and neural spine of the latter taxon are very similar to those of *Khurendukhosaurus* (see below).

The prominent supra- and infraglenoid tubercles on the scapulocoracoid for attachment of the triceps muscle tendon, suggest that the forelimbs could exert relatively large mechanical forces, perhaps related to a strongly amphibious habit. Representative remains of *Khurendukhosaurus orlovi* have been relatively well described and illustrated by Sigogneau-Russell and Efimov (1984).

Khurendukhosaurus bajkalensis Efimov, 1996

Holotype and locality. PIN 2234/201, right scapulocoracoid and dorsal rib; Kanon Ravine, western shore of Lake Gusinoe, Buryatia, Trans-Baikal, Russia.

Horizon. Ubukunskaya Svita, Neocomian, Lower Cretaceous.

Discussion. New choristoderan material from Russia, representing a small (less than 1 m) animal, has been described by Efimov (1996) as a new species of *Khurendukhosaurus*, *K. bajkalensis*. Only the holotype is known, however, and it is difficult to distinguish this species from *K. orlovi*, or indeed any plesiomorphic choristodere, because it is founded upon such a small amount of material. The animal's size and the proportions of its scapulocoracoid were thought to be distinctive (Efimov, 1996), but these can be considered to be variable characters within a species. The unique geographical and stratigraphical position of the specimen is by itself insufficient to allow erection of a new taxon, thus the validity of *K. bajkalensis* must remain tentative pending new material and a fuller study.

Irenosaurus Efimov, 1988
See Figures 19.8 and 19.9.

Diagnosis. A small choristodere of 1–1.5 m length, distinguished from typical champsosaurs by the relatively elongate vertebral centra. Differentiated particularly from *Tchoiria* by the lack of a prominent entepicondylar foramen in the humerus.

Irenosaurus egloni (Efimov, 1983)

Holotype and locality. PIN 3386/2, fragmentary postcranial skeleton; Hüren Dukh, Central-Gobi Aimag, Mongolia.

Horizon. Hühteeg Gorizont, Züünbayan Svita, Aptian, Lower Cretaceous.

Discussion. This monospecific taxon is known only from a partial postcranial skeleton, including most notably the humerus (Figure 19.8), from the Aptian of central Mongolia. Evans and Hecht (1993) show some scepticism regarding the validity of *Irenosaurus egloni*, suggesting that this taxon, originally described as a species of *Tchoiria* (Efimov, 1983), may show no greater variation from *Tchoiria namsarai* than do the

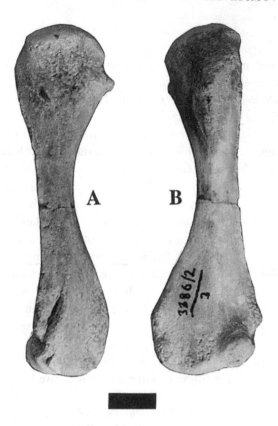

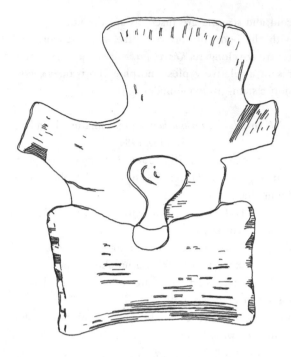

Figure 19.9. Dorsal vertebra of the holotype of *Irenosaurus egloni* (Efimov, 1983), PIN 3386/2, from the Aptian of Hüren Dukh, Mongolia, in left lateral view. 2.5× natural size.

Figure 19.8. Left humerus of the holotype of *Irenosaurus egloni* (Efimov, 1983), PIN 3386/2, from the Aptian of Hüren Dukh, Mongolia, in (A) oblique superior, and (B) inferior aspect. Scale bar = 10 mm.

several described species of *Champsosaurus* from each other. For example, while the humerus of *Irenosaurus* (Figure 19.8) seemingly lacks the prominent entepicondylar groove of *Tchoiria* (Efimov, 1988; Sigogneau-Russell and Efimov, 1984), this character is potentially ontogenetically controlled. However, the elongate vertebral centra and primitively shaped, rugose neural spines of *Irenosaurus* (Figure 19.9), are quite unlike those of *Tchoiria*, and rather suggest an affinity with *Khurendukhosaurus* known from the same locality as both *Irenosaurus* and *Tchoiria*. The supposedly well-formed articular ends of the humerus of *Irenosaurus* are surely individually and ontogenetically variable, while the overall morphology of the humerus is in keeping with that of plesiomorphic choristoderes,

most notably the probable stem-group member *Pachystropheus*.

Beyond the few irregularities in the shape of the humerus, no autapomorphies of *Irenosaurus* can be identified, in which case *Irenosaurus* may prove to be synonymous with one of the genera discussed above. The mixed remains of three putative genera in the fossil lake sediments of a single locality at Hüren Dukh present various problems of identification and taxonomy, and are only likely to be resolved by additional material.

Conclusion

The fossil record of Asian choristoderes is limited to the Lower Cretaceous (Neocomian to Albian) save an allusion to fragments in the Palaeocene of western Asia (Nesov, 1995). Their discovery and recognition, beginning in the 1970s with material collected by the

Soviet–Mongolian Palaeontological Expeditions, produced the first significant stratigraphical range extension of the Choristodera. Four choristoderan genera and five species have now been described from Mongolia and Russia, although the validity of some of these taxa is suspect.

At the beginning of the Cretaceous, tectonic activity associated with orogenesis in Siberia, Mongolia, and China resulted in the formation of a vast territory of structural depressions, often situated along the lines of deep faults. Against the backdrop of the humid palaeoclimate of these regions, such depressions became collecting basins for continental watersheds and hosted a broad range of aquatic and amphibious animals. Some of these faunas included choristoderes, most notably at Hüren Dukh in central Mongolia, where they are relatively common fossils and may have dominated the semi-aquatic fauna relative to the apparently less abundant crocodylians. Although the database is admittedly small, it suggests that only in the Late Cretaceous did the crocodylians become ascendant in Asia, while choristoderes disappeared regionally.

The ecological niche 'replacement' of choristoderes by crocodylians in the Late Cretaceous of Asia may have been in response to an increasingly arid climate and the destruction of many of the Asian lake systems. By contrast, choristoderes are common in the Late Cretaceous of North America and it is possible that the Asian aquatic systems may have contributed to an Asio-American faunal interchange (across Beringia) at the close of the Early Cretaceous. By the Palaeocene, choristoderes reappear in the western regions of Asia, perhaps migrating there from Europe along the shores of Tethys (Nesov, 1995), although the Early Tertiary choristoderes of Asia are known only from fragmentary remains.

The presence of significant quantities of as yet unprepared choristoderan remains in the museums of Moscow, St. Petersburg, and especially Ulaanbaatar, provides hope that in the near future our knowledge of Asian choristoderes will be greatly improved. In particular, their skeletal morphology and taxonomy should be examined anew, leading to a significant increase in the potential for phylogenetic analysis of the group as a whole. Ancilliary questions of taphonomy, palaeoecology and palaeobiogeography may also be addressed by the study of these collections.

Acknowledgements

The authors thank Mike Benton and David Unwin for the opportunity to contribute to this book. Susan Evans and David Gower kindly provided useful reviews of the typescript. G.W.S. thanks the Royal Society of London and the Natural Environment Research Council of the UK for financial support and the University of Bristol Department of Earth Sciences and the Museum of Natural History & Science at Cincinnati Museum Center for the use of their facilities.

References

Benton, M.J. 1985. Classification and phylogeny of the diapsid reptiles. *Zoological Journal of the Linnean Society* **84**: 97–164.

Brinkman, D.B. and Dong Z.-M. 1993. New material of *Ikechosaurus sunailinae* (Reptilia: Choristodera) from the Early Cretaceous Laohongdong Formation, Ordos Basin, Inner Mongolia, and the interrelationships of the genus. *Canadian Journal of Earth Sciences* **30**: 2153–2162.

Brown, B. 1905. The osteology of *Champsosaurus* Cope. *Memoir of the American Museum of Natural History* **9**: 1–26.

Carroll, R.L. and Currie, P.J. 1991. The early radiation of diapsid reptiles, pp. 354–424, in Schultze, H.-P. and Trueb, L. (eds.), *Origins of the Higher Groups of Tetrapods*. Cornell University Press: Ithaca.

Cope, E D. 1876. On some extinct reptiles and Batrachia from the Judith River and Fox Hills beds of Montana. *Proceedings of the Academy of Natural Sciences, Philadelphia, Paleontological Bulletin* **23**: 340–359.

—1884. The Choristodera. *The American Naturalist* **18**: 815–817.

Dollo, L. 1884. Première note sur le Simoedosaurien d'Erquelinnes. *Bulletin du Musée Royal d'Histoire Naturelle de Belgique* **3**: 151–186.

Efimov, M.B. 1975. [Champsosaurs from the Lower Cretaceous of Mongolia.] *Trudy Sovmestnoi Sovetsko-Mongol'skoi Paleontologicheskoi Ekspeditsii* 2: 84–93.

—1979. [*Tchoiria* (Champsosauridae) from the Early Cretaceous of Khamaryn Khural, MNR.] *Trudy Sovmestnoi Sovetsko-Mongol'skoi Paleontologicheskoi Ekspeditsii* 8: 56–57.

—1983. [Champsosaurs of Central Asia.] *Trudy Sovmestnoi Sovetsko-Mongol'skoi Paleontologicheskoi Ekspeditsii* 24: 67–75.

—1988. [Fossil crocodiles and champsosaurs of Mongolia and the USSR.] *Trudy Sovmestnoi Sovetsko-Mongol'skoi Paleontologicheskoi Ekspeditsii* 36: 1–108.

—1996. [Champsosaurid from the Lower Cretaceous of Buryatia.] *Paleontologicheskii Zhurnal* 1996: 122–123.

Erickson, B.R. 1972. The lepidosaurian reptile *Champsosaurus* in North America. *Monograph of the Science Museum of Minnesota (Paleontology)* 1: 1–91.

—1985. Aspects of some anatomical features of *Champsosaurus* (Reptilia: Eosuchia). *Journal of Vertebrate Paleontology* 5: 111–127.

—1987. *Simoedosaurus dakotensis*, new species, a diapsid reptile (Archosauromorpha: Choristodera) from the Paleocene of North America. *Journal of Vertebrate Paleontology* 7: 237–251.

Evans, S.E. 1988. The early history and relationships of the Diapsida, pp. 221–260, in Benton, M.J. (ed.), *The Phylogeny and Classification of the Tetrapods. Volume 1.* Oxford University Press: Oxford.

—1989. New material of *Cteniogenys* (Reptilia: Diapsida; Jurassic) and a reassessment of the phylogenetic position of the genus. *Neues Jahrbuch für Geologie und Paläontologie, Monatshefte* 1989: 577–589.

—1990. The skull of *Cteniogenys*, a choristodere (Reptilia: Archosauromorpha) from the Middle Jurassic of Oxfordshire. *Zoological Journal of the Linnean Society* 99: 205–237.

—1991. The postcranial skeleton of the choristodere *Cteniogenys* (Reptilia: Diapsida) from the Middle Jurassic of England. *Geobios* 24: 187–199.

—and Hecht, M.K. 1993. A history of an extinct reptilian clade, the Choristodera: longevity, Lazarus-taxa, and the fossil record. *Evolutionary Biology* 27: 323–338.

—and Manabe, M. 1999. A choristoderan reptile from the Lower Cretaceous of Japan. *Special Papers in Palaeontology*, 60, 101–119.

Fox, R.C. 1968. Studies of Late Cretaceous vertebrates. I.

The braincase of *Champsosaurus* Cope (Reptilia: Eosuchia). *Copeia* 1968: 100–109.

Gao, K. and Fox, R.C. 1998. New choristoderes (Reptilia: Diapsida) from the Upper Cretaceous and Palaeocene, Alberta and Saskatchewan, Canada, and phylogenetic relationships of Choristodera. *Zoological Journal of the Linnean Society*, 124: 303–353.

Gauthier, J., Kluge, A. and Rowe, T. 1988. Amniote phylogeny and the importance of fossils. *Cladistics* 4: 100–209.

Hecht, M.K. 1992. A new choristodere (Reptilia, Diapsida) from the Oligocene of France: an example of the Lazarus effect. *Geobios* 25: 115–131.

Hoffstetter, R. 1955. Rhynchocephalia, pp. 556–576, in Piveteau, J. (ed.), *Traité de Paléontologie*, 5. Masson et Cie: Paris.

Huene, E. von 1935. Ein Rhynchocephale aus dem Rhät (*Pachystropheus* n. g.). *Neues Jahrbuch für Mineralogie, Geologie und Paläontologie* 74: 441–447.

Nesov, L.A. 1995. [*The Dinosauria of North Eurasia: New Data on Composition, Complex, Ecology and Palaeobiogeography*.] St. Petersburg: Nauka.

Parks, W.A. 1927. *Champsosaurus albertensis*, a new species of rhynchocephalian from the Edmonton Formation of Alberta. *University of Toronto Studies, Geological Series* 23: 1–48.

Romer, A.S. 1956. *Osteology of the Reptiles*. Chicago: University of Chicago Press.

Russell, L.S. 1956. The Cretaceous reptile *Champsosaurus natator* Parks. *Bulletin of the National Museum of Canada* 145: 1–51.

Sigogneau-Russell, D. 1981a. Etude ostéologique du reptile *Simoedosaurus* (Choristodera). IIe partie: squelette postcranien. *Annales de Paléontologie (Vertébrés)* 67: 61–140.

—1981b. Présence d'un nouveau Champsosauridé dans le Crétace supérieur de Chine. *Comptes Rendus de l'Académie des Sciences, Paris* 292: 1–4.

—and Efimov, M.B. 1984. Un Choristodera (Eosuchia?) insolite du Crétacé Inférieur de Mongolie. *Paläontologische Zeitschrift* 58: 279–294.

—and Russell, D. 1978. Etude ostéologique du reptile *Simoedosaurus* (Choristodera). *Annales de Paléontologie (Vertébrés)* 64: 1–84.

Storrs, G.W. 1993. Terrestrial components of the Rhaetian (uppermost Triassic) Westbury Formation of southwestern Britain, pp. 447–451, in Lucas, S.J. and

Morales, M. (eds.) *The Nonmarine Triassic. New Mexico Museum of Natural History Science and Science Bulletin*, 3. New Mexico Museum of Natural History and Science: Albuquerque.

—1994. Fossil vertebrate faunas of the British Rhaetian (latest Triassic). *Zoological Journal of the Linnean Society* **112**: 217–259.

—1999. Tetrapods, pp. 223–238, in Swift, A. and Martill, D.M. (eds.) *Fossils of the Rhaetian Penarth Group. Field Guide to Fossils*, 9. The Palaeontological Association: Dorchester.

—, and Gower, D.J. 1993. The earliest possible choristodere (Diapsida) and gaps in the fossil record of semi-aquatic reptiles. *Journal of the Geological Society, London* **150**: 1103–1107.

—, Gower, D.J. and Large, N.F. 1996. The diapsid reptile *Pachystropheus rhaeticus*, a probable choristodere from the Rhaetian of Europe. *Palaeontology* **39**: 323–349.

Young, C.C. 1964. New fossil crocodiles from China. *Vertebrata PalAsiatica* **8**: 189–210.

20

Mesozoic crocodyliforms of north-central Eurasia

GLENN W. STORRS AND MIKHAIL B. EFIMOV

Introduction

Fossils of Mesozoic crocodylians *sensu lato* (Crocodyliformes) have been found in the territory of the former Soviet Union from Russia, the Ukraine, Kazakhstan, Uzbekistan, Kirgizstan and Tadzhikistan (Efimov, 1975, 1976, 1982a, b, 1988a, b). They are also an important component of vertebrate assemblages from Mongolia (Efimov, 1981, 1983, 1988a, b; Konzhukova, 1954; Mook, 1924; Osmólska, 1972). The oldest of these north and middle Eurasian crocodylians are Middle Jurassic in age, whereas the youngest Mesozoic forms are from the Maastrichtian. The most recent review of former Soviet and Mongolian crocodylians is by Efimov (1988b) and the fauna and geology of the Asian provinces has been reviewed by Rozhdestvenskii and Khozatskii (1967).

In this chapter, we discuss briefly the characters and status of the described taxa of fossil crocodylians from the Mesozoic deposits of the former Soviet Union and Mongolia. The localities of a number of these taxa are indicated in Figure 20.1. The described species and material are listed in Tables 20.1 and 20.2.

Repository abbreviations

AMNH, American Museum of Natural History, New York; GIN, Geological Institute, Academy of Sciences, Mongolian People's Republic, Ulaanbaatar; PIN, Paleontological Institute, Russian Academy of Sciences, Moscow; TsNIGRI, Central Scientific-Research Geological Exploration Museum, St. Petersburg; ZPAL, Institute of Paleobiology, Polish Academy of Sciences, Warsaw.

Systematic survey

DIAPSIDA Osborn, 1903
ARCHOSAURIA Cope, 1869
CROCODYLIFORMES Benton and Clark, 1988
Comments. The systematics and phylogeny of croco-dylians, or more correctly here, the Crocodyliformes *sensu* Clark (1986, 1994) and Benton and Clark (1988), remain a matter of some confusion. For instance, Benton and Clark (1988) consider that Protosuchia Mook, 1934, is most likely paraphyletic, whereas other authors, such as Wu *et al.* (1994) and Wu and Sues (1996) retain a monophyletic clade Protosuchia. Clark (1994) retains the Protosuchidae, but excludes the Mongolian *Gobiosuchus* from it as a plesiomorphic sister taxon. Efimov (1988b) considered *Gobiosuchus* to be part of Protosuchia, but recognized *Artzosuchus* and *Shartegosuchus* as 'notosuchians'. For convenience sake, *Gobiosuchus* and the other Mongolian 'protosuchians' discussed here are considered together, whether or not they are closely related within or without Protosuchia.

The relationships of mesosuchian-grade Mesoeu-crocodylia Whetstone and Whybrow, 1983, are partic-ularly problematic and no attempt to elucidate them is made here. Mesosuchia Huxley, 1875, is unquestion-ably a paraphyletic taxon, although many previous dis-cussions of northern Eurasian fossil crocodylians make use of this archaic systematic division. In this work, northern and central Eurasian 'mesosuchians' are dis-cussed in the framework of mesosuchian-grade Neosuchia following Benton and Clark (1988), and Thalattosuchia Fraas, 1901 (non-neosuchian Metasuchia Benton and Clark, 1988, are unknown from the former Soviet Union). Eusuchia Huxley, 1875,

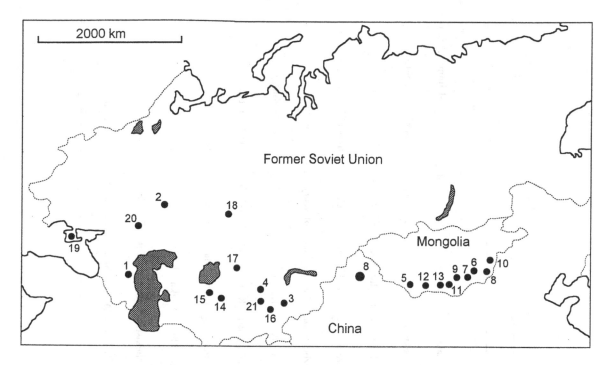

Figure 20.1. The main localities for north-central Eurasian Mesozoic crocodilians, as noted in text and tables. 1, Koisu; 2, Khvalynsk; 3, Tchanget; 4, Mihailovka; 5, Shar Teeg; 6, Bayan Zag; 7, Üüden Sair; 8, Amtgai; 9, Shireegiin Gashuun; 10, Khongil Tsav; 11, Nemegt; 12, Nogoon Tsav; 13, Ulaan Bulag; 14, Dzharakhuduk; 15, Sheichdzheili; 16, Kansai; 17, Shakh-Shakh; 18, Kushmurun; 19, Inkerman; 20, environs of Volgograd; 21, Kirkhudag: Dotted lines = international boundaries, solid lines = coastline.

is widely agreed to be monophyletic and the eusuchian taxa from Mongolia and the former Soviet Union are therefore dealt with in a single section below.

?PROTOSUCHIA Mook, 1934
SHARTEGOSUCHIDAE Efimov, 1988b

Comments. At the locality of Shar Teeg in southwestern Mongolia, thick sequences of sands and clays represent a gigantic lake basin of Tithonian age. The monsoonal character of the climate resulted in alternating wet and dry seasons (Efimov, 1988b). Numerous crocodilian remains are entombed in the shore and bottom sediments of the ancient lake at Shar Teeg, particularly in the restricted 'refugia' of deeper water basins that remained following lake retreat during dry spells. These fossils include the peculiar 'protosuchians', *Shartegosuchus asperopalatum*, and *Nominosuchus matutinus*.

Nominosuchus Efimov, 1996

Diagnosis. Small 'protosuchian' with total skull length estimated at 60 mm. Skull roof narrow and relatively low. Large supratemporal fenestrae. External nares divided by nasals. Small antorbital fossa at confluence of lachrymal, maxilla and jugal. Distinguished by large palpebrals, surface sculpturing of pterygoid and ectopterygoid, and 'choanae' divided anteroposteriorly by palatines.

Nominosuchus matutinus Efimov, 1996

Holotype and locality. PIN 4174/4 partial skull; Shar Teeg, Trans-Altai Gobi, southwestern Mongolia.

403

Table 20.1. Compilation of Mesozoic crocodyliform taxa based upon material from the former Soviet Union

Taxon	Holotype	Material	Locality	Svita	Stage
Kansajsuchus extensus Efimov, 1975	PIN No. 2399/301	right premaxilla	Tadzhikistan, Kansai	Yalovach	lower Santonian
Karatausuchus sharovi Efimov, 1976	PIN No. 2585/1	juvenile skeleton	Kazakhstan, Mikhailovka	Karabastau	Oxfordian–Kimmeridgian
Shamosuchus borealis (Efimov, 1975)	PIN No. 372/702	frontal/prefrontal	Uzbekistan, Dzharakhuduk	Bissekty	upper Turonian–Coniacian
Shamosuchus karakalpakensis Nesov et al.,1989	TsNIGRI 311/12457	frontal	Uzbekistan, Sheichdzheili	Khodzhakul	Cenomanian
Shamosuchus occidentalis Efimov, 1982a	PIN No. 327/721	partial rostrum	Uzbekistan, Dzharakhuduk	Bissekty	upper Turonian
Tadzikhosuchus kizylkumensis Nesov et al., 1989	TsNIGRI 331/12457	fragment left dentary	Uzbekistan, Dzharakhuduk	Bissekty	Turonian
Tadzikhosuchus macrodentis Efimov, 1982b	PIN No. 2399/457	fragment left dentary	Tadzhikistan, Kansai	Yalovach	lower Santonian
Tadzikhosuchus neutralis Efimov, 1988b	PIN No. 2399/458	fragment rt. dentary	Tadzhikistan, Kansai	Yalovach	lower Santonian
Thoracosaurus borissiaki Efimov, 1988b	TsNIGRI c. 3373 no.709	partial skull	Crimea Region, Inkerman	—	Maastrichtian
Turanosuchus aralensis Efimov, 1988a	PIN No. 2229/507	mandib. symphysis	Kazakhstan, Shakh–Shakh	Bostobe	lower Santonian
Zholsuchus procevus Nesov et al., 1989	TsNIGRI 381/12457	right premaxilla	Uzbekistan, Dzharakhuduk	Bissekty	Coniacian
Zhyrasuchus angustifroms Nesov et al., 1989	TsNIGRI 332/12457	frontal	Uzbekistan, Dzharakhuduk	Bissekty	Coniacian

Table 20.2. Compilation of *Mesozoic crocodyliform taxa based upon material from Mongolia*

Taxon	Holotype	Material	Locality	Horizon	Stage
Artzosuchus brachicephalus Efimov, 1983	GIN PST 10/23	partial skull and jaws	Üiden Sair	Baruungoyot	Campanian
Gobiosuchus kielanae Osmólska, 1972	ZPAL Mg.R 11/67	skull/partial skeleton	Bayan Zag	Baruungoyot	Campanian
Gobiosuchus parvus Efimov, 1983	GIN PST 10/22	skull and jaws	Üiden Sair	Baruungoyot	Campanian
Nominosuchus matutinus Efimov, 1996	PIN No. 4174/4	partial skull	Shar Teeg	Tsagaantsav	Tithonian
Shamosuchus ancestralis (Konzhukova, 1954)	PIN No. 551/21–1	skull roof fragment	Nemegt	Nemegt	lower Maastrichtian
Shamosuchus djadochtaensis Mook, 1924	AMNH 6412	partial skull and jaws	Bayan Zag	Baruungoyot	Campanian
Shamosuchus gradiliforms (Konzhukova, 1954)	PIN No. 554/1	skull and jaws	Shireegin Gashuun	Bayanshiree	Turonian–Santonian
Shamosuchus major (Efimov, 1981)	PIN No. 3726/501	partial skull	Khongil Tsav	Bayanshiree	Turonian–Santonian
Shamosuchus tersus Efimov, 1983	PIN No. 3141/501	partial skull	Nogoon Tsav	Nemegt	lower Maastrichtian
Shamosuchus ulanicus Efimov, 1983	PIN No. 3140/502	partial skull/skeleton	Ulaan Bulag	Nemegt	lower Maastrichtian
Shamosuchus ulgicus (Efimov, 1981)	PIN No. 3458/501	partial skull and jaws	Amtgai	Bayanshiree	Cenomanian–Turonian
Shartegosuchus asperopalatum (Efimov, 1988b)	PIN No. 4171/2	partial skull and jaws	Shar Teeg	Tsagaantsav	Tithonian
Sunosuchus shartegensis Efimov, 1988a	PIN No. 4171/1	partial skull and jaws	Shar Teeg	Tormkhon	Tithonian

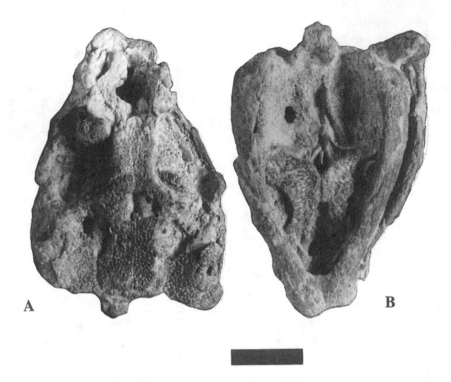

Figure 20.2. *Shartegosuchus asperopalatum* Efimov, 1988b, holotype skull (PIN 4174/2), from the Tithonian Tsagaantsav Gorizont of Shar Teeg, Mongolia, in (A) dorsal, and (B) palatal aspect. Scale bar equals 10 mm.

Horizon. Tsagaantsav (Ulaan Malgajtsky) Svita, Late Jurassic (Tithonian).

Discussion. This species, known from several partial specimens including two skulls, differs from traditional 'protosuchians' in the posterior position and divided nature of the "choanae". However, the anterior opening resembles the median palatal fenestra of *Shartegosuchus.* Efimov (1996), nevertheless, distinguishes *Nominosuchus* by the less posteriorly placed true choanae (on the palatine/pterygoid boundary).

Shartegosuchus Efimov, 1988b

Diagnosis. Small 'protosuchian' or 'notosuchian' with total body length up to 0.5 m. Small supratemporal fenestrae, broad skull roof. Distinguished from other crocodyliformes by the unusual sculptured surface of palate and three anteriorly positioned palatal fenestrae. Distinguished from typical Protosuchia by its posteriorly located choanae.

Shartegosuchus asperopalatum Efimov, 1988b
See Figure 20.2.

Holotype and locality. PIN 4171/2, partial skull and jaws of a juvenile animal; Shar Teeg, TransAltai Gobi, southwestern Mongolia.

Horizon. Tsagaantsav (Ulaan Malgajtsky) Svita, Late Jurassic (Tithonian).

Discussion. Shartegosuchus, known primarily from the deformed holotype of a juvenile individual, has been characterized by a combination of several unusual features (Efimov, 1988b). It is a small crocodyliform, with a skull reconstructed at about 40 mm in length. Extreme dorsoventral deformation prevents accurate assessment of the height of the skull. Antorbital fenestrae or depressions are present, and the skull table is broad. Most notably, like *Nominosuchus*, its palate is covered by fine sculpturing similar to that found on the skull roof, suggesting a possible osteodermal contribution to the surface of the palate. The anterior portion of the palate also contains three rounded and elongate fenestrae, the central one of which has been doubtfully considered to enter the course of the internal narial passage. The true choanae, apparently divided by a thin bony septum, occur posteriorly, at the back of short palatines and pterygoids, a condition clearly deviating from that of typical Protosuchia *sensu* Mook (1934), and hence rather more similar to mesosuchian-grade taxa such as *Notosuchus*. *Shartegosuchus*, like *Nominosuchus*, is most likely a stem-group crocodyliform and near sister taxon to Mesoeucrocodylia.

GOBIOSUCHIDAE Osmólska, 1972,

Comments. Alleged 'protosuchians' (Family Gobiosuchidae) are also known from the Upper Cretaceous of Mongolia (Efimov, 1983, 1988b; Osmólska, 1972). Like the Shartegosuchidae, these finds are rare, and to date are known from only four specimens.

Gobiosuchus Osmólska, 1972

Diagnosis. Small crocodyliforms with a unique suite of characters that include the seemingly complete secondary closure of the upper temporal fenestrae, pneumatization of the bones of the skull, the absence of mandibular fenestrae, and a highly developed osteodermal body armour, covering not only the spinal and abdominal areas, but also the proximal portions of the limbs.

Gobiosuchus kielanae Osmólska, 1972
See Figure 20.3.

Holotype and locality. ZPAL Mg.R 11/67, a skull and partial skeleton; Bayan Zag, Mongolia.
Horizon. Baruungoyot Svita, Cretaceous (Campanian).

Gobiosuchus parvus Efimov, 1983

Holotype and locality. GIN PST 10/22, skull and jaws; Üüden Sair, Mongolia.
Horizon. Baruungoyot Svita, Cretaceous (Campanian).
Discussion. Like *Shartegosuchus* and *Nominosuchus*, *Gobiosuchus kielanae* and *G. parvus* are small forms, and considered by Benton and Clark (1988) to be closely related to mesoeucrocodilians. *Gobiosuchus kielanae* and *G. parvus* are possibly conspecific.

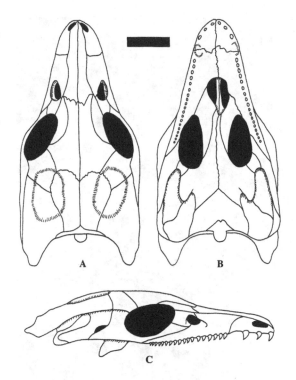

Figure 20.3. *Gobiosuchus kielanae* Osmólska, 1972, skull (ZPAL Mg.R 11/67), from the Cretaceous (Campanian) of Bayan Zag, Mongolia, in (A) dorsal, (B) palatal, and (C) lateral aspect (after Osmólska, 1972). Scale bar equals 10 mm.

ARTZOSUCHIDAE Efimov, 1983
Artzosuchus Efimov, 1983

Diagnosis. A small crocodyliform distinguished from other 'protosuchians/notosuchians' by its possession of a very short and low skull with lozenge-shaped supratemporal fenestrae and a jaw exhibiting a massive mandibular symphysis.

Artzosuchus brachicephalus Efimov, 1983

Holotype and locality. GIN PST 10/23, a partial skull and set of jaws; Üüden Sair, Mongolia.
Horizon. Baruungoyot Svita, Cretaceous (Campanian).
Discussion. This species apparently occurs as a separate taxon in the same Campanian deposits with *Gobiosuchus parvus* at Üüden Sair (Efimov, 1983, 1988b). *Artzosuchus*, however, is known only from an incomplete, unprepared skull. As for the taxa discussed above, it is unlikely that this poorly known species represents a true protosuchian of traditional usage.

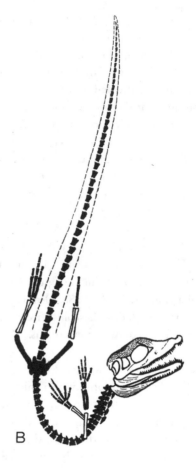

Figure 20.4. *Karatausuchus sharovi* Efimov, 1976, part of holotype skeleton (PIN 2585/1), from the Upper Jurassic of the Karatau Basin, Mikhailovka, Kazakhstan. (A) photograph, (B) line drawing to the same scale. Scale bar in centimetres.

MESOEUCROCODYLIA Whetstone and
Whybrow, 1983
NEOSUCHIA Benton and Clark, 1988
?ATOPOSAURIDAE Gervais, 1871
Karatausuchus Efimov, 1976
See Figure 20.4.

Diagnosis. A very small crocodyliform (approximately 160 mm long) with amphiplatyan vertebral centra, reduced dermal ossicles, and over ninety small, laterally compressed teeth.

Karatausuchus sharovi Efimov, 1976

Holotype and locality. PIN 2585/1, complete, but poorly preserved skeleton of a juvenile animal; Mikhailovka, Karatau Mountain Range, southern Kazakhstan.

Horizon. Karabastau Svita, Late Jurassic (Oxfordian or Kimmeridgian).

Discussion. Karatausuchus, from the calcareous shales of the Karatau lake, is one of the geologically oldest crocodyliforms from the former Soviet Union. *Karatausuchus* has been considered an atoposaur (Efimov, 1976, 1988b), but may be close to the paralligatorids (Efimov, 1996). The most recent analysis of Atoposauridae (Buscalioni and Sanz, 1988) considered *Karatausuchus* indeterminate 'with an imprecise position within the Metamesosuchia (*sensu* Buffetaut, 1982)'. As noted by Buscalioni and Sanz (1988), the features previously used for placement of *Karatausuchus* within Atoposauridae, such as small size, limb proportions, and reduced dermal armour (and implicitly, stratigraphic position) are all rather problematic.

Karatausuchus is known only from a single, poorly preserved, juvenile skeleton in part and counterpart. There are over ninety small, laterally compressed, *Alligator*-like teeth in the jaws, eight cervical vertebrae, seventeen dorsals, two sacrals and forty-six caudal vertebrae. All vertebrae are amphiplatyan in character. As in many specimens from Karatau, there is some indication of soft tissue preservation. In spite of, for example, the relatively large number of teeth for a brevirostrine taxon, the material might properly be considered nondiagnostic. Its lack of preserved anatomical detail, especially in the skull (Figure 20.4), and the obvious juvenile status of the specimen (exemplified by the extremely large orbits, relatively large head, and short rostrum) are troubling. Analysis of this specimen for taxonomic purposes must be suspect because of the possibility of a large amount of ontogenetic variation between it and any adult material recovered in the future. More data on this taxon must await additional specimens.

GONIOPHOLIDAE Cope, 1875
Sunosuchus Young, 1948
Diagnosis. A goniopholid crocodyliform with twinned anterior palatal openings, relatively small, circular, supratemporal fenestrae and a small maxillary (antorbital) depression. Distinguished from other goniopholids by its moderately elongate rostrum.

Sunosuchus shartegensis Efimov, 1988a
Holotype and locality. PIN 4171/1, partial skull and jaws; Shar Teeg, Mongolia.
Horizon. Tormkhon (Ulaan Malgajtsky) Svita, Late Jurassic (Tithonian).
Discussion. Sunosuchus shartegensis, from the basin of the ancient lake Shar Teeg in southwestern Mongolia, has been identified as a goniopholid by Efimov (1988a, b). It is a moderately longirostrine form, and presumed to be closely related to the Chinese species *Sunosuchus miaoi* Young, 1948. Young (1948) considered his new taxon to be a pholidosaur on the basis of its elongate rostrum, but Buffetaut (1986) considered the Chinese animal to be a member of the Goniopholidae.

Buffetaut (1986) also questioned the previously assigned Late Jurassic age of *S. miaoi*, rather preferring an estimate of Early Cretaceous. A third species, *S. thailandicus*, also from the Jurassic, was earlier recognized as an unusual longirostrine goniopholid (Buffetaut and Ingavat, 1980, 1984).

The condition of the palate and choanae of *Sunosuchus shartegensis* is 'mesosuchian' to the extent that the internal nares, although located rather posteriorly, are not found at the extreme posterior end of the palatines and pterygoids. However, additional twin openings also occur in the anterior portion of the palate as in *S. miaoi*. In this regard, the palate is similar to that of North American 'goniopholids' such as *Eutretauranosuchus* from the uppermost Jurassic Morrison Formation (Buffetaut 1986; Langston, 1973; Mook, 1967). *Sunosuchus shartegensis* has relatively small, circular, supratemporal fenestrae and a small maxillary depression, as does the Chinese *Sunosuchus* species, and the latter feature is again reminiscent of the antorbital depression of 'goniopholids'. The validity of a monophyletic Goniopholidae is, however, in doubt (Clark, 1994).

Kansajsuchus Efimov, 1975
See Figure 20.5A.
Diagnosis. A very large 'goniopholid' crocodyliform, up to 8 m long, with large, bicarinate, characteristically ribbed teeth.

Kansajsuchus extensus Efimov, 1975
Holotype and locality. PIN 2399/301, an isolated right premaxilla; Kansai, Fergana Basin, Tadzhikistan.
Horizon. Yalovach Svita, Late Cretaceous (lower Santonian).

Turanosuchus Efimov, 1988a
See Figure 20.5B.
Diagnosis. A 'goniopholid' with a flat and elongate mandibular symphysis and a very large fourth mandibular alveolus.

Turanosuchus aralensis Efimov, 1988a
Holotype and locality. PIN 2229/507, an isolated mandibular symphysis; Shakh-Shakh, southern Kazakhstan.

Figure 20.5. *Kansajsuchus extensus* Efimov, 1975, holotype right premaxilla (PIN 2399/301), from the lower Santonian Yalovach Svita of Kansai, Tadzhikistan, in (A) palatal aspect. *Turanosuchus aralensis* Efimov, 1988a, holotype mandibular symphysis (PIN 2229/507), from the lower Santonian Bostobe Svita of Shakh-Shakh, Kazakhstan, in (B) dorsal aspect. Scale bar in centimetres.

Horizon. Bostobe Svita, Late Cretaceous (lower Santonian).

Discussion. *Kansajsuchus*, *Turanosuchus*, and '*Zholsuchus*' from the Upper Cretaceous of Central Asia are potential 'goniopholids', but all are poorly known (Efimov, 1988b). As expected for traditional 'Mesosuchia', *Kansajsuchus* at least (on the basis of referred material), possessed amphicoelous vertebral centra, overlapping osteodermal scutes, and abdominal armour. *Kansajsuchus* is an especially large form (up to 8 m long) known only from collections of fragmentary material from the Kansai site, Fergana Basin, Tadzhikistan (Efimov, 1975). The teeth are large, bicarinate, and sometimes coarsely ribbed. The frontal bears transverse and longitudinal ridges and the inter-orbital distance is large.

Turanosuchus from Shakh-Shakh, southern Kazakhstan, is characterized by a very flat, elongate mandibular symphysis and an enlarged and prominent fourth mandibular alveolus (Efimov, 1988a). As in *Kansajsuchus*, the tooth crowns possess a characteristic enamel sculpturing. '*Zholsuchus*', based upon a fragment of premaxilla from Uzbekistan, (Nesov *et al.*, 1989) is insufficient material for diagnosis and identified here as a *nomen dubium*.

PARALLIGATORIDAE Konzhukova, 1954
Shamosuchus Mook, 1924

Diagnosis. Crocodyliform up to 4 m long with a broad rostrum of moderate length, a premaxillary pit for reception of the fourth mandibular tooth, large pterygoid flanges, no mandibular fenestra, platycoelous vertebral centra and well developed, non-overlapping osteodermal scutes.

Shamosuchus ancestralis (Konzhukova, 1954)
Holotype and locality. PIN 551/21–1, a skull roof fragment; Nemegt, Mongolia.
Horizon. Nemegt Svita, Late Cretaceous (lower Maastrichtian).

Shamosuchus borealis (Efimov, 1975)
Holotype and locality. PIN 372/702, isolated frontal and prefrontal fragment; Dzharakhuduk, Uzbekistan.
Horizon. Bissekty Svita, Late Cretaceous (upper Turonian or Coniacian).
Discussion. Both Nesov (1995 and Efimov (1988b) considered *S. borealis* to be the senior synonymn of *S. occidentalis* Efimov, 1982a, a species based upon a partial rostrum from the same svita and locality as *S. borealis.*

Shamosuchus djadochtaensis Mook, 1924
Holotype and locality. AMNH 6412, partial skull and jaws; Bayan Zag, Mongolia.
Horizon. Baruungoyot Svita, Cretaceous (Campanian).

Shamosuchus gradilifrons (Konzhukova, 1954)
See Figure 20.6.
Holotype and locality. PIN 554/1, isolated skull and jaws; Shireegiin Gashuun, Mongolia.
Horizon. Bayanshiree Svita, Late Cretaceous (Turonian or Santonian).

Shamosuchus karakalpakensis Nesov *et al.,* 1989
Holotype and locality. TsNIGRI 311/12457, isolated frontal; Sheichdzheili, Uzbekistan.
Horizon. Khodzhakul Svita, Late Cretaceous (Cenomanian).

Shamosuchus major (Efimov, 1981)
Holotype and locality. PIN 3726/501, partial skull; Khongil Tsav, Mongolia.
Horizon. Bayanshiree Svita, Late Cretaceous (Turonian or Santonian).

Shamosuchus tersus Efimov, 1983
See Figure 20.7.
Holotype and locality. PIN 3141/501, partial skull; Nogoon Tsav, Mongolia.

Horizon. Nemegt Svita, Late Cretaceous (lower Maastrichtian).

Shamosuchus ulanicus Efimov, 1983
Holotype and locality. PIN 3140/502, a partial skull and skeleton; Ulaan Bulag, Mongolia.
Horizon. Nemegt Svita, Late Cretaceous (lower Maastrichtian).

Shamosuchus ulgicus (Efimov, 1981)
See Figures 20.8 and 20.9.
Holotype and locality. PIN 3458/501, partial skull and jaws with associated dermal armour; Amtgai, Mongolia.
Horizon. Bayanshiree Svita, Late Cretaceous (Turonian or Santonian).
Discussion. Shamosuchus Mook, 1924 (= *Paralligator* Konzhukova, 1954 following Efimov, 1982a), represented by species up to 4 m long, is known from numerous specimens from the Upper Cretaceous of both Mongolia and Uzbekistan, and is thus the only crocodylian genus yet recognized from both Asian provinces. Over 30 localities in Mongolia alone have produced material attributed to this paralligatorid. Ten species of *Shamosuchus* have been named (some originally as *Paralligator*) from Mongolia and the former Soviet Union (Efimov, 1975, 1981, 1982a, 1983; Konzhukova, 1954; Mook, 1924; Nesov *et al.,* 1989), with an eleventh from China (Bohlin, 1953; Sun, 1958), although undoubtedly many of these are synonymous. Indeed, Nesov (1995) and Efimov (1988b) have synonymized *S. occidentalis* with its senior, *S. borealis.* Most new species descriptions have been based upon observed variations in the skull roof, palate, and lower jaw (Tables 20.1 and 20.2). There has been little study, however, of potential intraspecific variation or dimorphism in these populations, let alone the vagaries of their preservation. It is hoped that future work will address this outstanding problem that continues to cloud the specific content of the genus.

The generally broad, moderately lengthened rostrum of *Shamosuchus* contains a premaxillary pit for the reception of the fourth mandibular tooth, thus the alligatorine 'overbite' is developed in this genus

Figure 20.6. *Shamosuchus gradilifrons* (Konzhukova, 1954), holotype skull (PIN 554/1), from the Upper Cretaceous of Shireegiin Gashuun, Mongolia, in (A) lateral, (B) palatal, and (C) dorsal aspect. Scale bar in centimetres.

(Figures 20.6–20.8). Some species, at least, possessed squat rear teeth of crushing type, perhaps for use in masticating turtle shells. Large pterygoid flanges are present in all species. *Shamosuchus* is considered a near-sister taxon to Eusuchia by Benton and Clark (1988). Its well developed osteoderms (Figure 20.9) apparently occur in more than two longitudinal rows and do not overlap one another, as is also true for eusuchians. They do, however, also occur on the proximal and epipodial portions of the limbs. The verte-

brae are platycoelous and there is no mandibular fenestra.

THALATTOSUCHIA Fraas, 1901

Comments. Russian and Asian thalattosuchian fossils are, to date, largely fragmentary and have often been difficult to differentiate from the remains of other marine reptiles. Indeed, few strictly marine crocodyliforms have been reported from the territory of the

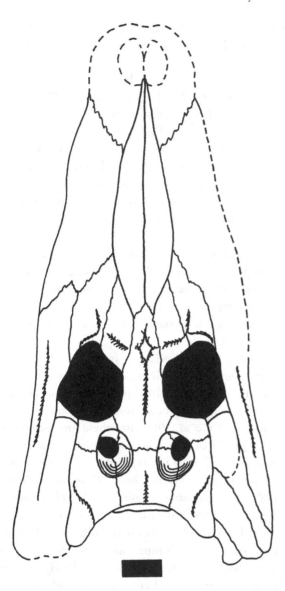

Figure 20.7. *Shamosuchus tersus* Efimov, 1983, skull (PIN 3141/501), from the Upper Cretaceous of Nogoon Tsav, Mongolia, in dorsal aspect. (After Efimov, 1983.) Scale bar equals 20 mm.

Figure 20.8. *Shamosuchus ulgicus* (Efimov, 1981), holotype skull (PIN 3458/501), from the Upper Cretaceous of Amtgai, Mongolia, in palatal aspect.

former Soviet Union (Storrs *et al.*, Chapter 11). The rare Russian Platform thalattosuchians, mostly if not exclusively Jurassic in age, and which represent the bulk of fossil crocodyliform material known from Russia itself, are largely undescribed.

Ochev (1981) reported a metriorhynchid vertebra and metatarsal V, possibly of *Dakosaurus*, from the

uppermost Jurassic or lowermost Cretaceous of Khoroshevsky Island (Khvalynsk, middle Volga region). An indeterminate thalattosuchian is known from the Kimmeridgian of the Moscow Region (Efimov and Chkhikvadze, 1987), while the partial remains of a thalattosuchian, as yet undescribed, has been found in the Volgian Jurassic of Gorodische near Ul'yanovsk (Storrs *et al.*, Chapter 11). *Poekilopleuron schmidti* was described from Cenomanian-age sedi-

Figure 20.9. *Shamosuchus ulgicus* (Efimov, 1981) fragment of dorsal dermal armour of the holotype (anterior at bottom) (PIN 3458/501), from the Upper Cretaceous of Amtgai, Mongolia. Scale bar in centimetres.

ments of the Kursk Region as a marine crocodylian (Kipriyanov, 1883), although Efimov and Chkhikvadze (1987) reinterpreted these remains as those of a dinosaur.

Teleosaurs, such as *Steneosaurus*, have been reported in Asia, in this case from the Jurassic (Aalenian) Karakhskaya Svita of Koisu, South Dagestan (Efimov, 1978, 1982a, 1988b). *Teleosaurus* itself is known, allegedly, from the uppermost Jurassic or lowermost Cretaceous of the Alaisky Mountains, Fergana Basin, Kirgizstan (Efimov and Chkhikvadze, 1987). The Chinese teleosaur genus *Peipehsuchus* has been reported from the Callovian of Changet (Fergana) and other localities (Nesov *et al.*, 1989), although the specimens are extremely fragmentary. Originally considered to be a pholidosaurid

(Steel, 1973), *Peipehsuchus* is clearly a teleosaurid as suggested by Buffetaut (1982) and shown by Li (1993).

EUSUCHIA Huxley, 1875
CROCODYLIDAE Cuvier, 1807
Tadzhikosuchus Efimov, 1982b
See Figure 20.10.

Diagnosis. Diplocynodontian crocodyliform with standard positioning of third and fourth mandibular teeth (i.e., immediately adjacent with no intervening space). Distinguished from *Diplocynodon* on the basis of aveolar shape and position.

Tadzhikosuchus macrodentis Efimov, 1982b

Holotype and locality. PIN 2399/457, symphysial portion of a fragmentary left dentary; Kansai, Tadzhikistan.

Horizon. Yalovach Svita, Late Cretaceous (lower Santonian).

Discussion. Tadzhikosuchus, from the Upper Cretaceous of Tadzhikistan and Uzbekistan, is a potential crocodylid with diplocynodontian affinities. Three species have been described (Efimov, 1982b, 1988b, Nesov *et al.*, 1989), but these are each based upon partial dentaries and realistically can not be differentiated. Thus the type species, *T. macrodentis*, should be given priority over '*T. neutralis*' Efimov, 1988b, from the same locality, and '*T. kizylkumensis*' Nesov *et al.*, 1989, from the Bissekty Svita (Turonian) of Dzharakhuduk, Uzbekistan. Like the Tertiary *Diplocynodon* of Europe and North America, *Tadzhikosuchus* is notable for the immediately adjacent positioning of the third and fourth mandibular teeth with no intervening space. Efimov (1982b) differentiates the two genera on the basis of subtle differences in alveolar shape and position, and implicitly, the disparity in their stratigraphic distribution. It remains to been seen whether or not such a differentiation will be supported by new material.

Zhyrasuchus Nesov *et al.*, 1989, from Dzharakhuduk, Uzbekistan also represents a possible Cretaceous crocodylid, but was described only on the basis of fragmentary remains from a variety of individuals. The species is supposedly characterized by a narrow interorbital bridge, elongate and narrow choanae, and

A

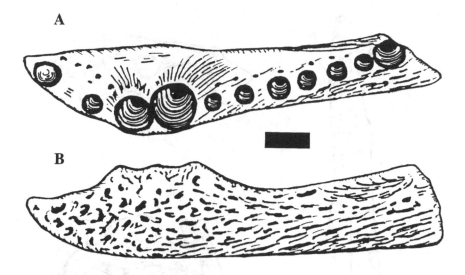

B

Figure 20.10. *Tadzhikosuchus macrodentis*, holotype dentary (PIN 2399/457), from the lower Santonian Yalovach Svita of Kansai, Tadzhikistan, in (A) dorsal, and (B) lateral aspect. (After Efimov, 1982b.) Scale bar equals 10 mm.

generally small size. On the basis of the poor material used to diagnose each genus, it remains unclear how *Zhyrasuchus* and *Tadzhikosuchus* are related, or indeed, whether or not they are synonymous.

THORACOSAURINAE Nopcsa, 1928
Thoracosaurus Leidy, 1852

Diagnosis. Longirostrine crocodylid with progressively tapering snout and no abrupt demarcation of rostrum and orbital/postorbital region. The long, slender nasals contact the premaxillae, but not the external nares. The long mandibular symphysis includes the splenial.

Thoracosaurus borissiaki Efimov, 1988b
See Figure 20.11.

Holotype and locality. TsNIGRI c. 3373 no.709, well preserved partial skull; Inkerman, Crimea Region.
Horizon. Details unknown, Late Cretaceous (Maastrichtian).
Discussion. Thoracosaurine remains are rare in Russia, but an undoubted species of *Thoracosaurus* has been described from the marine Maastrichtian of the Crimea (Inkerman Mines), near Sevastopol' (Borisyak, 1913; Ryabinin, 1946). This fossil, described by Borisyak (1913) as *T. macrorhynchus*, is represented by a well preserved skull (Figure 20.11). Steel (1973) subsequently assigned this material to a western European species, *T. scanicus*. Efimov (1988b), however, has referred the material to a new species, *T. borissiaki*. Because of the lack of a thorough description of the specimen, it is not yet possible to determine whether or not creation of this new species was warranted. Efimov (1988b) suggests that the moderate size of the species, the circular supratemporal fenestra and rounded lateral borders to the suborbital fenestra, and other meristic characters are significant. Efimov and Chkhikvadze (1987) also note a possible Maastrichtian thoracosaur from the Crimean village of Skalistoe, but no other details have been forthcoming, and a possible, fragmentary thoracosaur has also been found in the marine Cretaceous of Malaya Serdoba in the Penza Region (pers. communication L.S. Glikman).

Additional eusuchian record

The North American alligatorine *Brachychampsa* has been reported from locality 2 (Campanian) at Kirkuduk in western Asia (Nesov, 1995). If a genuine occurrence, this would represent an important

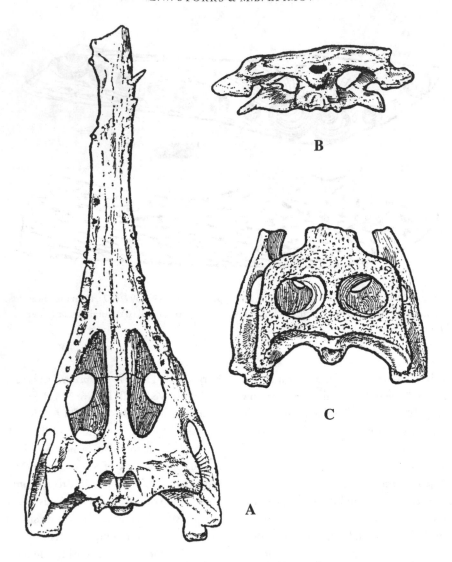

Figure 20.11. *Thoracosaurus borissiaki* Efimov, 1988b, skull (TsNIGRI c. 3373 no.709), from the Crimean Upper Cretaceous near Sevastopol', in (A) palatal, (B) occipital, and (C) posterior aspect. (After Borisyak, 1913.)

discovery with palaeozoogeographic implications. However, the material is problematic and requires more detailed examination prior to acceptance of this identification.

Discussion

In the Mesozoic, several regions present different crocodylian faunal complexes. In the Middle to Late Jurassic, the islands and north coast of the Tethys seaway (Russia to China) were home to at least two genera of thalattosuchian (*Steneosaurus* and ?*Dakosaurus*) (Efimov, 1978; Efimov and Chkhikvadze, 1987; Ochev, 1981). Similar habitats in the Maastrichtian (e.g. in the Simferopol' and Volgograd regions) are associated with the eusuchian *Thoracosaurus* (Borisyak, 1913; Ryabinin, 1946).

In the Late Jurassic, large, freshwater, intercontinental lakes of Kazakhstan and Mongolia (Mikhailovka and Shar Teeg, respectively) contained so-called 'protosuchians' (*Shartegosuchus* and *Nominosuchus*) and mesosuchian-grade mesoeucrocodylians (such as *Sunosuchus* and '*Karatausuchus*' from Shar Teeg, Mongolia and Karatau, Kazakhstan, respectively). Small, supposed protosuchians (*Gobiosuchus* and *Artzosuchus*, though see above) persisted in the Late Cretaceous ephemeral lakes and alluvial plains of Mongolia (Campanian sediments of Bayan Zag and Üüden Sair), in which are also found the paralligatorid *Shamosuchus*. *Shamosuchus* also occurs in several Late Cretaceous fully freshwater lake deposits of Mongolia (e.g. Nemegt, Ulaan Bulag and Amtgai).

Shamosuchus and other mesosuchian-grade taxa (*Kansajsuchus*, *Turanosuchus*, '*Zholsuchus*' are found in Late Cretaceous rocks representative of brackish, estuarine, deltaic and riverine systems of the old western coast of Central Asia, making *Shamosuchus* the one genus known from several realms. Early eusuchians (*Zhyrasuchus* and *Tadzhikosuchus*) are also found in these Central Asian fluviodeltaic deposits. Isolated occurrences of Mesozoic crocodylians in other parts of the former Soviet Union belong to less strictly defined paleogeographic provinces.

Conclusions

Mesozoic fossil crocodyliforms of the former Soviet Union and Mongolia have received only sporadic attention and study. Efimov (1975, 1976, 1981, 1982a, b, 1983, 1988a, b) has produced the largest and most recent body of work on these animals, but much remains to be done. Earlier studies by Borisyak (1913), Konzhukova (1954), Mook (1924), and Osmólska (1972) are now also dated. Nesov *et al.* (1989) studied principally fragmentary, and therefore problematic material. Many of the crocodyliform species described from northern and central Eurasia are, as a result, nondiagnosable or representative of previously known taxa. The problem of the synonymy of many of these species must be more fully addressed in future work. Many of these same comments can be applied also to the Tertiary crocodyliform faunas of the continent, subject matter outside the realm of this study, but reviewed previously by Efimov (1988b).

It is clear that the Central Asian countries and

Mongolia, in particular, have great potential for the production of crocodyliform specimens that may be useful in future taxonomic and phylogenetic investigations. The known Mesozoic associations of these regions preserve many apparently conservative lineages of mesosuchian-grade animals, and some unique forms as well. In particular, the unusual and so-called 'protosuchians' of the Jurassic and Cretaceous deposits deserve future attention, and all the Asian taxa require detailed redescription and illustration.

The Russian Platform has produced occasional marine occurrences and will likely continue to do so. This latter fauna bears a strong resemblance to that of western Europe, both in the Upper Jurassic and the uppermost Cretaceous, and suggests a prolonged opportunity for faunal interchange between the two regions as part of the wider Tethyan province (Efimov, 1988b). Unfortunately, little quality material has been collected from Russia and the likelihood that stratigraphically well constrained specimens will reach museum collections in the immediate future is not high. Nevertheless, here as well, there remains promise and potential for the longer term.

Acknowledgements

The authors thank Mike Benton and David Unwin for the opportunity to contribute to this book. James Clark and Eric Buffetaut aided with reviews of the typescript paper, although any existing errors remain our own. G.W.S. thanks the Royal Society of London and the Natural Environment Research Council of the UK for financial support, and the University of Bristol Department of Earth Sciences and the Museum of Natural History & Science at Cincinnati Museum Center for the use of their facilities.

References

Benton, J.M. and Clark, J.A. 1988. Archosaur phylogeny and the relationships of the Crocodylia, pp. 295–338, in Benton, M.J. (ed.), *The Phylogeny and Classification of the Tetrapods, Volume 1: Amphibians, Reptiles, Birds*, Clarendon Press: Oxford.

Bohlin, B. 1953. Fossil reptiles from Mongolia and Kansu. *Reports from the Scientific Expedition to the North-Western*

Provinces of China. The Sino-Swedish Expedition 6: 1–113.

Borisyak, A.A. 1913. Sur les restes d'un crocodile de l'étage supérieur du Crétacé de la Crimée. *Bulletin de l'Académie Impériale des Sciences, St. Petersbourg* 7: 555–558.

Buffetaut, E. 1982. Radiation évolutive, paléoécologie et biogéographie des crocodiliens mésosuchiens. *Société Géologique de France, Mémoires* 142: 1–88.

—1986. Remarks on the anatomy and systematic position of *Sunosuchus miaoi* Young, 1948, a mesosuchian crocodile from the Mesozoic of Gansu, China. *Neues Jahrbuch für Geologie und Paläontologie, Monatshefte* 11: 641–647.

—and Ingavat, R. 1980. A new crocodilian from the Jurassic of Thailand, *Sunosuchus thailandicus* n. sp. (Mesosuchia, Goniopholidae), and the palaeogeographical history of south-east Asia in the Mesozoic. *Geobios* 13: 879–889.

—and— 1984. The lower jaw of *Sunosuchus thailandicus* a mesosuchian crocodile from the Jurassic of Thailand. *Palaeontology* 27: 199–206.

Buscalioni, A.D. and Sanz, J.L. 1988. Phylogenetic relationships of the Atoposauridae (Archosauria, Crocodylomorpha). *Historical Biology* 1: 233–250.

Clark, J.A. 1986. *Phylogenetic Relationships of the Crocodylomorph Archosaurs.* Unpublished Ph.D. dissertation, University of Chicago: Chicago.

—1994. Patterns of evolution in Mesozoic Crocodyliformes, pp. 84–97, in Fraser, N.C. and Sues, H.-D. (eds.), *In the Shadow of the Dinosaurs.* Cambridge University Press: Cambridge.

Efimov, M.B. 1975. [Late Cretaceous crocodiles of Soviet Central Asia and Kazakhstan.]. *Paleontologicheskii Zhurnal* 9: 417–420.

—1976. [The oldest crocodile on the territory of the USSR.]. *Paleontologicheskii Zhurnal* 10: 115–117.

—1978. [Survey of the fossil crocodiles of the USSR.] [Abstract]. *Byulleten' Moskovskogo Obshchestva Ispytatelei Prirody* 53: 157.

—1981. [New paralligatorids from the Upper Cretaceous of Mongolia.]. *Trudy Sovmestnoi Sovetsko-Mongol'skoi Paleontologicheskoi Ekspeditsii* 15: 26–28.

—1982a. [New fossil crocodiles from the territory of the USSR.] *Paleontologicheskii Zhurnal* 2: 146–150.

—1982b. [A diplocynodontian crocodile from the Upper Cretaceous of Tadzhikistan.] *Paleontologicheskii Zhurnal* 4: 115–116.

—1983. [Review of fossil crocodilians from Mongolia.] *Trudy Sovmestnoi Sovetsko-Mongol'skoi Paleontologicheskoi Ekspeditsii* 24: 79–96.

—1988a. [On the fossil crocodilians from Mongolia and the USSR.] *Trudy Sovmestnoi Sovetsko-Mongol'skoi Paleontologicheskoi Ekspeditsii* 34: 82–86.

—1988b. [Fossil crocodiles and champsosaurs of Mongolia and the USSR.]. *Trudy Sovmestnoi Sovetsko-Mongol'skoi Paleontologicheskoi Ekspeditsii* 36: 1–108.

—1996. The Jurassic crocodylomorphs of inner Asia, pp. 305–310, in Morales, M. (ed.), *The Continental Jurassic,* Museum of Northern Arizona: Flagstaff.

—and Chkhikvadze, V.M. 1987. [Survey of the finds of fossil crocodiles in the USSR.] *Izvestiya Akademii Nauk Gruzinskoi SSR, Seriya Biologicheskaya* 13: 200–207.

Fraas, E. 1901. Die Meerkrokodile (*Thalattosuchia* n. g.), eine neue Sauriergruppe der Juraformation. *Jahreshefte des Vereins für vaterländische Naturkunde in Württemberg* 57: 409–418.

Huxley, T. 1875. On *Stagonolepis robertsoni*, and the evolution of the Crocodilia. *Quarterly Journal of the Geological Society of London* 31: 423–438.

Kipriyanov, W.A. 1883. Studien über die Fossilen Reptilien Russlands. III. Theil. Gruppe Thaumatosauria n. aus der Kreide-Formation und dem Moskauer Jura. *Mémoires de l'Académie Impériale des Sciences, St. Petersbourg* 31: 1–57.

Konzhukova, E.D. 1954. [New fossil crocodilians from Mongolia.]. *Trudy Paleontologicheskogo Instituta AN SSSR* 48: 171–194.

Langston, W. 1973. The crocodilian skull in historical perspective, pp. 263–284, in Gans, C. and Parsons, T.S. (eds.), *Biology of the Reptilia*, 4. Academic Press: London.

Li, J. 1993. [A new specimen of *Peipehsuchus teleorhinus* from the Ziliujing Formation of Daxian, Sichuan.] *Vertebrata PalAsiatica* 31: 85–94.

Mook, C.C. 1924. A new crocodilian from Mongolia. *American Museum Novitates* 117: 1–5.

—1934. The evolution and classification of the Crocodilia. *Journal of Geology* 42: 295–304.

—1967. Preliminary description of a new goniopholid crocodilian. *Kirtlandia* 2: 1–10.

Nesov, L.A. 1995. [*The Dinosauria of North Eurasia: New Data on Composition, Complex, Ecology and Palaeobiogeography.*] St. Petersburg: Nauka.

—Kaznyshkina, L.F. and Cherepanov, G.O. 1989. [Dinosaurs – Ceratopsidae and crocodiles from the

Mesozoic of Soviet Central Asia.], pp. 144–154, in *Trudy 33 Sessii Vsesoyuznogo Paleontologicheskogo Obshchestva.* St. Petersburg: Nauka.

Ochev, V.G. 1981. [Marine crocodiles in the Mesozoic of Povolzh'e (right bank of the Volga).] *Priroda* **1981**: 103.

Osmólska, H. 1972. Preliminary note on a crocodilian from the Upper Cretaceous of Mongolia. *Palaeontologia Polonica* **27**: 43–47.

Rozhdestvenskii, A.K. and Khozatskii, L.I. 1967. [Late Mesozoic terrestrial vertebrates from the Asiatic part of the USSR.], pp. 82–92, in Martinson, G.G. (ed.) [*Stratigraphy and Paleontology of Mesozoic and Palaeogene-Neogene Continental Deposits of the Asiatic Part of the USSR*]. St. Petersburg: Nauka.

Ryabinin, A.N. 1946. [New finds of fossil reptiles in the Crimea.]. *Priroda* **1946**: 65–66.

Steel, R. 1973. Crocodylia. *Handbuch der Paläoherpetologie* **16**: 1–116.

Sun, A.-L. 1958. A new species of *Paralligator* from Sungarian Plain. *Vertebrata PalAsiatica* **2**: 277–280.

Whetstone, K.N. and Whybrow, P. 1983. A cursorial crocodilian from the Triassic of Lesotho (Basutoland), South Africa. *Occasional Papers of the University of Kansas* **106**: 1–37.

Wu, X.-C., Brinkman, D.B. and Lu., J.-C. 1994. A new species of *Shantungosuchus* from the Lower Cretaceous of Inner Mongolia (China), with comments on *S. chuhsienensis* Young, 1961 and the phylogenetic position of the genus. *Journal of Vertebrate Paleontology* **14**: 210–229.

—and Sues., H.-D. 1996. Reassessment of *Platyognathus hsui* Young, 1944 (Archosauria: Crocodyliformes) from the Lower Lufeng Formation (Lower Jurassic) of Yunnan, China. *Journal of Vertebrate Paleontology* **16**: 42–48.

Young, C.C. 1948. Fossil crocodiles in China, with notes on dinosaurian remains associated with the Kansu crocodiles. *Bulletin of the Geological Society of China* **28**: 255–288.

21

Pterosaurs from Russia, Middle Asia and Mongolia

DAVID M. UNWIN AND NATASHA N. BAKHURINA

Introduction

Russia, Middle Asia and Mongolia form a large territory representing more than one sixth of the Earth's entire continental surface. Mesozoic deposits are widely distributed across this land mass and have yielded important and occasionally extensive remains of pterosaurs ranging in age from Middle Jurassic to latest Cretaceous (Bakhurina and Unwin, 1995a; Unwin *et al.*, 1997). The record is highly uneven, however, with most remains recovered from just a few localities often separated by large temporal gaps (Figure 21.1). This reflects the general situation with regard to the global pterosaur fossil record and is primarily a result of the relative fragility of the remains of these animals, which thus require exceptional conditions for their preservation.

Pterosaurs were first recognized at the end of the eighteenth century, from Late Jurassic deposits in Bavaria (Wellnhofer, 1991a), but were not reported in Russia and Middle Asia until the early twentieth century and in Mongolia only during the 1980s. At the time of writing (1999) pterosaurs have been recovered from five localities in Russia, about 12 in Middle Asia (mainly in Kazakhstan and Uzbekistan) and three in Mongolia (Bakhurina and Unwin, 1995a; Unwin *et al.*, 1997). At least seven species, each in a separate genus and representing at least five families of pterosaurs, have so far been reported (Bakhurina and Unwin, 1995a; Unwin *et al.*, 1997). Five of the seven species are known from reasonably complete skeletons, while the two remaining taxa, and all other records of pterosaurs, are based on isolated and often incomplete bones.

Despite their rarity, Russian and Asian pterosaurs are important for two reasons. First, many of the fossil remains have been collected from sequences that were deposited within a continental setting (Bakhurina and Unwin, 1995a, 1996). This is in contrast to much of the rest of the pterosaur record, which has generally been recovered from marginal marine or marine sediments (Wellnhofer, 1978, 1991a). Thus, although they are sparse, the Russian and Asian records provide important evidence of the evolutionary history of pterosaurs in continental environments. Second, some of the remains recovered, most notably those of *Sordes* and *Batrachognathus* from Karatau and the dsungaripterid from Mongolia, are exceptionally well preserved and provide unique insights into aspects of the anatomy, functional morphology and ecology of this group (Sharov, 1971; Bakhurina, 1986, 1988, 1989, Bakhurina and Unwin, 1992; Bakhurina, 1993; Unwin *et al.*, 1993; Unwin and Bakhurina, 1994; Bakhurina and Unwin, 1995a, 1995b, 1995c, 1995d, 1996, 1997; Unwin and Bakhurina, 1997; Unwin *et al.*, 1997).

Dedication

This chapter is dedicated to the late Dr Valerii Yu. Reshetov whose life was devoted to the JSMPE and to collections.

History of discovery

Russia

The first unequivocal pterosaur fossil to be found in Russia, the posterior half of a cervical vertebra of an

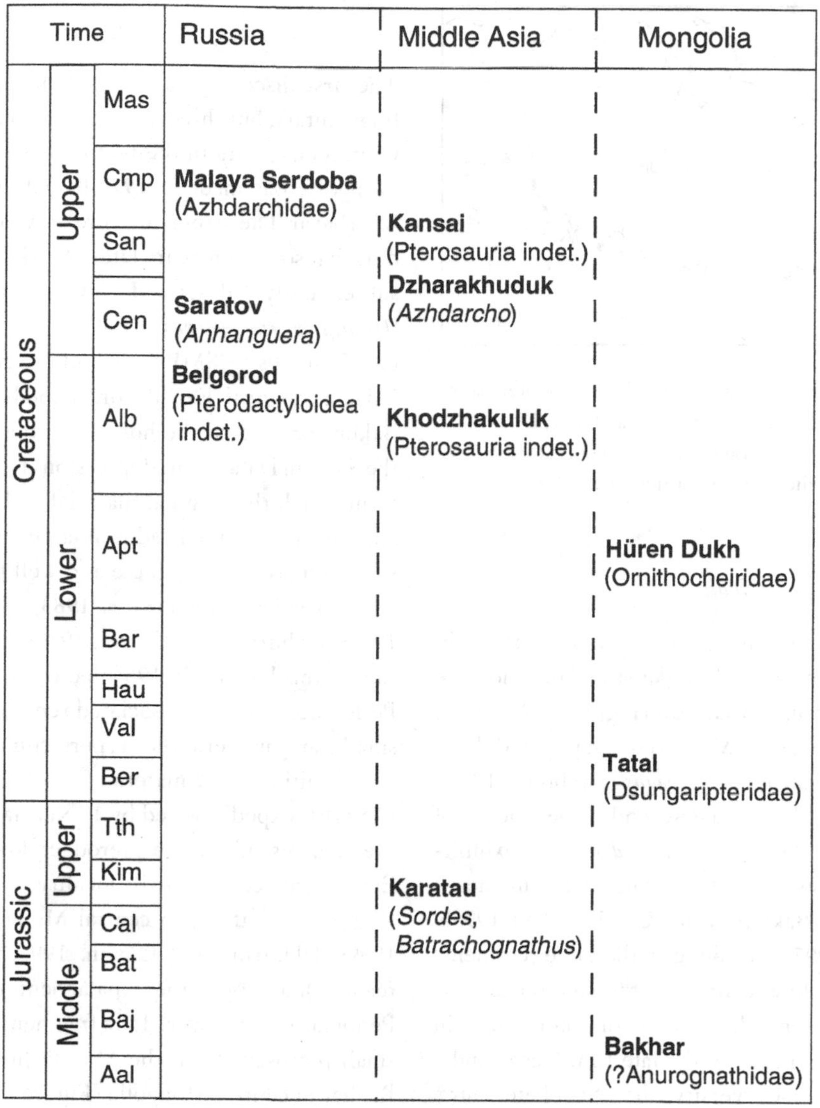

Time			Russia	Middle Asia	Mongolia
Cretaceous	Upper	Mas			
		Cmp	**Malaya Serdoba** (Azhdarchidae)		
		San		**Kansai** (Pterosauria indet.)	
		Cen	**Saratov** (*Anhanguera*)	**Dzharakhuduk** (*Azhdarcho*)	
	Lower	Alb	**Belgorod** (Pterodactyloidea indet.)	**Khodzhakuluk** (Pterosauria indet.)	
		Apt			**Hüren Dukh** (Ornithocheiridae)
		Bar			
		Hau			
		Val			
		Ber			**Tatal** (Dsungaripteridae)
Jurassic	Upper	Tth			
		Kim			
		Cal		**Karatau** (*Sordes, Batrachognathus*)	
	Middle	Bat			
		Baj			**Bakhar** (?Anurognathidae)
		Aal			

Figure 21.1. Stratigraphic distribution of important pterosaur localities (shown in bold) and taxa (in parentheses) from Russia, Middle Asia and Mongolia.

azhdarchid from Late Cretaceous marine sediments of the Volga region (Figure 21.2, site 1) was collected in 1911 by the geologist V.G. Khimenkov. Further isolated remains have occasionally been found in this region (listed in Bakhurina and Unwin, 1995a), but, with the exception of a short section of a mandibular symphysis (Khozatskii, 1995) they add little to our

knowledge of early Late Cretaceous pterosaurs. Elsewhere in Russia, a single incomplete humerus was recovered from the Late Jurassic of the Volga region and in the 1980s Nesov and associates found a few fragmentary remains in late Early to early Late Cretaceous deposits near Gubkin city in the Belgorod district (Nesov *et al.*, 1986).

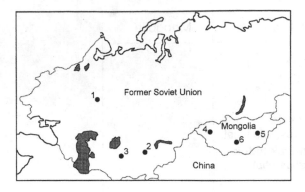

Figure 21.2. Geographic distribution of the main pterosaur localities in Russia, Middle Asia and Mongolia. (1) Lysaya Gora, (2) Karatau, (3) Dzharakhuduk, (4) Tatal, (5) Hüren Dukh, (6) Bakhar. Solid line = coastline, dotted line = international boundary.

Middle Asia

The first discovery of pterosaurs in Middle Asia, a thin slab bearing a semi-complete skeleton from the Late Jurassic of Karatau, Kazakhstan (Figure 21.2, site 2), was made in 1933 by M.A. Vedenyapin and later described as *Batrachognathus volans* (Ryabinin, 1948). Twenty years later a second specimen of *Batrachognathus* and a new taxon, *Sordes pilosus*, exhibiting evidence of soft tissues (Sharov, 1971; Unwin and Bakhurina, 1994; Bakhurina and Unwin, 1995a, 1995b, 1995c, 1995d, 1997) was found in the same sediments by A.G. Sharov, while searching for fossil insects.

The most recent discoveries of pterosaurs in Middle Asia were made by the late L.A. Nesov and a team from the University of St. Petersburg. Fragmentary, but well preserved remains of the Late Cretaceous pterosaur *Azhdarcho* were recovered during a series of expeditions to the locality of Dzharakhuduk in the Kyzylkum desert, Uzbekistan (Figure 21.2, site 3), in the late 1970s, the 1980s and the early 1990s (Nesov, 1984, 1989, 1990). A few additional remains were recovered from this locality by a joint Western-Russian-Kazakh expedition in 1997 (Archibald *et al.*, 1998). Nesov's team also found fragmentary remains of pterosaurs at other sites in the late Early and Late Cretaceous of the Middle Asian republics (see Bakhurina and Unwin, 1995a).

Mongolia

The first discovery of Mongolian pterosaurs, a few fragmentary, but three-dimensional bones, from Early Cretaceous strata in the region of Khovd in Western Mongolia was made in 1970 by V.F. Shuvalov and P. Khosbayar. These remains were only identified as pterosaurian some ten years later (Merkulova, 1980), and subsequently described under the name of '*Dsungaripterus parvus*' (Bakhurina, 1982). In the early 1980s JSMPE (Joint Soviet–Mongolian Palaeontological Expedition) expeditions led by N. Bakhurina located the horizon in the Tatal region of the Sangiin Dalai Nuur depression (Figure 21.2, site 4), from which the original material had been collected, and recovered much additional material, including some remarkably complete and well preserved skulls and lower jaws (Bakhurina, 1983, 1984, 1986, 1989, 1993; Bakhurina and Unwin, 1995a). A third expedition to this locality in 1988, led by B. Namsrai and A. Perle, recovered the associated remains of a relatively small dsungaripterid (Perle, pers. comm. 1993).

In addition to the material found at Tatal in 1981–2, a JSMPE expedition led by P. Narmandakh collected the remains of a large pterodactyloid from the late Early Cretaceous fossil locality of Hüren Dukh (Figure 21.2 site 5) in central Mongolia (Bakhurina, 1989; Bakhurina and Unwin, 1995a) and a JSMPE team led by the palaeoentomologist A.G. Ponomarenko recovered the fragmentary remains of a small pterosaur from the Middle Jurassic locality of Bakhar in central Mongolia (Figure 21.2, site 6) while searching for fossil insects (Bakhurina, 1989; Bakhurina and Unwin, 1995a).

Institutional abbreviations

GIN, Geological Institute, Mongolian Academy of Sciences, Ulaanbaatar; JSMPE, Joint Soviet Mongolian Palaeontological Expedition; MMNH, Mongolian Museum of Natural History, Ulaanbaatar; PIN, Palaeontological Institute, Russian Academy of Sciences, Moscow; TsNIGRI Central Institut of Geological Exploration, St. Petersburg; ZIN,

Figure 21.3. The holotype (PIN 52–2) of *Batrachognathus volans*, Ryabinin, 1948, from the Karabastau Formation of Karatau, Kazakhstan.

Collection of the Zoological Institute of the Russian Academy of Sciences, St. Petersburg, Russia.

Systematic review

Order Pterosauria Kaup, 1834
Family Anurognathidae Kuhn, 1937
Genus *Batrachognathus* Ryabinin, 1948

Type and only known species. Batrachognathus volans, Ryabinin, 1948.

Diagnosis. Batrachognathus is distinguished from *Anurognathus*, the only other genus in the Anurognathidae, by the greater number of teeth (at least 11, compared with 8 in *Anurognathus*), marked recurving of the tips of the teeth (straight in *Anurognathus*), and relatively short hind limbs (1.5–1.6 times the length of the humerus compared to >2.0 times the length of the humerus in *Anurognathus*) (Ryabinin, 1948; Bakhurina, 1988).

Batrachognathus volans, Ryabinin, 1948

Holotype. PIN 52–2, a well preserved skull and semi-complete postcranial skeleton (Ryabinin, 1948, pl. 1; fig. 8; Figure 21.3). Aulie, near Mikhailovka, Karatau ridge, Chimkent region, Kazakhstan. Karabastau Svita, Upper Jurassic (Oxfordian/Kimmeridgian: Dolu-denko and Orlovskaya [1976]).

Referred material. PIN 2585/4a. A heavily crushed skull and semi-complete postcranial skeleton associated with some evidence of soft tissues recovered from the holotype locality (Bakhurina and Unwin, 1995a, 1995b). This specimen lies adjacent to the holotype of *Sordes pilosus* on the same bedding plane: the only example, so far as we are aware, wherein representatives of two different genera of pterosaurs are preserved in direct association.

Diagnosis. As for the genus.

Comments. Batrachognathus was a relatively small pterosaur, about 0.75 m in wingspan and with a reconstructed skull length of 48 mm. The skull was short, broad and deep, but very lightly constructed, possibly kinetic, and with a large orbit (Bakhurina, 1988). The presence of short, peg-like, piercing teeth and the possibility of a large gape suggest an insectivorous diet, as has also been proposed for *Anurognathus* (Döderlein, 1923; Wellnhofer, 1975) and this is at least consistent with the very large numbers of fossil insects found in the Karabastau deposits (Sharov, 1968).

Bakhar anurognathid Bakhurina and Unwin, 1995a
Material. JSMPE (PIN), fragmentary wing bones of one individual. Bakhar, Bayanhongor Aimag, Central Mongolia. Bakhar Svita, Middle Jurassic (?Aalenian/Bajocian [Shuvalov, 1982]) (Bakhurina, 1989; Bakhurina and Unwin, 1995a, fig. 2).
Comment. This small, possibly juvenile, individual has an estimated wingspan of only 0.3–0.4 m. It shows some similarity to anurognathid pterosaurs, especially with respect to the morphology of the humerus, but further work is required to substantiate this systematic assignment.

Family Rhamphorhynchidae Seeley, 1870
Genus *Sordes* Sharov, 1971
Type and only known species. Sordes pilosus, Sharov, 1971
Diagnosis. Sharov (1971) cited numerous characters that supposedly differentiated *Sordes* from other pterosaurs, but many of these are not restricted to this taxon. Apomorphies that distinguish *Sordes* from other rhamphorhynchids include: only seven teeth per side in the upper jaw; and a pes digit four in which phalanx 4 exceeds the combined length of phalanges 1–3.
Comment. Sordes shows some similarity to the Late Jurassic form *Scaphognathus* and the clade that they form, the Scaphognathinae, appears to belong within the Rhamphorhynchidae (see Bakhurina and Unwin, 1995a, for discussion). This view has been challenged by Kellner (1996) who suggests that *Sordes* is a relatively primitive form that occupies a basal position within Pterosauria. Evidence in support of this hypothesis has yet to be published, however, and as *Sordes* exhibits a series of characters including: a skull that is more than three times longer than it is deep;

orbit larger than preorbital and nasal openings; premaxillae that separate the frontals anteriorly; a short metatarsal four; a mandibular symphysis; and loss of size dimorphism in the mandibular dentition, all of which unite *Sordes* with derived 'rhamphorhynchoids' (Unwin, 1995), we see no reason to adopt Kellner's proposal.

Sordes pilosus Sharov, 1971
Holotype. PIN 2585/3, almost complete skeleton with evidence of various types of soft tissues (Sharov, 1971, pl. 4; Bakhurina and Unwin, 1995a, 1995b; Unwin and Bakhurina, 1994). Aulie, near Mikhailovka, Karatau ridge, Chimkent region, Kazakhstan. Karabastau Svita, Upper Jurassic (Oxfordian/Kimmeridgian: Doludenko and Orlovskaya [1976]).
Referred material. Remains of another seven individuals (Sharov, 1971, pls 4, 5; Bakhurina, 1986, p. 33; Bakhurina and Unwin, 1995a, 1995b, 1995c, 1995d; Ivakhnenko and Korabel'nikov, 1987, figs. 261, 262; Unwin and Bakhurina, 1994, 1997; Figure 21.4).
Diagnosis. As for the genus.
Comments. Sordes was a medium-sized pterosaur with a wingspan of about 0.6 m and a skull length, in the holotype, of 80 mm. Remains of *Sordes* provide some of the best available evidence regarding wing-shape in pterosaurs. They show that the main wing membrane (cheiropatagium) was attached to the forelimb and to the hind limb as far as the ankle, and the presence of a cruropatagium stretched between the hind limbs and supported along its rear edge by the elongate fifth toes (Bakhurina and Unwin, 1995a; Unwin and Bakhurina, 1994). Other types of fossilized soft tissues include remains of the integument, which bear 'hair-like' structures in some regions, a tail flap, claw sheaths and foot webs (Sharov, 1971; Bakhurina and Unwin, 1992, 1995a, 1995b, 1995c, 1995d; Unwin *et al.*, 1993; Unwin and Bakhurina, 1994).

Suborder Pterodactyloidea Plieninger, 1901
Family Ornithocheiridae Seeley, 1870
'*Ornithocheirus* (?) sp.'
Material. ZIN PNT-S50–1, section of the mandibular symphysis lacking the rostral termination (Khozatskii,

Figure 21.4. The paratype (PIN 2470/1) of *Sordes pilosus* Sharov, 1971, from the Karabastau Formation of Karatau, Kazakhstan. Scale bar = 20 mm.

Figure 21.5. Section of the mandibular symphysis, in palatal view, of an ornithocheirid, cf. *Anhanguera* (ZIN.PNT-S50–1) from the early Late Cretaceous of Lysaya Gora, Saratov district, Russia. As preserved, the specimen is 90 mm in total length.

1995, fig. 2; Bakhurina and Unwin, 1995a; Figure 21.5). Lysaya Gora, near Proletarskii village, Saratov district, in the south European part of Russia (Glikman, 1953). Upper Cretaceous (Cenomanian).

Comments. The jaw fragment represents a medium to large-sized pterosaur with an estimated wingspan of 3–4 m. Although incomplete, the jaw exhibits a number of features, such as a rostral expansion containing relatively large teeth and an occlusal surface with a pronounced midline channel flanked by well developed ridges, that are typical of ornithocheirids (*Ornithocheirus* + *Anhanguera* + *Coloborhynchus*), and its assignment to this family by Khozatskii (1995) is accepted here. However, in *Ornithocheirus* (type species = *Ornithocheirus simus* Seeley, 1869) the jaws are straight in lateral view, rather than curved upward

anteriorly as in the Russian specimen. Curvature of the lower jaw does occur in another ornithocheirid, *Anhanguera*, and the Russian specimen is also similar in other respects to fossil remains assigned to species of *Anhanguera* from the Santana Formation of Brazil (e.g. Wellnhofer 1991b). Unfortunately, the incompleteness of the Saratov specimen prevents detailed comparisons, thus, for the present, we refer it to cf. *Anhanguera*.

Hüren Dukh ornithocheirid Bakhurina and Unwin, 1995a

Material. MMNH 100/30, rostral end of the upper jaw, almost complete lower jaw and most of the postcranial skeleton (Bakhurina, 1989; Bakhurina and

Figure 21.6. Anterior end of the mandibular symphysis, in right lateral view, of a large ornithocheirid from the late Early Cretaceous of Hüren Dukh, Mongolia. Scale bar = 10 mm.

Unwin, 1995a, fig. 12; Figure 21.6). Hüren Dukh, 60 km south west of Choir, middle Gobi region of Central Mongolia. Züünbayan Formation, Lower Cretaceous (Aptian–Albian) (Shuvalov, 1974).

Comments. The remains, amongst the most complete for a large Cretaceous pterosaur and a rare example of cranial and postcranial material preserved in association, represent a pterodactyloid of about 5.5 m in wingspan. This pterosaur exhibits a number of diagnostic features of the Ornithocheiridae (e.g. expanded jaw tips bearing three pairs of large, fang-like teeth, saggital bony crests located at the rostral ends of the jaws) and undoubtedly belongs in this family. Amongst ornithocheirids, the Hüren Dukh form shows closest similarity to species of *Anhanguera* and *Coloborhynchus*, but until the systematic status of these and other ornithocheirid genera has been reviewed we prefer not to assign the Mongolian ornithocheirid to any particular genus (Bakhurina and Unwin, 1995a).

Family Dsungaripteridae Young, 1964
'*Dsungaripterus parvus*' Bakhurina, 1982

Holotype. PIN 3953, fragments of the fore and hind limb bones of a single individual. Seventy kilometres north-north-east of lake Khar Us Nuur, Khovd Aimag, Western Mongolia. Tsagaantsav Svita, basal Early Cretaceous.

Referred material. Remains of at least 45 individuals, including juveniles (Figure 21.7) and GIN 100/31, a complete skull and low jaw (Ivakhnenko and Korabel'nikov, 1987, fig. 264; Wellnhofer, 1991a, p. 120) recovered from Tatal, Sangiin Dalai Nuur depression, Khovd Aimag, Western Mongolia. Upper part of the Tsagaantsav Svita (?Beriassian–Valanginian [Shuvalov and Trusova, 1976; Shuvalov, 1982]) (Bakhurina, 1983, 1984, 1986, 1993; Bakhurina and Unwin, 1995a, figs. 10, 11).

Comments. In 1982 Bakhurina established a new species of pterosaur, '*Dsungaripterus parvus*', on the basis of material (PIN 3953) recovered by Shuvalov and Khosbayar from western Mongolia in the 1970s. Further remains, collected by Bakhurina in the early 1980s. and including well preserved skull material (Bakhurina and Unwin, 1995a), were referred to this taxon and it was assigned to a new genus, '*Phobetor*' (Bakhurina, 1986), a name that we now know to be preoccupied (Bakhurina and Unwin, 1995a). Characters of the skull clearly distinguish the Tatal pterosaur from other pterodactyloids (Bakhurina, 1986; Bakhurina and Unwin, 1995a), but the holotype material (PIN 3953) appears to be indistinguishable from the corresponding bones of *Dsungaripterus* and *Noripterus*, and the validity of the name '*D. parvus*' seems doubtful.

The one complete skull is 360 mm in length and probably represents an individual of 2–2.5 m in wingspan. Most of the material collected appears to represent individuals of a similar size, though remains of specimens with estimated wingspans ranging from 1 to 4 m are present in the collections.

Figure 21.7. Skull remains, in left lateral view, of a juvenile dsungaripterid (MMNH) from the early Early Cretaceous of Tatal, Western Mongolia. Scale bar = 10 mm.

Family Azhdarchidae Nesov, 1984

Genus *Azhdarcho* Nesov, 1984

Type species. Azhdarcho lancicollis Nesov, 1984.

Diagnosis. According to Nesov 1984, p. 39, 'Spinous process in middle of tubular cervical vertebrae has the form of a weak crest, which is not approached by the lateral crests.' It is not clear that this feature distinguishes *Azhdarcho* from other pterosaurs and further work on the taxonomic status of this pterosaur is needed.

Azhdarcho lancicollis Nesov 1984

Holotype. TsNIGRI, LU-N 1/11915, anterior part of a mid-series cervical (Nesov, 1984, pl. 7, fig. 2). Dzharakhuduk (= Dzhyrakhuduk, Itemir, Beleuta), Navoi district of the Bukhara region, Uzbekistan. Lower and middle part of the Beleuta Svita, variously referred to by Nesov as the Taikarshin Beds (Nesov, 1981, 1984), or the Bissekty Svita (Nesov, 1990), Upper Cretaceous (Coniacian: Nesov and Roček, 1993; Archibald *et al.*, 1998).

Referred material. More than 40 fragmentary, disarticulated, but otherwise well-preserved bones representing parts of the jaws, vertebrae, pectoral and pelvic girdles and the fore and hind limbs (Nesov, 1984, pl. 7, figs. 1–11 and 13; 1986, pl. 2, fig. 1; Nesov and Yarkov, 1989, pl. 2, figs. 2–8; Bakhurina & Unwin, 1995a, fig. 13; Figure 21.8).

Diagnosis. Same as for the genus.

Comments. Azhdarcho exhibits derived features such as

the T-shaped cross-section of the second wing-phalanx, which unites it and other taxa, principally *Quetzalcoatlus* and *Zhejiangopterus*, in the family Azhdarchidae (Unwin and Lü, 1997). However, the remains of *Azhdarcho* have yet to be described in detail and its relationships to other azhdarchids are unclear.

Azhdarcho has been referred to as a giant pterosaur (Nesov and Roček, 1993), but most individuals are likely to have been only about 3–4 m in wingspan, although rare remains indicate the presence of animals with wingspans of up to 5–6 m. One or two tiny elements appear to represent very small individuals, perhaps less than 1 m in wingspan (Nesov, 1991).

Azhdarchidae genus and sp. indet.

Material. Posterior half of an elongate cervical vertebra (Bogolyubov, 1914; Bakhurina and Unwin, 1995a, fig. 14) from marine sediments near the village of Malaya Serdoba in what was the province of Saratov (now Penza district), Volga region, Russia. Late Cretaceous: Coniacian?–Santonian (Nesov, 1990) or possibly early Campanian in age (Glazunova, 1972).

Comments. The vertebra, now lost (Bakhurina and Unwin, 1995a), which represents an animal with an estimated wing span of about 3–4 m, is similar to the highly elongate, mid-series cervicals of *Quetzalcoatlus* and *Azhdarcho*, and there seem few grounds to doubt its assignment to Azhdarchidae (Nesov and Yarkov, 1989; Bakhurina and Unwin, 1995a). This specimen was made the holotype of '*Ornithostoma orientalis*' by

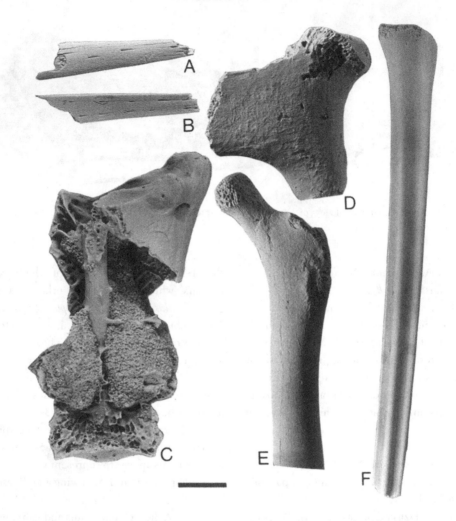

Figure 21.8. Fragmentary remains of *Azhdarcho lancicollis*, Nesov 1984, from the mid–Late Cretaceous of Dzharakhuduk, Uzbekistan. Section of mandibular symphysis in (A) right lateral and (B) occlusal view. Sixth (or possibly seventh) cervical in (C) dorsal view showing the ossified neural canal supported by fine bony trabeculae. Proximal end of left humerus in (D) dorsal view. Left femur in (E) anterior view. Left wing phalanx ii in (F) ventral view showing diagnostic longitudinal ventral ridge. Scale bar = 10 mm.

Bogolyubov (1914), but as it exhibits no features by which it can be distinguished from other azhdarchids the name must be considered a *nomen dubium*.

Pterosauria indet.

Indeterminate remains of pterosaurs have been reported from the Middle Jurassic Balabansai Formation of Northern Fergana in Kirgizstan (Nesov, 1990; Nesov *et al.*, 1987) and the Upper Jurassic of the Volga region of Russia (Khozatskii and Yur'ev, 1964). Fragmentary, indeterminate remains of pterosaurs have also been recovered from Early Cretaceous sediments at Klaudzin (= Kilodzhun) in the south-east part of the Fergana Valley, Kirgizstan, and at Khodzhakul and Scheikhdzheili, on the western part of the Sultanuvais ridge, in the south west Kyzylkum, Karakalpakia, Uzbekistan (Nesov *et al.*, 1987; Nesov, 1989, 1990).

Further indeterminate remains of pterosaurs have also been reported from the Upper Cretaceous localities of Shakh-Shakh, Baibishe and Buroinak on the Dzhusaly uplift in the Kyzyl-Orda district of the north-east Aral region, Kazakhstan (Nesov, 1984, 1990) and from the localities of Kansai, Zamuratscho and Kyzylpilial in the Kyzylbulag district of the north west part of the Fergana Valley, Tadzhikistan (Nesov, 1990).

Single pterosaur bones have been reported by Nesov (1990) from the Turonian of Khidzorut, in the Azizbeck region of Southern Armenia and by Bazhanov and Yeremin (1977) from Albian–Cenomanian beds near Kobyaki village, in the Kirsanov area of Tambov district, Russia. Nesov also collected a few poorly preserved pterosaur bones from late Early Cretaceous sediments exposed in quarries near Gubkin city in the Belgorod district of Russia (Nesov et al., 1986; Nesov, 1990). Finally, a few fragmentary indeterminate remains have also been collected from Late Cretaceous horizons in the south European part of Russia at the locality of Lysaya Gora, near Proletarskii village, Saratov district (Glikman, 1953), and from Luchiskina gorge, south west of Polunino village in the Dubovskii area of the Volgograd district (Nesov and Yarkov, 1989; Nesov, 1990).

Discussion

Pterosaurs from the former Soviet Union (FSU) and Mongolia span a considerable portion (110 Myr) of the known temporal range for this group (160 Myr) and represent more than half (five out of nine) of the main clades (Figure 21.9). We can gain some idea of the general significance of the taxa listed above by considering them within the general context of pterosaur evolutionary history.

Pre-Middle Jurassic pterosaurs are so far unknown from the FSU or Mongolia, though we suspect that they were present in this region during this interval since Late Triassic and Early Jurassic pterosaurs are well known from the western half of Eurasia (Wellnhofer, 1991a) and there would appear to have been few barriers to their early dispersal eastwards.

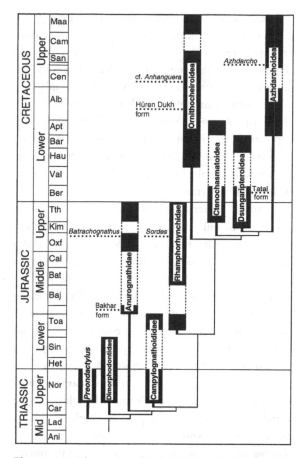

Figure 21.9. The stratigraphic distribution of major clades of pterosaur indicating the approximate position of important taxa from Russia, Middle Asia and Mongolia. The 'tree' used here is based on a synthesis of recent cladistic analyses of pterosaur relationships (see Unwin and Lü, 1997) and data on the stratigraphic distribution of taxa (from Wellnhofer, 1991a, updated to 1998) assigned to each of the nine major clades. Boxes with a solid boundary and black fill indicate the known fossil record, resolved to Stage level, for a particular clade; a dashed boundary represents an inferred record. Note that even if there is only a single record for a particular stage (e.g. *Anurognathus ammoni*) the taxon is assumed to span the entire stage length. Thick vertical lines indicate the maximum likely extension, backward in time, of particular clades, as inferred from the hypothesis of relationships proposed by Unwin (1995). Abbreviations: Alb, Albian; Ani, Anisian; Apt, Aptian; Bar, Barremian; Baj, Bajocian; Bat, Bathonian; Ber, Berriasian; Cal, Callovian; Cam, Campanian; Car, Carnian; Cen, Cenomanian; Hau, Hauterivian; Het, Hettangian; Kim, Kimmeridgian; Lad, Ladinian; Maa, Maastrichtian; Nor, Norian; Oxf, Oxfordian; San, Santonian; Sin, Sinemurian; Toa, Toarcian; Tth, Tithonian; Val, Valanginian.

Anurognathids, small to medium-sized, highly specialized, possibly insectivorous pterosaurs may have first appeared in the Late Triassic or Early Jurassic (Figure 21.9). The Mongolian and Kazakh records are important because, so far, they provide the only evidence to substantiate the predicted longevity of this clade. In addition, their recovery from continental deposits suggests that this group lived in terrestrial habitats and that the single specimen of *Anurognathus ammoni* found in the Solnhofen beds may represent an 'accidental' occurrence.

A similar situation might also pertain for *Sordes*, currently the only rhamphorhynchid known from the FSU or Mongolia. Examples of *Sordes* nearest known relative, *Scaphognathus*, are also exceptionally rare in the Solnhofen Plattenkalke (3 out of 300+ individuals [R. Kemp, pers comm., 1998]), though *Sordes* itself is relatively common in the Karatau deposits (8 out of 10 individuals recovered so far). By contrast, virtually all records of Jurassic rhamphorhynchines (*Dorygnathus* + *Rhamphocephalus* + *Nesodactylus* + *Rhamphorhynchus*) have been found in marginal marine or marine deposits. We hypothesize that rhamphorhynchines were aerial piscivores that lived predominantly in coastal environments, while scaphognathines (*Sordes* + *Scaphognathus*) inhabited continental environments and fed on fish caught in lakes and rivers. Notably, the dentition of scaphognathines is not specialized in the same way as that of rhamphorhynchines and it may be that their diet was not entirely confined to fish.

Ornithocheiroids (*Nyctosaurus* + '*Ornithodesmus*' + Ornithocheiridae + Pteranodontidae) have a good fossil record, for pterosaurs, and are known to have persisted from early in the Early Cretaceous through to the end of this period, though the group probably first arose in the Late Jurassic (Figure 21.9). These medium- to large-sized pterosaurs have been recovered from all continents, except Antarctica, and seem to have been efficient fliers (Bramwell and Whitfield, 1974). The Russian occurrence is of interest because it represents one of the youngest records for the Ornithocheiridae. This family is also known from the Cenomanian of England (Wellnhofer, 1978), but there are no reliable records from younger deposits. The Mongolian ornithocheirid is surprising because, unlike other fossil records for this family, virtually all of which have been found in marginal or fully marine deposits, this individual was recovered from a large, shallow, freshwater lake, located hundreds of kilometres from the nearest coastline. It is not clear, however, if this represents an 'accidental' occurrence, a migrant, or a resident (Bakhurina and Unwin, 1995a).

Dsungaripteroids (Germanodactylidae + Dsungaripteridae) are best known from the Early Cretaceous of Asia (Unwin *et al.*, 1997), and have also been reported from the Early Cretaceous of South America (Martill and Frey, pers comm., 1997) and the Late Jurassic of Europe (Buffetaut *et al.*, 1998) and East Africa (Galton, 1980; Unwin and Heinrich, 1999). The Tatal dsungaripterid is of particular interest because in some respects, for example, size, morphology and geological age, it forms an almost perfect intermediate between early dsungaripteroids such as *Germanodactylus* and the most derived taxon, *Dsungaripterus*. However, *Normannognathus* from the Late Jurassic of France, recently described by Buffetaut *et al.* (1998), does not fit comfortably within this sequence and suggests that dsungaripteroid evolutionary history may be more complex than previously thought.

The Azhdarchidae was first recognized by Nesov in the mid-1980s following his discovery of *Azhdarcho* which, until recently, was also the oldest certain record for this family. Cladistic analysis suggests that azhdarchids appeared before the end of the Early Cretaceous (Figure 21.9) and this seems to be supported by a report of a fossil remain from the Early Cretaceous Crato Formation of Brazil that exhibits at least one diagnostic azhdarchid feature (Martill and Frey, 1998). Following a string of new discoveries during the last ten years and reassignment of some previously described taxa, such as *Quetzalcoatlus* and *Arambourgiania*, to the Azhdarchidae, this family, now known from North and South America, Europe, Africa, Asia and possibly Australia (Unwin and Lü, 1997) appears to have been the most widespread and taxonomically diverse clade of Late Cretaceous pterosaurs.

Acknowledgements

We are grateful to the following for allowing access to specimens: A. Perle, Technical University, Ulaanbaatar; R. Barsbold, Geological Institute, Ulaanbaatar; and the late L. Nesov, University of St. Petersburg. We thank S. Powell (Bristol) and W. Harre (Berlin) for assisting with the figures and M.J. Benton (Bristol), E. Frey (Karlsruhe), D. Martill (Portsmouth) and P. Wellnhofer (Munich) for their helpful comments on earlier drafts of the MS. D.M.U. acknowledges the Royal Society and the Department of Earth Sciences, University of Bristol for their support.

References

Archibald, J.D., Sues, H.-D., Averianov, A.O., King, C., Ward, D.J., Tsaruk, O.A., Danilov, I.G., Rezvyi, A.S., Veretennikov, B.G. and Khodjaev, A. 1998. Precis of the Cretaceous paleontology, biostratigraphy and sedimentology at Dzharakuduk (Turonian?-Santonian), Kyzylkum Desert, Uzbekistan, pp. 21–27 in Lucas, S.G., Kirkland, J.I. and Estep, J.W. (eds.), *Lower and Middle Cretaceous terrestrial ecosystems. New Mexico Museum of Natural History and Science, Bulletin* **14**.

Bakhurina, N.N., 1982. [A pterodactyl from the Lower Cretaceous of Mongolia.] *Palaeontologicheskii Zhurnal,* **4**: 104–108.

—1983. [Early Cretaceous pterosaur localities in western Mongolia.] *Trudy Sovmestnoi Sovetsko-Mongol'skoi Paleontologicheskoi Ekspeditsii* **24**: 126–129.

—1984. [On the discovery of numerous remains of pterosaurs in the Early Cretaceous locality of Tatal, western Mongolia.] *Byulleten' Moskovskogo Obshchestva Ispytatelei Prirody, Otdel Geologicheskii* **59** (3): 130.

—1986. [Flying reptiles.] *Priroda* (1986), **7**: 27–36.

—1988. [On the first rhamphorhynchoid from Asia: *Batrachognathus volans* Ryabinin 1948, from Upper Jurassic beds of Karatau.] *Byulleten' Moskovskogo Obshchestva Ispytatelei Prirody, Otdel Geologicheskii* **63** (5): 132.

—1989. Flying reptiles from Mongolia: new information on morphology, systematics and palaeogeography of pterosaurs. *Basic research results of the Joint Soviet-Mongolian Palaeontological Expedition,* 1969–1988: 17–18, Moscow: Akademiya Nauk.

—1993. Early Cretaceous pterosaurs from western Mongolia and the evolutionary history of the Dsungaripteroidea. *Journal of Vertebrate Paleontology* **13** (Suppl. to 3): 24A.

—and Unwin, D M. 1992. *Sordes pilosus* and the function of the fifth toe in pterosaurs. *Journal of Vertebrate Paleontology* **12** (Suppl. to 3): 18A.

—and— 1995a. A survey of pterosaurs from the Jurassic and Cretaceous of the former Soviet Union and Mongolia. *Historical Biology* **10**: 197–245.

—and— 1995b. Taphonomy of pterosaurs from the Upper Jurassic lacustrine lithographic limestones of Karatau, Kazakhstan. *II International Symposium on Lithographic Limestones,* Lleida-Cuenca, Spain, Ediciones de la Universidad Autónoma de Madrid, 19–21.

—and— 1995c. A preliminary report on the evidence for 'hair' in *Sordes pilosus,* an Upper Jurassic pterosaur from Middle Asia, pp. 79–82, in Sun, A. and Wang Y. (eds.), *Sixth Symposium on Mesozoic Terrestrial Ecosystems and Biota.* Short Papers. China Ocean Press: Beijing.

—and— 1995d. The evidence for 'hair' in *Sordes* and other pterosaurs. *Journal of Vertebrate Paleontology* **15** (Suppl. to nb. 3): 17A.

—and— 1996. Pterosaurs from continental environments. *Journal of Vertebrate Paleontology* **16** (Suppl. to nb. 3): 20A.

—and—1997 Pterosaur 'hair'. *Journal of Morphology* **232**: 231.

Bazhanov, V.S. and Yeremin, A.V. 1977. [The first discovery of remains of reptiles in the Cretaceous beds of the Tambov district], pp. 20–21 in Darevskii, I.S. (ed.), *Problemy Gerpetologii* **4** (1977). Leningrad: Nauka Press.

Bogolyubov, N.N. 1914. [On the vertebra of a pterodactyl from the Upper Cretaceous beds of Saratoff Province.] *Ezhegodnik po Geologii i Mineralogii Rossii* **16**: 1–7.

Bramwell, C.D. and Whitfield, G.R. 1974. Biomechanics of *Pteranodon. Philosophical Transactions of the Royal Society of London* B **267**, 503–581.

Buffetaut, E., Lepage, J.-J. and Lepage G. 1998. A new pterodactyloid pterosaur from the Kimmeridgian of the Cap de la Hève (Normandy, France). *Geological Magazine* **135**: 719–722.

Döderlein, L. 1923. *Anurognathus Ammoni* ein neuer Flugsaurier. *Sitzungs-berichte der Bayerischen Akademie der Wissenschaften, mathematisch-naturwissenschaftliche Abteilung.* 117–164.

Doludenko, M.P. and Orlovskaya, E.R. 1976. Jurassic floras of the Karatau ridge, southern Kazakhstan. *Palaeontology* 19: 627–640.

Galton, P.M. 1980. Avian-like tibiotarsi of pterodactyloids (Reptilia: Pterosauria) from the Upper Jurassic of East Africa. *Paläontologische Zeitschrift* 54: 331–342.

Glazunova, A.E. 1972. [*Palaeontological basis for the stratigraphic subdivision of Cretaceous deposits in the Volga region*], 144 pp. Leningrad: Nedra Press.

Glikman, L.C. 1953. [Late Cretaceous vertebrates from the region of Saratov.] *Scientific Reports of Saratov State University* 38: 51–54.

Ivakhnenko, M.F. and Korabel'nikov, V.A. 1987. [*Life of the past world*], 253 pp. Moscow: Prosveshchenie Press.

Kaup, J. 1834. Versuch einer Einteilung der Säugetiere. *Isis.* 315pp. Jena.

Kellner, A.W.A. 1996. Pterosaur phylogeny. *Journal of Vertebrate Paleontology* 16 (Suppl. to nb. 3): 45A.

Khozatskii, L.I. 1995. [Pterosaur from the Cenomanian (Late Cretaceous) of Saratov.] *Vestnik Sankt-Peterburgskogo Universiteta, Seriya 3, Biologiya* 2 (2): 115–116.

—and Yur'ev, K.B. 1964. [Pterosauria], pp. 589–603 in Orlov, U.A. (ed.), *Osnovy Palaeontologii.* Moscow: Izdatel'stvo Nauka.

Kuhn, O. 1937. *Die fossilen Reptilien.* 121 pp. Berlin.

Martill, D.M. and Frey, E. 1998. A new pterosaur Lagerstatte in N.E. Brazil (Crato Formation: Aptian, Lower Cretaceous): preliminary observations. *Oryctos* 1: 79–85.

Merkulova, N.N. 1980. [The first discovery of pterosaurs in Mongolia.] *Byulleten' Moskovskogo Obshchestva Ispytatelei Prirody, Otdel Geologicheskii* 1980: 103–104.

Nesov, L.A. 1981. [Flying reptiles from the late Cretaceous of Kyzylkum.] *Paleontologicheskii Zhurnal* 1981 (4): 98–104.

—1984. [Upper Cretaceous pterosaurs and birds from Central Asia.] *Paleontologicheskii Zhurnal,* 1984 (1): 47–57.

—1989. [New discoveries of remains of dinosaurs, crocodiles and flying reptiles from the late Mesozoic of the USSR]. *Abstracts of the Seventh All Union Herpetological Conference, Kiev.* 173–174. Kiev: Naukova Dumka.

—1990. [Flying reptiles of the Jurassic and Cretaceous of the USSR and the significance of their remains for the reconstruction of paleogeographic conditions.]

Vestnik Leningradskogo Universiteta, Seriya 7, Geologiya, Geografiya 1990, 4 (28): 3–10.

—1991. [Giant flying reptiles of the family Azhdarchidae: I. Morphology and systematics]. *Vestnik Leningradskogo Universiteta, Seriya 7, Geologiya, Geografiya* 1991, 2 (14): 14–23.

—, Kaznyshkina, L.F. and Cherepanov, G.O. 1987. [Dinosaurs, crocodiles and other archosaurs from the late Mesozoic of Central Asia and their place in ecosystems]. *Abstracts of the thirty third session of the All-Union Paleontological Society, Leningrad:* 46–47.

—, Mertinene, R.A., Golovneva, L.B., Potapova, O.P., Sablin, M.B., Abramov, A.B., Bugaenko, D.V., Nalbandyan, L.A. and Nazarkin, M.V. 1986. [New discoveries of remains of ancient organisms in the Belgorod and Kursk districts], pp 124–131 in *Comprehensive study of biogeocoenoses of oak woods of forest steppe.* Leningrad: Leningrad University Press.

—and Roček, Z. 1993. Cretaceous anurans from central Asia. *Palaeontographica, Abteilung A* 226: 1–54.

—and Yarkov, A.A. 1989. [New birds from the Cretaceous and Palaeogene of the USSR and some remarks on the origin and evolution of the Class Aves], pp. 78–97 in Potapov, R.L. (ed.) [*Faunistic and ecological studies of Eurasian birds*], *Trudy Zoologicheskogo Instituta AN SSSR* 197.

Plieninger, F. 1901. Beiträge zur Kenntnis der Flugsaurier. *Paleontographica* 48: 65–90.

Ryabinin, A.N. 1948. [Remarks on a flying reptile from the Jurassic of the Karatau.] *Trudy Paleontologicheskogo Instituta AN SSSR* 15: 86–93.

Seeley, H.G. 1869. Index to the fossil remains of Aves, Ornithosauria and Reptilia in the Woodwardian Museum, Cambridge. *Proceedings of the Cambridge Philosophical Society* 3: 1–169.

—1870. *The Ornithosauria: An elementary study of the bones of Pterodactyles.* 130 pp. Cambridge.

Sharov, A.G. 1968. [*Jurassic insects from Karatau*], 252 pp. Moscow: Nauka Press.

—1971. [New flying reptiles from the Mesozoic of Kazakhstan and Kirgizstan.] *Trudy Paleontologicheskogo Instituta AN SSSR* 130: 104–113.

Shuvalov, V.F. 1974. [On the geological structure and age of the fossil localities Höövör and Hüren Dukh], pp. 296–304 in *Mesozoic and Cenozoic faunas and biostratigraphy of Mongolia. Trudy Sovmestnoi Sovetsko-Mongol'skoi Paleontologicheskoi Ekspeditsii* 1.

—1982. [Palaeogeography and historical development of

Mongolian lake systems in the Jurassic and Cretaceous], pp. 18–80 in Martinson, G.G. (ed.), *Mezozoiskie Ozernye Basseiny Mongolii.* Leningrad: Nauka.

—and Trusova, E.K. 1976. [New data on the stratigraphical position of late Jurasic and early Cretaceous conchostracans of Mongolia.] In *Paleontology and biostratigraphy of Mongolia, Trudy Sovmestnoi Sovetsko-Mongol'skoi Paleontologicheskoi Ekspeditsii* 3: 236–264.

Unwin, D.M. 1995. Preliminary results of a phylogenetic analysis of the Pterosauria (Diapsida: Archosauria), pp 69–72 in Sun, A. and Wang Y. (eds.), *Sixth Symposium on Mesozoic Terrestrial Ecosystems and Biota. Short Papers.* China Ocean Press: Beijing.

—and Bakhurina, N.N. 1994. *Sordes pilosus* and the nature of the pterosaur flight apparatus. *Nature* 371: 62–64.

—and— 1997. The significance of soft tissue preservation for understanding the palaeobiology of pterosaurs. *Journal of Morphology* 232: 332.

—, Bakhurina, N.N., Lockley, M.G., Manabe, M. and Lü, J. 1997. Pterosaurs from Asia. *Journal of the Palaeontological Society of Korea* 2: 43–65.

Unwin, D.M. and Heinrich, W.-D. 1999. On a pterosaur jaw from the Upper Jurassic of Tendaguru (Tanzania). *Mitteilungen Museum für Naturkunde Berlin, Geowissenschaftlichen Reihe* 2: 121–134.

—and Lü, J. 1997. On *Zhejiangopterus* and the relationships of pterodactyloid pterosaurs. *Historical Biology* 12: 199–210.

—, Martill, D.M. and Bakhurina, N.N. 1993. The structure of the wing membrane in pterosaurs. *Journal of Vertebrate Paleontology* 13 (Suppl. to 3): 61A.

Wellnhofer, P. 1975. Die Rhamphorhynchoidea der Oberjura-Plattenkalke Süddeutschlands. II: Systematische Beschreibung. *Palaeontographica*, A 148: 132–186.

—1978. *Handbuch der Paläoherpetologie. Teil 19, Pterosauria.* Stuttgart: Gustav Fischer Verlag.

—1991a. *The Illustrated Encyclopedia of Pterosaurs*, 192 pp. London: Salamander Books.

—1991b. Santana Formation pterosaurs, pp. 351–370 in Maisey, J.G. (ed.), *Santana Fossils.* Neptune City: T.F.H. Publications.

Young, C.C. 1964. On a new pterosaurian from Sinkiang, China. *Vertebrata PalAsiatica* 8: 221–255.

22

Theropods from the Cretaceous of Mongolia

PHILIP J. CURRIE

Introduction

Theropods were the most successful lineage of dinosaurs in the sense that they were amongst the first dinosaurs to appear more than 225 million years ago in Late Triassic times, and remained the dominant carnivores until the end of the Cretaceous. Because of their ancestral relationship to birds, it could even be said that they are the most successful group of air-breathing vertebrates today.

Non-avian theropods were never numerically as common as plant-eating dinosaurs, so it is not surprising that they are rare as fossils. Nevertheless, they were diverse and speciose during Mesozoic times. And the peak of their diversity, as presently understood, is represented by fossils from the Upper Cretaceous beds of Mongolia. Numerous sites have produced theropods from Neocomian to Maastrichtian stages, but are particularly strong in the Campanian to Maastrichtian Djadokhta, Baruungoyot and Nemegt 'Mongolian Land Vertebrate Ages' (Jerzykiewicz and Russell, 1991). With the possible exception of equivalent-aged beds in North America, no other region has produced so many fine specimens representing so many species.

Although Mongolia has some of the best Late Cretaceous dinosaur assemblages in the world, earlier intervals are not as well understood. Recent work by a joint Stanford University–Mongolian expedition to the western part of the country has shown the presence of Jurassic sauropods similar to those of northwestern China. This strongly suggests that Jurassic carnosaurs similar to *Monolophosaurus jiangi* (Zhao and Currie, 1993) and *Sinraptor dongi* (Currie and Zhao,

1993) from Xinjiang, will eventually be found in Mongolia. Dromaeosaurids, troodontids and ornithomimosaurs have already been found in Early Cretaceous beds of Mongolia, and provide some of the best information available on the ancestry of these groups.

Seven major theropod lineages (dromaeosaurids, oviraptorosaurs, therizinosauroids, troodontids, avimimids, ornithomimosaurs, and tyrannosaurids) lived during Cretaceous Mongolian times (Figure 22.1). All but the therizinosauroids and some families of oviraptorosaurs and ornithomimids were widely distributed in the Northern Hemisphere. The presence in North America of a possible Late Jurassic troodontid and Early Cretaceous dromaeosaurids and oviraptorosaurs suggests that these groups could have originated anywhere in the Northern Hemisphere. North American records of therizinosauroids (Currie, 1992) and avimimids are poor, but suggestive. The assignment to the Ornithomimosauria of the Late Jurassic *Elaphrosaurus* from Africa is completely unfounded (P. Makovicky, pers. comm. 1996). However, a possible ornithomimosaur from the Lower Cretaceous of Spain (Pérez-Moreno *et al.*, 1994) and reports of Lower Cretaceous ornithomimosaurs from Australia (Rich and Vickers-Rich, 1994) indicate that this clade may have originated and diversified somewhere other than Asia. Therizinosauroids and tyrannosaurids are the only two lineages for which a strong case can presently be made for central Asian origins. Even the latter has been questioned following the discovery of *Siamotyrannus* (Buffetaut *et al.*, 1996), although this is at least consistent with an Asian origin for tyrannosaurids.

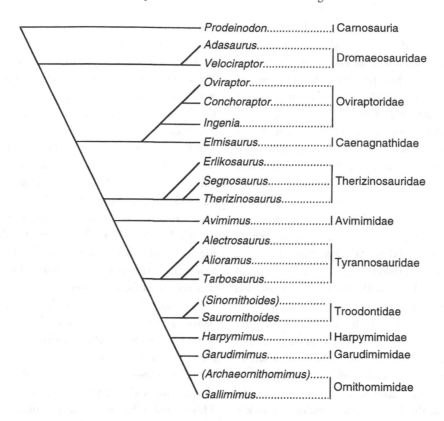

Prodeinodon.....................I Carnosauria
Adasaurus.........................
Velociraptor.....................| Dromaeosauridae
Oviraptor.........................
Conchoraptor....................| Oviraptoridae
Ingenia............................
Elmisaurus........................I Caenagnathidae
Erlikosaurus......................
Segnosaurus....................| Therizinosauridae
Therizinosaurus...............
Avimimus.........................I Avimimidae
Alectrosaurus...................
Alioramus.........................| Tyrannosauridae
Tarbosaurus.....................
(Sinornithoides)..............
Saurornithoides................| Troodontidae
Harpymimus....................I Harpymimidae
Garudimimus...................I Garudimimidae
(Archaeornithomimus)......
Gallimimus.......................| Ornithomimidae

Figure 22.1. Cladogram based on an analysis by Holtz (1994) showing the relationships of the best known Mongolian theropod genera. Genera of unknown affinities (*Bagaraatan*, *Deinocheirus*), and unnamed genera (specimens previously referred to as '*Oviraptor*' *mongoliensis*, 'undescribed giant dromaeosaur' and 'undescribed troodontid') are not included on the cladogram.

Regardless of whether any of the major theropod lineages originated in central Asia, Mongolian discoveries document theropod diversification better than anywhere else. Moreover, the beautiful preservation of so many specimens also permits more precise analysis of their relationships. In this review, each of the seven Mongolian theropod clades is considered, followed by some problematic fossils that represent valid taxa whose relationships are not yet clearly understood.

Systematic survey

Dromaeosauridae

Characteristics. The family Dromaeosauridae is generally subdivided into two subfamilies – Dromaeo- saurinae and Velociraptorinae. Recent discoveries by the American Museum of Natural History, and restudy of the type specimen of *Dromaeosaurus albertensis* (Currie, 1995) suggests that the two lineages could be separated at a higher taxonomic level (family or higher). Irrespective of the level of distinction, dromaeosaurine and velociraptorine theropods are more closely related to each other than either is to any other known theropod clade.

Dromaeosaurines tend to be more massive animals than velociraptorines of equivalent size. The only animal that can be assigned to this subfamily with certainty is *Dromaeosaurus albertensis* from Upper Cretaceous rocks of North America (Currie, 1995).

What we know about this genus is based almost entirely on the skull, which makes it difficult to compare with specimens that lack skulls. Although some Mongolian taxa have tentatively been referred to this subfamily (Paul, 1988b), this is not meaningful without good cranial material.

In 1989, a giant dromaeosaurid was discovered in Lower Cretaceous rocks of the southeastern Gobi (Perle *et al.*, 1999). *Achillobator* is from approximately the same time period as *Utahraptor* (Kirkland *et al.*, 1993) from the United States, and it is tempting to think that it might be related. Premaxillary tooth size indicates that *Utahraptor* is a velociraptorine, whereas serration size suggests that *Achillobator* might be a dromaeosaurine. Unfortunately, the absence of most of the postcranial skeleton of *Dromaeosaurus* makes these designations tentative.

Dromaeosaurids are easily distinguished from other theropods by many cranial and postcranial autapomorphies (Currie, 1995). These include a slender, T-shaped lacrimal; a T-shaped quadratojugal; a conspicuous lateral extension of the paroccipital process beyond the head of the quadrate; a broad, shallow, shelf-like retroarticular process with a vertical columnar process posteromedially; fusion of the interdental plates to each other and to the margins of the jaws; strongly angled intervertebral articulations in the cervical vertebrae; hyperelongated prezygapophyses in all but the most proximal caudals; hyperelongated anterior projections on all but the most proximal haemal arches; a retroverted pubis; confluence of the greater and lesser trochanters of the femur; and a highly specialized second digit of the foot bearing a sickle-shaped claw.

Record. The best record of dromaeosaurids, which were widespread in the northern hemisphere throughout most of the Cretaceous, comes from Mongolia. A possible dromaeosaurid has been reported from northern Africa (Rauhut and Werner, 1995), but better material is needed to confirm this. Four genera have been described to date, most of which are based on well-preserved material.

Velociraptor mongoliensis is the best known dromaeosaurid, having been described originally by Osborn in 1924 on the basis of a skull and partial skeleton from the Djadokhta beds of Bayan Zag (Figures 22.2A and 22.6A). One of the most remarkable dinosaur specimens ever discovered is a complete skeleton of *Velociraptor* from Tögrög preserved in association with a skeleton of *Protoceratops*. Although the association can be interpreted in many ways, the most likely explanation is that the *Velociraptor* had attacked the *Protoceratops* during a sandstorm (Jerzykiewicz *et al.*, 1993; Unwin *et al.*, 1995). Other specimens of *Velociraptor* have been reported from the Baruungoyot beds at Khulsan (Osmólska, 1980, 1982; Norell and Clark, 1992), and the Syuksyukskaya Svita of Kazakhstan (Nesov, 1995), although it has yet to be determined whether or not they represent the same species.

Barsbold (1983) gave a preliminary description of *Adasaurus mongoliensis* from the Nemegt site of Bügiin Tsav. The holotype includes a partial skull and parts of the skeleton. Other specimens include the best preserved dromaeosaurid pelvis (Barsbold, 1983) that clearly shows the retroverted pubis (Figure 22.5A). This animal is distinguished from other dromaeosaurids by the relatively small size of the ungual on the second pedal digit, and by unspecified features of the supporting metatarsal. Barsbold assigned this genus to the subfamily dromaeosaurinae, along with *Dromaeosaurus* and *Deinonychus*. However, the latter is clearly a velociraptorine (Paul, 1988a), so the criteria used to include *Adasaurus* in the dromaeosaurinae are suspect.

Hulsanpes perlei is based on an incomplete foot from the Baruungoyot Formation of Khulsan (Osmólska, 1982). Most of the similarities to a dromaeosaurid foot are plesiomorphic, and even its identification as a dromaeosaurid is uncertain. Chiappe and Norell believe this to be from another more speciose branch of the Maniraptora (Norell, pers. com.). The small size of the specimen suggests that metatarsal elongation may simply be a juvenile trait, and that negative allometry during ontogeny might have produced a metatarsal with proportions similar to adult *Velociraptor* specimens from the same locality.

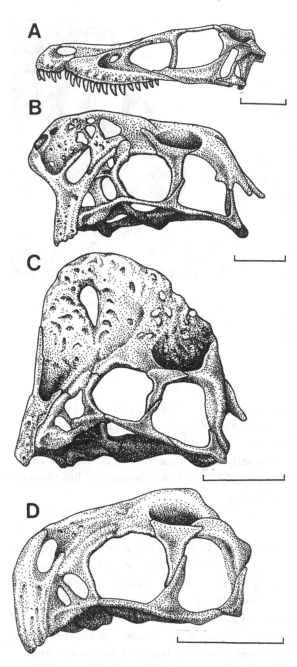

Figure 22.2. Skulls of Mongolian theropods. (A) *Velociraptor*, (B) *Oviraptor*, (C) '*Oviraptor*' *mongoliensis*, and (D) *Conchoraptor*. (A) after Paul (1988a), (B)–(D) after Barsbold *et al.* (1990). Scale bar = 50 mm.

Oviraptorosauria

As presently defined, the Oviraptorosauria includes at least two families – Oviraptoridae and Caenagnathidae. Barsbold (1983) has further subdivided the Oviraptoridae into the subfamilies Oviraptorinae and Ingeniinae. Until recently, caenagnathids were considered to have been restricted to North America. However, Currie *et al.* (1993) reported on the discovery of a caenagnathid, *Caenagnathasia martinsoni*, from the Late Cretaceous of Uzbekistan. Furthermore, it is now apparent that 'Elmisauridae' of Osmólska (1981) is the junior synonym of Caenagnathidae (Currie and Russell, 1988; Sues, 1994). Therefore, the Mongolian species *Elmisaurus rarus* is considered here as a caenagnathid.

Oviraptorosaurs are characterized by many autapomorphies including their toothless, birdlike skulls; loss of the intramandibular joint; fusion of the articular, surangular and coronoid; presence of an unusual jaw articulation with a prominent ridge on the articular; pneumatized vertebral centra, including those of the anterior caudals; and manual unguals with pronounced lips above the interphalangeal articulations.

Oviraptoridae

Characteristics. Mature oviraptorids seem to have ranged in length from one to four meters. Cranially (Figures 22.2B, C), they differ from caenagnathids in having deeper, shorter jaws, a higher, more anteriorly positioned external mandibular fenestra, and a process of the articular-surangular-coronoid ossification that invades the external mandibular fenestra. One of the most interesting differences is that oviraptorids lack the arctometatarsalian condition seen in caenagnathids (Figure 22.6C). The oviraptorid pectoral girdle has a relatively large, well-developed furcula, and the sternum is ossified. It is not known whether caenagnathids also had such a shoulder girdle.

Record. The first oviraptorid skeleton collected at Bayan Zag was associated with a nest of eggs, and was given the name *Oviraptor philoceratops*, which can be translated as 'egg seizer with a fondness for ceratopsian eggs' (Osborn, 1924). However, it has now been shown (Norell *et al.*, 1995; Dong and Currie, 1996) that this name is inappropriate in the sense that the eggs it was supposed to have been seizing were probably its own. Many oviraptorid skeletons with well-preserved skulls, including embryos (Norell *et al.*, 1994), have been discovered at Djadokhta beds sites in Mongolia and Inner Mongolia. Perhaps the most spectacular locality for oviraptorids is Ukhaa Tolgod, where more than twenty skeletons were discovered in 1993 and 1994 (Dashzeveg *et al.*, 1995). At present, it is not certain whether or not all of these specimens represent *Oviraptor*.

A mature specimen of *Oviraptor philoceratops* has a pneumatized crest over the snout anterior to the orbits (Figure 22.2B). The second and third fingers are subequal in length and each manual ungual has a distinctive dorsoposterior 'lip' that is lacking in the equivalent element in *Conchoraptor*.

A second species of *Oviraptor*, *O. mongoliensis*, was established by Barsbold (1986) on the basis of a well-preserved skull (Figure 22.2C) and partial skeleton from the Nemegt formation at Altan Uul. Subsequent work has suggested to Barsbold (pers. comm., 1996) that this species represents a genus distinct from *Oviraptor*. The crest of this animal is larger than that of *Oviraptor philoceratops*, and the parietal is incorporated into its construction. Although almost the same size as *O. philoceratops*, *O. mongoliensis* apparently has a more lightly built skeleton.

Specimens of *Conchoraptor gracilis* were recovered from Baruungoyot rocks at Hermiin Tsav. They are smaller animals than *Oviraptor philoceratops*, and lack any evidence of a crest (Figure 22.2D). The second and third digits of the hand are subequal in length as in *Oviraptor*, but each ungual lacks the well-developed 'lip' above the interphalangeal articulation.

In 1981, Barsbold set up the oviraptorid subfamily Ingeniinae. He subsequently (1986) elevated this to family level. At present, there is only one species, *Ingenia yanshini*, within this clade, represented by more than half a dozen skeletons from the Baruungoyot of Hermiin (Figures 22.5B and 22.6B). This a small, but relatively robust oviraptorosaur characterized by a

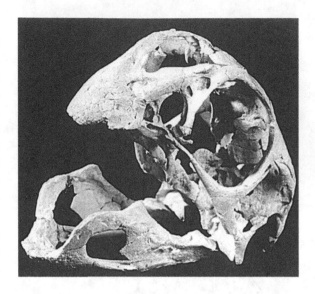

Figure 22.3. Anterolateral view of a beautiful oviraptorid skull in the collections of the Palaeontological Institute in Moscow (courtesy of P. Rich).

skull that apparently lacks a crest, and by a hand in which the first finger is longer and more powerful than the second and third fingers. The manual unguals are significantly longer than the corresponding penultimate phalanges, and lack posterodorsal lips above the articulations.

Caenagnathidae

Characteristics. Caenagnathids are more poorly understood than oviraptorids at present because of the incomplete nature of all skeletons. There appear to be two distinct lineages (which will be referred to here as caenagnathines and elmisaurines), one characterized by *Chirostenotes* from North America, and the other by *Elmisaurus* from both Mongolia and North America (Osmólska, 1981; Currie, 1989). Until recently, caenagnathines were known primarily from their lower jaws, which tend to be longer and lower than those of oviraptorids. Like oviraptorids, the braincase is highly pneumatized, although the basal tubera and basipterygoid processes are aligned vertically (Sues, 1994).

Most were 2–3 m in length at maturity, and had unfused, arctometatarsalian tarsometatarsi (Currie and Russell, 1988). No cranial material can be assigned with confidence to the Elmisaurinae, although the small size of *Caenagnathasia martinsoni* (Currie *et al.*, 1993) suggests that it may belong to this lineage. Elmisaurines were much smaller animals than caenagnathines, perhaps 1 m in length, but, nevertheless, had fused arctometatarsalian tarsometatarsi.

Record. Two caenagnathids are currently recognized from central Asia. *Caenagnathasia martinsoni* is from the Upper Turonian Bissekty Svita of Uzbekistan and possibly from the Bostobe Svita (probably Santonian) of Kazakhstan (Currie *et al.*, 1993). It is not unreasonable to think that its remains may also be found in Mongolia. Only dentaries have been identified to date, and these suggest that *Caenagnathasia* was a small animal that weighed less than 5 kg at maturity.

Elmisaurus rarus is based on partial skeletons from the Nemegt Formation at the Nemegt locality (Osmólska, 1981). These include hands, feet, and fragments of limb bones (Figure 22.6D). The hand looks remarkably similar to that of *Chirostenotes* from North America (Currie, 1990), but the more slender tarsometatarsus differs in being fused.

Therizinosauroidea (Alxasauridae, Therizinosauridae)

Although the first therizinosauroid specimens were found in the Irendabasu Formation of Inner Mongolia in 1923, there has been a lot of confusion about their relationships. The Irendabasu specimens were erroneously considered to belong to the tyrannosaurid *Alectrosaurus* (Mader and Bradley, 1989), *Therizinosaurus cheloniformis* from Mongolia was originally identified as a turtle (Maleev, 1954), and *Nanshiungosaurus brevispinus* from China was referred to the Sauropoda (Dong, 1979). In fact, they were not recognized as a distinct theropod taxon for 57 years (Perle, 1979), and their true nature remained obscure until much later (Russell and Dong, 1993a, b).

Many characters separate therizinosauroids from other theropods. These include small bulbous teeth with denticles aligned parallel to the longitudinal axis of each tooth; widely spaced cervical zygopophyses; elongate, highly pneumatic cervical vertebrae; tall neural arches in the anterior dorsal vertebrae; very broad hips; opisthopubic pelvis; deep preacetabular process of the ilium, which is strongly deflected outwards in *Segnosaurus* and *Nanshiungosaurus*; short postacetabular region of the ilium; relatively short metatarsus, less than a third the length of the tibia; and a functionally quadridactylous foot in which the proximal end of the first metatarsal reaches the tarsus.

Two families of therizinosauroids are recognized by Russell and Dong (1993a, b). Alxasauridae encompasses the more primitive, generally smaller therizinosauroids of the Lower Cretaceous, whereas therizinosaurids are more derived, Upper Cretaceous forms.

Alxasauridae

Characteristics. Alxasaurids are considered to be less derived than therizinosaurids in having teeth that extend to the front of the jaw, unfused cervical ribs, only five sacral vertebrae, a relatively small deltopectoral crest on the humerus, well developed ligament pits in the manual phalanges, and an elongate ilium with only moderate preacetabular expansion.

Record. At present, only one genus is included in the Alxasauridae. *Alxasaurus elesitaeiensis* is known from several partial skeletons collected from Lower Cretaceous (Albian?) strata in the Alxa (Alashan) Desert of Inner Mongolia. These remains suggest that it was a medium-sized dinosaur about four meters in length and 400 kg in weight. There are more teeth (40) in the dentary than there are in known therizinosaurids. Although its remains are unreported from Mongolia, its association with Early Cretaceous dinosaurs that are known from Mongolia suggests that it will be found north of the border.

Therizinosauridae

Characteristics. The generally larger therizinosaurids are distinguishable from alxasaurids in lacking teeth at the front of the mouth (Figure 22.4A), and in having

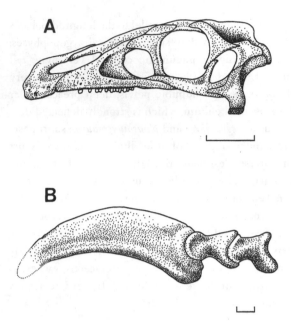

Figure 22.4. Fossil remains of Mongolian therizinosaurids. (A) Skull of *Erlikosaurus*, (B) phalanges of manual digit II of *Therizinosaurus*. (A) after Clark *et al.* (1994), (B) after Barsbold (1976). Scale bar = 50 mm for (A) and 10 mm for (B).

cervical ribs that are fused to the vertebrae, six sacral vertebrae, a large deltopectoral crest, shallow ligament pits in the manual phalanges (Figure 22.4B), and a shorter ilium with significant preacetabular expansion.

Therizinosaurids were a diverse assemblage of theropods that are relatively common at sites that seem to represent lake and river deposits. There are fewer teeth in therizinosaurids than there are in *Alxasaurus*, the anterior teeth having been lost and presumably replaced by a horny bill. This suggested to Barsbold and Perle (1980) that they may have been piscivorous, although Paul (1984) presented a strong case for herbivorous therizinosaurids.

Record. The holotype of *Erlikosaurus andrewsi* was recovered from Bayanshiree strata (Upper Cretaceous) of Baishin Tsav (Perle, 1981). Remains attributed to this animal have also been recovered from the Irendabasu beds of Inner Mongolia (Currie

and Eberth, 1993). (Note that the name has been spelled both as *Erlikosaurus* and '*Erlicosaurus*' by the original author and subsequent workers. The former spelling should be considered correct in that it was used first, and in that the animal is named after 'Erlik', a lamaist deity.) *Erlikosaurus* is smaller than most therizinosaurids, but the unguals are more trenchant (Figure 22.4B). It is the only therizinosaurid known from well preserved, well described cranial material (Clark *et al.*, 1994; Figure 22.4A). Notably, the premaxilla is edentulous, the maxillary teeth are inset from the side of the face, there are 31 dentary teeth, the external naris is relatively larger than those of other theropods, the parasphenoid-basisphenoid complex is highly pneumatic, there is a distinct depression around the otic region in the side of the braincase, and the coronoid bone has been lost from the lower jaw. Although some of these characters are similar in oviraptorosaurs, troodontids, ornithomimids, and other theropods, details allow one to distinguish *Erlikosaurus* easily from other taxa (Clark *et al.*, 1994). It is not yet known how widespread most of these characters are in other therizinosauroids.

The type specimen of *Enigmosaurus mongoliensis* consists of a relatively large pelvis from the Bayanshiree strata of Khar Hötöl. It is possible that this specimen might belong to *Erlikosaurus* (Barsbold, 1983), for which the pelvis is unknown.

Segnosaurus galbinensis is from Bayanshiree strata of Amtgai, Baishin Tsav, Khar Hötöl and Urilge Khudag in southeastern Mongolia (Barsbold and Perle, 1980), and from the Irendabasu formation of Inner Mongolia (Currie and Eberth, 1993). Specimens include the lower jaw (with 25 teeth) and much of the skeleton (Figure 22.5C, 22.6E). The front of the jaw is toothless and the anterior teeth are somewhat curved, whereas the posterior ones are smaller and straight.

Although Maleev (1954) originally thought the unguals of *Therizinosaurus cheloniformis* were from a sea turtle, Rozhdestvenskii (1970) and Osmólska and Roniewicz (1970) recognized that it was a theropod. A complete front limb and shoulder girdle was described by Barsbold (1976), and the hind limb was described

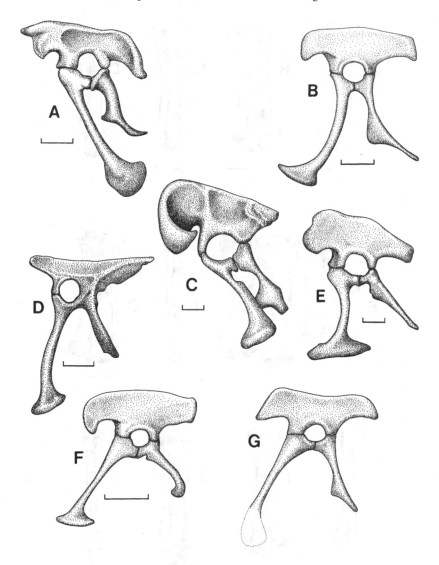

Figure 22.5. Pelvic girdles of Mongolian theropods. (A) *Adasaurus*, (B) *Ingenia*, (C) *Segnosaurus*, (D) *Avimimus*, (E) *Tarbosaurus*, (F) *Gallimimus*, (G) *Saurornithoides*. (A), (C) and (G) after Barsbold (1983), (B) after Barsbold *et al.* (1990), (D) after Kurzanov (1987), (E) after Maleev (1974), and (F) after Barsbold and Osmólska (1990). Scale bar = 50 mm for (A), (B), (D) and (F) and 20 mm for (C) and (E).

and figured by Perle (1982), but neither seemed to resolve the systematic position of this species. The discovery of *Alxasaurus* allowed Russell and Dong (1993a, b) to demonstrate an association between *Therizinosaurus* and 'segnosaurs'. *Therizinosaurus che-* *loniformis* remains, which consist mostly of parts of the front and hind limbs, have also been recovered from Nemegt beds near the Nemegt locality.

Therizinosaur remains are also found in Kazakhstan and Uzbekistan (Nesov, 1995), but are too incomplete

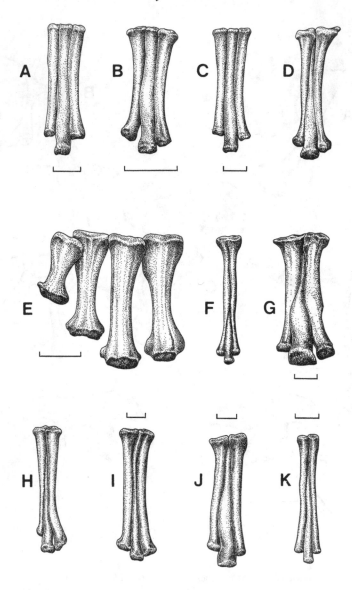

Figure 22.6. Metatarsi of Mongolian theropods. (A) *Velociraptor*, (B) *Ingenia*, (C) *Oviraptor*, (D) *Elmisaurus*, (E) *Segnosaurus*, (F) *Avimimus*, (G) *Tarbosaurus*, (H) *Tochisaurus*, (I) *Harpymimus*, (J) *Garudimimus*, and (K) *Gallimimus*. (A), (C) after Barsbold (1983), (B) after Barsbold *et al.* (1990), (D) after Osmólska (1981), (E) after Perle (1979), (F) after Kurzanov (1987), (G) after Maleev (1974), (H) after Kurzanov and Osmólska (1991), and (I), (J), and (K) after Barsbold and Osmólska (1990). Scale bar = 50 mm except for (E) and (G) = 100 mm.

for more specific identification at present. Recently, eggs with embryos from central China (Henan and Hubei provinces) have been referred to as therizinosaurs (Currie, 1996).

Avimimidae

Characteristics. Avimimids are small, turkey-sized theropods with unknown affinities. When discovered, *Avimimus portentosus* was identified as a bird (Kurzanov, 1981, 1987), though subsequent workers have tended to treat *Avimimus* as a small theropod. A short, deep premaxilla and the front of a lower jaw suggest that this animal was toothless. The braincase is inflated and rather birdlike, there is a reduced postorbital bar, and much of the skull roof is fused. The humerus is relatively long and slender, like that of a bird, but retains a theropod-like deltopectoral crest. The ulna supposedly has papillae for attachment of feathers, but these features are not distinct enough to be sure of their presence. There is a fused carpometacarpus. The hips are very broad, the ilium has a large intertrochanteric shelf (Figure 22.5D), the pubic canal is very broad, and the sacrum includes seven coossified vertebrae. The tarsometatarsus is coossified, but the distal ends of the metatarsals are separate (Figure 22.6F). The third metatarsal is constricted between the second and fourth metatarsals as in arctometatarsalian theropods including ornithomimosaurs, troodontids and caenagnathids. Other than the fact that the fifth metatarsal is included in the fused tarsometatarsus, this structure most closely resembles that of *Elmisaurus*. The vertebrae exhibit some similarities, such as the presence of hypapophyses on the anterior dorsals, to those of troodontids and ornithomimosaurs. Although reconstructed with a short tail by Kurzanov (1987), caudal vertebrae found in China, and the structure of the hips and femur suggest that *Avimimus* may have had a long, tapering tail. A number of isolated vertebrae, tarsometatarsi and unguals have been found in Upper Cretaceous strata of North America that closely resemble those of Mongolian avimimids (collections of Royal Tyrrell Museum of Palaeontology).
Record. *Avimimus portentosus* was recovered from several Djadokhta beds sites in southeastern and southwestern Mongolia. In addition to that, many isolated avimimid bones have been collected for more than 70 years from the Irendabasu Formation exposed near Erenhot in Inner Mongolia (Currie and Eberth, 1993), and in 1975 at Baishin Tsav (Kurzanov, pers. comm., 1991). Whether these two earlier occurrences represent the same species or not cannot be determined at this time.

Troodontidae

Characteristics. The Troodontidae is one of the most birdlike families of non-avian theropods. They were relatively gracile animals that were less than 3 m in length at maturity. The eyes are large, and because of the narrow snout and broad postorbital region, they face forward and have overlapping fields of view. Relative brain size is large, with that of the North American species *Troodon formosus* being the largest of any dinosaur presently known. Many of the skull bones are pneumatized, with air invading facial and palatal bones from the nasal region, and pneumatopores entering the braincase from the throat via the eustachian tube and the middle ear. The pneumatic parasphenoid has expanded into a bulbous, balloon-like structure. The teeth are relatively small, but are numerous and easily identified because of their relatively large, hook-like denticles. The premaxillary teeth are almost triangular in cross-section, whereas the teeth in the lower jaw are smaller than those in the upper jaw. Unlike other theropods, there are no interdental plates in the lower jaws, and the teeth are held in place by a ring of dental bone that wraps around a constriction between the root and crown of each tooth.

Postcranially, troodontids were highly adapted for a cursorial existence. In fact, limb proportions suggest that in terms of running speed they were probably second only to ornithomimids. Like dromaeosaurids, the second toe of the foot bore an enlarged raptorial claw that was kept off the ground to maintain its sharpness. Because only digits three and four contacted the ground, there are some unusual adaptations in the metatarsus. The second metatarsal, which supports the raised, raptorial claw, is reduced to a relatively thin bone. Like caenagnathids, ornithomimids, tyrannosaurids and several other types of Cretaceous theropods, the proximal part of the third metatarsal has been reduced to a small splint. Most of the weight of the animal therefore had to be borne by the fourth metatarsal, which is relatively larger in troodontids than in any other theropods.

Troodontids have many characterististics that make them one of the most easily defined clades in the Theropoda. However, defining their position amongst the theropods has not been as simple. Great importance was placed in the past on the presence in both dromaeosaurids and troodontids of the highly modified second digit of the foot with its enlarged claw. Because of this, the two families are usually included in a clade known as the Deinonychosauria (Colbert and Russell, 1969). However, fundamental differences in anatomical details suggest that this adaptation was attained independently (Currie and Peng, 1993). Similar changes in the second pedal digit seem to have also occurred in the Argentinian theropod *Noasaurus* (Bonaparte and Powell, 1980), and in at least one modern bird known as the seriema. A higher number of derived characters are shared by troodontids and ornithomimosaurs, which led Holtz (1994) to establish a clade that he called the Bullatosauria.

Record. The Troodontidae is best known from the Upper Cretaceous strata of Asia and North America, although it was clearly a well established family in Asia during Early Cretaceous times. The identification of teeth (*Koparion*) as those of troodontids from the Upper Jurassic Morrison Formation of the United States (Chure, 1994) would be more convincing if diagnostic postcranial elements are recovered. This is compounded by the fact that basal ornithomimosaurs and other theropods can have troodont-like teeth. However, given the highly derived nature of Early Cretaceous troodontids, it is not unlikely that the family traces its origins back into the Jurassic.

Numerous species of troodontids have been described from Mongolia and adjacent parts of China and Uzbekistan. The earliest, and incidentally the most complete, troodontid is *Sinornithoides youngi* (Russell and Dong, 1993a, b) from the Lower Cretaceous rocks of the Ordos Basin in Inner Mongolia. This was a small animal weighing only about 2.5 kg. Although it was a young animal, the degree of ossification and the relative proportions of the body suggest it was almost fully grown. A partial troodontid skeleton from the Lower Cretaceous Khamaryn Us locality of Mongolia (Barsbold *et al.,* 1987) is from a larger animal that may represent a distinct, unnamed species.

Saurornithoides mongoliensis was the first troodontid described from Asia (Osborn, 1924). The holotype was recovered from the Djadokhta beds at Bayan Zag, and specimens from equivalent aged beds are generally referred to this species. These include a small specimen from Bayan Mandahu in Inner Mongolia (Currie and Peng, 1993) that suggests that troodontid juveniles had disproportionately long, but slender metatarsals.

A second, somewhat larger species, *Saurornithoides junior*, was established on the basis of a lovely skull and partial skeleton from the Nemegt beds of Bügiin Tsav. In addition to size, this species has more teeth than *S. mongoliensis* (Barsbold 1974). *Borogovia gracilicrus* (Osmólska, 1987) is another troodontid collected from the Nemegt beds of Mongolia, but has only been reported so far from the Altan Uul IV locality. It is based on partial hind limbs, which include a distinctive second toe with a straight ungual. A third Nemegt genus and species of troodontid, *Tochisaurus nemegtensis*, was established on the basis of a metatarsus (Kurzanov and Osmólska, 1991; Figure 22.6H). Unfortunately, there is no significant overlap in the known specimens of these three Nemegt genera, which were recovered from the same geographic area, and it is conceivable that they all represent the same species.

Troodontid teeth and isolated cranial and postcranial bones from Iren Dabasu (Currie and Eberth, 1993) in Inner Mongolia and Jiayin in Heilongjiang, China, cannot be distinguished from *Saurornithoides* on the basis of size or morphology. A joint Mongolian–American expedition has recently recovered the skull of a new, as yet undescribed genus of troodontid (Novacek *et al.,* 1994) and troodontid teeth attributed to *Troodon asiamericanus* and *Pectinodon* have been described from the Khodzhakul Formation (Cenomanian) of Uzbekistan (Nesov, 1995).

Although *Archaeornithoides deinosauriscus* was described as a small, birdlike theropod, Elzanowski and Wellnhofer (1993) also speculated on the possibility that this animal, from the Djadokhta Formation of Bayan Zag, was a juvenile troodontid. The characters suggesting that it is not a troodontid include the presence of a wide palatal shelf and the absence of denticles on the teeth. However, new troodontid specimens from Montana show that these animals do have broad palatal shelves. Furthermore, the lack of denticles on the carina of the teeth of such a small animal is not surprising considering the fact that the teeth are less than half the size of the smallest known troodontid teeth. It is therefore quite possible that *Archaeornithoides deinosauriscus* might eventually be shown to be a very young specimen of *Saurornithoides mongoliensis*.

Although troodontid fossils clearly represent a diverse clade within the Cretaceous of Mongolia and neighbouring parts of China, the incompleteness of most specimens makes it difficult to determine evolutionary trends within the family. There is a tendency for the troodontid species to increase in size over time, and for them to increase the number of teeth. *Sinornithoides* and *Saurornithoides mongoliensis* have 18 maxillary teeth, *Saurornithoides junior* has 19–20, and the new, undescribed specimen evidently had at least 30.

Ornithomimosauria

The best fossils documenting the evolution and diversification of 'bird mimics' come from Mongolia. Three ornithomimosaur families are presently recognized – Harpymimidae, Garudimimidae and Ornithomimidae – and all of them are represented in Mongolia (Barsbold and Osmólska, 1990). *Pelecanimimus polyodon,* from the Lower Cretaceous of Spain, has also been identified as an ornithomimosaur (Pérez-Moreno *et al.,* 1994), but cannot be assigned to any of these three families.

Ornithomimosaurs are generally man-sized animals, although they have lightly built, birdlike heads and bodies. As in troodontids, the eyes are huge, and there is a bulbous parasphenoid. Most are toothless, a convergence with oviraptorids, and the jaws would have been encased by keratinous rhamphothecae. Toothed forms have either numerous, very small teeth (*Pelecanimimus*) or relatively few, poorly developed, peglike teeth (*Harpymimus*). As in dromaeosaurids, the premaxilla has a dorsoposterior process that excludes the maxilla from the narial opening. In ornithomimosaurs, however, this process is relatively longer, and separates the maxilla and the nasal to the level of the antorbital fossa. The lower jaws are slender and elongate, and the jaw articulation is in an anterior position ventral to the postorbital bar. The cervical vertebrae constitute about 40% of the length of the presacral vertebral column. Metacarpals II and III are almost the same length, and the first metacarpal is more than half of that length. In fact, with the exception of *Harpymimus*, the first metacarpal is usually only slightly shorter than either of the other two. The manual unguals are either weakly curved or straight, and have flexor tubercles that are more distally positioned than they are in other theropods. An ornithomimosaur ilium has an anteroventrally hooked process, the ischium is shorter than the pubis, and there is a wide pubic canal. The metatarsus/tibia ratio is higher than those of other theropods. The first toe is lost in all ornithomimosaurs except *Garudimimus*.

Ornithomimosaur remains are widely distributed in other parts of Asia, including China (Dong, 1992), Kazakhstan, Tadzhikistan, and Uzbekistan (Nesov, 1995).

Harpymimidae

Characteristics. Only one harpymimid has been found to date. There are teeth in the jaws of these animals, although they are not well-formed, appear to lack enamel, and may even have been covered over by a keratinous bill (Barsbold and Osmólska, 1990). The humerus is not twisted as it is in more advanced ornithomimosaurs, and the first metacarpal is not as elongate. The metatarsus (Figure 22.6I) seems to have been relatively shorter in that the length is only five times the width of the unit, compared with 7.5 to 9

times the length in ornithomimids. Perhaps more significant is the fact that the third metatarsal is not as constricted as it is in ornithomimids, and that it still separated the second and fourth metatarsals. However, there is some damage to the specimen (Osmólska, pers. com.) in this region, and this feature is uncertain.
Record. The only specimen of *Harpymimus okladnikovi* was recovered from the Aptian–Albian Shinekhudag beds of Dundgov'. It includes a skull and part of a skeleton, both of which need to be properly described before the systematic position of this species can be understood.

Garudimimidae

Characteristics. A single, well-preserved skull and its incomplete postcranial skeleton are all that are known of the Garudimimidae. Like more advanced ornithomimosaurs, there is a bulbous parasphenoid, and the jaws are edentulous. However, the postorbital region of the skull is relatively longer, and the jaw articulation is positioned posterior to the postorbital bar. The degree of constriction of the proximal end of the third metatarsal is intermediate between that of harpymimids and ornithomimids (Figure 22.6J). The first pedal digit has been retained, in contrast to ornithomimids where it is absent.
Record. The type specimen of *Garudimimus brevipes* was collected from the Cenomanian–Turonian Bayanshiree beds of Baishin Tsav. *Garudimimus* may actually be found in the same rocks that produce specimens of *Archaeornithomimus* (Currie and Eberth, 1993), a fact that could potentially lead to much confusion.

Ornithomimidae

Characteristics. Most ornithomimid genera from Mongolia are well represented by multiple skulls and skeletons. Each animal has metacarpals and fingers that are almost equal in length, a metatarsus that is more than two thirds the length of the tibiotarsus, a proximally pinched third metatarsal that permits proximal contact between metatarsals II and IV (Figure 22.6K), and loss of the first digit of the foot.
Record. The first ornithomimid described from central Asia was named *Ornithomimus asiaticus* by Gilmore

(1933), but was subsequently renamed *Archaeornithomimus asiaticus* by Russell (1972). This ornithomimid is poorly known, even though its remains occur in bonebeds in the Irendabasu Formation of Inner Mongolia, and thousands of partial skeletons and isolated bones have been collected (Currie and Eberth, 1993). The Irendabasu Formation is generally considered to be Cenomanian in age, but is best considered Early Senonian, and may ultimately prove to be as young as Campanian (Currie and Eberth, 1993).

Gallimimus bullatus is the best known ornithomimid, thanks to the recovery of several nearly complete skeletons with skulls from the Nemegt formation at Altan Uul, Bügiin Tsav, Nemegt, and Tsagaan Khushuu (Osmólska *et al.*, 1972). The youngest of these was only about 0.5 m high at the hips, while the largest was close to 2 m in the same dimension.

The Nemegt beds at Bügiin Tsav also produced the type and only specimen of *Anserimimus planinychus* (Barsbold, 1988). This partial skeleton is different from other Mongolian ornithomimids because of the powerful development of the deltoid crest of the humerus (which has never been illustrated), and because of peculiar, flattened unguals on the manus.

Tyrannosauridae

Tyrannosaurids are normally divided into two subfamilies – the Tyrannosaurinae, and the poorly understood Aublysodontinae. These large theropods are most readily characterized by their premaxillary teeth, which are D-shaped in cross-section, incisiform, and are smaller than most of the maxillary and dentary teeth. The cheek teeth are mediolaterally inflated, and can be subcircular in cross-section. This is correlated with increased tooth strength that reduced the chances of damage when bone was encountered during feeding. The nasals coossify in mature individuals, and their dorsal surfaces are rugose. In all species, the prominent nuchal crest extends double the height of the supraoccipital above the foramen magnum. Presacral vertebral centra are relatively shorter anteroposteriorly than those of other theropods. This, the reduction of the forelimbs,

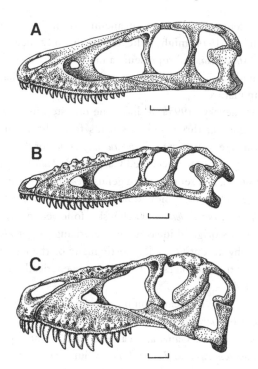

Figure 22.7. Skulls of Mongolian tyrannosaurids. (A) *Alectrosaurus*, (B) *Alioramus*, (C) *Tarbosaurus*. (A) after Perle (1977), (B) after Kurzanov (1976), and (C) after Maleev (1974). Scale bar = 50 mm for (A) and (B) and 100 mm for (C).

and the loss of all but manual digits I and II may be correlated with lightening of the front end of the skeleton. The legs are relatively long for such large animals, suggesting that they were fast movers. The feet are arctometatarsalian, with elongate metatarsals, the third one of which is proximally constricted.

Aublysodontinae

Characteristics. Aublysodontine tyrannosaurs seem to have all been medium sized theropods that grew to less than 5 m in length. Both *Aublysodon,* from North America, and *Alectrosaurus* (Figure 22.7A), from Central Asia, lack denticles (serrations) on their premaxillary teeth. In the latter, there are 17 maxillary, and 19 dentary teeth, which are higher numbers than counts for tyrannosaurine genera. The first two or three maxillary teeth are incisiform (Perle, 1977). The

teeth are narrower and more bladelike than those of their later, more specialized cousins. The skulls of these animals are relatively low and long, although this is, in part, a function of small size in that juvenile tyrannosaurines have similar cranial proportions. The dorsal surface of the fused nasal unit is smooth. The front limbs of *Alectrosaurus* are relatively large compared with advanced tyrannosaurids like *Tarbosaurus* (Perle, 1977), but the few measurements published resemble those of similar-sized individuals of *Gorgosaurus libratus.* Newly recovered specimens from Mongolia and Inner Mongolia suggest that there are many other postcranial characters, especially in the front limbs and hips, that distinguish *Alectosaurus* from other tyrannosaurids.

Record. Alectrosaurus olseni was described by Gilmore (1933) on the basis of parts of two different skeletons from the Iren Dabasu site near the modern city of Erenhot, Inner Mongolia. Perle (1977) and Mader and Bradley (1989) recognized that the robust arms that were supposed to belong to *Alectrosaurus* were in fact from a segnosaur, but that the hind limbs were unquestionably tyrannosaurid. The first specimens of this animal from Mongolia were described by Perle (1977) on the basis of cranial and postcranial material recovered from Baishin Tsav. Several partial, undescribed skeletons of *Alectrosaurus* collected from southeastern Mongolia are in the collections of the museum in Ulaanbaatar, and another new specimen was recently collected from Erenhot in China. *Aublysodon* and *Alectrosaurus* remains have been reported from Kazakhstan, Tadzhikistan and Uzbekistan (Nesov, 1995), although none of the specimens are complete enough for proper identification.

Tyrannosaurinae

Characteristics. Tyrannosaurines include some of the largest and most derived predators amongst the theropods, and large specimens of *Tarbosaurus* from Mongolia reached lengths of more than 12 m. Tyrannosaurines can be distinguished from aublysodontines in that they have serrated premaxillary teeth, have fewer than 17 maxillary teeth that are absolutely and relatively taller, have maxillary and dentary teeth that are labiolingually thicker, and have fused nasals with rugose dorsal surfaces.

Record. Numerous tyrannosaurines have been described from Mongolia since their discovery by the first Russian expedition to the Nemegt Valley in 1946. Tyrannosaurids are relatively common in this region. The Russians collected seven more or less complete skeletons in the 1940s, the Polish–Mongolian expeditions excavated at least three more, and at least six more skeletons are housed in Ulaanbaatar. Even today, there are reports of *Tarbosaurus* specimens being found and left in the field. Why there are so many of these large carnivores found in the Nemegt Formation is a matter for considerable speculation (Osmólska, 1980). Tyrannosaurid fossils recovered from Kazakhstan are sometimes referred to *Tarbosaurus* (Nesov, 1995), whereas teeth and poorly preserved remains of large tyrannosaurids from the Upper Cretaceous of the Heilongjiang (Amur) River of Russia and China, and other parts of China (Dong, 1992) are almost certainly attributable to *Tarbosaurus*.

Tyrannosaurs from Mongolia were originally described by Maleev (1955a, b, 1974) as *Tyrannosaurus bataar*, *Tarbosaurus efremovi*, *Gorgosaurus lancinator* and *Gorgosaurus novojilovi*. Rozhdestvenskii (1965) considered all Nemegt tyrannosaurs to be different growth stages of a single species, *Tarbosaurus bataar*. This has been generally accepted, although Carpenter (1992) used the type specimen of '*Gorgosaurus*' *novojilovi* to establish *Maleevosaurus novojilovi*. Most of the characters used to separate '*Maleevosaurus*' from specimens generally referred to as *Tarbosaurus* are differences in proportions that are ontogenetically controlled. Juvenile tyrannosaurs have lower, more elongate skulls than the adults, which means that the relative proportions of the fenestrae and individual bones (including the maxilla and dentary) go through some extreme changes. The moderate size of the lacrimal horn is not unexpected considering the well rounded, low nature of the dorsal surface of the lacrimal in *Tarbosaurus* (Figure 22.7C, 22.8). The jugal of '*Maleevosaurus*' appears to be very slender (Maleev, 1974), but it has also been damaged and is not complete. Although apparent fusion of the neural arch to

the centrum and the calcaneum to the astragalus (Maleev, 1974) might indicate that the only specimen of '*Maleevosaurus*' represents a mature individual, the scapula is not fused to the coracoid, suggesting that it is immature.

Olshevsky (1995a, b) has gone one step further in recognizing three tyrannosaurids from the Nemegt Basin. He resurrected *Tarbosaurus efremovi* for the 12 m long tyrannosaur, accepted Carpenter's *Maleevosaurus novojilovi*, and set up a third genus, *Jenghizkhan*, for the large (15 m) individual that Maleev called '*Tyrannosaurus*' *bataar*. Olshevsky followed many of Maleev's original ideas in characterizing *Jenghizkhan*, thereby accepting as diagnostic many of the features that other workers felt were ontogenetically controlled.

In my own research on ontogenetic series of specimens of *Gorgosaurus libratus* and *Daspletosaurus torosus* from Alberta, Canada, I can see trends that suggest *Maleevosaurus* and *Jenghizkhan* are junior synonyms of *Tarbosaurus*, as was proposed by Rozhdestvenskii (1965). In examining the many fine tyrannosaurid specimens in Moscow, Warsaw and Ulaanbaatar, I have never found differences significant enough to convince me that *Tarbosaurus bataar* should be subdivided. That does not mean that further research will not reveal convincing differences, but at the present time the most conservative approach is to accept only *Tarbosaurus bataar*.

There is one other tyrannosaurid from the Nemegt Formation. *Alioramus remotus* is from the Nogoon Tsav beds of the Ingenii Höövör valley (Kurzanov, 1976; Figure 22.7B). Based on a single specimen, this medium-sized tyrannosaurid is easily distinguished from *Tarbosaurus* by its higher tooth count and by a series of bumps on the nasals. The maxilla has 16, possibly 17 teeth, and the dentary has 18 teeth, compared with a maximum of 13 maxillary and 15 dentary teeth in *Tarbosaurus*. The bumps on the nasals are rather irregular, so it is possible that the number (five) will prove to be variable in other specimens of *Alioramus*. As pointed out by Kurzanov (1976), the skull is longer and lower than those of *Tarbosaurus*, *Albertosaurus*, *Daspletosaurus* and *Tyrannosaurus*. However, skull pro-

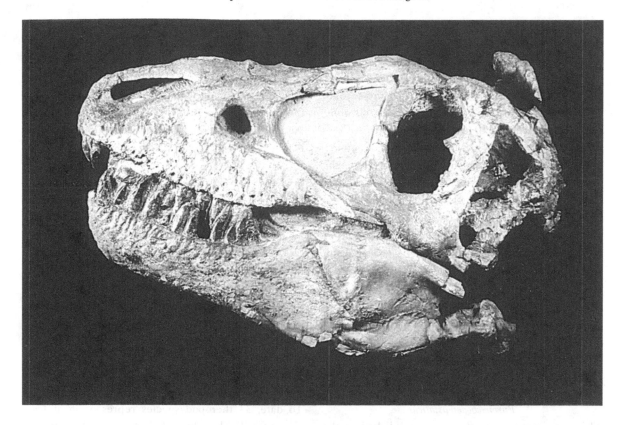

Figure 22.8. Skull of *Tarbosaurus* in left lateral view. (Courtesy of P. Rich.)

portions in tyrannosaurids are dependant on size and age, and juveniles of the other genera have skull proportions that are the same as those of *Alioramus*. The type specimen has a beautifully preserved braincase with a broad nuchal crest and a downturned occiput. These and other characters suggest that *Alioramus* is most closely related to *Tarbosaurus* in Asia, and to *Daspletosaurus* and *Tyrannosaurus* in North America.

Theropods of uncertain systematic position

Asiamericana asiatica

Nesov (1995) described small theropod teeth and jaws from the Bissekty Svita (Upper Turonian) of Uzbekistan and the Dabrazinskaya Formation (Santonian) of Kazakhstan as *Asiamericana*. Although these almost fish-like theropod teeth are distinctive, the systematic position of this theropod cannot be determined at this time.

Bagaraatan ostromi

An unusual medium-sized theropod described by Osmólska in 1996 is based on an incomplete skeleton from the Nemegt formation at the Nemegt locality. The type specimen of *Bagaraatan ostromi* includes a mandible with a shallow but massive dentary, and a fibula that is fused distally to both the tibia and the coossified astragalus and calcaneum. A more complete specimen is needed to determine the systematic position of this animal within the Theropoda.

Deinocheirus mirificus

The Deinocheiridae was erected to include only a single specimen of *Deinocheirus mirificus*, consisting of a remarkably large pair of front limbs, the shoulder girdle and assorted fragments (Osmólska and Roniewicz, 1970). The scapula is long and slender, the forelimbs are elongate, and the three fingers end in long, strong unguals. The relative proportions of the front limb elements are suggestive of ornithomimids (Osmólska and Roniewicz, 1970), with the manus being only slightly longer than the radius, and the first metacarpal being only 5% shorter than the second. The great length of the arms (close to 2.5 m) is more suggestive of therizinosaurs, however, which is one reason Barsbold (1976) included both types of animals in a new suborder that he called the Deinocheirosauria. This has not received widespread acceptance, and it is clear that the taxonomic position of *Deinocheirus* will not be resolved without more complete specimens. The single specimen of *Deinocheirus mirificus* was collected from Nemegt strata at Altan Uul III.

Embasaurus minax

Two vertebrae from the Lower Cretaceous of Kazakhstan were originally described as *Embasaurus* by Ryabinin (1931). They may be megalosaurid (Nesov, 1995).

Euronychodon asiaticus

The tooth genus *Euronychodon* was established on the basis of teeth from southern Europe, but Nesov (1995) proposed the species *E. asiaticus* for teeth he recovered from the Bissekty Svita (Upper Turonian) of Uzbekistan. The affinities of this small theropod are unknown.

Itemirus medullaris

A well-preserved braincase from the Turonian beds of the Kyzylkum Desert of Uzbekistan was described by Kurzanov (1976) as *Itemirus medullaris*. Although he originally referred to it as a carnosaur, it is much closer in all but one respect (the laterally excavated basipterygoid process) to a dromaeosaurid braincase (Currie, 1995). The Dzharakhuduk locality has produced many isolated theropod specimens, including dromaeosaurid teeth and bones, but without a more complete skeleton, *Itemirus* should not be assigned to the Dromaeosauridae.

Prodeinodon mongoliensis

Teeth from the Ondaisair and Öösh Formations (Early Cretaceous) of Mongolia have been referred to as *Prodeinodon mongoliensis* (Osborn, 1924). The teeth demonstrate that there was at least one species of large theropod in Mongolia at that time, but give no information on the type of theropod to which they might belong.

Shanshanosaurus huoyanshanensis

Olshevsky (1995b) has allied *Shanshanosaurus huoyanshanensis* from the Maastrichtian Subashi Formation of Xinjiang, China, with the aublysodontine tyrannosaurs because of its unserrated, incisiform, premaxillary teeth (Dong, 1977) and booted pubis. The reported presence of procoelic cervical vertebrae is incorrect, and re-examination of the type specimen (Currie and Dong, in prep.) suggests that it may be a juvenile *Tarbosaurus*.

Conclusions

To date, 33 theropod species representing at least eleven families have been described from Mongolia. It is doubtful whether all of these species are valid, but it is almost certain that at least 25 of them are (Table 22.1). An additional five theropods from neighbouring regions in China, Kazakhstan and Uzbekistan can be expected to be found eventually in Mongolia, and other species described from these regions, such as *Chilantaisaurus* and *Phaedrolosaurus* from China, may well turn out to be valid, and may also turn up in Mongolia. Furthermore, additional theropods will no doubt be discovered in Mongolia as the result of intensive collecting activity at established and newly discovered localities.

The diversity of Mongolian theropods gives us one of the best windows available on theropod evolution, including the origin of birds. Even though all of the small theropods discovered in Mongolia so far lived too late in time to have been bird ancestors, their superb preservation allows us to make detailed anatomical comparisons with birds. Furthermore, many of the Mongolian lineages seem to have originated

Table 22.1. *Theropods from Mongolia and adjacent regions. The first column lists the maximum number of species that have been proposed for central Asia, the second column is the most conservative interpretation of the first column, and the third gives the 'age' of the species according to Jerzykiewicz and Russell (1991)*

Achillobator giganticus	*Achillobator giganticus*	Lower Cret.
Adasaurus mongoliensis	*Adasaurus mongoliensis*	Nemegt
Alectrosaurus olseni	*Alectrosaurus olseni*	Bayanshiree
Alioramus remotus	*Alioramus remotus*	Nemegt
Alxasaurus elesitaeiensis	*	**China
Anserimimus planinychus	*Gallimimus bullatus*	Nemegt
Archaeornithomimus asiaticus	?*Archaeornithomimus*	**China
Archaeornithomimus bissektensis	*Archaeornithomimus bissektensis*	**Uzbekistan
Archaeornithoides deinosauriscus	*Saurornithoides mongoliensis*	Djadokhta
Asiamericana asiatica	*Asiamericana asiatica*	**Uzbekistan
Avimimus portentosus	*Avimimus portentosus*	Djadokhta
Bagaraatan ostromi	*Bagaraatan ostromi*	Nemegt
Borogovia gracilicrus	*Saurornithoides mongoliensis*	Nemegt
Caenagnathasia martinsoni	?*Elmisaurus*	**Uzbekistan
Conchoraptor gracilis	*Conchoraptor gracilis*	Baruungoyot
Deinocheirus mirificus	*Deinocheirus mirificus*	Nemegt
Elmisaurus rarus	*Elmisaurus rarus*	Nemegt
Enigmosaurus mongoliensis	*Erlikosaurus andrewsi*	Bayanshiree
Erlikosaurus andrewsi	*Erlikosaurus andrewsi*	Bayanshiree
Euronychodon asiaticus	*Euronychodon asiaticus*	**Uzbekistan
Gallimimus bullatus	*Gallimimus bullatus*	Nemegt
Garudimimus brevipes	*Garudimimus brevipes*	Bayanshiree
Harpymimus okladnikovi	*Harpymimus okladnikovi*	Shinekhudag
Hulsanpes perlei	*Velociraptor mongoliensis*	Baruungoyot
Ingenia yanshini	*Ingenia yanshini*	Baruungoyot
Itemirus medullaris	*Itemirus medullaris*	**Uzbekistan
Jenghizkhan bataar	*Tarbosaurus bataar*	Nemegt
Maleevosaurus novojilovi	*Tarbosaurus bataar*	Nemegt
Monolophosaurus jiangi	*	**China
'Oviraptor' mongoliensis	*Oviraptor mongoliensis*	Nemegt
Oviraptor philoceratops	*Oviraptor philoceratops*	Djadokhta
Prodeinodon mongoliensis	*Prodeinodon mongoliensis*	Öösh
Saurornithoides junior	*Saurornithoides mongoliensis*	Nemegt
Saurornithoides mongoliensis	*Saurornithoides mongoliensis*	Djadokhta
Shanshanosaurus huoyanshanensis	*	**China
Segnosaurus galbinensis	*Segnosaurus galbinensis*	Bayanshiree
Sinraptor dongi	*	**China
Sinornithoides youngi	?*Sinornithoides*	**China
Tarbosaurus efremovi	*Tarbosaurus bataar*	Nemegt
Therizinosaurus cheloniformis	*Therizinosaurus cheloniformis*	Nemegt
Tochisaurus nemegtensis	*Saurornithoides mongoliensis*	Nemegt
Troodon asiamericanus	*Saurornithoides mongoliensis*	**Uzbekistan
Undescribed troodontid	Undescribed troodontid	Djadokhta
Velociraptor mongoliensis	*Velociraptor mongoliensis*	Djadokhta

Notes:

A single asterix (*) means nothing has been found in Mongolia to indicate the presence of this species, although the probability of its discovery there is high. Double asterices (**) indicate the country of origin for species not presently known in Mongolia. A question mark refers to specimens already found in Mongolia that may eventually be identified as non-Mongolian genera.

during Jurassic times, and share with birds derived characters that would have been present in the ancestors of both these theropods and birds. Dromaeosaurids are considered by many experts to be the sister group of birds, but other authors have also made a case for a closer relationship between troodontids and birds. Several of the theropod families represented in Mongolia, including avimimids and caenagnathids, are so birdlike that they were originally identified as birds.

The large number of well preserved specimens from Mongolia also presents palaeontologists with opportunities to assess the morphological variation (individual, sexual and ontogenetic) of theropod species. Although such studies will inevitably lead to a reduction in the apparent diversity of Mongolian theropods, they will provide a much firmer foundation for understanding other aspects of dinosaurian biology, such as palaeoecology. Tyrannosaurids are one of the best examples from Mongolia of animals that can be used to assess variation because: (1) there are many well-preserved skulls and skeletons; (2) half-grown to adult individuals are known; and (3) their North American cousins have been used to demonstrate both ontogenetic and sexual variation. The large number of specimens and elaborate display crests on the skulls of some oviraptorids also make them prime candidates for such studies.

The fine preservation of Mongolian theropods has given science some of the best information on theropod behaviour through taphonomic studies. Examples include the predatory behaviour of *Velociraptor*, and the egg-laying and brooding behaviour of *Oviraptor* (Norell *et al.*, 1995). But there is still much to be learned. Why, for instance, are so many crested oviraptorids found at Ukhaa Tolgod?

Mongolia has clearly established itself as the 'Mecca' for specialists on theropod dinosaurs and an ever-increasing flow of exciting new discoveries is likely to insure this eminent position long into the future.

Acknowledgements

Over the years, I have been extended many courtesies by colleagues working on central Asian theropods, who have opened up their collections and freely discussed their research with me. These include Dong Zhiming (Beijing), Sergei Kurzanov (Moscow), Gene Gaffney and Mark Norell (New York), Halszka Osmólska (Warsaw), Altangerel Perle and Khishigjavyin Tsogtbaatar (Ulaanbaatar), Wei Zhengyi (Harbin), and the late Lev Nesov (St. Petersburg), and I am extremely grateful to them for all their help. In addition, I would like to thank Robert Bakker (Boulder), Ken Carpenter (Denver), Greg Paul (Baltimore) and Dale Russell (Raleigh) who have been a great source of inspiration and information. Michael Ryan (Drumheller) helped with translations. The paper greatly benefited from reviews by Thomas Holtz, Halszka Osmólska, and Dale Russell. The illustrations were prepared by Brenda Middagh.

References

Barsbold, R. 1974. Saurornithoididae, a new family of small theropod dinosaurs from central Asia and North America. *Palaeontologia Polonica* 30: 5–22.

—1976. [New data on *Therizinosaurus* (Therizinosauridae, Theropoda).] *Trudy Sovmestnoi Sovetsko-Mongol'skoi Paleontologicheskoi Ekspeditsii* 3: 76–92.

—1981. [Toothless carnivorous dinosaurs of Mongolia.] *Trudy Sovmestnoi Sovetsko-Mongol'skoi Paleontologicheskoi Ekspeditsii* 15: 28–39.

—1983. [Carnivorous dinosaurs from the Cretaceous of Mongolia.] *Trudy Sovmestnoi Sovetsko-Mongol'skoi Paleontologicheskoi Ekspeditsii* 19: 5–120.

—1986. [Theropod dinosaurs: Oviraptors], pp. 210–223 in *Herpetological Investigation in the Mongolian People's Republic. Collected Scientific Transactions of the Institute of Evolutionary Morphology of the U.S.S.R. Academy of Sciences.*

—1988. A new Late Cretaceous ornithomimid from the Mongolian People's Republic. *Paleontological Journal* 22: 124–127.

—Maryańska, T., and Osmólska, H. 1990. Oviraptorosauria, pp. 249–258, in Weishampel, D.B., Dodson, P. and Osmólska, H. (eds.), *The Dinosauria*. Berkeley: University of California Press.

—and Osmólska, H. 1990. Ornithomimosauria, pp. 225–244, in Weishampel, D.B., Dodson, P. and Osmólska, H. (eds.), *The Dinosauria*. Berkeley: University of California Press.

—, Osmólska, H. and Kurzanov, S.M. 1987. On a new troodontid (Dinosauria, Theropoda) from the Early Cretaceous of Mongolia. *Acta Palaeontologica Polonica* 32: 121–132.

—and Perle, A. 1980. Segnosauria, a new infraorder of carnivorous dinosaurs. *Acta Palaeontologica Polonica* 25: 185–195.

Bonaparte, J.F. and Powell, J.E. 1980. A continental assemblage of tetrapods from the Upper Cretaceous beds of El Brete, northwestern Argentina (Sauropoda-Coelurosauria-Carnosauria-Aves). *Mémoires de la Société Géologique de France* 139: 19–28.

Buffetaut, E., Suteethorn, V., and Tong, H. 1996. The earliest known tyrannosaur from the Lower Cretaceous of Thailand. *Nature* 381: 689–691.

Carpenter, K. 1992. Tyrannosaurids (Dinosauria) of Asia and North America, pp. 250–268, in Mateer, N.J. and Chen, P.J. (eds.), *Aspects of Nonmarine Cretaceous Geology, Proceedings of the First International Symposium of IGCP 245 Nonmarine Cretaceous Correlations, Urumqi, China, 1987*. Beijing: Ocean Press.

Chure, D.J. 1994. *Koparion douglassi*, a new dinosaur from the Morrison Formation (Upper Jurassic) of Dinosaur National Monument: The oldest troodontid (Theropoda: Maniraptora). *Brigham Young University Geological Studies* 40: 11–15.

Clark, J.M., Perle, A. and Norell, M.A. 1994. The skull of *Erlicosaurus andrewsi*, a Late Cretaceous 'segnosaur' (Theropoda: Therizinosauridae) from Mongolia. *American Museum Novitates* 3115: 1–39.

Colbert, E.H. and Russell, D.A. 1969. The small Cretaceous dinosaur *Dromaeosaurus*. *American Museum Novitates* 2380: 1–49.

Currie, P.J. 1989. The first records of *Elmisaurus* (Saurischia Theropoda) from North America. *Canadian Journal of Earth Sciences* 26: 1319–1324.

—1990. The Elmisauridae, pp. 245–248, in Weishampel, D.B., Dodson, P. and Osmólska, H. (eds.), *The Dinosauria*. Berkeley: University of California Press.

—1992. Saurischian dinosaurs of the Late Cretaceous of Asia and North America, pp. 237–249, in Mateer, N.J. and Chen, P.J. (eds.), *Aspects of Nonmarine Cretaceous Geology, Proceedings of the First International Symposium*

of *IGCP 245 Nonmarine Cretaceous Correlations, Urumqi, China, 1987*. Beijing: Ocean Press.

—1995. New information on the anatomy and relationships of *Dromaeosaurus albertensis* (Dinosauria: Theropoda). *Journal of Vertebrate Paleontology* 15: 576–591.

—1996. The great dinosaur egg hunt. *National Geographic Magazine* 189 (5): 96–111.

—and Eberth, D.A. 1993. Palaeontology, sedimentology and palaeoecology of the Iren Dabasu Formation (Upper Cretaceous), Inner Mongolia, People's Republic of China. *Cretaceous Research* 14: 127–144.

—, Godfrey, S.J. and Nesov, L. 1993. New caenagnathid (Dinosauria: Theropoda) specimens from the Upper Cretaceous of North America and Asia. *Canadian Journal of Earth Sciences* 30: 2255–2272.

—and Peng J.H. 1993. A juvenile specimen of *Saurornithoides mongoliensis* from the upper Cretaceous of northern China. *Canadian Journal of Earth Sciences* 30: 2224–2230.

—and Russell, D.A. 1988. Osteology and relationships of *Chirostenotes pergracilis* (Saurischia, Theropoda) from the Judith River (Oldman) Formation of Alberta, Canada. *Canadian Journal of Earth Sciences* 25: 972–986.

—and Zhao X.J. 1993. A new carnosaur (Dinosauria, Theropoda) from the Jurassic of Xinjiang, Peoples Republic of China. *Canadian Journal of Earth Sciences* 30: 2037–2081.

Dashzeveg, D., Novacek, M.J., Norell, M.A., Clark, J.M., Chiappe, L.M., Davidson, A., Mckenna, M.C., Dingus, L., Swisher, C. and Perle, A. 1995. Extraordinary preservation in a new vertebrate assemblage from the Late Cretaceous of Mongolia. *Nature* 374: 446–447.

Dong, Z.M. 1977. [On the dinosaurian remains from Turpan, Xinjiang.] *Vertebrata PalAsiatica* 15: 59–66.

—1979. [The Cretaceous dinosaur fossils in southern China], pp. 342–350 in *Mesozoic and Cenozoic Red Beds in Southern China*. Institute of Vertebrate Palaeontology and Palaeoanthropology and Nanjing Geological and Palaeontological Institute Science Press, Beijing.

—1992. *Dinosaurian Faunas of China*. Springer-Verlag, Berlin. 188 pp.

—and Currie, P.J. 1996. On the discovery of an oviraptorid skeleton on a nest of eggs at Bayan Mandahu, Inner Mongolia, People's Republic of China. *Canadian Journal of Earth Sciences* 33: 631–636.

Elzanowski, A. and Wellnhofer, P. 1993. Skull of *Archaeornithoides* from the Upper Cretaceous of Mongolia. *American Journal of Science* **293**: 235–252.

Gilmore, C.W. 1933. On the dinosaurian fauna of the Iren Dabasu Formation. *American Museum of Natural History, Bulletin* **67**: 23–78.

Holtz, T.R., Jr. 1994. The phylogenetic position of the Tyrannosauridae: implications for theropod systematics. *Journal of Paleontology* **68**: 1100–1117.

Jerzykiewicz, T., Currie, P. J., Eberth, D.A., Johnston, P.A., Koster, E.H. and Zheng J.J. 1993. Djadokhta Formation correlative strata in Chinese Inner Mongolia: an overview of the stratigraphy, sedimentary geology and paleontology and comparisons with the type locality in the pre-Altai Gobi. *Canadian Journal of Earth Sciences* **30**: 2180–2195.

—and Russell, D.A. 1991. Late Mesozoic stratigraphy and vertebrates of the Gobi Basin. *Cretaceous Research* **12**: 345–377.

Kirkland, J.I., Burge, D. and Gaston, R. 1993. A large dromaeosaur (Theropoda) from the Lower Cretaceous of eastern Utah. *Hunteria* **2** (10): 1–16.

Kurzanov, S.M. 1976. Braincase structure in the carnosaur *Itemirus* n. gen. and some aspects of the cranial anatomy of dinosaurs. *Paleontological Journal* **10**: 361–369.

—1981. [On some unusual theropods from the Upper Cretaceous in Mongolia.] *Trudy Sovmestnoi Sovetsko-Mongol'skoi Paleontologicheskoi Ekspeditsii* **15**: 39–50.

—1987. [Avimimidae and the problem of the origin of birds.] *Trudy Sovmestnoi Sovetsko-Mongol'skoi Paleontologicheskoi Ekspeditsii* **31**: 1–92.

—and Osmólska, H. 1991. *Tochisaurus nemegtensis* gen. et sp. n., a new troodontid (Dinosauria, Theropoda) from Mongolia. *Palaeontologia Polonica* **36**: 69–76.

Mader, B.J. and Bradley, R.L. 1989. A redescription and revised diagnosis of the syntypes of the Mongolian tyrannosaur *Alectrosaurus olseni*. *Journal of Vertebrate Paleontology* **9**: 41–55.

Maleev, E.A. 1954. [New turtle-like reptile in Mongolia.] *Priroda* **1954**: 106–108.

—1955a. [Gigantic carnivorous dinosaurs of Mongolia.] *Doklady AN SSSR* **104**: 634–637.

—1955b. [New carnivorous dinosaurs from the Upper Cretaceous of Mongolia.] *Doklady AN SSSR* **104**: 779–782.

—1974. [Gigantic carnosaurs of the family Tyrannosauridae.] *Trudy Sovmestnoi Sovetsko-Mongol'skoi Paleontologicheskoi Ekspeditsii* **1**: 132–191.

Nesov, L.A. 1995. [*Dinosaurs of northern Eurasia: new data on assemblages, ecology and palaeobiogeography.*] St. Petersburg, Izdatel'stvo Sankt-Peterburgskogo Univesiteta, 156 pp.

Norell, M.A. and Clark, J.M. 1992. New dromaeosaur material from the Late Cretaceous of Mongolia. *Journal of Vertebrate Paleontology* **12**: 45A.

—, Clark, J.M., Dashzeveg, D., Barsbold, R., Chiappe, L.M., Davidson, A.R., McKenna, M.C., Perle, A. and Novacek, M.J. 1994. A theropod dinosaur embryo and the affinities of the Flaming Cliffs dinosaur eggs. *Science* **256**: 779–782.

—, Clark, J.M., Chiappe, L.M. and Dashzeveg, D. 1995. A nesting dinosaur. *Nature* **378**: 774–776.

Novacek, M.J., Norell, M.A., McKenna, M.C. and Clark, J.M. 1994. Fossils of the Flaming Cliffs. *Scientific American* **271** (6): 60–69.

Olshevsky, G. 1995a. [The origin and evolution of the tyrannosaurids, Part 1.] *Kyoryugaku Saizensen* (*Dino Frontline*) **9**: 92–119.

—1995b. The origin and evolution of the tyrannosaurids, Part 2. *Kyoryugaku Saizensen* (*Dino Frontline*) **10**: 75–99.

Osborn, H.F. 1924. Three new theropods, *Protoceratops* zone, central Mongolia. *American Museum Novitates* **144**: 1–12.

Osmólska, H. 1980. The Late Cretaceous vertebrate assemblages of the Gobi Desert, Mongolia. *Mémoires de la Société Géologique de France* **139**: 145–150.

—1981. Coossified tarsometatarsi in theropod dinosaurs and their bearing on the problem of bird origins. *Palaeontologia Polonica* **42**: 79–95.

—1982. *Hulsanpes perlei* n.g. n.sp. (Deinonychosauria, Saurischia, Dinosauria) from the Upper Cretaceous Barun Goyot Formation of Mongolia. *Neues Jahrbuch für Geologie und Paläontologie, Monatshefte*, **1982**: 440–448.

—1987. *Borogovia gracilicrus* gen. et sp. n., a new troodontid dinosaur from the Late Cretaceous of Mongolia. *Acta Palaeontologica Polonica* **32**: 133–150.

—1996. An unusual theropod dinosaur from the Late Cretaceous Nemegt Formation of Mongolia. *Acta Palaeontologica Polonica* **41**: 1–38.

—and Roniewicz, E. 1970. Deinocheiridae, a new family of theropod dinosaurs. *Palaeontologia Polonica* **21**: 5–19.

—, Roniewicz, E. and Barsbold, R. 1972. A new dinosaur,

Gallimimus bullatus, n. gen. n. sp. (Ornithomimidae) from the Upper Cretaceous of Mongolia. *Palaeontologia Polonica* 27: 103–143.

Paul, G.S. 1984. The segnosaurian dinosaurs: relics of the prosauropod-ornithischian transition? *Journal of Vertebrate Paleontology* 4: 507–515.

—1988a. The small predatory dinosaurs of the mid-Mesozoic: the horned theropods of the Morrison and Great Oolite – *Ornitholestes* and *Proceratosaurus* – and the sickle-claw theropods of the Cloverly, Djadokhta and Judith River – *Deinonychus*, *Velociraptor* and *Saurornitholestes*. *Hunteria* 2 (4): 1–9.

—1988b. *Predatory dinosaurs of the world*. New York: Simon and Schuster.

Pérez-Moreno, B.P., Sanz, J.L., Buscalioni, A.D., Moratalla, J.J., Ortega, F. and Raskinn-Gutman, D. 1994. A unique multitoothed ornithomimosaur from the Lower Cretaceous of Spain. *Nature* 370: 363–367.

Perle, A. 1977. [On the first discovery of *Alectrosaurus* (Tyrannosauridae, Theropoda) from the Late Cretaceous of Mongolia.] *Problemy Mongol'skoi Geologii* 3: 104–113.

—1979. [Segnosauridae – a new family of theropods from the Late Cretaceous of Mongolia.] *Trudy Sovmestnoi Sovetsko-Mongol'skoi Paleontologicheskoi Ekspeditsii* 8: 45–55.

—1981. [A new segnosaurid from the Upper Cretaceous of Mongolia.] *Trudy Sovmestnoi Sovetsko-Mongol'skoi Paleontologicheskoi Ekspeditsii* 15: 50–59.

—1982. [On a new finding of the hind limb of *Therizinosaurus* sp. from the Late Cretaceous of Mongolia.] *Problemy Mongol'skoi Geologii* 5: 94–98.

—Norell, M. and Clark, J. 1999. *A new maniraptoran theropod – Achillobator giganticus (Dromaeosauridae) – from the Upper Cretaceous of Burkhant, Mongolia*. Ulaanbaatar: National University of Mongolia, Geology Department.

Rauhut, O.W.M. and Werner, C. 1995. First record of the family Dromaeosauridae (Dinosauria: Theropoda) in the Cretaceous of Gondwana (Wadi Milk Formation, northern Sudan). *Paläontologische Zeitschrift.* 69: 475–489.

Rich, T.H. and Vickers-Rich, P. 1994. Neoceratopsians and ornithomimosaurs: dinosaurs of Gondwana origin? *Research and Exploration* 10: 129–131.

Rozhdestvenskii, A.K. 1965. [Growth changes in Asian dinosaurs and some problems of their taxonomy.] *Paleontologicheskii Zhurnal* 1965: 95–109.

—1970. [Giant claws of enigmatic Mesozoic reptiles.] *Paleontologicheskii Zhurnal* 1970: 117–125.

Russell, D.A. 1972. Ostrich dinosaurs from the Late Cretaceous of western Canada. *Canadian Journal of Earth Sciences* 9: 375–402.

—and Dong Z. M. 1993a. A nearly complete skeleton of a troodontid dinosaur from the Early Cretaceous of the Ordos Basin, Inner Mongolia, China. *Canadian Journal of Earth Sciences* 30: 2163–2173.

—and— 1993b. The affinities of a new theropod from the Alxa Desert, Inner Mongolia, China. *Canadian Journal of Earth Sciences* 30: 2107–2127.

Ryabinin, A.N. 1931. [Two dinosaurian vertebrae from the Lower Cretaceous of Transcaspian Steppes.] *Zapiski Russkogo Mineralogicheskogo Obshchestva, Series 2*, 60: 110–113.

Sues, H.-D. 1994. New evidence concerning the phylogenetic position of *Chirostenotes* (Dinosauria: Theropoda). *Journal of Vertebrate Paleontology* 14: 48A.

Unwin, D.M., Perle, A. and Trueman, C. 1995. *Protoceratops* and *Velociraptor* preserved in association: evidence for predatory behaviour in dromaeosaurid dinosaurs? *Journal of Vertebrate Paleontology* 15: 57A–58A.

Zhao X.J. and Currie, P.J. 1993. A large crested theropod from the Jurassic of Xinjiang, People's Republic of China. *Canadian Journal of Earth Sciences* 30: 2027–2036.

23

Sauropods from Mongolia and the former Soviet Union

TERESA MARYAŃSKA

Introduction

Sauropods, a spectacular group of gigantic saurischians, are known from the Early Jurassic to the end of the Cretaceous from all continents, except Antarctica. An enormous number of genera (about 90) and species (over 150) of sauropods have been named and described, but most of them are based on imperfect, fragmentary material. Complete skulls and skeletons are rare, and this makes comparisons between different sauropod species difficult or impossible. For the same reason, the number of sauropod families varies over time.

In contrast to the long history of discovery and study of sauropods in Europe and North America which started in the nineteenth century, the first documented, undoubted sauropod material from Mongolia was found in 1922 by members of the Central Asiatic Expedition, organized by the American Museum of Natural History in New York. The first Mongolian sauropod genus and species, *Asiatosaurus mongoliensis*, was described by Osborn (1924). From that time, sauropod remains, primarily represented by isolated bones, have been discovered in many sites on the territory of Mongolia (cf. Kalandadze and Kurzanov, 1974; Gradziński *et al.*, 1977; Weishampel, 1990). The most important sauropod material from Mongolia was discovered by the Polish–Mongolian Paleontological Expedition in 1965 and by the Soviet–Mongolian Expedition in 1971 (see below). The first information on sauropod bones from the territory of the former Soviet Union was published by A.N. Ryabinin in the 1930s, and the first sauropod species from Kazakhstan, *Antarctosaurus jaxarticus*, was named by him in 1938.

Subsequently, many Cretaceous localities on the territory of Kazakhstan, Uzbekistan, Kirgizstan, Tadzhikistan (cf. Rozhdestvenskii and Khozatskii, 1967; Rozhdestvenskii, 1970, 1977) and Russia (Dmitriev and Rozhdestvenskii, 1968) have yielded sauropod remains, but most of them are represented by isolated bones. An exception is an undescribed, fragmentarily preserved postcranial skeleton of a Jurassic sauropod found by Russian palaeontologists in 1966 near Tashkumyr town in Kirgizstan (Rozhdestvenskii, 1969) and housed in the Paleontological Institute in Moscow. The most complete and comprehensive information on dinosaur remains from the territory of the former Soviet Union, with a list of localities, and some data on lithology and the age of local svitas, was recently published by Nesov (1995). According to him, isolated bones of Jurassic sauropods were discovered in five localities in Kirgizstan and one in Uzbekistan, bones of Early Cretaceous sauropods were recognized in four localities in Russia and Late Cretaceous sauropod remains are known from six localities on the territory of Uzbekistan, six in Kazakhstan, three in Kirgizstan, two in Tadzhikistan and one in Russia. In general, sauropods from Mongolia and the territory of the former Soviet Union are relatively poorly known, most of them are represented by incomplete and nondiagnostic bones, and there are no mass accumulations.

Among the five Cretaceous forms listed below, only three monospecific genera are represented by relatively well preserved and diagnostic material. They are: *Opisthocoelicaudia* Borsuk-Białynicka, 1977; *Nemegtosaurus* Nowiński, 1971, and *Quaesitosaurus* Bannikov and Kurzanov, 1983.

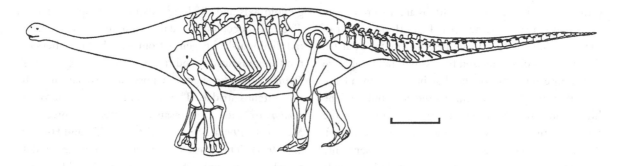

Figure 23.1. Reconstruction of the skeleton of *Opisthocoelicaudia skarzynskii*. Redrawn from Borsuk-Białynicka (1977). Scale bar = 1 m.

Institutional abbreviations are as follows: AMNH, American Museum of Natural History, New York; PIN, Paleontological Institute, Russian Academy of Sciences, Moscow; ZPAL, Institute of Paleobiology, Polish Academy of Sciences, Warsaw.

Systematic survey

The phylogenetic relationships of sauropods are not yet stabilized or generally accepted. For this reason, in this chapter, the conservative and basically familiar taxonomy of sauropods proposed by McIntosh (1990b) is utilized, and the assignment of *Opisthocoelicaudia* to the family Camarasauridae, and *Nemegtosaurus* and *Quaesitosaurus* to the diplodocid subfamily Dicraeosaurinae, as proposed by McIntosh (1990b), are also accepted.

Sauropodomorpha Huene, 1932
Sauropoda Marsh, 1887
Family Camarasauridae Cope, 1887
Subfamily Opisthocoelicaudiinae McIntosh, 1990b.
Genus *Opisthocoelicaudia* Borsuk-Białynicka, 1977
Type and only known species. Opisthocoelicaudia skarzynskii Borsuk-Białynicka, 1977
Holotype. ZPAL MgD-I/48, an almost complete postcranial skeleton (lacking skull and cervicals) discovered in 1965 by the Polish–Mongolian Paleontological Expedition to the Gobi Desert (Figure 23.1). The specimen was found in the Upper Cretaceous

(Campanian–Maastrichtian) Nemegt Formation of the locality of Altan Uul IV. (Gradziński *et al.*, 1969; Gradziński, 1970), in the Nemegt Basin, Gobi Desert, Mongolia. The holotype is presently housed in the Institute of Geology, Mongolian Academy of Sciences, Ulaanbaatar.

Referred material. To *O. skarzynskii* is also assigned a scapulocoracoid of a young individual (ZPAL MgD-I/25c).

Description. The most outstanding anatomical features of this sauropod are the relatively long vertebral centra of the dorsal and caudal vertebrae, the opisthocoelous caudal vertebrae and short tail. The dorsal vertebrae are characterized by distinctly opisthocoelous centra, comparatively low neural arches, and heavy transverse processes that are directed outwards. The neural spines do not form single elements, but along the dorsal region of the vertebral column they are divided into two stout elements, separated by a U-shaped cleft. The centra of the sacrals are coossified and bear low neural spines. The transverse process of the last sacral (caudosacral) vertebra is fused with the ilium and ischium. The centra of the perfectly preserved caudals are strongly opisthocoelous in the cranial half of the tail and none of the caudal centra have pleurocoels. Caudals 16 to 27 are amphiplatyan, and the more distal ones are biconvex. The caudal centra from 20 to 27 were probably coossified. The caudal neural arches are robust. The uniramous chevrons are not present beyond caudal 19. The pectoral girdle is massive and the forelimbs are also relatively stout and massive. Metacarpals I and II are almost of the same length and are longer than metacarpal III. The pelvis is characterized by a strong lateral flexure of the anterior wings of the iliac blade and by a large contribution of the ischium to the acetabulum. The hind limbs are robust. The fourth trochanter on the femur is situated below the middle length of the shaft. The stout, reduced pedal phalanges are present only in four digits and the phalangeal formula is 2:2:2:1:0. Such reduction of the pedal phalanges is not observed in other sauropods. The total height of the forelimb of *O. skarzynskii* is 1.87 m. The height of the hind limb is 2.46 m, and the humerus:femur ratio = 0.72.

Comments. The main anatomical features of *Opisthocoelicaudia* are shown in the reconstruction (Figure 23.1). The neck of this sauropod was probably rather short, as is documented by a reconstruction of the nuchal ligament presented by Borsuk-Białynicka (1977). The tail was held in a horizontal position during terrestrial locomotion, as is evidenced by the structure of the caudals: there are no wedge-shaped or downwardly flexed centra in the tail vertebrae, and the haemapophyses are firmly fused with the centra in caudals 6–17. The presence of opisthocoelous centra in the cranial half of the tail and some characters of the pelvis, such as the strongly deepened iliac section of the acetabulum, the outward bend of the ilia, and the fused pubic symphysis suggest, according to Borsuk-Białynicka (1977), that the tail in *Opisthocoelicaudia* could have served as a prop during occasional bipedal posture. If this posture was really possible for *Opisthocoelicaudia*, it may have been connected with some important part of its life activity, most probably with feeding.

The relationships of this genus are controversial. In the original description of *Opisthocoelicaudia*, the assignment of this genus to the Camarasauridae was discussed in detail. Camarasaurid features of this genus listed by Borsuk-Białynicka (1977) include: very low and divided spines in the dorsal vertebrae, low spines in the sacral region, simple transverse processes, uniramous chevrons, straight dorsal section of the vertebral column, and the small angle between the

glenoid axis and the long axis of the scapular blade. The opisthocoelous caudal vertebrae and very short tail distinguish this genus from all other members of the Camarasauridae and the structure of the limbs of *Opisthocoelicaudia* suggest a possible association with the Titanosauridae. The assignment of *Opisthocoelicaudia* to the Camarasauridae was accepted by McIntosh (1990a, b), Madsen *et al.* (1995), and Hunt *et al.* (1994). The presence of opisthocoelous caudal vertebrae was the basis for the proposal by McIntosh (1990b), of a new camarasaurid subfamily – Opisthocoelicaudiinae for this Mongolian genus. Subsequently, however, Upchurch (1994) removed this genus from the Camarasauridae and placed it as the sister taxon to the family Titanosauridae within the 'Titanosauroidea'.

Family Diplodocidae Marsh, 1884
Subfamily Dicraeosaurinae Janensch, 1944
Genus *Nemegtosaurus* Nowiński, 1971

Type and only known species. Nemegtosaurus mongoliensis Nowiński, 1971

Holotype. ZPAL MgD-I/9, an almost complete skull associated with the lower jaw, discovered in 1965 by the Polish–Mongolian Paleontological Expedition in the Upper Cretaceous (Campanian–Maastrichtian) Nemegt Formation at the Nemegt locality, Gobi Desert, Mongolia (Figure 23.2).

Referred material. A second, undescribed skull, probably of the same genus, is housed in the Geological Institute of the Mongolian Academy of Sciences, Ulaanbaatar.

Description. In its general characters the skull of *N. mongoliensis* exhibits a diplodocid type of structure, with an elongate snout and rostroventrally directed quadrate; chisel-like teeth are present only in the rostral portion of the jaws. The dental formula is 4:8/13. The external nares are situated far behind and most probably posterior to the level of the caudal end of the antorbital fenestra. The exact shape and size of the preorbital foramen is not known, but judging from preserved parts of the margin of this foramen, it was large and elongated. The infratemporal fossa seems to be narrower than in other diplodocids. The elongation

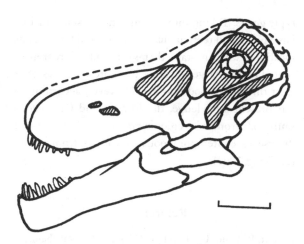

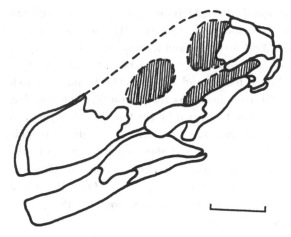

Figure 23.2. Reconstruction of the skull and mandible of *Nemegtosaurus mongoliensis.* Redrawn from Nowiński (1971). Scale bar = 100 mm.

Figure 23.3. Reconstruction of the skull and mandible of *Quaesitosaurus orientalis.* Redrawn from Kurzanov and Bannikov (1983). Scale bar = 100 mm.

of the basipterygoid processes, characteristic for diplodocids, is present in *Nemegtosaurus*, but the basipterygoid processes are stouter, shorter and not so rostrally directed as in some other diplodocids such as *Apatosaurus* and *Diplodocus* (Berman and McIntosh, 1978). The accessory fenestra in front of the antorbital fenestra, characteristic for some diplodocids, is not present in *Nemegtosaurus*. Another characteristic feature is a comparatively robust and long lacrimal that is strongly broadened in its dorsal portion. The lower jaw of *Nemegtosaurus* differs from that of other diplodocids in the length of the dentary, which in the Mongolian genus is evidently longer than the surangular–angular portion of the mandible. The mandibular symphysis in *Nemegtosaurus* is very weak.

Genus *Quaesitosaurus* Bannikov and Kurzanov, 1983
(in Kurzanov and Bannikov, 1983)
Type and only known species. Quaesitosaurus orientalis Bannikov and Kurzanov, 1983.
Holotype. PIN no. 3906/2, a partly damaged skull, with well preserved snout, occipital region and mandibles, was found by the Soviet–Mongolian Expedition in 1971 (Tsybin and Kurzanov, 1979) in Upper Cretaceous sediments of the Baruungoyot Svita at the

locality of Shar Tsav in south-eastern Gobi, Mongolia (Figure 23.3).
Description. In general the shape of the skull of *Q. orientalis* (Figure 23.3) is similar to that of *N. mongoliensis*, but differs in some structures. The snout of *Quaesitosaurus* seems to be broader than in *Nemegtosaurus*, and the squamosal is shorter and does not contact the quadratojugal. The basipterygoidal processes are stouter, and the parietal foramen is absent. The outstanding character of *Quaesitosaurus*, according to Kurzanov and Bannikov (1983), is the presence, in the basioccipital and basisphenoid, of a canal, leading from the hypophysis to the region beneath the occipital condyle. According to McIntosh (1990b), such a canal is absent not only in diplodocids, but also in other sauropods. Another feature characteristic for *Quaesitosaurus* is a comparatively large concavity on the caudal face of the quadrate, named by Kurzanov and Bannikov (1983) as the 'resonance' cavity. The chisel-like teeth, present only in rostral portion of the jaws, and the mandible are similar to those of *Nemegtosaurus*, but the maxillary tooth row is longer. The dental formula is 4:9/13.
Discussion. The assignment of *Nemegtosaurus* and *Quaesitosaurus* to the Dicraeosaurinae (Berman and

McIntosh, 1978; McIntosh, 1990a, b) is not accepted by some specialists. Calvo (1994), using characters of the tooth structure and feeding mechanism, referred *Nemegtosaurus* and *Quaesitosaurus* to the titanosaurids, whereas Barrett and Upchurch (1994) and Upchurch (1994) included both genera in the Nemegtosauridae, a new family erected by Upchurch for *Nemegtosaurus* and *Quaesitosaurus*. The Nemegtosauridae is considered, by Upchurch (1994), as a sister group to the dicraeosaurid–diplodocid clade within the 'Diplodocoidea'. Recently, however, Hunt *et al.* (1994) assigned *Nemegtosaurus* and *Quaesitosaurus* to the family Dicraeosauridae.

Other sauropods

The first sauropod from Mongolia, *Asiatosaurus mongoliensis*, was described by Osborn (1924) on the basis of just two teeth (AMNH 6264, the holotype and AMNH 6294, a paratype) from the Early Cretaceous Öösh Formation (= Hühteeg Svita of Aptian/Albian age according to Shuvalov (1973) and Weishampel (1990), or Öndörukhaa Svita of Valanginian to Late Neocomian age according to Shuvalov (1975), but see also Jerzykiewicz and Russell (1991)), at the locality of Ööshiin Nuur in the northern Gobi, Mongolia. Following McIntosh (1990b) this species is considered here as a *nomen dubium*.

The first sauropod named from the former Soviet Union, '*Antarctosaurus*' *jaxarticus* is a *nomen nudum*. This species was named, but not diagnosed or described, by Ryabinin (1939) on the basis of an isolated femur from an unknown locality in the Kyzylkum Desert, south Kazakhstan. According to Rozhdestvenskii (1969, 1971) this specimen comes from the Turonian–Santonian Dabrazinskaya Svita. This species is treated by McIntosh (1990a, b) as an unnamed titanosaurid.

Discussion

As mentioned above, the Jurassic and Cretaceous sauropods from Mongolia and the former Soviet Union are relatively rare, poorly known, and practically represented by only three reasonably good specimens, each assigned to a separate Late Cretaceous mono-

typic genus. Taking into account the presence of isolated bones and teeth in many Jurassic and Cretaceous localities, we can state that sauropods were represented on this large territory, by representatives of at least three sauropod families. However, according to our present knowledge, the Jurassic and Cretaceous sauropods from this region were not so abundant and diverse as contemporaneous sauropods from China (cf. Dong, 1992).

References

Barrett, P.M. and Upchurch, P. 1994. Feeding mechanisms of *Diplodocus*. *Gaia* **10**: 95–203.

Berman, D.S. and McIntosh, J.S. 1978. Skull and relationships of the Upper Jurassic sauropod *Apatosaurus* (Reptilia, Saurischia). *Bulletin of Carnegie Museum of Natural History* **8**: 5–35.

Borsuk-Białynicka, M. 1977. A new camarasaurid sauropod *Opisthocoeolicaudia skarzynskii* gen. n., sp. n. from the Upper Cretaceous of Mongolia. *Palaeontologia Polonica* **3**: 5–64.

Calvo, J.O. 1994. Jaw mechanics in sauropod dinosaurs. *Gaia* **10**: 183–193.

Dmitriev, G.A. and Rozhdestvenskii, A.K. 1968. [The bonebearing facies of the Upper Mesozoic lacustrinefluvial deposits], pp. 39–48 in *Mezozoiskie i Kainozoiskie ozera Sibiri*. Moscow: Izdatel'stvo Nauka.

Dong, Z. 1992. *Dinosaurian faunas of China*. Beijing: China Ocean Press and Berlin: Springer Verlag.

Gradziński, R. 1970. Sedimentation of dinosaur-bearing Upper Cretaceous deposits of the Nemegt Basin, Gobi Desert. *Palaeontologia Polonica* **21**: 147–229.

—, Kaümierczak, J. and Lefeld, J. 1969. Geographical and geological data from the Polish-Mongolian palaeontological expeditions. *Palaeontologia Polonica* **19**: 33–82.

—, Kielan-Jaworowska, Z., and Maryańska, T. 1977. Upper Cretaceous Djadokhta, Barun Goyot and Nemegt formations of Mongolia, including remarks on previous subdivision. *Acta Geologica Polonica* **2**: 281–318.

Hunt, A.P., Lockley, M.G., Lucas, S.G. and Meyer C.A. 1994. The global sauropod fossil record. *Gaia* **10**: 261–279.

Jerzykiewicz, T. and Russell, D.A. 1991. Late Mesozoic stratigraphy and vertebrates of the Gobi Basin. *Cretaceous Research* **12**: 345–377.

Kalandadze, N.N. and Kurzanov, S.M. 1974. [The Lower Cretaceous localities of terrestial vertebrates in Mongolia.] *Trudy Sovmestnoi Sovetsko-Mongol'skoi Paleontologicheskoi Ekspeditsii* 1: 288–295.

Kurzanov, S.M. and Bannikov, A.F. 1983. [A new sauropod from the Upper Cretaceous of Mongolia.] *Paleontologicheskii Zhurnal* 2: 90–96.

Madsen, J.H., McIntosh, J.S. and Berman, D.S. 1995. Skull and atlas-axis complex of the Upper Jurassic sauropod *Camarasaurus* Cope (Reptilia: Saurischia). *Bulletin of Carnegie Museum of Natural History* 31: 1–115.

McIntosh, J.S. 1990a. Species determination in sauropod dinosaurs with tentative suggestion for their classification, pp. 53–69 in Carpenter, K. and Currie, P.J. (eds.), *Dinosaur systematics*. Cambridge: Cambridge University Press.

—1990b. Sauropoda, pp. 345–401 in Weishampel, D.B., Dodson, P. and Osmólska, H. (eds.), *The Dinosauria*. Berkeley: California University Press.

Nesov, L.A. 1995. [*Dinosaurs of Northern Eurasia: new data on assemblages, ecology and palaeobiogeography.*] St Petersburg: Izdatel'stvo Sankt-Peterburgskogo Universiteta, 156 pp.

Nowiński, A. 1971. *Nemegtosaurus mongoliensis* n. gen., n. sp. (Sauropoda) from the Upper Cretaceous of Mongolia. *Palaeontologia Polonica* 25: 57–81.

Osborn, H.F. 1924. Sauropoda and Theropoda of the Lower Cretaceous of Mongolia. *American Museum Novitates* 128: 1–7.

Rozhdestvenskii, A.K. 1969. [*In search of dinosaurs in the Gobi.*] Moscow: Izdatel'stvo Nauka.

—1970. [Mesozoic and Cenozoic terrestrial vertebrate complexes of Central Asia and adjoining regions of Kazakhstan and their stratigraphic position], pp. 50–58 in *Biostratigraficheskie i paleofatsialnye issledova-niya i ikh prakticheskoe znachenie.* Leningrad: Izdatel'stvo Nedra.

—1971. [Investigations of the Mongolian dinosaurs and their role in the subdivision of the terrestrial Mesozoic.] *Trudy Sovmestnoi Sovetsko-Mongol'skoi Paleontologicheskoi Ekspeditsii* 1: 288–295.

—1977. The study of dinosaurs in Asia. *Journal of the Palaeontological Society of India* 20: 102–119.

—and Khozatskii, L.I. 1967. [Late Mesozoic land vertebrates from the Asiatic part of the U.S.S.R.], pp. 80–92 in Martinson, G.G. (ed.): *Stratigrafiya i paleontologiya mezozoiskikh i paleogen-neogenovykh kontinental'nykh otlozhenii aziatskoi chasti SSSR.* Leningrad: Izdatel'stvo Nauka.

Ryabinin, A.N. 1939. [Some results of the studies of the Upper Cretaceous dinosaurian fauna from the vicinity of the Station Sary-Agach, south Kazakhstan.] *Problemy Paleontologii* 4: 125–135.

Shuvalov, V.F. 1973. [Main stages of the development of the Mesozoic structures in central Mongolia]. *Trudy Instituta Geologii i Geofiziki Sibirskogo Otdeleniya AN SSSR* 173: 175–184.

—1975. [Mesozoic stratigraphy of central Mongolia.]. *Trudy Sovmestnoi Sovetsko-Mongol'skoi Paleontologicheskoi Ekspeditsii* 13: 50–112.

Tsybin, Y.I. and Kurzanov, S.M. 1979. [New data on Upper Cretaceous localities of vertebrates of Baishin Tsav region.] *Trudy Sovmestnoi Sovetsko-Mongol'skoi Paleontologicheskoi Ekspeditsii* 8: 08–111.

Upchurch, P. 1994. Sauropod phylogeny and palaeoecology. *Gaia* 10: 249–260.

Weishampel, D.B., 1990. Dinosaurian distribution, pp. 63–139 in Weishampel, D.B., Dodson, P. and Osmólska, H. (eds.), *The Dinosauria*. Berkeley: University of California Press.

24

Ornithopods from Kazakhstan, Mongolia and Siberia

DAVID B. NORMAN AND HANS-DIETER SUES

Introduction

The fossil record of ornithopods from Kazakhstan, Mongolia and Siberia is restricted to iguanodontians and the more derived hadrosaurids of the Cretaceous Period. This is in part an artefact of the fossil record, and also reflects some degree of isolation of the Asian landmass during mid-Mesozoic times.

Iguanodontians are known from the Early Cretaceous (Aptian/Albian) of Mongolia, and are anatomically similar to contemporary Northern Hemisphere forms (notably *Iguanodon*). More derived hadrosaurids appear in the Late Cretaceous of Asia (a fact which has already been documented in China through forms such as *Probactrosaurus*, *Gilmoreosaurus* and *Bactrosaurus*). Known hadrosaurids from Mongolia and Kazakhstan include forms representative of both the lambeosaurine and hadrosaurine radiations which have been documented in detail in North America. Some of the latest Late Cretaceous (Maastrichtian) forms from Asia: *Saurolophus angustirostris* Rozhdestvenskii, 1957, and *Procheneosaurus convincens* Rozhdestvenskii, 1968, are anatomically very similar to the North American hadrosaurids *S. osborni* and *Corythosaurus casuarius* respectively. Many of the Asian hadrosaurs so far described are based on very fragmentary or poorly preserved material and their status needs to be reassessed. In recent years new material has been discovered of ornithopods that appear to lie close to the base of the hadrosaurid clade.

Ornithopod dinosaurs are represented by a variety of medium-sized (5 m long) to large (10–13 m long) forms referable to the Iguanodontia from Cretaceous continental deposits of Mongolia and the territories of the former Soviet Union (Figure 24.1). To date, no skeletal remains of other well known groups of ornithopod dinosaurs, such as Heterodontosauridae and Hypsilophodontidae, have been recovered from these particular regions. This lack of representation is at least partially explained by the generally incomplete record of discoveries from this area of the World; for example basal hypsilophodontids are known to be present in the Jurassic of China (He and Cai, 1984) as well as Europe and North America (Galton, 1977) and in the Cretaceous of North America and Europe (Sues and Norman, 1990; Weishampel *et al.*, 1991). It should, however, also be noted that Jerzykiewicz and Russell (1991) regard the Early Cretaceous fauna of Mongolia (and by implication adjacent areas of China and the former Soviet Union) as being generally depauperate until Aptian/Albian times when there is supposed to have been a major phase of immigration of taxa derived from North America and Europe.

Rozhdestvenskii (1968, 1973, 1977) and Nesov (1995) have reviewed the history of discoveries of, and research on, dinosaurs from Mongolia and the territories of the former Soviet Union, and this account updates these earlier, and to some extent idiosyncratic, accounts.

Considerable confusion now exists concerning the systematics and phylogeny of the more derived ornithopods (Norman, 1998b, in prep., a). For the purposes of this review comments are confined to representatives of the euornithopod iguanodontian families Iguanodontidae (predominantly Early Cretaceous genera including: *Iguanodon, Ouranosaurus,* and *Lurdusaurus* from Niger (Taquet and Russell, 1999), *Altirhinus* (Norman, 1998a), *Probactrosaurus* (Norman,

Figure 24.1. Geographic location of sites from which ornithopods have been recovered in the Former Soviet Union and Mongolian People's Republic. 1, Shakh-Shakh; 2, Dzharakhuduk; 3, Alim Tau; 4, Syuk-Syuk; 5, Akkurgan; 6, Kyrk Khuduk; 7, Altan Uul II; 8, Northern Sair; 9, Nemegt N; 10, Baishin Tsav; 11, Hüren Dukh; 12, Khamaryn Khural; 13, Blagoveshchensk/Belye Kruchi; 14, Sinegorsk.

1998b, in prep., a) and possibly *Muttaburrasaurus* Bartholomai and Molnar, 1981), and the Late Cretaceous family Hadrosauridae (a diverse assemblage of highly derived and distinctive ornithopods characterized by a suite of cranial, dental and postcranial features).

Systematic survey

Suborder ORNITHOPODA Marsh, 1881
Infraorder IGUANODONTIA Dollo, 1888
Diagnosis. Expanded oral margin on the premaxilla; enlargement of external naris through the lateral flaring of the oral margin of the premaxillary beak, the rostral extension of the premaxilla and retreat caudally of the posterior narial margin; reduction of antorbital fenestra to form a small sub-circular opening bordered by the lacrimal and maxilla and an oblique channel behind (no antorbital fossa); the development of a denticulate margin on the predentary; dentary teeth have broad sub-rectangular lingual surfaces subdivided by two parallel vertical ridges;

marginal denticles of maxillary and dentary teeth form a mammillated ledge running from the labial to lingual side of the crown; metacarpals II–IV elongated and closely appressed; penultimate phalanges of digits II–IV of manus short and broad and capable of being hyperextended; ungual phalanx of digit II of manus elongate, twisted axially and flattened and larger than the ungual of digit III; ungual of digit three of manus is short, broad and hoof-shaped.

Family Iguanodontidae Cope, 1869
Genus *Iguanodon* Mantell, 1825
Type species. Iguanodon bernissartensis Boulenger, 1881.
Iguanodon bernissartensis Boulenger, 1881
Holotype. IRSNB 1535.
Diagnosis. Iguanodontid dinosaur of the Barremian/ Early Aptian, reaching a maximum body length of 11 m. Two free supraorbitals roof the upper margin of the orbit. Vertical tooth rows in maxilla reaching a maximum of 29 and in the dentary 25. No more than two teeth in each vertical tooth row. Oval foramen in the zygomatic arch located between the quadratojugal

463

and quadrate. Nasal bones neither thickened, nor strongly flexed dorsally. Sacrum composed of 8 fused vertebrae. Scapular blade little expanded distally. Intersternal ossification present. Forelimb long (70% of length of hindlimb) and stoutly constructed. Carpals co-ossified and bound by ossified ligaments. Phalangeal count of manus: 2,3,3,2,4. First phalanx of digit 1 of manus is a flat disk, the ungual is a relatively enormous conical spine. Manus very large (in proportion to the length of the forelimb), and stoutly constructed. Ilium has a broad brevis shelf caudally, and the caudal end tapers to a rounded point, the main body of the ilium is deep and laterally flattened, with a straight and slightly everted dorsal margin which is thickened above the ischial tuber; the rostral process of the ilium is triangular in cross-section and slightly downturned with a distally expanded tip. The rostral ramus of the pubis is laterally compressed, parallel-sided and expanded distally. Three distal tarsals are present. Metatarsal 1 is reduced to an obliquely oriented and transversely flattened splint lying against the medial surface of metatarsal 2.

Genus *Iguanodon* Mantell, 1825

Type species. I. bernissartensis Boulenger, 1881.
    I. orientalis Rozhdestvenskii, 1952 (junior subjective synonym)
Holotype. PIN 559/1.
Diagnosis. Specifically non-diagnostic.
Comments. Rozhdestvenskii (1952) described a maxilla and scapula (Figure 24.2A, B) from Early Cretaceous strata of Khamaryn Khural in the eastern Gobi Desert of Mongolia (Figure 24.1; location 12) and referred them to the European ornithopod taxon *Iguanodon* as a new species, *I. orientalis*. A recent review of the holotype material by Norman (1996) has demonstrated that it is non-diagnostic, and that that which is preserved is indistinguishable from the *Iguanodon bernissartensis*. *I. bernissartensis* is known from western Europe (England, France, Spain, Germany; see Norman, 1987; Norman and Weishampel, 1990; Norman, 1996). This new observation, if it proves to be correct, suggests a significant Northern Hemisphere range extension of this (Barremian/

Early Aptian) species, and would appear to accord with a model of dispersal from Europe to Asia during late Early Cretaceous times proposed by Jerzykiewicz and Russell (1991). This anatomical observation also provides a tentative biostratigraphic dating for the Khamaryn Khural locality.

Genus *Altirhinus* Norman, 1998a

Type species. A. kurzanovi Norman, 1998a.
Holotype. PIN 3386/8.
Diagnosis. Iguanodontid dinosaur of (conjecturally) Late Aptian/Early Albian age. Body size range up to 8 m and showing anatomical similarities to *I. bernissartensis*. Strongly dorsally flexed nasal bones with an elongate rostral process; internasal groove; ventrally deflected premaxillary beak; occluded antorbital fenestra; downturned rostral end of the dentary; deeper portions of the maxilla and dentary support three teeth in the vertical tooth row. Large, conical, but laterally compressed ungual phalanx of digit 1 of manus; carpus not co-ossified; digit IV of manus with a short broad ungual; ilium with a robust, downturned rostral process, a sinuous dorsal margin and strongly everted trochanteric ridge above the ischial tuber, brevis shelf absent; shaft of ischium straight and parallel-sided, but showing an axial twist.
Comments. In recent years some very distinctive skull (Figure 24.3) and associated parts of the postcranial skeleton of another ornithopod dinosaur have been recovered from the locality of Hüren Dukh (Figure 24.1; location 11) by joint Russian/Mongolian collaborative expeditions to Mongolia. This locality is of late Early Cretaceous age (Late Hühteegian 'age'; Jerzykiewicz and Russell, 1991). Biostratigraphic correlation, based purely on the derived nature of this ornithopod material when compared to that known from Europe, and the reassignment of the material formerly attributed to *I. orientalis*, is suggestive of a Late Aptian/Albian age.

The specimens collected to date have been referred to the now undiagnostic taxon *Iguanodon orientalis* (cf. Norman, 1985) in various exhibition catalogues and, most recently, in a review of non-hadrosaurian iguanodontians by Norman and Weishampel (1990).

A.

B.

C.

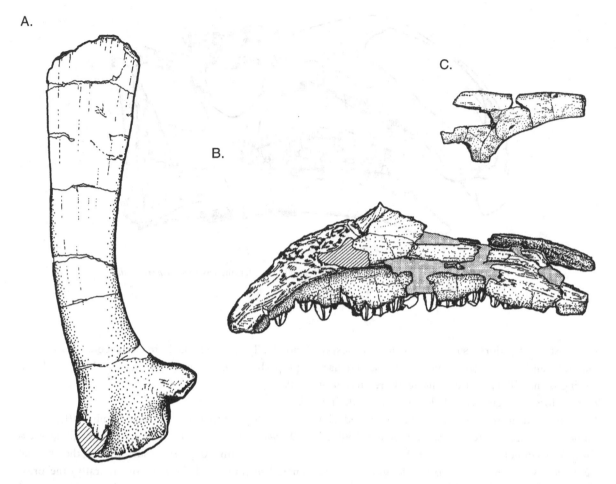

Figure 24.2. Anatomical details of *I. orientalis*. Left scapula (length 940 mm) in (A) lateral view. Right maxilla (length 440 mm) in (B) lateral view. Right nasal (length 145mm) in (C) lateral view. After Norman (1996).

Norman (1996) and Norman (1998) have shown that these specimens represent a new taxon that is characterized, most visibly, by a greatly dorsally enlarged nasal region (Figure 24.3). In many other respects the cranial and postcranial anatomy of this taxon is similar to other known late Early and early Late Cretaceous genera: *Iguanodon* (Norman, 1980; 1986), *Ouranosaurus* (Taquet, 1976) *Lurdusaurus* (Chablis, 1988; Taquet and Russell, 1999), *Probadros* (Head, 1998) and *Eolambia* (Kirkland, 1998); the poorly preserved and enigmatic ornithopod *Muttaburrasaurus* (Bartholomai and Molnar, 1981) may be of hypsilophodontian-dryomorphan grade.

Family Hadrosauridae Cope, 1869

Diagnosis. Close-packed tooth families which form 'dental batteries'; dentary and maxillary crowns which are narrow and subdivided by a median carina, enamel confined strictly to the median and lateral surfaces of the crowns respectively, and roots which are distinctively angular to accommodate adjacent teeth in the battery; a minimum of 3 replacement teeth per tooth position, and between one and three functionally occluding teeth per position; supraorbitals [palpebrals] fused to the dorsal orbital margin, rostral expansion of the jugal and elevation of the dorsal edge of the maxilla with consequent displacement of the antorbi-

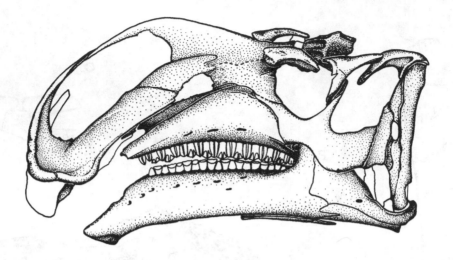

Figure 24.3. *Altirhinus kurzanovi* (holotype), previously identified as *Iguanodon orientalis*. (From Norman, 1998a).

tal fenestra to the dorsal surface of the maxilla; loss of quadrate and surangular foramina; mesiodistal narrowing of maxillary and dentary teeth, reduction of carpus, loss of metacarpal 1, loss of manus digit I; large, everted antitrochanter on the ilium, reduced pubic shaft, eight to ten sacral vertebrae. (Modified from Weishampel and Horner, 1990.)

Comments. A number of taxa belonging to the Hadrosauridae or duck-billed dinosaurs are known from the continental deposits of Late Cretaceous age in Kazakhstan, Mongolia, China and South-East Asia (Maryańska and Osmólska, 1981a, b). The precise stratigraphic age of many of these fossils remains to be established. Hadrosaurs appear to be less common and taxonomically diverse in Asia than in western North America (Weishampel and Horner, 1990).

Despite the difficulties associated with assignments of accepted and correlatable stratigraphic ages (which need to be resolved), the hadrosaurids of Asia are of considerable systematic and biogeographic importance; the anatomy of the early Late Cretaceous forms and, derived from this their systematics, suggest that they might have played a pivotal role in the origin and early diversification of a group of dinosaurs that was to become extremely numerically abundant and taxo-

nomically diverse in the Late Cretaceous (Norman, in prep., b, c; contradictory views are held by Head (1998)).

Subfamily Hadrosaurinae Lambe, 1918
Diagnosis. Transverse expansion of premaxillae, presence of a distinct depression surrounding the external narial opening, and (largely conjecturally) the presence of a mid-ventral groove on the posterior sacral vertebrae (Weishampel and Horner, 1990). To this might be added that the jugal suture with the maxilla is flush to the side of the maxilla and the jugal has a long rostral extension across the dorsal surface of the maxilla.

Genus *Saurolophus* Brown, 1912
Type species. S. osborni Brown, 1912.
Saurolophus angustirostris Rozhdestvenskii, 1957
(subjective validity)
Holotype. PIN 551/8.
Diagnosis. Skull narrow, especially across the snout; external narial opening short, its caudal border lies vertically above the first maxillary tooth; frontal with long vertical anterior process which provides a caudal prop to the lower half of the nasal crest; prefrontal

props the base of the nasal crest caudolaterally; rostral surface of nasal (within the crest) bears a longitudinal ridge or is covered with irregular, bony chambers (?); two supraorbitals identifiable, but fused in the orbital rim; jugal elongated rostrally into a sharp process wedged between maxilla and lacrimal; (lacrimal short compared to *S. osborni*); quadrate strongly bowed caudally; sacral neural spines perpendicular to long axis of sacrum, first two sacral ribs reinforce the pubo-iliac contact; scapula curved; radius shorter than humerus; postacetabular process of ilium broad and subrectangular; pes about a quarter of femur length (?); length ratios of: Mt III to femur = 0.27; pedal phalanx II-2 to pedal phalanx II-1 = 0.23 (from Maryańska and Osmólska, 1984: 132).

Comments. While this might seem an impressive list of diagnostic characters, the detailed cranial anatomy of the type species of *S. osborni* has never been described, and it will be clear to all workers on hadrosaurs that many of the anatomical characters listed for *S. angustirostris* may well fall within the normal range of intraspecific variability. The type species is wall-mounted in the ornithischian gallery at the American Museum of Natural History and this, along with comparable material from the collections, will need to be re-studied before the status of the Mongolian taxon can be assessed confidently. A simple examination of the skull of *S. angustirostris* described by Maryańska and Osmólska (1981a, b) (Figure 24.4A) confirms, as they suggested, that this is a juvenile (smaller size, relatively larger orbital cavity, less extensive development of the facial muzzle). Growth and differentiation of the skull to that seen in the example of *S. angustirostris* illustrated in Figure 24.4B seems to have produced a form virtually indistinguishable from the North American species (Figure 24.4C); immaturity of the smaller form (Figure 24.4A) may also account for the greater clarity in suture pattern between many of the facial bones that probably underlie differences in the descriptive accounts of the two currently recognized species.

The postcranial skeleton is typical of that of hadrosaurine ornithopods (notably the structure of the ischium, the shaft of which is straight and lacks the

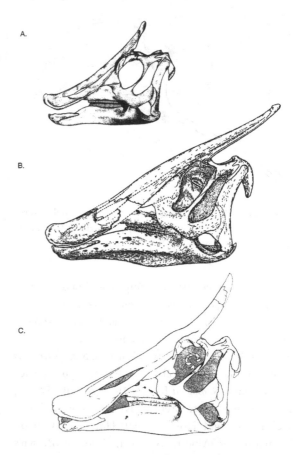

Figure 24.4. *Saurolophus* skulls in lateral view. (A) *S. angustirostris* small individual (after Weishampel and Horner, 1990). (B) *S. angustirostris* (after Rozhdestvenskii, 1957). (C) *S. osborni* (after Brown, 1912). Full-sized skulls approximately 1000 mm long.

distal expansion seen in lambeosaurines – see Figure 24.5). The pelvic girdle of *S. angustirostris* illustrated in Figure 24.7C was held by Maryańska and Osmólska (1984) to be specifically distinct from that of *S. osborni* particularly in details of the shape of the ilium and pubis. However, comparison with Rozhdestvenskii's 1957 illustration (Figure 24.5) of the same species would appear to offer a broadly comparable number of anatomical differences. For the moment we remain unconvinced by these apparent differences.

S. angustirostris is known from the skeletal remains of at least 15 individuals, including articulated cranial and postcranial material from the Nemegt Formation

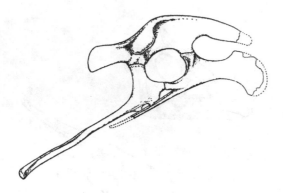

Figure 24.5. Pelvis of *Saurolophus angustirostris* in right lateral view; cf. Figure 24.7C. Full length approximately 1600mm. (From Rozhdestvenskii, 1957).

(Late Cretaceous: Campanian/Maastrichtian; Weishampel and Horner, 1990; Jerzykiewicz and Russell, 1991) of the Altan Uul and North Nemegt localities (Figure 24.1; locations 7 and 9), of the Nemegt Basin, Gobi Desert, Mongolia. *S. angustirostris* is very similar in general anatomy to the type-species of the genus, *S. osborni* Brown, 1912 (see also Brown, 1913), from the Late Cretaceous (Early Maastrichtian; Weishampel and Horner, 1990) Horseshoe Canyon Formation of Alberta (Canada). Both of Brown's papers represent brief descriptive announcements rather than full osteological descriptions of this species; and, in the absence of more detailed work, it has proved impossible (within the scope of this review) to resolve the taxonomy of these two species further.

For the present it is probably prudent to consider *S. angustirostris* as a distinct species of *Saurolophus*; however, the possibility that the Mongolian species is conspecific with the North American form *S. osborni* should be considered, and is a matter that needs to be resolved. This problem of taxonomic distinction of forms in the Late Cretaceous of Asia and western North America has a bearing on evolutionary modelling and biogeographical dispersal patterns, not to mention its significance for biostratigraphic correlations between localities in the western North American and Asian provinces. Similar taxonomic observations can be made on other faunal elements

such as the validity of the taxonomic distinctions claimed for *Tyrannosaurus* and *Tarbosaurus*, *Troodon* and *Saurornithoides*, *Deinonychus* and *Velociraptor*, and others.

Genus *Aralosaurus*

Type species. A. tubiferus Rozhdestvenskii, 1968.

Aralosaurus tubiferus Rozhdestvenskii, 1968 (*nomen dubium*)

Holotype. PIN 2229/1.

Diagnosis. No generically or specifically diagnostic characters can be defined for this taxon.

Comments. A partial skull of this form (Figure 24.6A) was collected from the Shakh-Shakh locality (Figure 24.1; location 1) on the eastern margin of the Dzhusalin uplift (Turonian; Rozhdestvenskii, 1974) in Central Kazakhstan. It comprises much of the skull roof, braincase and left side of the facial region and was described in some considerable detail by Rozhdestvenskii (1968). The strongest affinities of this species, judged by the similarity of the form of the skull roof, is with the hadrosaurines *Gryposaurus* (Figure 24.6B) and *Hadrosaurus*, which have been referred to as belonging to an infrasubfamilial grouping known as the gryposaurs (Weishampel and Horner, 1990).

This is a geographically important record of a gryposaur-type hadrosaurine from Asia. Further material will need to be obtained before the taxonomic validity of this species can be established; its current status is *nomen dubium*.

Subfamily Lambeosaurinae Parks, 1923

Diagnosis. The defining characters of this subfamily of hadrosaurids are: hollow supracranial crests; truncated rostral margin of the jugal bone; high angle >145° between the crown and root of the teeth of the lower jaw; high neural spines; a J-shaped and 'footed' ischium; and a midline ventral ridge on all the sacral vertebrae (Weishampel and Horner, 1990).

Genus *Barsboldia*

Type species. B. sicinskii Maryańska and Osmólska, 1981.

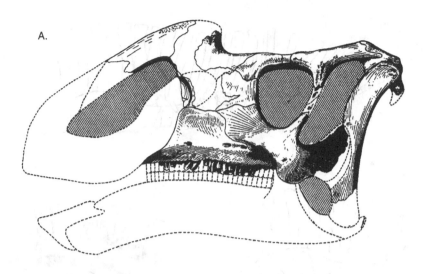

A.

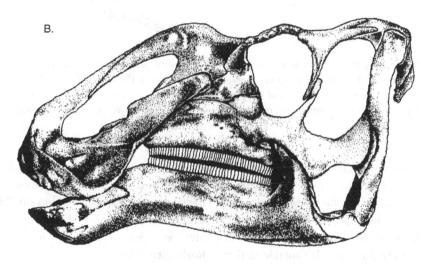

B.

Figure 24.6. Skulls of middle Asian hadrosaurs in lateral view. (A) *Aralosaurus tubiferus* (from Rozhdestvenskii, 1968), (B) *Gryposaurus notabilis* (from Weishampel and Horner, 1990). Original lengths approximately 600 mm.

Barsboldia sicinskii Maryańska and Osmólska, 1981
(*nomen dubium*)

Holotype. ZPAL MgD I/110.

Diagnosis. Hadrosaurid with very tall neural spines to its posterior dorsal, sacral and anterior caudal vertebrae.

Comments. This species is known only from incomplete postcranial remains from the Northern Sair (Figure 24.1; location 8) locality in the Nemegt Basin, Nemegt Formation (Maastrichtian) of the Gobi Desert, Mongolia.

The specimen (Figure 24.7A, B) comprises much of the posterior dorsal, sacral and anterior caudal series of vertebrae, the left ilium and both pubes, a number of ribs and ossified tendons, and fragments of the hindlimbs and elements of the pes. The neural spines

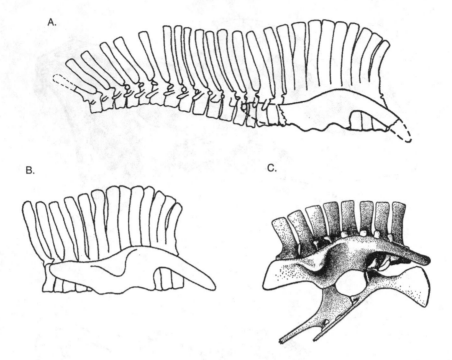

Figure 24.7. Postcranial remains of Mongolian hadrosaurs. (A) Ilium, sacrum and caudal vertebrae of *Barsboldia sicinskii* (from Maryańska and Osmólska, 1981b). (B) Ilium and vertebral column in lateral view of *B. sicinskii* (from Maryańska and Osmólska, 1981b). (C) Pelvis and sacral vertebrae in lateral view of *Saurolophus angustirostris* (from Maryańska and Osmólska, 1984).

are unusually tall (cf. the sacral vertebrae of the hadrosaurine *S. angustrostris*), even by lambeosaurine standards, being highest in the middle of the sacrum and decrease more rapidly in height anteriorly along the dorsal series than posteriorly. The tall neural spines of the anterior caudal vertebrae are also notable in that they bear club-shaped apices. The ventral midline of the sacrum is also marked by a keel, which has been claimed to be distinctive of lambeosaurines.

Genus *Jaxartosaurus*

Type species. J. aralensis Ryabinin, 1939.

Jaxartosaurus aralensis Ryabinin, 1939 (*nomen dubium*)

Holotype. PIN 5009/1.

Diagnosis. No specific or generically diagnostic features.

Comments. This taxon is based on a well-preserved skull roof (Figure 24.8) and braincase collected from Late Cretaceous strata (Coniacian; Rozhdestvenskii, 1974) of the Alim Tau site (Figure 24.1: location 3) in the Chuley region of the Chimkent/Tashkent area of eastern Kazakhstan. The holotype skull roof and braincase was described in great detail by Rozhdestvenskii (1968).

The skull roof shows very clearly the attachment area of the lambeosaurine hollow cranial crest; this takes the form of a very distinctive recess on the rostrodorsal surface of the frontals and causes modifications to the prefrontals such that their medial margins become elevated dorsally. However, the limited nature of the type material makes its taxonomic status dubious, even though, as will be mentioned below, other taxa have been referred to this species in recent years (Maryańska and Osmólska, 1981a, b; Weishampel and Horner, 1990).

A.

B.

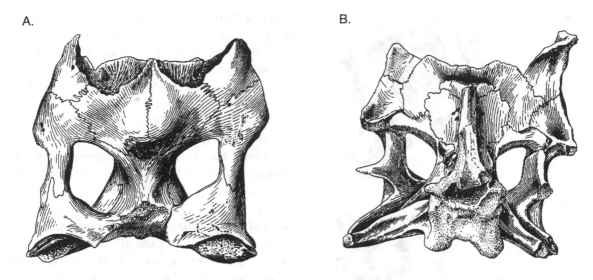

Figure 24.8. *Jaxartosaurus aralensis*, skull in (A) dorsal and (B) ventral view. (from Rozhdestvenskii, 1968).

Genus *Procheneosaurus* (Matthew, 1920)
Type species. P. praeceps Matthew, 1920.
Procheneosaurus convincens Rozhdestvenskii, 1968
(validity questionable)
Holotype. PIN 2230/1.
Diagnosis. A non-diagnosable corythosaur from the Late Cretaceous of Kazakhstan.
Comments. The nearly complete skeleton (lacking only the front of the skull and distal region of the tail) of *Procheneosaurus convincens* Rozhdestvenskii, 1968, was found at the Syuk-Syuk wells site (Figure 24.1; location 4) in the same general region as the type material of *Jaxartosaurus*, but at a different stratigraphic level: in the Dabrazinskaya Svita, referred to as Santonian–Campanian by Rozhdestvenskii in 1968 and later as Santonian by Rozhdestvenskii in 1974. The specimen is representative of a juvenile lambeosaurine hadrosaur (Figure 24.9).

Maryańska and Osmólska (1981a, b), following the work of Dodson (1975), suggested that a new generic name should be provided for this taxon. In its size and general anatomy *P. convincens* is most similar to the Canadian species *P. erectofrons* Parks, 1931 (reduced subjectively in synonymy by Dodson (1975) as a juvenile representative of the lambeosaurine *Corythosaurus*

casuarius). Weishampel and Horner (1990) condensed *P. convincens* with the taxon *J. aralensis* (see above); however, the basis for this taxonomic reassignment has never been made explicit. On the grounds of taxonomic stability, stratigraphic separation and geographic utility, associated with the differences in anatomy described by Rozhdestvenskii for the two species *P. convincens* and *J. aralensis* (differences in the shape of the occiput and layout of bones of the skull roof in these two specimens – even though *J. aralensis* is approximately twice the linear dimensions of *P. convincens*) it has been decided to retain the taxon at least temporarily, but record its status as dubious.

There is little doubt that this material is referable to a taxon within the infrasubfamilial 'corythosaur' group of lambeosaurines (including *Corythosaurus*, *Hypacrosaurus* and *Lambeosaurus*); it is, however, far from clear that any attempt should be made to directly synonymise this form with a particular genus such as *Corythosaurus*, as suggested recently by Nesov (1995). The presence in Kazakhstan and Mongolia of lambeosaurines is not doubted (cf. *Barsboldia*, *Jaxartosaurus*). What is less clear, however, is the stratigraphic horizons from which these remains have been recovered, or the extent to which they are comparable to better

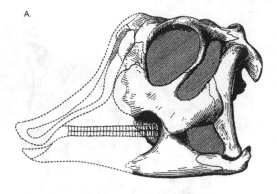

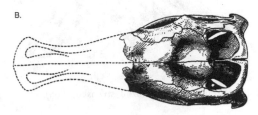

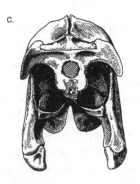

Figure 24.9. *Procheneosaurus convincens*, skull in (A) lateral, (B) dorsal view, and (C) occipital view. Original approximately 300 mm long (from Rozhdestvenskii, 1968).

known and defined taxa from elsewhere (primarily North America). This specimen, represented as it is by a nearly complete skeleton merits a thorough redescription of both the cranial and postcranial anatomy in order to try to resolve many of these problems of taxonomy and biogeography.

Genus *Nipponosaurus* Nagao, 1936
Type species. N. sachalinensis Nagao, 1936.
Nipponosaurus sachalinensis Nagao, 1936 (*nomen dubium*)
Holotype. Unknown.

Diagnosis. Currently non-diagnostic on the basis of the published material. This specimen comprises much of the skeleton of a hadrosaurid which is important biogeographically.

Comments. This taxon is based on a poorly preserved, incomplete skeleton, including fragments of the skull and much of the left hind limb, of a juvenile hadrosaur collected from Late Cretaceous (Senonian) strata in the former Kawakami coal mines (now Sinegorsk; Rozhdestvenskii, 1973) of southern Sakhalin (Figure 24.1; location 14).

The rarity of well-preserved lambeosaurine material in both Mongolia and adjacent areas of the former Soviet Union is particularly frustrating. The fact that *Procheneosaurus convincens* is reasonably well represented by skeletal remains from Kazakhstan, as is *Bactrosaurus* from China (Gilmore, 1933; Weishampel and Horner, 1986) show that lambeosaurines existed in this part of the world in the Late Cretaceous, and further specimens will surely be discovered. Detailed redescription of this material and comparison with related forms from Asia and North America is clearly needed. Work is currently in progress on this taxon (Weishampel, pers. comm., Dec. 1999).

Hadrosauridae incertae sedis or nomina dubia
Genus *Mandschurosaurus* Ryabinin, 1930
Type species. Trachodon amurense Ryabinin, 1925.
Mandschurosaurus amurensis (Ryabinin, 1925)
(*?incertae sedis*)

Diagnosis. The type material of this taxon is very poorly preserved and its distinction appears to rest upon the relatively low number of vertical tooth rows in the jaws (35) in addition to generalized hadrosaurid anatomical features.

Comments. Ryabinin (1930a) proposed *Mandschurosaurus* for the reception of *Trachodon amurense* Ryabinin, 1925. The holotype comprises the poorly preserved, disarticulated remains of a skeleton from Late Cretaceous conglomerates at Blagoveshchensk on the right bank of the Amur River (Nesov, 1995), opposite the villages of Kasatkino and Sagibovo, in Manchuria (Figure 24.1; location 13). The material described by Ryabinin (1930a) consists of an incomplete braincase, fragmentary dentaries, dorsal, sacral, and caudal vertebrae, two ribs, and a number of bones from the fore- and hind-limbs (Ryabinin, 1930a). The type material does not appear to be taxonomically informative and was regarded as a *nomen dubium* by Weishampel and Horner (1990).

In recent years there have been reports of new discoveries, either at the original site, or from adjacent localities (Bolotskii and Moiseenko, 1988) further along the Amur River (e.g. Belye Kruchi). Photographs of these finds suggest that more, and better preserved material has now been discovered. Ralph Molnar (pers. comm., 1997) also indicates that sufficient material has been found to be able to mount at least two skeletons of this hadrosaurid. Provided that the material proves to be associated, and is described in detail then we will be in a far better position to consider its status relative to other Asian hadrosaurids.

Saurolophus krischtofovici Ryabinin, 1930 (*nomen dubium*)
Diagnosis. Non-diagnostic hadrosaurid.
Comments. This taxon is based on an indeterminate fragment of an ischium from the same site in Manchuria as the specimen of *T. amurense* (Ryabinin, 1930b). This was regarded as a *nomen dubium* by Young (1958), Maryańska and Osmólska (1981a, b) and Weishampel and Horner (1990), and we have no grounds to disagree with this assignment.

Cionodon(?) *kysylkumense* Ryabinin, 1931 (*nomen dubium*)
Diagnosis. Non-diagnostic hadrosaurid.
Comments. This taxon is based on a dentary fragment, several vertebrae, and a proximal end of a tibia from the Late Cretaceous of Dzharakhuduk on the right bank of the Amu-Daria River in the Kyzylkum Desert, Kazakhstan (Figure 24.1; location 2). This material was referred to the genus *Thespesius* as *T. kysylkumense* by Steel (1969) but has since been declared to be non-diagnostic (Maryańska and Osmólska, 1981a, b; Weishampel and Horner, 1990).

Nesov (1995) has suggested that this material is referable to the lambeosaurine genus *Bactrosaurus* Gilmore, though there would appear to be no justification for this suggestion.

Bactrosaurus prynadai Ryabinin, 1939 (*nomen dubium*)
Diagnosis. Non-diagnostic hadrosaurid.
Comments. This taxon is based on dentaries and a partial maxilla of an indeterminate hadrosaur from the Late Cretaceous of Kyrk Khuduk, Kazakhstan (Figure 24.1; location 6). Rozhdestvenskii (1968) synonymized this taxon with *Jaxartosaurus aralensis*. This material is indeterminate and was regarded as a *nomen dubium* by Maryańska and Osmólska (1981a, b). We see no reason to dispute this assignment.

Orthomerus weberi Ryabinin, 1945 (*nomen dubium*)
Diagnosis. Non-diagnostic hadrosaurid.
Comments. Limb-bones of a hadrosaur referred to *Orthomerus weberi* by Ryabinin (1945) also noted by Rozhdestvenskii (1973) were collected from Late Maastrichtian marine deposits at Mt. Besh-Kosh on the Crimean peninsula in 1934. A recent review of this and related material can be found in Weishampel *et al.* (1990). The remains are not diagnostic beyond Hadrosauridae gen. et sp. indet.

Gilmoreosaurus (?) *atavus* Nesov, 1995 (*nomen dubium*)
Diagnosis. Non-diagnostic euornithopod teeth.
Comments. This taxon is based on a collection of 10 teeth. A single maxillary tooth of an ornithopod dinosaur is illustrated as the holotype (Nesov, 1995: pl. IX:1). The specimens were collected from Khodzhakul (middle–lower sections of the Khodzhakul Svita) in Karakalpakia (NW Uzbekistan); this site is dated as Late Albian. The holotype tooth does not belong to a hadrosaur: it has a relatively broad crown and the prominent keel which is typical of a less derived ornithopod. Reference to the genus *Gilmoreosaurus* is clearly incorrect, though this may well represent an interesting new taxon of ornithopod from Asia.

Gilmoreosaurus arkhangelskyi Nesov and Kaznyshkina, 1995 (in Nesov, 1995) (*nomen dubium*)
Diagnosis. Non-diagnosable assortment of ornithopod remains.
Comments. This taxon is based on a variety of ornithopod remains, mainly hadrosaurian jaw fragments,

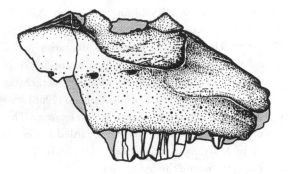

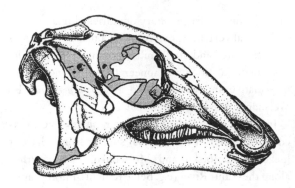

Figure 24.10. Posterior end of the right maxilla of the holotype of *Arstanosaurus akkurganensis*, Shilin and Suslov, 1982, in lateral view. Original has an overall length of 135 mm. From Norman and Kurzanov (1997).

Figure 24.11. Skull of a juvenile specimen, in right lateral view, of an unnamed hadrosaurid from Mongolia. Original 145 mm long. From Norman (in prep., b).

teeth and isolated cranial and postcranial elements collected from Dzharakhuduk (lower sections of the Bissekty Formation [Svita]) (Figure 24.1; location 2) in Uzbekistan, and dated by Nesov as Late Turonian. The material consists of cranial fragments, teeth, vertebrae (including a well preserved axis vertebra, and limb bones (Nesov, 1995: pls. VIII–XI). This taxon is claimed to be a dominant form of ornithopod from these beds in the Kyzylkum-PreAral region and its distribution is claimed to be from the Turonian–Coniacian of Uzbekistan and Tyul'keli (Kazakhstan).

Examination of the illustrations of these elements alone suggests that this is a mixed assortment of ornithopod remains, some of which are clearly hadrosaurid (indet) and others more basal ornithopods. Taxa erected on this quality of material are apt to cause confusion, but serve to register the existence of more ornithopod remains which need re-examination.

Arstanosaurus akkurganensis Shilin and Suslov, 1982
(*nomen dubium*)

Holotype. AAIZ 1/1.
Diagnosis. Non-diagnostic hadrosaurid.
Comments. This form was named on the basis of the posterior portion of a right maxilla (Figure 24.10) from the Late Cretaceous (?Santonian–Campanian) Bostobe Svita in the Kzyl-Orda Oblast' of the north-

eastern Aral region at the locality named Akkurgan, in central Kazakhstan (Figure 24.1; location 5). This specimen is not diagnostic, having been regarded as a *nomen dubium* by Weishampel and Horner (1990), but shares some features in common with *Bactrosaurus* from China (cf. Norman and Kurzanov, 1997).

Similarly, the distal portion of a left femur of an indeterminate large hadrosaur reported by Shilin and Suslov (1982) from the same site is uninformative.

Unnamed hadrosaurid taxa (Norman, in prep., b)
Comments. Additional material collected during the Soviet–Mongolian expeditions of the 1970s (Figure 24.11), includes well-preserved skull and skeletal remains of hadrosaurids which have been catalogued as referable to the genus *Arstanosaurus* in the collections of the Palaeontology Institute of the Russian Academy of Sciences, Moscow. This new and extremely informative material is being described and appears to be referable to a new species of hadrosaurid. Further new material which has an important bearing on this research is currently under study (Tsogtbaatar, pers. comm., 1999).

Conclusions

The fossil record of ornithopods in Mongolia and adjacent parts of the former Soviet Union is confined to Cretaceous exposures, and is of restricted utility, largely as a consequence of the poor preservation of material and the relatively low abundance of ornithopods at localities discovered to date. A notable exception is *Saurolophus angustirostris* whose remains have been discovered in some abundance at the 'Dragons' Grave' locality of Altan Uul II in the Gobi Desert.

Despite their dubious taxonomic status the remains that have been documented (Table 24.1), have the potential to be of considerable biogeographic and phylogenetic interest. Norman (1995, 1996, 1998a) has demonstrated that land-based dispersal of iguanodontid dinosaurs (*Iguanodon*) right across the Northern Hemisphere was possible during the latest Early Cretaceous (Barremian/Aptian) and may well demonstrate that dispersal, rather than vicariant patterns (as tentatively proposed by Milner and Norman, 1984) may be responsible for the distribution and evolutionary history of iguanodontid and hadrosaurid ornithopods. The appearance of derived iguanodontids in Asia during the Albian, is followed by some of the earliest hadrosaurids, both in Asia and the Cenomanian of Western North America (Head, 1998). This new material is likely to be supplemented by even earlier (Albian) material of basal hadrosaurids (Kirkland, pers. comm., 1998). It has become a tacit assumption among workers in this field that the early history and origins of hadrosaurids can be traced to Asia (Rozhdestvenskii, 1966); however, this view was partly challenged by Brett-Surman (1979) who noted that what appeared to be early hadrosaurids had been identified in South America as commented on more recently by Head (1998). The new discoveries in North America are challenging this long-held view and they also demand re-examination of the original material, which was only ever described in the briefest of detail (Norman, in prep., a, b).

Dating of the localities from which key taxa have been discovered is also a matter of considerable concern if the questions relating to the timing of appearance of taxa in different geographical regions and settings is to be correlated with systematic analyses. Currently, the stratigraphic age of many of the critical Asian localities is subject to doubt since much of the original dating was either based on comparative floral or faunal studies, or involves some circularity of argument.

Work over the next few years should throw considerable light on these important issues and it is hoped that this review shows the need for more detailed restudy of some of the outstanding specimens in Asian collections. Within the context of this review the following taxa: *Procheneosaurus convincens, Saurolophus angustirostris, Mandschurosaurus amurensis* and *Nipponosaurus sachalinensis*) need to be described in detail, and subjected to extensive comparative study (particularly with regard to the known taxa of North America) prior to the preparation of detailed systematic and palaeobiogeographic analyses. Of equal importance will be the establishment of geological investigation of the original localities (cf. Kordikova *et al.*, 1996), which needs to be undertaken in order that their stratigraphic ages can be more reliably established.

Acknowledgements

We would like to acknowledge the help of Sergei Kurzanov with work on the ornithopods of Mongolia and Kazakhstan during a research visit to the Paleontology Institute of the Soviet Russian Academy of Sciences, Moscow. We are grateful to Ralph Molnar (Queensland Museum) for helpful reviewer's comments, and to Jason Head (SMU, Texas) and Jim Kirkland (Dinamation International) for unpublished information on early hadrosaurids, and an anonymous reviewer – persistent errors in the manuscript are inevitably our responsibility. S. Oyun (Department of Earth Sciences, Cambridge) very kindly helped with the translation of sections of some Russian articles and Natasha Bakhurina corrected the errors in the bibliography. This work would not have been possible without the support of a study grant from the Royal Society of London and the Russian Academy of Sciences.

Table 24.1. *Tabulation of the ornithopod taxa reported from Mongolia and the former Soviet Union and their status following review in this article*

Taxon	Age	Locality	Material	Taxonomic source	Key reference	Status
Iguanodon orientalis	Barremian/ Aptian	Khamaryn Khural	Maxilla, premaxilla, nasal, scapula, ribs and vertebrae	Rozhdestvenskii, 1952	Norman, 1996	*Nomen dubium*
Altirhinus kurzanovi	Aptian/Albian	Hüren Dukh	Skull and postcranial elements	Norman & Weishampel, 1990	Norman, 1998	Valid
Saurolophus angustirostris	Campanian/ Maastrichtian	Altan Uul North Nemegt	Many skeletons and isolated elements	Rozhdestvenskii, 1957	Maryańska & Osmólska, 1981b; 1984	Validity is subjective
Aralosaurus tubiferus	Turonian	Shakh–Shakh	Skull roof, braincase and part of facial skeleton	Rozhdestvenskii, 1968	Rozhdestvenskii, 1968	*Nomen dubium*
Barsboldia sicinskii	Maastrichtian	Northern Sair	Dorsal, sacral and caudal vertebrae, part pelvis and other fragments	Maryańska & Osmólska, 1981a	Maryańska & Osmólska, 1981a	*Nomen dubium*
Procheneosaurus convincens	Santonian/ Campanian	Syuk–Syuk	Nearly complete skeleton	Rozhdestvenskii, 1968	Rozhdestvenskii, 1968	Validity questionable
Nipponosaurus sachalinensis	Senonian	Sinegorsk	Nearly complete skeleton	Nagao, 1936		*Nomen dubium*
Mandschurosaurus amurensis	Late Cretaceous	Amur River (Belye Kruchi)	Possibly several skeletons	Ryabinin, 1925	Bolotskii & Moiseenko, 1988	Incertae sedis
Saurolophus krischtofovici	Late Cretaceous	Amur River (Belye Kruchi)	Part ischium	Ryabinin, 1930a, b		*Nomen dubium*
Cionodon kysylkumense	Late Cretaceous	Dzharakhuduk	Fragments including a jaw, vertebrae and tibia	Ryabinin, 1931		*Nomen dubium*
Bactrosaurus prynadai	Late Cretaceous	Kyrk Khuduk	Jaw fragments	Ryabinin, 1939		*Nomen dubium*
Orthomerus weberi	Maastrichtian	Mt. Besh Kosh, Crimea	Limb fragments	Ryabinin, 1945		*Nomen dubium*
Gilmoreosaurus atavus	Late Albian	Khodzhakul	Teeth	Nesov, 1995	Nesov, 1995	Non-hadrosaur ornithopod
Gilmoreosaurus archangelskyi	Turonian	Dzharakhukuk	Assorted elements including teeth and postcranial	Nesov & Kaznyshkina, 1995 (in Nesov, 1995)		*Nomen dubium*
Arstanosaurus akkurganensis	Santonian/ Campanian	Akkurgan	Maxilla, isolated tooth and distal end of femur	Shilin & Suslov, 1982	Norman & Kurzanov, 1997	*Nomen dubium*
Un-named species	?Cenomanian	Baishin Tsav	Several skeletons including skulls	Norman & Kurzanov, 1997	Norman & Kurzanov, 1997	To be assigned

References

Bartholomai, A. and Molnar, R.E. 1981. *Muttaburrasaurus*, a new iguanodontid (Ornithischia: Ornithopoda) dinosaur from the Lower Cretaceous of Queensland. *Memoirs of the Queensland Museum* 20: 319–349.

Bolotskii, Y.L. and Moiseenko, V.G. 1988. [*Dinosaurs from the Amur River.*] Akademiya Nauk SSSR: Blagoveshchensk: 39pp.

Boulenger, G.A. 1881. Sur l'arc pelvien chez les dinosauriens de Bernissart. *Bulletins de l'Académie royal de Belgique* 1 (3me série (5)): 3–11.

Brett-Surman, M.K. 1979. Phylogeny and palaeobiogeography of hadrosaurian dinosaurs. *Nature* 277: 560–562.

Brown, B. 1912. A crested dinosaur from the Edmonton Cretaceous. *Bulletin of the American Museum of Natural History* 31: 131–136.

—1913. A new trachodont dinosaur, *Hypacrosaurus* from the Edmonton Cretaceous of Alberta. *Bulletin of the American Museum of Natural History* 32: 395–406.

Chablis, S. 1988. *Etude anatomique et systématique de Gravisaurus tenerensis n. gen et sp. (Dinosaurien, Ornithischien) du gisement de Gadoufaoua (Aptien de Niger).* Unpublished PhD thesis, Université de Paris VII.

Cope, E.D. 1869. Synopsis of the extinct Batrachia, Reptilia and Aves of North America. *Transactions of the American Philosophical Society* 14: 1–252.

Dodson, P. 1975. Taxonomic implications of relative growth in lambeosaurine hadrosaurs. *Systematic Zoology* 24: 37–54.

Galton, P.M. 1977. The ornithopod dinosaur *Dryosaurus* and a Laurasia-Gondwanaland connection in the Upper Jurassic. *Nature* 268: 230–232.

Gilmore, C.W. 1933. On the dinosaurian fauna of the Iren Dabasu Formation. *Bulletin of the American Museum of Natural History* 67: 23–78.

Hay, O.P. 1902. Bibliography and catalogue of the fossil Vertebrata of North America. *Bulletin of the United States Geological Survey* 179: 1–868.

He, X. and K. Cai. 1984. *[The Middle Jurassic dinosaurian fauna from Dashanpu, Zigong, Sichuan.]* Chengdu: Sichuan Scientific and Technical Publishing House.

Head, J.J. 1998. A new species of basal hadrosaurid (Dinosauria: Ornithischia) from the Cenomanian of Texas. *Journal of Vertebrate Palaeontology* 18: 718–738.

Jerzykiewicz, T. and Russell, D.A. 1991. Late Mesozoic stratigraphy and vertebrates of the Gobi Basin. *Cretaceous Research* 12: 345–377.

Kirkland, J.I. 1998. A new hadrosaurid from the Upper Cedar Mountain Formation (Albanian–Cenomanian) of Eastern Utah – the oldest known hadrosaurid (lambeosaurine?), pp. 283–295, in Lucas, S.G., Kirkland, J.I. and Estep, J.W. (eds.), *Lower and Middle Cretaceous Terrestrial Ecosystems.* Albuquerque, New Mexico Museum of Natural Hisory and Science.

Kordikova, E.G., Gunnell, G.F., Polly, P.D. and Kovrizhnykh, Y.B. 1996. Late Cretaceous and Paleocene vertebrate paleontology and stratigraphy in the Northeastern Aral Sea region, Kazakhstan. *Journal of Vertebrate Paleontology (Abstracts)* 16 (3): 46A.

Lambe, L.M. 1918. On the genus *Trachodon* of Leidy. *Ottawa Naturalist* 31: 135–139.

Mantell, G.A. 1825. Notice on the *Iguanodon*, a newly discovered fossil reptile, from the sandstone of Tilgate forest, in Sussex. *Philosophical Transactions of the Royal Society of London* 115: 179–186.

Marsh, O.C. 1881. Principal characters of American Jurassic dinosaurs. *American Journal of Science* Ser. 3, 21: 417–423.

Maryańska, T. and Osmólska, H. 1981a. First lambeosaurine dinosaur from the Nemegt Formation, Upper Cretaceous, Mongolia. *Acta Palaeontologica Polonica* 26: 243–255.

—and— 1981b. Results of the Polish–Mongolian palaeontological expeditions. Part IX. Cranial anatomy of *Saurolophus angustirostris* with comments on the Asian Hadrosauridae (Dinosauria). *Palaeontologia Polonica* 42: 5–24.

—and— 1984. Postcranial anatomy of *Saurolophus angustirostris* with comments on other hadrosaurs. *Palaeontologia Polonica* 46: 119–141.

Matthew, W.D. 1920. Canadian dinosaurs. *Natural History* 20 (5): 536–544.

Milner, A.R. and Norman, D.B. 1984. The biogeography of advanced ornithopod dinosaurs (Archosauria: Ornithischia) – a cladistic-vicariance model, in Reif, W.-E. and Westphal, F. (eds.), *Proceedings of the Third Symposium on Mesozoic Terrestrial Ecosystems.* Tübingen: Attempto Verlag.

Nagao, T. 1936. *Nipponosaurus sachalinensis*, a new genus and species of trachodont dinosaur from Japanese Saghalien. *Journal of the Faculty of Science, Hokkaido Imperial University* 3 (2)(Ser. IV): 185–220.

Nesov, L.A. 1995. [*Dinosaurs of northern Eurasia: new data about assemblages, ecology and palaeobiogeography.*] St. Petersburg: Izdatel'stvo Sankt-Peterburgskoi Universiteta, 156pp.

Norman, D.B. 1980. On the ornithischian dinosaur *Iguanodon bernissartensis* from Belgium. *Mémoires de l'Institut Royal des Sciences Naturelles de Belgique* **178**: 1–105.

—1985. *The illustrated encyclopedia of dinosaurs.* London, Salamander Books.

—1986. On the anatomy of *Iguanodon atherfieldensis* (Ornithischia: Ornithopoda). *Bulletin de l'Institut Royal des Sciences Naturelles de Belgique* **56**: 281–372.

—1987. A mass-accumulation of vertebrates from the Lower Cretaceous of Nehden (Sauerland), West Germany. *Proceedings of the Royal Society of London* B **230**: 215–255.

—1995. Ornithopods from Mongolia: new observations. *Journal of Vertebrate Paleontology (Abstracts)* **15** (3): 46A.

—1996. On Mongolian ornithopods (Dinosauria: Ornithischia). 1. *Iguanodon orientalis* Rozhdestvenskii, 1952. *Zoological Journal of the Linnean Society* **116** (3): 303–315.

—1998a. On Asian ornithopods (Dinosauria: Ornithischia). 3. A new species of iguanodontian. *Zoological Journal of the Linnean Society* **122**: 291–348.

—1998b. *Probactrosaurus* from Asia and the origin of hadrosaurs. *Journal of Vertebrate Paleontology* **18**: 66.

—(in prep., a). On Asian ornithopods (Dinosauria: Ornithischia). 4. Redescription of *Probactrosaurus gobiensis* Rozhdestvenskii, 1966.

—(in prep., b). On Asian ornithopods (Dinosauria: Ornithischia). 5. New hadrosaurids from Bainshin Tsav.

—(in prep., c). The systematics of the Iguanodontia and the origin of hadrosaurid ornithopods (Dinosauria: Ornithischia).

—and Kurzanov, S.M. 1997. On Asian ornithopods (Dinosauria: Ornithischia). 2. *Arstanosaurus akkurganensis* Shilin and Suslov, 1982. *Proceedings of the Geologists' Association* **108**: 191–199.

—and Weishampel, D.B. 1990. Iguanodontidae and related Ornithopoda, pp. 510–533, in Weishampel, D.B., Dodson, P. and Osmólska, H. (eds.), *The Dinosauria.* Berkeley: University of California Press.

Parks, W.A. 1931. A new genus and two new species of trachodont dinosaurs from the Belly River Formation of Alberta. *University of Toronto Studies in Geology Series* **31**: 1–11.

Rozhdestvenskii, A.K. 1952. [Discovery of iguanodonts in Mongolia.] *Doklady AN SSSR* **84**: (6): 1243–1246.

—1957. [Duck-billed dinosaur – *Saurolophus* from the Upper Cretaceous of Mongolia]. *Vertebrata PalAsiatica* **1** (3): 129–149.

—1966. [New iguanodonts from Central Asia. Phylogenetic and taxonomic interrelationships of late Iguanodontidae and early Hadrosauridae.] *Palaeontologicheskii Zhurnal* **1966** (3): 103–116.

—1968. [The hadrosaurs of Kazakhstan]. *Verkhnepaleozoiskie i mezozoiskie zemnovodnye i presmykayushchiesya SSSR.* Moscow, Izdatel'stvo Nauka: 97–141.

—1973. [Review of fossil reptiles in Russia.] *Paleontologicheskii Zhurnal* **1973** (2): 90–99.

—1974. [The history of dinosaur faunas in Asia and other continents and some problems of palaeogeography.] *Trudy Sovmestnoi Sovetsko-Mongol'skoi Paleontologicheskoi Ekspeditsii* **1**; 107–131.

—1977. The study of dinosaurs in Asia. *Journal of the Palaeontological Society of India* **20**: 102–119.

Ryabinin, A.N. 1925. [Discovery of the skeletal remains of *Trachodon amurense* sp. nov.] *Geologicheskii Komitet, Izvestiya* **45**: 1–12.

—1930a. [Review of the dinosaur fauna collected from the Amur River.] *Zapiski Rossiiskogo Mineralogicheskogo Obshchestva* **59**: 41–51.

—1930b. [*Mandschurosaurus amurensis* gen. et sp. nov. a new dinosaur from the Amur River.] *Russkoe Paleontologicheskoe Obshchestva, Monografy* **11**: 1–36.

—1931. [Dinosaur remains collected from the Amu-Daria River.] *Zapiski Rossiiskogo Mineralogicheskogo Obshchestva* **60**: 114–118.

—1939. [Review of the fossil fauna of Kazakhstan.] *Tsentral'nyi Nauchnoissledovatel'skii Geologo-razvedochnyi Institut Trudy* **118**: 1–40.

—1945. [Remains of dinosaurs from the Upper Cretaceous of Crimea.] *Materialy Vsesoyuzhogo Nauchnoissledovatel'skogo Geologorazvedochnogo Instituta, Paleontologiya i Stratigrafiya, Sbornik* **4**: 4–10.

Shilin, P.V. and Y.V. Suslov. 1982. [A hadrosaur from the northeastern Aral region.] *Palaeontologeskii Zhurnal* **1982** (1): 131–135 (translated version, pp. 132–136).

Steel, R. 1969. Ornithischia. *Handbuch der Paläoherpetologie* **15**: 1–84.

Sues, H.-D. and D.B. Norman. 1990. Hypsilophodontidae, *Tenontosaurus* and Dryosauridae, pp. 498–509 in Weishampel, D.B., Dodson, P. and Osmólska, H. (eds.), *The Dinosauria*. Berkeley: University of California Press.

Taquet, P. 1976. Géologie et paléontologie du gisement de Gadoufaoua (Aptien du Niger). *Cahiers de Paléontologie. Centre National de la Recherche Scientifique, Paris*. 1–191.

—and Russell, D.A. 1999. A massively-constructed iguanodont from Gadoufaoua, Lower Cretaceous of Niger. *Annales de Paleontologie* 85: 85–96.

Weishampel, D.B., Grigorescu, D. and Norman, D.B. 1991. The dinosaurs of Transylvania: Island biogeography in the Late Cretaceous. *National Geographic Research and Exploration* 7: 196–215.

—and Horner, J.R. 1986. The hadrosaurid dinosaurs from the Iren Dabasu fauna (People's Republic of China, Late Cretaceous). *Journal of Vertebrate Paleontology* 6: 38–45.

—and— 1990. Hadrosauridae, pp. 534–561 in Weishampel, D.B., Dodson, P. and Osmólska, H. (eds.), *The Dinosauria*. Berkeley: University of California Press.

Young, C.C. 1958. The dinosaurian remains of Laiyang, Shantung. *Palaeontologia Sinica C* 16: 53–138.

25

The fossil record, systematics and evolution of pachycephalosaurs and ceratopsians from Asia

PAUL C. SERENO

Introduction

Although little is known of the first half of their evolutionary history, margin-headed ornithischians (Marginocephalia) are represented by a remarkable array of small- and large-bodied species during the last 20 million years of the Mesozoic. Marginocephalians comprise two distinctive subgroups, pachycephalosaurs and ceratopsians, both characterized by a bony shelf that projects from the posterior skull margin. Pachycephalosaurs, as their group name suggests, have thickened the skull roof, the margins of which are ornamented with distinctive tubercles. Ceratopsians, by contrast, have extended the posterior shelf as a thin bony frill, which is often accompanied by one or more cranial horns.

Abundant in late Early and Late Cretaceous deposits in central Asia and western North America, marginocephalians are exceptionally rare earlier in the Cretaceous (Neocomian) and have never been found in deposits that are regarded with confidence as Jurassic in age. Marginocephalian origins, however, surely date back at least to the Early Jurassic, when they diverged from their sister group, the ornithopods.

In this chapter, the best-known Asian pachycephalosaurs and ceratopsians are reviewed and a general account of their osteology is presented. The phylogenetic relationships of all marginocephalians are considered. The biogeographic history of marginocephalians is particularly interesting, as it clearly involves a polar dispersal route across Beringia – a well-trodden passage that played a major role in the evolution of dinosaurs in the Northern Hemisphere during the Late Cretaceous (Sereno, 1997, 1999a).

Institutional abbreviations

AMNH, American Museum of Natural History, New York; GI, Geological Institute, Ulaanbaatar; GPI, Geologische-Paläontologisches Institut, Göttingen; IVPP, Institute of Vertebrate Paleontology and Paleoanthropology, Beijing; MIWG, Museum of the Isle of Wight (Geology), Sandown; MNHN, Muséum National d'Histoire Naturelle, Paris; MOR, Museum of the Rockies, Bozeman; NMC, National Museum of Canada, Ottawa; PAL, Institute of Paleobiology, Warsaw; PIN, Paleontological Institute, Moscow; TF, Department of Mineral Resources, Bangkok; TMP, Royal Tyrrell Museum of Palaeontology, Drumheller; UA, University of Alberta, Edmonton; USNM, National Museum of Natural History, Washington; YPM, Yale Peabody Museum, New Haven.

History of discovery

Excluding large-bodied ceratopsids from North America and *Psittacosaurus* and *Protoceratops* from Asia, marginocephalian fossils are generally rare and incomplete. The first relatively complete skeletal remains were discovered in Alberta and pertain to the pachycephalosaur *Stegoceras validus* (Gilmore, 1924; Sues and Galton, 1987) and the ceratopsian *Leptoceratops gracilis* (Brown, 1914). The type skull and skeleton of *Stegoceras* (UA 2), still the most complete pachycephalosaur skeleton from North America, revealed the peculiar anatomy of these bipedal ornithischians, known previously from isolated teeth and thickened skull caps. Shortly after the discovery of the first skeleton of *Leptoceratops* (AMNH 5205), three additional skeletons were discovered (Sternberg,

Table 25.1. *Age and known geographic range of pachycephalosaurs and ceratopsians*

Taxa	Age	Known geographic range
Pachycephalosauria		
Stenopelix valdensis	Barremian	central Europe
Wannanosaurus yansiensis	Campanian	eastern China
Goyocephale lattimorei	Campanian	southern Mongolia
Homalocephale calathocercos	Maastrichtian	southern Mongolia
Ornatotholus browni	Campanian	western North America
Yaverlandia bitholus	Barremian	western Europe
Stegoceras validus	Campanian	western North America
Prenocephale prenes	Maastrichtian	southern Mongolia
Tylocephale gilmorei	Campanian	southern Mongolia
Stygimoloch spinifer	Maastrichtian	western United States
Pachycephalosaurus wyomingensis	Maastrichtian	western United States
Ceratopsia		
Psittacosaurus	Barremian–early Aptian	China, Mongolia
Chaoyangsaurus youngi	latest Jurassic/Neocomian	northern China
Archaeoceratops oshimai	Neocomian	northern China
Leptoceratops gracilis	Maastrichtian	western Canada
Udanoceratops tschizhovi	Campanian	southern Mongolia
Bagaceratops rhozhdestvenskyi	Campanian	southern Mongolia
Protoceratops andrewsi	Campanian	southern Mongolia
Graciliceratops mongoliensis	Campanian	southern Mongolia
Montanoceratops cerorhynchus	Maastrichtian	western North America
Turanoceratops tardabilis	Cenomanian or Turonian	Uzbekistan
Ceratopsidae	Campanian–Maastrichtian	western North America

1951) that, likewise, constitute the most complete basal ceratopsian to date from North America.

Expeditions to the Gobi Desert in the 1920s by the American Museum in New York (see Chapter 12) and to northern China in the 1930s by the Palaeontological Institute in Uppsala brought to light much of what we currently know about marginocephalian diversity during the Late Cretaceous (Table 25.1). Numerous skulls and skeletons and the first well documented growth series were recovered for the basal ceratopsians *Psittacosaurus mongoliensis* (Osborn, 1923, 1924; Coombs, 1982; Sereno 1987, 1990a, b) and *Protoceratops andrewsi* (Granger and Gregory, 1923; Brown and Schlaikjer, 1940; Dong and Currie, 1993). Several new genera of pachycephalosaurs and basal ceratopsians were discovered by subsequent expeditions to Mongolia and northern China (Maryańska and

Osmólska, 1974, 1975; Perle *et al.*, 1982; Kurzanov, 1992; Dong and Azuma, 1997; see also Chapters 12 and 13).

Systematics of Asian marginocephalians

Taxonomic definitions

The utility of taxon names based on phylogenetic definitions has been explored by de Queiroz and Gauthier (1990, 1992). Node-based or stem-based phylogenetic definitions were applied to groups with living members to differentiate crown groups (node-based) from more inclusive groups (stem-based) that incorporate intervening extinct taxa.

Recently, this approach has been generalized to stabilize the phylogenetic meaning of widely used names

Table 25.2. *Unranked classification for marginocephalians. Taxa with node-based definitions are shown in bold, and those with stem-based definitions are shown in regular type (Sereno, 1997, 1998). This configuration of phylogenetic definitions specifies four node-stem triplets*

Marginocephalia
Pachycephalosauria
Pachycephalosauridae
Stegoceras
Pachycephalosaurinae
Ceratopsia
Neoceratopsia
Coronosauria
Protoceratopsidae
Ceratopsoidea
Ceratopsidae
Centrosaurinae
Ceratopsinae

for living or extinct clades (Sereno, 1997, 1998, 1999b). Taxa are defined with respect to one another using the same reference taxa to create stable node-stem triplets. The unranked classification used in this review is based on four node-stem triplets (Table 25.2).

Pachycephalosauria Maryańska and Osmólska, 1974
Definition. All marginocephalians closer to *Pachycephalosaurus* than *Triceratops* (Sereno, 1998)

Wannanosaurus yansiensis. *Wannanosaurus* is based on a partial skull (Figure 25.1) and several postcranial bones of one immature individual (IVPP V4447) with additional vertebrae and limb bones of a second individual found nearby (Hou, 1977). Like *Yaverlandia*, *Wannanosaurus* is a small pachycephalosaur, although the open sutures in the holotype cranium suggest that at maturity it may have reached a somewhat larger body size (contrary to Maryańska, 1990, p. 574, who remarked that the cranial sutures were obliterated). *Wannanosaurus* has a flat dorsal skull table with relatively large supratemporal fossae. The associated postcranial bones share several features with *Homalocephale* and *Stegoceras,* such as the short forelimb (humerus less than one-half femoral length), sigmoid-shaped humerus, and slender distal fibula. Diagnostic

features of *Wannanosaurus* include the low, fan-shaped dentary crowns with a marked median eminence on the lateral crown surface and the extreme flexure of the humerus (proximal and distal ends set at approximately 30° to one another).

Goyocephale lattimorei. *Goyocephale* is based on a relatively complete skeleton with a partial skull (GI SPS 100/1501) and is the best known of basal pachycephalosaurs (Perle *et al.,* 1982). It falls in the middle of the range of body size for pachycephalosaurs, similar to that of *Homalocephale, Stegoceras* and *Prenocephale.* Although Perle *et al.* (1982, p. 117) presented a lengthy diagnosis for *Goyocephale*, most of the listed features characterize other pachycephalosaurs as well. There are only a few features that are peculiar to *Goyocephale*, and these include the sinuous lateral margin of the skull as seen in dorsal view. The lateral margin is particularly prominent above the orbit where the two supraorbitals meet, resulting in an S-shaped edge as seen in dorsal view. In addition, the sternals in *Goyocephale* (Perle *et al.,* 1982, pl. 41) are more slender and gently curved than those in *Stegoceras* (UA 2).

Homalocephale calathocercos. *Homalocephale*, known from a partial skeleton and flat-headed skull (GI SPS 100/1201; previously listed as GI SPS 100/51, Maryańska and Osmólska, 1974), can be distinguished from other pachycephalosaurs by the crescent-shaped, ventrally deflected postacetabular process of the ilium. Other features, such as the sacral attachments to the ischium, may eventually prove to be diagnostic, but these are not preserved in any other pachycephalosaur.

Pachycephalosauridae Sternberg, 1945
Definition. *Stegoceras, Pachycephalosaurus*, their most recent common ancestor and all descendants (Sereno, 1998).

Prenocephale prenes. Based on a beautifully preserved cranium and partial postcranial skeleton (PAL MgD I/104; Figure 25.2), *Prenocephale* is currently the best known fully domed pachycephalosaur (Maryańska and Osmólska, 1974). The straight dorsal margin of the snout, which resembles that in *Goyocephale* (Perle *et*

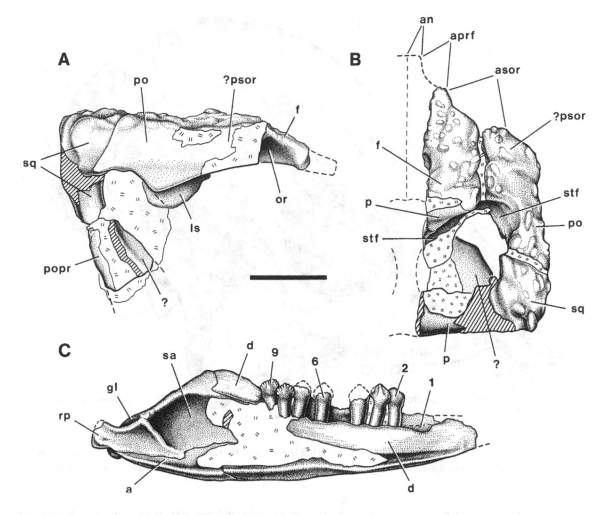

Figure 25.1. *Wannanosaurus yansiensis* (IVPP V4447), partial cranium in right lateral (A) and dorsal (B) views and left lower jaw in (C) medial view. Abbreviations: a, angular; an, articular surface for the nasal; aprf, articular surface for the prefrontal; asor, articulation for anterior supraorbital; d, dentary; f, frontal; gl, glenoid; ls, laterosphenoid; p, parietal; po, postorbital; popr, paroccipital process; psor, posterior supraorbital; rp, retroarticular process; sa, surangular; sq, squamosal; stf, supratemporal fossa; 1–9, position in tooth row. Scale bar equals 10 mm.

al., 1982), differs from the gently arched margin and shorter premaxilla in *Stegoceras* (Figure 25.3). In *Prenocephale*, the proximal end of the quadrate is tongue-shaped, and there is an unusual bulbous knob on the free dorsal margin of the quadratojugal (preserved on both sides). Aspects of cranial ornamentation, such as the unbroken line of tubercles that connect those on the postorbital with those on the jugal, may also be diagnostic for *Prenocephale*.

Tylocephale gilmorei. *Tylocephale*, known from a single weathered cranium and the posterior portion of the lower jaws (PAL MgD I/105; Figure 25.4), is a fully domed pachycephalosaur (Maryańska and Osmólska, 1974). As these authors noted, it differs in several details from *Prenocephale*, which it otherwise resembles quite closely. The dome and occiput are narrower, and the postorbital bar and quadrate are more slender (Maryańska and Osmólska, 1974, p. 51). The orbit was described as more elongate, but this may be the result

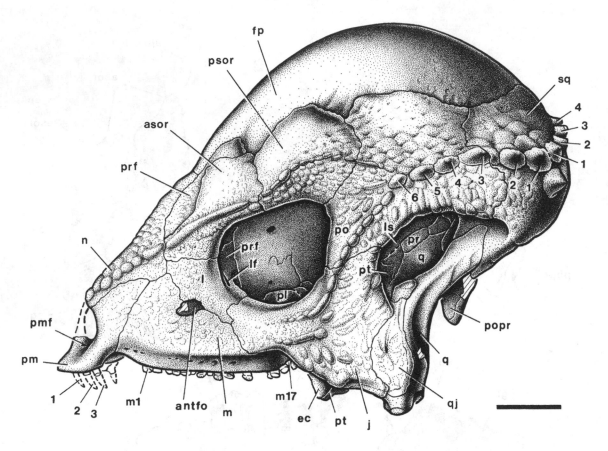

Figure 25.2. *Prenocephale prenes* (PAL MgD I/104), cranium in left lateral view. Abbreviations: as in Figures 25.1 and 25.3 and antfo, antorbital fossa; lf, lacrimal foramen; pmf, premaxillary foramen; pr, prootic; pt, pterygoid. Scale bar equals 30 mm.

of dorsoventral crushing of the cranium. The nodular ornamentation that characterizes the margins of the dome also differs from that in *Prenocephale. Tylocephale* has a dorsally upturned corner tubercle on the squamosal as in *Prenocephale*, but there are fewer tubercles medial to the corner tubercle in *Tylocephale*, probably only four on the right side and three on the left (contrary to Maryańska and Osmólska, 1974, fig. 1B4). In *Tylocephale* the ornamental tubercles on the postorbital bar are reduced compared to those in *Prenocephale*, but there are tubercles on the supraorbitals above the orbital margin that are absent in *Prenocephale*. A large oval depression, centred on the quadratojugal and preserved on both sides of the skull, may constitute a diagnostic feature of this pachycephalosaur. *Prenocephale* has a similar, although much smaller,

quadratojugal depression. In the lower jaw, the angular is ornamented with tubercles (contrary to Maryańska and Osmólska, 1974, p. 52), and the jaw articulation is positioned somewhat below the maxillary tooth row, as in other pachycephalosaurs. The dentary teeth, which are the best preserved, are characterized by a primary ridge and secondary ridges that extend down much of the crown surface. These crowns are easily distinguished from those of *Stegoceras*, in which the medial side of the crowns are dorsoventrally concave.

In summary, the skull and dentition of *Tylocephale* clearly indicate that it is a distinct, fully-domed pachycephalosaur that is similar, and quite possibly closely related, to *Prenocephale*. Diagnostic features include a narrow, deep skull and large fossa on the quadratojugal.

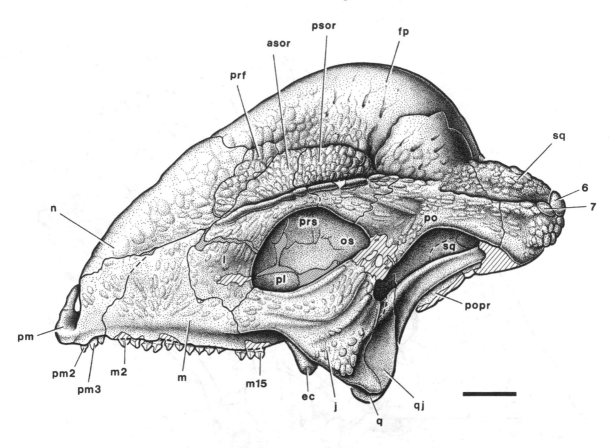

Figure 25.3. *Stegoceras validus* (UA 2), cranium in left lateral view. Abbreviations: as in Figure 25.1 and asor, anterior supraorbital; ec, ectopterygoid; fp, frontoparietal; j, jugal; l, lacrimal; m, maxilla; m1–17, maxillary tooth positions; n, nasal; os, orbitosphenoid; pl, palatine; pm, premaxilla; pm1–3, premaxillary tooth positions; prf, prefrontal; q, quadrate; qj, quadratojugal; 1–7, postorbital or squamosal tubercles. Scale bar equals 30 mm.

Ceratopsia Marsh, 1890

Definition. All marginocephalians closer to *Triceratops* than to *Pachycephalosaurus* (Sereno, 1998).

Psittacosaurus. Psittacosaurids, or 'parrot-beaked' dinosaurs, constitute a tightly knit group of species in the single genus *Psittacosaurus* (Figure 25.5), known only from Lower Cretaceous beds in China, Mongolia and Siberia (Sereno, 1987, 1990a, b; Eberth *et al.*, 1993). *P. mongoliensis* and *P. sinensis*, the former the larger and less derived of the two, are known from many skeletons, some with complete skulls. Two additional species, *P. xinjiangensis* and *P. meileyingensis* (Sereno and Chao, 1988; Sereno *et al.*, 1988), based on less complete material, have been described from China. The

former has a characteristic pyramidal jugal horn, and the latter has an extremely short skull that is nearly round in profile. Recently, two additional species have been described from Inner Mongolia, *P. neimongoliensis* and *P. ordosensis* (Russell and Zhao, 1996), which are extremely close to *P. mongoliensis* and *P. sinensis*, respectively. Finally, additional species have been described from China and Thailand (*P. mazongshanensis*, *P. sattayataki*), but their taxonomic status is questioned below.

The genus *Psittacosaurus* and the six species recognized here are diagnosed almost entirely on the basis of cranial characters. The deep and very short psittacosaurid snout, which constitutes less than 40% of skull length, most closely resembles that in the aberrant

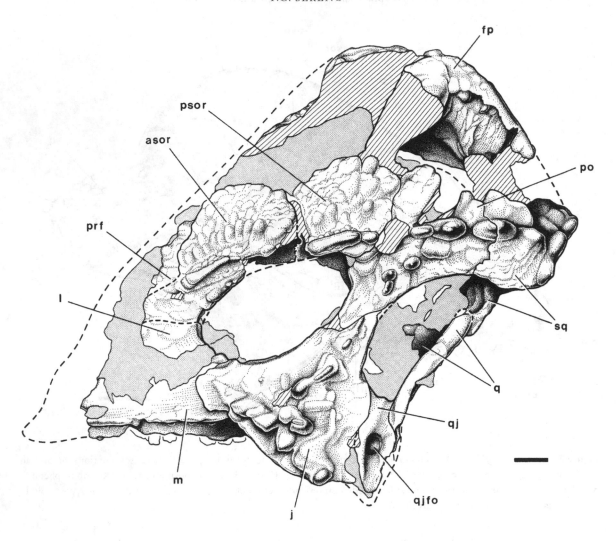

Figure 25.4. *Tylocephale gilmorei* (PAL MgD I/105), cranium in left lateral view. Abbreviations: as in Figures 25.1–25.3. Scale bar equals 10 mm.

theropod subgroup Oviraptoridae. The external naris is positioned very high on the snout, which is composed of the ceratopsian rostral bone and the broadly expanded premaxilla (Figure 25.5). The antorbital fenestra is closed and the antorbital fossa is absent. A small lateral depression is present on the maxilla in several species of *Psittacosaurus* and has been identified as a reduced antorbital fossa (Sereno, 1987; Sereno and Chao, 1988; Sereno *et al.*, 1988). This structure, however, is regarded here as a neomorphic depression unrelated to diverticulae of the cranial sinus system. Unlike any other dinosaurs, a section of the lateral wall of the lacrimal canal remains unossified in psittacosaurs; a foramen of variable size between the premaxilla and lacrimal exposes a section of the canal. Species differences are based primarily on cranial features, such as the shape of the jugal horn and length of the parietosquamosal frill.

The psittacosaurid postcranial skeleton is remarkably primitive compared to most other Cretaceous

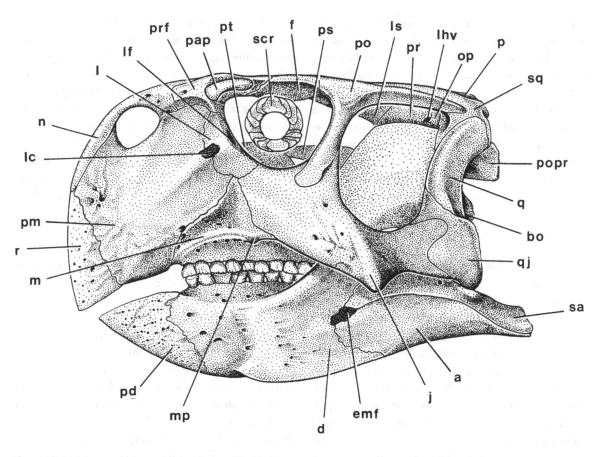

Figure 25.5. *Psittacosaurus mongoliensis*, skull reconstruction in left lateral view. Abbreviations: as in Figures 25.1–25.3 and bo, basioccipital; emf, external mandibular fenestra; lc, lacrimal canal; lhv, lateral head vein; mp, maxillary process; pap, palpebral; pd, predentary; ps, parasphenoid; r, rostral; sc, sclerotic ring. (From Sereno *et al.*, 1988.)

ornithischians. The relatively long and strongly built forelimb and the flattened manual unguals suggest that psittacosaurids may have been facultatively quadrupedal. Unlike later quadrupedal ceratopsians, however, the external digits of the manus are reduced or eliminated such that only digits I-III are functional.

Neoceratopsia Sereno, 1986

Definition. All ceratopsians closer to *Triceratops* than to *Psittacosaurus* (Sereno, 1998).

Chaoyangsaurus youngi. Described briefly by Zhao (1983, 1986), and in more detail by Zhao *et al.* (1999) from possible Jurassic beds in northern China,

Chaoyangsaurus is an intriguing basal ceratopsian intermediate in form between psittacosaurs and other neoceratopsians. Similar to a large psittacosaur in body size, *Chaoyangsaurus* is known only from the holotype specimen (IVPP V11527), which consists of a partial skull with lower jaws, several cervical vertebrae and a partial scapula and humerus. The rostral bone clearly establishes *Chaoyangsaurus* as a ceratopsian, and several features clearly link this early form with later neoceratopsians, including the narrow snout, strongly flared jugal arch, and pair of reduced, subconical premaxillary teeth (Zhao, 1983; Zhao *et al.*, 1999). As in other neoceratopsians, but unlike psittacosaurs and other outgroups, the skull appears to be quite large relative to girdle and appendicular bones, although more complete postcrania are needed for confirmation of this. The low, subtriangular maxillary and dentary crowns are primitive and resemble the condition in psittacosaurs. Likewise, the relatively broad proportions of the laterotemporal fenestra, absence of an epijugal ossification, substantial length of the postdentary elements of the lower jaw, and unfused condition of the anterior cervical vertebrae are plesiomorphic relative to other neoceratopsians. Unfortunately, the posterior portion of the dorsal skull roof and occiput are not preserved, so the presence and development of the parietosquamosal shelf is not known.

Archaeoceratops oshimai. Described recently from Early Cretaceous beds in Gansu Province, China, *Archaeoceratops* is known from two partial skeletons of relatively small size that include a relatively complete skull with lower jaws (Dong and Azuma, 1996, 1997; IVPP V11114, V11115). *Archaeoceratops* is clearly more advanced than *Chaoyangsaurus* on the basis of many features of the skull that closely resemble the condition in *Protoceratops* and other neoceratopsians, including the strong lateral crest on the jugal and the marked anteroposterior shortening of the laterotemporal region and postdentary elements of the lower jaw (Dong and Azuma, 1997, fig. 2). The antorbital fossa has a sharp rim and oval shape as in *Leptoceratops* and *Protoceratops* (contra Dong and Azuma, 1997, fig. 2A). The short parietosquamosal frill, low number of sacral vertebrae, and relatively long tapered tail establish *Archaeoceratops* as a very primitive neoceratopsian. Although reconstructed as a biped (Dong and Azuma, 1997, figs. 11 and 12), the pectoral girdle and forelimb are unknown, and the habitual posture of this early neoceratopsian cannot be reliably determined. Diagnostic features for the genus and species have yet to be identified, but may involve the dentition.

Udanoceratops tschizhovi. Recently described on the basis of a partial skull from Mongolia (Kurzanov, 1992), *Udanoceratops* has a skull length of approximately 0.6 m, which equals that of the largest specimens of *Protoceratops* (Brown and Schlaikjer, 1940, fig.

13). Distinguishing cranial features include an enlarged, oval external naris that, unlike other basal neoceratopsians, exceeds the orbit in maximum diameter. A depression on the posterolateral process of the premaxilla and an extremely deep and strongly arched lower jaw also distinguish this new neoceratopsian. *Udanoceratops* shares with *Leptoceratops* the strongly arched lower jaw and absence of premaxillary teeth, but differs from the latter in having straight tooth rows (Kurzanov, 1992, fig. 2b). The teeth are very similar to *Leptoceratops* and have enamel on both sides of the dentary crowns.

Coronosauria Sereno, 1986
Definition. *Protoceratops*, *Triceratops*, their most recent common ancestor and all descendants (Sereno, 1998).

Bagaceratops rozhdestvenskyi. Known from many specimens from the Hermiin Tsav red beds of the Baruungoyot Formation in Mongolia, *Bagaceratops* is second only to *Protoceratops* in the quantity of known remains, although the postcranial material has not been described in detail (Maryańska and Osmólska, 1975; Osmólska, 1986). As discussed below (see Problematic taxa), *Breviceratops* (Kurzanov, 1990) is regarded here as a junior synonym of *Bagaceratops*, and it is very likely that ?*P. kozlowskii* and *Bagaceratops rozhdestvenskyi* represent the same species.

Cranially and postcranially, *Bagaceratops* is similar to *Protoceratops* in nearly all details. The most outstanding differences in the cranium of *Bagaceratops* are an oval accessory fenestra between the premaxilla and maxilla and a coossified median nasal horn. The premaxillary–maxillary fenestra appears to decrease in size with maturity (Maryańska and Osmólska, 1975, fig. 9). The median nasal horn, which fuses early in growth and migrates posteriorly (Kurzanov, 1990), preserves traces of a median suture on its posterior aspect in immature individuals (PIN 3142/1; PAL MgD-I/125). Thus, the horn is composed of coossified processes of the nasals, as in centrosaurines (Gilmore, 1917), rather than a separate median ossification, as in some chasmosaurines.

Several other aspects of *Bagaceratops* appear to be

artefacts of preservation. The skull has been reconstructed with a short, unfenestrated frill, an antorbital fossa floored in part by the nasal, and a jugal without the accessory epijugal ossification (Maryańska and Osmólska, 1975, fig. 6). The frill in mature individuals, however, is fenestrated, as shown in additional specimens (Kurzanov, 1990, fig. 2; H. Osmólska, pers. comm.). The participation of the nasal in the antorbital fossa, a configuration not found in any ceratopsian, was reconstructed from a specimen that does not preserve this portion of the fossa (Maryańska and Osmólska, 1975, pls. 42 and 43). Attachment scars on the jugal and quadratojugal (PAL MgD-I/125) indicate that an epijugal, as large and prominent as in *Protoceratops*, is present in *Bagaceratops* (contrary to Dodson and Currie, 1990, p. 613). The squamosal–jugal contact occurs above the laterotemporal fenestra (contrary to Dodson and Currie, 1990, p. 613), but this contact is exposed only on the medial side of the postorbital (Maryańska and Osmólska, 1975, p. 158; Kurzanov, 1990, fig. 1). The reduction of this contact (and the posterior arching of the ascending ramus of the jugal) may characterize *Bagaceratops*.

Other features previously considered diagnostic for *Bagaceratops* – such as the low number of maxillary teeth (10) and straight margin of the lower jaw – are probably due to the immaturity of even the largest available specimens. All of these features occur in immature individuals of *Protoceratops* (Brown and Schlaikjer, 1940; Kurzanov, 1992). In addition, the absence of premaxillary teeth in *Bagaceratops* requires further documentation, given the poor preservation of the critical posterior margin of the premaxilla in all available specimens and the presence of premaxillary teeth in juvenile individuals (Dong and Currie, 1993).

Protoceratops andrewsi. Based on a splendid series of skeletons from hatchlings to adults, *Protoceratops* is the best known neoceratopsian (Brown and Schlaikjer, 1940). Generic and specific diagnoses for *Protoceratops*, nevertheless, do not include any derived features (e.g., Steel, 1969) because the skeleton is plesiomorphic in nearly all regards at the level of Neoceratopsia. Possible autapomorphies include the short lateral processes on the rostral, low tab-shaped processes on the frill margin (three on the squamosal and four or five on the parietal), parasagittal nasal prominences, and hoof-shaped pedal unguals (Figure 25.6).

Graciliceratops mongoliensis, n. gen., n. sp. Bohlin (1953) erected a new genus, *Microceratops*, with two new species on the basis of teeth, fragmentary jaws and assorted postcrania, much of which he believed to be from immature individuals. This fragmentary material came from two localities in different horizons in Gansu Province, China, the ages of which remain uncertain (Dong and Azuma, 1997). Bohlin (1953, p. 35) observed that the primary ridge in the dentary teeth in *Microceratops gobiensis* may be less prominent than in the closely related genus *Protoceratops*. No other diagnostic features were given, and it can be seen that the primary ridge varies in strength in the dentary crowns figured by Bohlin. Furthermore, the holotype dentary (Bohlin, 1953, fig. 14c) does not have any complete crowns and is now apparently lost (Z. Dong, pers. comm.). Young (1958, fig. 1B) referred an isolated maxilla and other small neoceratopsian material from Shansi Province, China to *M. gobiensis* on the basis of its small size. There appears to be no other basis for this referral. The second species, *Microceratops sulcidens*, is based on two isolated teeth, vertebrae and bones of the manus and pes (Bohlin, 1953, figs. 36–38, pl. II). The small size of this material is usually the only feature mentioned in taxonomic diagnoses (e.g., Steel, 1969). Dodson and Currie (1990, tab. 29.1) listed *M. sulcidens* as a junior synonym of *M. gobiensis*, but gave no reasons for this synonymy. Given the absence of any diagnostic features of the holotype material and the abundance of immature individuals at many Asian localities that have yielded ceratopsian remains, the genus *Microceratops* and the species *M. gobiensis* and *M. sulcidens* are regarded here as *nomina dubia*.

Maryańska and Osmólska (1975) referred an articulated skeleton (PAL MgD-I/156) from Shireegiin Gashuun in Mongolia to *Microceratops gobiensis*, ostensibly because of its small size relative to other basal neoceratopsians. Although no other reason was given, this referral has never been questioned. The primary

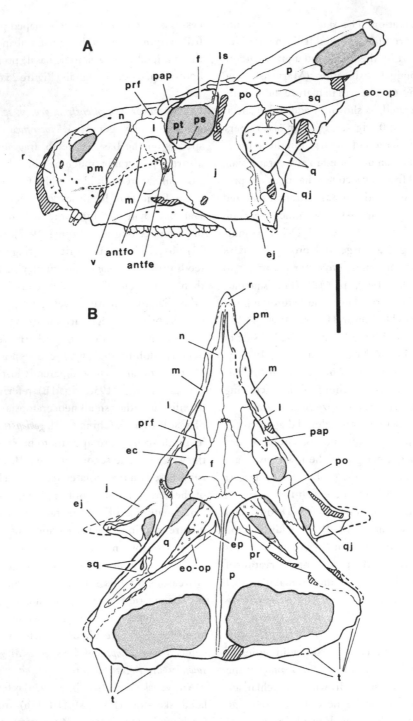

Figure 25.6. *Protoceratops andrewsi* (AMNH 6408), cranium in left lateral (A) and dorsal (B) views. Dashed outline shows position of vomer. Abbreviations: as in Figures 25.1–25.5 and antfe, antorbital fenestra; ej, epijugal; eo-op, exoccipital-opisthotic (fused); ep, epiotic; t, tab-shaped flange; v, vomer. Scale bar equals 50 mm.

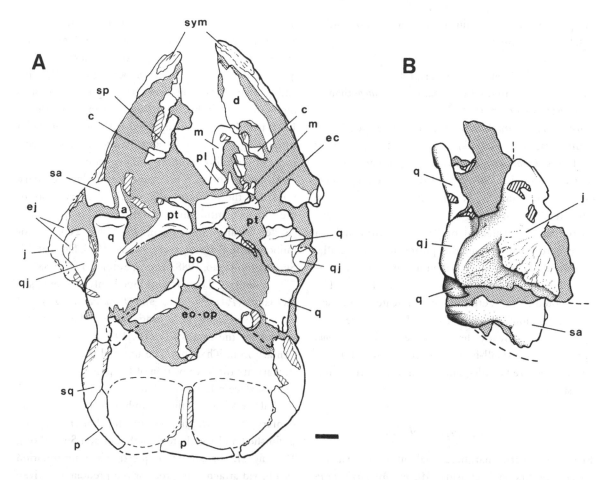

Figure 25.7. *Graciliceratops mongoliensis*, gen. nov., sp. nov. (A) Partial skull in ventral view and (B) posterior portion of skull in right lateral view showing inset articular surface for the epijugal. Abbreviations: as in Figures 25.1–25.6. Scale bar equals 10 mm.

ridge in the dentary crowns of this specimen, however, is as prominent as in other neoceratopsians. There is no basis, therefore, for referral of this specimen to *Microceratops* or to the species *M. gobiensis*. Because this skeleton is also from an immature individual (as shown by the disarticulated presacral neural arches and unfused sacral centra), its body size at maturity remains unknown and may well have equalled that of *Protoceratops*.

This skeleton is placed here in a new genus and species, *Graciliceratops mongoliensis* (*gracilis*, L. slender; *cerato-*, Gr. horn; *mongolia*, Mongolia; *-ensis*, L. place), characterized by the very slender median and poste-

rior parietal frill margins (Figure 25.7A) and high tibiofemoral ratio (1.2:1). The slender frill margins are very distinctive and much more delicate than reconstructed by Maryańska and Osmólska (1975, fig. 1). The frill extends quite far posterior to the occiput, and its lateral margins are formed by a well developed posterior process of the squamosal. Like *Bagaceratops*, the jugal and squamosal do not overlap extensively, as shown by well marked articular scars on the postorbital. A well demarcated scar on the jugal and quadratojugal indicates that a large epijugal was present (Figure 25.7B). The quadratojugal would have been exposed primarily in posterior view on the posterior aspect of the prominent jugal–epijugal horn, as in most other basal neoceratopsians.

Ceratopsoidea Hay, 1902

Definition. All coronosaurs closer to *Triceratops* than to *Protoceratops* (Sereno, 1998).

Turanoceratops tardabilis. Based on isolated teeth and cranial fragments of unknown association (Nesov *et al.*, 1989), *Turanoceratops* provides important evidence that two-rooted maxillary and dentary teeth, previously known only in North American ceratopsids, appeared first in much smaller Asian ceratopsians during the Cenomanian or Turonian. Broken horn cores (presumably from the postorbital), a maxilla, and predentary were described along with the two-rooted teeth (Nesov *et al.*, 1989, pl. 1, figs. 16–21). The two-rooted cheek teeth probably indicate an increase in packing along the tooth row, and a primitive tooth battery may already have evolved. *Turanoceratops* appears to be a small-bodied neoceratopsian. Further study is required to adequately characterize the genus and species.

Problematic marginocephalians

Five problematic marginocephalians from Asia are considered first and set aside. *Micropachycephalosaurus hongtuyanensis* (Dong, 1978; IVPP V5542) is based on fragmentary postcrania of discordant size. Much of the ilium (Dong, 1978, fig. 2) exists only as an impression in rock. Although this taxon has survived recent systematic review (Maryańska, 1990), no pachycephalosaurian features or autapomorphies are apparent in this material. *Micropachycephalosaurus* is here considered a *nomen dubium*.

Psittacosaurus sattayaraki, recently described on the basis of a partial dentary and possibly a fragment of the maxilla (TF 2449) from Cretaceous beds in Thailand, is regarded by the authors as 'clearly referable to the genus *Psittacosaurus*' (Buffetaut *et al.*, 1989, p. 370). The justification given by Buffetaut and Suteethorn (1992, pp. 803, 805) for the generic reference ('relatively deep and short dentary' and 'bulbous primary ridge and secondary denticles') and erection

of a new species ('small incipient ventral flange,' 'strongly convex' alveolar region of the dentary and 'five denticles on both sides of the primary ridge') is questionable. The association between the dentary and maxillary fragment, described as 'possibly belonging to the same individual' (Buffetaut and Suteethorn, 1992, p. 801), must be regarded with suspicion as no supporting evidence for association of these specimens, collected years apart, was presented.

All basal ceratopsians have short, deep dentary rami, and the primary ridge on the dentary tooth of the Thai ornithischian, such as it is preserved (Buffetaut and Suteethorn, 1992, fig. 2C), is not bulbous as in *Psittacosaurus* (Osborn, 1923, figs. 4 and 5; Sereno and Chao, 1988, fig. 5C; Sereno *et al.*, 1988, fig. 7D). A dentary flange is developed only in some psittacosaurs (*P. mongoliensis* and *P. meileyingensis*) and extends vertically as a ridge across the posterior portion of the ramus (Sereno, 1990a, b), unlike the dentary from Thailand. The unusual features of this dentary are the low position of the predentary attachment surface relative to the tooth row and the abrupt medial arching of the symphysial region of the dentary. The anterior end of the dentary appears unfinished and weathered (Buffetaut and Suteethorn, 1992, fig. 2E), bringing into question its interpretation as a broad attachment area for the predentary. Given the poor preservation of the dentary and its dubious association with the supposed maxillary fragment, assignment to a new genus of uncertain phylogenetic affinity is not warranted. '*Psittacosaurus*' *sattayaraki* is tentatively referred here to Ceratopsia, incertae sedis.

Psittacosaurus mazongshanensis, described recently from a skull and partial skeleton from Gansu Province, China (Xu, 1997; IVPP V12165), may represent a distinct species. The available description, however, does not establish that fact convincingly, as there are no clear diagnostic features that are absent in other psittacosaur species. Until the basis for this species is clarified, the partial skeleton is here referred to *Psittacosaurus*, incertae sedis, and the species *P. mazongshanensis* is regarded as a *nomen dubium*.

Asiaceratops salsopaludalis, described from disarticulated, fragmentary remains from Cenomanian or early

Turonian beds in Uzbekistan, is a small basal ceratopsian (Nesov *et al.*, 1989). If properly assigned to *Asiaceratops*, the unguals are pointed as in most basal ceratopsians and unlike the broader unguals in *Protoceratops*. The maxillary tooth rows in at least two individuals have nine teeth, which is fewer than in subadult specimens of *Protoceratops*. The dentition is indistinguishable from that in several other basal ceratopsians, and no other autapomorphies are apparent in the holotype (a left maxilla) or referred material. *Asiaceratops* therefore is regarded here as a *nomen dubium*.

Finally, Kurzanov (1990) recently transferred *Protoceratops kozlowskii* to a new genus, *Breviceratops*. Maryańska and Osmólska (1975) originally tentatively referred an immature holotype skull (PAL MgD 1/117) from the locality Khulsan to *Protoceratops* as a new species, *?P. kozlowskii*. The new referred material consists of five partial skulls of immature individuals (PIN 3142/1–5; Kurzanov, 1990, figs. 1 and 2) from a different locality (Hermiin Tsav) in the same formation (Baruungoyot). The taxonomic status of *?P. kozlowskii* and *Bagaceratops rozhdestvenskyi*, the latter based on material also collected at Hermiin Tsav, is complicated by the immaturity and incompleteness of many of the specimens.

The diagnostic features originally listed for *?P. kozlowskii* (Maryańska and Osmólska, 1975, pp. 143–144) are either present in juveniles or adults of other species or are difficult to assess. The position of the nasal–frontal suture above the orbit, for example, is not unique to *?P. kozlowskii* but rather characterizes *Protoceratops andrewsi* and juveniles of *Bagaceratops*. The supposed advanced characters in the postcranium, such as the stronger lateral flare of the iliac preacetabular process, are based on the very immature holotype skeleton and were not figured or photographed in a manner allowing comparison. There are no unique features linking *?P. kozlowskii* to the genus *Protoceratops*. The presence of a nasal horn, which occurs in specimens referred to this species (Kurzanov, 1990) and in *Bagaceratops*, cannot be determined in the material upon which the species was based. There is some indication that an accessory premaxilla–maxilla fenestra

may have been present in the holotype skull of *?P. kozlowskii* (Maryańska and Osmólska, 1975, pl. 50, fig. 1a), as also occurs in specimens later referred to this species (Kurzanov, 1990) and in *Bagaceratops*. The characters Kurzanov (1990) invoked to distinguish this species from *Bagaceratops* (larger size, higher and wider skull, and parietal fenestrae) are not valid given the better preserved material now known for *Bagaceratops*. The only apparent difference between *?P. kozlowskii* and *Bagaceratops* is the presence of premaxillary teeth in the former, but this can no longer be considered significant given the presence of premaxillary teeth in immature individuals of *Bagaceratops* (Dong and Currie, 1993, fig. 3). The subcylindrical premaxillary teeth in these juveniles appear to have been lost during growth in *Bagaceratops*, although the one adult skull with an intact ventral margin of the premaxilla is not complete posteriorly (PAL MgD-I/127; Maryańska and Osmólska, 1975, pl. 45, 1c). In summary, it seems very likely that all of the basal ceratopsian specimens from the Baruungoyot Formation pertain to a single species. The most appropriate name for that taxon is *Bagaceratops rozhdestvenskyi*, the holotype of which exhibits several diagnostic features. *Breviceratops* is regarded here as a junior synonym of *Bagaceratops*.

Phylogeny

In the following sections, previous work on the phylogeny of marginocephalians is reviewed, marginocephalian synapomorphies are re-examined, and the branching pattern within Pachycephalosauria and Ceratopsia is analyzed. The central phylogenetic issues to resolve are the affinities of two enigmatic early marginocephalians, *Stenopelix* (Schmidt, 1969; Sues and Galton, 1982; Sereno, 1987) and *Chaoyangsaurus* (Zhao, 1983; Zhao *et al.*, 1999), and the phylogenetic reality of the basal subgroups Homalocephalidae and Protoceratopsidae.

Marginocephalia

Traditional classification. After the description of the first relatively complete skull and skeleton (*Stegoceras validus*; Figure 25.3) by Gilmore (1924), early opinion presented two possibilities regarding the affinities of pachycephalosaurs (then termed 'troödonts'). Gilmore (1924) and others (Russell, 1932; Sternberg, 1933) regarded pachycephalosaurs as divergent ornithopods, and the Family Pachycephalosauridae was erected within Ornithopoda (Sternberg, 1945). Others linked pachycephalosaurs with ankylosaurs and were influenced by the downwardly curved shaft of the ischium and posterior extension of the palate (Romer, 1927, 1968) or by the armoured, akinetic condition of the skull (Nopcsa, 1929). Brown and Schlaikjer (1943, p. 146) sided with ornithopod origins, concluding that pachycephalosaurs shared 'a closer relationship to the Ceratopsia–Ornithopoda line than to the Stegosauria–Nodosauria group.' Besides noting similarities that are now clearly understood as plesiomorphic, they mentioned derived similarities shared with *Protoceratops*, such as grooved zygapophyseal articulations in the dorsal vertebrae and the downward curve of the ischial shaft. The former, now known in several pachycephalosaurs, is not present in *Protoceratops* or any other ceratopsian; the latter constitutes a potential synapomorphy as discussed below.

In summary, pre-cladistic notions of pachycephalosaurian ancestry were based as much on overall similarity as on the presence of shared derived characters. Bipedal ornithischians, such as pachycephalosaurs and psittacosaurs, were presumed to have evolved from a persistently primitive ornithopod stock (e.g., Romer, 1968; Steel, 1969; Galton, 1972; Thulborn, 1974) and were generally classified within Ornithopoda. More recently, pachycephalosaurs were removed from the Suborder Ornithopoda and accorded subordinal rank as Pachycephalosauria (Maryańska and Osmólska, 1975). Removal from Ornithopoda was not initiated on phylogenetic grounds, but rather was predicated upon the degree to which pachycephalosaurs were judged to have diverged from mainline ornithopods. Raising rank on the basis of morphologic distance, however, is an arbitrary phenetic decision (Sereno, 1990c), as arguments opposing such revision attest (Wall and Galton, 1979, p. 1185).

Pre-cladistic discussion of ceratopsian ancestry followed a similar pattern – an ambivalent relationship with Ornithopoda, the group believed to encompass the ancestral mainline of ornithischian evolution. Initially described as a 'pre-ceratopsian', *Protoceratops* was heralded as the bridge between ceratopsids and 'such primitive Jurassic Ornithopoda as *Hypsilophodon*' (Granger and Gregory, 1923, p. 4). The Family Protoceratopsidae, comprising *Protoceratops* and *Leptoceratops*, was later allied with Ceratopsidae within Ceratopsia (Gregory and Mook, 1925). Psittacosaurs, likewise, were originally described and classified as ornithopods (Osborn, 1923). Although Gregory (1927) outlined several ceratopsian features in the skull of psittacosaurs shortly after their initial discovery, many years elapsed before psittacosaurs were placed within Ceratopsia. The identification of the ceratopsian rostral bone, which was initially regarded as the premaxilla by Osborn (1923), played a key role in the recognition of psittacosaurs as basal ceratopsians (Romer, 1968; Maryańska and Osmólska, 1975).

Recent studies. Coombs (1979, p. 679) mentioned several features that unite pachycephalosaurs and ankylosaurids including the everted dorsal margin of the preacetabular process, ossification of an interorbital septum, 'tendency to close the supratemporal fenestra,' and 'armour-like texturing of the dorsal skull roof.' As discussed by Sues and Galton (1987, p. 36), these features fail to unite these groups because of problems of definition, homology, and distribution. One feature mentioned by Coombs (1979) – contact between the ilium and ischium on the anterior side of the acetabulum (i.e., exclusion of the pubis from the acetabular margin) – is a potential pachycephalosaur–ankylosaur synapomorphy. It is clearly manifest in both ankylosaurs (*Sauropelta*; YPM 541) and pachycephalosaurs (Maryańska and Osmólska, 1974) and is absent in all other ornithischians. This apomorphy, however, may not be present in the most primitive pachycephalosaur (see *Stenopelix* below) and is absent in thyreophoran outgroups to Ankylosauria as well. Thus, although character support for Marginocephalia is not overwhelming (as discussed below), opposing data is extraordinarily weak in the context of ornithischian phylogeny.

Sereno (1984, 1986) and Maryańska and Osmólska (1985) provided the first character evidence to establish a phylogenetic link between pachycephalosaurs and ceratopsians. Coining the name Marginocephalia for the combined clade, Sereno (1986) narrowed an initial list of nine proposed synapomorphies to four, which are further tailored here to three: (1) posterior extension of a parietosquamosal shelf that obscures the occiput in dorsal view of the skull; (2) median contact between the maxillae that excludes the premaxillae from the anterior margin of the internal nares; and (3) a short postpubic process that lacks the distal pubic symphysis.

The first two synapomorphies are unique among ornithischians. Sues and Galton (1987, p. 36) criticized the first synapomorphy because 'the parietosquamosal shelf of pachycephalosaurs shows no close resemblance to the frill of ceratopsians, which is characterized by transverse expansion of the parietal overhang.' The synapomorphy in question, however, concerns only the presence of a parietosquamosal shelf, not the relative composition of the shelf. The predominance of the parietal in forming the shelf was listed separately as a ceratopsian synapomorphy (Sereno, 1986), because the parietal in pachycephalosaurs and other ornithischians usually forms only a small proportion of the posterior margin of the skull roof. Dodson (1990, p. 562) remarked that the second synapomorphy is 'plesiomorphic for the group [Marginocephalia],' although no supporting evidence was cited. Marginocephalian outgroups, nevertheless, exhibit the plesiomorphic condition, in which the premaxillae form the anterior rim of the internal nares (e.g., *Hypsilophodon*, *Lesothosaurus*; Sereno, 1991). Sues and Galton (1987, p. 36) rejected the third synapomorphy because it is also present in Ankylosauria. The primitive condition (long postpubic process with distal symphysis), however, clearly obtains in more primitive armoured dinosaurs (thyreophorans), basal ornithopods, and the basal ornithischian *Lesothosaurus* and must be considered the outgroup condition for Marginocephalia. Other features, such as the ventral curvature of the ischial shaft, may eventually support Marginocephalia in a higher-level quantitative analysis. These features,

however, are particularly homoplastic – i.e., they are not uniformly present among marginocephalians and absent in outgroups.

Pachycephalosauria

Traditional classification. The only pre-cladistic phylogenetic tree of pachycephalosaurs shows an ancestral relationship between *Stegoceras* (*Troödon*) and *Pachycephalosaurus* (Brown and Schlaikjer, 1943, p. 148) – not surprising given that all other pachycephalosaurian genera have been described in the past 25 years.

Recent studies. The description of several new pachycephalosaurs from Mongolia (Maryańska and Osmólska, 1974; Perle *et al.*, 1982), China (Hou, 1977) and western North America (Giffen *et al.*, 1987) has opened the door to phylogenetic analysis. In the first cladogram of pachycephalosaurs, Sereno (1986) arranged five of the best known genera as a series of sister taxa to the large, fully domed, long-snouted genus *Pachycephalosaurus*. Three genera (*Wannanosaurus*, *Goyocephale*, *Homalocephale*) were positioned at basal nodes and comprise the so-called 'flat-headed' pachycephalosaurs. The domed genus *Stegoceras* occupied an intermediate position as sister taxon to two fully domed genera, *Prenocephale* and *Pachycephalosaurus*. Two familial names (Tholocephalidae, Domocephalinae) were proposed for subgroups in the analysis, but these are invalid because they are not based on existing genera. Although the phylogeny was based on 37 characters, only 13 apply to internal nodes on the cladogram, which reflects the very incomplete comparative information available for most pachycephalosaurs.

Sues and Galton (1987) presented an alternative phylogenetic arrangement, which divides pachycephalosaurs into 'flat-headed' (Homalocephalidae) and 'dome-headed' (Pachycephalosauridae) clades, following an earlier suggestion by Dong (1978). Maryańska (1990, fig. 27.5) followed Sues and Galton (1987), but (without explanation) altered the position of *Tylocephale* among pachycephalosaurids.

The character evidence listed by Sues and Galton (1987, p. 35) overlaps broadly with that in Sereno

(1986, pp. 243–244) with some notable exceptions. Thickening of the skull table characterizes all pachycephalosaurs known from cranial remains (Sereno, 1986). The only character evidence supporting the monophyly of 'flat-headed' pachycephalosaurs (Homalocephalidae) is the flat condition of the dorsal skull roof (as in Maryańska, 1990, p. 574). The growth series available for *Stegoceras* shows, however, that during growth the dome rises at the centre of an already thickened skull table (e.g., NMC 138; Lambe, 1918, pls. 1 and 2). The dome is surrounded by a broad, thickened, marginal shelf that is indistinguishable from that in 'flat-headed' forms. This strongly suggests that the 'flat-headed' condition is plesiomorphic within Pachycephalosauria (because it is also present early in growth in domed forms).

Within the 'flat-headed' group, Sues and Galton (1987) unite *Goyocephale* and *Homalocephale* (to the exclusion of *Wannanosaurus*) on the basis of the small size of the supratemporal fenestrae and the presence of squamosal tubercles. But these derived features are also present in all of the domed genera in their analysis (*Stegoceras*, *Tylocephale*, *Prenocephale*, *Pachycephalosaurus*) and characterize a more inclusive group of pachycephalosaurs (Sereno, 1986).

Much of the branching pattern in the domed clade outlined by Sues and Galton (1987) is based on an elaborate scenario for the evolution of the fully domed condition. A 'structural sequence from *Yaverlandia* to *Pachycephalosaurus*' is hypothesized, beginning with parasagittal frontal doming and followed by median frontal doming, frontal versus parietal doming, and ultimately frontoparietal doming (Sues and Galton, 1987; Maryańska, 1990). That *Majungatholus* – based on a thickened frontoparietal and braincase (MNHN MAJ4; Sues and Taquet, 1979; Sues, 1980) that pertains to an abelisaurid theropod – has been incorporated effortlessly into this sequence is telling. This scenario could be justified in a quantitative cladistic analysis only if it were coded as a single, ordered multistate character. Moreover, because this doming scenario is the only character evidence listed for several nodes within the domed clade (Sues and Galton, 1987), the implied *a priori* ordering of this character also specifies the structure of their cladogram.

As far as I can discern, several of the inferred stages in the development of the fully domed condition – such as the paired frontal thickenings in *Yaverlandia* (Galton, 1971) – do not occur in more than one taxon at any growth stage and therefore constitute autapomorphies. Other synapomorphies mentioned by Sues and Galton are based on incorrect information. *Tylocephale*, for example, has a distinct row of tubercles on the postorbital and a pair of supraorbital elements; there is no available character evidence to link *Tylocephale* and *Stegoceras* as closest relatives. Frontoparietal fusion, a synapomorphy used to link *Stegoceras* and fully domed pachycephalosaurs (Sues and Galton, 1987), is an informative synapomorphy, but is also present in *Yaverlandia*. In summary, no character evidence has been discovered to date that will support the monophyly of 'flat-headed' pachycephalosaurs. Domed pachycephalosaurs, on the other hand, have been viewed as a monophyletic subgroup (Sereno, 1986; Sues and Galton, 1987), in which the partially domed *Stegoceras* is the sister taxon to fully domed genera.

Stenopelix valdensis, based on the natural mold of a single postcranial skeleton (GPI 741–2) from the Early Cretaceous (Barremian) of Europe (inadvertently listed as Berriasian in age by Dodson, 1990, p. 563), has been regarded in recent studies as a basal pachycephalosaur (Maryańska and Osmólska, 1974; Sereno, 1987), a basal ceratopsian (Sues and Galton, 1982), and, most recently, the sister group to Pachycephalosauria plus Ceratopsia (Dodson, 1990, p. 563). Regarding the latter hypothesis, no supporting evidence was mentioned and it will not be considered further. Sues and Galton (1982, p. 188) also did not specify synapomorphies for their referral of *Stenopelix* to the Ceratopsia, stating only that such reference was based on the 'structure of the pelvic girdle, especially the form of the ilium and the reduced pubis.' However, I am not aware of any derived characters in the pelvic girdle shared by *Stenopelix* and ceratopsians or, for that matter, by ceratopsians alone. The downwardly curved preacetabular process of the ilium in *Stenopelix* occurs in several ornithischian subgroups (e.g., pachycephalosaurs), and the short prepubic process figured by Sues and Galton (1982, fig. 1A) for *Stenopelix* is half the length of the process as preserved in the natural mould (Schmidt, 1969, fig. 1).

The hypothesis of Maryańska and Osmólska (1974, pp. 48, 101), that *Stenopelix* shares a close relationship with pachycephalosaurs, deserves closer scrutiny. Two of the three characters listed to support this connection – tibia shorter than femur, and pubis excluded from the acetabulum – are not valid. Regarding the first, the tibia and femur are equal in length in *Stegoceras*, the only pachycephalosaur in which this can be measured. Other marginocephalians and a variety of marginocephalian outgroups, moreover, have very similar tibiofemoral ratios. Regarding the second feature, the pubis in *Stenopelix* clearly forms a significant portion of the acetabular margin, as observed by Sues and Galton (1987). The third feature mentioned by Maryańska and Osmólska (1974) – elongate anterior caudal ribs – is based, apparently, on the elongate posterior sacral ribs in the holotype skeleton (Sereno, 1987). This unusual feature, also present in pachycephalosaurs, is discussed below.

Present results. The following summary of pachycephalosaur phylogeny is based on an analysis of 41 characters in 12 species (Table 25.3; Figure 25.8; Appendix, Sereno, 1999a). The character data is derived for the most part from 37 synapomorphies listed in Sereno (1986, pp. 243–4). Some of these were omitted upon review; others were combined in the process of character coding; and several new characters have been added. One character, the position of a neomorphic process on the iliac blade (character 28), is not phylogenetically informative (because it cannot be polarized). Using Ceratopsia and Ornithopoda as successive outgroups, the analysis resulted in 15 most parsimonious trees (42 steps; consistency index, 0.95; retention index, 0.97; Fig. 25.8). These trees differ only in the position of two taxa that are based on frontoparietals – *Ornatotholus* and an undescribed dwarf pachycephalosaur from Alberta. *Ornatotholus* forms an unresolved trichotomy with *Homalocephale* and a group consisting of *Yaverlandia* and more derived, domed forms. The dwarf form belongs among fully domed pachycephalosaurs, but its more precise relations cannot be determined without additional information. Removal of *Ornatotholus* and the dwarf pachycephalo-

saur from the analysis results in a single tree involving nine pachycephalosaurs and that lacks any homoplasy (41 steps; consistency and retention indices, 1.0).

There is no available character evidence supporting the monophyly of flat-headed pachycephalosaurs. Using the framework phylogeny of nine pachycephalosaur genera mentioned above, six additional steps are required to maintain a clade of flat-headed pachycephalosaurs. Most of the additional homoplasy is introduced by *Wannanosaurus*, which is distinctly more primitive than other pachycephalosaurs. As discussed below, however, some of this apparent plesiomorphy may be attributable to the immaturity of the holotype.

Although the present analysis is nearly free of homoplasy and does not support the monophyly of flat-headed pachycephalosaurs, the most parsimonious arrangement is not particularly robust. Accepting trees two steps longer than the most parsimonious tree (42 steps) for the nine most complete taxa yields five trees, the strict consensus of which collapses the more advanced position of *Goyocephale* relative to *Homalocephale* and collapses most relationships among domed genera. The loss of structure is caused by the significant amount of missing data for most available taxa (approximately 50% or more in two-thirds of included genera).

In the following discussion, character numbers correspond to those tabulated in the Appendix, and synapomorphies are described at their least inclusive node (i.e., under delayed-transformation optimization).

Basal pachycephalosaurs. *Stenopelix* is positioned in this analysis as the most basal pachycephalosaur on the basis of three synapomorphies: (1) elongate posterior sacral ribs; (2) strap-shaped distal end of the scapular blade; and (3) distal expansion of the preacetabular process of the ilium. The peculiar elongate posterior sacral ribs (fourth to sixth) in *Stenopelix* and other pachycephalosaurs broaden by about 30% the transverse width of the posterior end of the sacrum (Maryańska and Osmólska, 1974). The strap-shaped scapular blade, preserved, but not yet described, in *Stenopelix* (Sereno, 1987) is very similar to that in *Stegoceras*, the only other pachycephalosaur in which

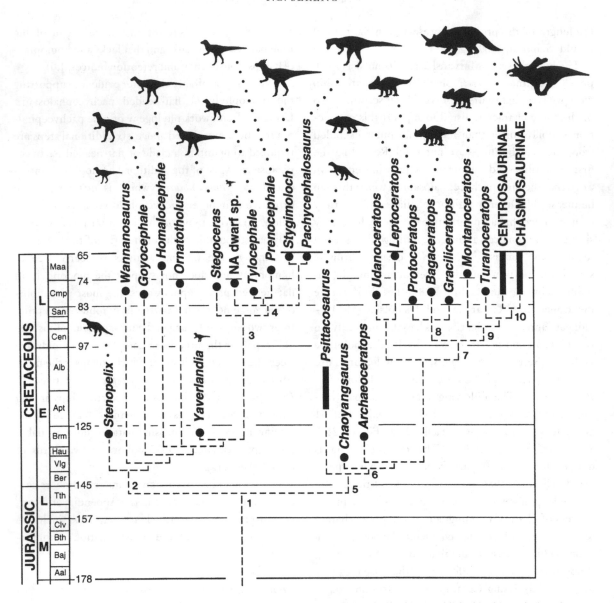

Figure 25.8. Calibrated phylogeny for marginocephalians based on cladistic relationships established in this analysis and recorded temporal ranges. The ages of *Chaoyangsaurus* and *Archaeoceratops* are uncertain, but probably lie somewhere between the latest Jurassic and the end of the Neocomian. Abbreviations: 1, Marginocephalia; 2, Pachycephalosauria; 3, Pachycephalosauridae; 4, Pachycephalosaurinae; 5, Ceratopsia; 6, Neoceratopsia; 7, Coronosauria; 8, Protoceratopsidae; 9, Ceratopsoidea; 10, Ceratopsidae.

this bone is known (Sues and Galton, 1987, fig. 10). A strap-shaped scapular blade also occurs in a few basal neoceratopsians (*Protoceratops;* Brown and Schlaikjer, 1940, fig. 26) and in heterodontosaurids (Santa Luca, 1980). The lobe-shaped expansion of the distal end of

the preacetabular process of the ilium in *Stenopelix* is broader than the base of the process by about 30%. This is very similar to the shape of the process in several pachycephalosaurs. Distal expansion of the preacetabular process occurs only rarely in other

Table 25.3. *Character-taxon matrix for Pachycephalosauria. (See Appendix for characters and character states.)*

		10			20			30			40	
ORNITHOPODA	00000	00000	00000	00000	00000	00X00	00000	000?0	0
CERATOPSIA	00000	00000	00000	00000	00000	00X00	00000	000?0	0
Stenopelix	111??	????0	0?0??	????0	?00??	??X00	0????	?????	?
Wannanosaurus	???11	11111	11100	0?0??	???0?	?????	?0000	0??00	?
Goyocephale	1?111	11111	11?11	11111	1110?	??00?	?0000	0??00	0
Homalocephale	1?11?	1111?	??111	11?11	11111	11111	10000	00000	0
Ornatotholus	???1?	??1??	???1?	?????	???1?	?????	?00?0	0????	?
Yaverlandia	???1?	1?0??	???1?	?????	?????	?????	?10?0	0????	?
Stegoceras	?1111	11111	11111	11111	11111	11111	11111	00000	0
NA dwarf sp.	???1?	??0??	???1?	?????	???1?	?????	?11?1	1????	?
Tylocephale	???1?	1111?	???11	1?1??	???11	?????	?1111	11100	0
Prenocephale	1?111	1111?	???11	11??1	?1111	11111	11111	11100	0
Stygimoloch	???1?	111??	???1?	1????	???1?	?????	?1111	1??11	1
Pachycephalosaurus	???1?	1111?	???11	11???	???11	?????	?1111	11011	1

ornithischians (e.g., *Centrosaurus*, Lull, 1933; *Kentrosaurus*, Galton, 1982).

All remaining pachycephalosaurs, including the diminutive *Wannanosaurus* (Figure 25.1), are united by a suite of cranial and postcranial synapomorphies. The classic cranial features of pachycephalosaurs are already evident and include a thickened frontal and parietal portion of the skull roof (4), broadened and flattened postorbital–squamosal bar (6), broad exposure of the squamosals on the occiput (7), two supraorbital elements forming the roof of the orbit lateral to the frontal (8), and an arched premaxillary–maxillary diastema (5) that very likely accommodated a dentary canine (Hou, 1977, fig. 1). Postcranial synapomorphies include the shortened forelimb (humerus less than 50% of femur) (10), bowed humeral shaft (11) with reduced deltopectoral crest (12), and slender midshaft of the fibula (13). Most of these postcranial synapomorphies can be verified as absent in *Stenopelix*.

Several synapomorphies link other 'flat-headed' pachycephalosaurs, in particular *Goyocephale* and *Homalocephale*, with more advanced forms. The supratemporal openings are reduced in size and the frontals are excluded from their margins (14). The skull is less kinetic, as evidenced by the broad postorbital–jugal bar (15) and the plate-shaped basal tubera (17).

Classic pachycephalosaur ornamentation is present with a linear row of at least five prominent tubercles on the posterior rim of the squamosal (16) and a smaller row of tubercles on the angular (18). Only rarely are the squamosal tubercles suppressed (Goodwin, 1990, fig. 14.5). In the most advanced pachycephalosaurs, such as *Pachycephalosaurus*, the squamosal tubercles are clumped. Diagnostic features in the girdles include shafted sternals (21), which resemble those in ankylosaurs and in advanced iguanodontians, and the presence of an unusual subtriangular process that projects medially from the dorsal margin of the iliac blade (23).

Several of these synapomorphies cannot be scored in *Wannanosaurus* because of the incompleteness of available material (IVPP V4447, V4447.1). The apparent immaturity of these remains (as suggested by the open sutures), moreover, may cast doubt on the interpretation of other features, such as the tubercle row on the angular, that may appear with age. Therefore, the position of *Wannanosaurus*, as the sister taxon to other pachycephalosaurs, is regarded here as tenuous.

Homalocephale appears to be more advanced than *Goyocephale* in two regards. In *Goyocephale* the parietal roof between the supratemporal fossae is smooth and transversely arched as in basal ceratopsians and ornithopods. In *Homalocephale* the parietal is flattened, broadened transversely, and textured (24), similar to other thickened portions of the skull roof, and the size of the supratemporal fossae is reduced (Perle *et al.*, 1982). Other cranial synapomorphies uniting *Homalocephale* and more advanced pachycephalosaurs include the highly derived pterygoquadrate processes, which project posteriorly above the palate (26), and the complete separation of subtemporal and occipital spaces by a flange of the prootic and basisphenoid (27) (the condition in *Goyocephale* remains unknown). In the postcranium, the medial process on the iliac blade, which is positioned above the acetabulum in *Goyocephale*, is located more posteriorly on the postacetabular process (28) and continues to the distal end of the process as a tapering flange (29).

In *Yaverlandia* and all other pachycephalosaurs, the frontals fuse early in ontogeny, completely obliterating the interfrontal suture internally and externally (32). The frontals also fuse to the parietal, although this suture is often visible on the roof of the braincase (common in *Stegoceras*). The holotype and only specimen of *Yaverlandia* (MIWG 1530) follows this pattern, with both the interfrontal and frontoparietal sutures fused externally, though the latter are still visible on the internal surface of the roof of the braincase.

Pachycephalosauridae, domed forms. *Stegoceras* bridges a morphological gap between flat-headed forms and those with a fully developed dome (Figure 25.3). In *Stegoceras* the coossified frontoparietal is strongly domed by upgrowth of vertical columns of bone (33), most of which occurs long after hatching. Doming of the frontoparietal on this scale appears to have evolved only once among pachycephalosaurs. In *Stegoceras* the dome never fully incorporates surrounding elements of the dorsal skull roof, which have deep columnar bone along their sutural contact with the frontoparietal, but which always maintain at least a narrow external shelf. This is true even in the oldest, most prominently domed individuals (see Goodwin, 1990, figs. 14.4 and 14.5). Other features that unite *Stegoceras* and fully domed forms include the closure, or near closure, of the supratemporal fossa (35) and the strong posterior displacement of the

parietal and squamosal over the occiput (34), as is best visualized in side view. Significant doming of the frontoparietal and closure of the supratemporal fossa may constitute correlated characters, yet they occur at different times during growth in *Stegoceras* (doming first, with closure of the fossa occurring very late in growth).

Pachycephalosaurinae, fully domed forms. In fully domed pachycephalosaurs, the bones that are sutured to the lateral and posterior aspects of the frontoparietal are fully incorporated into the vault of the dome (36). In the side view of the skull, tubercles occur only on the portion of the squamosal that projects away from the curve of the dome (Figure 25.2). The primitive parietosquamosal shelf, such as that in *Stegoceras*, extends posteriorly and laterally from the junction of the parietal, postorbital and squamosal – the remnant of the supratemporal fossa. No such shelf is present in fully domed pachycephalosaurs. In the top view of the skull, the posterior margin of the dome is vertical, or near vertical, and lacks any development of a posterior shelf.

This structural aspect of fully domed pachycephalosaurs has been confused by reference to the cluster of nodules on the squamosal in *Pachycephalosaurus* and *Stygimoloch* as a 'parietosquamosal' or 'squamosal' shelf (Galton and Sues, 1983; Maryańska, 1990). These nodules are attached to the back end of a fully domed skull that lacks any remnant of the original parietosquamosal shelf. This can be verified in specimens that lack the squamosals, as in the case of the disarticulated frontoparietal of *Stygimoloch* (Giffen *et al.*, 1987, fig. 3). In this specimen, the steep, shelfless profile of the dome is exposed even in this subadult individual. The ornamentation of the squamosal should not be confused with the primitive shelf that extends posteriorly from the supratemporal fossa.

The only other synapomorphy known to be shared by at least three fully domed forms (*Tylocephale, Prenocephale* and *Pachycephalosaurus*) is the establishment of a contact between the jugal and quadrate (37) (Figure 25.4). Although these bones approach each other in *Stegoceras* (Figure 25.3), they do not establish sutural contact. *Tylocephale* and *Prenocephale* share a unique oval fossa on the quadratojugal (38) that may indicate a close relationship (Figures 25.2 and 25.4).

Stygimoloch + Pachycephalosaurus, hypernoded forms. The large Maastrichtian pachycephalosaurs from western North America, *Pachycephalosaurus* and *Stygimoloch*, appear to be closely related. Both forms have a cluster of enlarged nodes on the squamosal (39), pronounced development of snout tubercles (41), and proportionately long snouts (40). A proportionately long, noded snout is preserved in *Pachycephalosaurus* and inferred for *Stygimoloch* on the basis of the low angle of the anterior end of the frontals and the presence of enlarged frontal nodes (Giffin *et al.*, 1987, p. 405, figs. 2, 3). The clumped configuration of nodes on the squamosal in these forms is also unique among pachycephalosaurs. Although it is difficult to establish a one-to-one correspondence, there are at least six or seven main nodes whose bases are in mutual contact in both *Pachycephalosaurus* and *Stygimoloch* (Sues and Galton, 1987). In the latter genus, three are extended as horn cores.

Ceratopsia

Traditional classification. Psittacosaurs, like pachycephalosaurs, were believed to have evolved from a central ornithopod stock and were originally classified within Ornithopoda (Osborn, 1923). Once the median, bill-supporting bone that capped the anterior end of the psittacosaur snout was properly identified as the ceratopsian rostral bone (Romer, 1956, 1968), psittacosaurs were allied with ceratopsians (Maryańska and Osmólska, 1975). Other small-bodied ceratopsians have been placed in the Family Protoceratopsidae, which was originally erected for *Protoceratops* (Granger and Gregory, 1923), but has served over the years as a repository for all small-bodied ceratopsians except psittacosaurs. The monophyly of the large-bodied forms within the Family Ceratopsidae has never been questioned.

Recent studies. The first cladistic analysis of basal ceratopsians not surprisingly placed *Psittacosaurus* as the outgroup to other ceratopsians, which were placed in

Neoceratopsia (Sereno, 1986). The arrangement of basal neoceratopsians, formerly classified within Protoceratopsidae, has been more controversial. Sereno (1986) argued that some protoceratopsids are more closely related to ceratopsids that others. In particular, *Leptoceratops* was regarded as more primitive, and *Montanoceratops* as more derived, than other protoceratopsids. Dodson and Currie (1990, p. 610, fig. 29.9), by contrast, favoured protoceratopsid monophyly, and presented a cladogram showing a fully resolved pectinate protoceratopsid clade. Although no evidence was given to support the branching sequence within the clade, three synapomorphies were mentioned to support the monophyly of the traditional Protoceratopsidae: a circular antorbital fossa, inclined parasagittal process of the palatine, and maxillary sinus.

A distinctly oval (rather than circular) antorbital fossa characterizes *Leptoceratops*, *Protoceratops*, *Bagaceratops*, and probably *Montanoceratops* (MOR 542). *Chaoyangsaurus* may also have an oval antorbital fossa, but only a portion of its margin is preserved. The principal difficulty with this synapomorphy is that available outgroups are difficult or impossible to score because the fossa is strongly reduced or absent. In *Psittacosaurus*, for example, there is no antorbital fenestra or fossa. The external depression on the maxilla (formerly identified as the antorbital fossa; Sereno *et al.*, 1988, fig. 5) does not communicate with the nasal cavity and is not homologous with the oval fossa in basal neoceratopsians. Among ceratopsids, chasmosaurines often retain at least a small antorbital fossa (Forster *et al.*, 1993, fig. 3). The dorsal margin of the fossa forms an arc across the maxilla and lacrimal and is not that different in shape, although less incised, from that in basal neoceratopsians. The posteroventral margin of the fossa, however, is straight. To conclude, the oval antorbital fossa may link basal neoceratopsians, but it is an ineffective character in a cladistic analysis because the plesiomorphic ceratopsian condition remains unclear.

The two remaining characters mentioned by Dodson and Currie (1990) to support protoceratopsid monophyly are difficult to justify. Osmólska (1986, p. 152) mentioned that the snout in both psittacosaurids and basal neoceratopsians was particularly deep and the angle of the palatine very steep. This is also true of ceratopsids, in which the palatine assumes a parasagittal orientation (Hatcher *et al.*, 1907, fig. 26). If Dodson and Currie (1990) meant to refer to the 'vertical transverse wing of the palatine,' this process appears to be fully developed in this manner only in *Protoceratops*, *Bagaceratops*, and ceratopsids; it is absent in *Leptoceratops* and *Psittacosaurus*, as noted by Osmólska (1986, p. 152). The distribution of the 'maxillary sinus' described by Osmólska is poorly known. Developed as a space above the tooth row, it communicates with the antorbital fossa, which communicates with the nasal cavity via the antorbital fenestra, suggesting that the 'maxillary sinus' may be a ramification of the nasal cavity (contrary to Osmólska, 1986, p. 154; Witmer, 1995). The distribution of this cavity among ceratopsians is poorly known.

Other evidence, such as the prominence of the wedge-shaped epijugal (24), may eventually be shown to support the traditional assemblage of protoceratopsids as a monophyletic clade. The jugal/epijugal crest is low in *Psittacosaurus*, *Chaoyangsaurus* and ceratopsids. A plate-shaped sagittal crest on the parietal (60) also links several basal neoceratopsians (*Leptoceratops*, *Protoceratops*, *Bagaceratops*) but is lacking in others (*Graciliceratops*, *Montanoceratops*).

Present results. The following summary of ceratopsian phylogeny is based on analysis of 72 characters in 10 ceratopsian genera (those reviewed above) and Ceratopsidae (see Appendix; Figure 25.8; Table 25.4). The character data are a modification and extension of synapomorphies listed in Sereno (1986, p. 244; 1990b, pp. 587–588). Using Pachycephalosauria and Ornithopoda as successive outgroups, the analysis yielded three most parsimonious trees differing only in the resolution of a trichotomy between *Protoceratops*, *Bagaceratops*, and *Graciliceratops* (consistency index, 0.86; retention index, 0.92). Accepting trees one step longer breaks the tenuous link between *Leptoceratops* and *Udanoceratops* and creates a trichotomy between these genera and Coronosauria. Protoceratopsidae *sensu stricto* (*Protoceratops*, *Bagaceratops*, *Graciliceratops*) collapses when trees two steps longer than the minimum are accepted.

Reconstituting the traditional Protoceratopsidae requires four extra steps; and eight are required if *Turanoceratops* is included in the family. Thus, these data show a decided preference for a paraphyletic arrangement of small-bodied neoceratopsians. Here, Protoceratopsidae is tentatively restricted to include only *Protoceratops*, *Bagaceratops*, and *Graciliceratops*, as discussed below. In the following discussion, character numbers correspond to those tabulated in the Appendix. When synapomorphies have an ambiguous location on the cladogram due to missing data or homoplasy, they are described under the least inclusive group that they could characterize (i.e. delayed-transformation optimization).

Ceratopsia. The monophyly of Ceratopsia is based exclusively on cranial synapomorphies, the most striking of which is the neomorphic rostral bone (1), a median, bill-supporting element sutured firmly to the tall and narrow anterior end of the snout (Figure 25.5). Other cranial features include broad, pointed jugals (3, 4), which give the skull a distinctly subtriangular shape in dorsal view (Gregory, 1927; Maryańska and Osmólska, 1975). The vaulted premaxillary palate (6) is deeply arched in psittacosaurs and narrower and more bird-like in neoceratopsians in contrast to the flat secondary palate that is present in *Lesothosaurus* (Sereno, 1991) and other ornithischians. The ventral process of the predentary has an unusually broad base (7) supporting the dentary symphysis.

The absence of ceratopsian postcranial synapomorphies reflects the conservative form of the postcranium in basal ceratopsians rather than missing information. In psittacosaurs the postcranium is remarkably primitive, differing only in minor ways from that in hypsilophodontids (Sereno, 1987). Basal neoceratopsians, likewise, exhibit few modifications in the postcranium. Except for some modification of the axial column, there is no major alteration of the postcranial skeleton among nonceratopsid ceratopsians.

Neoceratopsia. The discovery of *Chaoyangsaurus* (Zhao, 1983; Zhao *et al.*, 1999), the oldest known ceratopsian, has begun to bridge the substantial morphologic gap between psittacosaurs and neoceratopsians. Its linkage

Table 25.4. *Character-taxon matrix for Ceratopsia (see Appendix for characters and character states).*

	10	20	30	40	50	60	70	
ORNITHOPODA	00000 00000	00000 000X0	00000 000X0	00000 00000	00000 00000	00000 0000X	0X000	00
PACHYCEPHALOSAURIA	00000 00000	00000 X10X0	00000 00100	0000? 0?000	00000 00000	00000 00??0X	0X?00	00
Psittacosaurus	11XX0 ?1111	00000 X0000	00000 ?200?	00?00 ?2?00	10000 00000	00000 000X0	0X000	00
Chaoyangsaurus	??111 ?1111	00000 ??00?	?000? 00???	??000 00???	0000? ?2?0?	?2??? ?2?0?	0?000	0?
Archaeoceratops	??111 1?111	11111 ??00?	112?0 ?2???	?2?00 ?2?00	000?0 0000?	?000? ?200X	0?000	00
Leptoceratops	11XX1 11111	11111 11111	11111 11211	11111 11111	11100 00010	00001 0000X	0X000	00
Udanoceratops	1?XX? 1?XX?	11111 11111	?111? 1????	?21?? ?????	11100 ?????	????? ??0X	0??00	0?
Graciliceratops	??11? ?2??1	?2111 ?2111	?211? 1?21?	?21?? ?????	?201? ?????	?21? ?211?	11?0?	0?
Bagaceratops	11X11 ?2?11	?2111 ?2111	11211 11211	11111 11111	010?1 11121	?1?11 11100	11?0?	00
Protoceratops	11111 01111	?2111 ?2111	11211 11211	11111 11111	01011 11121	11111 11100	11100	00
Montanoceratops	11111 01111	?2111 ?2111?	11211 1?21?	?21?? ?????	01011 11121	11111 11100	11100	00
Turanoceratops	????? ?1???	?211? ?????	1?21? 1?21?	1???? ?????	????? ?????	????? ??1?	??11?	11
CERATOPSIDAE	1011? 11XX1	11111 11111	X2101 12111	11101 11100	10011 11121	11110 00011	12111	11

with later neoceratopsians is based on cranial synapomorphies alone. The subcylindrical, procumbent form of the premaxillary teeth in *Chaoyangsaurus* establishes this unusual tooth form as the plesiomorphic condition within Neoceratopsia. Similar premaxillary teeth are now known in *Archaeoceratops* (Dong and Azuma, 1997), *Protoceratops*, (Brown and Schlaikjer, 1940), at least one specimen of *Bagaceratops* (Dong and Currie, 1993), and a basal neoceratopsian from the Two Medicine Formation (Gilmore, 1939; USNM 13863).

The marked increase in the relative size of the skull (10), which measures (without the frill) as much as 20–30% of the length of the postcranial skeleton, characterizes neoceratopsians. *Chaoyangsaurus* appears to have a large skull relative to the preserved portions of the scapula. The keeled, pointed predentary (13), distally tapered ventral process of the predentary (14), and lack of a significant retroarticular process (15) constitute further links between *Chaoyangsaurus* and other neoceratopsians which can be scored as primitive in neoceratopsian outgroups. Other features, such as the keeled, pointed shape of the rostral bone (12), are less decisive because the rostral is a neomorphic bone; neoceratopsian outgroups that lack the rostral cannot be used to polarize characters involving this bone. Thus, potential neoceratopsian synapomorphies involving the rostral (11, 12) may also be regarded as plesiomorphic, with the condition in *Psittacosaurus* interpreted as derived.

Archaeoceratops + *Leptoceratops* + *Udanoceratops* + Coronosauria. Major modification of the ceratopsian skull is apparent in all neoceratopsians more advanced than *Chaoyangsaurus* (Zhao, 1983; Zhao *et al.*, 1999). The postorbital and supratemporal bars are broadened into strap-shaped struts (22), and the dorsal and particularly the ventral margins of the laterotemporal fenestra are shortened (23). Although the jugal is prominent in all ceratopsians, it forms a wedge-shaped process capped by the horn-covered epijugal (24) in neoceratopsians more advanced than *Chaoyangsaurus*. An epijugal was not described by Dong and Azuma (1997) in *Archaeoceratops*, but this bone is commonly disarticulated and lost in subadult individuals and was probably present in this early neoceratopsian. The

supratemporal region is reconfigured by the confluence of the supratemporal fossae in the midline (32) and the upward tilt of the posterior margin of the parietal (31). Modifications in the lower jaw include a cropping surface on the predentary (34), the participation of the splenial in the median symphysis (39), and major expansion of the coronoid process (37). The more closely packed dentition (20) in these neoceratopsians is characterized by the inset margin at the base of the maxillary and dentary crowns on their lateral and medial surfaces, respectively (17).

Leptoceratops + *Udanoceratops* + Coronosauria. Several cranial features unite *Leptoceratops*, *Udanoceratops* and coronosaurs, but nearly all of these have an ambiguous distribution because of missing data for *Archaeoceratops*. The most significant postcranial modifications among basal neoceratopsians involve the cervical and caudal vertebrae. The anterior three cervical vertebrae coalesce in *Leptoceratops* (NMC 1889) and more advanced neoceratopsians (41), and the neural spines of the mid-cervicals (third and fourth) are as tall as the axis (42). The distalmost caudal vertebrae have proportionately short centra, rudimentary neural arches and articulate with small chevrons (44). In *Psittacosaurus* and *Archaeoceratops*, by contrast, the distalmost caudals have cylindrical centra that lack neural arches and do not have associated chevrons. In *Leptoceratops* and Coronosauria, the mid-caudal vertebrae have particularly long neural spines and long chevrons, resulting in a 'leaf-shaped' tail in lateral view (45). *Psittacosaurus* (Sereno, 1987) and *Archaeoceratops* (Dong and Azuma, 1997, fig. 5) clearly lack these modifications, although the latter genus has been reconstructed with a leaf-shaped tail (Dong and Azuma, 1997, fig. 11).

Leptoceratops + *Udanoceratops*. A single synapomorphy suggests that *Leptoceratops* and *Udanoceratops* constitute a subgroup within Neoceratopsia. In several genera of basal neoceratopsians, the lower margin of the jaw is arched (*Leptoceratops*, *Udanoceratops*, *Bagaceratops*, *Protoceratops*), but in *Leptoceratops* and *Udanoceratops* the downward arching of the ventral margin (47) is pronounced and begins under the retroarticular process (48) rather than under the coronoid region. However, the material for *Udanoceratops* is very limited and the close relationship to *Leptoceratops* is supported only by this single feature.

Coronosauria. The synapomorphies that diagnose Coronosauria are located principally in the cranium. At least a rudimentary nasal horn is present (50), though least developed in *Protoceratops* (Brown and Schlaikjer, 1940, fig. 13B), and the supratemporal fossae are distinctly triangular (51) with long axes diverging posteriorly (52). In coronosaurs, enamel is present only on the lateral side of the maxillary crowns and medial side of the dentary crowns. In *Leptoceratops*, *Udanoceratops* (Kurzanov, 1992), and more basal ceratopsians, by contrast, enamel is present on both sides of the crowns.

The frill is particularly well developed (Figure 25.6). The parietal portion extends far posterior to the quadrate head (54), the distal portion of which has a sizable pair of fenestrae (55), one on each side of the midline. The fenestrae weaken the frill, which is frequently broken away along the anterior margin of these openings (as in the initial specimens of *Bagaceratops*; Maryańska and Osmólska, 1975). A distinct posterodorsal process of the squamosal, the frill process (59), forms much of the lateral margin of the frill. Except in chasmosaurines, the squamosal does not extend as far posteriorly as the parietal, and the posterolateral corners of the frill are rounded. In the anterior view of the skull (that encountered in display; Brown and Schlaikjer, 1940, pl. 6C), the frill forms a semicircular corona, from which the group name was derived. In *Montanoceratops* (TMP 82.11.1) the posterior extension of the parietal is less extreme, although the presence of sizable parietal fenestrae suggest that the parietal frill was longer than that in *Archaeoceratops* and *Leptoceratops*. The absence of a discrete frill process on the squamosal (AMNH 5464; TMP 82.11.1), however, introduces homoplasy. If coronosaurs split into protoceratopsid (*sensu stricto*) and ceratopsoid clades, as the data suggest as a whole, the frill process on the squamosal either was reduced in *Montanoceratops* or evolved independently in protoceratopsids (*sensu stricto*) and Ceratopsidae. I regard the former optimization (accelerated transformation with loss) as the more likely, given the derived form of the parietal (somewhat lengthened and fenestrated) in *Montanoceratops*.

Postcranial synapomorphies for Coronosauria are limited to the axial column and include the presence of a neomorphic element anterior to the atlas, the hypocentrum (56), and an increase to eight sacral vertebrae (57) with neural spines in mutual contact (58).

Protoceratopsidae (sensu stricto). Three synapomorphies suggest that *Graciliceratops*, *Protoceratops*, and *Bagaceratops* may constitute a monophyletic subgroup within Coronosauria. This subgroup, here referred to as Protoceratopsidae (*sensu stricto*), is characterized by a narrow strap-shaped paroccipital process (61), very small occipital condyle (62), and upturned dorsal margin of the predentary (63). Two of the three (*Protoceratops* and *Bagaceratops*) share a blade-shaped parietal sagittal crest (63) (Figure 25.6), but this is absent in *Graciliceratops* and present in at least one genus outside this subgroup (*Leptoceratops*).

Ceratopsoidea. *Montanoceratops* and ceratopsids share five synapomorphies, two of which are present in the poorly known central Asian species *Turanoceratops* (Nesov *et al.*, 1989). The anterior ramus of the squamosal is particularly deep in *Montanoceratops* (AMNH 5464; twice as long as deep) and plate-shaped in Ceratopsidae (22), which continues a trend in ceratopsians toward reduction of the laterotemporal fenestra. The nasal horn, which is quite well developed in *Montanoceratops* (AMNH 5464), is positioned over the external naris (65) rather than more posteriorly as in *Bagaceratops* and *Protoceratops*. Given that the nasal horn is a neomorph, however, the primitive position of the horn cannot be determined. It may be that the posteriorly positioned horn in protoceratopsids (*sensu stricto*) is derived, as suggested by its posterior migration during growth (Kurzanov, 1990). Two more decisive characters are present in the lower jaw in *Montanoceratops* and ceratopsids – the dentary ramus increases in depth toward its anterior end (66), a unique proportion among ornithischians, and the dentary teeth have very prominent primary ridges

(64). The latter can also be observed in *Turanoceratops* (Nesov *et al.*, 1989, pl. 1, fig. 16).

Turanoceratops + Ceratopsidae. The unique two-rooted cheek teeth (69) of ceratopsids, which lock together successive teeth in a vertical column, are also present in the recently discovered ceratopsoid *Turanoceratops* (Nesov, 1989, pl. 1, figs. 16 and 19). Despite the presence of two roots, there appears to be only two teeth in a vertical column in this ceratopsoid, as opposed to four or five in the much larger-bodied ceratopsids. Other aspects of the teeth in *Turanoceratops* are also advanced, including the sharp angle of the crown to the axis of the roots (70) and the reduction in height of the secondary ridges relative to the primary ridge (71). Broken horn cores (Nesov *et al.*, 1989, pl. 1, fig. 18) suggest that *Turanoceratops* had postorbital horns (72) as in ceratopsids. Postorbital horns have also recently been reported in an even more primitive neoceratopsian with single-rooted cheek teeth from approximately coeval deposits in western North America (Moreno Hill Formation; Childress, 1997). More complete and associated remains of these Cenomanian–Turonian forms will shed light on the initial stages of the evolution of the derived dental and cranial adaptations of ceratopsids.

Evolutionary trends

Body size

Both pachycephalosaurs and ceratopsians exhibit trends toward increasing body size, with maximum recorded body size (length) in each group appearing in the Maastrichtian (latest Cretaceous). The body size of the ancestral marginocephalian probably did not exceed 2 m, because known basal marginocephalians (*Stenopelix*, *Psittacosaurus*, *Chaoyangsaurus*) and basal members of marginocephalian outgroups (Ornithopoda, Thyreophora) have never exceeded this length.

Among pachycephalosaurs, moderate body size (2–3 m) probably evolved by the Early Cretaceous, when many pachycephalosaurs in this body size range

must have diverged. Large body size (6–8 m) was attained only among Maastrichtian pachycephalosaurs (*Stygimoloch*, *Pachycephalosaurus*) and presumably evolved some time in the Late Cretaceous. At least twice during the evolution of pachycephalosaurs, marked decrease in body size yielded some of the smallest ornithischians on record: *Yaverlandia* and an as yet undescribed North American species of similar size. These pachycephalosaurian dwarfs, represented by fully coossified skull caps of mature individuals, do not appear to form a clade, but rather seem to have evolved independently from ancestors of moderate body size. Not included here among dwarf pachycephalosaurians is the basal pachycephalosaur *Wannanosaurus*, the materials of which may be immature.

The trends described above are asymmetrical (McKinney, 1990). The range of body size increased over time, from a minimum skeletal length of about two metres in the Early Cretaceous to skeletons four or five times that length toward the end of the Late Cretaceous. In pachycephalosaurs, the body size range appears to have extended to smaller values as well, to skeletal lengths no greater than one metre. In both pachycephalosaurs and ceratopsians, the asymmetrical trend toward increase in body size is accretive, because species of moderate body size persisted alongside their larger cousins in the latest Cretaceous (Maastrichtian). In ceratopsians, increase in body size was also accretive, but unlike pachycephalosaurs, large-bodied species greatly outnumbered smaller species in the Maastrichtian. Mean body size for ceratopsians, therefore, increased more dramatically toward the end of the Cretaceous.

Doming of the skull roof

The extraordinary thickening of the skull roof in pachycephalosaurs occurred in several stages, according to the best estimate of the phylogenetic history of this group. First, the entire skull table was thickened; later, in one clade (Pachycephalosauridae), a dome arose composed principally of the frontal and parietal; finally, in one subgroup of that clade (Pachycephalosaurinae), the dome expanded to incorporate other bones of the skull table fully (Figures 25. 2 and 25.3). Excluded here are relatively minor proportional changes in the dome that characterize some genera.

The most remarkable fact about this trend in cranial thickening is that, despite body size evolution over more than an order of magnitude (*Yaverlandia* to *Pachycephalosaurus*), doming of the skull cap apparently occurred only once and was never reduced or eliminated. In this regard, the predominance in the data of characters pertaining to the skull roof is cause for concern, because of the potential to create an artificial transformation series. Nonetheless, there is no indication in available character evidence that a vaulted dome evolved more than once, that such a dome was ever later substantially reduced, or that it was ever subject to marked sexual dimorphism (contrary to Chapman *et al.*, 1981).

Extension of the frill

The evolution of the frill among ceratopsians followed a somewhat more complex course than the thickening of the skull roof among pachycephalosaurs. In psittacosaurs the short parietosquamosal shelf projects horizontally over the occiput (Figure 25.5). In the basal neoceratopsians *Archaeoceratops* and *Leptoceratops*, a posterodorsally inclined, transversely broadened frill has evolved, composed almost entirely of the parietal. The frill incorporates the squamosal laterally and becomes progressively more hyperextended in Late Cretaceous protoceratopsids (Figure 25.6) and ceratopsids, with the longest frills (relative to skull length) occurring among chasmosaurines.

The trend outlined above toward longer and broader frills is complicated by the presence in *Montanoceratops* of a short frill, composed almost entirely of the parietal. Other features of *Montanoceratops* clearly justify its derived position among neoceratopsians (such as the parietal fenestrae in the frill). Either the frill was shortened in *Montanoceratops*, with concomitant reduction in the participation of the squamosal, or the frill was lengthened independently in protoceratopsids (*sensu stricto*) and ceratopsids. Only the discovery of additional taxa can resolve this question.

Trophic adaptations

Pachycephalosaurian jaw morphology and tooth form appears to have undergone only superficial modification. The structure of the lower jaw remains primitive with the dentary forming no more than half of the lower jaw; the tooth row remains relatively loosely packed with spaces between adjacent crowns; tooth form remains primitive with triangular crowns and simple roots; and the dentary canine and associated diastema between the premaxillary and maxillary teeth – a derived condition present in basal pachycephalosaurs such as *Goyocephale* – is clearly maintained in fully domed forms such as *Prenocephale*.

The ceratopsian snout, jaws and teeth, by contrast, had undergone considerable transformation by the Late Cretaceous. Marked change in the form of the snout is present as early as the latest Jurassic or Early Cretaceous and clearly predates the angiosperm radiation (Figure 25.9). Moreover, the bird-like neoceratopsian snout, formed by a very narrow rostrum and pointed, upturned predentary, had also evolved by the earliest Cretaceous, as evidenced by *Chaoyangsaurus* (Figure 25.9). Other neoceratopsians show a more advanced condition of the lower jaw and tooth rows. The postdentary elements are reduced, a bevelled cropping edge is present on the predentary, the cheek teeth are more tightly packed, and the rate of tooth replacement is increased. The crowns of the cheek teeth, in addition, are taller than in psittacosaurs with enamel restricted to a single side. The discovery of *Archaeoceratops* in the Early Cretaceous of Asia demonstrates that many of these adaptations were established during the Early Cretaceous, when psittacosaurs with simple jaws and dentitions were more abundant. Most of these changes, likewise, significantly predate the rise of angiosperms toward the end of the Early Cretaceous (Figure 25.9).

Two-rooted cheek teeth must also have evolved before the end of the Early Cretaceous, given the presence of two-rooted teeth in the Cenomanian ceratopsian *Turanoceratops* (Figure 25.9). Dentary batteries are known only among ceratopsids from the Campanian and Maastrichtian of western North America and

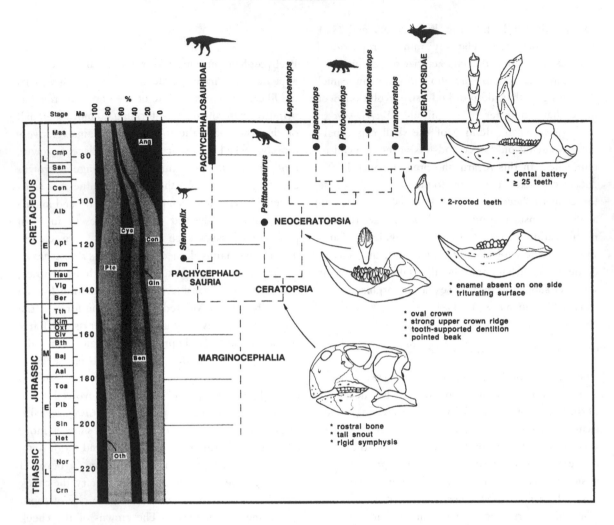

Figure 25.9. Calibrated phylogeny showing the temporal and phylogenetic origin of major cranial and dental features associated with herbivory in ceratopsians. Relative change in diversity (percent) in major plant clades is shown at left (based on Niklas, 1986). Abbreviations: ang, angiosperms; cyc, cycads; con, conifers; gin, ginkgophytes; oth, other; pte, pteridophytes.

might well be correlated with an increase in body size. Thus, there is a trend toward increased packing and replacement in the dentition which culminates in the tooth-supported dental batteries of ceratopsids. Almost identical trends occurred somewhat earlier in ornithopods, including the dominance of the dentary in the lower jaw, increased relative height of the tooth crowns and asymmetry of the enamel, and increased compaction and replacement of cheek teeth (Sereno, 1997).

Biochronology

Temporal calibration of the phylogeny provides insight into (1) major missing lineages that have left no fossil record and (2) the timing of cladogenic events. Three major missing lineages are apparent in the calibrated phylogeny of marginocephalians, the longest occurring before the earliest known marginocephalian. A missing lineage, possibly as long as 100 million years – one of the longest among major groups of dinosaurs – precedes the oldest marginocephalian (arguably *Chaoyangsaurus*), as evidenced by the appearance of the sister taxon to Marginocephalia (Ornithopoda) during the Early Jurassic (Figure 25.8). Given the relatively small body size of known basal marginocephalians, their predecessors may have been as small, or smaller, and less likely to have entered the fossil record. Other factors, such as habitat preference, may also may have contributed to the absence of fossil evidence for the early appearance of marginocephalians, because several small-bodied ornithopods are recorded during the early Jurassic. The origin of pachycephalosaurian and ceratopsian lineages may date back to the Early Jurassic, given the low number of derived features shared by both subgroups.

Other missing lineages, 40–50 million years in duration, are predicted for Late Cretaceous pachycephalosaurs and neoceratopsians. The discovery of the Early Cretaceous pachycephalosaur *Yaverlandia* and the neoceratopsians *Chaoyangsaurus* and *Archaeoceratops* identify major missing lineages preceding closely related genera (Figure 25.8). With the exception of *Stenopelix,* the most primitive pachycephalosaurs are Late Cretaceous in age, but must be the descendants of lineages that diverged early in the Cretaceous. Likewise, a long missing lineage precedes domed pachycephalosaurs (pachycephalosaurids). Among ceratopsians, a long missing lineage precedes all Late Cretaceous neoceratopsians, as established by the Early Cretaceous genus *Archaeoceratops* (Figure 25.8).

Not one of these missing lineages is associated with major structural modification. Among pachycephalosaurs, nearly all of the many unusual postcranial features (such as the ossified caudal tendons) are present in *Goyocephale*, and therefore must have evolved no later than the earliest Cretaceous (Figure 25.8). Likewise, among early ceratopsians, most structural change seems to have occurred after the divergence of *Chaoyangsaurus*. From available remains, *Archaeoceratops* appears to be very similar to primitive Late Cretaceous neoceratopsians such as *Protoceratops*, suggesting that the majority of the cranial modifications that characterize Late Cretaceous neoceratopsians had already evolved by the earliest Cretaceous,

though several of these features currently have an ambiguous temporal origin because of missing data for *Archaeoceratops*.

Biogeography

Except for two Early Cretaceous pachycephalosaurs, *Stenopelix* and *Yaverlandia*, and one ceratopsian, *Chaoyangsaurus*, marginocephalians are known exclusively from late Early and Late Cretaceous deposits in central Asia and western North America. Their phylogeny has direct bearing on their biogeographic history. Their limited biogeographic distribution, and the fact that no species has ever been found distributed across both areas, suggests a possible phylogenetic solution.

Previous hypotheses proposed a central Asian origin for the group in the Early Cretaceous, followed by a one-way dispersal event from Asia, across Beringia, to western North America in the Late Cretaceous for ceratopsians (Maryańska and Osmólska, 1975) and for other groups as well (Russell, 1993). This hypothesis predicts that, for each group distributed across these two areas, Asian taxa will compose a basal paraphyletic subgroup that, via a single, one-way dispersal event, gave rise to a monophyletic subgroup in western North America. Alternative biogeographic scenarios would be consistent with different phylogenetic patterns. If a given group were distributed initially across both central Asia and western North America and later divided by a Cretaceous vicariance event, for example, group members on each land mass would compose monophyletic sister clades. Or, if a given group had a more complicated biogeographic history, with multiple, bi-directional dispersal events across Beringia, an alternating pattern of Asian and western North American taxa would obtain.

The marginocephalian phylogeny presented here clearly favours the latter biogeographic scenario (Figure 25.10; Sereno, 1997, 1999a). Although Asian pachycephalosaurs and ceratopsians predominate at the basal end of their respective phylogenies, the alternating areal relationships among marginocephalians

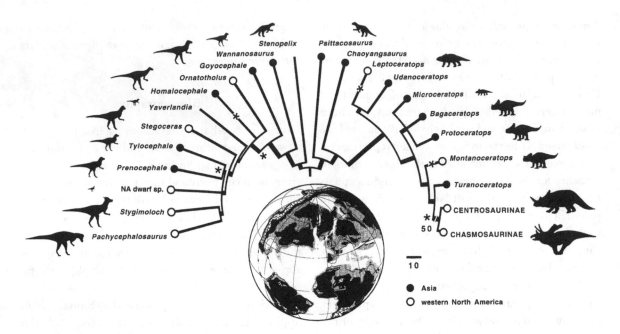

Figure 25.10. Polar sweepstakes dispersal by marginocephalians during the Cretaceous. Globe shows Maastrichtian (70 Ma) palaeogeography divided into orogenic belts (inverted Vs), lowlands (black), and shallow (grey) and deep seas (white). Cladogram for marginocephalians shows relative support for relationships based on the present analysis. Internal branch lengths are scaled to the number of synapomorphies under delayed character-state optimization (scale bar equals 10 synapomorphies; 50-synapomorphy ceratopsid branch shortened). Internal branch segments are filled (black) according to the number of unambiguous synapomorphies; open segments indicate ambiguous synapomorphies. Palaeogeographic distributions for marginocephalian genera are shown at branch tips (solid circles, Asia; open circles, western North America; blank, Europe). Six dispersal events (asterisks) across Beringia (in both directions) must be invoked to account for the palaeogeographic distributions shown by the cladogram (alternative positions for some of the dispersal events are possible). Increase in body size in pachycephalosaurs and ceratopsians is shown by body silhouettes (palaeogeographic projection courtesy of the Paleogeographic Atlas Project, University of Chicago).

requires a minimum of three dispersal events in two directions across Beringia in each group. Phylogenetic patterns suggesting bi-directional dispersal also occur in other Cretaceous dinosaurian groups with a similar bimodal distribution during the Late Cretaceous (hadrosaurids, ornithomimids, tyrannosaurids; Sereno, unpublished data).

The tectonic and palaeogeographic history of the north polar region during the Cretaceous is consistent with the emergence during the late Early–early Late Cretaceous of a high-latitude dispersal route. That route formed when the North Slope block collided with the East Siberian block, joining it with the western North American land mass (Worrall, 1991, fig.

8). Following the shoreline along a continuous, active trench, that route would have passed within five degrees of the paleopole during the Late Cretaceous. This polar passage, hidden from direct sunlight for six months of the year, may have functioned as a sweepstakes dispersal route (McKenna, 1973), periodically allowing passage of animals from one side to the other.

Acknowledgements

For the invitation to write this review, I would like to thank David Unwin and Michael Benton. I am indebted to Carol Abraczinskas for her skilled execution of the finished illustrations.

References

Bohlin, B. 1953. Fossil reptiles from Mongolia and Kansu. *Sino-Swedish Expeditions Publications* 37: 1–113.

Brown, B. 1914. *Leptoceratops,* a new genus of Ceratopsia from the Edmonton Cretaceous of Alberta. *Bulletin of the American Museum of Natural History* 33: 567–580.

Brown, B., and Schlaikjer, E.M. 1940. The structure and relationships of *Protoceratops. Annals of the New York Academy of Sciences* 40: 133–266.

—and— 1943. A study of the troödont dinosaurs with the description of a new genus and four new species. *Bulletin of the American Museum of Natural History* 82: 121–149.

Buffetaut, E., Sattayarak, N. and Suteethorn, V. 1989. A psittacosaurid from the Cretaceous of Thailand and its implications for the palaeogeographical history of Asia. *Terra Nova* 1: 370–373.

—and Suteethorn, V. 1992. A new species of the ornithischian dinosaur *Psittacosaurus* from the Early Cretaceous of Thailand. *Palaeontology* 35: 801–812.

Childress, J.O. 1997. A ceratopsian resurfaces out West. *Geotimes* 42: 12–14.

Coombs, W.P., Jr. 1979. Osteology and myology of the hindlimb in the Ankylosauria (Reptilia: Ornithischia). *Journal of Paleontology* 53: 666–684.

—1982. Juvenile specimens of the ornithischian dinosaur *Psittacosaurus. Palaeontology* 25: 89–107.

Chapman, R.E., Galton, P.M., Sepkoski, J.J. Jr., and Wall, W.P. 1981. A morphometric study of the cranium of the pachycephalosaurid dinosaur *Stegoceras. Journal of Paleontology* 55: 608–618.

de Queiroz, K., and Gauthier, J.A. 1990. Phylogeny as a central principle in taxonomy: Phylogenetic definitions of taxon names. *Systematic Zoology* 39: 307–322.

—and—1992. Phylogenetic taxonomy. *Annual Review of Ecology and Systematics* 23: 449–480.

Dodson, P. 1990. Marginocephalia. pp. 562–563 in Weishampel, D.B., Dodson, P. and Osmólska, H. (eds.), *The Dinosauria.* Berkeley: University of California Press.

—and Currie, P.J. 1990. Neoceratopsia, pp. 593–618 in Weishampel, D.B., Dodson, P. and Osmólska, H. (eds.), *The Dinosauria.* Berkeley: University of California Press.

Dong, Z. 1978. [A new genus of Pachycephalosauria from Laiyang, Shantung.] *Vertebrata PalAsiatica* 16: 225–228.

—and Y. Azuma 1996. *Dinosaurs from the Silk Road, China.* Guidebook to exhibition March 3 to May 26, 1996, 68 pp. Fukui Prefectural Museum, Japan.

—and—1997. On a primitive neoceratopsian from the Early Cretaceous of China, pp. 68–89 in Dong, Z. (ed.), *Sino-Japanese Silk Road Dinosaur Expedition.* Beijing: China Ocean Press.

—and Currie, P.J. 1993. Protoceratopsian embryos from Inner Mongolia, People's Republic of China. *Canadian Journal of Earth Sciences* 30: 2248–2254.

Eberth, D.A., Russell, D.A., Braman, D.R. and Deino, A.L. 1993. The age of the dinosaur-bearing sediments at Tebch, Inner Mongolia, People's Republic of China. *Canadian Journal of Earth Sciences* 30: 2101–2106.

Forster, C.A., Sereno, P.C., Evans, T.W. and Rowe, T. 1993. A complete skull of *Chasmosaurus mariscalensis* (Dinosauria: Ceratopsidae) from the Aguja Formation (Late Campanian) of west Texas. *Journal of Vertebrate Paleontology* 13: 161–170.

Galton, P.M. 1971. A primitive dome-headed dinosaur (Ornithischia: Pachycephalosauridae) from the Lower Cretaceous of England and the function of the dome in pachycephalosaurids. *Journal of Paleontology* 45: 40–47.

—1972. Classification and evolution of ornithopod dinosaurs. *Nature* 239: 464–466.

—1982. The postcranial anatomy of the stegosaurian dinosaur *Kentrosaurus* from the Upper Jurassic of Tanzania, East Africa. *Geologica et Palaeontologica* 15: 139–160.

—and Sues, H.-D. 1983. New data on pachycephalosaurid dinosaurs (Retilia: Ornithischia) from North America. *Canadian Journal of Earth Sciences* 20: 462–472.

Gilmore, C.W. 1917. *Brachyceratops,* a ceratopsian dinosaur from the Two Medicine Formation of Montana, with notes on associated fossil reptiles. *United States Geological Survey Professional Paper* 103: 1–45.

—1924. On *Troödon validus,* an orthopodous dinosaur from the Belly River Cretaceous of Alberta, Canada. *University of Alberta Bulletin* 1: 1–43.

—1939. Ceratopsian dinosaurs from the Two Medicine Formation, Upper Cretaceous of Montana. *Proceedings of the United States National Museum* 87: 1–18.

Giffin, E.B., Gabriel, D. and Johnson, R. 1987. A new pachycephalosaurid skull from the Cretaceous Hell Creek Formation of Montana. *Journal of Vertebrate Paleontology* 7: 398–407.

Goodwin, M.B. 1990. Morphometric landmarks of pachycephalosaurid cranial material from the Judith River Formation of northcentral Montana, pp. 189–201 in Carpenter, K. and Currie, P.J. (eds.), *Dinosaur Systematics: Approaches and Perspectives.* Cambridge: Cambridge University Press.

Granger, W. and Gregory, W.K. 1923. *Protoceratops andrewsi,* a pre-ceratopsian dinosaur from Mongolia. *American Museum Novitates* 72: 1–9.

Gregory, W.K. 1927. The Mongolian life record. *Science Monthly* 24: 169–181.

—and Mook, C.C. 1925. On *Protoceratops*, a primitive ceratopsian dinosaur from the Lower Cretaceous of Mongolia. *American Museum Novitates* 156: 1–9.

Hatcher, J.B., Marsh, O.C. and Lull, R.S. 1907. The Ceratopsia. *United States Geological Survey* 49: 1–198.

Hou, L. 1977. [A new primitive Pachycephalosauria from Anhui, China.] *Vertebrata PalAsiatica* 25: 198–202

Kurzanov, S.M. 1990. A new Late Cretaceous protoceratopsid from Mongolia. *Paleontological Journal* 1990: 85–91.

—1992. Gigantic protoceratopsid from the Upper Cretaceous of Mongolia. *Paleontological Journal* 26: 103–116.

Lambe, L.M. 1918. The Cretaceous genus *Stegoceras* typifying a new family referred provisionally to the Stegosauria. *Transactions of the Royal Society of Canada* 12: 23–26.

Lull, R.S. 1933. A revision of the Ceratopsia or horned dinosaurs. *Bulletin of the Peabody Museum of Natural History* 3: 1–175.

Maryańska, T. 1990. Pachycephalosauria, pp. 564–577 in Weishampel, D.B., Dodson, P. and Osmólska, H. (eds.), *The Dinosauria.* Berkeley: University of California Press.

—and Osmólska, H. 1974. Results of the Polish-Mongolian Palaeontological Expedition. Part V. Pachycephalosauria, a new suborder of ornithischian dinosaurs. *Palaeontologia Polonica* 30: 45–102.

—and—1975. Results of the Polish–Mongolian Palaeontological Expedition. Part VI. Protoceratopsidae (Dinosauria) of Asia. *Palaeontologia Polonica* 33: 133–182.

—and—1985. On ornithischian phylogeny. *Palaeontologia Polonica* 30: 137–150.

McKenna, M.C. 1973. Sweepstakes, filters, corridors, Noah's arks, beached Viking funeral ships in paleogeography, pp. 295–308 in Tarling, D.H. and Runcorn, S.K. (eds.), *Implications of Continental Drift to the Earth Sciences.* London and New York: Academic Press.

McKinney, M.L. 1990. Classifying and analysing evolutionary trends, pp. 28–58 in McNamara, K.J. (ed.), *Evolutionary Trends.* Tuscon: University of Arizona Press.

Nesov, L.A., Kaznyshkina, L.F. and Cherepanov, G.O. 1989. [Mesozoic ceratopsian dinosaurs and crocodiles of central Asia.] pp. 144–154 in Bogdanova, T.N. and Khozatskii, L.I. (eds.), *Theoretical and Applied Aspects of Modern Palaeontology.* Leningrad: Nauka Publishers.

Niklas, K.J. 1986. Largescale changes in animal and plant terestrial communities, pp. 383–405 in Raup, D.M. and Jablonski, D. (eds.), *Patterns and Processes in the History of Life.* Heidelberg: Springer-Verlag.

Nopcsa, F. 1929. Dinosaurierreste aus Siebenbürgen V. *Geologische Hungarica, Series Palaeontologische* 4: 1–76.

Osborn, H.F. 1923. Two Lower Cretaceous dinosaurs from Mongolia. *American Museum Novitates* 95: 1–10.

—1924. *Psittacosaurus* and *Protiguanodon:* Two Lower Cretaceous iguanodonts from Mongolia. *American Museum Novitates* 127: 1–16.

Osmólska, H. 1986. Structure of the nasal and oral cavities in the protoceratopsid dinosaurs. *Acta Palaeontologica Polonica* 31: 145–157.

Perle, A., Maryańska, T. and Osmólska, H. 1982. *Goyocephale lattimorei* gen. et sp. n., a new flat-headed pachycephalosaur (Ornithischia, Dinosauria) from the Upper Cretaceous of Mongolia. *Acta Palaeontologica Polonica* 27: 115–127.

Romer, A.S. 1927. The pelvic musculature of ornithischian dinosaurs. *Acta Zoologica* 8: 225–275.

—1956. *Osteology of the Reptilia.* Chicago: University of Chicago Press, 772 pp.

—1968. *Notes and Comments on Vertebrate Paleontology.* Chicago: University of Chicago Press, 304 pp.

Russell, D.A. 1993. The role of central Asia in dinosaur biogeography. *Canadian Journal of Earth Sciences* 30: 2002–2012.

—and X. Zhao 1996. New psittacosaur occurrences in Inner Mongolia. *Canadian Journal of Earth Sciences* 33: 637–648.

Russell, L.S. 1932. On the occurrence and relationships of the dinosaur *Troödon. Annual Magazine of Natural History* 9: 334–337.

Santa Luca, A.P. 1980. The postcranial skeleton of

Heterodontosaurus tucki (Reptilia, Ornithischia) from the Stormberg of South Africa. *Annals of the South African Museum* **79**: 159–211.

Schmidt, H. 1969. *Stenopelix valdensis* H. v. Meyer, der kleine Dinosaurier des norddeutschen Wealden. *Paläontologische Zeitschrift* **43**: 194–198.

Sereno, P.C. 1984. The phylogeny of Ornithischia: a reappraisal, pp. 219–226 in Reif, W. and Westphal, F. (eds.), *Third Symposium on Mesozoic Terrestrial Ecosystems, Short Papers*. Tübingen: Attempto Verlag.

—1986. Phylogeny of the bird-hipped dinosaurs (Order Ornithischia). *National Geographic Research* **2**: 234–256.

—1987. *The ornithischian dinosaur Psittacosaurus from the Lower Cretaceous of Asia and the relationships of the Ceratopsia*. Ph.D. dissertation, Columbia University, pp. 554.

—1990a. New data on parrot-beaked dinosaurs (*Psittacosaurus*), pp. 203–210 in Carpenter, K. and Currie, P.J. (eds.), *Dinosaur Systematics: Approaches and Perspectives*. Cambridge: Cambridge University Press.

—1990b. Psittacosauridae, pp. 579–592 in Weishampel, D.B., Dodson, P. and Osmólska, H. (eds.), *The Dinosauria*. Berkeley: University of California Press.

—1990c. Clades and grades in dinosur systematics, pp. 9–20 in Carpenter, K. and Currie, P.J. (eds.), *Dinosaur Systematics: Approaches and Perspectives*. Cambridge: Cambridge University Press.

—1991. *Lesothosaurus*, 'fabrosaurids,' and the early evolution of Ornithischia. *Journal of Vertebrate Paleontology* **11**: 168–197.

—1997. The origin and evolution of dinosaurs. *Annual Review of Earth and Planetary Science* **25**: 435–489.

—1998. A rationale for phylogenetic definitions with application to the higher-level taxonomy of Dinosauria. *Neues Jahrbuch für Geologie und Paläontologie Abhandlungen* **210**: 41–83.

—1999a. The evolution of dinosaurs. *Science* **284**: 2137–2147.

—1999b. Definitions in phylogenetic taxonomy: critique and rationale. *Systematic Biology* **48**: 329–351.

—and Chao, P. 1988. *Psittacosaurus xinjiangensis* (Ornithischia: Ceratopsia), a new psittacosaur from the Lower Cretaceous of northwestern China. *Journal of Vertebrate Paleontology* **8**: 353–365.

—, Chao, S., Cheng, Z. and Rao, C. 1988. *Psittacosaurus meileyingensis* (Ornithischia: Ceratopsia), a new psittaco-

saur from the Lower Cretaceous of northeastern China. *Journal of Vertebrate Paleontology* **8**: 366–377.

Steel, R. 1969. Ornithischia. *Handbuch der Paläoherpetologie* **15**: 1–84.

Sternberg, C.M. 1933. Relationships and habitat of *Troödon* and the nodosaurs. *Annual Magazine of Natural History* **11**: 231–235.

—1945. Pachycephalosauridae proposed for domeheaded dinosaurs, *Stegoceras lambei* n. sp., described. *Journal of Paleontology* **19**: 534–538.

—1951. Complete skeleton of *Leptoceratops gracilis* Brown from the Upper Edmonton member on the Red Dear River, Alberta. *Bulletin of the National Museum of Canada* **123**: 225–255.

Sues, H.-D. 1980. A pachycephalosaurid dinosaur from the Upper Cretaceous of Madagascar and its paleobiogeographical implications. *Journal of Paleontology* **54**: 954–962.

—and Galton, P.M. 1982. The systematic position of *Stenopelix valdensis* (Reptilia: Ornithischia) from the Wealden of north-western Germany. *Palaeontographica, Abteilung A* **178**: 183–190.

—and— 1987. North American pachycephalosaurid dinosaurs (Reptilia: Ornithischia). *Palaeontographica, Abteilung A* **198**: 1–40.

—and Taquet, P. 1979. A pachycephalosaurid dinosaur from Madagascar and a Laurasia-Gondwanaland connection in the Cretaceous. *Nature* **279**: 633–635.

Thulborn, R.A. 1974. A new heterodontosaurid dinosaur (Reptilia: Ornithischia) from the Upper Triassic Red Beds of Lesotho. *Zoological Journal of the Linnean Society* **55**: 151–175.

Wall, W.P., and Galton, P.M. 1979. Notes on pachycephalosaurid dinosaurs (Reptilia: Ornithischia) from North America, with comments on their status as ornithopods. *Canadian Journal of Earth Sciences* **16**: 1176–1186.

Witmer, L.M. 1995. Homology of facial structures in extant archosaurs (birds and crocodilians), with special reference to paranasal pneumaticity and nasal conchae. *Journal of Morphology* **225**: 269–327.

Worrall, D.M. 1991. Tectonic history of the Bering Sea and the evolution of Tertiary strike-slip basins of the Bering Shelf. *Geological Society of America Special Paper* **257**: 1–106.

Xu, X. 1997. A new psittacosaur (*Psittacosaurus mazongshanensis* sp. nov.) from Mazongshan area, Gansu

Province, China, pp. 48–67, in Dong, Z. (ed.), *Sino-Japanese Silk Road Dinosaur Expedition.* Beijing: China Ocean Press.

Young, C.C. 1958. The dinosaurian remains of Laiyang, Shantung. *Palaeontologia Sinica, New Series C* **16**: 1–138.

Zhao, X. 1983. Phylogeny and evolutionary stages of Dinosauria. *Palaeontologia Polonia* **28**: 295–306.

—1986. [Reptiles.] pp. 286–291 in Wang, S.E. *et al.* (eds.), *The Jurassic System of China.* Beijing: Geological Publishing House.

—Cheng, Z. and Xu, X. 1999. The earliest ceratopsian from the Tuchengzi Formation of Liaoning, China. *Journal of Vertebrate Paleontology.* **19**: 681–691.

APPENDIX

Character coding and distribution of character states are shown below for 41 characters in 12 pachycephalosaurian genera and 72 characters in 10 ceratopsian genera and Ceratopsidae. All characters in the pachycephalosaurian data set are binary; 6 of the 72 characters in the ceratopsian data set are three-state characters and the remainder are binary. One character in the pachycephalosaurian data set (character 28) and three characters in the ceratopsian data set (characters 12, 21, 67) are uninformative because the structures involved are neomorphic (i.e. impossible to polarize with outgroups). Character-state abbreviations: 0 = plesiomorphic state; 1, 2 = derived states; ? = not preserved or unknown; X = unknown as a result of transformation.

Pachycephalosauria

Characters and character states

Pachycephalosauria
1. Sacral rib length: subrectangular (0); strap-shaped (1).
2. Scapular blade, distal width: broad (0); narrow (1).
3. Preacetabular process, shape of distal end: tapered (0); expanded (1).

Wannanosaurus + other pachycephalosaurs
4. Frontal and parietal thickness: thin (0); thick (1).
5. Arched premaxilla–maxilla diastema, dentary canine: absent (0); present (1).
6. Postorbital–squamosal bar, form: bar-shaped (0); broad, flattened (1).

7. Squamosal exposure on occiput: restricted (0); broad (1).
8. Anterior and posterior supraorbital bones: absent (0); present (1).
9. Postorbital–squamosal tubercle row: absent (0); present (1).
10. Humeral length: more (0), or less than (1), 50% of femoral length.
11. Humeral shaft form: straight (0); bowed (1).
12. Deltopectoral crest development: strong (0); rudimentary (1).
13. Fibular mid-shaft diameter: 1/4 or more (0), or 1/5 or less (1), mid-shaft diameter of tibia.

Goyocephale + more derived pachycephalosaurs
14. Postorbital–parietal contact: absent (0); present (1).
15. Postorbital–jugal bar, shape: narrow (0); broad (1).
16. Squamosal tubercle row (5 to 7): absent (0); present (1).
17. Angular tubercle row: absent (0); present (1).
18. Basal tubera, shape: knob-shaped (0); plate-shaped (1).
19. Zygapophyseal articulations, form: flat (0); grooved (1).
20. Ossified interwoven tendons: absent (0); present (1).
21. Sternal shape: plate-shaped (0); shafted (1).
22. Iliac blade, lateral deflection of preacetabular process: weak (0); marked (1).
23. Iliac blade, medial tab: absent (0); present (1).

Homalocephale + more derived pachycephalosaurs
24. Parietal septum, form: narrow and smooth (0); broad and rugose (1).
25. Quadratojugal ventral margin, length: moderate (0); very short (1).
26. Pterygoquadrate rami, posterior projection of ventral margin: weak (0); pronounced (1).
27. Prootic-basisphenoid plate: absent (0); present (1).
28. Iliac blade, position of medial tab: above acetabulum (0); on postacetabular process (1).
29. Iliac blade, medial flange on postacetabular process: absent (0); present (1).
30. Ischial pubic peduncle, shape: transversely (0), or dorsoventrally (1), flattened.
31. Pubic body: substantial (0); reduced (1).

Yaverlandia + Pachycephalosauridae
32. Interfrontal and frontoparietal sutures: open (0); closed (1).

Pachycephalosauridae
33. Frontoparietal doming: absent (0); present (1).

34. Parietal–squamosal position relative to occiput: dorsal (0); posterodorsal (1).
35. Supratemporal opening: open (0); closed (1).

Pachycephalosaurinae

36. Frontoparietal doming, extent: incomplete (0), or complete (1), posteriorly and laterally.
37. Jugal–quadrate contact: absent (0); present (1).

Tylocephale + Prenocephale

38. Quadratojugal fossa: absent (0); present (1).

Stygimoloch + Pachycephalosaurus

39. Preorbital skull length: much less than (0), or subequal to (1), length from anterior orbital margin to posterior aspect of quadrate head.
40. Squamosal node cluster: absent (0); present (1).
41. Anterior snout nodes: absent (0); present (1).

Ceratopsia

Characters and character states

Ceratopsia

1. Rostral bone: absent (0); present (1).
2. Narial fossa, position: adjacent to (0), or separated by a flat margin from (1), the ventral margin of the premaxilla.
3. Jugal, lateral projection: chord from frontal orbital margin to extremity of jugal is less (0), or more (1), than minimum interorbital width.
4. Jugal (or jugal–epijugal) crest: absent (0); present (1).
5. Jugal infraorbital ramus, relative dorsoventral width: less (0), or subequal to or greater (1), than the width of the infratemporal ramus.
6. Premaxillary palate, form: flat (0); vaulted (1).
7. Predentary ventral process, width of base: less (0), or equal to or more (1), than half the maximum transverse width of the predentary.

Neoceratopsia

8. Premaxillary tooth number: 3 or more (0); 2 (1).
9. Premaxillary teeth, crown shape: recurved, transversely flattened (0); straight, subcylindrical (1).
10. Skull length (rostral–quadrate): 15% or less (0), or 20–30% (1), of length of postcranial skeleton.
11. Rostral anterior margin: rounded (0); keeled with point (1).
12. Rostral lateral processes: rudimentary (0); well developed (1).

13. Predentary anterior margin: rounded (0); keeled with point (1).
14. Predentary posteroventral process, shape: broader distally (0); narrower distally (1).
15. Retroarticular process length: long (0); very short or absent (1).

Archaeoceratops + Leptoceratops + Udanoceratops + Coronosauria

16. Edentulous maxillary/dentary margin, length: 2 (0), or 4 or 5 (1), tooth spaces.
17. Maxillary teeth, primary ridge development: low (0); prominent (1).
18. Maxillary/dentary primary ridge, position: near midline (0); offset posteriorly/anteriorly, respectively (1).
19. Maxillary/dentary teeth, packing: space between roots in adjacent teeth (0); no space between roots in adjacent tooth columns (1); no space between crowns within a tooth column (2).
20. Dentary tooth row, position of last tooth: anterior to (0), coincident with (1), or posterior to (2), the apex of the coronoid process.
21. Antorbital fossa shape: subtriangular (0); oval (1).
22. Postorbital and supratemporal bars, maximum width: narrow, bar-shaped (0); broad, strap-shaped (1); very broad, plate-shaped (2).
23. Infratemporal bar length: long, subequal to supratemporal bar (0); short, less than one-half supratemporal bar (1).
24. Jugal/epijugal crest: low (0); pronounced (1).
25. Quadrate shaft, anteroposterior width: broad (0); or narrow (1).
26. Predentary dorsal margin, form: sharp edge (0); bevelled cropping surface (1).
27. Dentary coronoid process, width and depth: narrow dentary process, low coronoid process (0); broad dentary process, moderately deep coronoid process (1); broad dentary process with distal expansion, very deep coronoid process (2).

Leptoceratops + Udanoceratops + Coronosauria

28. Maxillary/dentary crown, height: subequal to (0), or 1.5 times (1), maximum crown width.
29. Lateral maxillary/medial dentary crown base, form: convex (0), or inset (1), from root.
30. Jugal–squamosal contact above laterotemporal fenestra: absent (0); present (1).

31. Epijugal: absent (0); present (1).

32. Supratemporal fossae, relation: separated (0); joined in midline (1).

33. Posterior shelf composition: squamosal and parietal equal (0); squamosal dominant (1); parietal dominant (2).

34. Parietal shelf, inclination: horizontal (0); inclined posterodorsally (1).

35. Exoccipital–exoccipital contact below foramen magnum: absent (0); present (1).

36. Predentary surface between dentaries: absent (0); present (1).

37. Coronoid shape: strap-shaped (0); lobe-shaped (1).

38. Surangular eminence: absent (0); present (1).

39. Splenial symphysis: absent (0); present (1).

40. Axial neural spine, posterior margin: subtriangular (0); blade-shaped (1).

41. Cervicals 1–3, vertebral articulations: free (0); fused (1).

42. Cervicals 3–4, neural spine height: much shorter than (0), or subequal to (1), the axial neural spine.

43. Posteriormost caudals, neural spines and chevrons: absent (0); present (1).

44. Mid and posterior caudals, neural spine cross-section: subrectangular (0); oval (1).

45. Tail shape: tapering (0); leaf-shaped (1).

Leptoceratops + *Udanoceratops*

46. Premaxillary teeth: present (0); absent (1).

47. Dentary ventral margin, form: straight (0); curved (1).

48. Angular ventral margin, form: anterior portion (0), or nearly all of ventral margin (1), convex.

Coronosauria

49. Enamel distribution, maxillary/dentary teeth: both sides of crown (0); restricted to lateral/medial sides in maxillary/dentary teeth (1).

50. Nasal horn: absent (0); present (1).

51. Supratemporal fenestra, shape: oval (0); subtriangular (1).

52. Supratemporal fenestra, orientation of long axis: parasagittal (0); posterolaterally divergent (1).

53. Parietal width: subequal to (0), or much wider than (1), the dorsal skull roof.

54. Parietal posterior extension: as far posteriorly as (0), just posterior to (1), or far posterior to (2), the quadrate head.

55. Paired parietal fenestrae: absent (0); present (1).

56. Hypocentrum: absent (0); present (1).

57. Sacral number: 5 or 6 (0); 8 (1 dorsal, 1 caudal added) (1).

58. Sacral neural spines, mutual contact: absent (0); present (1).

Protoceratopsidae

59. Squamosal frill process: absent (0); present (1).

60. Sagittal crest, height: low and rounded (0); blade-shaped (1).

61. Paroccipital process, proportions: length is 2 (0), or 3 (1), times maximum depth of distal end.

62. Occipital condyle, size: large (0); small (1).

63. Predentary dorsal margin, inclination: horizontal (0); anterodorsally inclined (1).

Ceratopsoidea

64. Dentary teeth, primary ridge development: low (0); prominent (1).

65. Nasal horn position: posterior (0), or dorsal (1), to posterior margin of external nares.

66. Dentary ramus, position of maximum dorsoventral width: posterior (0); anterior (1).

67. Hypocentrum shape: wedge-shaped (0); U-shaped (1); ring-shaped (hemispherical occipital condyle) (2).

68. Mid cervical (C5–C7) neural spines, height: low (0); as high as dorsal neural spines (1).

Turanoceratops + Ceratopsidae

69. Maxillary/dentary teeth, root form: single (0); double (1).

70. Maxillary/dentary crowns, apical plane orientation: less (0), or more (1), than 45 degrees from the primary axis of the root.

71. Lateral maxillary/medial dentary crowns, secondary ridges: present (0); rudimentary or absent (1).

72. Postorbital horn: absent (0); present (1).

26

Armoured dinosaurs from the Cretaceous of Mongolia

TAT'YANA A. TUMANOVA

Introduction

Ankylosaurs, one of the main suborders of ornithischian dinosaurs, were distributed world-wide in the Mesozoic: their remains are found in Europe, Asia, North and South America, Australia and Antarctica (see Coombs and Maryańska (1990) for the most recent comprehensive review of the group). The earliest representatives of ankylosaurs are known from the Middle Jurassic (Galton, 1980; Dong, 1993), while the latest are Late Cretaceous (Maastrichtian) in age (e.g., Carpenter and Breithaupt, 1986). However, despite its wide distribution and representation by numerous fossil specimens, this group is still not well understood.

In the former USSR remains of ankylosaurs were found in two Lower Cretaceous sites, one in Yakutia (now the Republic Sakha) the other in Buryatia. They have also been recovered from various Upper Cretaceous horizons at 26 localities in Kazakhstan, Uzbekistan (Kyzylkum Desert) and Russia (Ryabinin,1939; Rozhdestvenskii, 1964; Nesov, 1995). These remains are represented by fragmentary material, usually disarticualated bones of the postcranial skeleton, armour, and, typically, teeth.

This article is devoted to the armoured dinosaurs of Mongolia, one of the most representative collections of ankylosaurs from Asia. The first discoveries of Mongolian ankylosaurs were made by the Central Asiatic Expedition of the American Museum of Natural History (1918–1930). Among large numbers of skeletons belonging to various dinosaurs from the highly fossiliferous locality of Bayan Zag, a skull and fragments of the postcranial skeleton of an ankylosaur

were collected and later described by Gilmore (1933) as *Pinacosaurus grangeri*. Abundant remains of ankylosaurs were excavated by the Palaeontological Expedition of the USSR Academy of Sciences and described by Maleev (1952a, b, 1954, 1956). Subsequently, new and well preserved ankylosaurs were discovered in the Upper Cretaceous of Mongolia by the Polish–Mongolian Palaeontological Expedition in the years 1963–1971 (Maryańska, 1969; 1971; 1977). Rich collections, including numerous interesting specimens of Upper Cretaceous ankylosaurs and the first Lower Cretaceous representatives of the group were excavated by the Soviet–Mongolian Expedition (now Mongolian–Russian Paleontological Expedition) during the past 20 years (Tumanova, 1977; 1987; 1993; Kurzanov and Tumanova, 1978). Material collected by these two expeditions and described by Maryańska (1977) and Tumanova (1987) are the principal sources for the generic diagnoses of the Mongolian ankylosaurs presented in this chapter.

The currently accepted systematic arrangement of ankylosaurs was developed by Coombs (1971, 1978a), who proposed that ankylosaurs should be divided into two families: Nodosauridae and Ankylosauridae. So far, all ankylosaurs found in Mongolia have been assigned to the family Ankylosauridae.

The oldest mongolian ankylosaur, *Shamosaurus scutatus* (see Figures 26.1–26.3), is the only member of the Ankylosauridae known so far from the Early Cretaceous (Albian–Aptian) of Asia. The construction of the skull of *Shamosaurus* is typical for ankylosaurids, although it also exhibits some plesiomorphic features. For this reason, *Shamosaurus* was placed in a new subfamily, Shamosaurinae (Tumanova, 1987). Owing to errors in the choice of characters, the genus *Saichania* was also included in this subfamily, but this classification of ankylosaurs at the suprageneric level has not been supported by other specialists (e.g. Coombs and Maryańska, 1990).

Upper Cretaceous ankylosaurs are common in all non-marine regional stratigraphic units. The oldest of these, the Bayanshiree Svita (Cenomanian–Turonian) contains remains of *Talarurus plicatospineus*, *Maleevus disparoserratus*, *Amtosaurus magnus* and possibly *Tsagantegia longicranialis* (see Figure 26.4). The succeeding unit, the Djadokhta Svita (?Late Santonian–Early Campanian), yielded *Pinacosaurus grangeri*, while the slightly younger Baruungoyot Svita (?Middle Campanian) produced *Saichania chulsanensis*. The youngest Mongolian ankylosaur, *Tarchia gigantea*, is known only from the Nemegt Svita (?Middle Campanian–Early Maastrichtian).

Repository abbreviations

AMNH, American Museum of Natural History, New York; GI SPS, Geological Institute, Section of Palaeontology and Stratigraphy, Mongolian Academy of Sciences, Ulaanbaatar; PIN, Palaeontological Institute, Russian Academy of Sciences, Moscow.

The anatomy of ankylosaurs

Ankylosaurs were usually large, reaching up to 7–8 m long, though some adults did not exceed 2.5 m in length. They were quadrupedal animals, heavily constructed, and with a broad body. Ankylosaur skulls are low, wide and rectangular in occipital view, with the long axis horizontal. The antorbital and supratemporal fenestrae are closed and the bones of the skull roof are covered by dermal plates which fuse with the underlying bones in adults, resulting in an even heavier, more massive skull. The special construction of the short neck, wherein the centra of the posterior cervical vertebrae have cranial and caudal faces of different height, enabled the huge weight of the heavy armoured head and dermal 'half-ring' to be supported.

Differences in the size of teeth of the two families (ankylosaurids have smaller teeth) and in the width of the muzzle led Carpenter (1982) to conclude that representatives of the two families employed different modes of feeding. Quite possibly, the nodosaurs, with their narrow muzzles, were more selective in their choice of vegetation, while the broad muzzled ankylosaurids were less so. A similar functional analogy can be found among African ungulates (Carpenter, 1982). In any case, an herbivorous mode of feeding for ankylosaurs seems to be clear, although their long, moveable tongue and some peculiarities in the mobility of the mandibles suggests the possibility that their diet also included insects and larvae (Nopcsa, 1928; Maryańska, 1977; Coombs and Maryańska, 1990).

The pectoral and pelvic girdles, like the limbs, were well adapted to redistributing the body weight. The scapula and coracoid were usually coossified, the ilium was widely expanded in the horizontal plane, and the acetabulum was imperforate. Indeed, the construction of the pelvic girdle is so consistent with quadrupedality that it is difficult to imagine transitional forms from a bipedal ancestor, though according to current models of the monophyletic origin of all ornithischians this must have been the condition of the ancestral form. The short massive limbs were situated beneath the body and moved mainly in a parasaggital plane. The forelimbs were shorter than the hind limbs resulting in a somewhat arch-like dorsal bend in the pelvic region. Coombs (1978b) concluded that ankylosaurs were mediportal, like the rhinoceros and hippopotamus, rather than graviportal like the elephant.

The armour of ankylosaurs consists of flat, keeled or spiked plates arranged in transverse and longitudinal rows. The plates, whose shape, size, height and symmetry vary in different parts of the armour, are surrounded by small, irregular, oval or tubercular ossicles. Distribution of the plates over the armour is also uneven. Usually, the keels of large plates are higher and sharper towards the flanks of the body and the direction of the plate tips differs in alternating longitudinal rows (Maryańska, 1977).

Unlike other thyreophorans, the armour elements of ankylosaurs did fuse to each other. The commonest constructions of the postcranial armour of ankylosaurs, and of Mongolian ankylosaurs in particular, are cervical and pectoral half-rings, each consisting of two bony layers. The deeper half-ring is constructed of fused plates, while the upper layer is formed from three pairs of sharply pointed large plates lying symmetrical to the mid-line. The arrangement of the armour elements follows the same general pattern for each ankylosaur family, but there are distinctive peculiarities in each species. For example, keeled plates with thick and dense walls occur in *Shamosaurus* (Figure 26.11A) and rib and groove ornamentation is found in *Talarurus*, while thin and strongly perforated spine walls are typcial of species of *Tarchia* (Figure 26.11B) and *Saichania*.

The peculiarly constructed tail of ankylosaurs, which could be used as a weapon of defence, is morphologically different in the representatives of both families. In Mongolian ankylosaurs, as in the group as a whole, the proximal part of the tail was flexible while the distal third consisted of a stiff rod formed from coossified posterior caudals in which long prezygapophyses and postzygapophyses overlapped each other up to the heavy, terminal tail club.

Systematic survey

Suborder Ankylosauria Osborn, 1923
Family Ankylosauridae Brown, 1908

Diagnosis (based on Coombs and Maryańska, 1990). Skull broad and triangular in dorsal view. Skull width more or less equal to length. Snout arches above the level of the postorbital skull roof. The caudal process of the premaxilla along the margin of the beak extends lateral to the most medial teeth. Paired sinuses in the premaxilla, maxilla, and nasal. The external nares are divided by a premaxillary septum with a ventral or lateral opening leading into a maxillary sinus. No ridge separating the premaxillary palate from the lateral maxillary shelf. The complex secondary palate twists the respiratory passage into a vertical S-shaped bend. The postorbital shelf, composed of postorbital and jugal bones, extends farther medially and ventrally than in nodosaurids. The lateral supraorbital

margin above the orbit is flat. The near-horizontal epipterygoid contacts the pterygoid and prootic. Dorsal surface of the skull roof covered by a large number of small scutes. Scute pattern poorly defined in the supraorbital region. Prominent, wedge-shaped caudolaterally projecting quadratojugal dermal plate. Large, wedge-shaped, caudolaterally projecting squamosal dermal plate. The infratemporal fenestra, paroccipital process and quadratojugal are hidden in lateral view by the united quadratojugal and squamosal dermal plates. There is a sharp lateral rim and low dorsal prominence for each lateral supraorbital element. Coronoid process lower and more rounded than in nodosaurids. Small teeth. Distal caudal vertebrae with centra partially or completely fused and with elongate prezygapophyses that are broad dorsoventrally and elongate postzygapophyses that are united to form a single dorsoventrally flattened, tongue-like process. Haemal arches of distal caudals elongate, with zygapophysis-like overlapping processes and elongate bases that contact to form a fully enclosed haemal canal. Fused sternal plates. Coracoid small relative to scapula. Scapular spine located on extreme anterior edge of scapular blade (see Figure 26.2). Acromion projects laterally and does not project above the dorsal margin of the scapular blade. Postacetabular process of ilium short. Ischium near vertical below acetabulum. Pubis reduced to nubbin and fused to other pelvic elements. Deltopectoral crest and transverse axis through distal humeral condyles in same plane. Fourth trochanter on distal half of femur. Terminal tail club consisting of large pair of lateral plates and two smaller plates. Ossified tendons surround distal caudal vertebrae. Keeled armour plates usually have deep internal hollow (Figure 26.11A).

Shamosaurus Tumanova, 1983

Type species. *Shamosaurus scutatus* Tumanova, 1983.
Holotype. PIN N 3779/2, complete skull and jaw; Khamaryn Us, Övörkhangai, Dornogov' Aimag; Hühteeg Svita, Lower Cretaceous (Aptian–Albian).
Referred material. In addition to the holotype a partial skull from the holotype locality (Tumanova, 1987) and

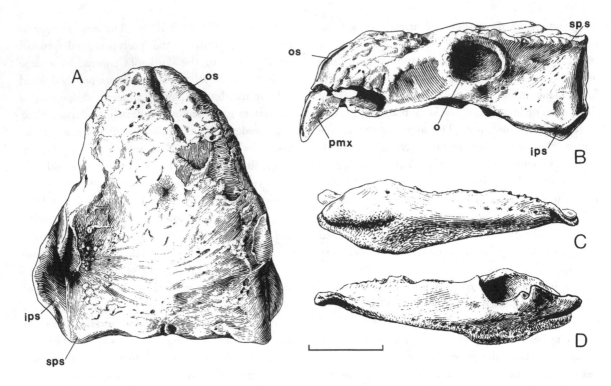

Figure 26.1. *Shamosaurus scutatus*, Tumanova, 1983. PIN 3779/2–1, holotype skull in (A) dorsal and (B) lateral view. Mandible in (C) right lateral and (D) internal view. After Tumanova, 1987, p. 24, fig. 4, p. 42, fig. 12. Abbreviations: ips, quadratojugal (lower postorbital) dermal plate; o, orbit; os, dermal plate; pmx, premaxilla; sps, squamosal (upper postorbital) dermal plate. Scale bar = 100 mm.

an undescribed mandible from Höövör are also referred to this taxon.

Description. The length of the body reaches about 7 m. The skull is up to 360 mm in length and 370 mm width. The dorsal surface of the of the skull was completely covered by osteodermal plates (Figure 26.1) bearing a sculpture pattern of small excrescences. Plates in the posterolateral corners of the skull do not develop into a scute and the quadratojugal and squamosal dermal plates do not hide the quadrate condylus. The anterior part of the snout is oval and narrow, and it is a little narrower than the distance between the posteriormost maxillary teeth. The posterior maxillary shelf is well developed and the ventral surfaces of the palatal bones are inclined laterally. The mandibular condyle of the quadrate is situated behind the posterior margin of the orbit. The occiput is inclined posteriorly and the occipital condyle is oriented ventrally. The quadrate

bones and paroccipital process are coossified and the ventral surface of the basioccipital is round and narrow. The walls of armour elements are thick and dense.

Talarurus Maleev, 1952

Type species. Talarurus plicatospineus Maleev, 1952.

Holotype. PIN N 557/91, occipital section of cranium with part of the skull roof. Bayan Shiree, Eastern Gobi desert; Bayanshiree Svita, Upper Cretaceous (Cenomanian-Turanian).

Referred material. A skull roof with occipital section (Kurzanov and Tumanova, 1978) and fragmentary remains of the postcranial skeletons of a number of individuals have been found at the same locality (Maleev, 1952a, b).

Description. The body length of this ankylosaur reached about 4–5 m. The skull (Figure 26.5) is about 240 mm long, and approximately 220 mm wide.

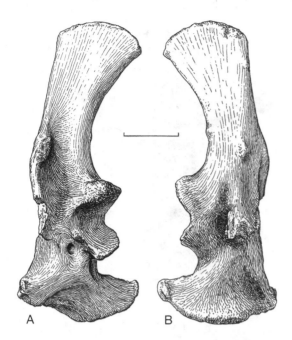

Figure 26.2. *Shamosaurus scutatus* Tumanova, 1983. PIN 3779/2–1, scapulocoracoid in (A) lateral and (B) medial view. Scale bar = 100 mm.

Pyramidal plates occur above and behind the orbits and the entire surface of the skull roof is covered by osteoderms ornamented with small tubercles (Figure 26.5A). The occipital surface is perpendicular to the skull roof and the paroccipital processes are directed a little posterolaterally. The occipital condyle is directed posteroventrally and there is no fusion of the quadrate bones and paroccipital processes. Anterior to the condyle the ventral surface of the basioccipital bears a medial eminence with depressions on both sides. The fundus of the brain cavity is level. The pectoral glenoid is deep and short and the humeral head is situated dorso-terminally. The manus is pentadactyl, the pes tetradactyl. Armour elements bear a 'furrow-rib' type of ornamentation and the tail-club is weakly developed.

Tsagantegia Tumanova, 1993
Type species. *Tsagantegia longicranialis* Tumanova, 1993.
Holotype (and only known specimen). GI SPS N 700/17,

skull; Tsagaan Teeg, Southeastern Gobi; Bayanshiree Svita, Upper Cretaceous (Cenomanian–Turonian).
Description. Large ankylosaur with a body up to 6–7 m long and a skull (Figure 26.4) about 300 mm long and 250 mm wide. The skull roof is covered by numerous small osteoderms with a weakly expressed relief. The upper postorbital spines are not developed and osteoderms do not expand above the occiput. The orbits are situated posterior to the middle of the skull length. The osteodermal ring around the orbit is separated from the other osteoderms by a prominent furrow. The orbit size is slightly reduced due to the presence of the osteodermal ring. The premaxillary beak is trapezoid and the anterior and posterior maxillary shelves are weakly developed. The medial part of the anterior wall of the pterygoid is inclined posteriorly and the basisphenoid and pterygoid are fused. The occipital surface is perpendicular to the skull roof and the lower side of each paroccipital process is directed a little medially, while the distal ends are bent slightly ventrally. The prootic, opisthotic and exoccipital bones are united, but with prominent borders between them. The occipital condyle is wide-oval and directed postero-ventrally. The quadrate and paroccipital process are coossified and the mandibular condyle of the quadrate is situated level with the posterior margin of the orbit, or further posteriorly. The ventral surface of the basioccipital bears a central depression, separated by gentle crests from the lateral depressions. The cingulum and lingulum of the maxillary teeth are separated by a vertical furrow.

Maleevus Tumanova, 1987
Type species. *Maleevus disparoserratus* (Maleev, 1952b)
Holotype. PIN N554/1, fragments of a right and left maxilla; Shireegiin Gashuun, Eastern Gobi Desert; Bayanshiree Svita, Upper Cretaceous (Cenomanian–Turonian).
Referred material. A fragment of a basicranium from the type locality (Figure 26.6).
Description. Anterior maxillary shelf weakly developed and originating from a point near the start of the tooth row. The maxillary teeth have a 'w'-shaped cingulum. The occipital condyle is almost hemispherical

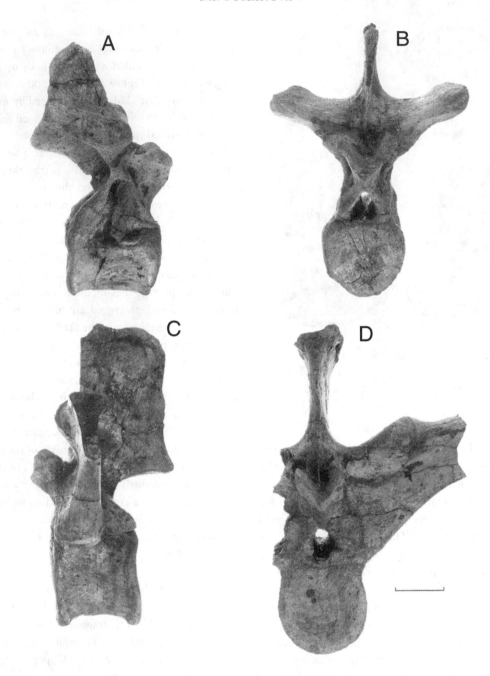

Figure 26.3. *Shamosaurus scutatus*, PIN 3779/2: A-B. Isolated dorsal vertebra in (A) lateral and (B) anterior view. Isolated dorsal vertebra with ankylosed rib in (C) lateral and (D) anterior views. Scale bar = 50 mm.

A

B

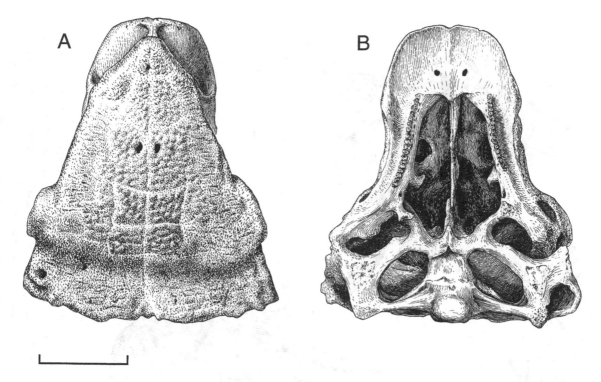

Figure 26.4. *Tsagantegia longicranialis* Tumanova, 1993. GI SPS 700/17, holotype skull in (A) dorsal and (B) palatal view. After Tumanova, 1993. Scale bar = 100 mm.

and ventrally directed. The ventral surface of the basicranium bears two weakly expressed longitudinal ridges which diverge anteriorly (Figure 26.6A). The medial depression upon the basicranium gradually transforms into a depression at the level of the sphenoccipital tubercles. The fundus of the brain cavity is almost level.

Comments. The holotype of this genus was first described by Maleev (1952a, b) under the name of *Syrmosaurus disparoserratus*. However, the type species of this genus, *Syrmosaurus viminicaudus*, proved to be a junior synonym of the genus *Pinacosaurus* and *S. disparoserratus* was assigned by Maryańska (1977) to the genus *Talarurus*. However, as the basicranium of *Syrmosaurus disparoserratus* is different from that of *Talarurus* and as the type material of *Talarurus* does not include the upper jaw, *S. disparoserratus* cannot be assigned to this genus and *Maleevus* was chosen as a replacement name.

Amtosaurus Kurzanov and Tumanova, 1978

Type species. Amtosaurus magnus Kurzanov and Tumanova, 1978.

Holotype (and only known specimen). PIN N 3780/2, brain case (Figure 26.7); Amtgai, Ömnögov', Gobi Desert; Bayanshiree Svita, Upper Cretaceous (Cenomanian–Turonian).

Description. This taxon is represented by a fragmentary basicranium. The occiput is high and the brain cavity is large. The occipital condyle is oval and directed posteroventrally. The ventral surface of the basioccipital bears two gentle longitudinal elevations situated symmetrically on either side of the medial depression (Figure 26.7A). The fundus of the brain cavity is slightly flexed anteriorly.

Pinacosaurus Gilmore, 1933

Type species. Pinacosaurus grangeri Gilmore, 1933.

Holotype. AMNH 6523, incomplete skull and maxilla,

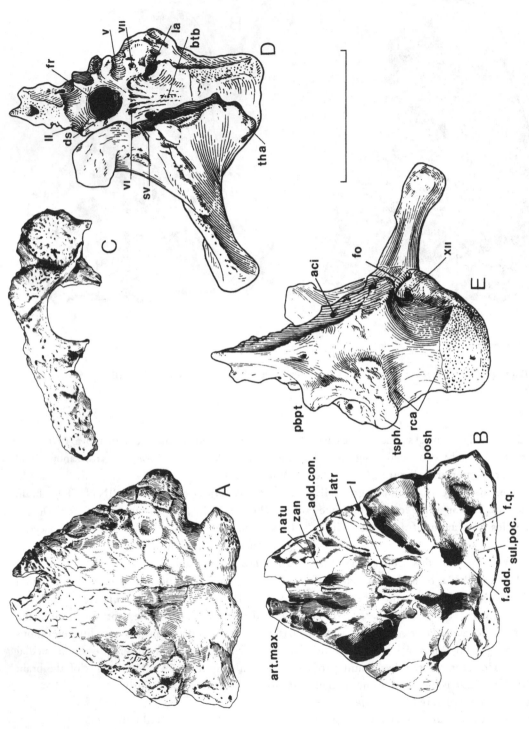

Figure 26.5. *Talarurus plicatospineus* Maleev, 1952. PIN 3780/1, partial skull roof in (A) dorsal, (B) palatal and (C) lateral view. Braincase in (D) dorsal and (E) ventral view. (After Tumanova, 1987, p. 27, fig. 5, p. 36, fig. 7.) Abbreviations: aci, arteria carotis interna; add. con, additus conchae; art. max, maxillary artery; btb, basis trabeculi basalis; ds, dorsum of the Turkish saddle; f. q, fossa for upper process of quadrate; fo, fenestra ovalis; fr, pituitary fossa; l, lagena; latr, lamina transversalis anterior; pbpt, basioccipital process; posh, postocular shelf; natu, nasoturbinals; rca, area for the origin of the M. rectus capiti anterior; sul. poc., sulcus for paroccipital process; sv, sulcus for vein; tha, tubera for neural arches of the first cervical vertebrae; tsph, sphenoccipital tubera; zan, zona annularis; I–XII, cranial nerves. Scale bar = 50 mm.

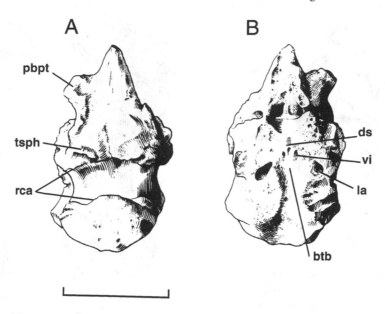

Figure 26.6. *Maleevus disparoserratus* (Maleev, 1952). PIN 554/2–1, partial braincase in (A) ventral and (B) dorsal view. After Tumanova, 1987, p. 39, fig. 10. For abbreviations see Figure 26.5. Scale bar = 50 mm.

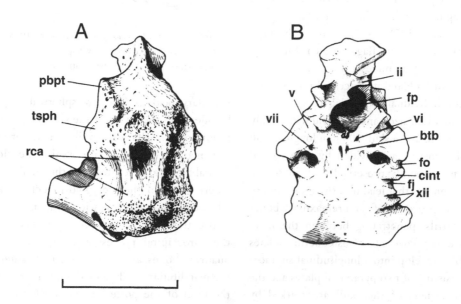

Figure 26.7. *Amtosaurus magnus* Kurzanov and Tumanova, 1978. PIN 3780/2, holotype braincase in (A) ventral and (B) dorsal view. After Tumanova, 1987, p. 38, fig. 9). Abbreviations as in Figure 26.5, and: cint, crista intervenestralis; fj, jugular foramen. Scale bar = 50 mm.

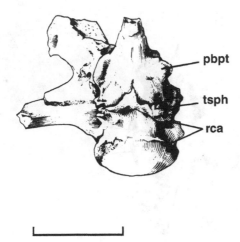

Figure 26.8. *Pinacosaurus grangeri* Gilmore, 1933. PIN 4043, braincase in ventral view. After Tumanova, 1987, p. 37, fig. 8. For abbreviations see Figure 26.5. Scale bar = 50 mm.

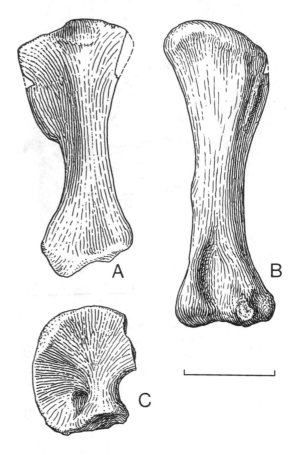

Figure 26.9. *Pinacosaurus grangeri* Gilmore, 1933. PIN 3144. Left humerus (A) in dorsal view. Right femur (B) and coracoid (C) in posterior view. Scale bar = 50 mm.

first cervical vertebra and some dermal scutes; Bayan Zag, Gobi Desert; Djadokhta Formation, Upper Cretaceous (?Late Santonian–Early Campanian).

Referred material. A well preserved skull and almost complete postcranial skeleton of a young individual, another complete postcranial skeleton (Maleev, 1952a, 1954) and fragmentary remains of postcrania and armour (Maryańska, 1971, 1977) all from the type locality; a basicranium from Baga Tariach (Kurzanov and Tumanova, 1978); an undescribed skull in a concretion from Shilt Uul; and undescribed fragmentary remains of a few individuals from Alag Teeg.

Description. A medium sized ankylosaurid with a body length of about 5 m and a skull 300 mm long and 340 mm wide. The nostrils are divided by a horizontal septum and in a superbly preserved skull there is a third pair of openings that lead into the premaxillary sinus. Ascending processes of the premaxillary bones divide the nostrils, penetrating between the nasal bones and are not covered by osteoderms. Plates above the orbits develop into a longitudinal supraorbital spine consisting of two pyramidal plates and the posterolateral corners of the skull are marked by upper and lower postorbital spines. Osteoderms do not overhang the medial part of the occiput, thus the occipital condyle is visible in dorsal view. The pre-

maxillary beak is quadrato-spherical in shape and broader than the width of the skull between the posteriormost maxillary teeth. Both anterior and posterior maxillary shelves are weakly developed, the palatal bones are elevated, and the medial part of the anterior wall of the pterygoid is inclined anteriorly. The occipital condyle is wide-oval and directed posteroventrally. The contact between the quadrate and the paroccipital process remained unossified. The quadrate bones are slightly inclined anteriorly and the mandibular condyle of the quadrate is located at the level of the posterior margin of the orbit. The ventral surface of the basisphenoid does not exhibit any marked relief (Figure 26.8). The postcranial skeleton (Figure 26.9) is relatively light, the limb bones are slender, the manus is pentadactyl and the pes tetradactyl.

Saichania Maryańska, 1977

Type species. Saichania chulsanensis Maryańska, 1977.

Holotype. GI SPS 101/151, skull with mandibles and the anterior part of the postcranial skeleton and armour in natural articulation; Khulsan, Gobi Desert; Baruungoyot Svita, Upper Cretaceous (?Middle Campanian).

Referred material. Fragments of a skull roof and armour from the type locality (Maryańska,1977) and an undescribed skull, mandibles and almost complete postcranial skeleton from Hermiin Tsav.

Description. Large ankylosaur with a body length up to 7 m, and a skull about 450 mm long and 480 mm wide. The skull roof exhibits prominent osteodermal plates and spinous postorbital osteoderms. The nostrils are large, terminally situated, and divided by a horizontal septum. A dorsal opening leads to a respiratory canal and a ventral one to a ventromedial canal which enters the premaxillary sinus. The premaxillary beak has a round-oval shape and is almost as broad as the distance between the posteriormost maxillary teeth. The palatine has strongly developed anterior and posterior maxillary shelves. The medial part of the anterior wall of the pterygoid is inclined anteriorly. The plane of the occiput is perpendicular to the skull roof and the paroccipital processes are low and perpendicular to the skull roof in their upper part, while their lower portion is deflected anteriorly. The occipital condyle is oval, weakly convex and directed ventrally. The quadrate and paroccipital process are coossified. The mandibular cotylus of the quadrate is located at the level of the middle part of the orbit. The anterior and posterior walls of the orbit are heavily ossified. The ventral surface of the basioccipital has no marked relief. The premaxillary bones are partially overlapped by osteoderms growing downward from the nasal bone. Strongly developed secondary plate-like intercostal ossifications occur along the lateroventral side of the trunk. The postcranial skeleton is extremely massive with strong ossification of the sternal complex. The manus is pentadactyl and the tail-club is large.

Tarchia Maryańska, 1977

Type species. Tarchia gigantea (Maleev, 1956).

Holotype. PIN 551/29, a series of caudal vertebrae, metacarpals and phalanges, and fragments of armour plates; Nemegt, Gobi Desert; Nemegt Svita, Upper Cretaceous (?Middle Campanian–Early Maastrichtian).

Referred material. Fragmentary remains of distal parts of tails and fragments of armour from the Nemegt Basin (Maryańska, 1977), an incomplete skull with skull roof, occiput and brain case from Khulsan (Maryańska, 1977) and a well preserved skull (Tumanova, 1977; 1987) with incomplete postcranium.

Description. This is the largest, but stratigraphically, the youngest Mongolian ankylosaur. The body length reached 8 m and the skull (Figure 26.10) was up to 400 mm long and 450 mm wide. The premaxillary part of the snout forms a rounded oval, and is as broad as the distance between the posteriormost maxillary teeth. The anterior and posterior maxillary shelves are well developed. The plane of the palatine bones is horizontally elevated and the anterior wall of each pterygoid is inclined forward. The occiput is inclined slightly posteriorly and the occipital condyle is brachy-oval with a slightly protruding joint surface that is directed posteroventrally. The paroccipital process is high, short, perpendicular to the plane of the skull roof and does not fuse to the quadrate. The mandibular condylus of the quadrate lies at the level of the posterior margin of the orbit. The ventral part of the basioccipital has no marked relief. The height of the foramen magnum exceeds its width, the brain cavity is very high and the openings for the cranial nerves are of large size. The pes is tetradactyl and the tail-club is large.

Comments. The type species of the genus *Tarchia*, *T. kielanae*, was established on the basis of an incomplete skull with skull roof, occiput and brain case (Maryańska, 1977). However, more comprehensive material has shown that *Dyoplosaurus giganteus* (Maleev, 1956) and *T. kielanae* belong to the same species: *Tarchia gigantea* (Maleev, 1956) (see Tumanova, 1977, 1987; Coombs and Maryańska, 1990).

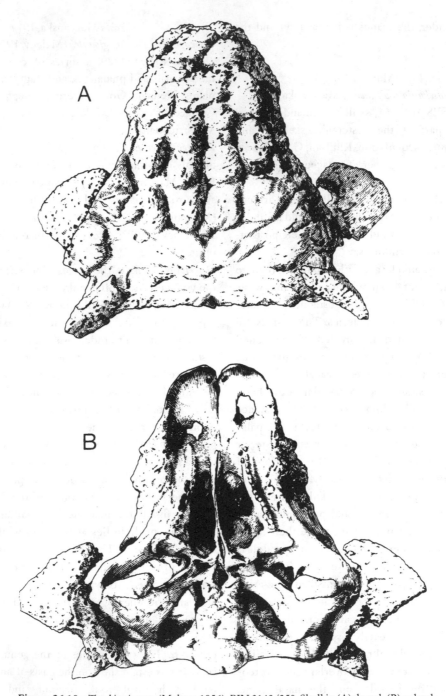

Figure 26.10. *Tarchia gigantea* (Maleev, 1956), PIN 3142/250. Skull in (A) dorsal, (B) palatal and (C) lateral view. Left mandible in (D) external and (E) internal view. Mandible in (F) dorsal view. After Tumanova, 1987, p. 19–21, fig. 3, p. 41, fig. 11. For abbreviations see Figure 26.5, and: art, articular; c, coronoid; f. meck., Meckelian foramen; f. sang, foramen supraangulare; os meck, Meckel's cartilage; pr, retroarticular process; sa, surangular. Scale bar = 100 mm.

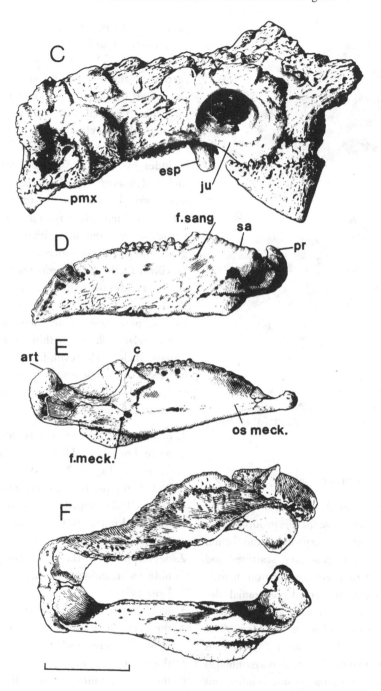

Figure 26.10. (*cont.*)

Figure 26.11. Different types of dermal scute surface: (A) *Shamosaurus scutatus* Tumanova, 1983, PIN 3779/2–1, holotype. (B) *Tarchia giantea* (Maleev, 1956), PIN 3142/250. Scale bar = 50 mm.

Discussion

At present, the fossil material of Mongolian ankylosaurs is the most representative for this group in Asia. Ankylosaur remains from the territory of the former Soviet Union (FSU) are fragmentary, scattered and because of the nature of their preservation: mainly teeth, armour and poorly preserved postcranial elements, they cannot be identified to the generic level.

The oldest ankylosaur so far known from Mongolia is *Shamosaurus* from the late Lower Cretaceous (Aptian–Albian). In the Upper Cretaceous assemblages from most of the regional stratigraphic subdivisions usually contain ankylosaurs. Thus they are known from the Cenomanian–Turonian (*Talarurus, Amtosaurus, Maleevus*) up to the Maastrichtian (*Tarchia*).

The recent discovery of the ankylosaur *Tianchiasaurus nedegoapeferima* (Dong, 1993) in the Middle Jurassic of China has greatly expanded the known time range of ankylosaurs from Asia. Surprisingly, however, although *T. nedegoapeferima* exhibits primitive features of the mandibles and postcranial skeleton, it has a small, flat, tail club that is characteristic for ankylosaurids. Thus all ankylosaurs from Asia seem to belong to the Ankylosauridae.

Besides the Asiatic genera, the Ankylosauridae includes two North American forms: *Euoplocephalus* and *Ankylosaurus*. *Euoplocephalus* is represented by numerous specimens and has the following characters: the external nares face rostrally and are divided by a vertical septum and the width of the beak is equal to or greater than the distance between the caudalmost maxillary teeth. *Ankylosaurus*, the youngest and largest ankylosaurid is distinguished by a strong expansion of the dermal armour in the nasal region that restricts the external nares to small circular openings.

Interrelationships within the Ankylosauridae remain unclear. However, by taking into account the older age of *Shamosaurus*, one can suggest the following general evolutionary transformation in the lineage *Shamosaurus–Ankylosaurus*:

1. quadratojugal and squamosal dermal plates become more horn-like. They are weakly expressed in *Talarurus* (Figure 26.5A, C), *Shamosaurus* (Figure 26.1 A, B), and in *Tsagantegia* (Figure 26.4 A), but are well developed and moderately pointed in *Tarchia* (Figure 26.10 A, C) and *Euoplocephalus*, and take the form of horns in *Ankylosaurus* and *Saichania*;
2. development of the quadratojugal plate tends to hide the mandibular condyle of the quadrate in lateral view;
3. widening of the premaxillary beak;
4. the plane of the orbits rotates forwards while the orbits shift posteriorly;
5. the tooth rows shorten;
6. there is some increase in skull kinetism because of the weakening of the basipterygoid joint: in *Shamosaurus* the pterygoids and basisphenoid were fused, but all isolated basicrania of *Talarurus*, *Maleevus* and *Amtosaurus* exhibit articular surfaces for the contact with the pterygoids.

Some morphological features of *Shamosaurus*, for example, fusion of the quadrate with the paroccipital process can be observed in later ankylosaurs such as *Tsagantegia* and *Saichania*. Often, however, the nature of this contact is rather different: the upper process of the quadrate reaches the skull roof anterior to the paroccipital process and contacts the latter to various degrees: close in *Talarurus*, and, probably, with ligaments in *Tarchia*.

Some features in the morphology of *Shamosaurus* can be considered as intermediate between Ankylosauridae and Nodosauridae. These include: a narrow premaxillary beak; immobile basipterygoidal contact; a quadrate condyle partially visible in lateral view; and a round, vertically oriented occipital condyle. These features suggest that these families are monophyletic and this is strongly supported by the discovery of Jurassic ankylosaurs in North America (Kirkland and Carpenter, 1994; Carpenter *et al.*, 1996). Study of the new Jurassic ankylosaurs may help to clarify phylogenetic relationships, both within families and at the family level.

Acknowledgments.

I am deeply grateful to Teresa Maryańska and Walter P. Coombs, Jr., for their careful reading of the manuscript, which has been greatly improved by their comments, and I thank Ken Carpenter for helpful information on new Jurassic ankylosaur specimens, and David Unwin for his useful comments during my work on this manuscript.

References

Carpenter, K. 1982. Skeletal and dermal armor reconstruction of *Euoplocephalus tutus* (Ornithischia: Ankylosauridae from the Late Cretaceous Oldman Formation of Alberta. *Canadian Journal of Sciences* **19**: 689–697.

—and Breithaupt, B. 1986. Latest Cretaceous occurrence of nodosaurid ankylosaurs (Dinosauria, Ornithischia) in western North America and the gradual extinction of the dinosaurs. *Journal of Vertebrate Paleontology* **6**: 251–257.

—Kirkland, J., Miles, C. and Cloward, K. 1996. Evolutionary significance of new Ankylosaurs (Dinosauria) from the Upper Jurassic and Lower Cretaceous, Western Interior. *Journal of Vertebrate Paleontology*, **19** (Suppl. to nb. 3): 25A.

Coombs, W.P., Jr. 1971. *The Ankylosauria*. Ph.D. thesis, Columbia University, New York, NY, 487 pp.

—1978a. The families of the ornithischian dinosaur order Ankylosauria. *Palaeontology* **21**: 143–170.

—1978b. Theoretical aspects of cursorial adaptations in dinosaurs. *Quarterly Review of Biology* **53**: 393–418.

—and Maryańska, T. 1990. Ankylosauria, pp. 456–483 in Weishampel, D.B., Dodson, P. and Osmólska, H. (eds.), *The Dinosauria*. University of California Press: Berkeley.

Dong, Z.M. 1993. An ankylosaur (Ornithischian dinosaur) from the Middle Jurassic of the Junggar Basin, China. *Vertebrata PalAsiatica* **31** (4): 258–266.

Galton, P.M. 1980. Armored Dinosaurs (Ornithischia: Ankylosauria) from the Middle and Upper Jurassic of England. *Geobios* **13**: 825–837.

Gilmore, C.W. 1933. Two new dinosaurian reptiles from Mongolia with notes on some fragmentary specimens. *American Museum Novitates* **679**: 1–20.

Kirkland, J.I. and Carpenter, K. 1994. North America's first pre-Cretaceous ankylosaur (Dinosauria) from the Upper Jurassic Morrison Formation of Western Colorado. *Brigham Young University Geology Studies* **40**: 25–42.

Kurzanov, S.M. amd Tumanova, T.A. 1978. [On the structure of the endocranium in some ankylosaurs from Mongolia.] *Paleontologicheskii Zhurnal* 1978: 90–96.

Maleev, E.A. 1952a. [A new family of armored dinosaurs from the Upper Cretaceous of Mongolia.] *Doklady AN SSSR* **87**: 131–134.

—1952b. [A new ankylosaur from the Upper Cretaceous of Asia.] *Doklady AN SSSR* **87**: 273–276.

—1954. [New turtle-like reptile in Mongolia.] *Priroda* **1954**: 106–108.

—1956. [Armoured dinosaurs from the Upper Cretaceous of Mongolia.] *Trudy Paleontologicheskogo Instituta AN SSSR* **62**: 51–91.

Maryańska, T. 1969. Remains of armored dinosaurs from the uppermost Cretaceous in Nemegt Basin, Gobi Desert. *Paleontologia Polonica* **21**: 22–34.

—1971. New data on the skull of *Pinacosaurus grangeri* (Ankylosauria). *Paleontologia Polonica* **25**: 45–53.

—1977. Ankylosauridae (Dinosauria) from Mongolia. *Paleontologia Polonica* **37**: 85–151.

Nesov, L.A. 1995. [*Dinosaurs of Nothern Eurasia: new data about assemblages, ecology and palaeobiogeography.*] St. Petersburg: Izdatel'stvo Sankt-Peterburgskogo Universiteta, 156 pp.

Nopcsa, F. 1928. Paleontological notes on reptiles. *Geologica Hungarica, ser. Paleont.* **1** (1): 1–84.

Rozhdestvenskii, A.K. 1964. [New data on the localities of dinosaurs in Kazakhstan and Middle Asia.] *Nauchnye Trudy Tashkent, Universiteta* **237**: 227–241.

Ryabinin, A.N. 1939. [The vertebrate fauna from the Upper Cretaceous of South Kazakhstan]. *Trudy*

Tsentral'nogo Nauchno-Issledovatel'skogo Geologo-Razvedsochnogo Instituta **118**: 1–40.

Tumanova, T.A. 1977. [New data on the ankylosaur *Tarchia gigantea.*] *Paleontologicheskii Zhurnal* **1977**: 92–100.

—1983. [The first ankylosaur from the Lower Cretaceous of Mongolia.] *Trudy Sovmestnoi Sovetsko-Mongol'skoi Paleontologicheskoi Expeditsii* **24**: 110–120.

—1987. [The armoured dinosaurs of Mongolia.] *Trudy Sovmestnoi Sovetsko-Mongol'skoi Paleontologicheskoi Expeditsii* **32**: 1– 80.

—1993. [A new armored dinosaur from South-Eastern Gobi.] *Paleontologicheskii Zhurnal* 1993, **27** (2): 92–98.

27

Mesozoic birds of Mongolia and the former USSR

EVGENII N. KUROCHKIN

Introduction

The first Mesozoic avian skeletal remains from Asia were found in Mongolia by the Polish–Mongolian Palaeontological Expedition (Elzanowski, 1974, 1976, 1981). Feathers from the Late Cretaceous of Kazakhstan were described by Bazhanov (1969) and Shilin (1977) and numerous fossil feathers from the Lower Cretaceous of Mongolia and East Siberia were collected by palaeoentomologists from PIN at the beginning of the 1970s. From the 1970s to the 1990s skeletal remains of various avian fossils were recovered from the Cretaceous of Mongolia by the Joint Russian–Mongolian Palaeontological Expedition and in the Cretaceous of Middle Asia by the late Lev Nesov as a result of his persistent exploration. Further remains of birds were found in recent years in Central Asia by the American–Mongolian expeditions (see Chapter 12) and it now appears that earlier American expeditions in the 1920s also recovered fossil birds, though they were not recognized as such until the 1990s (see Chiappe *et al.*, 1996, and refs therein).

Mesozoic birds from Mongolia and the former Soviet Union (FSU) are rare and usually fragmentary, but they provide some data regarding the early history of the group in this territory. In addition, many Mesozoic birds have recently been found in China (Chiappe, 1995). Together, these records provide the basis for analyses of Mesozoic avian assemblages in Asia (Kurochkin, 1995a). This chapter presents a summary of Mesozoic birds from the FSU and Mongolia, primarily discovered by Russian palaeontologists. Avian macro-taxonomy follows Kurochkin (1995c).

Institutional abbreviations

IVPP, Institute of Vertebrate Palaeontology and Palaeoanthropology, Beijing, China; IZASK, Institute of Zoology of the Kazakh Academy of Sciences, Alma-Aty, Kazakhstan; JRMPE, Joint Russian–Mongolian Palaeontological Expedition; MGI, Geological Institute of the Mongolian Academy of Sciences, Ulaanbaatar; PIN, Palaeontological Institute of the Russian Academy of Sciences, Moscow, Russia; PO, Collection of the Zoological Institute of the Russian Academy of Sciences, St. Petersburg, Russia; TsNIGRI, F.N. Chernyshev Central Museum for Geological Exploration, St. Petersburg, Russia; VPM, Volgograd Provincial Museum, Volgograd, Russia; ZPAL, Institute of Palaeobiology of the Polish Academy of Sciences, Warsaw, Poland.

Systematic description

Class Aves Linnaeus, 1758
Subclass Sauriurae Haeckel, 1866
Infraclass Enantiornithes Walker, 1981
Order Alexornithiformes Brodkorb, 1976
Family Alexornithidae Brodkorb, 1976

Contents. Gobipteryx Elzanowski, 1974; *Alexornis* Brodkorb, 1976; *Kizylkumavis* Nesov, 1984; *Zhyraornis* Nesov, 1984 (2 spp.); *Nanantius* Molnar, 1986 (2 spp.); *Sazavis* Nesov and Yarkov, 1989; *Neuquenornis* Chiappe and Calvo, 1994; *Lenesornis* Kurochkin, 1996.

Diagnosis. Cranial half of the synsacrum low and broad; synsacrum convex dorsally; ischium narrow; acrocoracoid and coracoidal process narrow and tapered; shaft of the coracoid strut-like and narrow; shaft of the coracoid with a deep and short depression on the dorsal side; scapular facet and scapular glenoid fused; ventral epicondyle of the humerus protrudes distad to a striking degree; wing digits clawless; shaft of the tibiotarsus thin; metatarsal III straight; metatarsals II–IV short and gracile.

Kizylkumavis Nesov, 1984

Type species. Kizylkumavis cretacea Nesov, 1984.

Diagnosis. Original diagnosis of Nesov (1984): olecranon fossa is narrow and displaced in the direction of the flexor process; flexor process is strongly projected distally; dorsal condyle very narrow and aligned at an angle of 65° to the longitudinal axis of the humeral shaft; ventral condyle is short, aligned almost transversely to the longitudinal axis of the humeral shaft, and only slightly projected in cranial aspect; intercondylar furrow runs slightly on the cranial side; brachial depression is not developed; a small groove is developed distal to a small ventral supracondylar tubercle on the cranial surface of the distal end, and aligned at an angle of 70° to the longitudinal axis of the shaft.

Comments. As I have not seen the specimen recently, the original diagnosis, which includes the generic characters of *Kizylkumavis*, as well as characters of the Enantiornithes and Alexornithiformes is presented here. The following characters form an emended diagnosis for this genus: distal end of the humerus is very wide in the dorsoventral direction; the ventral portion of the distal end of the humerus is remarkably enlarged; the dorsal condyle is broad; the intercondylar furrow is narrow; and the flexor process is strongly projected distally.

The humeri of *Kizylkumavis* and *Alexornis* of Mexico (Brodkorb, 1976) are similar in some respects. For example, they share: a remarkable distal projection of the ventral epicondyle, distal displacement of a shallow olecranal fossa, and an abrupt transition from the distal end to the shaft. However, they also exhibit some differences: the shape of the dorsal condyle (which is broad in *Kizylkumavis* and narrow and olive-shaped in *Alexornis*); a more spacious intercondylar furrow in *Alexornis*, and more distal projection of the flexor process in *Kizylkumavis*.

Kizylkumavis cretacea Nesov, 1984

Holotype. TsNIGRI 51/11915, distal fragment of a right humerus. Dzharakhuduk locality, Navoi District, Bukhara Province, Uzbekistan. Outcrop CBI-5a, Bissekty Svita (Coniacian).

Diagnosis. Same as for genus.

Comments. K. cretacea is a very small enantiornithine; the maximum width of the distal end of humerus is 5.1 mm. *K. cretacea* was the first member of the Enantiornithes to be described from the Old World (Nesov, 1984), although in the original description it was assigned to Aves *incertae sedis*. In spite of its fragmentary condition, there are no doubts as to the enantiornithine relationships of *K. cretacea* since it has no fossa for the *M. brachialis*, a transverse position of the dorsal condyle of the humerus, and an inclined position of the ventral condyle.

Zhyraornis Nesov, 1984

Type species. Zhyraornis kashkarovi Nesov, 1984.

Contents. Z. kashkarovi Nesov, 1984; *Z. logunovi* Nesov, 1992.

Diagnosis. Cranial portion of the synsacrum noticeably convex dorsally; cranial end of the synsacrum remarkably broad; synsacrum only slightly broadened across both sacral vertebrae; only one thoracic vertebra precedes two sacral vertebrae; caudal half of the synsacrum long and narrow; the two largest costal processes are inclined caudally; no longitudinal groove on the ventral side of the synsacrum.

Comments. There are four species of bird from Dzharakhuduk that are based on synsacra and assigned to the Ichthyornithiformes (Nesov, 1984, 1986, 1990, 1992b, d; Nesov and Yarkov, 1989; Nesov and Panteleev, 1993). Comparison of these remains with the synsacrum of *Nanantius valifanovi* Kurochkin, 1996 showed that they should be assigned to the Enantiornithes (Kurochkin, 1995c, 1996).

Zhyraornis kashkarovi Nesov, 1984

Holotype. TsNIGRI 42/11915, incomplete synsacrum, having at least seven coossified vertebrae and lacking the most caudal portion. Dzharakhuduk locality,

Navoi District, Bukhara Province, Uzbekistan. Outcrop CBI-4, Bissekty Svita (Coniacian).

Diagnosis. First vertebra on synsacrum expands gradually in the cranial direction; transverse processes on the second sacral vertebra are slightly marked on the dorsal surface; two pairs of the largest costal processes are slender and distinctly inclined lengthwise; the synsacrum is generally extended and narrow.

Comments. A thoracic vertebra (TsNIGRI 43/11915) from the Khodzhakulsai locality (Khodzhakul settlement, Karakalpakia, Uzbekistan; Khodzhakul Svita, Cenomanian) (Nesov, 1984, 1992b; Nesov and Borkin, 1983), regarded as an indeterminate Mesozoic bird by Nesov (1992d), a left scapula (TsNIGRI 44/11915) and the shaft of a left humerus (TsNIGRI 45/11915) from Dzharakhuduk (Nesov, 1984), were also assigned to *Z. kashkarovi*. The vertebra has a wide neural canal, deep lateral excavations on the centrum, and nearly flat cranial and caudal articular surfaces. It is comparable in size to the vertebrae of the synsacrum of *Z. kashkarovi*. The structure of the glenoid facet on the scapula and its acromion show a certain similarity to that of the enantiornithines, but this is not sufficient evidence for its assignment to *Z. kashkarovi*. The humerus shaft certainly is not enantiornithine, because its nutrient foramen is located in the typical position for neornithine taxa, whereas in the Enantiornithes the nutrient foramen occurs on the opposite side of the shaft.

Zhyraornis logunovi Nesov, 1992

Holotype. PO 4600, incomplete synsacrum, having at least five coossified vertebrae and lacking the most caudal portion. Dzharakhuduk locality, Navoi District, Bukhara Province, Uzbekistan. Outcrop CBI-5a, upper member of the Bissekty Svita (Coniacian).

Diagnosis. The first vertebra of the synsacrum expands abruptly in the cranial direction; the transverse processes of the second sacral vertebra are prominently marked on the dorsal surface; two pairs of the largest costal processes are thick; the costal processes of the second sacral vertebra are perpendicular to the sagittal plane; and the synsacrum is generally expanded and broadened.

Lenesornis Kurochkin, 1996

Type species. Lenesornis maltshevskyi (Nesov, 1986).

Diagnosis. The cranial portion of the synsacrum is only slightly convex dorsally, the cranial articular surface of the synsacrum is transversely elongated; the third and fourth vertebrae of the synsacrum possess the largest costal processes; the costal processes are at a right angle to the sagittal plane; and the ventral groove is wide and shallow.

Comments. This synsacrum was described as *Ichthyornis maltshevskyi* by Nesov (1986) in the family Ichthyornithidae (see also Nesov, 1992b, d). However, the similarity of this fossil to *Nanantius valifanovi* Kurochkin, 1996, and the abundance of postcranial remains of Enantiornithes at the Dzharakhuduk locality enabled it to be reidentified as an enantiornithine (Kurochkin, 1996). Because of a noticeable difference from *Zhyraornis*, it was assigned to a separate genus.

Lenesornis maltshevskyi (Nesov, 1986)

Holotype. PO 3434, cranial half of the synsacrum. Dzharakhuduk locality, Navoi District, Bukhara Province, Uzbekistan. Outcrop CBI-14, Bissekty Svita (Coniacian).

Diagnosis. Same as for genus.

Sazavis Nesov, 1989

Type species. Sazavis prisca Nesov, 1989.

Diagnosis. Small birds; distal end of the tibiotarsus is wide and its large medial condyle has a rounded dorsal margin in cranial aspect; intercondylar furrow is displaced laterally, therefore the lateral condyle is narrow. The diameter of the bone is strongly reduced dorsal to the distal end of the tibiotarsus; the ligamental tubercle on the cranial surface of the tibiotarsus is weak and located relatively distal.

Comments. As I have not seen the specimen recently, the original diagnosis, which includes the generic characters of *Sazavis*, as well as characters of the Enantiornithes and Alexornithiformes, is presented here. It should also be noted that in the original diagnosis lateral and medial aspects were confused, though these are corrected in the diagnosis given above. The following characters form an emended diagnosis for this genus: the transition from the shaft to the distal end of the tibiotarsus is abrupt; the medial condyle is nearly circular in cranial view; and the intercondylar fossa is somewhat medially displaced.

Sazavis was assigned, with some doubt, to the Alexornithidae (Nesov and Yarkov, 1989) or to the Enantiornithes (Nesov, 1992b, d). Assignment of *S. prisca* to enantiornithine birds is supported by the presence of a bulbous and enlarged medial condyle of the tibiotarsus, a small and transversely compressed lateral condyle, and a smooth tubercle in the centre of the ascending process.

Sazavis prisca Nesov, 1989

Holotype. PO 3472, distal fragment of the right tibiotarsus. Dzharakhuduk locality, Navoi District, Bukhara Province, Uzbekistan. Outcrop CBI-14, Bissekty Svita (Coniacian).

Diagnosis. Same as for genus.

Comments. S. prisca is another very small enantiornithine from the Dzharakhuduk locality. The width of the distal end of the tibiotarsus is 4.5 mm which is comparable in size with *Kizylkumavis cretacea* and a congeneric or conspecific relationship might be supposed. However, this cannot be certainly demonstrated because the remains are non-comparable.

Other alexornithiforms from the Kizylkum Desert

In various papers Nesov described other fragmentary avian fossils from the Dzharakhuduk locality in Bukhara Province, Uzbekistan. Some of these were assigned to the Enantiornithes indet. and some only to Aves indet. The remains of alexornithiforms are listed below.

Alexornithidae gen. indet.

A complete axis, PO 3473, was figured as avian by Nesov (1988, fig. 1: 3, 1992b, fig. 2: O-P) and then cited as similar to Gaviidae (Nesov, 1992d). This specimen shares, with the axis of *N. valifanovi*, a general craniocaudal elongation, poorly developed cranial articular facets, lateral extensions of the cranial articular facets,

a broad dorsal arch and a flat caudal lamina that is strongly projected caudally, a low dorsal process, and a shallow lateral excavation of the body. These characters allow assignment of PO 3473 to the Alexornithidae (Kurochkin, 1996). This specimen differs from the axis of *N. valifanovi* in being larger: the distance between the cranial and caudal articular surfaces is some 13.5 mm (measured from the figure in Nesov, 1988) but only 6.2 mm in *N. valifanovi*.

Two enantiornithine coracoids were described by Nesov and Panteleev (1993). Specimen PO 4819 is represented by the dorsal half of a coracoid with a narrow shaft, slight projection of the lateral margin, and shallow depression on the dorsal shaft, features characteristic for the Alexornithidae. Specimen PO 4818 is represented by the fragment of a shaft that shows a broadened sternal portion and a deep dorsal depression (Kurochkin, 1996). Because of its very fragmentary condition the family relationships of this bone is uncertain.

Enantiornithidae gen. indet.

The shaft of a right coracoid, TsNIGRI 56/11915, some 15 mm long (measured by the published figure in Nesov and Borkin, 1983), was provisionally assigned to birds (Nesov and Borkin, 1983) and later to the enantiornithines (Nesov and Panteleev, 1993). This specimen clearly belongs to the Enantiornithidae because it shows a distinctive deep dorsal depression, running far dorsally, and a convex lateral margin of the shaft (Kurochkin, 1996).

Alexornithiformes family indet.

'*Ichthyornis*' *minusculus* was based on a single thoracic vertebra, PO 3941 (Nesov, 1990). This specimen shares, with the synsacrum of *L. maltshevskyi*, an elliptical profile of the cranial articular surface and wide vertebral foramen. However, because of its fragmentary nature, '*I*'. *minusculus* can only be assigned to the Alexornithiformes fam. indet. (Kurochkin, 1996).

The proximal fragment of a left tarsometatarsus, PO 3394, was assigned to the Enantiornithidae by

Nesov and Borkin (1983) or to the Enantiornithes (Nesov, 1984, 1992b). The specimen shows the proximal fusion of the metatarsals and the enlargement of the medial cotyla. The depth of this proximal end is about 4 mm (measured from the figure in Nesov, 1992b). The specimen was correctly identified as an enantiornithine, and should be assigned to the Alexornithiformes on the basis of its very strongly reduced metatarsal IV (Kurochkin, 1996).

Discussion. In total, the Bissekty Svita of the Dzharakhuduk locality has produced 13 enantiornithines all based on fragmentary remains. Seven of them were described under species status here, but it is impossible to compare the other six fragments. Nevertheless, at least three small and two middle-sized enantiornithines might have existed concurrently in the Coniacian of Uzbekistan.

Family Alexornithidae Brodkorb, 1976
Gobipteryx Elzanowski, 1974
Type species. Gobipteryx minuta Elzanowski, 1974.
Amended diagnosis. Culmen straight and very thin above the nasal openings; rostral ends of the premaxilla and mandible flattened with rounded tips; mandibula with a low and short symphysis and thin rami; contact surfaces on the ventral margin of the premaxilla and maxilla and dorsal margin of the mandible flat; no distinct grooves on the lateral surface of the rostral portions of the premaxilla and mandible; dorsal mandibular margin distinctly elevated above the level of the lateral mandibular process; and large choanal fenestra located in the rostral area of the palatal shelf.
Comments. Originally, *Gobipteryx* was assigned to the Palaeognathae (Elzanowski, 1976, 1977), but later Martin (1983) attributed it to the Enantiornithes. There are no reliable arguments for assigning the skull portions of *Gobipteryx* to the Enantiornithes, since the skull remains of *Gobipteryx* are not known among undoubted enantiornithines. However, the bipartite mandibular articulation of the quadrate, anterior bifurcation of the pterygoid, the subsidiary palatal fenestra, and a hooked ectopterygoid (Elzanowski,

1976, 1977, 1995) might be features of the Enantiornithes, though at present these features are not known in described skull remains of this group which include the occipital cranium region of *Neuquenornis volans* from the Santonian–Coniacian of Argentina, rostral portions of the mandible and maxillary and palatal apparatus of *N. valifanovi* from the Late Campanian of Mongolia, and rostral portions of the mandible and maxillary and dorsal cranium of embryonic enantiornithines also from the Late Campanian of Mongolia.

Thus, there is only circumstantial evidence that *Gobipteryx* belongs to the Enantiornithes, since it does not have advanced characters in common with known enantiornithine birds. In addition *Gobipteryx* is characterized by some primitive characters of the quadrate, pterygoid and palate which are also unknown in other Enantiornithes.

Gobipteryx minuta Elzanowski, 1974
Holotype. ZPAL MgR-I/12, rostral portion of the skull. Khulsan locality, Nemegt Valley, South Gobi Desert, Mongolia. Baruungoyot Formation (Late Campanian), Late Cretaceous.
Referred material. Rostral portion of a skull, ZPAL MgR-I/32, from the same locality.
Diagnosis. Same as for genus.
Comments. G. minuta shows a gracile construction of the upper jaw and mandible and large rostral choanal fenestra bordered caudally by the palatines (Elzanowski, 1995). Following additional preparation of the holotype, Elzanowski (1995) discovered a small hooked ectopterygoid that has also been identified in both the Eichstätt and seventh specimen of *Archaeopteryx* (Wellnhofer, 1974; Elzanowski and Wellnhofer, 1996). The ectopterygoid is present in many theropods, and I evaluate this fact as a further confirmation of the relationship between the Enantiornithes and Archaeornithes and their assignment to a phylogenetic lineage separate from ornithurine birds. This and some other cranial synapomorphies support a theropodan origin for Sauriurae, but not all birds.

Nanantius Molnar, 1986

Type species. Nanantius valifanovi Kurochkin, 1996.

Contents. Nanantius eos Molnar, 1986 (Australia, Albian); *N. valifanovi* Kurochkin, 1997 (Mongolia, Late Campanian); *Nanantius* sp. (Australia, Albian).

Diagnosis. Maxillary and mandible short, high and stout; culmen very straight and thick above the nasal opening; the humerus has a curved shaft with a thin mid-portion; the shaft of the radius is bowed; the proximal phalanx of the major digit has a rectangular top; the tibiotarsus has a long and remarkably thin shaft that is bowed laterally; conspicuous intercotylar prominence on the caudal area of the proximal articular surface of the tibiotarsus; lateral area of the proximal articular surface of the tibiotarsus slopes distad; well-developed fibular crest reaches to the margin of the proximal articular surface; fibular crest located along the craniolateral edge of the shaft of the tibiotarsus; proximal origin of the fibular crest and top of the cranial cnemial crest united together; elongate depressions run along the cranial and caudal base of the fibular crest; medial condyle of the tibiotarsus transversely elliptical (not circular) in cranial view; fibula very short, flat and thin; metatarsals II–IV of a similar thickness.

Comments. In spite of the great geographical and temporal gaps between *N. valifanovi* from the Albian of Mongolia and *N. eos* from the Campanian of Australia (Molnar, 1986) it was decided to assign the Mongolian taxon to the same genus because of the great similarity in the derived characters of the tibiotarsus of the two forms.

Nanantius valifanovi Kurochkin, 1996

Holotype. PIN 4492–1, partial skeleton including portions of the skull, vertebrae, synsacrum, all bones of the shoulder girdle, pelvis, and most elements of the forelimb and hind limb (Figures 27.1 and 27.2). Hermiin Tsav locality, Trans-Altai Gobi Desert, South Gobi Aimag, Mongolia. Middle layers of the outcrop on the northern slope of Hermiin Tsav, Baruungoyot Formation (Late Campanian), Late Cretaceous.

Diagnosis. Shallow longitudinal groove on the ventral

side of the mandible; a shallow, broad axial groove on the ventral side of the synsacrum; a short fibular crest on the tibiotarsus that is approximately five times the transverse width of the proximal articular surface; position of the nutrient foramen on the cranial side of the tibiotarsus close to the distal extremity of the fibular crest; subtriangular cross-section and sharpened caudal edge of the proximal shaft of the tibiotarsus; transversely compressed lateral condyle of the tibiotarsus which projects markedly craniad in distal view; and metatarsal IV is shorter than metatarsal II.

Comments. The Hermiin Tsav locality, situated on the western border of the Trans-Altai Gobi Desert in Southern Mongolia, is famous for yielding numerous fossil eggs of birds, some of which contain embryos (Elzanowski, 1981; Chatterjee and Kurochkin, 1994; see below). The nearly complete holotype skeleton of *N. valifanovi* was found on the northern slope of Hermiin Tsav, about 4 km east of 'Bird's Hill', a locality for fossil avian eggs at the mouth of Hermiin Tsav. Some bones were collected in association including: rostral portions of the mandible and premaxillary; a portion of the palatal apparatus; some cervical vertebrae; part of the pelvis and synsacrum; the proximal end of the first phalanx of the major digit with the carpometacarpus; the distal end of the left femur and pelvis; the tarsometatarsus with some pedal phalanges; and other pedal phalanges. Many pieces of eggshell were also found in association with the bone remains of *N. valifanovi*. The eggshell has a laevisoolithid microstructure thus linking this type of eggshell to the Enantiornithes (Kurochkin, 1996; see Chapter 28).

N. valifanovi is distinguished from *N. eos* by the relatively longer fibular crest that extends distally, the greater transverse width of the proximal articular surface of the tibiotarsus, the presence of a nutrient foramen near the end of the fibular crest that is absent in *N. eos*, the sharpened caudal side of the proximal shaft, which is rounded in *N. eos*, a small tubercle at the centre of the ascending process which is located nearer to the top of this process in *N. eos*, greater cranial protrudence of the lateral condyle, a slightly compressed proximodistally medial condyle, which is

Figure 27.1. Holotype skeleton of *Nanantius valifanovi* Kurochkin, 1996 (PIN 4492–1). Scale bar = 20 mm.

more circular in *N. eos*, and a deeper and wider cranial intercondylar fossa. In additon, the tibiotarsus in *N. valifanovi* is about 30% longer than in *N. eos*.

The relationships of *N. valifanovi* to *Gobipteryx minuta* Elzanowski, 1974, need to be discussed here, because they were both found in the Baruungoyot Formation of the South Gobi. *G. minuta* is based on fragments of two skulls from the Khulsan locality located some 150 km east of Hermiin Tsav. This taxon shows some features in common with *N. valifanovi* in the configuration of the maxillary segment and in the presence of conspicuous rows of nutrient foramina in the maxillary and mandible. However, *N. valifanovi* differs from *G. minuta* in having more robust rostral

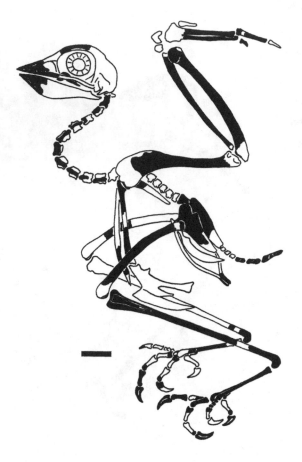

Figure 27.2. Restoration of the skeleton of *Nanantius valifanovi* Kurochkin, 1996. Scale bar = 10 mm.

ing the ability for powerful flexion of the digits. This is in contrast to the relatively small size of the trochlea of the left metatarsal I, and the phalanges of digit 1. The powerful anterior second and third toes with curved claws and weak hallux also suggest a climbing adaptation for the foot. The palatal elements and the rostral portions of the mandible and upper jaw have a very powerful and robust construction. This is, apparently, an adaptation for feeding on tough objects, perhaps fruits or seeds.

Family Avisauridae Brett-Surman and Paul, 1985
Contents. Enantiornis Walker, 1981; *Avisaurus* Brett-Surman and Paul, 1985; *Soroavisaurus* Chiappe, 1993.
Diagnosis. Deep fossa on the cranial surface of the scapula; scapular facet and scapular glenoid separated; ischium wide; tibiotarsus with a straight and robust shaft; metatarsal III with a strongly convex transversely dorsal surface; medial ridge of the trochlea on the metatarsal III projects markedly on the plantar side.

Enantiornis Walker, 1981
Type species. Enantiornis leali Walker, 1981
Contents. E. leali Walker, 1981 (Argentina, Maastrichtian); *E. walkeri* Nesov and Panteleev, 1993 (Uzbekistan, Coniacian); *E. martini* Nesov and Panteleev, 1993 (Uzbekistan, Coniacian).
Diagnosis. Shoulder end of the coracoid robust and short; shaft of the coracoid wide; acrocoracoid and coracoidal process stout; scapular acromion broad with an obtuse top; fossa or foramen on the cranial surface of the shoulder end of the scapula; scapular facet and glenoid facet of the scapula are separated; dorsal portion of the humeral head longer than the ventral portion; distinct cranial fossa in the cranial surface of the proximal end of the humerus; deltopectoral crest with a thin proximal base; distinct depression in the caudal surface of the proximal end of the humerus.
Comments. In the present state of knowledge of the Enantiornithes it is a difficult task to develop a generic diagnosis for *Enantiornis*, since almost all authors analyzed either the characters of the Enantiornithes or

portions of the upper jaw and mandible, sharper rostral ends of the upper jaw and mandible, thin and acute contact surfaces of the maxillary and mandible, deep grooves possessing nutrient foramina both in the upper jaw and mandible, nutrient foramina only in the anterior half of the maxillary rostrum, nutrient foramina which become larger caudally, and in the presence of an axial groove on the ventral side of the mandible.

The short tarsometatarsus and relatively short and powerful second and third pedal digits with strong and slightly curved claws of *N. valifanovi* reflect arboreal adaptations. The ungual phalanges are sturdy, similar in size and slightly curved, with symmetrical articular cotyles and well-developed flexor tubercles, signify-

separate species of this infraclass, beginning with Walker (1981) and including many other recent papers on the enantiornithines. In the diagnosis given above I attempted to select the advanced characters of *Enantiornis* which distinguish it from other enantiornithines also represented by the humerus and bones of the shoulder girdle.

Enantiornis walkeri Nesov and Panteleev, 1993
Holotype. PO 4825, dorsal fragment of the left coracoid. Dzharakhuduk locality, Navoi District, Bukhara Province, Uzbekistan. Outcrop CBI-5a, Bissekty Svita (Coniacian).
Diagnosis. Coracoidal process is stout, proximal portion of the shaft is gracile.

Enantiornis martini Nesov and Panteleev, 1993
Holotype. PO 4609, shoulder fragment of the right coracoid. Dzharakhuduk locality, Navoi District, Bukhara Province, Uzbekistan. Outcrop CBI-14, middle member of the Bissekty Svita (Coniacian).
Diagnosis. Coracoidal process is narrow latero-medially and the proximal portion of the shaft is stout.
Comments. The two shoulder portions of the coracoid, upon which the two species listed above are based, are of a very similar size (approximately 10 mm between the top of the acrocoracoid and the top of the coracoidal process: measured from published figures). However, they differ in the structures mentioned in the diagnoses. *E. walkeri* and *E. martini* are distinguished from *E. leali* by their much smaller size, narrower coracoidal process, shorter acrocoracoid, distal broadening of the proximal shaft, and by the narrower and more elongated coracoidal nerve foramen.

Enantiornithidae
Gurilynia Kurochkin, 1999a
Gurilynia nessovi Kurochkin, 1999a
Several wing and shoulder girdle bones of a large enantiornithid, *Gurilynia nessovi*, were collected by the JRMPE in South Mongolia in 1994 at the Guriliin Tsav locality, which is situated in the Bügiin Tsav depression, some 15 km to north of Altan Uul mountain, in the South Gobi Aimag. The sediments of this locality belong to the Late Cretaceous (Late Campanian–Early Maastrichtian) Nemegt Formation.

Gurilynia nessovi is characterized by the almost equal length of the ventral and dorsal portions of the humeral articular head, a very shallow cranial fossa in the middle of the proximal area of the cranial surface of the proximal end of the humerus, the thick base of the deltopectoral crest, a very shallow caudal depression in the caudal surface of the proximal end of the humerus, the pointed ventral termination of the processus coracoideus, and a remarkably thin coracoidal neck between the shoulder end and the shaft (Figure 27.3). The humerus also shows other specific characters, including the oval shape of the dorsal ligamental impression near the proximal base of the deltopectoral crest, the very wide and open angle between the ventral and dorsal portions of the articular head, a very stout, wide base that is not perforated by the axial foramen ventral tubercle, and the proximal end of the humerus is large and possesses a deep circular ventral ligamental impression on the ventrocranial angle of the ventral side (Kurochkin, 1999a)

The distal ends of a humerus and probably an ulna, radius and carpometacarpus also belong to this new form. It is a large bird, the width of the proximal end of the humerus being 26.2 mm and thus distinctly larger than *E. leali*.

The discovery of this large flying representative of the Enantiornithes increases their known diversity in the Late Cretaceous of Central Asia and shows that large enantiornithids inhabited Mongolia at the end of the Cretaceous, as well as North and South America.

Enantiornithine embryos from Mongolia
Elzanowski (1981) described embryonic avian skeletons in small elongate eggs from Hermiin Tsav which had previously been thought to be turtle or crocodile eggs. In all probability he assigned these embryos to *Gobipteryx minuta* Elzanowski, 1974, which was described from Khulsan. Martin (1983, 1995a), and later Elzanowski (1995) recognized all the fossil embryos as *G. minuta* and assigned them to the Enantiornithes. Later, two types of elongate avian fossil eggs, small and large, were found at Khulsan by

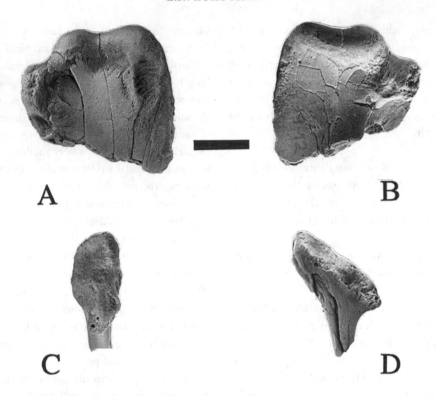

Figure 27.3. *Gurilynia nessovi* Kurochkin, 1999a, Nemegt Formation (Late Cretaceous) of Guriliin Tsav, Ömnögov' Aimag, Mongolia. Holotype PIN 4499-14, proximal end of the right humerus in (A) cranial and (B) caudal view. Paratype PIN 4499-13, shoulder end of the left coracoid in (C) dorsal and (D) lateral view. Scale bar = 10 mm.

the JRMPE. The small egg type is the same as that found at Hermiin Tsav (Mikhailov, 1995, 1996), but fossil embryos have not yet discovered in eggs from Khulsan.

The avian embryos from Hermiin Tsav, now in the collections of ZPAL and PIN (Figure 27.4), are beautifully preserved and show even the smallest ossified elements including, for example, the most proximal wing phalanges. This characteristic, as well as the complete ossification of the bones, from the ends across the shafts, differs fundamentally from the embryos of recent birds. In the latter case the process of ossification begins from centres in the ends of bones and in the shaft, so that cartilaginous insertions between these ossifications exist some weeks after hatching.

In earlier papers I concluded that two groups of birds were represented among the Hermiin Tsav embryos (Kurochkin, 1995a, c, 1996). However, after study of fossil material in the ZPAL collections I now agree that only one taxon is present and that it belongs to the enantionithines as originally proposed by Martin (1995a). Differences among the embryos (Kurochkin, 1996) represent different age stages, as was first noted by Elzanowski (1981). The structure of the eggshell is an important piece of evidence supporting the identification of the embryos as enantiornithine. The ratio between the spongy and mammillary layers, the ultrastructure, absence of asymmetry, and the greater thickness of the eggshell are all characters of enantiornithine eggs (Mikhailov, 1991, 1996). By contrast, the possible absence of the external layer of vertical crystals distinguishes the Hermiin Tsav eggshell from that of the Ratitae, Galliformes, and Anseriformes.

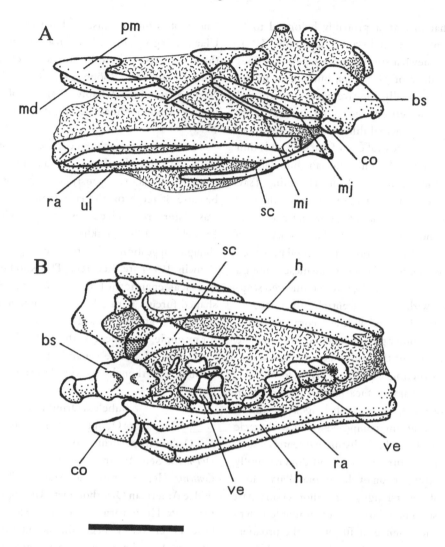

Figure 27.4. Fossil avian embryo (PIN 4492–3) from the Baruungoyot Formation (Late Cretaceous) of Hermiin Tsav, Ömnögov' Aimag, Mongolia. Left side (A) and posterior (B) view. Abbreviations: bs, basisphenoid; co, coracoid; h, humerus; md, mandible; mi, minor metacarpal; mj, major metacarpal; pm, premaxillary; ra, radius; sc, scapula; ul, ulna; ve, vertebrae. Scale bar = 5 mm.

However, the question still remains as to whether the Hermiin Tsav embryos can be assigned to *Gobipteryx minuta* from Khulsan. Specimen ZPAL MgR-1/34 (Elzanowski, 1981) is clearly enantiornithine, since it exhibits the boss and socket articulation between the coracoid and scapula, and other specialized postcranial characters of enantiornithines. In addition, this specimen has a long retroarticular process in the mandible, which is also found in *G. minuta* from Khulsan and appears to be characteristic for the Enantiornithes. However, a long retroarticular process is the only common feature for *Gobipteryx* and embryos. Elzanowski (1981) argued that the premaxillary of *Gobipteryx* and the embryo specimens were similar and that the latter probably belonged to *G. minuta*. Certainly, the embryonic specimen ZPAL MgR-1/88 is somewhat similar in its general outlines to the premaxillary of *G. minuta*, but differs in the larger and more cranially concave nasal aperture and in the sharper angle between the lateral surfaces of the premaxillary. The principal difference between them is the structure of the rostral end of the beak. It is thin and sharp in *Gobipteryx* from Khulsan, but flat and rounded in the embryos from Hermiin Tsav. Moreover, the exterior contact areas between the pre-maxilla and mandible in *Gobipteryx* are narrow while they are wide and flat in embryos. The embryos also show a double-headed quadrate with orbital process; a ventral flange on the rostral edge of the mesethmoid, a lateral groove on the mandible, a very elongated scapular acromion with a medioventral projection, no scapular labrum, a short and wide acrocoracoid, incorporation of the scapular glenoid and coracoid glenoid for the glenoid facet of the humerus, a merged scapulocoracoid construction, a proximal origin for the deltopectoral crest, a major metacarpal that is longer than the minor metacarpal, a manual phalangeal formula of 1–1–0, and a notarium composed of two thoracic vertebrae. They also exhibit distinctive features such as a small nasal aperture, a broad and dorsoventrally compressed rostral portion of the premaxillary, a long basal phalanx of the first digit, a very short cranial part of the ilium, a stout fibula which is of similar length to the tibia, and the absence of fusion of the proximal tarsalia (astragalo–calcaneus complex) to the tibia.

Thus, I consider, in contrast to Martin (1995a) and Elzanowski (1995), that assignment of the embryonic specimens ZPAL MgR-1/34, MgR-1/33, and MgR-1/88 to *Gobipteryx* or even to *G. minuta* cannot be justified and they together with the specimens in the PIN collections (Figure 27.7) must be assigned to a new taxon (Chatterjee and Kurochkin, 1994).

Order Euornithiformes Kurochkin, 1996
Family nov.

The partial skeleton of a small vertebrate was collected by the JRMPE at Kholboot in the Eastern area of the Mongolian Altai in the Bayankhongor Aimag.

Kholboot is located just 10 km west of the Khurilt Ulaan Bulag locality where *Ambiortus* was found and the sediments at these localities, which are currenty assigned to the Bööntsagaan Gorizont (possibly Neocomian – see Chapter 14) are of similar age, though Kholboot may be slightly younger (Sinitsa, 1993).

The Kholboot specimen consists of portions of the skull and some shoulder, wing and hind limb bones. Initially, it was erroneously identified as a pterosaur because of teeth on the jaws (Kurochkin, 1991), but was later recognized as avian (Unwin, 1993; Kurochkin, 1993; Bakhurina and Unwin, 1995), though suggestions of a relationship with *Ambiortus* (Unwin, 1993) are incorrect. The Kholboot specimen has clear enantiornithine characters including a V-shaped furcula with a long hypocleideum and metatarsals that are only fused proximally, and is similar, for example, in respect of the toothed jaws, to Lower Cretaceous Chinese enantiornithines, but distinguished from them by the very long tarsometatarsus.

Subclass Ornithurae
Infraclass Odontornithes Forbes, 1884
Order Hesperornithiformes Fürbringer, 1888
Diagnosis. See Martin (1984, p. 147).
Contents. Hesperornithidae Marsh, 1872; Baptornithidae American Ornithologists' Union, 1910.
Comments. Hesperornithidae and the genus *Hesperornis* have never really been diagnosed. Martin (1984) attempted to diagnose Hesperornithiformes and to this list one important character can be added: location of teeth in grooves.

Family Hesperornithidae Marsh, 1872
Contents. *Hesperornis* Marsh, 1872; *Parahesperornis* Martin, 1984; *Asiahesperornis* Nesov and Prizemlin, 1991.

Hesperornis Marsh, 1872
Type species. *Hesperornis regalis* Marsh, 1872.
Contents. *Hesperornis regalis* Marsh, 1872; *Hesperornis crassipes* Marsh, 1876; *Hesperornis gracilis* Marsh, 1976; *Hesperornis rossicus* Nesov and Yarkov, 1993.

Hesperornis rossicus Nesov and Yarkov, 1993

Holotype. VPM 26306/2, proximal portion of the right tarsometatarsus. Right shore of the Tzimlyanskoe Reservoir, Don river, between Rychkovo and the 278 km Station, Surovikinskii District, in the south-west of Volgograd Province, Central South Russia. Marine beds, *Belemnellocamax mamillatus* zone, upper zone of the Early Campanian.

Referred material. The shaft of a tarsometatarsus, an intermediate phalanx of the fourth pedal digit, a fragment of a thoracic vertebra, and a fragment of a cervical vertebra were all recovered from the same locality (Nesov and Yarkov, 1993). A fragment of the proximal end of a right tarsometatarsus from the Early Campanian of Ivö-Klack, Scone, Southern Sweden has also been assigned to this species (Nesov and Yarkov, 1993).

Diagnosis. The proximal articular surface of the tarsometatarsus has a very large transverse width and small dorsoplantar depth, the diagonal slant is strongly expressed, the lateral edge of the lateral cotyla exceeds the intercotylar prominence in proximal direction and the medial cotyla is located more distal in respect to the lateral cotyla.

Comments. *Hesperornis* is of the masculine gender, thus according to the *International Code of Zoological Nomenclature* (1985, article 32d, 33 (II), 34b) the original epithet *rossica* (Nesov and Yarkov, 1993) must be changed to *rossicus*.

H. rossicus is clearly hesperornithiform, but differs from *Hesperornis regalis* Marsh, 1880, from North America in the larger depth of the proximal articular surface, more proximal projection of the lateral cotyla and approximately 20% larger size of the tarsometatarsus.

Because of the proposed presence (see below) of a second species (*Hesperornis* sp.) in the same locality, the inclusion of the two vertebrae and the pedal phalanx in the referred material of *H. rossicus* is entirely arbitrary.

Hesperornis sp.

A fragment of the proximal end of a left tarsometatarsus from the same locality and same beds as the holo-

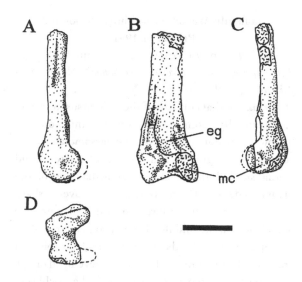

Figure 27.5. Right distal tibiotarsus of a small hesperornithiform from the Nemegt Formation (Late Cretaceous) of Tsagaan Khushuu, Ömnögov' Aimag, Mongolia. Lateral (A), cranial (B), medial (C), and distal (D) views. Abbreviations: eg, extensor groove; mc, medial condyle. Scale bar = 10 mm.

type of *H. rossicus* was assigned to *Hesperornis* sp. (Nesov and Yarkov, 1993). This specimen differs from *H. rossicus* in the more medial location of the intercotylar prominence and a larger ridge in the dorsal base of the intercotylar prominence (Nesov and Yarkov, 1993).

Hesperornithidae gen.

The distal portion of a tibiotarsus from the Nemegt beds of the Tsagaan Khushuu locality in the South Gobi Desert was announced as *Baptornis* sp. (Kurochkin, 1988). After comparison with all known hesperornithiforms in the collections of the Natural History Museum of Kansas University it was reidentified as representing a bird more closely related to *Parahesperornis*, on the grounds that it shows few differences between the lateral and medial condyles, no distal projection of the medial condyle, and the remarkable medial position of the extensor groove. The transverse width of the distal end of the tibiotarsus (Figure 27.5) is 11.7 mm, thus this was a small bird.

Subfamily Asiahesperornithinae Nesov and
Prizemlin, 1991

Asiahesperornis Nesov and Prizemlin, 1991

Type species. Asiahesperornis bazhanovi Nesov and
Prizemlin, 1991.

Diagnosis. Medial condyle of the tibiotarsus markedly
mediolaterally compressed, cranial intercondylar
furrow comparatively deep. The tarsometatarsus is
comparatively gracile and has parallel lateral and
medial sides. Both the lateral and medial crest on the
plantar side of the tarsometatarsus have a sharp
plantar margin and the flexor groove is shallow. The
dorsal facet in the middle of the tarsometatarsal shaft
is deep and narrow, and covered by a high dorsolateral
crest and the separate medial facet is developed on the
distal portion of the shaft. The base of the trochlea of
the fourth digit is much larger than the base of the
trochlea of the third digit and the fossa for metatarsal I
is very small and short.

Comments. The diagnosis given above is abstracted
from the author's original diagnosis and omits some
characters which appear to be characteristic for hesperornithiforms. However, further evaluation, via direct
comparison with remains of *Hesperornis* is required to
identify those characters that are derived for this new
taxon, especially in the context of some of Nesov's
remarks that several remains of *Asiahesperornis* can be
assigned to other taxa (see below). The following characters show that in any case the Kushmurun remains
are those of hesperornithiforms: the bones are heavily
constructed with well-expressed pachyostosis, a transversely compressed tarsometatarsus, and a strongly
developed fourth trochlea.

Asiahesperornis bazhanovi Nesov and Prizemlin, 1991
Holotype. IZASK 5/287/86a, shaft of a left tarsometa-
tarsus. Priozernyi Quarry, Kushmurun locality, near
the settlement of Kushmurun, Kustanaiskaya
Province, North Kazakhstan. Eginsai Svita (Latest
Santonian–Early Campanian).

Diagnosis. Same as for the genus.

Referred material. IZASK 5/287/86: shaft of a right tarsometatarsus, two thoracic vertebrae and a fragmentary distal portion of the tibiotarsus from the same

locality (Nesov, 1992c; Nesov and Prizemlin, 1991).

Comments. The distal portion of the right tibiotarsus
and the thoracic vertebra previously assigned to *A.
bazhanovi* were later illustrated as 'hesperornithiforms'
(see caption to fig. 5 of Nesov and Yarkov, 1993) and, in
the same caption, the proximal portion of the left tarsometatarsus was provisionally assigned to another
species.

The shaft of the tarsometatarsus of *Asiahesperornis
bazhanovi* exhibits a distinct transverse compression,
an acute medial plantar ridge, a strongly reduced facet
for metatarsal I, and a markedly craniocaudally compressed distal end of the tibiotarsus. The restored
length of the tarsometatarsus is 122 mm (Nesov and
Prizemlin, 1991). The hesperornithiform from
Kushmurun inhabited the Campanian Turgai Inferior
Seaway which ran from the Polar Ocean to the
Southern Ocean between Fennoscandia and Eurasia,
and was in some ways analogous to the Western
Inferior Seaway which divided North America in the
Campanian (Nesov and Prizemlin, 1991; Nesov,
1992a, c).

Family Baptornithidae American Ornithologists'
Union, 1910

Diagnosis. See Martin and Tate (1976).

Contents. Baptornis Marsh, 1877 (Coniacian, USA) and
Judinornis (Maastrichtian, Mongolia).

Comments. Judinornis Nesov and Borkin, 1983 does not
exhibit any of the characters listed in the diagnosis
given by Martin and Tate (1976). However, Martin
and Tate (1976) noted a small pit lying directly anterior to the diapophysis in the trunk vertebra of
Baptornis advenus and this is also present in *Judinornis
nogontsavensis* Nesov and Borkin, 1983. The flat ventral
side of the body is another character which might also
be apomorphic for the Baptornithidae and circular
pits in the articular surfaces of the centra of the thoracic vertebrae are also found in both *Baptornis* and
Judinornis.

Judinornis Nesov and Borkin, 1983

Type species. Judinornis nogontsavensis Nesov and
Borkin, 1983.

Diagnosis. The articular surfaces of the centrum of the thoracic vertebra are trapezoid-shaped and extend transversely. The ventral side of the centrum is distinctly narrowed in the middle but very broad caudally and the cranial zygapophyses are located close together on the centerline.

Contents. Type species only.

Judinornis nogontsavensis Nesov and Borkin, 1983
Holotype. PO 3389, incomplete thoracic vertebra. Nogoon Tsav locality, western area of the Trans-Altai Gobi, Bayankhongor Aimag, Southern Mongolia. Nemegt Formation, Late Cretaceous.

Diagnosis. Same as for genus.

Comments. Based on a single thoracic vertebra, *Judinornis nogontsavensis* was originally referred to the Charadriiformes (Nesov and Borkin, 1983). Later, it was referred to the Baptornithidae (Nesov, 1986, 1992a, b), but without any character evidence. *Judinornis nogontsavensis* is the first hesperornithiform to be recorded from the Old World. It is a middle-sized bird, the vertebral body of the type thoracic vertebra measuring 14.1 mm in length between the articular surfaces. The vertebra shows a very expanded caudal ventral surface of the body, narrow cranial zygapophyses, very deep pleurocoels, and transversely expanded cranial and caudal articular surfaces which are characteristic of hesperornithiforms.

Hesperornithiformes
Family nov.

Further evidence of the presence of small Hesperornithiformes in the interior water basins of the Mongolian Cretaceous was obtained from the Nemegt (Maastrichtian) beds of Bügiin Tsav in South Gobi. The fossil material consists of a large portion of a small tarsometatarsus, PIN 4491–8 (Figure 27.6). This specimen shows derived characters of the Hesperornithiformes, including an inclined cross-section of the tarsometatarsal shaft, the high proximal position of the second trochlea, and the proximal position of a facet for metatarsal I. In general, the tarsometatarsus is stout and short and the metatarsal shaft is transversely expanded. These characters distinguish

this bird as a separate taxon of small hesperornithiform. Two further remains (a cervical vertebra and the portion of a mandible) representing small hesperornithiforms were collected by the JRMPE in the Nemegt beds of Guriliin Tsav and Tsagaan Khushuu in Ömnögov' (South Gobi) Aimag. These fossils are also somewhat different from known representatives of this order.

The existence of small, possibly volant hesperornithiforms was first claimed by Nesov (1992a). This conclusion was based on his discovery, in North American museum collections, of several small bones of hesperornithiforms from the Late Campanian and Maastrichtian beds of Canada and the USA. Further records of small hesperornithiforms, from the Eginsai Svita (Latest Santonian–Early Campanian) of Kushmurun locality, in Kustanaiskaya Province, North Kazakhstan (see above), were subsequently published by Nesov (1992a, c).

Thus, small hesperornithiforms are found in the continental Maastrichtian of Mongolia and North America. The small, lightly constructed pneumatized bones of the hind limb support the possibility that they were volant.

Infraclass Neornithes Gadow, 1893
Parvclass Palaeognathae Pycraft, 1900

The monophyly of the Palaeognathae is supported by various morphological and molecular characters, as well as behavior and eggshell microstructure (Bock, 1963; Cracraft, 1986, 1988; Kurochkin, 1995b). At present, there is good evidence that the Palaeognathae radiated in the Paleocene and Eocene and evidence for this group has also been found in the Cretaceous of Central Asia and Europe (Kurochkin, 1995a, b).

Order Ambiortiformes Kurochkin, 1982
Family Ambiortidae Kurochkin, 1982

Amended diagnosis. Small flying birds; cervical vertebrae heterocoelous; sternum with a keel; a wide procoracoid process perpendicular to the shaft; scapular blade long and thin; ventral edge of the proximal end of the humerus strongly developed and with a distinct tubercle on its cranial surface; transverse groove is

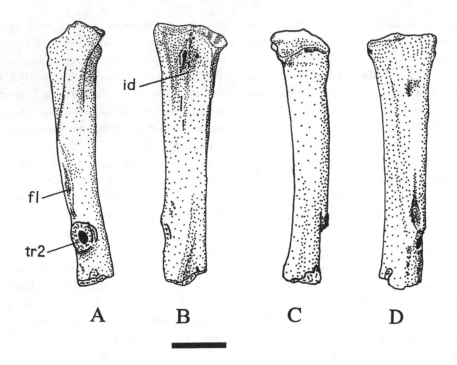

Figure 27.6. Left tarsometatarsus of a small, possibly volant hesperornithiform from the Nemegt Formation (Late Cretaceous) of Bügiin Tsav locality, Ömnögov' Aimag, Mongolia. Medial (A), cranial (B), lateral (C), and plantar (D) views. Abbreviations: fI, facet for metatarsal I; id, infracotylar depression; tr2, base of trochlea for digit 2. Scale bar = 10 mm.

short, fossa-like, and runs dorsoventrally; pneumotricipital fossa of the humerus not developed.

Contents. Ambiortus Kurochkin, 1982 (Neocomian, Mongolia) and *Otogornis* Hou, 1994 (Neocomian, China).

Comments. Otogornis genghisi Hou, 1994 is based on associated elements of the forelimb and shoulder girdle (holotype IVPP V 9607), together with some flight feathers, from the locality of Chabu Sumu, Otog Qi, Yikezhao-meng, in the Ordos Basin of Inner Mongolia, China. The mudstones which yielded the specimen belong to the Yijinhuoluo Formation of the Zidan Group and are earliest Cretaceous or possibly even Late Jurassic in age. *Otogornis* was first identified as an enantiornithid (Dong, 1993), but later assigned to Aves *incertae sedis* (Hou, 1994).

In 1995 I had an opportunity to investigate the holotype of *Otogornis genghisi* through the courtesy of Dr L. Hou and Z. Zhou. I found that *Ambiortus* and *Otogornis* shared a number of derived characters including: a thickened, three-edged acrocoracoid with an acute top; flat, wide humeral articular facet of the scapula; ventral position of a small, short, and oval articular head of the humerus; and a thin and long intermediate phalanx of the major wing digit. These characters provide evidence for a close relationship between *Ambiortus* and *Otogornis*, and for the assignment of *Otogornis* to the Ambiortidae (Kurochkin, 1999b).

The heterocoelous cervical vertebrae, U-shaped furcula, convex coracoidal cotyla in the scapula, and concave scapular cotyla in the coracoid are clear evidence that Ambiortidae belongs in Neornithes. Assignment of this family to Palaeognathae is based on the advanced condition of the strong ventrocaudal transverse processes of the cervical vertebrae; the well developed, dorsoventrally compressed scapular acromion with the tubercle or prominence on its dorsal side, and the projecting ventral edge of the proximal end of the humerus which bears a remarkable cranial tubercle with a centre pit on its cranial surface.

Ambiortus Kurochkin, 1982

Type species. Ambiortus dementjevi Kurochkin, 1982.

Diagnosis. Procoracoid process wide and long; scapular acromion long and dorsoventrally compressed; deep groove along lateral side of the scapula; scapular blade narrow; short, fossa-like groove cranial to the tubercle on the projecting ventral edge of the proximal end of the humerus; undivided capital groove in the proximal end of the humerus; metacarpals fused at the proximal end; intermediate phalanx of the major digit dorsoventrally compressed.

Comments. Ambiortus represents the earliest known stage in the evolution of neornithine birds. It indicates that early palaeognaths were keeled birds and good fliers. Comparison with the Lithornithiformes and the Ichthyornithiformes does not support the opinion of Martin (1991) and Elzanowski (1995) regarding the close relationships of *Ambiortus* to the Ichthyornithiformes.

Ambiortus dementjevi Kurochkin, 1982

Holotype. PIN 3790–271+, 3790–271–, and 3790–272, portion of the articulated left shoulder girdle, the left forelimb, and cervical and thoracic vertebrae (Figures 27.7 and 27.8). Khurilt Ulaan Bulag locality, Central Mongolian Altai Mountains, Bayankhongor Aimag. Bööntsagaan Gorizont, Neocomian, Lower Cretaceous. (The age of the Cretaceous shales and sandstones at Khurilt is disputed. According to the latest analysis (Sinitsa, 1993) the Khurilt beds were deposited between the Dundargalant Gorizont (Latest Jurassic) and the upper member of the Bööntsagaan Gorizont, which is of uncertain age, although, in places, the Khurilt beds are overlapped by the Khulsangol Svita (Aptian–Albian). The dating of the latter is based on lithofacies data, fossil fish, mollusks, and ostracods, thus the age of the Khurilt beds must be at least older than the Aptian. Palaeoentomologists

Figure 27.7. *Ambiortus dementjevi* Kurochkin, 1982. Holotype PIN 3790–271+ combined with a mould of PIN 3790–272. Scale bar = 10 mm.

and palaeobotanists consider the Khurilt beds as Late Neocomian (Zherikhin, 1978), Aptian (Krasilov, 1980, 1982; Dmitriev and Zherikhin, 1988), or just as the youngest insect assemblage among the three Lower Cretaceous assemblages of Central Mongolia (Ponomarenko, 1990). Based on geological data, Shuvalov (1982, and this volume) assigned the Khurilt and Kholboot beds to the Andaikhudag Formation that he dated as Hauterivian–Barremian. Kurochkin (1999b) further discusses the age of the Khurilt and Kholboot beds and possible correlations with the Chinese Jiufotang Formation, aged on the basis of absolute dates.

Diagnosis. Same as for the genus.

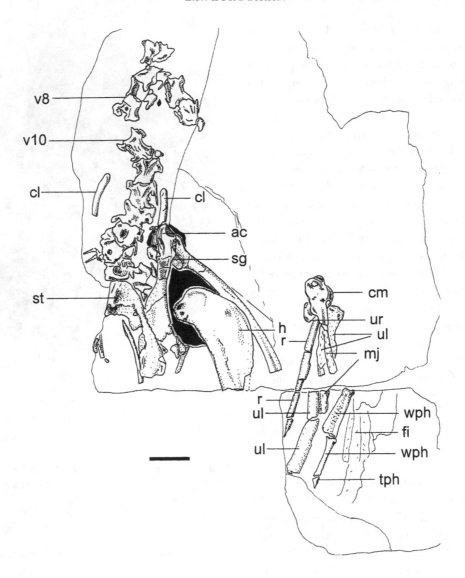

Figure 27.8. *Ambiortus dementjevi* Kurochkin, 1982. Holotype PIN 3790–271+ combined with a mold of PIN 3790–272. Abbreviations: ac, acrocoracoid; cl, clavicle; cm, carpometacarpus; fi, feather imprints; h, humerus; mj, major metacarpal; r, radius; sg, scapular glenoid; st, sternum; tph, terminal (ungual) phalanx of major digit; v8, eighth cervical vertebra; v10, tenth cervical vertebra; ul, ulna; ur, ulnare; wph, proximal and intermediate phalanges of major digit. Scale bar = 10 mm.

Comments. New preparation of *A. dementjevi* shows that the eight and tenth cervical vertebrae have heterocoelous not amphicoelous articular surfaces as previously reported (Kurochkin, 1985a, b). Further investigation also led to the discovery of a contact between the broken edges of the counterslab (PIN 3790–271) and the slab bearing the distal portion of the forelimb (PIN 3790–272). Thus the carpometacarpus, radius, and ulna on the main slab (PIN 3790–271+) extend to specimen PIN 3790–272. *Ambiortus dementjevi* shows some characters that confirm the primitive condition of this bird. The articular head of the humerus is oval and short, the bicipital crest and intumescence are absent, the pneumatic foramen and fossa are absent; the shaft of the radius is rounded in cross-section, the major and minor metacarpals are long, of the same length, and of similar thickness, the intermediate phalanx of the major digit is long and the major wing digit bears an ungual phalanx.

Parvclass Neognathae Pycraft, 1900

There are a few neognathous birds from the Cretaceous of Mongolia, Uzbekistan, and Russia. They are mainly represented by fragmentary remains and most have not yet been described. However, they provide important data on the distribution of neognaths in the Cretaceous and on the existence of some extant orders of birds at that time.

Order ? Gruiformes Bonaparte, 1854
Family indet.
Horezmavis Nesov, 1983

Type species. Horezmavis eocretacea Nesov, 1983.
Diagnosis. Medial cotyla of the tarsometatarsus inclined dorsally and located markedly more proximal than the lateral cotyla; intercotylar prominence low; dorsal infracotylar fossa deep and elongate; dorsomedial margin sharpened; tuberosity for insertion of *M. tibialis cranialis* short, high, and located in the proximal region of the fossa; the large vascular foramen on the lateral side and the impression for the ligamental attachment on the medial side are almost symmetrical with respect to the tuberosity; retinacula attachment located proximal to the ligamental attachment mentioned above, close to the dorsomedial margin; plantar crest relatively weak (Nesov and Borkin, 1983).

Horezmavis eocretacea Nesov, 1983

Holotype. PO 3390, proximal end of a left tarsometatarsus. Khodzhakul locality, outcrop CX-20, near to the north-western end of the Sultan-Uvais mountain ridge, Karakalpakia, Uzbekistan. Middle member of the Khodzhakul Svita (Late Albian).
Diagnosis. Same as for the genus.

Comments. H. eocretacea was assigned to the Gruiformes *sensu lato*, based on some (?) characters of the Ralli (Nesov and Borkin, 1983; Nesov, 1992d), but this needs to be confirmed. More recently, this taxon was erroneously assigned to the Enantiornithes by Martin (1995a). *Horezmavis* shows such characters of neognathous birds as a completely fused tarsometatarsus with dorsal infracotylar depression and an intercotylar prominence on the proximal articular surface. Relationships to any extant birds are difficult to establish because of the fragmentary condition of the material. However, *Horezmavis*, which was about the size of the extant *Gallinula chloropus* provides good evidence of the existence of neognathous birds in the latest Early Cretaceous.

Order Anseriformes Bechstein, 1804
Family Presbyornithidae Wetmore, 1926
Genus and species nov.
Unnamed taxon

There is a somewhat damaged, but complete tarsometatarsus of a presbyornithid from the Baruungoyot Formation at Üüden Sair, Ömnögov', Bulgan Sum, Mongolia (Kurochkin, 1988). This very small form, with a tarsometatarsus length of only 40.3 mm, is the only avian from the locality where the maniraptoran dinosaur *Avimimus* and the marsupial *Asiatherium* were discovered.

Order ? Pelecaniformes Sharpe, 1891
Family ? Fregatidae Garrod, 1891
Subfamily ? Limnofregatinae Olson, 1977
Volgavis Nesov and Yarkov, 1989

Type species. Volgavis marina Nesov and Yarkov, 1989.
Diagnosis. Tip of the mandible strongly ventrally deflected.
Comments. Volgavis was originally assigned, though with some doubt, to the Charadriiformes, then later determined as a member of the Limnofregatinae which belongs in the Pelecaniformes (Nesov, 1992d). This conclusion remains to be confirmed.

Volgavis marina Nesov and Yarkov, 1989
Holotype. PO 3638, rostral portion of the lower jaw

with both rami and a fragment of the surangular. Malaya Ivanovka locality, Dubovskii District, Volgograd Province, South Central Russia. Quartz-glauconite sands of Latest Maastrichtian or Danian.

Diagnosis. Same as for the genus.

Comments. The lower jaw fragment of *V. marina* is about 27 mm, as measured from figure 1–1a of Nesov and Yarkov (1989), and thus it was a small bird. The mandibular ramus exhibits the opening of the neuro-vascular canal on the medial side and some neurovascular foramina in a shallow groove on the lateral side. A ventrally deflected tip of the mandible also suggests a strongly hooked end of the upper jaw.

The beds that yielded this fossil, greenish, quartz-glauconite sands were originally considered to be Latest Maastrichtian (Nesov and Yarkov, 1989; Nesov, 1992b), but are now thought to be Danian (Palaeocene) (Nesov, 1988, 1992c).

Aves *incertae sedis*
Platanavis Nesov, 1992

Type species. *Platanavis nana* Nesov, 1992.

Diagnosis. Middle vertebrae of synsacrum heavily dorsoventrally compressed; vertebral foramen very spacious in the middle portion of the synsacrum; double ridge along ventral side of the synsacrum; pleurocoels very low and short; dorsal area of the middle vertebrae of the synsacrum broadened (Nesov 1992).

Platanavis nana Nesov, 1992

Holotype. PO 4601, fragment of the synsacrum consisting of two or three vertebrae. Dzharakhuduk locality, Navoi District, Bukhara Province, Uzbekistan. Outcrop CBI-5a, upper member of the Bissekty Svita (Coniacian).

Diagnosis. Same as for the genus.

Comments. *P. nana* represents a small bird with the following unique characters: the vertebrae of the synsacrum are strongly dorsoventrally compressed, the pleurocoels are deep and there is a doubled ridge on the ventral side of the synsacrum.

Family Kuszholiidae Nesov, 1992
Kuszholia Nesov, 1992

Type species. *Kuszholia mengi* Nesov, 1992.

Contents. Only the type species.

Diagnosis. Synsacrum wide; transverse process of the next to last vertebra on synsacrum strongly developed and stout; caudal pleurocoels small-sized, but deep; caudal articular surface large, wide and dorsoventrally compressed; postzygapophyses of the last vertebra on synsacrum very large; ventral groove especially deep in articulated areas of the centra; centrum of the third vertebra from caudal end heavily dorsoventrally compressed (Nesov, 1992).

Kuszholia mengi Nesov, 1992

Holotype. PO 4602, caudal portion of the synsacrum. Outcrop CBI-52, Dzharakhuduk locality, Navoi District, Bukhara Province, Uzbekistan. Upper member of the Bissekty Svita (Coniacian).

Referred material. Cranial portion of the synsacrum, PO 4623, from the same outcrop (Nesov, 1992d, plate IV, 5), and possibly some vertebrae from outcrops CBI-14 and CBI-57.

Diagnosis. Same as for the genus.

Comments. The holotype specimen (PO 4602) of *K. mengi* was first figured (fig. 2, 3a-3), under the number PO 3486, and identified, in the figure caption, as the synsacrum of a large ichthyornithid from outcrop CBI-52. *K. mengi* was a chicken-sized bird which had a stout synsacrum with an enlarged third pair of transverse processes. Originally placed in the Kuszholiidae *incertae sedis*, *Kuszholia* was later assigned to the Patagopterygiformes (Nesov and Panteleev, 1993), so far known only from the Coniacian–Santonian of Argentina. This hypothesis has yet to be verified, but it should be noted that the synsacra of *Kuszholia* and *Patagopteryx* are similar in that they both have an enlarged third pair of transverse processes and a convex ventral synsacrum. By contrast, *Patagopteryx* lacks the pleurocoels, a concave caudal articular surface, and a ventrally convex synsacrum, that are present in *Kuszholia*.

Fossil feathers

Subclass Praeornithes Rautian, 1978
Order Praeornithiformes Rautian, 1978
Family Praeornithidae Rautian, 1978
Praeornis Rautian, 1978
Praeornis sharovi Rautian, 1978

Holotype. PIN 2585/32, flight feather. Aulie (Mikhailovka) locality, Chimkent Province, Kazakhstan. Balabansai Svita, Late Jurassic.

Diagnosis. Relatively large bird, the size of a crow; edges of the barbs absolutely flat since they do not break up into barbules; barbs have pulp caps; outer and inner sides of barbs flattened and broadened, thus showing some similarity to vanes; the vanes located near to the dorsal side of the barb shaft, which is filled by the pulp caps; barbs form a complete vane; the number of the barbs in one centimetre is not more than four; the plane of the vane is twisted; the outer vane is narrower than the inner vane; the distal portion of the shaft of the flight feather is noticeably flexed in the horizontal plane; the pulp caps are large; pulp caps in the shaft are larger than ones in the barbs; ends of the barbs are clearly pointed.

Comments. This feather-like specimen (Figure 27.9) (Rautian, 1978) is possibly the earliest known record of a feather, but is doubted by some. Nesov (1992d) discussed the relationships of *Praeornis sharovi*, rejecting the designation of this specimen as avian, and supporting the conclusions of Bock (1986) who regarded this enigmatic fossil as a plant. New investigation of this specimen, using scanning electron microscopy (Glazunova *et al.*, 1991), supports its avian assignment, but the particular identity of this specimen remains unresolved.

Cretaceous feathers

A detailed record of fossil feathers from the former USSR was published by Nesov (1992b, d) and numerous fossil feathers from the Lower Cretaceous of Mongolia and Siberia were recorded and figured by Kurochkin (1985a, b, 1988).

There are a number of avian feathers from the

Figure 27.9. *Praeornis sharovi* Rautian, 1978. Holotype PIN 2585/32. Scale bar = 10 mm.

Lower Cretaceous of Western and Central Mongolia. Feather localities include: Hötöl, Khyra, Gurvan Ereen, Erdene Uul, Khurilt Ulaan Bulag, Kholboot, Shine Khudag, Böön Tsagaan, Altan Teel, Myangat, Andai Khudag and Ulaan Tolgoi. Among them, the fossiliferous deposits of the first three localities belong to the lowermost svitas of the Neocomian. In addition, the Lower Cretaceous localities of Baisa, Ust' Kara, Pad' Semen, and Turga in Transbaikalia, Russia, have yielded about three dozen feathers. Isolated feathers have been collected from the Late Cretaceous localities of Yantardakh (north of the Krasnoyarskii Krai, Khatanga River), Amka (Khabarovskii Krai, Okhotskii Region), Obetzautzii Creek (Magadan Province, Tenkinskii Region) in Russian Siberia, and Taldysai (Dzhezkazgan Province) which yielded *Cretaviculus sarysuensis* based on feather remains (Bazhanov, 1969; Shilin and Romanova, 1978) and Tulkeli (Kizylordin Province) (Shilin, 1977) in Kazakhstan.

Unfortunately, so far, Cretaceous feathers have provided little information on birds of the period. However, the specimens mentioned above are preserved in very different conditions. Some represent impressions of the feather structure, while others show different kinds of mineralization of the feathers. Most specimens represent small body contour feathers, but they do not show the microstructure of the distal barbules or microscopic barbicels which are of taxonomic significance in modern birds. Some specimens from Gurvan Ereen represent flight and tail feathers, and some are represented by plumules. One specimen from Baisa shows color patterns.

Fossil avian eggs

There are a number of fossil avian eggs and eggshell remains from the Upper Cretaceous of Mongolia, Kirgizstan and Uzbekistan (Chapter 28).

Non-avians

Class Reptilia Linnaeus, 1758
Theropoda Marsh, 1881
Maniraptora Gauthier, 1986
Family Parvicursoridae Karhu and Rautian, 1996
Mononykus (Perle *et al.*, 1993)

Type species. Mononykus olecranus (Perle *et al.*, 1993).

Holotype. MGI 107/6, posterior part of skull, most of the precaudal vertebrae, all four limbs, thoracic girdle and portion of ilium and pubis. Bügiin Tsav Locality, South Mongolia. Nemegt Formation, Upper Cretaceous (Maastrichtian?).

Referred material. A skull fragment and part of an articulated postcranial skeleton, MGI 100/99; Tögrögiin Shiree, South Mongolia; Djadokhta Formation, Upper Cretaceous (Campanian?).

Diagnosis. Edentulous maxilla; pronounced pectoral crest of humerus; single distal condyle of humerus; ventral tubercle of humerus pronounced; extremely short shafts of ulna and radius; very long olecranon process of ulna; carpometacarpus massive, short, quadrangular with no intermetacarpal space; manus digit 1 much larger than digits 2 and 3; claw of manus digit 1 robust; coracoid not expanded ventrally; sternal carina thick; one posterior dorsal vertebra biconvex and synsacrum procoelous; zygapophyses, costal fossa, and transverse process of anterior dorsal vertebrae on same level; sharply keeled posterior centra on synsacrum vertebrae; elongate haemal arches; ischium extremely slender; robust and horizontally projected antitrochanter; two cnemial crests on tibiotarsus; medial margin of ascending process of astragalus excavated by deep notch; metatarsal III limited to distal third, triangular in cross-section. This diagnosis is assembled from the original diagnoses in Perle *et al.*, (1993, 1994).

Comments. In recent years, there has been much discussion about the relationships of *Mononykus*. I do not consider it or its relatives to be avian but, because it has been assigned by some to Aves, it must be discussed here.

Cladistic analysis has suggested that *Mononykus* falls within Metornithes as a sister-group to Ornithothoraces (Norell *et al.*, 1993; Perle *et al.*, 1993, 1994; Chiappe *et al.*, 1996). *Archaeopteryx* is excluded from this evolutionary lineage and assigned to the Avialae *sensu* Gauthier (1986) placing *Archaeopteryx* as a sister-group to taxa including Maniraptora and traditional Aves, though in a recent paper Chiappe *et al.* (1996) changed Avialae to Aves for the taxon Metornithes + *Archaeopteryx*. At the same time, the cladistic approach utilizing the total group concept indicates theropodan relationships for *Mononykus* (Patterson, 1993a, b).

In spite of that, most palaeontologists using a morpho-phylogenetical concept of homology-analogy have found evidence of a non-avian, theropodan nature for *Mononykus* (Ostrom, 1994; Wellnhofer, 1994; Kurochkin, 1995c; Martin, 1995b; Zhou, 1995; Feduccia, 1996). The unambiguous relationships of *Mononykus* to birds was established on five (Perle *et al.*, 1993) or six (Chiappe *et al.*, 1996) characters: prominent ventral processes on cervicodorsal vertebrae, longitudinal and rectangular sternum, ossified sternal keel, ischium more than two-thirds of pubic length, carpometacarpus formed from fused distal carpals and metacarpals, prominent antitrochanter, and short fibula. All these characters can be explained as bipedal and digging adaptations (Zhou, 1995), and most are represented among different velociraptorine or maniraptoran theropods. Such characters as a single-headed quadrate, biconvex posterior dorsal vertebrae and opisthocoelous thoracic and cervical vertebrae, articulation of the cervical ribs at the same level as the cranial zygapophyses and transverse processes, an obtuse apex of the scapula that should be directed dorsad in natural articulation to the coracoid (the position of the scapula should be vertical, not horizontal as it was figured in the original reconstructions (Perle *et al.*, 1994 figs. 9 and 20)), a wide coracoid lacking an acrocoracoid, a single distal condyle of the humerus, femur longer than the metatarsus, and with a fourth trochanter, and unfused metatarsals with a proximally declined third metatarsal all clearly point to the theropodan nature of *Mononykus* (Kurochkin, 1995c). Indeed Ostrom (1994, p. 172), concluded that '*Mononykus* was not a bird, . . . it clearly was a fleet-fossorial theropod.'

Recently, Karhu and Rautian (1996) described a new maniraptoran, *Parvicursor remotus*, from the Late Cretaceous of Mongolia and discussed in detail the arctometatarsalian type of metatarsus. *Parvicursor* is extremely similar to *Mononykus* and both show the arctometatarsalian pes with a proximally declined third metatarsal. On this basis, Karhu and Rautian (1996) argued that arctometatarsalian theropods are not closely related either to Sauriurae, or to Ornithurae.

Discussion

Feather records show a wide distribution of birds from the earliest Cretaceous in the western part of Mongolia and in the southern part of Eastern Siberia. Late Cretaceous feather records confirm the existence of birds in Kazakhstan, Northern Siberia, Mongolia, and the Far East of Russia. Unfortunately, feathers reveal nothing regarding the taxonomic identity of these birds, but because of the wide distribution of downy and fine contour feathers they indicate the extensive distribution of birds with a warm-blooded physiology in the Early Cretaceous. These types of feathers may be associated with the origin of neognathous birds.

By contrast, the contour plumage for enantiornithine birds is problematic. Chinsamy *et al.* (1994, 1995) reported growth rings in the femora of Argentinian enantiornithines indicating cyclical growth of bones during life and slower growth rates than in extant birds. This also provides indirect evidence of their physiology, thus Chiappe (1995) and Chinsamy *et al.* (1995) have suggested that enantiornithines did not have endothermy or ectothermy, but some kind of intermediate level of metabolism. In any case, as this level of physiology is more similar to a reptilian mode of physiology, than an avian one, this may be sufficient to reject the idea of a good covering of contour feathers in the enantiornithines.

Osteological remains are more restricted than feathers, both in terms of their geological and geographical distribution, but they can be attributed to particular avian taxa. A small enantiornithine was present in the Early Cretaceous of Mongolia and during the late Late Cretaceous (Coniacian–Maastrichtian) small and middle-sized enantiornithines inhabited Mongolia and Uzbekistan. Their small size and long, curved pedal claws suggests that they may have been primarily arboreal birds.

Large, flightless hesperornithiforms are known from the Campanian of Russia and Kazakhstan, while small and possibly volant hesperornithiforms have been collected from the Santonian–Campanian beds of Kazakhstan and from the Maastrichtian beds of

Mongolia. This is supported by data from North America concerning the existence of previously unknown small representatives of hesperornithiforms in the interior basin of North America at the end of the Cretaceous.

The Palaeognathae is a separate lineage of neornithine birds that was represented in the Early Cretaceous of Mongolia by *Ambiortus*. In an adjoining region of China, *Otogornis*, a close relative of *Ambiortus*, has been reported from the earliest Cretaceous.

The Neognaths form a fourth group of Cretaceous birds. They were present in the Early Cretaceous as fossil records from Asia and Europe show. New data on birds from Liaoning in China also confirm the existence of neognathous birds in the Early Cretaceous (Hou *et al.*, 1996). In the Late Cretaceous Nemegt Formation undescribed remains from Mongolia indicate the existence of charadriiforms (Graculavidae), anseriforms (Presbyornithidae), pelecaniforms (a cormorant), and procellariiforms (an albatross) in this region. This is concordant with discoveries in North America and Antarctica. The fifth and last group of Asian Cretaceous birds, the Kuszholiidae, of uncertain relationships, is known only from the Late Cretaceous of Uzbekistan.

Most of these avian fossils are represented by fragmentary remains which mainly provide information on the distribution of Mesozoic birds. However, some of the Russian and Mongolian Cretaceous birds, such as *Ambiortus* and *Nanantius*, provide very important evidence of the mode of evolutionary processes in early birds. As I hypothesized (Kurochkin, 1985a, 1991), the known record of Mesozoic birds is more restricted than the true diversity of Mesozoic birds, and *Archaeopteryx* was not the direct ancestor for all later birds. New fossil data on Mesozoic birds (Chiappe, 1995; Feduccia, 1995; Hou *et al.*, 1995, 1996; Kurochkin, 1995c) confirm this assumption and these authors demonstrate a formerly unknown Mesozoic diversity of birds.

Data on Mesozoic birds has only appeared from Russia and Mongolia during the last few decades: this is in contrast to North America where Cretaceous birds have been collected and studied for more than a century. However, they provide good evidence for a greater avian diversity at higher taxonomic levels in the Mesozoic than in the Cenozoic. So, in the Cenozoic, only two evolutionary lineages: Palaeognathae and Neognathae of the infraclass Neornithes survived. During the Cretaceous, at least, there were five major phylogenetic lineages: the Enantiornithes, Hesperornithes, Ichthyornithes, Palaeognathae, and Neognathae.

Acknowledgements

I thank L. Hou and Z. Zhou for showing me the specimen of *Otogornis genghisi* and for discussion of this fossil, H. Osmólska and K. Sabath for the possibility to work on specimens of *Gobipteryx* and the enantiornithine embryos, A.V. Panteleev for additional data on birds from Dzharakhuduk, A.V. Mazin of the PIN laboratory of photography for the photographs of specimens, L. Martin for the idea of moulding specimen PIN 3790–272 and for its fabrication, and D. Unwin and two anonymous reviewers for critically reading this manuscript. This study was supported, in part, by the Institute of Vertebrate Palaeontology and Palaeoanthropology, Beijing, and by grants 95-05-16919d and 96-04-50822 of the Russian Fund for Fundamental Research.

1. The species of *Zhyraornis* listed on pp. 533–534 were assigned by Nesov (1984) to the Ichthyornithiformes, and subsequently to the Enantiornithes by Kurochkin (1996, this paper). However, Zhyraornithidae do not belong to the Enantiornithes, or to any other known fossil or living group of birds, as investigation of these remains in the TsNIGRI and in the PO, and discussion with A. Panteleev showed in February, 1998.
2. New genera and species of Enantiornithes based on coracoids mentioned on pp. 534–545 have been described by Panteleev (1998).

References

Bakhurina, N.N. and Unwin, D.M. 1995. A survey of pterosaurs from the Jurassic and Cretaceous of the Former Soviet Union and Mongolia. *Historical Biology* 10: 197–245.

Bazhanov, V.S. 1969. [On the record of a bird remain living in the Cretaceous in the USSR.] *Tezisy Dokladov XV Sessii Vsesoyuznogo Paleontologicheskogo Obshchestva*: 5–6. Leningrad.

Bock, W.J. 1963. The cranial evidence for Ratitae affinities. *Proceedings 13th International Ornithological Congress* 1: 39–54.

—1986. The arboreal origin of avian flight, pp. 57–72 in Padian, K. (ed.), *The Origin of Birds and the Evolution of Flight*. Memoirs of the California Academy of Sciences 8.

Brodkorb, P. 1976. Discovery of a Cretaceous bird, apparently ancestral to the order Coraciiformes and Piciformes (Aves: Carinatae). *Smithsonian Contributions to Paleobiology* 27: 67–73.

Chatterjee, S. and Kurochkin, E.N. 1994. A new embryonic bird from the Cretaceous of Mongolia. *Journal of Vertebrate Paleontology* 14 (3) Suppl.: 20A.

Chiappe, L.M. 1995. The first 85 million years of avian evolution. *Nature* 378: 349–355.

—and Calvo, J.O. 1994. *Neuquenornis volans*, a new Enantiornithes (Aves) from the Upper Cretaceous of Patagonia (Argentina). *Journal of Vertebrate Paleontology* 14: 223–246.

—, Norell, M.A. and Clark, J.M. 1996. Phylogenetic position of *Mononykus* (Aves: Alvarezauridae) from the Late Cretaceous of the Gobi Desert. *Memoirs of the Queensland Museum* 39 (3): 557–582.

Chinsamy, A., Chiappe, L.M. and Dodson, P. 1994. Growth rings in Mesozoic birds. *Nature* 368: 196–197.

—, —and— 1995. Mesozoic avian bone microstructure: physiological implications. *Paleobiology* 21: 561–574.

Cracraft, J. 1986. The origin and early diversification of birds. *Paleobiology* 12: 383–399.

—1988. The major clades of birds, pp. 339–361 in Benton, M.J. (ed.), *The Phylogeny and Classification of Tetrapods. Vol. 1. Systematics Association Special Volume 35A*. Oxford: Clarendon Press.

Dmitriev, V.Yu. and Zherikhin, V.V. 1988. [The changes of diversity in insect families on the data by method of the accumulated occurrences.] pp. 208–215 in Ponomarenko, A.G. (ed.), *Cretaceous biocenotic crisis and the evolution of insects*. Moscow: Nauka.

Dong, Zhi-Ming. 1993. A lower Cretaceous enantiornithine bird from the Ordos Basin of Inner Mongolia, People's Republic of China. *Canadian Journal of Earth Sciences* 30: 2177–2179.

Elzanowski, A. 1974. Preliminary note on the paleognathous bird from the Upper Cretaceous of Mongolia. *Palaeontologia Polonica* 30: 103–109.

—1976. Palaeognathous bird from the Cretaceous of Central Asia. *Nature* 264: 51–53.

—1977. Skulls of *Gobipteryx* (Aves) from the Upper Cretaceous of Mongolia. *Palaeontologia Polonica* 37: 153–165.

—1981. Embryonic bird skeletons from the Late Cretaceous of Mongolia. *Palaeontologia Polonica* 42: 1147–1179.

—1995. Cretaceous birds and avian phylogeny. *Courier Forshungsinstitut Senckenberg* 181: 37–53.

—and Wellnhofer, P. 1996. Cranial morphology of *Archaeopteryx*: evidence from the seventh skeleton. *Journal of Vertebrate Paleontology* 16: 81–94.

Feduccia, A. 1995. Explosive evolution in Tertiary birds and mammals. *Science* 267: 637–638.

—1996. *The Origin and Evolution of Birds*. New Haven: Yale Univ. Press. 420 pp.

Glazunova, K.P., Rautian, A.S. and Filin, V.R. 1991. [*Praeornis sharovi*: birds feather or plant leaf?] *Materialy 10 Vsesoyuznoi Ornitologicheskoi Konferentsii, Vitebsk, 17–20 Sentjabrya 1991*. Part 2, Book 1: 149–150. Minsk.

Hou, Lianhai. 1994. A Late Mesozoic bird from Inner Mongolia. *Vertebrata PalAsiatica* 32: 258–266.

—, Zhou, Zhonghe, Martin, L.D. and Feduccia, A. 1995. A beaked bird from the Jurassic of China. *Nature* 377: 616–618.

—, Martin, L.D., Zhou, Zhonghe and Feduccia, A. 1996. Early adaptive radiation of birds: evidence from fossils from Northeastern China. *Science* 274: 1164–1167.

International Code of Zoological Nomenclature. Third Edition, 1985. Russian translation, Leningrad: Nauka, 1988.

Karhu, A.A. and Rautian, A.S. 1996. [A New Family of Maniraptora (Dinosauria: Saurischia) from the Late Cretaceous of Mongolia.] *Paleontologicheskii Zhurnal* 4: 85–94.

Krasilov, V.A. 1980. [Fossil plants of Manlay.] *Trudy Sovmestnoi Sovetsko-Mongol'skoi Paleontologicheskoi Ekspeditsii* 13: 40–42.

—1982. Early Cretaceous flora of Mongolia. *Palaeontographica, Abteilung B*, 181: 1–43.

Kurochkin, E.N. 1985a. Lower Cretaceous birds from Mongolia and their evolutionary significance, pp.

191–199 in Il'ichev, V.D. and Gavrilov, V.M. (eds.), *Acta XVIII Congressus Internationalis Ornithologici*, v. I. Moscow: Nauka.

—1985b. A true carinate bird from Lower Cretaceous deposits in Mongolia and other evidence of early Cretaceous birds in Asia. *Cretaceous Research* 6: 271–278.

—1988. [Cretaceous birds of Mongolia and their significance for study of the phylogeny of class Aves.] *Trudy Sovmestnoi Sovetsko-Mongol'skoi Paleontologicheskoi Ekspeditsii* 34: 33–42.

—1991. [*Protoavis, Ambiortus* and other palaeornithological rarities.] *Priroda* 12: 43–53.

—1993. [*The principal stages in evolution of class Aves.*] Thesis on the scientific degree of the Doctor in Biological Sciences. Moscow: Palaeontological Institute, RAS. MS, 64 pp.

—1995a. The assemblage of the Cretaceous birds in Asia, pp. 203–208 in Sun, A. and Wang, Y. (eds.), *Sixth Symposium on Mesozoic Terrestrial Ecosystems and Biota, Short Papers*. Beijing: China Ocean Press.

—1995b. Morphological differentiation of the Palaeognathous and the Neognathous birds. *Courier Forschungsinstitut Senckenberg* 181: 79–88.

—1995c. Synopsis of Mesozoic birds and Early Evolution of class Aves. *Archaeopteryx* 13: 47–66.

—1996. *A new enantiornithid of the Mongolian Late Cretaceous and a general appraisal of the infraclass Enantiornithes (Aves).* Special Issue. 50 pp. Moscow: Palaeontological Institute.

—1999a. [A new large enantiornithid from the Upper Cretaceous of Mongolia (Aves, Enantiornithes).] *Trudy Zoologicheskogo Instituta, RAN* 277: 130–141.

—1999b. The relationships of the Early Cretaceous *Ambiortus* and *Otogornis* (Aves: Ambiortiformes). *Smithsonian Contributions to Paleobiology* 89: 275–284

Martin, L.D. 1983. The origin and early radiation of birds, pp. 291–338 in Brush, A.H. and Clark, G.A. Jr. (eds.), *Perspectives in ornithology*. Cambridge: Cambridge University Press.

—1984. A new hesperornithid and the relationships of the Mesozoic birds. *Transactions of the Kansas Academy of Science* 87 (3–4): 141–150.

—1991. Mesozoic birds and the origin of birds, pp. 485–540 in Schultze, H.-P. and Trueb, L. (eds.), *Origins of the higher groups of tetrapods, contrary and consensus*. Ithaca: Cornell University Press.

—1995a. The Enantiornithes: terrestrial birds of the Cretaceous. *Courier Forschungsinstitut Senckenberg* 181: 23–36.

—1995b. The relationship of *Mononykus* to ornithomimid dinosaurs. *Journal of Vertebrate Paleontology* 15 (3)Suppl.: 43A.

—and Tate, J., Jr. 1976. The skeleton of *Baptornis advenus* (Aves: Hesperornithiformes). *Smithsonian Contributions to Paleobiology* 27: 35–66.

Mikhailov, K.E. 1991. Classification of fossil eggshells of amniotic vertebrates. *Palaeontologia Polonica* 36: 193–238.

—1995. Systematic, faunistic and stratigraphic diversity of Cretaceous eggs in Mongolia: comparison with China, pp. 165–168 in Sun, A. and Wang, Y. (eds.), *Sixth Symposium on Mesozoic Terrestrial Ecosystems and Biota, Short Papers*. Beijing: China Ocean Press.

—1996. [Birds eggs in the Late Cretaceous of Mongolia.] *Paleontologicheskii Zhurnal* 1: 119–121.

Molnar, R.E. 1986. An enantiornithine bird from the Lower Cretaceous of Queensland, Australia. *Nature* 322: 736–738.

Nesov, L.A. 1984. [Upper Cretaceous pterosaurs and birds from central Asia.] *Paleontologicheskii Zhurnal* 1: 47–57.

—1986. [The first record of the Late Cretaceous bird *Ichthyornis* in the Old World and some other bird bones from the Cretaceous and Paleogene of Soviet Middle Asia.] *Trudy Zoologicheskogo Instituta, AN SSSR* 147: 31–38.

—1988. [New Cretaceous and Paleogene birds of Soviet Middle Asia and Kazakhstan and their environments.] *Trudy Zoologicheskogo Instituta, AN SSSR* 182: 116–123.

—1990. [Small *Ichthyornis* and others findings of the bird bones from the Bissekty Svita (Upper Cretaceous) of central Kyzylkum Desert.] *Trudy Zoologicheskogo Instituta, AN SSSR* 210: 59–62.

—1992a. [Flightless birds of meridional Late Cretaceous sea straits of North America, Scandinavia, Russia, and Kazakhstan as indicators of features of oceanic circulation.] *Byulleten' Moskovskogo Obshchestva Ispytatelei Prirody, Otdel Geologicheskii* 67 (5): 78–83.

—1992b. Mesozoic and Paleogene Birds of the USSR and their paleoenvironments. *Natural History Museum, Los Angeles County, Science Series* no. 36: 465–478.

—1992c. Russia, p. 13 in Mourer-Chauviré, C. (ed.), *Society of Avian Paleontology and Evolution, Information Letter*, 6, November 1992.

—1992d. [Record of the localities of the Mesozoic and Paleogene with avian remains of the USSR, and the description of new findings.] *Russkii Ornitologicheskii Zhurnal* 1 (1): 7–50.

—and Borkin, L.J. 1983. [New findings of bird bones from the Cretaceous of Mongolia and Soviet Middle Asia.] *Trudy Zoologicheskogo Instituta, AN SSSR* 116: 108–109.

—and Yarkov, A.A. 1989. [New Cretaceous-Paleogene birds of the USSR and some remarks about the origin and evolution of Class Aves.] *Trudy Zoologicheskogo Instituta, AN SSSR* 197: 78–97.

—and— 1993. [Hesperornithes in Russia]. *Russkii Ornitologicheskii Zhurnal* 2 (1): 37–54.

—and Panteleev, A.V. 1993. [On the similarity of the Late Cretaceous ornithofauna of South America and Western Asia.] *Trudy Zoologicheskogo Instituta, RAN* 252: 84–94.

—and Prizemlin, B.V. 1991. [A large advanced flightless marine bird of the order Hesperornithiformes from the Late Senonian of the Turgai Strait: The first finding of the group in the USSR.] *Trudy Zoologicheskogo Instituta, AN SSSR* 239: 85–107.

Norell, M., Clark, J. and Chiappe, L. 1993. Naming names. *Nature* 366: 518.

Ostrom, J.H. 1994. On the origin of birds and of avian flight, pp. 160–177 in Prothero, D.P. and Schoch, R.M. (eds.), *Major features of vertebrate evolution*. Knoxville: Univ. of Tennessee Press.

Panteleev, A.V. 1998. [New species of enantiornithines (Aves: Enantiornithes) from the Upper Cretaceous of Central Kyzylkum]. *Russkii Ornitologicheskii Zhurnal. Ekspress-vy.pvsk* 35: 3–15.

Patterson, C. 1993a. Bird or dinosaur? *Nature* 365: 21–22.

—1993b. Naming names. Patterson replies. *Nature* 366: 518.

Perle, A., Norell, M.A., Chiappe, L.M. and Clark, J.M. 1993. Flightless bird from the Cretaceous of Mongolia. *Nature* 362: 623–626.

—, Chiappe, L.M., Barsbold, R., Clark, J.M. and Norell, M.A. 1994. Skeletal morphology of *Mononykus olecranus* (Theropoda: Avialae) from the Late Cretaceous of Mongolia. *American Museum Novitates* 3105: 1–29.

Ponomarenko, A.G. 1990. [Insects and the Lower Cretaceous stratigraphy of Mongolia.] pp. 103–108 in Krasilov, V.A. (ed.), *Continental Cretaceous of the USSR*. Vladivostok: Far Eastern Branch of the USSR Academy of Sciences.

Rautian, A.S. 1978. [Unique feather from the beds of the Jurassic lake in Karatau ridge.] *Paleontologicheskii Zhurnal* 4: 106–114.

Shilin, P.V. 1977. [New record of the feather imprint of a Cretaceous bird in Kazakhstan.] p. 33 in Voinstvenskii, M.A. (ed.), *Sed'maya Vsesoyuznaya Ornitologicheskaya Konferentziya, Tezisy Dokladov*, Part I. Kiev: Naukova Dumka.

—and Romanova, E.V. 1978. [*Senonian floras of Kazakhstan*.] Alma-Ata: Kairat Publishing. 176 p.

Shuvalov, V.F. 1982. [Palaeogeography and the history of development of the Mongolian lake basins in the Jurassic and Cretaceous.] pp. 18–80 in Martinson, G.G. (ed.), *Mezozoiskie ozernye basseiny Mongolii*. Leningrad: Nauka.

Sinitsa, S.M. 1993. [Jurassic and Lower Cretaceous of Central Mongolia.] *Trudy Sovmestnoi Rossiisko-Mongol'skoi Paleontologicheskoi Ekspeditsii* 42: 1–238 p.

Unwin, D.M. 1993. Aves, pp.717–737 in Benton, M.J. (ed.), *The Fossil Record*. London: Chapman and Hall.

Walker, C.A. 1981. New subclass of birds from the Cretaceous of South America. *Nature* 292: 51–53.

Wellnhofer, P. 1974. Das fünfte Skelettexemplar von *Archaeopteryx*. *Palaeontographica, Abteilung* A185: 85–180.

—1994. New data on the origin and early evolution of birds. *Comptes Rendus de l'Académie des Sciences, Paris* 319: 299–308.

Zherikhin, V.V. 1978. [Formation and replacement of the Cretaceous and Cenozoic fauna complexes.] *Trudy Paleontologicheskogo Instituta, AN SSSR* 165: 1–198.

Zhou, Zhonghe. 1995. Is *Mononykus* a bird? *Auk* 112: 958–963.

28

Eggs and eggshells of dinosaurs and birds from the Cretaceous of Mongolia

KONSTANTIN E. MIKHAILOV

Introduction

Fossil eggs of dinosaurs and birds are now known from the Mesozoic of Central Asia, India, Southern Europe, South Africa, North America and South America (Hirsch, 1994; Mikhailov, 1997). Most of these eggs are from the Cretaceous, though there are some rare Triassic and Jurassic finds. At present Mongolia and China give the most complete picture of the remarkable diversity of dinosaurian eggs and also provide a great deal of information on their palaeoecology and taphonomy (Kurzanov and Mikhailov, 1989; Sabath, 1991; Thulborn, 1991; Mikhailov *et al.*, 1994). Most important in this respect has been the spectacular discoveries of dinosaur embryos in redbeds of both China and Mongolia (Norell *et al.*, 1994, 1995; Cohen *et al.*, 1995).

So far, Central Asia has yielded nine dinosaurian oofamilies containing 21 oogenera, and two avian oofamilies represented by three oogenera. Among these, twenty dinosaurian and four avian oospecies have been found in the southern part of Mongolia, in the territory of the Gobi Desert. In addition to the discoveries in China and Mongolia, fossil eggshells have also been reported from Central Asia including: eastern Kazakhstan (Zaisan Basin: Taizhuzgen River), western Kirgizstan (Naryn River Valley), central Kazakhstan and central Uzbekistan.

Most of the Central Asiatic dinosaurian oogenera occur in Mongolia and China. Judging from the preliminary descriptions and sketch-illustrations of Nesov and Kaznyshkin (1986), the materials from Kirgizstan belong to the same parataxa at the family and generic level, though their identification as oospe-

cies is not yet available. The materials from Uzbekistan and central Kazakhstan are very scarce and difficult to assess in parataxonomic terms, since most of the egg-remains have been found as shell scatterings. By contrast, whole and broken eggs, and nests with clutches of eggs are not uncommon in Mongolia.

Crocodilian eggs are not yet known from the Central Asia, while those of Testudinata, though found in Mongolia and possibly in Kirgizstan, are rare and have not yet been formally described. Some small, ovoid and ellipsoid forms from these regions have been proposed to be soft-shelled eggs of turtles and lizards (Sochava, 1969), but most of them, apart from some recently assigned to insects (Johnston *et al.*, 1996) are undoubtedly non-biological in origin (Mikhailov, 1991). These remains are therefore omitted from the following consideration.

All Mongolian egg fossils (eggs and their shells) are Cretaceous in age, and date mainly from the Upper Cretaceous. Only a few rare finds are known from the Lower Cretaceous of this region (Kurzanov and Mikhailov, 1989; Shuvalov, 1982). The main egg-bearing deposits are the Djadokhta and Baruungoyot Formations (Santonian–Campanian) and the Nemegt Formation (Maastrichtian). In the eastern part of the Mongolian Gobi some eggs and eggshells have been found in the Upper Cretaceous (Cenomanian–Santonian) Bayanshiree Formation (Figure 28.1).

This paper presents a synopsis of dinosaurian and avian egg-remains from the Cretaceous of Mongolia, that have now been studied in detail and described within a formal parataxonomy. Material from the republics of the former Soviet Central Asia, excluding those from the Taizhuzgen River sites, have only been

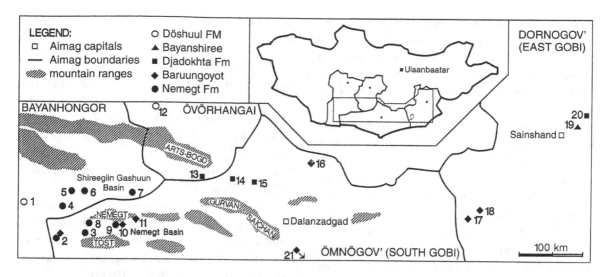

Figure 28.1. Cretaceous egg-bearing localities in the Mongolian Gobi Desert. The age of the localities is designated by the symbols of the formations. South-western group: 1, Dösh Uul; 2, Hermiin Tsav I and II; 3, Tsagaan Khushuu; 4, Khaichin Uul I; 5, Bügiin Tsav; 6, Guriliin Tsav; 7, Shireegiin Gashuun; 8, Altan Uul III; 9, Nemegt; 10, Khulsan; 11, Gilbent; 12, Builyastyn Khudag; 13, Üüden Sair; 14, Tögrögiin Shiree; 15, Bayan Zag; 16, Algui Ulaan Tsav. Eastern group: 17, Ih Shunkht; 18, Baga Mod Khudag; 19, Mogoin Ulaagiin Hets; 20, Baga Tariach (and Dariganga eastward); 21, Shilüüst Uul in southernmost Mongolia (Borzongiin Gobi); 22, Ukhaa Tolgod. Modified from Mikhailov *et al.*, (1994); drawn by Sabath.

subjected to a preliminary examination (Nesov and Kaznyshkin, 1986), and await formal description. Details of the discoveries in these regions can be found in Bazhanov (1961) and Nesov and Kaznyshkin (1986).

The discovery and study of fossil eggs in the Gobi Desert has a long history. In the early 1920s, some unusual elongate eggs (oogenera *Protoceratopsidovum* and *Elongatoolithus*) were found at Bayan Zag by the Central Asiatic Expedition of the American Museum of Natural History (Andrews, 1932; see also Chapter 12). These finds caused a sensation and, in the following decades, prompted many other new finds of fossil eggs all over the world. Recently, Carpenter *et al.* (1994) described some earlier unpublished materials collected by the American expedition. The Mongolian Palaeontological expeditions of the USSR Academy of Sciences (1946–1949), the Polish–Mongolian Palaeontological Expeditions (1963–1971), the Joint Soviet–Mongolian Palaeontological and Geological Expeditions (since 1969), and, finally, joint expeditions

of the Sino-Canadian Dinosaurian Project have significantly expanded the number and diversity of fossil materials collected in this region. The history of these discoveries is reviewed by Sochava (1969), Sabath (1991), Thulborn (1991), Jerzykiewicz *et al.* 1993, Mikhailov *et al.* (1994) and Carpenter *et al.* (1994). Recently, new finds, in particular those of embryos in eggs, have been made by the joint expedition of the American Museum of Natural History and the Mongolian Academy of Sciences (Norell *et al.*, 1994, 1995).

The investigation of the shell structure of fossil eggs from Mongolia was started by Van Straelen (1925) and continued by many other scholars. Sadov (1970) and Sochava (1969, 1971, 1972) contributed much to the description of the materials collected in the Gobi in the 1940s to 1960s and initiated the structural classification of fossil eggshells. Kurzanov and Mikhailov (1989), Sabath (1991), Mikhailov (1991, 1994a, b, 1996a, b), and Mikhailov *et al.* (1994) described new finds of fossil eggs and their shells from the Mongolian Gobi Desert and systematically orga- nized the entire egg collections held in the Palaeontological Institute of the Russian Academy of Sciences in Moscow and the Institute of Palaeobiology of the Polish Academy of Sciences in Warsaw.

A comparative review of the oospecies and some notes on their distribution are given below. With regard to the Mongolian ootaxa, this follows Mikhailov (1994a, b, 1996a, b), but also includes other sources (Sabath, 1991; Mikhailov et al., 1994). The general interpretation of the material from Kirgizstan is based on the descriptions and sketches given in Nesov and Kaznyshkin (1986). The distribution of ootaxa in China follows Zhao (1994) and Carpenter and Alf (1994). Diagnoses of the Mongolian oogenera and oospecies are summarized in Tables 28.1 and 28.2, and their particular diagnostic features are shown in Figures 28.2 and 28.3. Emended diagnoses of the oofa- milies and the location of Mongolian ootaxa in the overall diversity of the fossil record of Mesozoic eggs are given by Mikhailov (1997). Many general aspects of palaeo-oology, in particular those of fossil egg taphonomy and palaeobiology were cosidered by Hirsch (1994). Principles of fossil egg parataxonomy, in particular, the hierarchical ordering of ootaxa (oofamilies, oogenera, oospecies) and questions of nomenclature are discussed in detail by Mikhailov *et al.* (1996).

Institutional abbreviations

IVPP, Institute of Vertebrate Paleontology and Paleoanthropology, Beijing; PIN, Paleontological Institute, Moscow.

Systematic palaeontology

Veterovata (system of fossil vertebrate eggs and their shells)

Eggs with dinosauroid-spherulitic basic shell type

Oofamily Spheroolithidae Zhao, 1979
Comments. Association with embryos and neonates

suggests that eggs of this type were laid by hadrosaurs (Mikhailov, 1991; 1997; Mikhailov *et al.*, 1994).

Oogenus *Spheroolithus* Zhao, 1979
Sochava, 1969, plate 11: 4 and plate 12: 6–8; Mikhailov, 1991, plate 25: 2, 4; Mikhailov *et al.*, 1994, figs. 7.7A, 7.8A, B, E, H and 7.9A; Mikhailov, 1994b, plate 10: 4–6, plate 11; 2 and fig. 3; Mikhailov, 1997, pl. 6, plate 10: Fig. 8; text Fig. 18a.
Type oospecies. Spheroolithus (= Paraspheroolithus) irenen-sis Zhao and Jiang, 1974.
Comments. The name *Paraspheroolithus* is treated here as a junior synonym of *Spheroolithus* (Mikhailov, 1997).
Included oospecies. S. chianghiungtingensis (Zhao and Jiang, 1974); ?*S. megadermus* (Young, 1959); *S. maiasau-roides, S. tenuicorticus* (Mikhailov, 1994b).
Discussion. Eggshell scatterings assigned to the oogenus *Spheroolithus* are quite common in the Upper Cretaceous of Mongolia and possibly present in the collections from Kirgizstan. Complete eggs and, in particular, nests, are rare compared with China, and mostly known from the Guriliin Tsav and Shireegiin Gashuun localities of the Shireegiin-Gashuun Basin of southern Mongolia. In addition, one nest has been found at the locality of Shilüst Uul, Borzongiin Gobi (Figure 28.1).

Three *Spheroolithus* oospecies; *S. irenensis, S. tenuicor-ticus* and *S. maiasauroides* are known from Mongolia, and the first named oospecies is also common in eastern and north-eastern China. *S. maiasauroides* differs from all other Asiatic *Spheroolithus* oospecies in having a sculptured eggshell surface, similar to the eggs of *Maiasaura* from Montana. *S. tenuicorticus* differs from *S. irenensis* and the Chinese *S. chianghiungtingensis* in the range of eggshell thickness (Table 28.2).

Oofamily Ovaloolithidae Mikhailov, 1991
Comments. Taphonomic correlation with nests and bones suggest that this egg type may have been laid by hadrosaurs (Mikhailov, 1991; 1997; Mikhailov *et al.*, 1994).

Oogenus *Ovaloolithus* Zhao, 1979
Mikhailov, 1991, plate 27: 1, 4; Mikhailov *et al.*, 1994, figs. 7.8C, F, G, 7.9B, C and 7.10A, B; Mikhailov, 1994b,

Table 28.1. *Summary diagnoses of Mongolian oogenera*

Oogenera	Morphotype, pore system	Egg shape, egg size	Eggshell surface	Eggshell thickness (mm)
Spheroolithus	prolatospherulitic prolatocanaliculate	subspherical medium–size	sagenotuberculate: ornament or print	medium–thick (1–3)
Ovaloolithus	angustispherulitic rimocanaliculate	ellipsoid medium–size	sagenotuberculate: ornament or print	medium–thick (1–3)
Faveoloolithus	filispherulitic multicanaliculate	spherical large–size	not-sculptured	thick (2–3)
Dendroolithus	dendrospherulitic prolatocanaliculate	spherical–ellipsoid medium–size	not-sculptured or verrucous	thick (2–4)
Protoceratopsidovum	prismatic angusticanaliculate	elongate ellipsoid medium–size	linearituberculate or smooth	thin–medium (0.5–1.5)
Elongatoolithus	ratite (CL:ML = 2:1–3:1) angusticanaliculate	elongate ellipsoid medium–size	linearituberculate-discretituberculate	thin–medium (0.5–2.0)
Macroolithus	ratite (CL:ML = 2:1–3:1) angusticanaliculate	elongate ellipsoid–large–size	linearituberculate-discretituberculate	medium–thick (1–3)
Trachoolithus	ratite (CL:ML = 4:1) angusticanaliculate	?elongate ?medium–size	linearituberculate-discretituberculate	thin–medium (0.5–1.5)
Laevisoolithus	ratite (CL:ML = 1:1) angusticanaliculate	ellipsoid small–size	smooth	thin (0.5–1.0)
Subtiliolithus	ratite (CL:ML = 1:1–1:2) angusticanaliculate	? small–size	microtuberculate	thin (less than 1.0)
Gobioolithus	? (CL:ML = 2:1) angusticanaliculate	elongate ovoid small–size	smooth	very thin (less than 0.5)
Oblongoolithus	ratite (CL:ML = 1:1) angusticanaliculate	elongate ellipsoid medium–size	smooth	thin (less than 1.0)
Parvoolithus	?'spherulitic'?	? small–size	smooth	very thin (less than 0.5)

Abbreviation: CL:ML, the ratio of the thickness of the continuous layer to the mammillary layer in the eggshell of a ratite morphotype; only the standard range of eggshell thickness is shown.

Table 28.2. *Holotypes and quantitative characteristics of Mongolian oospecies*

Oospecies	Holotype	Egg-shape (El)	Egg size (cm)	Eggshell thickness (mm) −T1(T2)
Spheroolithus irenensis	IVPP ?V-733	1.1–1.2	7–8 × 8–10	1.1–2.2 (1.4–2.0)
Spheroolithus maiasauroides	PIN 4228-2	?1.2	7 × 9	1.0–1.6 (1.2–1.5)
Spheroolithus tenuicorticus	PIN 4476-4	?1.3	?	0.8–1.8 (1.0–1.3)
Ovaloolithus chinkangkouensis	IVPP ?V-732	1.2–?1.3	7–8 × ?8–10	1.4–3.0 (2.2–3.0)
Ovaloolithus dinornithoides	PIN 4231-1	1.4	7–8 × 10–11	1.1–1.8 (1.3–1.8)
Faveoloolithus ningxiaensis	IVPP V 4709	1.0	15–16.5 (d)	1.8–2.6
Dendroolithus verrucarius	PIN 3142-454	1.0	9–?12 (d)	1.4–4.3 (2.6–3.3)
Dendroolithus microporosus	PIN 4476-1	1.0–1.1	6 × 7	1.5–3.0 (2.0–2.7)
Protoceratopsidovum sincerum	PIN 614-58(1)	2.3–2.5	4–5 × 11–12	0.3–1.2 (0.6–0.7)*
Protoceratopsidovum minimum	PIN 4228-1	2.3–?2.5	4 × 10	0.3–0.7*
Protoceratopsidovum fluxuosum	PIN 3142-415	2.3–2.5	5–7 × 13–15	0.3–1.4 (0.6–0.7)*
Elongatoolithus frustrabilis	PIN 3143-126	2.0–2.2	6–7 × 15–17	0.8–1.5 (1.1–1.3)*
Elongatoolithus subtitectorius	PIN 3907-501	2.0–?	?similar	0.5–0.9 (0.7–0.8)
Elongatoolithus sigillarius	PIN 4216-403	2.0–2.2	6–7 × 15–17	0.3–1.1 (0.4–0.8)
Elongatoolithus excellens	PIN 522-402	2.2–?2.7	4 × 9–?11	0.3–0.9 (0.4–0.7)
Trachoolithus faticanus	PIN 4227-3	?2.2–?	?similar	0.3–0.5*
Macroolithus rugustus	IVPP ?V-2788	2.0–2.4	? × 18–?	0.8–1.5 (1.1–1.3)
Macroolithus mutabilis	PIN 4477-5	?	?larger	1.3–2.0 (1.5–1.8)
Laevisoolithus sochavai	PIN 2970-5	1.9	3–4 × 7	?0.5–0.6
Subtilioolithus microtuberculatus	PIN 4230-3	?similar	?similar	0.3–0.6 (0.3–0.4)
Gobioolithus minor	PIN 3142-422	1.8–2.0	2–2.4 × 3–4.6	0.1–0.2
Gobioolithus major	PIN 4478-1	1.8–2.0	2.6–3 × 5.3–7	0.2–0.4
Oblongoolithus glaber	PIN 3142-500/1	?2.0–?	?4 × ?9–11	0.3–0.7*
Parvoolithus tortuosus	PIN 4479	?	?2 × 4–5	0.07–0.1

Abbreviations: El, elongation of egg (egg length/egg width); d, diameter of spherical eggs; T1, complete range, including rare values, T2, standard range (both ranges through a single egg); asterisk (*) marks the range in the equatorial portion of elongated eggs (without polar values), measured without elements of the sculpture.

plate 11: 4–5; Mikhailov, 1997, plate 6: 3, and plate 10: 9; text Fig. 18B–D, text Fig. 19k.

Type oospecies. Ovaloolithus chinkangkouensis Zhao and Jiang, 1974.

Included oospecies. O. laminadermus (Zhao and Jiang, 1974); *O. tristriatus, O. mixtistriatus, O. monostriatus* (Zhao, 1979); *O. dinornithoides* (Mikhailov, 1994b).

Discussion. Like *Spheroolithus*, eggs and shell scatterings of *Ovaloolithus* are common in the Nemegt Formation (Maastrichtian) of Mongolia (Table 28.3), but rare in the earlier Baruungoyot and Djadokhta Formations (Santonian–Campanian). However, unlike *Spheroolithus*, this egg type is common in the Cenomanian– Campanian Bayanshiree Formation in the eastern part of Mongolia, but has yet to be found in

the Lower Cretaceous of this region. In Kirgizstan *Ovaloolithus* may occur in both the Upper Cretaceous and Lower Cretaceous (Nesov and Kaznyshkin, 1986).

Among Mongolian oospecies, *O. chinkangkouensis*, which is characteristic of the south-eastern part of Mongolia, is also common in eastern China, from which it has been primarily described (Zhao, 1994), though the stratigraphic correlation of this form in these two regions is not clear. Fine sagenotuberculate ornamentation and a hieroglyphic pore pattern clearly distinguish this species from *O. dinornithoides* which is wide spread in the southern part of the Mongolian Gobi. The latter, in turn, exhibits a particular (dinornithoid) pore pattern on the shell surface, and the eggshell is thinner (Table 28.2). The strati-

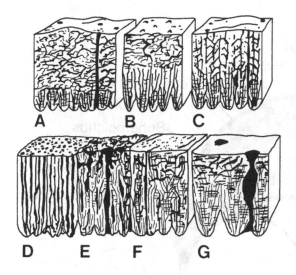

Figure 28.2. Microstructure of the shells of dinosaurian and avian eggs from Mongolia. A, Elongatoolithidae (ornithoid type, ratite morphotype); B, Laevisoolithidae (ornithoid type, ratite morphotype); C, Prismatoolithidae (dinosauroid-prismatic type, prismatic morphotype); spherulitic type: D, Faveoloolithidae (filispherulitic morphotype); E, Dendroolithidae (dendrospherulitic morphotype); F, Ovaloolithidae (angustispherulitic morphotype); G, Spheroolithidae (prolatospherulitic morphotype) (see also Table 28.1).

graphic distribution of these two oospecies in Mongolia is also different (Table 28.3). Unfortunately, comparison with other Chinese oospecies is not possible as their diagnoses are not available and type specimens seem not to have been indicated.

Oofamily Faveoloolithidae Zhao and Ding, 1976
Comments. Evidence from the palaeoecological setting of mass burials suggests that this egg type was laid by sauropods (Mikhailov, 1991; 1997; Mikhailov *et al.*, 1994).

Oogenus *Faveoloolithus* Zhao and Ding, 1976
Sochava, 1969, plate 11: 1 and plate 12: figs. 6–8; Sabath, 1991, plate 15: 1–3; Mikhailov, 1991, plate 24: 1–6; Mikhailov *et al.*, 1994, fig. 7.5A-C, D; Mikhailov, 1997, plate 12: 3; text. Fig. 16B, 19B, E, J.
Type and only oospecies. Faveoloolithus ningxiaensis Zhao and Ding, 1976.

Discussion. Scattered eggshell fragments of *F. ning-xiaensis* have recently been discovered at several localities in southern Mongolia (Table 28.3), but eggs and nests have been found only in three localities: Algui Ulaan Tsav, where they are most numerous, Ih Shunkht, and ?Dösh Uul (Figure 28.1). Unlike spheroolithid and ovaloolithid remains, *Faveoloolithus* eggshells are still not known from western Central Asiatic regions. *Faveoloolithus* differs from the Chinese oogenus *Youngoolithus* in the subspherical (versus strongly ellipsoid) egg-shape.

F. ningxiaensis is the largest egg from Central Asiatia and has a most peculiar arrangement of the pore system indicating an underground, chelonian type of incubation (Sochava, 1969; Sabath, 1991; Mikhailov *et al.*, 1994). In Mongolia, the eggs of this oospecies are always buried in red deposits and the eggshell is strongly impregnated by iron salts (red in colour). At Algui Ulaan Tsav complete nests and separate eggs of *F. ningxiaensis* are closely associated with the bones of sauropod dinosaurs (Sochava, 1969; Mikhailov *et al.*, 1994). However, sauropod eggs from other parts of the world including South America, India, and Southern Europe are different and have been assigned to a separate family, the Megaloolithidae (Mikhailov, 1991, 1997).

Oofamily Dendroolithidae Zhao and Li, 1988
Comments. Association with embryos shows that this egg type was lain by therizinosaurs (Cohen *et al.*, 1995; Mikhailov, 1997).

Oogenus *Dendroolithus* Zhao and Li, 1988
Sabath, 1991, plate 12: 1; Mikhailov, 1991, plate 24: 7; Mikhailov *et al.*, 1994, figs. 7.5E and 7.6 A–D; Mikhailov, 1994b, plate 10: 1–2; Mikhailov, 1997, plate 7: 3, and plate 10: 7, text fig. 16C, 19D, H.
Type oospecies. Dendroolithus wangdianensis Zhao and Li, 1988.
Included species. D. *verrucarius* (Mikhailov, 1994b), D. *microporosus* (Mikhailov, 1994b).
Discussion. Dendroolithid remains are characteristic for deposits of the Baruungoyot Formation in Mongolia (Figure 28.1, Table 28.3). Eggshell scatterings and transported shell debris are abundant in

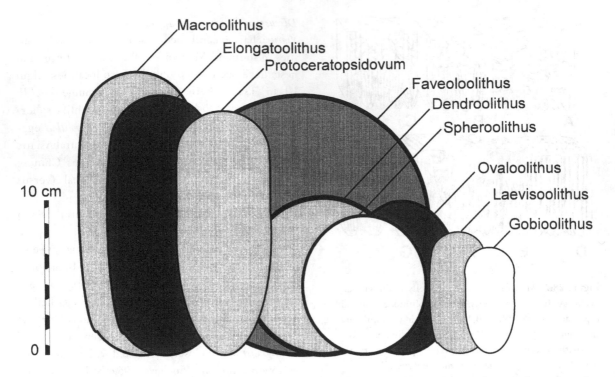

Figure 28.3. Size and shape of eggs from the Cretaceous of Mongolia. Shell thickness is shown in rough proportion to the scale. Regarding egg size, the oogeneric names are applied to the particular oospecies: *Macroolithus rugustus, Elongatoolithus frustrabilis, Protoceratopsidovum sincerum, Faveoloolithus ningxiaensis, Dendroolithus verrucarius, Spheroolithus irenensis, Ovaloolithus dinornithoides, Laevisoolithus sochavai* and *Gobioolithus major.*

several localities in the south of the country, but whole and broken eggs are sporadic and currently originate from just three localities: Hermiin Tsav, Gilbent, and Shilüüst Uul. The only nest is known from the last locality. This form is still not known from western Central Asiatic regions.

The Mongolian oospecies differ from *D. wangdianensis* of China in exhibiting a subspherical (versus slightly ellipsoid) shape and in their smaller size and more compact eggshell structure without large lacunae between the shell units: this is especially characteristic for *D. microporosus*. In contrast to *D. microporosus,* which is mostly smooth, *D. verrucarius* is the only oospecies of *Dendroolithus* that has a verrucous surface of the egg.

Dendroolithid eggshell, like that of *Faveoloolithus,* exhibits features suggestive of underground incubation in moist conditions.

Oofamily indet.
Oogenus *Parvoolithus* Mikhailov 1996b
Mikhailov, 1997, plate 8: 5.

Type and only oospecies. P. tortuosus Mikhailov 1996b.
Discussion. The oospecies *P. tortuosus* was established on the basis of a single incomplete small egg (25 × 40 mm), found in Upper Cretaceous red sediments of the ?Baruugoyot Formation at Khongil Tsav in the Ömnögov' Aimag, Mongolia. The eggshell structure is diagenetically altered, but in some parts of the egg, spherulitic organization of shell units in the lower half can be recognized. On this evidence *Parvoolithus* is tentatively identified as a dinosauroid-spherulitic shell type, but because of the incompleteness of the material and poor resolution of eggshell structure, this taxon cannot be attributed to any known family and is insufficient as a basis for establishing a new oofamily.

Table 28.3. *Stratigraphic distribution of oospecies in the Upper Cretaceous deposits of Mongolia*

Oospecies	Bayanshiree	Djadokhta	Baruungoyot	Nemegt
Spheroolithus irenensis				c (2)
Spheroolithus maiasauroides		r (2)		
Spheroolithus tenuicorticus			r (1)	
Ovaloolithus chinkangkouensis	c (2)			
Ovaloolithus dinornithoides		r (1)		c (3)
Faveoloolithus ningxiaensis	?K1		?c (5)	
Dendroolithus verrucarius			a (4)	
Dendroolithus microporosus			c (4)	
Protoceratopsidovum sincerum		a (3–4)	r (1)	
Protoceratopsidovum minimum		c (2)	r (1)	
Protoceratopsidovum fluxuosum		r (1)	c (2)	
Elongatoolithus frustrabilis		c (2)		
Elongatoolithus subtitectorius		r (1)		
Elongatoolithus sigillarius				r (1)
Elongatoolithus excellens				r (1)
Macroolithus rugustus				a (4)
Macroolithus mutabilis			r (1)	
Laevisoolithus spp.			c (2)	r (1)
Subtilioolithus spp.			a (5)	
Gobioolithus major		r (1)	r (2)	
Gobioolithus minor			a (3)	

The number of localities is given in brackets. Abbreviations: a, abundant; c, common; r, rare.

Eggs with dinosauroid-prismatic basic shell type
Oofamily Prismatoolithidae Hirsch, 1994
Oogenus *Protoceratopsidovum* Mikhailov, 1994a
Sabath, 1991, plates 11: 2–3, 16: 3, 17: 2–5 and 20: 12; Mikhailov *et al.*, 1994, figs. 7.11A–B, 7.12A–D, 7.13B, 7.14C and 7.15E; Mikhailov, 1994a, plate 9: 4 and figs. 2–3; Mikhailov, 1997, plate 8; 2–4, plate 10: 10, plate 12: 1–2, plate 13: 1 and plate 14: 3.
Comments. Association with remains of neonates and occurrence as mass-burials indicate that this egg type was laid by protoceratopsians (Mikhailov, 1991; 1997; Mikhailov *et al.*, 1994). The provisional assignment of *Prismatoolithus* from China to hypsilophodontid dinosaurs (Zhao and Li, 1993) may be incorrect and the same or similar eggs from the same locality (Bayan Mandahu) have been interpreted by Jerzykiewicz *et al.* (1993, p. 2189) as eggs of protoceratopsians.
Type oospecies. Protoceratopsidovum sincerum Mikhailov, 1994a.

Included species. P. minimum (Mikhailov, 1994a) and *P. fluxuosum* (Mikhailov, 1994a).
Discussion. In Mongolia, fossil eggs of the oogenus *Protoceratopsidovum* are one of the most characteristic remains from the deposits of the Djadokhta and Baruungoyot Formations (Table 28.3). Tiny eggshell fragments can be detected in most of the localities throughout the sedimentary layers, when they are intensively sampled, and this type of egg and the nests in which they are found dominate within egg assemblages in dinosaur nesting sites of this age (Table 28.4). These fossil eggs are also common in the equivalent deposits of the Chinese Gobi (Jerzykiewicz *et al.*, 1993).

The eggshell of both *P. sincerum* and *P. minimum* has a smooth surface and these species only differ from one another in egg size and shell thickness (Table 28.2). By contrast, *P. fluxuosum* exhibits fine linearituberculate ornamentation on the equatorial

Table 28.4. *Association of fossil eggs with dinosaurian and avian taxa at dinosaur nesting sites in Mongolia*

Predominant egg-fossils	Associated taxa
Mogoin Ulaagiin Hets (Döshuul Formation)	
Ovaloolithus chinkangkouensis	Lambeosaurines, hadrosaurines
Bayan Zag (Djadokhta Formation)	
Protoceratopsidovum sincerum	Protoceratopsians (?*Protoceratops*)
Elongatoolithus frustrabilis	Theropods (?*Velociraptor*, ?*Oviraptor*,)
Tögrögiin Shiree (Djadokhta Formation)	
Protoceratopsidovum sincerum	Protoceratopsians (*P. andrewsi*) (neonates)
Protoceratopsidovum minimum	Protoceratopsians (?*P. andrewsi*)
Elongatoolithus frustrabilis (one nest)	Theropods (?*Velociraptor*)
Hermiin Tsav (Baruungoyot Formation)	
Gobioolithus minor	Enantiornithids (embryos)
Subtiliolithus spp.	Enantiornithids (skeleton ?in the nest)
Protoceratopsidovum fluxuosum	Protoceratopsians (?*Breviceratops*)
Dendroolithus verrucarius	No association
Khulsan and Nemegt (red part) (Baruungoyot Formation)	
Gobioolithus minor	Enantiornithids (embryos)
Gobioolithus major	Enantiornithids (no remains)
Dendroolithus microporosus	No association
Algui Ulaan Tsav (?Cr1–?Cr2: Baruungoyot Formation)	
Faveoloolithus ningxiaensis	Sauropods
Khaichin Uul-1, Tsagaan Khushuu, Guriliin Tsav (Nemegt Formation)	
Macroolithus rugustus	Theropods (?*Tarbosaurus*)
Altan Uul I, II, III (Nemegt Formation)	
Ovaloolithus dinornithoides	Hadrosaurians (?*Saurolophus*)

part of the egg. This, however, can lead to confusion of this egg type with some species of the oogenus *Elongatoolithus* (oofamily Elongatoolithidae), in particular with *E. frustrabilis* which can be found at the same localities (Table 28.3). The most reliable diagnostic feature, in this case, is the eggshell microstructure in radial section (Figure 28.2). *Protoceratopsidovum* oospecies vary somewhat with regard to their distribution through the adjacent deposits of the Djadokhta-Baruungoyot continuum (Table 28.3) and in their correlation with particular taxa of protoceratopsians (Table 28.4; see Mikhailov, 1994a, for details).

At Bayan Zag and Tögrögiin Shiree eggs of *P. sincerum* are always found associated with skeletons of the ceratopsian dinosaur *Protoceratops andrewsi*. Incomplete eggs of *P. minimum* are found at Tögrögiin

Shiree suggest that some of the differences between this and the former oospecies might only reflect temporal changes in the populations of *P. andrewsi*. Eggs of *P. fluxuosum* possibly belong to another protoceratopsian dinosaur, namely *Breviceratops kozlovskii* (Mikhailov *et al.* 1994; Mikhailov 1994a).

Eggs with ornithoid basic shell type

Oofamily Elongatoolithidae Zhao, 1975
Comments. Recently, suggestions that eggs of this oofamily were those of theropod dinosaurs (Kurzanov and Mikhailov, 1989; Mikhailov *et al.*, 1994) have been confirmed by new material including an oviraptorid embryo within an elongatoolithid (?*Elongatoolithus*) egg collected at Ukhaa Tolgod by the Joint American–

Mongolian Palaeontological Expedition (Norell *et al.*, 1994, 1995). Unfortunately (as has frequently happened before) all similar elongate eggs with a slightly ridged ornamentation from the Djadokhta Formation, and in particular from Bayan Zag, have been alleged in these papers to be those of theropod dinosaurs. In fact, only part of them (those with ornithoid structure of the eggshell) belong to theropods while the others (with the prismatic-dinosauroid structure) were laid by protoceratopsian dinosaurs (see Mikhailov, 1994a).

The taphonomic settings and the arrangement of pore canals in elongatoolithid eggshells indicate semidry and dry incubation conditions and parental care of the nest (Mikhailov *et al.*, 1994)

Oogenus *Elongatoolithus* Zhao, 1975

Sabath, 1991, plates 13: 3–9, 14: 3, 16: 4 and 19: 1–4; Mikhailov, 1991, plates 28: 2 and 30: 1–2; Mikhailov *et al.*, 1994, fig. 7.13A, C; Mikhailov, 1994a, plate 9: 1–3, 5–6 and fig. 5; Mikhailov, 1997, plate 13: 2, and plate 14: 2.

Type oospecies. Elongatoolithus elongatus (Young, 1954).

Included species. E. andrewsi (Zhao, 1975), *E. magnus* (Zeng and Zhang, 1979); *E. frustrabilis* (Mikhailov, 1994a), *E. subtitectorius* (Mikhailov, 1994a), *E. sigillarius* (Mikhailov, 1994a), *E. excellens* (Mikhailov, 1994a); *Elongatoolithus* oosp. (Zeng and Zhang, 1979; Zhao *et al.*, 1991; Mikhailov, 1994a).

Discussion. Remains of *Elongatoolithus* type eggs are widespread in the Upper Cretaceous of Central Asia and are currently known from more than ten localities and three formations in southern Mongolia (Table 28.3), two svitas from Kirgizstan, and from more than twenty localities and sixteen formations in China. Though rarer, they are also present in the Lower Cretaceous of China (Zhao, 1993, table 1) and Kirgizstan.

All oospecies of *Elongatoolithus* are similar in their general morphology and shell microstructure and exhibit quite a large range of individual variation in terms of egg length and shell thickness (both within a clutch of eggs and within a single egg). Two of the four Mongolian oospecies, namely *E. frustrabilis* and *E. subtitectorius*, exhibit 'normal' linearituberculate ornamentation on the equatorial part of the egg. The two

remaining oospecies, *E. sigillarius* and *E. excellens*, exhibit the 'reversed' pattern (*sensu* Mikhailov, 1994a). Within each group some oospecies can be distinguished only by their eggshell thickness (Table 28.2).

Oogenus *Macroolithus* Zhao, 1975

Bazhanov, 1961, fig. 1; Sochava, 1969, plates 11: 6–8 and 12: 1–3; Mikhailov, 1991, plates 30: 3 and 31: 1–2; Mikhailov *et al.*, 1994, figs 7.14 and 7.15A–C; Mikhailov, 1994a, plate 10: 1–2; Mikhailov, 1997, plate 9: 1–4, plate 10: 11, and plate 14: 1.

Comments. In addition to egg size (Table 28.1), *Macroolithus* differs from *Elongatoolithus* in the higher maximal values of eggshell thickness on the upper pole of the egg and fewer lines of ridges and hillocks per 10 mm range (6–8 in *Macroolithus* versus 8–11 in *Elongatoolithus*) on the short diameter of the egg (Mikhailov, 1994a).

Type oospecies. Macroolithus rugustus (Young, 1965).

Included species. M. yaotunensis (Young, 1965); *M. mutabilis* (Mikhailov, 1994a).

Discussion. The three known oospecies differ in terms of their eggshell thickness (Table 28.2) and the range of variation of the surface ornamentation. Eggshell scatterings of *Macroolithus* (*M. rugustus*) are one of the most common fossil occurences in deposits from the Nemegt Formation of Mongolia (Table 28.3) and are also abundant in the Manrakskaya Svita of eastern Kazakhstan. The same oospecies is also common in the Maastrichtian of China. Another Mongolian oospecies, *M. mutabilis*, might be an aberrant form, but is common at the Ih Shunkht locality.

Oogenus *Trachoolithus* Mikhailov, 1994

Kurzanov and Mikhailov, 1989, fig. 12.1A–D; Mikhailov, 1991, plate 32: 1a, b; Mikhailov *et al.*, 1994, fig. 7.15D; Mikhailov, 1994a, plate 10: 3; Mikhailov, 1997, plate 9: 5–6.

Type and only known oospecies. Monotypic.

Discussion. Two large eggshell scatterings of this oospecies are currently known from the Lower Cretaceous type-locality Builyastyn Khudag in Mongolia. The remains were found in the Döshuul Formation, which is dated as Lower Cretaceous (Aptian). This oospecies differs from those of

Elongatoolithus and *Macroolithus* in some characteristics of its microstructure (such as the thickness ratio between the continuous and mammillary layers) and shell thickness (Tables 28.1 and 28.2).

Oofamily Oblongoolithidae Mikhailov, 1996b
Oogenus *Oblongoolithus* Mikhailov, 1996b
Mikhailov, 1997, plate 15: 5

Type and only known oospecies. *O. glaber* Mikhailov, 1996b.

Discussion. The single oospecies *O. glaber*, has been described from the Upper Cretaceous (?Baruungoyot Formation) Hermiin Tsav locality, of Ömnögov' Aimag, Mongolia. The short diameter of the eggs is less than 40 mm, and both poles are rather pointed. *Oblongatoolithus* differs from Elongatoolithidae in the absence of surface ornamentation. The eggshell microstructure is similar to that of Laevisoolithidae, but the egg shape is different and more similar to that of Elongatoolithidae.

Eggs with ornithoid basic shell type

Oofamily Laevisoolithidae Mikhailov, 1991

Comments. Association of a skeleton with a nest containing eggs shows that the egg type *Laevisoolithus* was laid by an enantiornithine bird (Mikhailov, 1996a; 1997; see also Chapter 27). Fossil eggs of the oogenus *Subtiloolithus* (Subtilioolithidae) are so similar to those of *Laevisoolithus* that Subtiliolithidae is treated as a junior synonym of Laevisoolithidae (Mikhailov, 1997).

Included oogenera (monotypic) and oospecies. Laevisoolithus sochavai and *Subtiliolithus microtuberculatus* (Mikhailov, 1991).

Mikhailov, 1991, plates 32: 2–4 and 33: 1–4; Mikhailov *et al.*, 1994, figs 7.16C, 7.17A, 7.18A–B; Mikhailov, 1996a, fig. 1a; Mikhailov, 1997, plate 10: land plate 15: 1–4; text-fig. 8E.

Discussion. The microstructure of the eggshell of *Laevisoolithus sochavai* (known from a single egg) and *Subtiliolithus microtuberculatus* is similar: they differ only in eggshell thickness and in that the shell surface of *L. sochavai* is absolutely smooth, while that of *S. microtuberculatus* has a sparse microtuberculation (Table 28.1).

These avian egg-fossils consisting of small parts of broken eggs and eggshells preserved in a sandy matrix are very common in deposits of the Baruungoyot Formation of southern Mongolia (Table 28.3). However, they have not yet been found in other regions of Central Asia.

Oofamily Gobioolithidae Mikhailov 1996a

Comments. The discovery of embryos within eggs show that this egg type was laid by enantiornithid birds (see Chapter 27).

Oogenus *Gobioolithus* Mikhailov 1996a

Mikhailov, 1991, plates 34: 1–4 and 35: 1–3; Mikhailov *et al.*, 1994, figs 7.16A–B and 7.17B; Mikhailov, 1996a, fig. 1b and 1c; Mikhailov, 1997, plate 10: 2–3 and plate 15: 6–7.

Type oospecies. *G. minor* (Mikhailov, 1996a).

Included species. *G. major* (Mikhailov, 1996a).

Discussion. Gobioolithid eggs, whole and broken, are characteristic remains from the deposits of the Baruungoyot Formation in southern Mongolia (Table 28.3), though they have not yet been found in other regions of Central Asia. Eggs of both oospecies are identical in shape and microstructure (Figures 28.2 and 28.3), but clearly differ in size and shell thickness (Table 28.2).

Fossil egg assemblages and their palaeoecological settings

The association of all particular ootaxa with skeletal remains, discovered at sites in which eggs and bones of dinosaurs and birds are buried together, are presented in Table 28.4. Three egg assemblages are typical for the territory of the Mongolian Gobi.

1. The 'hadrosaurian–theropodan' assemblage, characterized by eggshell scatterings of *Macroolithus rugustus* and *Ovaloolithus dinornithoides* (Table 28.3), comprises materials from the deposits of the Nemegt Formation (Maastrichtian) and their biostratigraphic equivalents.

2. The 'protoceratopsian–avian' assemblage, characterized by protoceratopsid eggs belonging to three

oospecies of the oogenus *Protoceratopsidovum*, and the eggs of small enantiornithid birds (two oospecies of the oogenus *Gobioolithus*), is found in deposits of the Djadokhta and Baruungoyot Formations (Santonian–Campanion).

3. Mass accumulations of fossil eggs of the oogenera *Protoceratopsidovum*, *Gobioolithus*, and *Faveoloolithus*, referred to in the literature as 'dinosaur nesting sites', occur mainly in the red deposits of the Baruungoyot and Djadokhta Formations (Table 28.4).

In addition, the yellow sands of the Djadokhta Formation at Tögrögiin Shiree yielded numerous eggs of protoceratopsians (*Protoceratopsidovum minimum* and *P. sincerum*) and the pink sands of the Döshuul Formation at Mogoin Ulaagiin Hets, produced strongly weathered nests belonging to ornithopod dinosaurs (*Ovaloolithus chinkangkouensis*).

Acknowledgements

I thank the editors of this volume for the opportunity to contribute to our current state of knowledge concerning fossilized eggs of dinosaurs and birds from Mongolia. Progress in this direction is mostly due to the selfless and highly successful labour of many generations of Russian, Mongolian, Polish and American participants of expeditions to the Mongolian Gobi.

References

Andrews, R.C. 1932 *The new conquest of Central Asia*. New York: American Museum of Natural History.

Bazhanov, V.S. 1961. [The first discovery of the dinosaurian eggshells in the U.S.S.R.] *Trudy Instituta Zoologii Kazakhskoi SSR* 15: 177–181.

Carpenter, K. and Alf, K. 1994. Global distribution of dinosaur eggs, nests and babies, pp. 15–30 in Carpenter, K., Hirsch, K.F. and Horner, J.R. (eds.), *Dinosaur Eggs and Babies*. Cambridge: Cambridge University Press.

—, Hirsch, K.F. and Horner, J.R. 1994. Introduction, pp. 1–11 in Carpenter, K., Hirsch, K.F. and Horner, J.R. (eds.), *Dinosaur Eggs and Babies*. Cambridge: Cambridge University Press.

Cohen, S., Cruickshank A., Joysey K., Manning T. and

Upchurch P. 1995. *The dinosaur egg and embryo project: exhibition guide*. Rock Art Publishing, Leicester, UK.

Hirsch, K.F. 1994. The fossil record of vertebrate eggs, pp. 269–294 in Donovan, S.K. (ed.). *The Palaeoecology of Trace Fossils*. Chichester: John Wiley and Sons. England.

Jerzykiewicz, T., Currie, P.J., Eberth, D.A., Johnston, P.A., Koster, E.H. and Zheng, J.J. 1993. Djadokhta Formation correlative strata in Chinese Inner Mongolia: an overview of the stratigraphy, sedimentary geology, and paleontology and comparisons with the type locality in the pre-Altai Gobi. *Canadian Journal of Earth Sciences* 30: 2180–2195.

Johnston, P.A., Eberth, D.A. and Anderson, P.K. 1996. Alleged vertebrate eggs from Upper Cretaceous redbeds, Gobi Desert, are fossil insect (Coleoptera) pupal chambers: *Fictovichus* new ichnogenus. *Canadian Journal of Earth Science* 33: 511–525.

Kurzanov, S.M. and Mikhailov, K.E. 1989. Dinosaur eggshells from the Lower Cretaceous of Mongolia, pp. 109–113 in Gillette, D.D. and Lockley, M.G. (eds.), *Dinosaur Tracks and Traces*. Cambridge: Cambridge University Press.

Mikhailov, K.E. 1991. Classification of fossil eggshells of amniotic vertebrates. *Acta Palaeontologica Polonica* 36: 193–238.

—1994a. [Eggs of theropod and protoceratopsid dinosaurs from the Cretaceous of Mongolia and Kazakhstan.] *Paleontologicheskii Zhurnal* 1994(2): 81–96.

—1994b. [Eggs of sauropod and ornithopod dinosaurs from the Cretaceous of Mongolia.] *Paleontologicheskii Zhurnal* 1994(3): 114–127.

—1996a. [The eggs of birds in the Upper Cretaceous of Mongolia.] *Paleontologicheskii Zhurnal* 1996(1): 119–121.

—1996b. [New genera of fossil eggs from the Upper Cretaceous of Mongolia.] *Paleontologicheskii Zhurnal* 1996(2): 122–124.

—1997. Fossil and Recent eggshell in amniotic vertebrates: fine structure, comparative morphology and classification. *Special Papers in Palaeontology* 56: 1–80.

—, Bray, E.S. and Hirsch, K.F. 1996. Parataxonomy of fossil egg remains (Veterovata): basic principles and applications. *Journal of Vertebrate Paleontology* 16: 763–769.

—, Sabath, K. and Kurzanov, S.M. 1994. Eggs and nests from the Cretaceous of Mongolia, pp. 88–115 in Carpenter, K., Hirsch, K.F. and Horner, J.R. (eds.),

Dinosaur Eggs and Babies. Cambridge: Cambridge University Press.

Nesov, L.A. and Kaznyshkin, M.N. 1986. [Discovery in the USSR. of strata with the egg remains of the Early Cretaceous and Late Cretaceous dinosaurs.] *Izvestiya Biologicheskikh Nauk* 9: 35–49.

Norell, M.A., Clark, J.M. and Dashzeveg, D. 1995. A nesting dinosaur. *Nature* 378: 774–776.

—, Clark, J.M., Dashzeveg, D., Barsbold, R., Chiappe, L.M., Davidson, A.R., McKenna, M.C., Perle, A. and Novacek, M.J. 1994. A theropod dinosaur embryo and the affinities of the Flaming Cliffs dinosaur eggs. *Science* 266: 779–782.

Sabath, K. 1991. Upper Cretaceous amniotic eggs from the Gobi Desert. *Acta Palaeontologica Polonica* 36: 151–192.

Sadov, I.A. 1970 [On the eggshell structure of extinct reptiles and birds.] *Paleontologicheskii Zhurnal* 1970(4): 88–91.

Shuvalov, V.E. 1982. [Paleogeography and history of the development of the lake systems of Mongolia in the Jurassic and Cretaceous.] pp. 18–80 in Martinson, G.G. (ed.). *Mezozoiskie ozernye basseiny Mongolii*. Leningrad: Nauka.

Sochava, A.V. 1969. [Dinosaur eggs from the Upper Cretaceous of the Gobi Desert.] *Paleontologicheskii Zhurnal* 1969(4): 517–527.

—1971. [Two types of egg shells in Cenomanian dinosaurs.] *Paleontologicheskii Zhurnal* 1971(3): 353–361.

—1972. [The skeleton of an embryo in dinosaur egg.] *Paleontologicheskii Zhurnal* 1972(4): 527–533.

Straelen, V. van 1925. The microstructure of the dinosaurian egg-shells from the Cretaceous beds of Mongolia. *American Museum Novitates* 173: 1–4.

Thulborn, R.A. 1991. The discovery of dinosaur eggs. *Modern Geology* 16: 113–126.

Young, C.C. 1954. Fossil reptilian eggs from Laiyang, Shantung China. *Scientia Sinica* 3: 505–522.

—1959. On a new fossil egg from Laiyang Shantung. *Vertebrata PalAsiatica* 3: 34–35.

—1965. [Fossil eggs from Nanhsiung Kwangtung and Kanchou Kiangsi.] *Vertebrata PalAsiatica* 9: 141–170.

Zeng, D.M. and Zhang J.J. 1979. [On the dinosaurian eggs from the Western Dongting Basin, Hunan.] *Vertebrata PalAsiatica* 17:77–84.

Zhao, Z.K. 1975. [The microstructure of the dinosaurian eggshells of Nanxiong Basin, Guandong Province.] *Vertebrata PalAsiatica* 13: 105–117.

—1979. [The advancement of researches on the dinosaurian eggs in China.] pp. 330–340 in IVPP and NGPI (eds.), [*Mesozoic and Cenozoic Red beds in Southern China*]. Peking: Science Press.

—1993. Structure, formation and evolutionary trends of dinosaurian eggshells, pp. 195–212 in Kabayashi, I., Mutvei, H. and Sahni, A. (eds.), *Structure, Formation and Evolution of Fossil Hard Tissues*. Tokyo: Tokai University Press.

—1994. Dinosaur eggs in China: on the structure and evolution of eggshells, pp. 184–203 in Carpenter, K., Hirsch, K.F. and Horner, J.R. (eds.), *Dinosaur Eggs and Babies*. Cambridge: Cambridge University Press.

—and Ding, S.R. 1976. [Discovery of the dinosaurian eggs from Alashanzuoqi (Alxa, Ningxia) and its stratigraphical meaning.] *Vertebrata PalAsiatica* 14: 42–45.

—and Jiang, Y.K. 1974 [=Chao, T.K. and Chiang, Y.K. 1974]. Microscopic studies on the dinosaurian eggshells from Laiyang, Shandong Province. *Scientia Sinica* 17: 73–83.

—and Li, R. 1988. [A new structural type of dinosaur eggs from Anly County, Hubei Province.] *Vertebrata PalAsiatica* 26: 107–115.

—1993. [First record of Late Cretaceous hypsilophodontid eggs from Bayan Mandahu, Inner Mongolia.] *Vertebrata PalAsiatica* 31: 77–84.

29

Mammals from the Mesozoic of Mongolia

ZOFIA KIELAN-JAWOROWSKA, MICHAEL J. NOVACEK, BORIS A. TROFIMOV AND
DEMBERLYIN DASHZEVEG

Introduction

Mongolia produces one of the world's most extraordinarily preserved assemblages of Mesozoic mammals. Unlike fossils at most Mesozoic sites, many of these remains are skulls, and in some cases these are associated with postcranial skeletons. By contrast, Mesozoic mammals at well-known sites in North America and other continents have produced less complete material, usually incomplete jaws with dentitions, or isolated teeth. In addition to the rich samples of skulls and skeletons representing Late Cretaceous mammals, certain localities in Mongolia are also known for less well preserved, but important, remains of Early Cretaceous mammals. The mammals from both Early and Late Cretaceous intervals have increased our understanding of diversification and morphologic variation in archaic mammals. Potentially this new information has bearing on the phylogenetic relationships among major branches of mammals.

Simpson (1925a) described the first Mesozoic mammal from Mongolia. This was a skull associated with fragments of the postcranial skeleton of a multituberculate, *Djadochtatherium matthewi*, collected in 1923 at Bayan Zag (known also as Shabarakh Usu) in the Gobi Desert, from rocks of the Upper Cretaceous Djadokhta Formation (Figure 29.1). The specimen was found by the Central Asiatic Expedition organized by the American Museum of Natural History in New York (1921–1930), and led by Roy Chapman Andrews (see Chapter 12). Principal field participants included Walter Granger who, with a small team, recovered eight additional mammal specimens in 1925. Gregory

and Simpson (1926) described these as placental (eutherian) insectivores. The deltatheroids originally included with the insectivores, more recently have been assigned to the Metatheria (Kielan-Jaworowska and Nesov, 1990). For about 40 years these were the only Mesozoic mammals known from Mongolia.

The next discoveries in Mongolia were made by the Polish–Mongolian Palaeontological Expeditions (1963–1971) initially led by Naydin Dovchin, then by Rinchen Barsbold on the Mongolian side, and Zofia Kielan-Jaworowska on the Polish side, Kazimierz Kowalski led the expedition in 1964. Late Cretaceous mammals were collected in three Gobi Desert regions: Bayan Zag (Djadokhta Formation), Nemegt and Khulsan in the Nemegt Valley (Baruungoyot Formation), and Hermiin Tsav, south-west of the Nemegt Valley, in the Red beds of Hermiin Tsav, which have been regarded as a stratigraphic equivalent of the Baruungoyot Formation (Gradziński *et al.*, 1977). The mammal collection made by these expeditions from all these sites contains about 170 specimens representing multituberculates and therians (Kielan-Jaworowska 1974, 1984c; Kielan-Jaworowska and Gambaryan, 1994; Gambaryan and Kielan-Jaworowska, 1995; Kielan-Jaworowska and Hurum, 1997, and references therein).

In 1968 the Soviet–Mongolian Geological Expedition found the skull of a multituberculate, *Buginbaatar transaltaiensis* at Khaichin Uul in the Bügiin Tsav region (referred to also as Bugin Cav), north-west of the Nemegt Basin (Kielan-Jaworowska and Sochava, 1969; Kielan-Jaworowska, 1974; Trofimov, 1975). The beds at Khaichin Uul are referred to the Upper Cretaceous and are equivalent

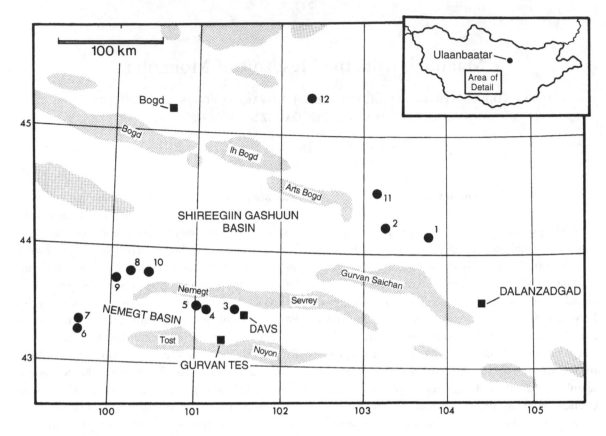

Figure 29.1. Diagrammatic map of mammal-bearing Mesozoic localities in Mongolia. (1) Bayan Zag, (2) Tögrögiin Shiree, (3) Ukhaa Tolgod, (4) Khulsan, (5) Nemegt, (6) Hermiin Tsav II, (7) Hermiin Tsav 1, (8) Bügiin Tsav, (9) Khaichin Uul, (10) Guriliin Tsav, (11) Üüden Sair, (12) Höövör.

to the Nemegt Formation (Gradziński *et al.*, 1977). Fossil mammals, however, were not found in the Nemegt Formation in the Nemegt Basin.

From 1969 until 1996, the Soviet–Mongolian Palaeontological Expeditions (SMPE), led by Rinchen Barsbold and Demberlyin Dashzeveg on the Mongolian side and by various Russian scientists, principally Valerii Reshetov, worked in Mongolia. They discovered an Early Cretaceous (Aptian or Albian) mammal site at Höövör (known also as Khovboor or Khoobur, and Guchin Us) in Guchinus Sum (county) in the Gobi Desert. This site, explored subsequently also by D. Dashzeveg (referred to further as D.D.) and the Mongolian Academy–American Museum Expeditions (see below) yielded numerous

eutherian and multituberculate mammals, more rare 'triconodonts' (now regarded as a polyphyletic group), very rare 'eupantotheres' (a paraphyletic group), symmetrodonts and aegialodontids (Belyaeva *et al.*, 1974; Dashzeveg, 1975, 1979, 1994; Trofimov, 1978, 1980; Dashzeveg and Kielan-Jaworowska, 1984; Kielan-Jaworowska *et al.*, 1987; Kielan-Jaworowska and Dashzeveg, 1989, 1998; Sigogneau-Russell *et al.*, 1992; Wible *et al.*, 1995). By contrast to Late Cretaceous mammals that were collected from the surface of outcrops and are represented by skulls, often associated with postcranial skeletons, the fossils from Höövör were collected using washing and screening techniques and consisted, by 1974, of about 500 isolated teeth and bones (Belyaeva *et al.*, 1974).

Another Early Cretaceous, though less rich, mammal locality was discovered by the SMPE at Khamaryn Us (= Gashuuny Khudag). situated southwest of the city of Sainshand in southeastern Mongolia. Only one mammal remain, a fragment of a dentary with m2 and m3 ('Triconodonta', possibly Amphilestidae) has been described (Reshetov and Trofimov, 1980) from this site.

The SMPE also collected Late Cretaceous mammals at various sites. Most important among these was the discovery at Üüden Sair, in beds corresponding to the Baruungoyot Formation, of the skull and postcranial skeleton of the first Late Cretaceous non-deltatheroid metatherian from Asia (Trofimov and Szalay, 1994; Szalay and Trofimov, 1996), and the skull of a large deltatheroidan in beds possibly equivalent to the Nemegt Formation, at Guriliin Tsav (= Gurlin Cav), north of the Nemegt Basin (Anonymous, 1983; Kielan-Jaworowska and Nesov, 1990; Szalay and Trofimov, 1996).

Finally, at the locality of Shar Teeg in the Trans-Altaian Gobi, the SMPE found the first and only known Jurassic mammal from Mongolia, a single tooth of a docodont that was initially identified as a symmetrodont (Tatarinov, 1994).

These samples were augmented by additional Late Cretaceous mammals collected by D.D. at Tögrögiin Shiree (= Tugrugeen Shireh, Tugrik) in beds equivalent to the Djadokhta Formation, and at other sites (Kielan-Jaworowska and Dashzeveg, 1978). A few Cretaceous mammals were found by other expeditions including the Italian–French–Mongolian Expedition in 1991 (Taquet, 1994) and by the Japanese–Mongolian Expeditions in 1993–1996 (Mahito Watabe, pers. comm. to Z.K.J.).

The American Museum of Natural History and the Institute of Geology of the Mongolian Academy of Sciences carried out ten palaeontological expeditions (1990–1999), under the guidance of Demberlyin Dashzeveg and Michael J. Novacek. This collaboration, known as the Mongolian Academy–American Museum Expeditions (MAE), has produced a spectacular collection of Late Cretaceous mammals from previously known and new localities, as well as large

samples of Early Cretaceous mammals from Höövör. The quantities of Late Cretaceous mammals recovered exceed in numbers the collections of all previous expeditions. For example, Ukhaa Tolgod, a new locality discovered in 1993, situated at the eastern part of the Nemegt Basin, east of Khulsan, has yielded over 800 mammal specimens in an unusually good state of preservation (Dashzeveg et al., 1995; Rougier et al., 1996a, b, 1997, 1998; Novacek et al., 1997).

During September of 1995 a small Mongolian–Polish team (Yo. Khand, H. Osmólska, T. Maryańska and K. Sabath) collected fossils at various localities in the Gobi Desert, including Ukhaa Tolgod where they recovered remains of nine multituberculates and one placental (Kielan-Jaworowska and Hurum, 1997; Kielan-Jaworowska, 1998).

It should also be noted that fossil mammals have been reported, but not described from Bayan Mandahu, a locality in northern China near the Mongolian–Chinese border, and thought to be equivalent in age to the Djadokhta Formation (Jerzykiewicz et al., 1993; Wang et al., 1995) About 50 skulls, three or four with incomplete skeletons, were tentatively identified as taeniolabidoid multituberculates, although one specimen was identified as the eutherian *Kennalestes*.

Gradziński et al. (1977) refined the descriptions of three Upper Cretaceous Gobi Desert formations. They suggested as 'best guesses' the following ages: Djadokhta Formation = ?upper Santonian and/or ?lower Campanian; Baruungoyot Formation = ?middle Campanian; Nemegt Formation = ?upper Campanian and ?lower Maastrichtian (see, however, Fox, 1978; Lillegraven and McKenna, 1986; Jerzykiewicz et al., 1993, and Chapters 14 and 15 for alternative age estimates). These assignments were based on fresh-water invertebrates and comparisons of dinosaurs and mammals with those from European and North American assemblages, essentially above the generic level. Unfortunately, without a context provided by palaeomagnetic, radiometric, or marine tie-ins, such correlations are tenuous at best.

The stratigraphic scheme of Gradziński et al. (1977) has been questioned by members of the Mongolian

Academy–American Museum Project. It appears from preliminary identifications of the fauna collected at Ukhaa Tolgod, that this locality provides a mixture of taxa known from both the Djadokhta and Baruungoyot formations (Novacek *et al.*, 1994; Dashzeveg *et al.*, 1995). On this basis it has been stated: 'Sampling at Ukhaa Tolgod, as well as ongoing work at other localities, blurs the distinction between Djadokhta assemblages and allegedly slightly younger Baruungoyot assemblages' (Dashzeveg *et al.*, 1995, p. 447). This statement contradicts the reported lack of overlap between taxa based on more than 200 mammal skulls, collected by previous expeditions from the Djadokhta and Baruungoyot formations and housed in the museums in New York, Warsaw, Moscow and Ulaanbaatar. The only mammal species previously cited as occurring in both formations is *Deltatheridium pretrituberculare*, stated by Kielan-Jaworowska (1975a) to be represented by different subspecies *(D. p. pretrituberculare* and *D. p. tardum)* in each formation, but demonstrated by Rougier *et al.* (1998) to be a single species.

Rougier *et al.* (1997) referred to the Ukhaa Tolgod beds as 'Djadokhta Formation equivalent?' and the mammal fauna yielded by this locality contains taxa mostly known from the Djadokhta Formation. In addition, these beds also contain new, undescribed taxa, as well as very rare specimens of *Chulsanbaatar* and *Nemegtbaatar*, previously known only from the Baruungoyot Formation and Red beds of Hermiin Tsav. However, a comprehensive analysis of the mammal assemblage from Ukhaa Tolgod has not been published, and the stratigraphic relationships of this section are the subject of ongoing analysis.

Aver'yanov (1997) described several mammal teeth from the Late Cretaceous Darbasa Formation of Alymatau in southern Kazakhstan. The age of this formation was established on the basis of marine invertebrates and shark teeth as early Campanian. Aver'yanov argued that the presence of the multituberculate *Bulganbaatar* in the Alymatau fauna, which occurs in the Djadokhta Formation and is not known from the Baruungoyot Formation suggests that the Alymatau fauna may be better correlated with that of the Djadokhta Formation. He also stated that the

Campanian is a marine stage established in Europe and divided on the basis of ammonites and other marine invertebrates into two substages. Therefore, the tripartite division of the Campanian in Asia and North America is inappropriate. Finally Aver'yanov (1997) stated: 'The evidence from this paper, although inconclusive as the tentative correlation presented above is on the generic level, supports in part the conclusion of Gradziński *et al.* (1977) in suggesting a somewhat earlier, perhaps early Campanian age for the Djadokhta Formation'.

In the present account we continue to recognize both the Djadokhta and Baruungoyot formations. We refer to the Djadokhta Formation as ?early Campanian and the Baruungoyot Formation as ?late Campanian, with the assignment of Ukhaa Tolgod pending faunal and stratigraphic analysis.

Abbreviations

I, C, P, M are used here for the upper incisors, canines, premolars and molars respectively and i, c, p, m, for the corresponding lower teeth. Mt I–V, metatarsals I–V. Institutions in which the collections are housed are abbreviated: AMNH, American Museum of Natural History, New York; PIN, Paleontological Institute of the Russian Academy of Sciences, Moscow; PSS, Institute of Geology, Section of Palaeontology and Stratigraphy, Academy of Sciences, Ulaanbaatar (abbreviated also as GI or GI PST); ZPAL, Institute of Paleobiology, Polish Academy of Sciences, Warsaw.

Systematic review

Subclass *incertae sedis*
Order Docodonta Kretzoi, 1958

Comments. Docodonts are a specialized group of very small mammals with molars that superficially resemble those of advanced Theria (Butler, 1988, and references therein; Sigogneau-Russell and Godefroit, 1997) and are known only from the Late Triassic, and the Middle and Late Jurassic. Lillegraven and Krusat (1991) regarded the docodonts as a sister group of all other

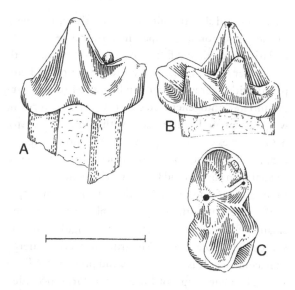

Figure 29.2. *Tegotherium gubini* Tatarinov, 1994, holotype right lower molar, PIN 4174/167, in (A) labial, (B) lingual and (C) occlusal views. Scale bar = 1 mm. (Drawn by L.P. Tatarinov.)

mammals, but Wible and Hopson (1993) and Wible *et al.* (1995) found evidence inconsistent with that conclusion and presented a tree (Wible and Hopson, 1993) in which the docodontid *Haldanodon* formed a sister taxon to triconodontids and other mammals to the exclusion of morganucodontids and *Sinoconodon*.

Family Docodontidae (Marsh, 1887)
Genus *Tegotherium* Tatarinov, 1994
Tegotherium gubini Tatarinov, 1994
See Figure 29.2.

Holotype (and only known specimen). PIN 4174/167, right lower molar. Shar Teeg, Trans-Altaia Gobi, Mongolia; Late Jurassic.

Description. The lower molar, which is 1.25 mm long, has three main cusps with an anterior basin situated between and in front of them. The principal cusp is the highest and placed posterolabially; it is conical and covered lingually with longitudinal ridges. On the margin of the crown there are two conical cusps. They are lower than the principal cusp and the posterior is situated opposite the principal cusp while the other is located more anteriorly. There are also two, poorly

defined cuspules, situated in front of the principal cusp and forming crenulations on the anterolabial margin of the anterior basin. There is a very small posterior cingulid. The tooth is similar to *Simpsonodon* from the Middle Jurassic of southern England (Kermack *et al.,* 1987), from which it differs in having three, rather than four main cusps (the anterolabial cusp of *Simpsonodon* being developed as two crenulations on the anterolabial margin of the anterior basin), a larger anterior basin, a less prominent posterior cingulum and a relatively higher principal cusp.

Comments. *Tegotherium gubini* is the only representative of the Docodonta in Mongolia. Tatarinov (1994) referred *Tegotherium* to the Symmetrodonta. He erected for it a monotypic family Tegotheriidae and a monotypic order Tegotheriidia. Hopson (1995) argued that *Tegotherium* was closely related to *Simpsonodon* from Oxfordshire (Kermack *et al.*, 1987), and that together with *Borealestes* they form a distinct clade of docodonts. We agree and we assign *Tegotherium* to the docodonts.

'Triconodonts' (= polyphyletic order Triconodonta Osborn, 1888)

Comments. Among Mesozoic mammals the 'triconodonts', especially the Morganucodontidae (assigned by Kermack *et al.*, 1973, to the order Morganucodonta), are most similar to the cynodonts. 'Triconodonta' are polyphyletic, however (e.g., Rowe, 1988; Rougier *et al.*, 1992; Wible and Hopson, 1993; Wible *et al.*, 1995; Kielan-Jaworowska, 1997), and are thus cited here in quotation marks. The Late Triassic–Early Jurassic Morganucodontidae and Megazostrodontidae, both of which retain a double jaw joint (Kermack *et al.*, 1973; Jenkins *et al.*, 1983) differ in many respects from the more modern, Late Jurassic–Early Cretaceous lineages Triconodontidae and Amphilestidae, which are characterized by a typical mammalian jaw joint (Jenkins and Crompton, 1979; Jenkins and Schaff, 1988).

Family Amphilestidae Osborn, 1888

Comments. Mills (1971) argued that the Amphilestidae would appear to belong to Pantotheria, as they exhibit a tooth morphology similar to that of *Kuehneotherium* and the 'obtuse-angled' symmetrodonts such as *Tinodon*. Jenkins and Crompton (1979), Jenkins and Schaff (1988) and Kielan-Jaworowska and Dashzeveg (1998) noted the similar occlusal pattern evident in symmetrodonts and amphilestids and the latter authors demonstrated that the interlocking mechanism of the lower molariforms in *Gobiconodon* resembles that of *Kuehneotherium* and *Tinodon*. Kielan-Jaworowska and Dashzeveg (1998) further stated that: 'The data presented in this paper give some support to Mills' idea on the therian affinities of the Amphilestidae although it cannot be excluded that the characters that unite the two groups may have developed in parallel'.

Subfamily Gobiconodontinae Chow and Rich, 1984
Comments. Jenkins and Schaff (1988) erected a new family, Gobiconodontidae (neglecting Gobiconodontinae Chow and Rich, 1984), to include *Gobiconodon* Trofimov, 1978, and, tentatively, *Guchinodon* Trofimov, 1978. Kielan-Jaworowska and Dashzeveg (1998) restored the subfamily status of Gobiconodontinae within Amphilestidae. The Gobiconodontinae share with known amphilestids the basic structure of the molars, but differ from them in small details of molar structure and in enlargement of the most mesial lower tooth. Kielan-Jaworowska and Dashzeveg considered these characters sufficient to warrant subfamilial, but not familial status. *Klamelia* Chow and Rich, 1984, assigned by its authors to the Gobiconodontinae, probably does not belong to this subfamily (Jenkins and Schaff, 1988).
Diagnosis. See Jenkins and Schaff (1988).

Genus *Gobiconodon* Trofimov, 1978
Synonym. Guchinodon Trofimov, 1978.
Type species. Gobiconodon borissiaki Trofimov, 1978.
Other species. G. hoburensis (Trofimov, 1978); *G. ostromi* Jenkins and Schaff, 1988.
Diagnosis. 'Very small to medium-sized (estimated skull length varies between 27 mm in *G. hoburensis* and 106 mm in *G. ostromi*) amphilestid 'triconodonts' with five rounded fossae on the palatal part of maxilla, situ-

ated close and slightly shifted posteriorly with respect to the corresponding upper molariform teeth. Five molariform teeth and five to six antemolariform teeth in the dentary. The i1 and c are semi-procumbent, p1–p3 with decreasing procumbency, p4 (disappearing in later ontogenetic stages in *G. ostromi*) vertical, with 3 cusps, molariform teeth with four or five cusps, m3 the largest. Main cusps in M3–M5 show incipient triangular pattern, with cusp A placed slightly more lingual than cusps B and C. Interlocking mechanism of lower molariforms of *Kuehneotherium* type, with cusp d of the anterior tooth fitting into embayment between small cusps e and f of the anterior cingulum of the succeeding molariform. Molariform teeth undergo replacement at least in *G. ostromi* and probably *G. borissiaki*. The main cusp a of lower molariforms occluded immediately in front of the distal margin of the corresponding upper molariform, between posterior cingulum, on which there is a small cusp D, and cusp C, rather than between cusps A and B, as in Morganucodontidae and Triconodontidae' (Kielan-Jaworowska and Dashzeveg, 1998).

Comments. Kielan-Jaworowska and Dashzeveg (1998) argued that *Guchinodon hoburensis* Trofimov, 1978 differs from *Gobiconodon borisiaki* Trofimov, 1978 in that it is almost twice as small, has a double-rooted p4, and by details of the position of the mental and infraorbital foramina. The dental formulae, arrangement of the teeth and molar structure are similar in both taxa, and there are only small differences in the relative heights of cusps b and c, resulting in a slightly different occlusal pattern. A specific rather than generic value was assigned to these differences and on this basis Kielan-Jaworowska and Dashzeveg (1998) regarded *Guchinodon* Trofimov, 1978, as a junior subjective synonym of *Gobiconodon* Trofimov, 1978, which we also accept here.

Gobiconodon borissiaki Trofimov, 1978
See Fig. 29.3.A.
Holotype. PIN 3101/09, incomplete right dentary. Höövör, Guchinus county, Gobi Desert, Mongolia; Höövör beds (Aptian or Albian).
Referred material. Several incomplete dentaries and

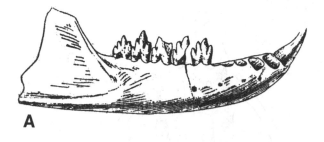

Figure 29.3. *Gobiconodon borissiaki* Trofimov, 1978, holotype, PIN 3101/09, incomplete right dentary in (A) labial view. *Guchinodon hoburensis* Trofimov, 1978, holotype, PIN 3101/24, incomplete right dentary in (B) labial view. Scale bars = 5 mm. (From Trofimov, 1978.)

fragments of maxillae with teeth from the type horizon and locality in the PIN and PSS collections. A dentary with broken teeth (in the collections of PIN) from the Early Cretaceous Shestakovo locality in Siberia was recently described by Mashchenko and Lopatin (1998).

Diagnosis. 'Medium sized *Gobiconodon*, estimated length of the skull about 48–50 mm. Differs from *G. ostromi* and *G. hoburensis* in dimensions, being approximately intermediate in size between these taxa. Differs from *G. hoburensis* and *G. ostromi* in having a single-rooted p4 (in *G. ostromi* it is incipiently double-rooted, either single-rooted or absorbed), which apparently does not disappear in adult ontogenetic stages, as characteristic of *G. ostromi* (but not of *G. hoburensis*). Differs from *G. ostromi*, apparently, in having more prominent cusps e and f on the anterior cingulum of m2–m5. The difference in size between i1 and c is smaller than in *G. ostromi*. Differs from *G. hoburensis* in having much more pronounced cusps b and c on lower molariform teeth, and cusp b on m1 placed lower. It shares these latter characters with *G. ostromi*' (Kielan-Jaworowska and Dashzeveg, 1998).

Gobiconodon hoburensis (Trofimov, 1978)
See Fig. 29.3B.

Holotype. PIN 3101/24, incomplete right dentary, Höövör, Guchinus county, Gobi Desert, Mongolia; Höövör beds (Aptian or Albian).

Referred material. Several incomplete dentaries and maxillae with teeth in the PIN and PSS collections, all from the type horizon and locality.

Diagnosis. 'Differs from *G. borissiaki* in being about 1.8 times smaller and in having a double-rooted p4; differs from *G. ostromi* in being almost 3 times smaller and in having p4 which apparently does not disappear during ontogeny. Differs from *G. borissiaki* and *G. ostromi* in having all the cusps (except cusp a) in lower molariform teeth less prominent, and cusp b in m1 situated relatively higher. The dentary is relatively wider mesially than in *G. borissiaki* and *G. ostromi*, which is related to a medial shift of i1 with respect to other teeth (it is more aligned in *G. borissiaki* and especially in *G. ostromi*). The three mental foramina are shifted slightly more anteriorly than in the two other species, especially *G. ostromi*. The palatal fossae are relatively deeper than in *G. borissiaki* (poorly known in *G. ostromi*). Differs from *G. borissiaki* in having the infraorbital foramen situated more anteriorly, above the P3–P4 embrasure, rather than above M1 (position not known in *G. ostromi*)' (Kielan-Jaworowska and Dashzeveg, 1998).

Subclass Allotheria Marsh, 1880
Order Multituberculata Cope, 1884

Comments. The Multituberculata were a very diverse group of Mesozoic and Early Tertiary mammals and are often referred to as the 'rodents of the Mesozoic'.

The multituberculate skull has been described by Granger and Simpson (1929), Simpson (1937), Kielan-

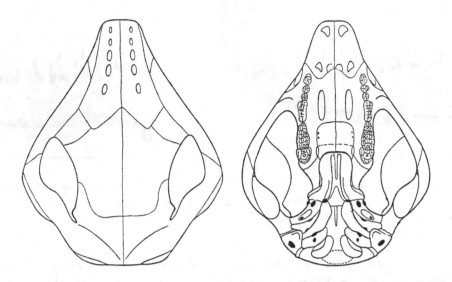

Figure 29.4. *Nemegtbaatar gobiensis* Kielan-Jaworowska, 1974. Diagrammatic reconstruction of the skull in dorsal (left) and ventral (right) views. The skull is about 40 mm long.

Jaworowska (1970, 1971, 1974), Kielan-Jaworowska and Dashzeveg (1978), Clemens and Kielan-Jaworowska (1979), Kielan-Jaworowska and Sloan (1979), Sloan (1979), Kielan-Jaworowska *et al.* (1986), Miao (1988, 1991, 1993), Hahn (1993, and references therein), Gambaryan and Kielan-Jaworowska (1995), Hurum (1994, 1998a, b), Râdulescu and Samson (1996), Kielan-Jaworowska and Hurum (1997), Rougier *et al.* (1997). The skull (Figures 29.4–29.6, 29.10–29.13) is dorsoventrally compressed, rather than laterally compressed as in therian mammals, and the postorbital process in most multituberculates is situated on the parietal rather than on the frontal as in therians. The anterior part of the orbit extends anteriorly, at least in all well preserved skulls of djadochtatherians (see below), as a pocket-like structure (orbital pocket) that has no floor, but is roofed dorsally and laterally (Gambaryan and Kielan-Jaworowska, 1995), in contrast to almost all therians. The zygomatic arches are stout, the jugal is placed on the medial wall of the zygomatic arch (Hopson *et al.*, 1989), the glenoid fossa is large and flat (rather than concave as in most therians), slopes backwards and stands out from the braincase. The cochlea is only slightly bent, the anterior lamina of the petrosal contributes largely to the lateral wall of the

braincase and there are three ear ossicles (Miao and Lillegraven, 1986; Miao, 1988) that are arranged as in modern mammals (Meng and Wyss, 1995; Hurum *et al.*, 1996; Rougier *et al.*, 1996b). The condylar process is low and forms at least part of the posterior margin of the dentary, which in most forms is semicircular. There is no angular process and the ventral margin of the dentary is inflected and forms the pterygoideus shelf for insertion of the pterygoid muscles.

The premolars and molars carry numerous cusps and, except in some Jurassic Paulchoffatiidae, they are of subequal height. In Jurassic taxa there are three pairs of upper incisors, but in later forms there are only two pairs, and one pair of lower incisors. More specialized conditions for multituberculates are represented by reduction in the number of premolars, presence of a diastema and cheek teeth that are longer anteroposteriorly and with a greater number of cusps. Multituberculates had a backward masticatory power stroke (Gingerich, 1977; Krause, 1982). In relation to this, all the masticatory muscles inserted more anteriorly than in other mammals, and the masseteric fossa and coronoid process are placed more anteriorly as well (Gambaryan and Kielan-Jaworowska, 1995).

The multituberculate brain (Figure 29.6) differs

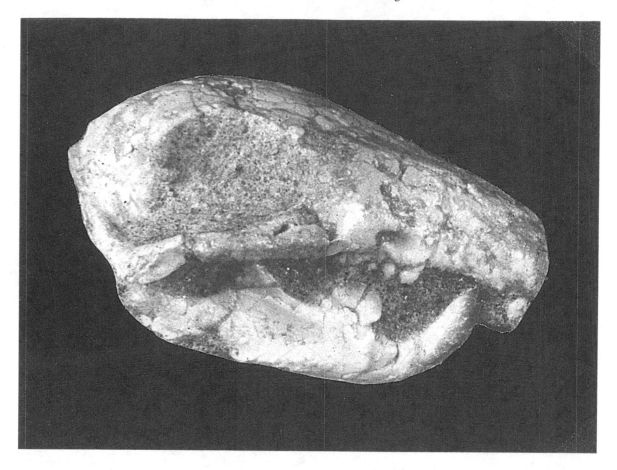

Figure 29.5. *Sloanbaatar mirabilis* Kielan-Jaworowska, 1970, holotype, ZPAL MgM-I/20, skull with lower jaws in lateral view. The skull is about 23 mm long. (From Kielan-Jaworowska, 1970.)

from that of primitive therian mammals in having a strongly expanded vermis, in lacking midbrain exposure on the dorsal side, and in the absence of the cerebellar hemispheres (Simpson, 1937; Kielan-Jaworowska, 1986). A similar pattern is known only in the triconodontid *Triconodon* (Simpson, 1927), indicating the possible relationship of multituberculates with the Triconodontidae, but not with other 'triconodonts' (Kielan-Jaworowska, 1997).

The multituberculate postcranial skeleton retains many characters that are primitive for mammals (Krause and Jenkins, 1983). For example, the interclavicles are present (Sereno and McKenna, 1995). The pes (Figure 29.7A, B) (Kielan-Jaworowska and Gambaryan, 1994) differs from that of all other

mammals in having a calcaneo-MtV contact and the MtIII abducted 30° from the longitudinal axis of the tuber calcanei (and the longitudinal axis of the body). However, the discovery of an almost complete triconodont skeleton by Ji *et al.* (1999, fig. 4) demonstrates the presence of a pes showing similar bone arrangements to that of multituberculates. This position of the pes would be ineffective in parasagittal limbs, where the main axis of the pes extends in a parasagittal plane. Similarly, the deep multituberculate pelvis with femoral adductors originating ventral to the acetabulum and the larger mediolateral rather than anteroposterior diameter of the tibia also indicate abduction of the hind limbs (Figure 29.7C). On this basis, Kielan-Jaworowska and Gambaryan (1994)

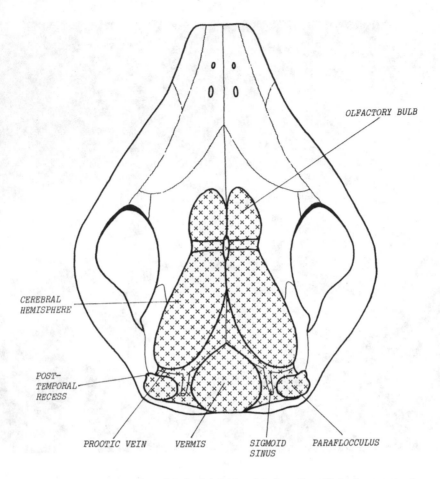

OLFACTORY BULB

CEREBRAL
HEMISPHERE

POST-
TEMPORAL
RECESS

PROOTIC VEIN   VERMIS   SIGMOID   PARAFLOCCULUS
                            SINUS

Figure 29.6. Reconstruction of the brain of the multituberculate *Chulsanbaatar vulgaris*.
The skull is about 20 mm long. (Modified from Kielan-Jaworowska, 1986.)

reconstructed the multituberculate stance as sprawling (Figure 29.8) and argued that the structure of the
lumbar vertebrae, with long transverse and high
spinous processes shows that multituberculates were
adapted for asymmetrical gaits with steep jumps.

Sereno and McKenna (1995) demonstrated that in
the Late Cretaceous Mongolian taxon cf. *Bulganbaatar*
the humerus shows only a small degree of torsion
(15°) and on this basis concluded that multituberculate forelimbs operated in a parasagittal fashion.
However, the humerus of cf. *Bulganbaatar* differs considerably from that of *?Lambdopsalis,* a Palaeocene
multituberculate from China in which the torsion
approaches 38° and from that of *Kryptobataar*, from the

Late Cretaceous of Mongolia, in which humeral
torsion is 31° (Kielan-Jaworowska, 1998). Gambaryan
and Kielan-Jaworowska (1997) recognized several features characteristic of the humeri that indicate a primarily sprawling stance. The most important of these
is the presence of prominent radial and ulnar condyles, and absence of a trochlea, traits found in parasagittal forms. Torsion of the humerus occurs in
terrestrial tetrapods with abducted forelimbs, that use
symmetrical diagonal gaits, but not in anurans, which
have abducted forelimbs, but use asymmetrical jumps,
and, with the exception of Chrysochloridae, not in
fossorial therians with a sprawling or semi-sprawling
stance. Therefore Gambaryan and Kielan-Jaworowska

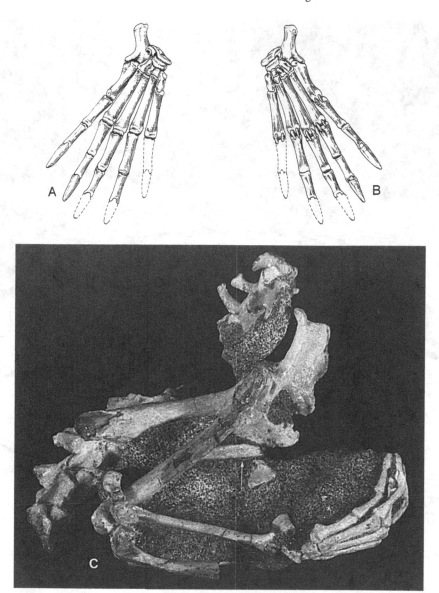

Figure 29.7. *Kryptobaatar dashzevegi* Kielan-Jaworowska, 1970, ZPAL MgM-I/41.
Reconstruction of the right pes in (A) dorsal and (B) plantar views (based on the specimen
in C). Pelvic girdle and hind limbs in (C) left lateral view. The arrow points to the epipubic
bone. × 2.5. (Modified from Kielan-Jaworowska and Gambaryan, 1994.)

(1997) argued that the lack of torsion is not always indicative of parasagittalism. They further argued that, as there is no trace of even an incipient trochlea in any known multituberculate (the trochlea made its appearance in therians possibly during the Late Jurassic), the idea that parasagittalism occurred in mammalian evolution in common ancestors of therians and multituberculates does not hold.

In the multituberculate pelvic girdle, epipubic ('marsupial') bones are present (Kielan-Jaworowska,

Figure 29.8. Reconstruction of the posture of the multituberculate *Nemegtbaatar*. (Modified from Kielan-Jaworowska and Gambaryan, 1994.)

1969b; see also Figure 29.7C). The pelvis is deep and very narrow, with a small ischial arc and a process-like ischial tubercle, and there is a long ischiopubic symphysis with a ventral ridge. The length of this ridge and the degree of fusion indicate that the pelvis was rigid and could not have spread apart during parturition. The profile of the passage within the ischial arc is triangular and the space available for the passage of an egg is so small, that any known cleidoic egg could not have passed through it. On this basis, Kielan-Jaworowska (1979) concluded that multituberculates may have been viviparous.

The origin and relationships of multituberculates to other mammals is controversial. In recent analyses, multituberculates have been regarded as: (1) a sister taxon of the Theria (e.g., Rowe, 1988, 1993; Lucas and Luo, 1993; Sereno and McKenna, 1995; Ji *et al.*, 1999); (2) a sister taxon of Monotremata + Theria (e.g., Miao, 1991; Wible, 1991; Lillegraven and Hahn, 1993); (3) a sister taxon of the Monotremata, both groups together being a sister taxon of the Theria (e.g., Kielan-Jaworowska, 1971; Kemp, 1983 – tentatively; Wible and Hopson, 1993; Meng and Wyss, 1995); (4) a sister taxon of all other mammals (e.g., McKenna, 1987; Miao, 1993; Kielan-Jaworowska and Gambaryan, 1994).

Information on Late Cretaceous Mongolian multituberculates has an important bearing on arguments regarding the relationships of multituberculates to other mammals and mammalian phylogenetics (Novacek, 1990). Data on the structure of the multituberculate skull, the brain (as inferred from endocranial casts), the ear region and the postcranial skeleton are based primarily on the rich collections of Late Cretaceous multituberculates from Mongolia, only a part of which have been described so far (Kielan-Jaworowska, 1970, 1971, 1974, 1986, 1994, 1998; Kielan-Jaworowska et al., 1986; Wible and Hopson, 1993; Hurum, 1994, 1998a, b; Kielan-Jaworowska and Gambaryan, 1994; Novacek et al., 1994; Dashzeveg et al., 1995; Sereno and McKenna, 1995; Gambaryan and Kielan-Jaworowska, 1995, 1997; Rougier et al., 1996b, 1997; Kielan-Jaworowska and Hurum, 1997).

Material of purported multituberculates from the Early and Middle Jurassic consists of a few fragmentary teeth known only from Europe (Freeman, 1979; Hahn et al., 1987; Kermack, 1988, and references therein). These teeth may be those of multituberculates, but this cannot be unequivocally demonstrated. However, P.M. Butler (pers. comm., 1999) informed us that in the collection of isolated teeth from Kirtlington, Oxfordshire, housed in the Natural History Museum, London, there are isolated lower and upper fourth premolars that belong to multituberculates. The next unequivocal multituberculates are from the Kimmeridgian of Portugal (Hahn, 1969, 1993, and references therein).

The South American Gondwanatheria have been tentatively assigned to 'Plagiaulacoidea' (e.g., by Krause et al., 1992), but according to Pascual et al. (1999) should be classified as Mammalia incertae sedia. Gondwanatheria include two families, the brachyodont Ferugliotheriidae and the advanced hyspodont Sudamericidae. Pascual et al. (1999) described a

dentary of a Palaeocene sudamericid Sudamerica, with two molariform teeth and two more molar loci posterior to them. Sudamerica apparently had four molariform teeth, a condition unknown for multituberculates. We agree with Pascual et al. (1999) that Gondwanatheria may not belong to Multituberculata, but we believe that some fossils from the Late Cretaceous Los Alamitos Formation of Argentina, tentatively attributed to Ferugliotheriidae, might be multituberculates. These are: upper premolars, with few conical cusps, similar to those of multituberculates (described by Krause et al., 1992), and a dentary with an alveolus for an incisor, a diastema, and a blade-like premolar with ridges (?p4), similar to those in multituberculates (described by Kielan-Jaworowska and Bonaparte, 1997).

In this review we regard Multituberculata Cope, 1884 as an order within the subclass Allotheria Marsh, 1880, and we adopt the following systematic arrangement.

1. Suborder 'Plagiaulacoidea' Simpson, 1925a, Hahn, 1969. This contains seven families: Paulchoffatiidae, Pinheirodontidae, Allodontidae, Plagiaulacidae, Eobaataridae, Zofiabaataridae, Albionbaataridae, and the genus Glirodon, assigned to 'Plagiaulacoidea' incertae sedis. 'Plagiaulacoidea' are known from the ?Middle Jurassic to Early Cretaceous of the Northern Hemisphere.

2. Suborder incertae sedis, family Arginbaataridae. Aginbaataridae are known only from the Early Cretaceous of Mongolia.

3. Suborder Cimolodonta McKenna, 1975, with three infraorders: Djadochtatheria, Taeniolabidoidea (limited to a sole family Taeniolabidae), and Ptilodontoidea; four families left in an infraorder incertae sedis: Cimolomyidae, Eucosmodontidae, Microcosmodontidae, and Kogaionidae; and several incompletely known genera, left in an infraorder and family incertae sedis. Cimolodonta are known from the Early Cretaceous to Eocene of the Northern Hemisphere.

4. ?Multituberculata incertae sedis – here we assign, tentatively, the multituberculate-like upper pre-

molars and a dentary with a premolar (?p4), that have been tentatively referred to the gondwanatherian genus *Ferugliotherium*. These fossils are known only from the Late Cretaceous of Argentina.

For further data the reader is referred to Simpson (1928, 1929), Sloan and Van Valen (1965), Hahn (1969, 1993, and references therein), Clemens and Kielan-Jaworowska (1979), Archibald (1982), Kielan-Jaworowska *et al.* (1987), Bakker (1992), Krause *et al.* (1992), Kielan-Jaworowska and Ensom (1992), Eaton (1995), Kielan-Jaworowska and Bonaparte (1996), Sigogneau-Russell (1991a), McKenna (1975), Rådulescu and Samson (1996), Rougier *et al.* (1997), Engelmann and Callison (1999), Hahn and Hahn (1999), and Fox (1999).

Suborder 'Plagiaulacoidea' Simpson, 1925a, Hahn, 1969

Comment. Simmons (1993) and Rougier *et al.* (1997) argued that 'Plagiaulacoidea' are paraphyletic; we follow this opinion and therefore cite this name in quotation marks. The subordinal name Plagiaulacoidea (Simpson, 1925a) has been commonly used for many years. However, the *International Commission on Zoological Nomenclature* (1999) recommends using the suffix 'oidea' for superfamilies. Therefore this name should be replaced by another one, adequate for the suborder. McKenna and Bell (1997) chose not to use 'Plagiaulacoidea' because this taxon is paraphyletic, using instead the family name Plagiaulacidae Gill, 1972, in which they included paulchoffatiid, allodontid, and some plagiaulacid genera. We do not support this procedure, believing that some 'plagiaulacoid' families are monophyletic.

'Plagiaulacoidea' are a grade group characterized by four blade-like lower premolars (three in advanced forms) that are rectangular in lateral view. These characters are apomorphic for Multituberculata, but plesiomorphic within the group.

Family Eobaataridae Kielan-Jaworowska *et al.*, 1987
Included genera. *Loxaulax* Simpson, 1928; *Eobaatar* Kielan-Jaworowska *et al.*, 1987; the poorly known *Monobaatar* Kielan-Jaworowska *et al.*, 1987, currently

assigned to subfamily *incertae sedis* may also belong here.
Geographical and stratigraphical range. Early Cretaceous of Europe and Asia.
Diagnosis (emended after Kielan-Jaworowska and Ensom, 1992). Eobaataridae differ from most Plagiaulacidae in having a limited enamel band on the lower incisor and share this character with *Glirodon* Engelmann and Callison, 1999 and many cimolodontans. The m1 and m2 are asymmetrical and shorter lingually than labially. The P5 lacks labial cuspules and M1 has a posterolingual wing that is more prominent than in the Plagiaulacidae. Eobaataridae differ from Paulchoffatiidae, Plagiaulacidae, and Ptilodontoidea in having gigantoprismatic enamel, and share this character with Arginbaataridae and most Cimolodonta.
Comments. Eobaataridae are morphologically intermediate between some Late Jurassic Plagiaulacidae and the Cretaceous–Tertiary Cimolodonta. Eobaataridae have three lower premolars, which are relatively smaller than those in the Jurassic Plagiaulacidae (which have three to four lower premolars), and the p4 is relatively large, parallel-sided and does not overhang the p3 (it does so primitively in Eucosmodontidae, Djadochtatheria, and Ptilodontoidea). There is no row of accessory basal cusps on p4 (a characteristic of the Plagiaulacidae), but the posterior basal cusp is present. There are five upper premolars (as in the Plagiaulacidae).

Genus *Eobaatar* Kielan-Jaworowska *et al.*, 1987
Type species. Eobaatar magnus Kielan-Jaworowska, Dashzeveg and Trofimov, 1987.

Eobaatar magnus Kielan-Jaworowska *et al.*, 1987
See Figure 29.9A.
Holotype. PIN 3101/57, left p4. Höövör, Guchinus county, Gobi Desert, Mongolia. Höövör beds (Aptian or Albian).
Referred material. About 20 isolated lower and upper teeth in the PIN and PSS collections. See Kielan-Jaworowska *et al.* (1987, p. 6) for a description of the method of matching isolated teeth.
Description. Estimated skull length is about 30 mm.

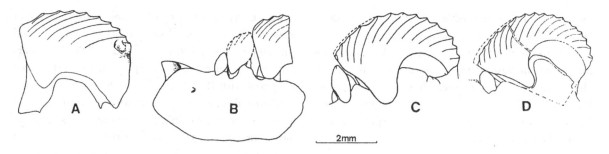

Figure 29.9. Comparison of p4 and adjacent teeth of Early Cretaceous multituberculates from Höövör, Mongolia. (A), *Eobaatar magnus*, holotype, PIN 3101/57, left p4. (B), *Eobaatar minor*, holotype, PIN 3101/70, fragment of the right lower jaw (reversed), with anterior part of p4, p2 and alveolus for incisor, p3 reconstructed. (C), *Arginbaatar dimitrievae*, holotype, PIN3101/51, right p2–p4 (reversed), preserved in lower jaw. (D), *Arginbaatar dimitrievae*, PSS 10–12, left p4 showing older ontogenetic stage than the specimen in Fig. C; p4 is rotated over reduced p3, while p2 has disappeared. (Modified from Kielan-Jaworowska *et al.*, 1987.)

The p4 is rectangular in labial view, 3.1–3.6 mm long, with 9–10 serrations and bearing ridges and a V-shaped groove above the basal cusp. There are 4:2 cusps on m1 and 3:2 on m2. The second labial cusp on m1 is larger than the others. M1 has 3–4 cusps and a small crescentic posterolingual wing. M2 has 1:3:3 cusps and a straight anterior margin. The lower molars are ornamented with comma-shaped grooves, while the upper premolars and M1 are ornamented with prominent striae. In m2 only, the anterolabial wing and a middle medial cusp are ornamented with grooves and striae.

Eobaatar minor Kielan-Jaworowska, Dashzeveg and Trofimov, 1987
See Figure 29.9B.

Holotype. PIN 3101/70, incomplete right dentary with alveolus for incisor, p2, the roots and crown fragments of p3 and anterior part of p4. Höövör, Guchinus somon, Gobi Desert, Mongolia; Höövör beds (Aptian or Albian).

Referred material. An incomplete p4, PSS 10–23, from the type horizon and locality, is also tentatively assigned to this species.

Description. *E. minor* is smaller than *E. magnus* and the estimated length of the skull is about 10 mm. There is a short diastema between the incisor and p2. The p4 is distinctly lower and apparently shorter anteroposteri-orly than the p4 in *E. magnus*. The ridges on p4 are weaker and more closely arranged than in *E. magnus*.

Eobaatar sp.
Description. In addition to the specimens assigned to *E. magnus* and *E. minor*, Kielan-Jaworowska *et al.* (1987) described two incomplete lower incisors as *Eobaatar* sp. *a* and *Eobaatar* sp. *b*. These incisors differ from those of most 'Plagiaulacoidea', except for the North American *Glirodon* (see Engelmann and Callison, 1990) from the Morrison Formation, in having a limited enamel band, but share this character with numerous Cimolodonta.

Subfamily *incertae sedis*
Genus *Monobaatar* Kielan-Jaworowska *et al.*, 1987
Type species by monotypy. Monobaatar mimicus Kielan-Jaworowska *et al.*, 1987.

Monobaatar mimicus Kielan-Jaworowska *et al.*, 1987
Holotype. PIN 3101/65, incomplete left maxilla with P2–P4, broken alveolus for P1 and anterior alveolus for P5. Höövör, Guchinus county, Gobi Desert, Mongolia; Höövör beds (Aptian or Albian).
Referred material. The following specimens in the PSS collection and all from the type horizon and locality are assigned to this species: a fragmentary left maxilla with P1 and P2, a left P4, and possibly a left M2.

Description. The estimated length of the skull is about 20 mm. There is a single infraorbital foramen positioned above the P3–P4 embrasure. The posterior margin of the base of the zygomatic arch is situated above the P4–P5 embrasure. There are 3:4 cusps on P4, the second cusp of the labial row being the largest. M2 has a weakly sigmoid anterior margin and 1:2:3 relatively robust cusps.

Suborder *incertae sedis*
Family Arginbaataridae Hahn and Hahn, 1983
Comment. This is a monotypic, highly specialized family and includes only a monotypic genus *Arginbaatar* Trofimov, 1980 (see Trofimov, 1980; Kielan-Jaworowska *et al.*, 1987; Kielan-Jaworowska and Ensom, 1992). The presence of five upper premolars is characteristic for members of the 'Plagiaulacoidea'. We place Arginbaataridae in a suborder *incertae sedis*. The only taxon that shows some similarities to *Arginbaatar* is the Late Cretaceous Mongolian *Nesovbaatar*, described below.

Genus *Arginbaatar* Trofimov, 1980
Type species by monotypy. Arginbaatar dimitrievae Trofimov, 1980.

Arginbaatar dimitrievae Trofimov, 1980
See Figure 29.9C, D.
Holotype. PIN 3101/49, incomplete right dentary with p2–m1 and an incisor apparently belonging to the left dentary of the same specimen, found in association. Höövör, Guchinus county, Gobi Desert, Mongolia; Höövör beds (Aptian or Albian).
Referred material. A. dimitrievae is the most common multituberculate species in the Höövör Beds and, in addition to the holotype specimen, is represented in the PIN collections by 17 fragmentary dentaries or lower teeth, and seven maxillae with teeth, and in the PSS collection by nine dentaries with teeth or isolated teeth, and four maxillae with teeth, all from the type horizon and locality (see Kielan-Jaworowska *et al.*, 1987, for methods of reconstruction).
Description. The estimated length of the skull is about 20 mm. There are two infraorbital foramina, three

lower and five upper premolars, and possibly an upper canine. The p2 is peg-like. The p3 is double-rooted and its length equal to 1/7 that of p4. The most characteristic feature is a fan-shaped p4, without a basal cusp and with 15–18 serrations. The enamel on p4 is limited and covers only the anterior and posterodorsal parts of the crown. In early ontogenetic stages the posterior part of p4 is obscured by bone. The p4 rotated anteroventrally during ontogeny, over the worn p2 and p3, which disappeared in later ontogenetic stages. The anterior root of p4 is smaller than the posterior and both roots are open. The m1 is small, 2.7 times shorter than p4, with 3:2 cusps and a sigmoid anterior margin. P4 is similar to P5 and both have two rows of cusps. M1 is relatively small with 3:4 cusps and a small posterolingual wing. M2 has a strongly sigmoid anterior margin, a small anterolabial cusp, a cusp formula 1:2:3, and is relatively slender with respect to the width. Cusps on the upper and lower molars are widely separated from each other, the premolar cusps bear weak striations and the enamel is gigantoprismatic.

Suborder Cimolodonta McKenna, 1975
Infraorder Djadochtatheria Kielan-Jaworowska and Hurum, 1997 (new rank)
Comments. The multituberculate assemblages from the Djadokhta and Baruungoyot Formations, the Red beds of Hermiin Tsav, the Ukhaa Tolgod beds, and horizons equivalent to the Nemegt Formation have yielded twelve described and several undescribed species, eleven of which are assigned to Djadochtatheria, while one, *Buginbaatar transaltaiensis*, is tentatively assigned to the Cimolodontidae, suborder *incertae sedis*. The idea that most Mongolian Late Cretaceous multituberculates belong to a monophyletic group, that falls outside the Cimolodonta, has been reached independently by Rougier *et al.* (1997) and Kielan-Jaworowska and Hurum (1997). The latter authors proposed the suborder Djadochtatheria for these taxa. In this review we place the infraorder Djadochtatheria (new rank) in the suborder Cimolodonta.
Diagnosis. (modified from Kielan-Jaworowska and

Hurum, 1997, see also character analysis in Rougier *et al.*, 1997). Multituberculates with skull length varying between 20 and 70 mm, dental formula 2/1, 0/0, 3–4/2, 2/2, unicusped I2 and double-rooted upper premolars. Additional synapomorphies include: U-shaped fronto-parietal suture; no frontal-maxilla contact; and an edge between the palatal and lateral walls of the premaxilla. It is also possible that in djadochtatherians the postglenoid region of the brain-case is longer in respect to the skull length than in other multituberculates. A large, roughly rectangular facial surface of the lacrimal, exposed on the cranial roof and separating the frontal from the maxilla is present in all djadochtatherians, but may be a plesio-morphic feature. The frontals are extensive, pointed together anteriorly in the midline and deeply inserted between the nasals. Djadochtatheria differ from all other multituberculates except *Stygimys* and *Meniscoessus* in having I3 placed on the palatal part of premaxilla. Members of this taxon differ from 'Plagiaulacoidea' in having two upper incisors rather than three, a single-cusped I3, and two lower premo-lars, rather than three or four. Further differences include the presence of only one basal cuspule on p4, rather than a row of cuspules and the arcuate rather than rectangular shape of p4 (secondarily subtrape-zoidal in *Catopsbaatar*). Djadochtatherians share with Taeniolabidoidea, Cimolomyidae, *Glirodon* (Engelmann and Callison, 1999), and *Eobaatar* (Kielan-Jaworowska *et al.*, 1987) the presence of a limited enamel band on the lower incisor, and also share, with most Taeniolabidoidea, Cimolomyidae and a few *incertae sedis* multituberculate genera, gigan-toprismatic enamel. Djadochtatherians differ from the Taeniolabidae in having two lower premolars instead of one, an arcuate rather than triangular p4 (secon-darily subtrapezoidal in *Catopsbaatar)*, three or four upper premolars rather than one, and a smaller number of cusps on the upper and lower molars. They share with the Eucosmodontidae a low p4, which does not protrude dorsally above the level of the molars, but differ from them in having a smaller number of serrations (up to nine) on p4 and a smaller number of cusps on the lower and upper molars. They differ from

the Ptilodontoidea in having a more robust lower incisor, with limited enamel band, and in having a smaller p4 which does not protrude dorsally over the level of the molars (p4 strongly protrudes in Ptilodontoidea).

The North American taxa tentatively assigned here, by Rougier *et al.* (1997) and Kielan-Jaworowska and Hurum (1997), probably do not belong to Djadochtatheria.

Stratigraphic and geographical range. Late Cretaceous Djadokhta and Baruungoyot Formations (?early and ?late Campanian), Red beds of Hermiin Tsav, and Ukhaa Tolgod beds, Gobi Desert, Mongolia, and early Campanian of southern Kazakhstan (Aver'yanov, 1997). As the North American taxa (see Jepsen, 1940) are only tentatively assigned, we restrict the strati-graphic and geographical ranges to Asian localities.

Family Sloanbaataridae Kielan-Jaworowska, 1974
Genus *Sloanbaatar* Kielan-Jaworowska, 1970 (the only genus included).

Type species by monotypy. Sloanbaatar mirabilis Kielan-Jaworowska, 1970.

Sloanbaatar mirabilis Kielan-Jaworowska, 1970
See Figure 29.5.

Holotype (and only known specimen). ZPAL MgM-I/20, a complete skull associated with both dentaries and fragments of the postcranial skeleton. Bayan Zag, Gobi Desert, Mongolia; Djadokhta Formation (?early Campanian).

Description. The skull is about 25 mm in anteroposter-ior length, with a narrow and rectangular snout, zygo-matic arches that are very strongly expanded laterally, and a narrow and long glenoid fossa. There is one pair of vascular foramina on the nasals and two pairs of palatal vacuities. The lower margin of the dentary is arranged at an angle of 28° to the occlusal level of the molars (rather than 11–14° as in most djadochtathe-rians, except *Catopsbaatar*), and the condylar process faces upwards and is situated above the level of the molars. The coronoid process is relatively short and flares laterally – a character found also in *Kamptobaatar* and *Nesovbaatar*. The cusp formula for P4 is 2:5; M1,

4:4:ridge; M2, 1:2:3; p4, 8 serrations and basal cusp; m1, 4:3; m2, 2:2.

Family Djadochtatheriidae Kielan-Jaworowska and Hurum, 1997

Type genus. Djadochtatherium Simpson, 1925a.

Included genera. Djadochtatherium Simpson, 1925a; *Kryptobaatar* Kielan-Jaworowska, 1970; *Catopsbaatar* Kielan-Jaworowska, 1994; *Tombaatar* Rougier *et al.,* 1997.

Diagnosis (modified from Kielan-Jaworowska and Hurum, 1997). Djadochtatheriidae differ from all other multituberculates (and all other mammals) in having a subtrapezoidal snout in dorsal view, with a wide anterior margin and lateral margins confluent with the zygomatic arches rather than incurved in front of the arches. They differ from other members of the Djadochtatheria in having a snout that extends for 50% or more of the skull length (rather than for less than 49%). They share with *Chulsanbaatar* two pairs of vascular foramina on the nasals and with *Chulsanbaatar, Kamptobaatar,* and Taeniolabidae the lack of palatal vacuities.

Genus *Kryptobaatar* Kielan-Jaworowska, 1970

Synonyms. Gobibaatar Kielan-Jaworowska, 1970, and *Tugrigbaatar* Kielan-Jaworowska and Dashzeveg, 1978.

Diagnosis. The smallest member of the Djadochtatheriidae (skull length 25–32 mm). This genus is most similar to *Djadochtatherium,* with which it shares an arcuate p4 (rather than trapezoidal as in *Catopsbaatar*). The p4 in *Kryptobaatar* is relatively longer than in *Djadochtatherium* and has eight serrations (unknown in *Djadochtatherium*). *Kryptobaatar* differs from *Djadochtatherium* and *Catopsbaatar* in having a relatively smaller facial surface of the lacrimal and a shorter postorbital process. It differs from *Catopsbaatar* and *Tombaatar* in having four upper premolars, and from *Catopsbaatar* in that the inner ridge of the M1 extends for less than a half, or a half of the tooth length. It shares a short inner ridge of the M1 with *Chulsanbaatar, Sloanbaatar, Tombaatar,* and *Paracimexomys* from North America (Lillegraven, 1969; Archibald, 1982). It differs from *Tombaatar* in having M1 cusp formula of 4–5:4:3–5

rather than 5:5:2, and from *Djadochtatherium* and *Catopsbaatar* in having a shorter snout (but longer than in non-djadochtatheriid djadochtatherians), a relatively less robust lower incisor, and less prominent masseteric and parietal crests.

Type species. Kryptobaatar dashzevegi Kielan-Jaworowska, 1970.

Other species. Kryptobaatar saichanensis (Kielan-Jaworowska and Dashzeveg, 1978).

Kryptobaatar dashzevegi Kielan-Jaworowska, 1970
See Figures 29.7 and 29.10.

Holotype. ZPAL MgM-I/21, rostrum associated with right and left incomplete dentaries. Bayan Zag, Gobi Desert, Mongolia; Djadokhta Formation (?early Campanian).

Referred material. Twenty one specimens (incomplete skulls often associated with dentaries) from the Djadokhta Formation at Bayan Zag in the ZPAL collection, and five specimens from Tögrög and more than 150 specimens from Ukhaa Tolgod in the PSS collections.

Description. The length of the skull reaches 32 mm. The facial surface of the lacrimal is relatively smaller than in *Catopsbaatar* and the postorbital process much shorter than in *Djadochtatherium* and *Catopsbaatar.* The cusp formula for P4 is 2:4; M1, 4–5:4:3–5; M2, 1:2:3. The p4 has seven ridges and a basal cusp; m1, cusps 4:3; and m2, 3:2. The angle between the lower margin of the dentary and the occlusal surface of the molars is about 11°. An unusually well preserved pelvic girdle and hind limbs are also known for *Kryptobaatar* (Figure 29.7C).

Kryptobaatar saichanensis (Kielan-Jaworowska and Dashzeveg, 1978)

Holotype (and only known specimen). PSS 8–2 PST, a damaged skull associated with both dentaries and an incomplete postcranial skeleton. Tögrög, Gobi Desert, Mongolia; Tögrög beds, equivalent to the Djadokhta Formation (?early Campanian).

Description and comments. K. saichanensis differs from *K. dashzevegi* in having a slightly longer M1, and 4:2 cusps in m2 (3:2 in *K. dashzevegi*). In addition, the palatal part

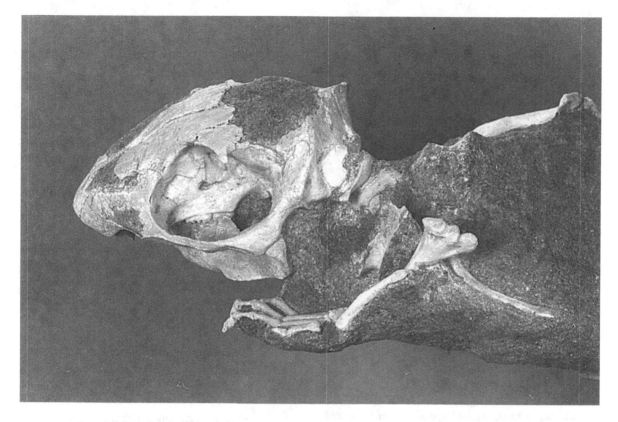

Figure 29.10. *Kryptobaatar dashzevegi* Kielan-Jaworowska, 1970, PSS MAE 101, skull associated with both lower jaws and incomplete postcranial skeleton, Ukhaa Tolgod, Gobi Desert. × 3.

of the premaxillary–maxillary suture occurs in front of P1 rather than on the level of P1, and there seems to be a smaller glenoid fossa, placed more laterally, on a longer stem. There is also a small foramen of unknown function in the middle of the palatal part of the premaxilla, though a similar foramen occurs on one side of the premaxilla in a skull of ?*K. dashzevegi* from Ukhaa Tolgod, and thus this cannot be a diagnostic character for *K. saichanensis*. In the dentary, distinctions consist of the presence of a smaller mandibular condyle, facing more dorsally, with a longer and more laterally directed ascending ramus.

Djadochtatherium Simpson, 1925a

Type species by monotypy. Djadochtatherium matthewi Simpson, 1925a.

Djadochtatherium matthewi Simpson, 1925a

Holotype. AMNH 20440, rostral part of the skull associated with both dentaries and fragments of the postcranial skeleton. The upper and lower molars are not preserved and right p4 is badly damaged. Bayan Zag, Gobi Desert, Mongolia; Djadokhta Formation (?early Campanian).

Referred material. An incomplete right dentary from Ukhaa Tolgod, GI 5/301, and a skull (not described) found by the Japanese–Mongolian Expedition at Tögrög in 1994.

Description. Skull length is about 50 mm and the m1 cusp formula is 4:3. *D. matthewi* differs from *Kryptobaatar* in being distinctly larger, but shares with *Kryptobaatar* and *Catopsbaatar* a similar shape of the snout. It shares with *Kryptobaatar* an arcuate p4 and the

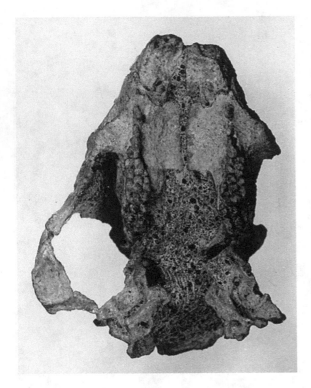

Figure 29.11. *Catopsbaatar catopsaloides* (Kielan-Jaworowska, 1974), holotype, ZPAL MgM-I/78, skull in ventral view. Red Beds of Hermiin Tsav, Hermiin Tsav II. × 1.4. (Modified from Kielan-Jaworowska, 1974.)

presence of four upper premolars, though in this respect it differs from *Catopsbaatar* and *Tombaatar*, which have three upper premolars. It shares with *Catopsbaatar* a very long postorbital process and a robust lower incisor.

Genus *Catopsbaatar* Kielan-Jaworowska, 1994
Type species by monotypy. Catopsbaatar catopsaloides (Kielan-Jaworowska, 1974).

Catopsbaatar catopsaloides (Kielan-Jaworowska, 1974)
See Figure 29.11.
Holotype. ZPAL MgM-I/78, an almost complete skull associated with both dentaries. Hermiin Tsav II, Gobi Desert, Mongolia; Red beds of Hermiin Tsav (?late Campanian).

Referred material. Two skulls from Hermiin Tsav, one of which is associated with incomplete dentaries, and a left m2 with a fragment of a dentary from Khulsan, all in the ZPAL collection.

Description. The length of the skull is about 60 mm. *C. catopsaloides* is similar to *Djadochtatherium* in the shape of the skull, but differs from it and from other djadochtatherians in having a very long snout and an orbit situated far posteriorly. The glenoid fossa is wide and roughly oval in shape. *C. catopsaloides* differs from *Djadochtatherium* in having only three upper premolars (the P2 is absent), and a less arcuate and distinctly smaller p4. The cusp formula is: P4, 5:1; M1, 5:5:4; M2, 2:3:3; p3 present, p4 small, with 3 cusps and basal cusp, no ridges; m1, 4:4; m2, 2:2. By contrast to most djadochtatherians and eucosmodontids, the condylar process is situated above the level of the molars and the coronoid process is relatively large.

Genus *Tombaatar* Rougier *et al.*, 1997
Type species by monotypy. Tombaatar sabuli Rougier *et al.*, 1997.

Tombaatar sabuli Rougier *et al.*, 1997
Holotype (and only known specimen). PSS-MAE 122A, a fragmentary rostrum and anterior portion of the braincase, with most of the dentition. Ukhaa Tolgod, Nemegt Basin, Gobi Desert, Mongolia; Ukhaa Tolgod beds (?early Campanian).

Referred material. Additional specimens, possibly referable to *Tombaatar*, have recently been recorded at Ukhaa Tolgod by the MAE.

Description. This is a large djadochtatheriid, generally similar to *Djadochtatherium* and *Catopsbaatar*. It shares with *Catopsbaatar* the absence of P2 and differs from all djadochtatherians in having a cusp formula of M1 4:5:2. The I3 alveolus is formed by both the premaxilla and maxilla, and there is a very prominent postpalatine torus. The lower jaw is not known.

Family *incertae sedis*
Genus *Bulganbaatar* Kielan-Jaworowska, 1974
Type species. Bulganbaatar nemegtbaataroides Kielan-Jaworowska, 1974.

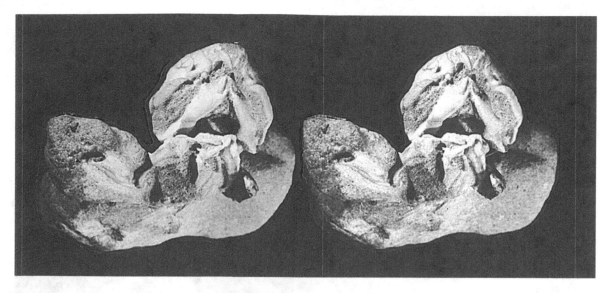

Figure 29.12. *Bulganbaatar nemegtbaataroides*, PSS MAE 103, skull associated with incomplete postcranial skeleton. Djadokhta Formation, Bayan Zag. Stereo-pair. × 1.5. (From Sereno and McKenna, 1995.)

Other species. ?*Bulganbaatar* sp. This taxon is based on a P4 from the early Campanian of southern Kazakhstan (Aver'yanov, 1997).

Bulganbaatar nemegtbaataroides Kielan-Jaworowska, 1974
See Figure 29.12.

Holotype. ZPAL MgM-I/25, damaged rostrum with teeth. Bayan Zag, Gobi Desert, Mongolia; Djadokhta Formation (?early Campanian).

Referred material. In addition to the holotype, there is a complete skull with lower jaws, shoulder girdle and almost complete forelimbs, PSS-MAE-103, also from Bayan Zag (Sereno and McKenna, 1995).

Description. B. nemegtbaataroides is similar to *Nemegtbaatar* in having lateral margins of the snout that are strongly incurved in front of the zygomatic arches (they are not incurved in Djadochtatheriidae) and one pair of palatal vacuities. The cusp formula is: P4, 2:5; M1, 5:5:3; M2, 1:2:2. There are more cusps on M1 than in *Kryptobaatar*, *Chulsanbaatar* and *Sloanbaatar*, and the M1 has a long lingual ridge, in which it resembles *Nemegtbaatar. B. nemegtbaataroides* differs from *Nemegtbaatar* in being smaller and in having a smaller number of cusps on the P4 and upper molars.

Genus *Nemegtbaatar* Kielan-Jaworowska, 1974
Type species by monotypy. Nemegtbaatar gobiensis Kielan-Jaworowska, 1974.

Nemegtbaatar gobiensis Kielan-Jaworowska, 1974
See Figures 29.4 and 29.8.

Holotype. ZPAL MgM-I/81, an almost complete dorsoventrally compressed skull, associated with both dentaries and most of the postcranial skeleton. Hermiin Tsav II, Gobi Desert, Mongolia; Red beds of Hermiin Tsav (?late Campanian).

Referred material. This species is known from several well preserved specimens in the ZPAL and PIN collections, from Hermiin Tsav II, Khulsan and Nemegt localities.

Description. The skull is 40 mm long thus *N. gobiensis* is larger than *Kryptobaatar*, but smaller than other djadochtatheriid taxa. This species differs from other members of the Djadochtatheriidae in having a differently shaped snout, in which the lateral margins are incurved in front of the zygomatic arches, a feature it shares with other djadochtatherians. The anterior part of the snout is narrow and elongated. Further differences include the presence of one pair of palatal vacuities and five pairs of vascular foramina on the nasals. *Nemegtbaatar* differs from all other djadochtatherians in respect of the cusp formula which is: P4, 3:6:1; M1, 6:7:4; M2, 1:3:2 (Figure 29.4); p4 with 7 ridges and basal cusp; m1, 5:4; m2, 3:2. The taxon closest to *Nemegtbaatar* is *Bulganbaatar*, which, however, is smaller and has fewer cusps on P4 and on the upper molars. Well preserved postcranial skeletons of *Nemegtbaatar* and *Kryptobaatar* enabled Kielan-Jaworowska and Gambaryan (1994) to reconstruct multituberculate postcranial musculature and demonstrate an abducted limb posture (Figure 29.8).

Genus *Chulsanbaatar* Kielan-Jaworowska, 1974
Type species by monotypy. Chulsanbaatar vulgaris Kielan-Jaworowska, 1974.

Chulsanbaatar vulgaris Kielan-Jaworowska, 1974
See Figure 29.6.

Holotype. ZPAL MgM-I/139, a partly damaged skull associated with left and incomplete right dentaries. Khulsan, Gobi Desert, Mongolia; Baruungoyot Formation (?late Campanian).

Referred material. This is the commonest species in the Baruungoyot Formation and in the Red beds of Hermiin Tsav, at Khulsan, Nemegt and Hermiin Tsav II, and is represented in the ZPAL collection by some 36 specimens consisting of skulls often associated with dentaries and fragments of the postcranial skeleton.

Description. Chulsanbaatar is the smallest representative of the Djadochtatheria and the skull length varies around 18–22 mm. Like other members of the Djadochtatheriidae it lacks the palatal vacuities and has two pairs of vascular foramina on the nasals, but differs from this family in being distinctly smaller and in having anterior margins of the skull that incurve in front of the zygomatic arches, and a rectangular snout. It also differs from other djadochtatherians in respect of the cusp formula, which is: P4, 2:6; M1, 4:5:ridge; M2, 1:2:2; p4 has 7 ridges and basal cusp; m1, 4:3; m2, 2:2.

Genus *Kamptobaatar* Kielan-Jaworowska, 1970
Type species by monotypy. Kamptobaatar kuczynskii Kielan-Jaworowska, 1970.

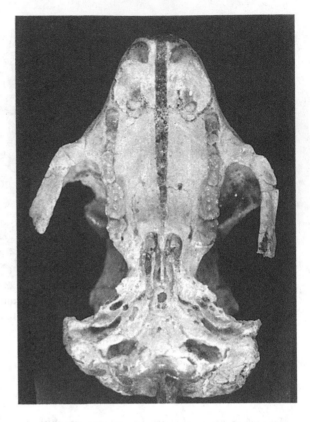

Figure 29.13. *Kamptobaatar kuczynskii* Kielan-Jaworowska 1970, holotype, ZPAL MgM-I/33, skull in ventral view. Djadokhta Formation, Bayan Zag. × 5. (From Kielan-Jaworowska, 1971.)

Kamptobaatar kuczynskii Kielan-Jaworowska, 1970
See Figure 29.13.

Holotype. ZPAL MgM-I/33, skull with broken zygomatic arches of a juvenile individual. Bayan Zag, Gobi Desert, Mongolia; Djadokhta Formation (?early Campanian).

Referred material. Three incomplete skulls in the ZPAL collection, two of which are associated with dentaries.

Description. The snout in *Kamptobaatar* is incurved in front of the zygomatic arches, with the anterior part of the arches directed roughly transversely, or posterolaterally. The anterior part of the snout is relatively wide, rectangular and blunt. The glenoid fossa is approximately half-oval, and oriented anterolaterally. There are asymmetrical vascular foramina on the nasals, no palatal vacuities, and the foramen ovale is divided into five foramina (two in other multituberculates). The dentary is relatively short and the coronoid process is low and triangular, flaring laterally as in *Sloanbaatar* and *Nesovbaatar*. The mandibular condyle, which is situated at the top of the rounded posterior edge of the dentary, faces dorsally, not posterodorsally as in most djadochtatherians. The cusp formula is: P4, 3:5–6; M1, 5:5:ridge; M2, 1:2:3; p4 with 7 serrations with ridges; m1, 4:3; m2 unknown.

Genus *Nesovbaatar* Kielan-Jaworowska and Hurum, 1997

Type species by monotypy. Nesovbaatar multicostatus Kielan-Jaworowska and Hurum, 1997.

Nesovbaatar multicostatus Kielan-Jaworowska and Hurum, 1997

Holotype. ZPAL MgM-I/103, both dentaries with dentition (incisors broken). Hermiin Tsav II, Gobi Desert, Mongolia; Red beds of Hermiin Tsav (?late Campanian).

Description. The specimen is a juvenile, the length of the dentary, measured from the base of incisor to the posterior end of the condyle, is 13.5 mm. *Nesovbaatar* differs from *Sloanbaatar* in the structure of the p4 and in its smaller size, but shares with it and with *Kamptobaatar*, a relatively small coronoid process that flares laterally. It shares with *Sloanbaatar*, *Kamptobaatar*, *Djadochtatherium* and *Catopsbaatar* the condyle placed above the occlusal level of the molars and facing dorsally. It differs from all the djadochtatherians in having a larger p4 with 10 cusps and 9 serrations, 8 of which are provided with weak ridges. It resembles *Sloanbaatar* in having 4:3 cusps on m1, but differs from it in exhibiting an angle of 18° between the lower margin of the dentary and the occlusal level of the molars (28° in *Sloanbaatar*), and 3:2 cusps on m2, rather than 2:2. The lower molars are similar in size and cusp formulae to those of *Chulsanbaatar*. The p4 resembles that of the Early Cretaceous Mongolian form *Arginbaatar* in being fan-shaped, but differs from it in being less vaulted, having a smaller number of serrations with ridges, and being completely covered with enamel.

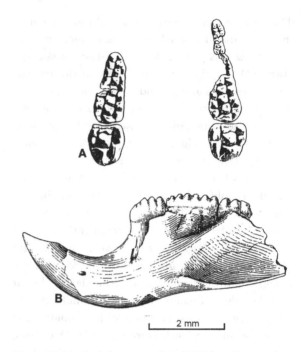

Figure 29.14. *Buginbaatar transaltaiensis* Kielan-Jaworowska and Sochava, 1969, PIN 3487. (A), Right M1 and M2 and left P4, incomplete M1 and M2 in occlusal view. (B), Left lower jaw in labial view. (Modified from Trofimov, 1975.)

Infraorder *incertae sedis*
Family ?Cimolomyidae Marsh, 1889
Genus *Buginbaatar* Kielan-Jaworowska and Sochava, 1969

Type species by monotypy. Buginbaatar transaltaiensis Kielan-Jaworowska and Sochava, 1969.

Buginbaatar transaltaiensis Kielan-Jaworowska and Sochava, 1969
See Figure 29.14.

Holotype. PIN 3487–1 and 3487–2, incomplete maxillae with P4–M2 and both incomplete dentaries with teeth. Khaichin Uul I, in the region of Bügiin Tsav, Transaltaian Gobi, Mongolia; beds equivalent to the Nemegt Formation (?early Maastrichtian).

Description. The dental formula is ?1/1, 0/0, ?1/1, 2/2, but only P4 of the upper premolars has been preserved. The cusp formulae are: P4, 3:6:1; M1, 7:8:6; M2, 2:3:2; p4 has 4 cusps with ridges; m1, 5:6; and m2, 3:3.

The lower incisor is very robust and thus similar to those in the Taeniolabidae, *Djadochtatherium* and *Catopsbaatar*. As argued by Kielan-Jaworowska and Hurum (1997), the apparent absence of the anterior upper premolars and p3, and the enlarged upper and lower molars with numerous cusps place *Buginbaatar* outside Djadochtatheria. *Buginbaatar* was first assigned to the ?Cimolomyidae and then to the Eucosmo-dontidae (Kielan-Jaworowska and Sochava, 1969; Trofimov, 1975). Hahn and Hahn (1983) established for it a subfamily Buginbaatarinae within the Eucosmodontidae. The apparent lack of anterior upper premolars and a rectangular rather than arcuate p4 suggest that *Buginbaatar* is not an eucosmodontid (Jepsen, 1940; Sloan and Van Valen, 1965), and the elongate multi-cusped P4 and rectangular p4 place it outside the Taeniolabidae. Its similarity to *Meniscoessus* suggests affiliation with Cimolomyidae (Clemens and Kielan-Jaworowska, 1979; Archibald, 1982) as suggested by Kielan-Jaworowska and Sochava (1969, see also Rougier *et al.*, 1997: fig. 6C). However, McKenna and Bell (1997) placed it in Taeniolabidae.

Subclass Theria Haswell and Parker, 1897

Comments. Simpson (1925b) erected the order Symmetrodonta, which he assigned to the subclass Theria Haswell and Parker, 1897. This assignment was subsequently followed by most authors. Recently, however, Rowe (1988), Szalay (1993a, b), Trofimov and Szalay (1994), Hopson (1994), and others have limited Theria to the marsupials, placentals and their last common ancestor, excluding even some forms with tribosphenic molars, for example, the 'tribotheres' of Butler (1978, 1990). Hopson (1994, p. 208) coined the informal term 'holotheres': '. . . to refer to the entire group of mammals characterized by a 'reversed triangles' molar pattern'. The senior author (but not M.J.N.) considers this term redundant and a source of confusion, as 'holotheres' are the same as Theria of Simpson (1925a, b), while the Theria of Rowe (1988) is the same as Tribosphenida McKenna, 1975. We follow Simpson and use the term Theria (contra the opinion of M.J.N.) to include Symmetrodonta and 'eupan-totheres'.

Order Symmetrodonta Simpson, 1925b

Comments. The order Symmetrodonta includes the most primitive Theria, described representatives of which are mostly known only from incomplete maxil-lae and dentaries with teeth, although a skull and almost complete skeleton of a symmetrodont *Zhangheotherium quinquecuspidens* Hu *et al.*, 1997, has been found in the Lower Cretaceous beds of Liaoning Province, China (Li *et al.*, 1995; Hu *et al.*, 1997). According to Hu *et al.* (1997) *Zhangheotherium* lacks the more parasagittal posture of the forelimb found in most living therian mammals. The authors also con-cluded that archaic therians, such as symmetrodonts, retained a primitive feature of non-therian mammals: a finger-like promontorium (possibly with an uncoiled cochlea). Symmetrodonts, shrew-sized mammals with postcanine teeth characterized by a triangular arrangement of the three main cusps and absence of an angular process in the dentary, are known from the Late Triassic through to the Late Cretaceous.

Family Amphidontidae Simpson, 1925b

Comment. Prothero (1981) erected for this family a sublegion Amphidontoidea, but we follow the division of Fox (1985), who treated the Amphidontidae as a family. Trofimov (1997) replaced his genus *Gobiodon* Trofimov, 1980 (preoccupied by *Gobiodon* Bleeker, 1856 – Teleostei, Perciformes, Gobiidae) by *Gobiotheriodon* Trofimov, 1997.

Genus *Gobiotheriodon* Trofimov, 1997
See Fig. 29.15.

Type species by monotypy. Gobiotheriodon infinitus (Trofimov, 1980).

Holotype. PIN 3101/50, almost complete right dentary with the last three molars and alveoli for six remaining teeth. Höövör, Guchinus county, Gobi Desert, Mongolia; Höövör beds (Aptian or Albian).

Referred material. A fragment of a maxilla with M3 and an isolated molar, both in the PIN collections, have also been assigned to this taxon.

Description. The dental formula, known only for the lower dentition, was described by Trofimov (1980) as: 3, 1, 4, 5. The three main cusps on the lower molars

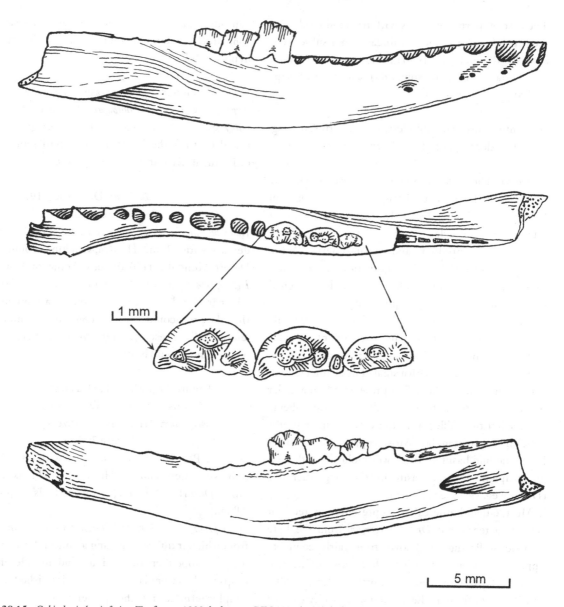

Figure 29.15. *Gobiotheriodon infinitus* Trofimov, 1980, holotype, PIN 3101/80, right lower jaw in lateral, occlusal and medial views. Höövör beds, Höövör. (From Trofimov, 1980.)

form an angle which decreases posteriorly: this angle is strongly obtuse in m5, but forms almost a right angle in m3; there are also two small additional cuspules, anterior and posterior.

Comments. Gobiotheriodon differs from *Amphidon* in having a well developed paraconid and metaconid on

all three molars that are preserved in the holotype specimen.

An isolated petrosal bone from the Early Cretaceous Höövör locality was described by Wible *et al.* (1995). As a result of their cladistic analysis, two possible allocations of this specimen were proposed.

The first supported affinities with triconodontids, and the second with the therian lineage. In a subsequent analysis (Rougier *et al.*, 1996a), the second hypothesis, i.e. therian (Prototribosphenida) affinity was supported.

'Eupantotheres' (a paraphyletic taxon, corresponding to the order Eupantotheria Kermack and Mussett, 1958)

Comments. The order Eupantotheria was erected by Kermack and Mussett (1958) to replace part of Pantotheria Marsh, 1880; Kermack and Mussett retained the infraclass Pantotheria, to which they assigned Symmetrodonta and Eupantotheria. We use the term 'eupantotheres' in quotation marks, since McKenna (1975) and Prothero (1981) have argued that 'Eupantotheria' is a paraphyletic group.

'Eupantotheres' are known from the Middle Jurassic to the Late Cretaceous. They differ from symmetrodonts in having an angular process on the dentary, lower molars with a small, but distinct talonid and wider upper molars. The presence of an angular process links them to mammals with tribosphenic molars, but they differ from these in lacking a protocone on the upper molars. Most 'eupantotheres' do not have a talonid basin on the lower molars, but there is an incipient talonid basin in the Arguitheriidae (Dashzeveg, 1994; see also below).

Most 'eupanthotheres' are represented by teeth or jaws with teeth, but *Drescheratherium* from the Late Jurassic of Portugal is known from fairly complete upper and lower jaws with teeth. *Henkelotherium*, also from the Early Cretaceous of Portugal (Krebs, 1991) and *Vincelestes* from the Early Cretaceous of Argentina (Rougier, 1993) are known from skulls associated with postcranial skeletons, but most 'eupantotheres' are represented only by teeth or jaws with teeth. *Vincelestes* is regarded as a sister taxon of the Tribosphenida McKenna, 1975 (Rougier, 1993; Rowe, 1993).

Family Arguimuridae Dashzeveg, 1994

Comments. Arguimuridae, according to Dashzeveg (1979, 1994), are close to Amphitheriidae and Paurodontidae, but distinguished by possession of a comparatively large talonid with a well-developed hypoconulid.

Genus *Arguimus* Dashzeveg, 1979

Type species. *Arguimus khosbajari* Dashzeveg, 1979.
Other species. Dashzeveg (1994) also assigned the so-called Porto Pinherio molar from the Kimmeridgian of Portugal (Krusat, 1969) to *Arguimus*.

Arguimus khosbajari Dashzeveg, 1979
See Figure 29.16B.

Holotype (and only known specimen). PSS 10–15, left dentary with p3–m2. Höövör, Guchinus county, Gobi Desert, Mongolia; Höövör beds (Aptian or Albian).
Description. The fifth lower premolar is molariform. The paraconid and metaconid on m2 are much lower than the protoconid, the hypoconulid is prominent, an 'entoconid' is present, and the hypoconid is indistinct. There is no talonid basin.

Family Arguitheriidae Dashzeveg, 1994
Genus *Arguitherium* Dashzeveg, 1994
Arguitherium cromptoni Dashzeveg, 1994
See Figure 29.16A.

Holotype. PSS 10–31, right dentary with p4–m1 – the only specimen known. Höövör, Guchinus county, Gobi Desert, Mongolia; Höövör beds (Aptian or Albian).
Description. *Arguitherium* is close to the Peramuridae, from which it differs in having a non-molariform p5, a less developed cristid obliqua and undifferentiated cusps on the talonid of m1. It is distinguished from the Amphitheriidae by the presence of an incipient talonid basin on the molars and a better developed cingulum on the labial side of m1.

Legion Tribosphenida McKenna, 1975
Order Aegialodontia Butler, 1978

Comments. A type of molar dubbed the tribosphenic molar, first occurs in Early Cretaceous Theria. The upper molars of the tribosphenic type differ from those of 'eupantotheres' in that they are wider and

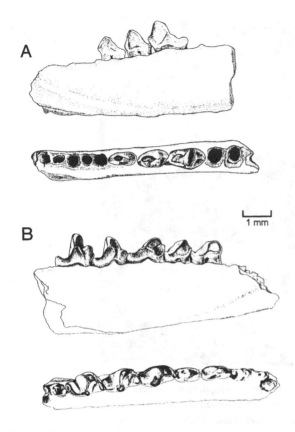

Figure 29.16. *Arguitherium cromptoni* Dashzeveg, 1994, holotype, PSS 10–31, right lower jaw with p4–m1 in (A) lingual and occlusal views. *Arguimus khosbajari* Dashzeveg, 1979, holotype, PSS 10–15, left lower jaw with p3–m2 in (B) lingual and occlusal views. (From Dashzeveg, 1994.)

possess a lingual cusp named the protocone (this is absent, apparently with one exception, in 'eupantotheres'). The three main cusps of the tribosphenic upper molar are arranged in a triangle (trigon) and are called the protocone, metacone and paracone. There are also accessory cusps (styles) situated on the labial margin of the triangular upper molar. The tribosphenic lower molars differ from those of 'eupantotheres' (except for *Arguitherium*) in having a talonid basin, which received the protocone of the upper molar. On the margin of the talonid there are usually three cusps: a hypoconid, a hypoconulid and an ento-

conid, though only two of these – the hypoconid and the hypoconulid – occur in some early forms. Tribosphenic molars are characteristic of Cretaceous and most early Tertiary therians, as well as of some extant therians.

Kermack *et al.,* (1965) described *Aegialodon dawsoni* from the lower Wealden (Valanginian) of southeastern England, based on a single tiny tribosphenic lower molar with a small, but basined talonid. This tooth was believed at that time to be the oldest known tribosphenic molar and *Aegialodon* was regarded by these authors as an ancestor of all mammals with tribosphenic molars. Subsequently, Kermack (1967) erected the family Aegialodontidae for this genus. Dashzeveg (1975) described another aegialodontid lower molar from the Aptian or Albian of Höövör (Guchinus county) in Mongolia, and named it *Kielantherium gobiensis.* Butler (1978) erected the order Aegialodontia (including Aegialodontidae, Kermackiidae, Deltatheridiidae and, tentatively, *Potamotelses*) within his infraclass Tribotheria. Fox (1980) excluded the Kermackiidae and included Picopsidae in the Aegialodontia, while Kielan-Jaworowska and Nesov (1990) excluded the Deltatheridiidae, arguing that deltatheroidans are metatherians. Subsequently, Butler (1990) accepted the metatherian nature of the Deltatheroida, withdrew his infraclass Tribotheria and restricted Aegialodontia to *Aegialodon* and *Kielantherium*, an action tentatively accepted here.

Family Aegialodontidae Kermack, 1967
Genus *Kielantherium* Dashzeveg, 1975
Type species by monotypy. Kielantherium gobiensis Dashzeveg, 1975.

Kielantherium gobiensis Dashzeveg, 1975
See Figure 29.17.
Holotype. PSS 10–4, right ?m2. Höövör, Guchinus county, Gobi Desert, Mongolia; Höövör beds (Aptian or Albian).
Referred material. PSS 10–16, an incomplete right dentary with four molars, alveoli or roots of four double-rooted premolars, and one broken alveolus for

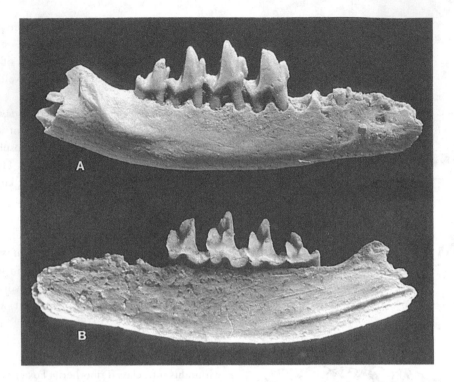

Figure 29.17. *Kielantherium gobiensis* Dashzeveg, 1975, PSS 10–16, right incomplete lower jaw with four molars and eight alveoli or roots of four double-rooted premolars and one broken alveolus for one more premolar or canine in (A) labial and (B) medial view. × 6. (From Dashzeveg and Kielan-Jaworowska, 1984).

another premolar or the canine (Dashzeveg and Kielan-Jaworowska, 1984).

Description. In *Kielantherium* the trigonid is larger than the talonid, the protoconid is the highest cusp, and the paraconid is higher than the metaconid, as is characteristic of the Deltatheroida. The talonid basin is narrow, with a hypoconid and hypoconulid; the entoconid is not developed.

Comments. The Aegialodontia are known only from lower molars. The oldest tribosphenic molar is possibly a talonid from the Purbeck Limestone Group (?early Barriasian) of England (Sigogneau-Russell and Ensom, 1994). *Aegialodon* (Kermack *et al.*, 1965) from the Valanginian of Sussex is younger. Similarly *Tribotherium africanum* and other 'tribotheres' from the Berriasian of Morocco (Sigogneau-Russell, 1991b, 1995) may be slightly younger. The upper, and tentatively assigned lower molars, of *Tribotherium* are

different from those of *Prokennalestes*, *Kielantherium* (of which only the lower molars are known), and *Deltatheridium*.

Metatherian–eutherian dichotomy

Comments. For many years it has been generally accepted that metatherians and eutherians differentiated from a common ancestor during the Early Cretaceous, and that metatherians diversified in North or South America, while eutherians diversified in Asia. It was presumed that this vicariance resulted from the appearance of marine barriers between the main land-masses, which restricted intercontinental exchange (e.g., Lillegraven, 1969, 1974). At that time it was also believed that metatherians did not reach Asia. However, palaeontological discoveries of subsequent years have challenged this hypothesis.

A family of carnivorous therians, the Deltatheridiidae Gregory and Simpson, 1926 (see below), known originally from the Late Cretaceous of Asia, was for a long time regarded as an important group of placental carnivores. However, Butler and Kielan-Jaworowska (1973) showed that this group has a marsupial-like postcanine dental formula. Accordingly, Kielan-Jaworowska and Nesov (1990) assigned the Deltatheroida Kielan-Jaworowska, 1982 to the Metatheria. Szalay and Trofimov (1996) described the skull and postcranial skeleton of a metatherian, *Asiatherium reshetovi*, from the Late Cretaceous of Mongolia.

The oldest known marsupials are Cenomanian in age (Cifelli and Eaton, 1987), the deltatheroidans currently date from the late Turonian (Eaton and Cifelli, 1988), while the oldest eutherians are from the ?late Aptian or ?early Albian of Mongolia (Kielan-Jaworowska and Dashzeveg, 1989). These and other data, discussed under Aegialodontia, suggest that the marsupial–placental dichotomy occurred earlier than previously thought, possibly during the latest Jurassic.

Theria *incertae sedis*
Genus *Hyotheridium* Gregory and Simpson, 1926
Type species by monotypy. Hyotheridium dobsoni Gregory and Simpson, 1926.

Hyotheridium dobsoni Gregory and Simpson, 1926
Holotype. AMNH 21702, a single, damaged rostrum with anterior parts of both dentaries. Bayan Zag, Gobi Desert, Mongolia; Djadokhta Formation (?early Campanian).
Description and comments. H. dobsoni is very poorly known. It has large upper and lower canines and three premolars, which might indicate deltatheroidan affinities (see below). As the number of molars and their occlusal surfaces are not known, we prefer to classify it as Theria *incertae sedis*.

Infraclass Metatheria Huxley, 1880
Order Deltatheroida Kielan-Jaworowska, 1982
Comments. We follow Kielan-Jaworowska and Nesov (1990) in assigning Deltatheroida Kielan-Jaworowska, 1982, to the Metatheria, in agreement with Marshall

and Kielan-Jaworowska (1992), and Trofimov and Szalay (1994), Nesov (1997) and Rougier *et al.* (1998), but *contra* Cifelli (1993).

Deltatheroida are an order of relatively large (by Mesozoic standards), carnivorous mammals, the skull length of which ranges between 40 and 70 mm. Deltatheroidans retain several characters that are plesiomorphic for therians. For example, the nasals are expanded posteriorly and there is a long jugal that contributes to the glenoid fossa. They also have an incipient alisphenoid bulla, a feature regarded by Kielan-Jaworowska and Nesov (1990) as a metatherian trait, though it is absent in the South American Palaeocene marsupials *Pucadelphys* and *Mayulestes* (Muizon, 1994).

Marsupial characters (Rougier *et al.*, 1998) include: a premaxilla with a posteriorly directed process that reaches the alveolus for the canine; and a dentary with a shelf-like medially directed process. The petrosal is similar to those attributed to metatherians from the Late Cretaceous of North America (Wible, 1990) and shows two major metatherian synapomorphies: (1) a marked reduction or absence of the stapedial arterial system; and (2) a small, horizontally directed prootic canal connected to the postglenoid venous system. The dental formula: 4/3, 1/1, 3/3, 4/4 (as established for *Deltatheridium*), is marsupial-like and there is a sharp morphological break between the molars and premolars, which are not molariform, as is characteristic of metatherians. The most distinctive similarity between *Deltatheridium* and living marsupials is the tooth replacement pattern, characterized by replacement of a single tooth (last premolar). Unlike marsupials, the hypoconulid and entoconid on the lower molars are not approximated. The upper molar stylar shelf is, relatively speaking, the widest among tribosphenid mammals.
Stratigraphic and geographical range. Late Cretaceous of the Northern Hemisphere (Gregory and Simpson, 1926; Fox, 1974; Kielan-Jaworowska, 1975a; Kielan-Jaworowska and Nesov, 1990; Marshall and Kielan-Jaworowska, 1992; see also Cifelli, 1993, for review).
Comment on systematics. Kielan-Jaworowska and Nesov (1990) divided the order Deltatheroida Kielan-

Jaworowska, 1982, into two famillies. Deltatheridiidae Gregory and Simpson, 1926, and Deltatheroididae Kielan-Jaworowska and Nesov, 1990. Deltatheridiidae are characterized by lack of palatal vacuities and three upper molars (rather than the usual four), while Deltatheroididae are characterized by the presence of palatal vacuities and four upper molars. Rougier *et al.* (1998) demonstrated on the basis of better preserved material of *Deltatheridium* from Ukhaa Tolgod in Mongolia that this genus, and in consequence Deltatheridiidae, have four upper molars. Thus the only difference between the two families would be the presence or absence of palatal vacuities. As in other groups of mammals, however, the presence of palatal vacuities is a generic, rather than family character, and we regard Deltatheroididae Kielan-Jaworowska and Nesov, 1990, as a junior subjective synonym of Deltatheridiidae Gregory and Simpson, 1926.

Nesov (1985) proposed Sulestinae as a subfamily of the Deltatheridiidae, but McKenna and Bell (1997) synonymized this subfamily with Deltatheridiidae Gregory and Simpson, 1926. As we have not found characters that would differentiate the Sulestinae and a nominal subfamily Deltatheridiinae at the subfamily level, we follow McKenna and Bell in this respect.

Family Deltatheridiidae Gregory and Simpson, 1926
Included genera. Deltatheridium Gregory and Simpson, 1926; *Deltatheroides* Gregory and Simpson, 1926; *Sulestes* Nesov, 1985; *Deltatheroides*-like mammals (Fox, 1974); and a deltatheroidan skull from Guriliin Tsav (Anonymous, 1983; Szalay and Trofimov, 1996, fig. 22). *Stratigraphic and geographical range.* Late Cretaceous of Mongolia and North America.

Genus *Deltatheridium* Gregory and Simpson, 1926
Type species by monotypy. Deltatheridium pretrituberculare Gregory and Simpson, 1926.

Deltatheridium pretrituberculare Gregory and Simpson, 1926
See Figure 29.18.
Synonym. Deltatheridium pretrituberculare tardum Kielan-Jaworowska, 1975.

Holotype. AMNH 21705, a damaged rostrum associated with both dentaries. Bayan Zag, Gobi Desert, Mongolia; Djadokhta Formation (?early Campanian).
Referred material. Five incomplete skulls from the Baruungoyot Formation and Red beds of Hermiin Tsav in the ZPAL collection, several skulls from Ukhaa Tolgod in the PSS collection.
Description. Deltatheridium pretrituberculare is a relatively small carnivorous mammal (skull length about 40 mm long), with shortened snout, no palatal vacuities, and a lacrimal with a large facial wing. The stylar shelf on the molars is very large, with a deep ectoflexus, paracone and metacone placed in the middle of the tooth width, and large, convex conules. The protocone is small and low. The p1 is single-rooted, while p2 and p3 are double-rooted with basal cusps. In lower molars the protoconid is tall, the metaconid is smaller than the paraconid, and the talonid is transversely narrow.

Kielan-Jaworowska (1975a) erected a new subspecies *D. pretrituberculare tardum*, based on five specimens from the Baruungoyot Formation and Red beds of Hermiin Tsav. As demonstrated by Rougier *et al.* (1998), the apparent differences between *D. p. tardum* and the nominal subspecies are due to the state of preservation and do not merit distinction at the subspecies level.

Genus *Deltatheroides* Gregory and Simpson, 1926
Type species by monotypy. Deltatheroides cretacicus Gregory and Simpson, 1926

Deltatheroides cretacicus Gregory and Simpson, 1926
Holotype. AMNH 21700, a damaged snout. Bayan Zag, Gobi Desert, Mongolia; Djadokhta Formation (?early Campanian).
Referred material. ZPAL MgM-I/29, a left dentary from the same formation and locality as the holotype.
Description. The skull and dentition in the holotype specimen are badly damaged. Individual teeth in ZPAL MgM-I/29 are about 1.2–1.35 times greater than those of *Deltatheridium pretrituberculare* ZPAL MgM/–I/91 (see Kielan-Jaworowska, 1975a, table 1), but do not differ much in morphology. Owing to the poor state of

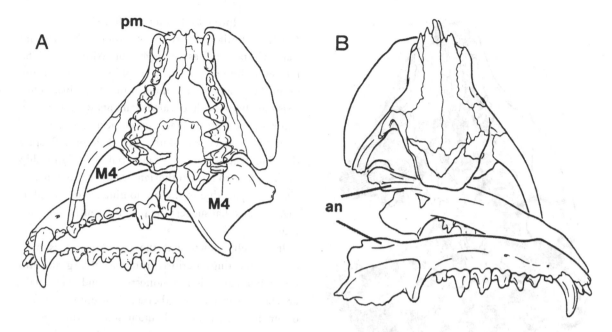

Figure 29.18. *Deltatheridium pretrituberculare* Gregory and Simpson, 1926, PSS-MAE 133, rostrum and dentaries. Palate in occlusal view and dentaries in oblique dorsal view (A). Rostrum in dorsal view and dentaries in ventral oblique view, showing the inflected angle (B). × 2.2 (From Novacek *et al.*, 1997.) Abbreviations: an, angle; M4, fourth molar; pm, premaxilla.

preservation of the holotype specimen it cannot be stated, with any certainty, whether the palatal vacuities, which are present in the Guriliin Tsav skull, occur also in *Deltatheroides*. The differences between the two Mongolian deltatheroidan genera *Deltatheridum* and *Deltatheroides* are not clear, and in the future it may be shown that the two genera are congeneric.

Comments. The beautifully preserved, almost complete, so-called deltatheroidan skull from Guriliin Tsav (Anonymous, 1983; Kielan-Jaworowska and Nesov, 1990; Szalay and Trofimov, 1996, fig. 22) from beds corresponding to the Nemegt Formation awaits description. It is the largest known deltatheroidan with distinct palatal vacuities.

Order Asiadelphia Trofimov and Szalay, 1994
Family Asiatheriidae Trofimov and Szalay, 1994
Genus *Asiatherium* Trofimov and Szalay, 1994
Type species by monotypy. Asiatherium reshetovi Trofimov and Szalay, 1994.

Asiatherium reshetovi Trofimov and Szalay, 1994
See Figure 29.19.

Holotype (and only known specimen). PIN 3907, a crushed skull with both dentaries and an almost complete postcranial skeleton. Üüden Sair, Gobi Desert, beds equivalent to the Baruungoyot Formation (?late Campanian).

Description. A. reshetovi has three premolars and four molars with a sharp morphological break between the last premolar and first molar. The stylar cusps are poorly developed and the conules and pre- and post-cingula are present. Szalay and Trofimov (1996, p. 474) stated: 'the closely twinned hypoconulid and entoconid, correlated with a (relative) hypertrophy of the metacone, an alisphenoid component to the bulla (possibly an independently derived trait), oval (not elliptical) fenestra vestibuli, and an elliptical fenestra cochleae, along with other unmistakenly marsupial-like characters (and therian as well as pretherian) postcranial taxonomic properties, all attest the noneutherian status of *Asiatherium*.' Further analy-

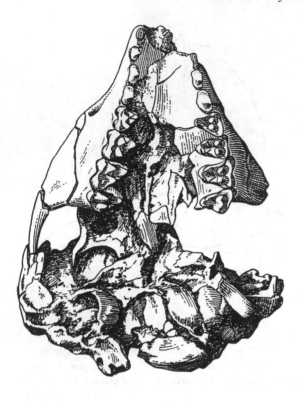

Figure 29.19. *Asiatherium reshetovi* Trofimov and Szalay, 1994, holotype, PIN 3907, drawing of the palatal view of the skull from beds equivalent in age to the Baruungoyot Formation, Üüden Sair, Southern Mongolia. (From Trofimov and Szalay, 1994.) Scale bar, 5 mm.

sis, including comparisons with new deltatheroidan material from Ukhaa Tolgod is required.

Comments. Aver'yanov and Kielan-Jaworowska (1999) described additional material of *Marsasia* Nesov, 1997, from the Coniacian Bissekty Formation of Uzbekistan, and assigned it tentatively to Asiadelphia. They suggested that the phylogenetic position of *Marsasia* might lie between the Albian form *Kokopellia* (Cifelli and Muizon, 1997) and the Campanian form *Asiatherium*, thus supporting Cifelli and Muizon's conclusion regarding the marsupial affinities of *Asiatherium*.

Infraclass Eutheria Gill, 1872

Comments. Eutherians are known from both Early and Late Cretaceous assemblages of Mongolia. The former represent possibly the oldest record of the group, but the material representing these forms consists largely of teeth, or jaw fragments with teeth. By contrast, the Late Cretaceous eutherians are represented by spectacular skulls and skeletons. Despite this abundance of excellent material, the relationship of these taxa to later eutherian orders is problematic. This is in part due to the predominance of morphological traits that suggest the primitive eutherian condition.

It is well known that the epipubic (marsupial) bones, projecting anteriorly from the pelvic girdle into the abdominal region in monotremes and marsupials, are absent from extant eutherians. The epipubic bones occur also in tritylodontids and in some extinct groups of early mammals: multituberculates (Kielan-Jaworowska 1969b) and symmetrodonts (Hu *et al.* 1997). Kielan-Jaworowska (1975c) demonstrated that the Late Cretaceous eutherian *Barunlestes* (family Zalambdalestidae) has a triangular fossa on the anterior margin of the acetabular branch of the pubic bone, which she interpreted as the attachment area for the epipubic bone. She suggested that epipubic bones might have been present in all early mammals and their disappearance in the Tertiary Eutheria may be connected with the gradual evolution of a prolonged internal gestation period. Kielan-Jaworowska's prediction has been confirmed by the discovery of epipubic bones in symmetrodonts (Hu *et al.* 1997) and especially by the discovery by Novacek *et al.* (1997) of epipubic bones in two lineages of Late Cretaceous eutherians from Mongolia: a zalambdalestid – cf. *Zalambdalestes* and an asioryctid, *Ukhaatherium nessovi* Novacek *et al.*, 1997.

The question of which is the oldest eutherian mammal is still debated. *Ausktribosphenos*, allegedly an Early Cretaceous placental from Australia (Rich *et al.*, 1997), is, in our opinion, not a eutherian (see also Kielan-Jaworowska *et al.*, 1998). *Prokennalestes*, described below, is a likely candidate and is contemporaneous with *Slaughteria* from the Albian of Texas (Marshall and Kielan-Jaworowska, 1992, but see also Butler, 1978, 1990), and *Montanalestes* from the Cloverly Formation (Aptian or Albian), see Cifelli (1999).

Comments on systematics. There is a dearth of derived features that can be utilized to indicate the affinities of Mongolian Cretaceous eutherians (Novacek, 1992, 1993; see also McKenna, 1975; Novacek, 1986b; and MacPhee and Novacek, 1993, for reviews). Consequently, there is a lack of consensus on the ordinal assignments of early eutherians in general (see systematic arrangements proposed by e.g., Carroll, 1988; McKenna and Bell, 1997; Novacek *et al.*, 1997; Nesov, 1997, and many others). Below we present a concise summary of selected opinions regarding the systematics of Mongolian eutherians.

Szalay and McKenna (1971) erected the order Anagalida to which they assigned the Mongolian Late Cretaceous Zalambdalestidae. McKenna (1975) erected the cohort Epitheria, including a new magnorder Erontheria, and a new superorder Kennalestida. He regarded two Late Cretaceous Mongolian genera *Kennalestes* and *Asioryctes* (assigned to Palaeoryctinae) as the earliest ernotheres. He also referred to early ernotheres the Late Cretaceous Mongolian zalambdalestids, coeval with anagalids and several other groups of early eutherians.

Kielan-Jaworowska (1981) erected Kennalestidae to include *Kennalestes* and tentatively '*Prokennalestes*' (cited at that time as a *nomen nudum*), and assigned her new family to Proteutheria Romer, 1966. She assigned to the same suborder the Palaeoryctidae, and erected within it Asioryctinae to include the sole genus *Asioryctes*. Nesov (1985) erected the suborder Mixotheridia (within Proteutheria), and Nesov *et al.* (1994) assigned the Zalambdalestidae to this taxon, and was followed in this respect by Nesov (1997). In this last paper Nesov placed Kennalestidae and Otlestidae in Proteutheria.

Kielan-Jaworowska and Dashzeveg (1989) classified *Prokennalestes* within Otlestidae, superfamily Kennalestoidea (as a new rank assigned to Kennalestidae),

order Proteutheria, and were followed in this by Sigogneau-Russell *et al.* (1992).

Novacek *et al.* (1997) placed *Asioryctes* and a new closely related genus, *Ukhaatherium*, in the Asioryctidae Kielan-Jaworowska, 1981, and assigned this family to Asioryctitheria (a suborder?) which was identified as Eutheria *incertae sedis*. According to these authors Kennalestidae (containing *Kennalestes*), are closely related to *Asioryctes* and also belong to Asioryctitheria.

McKenna and Bell (1997) assigned *Prokennalestes* and *Kennalestes* to Gypsonictopidae, in the superorder Leptictida McKenna, 1975, and placed *Asioryctes* in the magnorder Epitheria McKenna, 1975 (attributing a new rank to this taxon). In this review we assign Cretaceous Mongolian eutherian genera to two orders: Asioryctitheria Novacek *et al.*, 1997 and Anagalida Szalay and McKenna, 1971.

Order Asioryctitheria Novacek, Rougier, Wible, McKenna, Dashzeveg, and Horovitz, 1997
Family Kennalestidae Kielan-Jaworowska, 1981
Genus *Prokennalestes* Kielan-Jaworowska and Dashzeveg, 1989

Type species. Prokennalestes trofimovi Kielan-Jaworowska and Dashzeveg, 1989.

Other species. Prokennalestes minor Kielan-Jaworowska and Dashzeveg, 1989.

Prokennalestes trofimovi Kielan-Jaworowska and Dashzeveg, 1989
See Figure 29.20.

Holotype. PSS 10–6, posterior part of the right dentary with coronoid and condylar processes, m2 and m3. Höövör, Guchinus county, Gobi Desert, Mongolia; Höövör beds (Aptian or Albian).

Referred material. Eighteen dentaries and maxillae with teeth in the PSS collection and a large number of as yet undescribed specimens in the PIN collection.

Description. P. trofimovi is a small mammal, the estimated skull length being 24–27 mm. It has a labial mandibular foramen and remnants of the coronoid bone and the Meckelian groove. There are five pre-

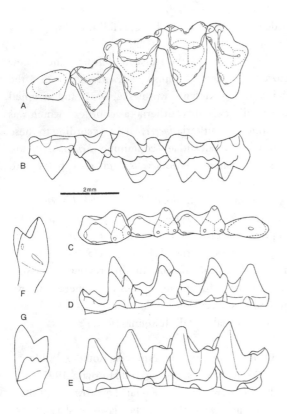

Figure 29.20. *Prokennalestes trofimovi* Kielan-Jaworowska and Dashzeveg, 1989, reconstruction of the dentition. P4–M3 in (A) occlusal and (B) labial view; p5–m3 in (C) occlusal, (D) lingual and (E) labial view; m2 in (F) anterior and (G) posterior view. (From Kielan-Jaworowska and Dashzeveg, 1989.)

molars and three molars. P5 is semimolariform without a metacone. The molars have three cusps in the parastylar region: the paracone is larger than the metacone, the conules are unwinged, and there is no pre- and postcingula. In the lower molars there is a 3-cusped talonid that is narrower than the trigonid (Figure 29.20). The dental formula in these specimens supports the notion that primitively eutherians had five premolars (McKenna, 1975; Novacek, 1986a).

Prokennalestes minor Kielan-Jaworowska and Dashzeveg, 1989

Holotype. PSS 10–7a, fragment of the posterior part of the left dentary with m2; locality and horizon as for *P. trofimovi*.

Referred material. Eighteen dentaries and maxillae with teeth in the PSS collection and a large number of as yet undescribed specimens in the PIN collection.

Description. *P. minor* differs from *P. trofimovi* in being distinctly smaller and in that the Meckelian groove is only slightly flexed.

Comment. As noted by Kielan-Jaworowska and Dashzeveg (1989; see also Sigogneau-Russell *et al.*, 1992), *P. trofimovi* and *P. minor* differ only in size and may be sexual morphs of the same species.

Eutheria gen. et sp. indet.

Material. In the PSS collections from Höövör there is a heavily worn undescribed lower molar 2.8 mm long. This specimen shows that in addition to the minute *Prokennalestes*, another, somewhat larger eutherian taxon, occurred in this assemblage.

Family Kennalestidae Kielan-Jaworowska, 1981
See Figures 29.21 and 29.22.B.
Genus *Kennalestes* Kielan-Jaworowska, 1969
Type species by monotypy. Kennalestes gobiensis Kielan-Jaworowska, 1969.

Kennalestes gobiensis Kielan-Jaworowska, 1969
Holotype. ZPAL MgM-I/3, anterior part of the skull associated with both dentaries. Bayan Zag, Gobi Desert, Mongolia; Djadokhta Formation (?early Campanian).

Referred material. Five fragmentary skulls and one complete juvenile skull in the ZPAL collection; except for an atlas the postcranial skeleton is not known.

Description. The skull length of adult individuals is about 26 mm. The dental formula is 4/3, 1/1 4/4, 3/3, but, apparently, there are five premolars in juvenile specimens. The upper and lower canines are double-rooted. P1 and P2 are small with minute diastemae in front, behind and between them. P3 is the strongest tooth in the postcanine series, while P4 is partly

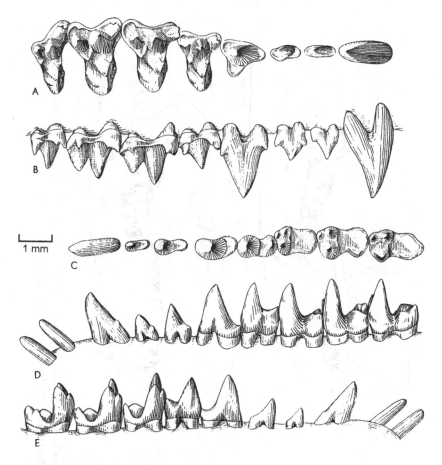

Figure 29.21. *Kennalestes gobiensis* Kielan-Jaworowska, 1969. Upper canine and cheek teeth in (A) occlusal and (B) labial views. Lower canine and cheek teeth in (C) occlusal view. Lower dentition in (D) labial and (E) lingual views. (From Kielan-Jaworowska, 1969a.)

molariform, but has only an incipient metacone. The molars have a large parastylar area with three cusps and the preparastyle characteristic of *Prokennalestes* is present. The conules are winged and there are pre- and postcingula. The dentary is slender and the angular process is slightly inflected. The snout is tubular and the nasals are expanded posteriorly. The structure of the basicranial region is discussed below together with the description of *Asioryctes*.

Comments. In terms of its cranial anatomy *Kennalestes* is a very primitive Mongolian Cretaceous mammal. Its suggested alliance with North American Cretaceous and Early Tertiary Leptictidae (Kielan-Jaworowska *et al.*, 1979) has been questioned (Fox, 1976; Novacek, 1986b),

and the problem is in need of further study. *Kennalestes* has a number of features, including the conservative construction of its tympanic region, and the presence of five premolars in the dentaries of some juvenile individuals (see McKenna, 1975; Kielan-Jaworowska *et al.*, 1979), that bear on arguments concerning the transformation of important mammalian features.

Family Asioryctidae Kielan-Jaworowska, 1981 (new rank)

Genus *Asioryctes* Kielan-Jaworowska, 1975b

See Figures 29.22A and 29.23.

Type species by monotypy. Asioryctes nemegetensis Kielan-Jaworowska, 1975b.

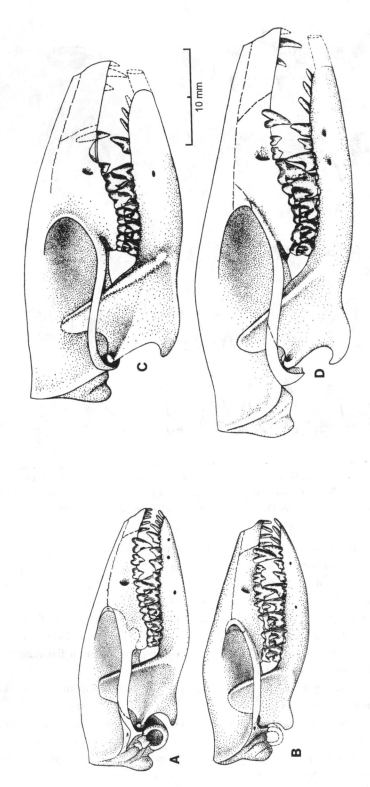

Figure 29.22. Comparison of reconstructed skulls of four Late Cretaceous eutherian mammals from Mongolia all shown in lateral view. (A), *Asioryctes nemegetensis*, (B), *Kennalestes gobiensis*, (C), *Barunlestes butleri*, (D), *Zalambdalestes lechei*. (Modified from Kielan-Jaworowska, 1975b.)

10 mm

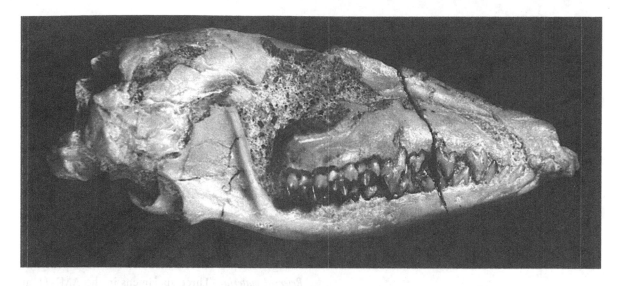

Figure 29.23. Skull of *Asioryctes nemegetensis* Kielan-Jaworowska 1975, ZPAL MgM-I/98, in right lateral view. Red beds of Hermiin Tsav, Hermiin Tsav II, Gobi Desert. ×4. (From Kielan-Jaworowska, 1975b.)

Asioryctes nemegetensis Kielan-Jaworowska, 1975
Holotype. ZPAL MgM-I/56, almost complete skull with both dentaries in occlusion, atlas and axis. Nemegt, Nemegt Basin, Gobi Desert, Mongolia; Baruungoyot Formation (?late Campanian).
Referred material. Ten specimens including skulls or fragmentary skulls with dentaries, and, in one case, associated with an incomplete postcranial skeleton (Kielan-Jaworowska, 1975b, 1977, 1981) in the ZPAL collection from the localities of Nemegt, Khulsan and Hermiin Tsav II, in the Gobi Desert.
Description. Skull length is about 30 mm. The dental formula is 5/4, 1/1, 4/4, 3/3. In lateral view the teeth show the same arrangement as in *Kennalestes*, but the molars differ considerably in being strongly elongated transversely. On the molars there is no pre- and postcingula, the paracone and metacone are situated more labially than in *Kennalestes* and are connate at their bases. The metacone is much shorter than the paracone and the paraconule is larger than the metaconule. The lower molars differ from those in *Kennalestes* in having a smaller paraconid and a trigonid that is shorter in relation to the talonid. The jugal is deeper than in *Kennalestes* and its anterior portion is deep and

meets the maxilla with a sigmoid suture. The coronoid process is larger than in *Kennalestes* and the angular process is similarly inflected.

The postcranial skeleton, which is partly known, shows that the pollex and hallux were not opposable, which together with sedimentological data indicate terrestrial (not scansorial) habits (Kielan-Jaworowska, 1977).

Comments. Kielan-Jaworowska (1981) regarded the following features of the skull structure, characteristic for both *Kennalestes* and *Asioryctes*, as symplesiomorphic therian character states: inclination of the occipital plate forwards from the condyles; development of a basisphenoid wing homologous to the basipterygoid process; a foramen rotundum that is confluent with the sphenorbital fissure; an ectotympanic inclined 45° to the horizontal; medial internal carotid and stapedial arteries present; no entotympanic; a long jugal; a subsquamosal foramen; no paroccipital process, and a medial inflection of the angular process.

Asioryctes has been assigned to the Asioryctinae within Palaeoryctidae (Kielan-Jaworowska, 1975b, 1981; Kielan-Jaworowska *et al.*, 1979), a family allied by some with soricomorph insectivorans (Lillegraven *et al.*, 1981). Nonetheless, as noted above, many aspects of *Asioryctes*, most notably the postcranial skeleton, are extremely primitive (Kielan-Jaworowska, 1977) and even defy a clear association with other selected eutherian clades (Novacek, 1980). Moreover, several features of the anterior tympanic roof and the shape of the postglenoid process do not (*contra* Kielan-Jaworowska, 1981) necessarily point to a close relationship with Early Tertiary North American palaeoryctids (MacPhee and Novacek, 1993).

Genus *Ukhaatherium* Novacek *et al.*, 1997
Type species by monotypy. Ukhaatherium nessovi Novacek, Rougier, Wible, McKenna, Dashzeveg, and Horovitz, 1997.

Ukhaatherium nessovi Novacek *et al.*, 1997
Holotype. PSS-MAE 102, skull with nearly complete skeleton, Ukhaa Tolgod, Nemegt Basin, Gobi Desert, Mongolia; ?Djadokhta Formation (?early Campanian). *Referred material.* PSS-MAE 103–106, skulls and skeletons found in association with the holotype; PSS-MAE 110, skull with lower jaws articulated, but lacking anterior snout and anterior mandible; PSS-MAE 111, complete skull with lower jaws in articulation.
Description. Ukhaatherium is united with *Kennalestes* and *Asioryctes* in the Asioryctitheria by the following characters: drainage for the postglenoid vein within rather than posterior to the glenoid buttress which is developed medially into an entoglenoid process; distinct interfenestral ridge on the promontorium; and large piriform fenestra. *Ukhaatherium* and *Asioryctes* both have: P2 (second upper premolar) smaller than P1; upper molars more strongly elongated transversely, lacking pre- and postcingula; occipital exposure of the mastoid rectangular in outline (triangular in *Kennalestes*); and a large lower foramen on the occipital exposure of the mastoid. *Ukhaatherium* differs from *Asioryctes* in having an enlarged upper canine with a single root (smaller and double-rooted in the latter); a less robust P3 with a less salient paracone; only two mental foramina in the dentary; and a smaller facial process of the lacrimal. See Figure 29.24.

Order Anagalida Szalay and McKenna, 1971
Family Zalambdalestidae Gregory and Simpson, 1926
Genera included. Zalambdalestes Gregory and Simpson, 1926, *Barunlestes* Kielan-Jaworowska, 1975b, *Alymlestes* Aver'yanov and Nesov, 1995.

Genus *Zalambdalestes* Gregory and Simpson, 1926
Type species by monotypy. Zalambdalestes lechei Gregory and Simpson, 1926.

Zalambdalestes lechei Gregory and Simpson, 1926
See Figures 29.22D and 29.25.
Holotype. AMNH 21708, a damaged skull with a large part of the left dentary. Bayan Zag, Gobi Desert, Mongolia; Djadokhta Formation (?early Campanian).
Referred material. Three specimens in the AMNH, at least seven in the PSS, collected by recent MAE expeditions (some of which are still unprepared), 12 in the ZPAL, and one in the PIN collections. These include complete skulls and parts of the postcranial skeleton.
Description. The skull (Kielan-Jaworowska 1984a) is up to 50 mm long, strongly constricted in front of P1, and elongated into a very long, narrow snout. The zygomatic arches are slender and strongly expanded laterally, and the lambdoidal crests are prominent. The dental formula is: 3/3, 1/1, 4/4, 3/3. I1 was apparently small, I2 enlarged, caniniform, and I3 small, with a long diastema between I3 and C. The upper canine is very large, double-rooted, and placed some distance behind the premaxillary–maxillary suture. P1 and P2 are small, P3 is the tallest of all the teeth with a spur-like protocone, while P4 has a protocone developed as in molars, but no metacone. The upper molars lack cingula and are strongly elongated transversely, with incipient conules and a small stylar shelf. M3 is very small with respect to M1 and M2. The i1 is very large and procumbent, but procumbency decreases between i2 and p2. The c is single-rooted. The p4 has a three-cusped trigonid and a talonid without basin. The lower molars have trigonids narrower than the talonids, and the paraconid and metaconid are connate at their bases. The paraconid is very small, the protoconid is the highest, and the talonid is strongly basined.

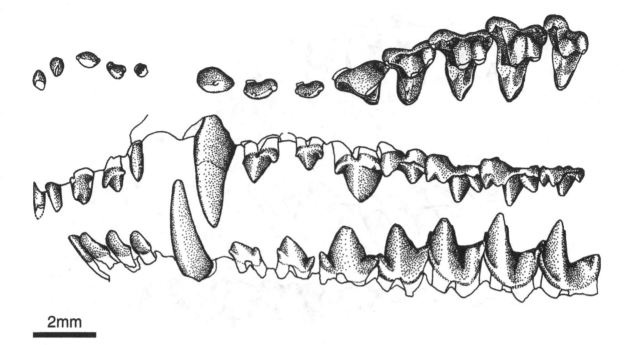

Figure 29.24. Dentition of *Ukhaatherium nessovi* (PSS-MAE 102). Occlusal (top) and labial (middle) views of the left upper dentition; I1–5, C, P1–4, M1–3. Labial view of the lower dentition (bottom); i1–4, c, p1–4, m1–3. Lower dentition is a composite of the right i1–3 and c, and the left i4, p1–4, and m1–3.

The structure of the braincase and postcranial skeleton will be discussed together with *Barunlestes*.

Genus *Barunlestes* Kielan-Jaworowska, 1975b
Type species by monotypy. Barunlestes butleri Kielan-Jaworowska, 1975b.

Barunlestes butleri Kielan-Jaworowska, 1975b
See Figures 29.22C and 29.26.
Holotype. ZPAL MgM-I/77, a damaged skull with both dentaries and a large part of the postcranial skeleton. Khulsan, Nemegt Basin, Gobi Desert, Mongolia; Baruungoyot Formation (?late Campanian).
Referred material. Five specimens in the ZPAL, one in the PIN and one in the PSS collections, representing incomplete skulls.
Description. Barunlestes differs from *Zalambdalestes* in having a shorter and somewhat more robust skull (about 35 mm long), a single-rooted upper canine, and only three upper premolars (the P2 is lacking). In

addition, the dentary is deeper and there is a higher coronoid process with a powerful coronoid crest, equipped with a knob-like projection and a medial prominence.

Although *Zalambdalestes* and *Barunlestes* differ in dental formulae and the shape of the skull and dentary, their braincase structure is closely similar (Kielan-Jaworowska and Trofimov, 1980; Kielan-Jaworowska, 1984a; Novacek *et al.*, in prep.). The zalambdalestid braincase (Figure 29.26) is more inflated and the mesocranial region is shorter than in *Kennalestes* and *Asioryctes*. The occipital plate is inclined forwards from the condyles, the maxilla extends backwards along the choanae, and the presphenoid has a prominent median process. There is a large pterygoid process of the basisphenoid and the postglenoid process extends only along the medial part of the cupola-like glenoid fossa. The promontorium is flattened and lacks definite grooves for trans-promontorian arteries, the carotid foramen is placed

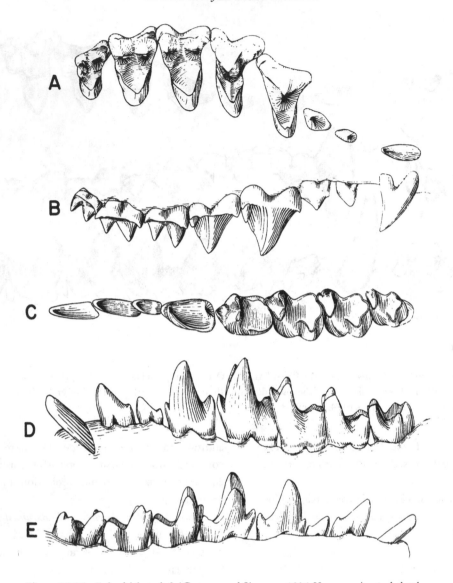

Figure 29.25. *Zalambdalestes lechei* Gregory and Simpson, 1926. Upper canine and cheek teeth in (A) occlusal and (B) labial views. Lower canine and cheek teeth in (C) occlusal, (D) labial and (E) lingual views. × 6. (From Kielan-Jaworowska, 1969a.)

medially, and there is a foramen arteriae stapediae and sulcus arteriae stapediae, but no sulcus arteriae promontorii. This arrangement shows that the carotid arteries, the main channels supplying blood to the brain, entered the skull along the midline rather than at the sides, as they do in most living mammals.

The postcranial skeleton (Figure 29.27), parts of

which are preserved in *Zalambdalestes* and parts in *Barunlestes*, was probably similar in all zalambdalestids and has been reconstructed on evidence of both genera (Kielan-Jaworowska, 1978). This skeleton shows a mosaic of primitive and advanced characters. The axis has a very long spinous process, but the thoracic vertebrae bear only short spinous processes. The

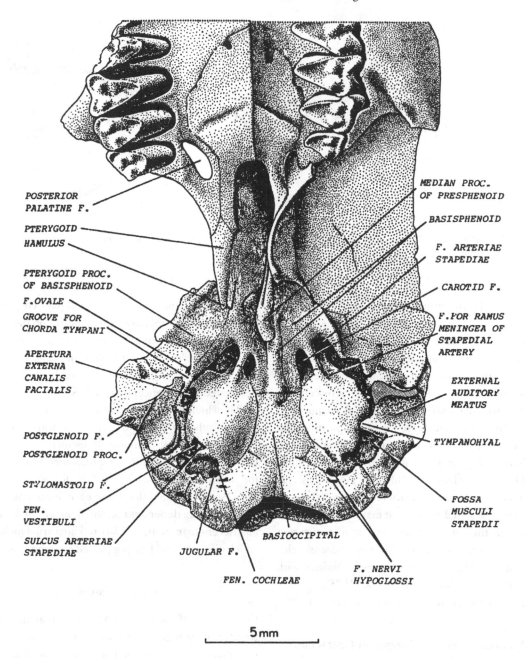

Figure 29.26. *Barunlestes butleri* Kielan-Jaworowska, 1975, PIN 3142–701, ventral view of the braincase. Red beds of Hermiin Tsav, Hermiin Tsav II, Gobi Desert. Abbreviations: F., foramen; FEN., fenestra; PROC, process. (From Kielan-Jaworowska and Trofimov, 1980.)

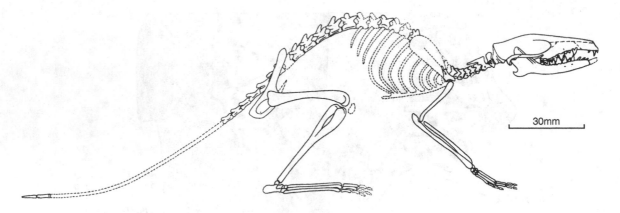

Figure 29.27. *Zalambdalestes lechei* Gregory and Simpson, 1926. Reconstruction of the postcranial skeleton, partly based on *Barunlestes*. (From Kielan-Jaworowska, 1978.)

tibia and fibula are strongly fused, and a calcaneal fibular facet is lacking, but the tibial trochlea on the astragalus is well developed. The hind limbs, especially the metatarsals, are very long. It has been presumed that the locomotion and mode of life of the zalambdalestids was similar to that of macroscelidids, that is quadrupedal walking, running and jumping.

The zalambdalestids are intriguing in some respects. Despite a number of curious specializations, including greatly enlarged lower incisors and elongated hindlimbs, zalambdalestids were given the status of Proteutheria *incertae sedis* (Kielan-Jaworowska *et al.*, 1979), with the added notion that many of the skeletal features of this group resemble (in a convergent fashion) modern, saltatorial macroscelideans (elephant shrews). Others have suggested affinities with lagomorphs or rodents (McKenna, 1975), a view rejected, however, by Kielan-Jaworowska *et al.* (1979).

Endocranial casts of Mongolian Cretaceous eutherian mammals

The endocranial casts that are partly or entirely preserved in four out of five eutherian genera discussed above (Kielan-Jaworowska 1984b, 1986, and references therein), are the oldest known eutherian endocasts (Figure 29.28). They indicate a primitive therian lissencephalic brain with very large olfactory bulbs and cerebral hemispheres widely separated posteriorly. There is a large midbrain exposure on the dorsal side and a comparatively short and wide cerebellum with well developed cerebellar hemispheres. This construction belongs to the type of brain structure designated eumesencephalic by Kielan-Jaworowska (1986), which is very different from that characteristic of multituberculates and Triconodontidae. Kielan-Jaworowska (1984b) tentatively estimated encephalization quotients of 0.36 for *Kennalestes gobiensis*, 0.56 for *Asioryctes nemegetensis* and 0.70 for *Zalambdalestes lechei*. She concluded that these early eutherians were probably more dependent on smell than most Tertiary and Recent mammals, and favored nocturnal niches in which olfaction and hearing played an important role.

Comparisons

Comparison of Mongolian Mesozoic mammals with those of other regions is limited because Late Triassic mammals have not been found in Mongolia, while those from the Jurassic are represented only by a single tooth of a docodont (Tatarinov, 1994; see also section on the Docodonta above). The record of Cretaceous Mongolian mammalian assemblages is more complete. The Cretaceous mammals of Gondwana developed in isolation from those of Holarctica, and do not invite a close comparison (see

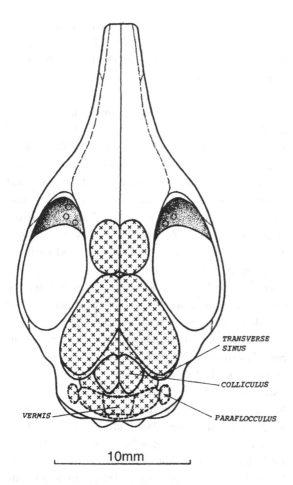

Figure 29.28. *Kennalestes gobiensis* Kielan-Jaworowska 1969. Reconstruction of the endocranial cast. (From Kielan-Jaworowska, 1986.)

Bonaparte and Kielan-Jaworowska, 1987; Bonaparte, 1990; and Sigogneau-Russell, 1991a, b, 1995; Rich *et al.*, 1997; Kielan Jaworowska *et al.*, 1998 and Flynn *et al.*, 1999 for reviews and references). Thus, we compare Mongolian Cretaceous mammal assemblages with those from other areas of the Holarctica. As the Cretaceous mammals of Europe are hardly known, comparisons are largely confined to North American and southwestern Asian taxa.

The Early Cretaceous Mongolian mammals are known only from Aptian or Albian Höövör beds, at Höövör, in the Gobi Desert and a single specimen of a 'triconodont' from southeastern Mongolia (Reshetov and Trofimov, 1980). A rich mammalian assemblage encountered at Höövör consists of 'triconodonts', multituberculates, symmetrodonts, 'eupantotheres', aegialodontids, and eutherians. Höövör mammals, except for the 'triconodont' *Gobiconodon*, and the multi-tuberculate *Eobaatar*, are endemic at the generic level. *Gobiconodon* has been reported in the North American Cloverly Formation (Jenkins and Schaff, 1988) and in the Early Cretaceous of Siberia (Maschenko and Lopatin, 1998) and is represented in Asia and North America by different species. *Eobaatar* also occurs in the Barremian of Spain, where it is represented by *E. hispanicus* Hahn and Hahn, 1992, based on isolated teeth.

Out of eight mammal families or subfamilies represented at Höövör, three are endemic to Mongolia. These are: the highly specialized multituberculate family Arginbaataridae and the two 'eupantotherian' families Arguimuridae and Arguitheriidae. The latter two families are poorly known, however, and based on genera represented by incomplete dentaries. It cannot be excluded that when better known, the type genera of these families may be assigned to other families, known from other regions.

Five mammalian families from Höövör are known also from other regions. The 'triconodont' family Gobiconodontidae occurs in Mongolia, Siberia, North America, and in the Middle or Late Jurassic of north-western China (Chow and Rich, 1984). The 'plagiaula-coid' family Eobaataridae contains, in addition to two Mongolian genera, *Eobaatar* (known from Mongolia and Spain, see Hahn and Hahn, 1992), and the tentatively assigned *Monobaatar*, also *Loxaulax* from the Valanginian (Wealden) of Great Britain. The symmetrodont family Amphidontidae comprises *Amphidon* from the latest Jurassic Morrison Formation of North America, *Gobiodon* from Mongolia, *Manchurodon* from the ?Early Cretaceous of China (Yabe and Shikama, 1938) and *Nakunodon* from the Early Jurassic Kota Formation of India (Yadagiri, 1985). The aegialodontian family Aegialodontidae comprises the Mongolian *Kielantherium* and *Aegialodon* from the Valanginian of Great Britain. To the eutherian family Otlestidae, in addition to the Mongolian form *Prokennalestes*, belongs also *Otlestes* from the early Cenomanian of Uzbekistan.

The mammalian fauna from the Antlers Formation (Albian) of Texas is known almost exclusively from isolated teeth, which do not invite a close comparison with Höövör mammals (see Butler, 1978, and references therein). Multituberculates from the same formation, although found half a century ago, have not been described, except for an abstract (Krause *et al.*, 1990). Multituberculates from the late Albian Cedar Mountain Formation of Utah are represented by several species of *Paracimexomys* and these teeth are of uncertain affinity. Eaton and Nelson (1991) assigned *Paracimexomys* to ?Ptilodontoidea and concluded (p. 11): 'Comparison of the material from Asia and Utah provided no evidence of multituberculate exchange between Asia and North America during the late Early Cretaceous'.

Mammals from the early part of the Late Cretaceous have not been recorded in Mongolia. Mammalian assemblages from this interval are rare and generally poorly known, although in the past two decades new discoveries have been made in North America and Asia. North American mammals from Utah were described by Cifelli and Eaton (1987), Eaton and Cifelli (1988), Cifelli (1990, and references therein) and Eaton (1995).

Nesov published a series of papers with descriptions of mammalian assemblages from the vast territory of southwestern Asia (Uzbekistan, Kazakhstan and Tadzhikistan, traditionally referred to in the Soviet literature as 'Middle Asia'). The Cretaceous beds encountered there range possibly from Berriasian to Maastrichtian, but mammals are known only from Late Albian to the Campanian (summarized by Nesov, 1985; Nesov and Kielan-Jaworowska, 1991; Nesov *et al.*, 1994; personal communication from A.O. Aver'yanov, see also Chapter 30).

Late Cretaceous mammal assemblages of Holarctica are much better known. In Mongolia numerous Late Cretaceous mammals occur in the Djadokhta and Baruungoyot Formations, which are of ?early and ?late Campanian age. In the younger Nemegt Formation mammals have not been found,

but the so called deltatheroidan skull from Guriliin Tsav and the multituberculate *Buginbaatar* from Bügiin Tsav region were found in beds possibly equivalent to the Nemegt Formation.

In North America the Late Cretaceous (especially Campanian–Maastrichtian) mammal faunas are rich and diversified (see Lillegraven *et al.*, 1979, and Cifelli, 1990, for reviews and references). While Late Cretaceous mammals from Mongolia are represented by multituberculates, eutherians, deltatheroidans, and endemic asiadelphians, in North America, these groups co-occur with 'triconodonts' and symmetrodonts (Fox, 1969, 1976, 1985). Thus in North America 'triconodonts' and symmetredonts survived until the Campanian.

During the Late Cretaceous, the North American multituberculate assemblages were dominated by the Ptilodontoidea, the Taeniolabidoidea making their appearance only at the very end of the Cretaceous period (during the Maastrichtian). By contrast, representatives of Ptilodontoidea have not been found in Asia (Kielan-Jaworowska, 1980). Until recently, Mongolian Late Cretaceous multituberculates were assigned to the Taeniolabidoidea. Recent analyses (Rougier *et al.*, 1997; Kielan-Jaworowska and Hurum, 1997) demonstrated, however, that all the Campanian Mongolian multituberculates belong to a separate monophyletic clade, for which Kielan-Jaworowska and Hurum (1997) erected the suborder Djadochtatheria, regarded by us as an infraorder within the suborder Cimolodonta. Thus the Campanian multituberculate assemblage of Mongolia is very different from that of North America at higher taxonomic levels. The only Late Cretaceous Mongolian multituberculate taxon not assigned to Djadochtatheria is *Buginbaatar*, which is tentatively attributed to the ?Cimolomyidae. This taxon occurs in beds possibly equivalent to the Nemegt Formation and of ?early Maastrichtian age. The presence of rare putative djadochtatherians in North America is doubtful.

The composition of the Late Cretaceous North American and Mongolian therian assemblages is also distinctly different. While marsupials prevail in North

America and eutherians are more common only towards the end of the Cretaceous, in Mongolia the marsupials are rare. Mongolian Metatheria are represented by *Asiatherium* and *Marsasia*, assigned to the endemic order Asiadelphia, and by representatives of a more common order Deltatheroida also known from the Cretaceous of North America (Cifelli, 1993).

Only five Late Cretaceous Mongolian eutherian genera (all of which are monotypic) have been described: *Kennalestes*, *Asioryctes*, *Ukhaatherium*, *Zalambdalestes*, and *Barunlestes*. All are endemic to Mongolia although Zalambdalestidae are known from Uzbekistan and Kazakhstan (Aver'yanov and Nesov, 1995; Aver'yanov, 1997). Assignment of these taxa to orders is debated (e.g., McKenna and Bell, 1997; Novacek *et al.*, 1997) and here we placed them in the orders Asioryctitheria and Anagalida.

Late Cretaceous mammals from other areas of Asia are of particular interest for comparisons with Mongolian assemblages. The closest are those from Kazakhstan (Aver'yanov and Nesov, 1995; Aver'yanov, 1997, and references therein). Aver'yanov (1997) demonstrated that the Kazakhstan assemblage contains, at the family level, taxa from the Turonian–Coniacian Bissekty Formation of Uzbekistan (Nesov, 1990) and from the Campanian of Mongolia. Among the mammal families discovered in Kazakhstan, the 'Zhelestidae' (Archibald, 1996a; Nesov *et al.*, 1998) are known from Uzbekistan and Japan (Setoguchi *et al.*, 1999) and the Deltatheridiidae occur both in Mongolia and Uzbekistan. Zalambdalestidae are known from Mongolia and Kazakhstan (Aver'yanov, 1997). The single zalambdalestid *?Zalambdalests nymbulakensis* Nesov, 1985, reported by Nesov from the Coniacian of Uzbekistan is a junior synonym of *Sorlestes budan* Nesov, 1985 (see Nesov *et al.*, 1994, and references therein). One multituberculate genus, *Bulganbaatar*, represented in Kazakhstan by a single tooth, is characteristic for the Mongolian Djadokhta Formation.

The assemblages of the Late Cretaceous southwestern Asian mammals (Uzbekistan and Tadzhikistan) differ markedly from those of Mongolia. The southwestern Asian faunas are rich in eutherian taxa, and also include deltatheroidans (Kielan-Jaworowska and Nesov, 1990), while multituberculates are extremely rare. Kielan-Jaworowska and Nesov (1992, p. 2) stated: 'Multituberculates were apparently very rare on southwestern coastal plains of Cretaceous Asia. The latest Albian, Early Cenomanian and Early Santonian–Middle Campanian mammals of these regions are represented only by eutherians while multituberculates are not known (Nesov & Kielan-Jaworowska, 1991). Only the Coniacian of Uzbekistan has yielded uncontested multituberculates, which represent not more than one percent of the known mammalian specimens. This contrasts with the Late Cretaceous assemblages from Mongolia and North America (Lillegraven *et al.*, 1979, and references therein) where multituberculate specimens represent 50%–70% of the mammalian assemblages, and in the Hell Creek (Sloan & Van Valen, 1965) and Lance Formations (Krause, 1986) they are even more numerous'.

Perhaps the most intriguing feature of these southwestern Asian Late Cretaceous mammal assemblages is the occurrence in the Bissekty Formation (Coniacian) of eutherians that have ungulate affinities. These fossils, plus possibly some others from North America and Europe, comprise the family 'Zhelestidae' (Nesov, 1985; Nesov and Kielan-Jaworowska, 1991). Archibald (1996a) argued that the 'Zhelestidae' are paraphyletic and therefore cited it in quotation marks. In the same paper Archibald erected the supergrandorder Ungulatomorpha, which comprises the 'Zhelestidae' and all later Ungulata. The rich fauna of southwestern Asian Late Cretaceous 'zhelestids' has been described in detail by Nesov *et al.* (1998).

Nesov attempted to explain the differences between the Late Cretaceous mammal assemblages of southwestern Asia and Mongolia, using palaeobiogeographic and palaeoclimatologic data (see Nesov and Kielan-Jaworowska, 1991; Nesov *et al.*, 1994, and Nesov *et al.*, 1998, for summaries). It is generally accepted (see Jerzykiewicz and Russell, 1991, and Jerzykiewicz *et al.*, 1993) that the Mongolian Campanian Djadokhta and Baruungoyot formations were deposited in inland areas with semi-arid and arid climates. The Upper Cretaceous deposits of south-western Asia (especially the best known Coniacian Bissekty Formation) were deposited, as argued by Nesov, on low coastal plains in semi-humid subtropical conditions and the same environment was characteristic for most Upper Cretaceous formations of North America (Archibald 1996b). In the Late Cretaceous vertebrate assemblages from southwestern Asia and the western part of North America there were (in addition to mammals), sharks, bony fishes, amphibians, turtles, and crocodilians, all of which are much less common or absent in Mongolian sites.

As stated above, multituberculates were very common during the Late Cretaceous of Mongolia and western North America (although represented on both continents by different suborders), while they were very rare in southwestern Asia (Kielan-Jaworowska and Nesov, 1992). Nesov *et al.* (1998) hypothesized that the 'zhelestids' and multituberculates were ecological competitors. According to these authors the lack of multituberculates in Upper Cretaceous deposits of southwestern Asia may be explained by the diversification of the 'zhelestids' (the first eutherians adapted to herbivorous niches) which ecologically replaced the multituberculates. Nesov *et al.* (1998) also argued that the 'zhelestids' flourished in western Asia during the Coniacian, preceding by some 20 million years the first appearance of the archaic ungulates in North America.

Although the Late Cretaceous Asian mammal assemblages are less diversified than those from the western North America and southwestern Asia, they are of great value because of their unparalleled preservation. While most Mesozoic mammal sites in the world yield isolated teeth, dentaries and maxillae with teeth, complete skulls with postcranial skeletons are only very rarely found. By contrast, Mongolian Late Cretaceous mammals are represented, as a rule, by entire skulls, often associated with postcranial skeletons.

Most of the Late Cretaceous multituberculates belong to a different infraorder than those from other areas, but they provided very important anatomical information on the multituberculate structure as a whole. Moreover, the general debate on the phylogenetic position of multituberculates among mammals is also largely based on the anatomical data provided by the Late Cretaceous Mongolian multituberculates.

The eutherian Late Cretaceous Mongolian mammals, although currently represented by only five taxa, also provide important data on the skull, brain structure, and the postcranial skeleton of early eutherians. The Cretaceous marks the juxtaposition of several archaic Mesozoic lineages with the emergence of clades related to extant mammals. In many cases the taxa from Mongolia provide the only evidence, beyond teeth, for these critical branches. There are, however, many ambiguities remaining with respect to phylogenetic patterns. It is hoped that newly discovered material of Late Cretaceous Mongolian mammals, as well as ongoing anatomical and phylogenetic analysis will improve this picture.

References

Anonymous. 1983. [*Palaeontological Institute of the USSR Academy of Sciences*]. Vneshtorgizdat, Izdatel'stvo No. 3496. 35 pp.

Archibald, J.D. 1982. A study of Mammalia and geology across the Cretaceous-Tertiary boundary in Garfield County, Montana. *University of California Publications in Geological Sciences* 122: 1–286.

—1996a. Fossil evidence for a Late Cretaceous origin of 'hoofed' mammals. *Science* 272: 1150–1153.

—1996b. *Dinosaur Extinction and the End of an Era: what the Fossils Say.* New York: Columbia University Press, XV + 237 pp.

Aver'yanov, A.O. 1997. New Late Cretaceous mammals of southern Kazakhstan. *Acta Palaeontologica Polonica* 42: 243–256.

—and Kielan-Jaworowska, Z. 1999. Marsupials from the Late Cretaceous of Uzbekistan. *Acta Palaeontologica Polonica* 44:71–81.

—and Nesov, L.A. 1995. A new Cretacous mammal from the Campanian of Kazakhstan. *Neues Jahrbuch für Geologie und Paläontologie, Monatshefte* 1995: 65–74.

Bakker, R.T. 1992. Zofiabaataridae, a new family of multituberculate mammals from the Breakfast Bench fauna at Como Bluff. *Hunteria* 2: 24.

Belyaeva, E.I., Trofimov, B.A. and Reshetov, V.J. 1974. [General stages in evolution of late Mesozoic and early Tertiary mammalian faunas in Central Asia.] pp. 19–45. *Trudy Sovmestnoi Sovetsko-Mongol'skoi Paleontologicheskoi Ekspeditsii* 1: 19–45.

Bonaparte, J.F. 1990. New Late Cretaceous mammals from the Los Alamitos Formation, Northern Patagonia. *National Geographic Research* 6: 63–93.

—and Kielan-Jaworowska, Z. 1987. Late Cretaceous dinosaur and mammal faunas of Laurasia and Gondwana, pp. 24–29 in Currie, P.J. and Koster, E.H. (eds.), *Fourth Symposium on Mesozoic Terrestrial Ecosystems. Short Papers. Occasional Papers of the Tyrrell Museum of Palaeontology, Drumheller* 3.

Butler, P.M. 1978. A new interpretation of the mammalian teeth of tribosphenic pattern from the Albian of Texas. *Breviora* 446: 1–27.

—1988. Docodont molars as tribosphenic analogues (Mammalia, Jurassic), pp. 329–340 in Russell, D.E., Santoro, J.P. and Sigogneau-Russell, D. (eds.), *Teeth Revisited: Proceedings of the VIIth International Symposium on Dental Morphology. Mémoirs du Muséum National d'Histoire Naturelle, Paris* 53.

—1990. Early trends in the evolution of tribosphenic molars. *Biological Reviews* 65: 529–552.

—and Kielan-Jaworowska, Z. 1973. Is *Deltatheridium* a marsupial? *Nature* 245: 105–106.

Carroll, R.L. 1988. *Vertebrate paleontology and evolution.* New York: W.H. Freeman and Company. xiv+698pp.

Chow, M. and Rich, C. 1984. A new triconodontan (Mammalia) from the Jurassic of China. *Journal of Vertebrate Paleontology* 3: 226–231.

Cifelli, R.L. 1990. Cretaceous mammals of southern Utah. IV. Eutherian mammals from the Wahweap (Aquilan) and Kaiparowits (Judithian) formations. *Journal of Vertebrate Paleontology* 10: 343–360.

—1993. Early Cretaceous mammal from North America and the evolution of marsupial dental characters. *Proceedings of the National Academy of Sciences, USA* 90: 9413–9416.

—1999. Tribosphenic mammal from the North American Early Cretaceous. *Nature* 401: 363–366.

—and Eaton, J.G. 1987. Marsupial from the earliest Late Cretaceous of Western US. *Nature* 325: 520–522.

—and Muizon, C. de 1997. Dentition and jaw of *Kopellia juddi*, a primitive marsupial or near-marsupial from the Medial Cretaceous of Utah. *Journal of Mammalian Evolution* 4: 241–258.

Clemens, W.A. and Kielan-Jaworowska, Z. 1979. Multituberculata, pp. 99–149 in Lillegraven, J.A., Kielan-Jaworowska, Z. and Clemens, W.A. (eds.), *Mesozoic Mammals: The First Two-Thirds of Mammalian History.* Berkeley: University of California Press.

Dashzeveg, D. 1975. New primitive therian from the Early Cretaceous of Mongolia. *Nature* 256: 402–403.

—1979. *Arguimus khosbajari* gen. n., sp. n. (Peramuridae, Eupantotheria) from the Lower Cretaceous of Mongolia. *Acta Palaeontologica Polonica* 24: 199–204.

—1994. Two previously unknown eupantotheres (Mammalia, Eupantotheria). *American Museum Novitates* 3107: 1–11.

—and Kielan-Jaworowska, Z. 1984. The lower jaw of an aegialodontid mammal from the Early Cretaceous of Mongolia. *Zoological Journal of the Linnean Society* 82: 217–227.

—, Novacek, M.J., Norell, M.A., Clark, J.M., Chiappe, L.M., Davidson, A., McKenna, M.C., Dingus, L., Swisher, C. and Altangerel, P. 1995. Extraordinary preservation in a new vertebrate assemblage from the Late Cretaceous of Mongolia. *Nature* 374: 446–449.

Eaton J.G. 1995. Cenomanian and Turonian (early Late Cretaceous) multituberculate mammals from southwestern Utah. *Journal of Vertebrate Paleontology* 15: 761–784.

—and Cifelli, R.L. 1988. Preliminary report on Late Cretaceous mammals of the Kaiparowits Plateau, southern Utah. *Contributions to Geology, University of Wyoming* 26: 45–55.

—and Nelson, M.E. 1991. Multituberculate mammals from the Lower Cretaceous Cedar Mountain Formation, San Rafael Swell, Utah. *Contributions to Geology, University of Wyoming* 29: 1–12.

Engelmann, G.F. and Callison, G. 1999. *Glirodon grandis*, a new multituberculate mammmal from the Upper Jurassic Morrison Formation. In Gillette, D.D. (ed.), *Vertebrate Paleontology in Utah. Utah Geological Survey Publication* 99-1: 161–178.

Flynn, J.J., Parrish, J.M., Rakotosamimanana, B., Simpson, W.F. and Wyss, A.R. 1999. A middle Jurassic mammal from Madagascar. *Nature* 401: 57–60.

Fox, R.C. 1969. Studies of Late Cretaceous vertebrates. III. A triconodont mammal from Alberta. *Canadian Journal of Zoology* 47: 1253–1256.

—1974. *Deltatheroides*-like mammals from the Upper Cretaceous of North America. *Nature* 249: 392.

—1976. Additions to the mammalian local fauna from the

Upper Milk River Formation (Upper Cretaceous), Alberta. *Canadian Journal of Earth Sciences* **13**: 1105–1118.

—1978. Upper Cretaceous terrestrial vertebrate stratigraphy of the Gobi Desert (Mongolian People's Republic) and western North America, pp. 577–594 in Stelck, C.R. and Chatterton, D.E. (eds.), *Western and Arctic Canadian Biostratigraphy. Geological Association of Canada, Special paper* 18.

—1980. *Picopsis pattersoni*, n. gen. and sp., an unusual therian from the Upper Cretaceous of Alberta, and the classification of primitive tribosphenic mammals. *Canadian Journal of Earth Sciences* 17: 1489–1498.

—1985. Upper molar structure in the Late Cretaceous symmetrodont *Symmetrodontoides* Fox, and a classification of the Symmetrodonta (Mammalia). *Journal of Paleontology* 59: 21–26.

—1999. The monophyly of the Taeniolabidoidea (mammalia: Multituberculata), p. 26. *In* Leanza, H.A. (ed.) *VII International Symposium on Mesozoic Terrestrial Ecosystems. Abstracts.* Buenos Aires, 64 pp.

Freeman, E.F. 1979. A Middle Jurassic mammal bed from Oxfordshire. *Palaeontology* 22: 135–166.

Gambaryan, P.P. and Kielan-Jaworowska, Z. 1995. Masticatory musculature of Asian taeniolabidoid multituberculate mammals. *Acta Palaeontologica Polonica* 40: 45–108.

—and— 1997. Sprawling *versus* parasagittal stance in multituberculate mammals. *Acta Palaeontologica Polonica* 42: 13–44.

Gingerich, P.D. 1977. Patterns of evolution in the mammalian fossil record, pp. 469–500 in Hallam, A. (ed.), *Patterns of Evolution.* Amsterdam: Elsevier Scientific Publishing Company.

Gradziński, R., Kielan-Jaworowska, Z. and Maryańska, T. 1977. Upper Cretaceous Djadokhta, Barun Goyot and Nemegt formations of Mongolia, including remarks on previous subdivisions. *Acta Geologica Polonica* 27: 281–318.

Granger, W. and Simpson, G.G. 1929. A revision of the Tertiary Multituberculata. *Bulletin of the American Museum of Natural History* 56: 601–676.

Gregory, W.K. and Simpson, G.G. 1926. Cretaceous mammal skulls from Mongolia. *American Museum Novitates* 225: 1–20.

Hahn, G., 1969. Beiträge zur Fauna der Grube Guimarota Nr.3. Die Multituberculata. *Palaeontographica, Abteilung* A 133: 1–100.

—1993. The systematic arrangement of the Paulchoffatiidae (Multituberculata) revisited. *Geologica et Palaeontologica* 27: 201–214.

—and Hahn, R. 1983. *Multituberculata*, in Westphal, F. (ed.), *Fossilium Catalogus, I Animalia*, Pars 127, Amsterdam, Kugler Publications: 409 pp.

—and— 1992. Neue Multituberculaten-Zähne aus der Unter-Kreide (Barremium) von Spain (Galve und Uña). *Geologica et Palaeontologica* 26: 143–162.

—and— 1999. Pinheirodontidae n. fam. (multituberculata) (mammalia) aus der tiefen Unter-Kreide Portugals. *Palaeontographica, Abteilung, A,* 253: 1–146.

—Lepage, J.C. and Wouters, G. 1987. Ein Multituberculaten-Zahn aus der Ober-Trias von Gaume (S-Belgien). *Bulletin de la Société Belge de Géologie* 96: 39–47.

Hopson, J.A. 1994. Synapsid evolution and the radiation of non-therian mammals, pp. 190–219 in Prothero, D.R. and Schoch, R.M. (eds.), *Major Features of Vertebrate Evolution. Short Courses in Paleontology* 7: 190–219.

—1995. The Jurassic mammal *Shuotherium dongi*: 'pseudotribosphenic therian', docodontid or neither? *Journal of Vertebrate Paleontology* 15, Supplement to No. 3: 36A.

—, Kielan-Jaworowska, Z. and Allin, E.F. 1989. The cryptic jugal of multituberculates. *Journal of Vertebrate Paleontology* 9: 201–209.

Hu, Y., Wang, Y., Luo, Z. & Li, Ch. 1997. A new symmetrodont mammal from China and its implications for mammalian evolution. *Nature* 390: 127–142.

Hurum, J.H. 1994. The snout and orbit of Mongolian multituberculates studied by serial sections. *Acta Palaeontologica Polonica* 39: 181–222.

—1998a. The braincase of two Late Cretaceous Asian multituberculates studied by serial sections. *Acta Palaeontologica Polonica* 43: 21–52.

—1998b. The inner ear of two Late Cretaceous multituberculate mammals and its implications for multituberculate hearing. *Journal of Mammalian Evolution* 5: 65–94.

—Presley, R. and Kielan-Jaworowska, Z. 1996. The middle ear in multituberculate mammals. *Acta Palaeontologica Polonica* 41: 253–275.

International Commission on Zoological Nomenclature. 1999. *International Code of Zoological Nomenclature. Fourth Edition.* International Trust for Zoological nomenclature, London. The Natural History Museum, London, xxix + 306 pp.

Jenkins F.A. Jr., and Crompton, A.W. 1979. Triconodonta,

pp. 74–90 in Lillegraven, J.A., Kielan-Jaworowska, Z. and Clemens, W.A. (eds.), *Mesozoic Mammals: The First Two-Thirds of Mammalian History*. Berkeley, University of California Press.

—Crompton, A.W. and Downs, W.R. 1983. Mesozoic mammals from Arizona: new evidence on mammalian evolution. *Science* 222: 1233–1235.

—and Schaff, C.R. 1988. The Early Crataceous mammal *Gobiconodon* (Mammalia, Triconodonta) from the Cloverly Formation in Montana. *Journal of Vertebrate Paleontology* 6: 1–24.

Jepsen, G.L. 1940. Paleocene faunas of the Polecat Bench Formation, Park County, Wyoming. *Proceedings of the American Philosophical Society* 83: 217–341.

Jerzykiewicz, T., Currie, P.J., Eberth, D.A., Johnston, P.A., Koster, E.H. and Zheng, J.-J. 1993. Djadokhta Formation correlative strata in Chinese Inner Mongolia: an overview of the stratigraphy, sedimentary geology, and paleontology and comparisons with the type locality in the pre-Altai Gobi. *Canadian Journal of Earth Sciences* 30: 2180–2195.

—and Russell, D.A. 1991. Late Mesozoic stratigraphy and vertebrates of the Gobi Basin. *Cretaceous Research* 12: 345–377.

Ji, Q., Luo, Z. and Ji, S. 1999. A Chinese triconodont mammal and mosaic evolution of the mammalian skeleton. *Nature* 398: 326–330.

Kemp, T.S. 1983. The relationships of mammals. *Zoological Journal of the Linnean Society* 77: 353–384.

Kermack, K.A. 1967. The Aegialodontidae – a new family of Cretaceous mammals. *Proceedings of the Geological Society of London* 1640: 146.

—1988. British Mesozoic mammal sites. *Special Papers in Palaeontology* 40: 85–93.

—, Lee, A.J., Lees, P.M. and Mussett, F. 1987. A new docodont from the Forest Marble. *Zoological Journal of the Linnean Society* 89: 1–39.

—, Lees, P.M. and Mussett, F. 1965. *Aegialodon dawsoni*, a new tuberculosectorial tooth from the Lower Wealden. *Proceedings of the Royal Society of London, series B* 162: 535–554.

—and Mussett, F. 1958. The jaw articulation of the Docodonta and the classification of Mesozoic mammals. *Proceedings of the Royal Society of London, series B* 148: 204–215.

—, Mussett, F. and Rigney, H.W. 1973. The lower jaw of *Morganucodon. Zoological Journal of the Linnean Society* 53: 87–175.

Kielan-Jaworowska, Z. 1969a. Preliminary data on the Upper Cretaceous eutherian mammals from Bayan Dzak, Gobi Desert. *Palaeontologia Polonica* 19: 171–191.

—1969b. Discovery of a multituberculate marsupial bone. *Nature* 222: 1091–1092.

—1970. New Upper Cretaceous multituberculate genera from Bayan Dzak, Gobi Desert. *Palaeontologia Polonica* 21: 35–49.

—1971. Skull structure and affinities of the Multituberculata. *Palaeontologia Polonica* 25: 5–41.

—1974. Multituberculate succession in the Late Cretaceous of the Gobi Desert (Mongolia). *Palaeontologia Polonica* 30: 23–44.

—1975a. Evolution of the therian mammals in the Late Cretaceous of Asia. Part I. Deltatheridiidae. *Palaeontologia Polonica* 33: 103–132.

—1975b. Preliminary description of two new eutherian genera from the Late Cretaceous of Mongolia. *Palaeontologia Polonica* 33: 5–16.

—1975c. Possible occurrence of marsupial bones in Cretaceous eutherian mammals. *Nature* 255: 698–699.

—1977. Evolution of the therian mammals in the Late Cretaceous of Asia. Part II. Postcranial skeleton in *Kennalestes* and *Asioryctes. Palaeontologia Polonica* 37: 65–84.

—1978. Evolution of the therian mammals in the Late Cretaceous of Asia. Part III. Postcranial skeleton in Zalambdalestidae. *Palaeontologia Polonica* 38: 5–41.

—1979. Pelvic structure and nature of reproduction in Multituberculata. *Nature* 277: 402–403.

—1980. Absence of ptilodontoidean multituberculates from Asia and its palaeogeographic implications. *Lethaia* 13: 169–173.

—1981. Evolution of the therian mammals in the Late Cretaceous of Asia. Part IV. Skull structure in *Kennalestes* and *Asioryctes. Palaeontologia Polonica* 42: 25–78.

—1984a. Evolution of the therian mammals in the Late Cretaceous of Asia. Part V. Skull structure in Zalambdalestidae. *Palaeontologia Polonica* 46: 107–117.

—1984b. Evolution of the therian mammals in the Late Cretaceous of Asia. Part VI. Endocranial casts of eutherian mammals. *Palaeontologia Polonica* 46: 157–171.

—1984c. Evolution of the therian mammals in the Late Cretaceous of Asia. Part VII. Synopsis. *Palaeontologia Polonica* 46: 173–183.

—1986. Brain evolution in Mesozoic mammals.

Contributions to Geology, University of Wyoming, Special Paper **3**: 21–34.

—1994. A new generic name for the multituberculate mammal '*Djadochtatherium*' *catopsaloides*. *Acta Palaeontologica Polonica* **39**: 134–136.

—1997. Characters of multituberculates neglected in phylogenetic analyses of early mammals. *Lethaia* **29**: 249–266.

—1998. Humeral torsion in multituberculate mammmals. *Acta Palaeontologica Polonica* **43**: 131–1324.

—and Bonaparte, J.F. 1996. Partial dentary of a multituberculate mammal from the Late Cretaceous of Argentina and its taxonomic implications. *Revista del Museo Argentino de Ciencias Naturales 'Bernardino Rivadavia'* N.S. **145**: 1–9.

—Bown, T.M. and Lillegraven, J.A. 1979. Eutheria, pp. 221–258, in Lillegraven, J.A., Kielan-Jaworowska, Z. and Clemens, W.A. (eds.), *Mesozoic mammals: The first two-thirds of mammalian history.* Berkeley, University of California Press.

—Cifelli, R.L. and Luo, Z. 1998. Alleged Cretaceous placental from down under. *Lethaia* **31**: 267–268.

—and Dashzeveg, D. 1978. New Late Cretaceous mammal locality and a description of a new multituberculate. *Acta Palaeontologica Polonica* **23**: 115–130.

—and— 1989. Eutherian mammals from the Early Cretaceous of Mongolia. *Zoologica Scripta* **18**: 347–355.

—and— 1998. Early Cretaceous amphilestid ('triconodont' mammals from Mongolia. *Acta Palaeontologica Polonica* **43**: 413–438

—,— and Trofimov, B.A. 1987. Early Cretaceous multituberculates from Mongolia and a comparison with Late Jurassic forms. *Acta Palaeontologica Polonica* **32**: 3–47.

—and Ensom, P.C. 1992. Multituberculate mammals from the Upper Jurassic Purbeck Limestone Formation of Southern England. *Palaeontology* **35**: 95–126.

—and Gambaryan, P.P. 1994. Postcranial anatomy and habits of Asian multituberculate mammals. *Fossils and Strata* **36**: 1–92.

—and Hurum, J.H. 1997. Djadochtatheria, a new suborder of multituberculate mammals. *Acta Palaeontologica Polonica* **42**: 201–242.

—and Nesov, L.A. 1990. On the metatherian nature of the Deltatheroida, a sister group of the Marsupialia. *Lethaia* **21**: 1–10.

—and— 1992. Multituberculate mammals from the Cretaceous of Uzbekistan. *Acta Palaeontologica Polonica* **37**: 1–17.

—, Presley, R. and Poplin, C. 1986. The cranial vascular system in taeniolabidoid multituberculate mammals. *Philosophical Transactions of the Royal Society of London, series* B **313**: 525–602.

—and Sloan, R. 1979. *Catopsalis* (Multituberculata) from Asia and North America and the problem of taeniolabidid dispersal in the Late Cretaceous. *Acta Palaeontologica Polonica* **24**: 187–197.

—and Sochava, A.V. 1969. The first multituberculate from the uppermost Cretaceous of the Gobi Desert (Mongolia). *Acta Palaeontologica Polonica* **14**: 355–371.

—and Trofimov, B.A. 1980. Cranial morphology of the Cretaceous eutherian mammal *Barunlestes*. *Acta Palaeontologica Polonica* **25**: 167–185.

Krause, D.W. 1982. Jaw movement, dental function, and diet in the Paleocene multituberculate *Ptilodus*. *Paleobiology* **8**: 265–281.

—1986. Competitive exclusion and taxonomic displacement in the fossil record: the case of rodents and multituberculates in North America. *Contributions to Geology, University of Wyoming Special Paper* **3**: 95–117.

—and Jenkins, F.A. Jr. 1983. The postcranial skeleton of North American multituberculates. *Bulletin of the Museum of Comparative Zoology* **150**: 199–246.

—Kielan-Jaworowska, Z. and Bonaparte, J.F. 1992. *Ferugliotherium* Bonaparte, the first known multituberculate from South America. *Journal of Vertebrate Paleontology* **12**: 351–376.

—Kielan-Jaworowska, Z. and Turnbull, W.D. 1990. Early Cretaceous Multituberculata (Mammalia) from the Antlers Formation, Trinity Group, of north-central Texas. *Journal of Vertebrate Paleontology* **10**, Supplement to no. 3: 109A.

Krebs, B. 1991. Das Skelett von *Henkelotherium guimarotae* gen. et sp. nov. (Eupantotheria, Mammalia) aus dem Oberen Jura von Portugal. *Berliner Geowissenschaftliche Abhandlungen, Abteilung* A **133**: 1–121.

—1998. *Drescheratherium actum* gen. et sp. nov., ein neuer Eupantotherier (Mammalia) aus dem Oberen Jura von Portugal. *Berliner Geowissenschaftliche Abhandlungen, Abteilung* E **28**: 91–111.

Krusat, G. 1969. Ein Pantotheria-Molar mit dreispitzigen Talonid aus dem Kimmeridge von Portugal. *Paläontologische Zeitschrift* **43**: 52–56.

Li, Ch., Wang, Y., Hu, Y. and Zhou, M. 1995. A symmetrodont skeleton from the Late Jurassic of Western Liaoning, China, p. 233 in Sun, A. and Wang, Y. (eds.), *Sixth Symposium on Mesozoic Terrestrial Ecosystems and Biota. Short Papers.* Beijing, China Ocean Press.

Lillegraven, J.A. 1969. Latest Cretaceous mammals of upper part of Edmonton Formation of Alberta, Canada and review of marsupial-placental dichotomy in mammalian evolution. *The University of Kansas, Paleontological Contributions* 50 (Vertebrata 12): 1–122.

—1974. Biogeographical considerations of the marsupial-placental dichotomy. *Annual Review of Ecology and Systematics* 5: 163–283.

—and Hahn, G. 1993. Evolutionary analysis of the middle and inner ear of Late Jurassic multituberculates. *Journal of Mammalian Evolution* 1: 47–74.

—, Kielan-Jaworowska, Z. and Clemens, W.A. (eds.) 1979. *Mesozoic Mammals. The First Two Thirds of Mammalian History.* Berkeley: University of Califfornia Press, 311 pp.

—, Krusat, G. 1991. Cranio-mandibular anatomy of *Haldanodon exspectatus* (Docodonta; Mammalia) from the Late Jurassic of Portugal and its implications to the evolution of mammalian characters. *Contributions to Geology, University of Wyoming* 28: 39–138.

—and McKenna, M.C. 1986. Fossil mammals from the 'Mesaverde' Formation (Late Cretaceous, Judithian) of the Bighorn and Wind River Basins, Wyoming, with definition of Late Cretaceous North American land-mammal 'Ages'. *American Museum Novitates* 2840: 1–68.

—McKenna, M.C. and Krishtalka, L. 1981. Evolutionary relationships of Middle Eocene and younger species of *Centetodon* (Mammalia, Insectivora, Geolabididae) with a description of the dentition of *Ankylodon* (Adapisoricidae). *Contributions to Geology, University of Wyoming* 45: 1–115.

Lucas, S.G. and Luo, Z. 1993. *Adelobasileus* from the Upper Triassic of West Texas: the oldest mammal. *Journal of Vertebrate Paleontology* 13: 309–334.

MacPhee, R.D.E. and Novacek, M.J. 1993. Definition and relationships of Lipotyphla, pp. 13–31 in Szalay, F.S., Novacek, M.J. and McKenna, M.C. (eds.), *Mammal phylogeny: Placentals.* New York: Springer-Verlag.

McKenna, M.C. 1975. Toward a phylogenetic classification of the Mammalia, pp. 21–46 in Luckett, W.P. and Szalay, F.S. (eds.) *Phylogeny of the Primates.* New York: Plenum Press.

—1987. Molecular and morphological analysis of high-level mammalian interrelationships, pp. 55–95 in Patterson, C. (ed.) *Molecules and Morphology in Evolution: Conflict or Compromise?* Cambridge: Cambridge University Press.

—and Bell, S.K. 1997. *Classification of Mammals Above the Species Level.* New York: Columbia University Press.

—, Kielan-Jaworowska, 2. and Meng, J. 2000. Earliest eutherian mammal skull from the Late Cretaceous (Coniacian) of Uzbekistan. *Acta Palaeontologica Polonica* 45: 1–54.

Marshall, L.G. and Kielan-Jaworowska, Z. 1992. Relationships of the dog-like marsupials, deltatheroidans and early tribosphenic mammals. *Lethaia* 25: 361–374.

Maschenko, E.N. and Lopatin, A.V. 1998. First record of an Early Cretaceous triconodont mammal in Siberia. *Bulletin de l'Institut Royal des Sciences Naturelles de Belgique, Sciences de la Terre* 68: 233–236.

Meng, J. and Wyss, A. 1995. Monotreme affinities and low-frequency hearing suggested by multituberculate ear. *Nature* 377: 141–144.

Miao, D. 1988. Skull morphology of *Lambdopsalis bulla* (Mammalia, Multituberculata) and its implications to mammalian evolution. *Contributions to Geology, University of Wyoming* Special Paper 4: 1–104.

—1991. On the origins of mammals, pp. 579–597 in Schultze, H.P. and Trueb, L. (eds.), *Origins of the Higher Groups of Tetrapods: Controversy and Consensus,* Ithaca: Cornell University Press.

—1993. Cranial morphology and multituberculate relationships, pp. 63–74 in Szalay, F.S., Novacek, M.J. and McKenna, M.C. (eds.), *Mammal Phylogeny: Mesozoic Differentiation, Multituberculates, Monotremes, Early Therians, and Marsupials.* New York: Springer Verlag.

—and Lillegraven, J.A. 1986. Discovery of three ear ossicles in a multituberculate mammal. *National Geographic Research* 2: 500–507.

Mills, J.R.E. 1971. The dentition of *Morganucodon*, pp. 29–63 in Kermack, D.M. and Kermack, K.A. (eds.), *Early Mammals. Zoological Journal of the Linnean Society of London* 50 Supplement 1.

Muizon, de, Ch. 1994. A new carnivorous marsupial from the Paleocene of Bolivia and the problem of marsupial monophyly. *Nature* 370: 208–211.

Nesov, L.A. 1985. [New mammals from the Cretaceous of Kyzylkum.] *Vestnik Leningradskogo Universiteta* 17: 8–18.

—1990. [Small ichthyornithiform bird and other bird remains from the Bissekty Formation (Upper Cretaceous) of Central Kyzylkum Desert]. *Trudy Zoologicheskogo Instituta AN SSSR* 21: 59–62.

—1997. [*Cretaceous Non-Marine Vertebrates of Northern Eurasia.*] (Edited by Golovneva, L.B. and Aver'yanov, A.O., St. Petersburg: Institute of the Earths Crust, St. Petersburg University, 218 pp.

—, Archibald, J.D. and Kielan-Jaworowska, Z. 1998. Ungulate-like mammals from the Late Cretaceous of Uzbekistan and a phylogenetic analysis of Ungulatomorpha, pp. 40–88 in Beard, C. and Dawson, M. (eds.), *The Dawn of the Asian Mammals. Bulletin of the Carnegie Museum of Natural History.*

—and Kielan-Jaworowska, Z. 1991. Evolution of the Cretaceous Asian therian mammals, pp. 51–52 in Kielan-Jaworowska, Z., Heintz, N. and Nakrem, H.A. (eds.), *Fifth Symposium on Mesozoic Terrestrial Ecosystems and Biota. Extended Abstracts. Contributions from the Palaeontological Museum, University of Oslo* 364.

—, Sigogneau-Russell, D. and Russell, D.E. 1994. A survey of Cretaceous tribosphenic mammals from Middle Asia (Uzbekistan, Kazakhstan and Tajikistan), of their geological setting, age and faunal environment. *Palaeovertebrata* 23: 51–92.

—and Trofimov, B.A. 1979. [Ancient insectivores from the Cretaceous of Uzbek SSR.] *Doklady AN SSSR* 247: 952–955.

Novacek, M.J. 1980. Cranioskeletal features in tupaiids and selected Eutheria as phylogenetic evidence., pp. 35–94 in Luckett, W.P. (ed.), *Comparative Biology and Evolutionary Relationships of Tree Shrews.* New York: Plenum Press.

—1986a. The primitive eutherian dental formula. *Journal of Vertebrate Paleontology* 6: 191–196.

—1986b. The skull of leptictid insectivorans and the higher-level classification of eutherian mammals. *Bulletin of the American Museum of Natural History* 183: 1–111.

—1990. Morphology, paleontology, and the higher clades of mammals. *Current Mammalogy* 2: 507–543.

—1992. Mammalian phylogeny: shaking the tree. *Nature* 356: 121–125.

—1993. Patterns of diversity in the mammalian skull, pp. 438–545 in Hanken, J. and Hall, B.K. (eds.), *The Skull.*

Volume 2: Patterns of Structural and Systematic Diversity. Chicago: The University of Chicago Press.

—Norell, M., McKenna, M.C. and Clark, J. 1994. Fossils of the Flaming Cliffs. *Scientific American* December, 1994: 60–69.

—Rougier, G.W., Wible, J.R., McKenna, M.C., Dashzeveg, D. and Horovitz, I. 1997. Epipubic bones in eutherian mammals from the Late Cretaceous of Mongolia. *Nature* 389: 483–486.

Pascual, R., Goin, F.J., Krause, D.W., Oritz-Jaureguizar, E. and Carlini, A.A. 1999. The first gnathic remains of Sudamerica: Implications for gondwanthere relationships. *Journal of Vertebrate Paleontology* 19: 373–382.

Prothero, D. 1981. New Jurassic mammals from Como Bluff, Wyoming, and the interrelationships of non-tribosphenic Theria. *Bulletin of the American Museum of Natural History* 167: 280–325.

Rădulescu, C. and Samson, P.-M. 1996. The first multituberculate skull from the Late Cretaceous (Maastrichtian) of Europe (Hațeg Basin, Romania). *Anuarul Institutului Geologic al Romaniei* 96, Supplement 1 (Abstracts): 177–178.

Reshetov, V.Y. and Trofimov, B.A. 1980. [The main stages in the development of Asian mammals,] pp. 114–121 in Sokolov, B.S. (ed.), *Palaeontology. Stratigraphy.* International Geological Congress, 26 session. Moskva, Nauka.

Rich, T.H., Vickers-Rich, P., Constantine, A., Flannery, T.A., Kool, L., and von Klaveren, N. 1997. A tribosphenic mammal from the Mesozoic of Australia. *Science* 278: 1438–1442.

Rougier, G.W. 1993. *Vincelestes neuquenianus Bonaparte (Mammalia, Theria) un primitivo mamífero del Crétacico Inferior de la Cuenca Neuquina.* Unpublished Ph.D. Thesis, Universidad Nacional de Buenos Aires, Facultad de Ciencias Exactas y Naturales. Buenos Aires. 720 pp.

—, Novacek, M.J. and Dashzeveg, D. 1997. A new multituberculate from the Late Cretaceous locality Ukhaa Tolgod, Mongolia. Considerations on multituberculate interrelationships. *American Museum Novitates* 3191: 1–26.

—, Wible, J.R. and Hopson, J.A. 1992. Reconstruction of the cranial vessels in the Early Cretaceous mammal *Vincelestes neuquenianus:* implications for the evolution of mammalian cranial system. *Journal of Vertebrate Paleontology* 12: 188–216.

—, Wible, J.R. and Hopson, J.A. 1996a. Basicranial anatomy and relationships of *Priacodon fruitaensis* (Triconodontidae, Mammalia) from the Late Jurassic of Colorado. *American Museum Novitates* **3183**: 1–38.

—, Wible, J.R. and Novacek, M.J. 1996b. Middle-ear ossicles of the multituberculate *Kryptobaatar* from Mongolian Late Cretaceous: implications for mammaliamorph relationships and evolution of the auditory apparatus. *American Museum Novitates* **3187**: 1–43.

—, Wible, J.R. and Novacek, M.J. 1998. Implications of *Deltatheridium* specimens for early marsupial history. *Nature* **396**: 459–463.

Rowe, T. 1988. Definition, diagnosis, and origin of Mammalia. *Journal of Vertebrate Paleontology* **8**: 241–264.

—1993. Phylogenetic systematics and the early history of mammals, pp. 129–145. in Szalay, F.S., Novacek, M.J. and McKenna, M.C. (eds.), *Mammal Phylogeny: Mesozoic Differentiation, Multituberculates, Monotremes, Early Therians, and Marsupials.* New York: Springer-Verlag.

Sereno, P. and McKenna, M.C. 1995. Cretaceous multituberculate skeleton and the early evolution of the mammalian shoulder girdle. *Nature* **377**: 144–147.

Setoguchi, T., Tsubamoto, T., Hanamura, H. and Hachiya, K. 1999. An early Late Cretaceous mammal from Japan, with reconsideration of the evolution of tribosphenic molars. *Paleontological Research* **3**: 18–28.

Sigogneau-Russell, D. 1991a. First evidence of Multituberculata (Mammalia) in the Mesozoic of Africa. *Neues Jahrbuch für Geologie und Paläontologie, Monatshefte* **1991**: 119–125.

—1991b. Découverte de premier mammifère tribosphenique du Mésozoïque africain. *Comptes Rendus de l'Académie des Sciences, Paris* **313**: 1635–1640.

—1995. Further data and reflexions on the tribosphenid mammals (Tribotheria) from the Early Cretaceous of Morocco. *Bulletin du Museum national d'Histoire naturelle, Paris* 4 sér. **16**: 291–312.

—Dashzeveg, D. and Russell, D.E. 1992. Further data on *Prokennalestes* (Mammalia, Eutheria *inc. sed.*) from the Early Cretaceous of Mongolia. *Zoologica Scripta* **21**: 205–209.

—and Ensom, P. 1994. Découverte, dans le Group de Purbeck (Berriasien, Angleterre), du plus ancien témoignage de l'existence de mammifères tribospheniques. *Comptes Rendus de l'Académie des Sciences, Paris* **319**, série II: 833–838.

—and Godefroit, P. 1997. A primitive docodont (Mammalia) from the Upper Triassic of France and the possible therian affinities of the order. *Comptes Rendus de l'Académie des Sciences, Paris* **324**: 135–140.

Simmons, N.B. 1993. Phylogeny of Multituberculata, pp. 146–164, in Szalay, F.S., Novacek, M.J. and McKenna, M.C. (eds.), *Mammal Phylogeny: Mesozoic Differentiation, Multituberculates, Monotremes, Early Therians and Marsupials.* New York: Springer Verlag.

Simpson, G.G. 1925a. A Mesozoic mammal skull from Mongolia. *American Museum Novitates* **201**: 1–11.

—1925b. Mesozoic Mammalia III: Preliminary comparison of Jurassic mammals except multituberculates. *American Journal of Science* **10**: 559–569.

—1927. Mesozoic Mammalia, IX: The brain of Jurassic mammals. *American Journal of Science* **14**: 259–268.

—1928. *A Catalogue of the Mesozoic Mammalia in the Geological Department of the British Museum.* London: British Museum (Natural History), 215 pp.

—1929. American Mesozoic Mammalia. *Memoirs of the Peabody Museum of Yale University* **3**: 1–235.

—1937. Skull structure of the Multituberculata. *Bulletin of the American Museum of Natural History* **73**: 727–763.

Sloan, R.E. 1979. Multituberculata, pp. 492–498 in Fairbridge, R.W. and Jablonski, D. (eds.), *The Encyclopedia of Paleontology.* Stroudsburg; Dowden, Hutchison and Ross, Inc.

—and Van Valen, L. 1965. Cretaceous mammals from Montana. *Science* **148**: 220–227.

Szalay, F.S. 1993a. Pedal evolution of mammals in the Mesozoic: tests for taxic relationships, pp. 108–128 in Szalay, F.S., Novacek, M.J. and McKenna, M.C. (eds.), *Mammal Phylogeny: Mesozoic Differentiation, Multituberculates, Monotremes, Early Therians, and Marsupials.* New York: Springer Verlag.

—1993b. *Evolutionary History of the Marsupials and an Analysis of the Osteological Characters.* New York: Cambridge University Press: 12+481 pp.

—and McKenna, M.C. 1971. Beginning of the age of mammals in Asia: the Late Paleocene Gashato fauna, Mongolia. *American Museum of Natural History Bulletin* **144**: 269–318.

—and Trofimov, B.A. 1996. The morphology of Late Cretaceous *Asiatherium*, and the early phylogeny and paleobiology of Metatheria. *Journal of Vertebrate Paleontology* **16**: 474–509.

Taquet, P. 1994. *L'Empreinte des Dinosaures.* Paris, Editions Odile Jacob, 363 pp.

Tatarinov, L.P. 1994. [On an unusual mammalian tooth from the Mongolian Jurassic]. *Paleontologicheskii Zhurnal* 1994: 97–105.

Trofimov, B.A. 1975. [New data on *Buginbaatar* Kielan-Jaworowska et Sochava, 1969 (Mammalia, Multituberculata) from Mongolia.] *Trudy Sovmestnoi Sovetsko-Mongol'skoi Paleontologicheskoi Ekspeditsii* 2: 7–13.

—1978. [The first triconodonts (Mammalia, Triconodonta) from Mongolia]. *Doklady AN SSSR* 243: 213–216.

—1980. [Multituberculata and Symmetrodonta from the Lower Cretaceous of Mongolia.] *Doklady AN SSSR* 251: 209–212.

—1997. A new generic name *Gobiotheriodon* for a symmetrodont mammal *Gobiodon* Trofimov, 1980. *Acta Palaeontologica Polonica* 42: 496.

—and Szalay, F.S. 1994. New Cretaceous marsupial from Mongolia and the early radiation of Metatheria. *Proceedings of the National Academy of Sciences, USA* 91: 12569–12573.

Wang, Y., Hu, Y., Zhou, M. and Li, C. 1995. Mesozoic mammal localities in Western Liaoning, northeast China, pp. 221–227 *in* Sun, A. and Wang, Y. (eds.) *Sixth Symposium on Mesozoic Terrestrial Ecosystems and Biota. Short papers*, Ocean Press, Beijing.

Wible, J.R. 1990. Petrosals of Late Cretaceous marsupials from North America, and a cladistic analysis of the petrosal in therian mammmals. *Journal of Vertebrate Paleontology* 10: 183–205.

—1991. Origin of Mammalia: the craniodental evidence reexamined. *Journal of Vertebrate Paleontology* 11: 1–28.

—and Hopson, J.A. 1993. Basicranial evidence for early mammal phylogeny, pp. 45–62 in Szalay, F.S., Novacek, M.J. and McKenna, M.C. (eds.), *Mammal Phylogeny: Mesozoic Differentiation, Multituberculates, Monotremes, Early Therians, and Marsupials.* New York: Springer Verlag.

—, Rougier, G.W., Novacek, M.J., McKenna, M.C. and Dashzeveg, D. 1995. A mammalian petrosal from the Early Cretaceous of Mongolia: implications for the evolution of the ear region and mammaliamorph relationships. *American Museum Novitates* 3149: 1–19.

Yabe, H. and Shikama, T. 1938. A new Jurassic Mammalia from south Manchuria. *Japanese Journal of Geology and Geography* 14: 353–357.

Yadagiri, P. 1985. An amphidontid symmetrodont from the Early Jurassic Kota Formation, India. *Zoological Journal of the Linnean Society* 85: 411–417.

30

Mammals from the Mesozoic of Kirgizstan, Uzbekistan, Kazakhstan and Tadzhikistan

ALEXANDER O. AVERIANOV[1]

Introduction

In the vast territory of Russia, Mesozoic continental deposits are rare. Thus, Russian vertebrate palaeontologists extended their search for Mesozoic mammals to territories surrounding the former Soviet Union, and beyond. The great Russian palaeontologist Vladimir Kovalevskii (1842–1884) was interested in the problem of the origin of ungulates. As the oldest ungulates known at that time were from what we now call the Palaeocene, he started, in 1873, to search for their primitive ancestors in Cretaceous continental deposits of southern France (Kovalevskii, 1950, p. 217), but without positive results. Purported remains of a Cretaceous mammal (specifically, a sea cow 'Halicore maximovitschi') were reported at that time by A.S. Rogovich (1875) from the 'upper green sandstone of the Cretaceous Formation' outcropping near Kanev, Ukraine, but these remains belong, in fact, to Eocene cetaceans (personal observation). Rogovich clearly understood the importance of mammals from the 'Secondary System' and tried to search for their Mesozoic remains, but his efforts were in vain.

For a long time Mesozoic mammals were totally unknown from the vast territory of the former Soviet Union. This gap was emphasised by the late Professor Yurii A. Orlov (1964), who believed that success in the search for Mesozoic mammals might be achieved by using screening techniques. The first discovery that filled this gap was a fragment of an edentulous dentary from the Cretaceous red beds near Zhalmaus Well in Kyzyl-Orda Province, Kazakhstan (Bazhanov, 1972) (Figure 30.1). This fragment was found by F.E. Vetrov during the screen-washing of 25 m^3 of sediment in the summer of 1962. Vetrov was a member of an expedition of the Laboratory of Paleobiology of the Institute of Zoology, in the Academy of Sciences of Kazakhstan led by T.N. Nurumov. The locality that yielded this dentary is close to Zhalmaus Well (not Baibolat Well as was originally indicated) in the lower part of the Bostobe Svita (not Beleuty Svita), and thus its age is most probably Santonian rather than Coniacian (see Nesov and Khisarova, 1988; Nesov et al., 1994b). In the description of this specimen, named Beleutinus orlovi by Bazhanov (1972) the slope of the coronoid process was considered as the 'diastema' between the canine and premolars and thus the orientation of the dentary and tooth positions were mistakenly reversed (Lillegraven et al., 1979, p. 37).

The discovery of Beleutinus orlovi suggested the possibility of finding more complete remains of Mesozoic mammals in Kazakhstan and adjacent areas. By the time that the Soviet–Mongolian Palaeontological Expeditions started their work in 1969, however, the palaeontological fascination with Mesozoic mammals had shifted to Mongolia (see Chapter 29). This is the primary reason why Kazakhstan and the so-called neighbouring areas of 'Middle Asia' were almost totally neglected by most Soviet palaeontologists. A notable exception was Dr Lev A. Nesov (1947–1995) of Leningrad State University (now St. Petersburg University). With considerable enterprise, unusual tenacity, and great effort, he and his students assembled the first collection of Mesozoic mammals from remote deserts of the former Soviet 'Middle Asia'.

[1] Alexander Aver'yanov prefers the transliteration Averianov for his name.

Figure 30.1. Diagrammatic map of mammal-bearing Mesozoic localities in Kazakhstan, Uzbekistan, Tadzhikistan and Kirgizstan: 1, Kalmakerchin. 2, Khodzhakul. 3, Khodzhakulsai. 4, Sheikhdzheili. 5, Chelpyk. 6, Ashchikol. 7, Dzharakhuduk. 8, Kansai. 9, Zhalmauz. 10, Alymtau.

Nesov began his work in the 1970s as a specialist on Mesozoic turtles. He travelled across 'Middle Asia' and Kazakhstan collecting turtles as well as remains of other vertebrates including dinosaurs, crocodiles, chondrichthyan and osteichthyan fishes. The turning point in his scientific career was the discovery of a bone of a frog in the upper Cretaceous deposits at Dzharakhuduk (central Kyzylkum Desert, Uzbekistan), in the twilight of the last day of the field season of 1977 (Averianov, 1996). Before this discovery, anuran remains were not known from the Cretaceous of the USSR. Impressed by this discovery, Nesov moved from the study of turtles to work on more poorly known groups of Cretaceous vertebrates.

The recovery of amphibians and rare reptiles (lizards and pterosaurs) from the Cretaceous of 'Middle Asia' showed that there was also some possibility of finding Cretaceous mammals, since all these groups co-occur in similar vertebrate assemblages in North America. Encouraged by these correspondences, Nesov concentrated on the search for Cretaceous mammals. The first discoveries were made by Nesov in 1978 in the central and southwestern Kyzylkum Desert and he subsequently organised and led twelve further expeditions between 1979 and 1994. On each expedition, Nesov was accompanied by one to seven participants, usually students from Leningrad State University. Because of limited finances, once the expedition reached Khodzheili or Tashkent by train from St. Petersburg, the group relied on local transportation such as buses, or sometimes even by hitch-hiking. The expedition members carried all their equipment and part of their food. Upon reaching the desert, they made their way on foot, using rucksacks and pushcarts for the transport of food, water, and equipment.

Nesov and his team recovered and described a rich collection of Cretaceous mammals including, among others, previously unknown assemblages from the early half of the Late Cretaceous. The stratigraphical and geographical scope of the material collected ranged from the Albian and lower Cenomanian of southwestern Kyzylkum, through the upper Turonian and Coniacian of central Kyzylkum (Uzbekistan), to the Santonian of Kansai (Tadzhikistan) (Nesov and Trofimov, 1979; Nesov and Gureev, 1981, 1982; Nesov, 1982a, b, 1984, 1985a, b, 1986, 1987, 1989, 1990a, 1993, 1997; Nesov and Khisarova, 1988; Kielan-Jaworowska and Nesov, 1990, 1992; Nesov and Kielan-Jaworowska, 1991; Nesov et al., 1994b; Nesov et al., 1995, 1998; Archibald, 1996). This collection contains about 500 specimens of Cretaceous mammals including isolated teeth, fragmentary dentaries and maxillae with and without teeth, petrosals, postcranial bones, and one skull. In addition, Nesov and co-workers collected fossil material from the early Tertiary (Palaeocene), which is not discussed in this account (Nesov, 1986, 1987).

In 1988 and 1991 two new Cretaceous mammal localities were discovered in Kazakhstan. The holotype of *Sorlestes kara* was found by local geologists at Ashchikol, and the holotype of *Alymlestes kielanae* was recovered from Alymtau by a small team led by A. Averianov (Averianov and Nesov, 1995). Fragmentary mammal remains were also found in the Upper Jurassic of Kirgizstan (Kalmakerchin) by the Russian–Norwegian Expedition led by Nesov in 1992 (Nesov et al., 1994a). In 1994 a joint Russian–American–Uzbekistanian Expedition led by Nesov and J.D. Archibald worked in the central and southwestern Kyzylkum Desert, Uzbekistan.

Beginning in 1979 Nesov published a number of preliminary descriptions of Middle Asian Cretaceous mammals. In 1990 he invited a number of foreign palaeontologists to join him in his efforts. Primary among these were: Z. Kielan-Jaworowska, J.D. Archibald, D. Sigogneau-Russell, D. Russell, and M.C. McKenna. His untimely death in 1995 left parts of his mammal collection still undescribed.

Abbreviations

Capital and lower case letters, I/i (incisor), P/p (premolar) and M/m (molar), refer to upper and lower teeth, respectively. Institutions in which the collections are housed are: Institute of Zoology, Kazakhstan Academy of Sciences, Almaty (IZK); Chernyshev's Central Museum of Geological Exploration, St. Petersburg (TsNIGRI); Zoological Institute, Russian Academy of Sciences, St. Petersburg (ZIN).

Middle Asian mammal localities

A number of Middle Asian Mesozoic fossil localities (Figure 30.1) have yielded the remains of mammals. A list of the most important specimens is given in Table 30.1. Below there is a list of localities with references to their location and dating and some comments on mammalian specimens.

Jurassic

1. *Kalmakerchin.* This village is located on the Kokart River, in the foothills of the Fergana Range, in the northeastern part of the Fergana Depression, Kirgizstan. The mammal-bearing level is in the red-coloured part of the Balabansai Svita, which outcrops along the Kokart River in the vicinity of Kalmakerchin. In Nesov et al., 1994a, pp. 320, 322, this mammal locality was erroneously cited as 7 km NW of Tashkumyr.

The mammal remains were found with bones of salamanders belonging to the family Karauridae, which is known only from the Jurassic. The level yielding these remains is overlain by layers containing a leaf flora that suggests a Middle Jurassic age (Burakova and Fedorov, 1989). Above the plant-bearing levels there are layers with remains of ptycholepid palaeoniscoids

Table 30.1. *List of most important mammalian specimens from late Mesozoic of Kirgizstan, Uzbekistan, Kazakhstan and Tadzhikistan*

Locality	Stratigraphic division	Site	Age	Specimen	Collection number	Identity	Main references
Kalmakerchin	Balabansai Sv.	KUG-1	Bathonian	upper molar	ZIN 79026	?Docodonta indet.	Nesov *et al.*, 1994a, fig. 4B–D
Kalmakerchin	Balabansai Sv.	KUG-1	Bathonian	ulna	ZIN 79027	Mammalia indet.	Nesov *et al.*, 1994a, fig.5
Khodzhakul	l.-m. part of Khodzhakul Sv.	SKH-20	late Albian	e.d.f.	2/11658	Theria indet.	Figure 30.3I; Nesov, 1984, fig. v
Khodzhakul	l.-m. part of Khodzhakul Sv.	SKH-20	late Albian	maxilla with M2–3	2/12176, holotype	*Bobolestes zenge*	Figure 30.3A, B; Nesov, 1985a, tab. 1, fig. 1; 1993, fig. 5, 1; Nesov *et al.*, 1994b, pl. 2, fig. 3
Khodzhakul	l.-m. part of Khodzhakul Sv.	SKH-20	late Albian	e.m.f.		Mammalia indet.	Nesov and Kielan-Jaworowska, 1991, fig. 1
Khodzhakul	l.-m. part of Khodzhakul Sv.	SKH-20	late Albian	tooth fragment with thick enamel	75/12455	Mammalia indet.	Nesov, 1993, fig. 3, 1
Khodzhakulsai	u. part of Khodzhakul Sv.	SKH-5	e. Cenomanian	e.d.f.		Theria indet.	
Sheikhdzheili	u. part of Khodzhakul Sv.	SSHD-8	e. Cenomanian	fused cervicals 2–3	6/11758, holotype	*Oxlestes grandis*	Figure 30.3N; Nesov, 1981, fig. 9, 23; 1982a, tabl. 1, fig. 1; Nesov *et al.*, 1994b, pl. 1, fig. 5
Sheikhdzheili	u. part of Khodzhakul Sv.	SSHD-8	e. Cenomanian	dentary with p4–5 m1–3	7/12176, holotype	*Otlestes meiman*	Figure 30.3G, O; Nesov, 1985a, tabl. 1, fig. 13; 1993, fig. 5, 2; Nesov *et al.*, 1994b, pl. 3
Sheikhdzheili	u. part of Khodzhakul Sv.	SSHD-8	e. Cenomanian	dentary with p5 and m3	9/12176	*Otlestes meiman*	Figure 30.3H; Nesov, 1985a, tabl. 1, fig 12
Sheikhdzheili	u. part of Khodzhakul Sv.	SSHD-8	e. Cenomanian	maxilla with M1–2	8/12176	*Otlestes meiman*	Figure 30.3C; Nesov, 1985a, tabl. 1, fig. 4; 1993, fig. 5, 3

Locality	Stratigraphic unit	Section	Age	Element	Specimen no.	Taxon	References
Sheikhdzheili	u. part of Khodzhakul Sv.	SSHD-8	e. Cenomanian	m1	26/12176	'Zhelestidae' gen. et sp. nov.	Figure 30.3D, E; Nesov, 1985a, tabl. 1, fig. 3; Nesov et al., 1994b, pl. 7, fig. 2
Sheikhdzheili	u. part of Khodzhakul Sv.	SSHD-8	e. Cenomanian	e.d.f.	1/11658	Eutheria indet.	Nesov, 1984, fig. a, b; Nesov et al., 1994b, pl. 7, fig. 3
Sheikhdzheili	u. part of Khodzhakul Sv.	SSHD-8	e. Cenomanian	l. pm.		Theria indet.	Figure 30.3F; Nesov, 1985a, tabl. 1, fig. 11
Sheikhdzheili	u. part of Khodzhakul Sv.	SSHD-8a	e. Cenomanian	parietal		*Oxlestes*? sp.	Figure 30.3J, K; Nesov, 1985a, tabl. 2, fig. 1; Nesov et al., 1994b, pl. 7, fig. 1
Chelpyk	u. part of Khodzhakul Sv.	STCH-1	e. Cenomanian	calcaneus		Theria indet.	Figure 30.3L, M; Nesov, 1985a, tabl. 2, fig. 2
Ashchikol	grey colored unnamed beds	STCH-1	e. Turonian	dentary with p5 m1–3	101/12455, holotype	*Sorlestes kara*	Figure 30.4L, M; Nesov, 1993, fig. 1, 1; Nesov et al., 1994b, pl. 7, fig. 4
Dzharakhuduk	l. part of Bissekty Sv.	CDZH-17a	late Turonian	dentary with c1 and m1	1/11758, holotype	*Daulestes kulbeckensis*	Figure 30.4A, B; Nesov and Trofimov, 1979, fig. 1; Nesov, 1981, fig. 10, 18; 1982a, tabl. 1, fig. 4; Nesov et al., 1994b, pl. 1, fig. 2
Dzharakhuduk	l. part of Bissekty Sv.	CDZH-17a	late Turonian	canine	5/11758	Deltatheroida indet.	Nesov, 1981, fig. 9, 22; 1982a, tabl. 1, fig. 8
Dzharakhuduk	l. part of Bissekty Sv.	CDZH-17a	late Turonian	dentary with m2	8/11758, holotype	*Taslestes inobservabilis*	Figure 30.4E; Nesov, 1982a, tabl. 1, fig. 4; Nesov et al., 1994b, pl. 1, fig. 4
Dzharakhuduk	l. part of Bissekty Sv.	CDZH-17a	late Turonian	M1	5/12176, holotype	*Kulbeckia rara*	Figure 30.4J; Nesov, 1993, fig. 4, 8
Dzharakhuduk	l. part of Bissekty Sv.	CDZH-17a	late Turonian	dentary with p4, 5 m1, 2	4/12176, holotype	*Aspanlestes aptap*	Figure 30.4C, D; Nesov, 1985a, tabl. 2, fig. 11; Nesov et al., 1994b, pl. 4, fig. 1; Archibald, 1996, fig. 2F
Dzharakhuduk	l. part of Bissekty Sv.	CDZH-17a	late Turonian	M1	38/12000	cf. *Zhelestes* sp.	Figure 30.4I; Nesov, 1985b, tabl. 2, fig. 5
Dzharakhuduk	l. part of Bissekty Sv.	CDZH-17a	late Turonian	u. molar	23/12176	'Zhelestidae' indet.	Figure 30.4G, H; Nesov, 1985a, tabl. 2, fig. 10; Nesov et al., 1994b, pl. 7, fig. 5

Table 30.1. (*cont.*)

Locality	Stratigraphic division	Site	Age	Specimen	Collection number	Identity	Main references
Dzharakhuduk	l. part of Bissekty Sv.	CDZH-17a	late Turonian	e.d.f.	14/11758	Theria indet.	Nesov, 1982a, tabl. 1, fig. 5
Dzharakhuduk	l. part of Bissekty Sv.	CDZH-17a	late Turonian	e.d.f.	18/11758	Theria indet.	Nesov, 1982a, tabl. 1, fig. 7
Dzharakhuduk	l. part of Bissekty Sv.	CDZH-17a	late Turonian	ulnae		Theria indet.	Figure 30.4K; Nesov, 1985a, tabl. 2, fig. 8
Dzharakhuduk	l. part of Bissekty Sv.	CDZH-17a	late Turonian	dentary with erupting canine	3/11658	*Aspanlestes aptap*	Figure 30.4F; Nesov, 1984, fig. g, d
Dzharakhuduk	m. part of Bissekty Sv.	CBI-4	Coniacian	M1	4/12455	Eutheria gen et sp. nov.	Nesov, 1987: tabl. 1, fig. 4; Nesov et al., 1994b, pl. 7, fig. 6
Dzharakhuduk	m. part of Bissekty Sv.	CBI-4	Coniacian	dentary with m2	72/12455	*Kennalestes* sp. nov.	Figure 30.5R; Nesov, 1993, fig. 5
Dzharakhuduk	m. part of Bissekty Sv.	CBI-4b	Coniacian	dentary with m1, 2	1/12176, holotype	*Kumlestes olzha*	Figure 30.5A, B; Nesov, 1985a, tabl. 2, fig. 15; Nesov et al., 1994b, pl. 2, fig. 2
Dzharakhuduk	m. part of Bissekty Sv.	CBI-4b	Coniacian	maxilla with M1–2	35/12000: holotype	*Sulestes karakshi*	Figure 30.5K; Nesov, 1985a, tabl. 3, fig. 15; Nesov, 1985b, tabl. 2, fig. 1; Kielan–Jaworowska and Nesov, 1990, figs. 3 and 4A; Nesov et al., 1994b, pl. 4, fig. 3
Dzharakhuduk	m. part of Bissekty Sv.	CBI-4b	Coniacian	frontal	39/12000	?Deltatheridia	Figure 30.5N; Nesov, 1985b, tabl. 2, fig. 6
Dzharakhuduk	m. part of Bissekty Sv.	CBI-4b	Coniacian	e.d.f.		?Marsupialia	Figure 30.5L, M; Nesov, 1985a, tabl. 2, fig. 5
Dzharakhuduk	m. part of Bissekty Sv.	CBI-4b	Coniacian	dentary with m2	36/12000, holotype	?*Zalambdalestes mynbulakensis*	Figure 30.5E, F; Nesov, 1985b, tabl. 2, fig. 2; Nesov et al., 1994b, pl. 5, fig. 2
Dzharakhuduk	m. part of Bissekty Sv.	CBI-4b	Coniacian	femur		Theria indet.	Figure 30.5Q; Nesov, 1985a, tabl. 1, fig. 5
Dzharakhuduk	m. part of Bissekty Sv.	CBI-4b	Coniacian	squamosal		Mammalia indet.	Nesov, 1985a, tabl. 2, fig. 6
Dzharakhuduk	m. part of Bissekty Sv.	CBI-4v	Coniacian	e.d.f.	16/11758	?Deltatheridia	Nesov, 1982a, tabl. 2, fig. 7

Locality	Stratigraphic unit	Section	Age	Element	Specimen number	Taxon	References
Dzharakhuduk	m. part of Bissekty Sv.	CBI-4v	Coniacian	M1	7/11758, holotype	*Sailestes quadrans*	Nesov, 1981, fig. 11, 29; 1982a, tabl. 2, fig. 8; 1993, fig. 5, 4; Nesov *et al.*, 1994b, pl. 1, fig. 6
Dzharakhuduk	m. part of Bissekty Sv.	CBI-4v	Coniacian	M3	12/12176, holotype	*Bulaklestes kezbe*	Nesov, 1985a, tabl. 3, fig. 6; 1993, fig. 4, 10; Nesov *et al.*, 1994b, pl. 4, fig. 2
Dzharakhuduk	m. part of Bissekty Sv.	CBI-4v	Coniacian	dentary with dp2–4		*?Aspanlestes aptap*	Figure 30.5, 3; Nesov, 1985a, tabl. 2, fig. 17
Dzharakhuduk	m. part of Bissekty Sv.	CBI-4v	Coniacian	dentary with m1–3	12/11758, holotype	*Kumsuperus avus*	Figure 30.5C; Nesov, 1984, fig. e, zh, z; 1982a, tabl. 2, fig. 2; Nesov *et al.*, 1994b, pl. 2, fig. 1
Dzharakhuduk	m. part of Bissekty Sv.	CBI-4v	Coniacian	m1?	37/12000	'Zhelestidae' indet.	Nesov, 1985b, tabl. 2, fig. 3
Dzharakhuduk	m. part of Bissekty Sv.	CBI-4v	Coniacian	u. pm.		'Zhelestidae' indet.	Nesov, 1985a, tabl. 3, fig. 1
Dzharakhuduk	m. part of Bissekty Sv.	CBI-4v	Coniacian	molar trigonid		'Zhelestidae' indet.	Figure 30.5G; Nesov, 1985a, tabl. 3, fig. 4
Dzharakhuduk	m. part of Bissekty Sv.	CBI-4v	Coniacian	e.d.f.	15/11758	Eutheria indet.	Nesov, 1982a, tabl. 2, fig. 3
Dzharakhuduk	m. part of Bissekty Sv.	CBI-4v	Coniacian	e.d.f.		Eutheria indet.	Nesov, 1985a, tabl. 3, fig. 3
Dzharakhuduk	m. part of Bissekty Sv.	CBI-4v	Coniacian	femur		Eutheria indet.	Nesov, 1982a, tabl. 2, fig. 1
Dzharakhuduk	m. part of Bissekty Sv.	CBI-4v	Coniacian	l. pm.	9/11758	Eutheria indet.	Nesov, 1985a, tabl. 3, fig. 2
Dzharakhuduk	m. part of Bissekty Sv.	CBI-4v	Coniacian	ilium	10/11758	Theria indet.	Nesov, 1982a, tabl. 2, fig. 4
Dzharakhuduk	m. part of Bissekty Sv.	CBI-4v	Coniacian	ischium	11/11758	Theria indet.	Nesov, 1982a, tabl. 2, fig. 5
Dzharakhuduk	m. part of Bissekty Sv.	CBI-4v	Coniacian	incisor		Mammalia indet.	Figure 30.5O, P; Nesov, 1985a, tabl. 2, fig. 3; Nesov *et al.* 1994a, pl. 7, fig. 7
Dzharakhuduk	m. part of Bissekty Sv.	CBI-14	Coniacian	p4	100/12455, holotype	*Uzbekbaatar kizylkumensis*	Figure 30.6A–D; Kielan-Jaworowska and Nesov, 1992, fig. 1; Nesov, 1993, fig. 1, 2; 1995, tabl. 11, fig. 5
Dzharakhuduk	m. part of Bissekty Sv.	CBI-14	Coniacian	incisor?	102/12455	*Uzbekbaatar kizylkumensis*	Kielan-Jaworowska and Nesov, 1992, fig. 4A, B

Table 30.1. (*cont.*)

Locality	Stratigraphic division	Site	Age	Specimen	Collection number	Identity	Main references
Dzharakhuduk	m. part of Bissekty Sv.	CBI-14	Coniacian	e.d.f.	101/12455	*Uzbekbaatar kizylkumensis*	Kielan-Jaworowska and Nesov, 1992, fig. 4D–F
Dzharakhuduk	m. part of Bissekty Sv.	CBI-14	Coniacian	femur	104/12455	*Uzbekbaatar kizylkumensis*	Kielan-Jaworowska and Nesov, 1992, fig. E–G
Dzharakhuduk	m. part of Bissekty Sv.	CBI-14	Coniacian	humerus	103/12455	*Uzbekbaatar kizylkumensis*	Kielan-Jaworowska and Nesov, 1992, fig. 5A–B
Dzharakhuduk	m. part of Bissekty Sv.	CBI-14	Coniacian	m1	5/12455	*Sulestes* sp.	Figure 30.6M, N; Nesov, 1987, tabl. 1, fig. 5; Kielan-Jaworowska and Nesov, 1990, fig. 1, fig. 2A–E
Dzharakhuduk	m. part of Bissekty Sv.	CBI-14	Coniacian	m2?	40/12455, holotype	*Deltatheroides kizylkumensis*	Figure 30.6E; Nesov, 1993, fig. 4, 1
Dzharakhuduk	m. part of Bissekty Sv.	CBI-14	Coniacian	m3	41/12455	*Deltatheroides kizylkumensis*	Nesov, 1993, fig. 4, 2
Dzharakhuduk	m. part of Bissekty Sv.	CBI-14	Coniacian	e.m.f.	42/12000	?*Deltatheroides kizylkumensis*	Figure 30.6, F; Nesov, 1993, fig. 2, 2
Dzharakhuduk	m. part of Bissekty Sv.	CBI-14	Coniacian	M1?	6/12455	?*Kulbeckia* sp.	Figure 30.6, J; Nesov, 1987, tabl. 1, fig. 6; 1995, tabl. 2, fig. 22
Dzharakhuduk	m. part of Bissekty Sv.	CBI-14	Coniacian	M3	54/12455	?*Kulbeckia* sp.	Figure 30.6O, P; Nesov, 1993, fig. 3, 2, fig. 4, 9
Dzharakhuduk	m. part of Bissekty Sv.	CBI-14	Coniacian	dentary with m2, 3	67/12455, holotype	*Paranyctoides aralensis*	Figure 30.7K; Nesov, 1993, fig. 2, 5
Dzharakhuduk	m. part of Bissekty Sv.	CBI-14	Coniacian	M1	3/12455	*Aspanlestes aптap*	Figure 30.6K, L; Nesov, 1987, tabl. 1, fig. 3; Nesov and Kielan-Jaworowska, 1991, fig. 1
Dzharakhuduk	m. part of Bissekty Sv.	CBI-14	Coniacian	dentary with m1, 2	6/12176	*Aspanlestes aптap*	Figure 30.6G, H; Nesov, 1985a, tabl. 2, fig. 7
Dzharakhuduk	m. part of Bissekty Sv.	CBI-14	Coniacian	maxilla with P5 M1, 2	1/12455, holotype	?*Zhelestes bezelgen* (= *Aspanlestes aптap*)	Figure 30.7B; Nesov, 1987, tabl. 1, fig. 1; Nesov *et al.*, 1994b, pl. 6, fig. 1; Archibald, 1996, fig. 2E

Locality	Member	Horizon	Stage	Element	Specimen	Taxon	Figure/Reference
Dzharakhuduk	m. part of Bissekty Sv.	CBI-14	Coniacian	maxilla with M2	68/12455	*Aspanlestes aptap*	Figure 30.7C; Nesov, 1993, fig. 2, 3
Dzharakhuduk	m. part of Bissekty Sv.	CBI-14	Coniacian	M2	103/12455	*Aspanlestes aptap*	Nesov, 1993, fig. 5, 5
Dzharakhuduk	m. part of Bissekty Sv.	CBI-14	Coniacian	dentary with m2	3/12176, holotype	*Sorlestes budan*	Figure 30.7H, I; Nesov, 1985a, tabl. 2, fig. 13; Nesov et al., 1994b, pl. 1, fig. 7
Dzharakhuduk	m. part of Bissekty Sv.	CBI-14	Coniacian	dentary with c and pm	15/12953	*Sorlestes budan*	Figure 30.6X
Dzharakhuduk	m. part of Bissekty Sv.	CBI-14	Coniacian	maxilla with P4, 5 M1–3	70/12455, holotype	*Parazhelestes robustus*	Figure 30.7A; Nesov, 1993, fig. 2, 1
Dzharakhuduk	m. part of Bissekty Sv.	CBI-14	Coniacian	M1	20/12953	*Parazhelestes robustus*	Figure 30.6R
Dzharakhuduk	m. part of Bissekty Sv.	CBI-14	Coniacian	M1	11/12953	*Parazhelestes* sp. nov.	Figure 30.6Q; Archibald, 1996, fig. 2G
Dzharakhuduk	m. part of Bissekty Sv.	CBI-14	Coniacian	maxillae with P5 M1, 2	11/12176	*Parazhelestes* sp. nov.	Nesov, 1985a, tabl. 3, fig. 5; Nesov et al, 1994b, pl. 7, fig. 8; Archibald, 1996, fig. 2H
Dzharakhuduk	m. part of Bissekty Sv.	CBI-14	Coniacian	M2	2/12455	'Zhelestidae' gen. et sp. nov.	Figure 30.6S; Nesov, 1987, tabl. 1, fig. 2; 1993, fig. 1, 3; 1995, tabl. 11, fig. 6; Archibald, 1996, fig. 2D
Dzharakhuduk	m. part of Bissekty Sv.	CBI-14	Coniacian	m3	18/12953	cf. 'Zhelestidae' gen. et sp. nov.	Figure 30.6T
Dzharakhuduk	m. part of Bissekty Sv.	CBI-14	Coniacian	M2	104/12455	'Zhelestidae' indet.	Nesov, 1993, fig. 5, 6
Dzharakhuduk	m. part of Bissekty Sv.	CBI-14	Coniacian	e.d.f.	23/12953	'Zhelestidae' indet.	Figure 30.7J
Dzharakhuduk	m. part of Bissekty Sv.	CBI-14	Coniacian	dentary with m2–3	69/12455	Eutheria indet.	Nesov, 1993, fig. 2, 4
Dzharakhuduk	m. part of Bissekty Sv.	CBI-14	Coniacian	humerus	7/12455	Theria indet.	Figure 30.6, I; Nesov, 1985a, tabl. 2, fig. 9
Dzharakhuduk	m. part of Bissekty Sv.	CBI-14	Coniacian	femur	7/12455	Theria indet.	Figure 30.6U, V; Nesov, 1987, tabl. 1, fig. 7
Dzharakhuduk	m. part of Bissekty Sv.	CBI-14	Coniacian	fused tibia and fibula	8/12455	Theria indet.	Figure 30.6W; Nesov, 1985a, tabl. 1, fig. 8
Dzharakhuduk	m.-u. part of Bissekty Sv.	CBI-17, on the level of CBI-5a	Coniacian	maxilla with P2–5 M1–3	10/12176, holotype	*Zhelestes temirkazyk*	Figure 30.7D, E; Nesov, 1985a, tabl. 3, fig. 14, Nesov et al., 1994b, pl. 5, fig. 1

Table 30.1. (*cont.*)

Locality	Stratigraphic division	Site	Age	Specimen	Collection number	Identity	Main references
Dzharakhuduk	m.-u. part of Bissekty Sv.	CBI-5	Coniacian	skull	ZIN C.79066	*Taslestes*? sp. nov.	Figure 30.8A, B; Nesov, 1993, fig. 1, 4; 1995, tabl. 11, fig. 7
Dzharakhuduk	m.-u. part of Bissekty Sv.	CBI-5a	Coniacian	M2	52/12455, holotype	*Kulbeckia kulbecke*	Figure 30.5H–I; Figure 30.8C, D; Nesov, 1993, fig. 3, 3, fig. 4. 3
Dzharakhuduk	m.-u. part of Bissekty Sv.	CBI-5a	Coniacian	m1?	60/12455	*Kulbeckia kulbecke*	Figure 30.8O, P; Nesov, 1993, fig. 3, 4, fig. 4, 4
Dzharakhuduk	m.-u. part of Bissekty Sv.	CBI-5a	Coniacian	m1?	53/12455	*Kulbeckia kulbecke*	Nesov, 1993, fig. 4, 5
Dzharakhuduk	m.-u. part of Bissekty Sv.	CBI-5a	Coniacian	m2	102/12455	*Kulbeckia kulbecke*	Nesov, 1993, fig. 4, 6
Dzharakhuduk	m.-u. part of Bissekty Sv.	CBI-5a	Coniacian	m1?	74/12455	Eutheria indet.	Nesov, 1993, fig. 3, 5
Dzharakhuduk	u. part of Bissekty Sv.	CBI-7a	Coniacian	l. pm	78/12455	Eutheria indet.	Nesov, 1993, fig. 4, 11
Dzharakhuduk	u. part of Bissekty Sv.	CBI-7a	Coniacian	femur		Eutheria indet.	Figure 30.7F; Nesov, 1985a, tabl. 3, fig. 13
Kansai		FKA-7a	l. Santonian	M1	9/12455, holotype	*Kulbeckia kansaica*	Figure 30.8E–G; Nesov, 1987, tabl. 1, fig. 9, 1993, fig. 4, 7; Nesov *et al.*, 1994b, pl. 7, fig. 9
Kansai		FKA-7a	l. Santonian	M1?	73/12455	*Kulbeckia* sp.	Figure 30.8H, I; Nesov, 1993, fig. 3, 6
Zhalmaus	l. part of Bostobe Sv.		Santonian	dentary with damaged molars	IZK 1–751/ III-1962, holotype	*Beleutinus orlovi*	Figure 30.8J–L; Bazhanov, 1972, figs. 1–5; Nesov, 1987, tabl. 1, fig. 10; Nesov *et al.*, 1994b, pl. 1, fig. 1
Zhalmaus	l. part of Bostobe Sv.		Santonian	cervical	IZK 3679/ III-III-62	Mammalia indet.	Nesov and Khisarova, 1988, fig. 7
Alymtau	l. Darbasa Sv.	ALT-1 ("Grey Mesa")	l. Campanian	m1	ZIN C.78332, holotype	*Alymlestes kielanae*	Figure 30.8M, N; Nesov *et al.*, 1994b, pl. 6, fig. 2; Aver'yanov and Nesov, 1995, figs. 3 and 4

Abbreviations: e., early; e.d.f., edentulous dentary fragment; e.m.f, edentulous maxillary fragment; l., lower; m., middle; pm, premolar; Sv., Svita; u, upper. The numbered specimens belong to the TsNIGRI collection, if not otherwise indicated.

(unknown from the Cretaceous), sharks, and turtles of Jurassic age (Nesov *et al.*, 1994a).

Cretaceous

2. Khodzhakul. This locality occurs in cliffs north of the dried lake of Khodzhakul, at the northern extremity of the Sheikhdzheili ridge, which is west- northwest from the Sultan Uvais Range, Karakalpakistan, Uzbekistan. Locality SKH-20 is in the lower or middle part of the Khodzhakul Svita and dated as late Albian on the basis of the sharks *Paraisurus macrorhiza* and *Eoanacorax dalinkevichiusi*, neither of which is known from deposits younger than the late Albian (Nesov *et al.*, 1994b; Nesov, 1995). Mammals were first found here in 1978.

3. Khodzhakulsai. Localities SKH-5 and SKH-5a occur in the upper part of the Khodzhakul Svita, which outcrops in the ravine of Khodzhakulsai near the Khodzhakul hole, Karakalpakia, Uzbekistan. The upper part of the Khodzhakul Svita was deposited during a regressive phase between the marine transgressions in the late Albian and late Cenomanian–early Turonian. Based on this and the recovery of the shark *Hybodus nukusensis* and the turtle *Ferganemys itemirensis*, it is dated as early Cenomanian (Nesov *et al.*, 1994b; Nesov, 1995). Mammals were found here in 1979.

4. Sheikhdzheili. Localities SSHD-8 and SSHD-8a occur in the upper part of the Khodzhakul Svita of Sheikhdzheili Ridge, Karakalpakistan, Uzbekistan. They are the same age as those at Khodzhakulsai. Mammals have been found here since 1979. Besides remains mentioned in Table 30.1, a series of as yet undescribed teeth, jaw fragments and postcranial bones is also known from this locality.

5. Chelpyk. Locality STCH-1 occurs in the upper part of the Khodzhakul Svita at the base of Chelpyk Hill, north of the Sultan Uvais Range, Karakalpakistan, Uzbekistan. This locality is also the same age as those at Khodzhakulsai. Mammal bones were found here in 1980.

6. Ashchikol. In 1988 the holotype of *Sorlestes kara* (Figure 30.4L, M) was found in a drilling core from a depth of about 500 m. The age for sediments in this core is lower Turonian based upon spores and pollen (Nesov, 1993; Nesov *et al.*, 1994b).

7. Dzharakhuduk. This series of localities occurs in the Bissekty Svita, outcropping above cliffs stretching about 9 km to the east from the Dzharakhuduk settlement (near Bissekty and Kulbeke wells, Figure 30.2) in the Mynbulag Depression of the central Kyzylkum Desert, Uzbekistan. There is one possible upper Turonian and several Coniacian levels in the Bissekty Svita that yield mammal remains.

7a. Possible upper Turonian level. Localities CDZH-17a, CDZH-17g and CDZH-25 are located in the lower part of the Bissekty Svita, which is thought, on the basis of an anacoracid and sharks to be late Turonian in age (Nesov, 1993; Nesov *et al.*, 1994b; Nesov *et al.*, 1998). Besides specimens listed in Table 30.1, a hystricognathous dentary without teeth from an undescribed multituberculate, other teeth, jaw fragments and postcranial bones were found at this level.

7b. Coniacian; localities CBI-4, CBI-4a, CBI-4b, CBI-4v and others. The dating of these localities is based on the remains of sharks, including *Ptychocorax aulaticus* and a small *Squalicorax* sp., that have been found in association with Coniacian invertebrates and turtles in marine sediments of Tadzhikistan and Turkmenistan (Nesov, 1993; Nesov *et al.*, 1994b; Nesov *et al.*, 1998). The holotype of *Cretasorex arkhangelskyi* (TsNIGRI 2/11758: Nesov and Gureev, 1981, fig. 1; Nesov, 1981, fig. 11, 30; 1982, tabl. 2, fig. 6; 1985b, tabl. 2, fig. 4; Nesov *et al.*, 1994b, pl. 1, fig. 3) was found in 1979, but most probably is late Cenozoic in age (see Nesov, 1993; Nesov *et al.*, 1994b). The majority of remains from this level were collected during intensive dry screening at locality CBI-4v in 1979 and 1980. In the latter year 12 tons of matrix were sifted. Further remains were found in 1989. In addition to specimens listed in Table 30.1, several dozen other undescribed remains come from this level.

Figure 30.2. Cliffs exposing upper Cretaceous deposits near Dzharakhuduk settlement. Photo by L.F. Kaznyshkina, 1979.

7c. Coniacian, localities CBI-14, CBI-14a, CBI-50 and others. Intensive surface collecting was carried out at locality CBI-14 in 1987 and 1989, yielding the majority of mammalian material from Dzharakhuduk. A dentary with m3, more than 100 teeth, jaw fragments and postcranial bones of therian mammals as yet undescribed from this locality are not included in Table 30.1.

7d. Coniacian, locality CBI-17. The holotype of *Zhelestes temirkazyk* (Figure 30.7D, E) came from the ravine containing site CBI-17, from the top deposits approximately at CBI-5a level.

7e. Coniacian, localities CBI-5 and CBI-5a. Besides specimens listed in Table 30.1, several undescribed postcranial elements were found at these localities.

7f. Coniacian, locality CBI-7a. A few mammalian specimens are known from this level (Table 30.1), including several dentary fragments.

8. Kansai. Mammals remains have been recovered from the lower bone bearing horizon in the Yalovach Svita at locality FKA-7a which is located 3 km east of Kyzylbulag settlement, near Kansai village, Tadzhikistan. The age of the beds was established on the basis of molluscs, rays, fishes and turtles as lower Santonian (Nesov, 1993; Nesov *et al.*, 1994b). Two mammalian teeth were found here in 1984 (Table 30.1).

9. Zhalmaus. Mammal remains have been found in the lower part of the Bostobe Svita, dated on the basis of rays and turtles as Santonian (Nesov and Khisarova,

1988; Nesov, 1993) at Zhalmaus Well, near Baibolat Well, and close to the Shakh-Shakh cliffs, north-east of the Aral Sea, in Kyzyl-Orda Province, Kazakhstan. Here, in 1962, was found the holotype of *Beleutinus orlovi* (Figure 30.8J-L). In 1982 the centrum of a mammalian cervical vertebra was found by Nesov in material screen-washed in 1962.

10. *Alymtau.* The locality of ALT-1 ('Grey Mesa') is located on the northern slope of the Alymtau Range in southern Kazakhstan. Beds exposed here belong to the lower Darbasa Formation which is dated as lower Campanian according to the shark and tetrapod assemblage and its position within the section (Nesov, 1993; Aver'yanov and Nesov, 1995). The holotype of *Alymlestes kielanae* (Figure30.8M, N) was found in 1991. Seven additional mammalian specimens, including a multituberculate (*Bulganbaatar* sp.) and a deltatheroidan (*Deltatheridium nessovi*), were found here in 1996 (Averianov, 1997).

Additional localities yielding Mesozoic mammals include Madygen in Kirgizstan and Yantardakh in northern Russia. From the former locality the skeleton of a mammal-like reptile, morphologically very close to mammals has been reported from beds of upper Triassic age (Tatarinov, 1980). The second locality yielded mammalian hair in amber of Santonian age (Nesov *et al.*, 1994b).

Systematic review

Docodonts

Docodonts are known from the Middle and Upper Jurassic of Europe and North America. In Asia, one tooth has been found in the Upper Jurassic of Mongolia (see Chapter 29). A heavily damaged upper tooth, similar in general outlines to docodont molars, was found in Middle Jurassic deposits at Kalmakerchin in Kirgizstan (Nesov *et al.*, 1994a). The fragment of an ulna of a relatively large mammal found together with this tooth may also belong to a docodont. If this attribution is correct, it may indicate, together with the Mongolian discovery, an as yet unknown abundance of docodonts in Mid–Late Jurassic mammal assemblages of Asia.

Multituberculates

Multituberculates, the dominant group of Cretaceous mammals in Mongolia and adjacent areas of Central Asia (see Chapter 29), are very rare in Cretaceous mammal assemblages of western Asia. They have been found in late Turonian and Coniacian deposits of Uzbekistan and early Campanian deposits of Kazakhstan. In Uzbekistan their remains constitute one percent of all mammalian material (Kielan-Jaworowska and Nesov, 1992; Nesov, 1993).

Turonian multituberculates may be represented by an undescribed edentulous dentary with the alveolus of an unreduced p3, a relatively small p4 and molars, and showing a tendency to hystricognathy. Coniacian multituberculates are represented by five fragments, among which a small p4 is the most informative. This tooth (Figure 30.6A–D), the holotype of *Uzbekbaatar kizylkumensis* Kielan-Jaworowska and Nesov, 1992, has an arcuate crown without a row of posterobuccal cusps, and a small number (9) of serrations. The structure of this p4 suggests the presence of an unreduced p3. This taxon was identified by Kielan-Jaworowska and Nesov (1992) as a non-specialized cimolodontan, but was not assigned to a particular family or infraorder.

Recently, an isolated P4 referred to *Bulganbaatar* sp. was found among seven additional mammalian specimens from the Alymtau assemblage (Averianov, 1997).

Therians

Therians form the majority of mammals that lived on the western margin of the ancient Asian continent in the Cretaceous. The bulk of these therians are eutherians, but remains attributed to deltatheroidans and some enigmatic marsupial-like forms have also been found.

?Marsupial-like forms

The fragment of a left dentary from the locality of CBI-14 in Coniacian beds of Dzharakhuduk (Uzbekistan) bears the last molar, which has a large paraconid and twinned hypoconulid and entoconid and thus shows marsupial-like characters. This specimen is currently under study by Professor Zofia Kielan-Jaworowska. Two fragmentary edentulous dentaries, each with a strongly inflected internal angular process (Figure 30.5L, M), from the Coniacian localities of CBI-51 and CBI-4b at Dzharakhuduk, also show marsupial-like characters. The dental formulae of the better preserved specimen has been interpreted as p1–3, m1–4 (Nesov, 1997).

Deltatheroidans

Deltatheroidans are large (by Mesozoic standards) carnivorous mammals, and may be the sister group of the Metatheria (Kielan-Jaworowska and Nesov, 1990). In western Asia they have been found in the late Turonian and Coniacian of central Kyzylkum (Uzbekistan), where they are represented by two subfamilies of the family Deltatheridiidae: the monotypic Sulestinae and the Deltatheridiinae. Both are close to the Mongolian forms.

The Sulestinae are endemic to western Asia and represented by *Sulestes karakshi* Nesov, 1985b, from the Coniacian site CBI-4b at Dzharakhuduk, Uzbekistan (Figure 30.5K) and *Sulestes* sp. from CBI-14 at Dzharakhuduk (Figure 30.6M, N). The upper molars of *Sulestes* differ from those of Mongolian deltatheroidans mainly in the presence of well-developed cuspules on the ectoflexus and on the metastylar region. The morphology of the molars, with their relatively long protocone region, wide talonid, and low trigonid, which is more similar to the condition of marsupials than deltatheroidans, also distinguishes *Sulestes*.

'*Deltatheroides*' *kizylkumensis* Nesov, 1993 is represented by lower molars (Figure 30.6E) and an edentulous maxilla (Figure 30.6F) from site CBI-14 at Dzharakhuduk. This species was originally assigned to the Deltatheroididae (Nesov, 1993), but subsequently Nesov (1997) assigned it to a new genus within the Deltatheridiidae. This taxon differs from *Sulestes* in its larger size, the sharper cutting edges of the molars, and in the less developed talonid. It differs from the Mongolian deltatheridians in having a relatively larger upper canine, two-rooted P1, and a lower position of the infraorbital foramen.

Deltatheridium known from the early Campanian Alymtau assemblage (Averianov, 1997), is represented by a partial M2 that retains some primitive traits in molar morphology.

The early Cenomanian *Oxlestes grandis* Nesov, 1982a, known from a fused cervical centra (Figure 30.3N), and, less certainly, fragments of a canine and parietal (Figure 30.3J, K) from site SSHD-8 at Sheikhdzheili in Karakalpakia, might be a deltatheroidan (Nesov *et al.*, 1994b). This large, apparently carnivorous, mammal possibly preyed on small protoceratopsid dinosaurs such as *Asiaceratops* (Nesov, 1982b). The systematic position of this form remains uncertain.

Eutherians

Eutherians, or placental mammals, are the dominant mammal group in the Cretaceous of the western part of ancient Asia. They occur as early as the late Albian, where they are represented by *Bobolestes zenge* Nesov, 1985a, known only from the holotype, a maxilla with the last two molars (Figure 30.3A, B). Originally, *Bobolestes* was placed in the Pappotheriidae (Nesov, 1985a), a family that is now considered to be close to the metatherian–eutherian dichotomy (Butler, 1990). Pappotheridians retain more primitive dental characters than other placentals, and probably had four molars. *Bobolestes* is somewhat more advanced than pappotheriids in having winged conules and a more reduced parastylar region. Subsequently *Bobolestes* was placed by Nesov (1989, 1993) in its own family, the Bobolestidae.

Cenomanian eutherians are represented by *Otlestes meiman* Nesov, 1985, based on lower and upper jaw fragments (Figure 30.3C, G, H, O) from Sheikhdzheili, Karakalpakia. The dental morphology

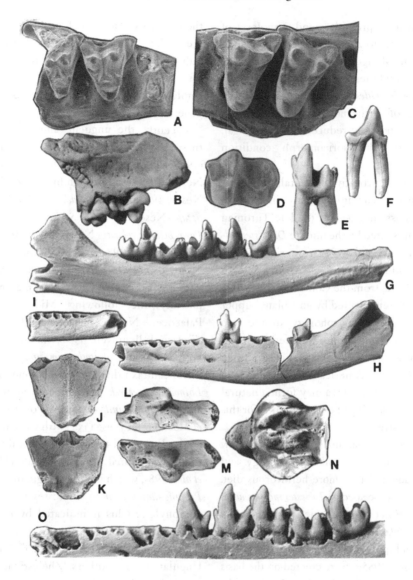

Figure 30.3. Mammal remains from the upper Albian and lower Cenomanian of Southwestern Kyzylkum, Uzbekistan (A, C, from Nesov, 1993; B, E–H, J–M, O, from Nesov, 1985a; I, from Nesov, 1984; N, from Nesov, 1982a). A and B fragment of right maxilla with M2–3, TsNIGRI 2/12176, SKH-20, the holotype of *Bobolestes zenge*, in occlusal (A) ($\times$13) and labial (B) ($\times$10) views. C. left maxilla with M1 and 2, TsNIGRI 8/12176, SSHD-8, attributed to *Otlestes meiman* ($\times$13). D and E, left lower molar of 'Zhelestidae' gen. et sp. nov. (Nesov, 1997), TsNIGRI 26/12176, SSHD-8, in occlusal (D) ($\times$13) and labial (E) ($\times$8) views. F, lower premolar of therian, SSHD-8, lateral view ($\times$10). G and O, left dentary with p4–5 m1–3, TsNIGRI 7/12176, SSHD-8, the holotype of *Otlestes meiman*, in lingual (G) ($\times$8) and labial (O) ($\times$10) view. H, left dentary with p5 and talonid of m3 of *Otlestes meiman*, TsNIGRI 9/12176, SSHD-8, labial view ($\times$6). I, edentulous dentary fragment, TsNIGRI 2/11658, SKH-20, labial view ($\times$5). J and K, parietal, possibly of *Oxlestes*, SSHD-8a, in dorsal (J) and ventral (K) views ($\times$2). L and M, calcaneus, STCH-1, in dorsal (L) and lateral (M) views ($\times$5). N, vertebral body of C2 and C3, TsNIGRI 6/11758, SSHD-8, the holotype of *Oxlestes grandis*, ventral view ($\times$3).

and dental formula (5 premolars and 3 molars) of *Otlestes* is very similar to that of the Aptian–Albian *Prokennalestes* from Mongolia, and both taxa were assigned to the family Otlestidae (Kielan-Jaworowska and Dashzeveg, 1989). *Otlestes* may belong to another evolutionary line of eutherians, as it differs from *Prokennalestes* in the possibly reduced or absent jugal, which approximates the soricomorph condition (Nesov, 1997).

In the late Turonian, primitive placentals of palaeo-ryctoid appearance occur infrequently in western Asian mammalian assemblages. At the late Turonian level, they are represented by the minute *Daulestes kulbeckensis* Trofimov and Nesov, 1979, from site CDZH-17a at Dzharakhuduk (Figure 30.4A, B). *Sailestes quadrans* Nesov, 1982a, from the Coniacian site CBI-4 at Dzharakhuduk is represented by an isolated upper molar, similar in dental morphology to the late Palaeocene *Bustylus* from France (Gheerbrant and Russell, 1991) and the Campanian?–Maastrichtian *Bistius* from North America (Clemens and Lillegraven, 1986). All these taxa may form a natural group, but so far they are insufficiently known for this to be clearly demonstrated.

In the late Turonian mammals appear that show additional modifications in dental morphology and were apparently adapted to a more herbivorous diet. The genera *Taslestes* (type species *Taslestes inobservabilis* Nesov, 1982a), *Sorlestes* (type species *Sorlestes budan* Nesov, 1985a), *Aspanlestes* (type species *Aspanlestes aptap* Nesov, 1985a), and possibly *Kumlestes* (type species *Kumlestes olzha* Nesov, 1985a) were erected on the basis of lower dentitions and assigned to the suborder (subsequently order) Mixotheridia (Nesov, 1985a).

Mixotheridia Nesov, 1985a is characterized by a mosaic of marsupial and eutherian characters of the lower molars: the paraconid is reduced (a eutherian feature) and the hypoconulid approximates the entoconid (a marsupial feature). In this respect, mixotheridians approximate to the Recent tupaiids (Scandentia). These mammals, whose upper dentition is characterized by an incipient or moderately developed cingula (Figure 30.7A–E, G), were assigned by Nesov (1985a, 1987) to the subfamily Zhelestinae in

the Kennalestidae, though Nesov (1993) later considered them as a separate family, Zhelestidae. Subsequently, Nesov (1993) recognized that both lower and upper dentitions belonged to the same or similar animals, which led to some confusion at the generic level.

Recently, the ungulate affinities of a number of these taxa has been examined in detail by Nesov *et al.* (1998). These authors conclude that *Kumsuperus avus* Nesov, 1984, nomen dubium, *Zhelestes temirkazyk* Nesov, 1985a, *Sorlestes budan* Nesov, 1985a, *Aspanlestes aptap* Nesov, 1985a, *Sorlestes kara* Nesov, 1993, *Parazhelestes robustus* Nesov, 1993, *Parazhelestes* sp. nov. Nesov *et al.* (1998), and 'Zhelestidae' gen. et sp. nov. Nesov *et al.* (1998), share an ancestry that is not shared with any other late Cretaceous mammals, except for possibly the following Mid-Campanian– Early Palaeocene North America taxa: *Gallolestes pachymandibularis* Lillegraven, 1976, *Alostera saskatchewanensis* Fox, 1989, and *Avitotherium utahensis* Cifelli, 1990. European taxa that may also belong in this group are *Lainodon orueetxebarriai* Gheerbrant and Astiba, 1994, and *Labes quintanillensis* Sig in Pol *et al.*, 1992. Nesov *et al.* (1998) refer these taxa (with some doubt in the case of the European forms) to the family Zhelestidae. A phylogenetic analysis by Archibald (1996) and Nesov *et al.* (1998), which includes ungulates (represented by *Protungulatum donnae*), indicates that 'Zhelestidae' is paraphyletic (this is indicated by quotation marks) because some 'zhelestid' taxa, notably a new form (Nesov *et al.*, 1998) share a more recent ancestry with Ungulata than do others. 'Zhelestidae' plus Ungulata were included in a new supraordinal taxon, Ungulatomorpha (Archibald, 1996; Nesov *et al.*, 1998). The 'zhelestids' are the most common mammals in the Coniacian mammal assemblages of Uzbekistan.

One of the most important fossils found in the Coniacian of Uzbekistan (site CBI-5a), in the Bissekty Formation at Dzharakhuduk, is an almost complete skull of a juvenile mammal, associated with both lower jaws. Most of the dentition is preserved, as are the ear ossicles, but the posterior end of the skull is broken and distorted (Figure 30.8A, B). This may be the oldest known skull of an eutherian mammal and is possibly

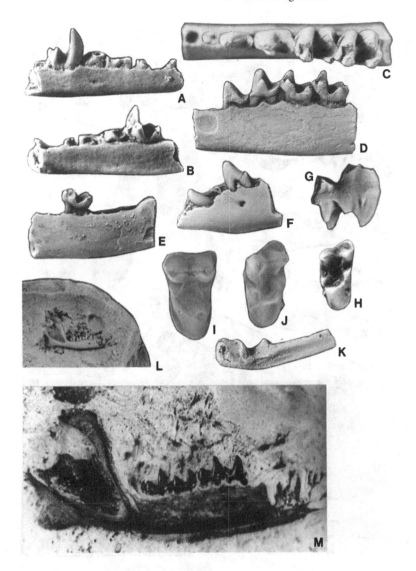

Figure 30.4. Mammal remains from the Turonian of Uzbekistan and Kazakhstan (A, B, E, from Nesov, 1982a; C, D, G, H, K, from Nesov, 1985a; F, from Nesov, 1984; J, L, from Nesov, 1993). A and B, right dentary fragment with c1 and m1, TsNIGRI 1/11758, CDZH-17a, the holotype of *Daulestes kulbeckensis*, in labial (A) and lingual (B) views (×10). C and D, right dentary fragment with p4–5 m1–2, TsNIGRI 4/12176, CDZH-17a, the holotype of *Aspanlestes aptap*, in occlusal (C) (×8) and lingual (D) (×6) views. E, right dentary fragment with m2, TsNIGRI 8/11758, CDZH-17a, the holotype of *Taslestes inobservabilis*, lingual view (×8). F, left dentary fragment with erupting c1 and a premolar of *Aspanlestes aptap*, TsNIGRI 3/11658, CDZH-25, labial view (×10). G and H, left M1 or M2 of 'Zhelestidae'? indet., TsNIGRI 23/12176, CDZH-17a, posterior (G) and occlusal (H) views (×10). I, right M1 of cf. *Zhelestes* sp., TsNIGRI 38/12000, CDZH-17a, occlusal view (×13). J, left M1, TsNIGRI 5/12176, CDZH-17a, the holotype of *Kulbeckia rara*, occlusal view (×13). K, ulna, CDZH-17a, lateral view (×4). L and M, right dentary with p5, m1–3 from a drilling core, TsNIGRI 101/12455, Ashchikol, the holotype of *Sorlestes kara*, labial view (L) (×1) and (M) (×4).

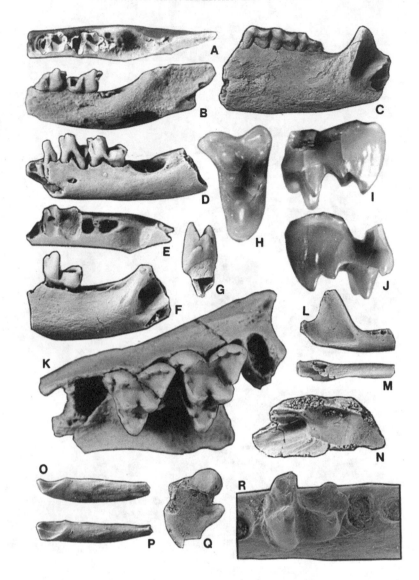

Figure 30.5. Mammal remains from the Coniacian of Central Kyzylkum, Uzbekistan (A, B, D, G, L, M, O–Q, from Nesov, 1985a; C, from Nesov, 1984; E, F, K, N, from Nesov, 1985b; H–J, R, from Nesov, 1993). A and B, left dentary fragment with m1–2, TsNIGRI 1/12176, CBI-4b, the holotype of *Kumlestes olzha*, in occlusal (A) and labial (B) views (×10). **C**, left dentary fragment with m1–3, TsNIGRI 12/11758, CBI-4, the holotype of *Kumsuperus avus*, labial view (×4.3). D, left dentary fragment with dp2–4, possibly of *Aspanlestes aptap*, CBI-4v, labial view (×8). E and F, left dentary fragment with m2, TsNIGRI 36/12000, CBI-4b, the holotype of *?Zalambdalestes mynbulakensis* (possible junior synonym of *Sorlestes budan*), in occlusal (E) and labial (F) view (×6). G, molar trigonid of 'Zhelestidae' indet., CBI-4v, anterior view (×8). H–J, left M2, TsNIGRI 52/12455, CBI-5a, the holotype of *Kulbeckia kulbecke*, in occlusal (H), posterior (I) and anterior (J) views (×15). K, left maxilla fragment with M1–2, TsNIGRI 35/12000, CBI-4b, the holotype of *Sulestes karakshi*, occlusal view (×10.5). L and M, right dentary fragment, CBI-4b, in labial (L) and ventral (M) view (×2.5). N, frontal, TsNIGRI 39/12000, CBI-4b, ventral view (×3). O and P, incisor, CBI-4b, in lateral (O) and posterior (P) view (×10). Q, proximal part of femur, CBI-4b, posterior view (×4). R, left dentary fragment with m2 of *Kennalestes* sp. nov. (Nesov, 1997), TsNIGRI 72/12455, CBI-4, occlusal view (×26).

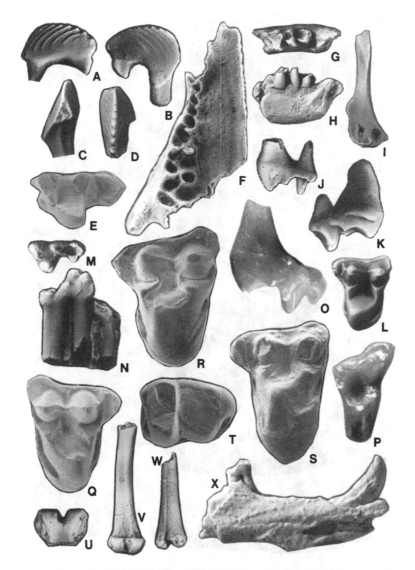

Figure 30.6. Mammal remains from sites CBI-4 (Q), CBI-14 (A–N, R–X) and CBI-5a (O, P) (Coniacian) of central Kyzylkum, Uzbekistan (A–F, O, P, from Nesov, 1993; G–I from Nesov, 1985a; J–N, U–W, from Nesov, 1987). A–D, left p4, TsNIGRI 100/12455, the holotype of *Uzbekbaatar kisylkumensis*, in lingual (A), labial (B), anterior (C) and occlusal (D) views (×9). E, left m2?, TsNIGRI 40/12455, the holotype of *Deltatheroides kizylkumensis*, occlusal view (×13). F, edentulous right maxilla, possibly of *Deltatheroides kizylkumensis*, TsNIGRI 42/12000, ventral view (×2.5). G and H, left dentary fragment with incomplete m2 and m3 of *Aspanlestes aptap*, TsNIGRI 6/12176, in occlusal (G) and labial (H) views (×5). I, right humerus, anterior view (×2). J, right M1? of ?*Kulbeckia* sp., TsNIGRI 6/12455, posterior view (×10). K and L, left M1 of *Aspanlestes aptap*, TsNIGRI 3/12455, in posterior (K) and occlusal (L) views (×10). M and N, left m1 of *Sulestes* sp., TsNIGRI 5/12455, in occlusal (M) and lingual (N) views (×8.8). O and P, right M3 of ?*Kulbeckia* sp., TsNIGRI 54/12455, in posterior (O) and occlusal (P) view (×15). Q, right M1 of *Parazhelestes* sp. nov. (Nesov *et al.*, 1998), TsNIGRI 11/12953, occlusal view (×13). R, left M1 of *Parazhelestes robustus*, TsNIGRI 20/12953, occlusal view (×13). S, left M2 of 'Zhelestidae' gen. et sp. nov. (Nesov *et al.*, 1998), TsNIGRI 2/12455, occlusal view (×12). T, left m3 of cf. 'Zhelestidae' gen. et sp. nov. (Nesov *et al.*, in press), TsNIGRI 18/12953, occlusal view (×13). U and V, distal part of femur, TsNIGRI 7/12455, in distal (U) (×3) and posterior (V) (×2) views. W, distal part of fused tibia and fibula, TsNIGRI 8/12455 (×3). X, left dentary fragment with c1 and a premolar of *Sorlestes budan*, TsNIGRI 15/12953, lingual view (×5).

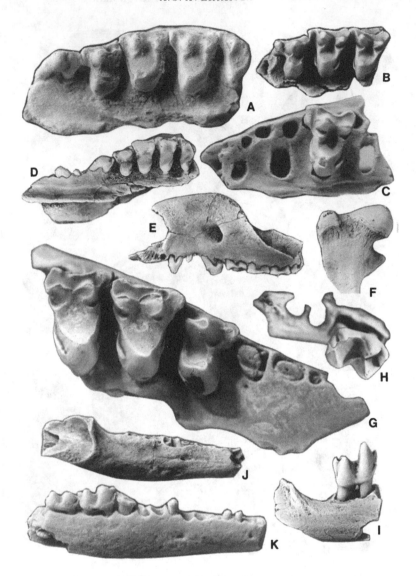

Figure 30.7. Mammal remains from the Coniacian of central Kyzylkum, Uzbekistan (A, C, K, from Nesov, 1993; B, from Nesov, 1987; D–I, from Nesov, 1985a). All specimens, except D, E and F, from site CBI-14. A, left maxilla with P4–5 M1–3, TsNIGRI 70/12455, the holotype of *Parazhelestes robustus*, occlusal view (×6). B, left maxilla with P5, M1–2, TsNIGRI 1/12455, the holotype of ?*Zhelestes bezelgen* (junior synonym of *Aspanlestes aptap*), occlusal view (×6.8). C, left maxilla with M2 of *Aspanlestes aptap*, TsNIGRI 68/12455, occlusal view (×8). D and E, left maxilla with P2–5 M1–3, TsNIGRI 10/12176, the holotype of *Zhelestes temirkazyk*, found in the ravine at site CBI-17, but on the level of site CBI-5a, in occlusal (D) and labial (E) view (×3). F, proximal part of femur, CBI-7a (×3). G, right maxilla with P5 M1–2 of *Parazhelestes* sp. nov. (Nesov *et al.*, 1998), TsNIGRI 11/12176, occlusal view (×10). H and I, right dentary fragment with m2, TsNIGRI 3/12176, the holotype of *Sorlestes budan*, in occlusal (H) (×8) and labial (I) (×6) views. J, right edentulous dentary of 'Zhelestidae' indet., TsNIGRI 23/12953, labial view (×2.5). K, right dentary fragment with m2–3, TsNIGRI 67/12455, the holotype of *Paranyctoides aralensis*, labial view (×8).

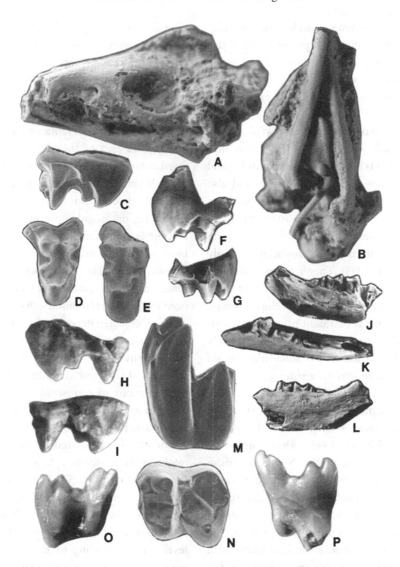

Figure 30.8. Late Cretaceous mammal remains from Uzbekistan, Tajikistan and Kazakhstan (C–E, H, I, O, P, from Nesov, 1993; F, G, J, L, from Nesov, 1987; M, N, from Averianov and Nesov, 1995). A and B, eutherian skull, ZIN C.79066, CBI-5, in lateral (A) and ventral (B) view (×6). C and D, left M2, TsNIGRI 52/12455, CBI-5a, the holotype of *Kulbeckia kulbecke*, in posterior (C) and occlusal (D) views (×13). E–G, left M1, TsNIGRI 9/12455, FKA-7a, the holotype of *Kulbeckia kansaica*, in occlusal (E) (×13), anterior (F) (×10) and posterior (G) (×10) views. H and I, left M1?, TsNIGRI 73/12455, FKA-7a, in anterior (H) and posterior (I) views (×15). J–L, right dentary fragment with heavily worn teeth, IZK 1–751/III-1962, Zhalmaus, the holotype of *Beleutinus orlovi*, in labial (J) (×3.2), occlusal (K) (×4.1) and lingual (L) (×3.2) views. M and N, left m1, ZIN C78332, ALT-1, the holotype of *Alymlestes kielanae*, in labial (M) and occlusal (N) views (×15). O and P, right m1? of *Kulbeckia kulbecke*, TsNIGRI 60/12455, CBI-5a, in labial (O) and lingual (P) views (×15).

congeneric with *Taslestes* (McKenna, Kielan-Jaworowska, Meng and Nesov, in preparation). If this assignment is correct, *Taslestes* should be removed from the Mixotheridia, although the holotype of *T. inobservabilis* is consistent with the diagnosis of Mixotheridia.

The rather peculiar mammals *Kulbeckia kulbecke* Nesov, 1993 and *Kulbeckia rara* Nesov, 1993 from sites CBI-5a and CDZH-17a at Dzharakhuduk, and *Kulbeckia kansaica* Nesov, 1993 from Kansai and also found in the Turonian and Coniacian of Uzbekistan (Figure 30.4J; Figure 30.5H–J; Figure 30.6O, P; Figure 30.8C, D) and Santonian of Tadzhikistan (Figure 30.8F–I), are united in the monotypic family Kulbeckiidae (Nesov, 1993). They are represented by isolated teeth characterized by a narrow stylar shelf, sharp and tall conules with well-developed wings, high protocone, vestigial or absent pre- and postcingula, paraconid close to the metaconid, a moderately reduced, deep talonid basin, and spine-like talonid tubercles. Kulbeckiidae may be a sister group of the taxon Zhelestidae + Ungulata (Nesov, 1997).

One dentary fragment from the Coniacian of Dzharakhuduk (Figure 30.7K), the holotype of *Paranyctoides aralensis* Nesov, 1993, with alveoli for five premolars, was attributed to the North American genus *Paranyctoides* (Fox, 1979, 1984), which is sometimes placed in the lipotyphlan family Nyctitheriidae.

In addition to the kulbeckiids, Santonian mammals of western Asia are represented possibly by Zalambdalestidae, if *Beleutinus orlovi* from Kazakhstan (Figure 30.8J–L) belongs to this group. Beginning in the Santonian, this group apparently became dominant in the western part of the Asian continent. Three fragments, including the holotype of *Alymlestes kielanae* Aver'yanov and Nesov, 1995 (Figure 30.8M, N), represent Campanian zalambdalestids (Averianov, 1997). In dental morphology, *Alymlestes* is somewhat more derived than the Mongolian *Zalambdalestes* and *Barunlestes*, having a more developed unilateral hypsodonty and a higher tooth crown. Structurally, the molar of *Alymlestes* is very close to the initial condition for Lagomorpha. Possibly, zalambdalestids were a more numerous and diverse group than previously

thought, based on the Mongolian genera, and some members were close to the ancestral stock for Lagomorpha and Glires.

Conclusions

Although mammalian remains from the Cretaceous of the western part of the ancient Asian continent are fragmentary in comparison to central Asian records, they provide important information about the origin and early diversification of some mammalian orders. In this region, ungulatomorph 'zhelestids' and other placentals dominated, whereas in central Asia the most abundant mammals were multituberculates (Nesov and Kielan-Jaworowska, 1991). This may reflect different climatic and environmental conditions in these parts of ancient Asia.

The Coniacian deposits of Uzbekistan contain vertebrate assemblages consisting of numerous disarticulated freshwater amphibians and fish, brackish and saltwater sharks and actinopterygians, together with rare remnants of terrestrial lizards, mammals, dinosaurs, pterosaurs, and small birds. Nesov (1990b), Nesov and Kielan-Jaworowska (1991) and Roček and Nesov (1993) speculated on how these associations might have been formed. As put by Nesov and Kielan-Jaworowska (1991, p. 52): 'The fossils have been preserved only in channels linking brackish water and fresh water basins. The channels show dual current directions resulting mostly from winds. A rise in water level caused flooding of the flat coastal plains and many terrestrial animals drowned. The influx of brackish or salt water (especially unoxygenated water from deep parts of the channels) into freshwater basins and swamps killed the amphibians and fish. Flooding of the land after heavy rains caused the death of sharks and actinopterygians living in brackish or saline basins. As the multituberculates did not live in the swamps or on wet coastal plains, and archaic ungulates occupied herbivorous mammal niches, multituberculate remains are very rare.'

The Djadokhta and Baruungoyot formations of Mongolia have also yielded mammals (Chapter 29), but these and other taxa including dinosaurs and their eggs, lizards, and birds were buried in eolian sand (Jerzykiewicz *et al.*, 1993) and probably lived under rather different conditions: dry steppes and semi-deserts far away from the sea. The therian Late Cretaceous assemblage of Uzbekistan, was more diversified than that of the Gobi Desert, and perhaps this reflects the rather more favourable subhumid, subtropical conditions in which they lived.

Acknowledgements

I am indebted to the late Dr Lev A. Nesov for the opportunity to collect and study Cretaceous mammal remains from 'Middle Asia'. I am very grateful to Professor Zofia Kielan-Jaworowska, Professor J. David Archibald, and Dr Richard C. Fox for reading and improving the text, and to Dr Lena B. Golovneva and Miss Alla A. Nesova for access to photographs from Lev Nesov's archive.

Note added in proof. While this paper was in preparation, four new Mesozoic mammal localities were discovered on the territory of the former USSR.

1. A sediment-filled fissure containing clays assigned to the Meshcherskii Gorizont of the Moskvoretskaya Svita (Middle Jurassic: upper Bathonian) at the locality of Peski, about 100 km sout-east of Moscow, Moscow Province, central Russia, yielded an almost complete femur of a morganucodontid mammal (Novikov *et al.* 1998, 1999).

2. Sediments of the Mogoito Member of the Murtoi Svita (Lower Cretaceous: upper Barremian–lower Aptian) at the locality of Mogoito, on the western coast of the Gusinoe Ozero [Goose Lake], Buryatiza, western Transbaikalia, Russia (N 51°12'03", E 106°17'06"), yielded three molars (M1, M2 and m3) of the oldest eutherian mammal. This was first determined as *Prokennalestes* sp. n. (Averianov and Skutschas, 1999, in press a), but was subsequently referred to Kennalestoidea gen. et sp. n. (Averianov and Skutschas, 1999, in press b).

3. Jaw fragments representing *Gobiconodon borissiaki* Trofimov, 1978, up to two other species of *Gobiconodon* (one of which is new), and a possible

symmetrodont have been reported (Mashchenko and Lopatin, 1998; Mashchenko, 1999a, b; E.N. Mashchenko, pers. comm.) from the Ilekskaya Svita (Lower Cretaceous: ?Albian) of Shestakovo locality, on the right bank of the Kiya River, a tributary of the Chulym River in Kemerovskaya Province, Western Siberia, Russia (N 55°54'12", E 87°57'28").

4. Isolated mammal teeth and tooth fragments representing *Uzbekbaatar wardi* Averianov, 1999, an M3 identified as cf. *Parazhelestes* sp., a fragment of an upper molar of a deltatheroid, and other remains (Averianov 1999, in press) have been recovered from marine sediments of the Aitym Svita (Late Cretaceous: ?Santonian) in Aitym (locality CBI-117, the 'shark locality') at Dzharakhuduk in the Central Kyzylkum Desert of Uzbekistan (N 42°07'24", E 62°39'29").

In addition, considerable new data on Mesozoic mammals were gathered during fieldwork by joint Uzbek–Russian–British–American–Canadian expeditions in 1997–1999. The new records come from the Bissekty Svita (upper Turonian – ?Coniacian) at the locality CBI-14, at Dzharakhuduk, in central Kyzylkum, and from the upper Khodzhakul Svita (lower Cenomanian) at the locality of SSHD-8a at Sheikhdzheili, in south-western Kyzylkum (Archibald *et al.*, 1998, 1999). A new record of the marsupial *Marsasia* sp, consisting of a dentary fragment with m3, from locality CBI-14, was reported by Averianov and Kielan-Jaworowska (1999).

References

Archibald, J.D. 1996. Fossil evidence for a late Cretaceous origin of 'hoofed' mammals. *Science* **272**: 1150–1153.

—, Sues, H.-D., Averianov, A., Danilov, I., Rezvyi, A., Ward, D., King, C., and Morris, N. 1999. New paleontologic, biostratigraphic, and sedimentologic results at Dzhara Kuduk (U. Cret.), Kyzylkum Desert, Uzbekistan. *Journal of Vertebrate Paleontology* **19**, suppl. to No. 3: 29A–30A.

—Sues, H.-D., Averianov, A., King, C., Ward, D.J., Tsaruk, O.I., Danilov, I.G., Rezuyi, A.S. Veretennikov, B.G. and Khodjaev, A. 1998. Précis of the paleontology, biostratigraphy, and sedimentology at Dzharakuduk (Turonian?–Santonian), Kyzylkum Desert, Uzbekistan, pp. 21–28 in Lucas, S., Kirkland, J.I. and Estep, J.W. (eds.) *Lower and Middle Cretaceous Terrestrial Ecosystems. Bulletin of the New Mexico Museum of Natural Hisotry and Science* **14**.

Averianov, A. 1996. [Lev Aleksandrovich Nesov (1947–1995).] *Russian Journal of Herpetology* **3**: 105–106.

—1997. New late Cretaceous mammals of southern Kazakhstan. *Acta Palaeontologica Polonica* **42**(2): 243–256.

—1999. A new species of multituberculate mammal *Uzbekbaatar* from the Late Cretaceous of Uzbekistan. *Acta Paleontologica Polonica* **44**: 301–304.

—(in press). A 'zhelestid' mammal from the Upper Cretaceous Aitym Formation in Uzbekistan. *Acta Paleontologica Polonica.*

—and Kielan-Jaworowska, Z. 1999. Marsupials From the Late Cretaceous of Uzbekistan. *Acta Paleontologia Polonica* **44**: 71–81.

—and Nesov, L.A. 1995. A new Cretaceous mammal from the Campanian of Kazakhstan. *Neues Jahrbuch für Geologie und Paläontologie, Monatshefte* **1995**: 65–74.

—and Skutschas, P.P. 1999. The oldest eutherian mammal, p. 6 in Orlov, V.N. (ed.) *VI Congress of Theriological Society.* Theriological Society. Moscow.

—and—(in press a). A eutherian mammal from the Early Cretaceous of Russia and biostratigraphy of the Asian Early Cretaceous vertebrate assemblages. *Lethaia.*

—and—(in press b). [The oldest placental mammal]. *Doklady RAN.*

Bazhanov, V.S. 1972. [First Mesozoic Mammalia (*Beleutinus orlovi* Bashanov) from the USSR.] *Teriologiya* **1**: 74–80.

Burakova, L.T. and Fedorov, P.V. 1989. [On the age of the lower part of the red-coloured layers in the upper Kugart River.] *Vestnik Leningradskogo Universiteta* **7**: 67–70.

Butler, P.M. 1990. Early trends in the evolution of tribosphenic molars. *Biological Reviews* **65**: 529–552.

Cifelli, R.L. 1990. Cretaceous mammals of southern Utah. IV. Eutherian mammals from the Wahweap (Aquilan) and Kaiparowits (Judithian) Formations. *Journal of Vertebrate Paleontology* **10**: 346–360.

Clemens, W.A. and Lillegraven, J.A. 1986. New Late Cretaceous, North American advanced therian mammals that fit neither the marsupial nor eutherian molds. *Contributions to Geology, University of Wyoming, Special Paper* **3**: 55–85.

Fox, R.C. 1979. Mammals from the Upper Cretaceous Oldman Formation, Alberta. III. Eutheria. *Canadian Journal of Earth Sciences* **16**: 114–125.

—1984. *Paranyctoides maleficus* (new species), an early eutherian mammal from the Cretaceous of Alberta. *Carnegie Museum of Natural History, Special Publication* **9**: 9–20.

—1989. The Wounded Knee local fauna and mammalian evolution near the Cretaceous-Tertiary boundary, Saskatchewan, Canada. *Palaeontographica, Abteilung A* **208**: 11–59.

Gheerbrant, E. and Astiba, H. 1994. Un nouveau mammifer du Maastrichtien de Laa (Pays Basque espagnol). *Comptes Rendus de l'Académie des Sciences*, serie II **318**: 1125–1131.

—and Russell, D.E. 1991. *Bustylus cernaÿsi* n. g., n. sp., nouvel adapisoriculide (Mammalia, Eutheria) paleocene d'Europe. *Geobios* **24**: 467–481.

Jerzykiewicz, T., Currie, P.J., Eberth, D.A., Johnson, P.A., Koster, E.H. and Zheng, J.-J. 1993. Djadokhta Formation correlative strata in Chinese Inner Mongolia: an overview of the stratigraphy, sedimentary geology, and paleontology and comparisons with the type locality in the pre-Altai Gobi. *Canadian Journal of Earth Sciences* **30**: 2180–2195.

Kielan-Jaworowska, Z. and Dashzeveg, D. 1989. Eutherian mammals from the Early Cretaceous of Mongolia. *Zoologica Scripta* **18**: 347–355.

—and Nesov, L.A. 1990. On the metatherian nature of the Deltatheroida, a sister group of the Marsupialia. *Lethaia* **23**: 1–10.

—and— 1992. Multituberculate mammals from the Cretaceous of Uzbekistan. *Acta Palaeontologica Polonica* **37**: 1–17.

Kovalevskii, V.O. 1950. [*The complete works*] 1. 478pp. Moskva: Izdatel'stvo Akademii Nauk SSSR, 478 pp.

Lillegraven, J.A. 1976. A new genus of therian mammal from the late Cretaceous 'El Gallo Formation', Baja California, Mexico. *Journal of Paleontology* **50**: 437–443.

—Kielan-Jaworowska, Z. and Clemens, W.A. (eds.) 1979. *Mesozoic Mammals. The first two thirds of mammalian history.* Berkeley: Univ. Calif. Press, 311 pp.

Mashchenko, E.N. 1999a. [The first record of a Mesozoic mammal in Siberia.] *Byulleten' Moskovskogo Obshchestva Ispytatelei Prirody, Otdel Geologicheskii* **74**(1): 80.

—1999b. [A mesozoic triconodont in the territory of Siberia.] *Priroda* **7**: 52–53.

—and Lopatin, A.V. 1998. First record of an Early Cretaceous triconodont mammal in Siberia. *Bulletin de l'Institute Royal des Sciences Naturelles de Belgique, Sciences de la Terre* **68**: 233–236.

Nesov, L.A. 1981. [Cretaceous salamanders and frogs of Kyzylkum Desert.] *Trudy Zoologicheskogo Instituta AN SSSR* **101**: 57–88.

—1982a. [The ancient mammals of the USSR.] *Ezhegodnik Vsesoyuznogo Paleontologicheskogo Obshchestva* **25**: 228–242.

—1982b. [Late Cretaceous mammal assemblages of Middle Asia.] *III S'ezd Vsesoyuznogo Teriologicheskogo obshehestva, Tezisy Dokladov* **1**: 59–60.

—1984. [On some remains of mammals in the Cretaceous deposits of the Middle Asia.] *Vestnik Zoologii* **2**: 60–65.

—1985a. [New mammals from the Cretaceous of Kyzylkum.] *Vestnik Leningradskogo Universiteta* **17**: 8–18.

—1985b. [Rare bony fishes, terrestrial lizards and mammals from the zone of estuaries and coastal plains of the Cretaceous of Kyzylkum.] *Ezhegodnik Vsesoyuznogo Paleontologicheskogo Obshchestva* **28**: 199–219.

—1986. [Late Mesozoic and early Paleogene mammals of the USSR.] *IV S'ezd Vsesoyuznogo Teriologicheskogo obshchestva, Tezisy Dokladov* **1**: 23–24.

—1987. [Results of search and study of Cretaceous and early Paleogene mammals on the territory of the USSR.] *Ezhegodnik Vsesoyuznogo Paleontologicheskogo Obshchestva* **30**: 199–218.

—1989. [Mammals of the first half of late Cretaceous of Asia.] *Operativno-informatsionnye Materialy k I Vsesoyuznomu Soveshchaniyu po Paleoteriologii.* 45–47.

—1990a. [Deltatheroids, early placental and multituberculate mammals of the coastal plains of late Cretaceous of the western Middle Asia.] *V S'ezd Vsesoyuznogo Teriologicheskogo Obshchestva, Tezisy Dokladov* **1**: 22–23.

—1990b. [Small ichthyornithiform bird and other bird remains from Bissekty Formation (upper Cretaceous) of Central Kyzylkum Desert.] *Trudy Zoologicheskogo Instituta AN SSSR* **210**: 59–62.

—1993. [New Mesozoic mammals from Middle Asia and Kazakhstan and comments on the evolution of coastal plain Cretaceous theriofaunas.] *Trudy Zoologicheskogo Instituta RAN* **249**: 105–133.

—1995. [*Dinosaurs of Northern Eurasia: new data about assemblages, ecology and paleobiogeography.*] St. Petersburg: Izdatel'stvo Sankt-Peterburgskogo Universiteta, 156 pp.

—1997. [*Cretaceous nonmarine vertebrates of Northern Eurasia.*] St. Petersburg: Izdatel'stvo Sankt-Peterburgskogo Universiteta, 218 pp..

—, Archibald, J.D. and Kielan-Jaworowska, Z. 1995. Ungulate-like mammals from the late Cretaceous of Uzbekistan and a rediagnosis of Ungulata. *Journal of Vertebrate Paleontology* **15**: 58A.

—, Archibald, J.D. and Kielan-Jaworowska, Z. 1998. Ungulate-like mammals from the late Cretaceous of Uzbekistan and a phylogenetic anaylsis of Ungulatomorpha. *Bulletin of the Carnegie Museum of Natural History* **34**: 40–88.

—and Gureev, A.A. 1981. [The find of a jaw of the most ancient shrew in the upper Cretaceous of the Kyzylkum Desert.] *Doklady AN SSSR* **257**: 1002–1004.

—and— 1982. [On the time of appearance of some morphological characters of Soricidae.] *III S'ezd Vsesoyuznogo Teriologicheskogo Obshchestva, Tezisy Dokladov* **1**: 60–61.

—and Khisarova, G.D. 1988. [New data about upper Cretaceous vertebrates from Shakh-Shakh and Baibolat localities (North-Eastern Aral area).] *Materialy po istorii fauny i flory Kazakhstana* **10**: 5–14.

—and Kielan-Jaworowska, Z. 1991. Evolution of the Cretaceous Asian therian mammals. Fifth Symposium on Mesozoic Terrestrial Ecosystems and Biota. Extended Abstracts. *Contributions from the Paleontological Museum, University of Oslo* **364**: 51–52.

—, Kielan-Jaworowska, Z., Hurum, J.H., Averianov, A.O., Fedorov, P.V., Potapov, D.O. and Froyland, M. 1994a. First Jurassic mammals from Kirghizia. *Acta Palaeontologica Polonica* **39**: 315–326.

—, Sigogneau-Russell, D. and Russell, D.E. 1994b. A survey of Cretaceous tribosphenic mammals from Middle Asia (Uzbekistan, Kazakhstan and Tajikistan), of their geological setting, age and faunal environment. *Palaeovertebrata* **23**: 51–92.

—and Trofimov, B.A. 1979. [The oldest insectivore of the Cretaceous of the Uzbek SSR.] *Doklady AN SSSR* **247**: 952–954.

Novikov, I.V., Lebedev, O.A. and Alifanov, V.R. 1998. New Mesozoic vertebrate fossil sites of Russia, p. 58, in Jagt, J.W.M., Lambers, P.H., Mulder, E.W.A. and

Schulp, A.S. (eds), *Third European Workshop on Vertebrate Palaeontology*, Maastricht, 6–9 May 1998.

—, Alifanov, V.R. Lebedev, O.A. and Lavrov, A.V. 1999. [New data on Mesozoic vertebrate localities in Russia], p. 184, in *IV International Conference 'New Ideas in Earth Sciences'*, Abstracts 1. Moscow.

Orlov, Y.A. 1964. [On one task of vertebrate paleontology in the USSR.] *Paleontologicheskii Zhurnal* 1: 131–132.

Pol, C, Buscalioni, A.D., Carballeira, J., Frances, V., Lopez Martinez, N., Marandat, B., Moratalla, J.J., Sanz, J.L., Sig, B. and Villate, J. 1992. Reptiles and mammals from the late Cretaceous new locality Quintanilla del Coco (Burgos Province, Spain). *Neues Jahrbuch für Geologie und Paläontologie, Abhandlungen* **184**, 279–314.

Roček, Z. and Nesov, L.A. 1993. Cretaceous anurans from Central Asia. *Palaeontographica, Abteilung A* **226**: 1–54.

Rogovich, A.S. 1875. [Note on the localities of bones of fossil mammal animals in southwestern Russia.] *Zapiski Kievskogo Obshchestva Estestvoispytatelei* **4**, 1: 33–45.

Tatarinov, L.P. 1980. [Towards a prehistory of mammals.] pp. 103–114 in Sokolov, B.S. (ed.), *Paleontology. Stratigraphy* (International Geological Congress, 26 session). Moscow: Nauka.

INDEX

The index includes mainly proper names (people, places, stratigraphic units, taxa). Page numbers for illustrations are indicated in **bold** type. The index was compiled by M.J.B. and D.M.U., with considerable assistance from Natasha Bakhurina.

Abdarain Nuur **280**
Academy of Sciences of the USSR/of Russia 1, 226, 232, 235–7, 240–1, 256
Acanthostega 38
Acanthostegidae 35
Achillobator 436
Acontias 379
Actinopterygii 178, 648
Adamisauridae 376, 383–6
Adamisaurus 376, **376**
Adasaurus 270, **435**, 436, **441**, 451
Admetophoneus 101
Admiralteistva peninsula 172
Admiralteistva Svita 172
Adocidae 310, 320–2, 331–3, 334–5, 359–61
Adocoides 333, 335–6, **335**, 338
Adocus 321, 332–4, 336, 338, 359–61
Adzhat River 195
Aegialodon 599, **600**, 615
Aegialodontia 598–601, 615
Aegialodontidae 574, 599, 600, 615
Aeolian dune deposits **288**
Aetosauria 140, 181
Agamidae 369, 371, 373, 383–4
Aginbaataridae 585
Aguinbaatar 262
Aikino district 143, 153
Aitym Svita 649
Akbatyrovo mines 6, 42
Akbulak district 106, 144, 145
Akkurgan **463**, 474, 476
Aktyubinsk Province 100
Alag Teeg locality 241–3, **242**, 526

Alaiskii Ridge 204, 414
Alamosemys 333
Alamyshyk Svita 205, 321, **326**, **330**, **332**
Albanerpetontidae 306–7
Alberta Basin **281**
Albertosaurus 448
Albionbaataridae 585
Alectrosaurus **435**, 439, 447, **447**, 451
Alegeinosaurus 45
Alekseichik, A.N. 236
Alexornis 533–4
Alexornithidae 533–40
Alexornithiformes 533–44
Alf, R. 562
Algui Ulaan Tsav locality 237, 243, **245**, 256, 264, **265**, **267**, **280**, **561**, 565, 568
Alifanov, V.R. 368–9, 372–5
Alim Tau locality **463**, 470
Alioramus **435**, **447**, 448–9, 451
Alligatoridae 409
Alligatorinae 415
Allodontidae 585–586
Allotheria 579–96
Alma Aty (*see* Almaty)
Almaty 63, **628**
Almaty, Institute of Zoology, Kazakhstan Academy of Sciences 310, 533, 627, 629
Alostera 642
Altai (*see* Mongol Altai)
Altai Sum 245
Altan Teeli 553
Altan Uul II locality 304, **463**, 475
Altan Uul III locality 450, **561**

Index

Altan Uul IV locality **280**, 444, 457
Altan Uul locality 237–9, 260, **265**, 266, **267**, 268–9, 284, 331, 476, 568
Altan Uul ridge 237–9, 438, 446, 468, 541
Altanteeli locality 239, 271
Altanteius 374
Altanulia 304
Altanuul Formation/Svita 257
Altirhinus 462, **466**
Alxa (Alashan) Desert 439
Alxasauridae 439
Alxasaurus 439–41, 451
Alymlestes 610, 629, 639, 648
Alymtau 576, **628**, 629, 639–40
Amalitskii collection **78**
Amalitskii, V. P. 2–5, **4**, 77–80, 103
Amalitzkia 94
Ambiortidae 547–51
Ambiortiformes 547–51
Ambiortus 544, 548–50, **549**, **550**, 556
Ambystomatoidea 305–6
American Museum of Natural History (AMNH) 211–25, 233, 235, 369, 402, 435, 456–7, 467, 480–1, 518, 573, 575–6, 610
   Central Asiatic Expeditions 211–25, 235–6, 256, 309–10, 444, 481, 517, 533, 561, 569, 573
American Ornithologists' Union 544, 546
Amida 268
Amka 554
Ammonites 129, 132, 576
Amniota 37–8, 61, 162, 165, 174
Amphibamidae 36
Amphibia
   Central Asia 297–308, 628, 648
   Mesozoic 246, 297–308, 628, 648
   Mongolia 246, 297–8, 301–2, 304, 618
   Permian 17–26
   Permo-Triassic 1–13, 35–70, 125, 135
Amphidon 597, 615
Amphidontidae 596–8, 615
Amphidontoidea 596
Amphilestidae 575, 577–9
Amphisbaenia 373–4, 382
Amphitheriidae 598
Amtgai 243, 250, 331, **335**, **403**, 405, 411, **413**, **414**, 417, 440, 523
Amtosaurus 518, 523, **525**, 530

Amu-Daria River 473
Amur River 199, 472, 473, 476
Amyda 345, **345**
Anagalida 605, 610–14, 617
Ananinskoe locality **121**
Anapsida 6, 205
Anatolemys 205, 327, **329**, 332
Anchaurosaurus 369
Andai Khudag locality 236, 238, 257, 259, 262, **265**, 553
Andaikhudag Formation/Svita **243**, 256–7, 259–60, **260**, 282, **283**, 549
Andreevka village 145, 167
Andrews, R.C. 211–25, **218**, **219**, **223**, **224**, 227, 230–2, 236, 573
Angelosaurus 23
Angiosperms 249, 272, 507
Anguidae 382–6
Anguimorpha 203, 380–3
Anguis 382
Anguoidea 380
Angusaurus 50, 52, **52**, **122**, **123**, 130
Anhanguera 425–6, **426**, **429**
Ankylosauria 233, 237, 242, 244, 246–7, 249, **252**, 268, 494–5, 499, 517–32
Ankylosauridae 519–31
Ankylosaurus 530
Anna 110
Annatherapsidinae 110
Annatherapsidus 26, **109**, 110
Annemys 314–15, **314**, **317**, 361
Anomodontia 18, 20–1, 23–30, **28**, **29**, 87–8, 102–8, **104**, 123, **124**, 135
Anomoiodon 165
Anoplosuchus 20, 96
Anosteira 344
Anosteirinae 343–4, 361
Anseriformes 542, 551, 556
Anserimimus 446, 451
Antarctosaurus 456, 460
Antecosuchus 112, **122**, **124**
Anteosauria 20, 23–4, 27
Anteosauridae 95, 96, **99**
Anteosaurus 23
Anthecosuchus 134
Anthodon 72, **72**, 73, 75, 76
Anthracosauria 37–8, 60–70, 125, 128, 191
Anthracosauroideae 60, 62, 63–4

Anthracosauromorpha 35, 60–70, 123, **123**

Anthracosaurus 61

Anura 35, 37, 39, 297, 300–4, 628

Anurognathidae 423–4, **429**, 430

Anurognathus 423–4, **429**, 430

Apatheon 45

Apatosaurus 459

Aral Sea 268, 297, 429, 474, 639

Aralobatrachus 303–4

Aralosaurus 468, **469**, 476

Arambourgiania 430

Aranetsia 54, **122**, **123**, 132

Aratrisporites 132

Araucarians 272

Archaeoceratops 481, 488, **498**, 503–5, 507, 509, 515

Archaeopteryx 537, 554, 556

Archaeornithes 537

Archaeornithoides 445, 451

Archaeornithomimus **435**, 446, 451

Archaeosyodon 20, 95, **99**, 100

Archegosauridae 20, 39, 42–3

Archegosauroidea 12, 36, 39, 40, 42–4

Archibald, J.D. 617, 629, 642

Archives of the Russian Academy of Sciences (ARAS) 235, 237–40

Archosauria
  Cretaceous 402
  Jurassic 204
  Triassic 5, 10–12, 26, 121, 124–5, **124**, 134–5, 140–59, **141**, **157**, 182–4

Archosauromorpha 174, 390

Archosaurus 11, 12, 26, **141**, 142–3, **142**, 156

Arctognathus 93

Ardeosauridae 377

Ardeosaurus 377, 378

Ardyn Ovoo 236

Arginbaatar **587**, 588, 595

Arginbaataridae 585–8, 615

Arguimuridae 598, 615

Arguimus 598, **599**

Arguitherium 599, **599**

Arguniella 259

Ariekanerpeton 63

Arkhangel'sk Province 22–3, 25, 64, 78, 91, 94, 110, 114, 167, 173–4

Arkhangel'skii, M.S. 200, 203

Armenia, pterosaur 429

Arretosauridae 383–4

Arstanosaurus 268, 474, **474**

Arts Bogd Ridge 245, 264, 266, 270, 272, **280**, **561**, **574**

Artzosuchidae 407

Artzosuchus 402, 405, 407, 417

Arvayheer **238**, **244**, **250**, **265**, **280**

Ashchikol **628**, 629, 631, 637, **643**

Asiaceratops 492, 493, 640

Asiachelys 325, 326, **327**

Asiadelphia 603–4, 616–17

Asiahesperornis 544, 546

Asiahesperornithinae 546

Asiamericana 449, 451

Asiatheriidae 603–4

Asiatherium 248, 551, 601, 603–4, **604**, 617

Asiatoceratodus 178

Asiatosaurus 456, 460

Asioryctes 605, 607, **608**, 609–11, **609**, 614, 617

Asioryctidae 604–5, 607–10

Asioryctinae 605

Asioryctitheria 605–10, 617

Aspanlestes 631–5, 642, **643**, **644**, **645**, **646**

Aspideretes 322, 354

Astashikha I locality 105, **121**

Astashikhian Member 126

Astrakhanovka **126**

Atoposauridae 408

Aublysodon 447

Aublysodontinae 446–7, 450

Auerbakh, I.B. 5

Aulie (Mikhailovka) locality 423–4, 553

Ausktribosphenos 604

Australochelys 359

Australosyodon 27, 97

Averianov, A.O. (also Aver'yanov) 306–7, 628–9, 649

Aversor 63

Aver'yanov 576, 604, 617

Aves, Cretaceous 184, 217, 219, 233, 239–40, 243, 248–50, 292, 434, 443, 450, 452, 533–60, 648–9

Avesuchia 140, 150–1, 156–7, **157**

Avialae 554

Avimimidae 244–7, 434, **435** 443, 452

Avimimus **435**, **441**, **442**, 443, 451, 551

Avisauridae 540–1

Avisaurus 540

Avitotherium 642

Axestemys 322, 346

Axitectum 68, **68**, **122**, **123**, 125, 131
Azerbaijan
  ichthyosaur 188, 203
  mosasaur 203
Azhdarchidae 421, 427, 430
Azhdarcho 422, 427, **428**, **429**, 430
Azhdarchoidea **429**
Azi Molla II **121**
Azizbeck region 429
Azuma 504

Bactrosaurus 462, 472–4, 476
Badmayapov, T. 214
Baena 346
Baenidae 322, 359, 360, 361
Baga Mod Khudag **561**
Baga Tariach **265**, 268, 526
Baga Zos Nuur (lake) 257, 272
Bagaceratops 233, 481, 488–9, 491, 493, **498**, 501–5, **508**,
  510
Baganuur locality 260, 272
Bagaraatan **435**, 449, 451
Bagazosnuur Formation/Svita 257
Baibishe 429
Baibolat well 627
Bain Chire (*see* Bayan Shiree)
Bainguidae 382
Bainguis 382
Bairdestheria 259–60
Baisa 554
Baishin Tsav depression
  dinosaurs 440, 443, 446–7, **463**
  geology 260, **265**, 266, 268, 270–1, **280**
  historical 243, 250, **251**
  turtles 331, **341**
Baissocovixa 258
Baisun 196
Bakhar locality **421**, 422, **422**, 424, **429**
Bakhar Svita 424
Bakhar Uul **280**
Bakhurina, N.N. 245, 258, 422, 424, 426
Balabansai Svita 299, 307, **313**, 428, 553, 629, 632
Balinski, A. 233
Balkhash Lake 297
Bambuu Khudag locality 268, **280**, 309, 331, **356**
Bannikov, A.F. 459
Bannovka 201
Baptornis 545–6

Baptornithidae 544, 546–7
Barasaurus 82, 160
Barghusen, H.R. 86, 87, 95, 110–11, 114
Barrett, P. 460
Barsbold, R. 232–3, 239–42, **240**, **241**, 245, 271, 273, 436–8,
  440, 450, 573–4
Barsboldia 468, **470**, 471, 476
Barungoia 375, **375**
Barunlestes 604, **608**, 610–12, **613**, **614**, 617, 648
Baruun Bayan cliffs 256–7, 264–5, 272, **280**
Baruunbayan Formation/Svita 257, 261, 263–6, **265**, **267**,
  272
Baruun Goyot 257, **265**, **286**, 331, 369, 376, 405, 434, 438,
  451, 567
Baruungoyot Formation/Svita
  amphibians 301
  birds 537–9, **543**, 551, 560, **561**, 564–8, 570–1
  crocodilians 407, 411
  dinosaurs 436, 438, 459, 488, 493, 518, 527
  geology 256–7, 261, **265**, 266, 268–71, **269**, 273–4, 279,
    283, 284, **285**, 287, **288**, 289, 292
  historical 233, **240**, **242**, 249, 250, **251**, **252**
  lizards 368–9, 383, 385
  mammals 573, 575–6, 588–90, 594, 602–3, **604**, 609, 611,
    616, 617, 648
  turtles 333, 339, 342–3, **342**, **343**, 355, **356**
Baruunurt **238**, **244**, **250**, **265**
Bashkir Mines 20
Bashkirosaurus 43
Bashkortostan Republic 1, 6, 8, 19–22, 89, 96–7, 100,
  102–3, 106–8, 121, **121**, 132, 152, 154
Bashkyroleter 63
Basilemys 233, 322, 339–40, 361
Batagurinae 353
Batrachognathus 420, 422–3, **423**, **429**
Batrachomorpha 37, 297
Batrachosauria 6, 8
Batrachosauroididae 306
Batrachosuchoides 54, **122**, **123**, 131
Battail, B. 86–7, 112, 115, 134
Bauriamorpha 108
Bauriidae 112–13, 134
Bauriinae 112
Baurioidea 109, 111–13
Bayandalai Sum 268
Bayankhongor Aimag **238**, **244**, **250**, **265**, **280**, 424, 544,
  547, 549, **561**
Bayan Mandahu locality 273, **280**, 284, **290**, 444, 575

Bayan Mandahu basin 281

Bayan Shiree locality 227, 229, 237, 256–7, **265**, 268, **280**, 331, **345**, 405, 451, 518, 520, **561**, 567

Bayanshiree Formation/Svita
  birds 560, 564
  crocodilians 411
  dinosaurs 440, 446, 520–1, 523
  geology 256–7, 261, **265**, 266–8, **267**, 273–4, 282–3, **283**, 287, 289
  historical 250
  turtles 332–3, 335–6, **335**, **337**, 339–40, **341**, 344–5, **345**, 355

Bayan Tsagaan Gobi **280**

Bayan Zag locality
  birds 561, **561**, 568–9
  crocodilians **403**, 405, 407, **407**, 411, 417
  dinosaurs 436–7, 444–5, 517, 526
  geology 255–6, 262, **265**, 266, 268–9, **280**, 284, **284**
  historical 216–24, **217**, **220**, **221**, **222**, **223**, **224**, 227–9, 231–3, 236–7, 239, 241
  lizards 375–6, 379–80, 382
  mammals 573, **574**, 590–1, 593–4, **593**, **594**, 601–2, 606, 610
  turtles 309, 331

Bayanzag Formation/Svita 268

Baybishe 307

Bayn Dzak (*see* Bayan Zag)

Baynshin (*see* Baishin)

Bayn Shire (*see* Bayan Shiree)

Bazhanov, V.S. 533, 561, 627

Beaufort Series 17

Beger Nuur depression 239

Beijing 216

Beijing, Institute of Vertebrate Palaeontology and Palaeoanthropology (IVPP) 309, 480, 533, 556, 562

Bekovo District, **194**

Belebeiskaya Svita 43, 45

Belebey 6, 12, 18, **18**, **19**, 21–3, 43, 45, 89

Belebey 12, 21–2, **22**

Belebey fauna 6, 9, 21–2

Belemnitella mucronata Beds 203

Beleuta Svita 427, 627

Beleutinus 627, 639, **647**, 648

Belgorod **421**

Belgorod district 201, 205, 421, 429

Bell, S.K. 586, 596, 602, 605

Belyaeva, E.I. 236

Belyaevka district 94

Belye Kruche **463**, 473, 476

Benthosphenus 11, 49, 129

Benthosuchidae 12, 40, 49–51, 128–131

Benthosuchus 6, 49, **49**, **122**, **123**, **124**, 125, 128

Benthosuchus fauna/grouping 120, 122–4, 128–9

Benthosuchus-Wetlugasaurus fauna 39–40, 120–30, **122**, 130–1

Benton, M.J. 156, 181–2, 407

Berdyanka I locality **11**, 106–7, 112, 150

Berdyanka II locality 112, 144, 152

Berdyanka River 10, 48, 134, 153, 169

Berezniki **121**, 196

Bereznikovskaya Svita 129

Berezovye Polyanki 20

Beringia 480, 509, 510, **510**

Berkey, C.P. 214, 235–6, 259, 266

Beshtyube Svita 205, 306–7, **344**

Biarmica 63

Biarmosaurus 89

Biarmosuchia 20–2, 30, 86–92

Biarmosuchidae 86, 88, 89

Biarmosuchoides 90

Biarmosuchus 20, 23, 89, **90**

Biasala 203

Bird's Hill 538

Birds (*see* Aves)

Bishara 307

Bishkek **628**

Bissektia 304

Bissekty Svita
  amphibians 301–4, 306, 307
  birds 534–7, 541, 552
  crocodilians 411, 414
  dinosaurs 439, 449–50, 474
  mammals 604, 617–18, 631–5, 637, 642, 649
  pterosaurs 427

Bissekty well 404, 637

Bistius 642

Bivalves 134, 178, 287

Blagoveshensk **463**

Blomia 122, 124, 126, **141**, 144

Blomosaurus **122**, **124**, 126

Blyumental' ravine 49

Blyumental' 3 locality 25, 114

Bobolestes 630, 640, **641**

Bobolestidae 640

Bock, W. 553

Bogdinskaya Svita 130
Bogdo Mountain 132
Bogolyubov, N.N. 188, 195–6, 200, 428
Bohlin, B. 310, 489
Bol'shoe Linovo 114
Bolosauridae 12, 20–2, 27
Bolosaurus 12, 20
Bolotskii 473
Bolshaya Sludka **121**
Bolshaya Synia depression **121**, 133, 195
Bolshaya Synya River 68, **69**
Bolshoe Bogdo 5, 52, **121**
Bolshoi Kityak 96
Bolshoi Yushatyr' River 48
Bombina 304
Boonstra, L.D. 100
Böön Tsagaan locality 249, **280**, 321, 553
Bööntsagaan Gorizont 544, 549
Boremys 359
Borealestes 577
Boreopelta 11, 39, 55, **55**, 129
Boreopricea **122**, **124**, 130, 174
Borisoglebskaya, M. **246**
Borisyak, A.A. 5, 235–6, 415
Borogovia 444, 451
Borshchevka 51
Borskoi district 155
Borsuk-Białynicka 232, 368–9, 372, 377, 380, 383, 458
Borzongiin Gobi 268–9, **280**, **561**, 562
Bostobe Svita 268, 307, 404, 410, **410**, 439, 474, 627, 637–8
Bozeman, Museum of the Rockies 480
Brachigrapta 259
Brachychampsa 415
Brachyopidae 36, 40, 45, 53–4, 131, 297–9
Brachyopoidea 11, 26, 35–6, 39, 40, 52–4, 125
Brachypterygius 197, 203
Bradysaurus 24, 72, **72**, 75, 80, 82
Bragin, M. **240**
Branchiosauridae 36, 45
Brett-Surman, M.K. 475
Breviceratops 488, 493, 568
Brink, A.S. 112
Brinkman, D.B. 313, 317, 322, 326, 345, 395
Brithopodidae 88, 96, 97
Brithopus 1, 20, 96–7
British Museum (Natural History), London 86
Brontotheres 235, 246
Broomicephalus 93

Brown, B. 217, **219**, 230, 468, 494, 505
Buckantaus 375
Buffetaut, E. 183, 300, 409, 430, 492
Bügiin Tsav depression 541
Bügiin Tsav locality
    birds 547, **548**, 554, **561**
    dinosaurs 436, 444, 446
    geology **265**, 266, 268–9, **280**
    historical 239
    lizards 369
    mammals 573, **574**, 595, 616
    turtles 331
Buginbaatar 573, 588, 595–6, **595**, 616
Buginbaatarinae 596
Builyastyn Khudag **561**, 569
Builyastyn Formation/Svita 256
Buinsky Mine 191, **192**
Bukhara Province 427, 534–6, 541, 552
Bukobaja 48, **122**, **123**, 135
Bukobay Gorizont
    anthracosaurs 68, **69**
    archosaurs **141**, 149, 152–3
    geology 120, 122–4, **122**, 132, 134–5
    synapsids 108
    temnospondyls 48, 54
Bukobay I–V localities 48, **121**
Bukobay I locality 149, 152
Bukobay V locality 149, 153
Bukobay VI locality 149
Bukobay VII locality 149
Bukobay Svita 134
Bulaklestes 633
Bulgan Sum **238**, **244**, **250**, **265**, 266, 551
Bulganbaatar 576, 582, 592–4, **593**, 617, 639
Bulganemys 339, 342, **342**
Bulgant Formation/Svita 256
Bullatosauria 444
Buntsandstein 128, 132
Burnetia 25, 91
Burnetiidae 86, 89, 90, 92, 94
Buroinak 429
Burtensia **122**, **123**, 131, 161, 165, 167, **169**
Buryatia 321, 324, 390, 396–7, 517, 640, 649
Buscalioni, A. 408
Bustylus 642
Butler, P.M. 585, 596, 599, 601
Buur River 55
Buylyasutuin (*see* Builyastyn)

Buzulukia 63
Bystrov, A.P. 6, 7, 93
Bystrowiana 67, **67**, 68
Bystrowianidae 61–2, 66, 125, 128
Byzovaya **121**
Byzovskaya Svita 129

Caenagnathasia 437, 439, 451
Caenagnathidae **435**, 437–9, 443, 452
Caenagnathinae 438–9
Callenosaurus **122, 124**
Calleonasus 103, 108
Callopistes 376
Calvo, J.O. 460
Camarasauridae 457–8
Campylognathoididae **429**
Candelaria 162
Cape Nikolaya 172
Capitosauridae 39, 40, 45–8, 299
Capitosauroidea 10–11, 36, 39, 40, 45–8, 132–3, 297–300
Capitosaurus 120
Captorhinidae 8, 18–20, 27, 71, 160
Carettochelyidae 320, 322, 332, 335, 339, 343–4, 359, 361
Carnegie Museum, Pittsburg 215
Carnosauria 233, 237, 434, **435**, 450
Carpenter, K. 448, 518, 531, 561–2
Carter, Jimmy (sobaka) **8**
Carusia 377, 378, **378**
Carusiidae 377–8, 383–6
Caseidae 8, 17, 20, 23
Caspian basin 39, 48, 54, 132, 140
Catopsbaatar 590–2, **592**, 595–6
Caudata 37, 305–7
Cedrus 272
Central Asia
  amphibians 297, 299, 301
  birds 533, 547, 560, 566, 569, 570
  choristoderes 390
  crocodilians 410, 417
  dinosaurs 439, 446–7, 480
  geology 273, 287
  historical 212, 224, 229, 235, 248
  lizards 368, 381, 386
  mammals 639
  marine reptiles 204
  turtles 309–10, 312–13, 321–2, 333–4, 345, 349, 357, 360–2
Central Asia Fold Belt **280**

Central Asiatic Expeditions of the AMNH (CAE)
  birds 533, 561, 569
  dinosaurs 444, 456, 481, 517
  geology 256
  historical 211–25, **214, 216, 217, 218, 219, 220, 222, 226**, 227–8, 230, 235–8
  mammals 573
  turtles 309–10
Central Cliffs **285**
Central Geological Museum (*see* St. Petersburg)
Central Gobi Aimag (*see* Dundgov' Aimag)
Central Sair **286**
Centrosaurinae 482, 488, **498, 510**
Centrosaurus 499
Cephalerpeton 161
Ceratodus 131, 132
Ceratopsia 218, 220, 230, 480–2, 485–94, **498**, 499, 501–10, **508, 510**, 515–16
Ceratopsidae 481–2, 494, **498**, 501–3, 505–7, **508**, 514, 516
Ceratopsinae 482
Ceratopsoidea 482, 492–3, **498**, 505–6, 516
Cetacea 627
Chabu Sumu 548
Chaikovskii, A.P. 236
Chalcosaurus 63
Chalishevia **122, 124**, 135, **141**, 145, 149–50, **149**, 156
Chamaeleognathus 370–1
Chamaeleonidae 369, 371–2
Chamopinae 373
Champsosauridae 243, 263, 390–401
Champsosaurs (*see* Champsosauridae; Choristodera)
Champsosaurus 391, 395, 397–8
Changet 414
Changetisaurus 378
Chaoyangsaurus 481, 487–8, **498**, 501–4, 506–7, 509, **510**
Chapayevka River 50
Chapman, F. 215
Charadriiformes 547, 551, 556
Charig, A.J. 142, 145, 147–8
Charkabozhskaya Svita 46, 129, 174
Charkov region 196
Charophyte algae 264, 265, 270–3
Chasmatosaurus 143, 145
Chasmatosuchus **122, 124**, 125, 129–30, 140, **141**, 142, 144–5, 147–8, 155
Chasmosaurinae 482, 488, **498**, 501, 507, **510**
Chelonia (*see* Testudines)
Cheloniamorpha 62–3

Chelonioidea 332
Chelospharginae 205
Chelpyk Hill 306, **628**, 631, 637
Chelydridae 350, 361
Chelydroidea 350, 359, 361
Chelydropsis 352
Chengyuchelyidae 312, 359
Chengyuchelys 312
Cherepanka locality 155
Cherkassia 203
Cherminotus 380, **381**
Cherminsaurus 373
Chernyi Bor **121**
Chernyshev's Central Museum of Geological Exploration
　　(*see* St. Petersburg)
Chia-yü-kuan locality 325, 333, 354
Chiappe, L. 436, 533, 554–5
Chilantaisaurus 450
Chilingosaurus 375
Chimkent Province 470, 553
China
　amphibians 299
　archosaurs (Triassic) 143, 149–50
　birds 544, 548–9, 556
　choristoderes 390, 395, 399
　Cretaceous geology 259, 271, 273, 280
　crocodilians 411, 416
　dinosaurs 434, 439–41, 443–8, 450–1, 462, 466, 472, 474,
　　481–3, 485–9, 492, 495, 530
　eggs 560, 562, 564–7, 569
　historical 214, 235–6, 239
　lizards 369–72, 376, 378
　mammals 575, 582, 596, 615
　Permian faunas 22, 23
　procolophonoids 162, 165
　synapsids 105, 112
　Triassic tetrapods 143, 149–50, 162, 165
　turtles 309–31, **318**, **319**, 319, **320**, **328**, 333–4, 337,
　　339–46, 349, 354, 359
Chingshan Formation 329
Chinsamy, A. 555
Chirostenotes 438–9
Chkalov 195, 196
Chkhikvadze, V.M. 273, 311, 327–8, 330, 333, 350, 354, 415
Choibalsan depression **238**, **244**, **250**, 260, **265**, 271
Choibalsan series 259
Choir 262, 426

Choir depression 262
Chondrichthyes 273, 576, 618, 628, 637, 639, 648–9
Chondrostei 258
Choristodera 390–401, **391**
Chroniosaurus 64, **66**
Chroniosuchia 8, 12, 60–9, 125, 128, 131, 135
Chroniosuchida 62
Chroniosuchidae 25–6, 62, 64
Chroniosuchus 64, **65**
Chrysochloridae 582
Chthamaloporus 20, 101
Chthonosauridae 110
Chthonosaurus 110
Chudinov, B.M. 236
Chudinov, P. K. 8–11, **8**, 18–19, **19**, **21**, **22**, 83–4, 88, 95, 97,
　　100–2, 161, 239, **240**, 241
Chukotka 199
Chuley region 470
Chulsanbaatar 576, **582**, 590, 593–5
Chulym River 649
Chuvashia Republic 189, 191, **192**, 195
Cifelli, R.L. 601, 604, 616
Cimoliasauridae 196, 205
Cimoliasaurus 188, 190–1
Cimolodonta 585–96, 616, 639
Cimolodontidae 588
Cimolomyidae 585, 595–6, 616
Cionodon 473, 476
Cis-Aral region 297, 307
Cis-Caspian depression 5, 121, **121**, 130, 132, 133, 135
Cis-Caucasus 129
Cis-Urals (Cisuralia)
　anthracosaurs 60, 62
　archosaurs 142
　historical 1, 2, 6, 8–12
　marginal trough 120–1, **121**, 125, 128–9, 132–3, 135, 140
　marine reptiles 195–6
　Permian 18–19, **18**
　procolophonoids 161
　synapsids 97
　temnospondyls 39, 40, 48, 54
　Triassic 120–1, **127**, 128–35, **131**, **133**
Cladistic analysis
　Anomodontia 27–30
　Archosauria 156–7
　Ceratopsia 501–6, 515–16
　Marginocephalia 493–516

Pachycephalosauria 495–501, 514–15
Pareiasauria 72, 82–3
Pterosauria 429
Clamorosaurus 19, 41, **41**
Clark, J. 148, 156, 407
Clidastes 203
Cliorhizodon 97
Cnemidophorus 376
Coal 272
Coelodontognathus **122**, **123**, 131, 161, 174–5
Collidosuchus 2
Coloborhynchus 425–6
Colosteidae 35
Colpodontosaurus 381
Colymbosaurus 188–9, 195
COMECON 232
Como Bluff 211, 215
Conchoraptor **435**, **437**, 438, 451
Conchostraca 128, 134, 258–60, 262–4, 266, 270–1, 273
Conicodontosaurus 375
Conifers 272
Conodonts 129
Contogenys 378
Contritosaurus **122**, **123**, 125, 161–4, **162**, **163**, **164**, 175
Coombs, W.C. 494, 517–19, 527
Cope, E.D. 211, 224, 232
Copper Sandstones (Upper Permian) 1–3, 6, 19, 20, 42, 63, 86, 89, 96–7, 100
Coprolites 258, 263
Cordaites 237
Corixid insects 258
Coronosauria 482, 488–93, **498**, 502, 504–6, 515–16
Corythosaurus 462, 471
Cotylorhynchus 23
Cotylosauria 6, 9, 18
Council of Ministers of the USSR 236
Crailsheim 134
Craspedites 187, 200
Crassigyrinidae 35, 37
Crato Formation 430
Creberidentat 373
Cretaceous
  amphibians 297–308
  birds 533–59
  dating of, in Mongolia 287
  mammals 573–652
  Mongolia 279–96

palaeoenvironments 287–93
  reptiles 309–532
Cretaceous/Tertiary boundary 287, 361, 383
Cretagama 370–1
Cretasalia 302, 303
Cretasorex 637
Cretaviculus 554
Crimea 189, 197, 203, 205, 415, 473, 476
Crocodylia
  Cretaceous 223, 237–8, 246–7, 263, 268–70, 292, 298, 360, 395, 399, 402–19, 560, 618, 628
  Jurassic 187, 204–5
Crocodylidae 205, 414–15
Crocodyliformes 204–5, 390, 402–19
Crocodylomorpha 140, 150
Crompton, A.W. 578
Crustacea 178
Cryptoclididae 195–6, 205
Cryptoclidus 188–9
Cryptodira 346, 359
Cteniogenys 397
Ctenochasmatoidea **429**
Currie, P.J. 437, 439, 443, 450, 489, 501–2
Cyclotosauria 36
Cyclotosauridae 135, 299
Cyclotosaurus 48, **122**, **123**
Cymbospondylus 199
Cynodontia 24–6, 30, 88, 108, 112–15, **113**, **114**, 178, 577
Cynognathus Zone 150, **157**
Cypridea 259, 269

Dabrazinskaya Svita 449, 460, 471
Dagestan, fossil crocodilian 204, 414
Dakosaurus 204, 413, 416
Dalanshandkhudag Formation/Svita 297–8
Dalanzadgad 237, **238**, **244**, **250**, 263, **265**, 268, **280**, **561**, **574**
Daonella Shales 199
Darbasa Svita 576, 637, 639
Darbi Somon (*see* Darvi Sum)
Darchansaurus 373, **374**
Dariganga **561**
Darvi Sum 258–9, **265**
Dashankou 22, 23
Dashzeveg, D. 232, 240, 242, 245, **246**, 574–6, 578, 598–9, 605
Daspletosaurus 448–9

Dating, radiometric 259, 268–9, 274
Daulestes 631, 642, **643**
Davletkulia 12, 22
Davs settlement 268, **574**
de Queiroz, K. 481
Deinocheiridae 247, 270, 449
Deinocheirosauria 450
Deinocheirus 233, 270, **435**, 449–51
Deinonychosauria 444
Deinonychus 436, 468
Delgerekhu Uul **280**
Delnov, N.I. 236
Deltatheridia 632, 640
Deltatheridiidae 599, 601–3, 617, 640
Deltatheridiinae 640
Deltatheridium 576, 600–3, **603**, 639–40
Deltatheroida 575, 599–604, 616–17, 631, 639–40
Deltatheroides 269, 602–3, 634, 640, **645**
Deltatheroididae 573, 602, 640
Deltavjatia 24–5, 71–6, **72**, **73**, **74**, 79, 81–2
Demsk Mines 20
Dendroolithidae 565–6, **565**
Dendroolithus 563–8, **566**
Dermatemydidae 262, 310, 331, 333, 339, 350, 352, 361
Dermatemys 333, 350
Dermochelys 346
Desmatochelyidae 205
Detskii (Children's) Sanatorium area 200–1
Deuterosauridae 101
Deuterosaurus 2, 96, 100
Devitcha 196
Devyatkin, E.V. 240, **246**
Diadectomorpha 71
Diapsida 71, 129–30, 178, 181, 184, 188, 390, 402
Dicraeosauridae 460
Dicraeosaurinae 457–9
Dicrodon 376
Dicynodon 24, 26, 28, **29**, 34, 103, 105, **105**
Dicynodontia 3, 9–12, **10**, 21, 24–30, **28**, **29**, 87–8, 102–8, **105**, **106**, **107**, **108**, 126–8, 132, 134–5
Dicynodontidae 105
Dicynodontinae 105
Dimorphodontidae **429**
Dinocephalia 1–2, 6–7, 10, 18, 20, 23–4, 26–7, 29, 39, 83, 87–8, 91, 94–102, **95**, **96**, **98**, **99**, **101**
Dinocephalian fauna/complex 11, 18–19, 23–4
Dinosaur National Monument 211

Dinosauria
  eggs 219–22, 224, **224**, 228, 231, **245**, 247, **267**, 560–72
  embryos 560, 568
  Kazakhstan 441, 443, 450, 456, 460, 462–3, 468, 470–4, 476, 517
  Kirgizstan 456
  Mongolia, finds 298, 390, 414, 628, 648
  Mongolia, geology 256, 258–60, 262–4, 266, 268–70, 272–3, 282–3, 289–90, 293
  Mongolia, historical 216–17, 219, 224–5, 229–31, **230**, 236–9, 245, 250, 575
  Mongolia, overviews 434–530
  nests 437, 560, 565, 571
  Russia 448, 463, 472–3, 476, 517
  Tadzhikistan 445, 447, 456
  Uzbekistan 441, 443, 449–51, 456, 463, 473–4, 476, 481, 492–3, 517
Dinosauromorpha 140, 145, 156, **157**
Dinosaurus 97, 102
Diplocynodon 414
Diplodactilidae 380
Diplodocidae 458–60
Diplodocoidea 460
Diplodocus 459
Diploglossa 380
Dipnoi 120, 130–1, 132, 178, 298
Discoglossidae 300, 301, 303, 304
Discoglossus 300, 304
Discosauriscidae 12, 62–3
Discosauriscus 63
Dissorophidae 20, 36, 39, 44–5
Dissorophoidea 35–6, 44–5
Djadochtatheria 580, 585, 586, 588–96, 616
Djadochtatheriidae 590–4
Djadochtatherium 269, 573, 590–1, 595–6
Djadokhta Formation/Svita
  amphibians 304
  birds 554, 560, **561**, 567–9, 571
  dinosaurs 436, 438, 443–5, 518, 526
  geology 256, 266, 268–9, 273, 279, 283–4, **283**, **284**, 287, **288**, 289, **290**
  historical 233
  lizards 369, 372, 376, 383, 385
  mammals 573, 575–6, 588–91, 593–4, **593**, **594**, 601–2, 606, 610, 616–17, 646
Djadokhta locality 292, 331, 333, 368–9, 371, 376, 383, 434, 451, 567, 648

Dmitrieva, E. 242
Dobruskina, I.A. 129, 135, 177
Docodonta 575–7, 614, 630, 639
Docodontidae 577
Dodson, P. 471, 489, 501–2
Doleserpetontidae 36
Dolichosauria 380
Doliosauriscus 23, 98, **99**
Doliosaurus 98
Dollapa 47
Dollo, L. 204
Dollosaurus 203, 204, **204**
Dombrovski, B.S. 236
Domocephalidae 495
Domocephalinae 495
Don River 47, 203, 545
Don River basin 121, 130–2, 171, 174, 189–90, 196, 203
Donets River 203
Dong, Z.-M. 375, 395, 439, 441, 460, 495, 504, 530
Dongosuchus **122**, **124**
Dongsheng Formation 282, **283**
Dongusaurus 112, **122**, **124**, 134
Dongusia **122**, **124**, **141**, 145
Dongusuchus 134, **141**, 151, 153–4, **154**, 156
Donguz Gorizont
    archosaurs **141**, 144–5, 148–50, 152–3, 156
    geology 120, 122–4, **122**, 132–5, **133**
    procolophonoids 169, 175
    synapsids 105–8, 112, 115
    temnospondyls 48, 54
Donguz I locality 105, 107, 112, 145, 149–50, 153
Donguz VI locality **66**, 164
Donguz IX locality 155
Donguz X locality **121**
Donguz XII locality 154
Donguz River 10, 54, 65, 100, 105, 164
Donguz Svita 133, **133**
Doniceps **122**, **124**
Donskaya Luka locality **121**, 155–6
Dornogov' Aimag (= Eastern Gobi) 236, 238, **246**, **247**, **335**, **337**, **341**, 395, **396**, 519, **561**
Dorognathus 134
Dorosuchus **122**, **124**, **141**, 150, **151**, 156
Dorsetisauridae 379, 382, 384–6
Dorsetisaurus 382
Dorsetochelys 359
Dorsoplanites 187, 191, **192**, 200

Dorygnathus 430
Dösh Uul locality 257, 264, 271–2, **280**, **561**, 565
Dösh Uul II locality **331**
Döshuul Formation/Svita
    birds **561**, 568–9, 571
    geology 257, 262, 265, 282, **283**
    turtles **323**, **325**, 329, **331**
Douglass, E. 211
Dovchin, N. 232, 573
Dracochelys 310, 327, **328**, 359
Dragon's Tomb locality 229, 238, 475
Drescheratherium 598
Dromaeosauridae 270, 434–6, **435**, 443–5, 450–2
Dromaeosaurinae 435–6
Dromaeosaurus 435–6
Dromasauria 24, 103
Dromotectum 68, **68**, **122**, **123**
Drumheller (*see* Royal Tyrrell Museum of Palaeontology)
Dryomorpha 465
Dsungaripteridae 420, 422, 426, **427**, 430
Dsungaripteroidea **429**, 430
Dsungaripterus 258, 422, 426, 430
Dubeikovskii, S.G. 195
Dubovka I locality 90
Dubovskii area 429, 552
Dundargalant Gorizont 549
Dundgov' Aimag (= Central Gobi) **243**, **245**, 312, **317**, 327, 391, **393**, 397, 399, 446
Dush Uul (*see* Dösh Uul)
Dushanbe **628**
Dvinia 26, 113, **113**, **114**
Dviniidae 113
Dvinosauridae 52–3
Dvinosaurus 5, 24–5, 39, 53, **53**
Dyoplosaurus 527
Dyugadyak River Basin 197
Dzamyn (*see* Zamyn)
Dzergen (*see* Zereg)
Dzhadochtosaurus 375
Dzhailyau-Cho 177
Dzharakhuduk localit
    amphibians 301–3, 305–7
    birds 534–7, 541, 552
    crocodilians **403**, **404**, 411, 414, **421**, 422, **422**, 427, **428**
    dinosaurs 450, **463**, 473–4
    mammals 628, **628**, 631–5, 637–8, **638**, 640, 642, 648–9
    turtles **338**, **355**

Dzhezkazgan Province 554
Dzhibkhalantu (*see* Javkhlant)
Dzhirgalantuin (*see* Jargalantyn)
Dzhungarian depression 313, 316, 319
Dzhusalin uplift 429, 468
Dzhyrakuduk (*see* Dzharakhuduk)
Dzun Bayan (*see* Züün Bayan)
Dzurumtai (*see* Zurumtai)

East Africa 131, 134, 143, 145, 154
East European Platform 120, 125, 132
Eastern Europe 132, 135, 161, 175
Eastern Gobi (*see* Dornogov' Aimag)
Eaton, T.H. 616
Eberth, D. 443
Edaxosaurus 103, 108, **122, 124**
Efimov, M. B. 397, 402, 411, 414–15, 417
Efremov, I.A.
    Cretaceous of Mongolia 226–31, **227, 228, 231**, 236–9,
        238, 309–11, 346
    Permo-Triassic researches 5–10, **6**, 12, 18–19, **21**, 76, 83,
        89, 93–4, 97, 100–2, 120–1, 140
Eggs
    bird 538, 542, 554, 560–72
    crocodilian 541, 560
    dinosaur 219–22, 223–4, **224**, 231, 236, **245**, 247, 263–4,
        266, **267**, 269, 272–3, 560–72
    enantiornithine 542
    fossil 217, 219–20, 223, 228, 232, 244, 247, 270, 443, 538,
        542, 547, 554, 570, 649
    sauropod 565
    turtle 264, 541, 560
Eginsai Svita 546, 547
Eglon, J. 237
Ehrmaying Formation 150
Eichstaettisauridae 376–7, 383–6
Eichstaettisaurus 377
Eichstätt 537
Eichwald, E.I. von 2
Elaphrosaurus 434
Elasmosauridae 189, 191, 195–6, 205
Elasmosaurus 188–9
Elatosaurus 103, 108, **122, 124**, 135
Elephantosaurus 103, 107, **122, 124**
Elephant shrews 614
Elginia 72, **72**, 79, 83
Elkemys 354

Elliotsmithia 17, 27
Elmisauridae 437–9
Elmisaurus **435**, 437–9, **442**, 443, 451
Elongatoolithidae **565**, 568–70
Elongatoolithus 561, 563–4, **566**, 567–9
Elshanka **127**
Elton Gorizont 133
Elva Vymskaya **121**
Elzanowski, A. 445, 533, 537, 541–4, 549
Embasaurus 450
Embolomeri 61–3
Embrithosaurus 24, **72**, 82
Embryos
    bird 240, 244, 538, 541–4, **543**
    dinosaur 560, 562, 565, 570
    enantiornithine 537, 541–4, 556
    oviraptorid 568
    theropod 438, 443
Emeroleter 63
Emydidae 349, 354, 359, 362
Enantiornis 540–1
Enantiornithes 248, 533–44, 551, 555–6
Enantiornithidae 536, 541, **542**, 548, 568, 570–1
    eggs 568, 470–1
Enantiornithiformes 544
Energosuchus **122, 124**, 135, **141**, 148, 151, **152**, 153
Enigmosauridae 244
Enigmosaurus 440, 451
Ennatosaurus 23, 27, 86
Enosuchus 23, 24, 64
Eoanacorax 637
Eobaatar 585–7, **587**, 590, 615
Eobaataridae 585–8, 615
Eocyclotosaurus community 132
Eodicynodon 23, 28, **29**, 34
Eodicynodon–Tapinocaninus Assemblage Zone 27
Eodiscoglossus 304
Eogyrinidae 63
Eoherpetontidae 62
Eolacertilia 382
Eolambia 465
Eopelobates 301, 304
Eoscapherpeton 305–7
Eoscapherpetontinae 305, 307
Eosuchia 390
Eotheriodontia 6, 10, 88–95
Eotitanosuchia 87, 88, 91–2, **92**

Eotitanosuchidae 91

Eotitanosuchus 20, 87, 88, 91, **92**

Eoxanta 379–80

Ephemeropsis 259

Epipubic bone **583**

Epitheria 605

Epivirgatites 187, 200

Erbaeva, M. 241

Erdene Tsogtyn Gobi gorge 263, 272

Erdene Uul mountains 260, 263, **280**, 553

Erdenetosaurus 373

Erenhot locality **280**, 283, 333, 443, 447

Eretmosaurus 188, 189

Ergil Ovoo locality 236–8, 243, 309

Ergiliin Zoo locality 236, 243, **246**, **247**, **248**, 333, 384

Ergiliinzoo Formation/Svita **247**, **248**

Erlikosaurus **435**, 440, **440**, 451

Erontheria 605

Eryopidae 41

Eryopoidea 8, 19, 39–42

Eryosuchus 10, 11, **11**, 40, 47, **48**, 120, **122**, **123**, **124**, 133

Eryosuchus fauna 40, 103, 122–4, **122**, 132–5

Erythrosuchidae **9**, 131–2, 135, 140, 142, 145, 147–9, 153, 156, **157**

Erythrosuchus **122**, **124**, 134, 145, 147–9

Erzovka 44

Estemmenosuchidae 20–1, 23–4, 27, 95, 101–2

Estemmenosuchus 20, 95–6, **95**, **96**

Estes, R. 373, 375–7

Estesia 380

Estesina 304

Etosha 292

Eublepharidae 380

Euchambersiidae 25–6, 109–10

Eucosmodontidae 585–6, 590, 596

Eucryptodira 333

Euoplocephalus 530

Euornithiformes 544

Euornithopoda 462, 473

Eupantotheria 574, 596, 598–9, 615

Euparkeria 150, **157**

Euparkeriidae 12, 134, 140, 150–1, 156

Eupelycosauria 27

Eureptilia 82

Euronychodon 450–1

European Russia **121**, **123**, 125, 129, 135, 142, 155, 161

Eurosaurus 97, 102

Eusuchia 402–3, 412, 414–17

Eutheria 224, 248, 262, 573–5, 575, 596, 600–1, 604–15, **608**, 631–3, 635, 637, 639–49, **647**

Eutherocephalia 109–13

Eutretauranosuchus 409

Evans, S.E. 395, 397

Exilisuchus **122**, **124**, 130, **141**, 144

Expeditions

American–Mongolian 233, 561, 574–6

Central Asiatic (CAE; AMNH) 211–25, 235–6, 256, 309–10, 444, 481, 517, 533, 561, 569, 573

Italian–French–Mongolian 575

Japanese–Mongolian 575, 591

Joint Polish–Mongolian, Palaeontological 231–3, 239–40, 256, **288**, 301, 311, 368, 448, 456, 457, 458, 517, 533, 561, 573, 575

Joint Russian–Mongolian, Palaeontological (JRMPE) 233, 246–51, 309, 368, 533, 541–2, 547

Joint Soviet–Mongolian, Palaeontological (JSMPE) 225–30, 235–54, **238**, **244**, **247**, **250**, 256, 298, 304, 309, 311, 345–6, 368, 399, 420, 422, 424, 456, 459, 464, 474, 517, 533, 561, 573–4, 627

Sino-Canadian Dinosaur Project 310

Swedish–Chinese 309–10

Ural–Dvina 6

Volga–Kama 6

Ezhovo (Ocher) fauna 19, **19**, 20–1, 24

Ezhovo **18**, 19, 20, 23, 42, 44, 89, 91, 95–6, 101, 103

F.N. Chernyshev Central Museum for Geological Exploration (*see* St. Petersburg)

Faustovo **121**

Faveoloolithidae 565, **565**

Faveoloolithus 264, 563–6, **566**, 567–8, 571

Feathers, fossil 248–9, 533, 553–5

Fedorovka River 68, **121**, 167

Fedorovskaya Svita 54, 130

Fedorovskian Gorizont **121**, 131–2

Fergana basin (valley)

amphibians 297, 299, 304, 306, 307

crocodilians 409–10, 414

geology 177–8, 268

historical 12

lizards 379

mammals 627, 629

marine reptiles 204–5

pterosaurs 428–9

Fergana basin (valley) (*cont.*)
Triassic diapsids 177–86
turtles 310, 312–13, **313**, 320–1, **326**, 327, **330**, **332**, 333, 337
Ferganemys 320–1, **330**, 332–3, 336–7, 637
Ferganobatrachus 298–9
Ferugliotheriidae 585
Ferugliotherium 586
Fighting dinosaurs 233, 436
Fischer von Waldheim, G.F. 2, 191
Fish shales 259
Fishes 194, 237–8, 239, 243, 245–6, 249, 259–60, 360, 549, 638, 648
Flaming Cliffs (*see also* Bayan Zag) 216–24, **222**, 231, 233, 256, **284**, 256
Flaviagama 370, 371
Flaviagaminae 371
Flerov, K.K. 237, 241
Fox, R.C. 306, 375, 596, 599
Franz-Josef Land 195
Fregatidae 551–2
Frey, D. 430
Frogs (*see* Anura)
Frolkin, N. **245**

Gaffney, E. 322, 326, 333, 335, 339, 352, 358
Galechirus 28, 103
Galeopidae 103
Galeops 25, 27–30, **28**, **29**, 34, 103
Galepus 28, 103
Galesauridae 25, 115, 131
Galliformes 542
Gallimimus 233, 270, **435**, **441**, **442**, 446, 451
Gallinula 551
Gallolestes 642
Galton, P.M. 430, 494–7
Gam locality **121**, 153
Gambaryan, P.P. 581, 594
Gamosaurus **122**, **124**, 131, **141** 143
Gamskaya Svita 47, 130, 132, 169
Gamskian Gorizont 121, 131–2
Gans, C. 179–81
Gansu Province (China) 309, 316, 325, 329, 333, 354, 488–9, 492
Gao, K. 371–2, 374–5
Gardner, J.D. 306, 307
Garjainia 9, **122**, **124**, 131–2, **141**, 142, 145, **146**, 147–9, 156

Garjainiidae 147
Garudimimidae 244, **435**, 445–6
Garudimimus **435**, **442**, 445, 446, 451
Gashato locality 236, 237
Gashuuny Khudag 243, 249, 575
Gastropods 271, 282
Gauthier, J. 481
Gaviidae 536
Gecatogomphius 20, 27
Gekkonidae 380, 383–6
Gekkota 377–8, 380
Geological Institute (*see* Ulaanbaatar)
Geologische-Paläontologisches Institut, Göttingen 480
Georgia 194, **194**
Georgiasaurus 188–9, 194, **194**, 196
Gephyrostegidae 61–4
Germanic basin 128, 132, 134–5, 177
Germanodactylidae 1, 430
Germanodactylus 430
Gerontoseps 375
Getmanov, S.N. 12
Gilbent Ridge 237, 239, 268, 566
Gilmore, C.W. 309, 368, 372, 376, 447, 494, 517
Gilmoreosaurus 462, 473, 476
Ginkgos 272
Gissar Mountains 196
Gladidenagama 371
Glikman, L.C. 429
Glires 648
Glirodon 585–7, 590
Globaura 376–7, **377**
Glyptops 359
Gnathorhiza 125, 131, 132
Gobekko 380
Gobi Altai (*see* Gov'altai)
Gobi Basin **280**, **281**, **283**
Gobi Desert
amphibians 297, 301, 303–4
birds 537, 539, 545, 547, 560–2, 567, 570
choristoderes 391
dinosaurs 436, 457–60, 464, 468–9, 475, 481, 520–1, 523, 526–7
geology 256, **263**, 272–3, 280–2, 287
historical 212–13, **214**, 215, 217, 227, 230, 236–8
lizards 368, 372, 376, 382–3
mammals 573, 575, 578–9, 586, 588–96, 598–9, 601–3, 605–6, 609–10, 615, 649
turtles 327, 333, 335, 340, 342–3, 345

Gobi Downwarp 272
Gobiates 300–2, **300**, 304
Gobiatidae 300–1, 304
Gobiatoides 300–1, 304
Gobibaatar 590
Gobiconodon 262, 578–9, **579**, 615, 649
Gobiconodontidae 578, 615
Gobiconodontinae 578–9
Gobiderma 381
Gobiidae 596
Gobinatus 374
Gobiocypris 269
Gobiodon 262, 596, 615
Gobioolithidae 570
Gobioolithus 563–4, **566**, 567–8, 570–1
Gobiops 297–300, **298**, **299**
Gobiosuchidae 407
Gobiosuchus 402, 405, 407, **407**, 417
Gobiotheriodon 596–7, **597**
Gobipteryx 240, 533, 537, 539, 541, 543–4, 556
Golyusherma complex 18, 44, 101
Gomphodont cynodonts 115, 134
Gondwana 128–9, 614
Gondwanatheria 585–6
Gonioglyptus 125, 128
Goniopholidae 409
Goodwin, B.C. 500
Goose Lake 649
Gorbatovo **121**
Gordonia 105
Gorgonops 28, **29**, 34
Gorgonopsia 3, 5, 24–6, 28–30, 87–8, 90, 92–4, **93**, 108, 128
Gorgonopsidae 87, 92
Gorgonopsinae 93
Gorgosaurus 447, 448
Gorizont concept xvii–xviii
Gorka 20
Gorky 8
Gorky batrachosaur complex 8
Gorky I locality **66**
Gornyi 192, **193**, 200
Gorodishche 195, 200, 204, 233, 413
Gov'altai 235, **258**, 260, **260**, 264
Gower, D.J. **126**, **131**, 144, 148, 150–1, 156
Goyocephale 233, 481–2, 495–7, **498**, 499, 500, 507, 509, **510**, 514
Graciliceratops 481, 489, **491**, **498**, 502–3, 505
Graculavidae 556

Gradziński, R. 232, 256, 279, 284, 575–6
Granger, W. 214–25, **217**, **218**, **219**, **219**, **221**, 230, 232, 235, 279, 309, 573
Gravemys 349, 355, **356**
Greenland 9, 128
Greererpeton 38
Gregory, W.K. 218, 494, 573
Gromov, V.I. 237
Gruiformes 551
Gryposaurus 468, **469**
Guandun Formation 333, 339, 343, 354
Guberlin Mountains 196
Gubin, Yu.M. 12, 245
Gubkin city 201, 421, 429
Guchinodon 262, 578, **579**
Guchinus Sum locality
  amphibians 304
  geology 262, **265**, **280**
  historical 242
  mammals 574, 578–9, 586, 588, 596, 598–9, 605
Gui Suin Gobi depression 239
Guriliin Tsav locality
  dinosaurs 541, **542**, 547, 562
  eggs 568
  geology **280**
  historical 243
  lizards 369
  mammals **574**, 575, 602–5, 616
  turtles 331
Gurilynia 541
Gurmaya Hills **3**
Gurvan Ereen ridge 249, 257–8, **265**, 271, 553–4
Gurvanereen Formation/Svita 257–9
Gurvan Saichan **280**, **561**, **574**
Gurvan Tes **574**
Gurvansaurus 375
Gusinoe Ozero 649
Gymnophiona 37–8
Gymnophthalmidae 380
Gypsonictopidae 605

Hadrokkosaurus 39
Hadrosauria 568
Hadrosauridae 229–30, 237, 239, 247, 268, 462–3, 465–76, **474**, 510, 562
  eggs 568, 570–1
Hadrosaurinae 462, 466–8, 470, 568
Hadrosaurus 468

Hahn, G. 596
Haichemydidae 332, 357–9, 361, 362
Haichemys 357, 358, **358**
Halazhaisuchus 150
Haldanodon 577
Halicore 627
Halstead, L.B. 183, 191, 193
Hanbogd (*see* Khanbogd)
Hanbogdemys 339–40, **341**, 342
Hangai (*see* Khangai)
Hangaiemys 321–5, **323**, **325**, 327, 361
Hardegsen beds 132
Harpymimidae 244, **435**, 445–6
Harpymimus **435**, **442**, 445–6, 451
Hartmann-Weinberg, A.P. 5, 73–81, 93
Haubold, H. 183
Haughton, S.H. 112
Hecht, M.K. 395, 397
Heilongjiang (Amur) River 444, 448
Heinrich, W.-D. 430
Heishanemys 328
Helodermatidae 382
Henan Province 443
Henkelotherium 598
Hentii mountains 271, **280**
Herlen Gol **280**
Hermiin Tsav locality
    amphibians 301, 303
    birds 538–9, 541–4, **543**, 566, **566**, 570
    dinosaurs 438, 488, 527
    geology **265**, 268–9, **280**, 284
    historical 233, 243, 250
    lizards 369–71, 373–4, 376–7, 379–80
    mammals 573, **574**, 576, 588, 590, 592–5, **592**, 602, **609**,
        613
    turtles 356
Hermiin Tsav I locality **561**
Hermiin Tsav II locality **561**, **574**, 592–5, **592**, 609, **609**,
    613
Hesperornis 544–6
Hesperornithes 556
Hesperornithidae 248, 544–6
Hesperornithiformes 544–7, **545**, **548**, 555–6
Heterodontosauridae 462, 498
Hetsüü Tsav beds **247**
Hexacynodon 26, 30, 109–10
Hipposauridae 86, 89
Hirayama, R. 353

Hirsch, K. 560, 562
Hobur (*see* Höövör)
Hoburogekko 380
Hodzhakulia 382
Hodzhakuliidae 382, 384–6
Hofmeyriidae 109
Holotheria 596
Holtz, T. **435**, 444
Homalocephale 233, 481–2, 495–7, **498**, 499, 500, **510**, 514
Homalocephalidae 493, 495–6
Hongilemys 349, 354, **356**
Hooker Island 195
Höövör locality
    amphibians 304
    dinosaurs 520
    geology 262, 264, **265**, 271
    historical 233, 242–3, **243**, 248–9
    lizards 369, 375, 378, 382, 385–6
    mammals 574–5, **574**, 578–9, 586, **587**, 588, 596–9, **597**,
        605–6, 615–16
    turtles 309, 321–2, **323**, **325**
Hoplocercidae 371–2, 383–6
Hopson, J.A. 86–8, 95, 103, 105, 108, 110–12, 114–15, 577,
    596
Horezmavis 551
Horezmia 306
Horner, J.R. 471, 473
Hötöl locality 553 (*see also* Khar Hötöl)
Hou, L. 371–2, 374
Hovd (*see* Khovd)
Hu, Y. 596
Huachi Formation 282, **283**
Huang He (Yellow river) **280**
Huanhe Formation 282, **283**
Hubei Province 443
Huene, F. von 2, 140, 142, 145, 148
Hughes, B. 142
Hühteeg locality 257, 262–4, 392
Hühteeg Formation/Svita 257, 260–5, 271, 369, 460, 464,
    519
Hühteeg Gorizont 260–5, **265**, 274, 321, 391, 395, 397
Hui-hui-p'u 333
Hulsanpes 436, 451
Hungary 265
Hunt, A. 458, 460
Hüren Dukh locality
    choristoderes 390–1, **391**, 392, **393**, 396–9, **396**, **398**
    dinosaurs **463**, 464, 476

Index

geology 262, 264, **265**, 271
  historical 243, 249
  pterosaurs 421, 422, **422**, 425–6, **426**, **429**
  turtles 321, **325**, **327**
Hürmen Sum 263
Hurum, J.H. 588, 590, 596, 616
Huxley, T.H. 184
Hybodus 637
Hyotheridium 601
Hypacrosaurus 471
Hypocentra 61
Hypsilophodon 494–5
Hypsilophodontidae 462, 465, 567
Hypsodonty 585
Hypsognathus 83, 165

Ibresinsky district 191, **192**
Ichthyornis 535–6
Ichthyornithes 556
Ichthyornithidae 535
Ichthyornithiformes 534, 549, 556
Ichthyosauria 187, 189, 196–203, 205
Ichthyosauridae 199, 200, 205
Ichthyosaurus 197, 199, 200–1, 203
Ichthyostegidae 35
Ictidorhinidae 86, 89, 92
Ictidosuchopsidae 111–12
Iganii River Basin 197
Igua 369
Iguana 374
Iguania 369–73, 385
Iguanidae 369, 384
Iguanodon 219, 262–4, 462–5, **465**, **466**, 475–6
Iguanodontia 230, 462–76, 499
Iguanodontidae 244, 462–5
Ih Bayan Uul mountains 238
Ih Bogd **574**
Ih Ereen mountain 262, 263, **265**
Ih Shunkht locality **561**, 565, 569
Ih Zos Nuur locality 260
Ihes Nuur locality 271
Ikechosaurus 390–3, 395, **396**, 397
Ikh (*see* Ih)
Ikhe Dzosu Nuur (*see* Ih Zos Nuur)
Il'inskoe 7, 78, 93
Ilekskaya Svita 649
Ilovlyanskii district 155
Inder Gorizont 54, 133, 135

Inder Lake 5, 54, 132, 135
India 105, 128, 615
Inflectosaurus 52, **122**, **123**, 131–2
Ingenia **435**, 438, **438**, **441**, **442**, 451
Ingenii Höövör depression 268, 271–2, 448
Ingenii Tsav locality **265**, **270**, **280**
Ingeniinae 437–8
Inkerman Mines 205, **403**, 404, 415
Inner Mongolia (northern China)
  birds 548
  choristoderes 395
  dinosaurs 438–40, 443–4, 446–7, 485
  geology 280, **280**, 284
  historical 225, 235
  lizards 369–71, 378
  turtles 309–10, 316, **319**, 325, 333
Inostrancevia **4**, 5, 25–6, 93–4, **93**
Inostranceviidae 26
Inostranceviinae 93–4
Insectivora 262, 573
Insects 178, 243, 246, 249, 259, 263, 298
Institute of Geology (*see* Ulaanbaatar)
Institute of Paleobiology (*see* Warsaw)
Institute of Vertebrate Paleontology and
  Paleoanthropology (*see* Beijing)
Institute of Zoology, Kazakhstan Academy of Sciences (*see*
  Almaty)
Insulophon **122**, **123**, 130, 161, 174–5
Inta fauna 12, 19, 20
Inta River 8, **18**, **19**, 20, 41, 63, 112
Intasuchidae 36, 39, 41
Intasuchus 19, 41, **42**
International Code of Zoological Nomenclature 372
Intinskaya Svita 41, 63
Irdyn Manga 333
Iren Dabasu locality 216–17, **217**, 224, 256, 444, 447
Iren Dabasu lake 235
Irendabasu Formation/Svita 283, 333, 439–40, 443, 446
Irenosaurus 390, 392, 397, 398, **398**, **399**
Isheevo 6, 8, 18–20, **18**, **19**, 23–4, 44 97, 99, 102, 104, 110
Isheevo Dinocephalian Complex 6, 19, 23–4, 103
Isodontosauridae 372–3, 383–6
Isodontosaurinae 372
Isodontosaurus **3**, 372–3
Italian–French–Mongolian Expedition 575
Itemir 427
Itemirella 303
Itemirus 450–1

Ivakhnenko, M.F. 11–12, **19**, 20, 23, 29, 71, 75–6, 80–1, 83, 89, 91, 95, 97, 100–2, 161–3, 165, 167, 174
Ivantosaurus 20, 87, 91
Ivö-Klack locality, Sweden 545

Jaikosuchus **122**, **124**, 131, **141**, 142, 151, **152**, 155
Japan 233, 617
Japanese–Mongolian Expedition 575, 591
Jargalantyn Gol 272, **280**
Jargalantyn mountains 239, 272
Jarilinus 65
Jastmelchyi 315
Javkhlant Formation/Svita 256
Javkhlant Uul 268, **280**
Jaxartosaurus 470–1, **471**, 473
Jenghizkhan 448, 451
Jenkins, F.A. 578
Jerzykiewicz, T. 279, 283–4, 339, 460, 462, 567, 649
Jiayin 444
Jibhalanta (*see* Javkhlant)
Jingchuang Formation 282, **283**
Jiufotang Formation 549
Johnson, A. 217, **218**
Joint Polish–Mongolian Palaeontological Expedition 231–3, 239–40, 256, **288**, 301, 311, 368, 448, 456, 457, 458, 517, 533, 561, 573, 575
Joint Russian–Mongolian Palaeontological Expedition (JRMPE) 233, 246–51, 309, 368, 533, 541–2, 547
Joint Soviet–Mongolian, Palaeontological Expedition (JSMPE)
  amphibians 298, 304
  birds 533, 561
  dinosaurs 456, 459, 464, 474, 517
  geology 256
  historical 225–30, 235–54, **238**, **244**, **247**, **250**
  lizards 368, 399
  mammals 573–5, 627
  pterosaurs 420, 422, 424
  turtles 309, 311, 345–6
Journal titles xxix–xxxiv
Judinornis 546, 547
Jugosuchus 64
Jurassic
  crocodilians 402–6, 408–9, 412–14, 416
  marine reptiles 187–210
  pterosaurs 420–4, 428–30
  Russia 187–210
  sauropod dinosaurs 456

Jushatyria **122**, **124**, 135, **141**, 151–3
Jurung-Tumus Peninsula 195
Juul, L. 152, 153

Kaisen, P. 217, **218**
Kalahari Desert 289, 291, **291**
Kalandadze, N.N. 11–12, 153
Kalgan 212, 214–15
Kallokibotion 346, 349, 359, 362
Kalmakerchin locality **628**, 629–30, 639
Kama River basin 101, 189, 195–6
Kamacops 44
Kamennyi Yar ravine 50, 97
Kamptobaatar 590, 594–5, **594**
Kanev locality 627
Kannemeyeriidae 9, 10, 134–5
Kannemeyeriinae 105, 108
Kannemeyeriini 106
Kannemeyeroidea 132, 134–5
Kanon Ravine 397
Kansai locality
  amphibians 304, 306
  crocodilians **403**, 404, 409–10, **410**, 414, **421**, 429
  mammals **628**, 629, 637–8, 648
  turtles 333
Kansajsuchus 404, 409–10, **410**, 417
Kapes **122**, **123**, 131–2, 134, 161, 165, 168–70, **170**, **171**, 172, 175
Kara Sea 196
Kara-Bolla-Kantemir **121**
Karabastau 404, 424
Karabastau Svita 305, 408, 423, **423**, 424, **425**
Karagachka locality **121**, 148
Karagachka River 49, 106, 108, 115
Karakalpakia
  amphibians 304, 306
  birds 535, 551
  dinosaurs 473
  mammals 637, 640
  marine reptiles 203, 205
  pterosaurs 428
  turtles 310, 319, 327, 333, 344, **344**
Karakhskaya Svita 204, 414
Karatau lake 297, 408–9, **408**, 417
Karatau ridge 305, 408, 420, **421**, 422–4, **422**, **423**, **424**, 430
Karatausuchus 404, 408–9, **408**, 417
Karauridae 305, 629

Karauroidea 305
Karaurus 305, **305**
Kargala Mines 2, 6, **18**, 24, 63, 97, 100–1, 104
Karhu, A.A. 555
Karoo Basin 17, 27
Karpinskiosauridae 62–3
Karpinskiosaurus 25, 63
Karpogory **18**, 23
Kasatkino 472
Kashpir 187, 200
Kasyanovtsky locality 164
Kawakami coal mines 472
Kawinga Formation 143
Kayentachelys 349, 358–9
Kazakhstan
  amphibians 297, 305, 307
  birds 533, 546–7, 553–5
  crocodilians 402, 404, 408–10, **408**, **410**, 416–17
  dinosaurs 436, 439, 441, 443, 445, 447–50, 456, 460,
    462–3, 466, 468, 470–4, 476, 517
  eggs 560–1, 569
  historical 12, 235
  lizards 378
  mammals 576, 590, 593, 616–17, 627–8, **628**, 639, **643**,
    647, 648
  mosasaur 203
  plesiosaurs 188, 195–6, 199
  pterosaurs 420–5, **423**, **425**, **428**, 429–30
  synapsids 100, 132, 135
  turtles 310, 320, 322, 332–3, 346, 354, 358
Kazan' University 2
Kazan', Geology and Mineralogy Museum 188
Kazmierczak, J. 232
Kaznyshkin, M.N. 310, 562
Kaznyshkina **638**
Kemerovskaya Province 649
Kemp, T.S. 87, 88, 108, 430
Kennalestes 233, 262, 575, 605–7, **607**, **608**, 609–11, 614,
  615, 617, 632, **644**
Kennalestidae 605–7, 642
Kennalestoidea 605, 649
Kentrosaurus 499
Kermack, K.A. 577, 598–9
Kermackiidae 599
Keros locality **121**
Keryamayol 135
Khabarovskii Krai 554
Khaichin Uul 244, 248, **280**, 331, **358**, 563, 568, 573

Khaichin Uul I locality **561**, 595
Khaichin Uul II locality 242, 384
Khaichin Uul II–V localities 243
Khamaryn Khural locality 237, 262–4, **265**, 390, **391**, 392,
  395, **396**, **463**, 464, 476
Khamaryn Us locality 237, 243, 249, 444, 519, 575
Khanbogd Sum 268
Khand, E. 241, 575
Khangai mountains 259, 260, **265**, 271, 279–80, **280**
Khangil **280**
Khar Hötöl locality 233, 237, 239, 264, 271, **280**, 282, 331,
  440
Khar Hötöl Uul 256, 263, **265**, 268, 272
Khar Khutul (*see* Khar Hötöl)
Khar Us Nuur locality 426
Khara Khutul (*see* Khar Hötöl)
Khashaat locality **280**, **284**
Khatanga River 554
Khei-Yaga River basin 54, **121**, 172–3
Khentei (*see* Hentii)
Khermiin (*see* Hermiin)
Khetzu (*see* Hetsüü)
Khidzorut 429
Khimenkov, V.G. 421
Khoboor (*see* Höövör)
Khodzhakul Formation/Svita
  amphibians 304, 306–7
  birds 535, 551
  crocodilians 411
  dinosaurs 445, 473
  mammals 630–1, 637, 649
  marine reptiles 205
  turtles **344**
Khodzhakul lake 205, 306, 390, 637
Khodzhakul locality 204, **391**, 404, **421**, 428, 473, 535, 551,
  628, 630, 637
Khodzhakul'sai ravine **344**, 535, **628**, 630, 637
Khodzheili 628
Khoer (*see* Khoyor)
Kholboot 249, 256, 544, 553
Kholboot Formation/Svita 256, 549
Kholboot Gol locality **260**
Kholboot Sair locality 260
Kholbotu (*see* Kholboot)
Khongil Tsav **252**, **265**, 268, **280**, **356**, **403**, 405, 411, 566
Khooldzin plateau 235
Khoolsun (*see* Khulsan)
Khoroshevskii Island 204, 413

Khosbayar, P. 240, 422, 426
Khovboor (*see* Höövör)
Khovd **238**, 239, **244**, **250**, **265**
Khovd Aimag 426
Khoyor Zaan locality 384
Khozatskii, L.I. 310–11, 322, 325, 402, 425
Khudiakovia 197
Khukhtesk (*see* Hühteeg)
Khukhtyk (*see* Hühteeg)
Khulchin Uul **574**
Khulsan Gol 257, **280**
Khulsan locality
   birds 537, 539, 541, 543–4, **561**, 568
   dinosaurs 436, 493, 527
   geology **265**, 268–9, **280**, 284, **285**, **288**
   historical 233, 240
   lizards 369, 373, 375–6, 379–81
   mammals 573, **574**, 575, 592, 594, 609, 611
   turtles 331, 343
Khulsangol Formation/Svita 257, 259–60, 262–4, **263**,
   282, **283**, **325**, **327**, 549
Khulsyn (*see* Khulsan)
Khunnuchelys 333, 359
Khuren Dukh (*see* Hüren Dukh)
Khurendukhosaurus 390, 392, 396–8, **396**
Khurilt Ulaan Bulag locality 544, 549, 553
Khurmen (*see* Hürmen)
Khutoolyin (*see* Hötöl)
Khvalynsk **403**, 413
Khyra 553
Kielan-Jaworowska, Z. 232–3, 239, 311, 573, 576, 578, 581,
   584, 587–90, 594, 596, 599, 601–2, 604–5, 609, 614,
   616–17, 639, 648–9
Kielantherium 599, 600, **600**, 615
Kilodzhun (*see* Kylodzhun)
King, G.M. 87, 95, 105, 108
Kinosternoidae 335
Kipriyanov, W.A. 188, 195–6, 201
Kirghizia (*see* Kirgizstan)
Kirgizemys 205, 321–2, 324, **326**
Kirgizstan
   amphibians 297, 299, 305, 307
   birds 554, 560, 562, 564, 569
   crocodilians 204–5, 402, 414
   dinosaurs 456
   eggs 560, 562, 564, 569
   historical 12
   lizards 379

   mammals **628**, 629, 639
   pterosaurs 420–1, 428
   Triassic reptiles 177–86
   turtles 205, 310, 312–13, **313**, 320, 324–6, **326**, 330, 332,
   332
Kirkhuduk **403**, 415
Kirkland, J. 531
Kirov Province
   anthracosaurs 64, 68, **68**
   historical 7, 12
   marine reptiles 203
   pareiasaurs 74
   Permian faunas 20, 25
   procolophonoids164, 167
   synapsids 91, 96, 101, 104, 111, 114
Kirpichnikov, A.A. 237
Kirsanov region 429
Kitching, J. 112, 115
Kiya River 649
Kizylkum (*see* Kyzylkum)
Kizylkuma 303, 304
Kizylkumavis 533, 534, 536
Kizylkumemys 322, 332, 344–5, **344**
Kizylordin Province 554
Klamelia 578
Klaudzin 428
Klyuchevskii mine 45
Kobdo (*see* Khovd)
Kobyaki village 429
Kocherzhenko, E. **246**
Koiloskiosaurus 165
Koinia 44
Koisu **403**, 414
Kokart River 629
Kokartus 305
Kokopellia 604
Kolguev Island 161, 174
Koltaevo localities **10**, 48, **121**
Koltaevo I locality 54
Koltaevo II locality 106–8, 149, **133**, 154
Koltaevo III locality 108, 149, 152–3
Kolymskii 196
Komatosuchus 48, **122**, **123**, 132
Komi Republic 45, 63, 68, **69**, 112, 143, 153, 155, 169
Konaki locality **68**
Konoplyanka River 195–6, 203
Konzhukova, E.D. 8
Konzhukovia 43, **44**

Kopanskaya Svita 46, 49, 125, **126, 127, 131**

Koparion 444

Kordikova, E. 310

Kormitsa River 50

Korneevskoe 50

Korotaikha depression 133, 173

Kostroma region 195

Kotel'nich complex/fauna 24–5, 103

Kotel'nich locality 2, 7, 12, 18–19, **18**, 21–2, **21**, 24–5, 74, 76, 90, 104, 111

Kotel'nich Museum 71

Kotlas district 25, 94, 110, 113

Kotlas rail station 4

Kotlassia 5, 25, 63

Kotlassiomorpha 62

Kovalevskii, V.O. 627

Kowalski, K. 232, 573

Kozlat'yevo **121**

Kramarenko, N.N. 241–2

Krasnie Baki village 164

Krasnoarmeisk district 201

Krasnokamenskaya Svita 133

Krasnoyarskii Krai, 554

Krasnye Pozhni **121**

Kriushi village 187

Krusat, G. 577

Krymskii locality 22

Krymskoe 12

Kryptobaatar 582, **583**, 590–1, **591**, 593–4

Kubiak, H. 232

Kudanga **121**

Kuehneotherium 578

Kugart basin 305

Kuhn, O. 76, 377

Kulbeckia 631, 634, 637, **643, 644, 645, 647**, 648

Kulbeckiidae 648

Kulbeke well 637

Kulczycki, J. 232

Kulicki, C. 233

Kultchitskii, J. 239

Kumanskaya Svita 129

Kumertau district 152, 154

Kumlestes 632, 642, **644**

Kumsuperus 633, 642, **644**

Kupletskii, B.M. 236

Kuramin Ridge 63

Kurochkin, E.N. 241–3, **246**, 533, 549

Kursk Osteolite 201

Kursk region 189, 195–6, 201, 204, 414

Kurzanov, S.M. 242, **242**, 245, **252**, 443, 448, 450, 459, 493

Kushmurun locality **403**, 546–7

Kustanaiskaya Province 546–7

Kuszholia 552

Kuszholiidae 552, 556

Kutorga, S.S. 1, 97

Kutulukskaya Svita 64, 114

Kutyrkin, V. **246**

Kuzikov, I. **246**

Kuznetsov, V.V. 310

Kylodzhun locality 321, **326, 330, 332**, 428

Kyrk Khudag **463**, 473, 476

Kyzyl-Orda Province 429, 474, 627, 639

Kyzylbulag district 429, 638

Kyzylkum Desert
  amphibians 297, 300, 301–7
  birds 536
  dinosaurs 450, 460, 473–4, 517
  mammals 628–9, 637, 640, **641, 644, 645, 646**, 649
  marine reptiles 204–5
  pterosaurs 422, 428
  turtles 310, 326, 333, 337, **338**, 349, 354

Kyzylpilial locality 429

Kyzylsu River 305

Kzyl-Oba River, 167

Kzyl-Sai II 2 locality 145

Kzyl-Sai III 2 locality 144

Kzyl-Sai ravine 47, 55, 167

Kzylsaiskaya Svita 129, 173

Labes 642

Labyrinthodontia 35, 37–9, 60–1, 120

Lacertidae 383–4

Lacertilia (*see* Sauria)

Laevisoolithidae **565**, 570

Laevisoolithus 563–4, **566**, 567, 570

Lagomorpha 614, 648

Lainodon 642

Lake Baskunchak 5

Lake Gusinoe locality 390, **391**, 392, 396–7

Lake Tanganyika 292

Lakes Valley 236, 238

Lamawan Formation 282, **283**

Lambdopsalis 582

Lambeosaurinae 462, 467–8, 470–3, 568

Lambeosaurus 471

Lang Shan massif **280**, 281

Lanthaniscus 23, 63
Lanthanosuchidae 24, 27, 63, 82
Lanthanosuchus 23, 63
Lanthanotidae 380
Laolonghuoze locality **319, 319, 327**
Lapinsk region 196
Laptev Sea 195
Late Dinocephalian (Isheevo) fauna 23–4
Latonia 300
Laurussia 17, 27
Lazarussuchus 397
Lebedeva, Z.A. 236
Lebedinsk mine 201, 205
Lee, M.S.Y. 24, 71, 160
Lefeld, J. 232
Lehe Formation 282, **283**
Leiopelmatidae 301
Lena River basin 195
Lenesornis 533, 535
Leningrad (*see* St. Petersburg)
Leninskoe district 114
Lepidosauria 71, 142, 181
Lepospondyli 37–9
Leptictida 605
Leptictidae 607
Leptoceratops 480, 488, 494, **498**, 501–5, 507, **508, 510,**
    515–16
Leptochamos 373
Leptoglossa 373
Leptopleuron 165
Leptopleuroninae 161, 165, 168
Leptopterygidae 200–2, 205
Leptoropha 63, 84
Leptorophidae 62–3, 84
Lesothosaurus 495, 502
Lestanshor Creek 173
Lestanshoria **122, 123,** 130, 161, 168, **172,** 173, 175
Lestanshorskaya Svita 54, 130, 173
Leutkesaurus 188
Liaoning Province 316, **320,** 556, 596
Lillegraven, J.A. 577
Limnocyrena 259
Limnofregatinae 551–2
Lin-Ho **280**
Lindholmemydidae 322, 331–2, 349–50, 352–4, 359–62
Lindholmemydinae 350
Lindholmemys 269, 332, 349–50, 354–5, **355**

Linovo 25
Liodon 203
Liopleurodon 188–9, 191, **192,** 195
Lipotyphla 648
Lipovaya Balka 47, 171, 174
Lipovo village 164
Lipovskaya Svita 47, 52, 171, 174
Liska River basin 196, 203
Liskun, I. **246**
Lissamphibia 37, 306
Litakis 382
Lithornithiformes 549
Lizards (*see* Sauria)
Logachevka 51
Longisquama 12, 177, **178,** 182–4, **182**
Longisquamidae 182–4
Loo 238
Lopatin, A.V. 579, 615, 649
Lopatino locality **121**
Loxaulax 586, 615
Loxommatidae 35, 37
Lozovskiy, V.R. 10, 128–9, 131
Lü, J. 430
Luchiskina gorge 429
Luetkesaurus 189
Lugansk region 203
Luk'yanov, I. **240, 242**
Luohadong Formation 282, **283**
Lurdusaurus 462, 465
Luza River basin 45, **121,** 130
Luzocephalus 39, 45, **46, 122, 123,** 125, 128
L'vov 203
Lyailyakskii district 177
Lyapin region 196
Lycopsida 128–9
Lycoptera 259
Lydekkerinidae 39, 45, 125, 128
Lysaya Gora **422,** 425, **426,** 429
Lysogorsk district 196
Lystrosaurini 105
Lystrosaurus 11, 12, 55, 103, 105, **106, 107, 122, 124,** 126–8
Lystrosaurus-lydekkerinid episode 128
Lystrosaurus/Procolophon assemblage zone 161

Macrobaena 322–4, 327, 360–1
Macrobaenidae 310, 315, 317–18, 321–4, 327, 331–2, 353,
    359–62

Macrocephalosauridae 373–6, 383–6
Macrocephalosaurus 373
Macroleter 22–3, 27, 63
Macroolithus 563–4, **566**, 567–70
Macrophon **123**, 161, 165, 168–9, **169**, **170**
Macropterygius 199
Macroscelidia 614
Mader, B. 447
Madsen, J. 458
Madygen locality 12, 177–86, 639
Madygen Svita 177–8, **178**, 183, 307
Madygenia 178
Magadan Province 199, 554
Maiasaura 562
Majungatholus 496
Makar'ev district 49
Makgadikgadi 292
Makulbekov, N.M. 245
Malaya Ivanovka locality 552
Malaya Kinel River 44
Malaya North Dvina River 64
Malaya Serdoba 187, 195–6, 203, 205, 415, **421**, 427
Malecki, J. 232
Maleev, E. A. **227**, 237, 440, 448, 517, 523
Maleevosaurus 448, 451
Maleevus 518, 521, 523, **525**, 530
Malmyzh district **18**, 96
Malokinel'skaya Svita 65
Malutinisuchus **122**, **124**, 135
Malye Undory 199
Malyi Uran River **18**, 23, 43, 100
Mamadysh district 43
Mammalia
  historical 216, 223–4, 232, 235–9, 243, **243**, 245, 248, 250
  geology 262, 264, 266, 269–70, 273, 283, 287
  Kazakhstan 627–8, **628**, 639, **643**, **647**, 648
  Kirgizstan **628**, 629, 639
  Mongolia 223–4, 232, 573–626
  origins 86, 88
  Tadzhikistan **628**, 629, 637–8, **647**, 648
  Turkmenistan 637
  Ukraine 627
  Uzbekistan **628**, 629, 637, 639–40, **641**, 642, **643**, **644**, **645**, **646**, **647**, 648–9
Mammal-like reptiles (*see* Synapsida)
Manchuria 472–3
Manchurochelys 310, 316, 317, 319, 320, **320**, 321

Manchurodon 615
Mandalgov' Aimag **238**, **244**, **250**, 260, **265**
Mandschurosaurus 472, 475–6
Mangyshlak Peninsula 11, **121**, 132
Maniraptora 436, 554–5
Manjurochelys 315
Manlai Lake 249, 287
Manrakskaya Svita 569
Marginocephalia 480–516, **498**, **508**
Marine reptiles, Mesozoic 187–210
Marinov, N.A. 236
Markovka village 172
Marsasia 604, 617, 649
Marsh, O.C. 211, 224, 232
Marshall, L.G. 601
Marsupialia 248, 596, 601, 617, 632, 639–40, 642, 649
Martill, D.M. 430
Martin, L.D. 537, 541–2, 544, 546, 549, 551
Martinson, G.G. 240, 256
Marxism xviii–xix
Maryańska 232, 467, 471, 473, 483, 489, 491, 493–5, 497, 505, 517, 519, 523, 527, 575
Mashchenko, E.N. 579, 615, 649
Masteksayan Gorizont 135
Mastodonsauridae 36, 40, 48, 135, 299
Mastodonsaurus 8, 10, 40, 48, 120, **123**, **124**, 135
Mastodonsaurus fauna 40, 103, 122–4, **122**, 132, 134–5
Matthew, W.D. 211, 215, 309
Mayulestes 601
Mazin, J.-M. 199
McDowell, S.B.J. 359
McGowan, C. 199, 201–2
McIntosh, J.S. 457–60
McKenna, M.C. 582, 586, 596, 598, 602, 605
Mechet' ravine 50, 68, **68**
Mechet' II locality **121**, 155
Medvedkovo 64
Megalanidae 380
Megaloolithidae 565
Megalosauridae 450
Megazostrodontidae 577
Meiolania 346, 359
Meiolanidae 346, 362
Melanopelta 54, **122**, **123**
Melosauridae 20, 23, 39, 43–4
Melosaurus 43, 44
Mendrez, C.H. 110

Mengyin Formation **318**, 329

Meniscoessus 590, 596

Mesenosaurus 6, 23, 27, 86

Meshcherskii Gorizont 649

Meshcheryakovka locality 167

Mesochara 264, 265

Mesochelys 359

Mesodira 195

Mesoeucrocodylia 402, 407–17

Mesolanistes 271

Mesosauria 71, 161

Mesosuchia 402, 410, 417

Metamesosuchia 408

Metasuchia 402

Metatheria 573, 575, 600–4, 617, 640

Metoposauridae 36

Metornithes 554

Metriorhynchidae 204–5

Meyer, H. von 2

Meylan, P. 322, 326, 333, 335, 339, 352

Mezen' Group 22

Mezen' locality **18**, **19**, **121**

Mezen' River basin 6, 18–19, 22–4, 89, 91, 130, 155, 173

Mezen' syncline 121, **121**, 129–30, 132

Mezen'–Belebey cotylosaur complex 6, 9, 22–3

Mezhog locality 153

Microceratops 333, 489, **510**

Microcnemus **122**, **124**, 126, 129–30, 142

Microcosmodontidae 585

Micropachycephalosaurus 492

Microphon 161, 164–5, **165**, 175

Microsauria 37–8

Microsyodon 95, 101

Microtheledon 165

Middle Asia 310–11, 324, 334, 339, 344–5, 349, 354, 370, 372, 375, 382, 420, 422, 616, 627–8

Middle Gobi (*see* Dundgov' Aimag)

Mikhailov, K.E. **252**, 542, 560, 562

Mikhailovka village 153, 305, **403**, 404, 408, **408**, 416, 423, 424

Millerettidae 71, 82

Milner, A.R. 37, 306

Mimeosaurus 371, 372, **372**

Mimobecklesiosaurus 378

Minikh, M.G. 130

Miospores 130, 132, 134–5

Mishakovskaya **121**

Mixosauridae 197, 199, 205

Mixosaurus 199

Mixotheridia 605, 642, 648

Młynarski, M. 311, 339, 350

Mlynarskiella 333, 336

Mnemeiosaurus 100

Mnevniki 195

Mogoin Ulaagiin Hets **561**, 568, 571

Mogoito Member 649

Moiseenko, V.G. 473

Moldavia 203

Molluscs 258–9, 262–6, 268–70, 272–3, 282, 287, 549, 638

Molnar, R. 473

Molybdopygus 96

Mongol Altai Mountains 238, **238**, **244**, **250**, **265**, 280, **280**, 544, 549

Mongolemys 262, 311, 322, 349–50, **351**, 352–5, **353**, 360, 362

Mongolia

  amphibians 297–8, 301–2, 304

  ankylosaurs 517–32

  archosaurs 402–559

  birds 533–59

  ceratopsians 480–516

  choristoderes 390–401

  Cretaceous faunas 211–624

  Cretaceous geology 256–96

  crocodilians 402–19

  diapsids 368–559

  dinosaurs 434–559

  dinosaur expeditions 211–55

  eggs, dinosaur and bird 560–72

  expeditions, history 211–55, **214**, **217**, **222**, **227**, **230**

  geology 256–96

  history of research 211–55

  lizards 368–89

  locations xxiii–xxviii

  mammals, Mesozoic 223–4, 232, 573–627, 639, 642

  ornithopods 462–79

  pachycephalosaurs 480–516

  palaeobiogeography 359–62, 385–6, 509–10

  palaeoclimates 271–4, 287–93, 399

  palaeoenvironments 271–4, 287–93, 399

  palaeogeography 271–4, 281

  pterosaurs 420–33

  reptiles, Mesozoic 309–626

  sauropods 456–61

sedimentology, Cretaceous 279–96
stratigraphic units xxiii–xxviii
stratigraphy of Cretaceous 256–71, 279–96
theropods 434–55
turtles 309–67
Mongolian
    alphabets xxii–xxiii
    geologists xxv–xxxix
    language xxii–xxiii
    palaeontologists xxxv–xxxix
Mongolian Academy of Sciences (MAS) 232, 235, 239–40, 251, **317**
Mongolian Museum of Natural History (*see* Ulaanbaatar)
Mongolian–Polish Palaeontological Expedition (*see* Joint Polish–Mongolian Palaeontological Expedition)
Mongolian–Russian Palaeontological Expedition (*see* Joint Russian–Mongolian Palaeontological Expedition)
Mongolochamopidae 374–5, 383–6
Mongolochamops 374
Mongolochelyidae 312, 332, 346, 349, 359–60, 362
Mongolochelys 311, 346, **347, 348**, 349, 362
Monobaatar 586–7, 615
Monolophosaurus 434, 451
Mononykus 233, 554–5
Monotremata 584
Montanalestes 605
Montanoceratops 481, **498**, 501–3, 505, 507, **508, 510**
Moody, R.T.J. 371–2
Mook, C.C. 218
Mordovo 187
Morganucodonta 577
Morganucodontidae 577–8, 649
Moroznitsa 23
Morris, F.K. 214, **214, 218**, 236, 259, 266
Morrison Formation 409, 444, 587, 615
Morunasius 370, 371
Mosasauria 187, 203–4
Mosasauridae 187, 203–4, 205, 380, 383
Mosasaurinae 205
Mosasaurus 203
Moschops 23, 101
Moschorhinidae 110
Moschorhinus 28, 34
Moschowhaitsia 26, 110–11, **111**
Moschowhaitsiinae 110
Moscow 195, 199, 225, 230, 232, 236–8, 245, 360, 399, 448, 576

Moscow Geological Prospecting Institute 188
Moscow, Palaeontological Institute (PIN) 5–12, 35, 71, 74, 80, 86, 140–1, 161, 177, 188, 225–6, 235–55, 297, 309, 369, 391, 402, 422, 456–7, 474, 480, 517–18, 533, 562, 576, 579, 586, 588, 610–11
Moscow Province 187, 189, 195, 204, 300, 413, 649
Moscow River basin 187, 195–6
Moscow Syncline 121, **121**, 125, 128–30
Moscow University 242
Moskovoretskaya Svita 649
Mt. Besh-Kosh 473, 476
Mugai River 195
Muizon, C. de 604
Multituberculata 248, 262, 573–6, 579–96, **582, 584, 587,** 604, 614–18, 637, 639, 648
Muraenosaurus 188–90, 195
Muraptalovo 49
Murchison, R.I. 2, **3**
Muren **238, 244, 250, 265**
Murtoi Svita 649
Murzaev, E.M. 236
Museum für Naturkunde, Humboldt Universität, Berlin 35
Museum of the Isle of Wight (Geology) (*see* Sandown)
Museum of the Rockies, Bozeman 480
Muséum National d'Histoire Naturelle, Paris 480
Mussett, F. 598
Mutovino 65
Muttaburrasaurus 463, 465
Myangat Sum 249, 259, 553
Myctosuchus 104
Mylva River **19**, 20
Mynbulag depression 637
Mynbulakia 306
Myocephalus 165
Myopterygius 197, 201–2, **202**

Nadkrasnokamenskaya Svita 68, **69**, 135
Nakunodon 615
Nalaih 260
Namsrai, T.N. 422
Namsray, G. **240**, 240, **241, 242**, 533–5, 538, **539, 540**, 556
Nanhsiung Group 333, 339, 343
Nanhsiungchelyidae 322, 331–3, 335, 339, 341, 359, 361
Nanhsiungchelys 333, 339, 343
Nanhsiungosaurus 333
Nanocynodon 25, 114, **115**

Nanoparia 72
Nanshiungosaurus 439
Nanxiong Formation 333, 339
Naran Bulag locality, 237, 239, 243–4, **280**, 334
Naranbulag Formation/Svita 327, 349, 357
Naran Gol section 287
Narmandakh, P. **240**, 241, 309, 311, 340, 422
Naryn River valley 560
National Geographic Society **18**
National Museum of Canada, Ottawa 480
National Museum of Natural History, Washington 480
Natural History Museum, Kansas University 545
Natural History Museum, London 145, 154
Navoi District 534, 535, 536, 541, 552
Navoloki 93
Necrosauridae 381–6
Nectridea 37
Nelson, M.E. 616
Nemegt basin
  dinosaurs 448, 457, 468–9, 527
  geology **280**, 281, 284, **285**, 287, 292
  historical 229–31, **230**, 233, 239
  mammals 573–5, 592, 609–11
Nemegt Formation/Svita
  amphibians 304
  birds 541, **542**, 545, **545**, 547, **548**, 554, 556, 560, **561**, 564, 566–70
  crocodilians 411
  dinosaurs 438–9, 441, 444, 446, 448–50, 457–8, 467, 469, 518, 527
  geology 256–7, **265**, 266, 270–1, **270**, 279, **283**, 284, **286**, 287, 289, 292
  historical **245**
  lizards 369, 375, 380–1, 383, 385
  mammals 574–5, 588, 595, 603, 616
  turtles 311, 331–3, 346, **347**, 349, **351**, 353, **353**, 355, **356**, 357–8, **358**, 362
Nemegt locality
  birds 537, **561**, 567, 568
  crocodilians **403**, 405, 411, 417
  dinosaurs 434, 436, 439, 441, 446, 448–9, 451, 458, **463**, 468, 476, 527
  geology 256–7, **265**, 266, 268, **269**, **280**, 284, **286**, 292
  historical 229–33, 237–9
  lizards 369, 377, 383
  mammals 573, **574**, 593, 594, 609
  turtles 331, 343, **347**, **353**
Nemegtbaatar 576, **580**, **584**, 593–4

Nemegtosauridae 460
Nemegtosaurus 233, 456–60, **459**
Nenetskii National District 173
Neoceratopsia 482, 487–93, 498, **498**, 501–9, **508**, 515–16
Neodiapsida 390
Neognathae 551–2, 555–6
Neopliosaurus 188–9
Neoprocolophon 162
Neorachitome Fauna 12
Neorachitomi 6, 10, 120
Neornithes 547–52, 556
Neosuchia 402, 408–17
Nesodactylus 1 430
Nesov, L.A.
  amphibians 297, 300–1, 304, 306–7
  birds 533–6, 547, 552, 553, 560, 562
  crocodilians 411
  dinosaurs 441, 447, 449–50, 456, 462, 471, 473
  lizards 370, 375
  mammals 599, 601–2, 605, 616–18, 627–9, 639–40, 642, **643**, **644**, **645**, **646**, **647**, 648
  marine reptiles 196, 199, 201
  pterosaurs 421–2, 427, 429–30
  turtles 310–11, 313, 322, 325, 328–30, 332–3, 337, 344–5
Nesovbaatar 588, 590, 595
Nesovemys 328
Nest, dinosaur 219–20, 223, 233, 283, 437, 565, 567, 571
Neuquenornis 533, 537
Neurankylus 346, 359
Niaftasuchidae 89, 91
Niaftasuchus 23, **90**, 91
Nikitin, S.N. 120
Nilgin (*see* Nyalga)
Ningjiagou locality **318**
Nipponosaurus 472, 475–6
Niuksenitia 93, 94
Niz'ma **121**
Nizhneustinskaya Svita 23
Nizhnii Novgorod Province 8, 26, 65, **66**, 68, **68**, 105, 110, 143–4, 162, 164
Noasaurus 444
Nochelesaurus **72**, 82
Nodosauridae 494, 517, 519, 531
Nogoon Tsav Formation/Svita 271, 448
Nogoon Tsav gorge 243, 264
Nogoon Tsav locality
  birds 547
  crocodilians **403**, 405, 411, **413**

geology **265, 270, 280**
historical 243
turtles 309, **315**, 331
Nominosuchus 403, 406–7, 417
Nopcsa, F. von 2, 100
Norell, M. 436
Noripterus 426
Norman, D.B. 464–5, 475
Normannognathus 430
North Dvina Commission 5
North Dvina excavations 3–5
North Dvina Gallery 5
North Dvina pareiasaur complex 7, 8
North Dvina River basin 3–5, **4**, 7, 8, 12, 25, 60, 78–81, 94, 103, **121**
North Korea Block **280**
North Mongolian Uplift 272
Northern Cliffs **285**
Northern Sair 286, 289, **463**, 469, 476
Nothogomphodon 111, 112, **122, 124**, 134
Nothogomphodontinae 112
Nothosauria 2, 124, 195
Nothosauriformes 195
Nothosaurus **122, 122**, 195
Notobatrachus 301
Notosuchia 402, 406–7
Notosuchus 407
Notosyodon 100
Novacek, M.J. 575, 604–5
Novaya Bedenga 202
Novaya Zemlya Archipelago 172
Novikov, I.V. 12
Novo-Alexandrovka locality 153
Novo-Nikolskoe 105
Novokhatskii, I.P. 196
Novosergievka district 90
Novozhilov Hills **248**
Novozhilov, N.I. 8, 191–4, **192, 193**, 230, **231**, 237
Nowiński, A. 232–3
Noyon Sum 237, 246, 263, 298, **574**
Noyonsum Formation/Svita 297
Nukusaurus 306–7
Nurumov, T.N. 627
Nyadeitinskaya Svita 133
Nyafta 91
Nyalga depression 260, 271
Nycteroleter 6, 21–3, 63, 83
Nycteroleteridae 22, 24–5, 27, 63, 82–3

Nycteroleterina 63
Nyctiboetus 64
Nyctiphruretidae 82, 160
Nyctiphruretus 6, 23, 27, 83
Nyctitheriidae 648
Nyctosaurus 430
Nyuksenitsa district 93

Obetzautzii Creek 554
Oblongatoolithus 570
Oblongoolithidae 570
Oblongoolithus 563, 564, 570
Obruchev, V.A. 212, 235
Obshchii Syrt region 121, **131**, 155
Ocher (Ezhovo) fauna 9–10, 18–21, 103
Ocher district 9, 20–1, 89, 91, 95, 96, 101, 103
Ochev, V.G.
    archosaurs 140, 142, 144, 147–9, 152–3
    historical 10–11, **10, 11**
    marine reptiles 188, 194–5, **194**, 195, 199
    Mongolia 242
    procolophonoids 161
    Triassic biostratigraphy 129, 132–3, **133**
Odontornithes 544–7
Oka River basin 11
Okavango Delta **291**
Okavango Oasis 280, 291–2, **291**
Okhotsk Sea 199
Okhotskii region 554
Okunevo locality 54, 68, **121**, 167
Olenek River basin 11
Ölgii Hiid locality **238, 244, 250, 265**
Ologoy Ulaan Tsav (*see* Algui Ulaan Tsav)
Olsen, G. 217, **218**, 219, **223**
Olshevsky, G. 448, 450
Olson, E.C. 9, 17, 20–1, 23, 88–9, 100
Ömnögov' Aimag (= South Gobi)
    amphibians 298, 303–4
    birds 538, 541, **542, 543, 545**, 547, **548**, 551, **561**, 566, 570
    dinosaurs 523
    geology 268
    historical **240, 242, 249, 251, 252**
    turtles **342**, 346
Omolon Massif 199
Ondai Sair locality 224, 256, 262, 309 (*see also* Andai Khudag locality)
Ondaisair Formation/Svita 259, 282, **283**, 450

Index

Öndörshil Sum 260
Öndör Ukhaa 257, **258**
Öndörukhaa Formation/Svita 257–9, **258**, 262, 460
Ongon locality **217**, 224
Ongon Ulaan Uul 256, **280**
Önjüül 312, **317**
Öösh Basin **217**, **226**, 256
Öösh Formation/Svita 450, 460
Öösh locality 236, 256
Ööshiin Nuur locality 224, **224**, 236, 238–9, **263**, **265**, **280**, 451, 460
Ophthalmosaurus 197, 199, 200
Opisthocoelellus 300
Opisthocoelicaudia 233, 456–8, **457**
Opisthocoelicaudiinae 457–8
Ordos basin 280–2, **281**, **283**, 287, **319**, **327**, 444, 548
Ordos block **280**
Ordosemys 325–6, **327**
Orenburg **3**, **7**, 195–6
Orenburg Province
    anthracosaurs 63–5, **65**, **66**, 68
    archosaurs 142, 144–5, 147–50, 152–3, 155, 164, 167, 169, 172
    historical 1, 8–10, 12
    marine reptiles 189–90, 203
    Permian biostratigraphy 22–4, 25
    synapsids 90, 94, 97, 100–1, 103–10, 112, 114–15
    temnospondyls 42, 44
    Triassic biostratigraphy 121, **121**, **126**, **127**, **131**, 132
Orenburgia **122**, **123**, 130–1, 161, 165, 168–70, **169**, **171**, 173–5
Orkhon river 260
Orlov, Yu.A. 6, 226, 236–7, 240–1, 627
Ornatotholus 481, 497, **498**, 499, **510**
Ornithischia 480, 487, 492, 494–5, 499, 502, 505, 517–18
Ornithocheiridae 424–6, **426**, 430
Ornithocheiroidea **429**, 430
Ornithocheirus 424–5
Ornithodesmus 430
Ornithodira 181–3
Ornithomimidae 270, 434, **435**, 443, 445–6, 450, 510
Ornithomimosauria 229, 233, 268, 434, 444–6
Ornithomimus 446
Ornithopoda 462–80, **463**, 494–5, 497, 499–503, 506, 508–9, 571
Ornithostoma 427
Ornithosuchidae 140
Ornithothoraces 554

Ornithurae 537, 544–52, 555
O'Rourke, J.E. xviii, 282
Ors'yu **121**
Orsk 203
Orthomerus 473, 476
Orthopus 97
Osborn, H.F. 211–12, **218**, 230, 436, 456, 460, 494
Osh (Mongolia; *see* Öösh)
Osh Province, Kazakhstan 177
Oshi Nuur (*see* Ööshiin Nuur)
Osmólska, H. 232, 439–40, 449, 467, 471, 473, 483, 489, 491, 493–5, 497, 501–2, 505, 575
Osteichthyes 618, 628
Osteopygis 310, 325
Ostracoda 128, 258–60, 264, 269–73, 282, 549
Otlestes 616, 630, 640, **641**, 642
Otlestidae 605, 615, 642
Otog Qi 548
Otogornis 548, 556
Otschevia 197, 202–3
Otsheria 20, **21**, 24, 27–8, **28**, **29**, 30, 34, 103, **104**
Otsheriidae 103
Ottawa, National Museum of Canada 480
Otter Sandstone Formation 170
Oudenodon 105
Ouranosaurus 462, 465
Outer Mongolia (*see* Mongolia)
Ovaloolithidae 562, 564–5, **565**
Ovaloolithus 562–4, **566**, 567–8, 570–1
Övdög Khudag locality 260, 272
Oviraptor 223, 233, 269, 283, **435**, 437–8, **437**, **442**, 451–2, 568
Oviraptoridae 233, 244, 247, **435**, 437–8, 445, 452, 486, 568
    eggs 568–9
Oviraptorinae 437
Oviraptorosauria 434, 437–9
Övörkhangai Aimag **243**, 262, 304, 519, **561**
Owen, R. 2
Owenetta 82, 160
Owenettidae 160–1
Oxemys 205
Oxia 382
Oxlestes 630–1, 640, **641**
Oxynaia 264, 265
Ozink Mine 191

Pachycephalosauria 233, 240, 480–4, 493–503, **498**, 506–10, **508**, **510**, 514–15

Pachycephalosauridae 482–4, 494–6, **498**, 500–1, 506, **508**, 509, 514–15
Pachycephalosaurinae 482, **498**, 500–1, 506, 515
Pachycephalosaurus 481–2, 485, 495–6, **498**, 499–501, 506–7, **510**, 515
Pachystropheus 397–8
Pad' Semen 554
Pakistan 128
Palaeobiogeography, Mongolia, Cretaceous
  ankylosaurs 530–1
  crocodilians 416–17
  lizards 385–6
  marginocephalian dinosaurs 509–10
  turtles 359–62
Palaeobiological Institute (*see* Warsaw)
Palaeognathae 537, 547–51, 556, 568
Palaeontological Institute (*see* Moscow)
Palaeontological Institute, Uppsala 481
Palaeoryctidae 605, 609–10
Palaeoryctinae 605
Palaeotrionyx 346
Palaeozoological Institute, St. Petersburg 5
Paleogeographic Atlas Project, University of Chicago **510**
Paleosaniwa 381
Paleosols 282, 287, 289
Palynology 287
Panchen, A.L. 38
Pangaea 27, 115
Panik creek 51
Panteleev, A.V. 536
Pantotheria 248, 577, 598
Paoetodon 165
Pappotheriidae 640
Parabenthosuchus 125
Parabradysaurus 83, 102
Paracimexomys 590, 616
Parahesperornis 544–5
Paraisurus 637
Paralligator 411
Paralligatoridae 408, 410–11, 417
Paramacellodidae 258, 378–9, 384–6
Parameiva 374
Paranyctoides 634, **646**, 648
Paraophthalmosaurus 197–8, 200
Paraplastomenus 322, 346
Parareptilia 17, 22–3, 30, 62–3, 160–1
Parasaniwidae 381–2
Parasaurus **72**, 83

Paraspheroolithus 562
Parasuchia 140
Parathalassemys 326
Paravaranidae 382–6
Paravaranus 382–3, 385
Parazhelestes 635, 642, **645**, **646**, 649
Pareiasauria 3, **4**, 5, 6–7, 12, 24–7, 71–85, **73**, 102, 128
Pareiasaurian complex (zone IV) 6–7, 8, 11, 18
Pareiasaurian–gorgonopsian fauna 18, 24–6
Pareiasaurida 63
Pareiasauridae 71–85
Pareiasaurus 71, **72**, 76–7, 79–81
Pareiasuchus 72–6, **72**
Pareiosaurus 77, 80
Paris, Muséum National d'Histoire Naturelle 480
Parotosuchus 10–11, 40, 46–8, **47**, 120, **122**, **123**, **124**, 131–2
Parotosuchus fauna 40, 122–4, **122**, 130–2, 134
Parotosuchus–Trematosaurus amphibian assemblage 132
Parrington, F.R. 143
Parrish, J.M. 148, 150–1, 156
Parshino **121**
Partridge, D. **127**
Parvicursor 555
Parvicursoridae 554–5
Parviderma 381
Parvoolithus 563–4, 566
Pascual, R. 585
Patagopterygiformes 552
Patagopteryx 552
Patranomodon 27–8, **28**, **29**, 34, 103
Paul, G.S. 440
Paulchoffatiidae 580, 585–6
Paurodontidae 598
Pechora River basin
  anthracosaurs 60, 68
  archosaurs 155
  historical 8, 12
  marine reptiles 195, 199, 203
  Permian biostratigraphy 19
  temnospondyls 41, 48, 55
  Triassic biostratigraphy 121, 130, 132–3
Pechora Syncline 121, **121**, 129, 135
Pectinodon 444
Peipehsuchus 414
Peishanemydidae 328
Peishanemys 310, 328, 329, 331, **331**
Pelecaniformes 551–2, 556
Pelecanimimus 445

Pelobatidae 300–1, 303–4
Peloneustes 188, 190, 193, 195
Peltobatrachidae 131
Peltochelys 339
Peltostegidae 55
Pelycosauria 2, 17, 86
Peneteius 376
Peng, J.-H. 313, 317, 322, 326
Penza Region 187, 189, 190, 194–6, **194**, 203, 205, 415, 427
Penza Regional Local History Museum 188
People's Republic of Mongolia (*see* Mongolia)
Peramuridae 598
Perciformes 596
Perle, A. 233, **240**, 241, 422, 439–41, 447, 482, 554
Perm' Province 1, 9, 20, 89, 91, 95–7, 101, 103
Perm' University 242
Permian
    amphibians 35–70
    palaeobiogeography 26–30
    reptiles 71–119
    stratigraphy 17–26
    temnospondyl faunas 39
    vertebrates, history of study 1–16
Permo-Triassic vertebrates
    history of study 1–16
Permocynodon 113–14, **113**, **114**
Permogor'e **121**
Perovka 9, 47, 105
Pervushoviasaurus 201
Peski 649
Petropavlovka **7**, **121**, **131**, 169
Petropavlovskaya Svita 46–7, 55, 112, 130, **133**, 148, 167
Phaanthosaurus **122**, **123**, 125, 161–3, **162**, **163**, 165, 175
Phaedrolosaurus 450
Phobetor 426
Pholiderpeton 38
Pholidosauridae 409, 414
Phreatophasma 20
Phreatosaurus 20, 102
Phreatosuchidae 20
Phreatosuchus 20, 102
Phrynosoma 369
Phrynosomatidae 369, 383–6
Phrynosomimus 370–1
Phthinosauridae 89
Phthinosaurus 6, 22, 89
Phthinosuchia 87–9
Phthinosuchidae 88–9

Phthinosuchus 89, **90**
Phylogenetic analysis (*see* Cladistic analysis)
Phylogenetic nomenclature 481–2
Picea 272
Picopsidae 599
PIN (*see* Moscow, Palaeontological Institute)
Pinacosaurus 223, 232, 283, 292, 517–18, 523, **526**
Pinega River 23
Pingfengshan **313**
Pinheirodontidae 585
Pinus 272
Piramicephalosaurus 374
Pistosauridae 205
Pistosaurus 195
Pizhmo-Mezen' River 167, 173
Pizhmomezen'skaya Svita 129, 173
Placental mammals (*see* Eutheria)
Placeriini 108
Placodontia 2
Plagiaulacidae 262, 585–6
Plagiaulacoidea 585–8, 590, 615
Plagiosauria 10, 11
Plagiosauridae 40, 54–5, 131–4, 298
Plagiosauroidea 10–11, 35–6, 40, 45, 54–5, 135
Plagioscutum 54, **55**, **122**, **123**, 133, 135
Plagiosternum 54, **55**, **122**, **123**, 133–5
Plagiosuchus 134
Planiplastron 358
Plants 135, 178, 239, 243, 246, 249, 250, 287
Platanavis 552
Platecarpus 203
Platychelys 360
Platynota 380
Platyoposauridae 21
Platyoposaurus 42, **43**, 44
Platyops 2
Platypeltis 346
Platypterygius 197, 200–2, **202**
Platysternidae 330–1, 350, 352, 358, 361–2
Platysternini 352
Platysternon 350, 352, 357
Ples 52, **121**, 195
Plesiochelyidae 310, 322, 350, 359–61
Plesiochelys 313, 320, 321, 333, 337, 350, 359
Plesiosauria 187–96, 205
Plesiosaurus 188–9, 191, 197, 203
Pleurocentra 61
Pleurodira 359–61

Pleurodontagama 371–2
Pleurodontagaminae 372
Pleurokinetism 36
Pleuromeia 128–9
Pleurosternidae 315, 318, 349, 359–60, 362
Pleurosternon 359
Plioplatecarpinae 203–5
Plioplatecarpus 203
Pliosauridae 189–96, 205
Pliosaurus 188–93, **192**, **193**, 195
Plutoniosaurus 197, 201
Podolia 195
Podopterygidae 178–82
Podopteryx 12, 177–82
Podsaraitsa **121**
Poekilopleuron 204, 413
Poland 232, 311
Poldarsa 66
Polish–Mongolian Palaeontological Expedition (*see* Joint
    Polish–Mongolian Palaeontological Expedition)
Polrussia 369, **370**
Polubotko, I.V. 199
Polunino village 429
Polycotylidae 189–90, 194–6, 205
Polycotylus 188, 190, 196
Polycryptodira 322
Polyglyphanodontinae 376
Polyptychodon 188–9, 195
Ponomarenko, N.G. **240**, 241, 422
Poposauridae 140
Porosteognathus 109–10
Portlandemys 359
Potamolestes 599
Povolzh'e 187, 191, 203
Praeornis 553, **553**
Praeornithes 553
Praeornithidae 553
Praeornithiformes 553
Pravoslavlev, P.A. 5, 94, 188, 195
Pravoslavlevia 26, 94
Pre-Amur 354
Pre-Urals (*see* Cis-Urals)
Pre-Aral region 474
Prenocephale 233, 481–4, **484**, 495–6, **498**, 499, 500, 507,
    510, 515
Preondactylus **429**
Presbyornithidae 551, 556
Prestosuchidae 151

Prestosuchus 153
Priozernyi quarry 546
Priscagama 369–72, **371**
Priscagamidae 369–72, 383–6
Priscagaminae 369, 371–2
Prismatoolithidae **565**, 567–8
Prismatoolithus 567
Pristerodon 28, **29**, 34
Pristerodontioidea 105–8
Pristerognathidae 109
Pristerognathus 28, **29**, 34
Pristerosauria 109–10
Priural'sk district 196
Probactrosauria 321
Probactrosaurus 462
Proburnetia **21**, 25, 90–1, **90**
Procellariiformes 556
Procerobatrachus 303
Procheneosaurus 462, 471–2, **472**, 475–6
Procolophon 161–2, 165
Procolophonia 11–12, 26
Procolophonidae 7, 8, **9**, 11–12, 18, 26, 82–3, 121, 123, **123**,
    125, 128–32, 134, 160–76
Procolophoninae 128, 160–2, 165–74
Procolophonoidea 71, 160–76
Procolophonomorpha 71, **71**, 82
Procynosuchidae 25, 114–15
Procynosuchus 114
Prodeinodon **435**, 450–1
Prodenteia 374
Proelginia 76, **77**, 81
Proganochelys 359–60
Prognathodon 203–4
Prohadros 465
Prokennalestes 600, 604–7, **606**, 616, 642, 649
Prolacertiformes **9**, 125–6, 130–1, 135, 142, 177–84
Proletarskii village 425, 429
Pron'kino 8, 64, **65**, 110
Proplatynotia 381, **381**
Prosauropoda 140
Prosirenidae 307
Proteida 306
Proterochampsidae 140, 156, **157**
Proterogyrinidae 61
Proterosuchia 9, 125, 129–30, 134, 140, 144, 147, 151
Proterosuchidae 128, 140–5, 147, 153, 155–6
Proterosuchus 143, **157**
Proteutheria 605, 614

Prothero, D. 596, 598
Prothoosuchus 50, **122**, **123**, 130
Protiguanodon 259
Protoceratops 218–25, **221**, **222**, **223**, 228, 231, 233, 236, **241**,
    244, **249**, 269, 283, **288**, 292, 333, 436, 480–1, 488–9,
    490, 491–4, 498, **498**, 501–5, **508**, 509, **510**, 568
Protoceratopsidae 247, 482, 493–4, **498**, 501–2, 505, 507,
    516, 567–71, 640
  eggs 567–8, 570–1
Protoceratopsidovum 561, 563, **566**, 567–8, 571
Protorosauria 181
Protorothyrididae 71, 161
Protorothyris 160
Protosirenidae 306
Protostegidae 205
Protosuchia 402–7, 417
Protosuchidae 402
Prototeius 375
Prototribosphenida 598
Protungulatum 642
Pseudanosteira 344
Pseudochrysemys 362
Pseudograpta 260
Pseudohyria 268
Pseudosuchia 177, 181, 183
Psittacosauridae 247, 263, 321, 485–8, 492, 494, 501, 507
Psittacosaurus 236, 243–4, 249, 258–9, 262–3, 480–1, 485–7,
    487, 492, **498**, 501–4, 506, **508**, **510**
Pteranodontidae 430
Pterodacyloidea 424–8
Pterosauria 140, 177, 181, 245, 248, 258, 321, 390, 420–33,
    421, **429**, 544, 628, 648
Ptilodontoidea 585–6, 590, 616
Ptychocorax 637
Ptycholepidae 629
Pucadelphys 601
Pugachev district 192, **193**, 200
Pugachev Regional Museum 188
Pumpelly, R. 212
Purly locality 26, 53, 143
Purly–Vyazniki faunal assemblage 26
Pygopodidae 380

Qianshanosaurus 373
Qigu Formation **313**
Quaesitosaurus 456–7, 459–60, **459**
Quetzalcoatlus 427, 430
Quinling **280**

Rabidosaurus 103, 106, **108**, **122**, **124**
Rachkovskii, I.P. 236
Radiometric dating 259, 265, 268–9, 274
Radkevich, N. **240**, **242**, **246**
Ralli 551
Raphanodon 63
Rassypnaya locality 111, 142, 147, 155
Rassypnoe 9
Ratitae 542
Rauisuchia 129, 150–1, 153, 181
Rauisuchidae 11–12, 130–2, 134–5, 140, 145, 150–6
Rautian, A.S. 555
Rays 638
Red Deer River 217, 230
Red Walls **286**
Reig, O.A. 142, 145, 147
Reptilia
  Cretaceous 246, 273, 292, 307–530, 554
  Jurassic 187–210
  marine 187–210
  Permo-Triassic 1–34, 71–186
Reptiliomorpha 37
Reshetov, V.Yu. 241–2, **242**, 245, 420, 574
Reshma locality **121**
Rhadiodromus 9, 103, 105, **107**, **122**, **124**
Rhamphocephalus 430
Rhamphorhynchidae 424, **429**, 430
Rhamphorhynchinae 430
Rhamphorhynchus 430
Rhinesuchus 39
Rhinocerocephalus 105
Rhinoceros 518
Rhinocypris 269
Rhinodicynodon 103, 107, **108**, **122**, **124**, 134
Rhinosauriscus 190–1
Rhinosaurus 188, 190–1
Rhipaeosauridae 63, 83–4
Rhipaeosaurus 6, 22, 63, 83
Rhipidistia 36
Rhizoliths 289
Rhopalodon 2, 97, 102
Rhopalodontidae 89, 102
Rhynchocephalia 390
Rhytidosteidae 11, 36, 40, 55, 131–2
Rhytidosteoidea 39, 55
Rhytidosteus 11, 40, 55, **122**, **123**, 131
Riabininus 19
Richthofen, F.F. von 212, 235

Rjabininus 12
Roček, Z. 300, 301, 304, 648
Rodents 614
Rogovich, A.S. 627
Romer, A.S. 87–8, 494
Roniewicz, E. 440
Rossypnaya 46
Rossypnoye **121**
Rougier, G.W. 576, 586, 588, 590, 596, 601–2
Royal Tyrrell Museum of Palaeontology, Drumheller 443,
    480
Rozhdestvenskii, A.K. 191, 226–32, **227**, **228**, **231**, 237,
    239, 273, 402, 440, 448, 456, 460, 462, 464, 467–8,
    470–1, 473
Rozhdestvenskii mine 42
Rubidge, B.S. **28**, 103, 105
Rubidgea 94
Rubidgeinae 93
Russell, D.A. 204, 385, 439, 441, 460, 462
Russia
  amphibians, Permo-Triassic 17–59
  anthracosaurs 60–70
  archosaurs, Permo-Triassic 140–59
  birds 530, 545, 551–2, 554–6
  choristoderes 390, 392, 396–7, 399
  crocodilians 402, 404, 412–16
  dinosaurs 448, 456, 463, 472–3, 476, 517
  mammals 627, 639, 649
  marine reptiles 187–208
  pareiasaurs 71–86
  Permian biostratigraphy 17–34
  Permo-Triassic, historical 1–16
  procolophonoids 160–76
  pterosaurs 420–1, 424–5, 427–9
  reptiles, Permo-Triassic 17–34, 71–187
  synapsids, Permo-Triassic 86–119
  temnospondyls 35–59
  therapsids, Permo-Triassic 86–119
  Triassic biostratigraphy 120–39
  turtles 310, 321, 324–5, 360
Russian
  alphabet xvii
  geologists xxv–xxxix
  journal titles xxix–xxxiv
  palaeontologists xxxv–xxxix
  stratigraphic system xvii–xix
  stratigraphic units xx–xxi
  transliteration xvi–xvii

Russian Academy of Sciences (*see* Academy of Sciences)
Russian Arctic 174
Russian Far East 129
Russian platform 140, 187, 413, 417
Russian–Mongolian Palaeontological Expedition (*see* Joint
    Russian–Mongolian Palaeontological Expedition)
Russkaya River 199
Russkii Island 49, 199
Ryab' **121**
Ryabinin, A.I. 5, 195, 310, 422, 450, 456, 460, 472–3, 517
Ryabinskian Member 105, 127
Ryazan' region 195
Rybinsk 12
Rybinskaya Svita 49, 128
Rybinskian Gorizont 49–51, 68, **68**, 120, 122–4, **122**,
    128–30, **141**, 142, 155, 167–8, 175
Rychkov, P.I. 1

Sabath, K. 575
Saevesoederberghia 303
Sagibovo 472
Saichania 233, 518, 527, 530–1
Sailestes 633, 642
Sainsar Bulag locality 246
Sainshand locality
  eggs **561**
  geology 256, 263, **265**, **280**, 282
  historical 237, 239, **238**, 239, 243, **244**, 249, **250**
  mammals 575
Sainshand Formation/Svita 256, 263, 268, 282, **283**, 287
Sainshandia 268
St. Petersburg 311, 399
St. Petersburg, F.N. Chernyshev Central Museum for
    Geological Exploration (TsNIGRI) 35, 161, 188,
    309–10, 402, 422, 533, 629
St. Petersburg Naturalists' Society 4
St. Petersburg School of Mines/Mining Institute 2, 86
St. Petersburg University 297, 422, 627
St. Petersburg, Zoological Institute 188, 297, 309–10, 423,
    533, 629
Saitsev, A. **246**
Sakha Republic 517
Sakhalak (*see* Sakhlag)
Sakhalin Island 189, 195, 203, 472
Sakhlag Uul **263**
Sakmara River **7**, 63, 169
Salamanders (*see* Caudata)
Salarevskaya Svita 64

Samara Province **68**, 187, 195, 203
Samara River basin 68, 128, 172
Samaria **122**, **123**, 130, 161, 168, 170, 172–3, **172**, 175
Sanchuansaurus **72**, 83
Sandown, Museum of the Isle of Wight (Geology) 480
Sangiin Dalai Nuur depression 258, **265**, 422, 426
Saniwa 380
Saniwides 380
Santagulova Mines **18**, **19**, 20
Santana Formation 425
Sanz, J.-L. 408
Sarafanikha site 164
Saratov 190, **193**, 196, 199, 201, 203, **421**, **426**, 427
Saratov Province 187, 191–2, 196, 200–1, 203, 425, 429
Saratov State University (SGU) 10–12, **10**, 35, 86, 140–1, 161, 188, 242
Saratov–Bristol Expedition to the South Urals 7, **8**, **10**, **126**, **127**, **131**, **133**
Saray-Gir 22
Sarmatosuchus **122**, **124**, 134, **141**, 144, **145**, 148, 156
Sarykamyshsai locality 299, **313**
Sauria (lizards) 233, 239, 242–3, 248, 250, 269, 270, 368–89, 626, 646, 647
Saurichthys 178
Saurischia 456
Sauriurae 533–44, 555
Sauroctonus 26, 93, **93**
Saurolophus 229–30, 238–9, 244, 462, 466, **467**, 468, **468**, **470**, 473, 475–6, 568
Sauropelta 494
Sauropoda 233, 237, 239, 243, 249, 262–3, 434, 439, 456–61, 565, 568
   eggs 568
Sauropodomorpha 457–60
Sauropterygia 124, **124**, 187–96
Saurornithoides 223, **435**, **441**, 444–5, 451, 468
Saurosuchus 153
Savel'evsk oil shale mines 192, **193**, 200
Sayn Shand (*see* Sainshand)
Sazavis 533, 535, 536
Scalenodon 115, **122**, **124**, 134
Scalopognathidae 112
Scalopognathus 112, **122**, **124**, 129
Scaloposauria 110
Scandentia 642
Scanisaurus 188, 190
Scapherpetontidae 305

Scaphognathinae 424, 430
Scaphognathus 424, 430
Sceloporini 369
Sceloporus 369
Schaff, C. 578
Scharschengia **122**, **124**, 129–30
Scheikhdzheili (*see* Sheikhdzheili)
Schlaikjer, E.M. 494, 505
Schukino 195–6
Scincidae 380
Scincomorpha 373–80
Scleromochlus 181
Sclerosaurus 82
Scotiophryne 300, 304
Scutemys 328
Scutosaurus **4**, 5, 24–5, 71–2, **72**, 74–83, **77**, **78**, **79**
Scylacops 26, 93
Scylacosauridae 26, 109
Scylacosaurus 110
Scylacosuchus 29, 109, **109**
Scythosuchus 156
Scytophyllum flora 134–5, 177
Sea cow 627
Sechuan (*see* Sichuan)
Seeley, H.G. 2, 100
Segnosauridae (*see* Therizinosauridae)
Segnosaurus **435**, 439–40, **441**, **442**, 451
Semigorye **121**
Semin Ravine 26, 93
Sennikov, A.G. 12, **19**, 134, 140, 142–5, 147–50, 152–5
Serdoba River 187
Sereno, P.C. 156, 495–7, 501, 582
Serov region 203
Serpentes 204–5
Sevastopol' 205, 415, **416**
Severnii Island 172
Severodvinskian Gorizont **19**, 24–6, 65–6, **66**, 74, 93, 109, 164, 175
Seversk Sandstone 201, **202**
Sevkhul Khudag **246**, **247**
Sevrei ridge 268, **574**
Seymouriamorpha 23, 25, 35, 37–8, 60–2, 64, 83–4
Seymourida 62
Seymouriidae 190–1
Shabarakh Us (= Bayan Zag) 216–24, **217**, **220**, **221**, **223**, **224**, 227–9, 231, 266, 573
Shachemydinae 320, 332, 337

Shachemys 332–3, 337–9, **338**

Shagayn Teeg **280**

Shakh-Shakh locality 307, 333, **403**, 404, 409–10, **410**, 429, **463**, 468, 476, 639

Shakhun'ya district 110

Shalneva, Z. **246**

Shamosaurinae 517

Shamosaurus 517, 519, **520**, **521**, **522**, 530–1, **530**

Shamosuchus 268, 404–5, 410–12, **412**, **413**, **413**, **414**, 417

Shandong 310, 316, **318**, 322, 329

Shanh Sum 260

Shanshanosaurus 450–1

Shansi Province 489

Shansiodontidae 134

Shansiodontini 107

Shansisaurus **72**

Shansisuchus 148–50, 156

Shar Teeg locality
    amphibians 298
    crocodilians 403, **403**, 405–6, **406**, 409, 416–17
    historical 245, 250
    mammals 575, 577
    turtles 309, 312, **314**, **316**

Shar Tsav locality 250, 459

Shariliin Formation/Svita 256

Sharks (*see* Chondrichthyes)

Sharov, A.G. 12, 177, 180–1, 183–4, 422, 424

Sharovipterygidae 178–82

Sharovipteryx 12, **13**, 177–82, **178**, **179**, 184

Sharovisaurus 378

Shartegemys 315, **316**

Shartegosuchidae 403, 407

Shartegosuchus 402–3, 405–6, **406**, 407, 417

Sharzhenga River 167

Shastasauridae 199, 205

Shastasaurus 190, 196–7, 199

Sheikhdzheili ridge **403**, 404, 411, 428, **628**, 630, 631, 637, 640, 649

Sheshminskian Gorizont 41

Shestakovo locality 579, 649

Shihtienfenia **72**

Shikhovo-Chirki localities 22, 43

Shilikha locality **121**

Shilikhinskaya Svita 128

Shilin 533

Shiljust (*see* Shilüüt)

Shilt Uul locality 526

Shilüüt Uul locality **561**, 562, 566

Shine Khudag locality 257, 259–60, **260**, **265**, **337**, 451, 553

Shinekhudag Gorizont 257, 259–60, 271, 321, 446

Shineusemys 333, 336–7, **337**

Shinisauridae 382

Shinisauroides 377, 378

Shin Khuduk (*see* Shine Khudag)

Shireegiin Gashuun basin 271, **280**, 281, 283, 292, 562

Shireegiin Gashuun locality
    crocodilians **403**, 405, 411, **412**
    dinosaurs 489, 521
    eggs **561**, 562
    geology 268, **280**
    historical 237
    mammals **574**

Shirilin (*see* Shariliin)

Shishkin, M.A. 10–11, 81, 128–9, 141, 147, 241

Shteinberg, S.M. **353**

Shulaevka locality 46

Shuvalov, V.F. 240, 256, **258**, **260**, 273, 422, 426, 460

Shyrendyb, Prof. 232

Siamotyrannus 434

Siberia 39, 195–6, 258, 280, 399, 462, 485, 510, 533, 553–5, 579, 615, 649

Siberian Block **280**

Sichuan Province 299, 310, 312–13, 316, 320, 337, 354

Sigogneau-Russell, D. 86–90, 92–3, 397

Silphedestidae 112

Silphedosuchus **9**, 111–12, **111**, **122**, **124**, 131

Simbirskiasaurus 197, 201

Simbirskites Clay 201

Simbirtsit Industrial Works 188

Simferopol' region 416

Simmons, N.B. 586

Simoedosauridae 395

Simoedosaurus 391, 393, 395

Simolestes 188

Simoliophidae 205

Simpson, G.G. 573, 596

Simpsonodon 577

Simukov, A.D. 236

Sinaspideretes 320, 344

Sinegorsk **463**, 472

Sinemydidae 262, 310, 312, 316–17, 319, 321–3, 331, 359–60, 362

Sinemys 310, 316–18, **318**, **319**, **319**, 359

Sinobrachyops 299
Sino-Canadian Dinosaur Project 310, 561
Sinochelyidae 321, 327–31, 361
Sinochelys 310, 327, 328, 329
Sinoconodon 577
Sinokannemeyeriini 105
Sinornithoides **435**, 444–5, 451
Sinraptor 434, 451
Sinzyan 313
Skalistoe village 205, 415
Skin, fossilized 238–9
Slaughteria 605
Slavoia 379–80, **379**
Slavoiidae 379, 383–6
Sloanbaatar **581**, 590, 593, 595
Sloanbaataridae 589–90
Sludkian Gorizont 45, 48–52, 120, 122–4, **122**, 129–30, **141**, 144, 173–5
Sludkinskaya Svita 129
Smolensk region 203
Snakes (*see* Serpentes)
Sobolevskii, Captain 1
Sochava, A.V. 240, 268, 560, 596
Socognathus 375
Soilensk mine 201, 205
Sokolki 4, 12, **18**, 19, 25–6, 53, 94, 110, 113
Sokolki complex/fauna 25–6, 103
Sol'-Iletsk district 10, 100, 105, 107, 144–5, 148–50, 152–3
Sol'-Iletsk phenomenon 10
Solnhofen 430
Solnhofia 359
Sordes 420, 422–4, **425**, **429**, 430
Soricomorpha 609
Sorlestes 617, 629, 631, 635, 637, 642, **643**, **644**, **645**, **646**
Soroavisaurus 540
Sorochinsk district 110
Soroka River 172
South Africa 17–19, 25–7, 82, 89, 91–3, 97, 102–3, 105, 110, 114, 131, 143, 147, 150, 161–2, 165
South African Museum, Cape Town 80
South China Block **280**
South Gobi (*see* Ömnögov' Aimag)
South Korea Block **280**
Southern Monadnocks **285**
Soviet–Mongolian Geological Expedition 298
Soviet–Mongolian Palaeontological Expedition (*see* Joint Russian–Mongolian Palaeontological Expedition)

Spasskoe I locality 144
Spasskoe locality 53, 68, **68**, **121**, 162
Spasskoe-Semenovskoe **121**
Speetoniceras 187, 201
Spencer, P.S. 71, 160–1, 167
Sphenacodontidae 17
Sphenosiagon 375
Spheroolithidae 562, 565, **565**
Spheroolithus 562–4, **566**, 567
Sphodrosaurus 165
Špinar, Z. 300
Spondylolestes 161, 162
Spondylolestinae 125, 128, 161–2, 165, 167
Spondylosaurus 188, 190
Sporomorphs 128, 135
Squalicorax 637
Squamata 203, 368–89
Srednyaya Makarikha 112
Stagonolepididae 152
Stahleckeria 108
Stahleckeriini 107
Stahlekeriidae 134
Stanford University–Mongolian expedition 434
Stanislavsk 201
Staritskaya Svita 128
Stegoceras 480–4, **485**, 493, 495–7, **498**, 499, 500, **510**
Stegosauria 229, 494
Steneosaurus 204, 414, 416
Stenopelix 481, 493–4, 496–9, **498**, 506, **508**, 509, **510**
Stenopterygiidae 202–3
Sterlitamak district 43, 97
Stichopterus 258
Storrs, W.G. **126**, **131**, 195
Straelen, V. van 561
Stratigraphy
    Cretaceous of Mongolia 279–96
    international system xviii–xix
    Mongolian units xxiii–xxviii
    Permian of Russia 17–26
    Russian system xvii–xix, 279
    Russian units xx–xxi
    Triassic of Russia 120–39
Stretosaurus 191
Stromatolites 263
Strongylokrotaphus 188, 190–3, **193**
Stuckenberg, A.A. 3
Stygimoloch 481, **498**, 499–501, 506, **510**, 515
Stygimys 590

Index

Subashi Formation 333, 450
Subtilioolithidae 570
Subtilioolithus 563, 564, 567, 568, 570
Suchia 156, **157**
Suchonica 66
Sudamerica 585
Sudamericidae 585
Sues, H.-D. 142, 147, 148, 161, 167, 494–7
Suhbaatar **238**, **244**, **250**, **265**
Sukhani Gobi **280**
Sukhanov, V. 241, 311, 340
Sukhona River 93
Sukhonskaya Svita 44, 93
Sulestes 632, 634, 640, **644**, **645**
Sulestinae 602, 640
Sulimski, A. 232–3, 374, 376, 380
Sultan Uvais range **344**, 428, 551, 637
Suminia 21, 25, 28, **28**, **29**, 30, 34, 103, **104**
Sun A. 112
Sundokovo village 187
Sunosuchus 405, 409, 417
Surkov, M.V. **8**, **133**
Surovikinskii District 545
Sushkin, P.P. 5
Suteethorn, V. 492
Svita concept xvii, xviii, 279
Swedish–Chinese Expeditions 309–10
Sychevskaya, E.K. 242, 245
Symmetrodonta **243**, 248, 262, 298, 574–5, 577–8, 596–8,
   604, 615–16, 649
Synapsida 1–3, 6, 9–12, 17, 20, 23, 24–30, 71, 86–119, 174,
   639
Syndyodosuchus 19, 41
Synesuchus 68, **69**, **122**, **123**, 135
Syninskaya Svita 135
Synya locality **121**
Syodon 1, 23, **96**, 97, 100
Syodontidae 97
Syrmosaurus 229, 269, 523
Sytin, M. **246**
Syuk-Syuk wells locality **463**, 471, 476
Syuksyukskaya Svita 436
Syzran' 199, 200
Szalay, F.S. 601, 603, 605
Szaniawski, H. 233

Taas-Krest River 195
Tadzhik depression 196

Tadzhikistan
   amphibians, Jurassic–Cretaceous 297, 304, 306
   anthracosaurs 63
   crocodilians 402, 404, 409–10, **410**, 414–15, **415**
   dinosaurs 445, 447, 456
   historical 12
   mammals 616–17, **628**, 629, 637–8, **647**, 648
   pterosaurs 429
   turtles 310, 327, 333
Tadzhikosuchus 404, 414–15, **415**, 417
Taeniolabidae 585, 590, 596
Taeniolabidoidea 575, 585, 590, 616
Taikarshin Beds **338**, **355**, 427
Taizhuzgen River 560
Talarurus 268, 518–20, 523, **524**, 530–1
Taldysai 554
Talkhabskaya Svita 196
Tambov district 203, 429
Tamtsag depression 268, 271
Tan Lu fault **280**
Taphonomy 7, 238, 273
Tapinocaninus 23, 102
Tapinocephalia 89, 91, 95, 102
Tapinocephalidae 23–4, 27, 95, 101, **101**, **102**
Tapinocephalinae 95
Tapinocephalus 23, 27
Tapinocephalus Zone 24, 82
Tarasovo **121**
Tarbosaurus 233, 238, 244, 270, **435**, **441**, **442**, 447–50, **447**,
   449, 451, 468
Tarchia 518–19, 527, **528**, **529**, 530–1, **530**
Tariat Uul 259, **280**
Tarlo, L.B.H. (*see* Halstead, L.B.)
Tartarovo 196
Taryatu (*see* Tariat)
Tashkent 470, **628**, **628**
Tashkumyr 299, 305, 307, 456, 629
Taslestes 631, 637, 642, **643**, 648
Tatal locality **421**, 422, **422**, 426, **427**, **429**, 430
Tatal Gol locality 233, 236, 238–9
Tatal Yavar locality **265**
Tatarinov, L.P.
   archosaurs 140, 142–3, 145, 147–9
   diapsids 178–9, 181
   Mongolia 242, 300
   synapsids 11, 89, 93, 94, 108, 110–14
Tatarstan Republic (Tataria) 7, 23, 64, 93, 97, 99, 102, 104,
   110, 187

Tate, J., Jr. 546
Tavolzhanka River 68
Taxodium 237
Tbilisi 311
Tchanget **403**
Tchangtoomoor, A. **240, 246**
Tchingisaurus 375
Tchoiria 390–3, **393, 394**, 395, **395**, 397–8
Tebshiin (*see* Tevshiin)
Teel Ulaan Uul **265**, 268
Tegaimiao Formation 282, **283**
Tegotheriidae 577
Tegotheriidia 577
Tegotherium 577, **577**
Teiidae 373–6, 383–6
Teiinae 373
Teius 376
Teleosauridae 204–5, 414
Teleosaurus 204, 414
Teleostei 596
Telmasaurus 380
Tel Ulan Ula (*see* Teel Ulaan Uul)
Telyanino **121**
Temnodontosaurus 199
Temnospondyli 2, 5, **7**, 8, **9**, 10–12, **11**, 19–21, 23, 25, 26,
    35–61, 123, **123**, 125, 128, 131–4, 246, 297–300, 360
Tenkinskii region 554
Teryukhan **121**
Teryutekhskaya Svita 55, 129
Testudines
    Central Asia 309–67
    China 309–31, 333, 337, 339–43, 354
    Cretaceous 320–58, 360–2
    Jurassic 312–20, 359–60
    Kazakhstan 310, 320, 322, 332–3, 346, 354, 358
    Kirgizstan 205, 310, 312–13, **313**, 320, 324–6, **326**, 330,
        332, **332**
    marine 187, 205
    Mongolia 223, 233, 237–9, 247, 262–4, 266, 268–70, 273,
        309–67, 439, 541, 560, 618
    Palaeogene 361–2
    relationships 38, 71, **72**, 82
    Russia 310, 321, 324–5, 360
    Tadzhikistan 310, 327, 333
    Ukraine 628, 637–8
    Uzbekistan 205, 310, 319, 321, 326–7, 333, 337–8, **338**,
        344, **344**, 349, 354–5, **355**
Testudinidae 330, 339, 359

Testudinoidea 330, 335, 350, 352, 359, 361
Tethys seaway 399, 416–17
Tetrapoda 1, 37, 39, 128, 130, 132, 134–5
Tetyushi 7, 26
Tevsh Formation/Svita 256, 259, **283**
Tevshiin Gobi gorge 260
Thailand 300, 359, 485, 492
Thalassemydidae 320, 326, 360
Thalattosuchia 204, 402, 412–14, 416
Thaumatosaurus 188, 190, 195
Thecodontia 145, 155, 181
Thecodontosaurus 140
Thelegnathus 165
Therapsida 1–3, 6, 9–12, 17, 20–1, **21**, 23–30, 86–119, **90**,
    115, 125, 128, 131, 134
Theria 233, 573, 577, 580, 583–5, 596–614, 616, 630–3, 635,
    638–48
Theriodontia 2, 6, **9**, 10, 108, 121, **124**, 128–9, 131, 134
Therizinosauridae 244, 247, 268, **435**, 439–43, **440**, 447,
    565
Therizinosauroidea 434, 439–43
    eggs 565
Therizinosaurus 270, **435**, 439–41, **440**, 451
Therocephalia 24–6, 28–30, 88, 108–13, **109**, **111**
Theropoda 184, 229–30, 233, 236–8, 244, 258, 262, 263,
    434–55, **435, 437, 441, 442**, 537, 554–5, 568–9
    eggs 568, 570
Thespesius 473
Tholocephalidae 495
Thoosuchinae 128
Thoosuchus 50, **50, 122, 123**, 128
Thoracosaurinae 204–5, 415
Thoracosaurus 205, 404, 415–16, **416**
Thrinaxodontidae 114–15
Thung Song locality 300
Thyreophora 494–5, 506, 518
Tianchiasaurus 530
Tichvinskia **122, 123**, 28, 130–1, 161, 165, **166**, 167–72,
    167, 168, 169, 175
Tienfucheloides 316, 319
Tienfuchelys 316
Tikhvinskoe 12, 49
Timanophon **122, 123**, 130, 161, 167, **172**, 173, 175
Timan–North Urals 12
Tinodon 578
Tirolites 132
Titanophoneus **22**, 23, 28, **29**, 34, 97, **98**, 99, **99**
Titanosauridae 458, 460

Titanosauroidea 458
Titanosuchia 95
Titanosuchidae 95, 97, 100, 101
Titanosuchinae 95
Tlaanbaatar, Geological Institute 518
Tochisaurus **442**, 444, 451
Tögrög locality **280**, **288**, 436, 575, 590–1
Tögrög Bulag locality 268, 289
Tögrögiin Shiree locality
　birds 554, **561**, 568, 571
　geology **265**, 269
　historical 233, 240–1, **240**, **241**, 243, **249**
　lizards 369–70, 375
　mammals **574**, 575
Tokosauridae 63
Tokosaurus 63
Tombaatar 590, 592
Tomb of the Dragons **230**
Tomurtogoo, T. 240
Toogreek (*see* Tögrög)
Tormkhon Formation/Svita 256, **280**, 405, 409
Tost ridge **561**, **574**
Toxochelyidae 205, 322–3, 361
Trachodon 230, 472
Trachoolithus 563, 564, 569
Trans-Altai Gobi 236, 238, 250, 256, 260, 264, **270**, 271,
　280, 298, 312, **314**, 315, **316**, 329, **331**, 336, 343, 346,
　353, 355, **356**, 357–8, **358**, 403, 406, 538, 547, 575,
　577, 595
Trans-Baikal region (Transbaikalia) 258–9, 310, 390,
　396–7, 554, 649
Transliteration
　Mongolian xxii–xxiii
　Russian xvi–xxi
Trautschold, H. 195
Traversodontidae 112, 115
Traversodontoides 112
Trematosauridae 40, 52, 128–31
Trematosauroidea 35, 48–52, 125
Trematosaurus 52, **122**, **123**, 132
Trematotegmen 51, **122**, **123**, 128
Triassic
　amphibians 35–70
　China 143, 149–50, 162, 165
　faunas 120–39
　Kirgizstan 177–86
　palaeoenvironments 120–39
　reptiles 86–119, 140–86

Russia 120–39
　sedimentology 120–39
　stratigraphy 120–39
　temnospondyl faunas 39–40
　tetrapod faunas 120–39
　vertebrates, history of study 1–16
Triassuridae 307
Triassurus 178, 307
Tribosphenida 596, 598–614
Tribotheria 596, 599, 600
Tribotherium 600
Triceratops 230, 482, 485, 487–8, 492
Triconodon 581
Triconodonta **243**, 248, 262, 574–5, 577–9, 581, 598,
　615–16
Triconodontidae 577–8, 581, 598, 614
Trilophosauridae 123, **123**, 131, 174
Trimerorhachoidea 39
Trinacromeriidae 193
Trionychia 335
Trionychidae 310, 320, 322, 332–3, 335, 339, 345–6, 359,
　361
Trionychoidae 335, 339
Trionychoidea 335, 339, 352, 359
Trionyx 311, 322, 332, **332**, 346
Tritylodontidae 604
Trofimov, B.A. 239, 596, 601, 603
Troodon 443, 444, 451, 468, 495
Troodontidae 233, 434, **435**, 443–5, 451–2, 494
Tropidostoma 24, 103
Tryphosuchus 44
Tsagaan Gol locality 257, **265**
Tsaagangol Formation/Svita 257, 260–1
Tsagaan Khushuu locality
　birds 545, **545**, 547, **561**, 568
　dinosaurs 446
　geology **280**
　historical 238–9, 243, 248
　lizards 369, 373, 384
　turtles 309, 331
Tsagaan Teeg locality 521
Tsagaan Tsav well 257, 259, 405
Tsagaantsav Formation/Svita 256–9, 261–2, 271, 274, 321,
　406, 426
Tsagaantsav Gorizont 258–9, **258**, 261, **265**, 271, 321, **406**
Tsagaan Uul locality 238–9
Tsagantegia 518, 521, **523**, 530–1
Tsakhir Uul **280**

Tsakhiurt 243
Tsaotanemys 310, 349, 354
Tsaregradskii, V. **204**
Tsast Bogd Mountains 239
Tsel Bulag **280**
Tsastoo Bogdo (*see* Tsast Bogd)
Tsentralny Nauchno-Issledovatelskii Geologo-
　　Razvedochnyi Muzei, St. Petersburg (TsNIGRI) 35,
　　161, 188, 309–10, 402, 422, 533, 629
Tsetsen Uul 260
Tsetserleg **238**, **244**, **250**, **265**
Tsogtovoo Sum 259, **280**
Tsulutu Uul **280**
Tsyl'ma River 46, 51
Tsyl'ma tetrapod complex 12
Tsylmosuchus **122**, **124**, 129–31, 140, **141**, 142, 151, **152**,
　　154–5
Tugrig (*see* Tögrög)
Tugrigbaatar 590
Tugrikin (*see* Tögrögriin)
Tugulo Series **328**
Tulerpeton 60
Tulkeli locality 554
Tumanova, T.A. 517, 527
Tunguska River 45
Tungussogyrinus 45
Tupaiidae 642
Tupilakosauridae 36, 39, 53, 128
Tupilakosaurus 9, 11, 53, **53**, **122**, **123**, **124**, 125, 127–8
Tupilakosaurus assemblage 103
Tupilakosaurus–Luzocephalus assemblage 120, 122–8
Tupinambinae 373
Turanoceratops 481, 492, **498**, 502–3, 505–7, **508**, **510**
Turanosuchus 404, 409–10, **410**, 417
Turazhol village 100
Turfan depression 333
Turfanosuchus 150
Turga locality 554
Turgai Interior Seaway 206, 546
Turkmenistan, mammals 637
Turtles (*see* Testudines)
Tushilge (*see* Tüshleg)
Tüshleg 237, 390
Tüshleg Uul 259–60, **280**, **391**
Tuul Gol **280**
Tverdokhlebov, V.P. **7**, 10, 134, 242
Tverdokhlebova, G.I. 12

Twelvetrees, W.H. 2
Tylocephale 233, 481, 483–4, **486**, 495–6, **498**, 499, 500, **510**,
　　515
Tylosaurinae 205
Tylosaurus 203
Tyrannosauridae 231–2, 270, 434, **435**, 439, 444, 446–9,
　　447, 452, 510
Tyrannosaurinae 446–9
Tyrannosaurus 229, 448–9, 468
Tyul'keli 474
Tzimlyanskoe Reservoir 545
Tzybin, Yu. 242
Tzylma locality **121**

Ubukunskaya Svita 392, 397
Ubur Khangai (*see* Övörkhangai)
Udan Sayr (*see* Üüden Sair)
Udanoceratops 481, 488, **498**, 502–5, **510**, 515–16
Udmurt Republic 101–2
Ufa 1
Uilgan Formation/Svita 257
Uilgan River 257, 259
Ukhaa Tolgod locality
　dinosaurs 438, 452
　eggs **561**, 568
　geology **265**, 268, 273
　historical 233
　mammals **574**, 575–6, 588, 590–2, **591**, 602, 604, 610
　turtles 331
Ukhaatherium 604–5, 610, **611**, 617
Ukhta region 199
Ukraine
　crocodilian 402
　mammal 627
　marine reptiles 195, 203, **204**
Ulaanbaatar 212, 214–16, 227, 231–2, 237, **238**, 239, **244**,
　　245–6, 250, **250**, **265**, **280**, 399, 447–8, **561**, **574**
Ulaanbaatar, Geological Institute 245, 251, 309, 402, 422,
　　457–8, 480, 533, 575–6, 579, 586–8, 590, 593, 602,
　　610–11
Ulaanbaatar, Mongolian Museum of Natural History 422
Ulaanbaatar, State Museum 245, 250
Ulaan Bulag locality **403**, 405, 411, 417
Ulaandel Formation/Svita 257, 263
Ulaandel Uul 239, 257
Ulaangom **238**, **244**, **250**, **265**
Ulaan Nuur basin 281–2

Ulaan Öösh locality 237, 256, 264, **265**, **280**
Ulaan Sair (*see* Üüden Sair)
Ulaan Tolgoi locality **280**, 553
Ulaanush (*see* Ulaan Öösh)
Ulan Bator (= Ulaanbaatar)
Ulan Osh (*see* Ulaan Öösh)
Ulemica 21, 24, 27–8, **28**, **29**, 30, 34
Ulemosauridae 102
Ulemosaurus **22**, 23–4, 100–1, **101**, **102**
Ulgii (*see* Ölgii)
Ultan Tsav Uul **280**
Ulugei (*see* Ölgii)
Ul'yanovsk Province 187–90, 195–7, 199–202, 304, 413
Ulyasutai town **238**, **244**, **250**, **265**
Ulzii (*see* Ölgii)
Umnogovi (*see* Ömnögov')
Underhaan **238**, **244**, **250**, **265**
Undershil (*see* Öndörshil)
Under Ukhaa (*see* Öndörukhaa)
Underukhiin (*see* Öndörukhaa)
Undjulemys 315, **317**
Undorosaurus 197–8, 200
Undory town 187–8
Undory Palaeontological Musuem 188
Undur Ukha (*see* Öndör Ukhaa)
Undzhul (*see* Önjüül)
Ungulata 617, 627, 642
Ungulatomorpha 617, 642
University of Alberta, Edmonton 480
Unwin, D.M. 424, 430
Unzha River 195
Upchurch. P. 458, 460
Uppsala, Palaeontological Institute 481
Urad Houqi **280**
Ural–Dvina Expedition 6
Uralerpeton 65, 66
Ural Mountains 1, **3**, **9**, **131**, 133, 161, 196, 203
Uralocynodon 25, 114
Uralokannemeyeria 103, 106, **107**, **122**, **124**, 134
Uralosaurus **141**, 145, 148–9, 156
Uralosuchus 44
Ural River basin 10, 40, 44, 60, 65, 94, 112, 125, 130, 132, 134, 167, 169, 203
Ural–Samara Basin 121, 125, 128–9
Ural'sk 195
Uraniscosaurus 100
Uranoceratops 516

Urbanek, A. 233
Urga (= Ulaanbaatar)
Urilge Khudag locality 440
Urodela 178, 297, 304, 307
Uromasticidae 383–4
Uromastyx 373
Urzhumian Gorizont 110
Ust' Kara 554
Ustmylian Gorizont
    archosaurs **141**, 142, 155
    biostratigraphy 120, 122–4, **122**, 129–30
    procolophonoids 172–3, 175
    temnospondyls 49, 51
Ust'-Tsyl' ma district, 155
Us'va 19, 20
Utahraptor 436
Utegenia 63
Üüden Sair locality
    amphibians 304
    birds 551, **561**
    crocodilians **403**, 405, 407, 417
    geology **265**, 266
    historical 245, 248
    mammals **574**, 575, 603, **604**
Uvdeg (*see* Övdög)
Uverhangai (*see* Övörkhangai)
Uzbekbaatar 633, 634, 639, **645**, 649
Uzbekistan
    amphibians 297, 300–7
    birds 530, 534–7, 540–1, 551–2, 554–6
    choristoderes 390
    crocodilians 402, 404, 410–11, 414
    dinosaurs 437, 439, 441, 443–5, 447, 449–51, 456, 463, 473–4, 476, 481, 492–3, 517
    eggs 560
    lizards 382
    mammals 604, 616–17, **628**, 629, 637, 639–40, **641**, 642, **643**, **644**, **645**, **646**, **647**, 648–9
    marine reptiles 188, 196, 205
    pterosaurs 420–2, 427–8
    snake 204
    turtles 205, 310, 319, 321, 326–7, 333, 337–8, **338**, 344, **344**, 349, 354–5, **355**

Vaimos **121**
Vakhnevo locality 49, **121**, 167
Varanidae 380–1, 383–6

Varanoidea 380, 382
Varanopseidae 17, 23, 27
Varanus 380, 382
Vashka locality **121**
Vasil'ev, V.G. 256
Vedenyapin, M.A. 422
Velikoretskoe **121**
Velociraptor 223, 233, 269, 283, **435**, 436, **437**, **442**, 451–2,
    468, 568
Velociraptorinae 435–6
Venyukov, P.N. 2
Venyukovia 2, 21, 23–4, 27–8, **28**, 103–5, **104**
Venyukoviamorpha 103
Venyukoviidae 10, 104–5
Venyukovioidea 28–9, **29**
Vernadskii State Geological Museum, Moscow 188
Vertebrae, diplospondylous 62
Vertebrae, gastrocentrous 61
Vertexia 128
Verzilin, N.N. 273, 330
Veterovata 562
Vetluga River basin 8, 11, 68, 105, 120, 125–6, 128–9, 144,
    162, 164
Vetlugian Supergorizont 112, 120–30, **122**, 140, **141**, 142
Vetrov, F.E. 627
Viatkosuchus 111
Vieraella 301
Vilyui River 189, 196
Vincelestes 598
Vinnitsa locality 203
Vinogradov, G. **252**
Vitalia **122**, **123**, 131, 161, 165, 174
Vjushkovia 131, **141**, 145, **146**, 147–9, **147**
Vjushkovisaurus 11, **124**, 134, **141**, 151–2, **152**
Vladimir Province 12, 26, 66–7, **67**, 111, 142
Vladivostock, Far-Eastern University 236
Vokhmian Gorizont
    anthracosaurs 68, **68**
    archosaurs **141**, 142, 144
    biostratigraphy 120, 122–9, **122**, **126**, **127**, **131**
    procolophonoids 162, 164, 175
    synapsids 105
    temnospondyls 45, 53
Vokhminskaya Svita 125
Volga–Kama Expedition 6
Volga region 413, 421, 427–8
Volga River basin 12, 187, 191, 196–7, 200–5

Volga–Ural Anticline 121, **121**, 125, 128–30
Volgavis 551
Volgian oil shales 187, **192**, **193**
Volgograd Province 47, 155, 171, 174, 196, 199, 203, **403**,
    416, 429, 545, 552
Volgograd Provincial Museum, Volgograd, 533
Vologda Province 49, 65–6, 93, 167, 195
Vologdin, A. 241
Vonhuenia **122**, **124**, 126, **141**, 142, 144
Voronezh region 190, 196
Voroshilovgrad region, 203, **204**
Voskresensk locality 187
Vrag locality **121**
Vshivtsevo locality **121**
Vyatka River basin 7, 19, 20, 24, 48, 54, 68, 74, 91, 111, 125,
    130, 167, 195–6
Vyatskian Gorizont **19**, 25–6, 63–6, **65**, **66**, **67**, 78, 94,
    110–11, 114, **141**, 143
Vyazniki 1 locality **18**, **67**, 111, 142
Vyazniki 2 locality 2, 26, 53, 66–7, 111
Vybor River 49, **121**
Vyborosaurus 49, **122**, **123**, 130
Vychegda River basin 130, 143, 153, 169
Vym' River 44–5
Vytshegdosuchus **122**, **124**, 131–2, **141**, 151, 153–4, **154**,
    156
V'yushkov, B.P. 8, 9, 12, 18, 93, 108, 113, 120–1, 161

Walker, C.A. 541
Wangenheim von Qualen, F. 1
Wangisuchus 150
Wannanosaurus 481–2, **483**, 495–7, **498**, 499, 506, **510**, 514
Warsaw 232, 448, 576
Warsaw, Institute of Paleobiology (ZPAL) 232, 297, 369,
    402, 457, 480, 533, 562, 577, 590, 592, 594, 602,
    610–11
Washington, National Museum of Natural History 480
Watongia 92
Watongiidae 92
Watson, D. M. S. 79–80, 87
Weishampel, D.B. 464, 471–3
Welles, S.P. 188
Wellnhofer, P. 430, 445
Wetlugasaurus 45, **46**, 48, 121, **122**, **123**, **124**, 125, 129–31
Wetlugasaurus faunal grouping 122–4, 129–30
Whaitsiidae 26, 109, 110
Whaitsiinae 110

Wible, J.R. 577
Wilkinson, M. 156
Wiman, C. 310, 328
Witmer, L. 184
Wuerho district **328**

Xantusiidae 379
Xenosauridae 382, 384–6
Xihaina 369
Xilousuchus 150
Xinjianchelyidae 360
Xinjiang Province 105, **313**, **328**, 333, 434, 450
Xinjiangchelyidae 312–15, 318, 321–2, 331, 349–50, 353, 359–60, 362
Xinjiangchelys 312, **313**, 315

Yablonovy **121**
Yagaan Khovil 331, **342**
Yakovlev, N.N. 140
Yakovlev, V. 242
Yakutia 199, 517
Yale Peabody Museum, New Haven 480
Yalovach Svita 268, 304, 306, 354, 404, 409, **410**, 414, **415**, 637–8
Yantardakh locality 554, 639
Yarenga River 52
Yarengia 52, **121**, **122**, **123**, 131
Yarengiidae 40, 51–2, 131
Yarenskian Gorizont
  anthracosaurs 68
  archosaurs **141**, 142, 144–5, 147–8, 153–6
  biostratigraphy 120–4, **122**, 130–2, **131**
  procolophonoids 167, 169, 171, 174–5
  synapsids 112
  temnospondyls 52
Yarilinus **66**
Yarkov, A.A. 552
Yaroslavl' 195
Yasykovia 198, 200
Yaverlandia 481–2, 496–7, **498**, 499, 500, 506–7, 509, **510**, 514
Yaxartemys 320
Yeh H.-K. 310, 313, 337, 343
Yellow River valley 279, 280, 282
Yiginholo Formation 282, **283**
Yijinhuoluo Formation 548
Yijun Formation **283**

Yikezhao-meng 548
Yin Shan **280**
Yixian Formation **320**
Young, C.C. 142, 147–8, 409
Youngoolithus 565
Yumenemys 310, 346
Yunatov, A.A. 236
Yushatyr Svita 134
Yuza I, II localities **121**

Zaaltai (*see* Trans-Altai)
Zaisan basin 333, 560
Zakhar'evsk Mine 201
Zalambdalestes 233, 262, 269, 604, **608**, 610–12, **612**, 614, **614**, 617, 632, **644**, 648
Zalambdalestidae 604–5, 610–14, 617, 648
Zamuratscho 429
Zamyn Khond locality 331, 343, **343**, 369
Zangerl, R. 328
Zangerlia 339–40, 342–3, **343**
Zanul'e locality **121**
Zaocishan locality **320**
Zaplavnoe locality **121**
Zatolokino locality **194**
Zatrachydidae 40
Zavolzh'e 187
Zavrazhe 94
Zereg depression 239, 257, 271
Zereg Formation/Svita 257, 260, 263
Zhaksy-Karagala River 100
Zhalmaus well 627, **628**, 637–9, **647**
Zhangheotherium 596
Zhangjiakou 212
Zhao Z.K. 487, 562
Zhegallo, V.I. 241–2, **246**
Zhejiangopterus 427
Zhelestes 631, 634, 636, 638, 642, **643**, 646
Zhelestidae 617–18, 631, 633, 635, **641**, 642, **643**, **644**, 645, **646**, 648
Zhelestinae 642
Zhelubovskii, Yu.S. 236
Zheshart locality 47, **121**, 143, 153, 169
Zhidan Group 282, 287
Zhigansk 196
Zholsuchus 404, 410, 417
Zhuravlev, K.I. 194
Zhyraornis 533–5, 556

Index

Zhyraornithidae 556
Zhyrasuchus 404, 414–15, 417
Zidan Group 548
Znamenskoye **121**
Zofiabaataridae 585
Zones V–VII (Permo-Triassic) 120
Zopherosuchus 20, 96

Zubovskoe 45, **121**
Zurumtai depression 271
Züün Bayan 256, 264, **265**, 268, 271–2
Züünbayan Formation/Svita 243, 249, 256, 260, 391, 395, 397, 426
Zygoramma 333
Zygosaurus 2, 45